Springer Geology

The Springer Geology series seeks to publish a broad portfolio of scientific books, aiming at researchers, students, and everyone interested in geology. The series includes peer-reviewed monographs, edited volumes, textbooks, and conference proceedings. It covers the entire research area of geology including, but not limited to, economic geology, mineral resources, historical geology, quantitative geology, structural geology, geomorphology, paleontology, and sedimentology.

More information about this series at http://www.springer.com/series/10172

Bernard Charlier • Olivier Namur • Rais Latypov
Christian Tegner
Editors

Layered Intrusions

Springer

Editors
Bernard Charlier
University of Liège
Sart Tilman
Belgium

Rais Latypov
University of the Witwatersrand
Johannesburg
South Africa

Olivier Namur
University of Hannover
Hannover
Germany

Christian Tegner
Aarhus University
Aarhus
Denmark

ISBN 978-94-017-9651-4
ISBN 978-94-017-9652-1 (eBook)
DOI 10.1007/978-94-017-9652-1

Library of Congress Control Number: 2015932152

Springer Dordrecht Heidelberg New York London

Printed on acid-free paper

Springer is part of Springer Science+Business Media (www.springer.com)

Preface

Layered intrusions have received continuous interest since the publication of the treatise on 'Layered Igneous Rocks' by Lawrence Wager and Malcolm Brown, updated in books edited by Ian Parsons in 1987 and Grant Cawthorn in 1996. The study of these fossilized magma chambers keep inspiring a number of scientists with a range of interests including petrology and igneous differentiation, geochronology, geochemistry, mineralogy, rock textures and fabric, fluid dynamics, and ore deposits. The goal of this book is to further our understanding of magma chamber processes and crystal-liquid relationship during magma cooling magma. Physical and chemical processes are now better quantified thanks to the development analytical and computing tools such as compositional mapping, 3D X-ray computed tomography, in situ analyses for trace elements and isotopes, development of new experimental facilities, and progress in instrument sensitivity.

The book is subdivided into two parts. The first includes reviews and new views on chronological, textural, mineralogical, geochemical, and magnetic characteristics of layered igneous rocks. The second part reviews recent progress in the study of layered intrusions. A newcomer on the layered intrusions scene is the Panzhihua intrusion (SW China) that has been intensively studied recently. Reviews of recent findings for Sept Iles, Bushveld, Kiglapait, Ilímaussaq, and layered rocks in ophiolites are also presented. Interest in layered intrusions is also driven by their natural resources. Many intrusions host world-class ore bodies of chromium, platinum group elements (PGE), vanadium, titanium and phosphorous. Ore-forming processes and important deposits associated with layered intrusions are described and their origin is discussed.

The objective of this book is to outline the most recent ideas and challenges in the study of layered igneous bodies. It has also the purpose to aid in teaching, and to encourage new studies to tackle major issues in the understanding of magma chamber processes and associated ore-forming processes.

The book has benefited from detailed comments by a many reviewers, who are greatly acknowledged: Tom Andersen, Lewis Ashwal, Olivier Bolle, Alan Boudreau, Georges Ceeleneer, Kevin Chamberlain, Jean-Clair Duchesne, Bernard Henry, Michael Higgins, Clément Ganino, Lotte Melchior Larsen, Johan Lissenberg, Wolfgang Maier, Edmond Mathez, Iain McDonald, Jim Mungall, Richard Naslund, Troels Nielsen, Brian O'Driscoll, Ariel Provost, James Roberts, Brian Robins, Jill VanTongeren, Richard Wilson, and other anonymous referees.

Contents

Eric C. Ferré Department of Geology, Southern Illinois University at Carbondale, Carbondale, IL, USA

Bélinda Godel CSIRO Mineral Resources Flagship, Kensington, WA, Australia

Michael D. Higgins Sciences de la Terre, Université du Québec à Chicoutimi, Chicoutimi, Québec, Canada

Marian B. Holness Department of Earth Sciences, University of Cambridge, Cambridge, UK

Rais Latypov School of Geosciences, University of the Witwatersrand, Wits, South Africa

Craig Magee Department of Earth Science and Engineering, Imperial College, London, UK

Gregor Markl Mathematisch-Naturwissenschaftliche Fakultät, FB Geowissenschaften, Universität Tübingen, Tübingen, Germany

Michael A. W. Marks Mathematisch-Naturwissenschaftliche Fakultät, FB Geowissenschaften, Universität Tübingen, Tübingen, Germany

D. Morata Departamento de Geología & Andean Geothermal Center of Excellence (CEGA, Fondap-CONICYT), Facultad de Ciencias Físicas Matemáticas, Universidad de Chile, Santiago, Casilla, Chile

Stearns A. Morse Department of Geosciences, University of Massachusetts, Amherst, MA, USA

James E. Mungall Department of Earth Sciences, University of Toronto, Toronto, ON, Canada

Olivier Namur Institute of Mineralogy, University of Hannover, Hannover, Germany

Troels F. D. Nielsen Geological Survey of Denmark and Greenland, Copenhagen, Denmark

Brian O'Driscoll School of Physical and Geographical Sciences, Keele University, Keele, UK; The University of Manchester School of Earth, Atmospheric and Environmental Sciences (SEAES), Manchester, UK

Kwan-Nang Pang Institute of Earth Sciences, Academia Sinica, Taipei, Taiwan, China; Department of Geosciences, National Taiwan University, Taipei, Taiwan, China

Part I
Formation and Evolution of Cumulate Rocks

Thomas Peacock Department of Mechanical Engineering, Massachusetts Institute of Technology, Cambridge, MA, USA

James S. Scoates Department of Earth, Ocean & Atmospheric Sciences, Pacific Centre for Isotopic and Geochemical Research, University of British Columbia, Vancouver, BC, Canada

J. Gregory Shellnutt Department of Earth Sciences, National Taiwan Normal University, Taipei, Taiwan, China

Carl J. Spandler School of Earth and Environmental Sciences, James Cook University, Townsville, QLD, Australia

Carl T. E. Stevenson School of Geography, Earth and Environmental Sciences, University of Birmingham, Birmingham, UK

Atsushi Toramaru Department of Earth and Planetary Sciences, Kyushu University, Fukuoka, Japan

Jacqueline Vander Auwera Department of Geology, University of Liege, Liege, Belgium

Ilya V. Veksler Department of Mineralogy and Petrology, Technical University Berlin, Berlin, Germany; Helmholtz Centre Potsdam, GFZ German Research Centre for Geosciences, Potsdam, Germany; Department of Geology, Perm State University, Perm, Russia

Ron H. Vernon Department of Earth and Planetary Sciences and National Key Centre for GEMOC, Macquarie University, Sydney, Australia

Corey J. Wall Department of Earth, Ocean & Atmospheric Sciences, Pacific Centre for Isotopic and Geochemical Research, University of British Columbia, Vancouver, BC, Canada

Mei-Fu Zhou Department of Earth Sciences, The University of Hong Kong, Hong Kong, China

Chapter 1
Geochronology of Layered Intrusions

James S. Scoates and Corey J. Wall

Abstract Layered intrusions crystallize mainly from basaltic magma to form large bodies of igneous rocks that exhibit prominent layering and they preserve stunning rock records of the processes by which magma evolves in crustal magma chambers. These intrusions contain world-class deposits of chromium, platinum group elements (PGE), and vanadium, metals that are vital to industry and society in general. Despite their scientific and practical importance, precise age constraints are lacking for many layered intrusions, and geochronological frameworks linking crystallization and cooling ages for the most part do not exist. This has resulted in critical knowledge gaps related to their origin and formation. This chapter provides an overview of dating methods (U–Th–Pb, $^{40}Ar/^{39}Ar$) and mineral chronometers (e.g., zircon, baddeleyite, rutile, apatite, titanite) potentially present in layered intrusions that is coupled with field, textural, and petrographic criteria for targeting sample selection to allow for the successful implementation of geochronologic studies of layered mafic-ultramafic rocks of any age. As an application, we demonstrate how the thermal history of the Bushveld Complex is documented by mineral ages from samples of the PGE-rich Merensky Reef. High-precision U–Pb zircon ages, involving pretreatment of zircon by the chemical abrasion (annealing and leaching) or CA-TIMS technique, for two samples separated by >300 km are indistinguishable from each other (2056.88 ± 0.41 Ma, Eastern Limb; 2057.04 ± 0.55 Ma, Western Limb; uncertainty reported as 2s) confirming synchronous crystallization of this horizon at near-solidus conditions across the intrusion. Rapid cooling (~125 °C/Ma) down to temperatures of ~400–450 °C is defined by U–Pb rutile ages from the same samples (2052.96 ± 0.61 Ma, 2053.0 ± 2.7 Ma) and a regional hydrothermal event is signaled in $^{40}Ar/^{39}Ar$ biotite ages (1999 ± 10 Ma, 2002 ± 10 Ma). The geochronology of layered intrusions, where magma differentiation processes are captured in a wide range of rock textures and structures, represents an essential tool for assessing the evolution of mafic magmatism in the Earth's crust.

J. S. Scoates (✉) · C. J. Wall
Department of Earth, Ocean & Atmospheric Sciences, Pacific Centre for Isotopic and Geochemical Research, University of British Columbia, Vancouver, BC V6T-1Z4, Canada
e-mail: jscoates@eos.ubc.ca

C. J. Wall
e-mail: cwall@eos.ubc.ca

B. Charlier et al. (eds.), *Layered Intrusions,* Springer Geology,
DOI 10.1007/978-94-017-9652-1_1

Keywords Zircon · Baddeleyite · Cumulates · Bushveld · U–Pb dating

Introduction

For over a century, mafic-ultramafic layered intrusions have fascinated geologists with their detailed records of crystal accumulation and growth that reflect the interplay of a wide variety of magmatic, and post-magmatic, processes in crustal magma chambers (Harker 1904; Bowen 1928; Wager and Brown 1967; Parsons 1987; Cawthorn 1996). Layered intrusions crystallize predominantly from basaltic magma and form sill-like bodies of plutonic igneous rocks that exhibit prominent layering (Irvine 1982) (e.g., Muskox layered intrusion, Fig. 1.1). They consist of cumulates (e.g., Figs. 1.1b, c and d), rocks formed by crystal accumulation, and are the repositories of crystals fractionated from magma. As such, layered intrusions play an important role in illustrating compositional diversity in magmas and they represent a critical link in the evolution of basaltic magma from partial melting in the mantle to differentiation in crustal reservoirs to eruption, in many cases as voluminous and extensive flood basalts. These intrusions are integral components of Earth's continental crust throughout geological time and some host world-class ore bodies of chromium, platinum group elements (PGE), and vanadium (Cawthorn et al. 2005). Mafic layered intrusions may even be exposed on the surface of Mars at Columbia Hills and thus potentially serve as a model for early crustal development of planets (Francis 2011). Accurate and precise ages of rocks from different parts of layered intrusions will define how they are related to each other temporally as well as genetic relations between layered intrusions and their associated ore bodies, links to mafic dike swarms and volcanic rocks within large igneous provinces, and even potential environmental impacts. Yet surprisingly, even the classic mafic layered intrusions in petrology (e.g., Stillwater, Bushveld, Skaergaard) do not yet have robust geochronological frameworks.

In the past decade, major advances in sample pretreatment, instrument sensitivities, and data reduction protocols for U–Th–Pb and $^{40}Ar/^{39}Ar$ geochronology have led to significantly improved precision and accuracy of ages. These advances include development of the combined annealing and partial dissolution analysis of zircon or "chemical abrasion" method of Mattinson (2005) for removing zircon domains that have lost lead, establishment of the "EARTHTIME Initiative" with the goal of improving the precision of U–Pb geochronology through the use of widely available high-purity tracers, synthetic standard solutions, and software systems for data-handling (Schmitz and Schoene 2007; Bowring et al. 2011), and improved accuracy of $^{40}Ar/^{39}Ar$ ages through new calibrations of decay constants and standard ages (Kuiper et al. 2008; Smith et al. 2010; Renne et al. 2010). Combined with the recognition that zircon and other mineral chronometers (e.g., baddeleyite, apatite, rutile, titanite, biotite) can be successfully separated from mafic-ultramafic rocks (e.g., Scoates and Chamberlain 1995; Schwartz et al. 2005; Scoates and Friedman 2008; Grimes et al. 2009; Morisset et al. 2009), it is now possible to date multiple samples from individual layered intrusions allowing for the assessment of variations in their crystallization ages and cooling histories.

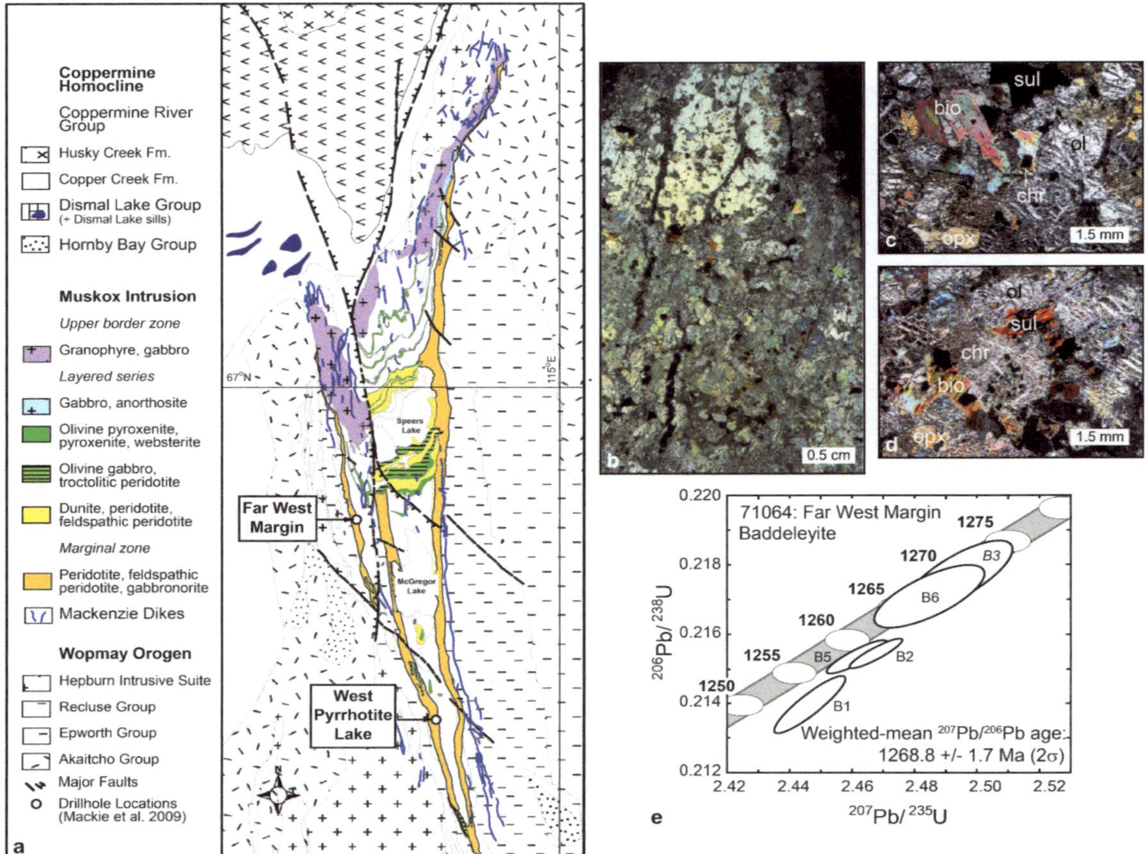

Fig. 1.1 Composite figure showing the application of U–Pb geochronology in determining the age of crystallization of

This contribution begins by providing an overview of the major dating techniques and mineral chronometers available to researchers on layered intrusions. Using the Archean Stillwater Complex (Montana, USA) as a test case, we then outline a field and petrologic approach for a sampling strategy to identify rock types and textures in layered intrusions that are most likely to yield dateable minerals for geochronology. This is followed by an application to dating of the Merensky Reef, host to one of the world's largest concentrations of platinum group elements, from the giant Bushveld Complex in South Africa. Two samples of the Merensky Reef, separated by > 300 km, each yielded multiple mineral chronometers (zircon, rutile, apatite, biotite) that allow for a detailed investigation of their crystallization ages, including a revision of the age reported in Scoates and Friedman (2008), and subsequent cooling path over 50–60 million years in the Paleoproterozoic using combined U–Pb and $^{40}Ar/^{39}Ar$ isotopic systematics. Petrographic relations among dated accessory minerals and rock-forming minerals in the Merensky Reef, coupled with cathodoluminescence imaging of the internal structure of zircon, are fundamentally imperative to interpreting the geochronological results. Finally, we conclude with an overview on future directions in the geochronology of layered intrusions and identify some of the major analytical and conceptual advances that will significantly improve our understanding of the timing of emplacement and geochemical evolution of these remarkable igneous bodies.

Layered Intrusions Through Time

Layered intrusions are present in crustal sequences of the Earth's earliest greenstone belts, they are common in Archean and Proterozoic terranes, and they may be found associated with young Cenozoic continental and oceanic flood basalts (Table 1.1). The oldest terrestrial layered igneous bodies may be the metamorphosed ultramafic sills, 5–30 m wide, which occur within the ca. 3.8 Ga Nuvvuagittuq greenstone belt

---

an ultramafic rock from the marginal zone of the Muskox layered intrusion, Nunavut, Canada. **a** Geologic map of the Muskox intrusion showing the location of the Far West Margin and West Pyrrhotite Lake drill holes in the marginal zone studied by Mackie et al. (2009) (map after Hulbert 2005, based on the map of Smith 1962). **b** Thin section scan (XPL) of 71064c, a sulphide-bearing, chromite-rich peridotite (sample 71064; 40.3 wt% MgO anhydrous) from near the *top* of the Far West Margin drill hole (28 m depth, ~ 89 m from contact) that was targeted for dating based on the presence of abundant interstitial biotite (but no interstitial plagioclase). The scan shows several large, cm-scale, poikilitic orthopyroxene crystals including mm-sized olivine (pervasively serpentinized); prominent fractures are filled with secondary magnetite. **c** Photomicrograph showing coarse sulphide (pyrrhotite, pentlandite, chalcopyrite), euhedral chromite, serpentinized olivine, and late biotite overgrowths (note zoning in biotite). **d** Photomicrograph showing interstitial biotite overgrowths around sulphide and chromite in a matrix of serpentinized olivine and poikilitic orthopyroxene. **e** Concordia diagram showing U–Pb geochronological results for baddeleyite from peridotite sample 71064; individual fractions are represented by a 2s error ellipse and labeled B1, B2, etc.; shaded grey band shows the error bounds of the concordia curve taking into consideration uncertainty in the U decay constants. Abbreviation: *Fm* Formation

Table 1.1 Summary of ages of representative layered intrusions through geological time

Intrusion	Region	Location	Era/Period	Age(s)[a]	Method[b]	Rock Type[c]	Reference
Ujaraaluk	Nuvvuagittuq	Québec	Hadean	4406 + 14/− 17 Ma	$^{146}Sm-^{142}Nd$ isochron	Meta-mafic/ultramafic rocks	O'Neil et al. (2012)
Akilia	Itsaq, Isua	Greenland	Eoarchean	3811 ± 4 Ma (min.)	U–Pb zircon SHRIMP	Enclaves in tonalite	Nutman et al. (1996)
Stella		South Africa	Mesoarchean	3033.5 ± 0.3 Ma	U–Pb zircon ID-TIMS	Gabbro	Maier et al. (2003)
Fiskenaesset		Greenland	Mesoarchean	2973 ± 28 Ma	$^{147}Sm-^{143}Nd$ errorchron	Mafic-ultramafic rocks	Polat et al. (2010)
Windimurra	Yilgarn Craton	Australia	Mesoarchean	2813 ± 3 Ma	U–Pb zircon ID-TIMS	Rhyolite in roof pendant	Ivanic et al. (2010)
Bird River	Manitoba	Canada	Neoarchean	2743.0 ± 0.5 Ma	U–Pb zircon CA-TIMS	Gabbro	Scoates and Scoates (2013)
Mulcahy Lake	Ontario	Canada	Neoarchean	2733.2 ± 1.0 Ma	U–Pb zircon ID-TIMS	Mt-rich gabbro	Morrison et al. (1985)
Stillwater	Montana	USA	Neoarchean	2701 ± 8 Ma	$^{147}Sm-^{143}Nd$ isochron	Gabbronorite	DePaolo and Wasserburg (1979)
				2704 ± 1 Ma	U–Pb zircon ID-TIMS	Pegmatitic anorthosite	Premo et al. (1990)
				2709.05 ± 0.85 Ma	U–Pb zircon CA-TIMS	Pegmatitic troctolite	Wall et al. (2010)
Great Dyke		Zimbabwe	Neoarchean	2575.9 ± 1.0 Ma	U–Pb zircon ID-TIMS	Orthopyroxenite	Oberthür et al. (2002)
Lukkulaisvaara	Karelia	Russia	Paleoproterozoic	2442.1 ± 1.4 Ma	U–Pb zircon ID-TIMS	Gabbroic pegmatite	Amelin et al. (1995)
Kemi	Elijfirvi	Finland	Paleoproterozoic	2.44 Ga	U–Pb zircon ID-TIMS	Gabbro	Patchett et al. (1981)
Bushveld		South Africa	Paleoproterozoic	2056.88 ± 0.41 Ma	U–Pb zircon CA-TIMS	Pegmatitic orthopyroxenite	This study

Table 1.1 (continued)

Intrusion	Region	Location	Era/Period	Age(s)[a]	Method[b]	Rock Type[c]	Reference
Fox River	Manitoba	Canada	Paleoproterozoic	1882.9 ± 1.5 Ma	U–Pb zircon ID-TIMS	Gabbroic pegmatite	Heaman et al. (1986)
Panton	East Kimberley	Australia	Paleoproterozoic	1856 ± 2 Ma	U–Pb zircon SHRIMP	Anorthosite	Page and Hoatson (2000)
Poe Mountain	Laramie	Wyoming, USA	Mesoproterozoic	1434.4 ± 0.6 Ma	U–Pb baddeleyite ID-TIMS	Anorthosite	Scoates and Chamberlain (1995)
Kiglapait	Nain, Labrador	Canada	Mesoproterozoic	1305 ± 22 Ma	^{147}Sm–^{143}Nd isochron	Ferrodiorite	DePaolo (1985)
Muskox	Nunavut	Canada	Mesoproterozoic	1269.4 ± 1.1 Ma	U–Pb baddeleyite ID-TIMS	Feldspathic peridotite	French et al. (2002)
				1268.8 ± 1.7 Ma	U–Pb baddeleyite ID-TIMS	Chr-bearing peridotite	Mackie et al. (2009)
Ilimaussaq	Gardar Province	Greenland	Mesoproterozoic	1160 ± 5 Ma	U–Pb baddeleyite ID-TIMS	Augite syenite	Krumrei et al. (2006)
Duluth	Duluth Complex	Minnesota, USA	Mesoproterozoic	1099.1 ± 0.2 Ma	U–Pb zircon ID-TIMS	Anorthosite	Schmitz et al. (2003)
Sonju Lake	Duluth Complex	Minnesota, USA	Mesoproterozoic	1096.1 ± 0.8 Ma	U–Pb baddeleyite ID-TIMS	Apatite ferrodiorite	Paces and Miller (1993)
Bjerkreim–Sokndal	Rogaland	Norway	Neoproterozoic	932 ± 5 Ma	U–Pb zircon ID-TIMS	Quartz mangerite	Pasteels et al. (1979)
Sept Iles	Québec	Canada	Neoproterozoic	565 ± 4 Ma	U–Pb zircon ID-TIMS	Granophyre	Higgins and van Breemen ((1998)
Fongen-Hyllingen	Caledonides	Norway	Silurian	437.8 ± 2.3 Ma	U–Pb zircon ID-TIMS	Monzonite	Nilsen et al. (2007)

Table 1.1 (continued)

Intrusion	Region	Location	Era/Period	Age(s)[a]	Method[b]	Rock Type[c]	Reference
Hongge	Emeishan	China	Permian	259.3 ± 1.3 Ga	U–Pb zircon ID-TIMS	Ilm-rich gabbro	Zhong and Zhu (2006)
Dufek	Ferrar	Antarctica	Jurassic	183.9 ± 0.3 Ma	U–Pb zircon ID-TIMS	Granophyre	Minor and Mukasa (1997)
Duke Island	Alaska	USA	Cretaceous	108 ± 1 Ma	U–Pb zircon ID-TIMS	Hbl-plag pegmatite	Saleeby (1992)
Rum	Inner Hebrides	Scotland	Paleocene	60.53 ± 0.08 Ma	U–Pb zircon ID-TIMS	Feldspathic pegmatite	Hamilton et al. (1998)
Skaergaard	East Greenland	Greenland	Eocene	55.40 ± 0.14 Ma	$^{40}Ar/^{39}Ar$ hornblende	Granophyre	Hirschmann et al. (1997)
				55.960 ± 0.018 Ma	U–Pb zircon CA-TIMS	Ferrodiorite	Wotzlaw et al. (2012)
Kap Edvard Holm	East Greenland	Greenland	Eocene	48.8 ± 0.2 Ma	$^{40}Ar/^{39}Ar$ hornblende	Gabbroic pegmatite	Tegner et al. (1998)
Val	Kerguelen	Indian Ocean	Oligocene	24.25 ± 0.15 Ma	U–Pb zircon ID-TIMS	Gabbro	Scoates et al. (2007)

[a] Uncertainty reported as 2s; all ages reported as Ma, except where there is insufficient precision (Ga)

[b] Abbreviations: SHRIMP = sensitive high-resolution ion microprobe, *ID-TIMS* isotope dilution-thermal ionization mass spectrometry, *CA-TIMS* chemical abrasion-thermal ionization mass spectrometry

[c] Abbreviations: *Mt* magnetite, *Chr* chromite, *Ilm* ilmenite, *Hbl* hornblende

of the Superior Province, Canada (O'Neil et al. 2007, 2012). The sill interiors are composed predominantly of serpentine and talc, however, they also contain amphibole-rich layers and layers of metamorphic orthopyroxene that represent original pyroxene cumulate horizons (O'Neil et al. 2007). These ultramafic sills are geochemically correlative with the Ujaraaluk unit amphibolites of volcanic protolith (O'Neil et al. 2011) for which a ^{146}Sm–^{142}Nd isochron age of 4406 + 14/− 17 Ma has been determined (O'Neil et al. 2012) (Table 1.1). Neoarchean layered intrusions hosting stratiform chromitites occur in the 3.81 Ga deformed layered anorthosite-ultramafic intrusions in the Ivîsartoq region of West Greenland (Rollinson et al. 2002) and Paleoarchean komatiitic sills with chromitite layers are recognized in the Zimbabwe craton (Prendergast 2008). Mesoarchean stratiform chromitites and layered intrusions are found in India (Mondal et al. 2006), South Africa (Maier et al. 2003), Greenland (Fiskenaesset, Polat et al. 2010), and western Australia (Ivanic et al. 2010). Layered intrusions are notable in the Neoarchean rock record from ca. 2700–2750 Ma, especially in the Superior Province of Canada, and also include the Stillwater Complex in the Wyoming Province of southwestern Montana (Hess 1960; Jackson 1961; McCallum 1996). The Bushveld Complex, the world's largest layered intrusion and host to remarkable resources of chromium, platinum group elements, and vanadium (Eales and Cawthorn 1996; Cawthorn et al. 2005; Cawthorn (Chap. 12, The Bushveld Complex), is one of a number of major Paleoproterozoic intrusions that are also exemplified by mafic intrusions in the eastern Baltic (Fennoscandian) Shield (Amelin et al. 1995), the Kemi intrusion of northern Finland with its major chromite deposit (Alapieti et al. 1989), the 250 km-long Fox River Sill in northeastern Manitoba (Scoates 1990), and numerous layered mafic-ultramafic intrusions in the East Kimberley of western Australia (Hoatson and Blake 2000). Mesoproterozoic layered intrusions are typically related to either crustal extension and rifting (e.g., 1.27 Ga Muskox intrusion, Mackenzie large igneous province, northern Canada, Fig. 1.1; 1.1 Ga Duluth complex, Keweenawan or Midcontinent Rift, Minnesota; 0.56 Ga Sept Iles intrusion, eastern Québec) or with Proterozoic anorthosite plutonic suites (e.g., 1.43 Ga Poe Mountain intrusion, Laramie; 1.3 Ga Kiglapait intrusion, Nain; ca. 0.93 Ga Bjerkreim–Sokndal intrusion, Rogaland) (Table 1.1). The Phanerozoic record of layered intrusions, including the 56 Ma Skaergaard intrusion in East Greenland (Wager and Brown 1967; Wotzlaw et al. 2012), is dominated by those that are part of large igneous provinces (e.g., Emeishan, Ferrar, North Atlantic) (Table 1.1). This association reflects their emplacement during large-scale crustal rifting and outpouring of voluminous flood basalt sequences and development of high-level magma reservoirs as staging chambers prior to the eruption of flood basalts on oceanic islands (e.g., 24 Ma Val gabbro, Kerguelen Archipelago, southern Indian Ocean; Scoates et al. 2007).

Cumulates and Geochronology

Layered intrusions consist of cumulates (Fig. 1.2), rocks that began by forming on magma chamber floors, walls, and roofs as a framework of touching minerals, cumulus crystals or "primocrysts", with interstitial melt or intercumulus melt filling

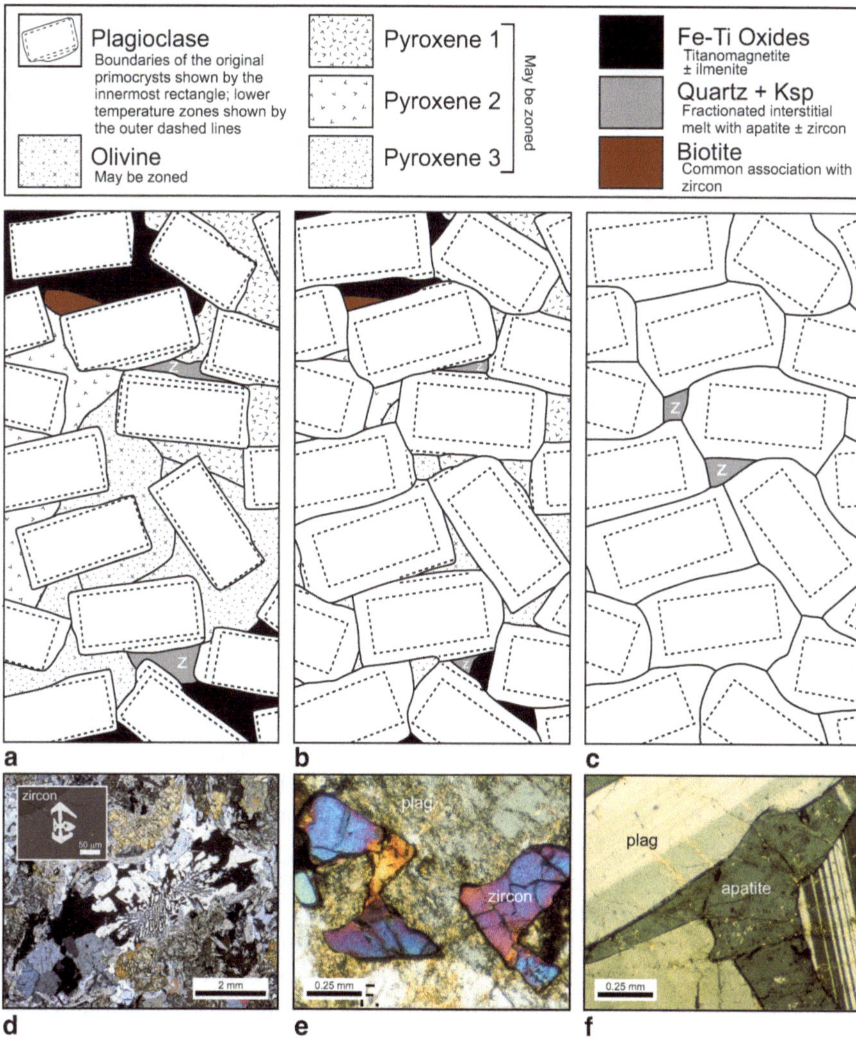

Fig. 1.2 Schematic diagrams and photomicrographs showing the textural setting of dateable accessory minerals in different types of plagioclase cumulates in layered intrusions. Diagrams are after Wager et al. (1960). **a** Plagioclase cumulate with abundant poikilitic pyroxene, olivine, and Fe–Ti oxides, and interstitial *patches* of quartz/K-feldspar that represent a component of the most fractionated interstitial melt; zircon (and apatite) is preferentially associated with these interstitial residual areas and with late biotite overgrowths. **b** Plagioclase cumulate with minor amounts of interstitial pyroxene, olivine, and Fe–Ti oxides, and small interstitial *patches* of fractionated interstitial melt (quartz/K-feldspar). **c** Pure plagioclase cumulate (anorthosite) with cumulus and interstitial plagioclase; anorthosites generally make excellent target rocks for dating. **d** Photomicrograph of a large granophyre clot (quartz/K-feldspar) within subophitic gabbronorite of the marginal zone, Muskox intrusion (West Pyrrhotite Lake, sample 71139). The clot contains abundant acicular apatite and minor hopper (skeletal) zircon shown in the backscatter electron image inset. **e** Photomicrograph of interstitial (poikilitic) zircon in layered leucogabbro of the Poe Mountain intrusion, Laramie anorthosite complex, Wyoming. **f** Photomicrograph of interstitial apatite in layered leucogabbro of the Poe Mountain intrusion, Laramie anorthosite complex, Wyoming (sample PM310)

the pore spaces (Wager et al. 1960; Wager and Brown 1967; Irvine 1982; Hunter 1996). From the perspective of geochronology, the evolution of this interstitial melt and its potential to saturate in the U–Th–Pb-bearing minerals (e.g., zircon, baddeleyite, apatite) and K-bearing minerals (e.g., biotite, amphibole) that can be dated by modern, high-precision geochronological techniques is of critical importance (Figs. 1.2a, b and c). Crystal frameworks in slowly cooled intrusions are characterized by an initial high porosity and permeability (e.g., Jerram et al. 2003) that allow exchange of melt from the crystal mush to the main body of magma by compositional convection or by diffusion (e.g., Morse 1986). Elimination of this porosity occurs by consolidation or densification of the crystal mush (e.g., Wager and Brown 1967; Irvine 1982; Hunter 1996). This may involve a variety of processes, including (1) continued crystallization and growth of cumulus crystals from interstitial melt, (2) crystallization of new minerals from evolved interstitial melt that grow as poikilitic or subpoikilitic crystals surrounding cumulus crystals, (3) crystallization of remaining melt as zoned overgrowths on cumulus crystals, as pockets of minerals that crystallize from fractionated melt such as quartz, alkali feldspar, biotite ± zircon/baddeleyite/apatite, or as mono-mineralic rims on cumulus crystals that represent the presence of grain boundary liquid during the final stages of crystallization of mafic intrusions (e.g., Holness et al. 2007), and (4) mechanical compaction. Some adcumulates, cumulates with only minor postcumulus material (Irvine 1987), may also have formed directly on the surface of the cumulate pile (Campbell et al. 1983; Campbell 1987). Crystal ageing (Boudreau 2011) and textural coarsening (Higgins 2011; Higgins (Chap. 3, Quantitative textural analysis of rocks in layered mafic intrusions)) may also modify the final texture of a cumulate.

The ability to successfully date a layered intrusion depends on selecting appropriate samples that have a high probability of containing extractable quantities of minerals that can be used as chronometers. For reasons outlined in subsequent sections, zircon ($ZrSiO_4$) is the most important mineral that can be dated to high precision and accuracy from layered intrusions. Not only is zircon found in mafic or ultramafic cumulates from layered intrusions (e.g., Premo et al. 1990; Scoates and Friedman 2008; Scoates and Scoates 2013), but it also occurs in a wide range of other mafic cumulates, including gabbros from mid-ocean ridges (e.g., Grimes et al. 2009; Schmitt et al. 2011) and plagioclase-rich rocks from Proterozoic anorthosite plutonic suites (e.g., McLelland and Chiarenzelli 1990; Scoates and Chamberlain 1995, 2003). Experimental results demonstrate that zircon cannot crystallize directly from basaltic liquids because unrealistically high Zr concentrations (> 5000 ppm) in the melt would be required (Boehnke et al. 2013). Thus, zircon that is found in mafic-ultramafic rocks, and that is not inherited as pre-existing crystals from other rocks, must have crystallized from late-stage, evolved melts probably at near-solidus temperatures (e.g., Scoates and Chamberlain 1995; Grimes et al. 2009; Boehnke et al. 2013). If zircon and other U–Th–Pb-bearing accessory minerals crystallize from liquid that is interstitial to cumulus grains and is trapped (i.e., the permeability of the consolidating cumulate goes to zero), these accessory phases should be directly associated with other lower temperature minerals that will saturate (e.g., quartz, alkali feldspar, magnetite/ilmenite, biotite; Fig. 1.2d) (Meurer and Meurer

2006). In contrast, if zircon is not associated with these near-solidus minerals or occurs as poikilitic overgrowths (Fig. 1.2e), it did not crystallize from a trapped liquid, but rather from an evolved liquid undergoing expulsion from the semi-consolidated cumulate by compaction (e.g., Meurer and Boudreau 1998; Meurer and Meurer 2006). In this contribution, textural observations of the in situ relationship between zircon and other minerals in mafic-ultramafic cumulates that are fundamental in constraining the relative timing of zircon crystallization will be addressed in the context of zircon from the Merensky Reef of the Bushveld Complex.

Dating Methods for Layered Intrusions

There are number of excellent and comprehensive references on geochronological techniques that should be a starting point for any researcher interested in the dating of layered intrusions. These include the book by McDougall and Harrison (1999) on $^{40}Ar/^{39}Ar$ geochronology, volume 53 of the Reviews in Mineralogy and Geochemistry entitled "Zircon" published by the Mineralogical Society of America (Hanchar and Hoskin, editors), which includes key chapters on the history of U–Th–Pb geochronology, analytical methods, and a very useful atlas of zircon textures (Corfu et al. 2003), and the May 2013 issue of Elements focused on "One Hundred Years of Geochronology" (Condon and Schmitz, guest editors). The Second Edition of the Treatise on Geochemistry also contains an exceptional overview on "U–Th–Pb Geochronology" by Schoene (2014). In the section below, the most commonly used dating approaches, isotopic systems, analytical techniques and protocols, and mineral chronometers are reviewed in the context of their application to layered intrusions.

Mineral-Whole Rock Isochrons

Due to elemental redistribution during post-crystallization hydrothermal alteration and metamorphism in ancient (Proterozoic to Archean) mafic and ultramafic rocks, obtaining precise and accurate dates of layered intrusions in the 1960s and 1970s was a challenge using conventional Rb–Sr, K–Ar, and U–Th–Pb methods. This changed with the pioneering application of Sm–Nd isotopes to the Stillwater Complex by DePaolo and Wasserburg (1979) who reported a precise age of 2701 ± 8 Ma for a gabbronorite from just below the J-M Reef (West Fork Adit). This age was based on the isochron relationship defined by the analyses of the whole rock and separates of plagioclase, clinopyroxene, and orthopyroxene (Fig. 1.3a). A similar U–Pb zircon-baddeleyite age of 2704 ± 1 Ma was determined from the Lower Banded Series of the Stillwater Complex by Premo et al. (1990) and more recent dating has revealed that most of the Lower Banded Series crystallized at 2709 Ma (Wall et al. 2010).

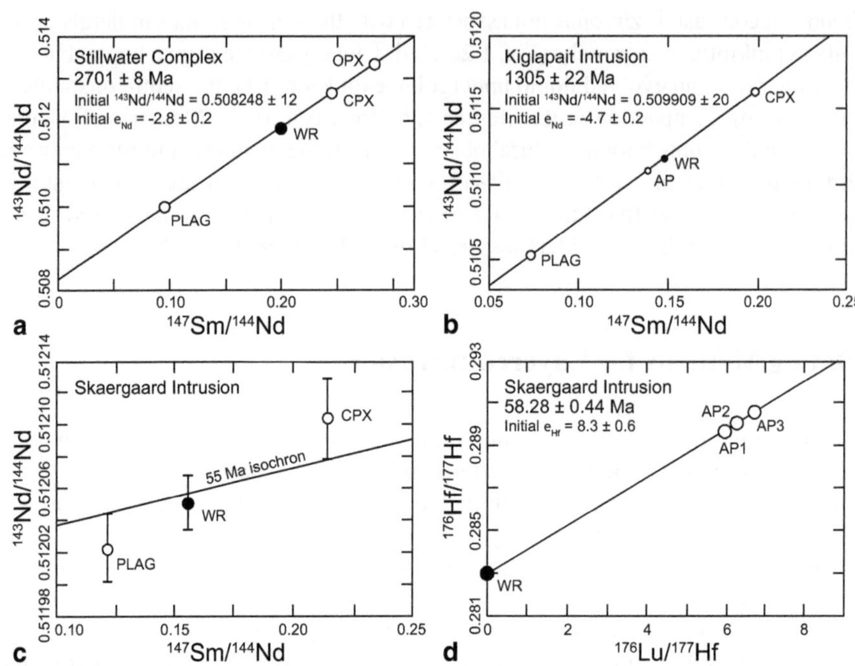

Fig. 1.3 Isochron plots showing mineral–whole rock isotopic results for representative layered intrusions. **a** Stillwater Complex: $^{147}Sm/^{144}Nd$ vs. $^{143}Nd/^{144}Nd$ plot showing analytical results of the whole rock, plagioclase, clinopyroxene, and orthopyroxene from a gabbronorite (STL-100) from the West Fork of the Stillwater River (top of Gabbronorite II in the Lower Banded Series) that define an isochron with an age of 2701 ± 8 Ma (modified from DePaolo and Wasserburg 1979). **b** Kiglapait intrusion: $^{147}Sm/^{144}Nd$ vs. $^{143}Nd/^{144}Nd$ plot showing analytical results for the whole rock, plagioclase, apatite, and clinopyroxene from a ferrodiorite (3379) from the *Upper* Zone at the 98.6 PCS level (modified from DePaolo 1985) that define an isochron age of 1305 ± 22 Ma. **c**. Skaergaard intrusion: $^{147}Sm/^{144}Nd$ vs. $^{143}Nd/^{144}Nd$ plot showing analytical results of the whole rock, plagioclase, and clinopyroxene from a ferrogabbro (SK-80) from *Upper* Zone c, 2540 m above the base of the Layered Series that appear to show isotopic heterogeneity between minerals with a 55 Ma reference isochron only just barely intersecting the error bars of the analyses for co-existing plagioclase and clinopyroxene (modified from Stewart and DePaolo 1990). **d** Skaergaard intrusion: $^{176}Lu/^{177}Hf$ vs. $^{176}Hf/^{177}Hf$ plot showing analytical results for the whole rock and three apatite fractions from an apatite-rich ferrogabbro (GGU-348881) from *Upper* Zone b that define a precise Lu–Hf isochron age of 58.28 ± 0.44 Ma (modified from Barfod et al. 2003). Abbreviations: *PLAG* plagioclase; *WR* whole rock; *CPX* clinopyroxene; *OPX* orthopyroxene; *AP* apatite (AP1, AP2, AP3 indicate different analyses of apatite)

Isochron diagrams compare the number of parent atoms in a radioactive decay scheme to the number of daughter atoms where the isotopes of the parent-daughter atoms are normalized to a stable isotope in the system of interest (e.g., Rb–Sr method: $^{87}Sr/^{86}Sr$ vs. $^{87}Rb/^{86}Sr$; Sm–Nd method: $^{143}Nd/^{144}Nd$ vs. $^{147}Sm/^{144}Nd$) (Faure and Mensing 2005; White 2013). The slope of the line passing through the data for the analyzed samples, with variable parent-daughter concentration ratios, is related to the age of the system and is called an isochron (Fig. 1.3). The intercept of the isochron is equal to the initial ratio, which is useful for fingerprinting magma sources and contamination. The isochron method of dating requires that all samples or minerals dated have the same initial isotopic ratio and crystallized at the same time. Precise isochron dating is critically dependent on analyzing minerals and whole rocks from the same samples with a wide range of parent-daughter concentration ratios (e.g., Rb/Sr, Sm/Nd, Lu/Hf). Deviations from the isochron may indicate open-system behavior during alteration or metamorphism and this leads to the possibility of defining secondary isochrons that date the approximate timing of these events. For example, in the Stillwater Complex, the $^{87}Rb/^{86}Sr$ isotopic systematics of plagioclase, clinopyroxene, and orthopyroxene from the gabbronorite that yielded the precise Sm–Nd whole rock-mineral isochron do not define a straight line in an isochron plot and, instead, lie closer to a 2.2 Ga isochron (DePaolo and Wasserburg 1979). These disparate results are consistent with differences in the relative mobility of the two parent-daughter systems during low-grade metamorphism with the Sm–Nd isotope systematics remaining unaffected.

The use of mineral-whole rock isochrons, especially Sm–Nd isochron dating, has been successfully applied to constraining the ages of crystallization and other events in a number of important layered intrusions (e.g., Stillwater, Kiglapait). Even though the uncertainty of these ages can be high (up to several percent), they are useful for broadly characterizing temporal relationships between intrusions and host rocks and for assessing regional tectonic controls on magmatism. DePaolo (1985) reported a Sm–Nd isochron age of 1305±22 Ma for a ferrodiorite from the Upper Zone of the Kiglapait intrusion, Labrador, based on the analytical results from plagioclase, apatite, the whole rock, and clinopyroxene (Fig. 1.3b). For the same sample, it was not possible to determine a precise Rb–Sr isochron due to the limited range of Rb/Sr in the minerals. For the Muskox intrusion (Fig. 1.1a), part of the giant Mackenzie large igneous province that also includes the Coppermine River flood basalts and the Mackenzie dike swarm in northern Canada, a Sm–Nd isochron (plagioclase, whole rock, clinopyroxene, orthopyroxene) of 1258±40 Ma was determined for a gabbronorite from Cyclic Unit 22 near the top of the cumulate sequence (Stewart and DePaolo 1996). This age, while relatively imprecise, is in agreement with the higher precision U–Pb baddeleyite dating of the Muskox intrusion (e.g., 1270±4 Ma, LeCheminant and Heaman 1989; 1268.8±1.7 Ma, Mackie et al. 2009). Intriguingly, Stewart and DePaolo (1996) also reported an Sm–Nd isochron age of 1822±165 Ma for a sample of roof zone breccia in the Muskox intrusion, thus providing evidence that this material actually represents a stoped piece of surrounding wall rock. All Rb–Sr isochrons in the Muskox intrusion are young (1000–1139 Ma) due to disturbance of the Sr isotopic systematics during a ca. 1000 Ma metamorphic event (Stewart and DePaolo 1996).

When applied to the Skaergaard intrusion of East Greenland, mineral-whole rock isochron results provide both crystallization age information (i.e., data from analyses of individual samples lie along 55 Ma isochrons) and evidence for resetting due to metamorphism and partial melting from the nearby Basistoppen intrusion (Stewart and DePaolo 1990), despite showing limited variations in Rb/Sr and Sm/Nd and in initial $^{87}Sr/^{86}Sr$ and $^{143}Nd/^{144}Nd$. Importantly, the Sm–Nd isotopic systematics of a sample from Upper Border Zone c, where the 55 Ma isochron barely intercepts the error bars of the analyses for plagioclase and clinopyroxene (Fig. 1.3c) and where the Rb–Sr isochron is coherent, suggested that isotopic heterogeneity between minerals in cumulates may need to be considered (e.g., incomplete equilibration between cumulus crystals and interstitial melt derived from another part of the intrusion). Evidence for mineral-scale isotopic heterogeneity in the Skaergaard intrusion was confirmed by McBirney and Creaser (2003) and there is now an increasing number of studies from the Bushveld Complex that document initial Sr–Nd isotopic differences between plagioclase and orthopyroxene (Prevec et al. 2005; Chutas et al. 2012; Roelofse and Ashwal 2012).

A promising approach to dating layered intrusions involves utilizing the ^{176}Lu –^{176}Hf isotopic systematics of apatite, which is a relatively common accessory mineral in mafic-ultramafic rocks. Apatite preferentially incorporates rare earth elements (REE) and contains little Hf (typically $\ll$1 ppm) (e.g., Fujimaki 1986). Barfod et al. (2003) presented a method for the separation of Lu and Hf from apatite that allows for rapid sample processing prior to analysis by multiple collector-inductively coupled plasma-mass spectrometry (MC-ICP-MS). When applied to a sample from Upper Zone b of the Skaergaard intrusion, isotopic data from three apatite fractions and the whole rock yield an isochron with a precise Lu–Hf age of 58.28 ± 0.44 Ma (Barfod et al. 2003: decay constant of ^{176}Lu of 1.869×10^{-11} year–^{1} from Scherer et al. 2001) (Fig. 1.3d). The Lu–Hf isochron is older than the recently published U–Pb zircon age of 55.960 ± 0.018 Ma (Wotzlaw et al. 2012), perhaps the result of disturbance of the original Lu–Hf systematics, but nonetheless demonstrates the potential for determining precise ages of layered intrusions from apatite when zircon is not present in the samples of interest.

U–Th–Pb Geochronology

The U–Th–Pb dating method has been a longstanding geochronological tool for the last 100 years, with significant improvements being made in the last 50 years, and has been used to determine the ages of U- and Th-bearing accessory minerals in rocks representing the span of Earth history (e.g., Parrish and Noble 2003; Mattinson 2013; Schoene 2014), including the vast majority of dated layered intrusions (Table 1.1). U–Th–Pb geochronology is based on the decay (parent to daughter) of the long-lived radioactive isotopes ^{238}U to ^{206}Pb, ^{235}U to ^{207}Pb, and ^{232}Th to ^{208}Pb where the Pb is referred to as radiogenic (i.e., Pb*); a fourth Pb isotope, ^{204}Pb, is stable and has no radioactive parent. From each of these decay systems, an independent age can be determined following isotopic measurements of the mineral of interest (e.g., zircon, monazite, baddeleyite, rutile) with knowledge of the decay constants of ^{238}U, ^{235}U, and ^{232}Th, an assumed present-day ^{238}U/^{235}U isotopic composition (Condon et al. 2010; Hiess et al. 2012), and the initial or common Pb present (i.e., Pb_c), which is the nonradiogenic Pb that occurs in the mineral upon crystallization or Pb introduced during sample processing in the laboratory (e.g., Dickin 2005; Faure and Messing 2005). A ^{207}Pb/^{206}Pb date can be calculated by combining

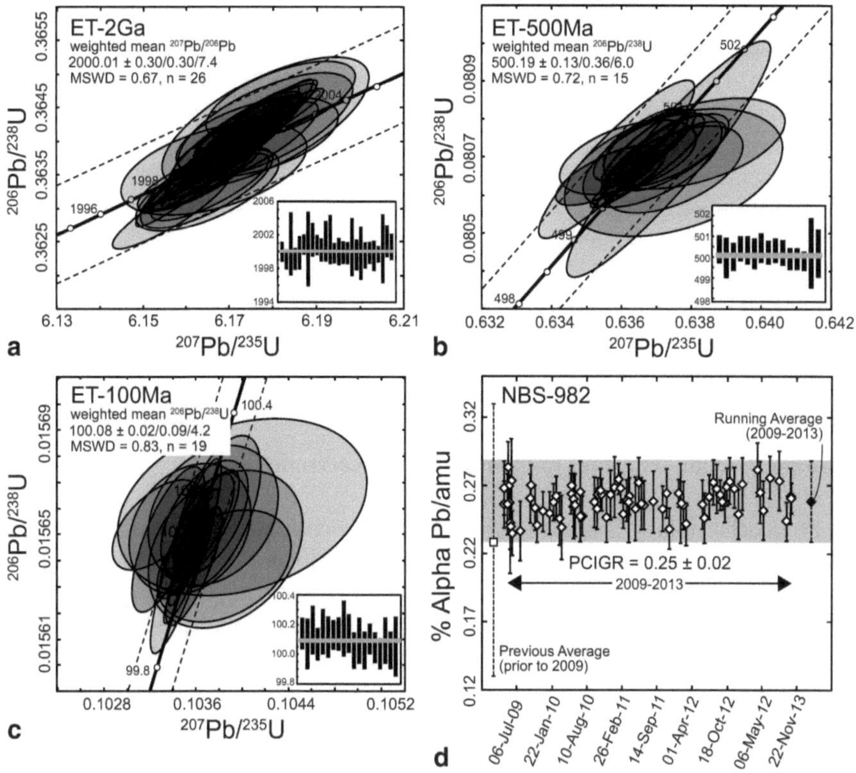

Fig. 1.4 Plots showing results for standard analyses at PCIGR (UBC) since early 2009. **a** Concordia diagram showing U–Pb results for the EARTHTIME 2 Ga synthetic solution (ET-2 Ga). **b** Concordia diagram showing U–Pb results for the EARTHTIME 500 Ma synthetic solution (ET-500 Ma). **c** Concordia diagram showing U–Pb results for the EARTHTIME 100 Ma synthetic solution (ET-100 Ma). The U–Pb errors are reported in the ± X/Y/Z format of Schoene et al. (2006) with internal error in the absence of all systematic errors (± X), tracer calibration error (± Y), and uncertainty due to decay-constant errors (± Z), which is used for comparison with other isotopic geochronometers (e.g., $^{40}Ar/^{39}Ar$). The *dashed* lines indicate the error bounds of the Concordia curve. The inset shows the $^{207}Pb/^{206}Pb$ ages for individual analyses for ET-2 Ga in panel **a** and the $^{206}Pb/^{238}U$ ages for panels **b** and **c**; the *grey horizontal* lines correspond to the weighted mean ages for each synthetic solution. **d** Fractionation factor, alpha (%/amu), from measurements of $^{208}Pb/^{206}Pb$ for 66 analyses of the NBS-982 Pb standard over nearly 5 years (2009–2013). The *dark grey* band shows the average value (0.25 ± 0.02, 2s) compared the previous value (0.23 ± 0.05, 2s) used at PCIGR

the daughter isotope determinations from the two uranium decay systems. For visualization, U–Pb results are most commonly presented on a Wetherill concordia diagram (Wetherill 1956), which plots $^{207}Pb^*/^{235}U$ vs. $^{206}Pb^*/^{238}U$ from the same analyses (e.g., Figs. 1.1e, 1.4a, b, and c). The concordia curve represents the locus of points of equal age from the two different uranium decay schemes and is non-linear due to the different half-lives of ^{238}U (4.468×10^9 years or 4.468 Ga) and ^{235}U (7.038×10^8 years or 703.8 Ma) (Jaffey et al. 1971). Concordant U–Pb results (i.e.,

$^{207}Pb^*/^{235}U$ age = $^{206}Pb^*/^{238}U$ age) for an analysis indicate that the sample (i.e., mineral) has remained a closed system since the time of formation (see analyses B3 and B6 on Fig. 1.1e). U–Pb results that do not lie on concordia are referred to as being discordant (see analyses B2, B5, and B1 on Fig. 1.1e) and result from open-system processes at some point after formation. This can include a variety of processes, but is most commonly attributed to Pb loss and mixing of materials with different ages (e.g., inheritance of older grains, metamorphic overgrowths). Background information on the most commonly used analytical methods (ID-TIMS, SIMS, LA-ICP-MS), pretreatment protocols for zircon analysis, and community-wide innovations spurred on by EARTHTIME is provided in the following sub-sections because of the importance of U–Th–Pb geochronology in determining the ages of layered intrusions.

Isotope Dilution-Thermal Ionization Mass Spectrometry

U–Pb geochronology using isotope dilution-thermal ionization mass spectrometry (ID-TIMS) is by far the most precise analytical technique with average precisions now reported from most laboratories on single zircons of ~0.2 % of the reported date and some laboratories achieving significantly higher precision (<0.1 %) (Schoene 2014). Comprehensive summaries of the history and application of TIMS to U–Th–Pb geochronology can be found in Davis et al. (2003) and Parrish and Noble (2003). Briefly, in ID-TIMS, minerals selected for dating after standard mineral separation procedures are spiked with a tracer solution, dissolved in Teflon vessels using HF or HCl, and U and Pb are chemically separated from other elements using ion exchange chromatography (Krogh 1973; Davis et al. 2003; Parrish and Noble 2003). Addition of a mixed U–Pb isotopic tracer of known composition and quantity, or isotope dilution, allows for correction of elemental fractionation during analysis on a mass spectrometer and determination of precise isotopic ratios as well as elemental abundances. The separated sample containing U and Pb after column chemistry is placed on a metal filament (e.g., Re), which is heated under vacuum in the source of the TIMS to ionize both U and Pb with the ions then accelerated into and analyzed by a magnetic sector mass spectrometer (Parrish and Noble 2003; Schoene 2014). The method is highly accurate and precise with very stable beams and measurement times of 2–3 h per sample, which are substantially longer than those for in situ techniques described below (e.g., SIMS, LA-ICP-MS). The vast majority of the reported ages for layered intrusions worldwide are based on the ID-TIMS methods for U–Th–Pb geochronology (Table 1.1).

Pretreatment Methods for Zircon

Zircon ($ZrSiO_4$) is the most widely used U–Th–Pb-bearing mineral in geochronology (e.g., Davis et al. 2003; Parrish and Noble 2003, and references therein) due to its presence in a wide range of igneous, metamorphic, and sedimentary rocks, to its stability under crustal and upper mantle conditions, and to its high closure temperature for diffusion of Pb (>1000 °C, Cherniak and Watson 2000). Zircon is, however, susceptible to radiation damage (e.g., Holland and Gottfried 1955; Ewing et al. 2003). The crystal structure can become metamict through self-irradiation from alpha recoil during the decay of U and Th, which produces defects in the crystal lattice and amorphous material (Rios et al. 2000). Such amorphous zones can be linked along microcracks to form passages that allow Pb (the daughter product of U and Th) to be lost from the crystal during interaction with hydrothermal or metamorphic fluids (e.g., Rios et al. 2000; Geisler et al. 2003; Nasdala et al. 2005), resulting in significant discordance of U–Pb results and young apparent U–Pb ages. A major issue in U–Pb geochronology has long been the identification and removal of Pb-loss zones in zircon (e.g., Tilton et al. 1957; Silver and Deutsch 1963; Krogh and Davis 1974). The technique of Krogh (1982a, b) allowed for removal of the outer, U+Th-rich layer of zircon crystals by air abrasion and for the determination of precise, concordant U–Pb results after careful grain selection. The more recently developed chemical abrasion technique (CA-TIMS) of Mattinson (2005) allows for removal of zircon domains that have lost Pb from the interior of grains and the analysis of low-U+Th, closed system residues with both concordant results and ages with significantly improved accuracy. For low to moderately damaged zircon, annealing and leaching is effective at removing altered discordant domains in zircon and the technique allows for determination of ages with a high degree of concordance. For highly damaged zircon, especially high-U Precambrian zircon, it may not be possible to fully restore primary U–Pb ages under any annealing conditions (Das and Davis 2010). The CA-TIMS technique provides very robust results for determining ages of layered intrusions to high precision and accuracy, although relatively few studies have reported U–Pb zircon CA-TIMS ages to date (Table 1.1).

Isotope Tracers, Synthetic Solutions, and Standards

The "EARTHTIME Initiative", initiated in 2003 with the goal of sequencing Earth history throughout geologic time by producing a highly resolved, calibrated geologic timescale, has been instrumental in spearheading significant improvements in high-precision geochronology (Bowring et al. 2011; Schmitz and Kuiper 2013). One of the main goals of EARTHTIME is to stimulate cooperation and intercalibration of U–Pb and $^{40}Ar/^{39}Ar$ laboratories worldwide through the use of common enriched isotope tracer solutions, or spikes, and synthetic Pb–U isotope solutions. A new $^{205}Pb-^{233}U-^{235}U$ spike (ET535) was made available to U–Pb labs, including the Pacific Centre for Isotopic and Geochemical Research (PCIGR) at the University of British Columbia (Vancouver, Canada) where the analyses in this study were conducted (Table 1.2), provided they met certain criteria following an international quorum with the different labs analyzing a range of reference materials. Another tracer that is available from EARTHTIME is the $^{202}Pb-^{205}Pb-^{233}U-^{235}U$ (ET2535) spike, which allows for in-run correction of instrumental Pb fractionation instead of relying on an external correction from repeat analyses of a Pb standard (e.g.,

Table 1.2 U–Pb geochronologic analytical techniques and protocols at PCIGR (UBC)

Variable	Pre-2008	Post-2008
Mass spectrometer	VG54R single collector, analog Daly	VG354S single collector, Sector 54 electronics, analog Daly
Pyrometer	Analog optical	Digital optical
Data collection	Peak-hopping on Daly, RL Armstrong (unpubl.)	Micromass Sector 54 IR software; employs Dodson (1978) 2nd-degree interpolation (release 3.977)
Spike	Parrish (1987) ^{205}Pb–$^{233-235}$U spike	EARTHTIME 535 ^{205}Pb–$^{233-235}$U spike
Pb standard	NBS-982	NBS-982
Reference	R33 zircon + others	EARTHTIME 100Ma, 500Ma, 2Ga synthetic solutions
Pb blanks	1.4–9 pg	0.4–1.7 pg
Data reduction	HTB—Roddick (1987)	Tripoli + U-Pb_Redux (Bowring et al. 2011) & UPbr (Schmitz and Schoene 2007)

NBS-981) and thus improved precision of ^{207}Pb/^{206}Pb ages. Using the same isotopic tracer allows for the direct comparison of analyses between labs without having to propagate errors related to spike calibrations and thus effectively eliminates inter-laboratory bias due to tracer calibration.

Another major component of EARTHTIME intercalibration process was the production of synthetic U-Pb solutions with concordant isotopic compositions and elemental ratios (corresponding to 10 Ma, 100 Ma, 500 Ma, and 2 Ga) for distribution between various U–Pb labs to monitor long-term mass spectrometer performance (Condon et al. 2007; Schmitz and Kuiper 2013). The solutions were prepared by iterative mixing of natural uranium (SRM 950a), ^{206}Pb (SRM 991) and ^{207}Pb (metal from Cambridge Isotopes) solutions. Determination of the ^{206}Pb –^{238}U and ^{207}Pb –^{235}U ratios is being achieved via calibration against the ET535 tracer. As these solutions are pure U and Pb, they do not require any chemical purification, thus the U–Pb loads can be prepared by mixing in a pre-cleaned beaker, which means that exposure to laboratory blank and effort in preparing loads is greatly reduced compared to using a mineral standard. These solutions have the potential to replace natural mineral standards (e.g., R33, Temora) for ID-TIMS, allowing for a more accurate evaluation of mass spectrometry behaviour because post-crystallization Pb-loss and natural compositional variation observed in zircon are not factors. PCIGR has been analyzing three synthetic U–Pb solutions since 2009 with the following results: ET-2 Ga, weighted mean ^{207}Pb/^{206}Pb date = 2000.01 ± 0.30 Ma (uncertainty quoted at 2s with internal errors); ET-500 Ma, weighted mean ^{206}Pb/^{238}U date = 500.19 ± 0.13 Ma; ET-100 Ma, weighted mean ^{206}Pb/^{238}U date = 100.08 ± 0.02 Ma (Figs. 1.4a, b, and c).

Additional outgrowths of EARTHTIME include the development of common, open-source software platforms or cyber infrastructure in data collection and analysis (Tripoli and U–Pb_Redux; Bowring et al. 2011; McLean et al. 2011) and derivation of isotope ratios, errors, and error correlations for U–Pb geochronology using the mixed ET535 (or ET2535) tracer (Schmitz and Schoene 2007), which are now in use at PCIGR. Other areas of potential analytical improvement include the characterization of Pb fractionation during thermal ionization mass spectrometry, which is the largest source of uncertainty in $^{207}Pb/^{206}Pb$ ages when analyzing Archean zircon grains due to the high amounts of Pb per analysis (can be over 1 ng). Over the past 5 years at PCIGR, we have characterized the Pb fractionation value of the commonly used Pb standard NBS-982 based on the $^{208}Pb/^{206}Pb$ and $^{207}Pb/^{206}Pb$ isotopic ratios measured on a VG354S single collector mass spectrometer with Sector 54 electronics and an analog Daly (Table 1.2). The calculated Pb fractionation value for 66 analyses is 0.25 ± 0.03 %/amu (2s, error based on $^{208}Pb/^{206}Pb$ as these are the most abundant isotopes for NBS-982) (Fig. 1.4d), which means that uncertainties as low as 0.02 % are now achievable provided all other sources of uncertainty (e.g., Pb blanks) are kept at a minimum (Table 1.2). An additional way to reduce analytical uncertainty related to Pb fractionation, even in sub-nanogram sample loads, is through the use of a double Pb spike (e.g., ^{202}Pb–^{205}Pb, Amelin and Davis 2006) or the mixed ET2535 tracer. Collectively, these improvements provide the opportunity to discriminate different magmatic episodes in the crystallization sequences of layered intrusions.

In Situ Methods (SIMS, LA-ICP-MS)

In addition to high-precision TIMS analysis using the chemical abrasion pretreatment technique, modern U–Th–Pb geochronology has been revolutionized by the advent of high-spatial resolution isotope analyses by secondary ionization mass spectrometry (SIMS) and laser ablation inductively coupled plasma mass spectrometry (LA-ICP-MS) (e.g., Ireland and Williams 2003; Nemchin et al. 2013). These in situ techniques allow for the determination of isotopic ratios from within single zones of crystals at scales of tens of microns that can target areas with distinctive textural, chemical, and isotopic signatures. The number of minerals than can be analyzed for U–Th–Pb geochronology by in situ methods is rapidly expanding and includes zircon, monazite, xenotime, apatite, titanite, baddeleyite, rutile, pervokskite, and allanite.

In a SIMS instrument, or ion microprobe (e.g., SHRIMP, CAMECA), a beam of high-energy primary ions (e.g., oxygen) is focused onto a polished target sample and a small amount of material is sputtered, typically with dimensions of 10–50 microns in diameter and less than 5 microns depth. Secondary ions are produced during sputtering and they are analyzed in the mass spectrometer (Ireland and Williams 2003). The SIMS technique is effectively non-destructive and ideal for analyzing complexly zoned zircon and for distinguishing the ages of inherited cores and magmatic and metamorphic overgrowths. The SHRIMP, or sensitive high-resolution ion microprobe, was specifically designed for the U–Th–Pb geochronology of zircon and the first results were reported in the early 1980s (e.g., Williams et al. 1983; Compston et al. 1984). In a LA-ICP-MS instrument, high intensity photons or laser radiation of uniform wavelength (e.g., 213- and 193-nm), phase, and polarization interact with a solid sample, which results in vaporization and ablation of atoms, ions, molecules, melt, and solid particles (e.g., Kosler and Sylvester 2003; Chang et al. 2006; Frei and Gerdes 2009; Fisher et al. 2010). This material is then introduced into an ICP-MS and isotope ratios analyzed by mass spectrometry. Compared to SIMS, the LA-ICP-MS technique involves destruction and removal of more material with typical spot sizes of 30–60 microns and depths of 10–20 microns, although recent studies are demonstrating successful application of single shot laser ablation analysis with small spot sizes (down to 6 microns) (Cottle et al. 2009, 2012). The availability, lower capital cost of the instrumentation compared to TIMS and SIMS, and the capability of obtaining ages from 100s of unknowns in a single analytical session has resulted in widespread application of U–Pb dating by LA-ICP-MS over the past 5–10 years (Fisher et al. 2010).

For geochronologic applications, both SIMS and LA-ICP-MS require comparison of measured values from a sample with those from matrix-matched reference materials that have been previously determined, typically from ID-TIMS or solution multiple collector-ICP-MS. As a result, the precision and accuracy of the analyses is limited to the precision and accuracy of the elemental ratios of interest in the reference materials being used with achievable precision typically not better than 2 % at the 2s level (Nemchin et al. 2013). Analyses by in situ techniques must also be designed to deal with changes in the measured isotopic ratios with time (e.g., sputtering-related changes in secondary ion emissions by SIMS, Ireland and Williams 2003; laser-induced fractionation of U relative to Pb by LA-ICP-MS, Fisher et al. 2010). The lack of chemical preparation, the ability to rapidly collect data, and the capability of determining age variations within mineral grains and structurally complex minerals make in situ analyses an essential dating technique where the scale of mineral (and age) heterogeneity is of concern. These techniques have as yet seen little application in dating the age of layered intrusions (Table 1.1) given the relatively high uncertainty compared to ID-TIMS (~2 % vs. <0.1 %) and typically simple zircon morphologies encountered (see below).

$^{40}Ar/^{39}Ar$ Geochronology

The $^{40}Ar/^{39}Ar$ dating method is a widely applicable and precise method of geochronology, with relative precision better than 0.1 % in many cases, that has been used to calibrate Cenozoic time scales, determine the age of emplacement of volcanic rocks, and constrain the cooling path of plutonic and metamorphic rocks (e.g., Renne et al. 1998; McDougall and Harrison 1999; Renne et al. 2010). $^{40}Ar/^{39}Ar$ dating, based on the ^{40}K–^{40}Ar method (i.e., decay of ^{40}K to ^{40}Ar), is applied to K-bearing minerals or rocks in which potassium is an essential structural constituent (e.g., micas, feldspars) or present in limited amounts (e.g., amphibole, <1 % K) (Harrison and Zeitler 2005). An $^{40}Ar/^{39}Ar$ date requires calculation of a $^{40}K/^{39}Ar$ ratio from isotopic measurements after a sample has been irradiated with fast neutrons to transform a proportion of the ^{39}K atoms to ^{39}Ar. A standard sample of known age (e.g., 28.305 ± 0.036 Ma Fish Canyon sanidine, Renne et al. 2010) is irradiated together with the unknown sample and the age of the unknown is derived by comparison with the $^{40}Ar/^{39}Ar$ of the standard (Harrison and Zeitler 2005). Systematic errors in the $^{40}Ar/^{39}Ar$ method arise from imprecise K and Ar isotopic data for standards and from uncertainties in ^{40}K decay constants (Renne et al. 1998; Min et al. 2000). When data pairs of $^{206}Pb/^{238}U$ and conventional $^{40}Ar/^{39}Ar$ ages are compared from the same samples, a systematic bias of older U–Pb ages is observed ranging from > 1.5 % for young rocks (Phanerozoic) to 0.5 % for rocks as old as 2 Ga (e.g., Renne et al. 2010, 2011). It is now possible to reconcile the two isotopic systems (Smith et al. 2010) by using the revised age for the Fish Canyon standard (Kuiper et al. 2008).

$^{40}Ar/^{39}Ar$ geochronology has been most successfully applied to young (Phanerozoic) intrusions, such as the Paleogene mafic layered intrusions related to the North Atlantic Igneous Province (e.g., Skaergaard, Kap Edvard Holm, Rum), where post-crystallization hydrothermal alteration and metamorphism have not resulted in open-system behavior. Hirschmann et al. (1997) reported the first precise date for the Skaergaard intrusion based on hornblende (55.48 ± 0.30 Ma) and biotite (55.40 ± 0.14 Ma) from a Transgressive granophyre from Upper Zone a of the Layered Series (Table 1.1) and used these cooling ages to evaluate the cooling history of the intrusion and the timing of the Skaergaard intrusion relative to eruption of the East Greenland basalts. Hamilton et al. (1998) reported the first precise dates from the Rum intrusion (Isle of Rum, Hebrides) from combined U–Pb zircon (60.53 ± 0.08 Ma: marginal alkaline segregation) (Table 1.1) and $^{40}Ar/^{39}Ar$ phlogopite (60.1 ± 1.0 Ma: gabbro pegmatite) systematics. These results indicated rapid cooling of the Rum intrusion and were used to calculate magma production rates for the Skye volcanic centre comparable to those for magmatism above upwelling mantle plumes such as Hawaii. $^{40}Ar/^{39}Ar$ geochronological results from Tegner et al. (1998) for plagioclase, biotite, amphibole, and whole rocks from a wide range of East Greenland layered intrusions document their ages of emplacement allowing for the recognition of discrete mantle melting episodes, the timing of coeval flood basalt magmatism and continental breakup, and even the potential links between massive outpourings of basalt and environmental impacts (e.g., Paleocene-Eocene Thermal Maximum, Storey et al. 2007).

U–Th–Pb-bearing Minerals in Layered Intrusions

U–Pb geochronology involves determination of ages for U–Th–Pb-bearing minerals (e.g., zircon, baddeleyite, monazite, titanite) to constrain timing of a wide range of geological events from magma crystallization to metamorphism to hydrothermal activity throughout all of geological history (e.g., Heaman and Parrish 1991; Davis et al. 2003; Parrish and Noble 2003). These accessory minerals are characterized by variable U concentrations, the amount of common Pb incorporated during crystallization, and closure temperatures (Table 1.3). Closure temperature is a complex function involving diffusivity of the element of interest (i.e., Pb), effective grain radius, which may not be the same as grain size, and cooling rate (Dodson 1973). For the U–Th–Pb-bearing minerals typically encountered in layered intrusions

Table 1.3 Geochemical characteristics of dateable accessory minerals in layered intrusions

Mineral	Chemical Formula	Closure Temperature (°C)	U Concentration (ppm)	Common Pb (ppm)	References
Zircon	$ZrSiO_4$	>950	1 to >10,000	<2	Cherniak and Watson (2000, 2003); Belousova et al. (2002)
Baddeleyite	ZrO_2	>950	1 to >3000	<2	Heaman and LeCheminant (1993, 2000)
Monazite	$(Ce,La,Nd,Th)PO_4$	>750	100–1000	<2	Cherniak (2010)
Titanite	$CaTiSiO_5$	600–650	20–1000	0.2–10	Frost et al. (2000); Cherniak (2010)
Apatite	$Ca_5(PO_4)_3(F,Cl,OH)$	425–500	8–100	<5–800	Cherniak (2000); Chamberlain and Bowring (2000)
Rutile	TiO_2	400–450	<<1–400	<2–100	Schmitz and Bowring (2003)

(Table 1.3), closure temperatures range from ~1000 °C down to 400 °C, which allows for the possibility of determining integrated crystallization and cooling ages. Key geochronological characteristics of the most common, and not so common, U–Th–Pb-bearing minerals that may be present in layered intrusions and indicate both successful and potential dating applications are briefly outlined below.

Zircon

Zircon, $ZrSiO_4$, a common accessory mineral occurring in a wide variety of terrestrial rocks (igneous, metamorphic, sedimentary) and extraterrestrial rocks (meteorites, lunar rocks, tektites), is one of the most widely used minerals for geochronology. Zircon is a refractory mineral and easily separated from crushed rock samples due to its high density (ρ=4.6–4.7 g/cm^3) and hardness (H=7.5). Zircon incorporates a wide range of trace elements with large ionic radii, including the heavy rare earth elements, Nb, Ta, Hf, and the actinides, U and Th (Belousova et al. 2002; Finch and Hanchar 2003). A fundamental aspect for geochronological applications is the incorporation of U (typical range of 10–1000 ppm) and Th (1–100 ppm) into the zircon structure during crystallization and the preferential exclusion of Pb (Parrish and Noble 2003) (Table 1.3) such that the Pb accumulated in zircon over time from the radioactive decay of parent U and Th isotopes is considered as radiogenic Pb (i.e.,

Pb*). The closure temperature for Pb diffusion in zircon is in excess of 950–1000 °C (Cherniak 1993, 2010; Cherniak and Watson 2000, 2003), which corresponds to the temperature range estimated for near-solidus conditions in layered intrusions (e.g., Lindsley et al. 1969; Morse 2008) based on experimental and field-based studies. Lead diffusion in crystalline zircon is so slow at geologically reasonable temperatures in the crust that Pb-loss cannot be attributed to volume diffusion through its crystal lattice, but rather is related to more rapid diffusion in metamict zircon that has been radiation-damaged during decay of U+Th through the release of a low-energy, but heavy, alpha-recoil nucleus (daughter isotope) or a high-energy, and light, alpha-particle (He nucleus) (Cherniak and Watson 2003; Geisler et al. 2007). Primary magmatic zircon can also be replaced by secondary hydrothermal zircon with significant elemental gains and losses, including both U and Pb (Ewing et al. 2003; Geisler et al. 2003, 2007; Nasdala et al. 2005) with important implications for zircon geochronology. Fortunately, careful grain examination during selection under a binocular microscope coupled with imaging by cathodoluminescence and backscattered electron techniques (e.g., Corfu et al. 2003) can be used to identify zircon that may not record primary crystallization ages. As noted previously, zircon is the mineral of choice for high-precision geochronology of layered intrusions (Table 1.1) given its relative abundance amongst all dateable accessory minerals, high U/Pb ratio upon crystallization, and high closure temperature similar to the temperatures of crystallization of zircon from late-stage, fractionated interstitial melt in mafic-ultramafic cumulates. In addition, heavy mineral separates from mafic-ultramafic rocks tend to be dominated by a single population of zircon without evidence for inheritance of zircon from earlier intrusive phases and are characterized by relatively simple zoning patterns. These grains do not typically record the complex, multi-stage crystallization histories that are so commonly preserved in zircon found in granitoids and other felsic rocks.

Baddeleyite

Baddeleyite, ZrO_2, occurs in many terrestrial and lunar rocks, and some meteorites, and is the mineral with the highest potential for U–Pb dating in mafic dikes and high-level mafic sills (Heaman and LeCheminant 1993). In mafic rocks, it is found in the late-stage, chemically fractionated regions with K-feldspar, mica, amphibole, and small amounts of quartz (Heaman and LeCheminant 1993). Baddeleyite is ideal for U–Pb dating yielding concordant high-precision ages due to the abundant U incorporated into the crystal lattice (generally <1500 ppm, but up to 3000 ppm), negligible initial common Pb, and an apparent high U–Pb closure temperature similar to that of zircon (Heaman and LeCheminant 1993, 2000; Lumpkin 1999) (Table 1.3). Baddeleyite has become a major tool for dating mafic rocks, especially dike swarms associated with large igneous provinces, and the resultant ages provide important constraints on supercontinent reconstructions and mantle dynamics (e.g., Ernst and Buchan 2001; Söderlund et al. 2010). The newly developed in situ U–Pb SIMS technique for dating micro-baddeleyite needles (<10–40 μ long by a few microns wide), which are difficult to recover with conventional concentration techniques, now allows for non-destructive dating with precisions ranging from 0.1 % for Precambrian rocks to 3–7 % for Phanerozoic rocks (Chamberlain et al. 2010; Schmitt et al. 2010). Post-crystallization Pb loss and intergrowth of baddeleyite and zircon resulting in discordant results has been documented (Heaman and LeCheminant 1993; Rioux et al. 2010). Recent advances in multi-step digestion have been used successfully to isolate magmatic baddeleyite from secondary zircon inter- and overgrowths so that igneous and subsequent hydrothermal or metamorphic events can now be resolved (Rioux et al. 2010). Discordance in baddeleyite may also involve several isotope-specific mechanisms, including (1) ^{206}Pb deficit due to disequilibrium in Th/U (i.e., low Th in baddeleyite), (2) ^{207}Pb excess from protactinium (Amelin and Zaitsev 2002), and (3) preferential loss of ^{206}Pb from radon loss (Davis and Sutcliffe 1985; Heaman and LeCheminant, 2000). Application of U–Pb baddeleyite dating to layered intrusions is still relatively restricted (e.g., Poe Mountain, Scoates and Chamberlain 1995; Muskox intrusion, Mackie et al. 2009, Fig. 1.1; Ilimaussaq, Krumrei et al. 2006; Sonju Lake, Paces and Miller 1993) as zircon is the principal Zr-bearing phase recovered and dated from layered intrusions worldwide (Table 1.1).

Apatite

Apatite, $Ca_5(PO_4)_3(F,Cl,OH)$, is a common accessory mineral in many igneous rocks, including mafic-ultramafic rocks in layered intrusions (e.g., Poe Mountain intrusion, Fig. 1.2f) where it ranges in composition from fluorapatite-hydroxylapatite in most cumulates to anomalous chlorapatites in the sequences below the PGE reefs in both the Stillwater and Bushveld complexes (e.g., Boudreau et al. 1986; Boudreau and Hoatson 2004). Apatite is a relatively low-U mineral (< 100 ppm), with lattice-bound U and Pb, and closure temperatures to Pb diffusion in the range of 425–500 °C (Cherniak et al. 1991; Chamberlain and Bowring 2000) (Table 1.3), depending on diffusion radii and cooling rates. These characteristics make apatite an important thermochronometer for investigating the cooling history of layered intrusions, however, there are as yet few studies that exploit its potential. Because apatite can accommodate significant amounts of initial Pb in its crystal structure, the ratio of radiogenic Pb to common Pb can be very low and thus the accuracy and precision of U–Pb apatite ages requires careful selection of initial Pb isotopic compositions in age calculations (e.g., co-existing feldspar) (Chamberlain and Bowring 2000).

Rutile

Rutile, TiO_2, is the most common naturally occurring titanium dioxide polymorph and is found in a wide range of rocks as an accessory mineral, including granitoids, metamorphic rocks, mantle rocks, and meteorites (Meinhold 2010). Rutile is a major host for Nb, Ta, and other high field strength elements and it contains sufficient U (up to 100 ppm) making precise U–Pb geochronology possible (e.g., Mezger et al. 1989; Luvizotto et al. 2009). However, rutile can incorporate significant non-radiogenic or common Pb during crystallization (Table 1.3). This common Pb must be corrected for either by analyzing a coexisting leached feldspar for common Pb (Mezger et al. 1989), using a Pb growth model (Stacey and Kramers 1975), or using the measured $^{238}U/^{206}Pb$ and $^{207}Pb/^{206}Pb$ ratios (Tera and Wasserburg 1972). The closure temperature to Pb diffusion in rutile is variable. Experimental Pb diffusion data for rutile from Cherniak (2000, 2010) indicate relatively high temperatures (600–700 °C), whereas field-based studies indicate lower temperatures (400–450 °C) when compared to co-existing U–Pb and $^{40}Ar/^{39}Ar$ chronometers (Mezger et al. 1989; Schmitz and Bowring 2003).

In igneous rocks, rutile is most commonly found in granites and associated quartz veins, pegmatites, carbonatites, kimberlites, and metallic ore deposits (Meinhold 2010). Rutile is rarely reported in layered intrusions. It has however been noted throughout the Critical Zone of the Bushveld Complex (Cameron 1979), mostly associated with chromite and as inclusions in plagioclase and orthopyroxene. It is present in bronzitites, orthopyroxenites, of the Great Dyke (Oberthür et al. 2002), where interstitial quartz, K-feldspar, and phlogopite are present, the latter containing inclusions of coarse rutile. We have observed that rutile is a common accessory in many ultramafic and mafic cumulates from both the Stillwater Complex and Bushveld Complex. U–Pb geochronological studies of rutile from layered intrusions are rare. Metamorphic rutile ages (~650 Ma) are reported for the Niquelandia layered intrusion in central Brazil (Ferreira-Filho et al. 1994). To our knowledge, only two studies provide U–Pb rutile dates for magmatic rutile from layered intrusions: (1) a 2575.0 ± 1.0 Ma age from sulphide-bearing bronzitites of the Main Sulphide Zone of the Great Dyke, which overlaps within uncertainty the U–Pb zircon age of 2575.9 ± 1.0 Ma from the same sample (Oberthür et al. 2002), and (2) a 2055.0 ± 3.9 Ma age from the Merensky Reef of the Bushveld Complex, where rutile occurs mainly as inclusions within or overgrowths on chromite, especially in the chromitite stringers that bound the reef, and also as acicular grains with biotite (Scoates and Friedman 2008).

Titanite

Titanite, $CaTiSiO_5$, is a U–Pb-bearing mineral that is commonly found in mafic and calc-silicate metamorphic rocks and in relatively oxidized igneous rocks (Frost et al. 2000). Titanite is useful for U–Pb dating as U concentrations in igneous rocks vary from 10s to 100s of ppm, which combined with relatively low 0.2–10 ppm common Pb (Table 1.3) allows for the possibility of determining high-precision U–Pb ages (Frost et al. 2000). The closure temperature to Pb diffusion in titanite is high (>600 °C; Cherniak 1993, 2010) making it a valuable chronometer for reconstructing the high-temperature cooling history of intrusions. U–Pb titanite dating has been successfully applied to ancient and young felsic rocks, ranging from Archean granitoids (Corfu et al. 1985) to the ca. 28 Ma Fish Canyon Tuff (Hemming and Rasbury 2000; Chew et al. 2014). There are as of yet few published studies using U–Pb titanite for dating layered intrusions, although titanite may be a relatively common mineral in late-stage granophyric segregations that occur locally in many intrusions (e.g., Stillwater granophyres, Wall 2009). A notable exception is the Bushveld Complex where Buick et al. (2001) reported a U–Pb titanite age of 2058.9 ± 0.8 Ma (weighted mean $^{207}Pb/^{206}Pb$ age of three multigrain fractions) from a hydrothermally retrogressed calc-silicate xenolith in the Upper zone of the Bushveld Complex. Assuming that fluid flow and titanite growth in the xenolith were directly related to Bushveld magmatism, they proposed that this metamorphic titanite age restricted crystallization of the layered mafic-ultramafic rocks of the Bushveld Complex to the interval of 2059–2061 Ma.

Other Minerals

Two other minerals that have been identified in mafic-ultramafic rocks may also be potential U–Pb chronometers. Zirconolite, $CaZrTi_2O_7$, is a rare U-bearing oxide mineral that has been reported in lunar basalts and meteorites, in silica undersaturated alkaline rocks, in mafic dikes, and in a few layered intrusions (Heaman and LeCheminant 1993; Rasmussen and Fletcher 2004; Zaccarini et al. 2004). Zirconolite from mafic intrusions in Australia typically occurs as euhedral laths and needles (<200 microns long and <20 μ wide), is light tan to dark brown to opaque, and contains several wt% $UO_2 + ThO_2$ and up to 15 wt% combined REE (Rasmussen and Fletcher 2004). U–Pb SHRIMP dating of zirconolite from mafic and lunar rocks shows that it typically contains minimal common Pb and can produce precise and tightly grouped $^{207}Pb/^{206}Pb$ dates due largely to the high ionization efficiency for Pb^+(Rasmussen and Fletcher 2004; Norman and Nemchin 2014). Loveringite, $(Ca,Ce)(Ti,Fe,Cr,Mg)_{21}O_{38}$, a U-bearing Ca-end-member of the crichtonite group, was discovered in orthopyroxene-rich cumulates of the Jimberlana intrusion, Western Australia (Gatehouse et al. 1978), and has been found in gabbroic cumulates from ophiolites (Cabella et al. 1997) and in some layered intrusions in Finland and Russia (e.g., Koitelainen, Tarkian and Mutanen 1987; Penikat, Alapieti 1982; Barkov et al. 1996). In the Jimberlana intrusion, loveringite is black, occurs as rare anhedral to needle-like grains (50 × 100 microns), is closely associated with quartz, K-feldspar, or phlogopite, and contains 0.18–0.34 wt% UO_2 and 0.09–0.53 wt% ThO_2 (Campbell and Kelly 1978). There are as yet no published U-Pb geochronological determinations of loveringite. It is possible that both zirconolite and loveringite are more widespread in layered intrusions than their documented occurrences suggest, however, their very fine grain size, trace abundance, and morphological similarities to rutile and baddeleyite to date hinder their routine separation and identification.

Sampling Strategy for Dateable Accessory Minerals in Layered Intrusions

Previous Geochronologic Investigations of the Stillwater Complex

Using the Stillwater Complex as a test case, we describe the criteria to successfully identify samples in mafic-ultramafic cumulates in layered intrusions that are likely to yield small quantities of dateable accessory minerals. The Stillwater Complex is a large fragment of a mafic-ultramafic layered intrusion in the Beartooth Range (Montana, USA) of the Archean Wyoming Province (Fig. 1.5a) and represents a crystallized sub-volcanic magma chamber that fed voluminous volcanic rocks (Hess 1960; Jackson 1961; McCallum, 1996). From base-to-top, the intrusion is subdivided into the thin Basal Series, the 2000 m-thick Ultramafic Series with a lower Peridotite Zone and upper Bronzitite Zone, and the overlying Banded Series, which consists of the Lower Banded Series (host to the PGE-rich J-M Reef horizon), Middle Banded Series, and Upper Banded Series (Fig. 1.5b). Determining the precise age of crystallization of the Stillwater Complex has been the focus of studies since the late 1960s (e.g., Fenton and Faure 1969; Nunes and Tilton 1971), but has been difficult to achieve. As noted above, DePaolo and Wasserburg (1979) produced a relatively precise internal Sm–Nd isochron for mineral-whole rock analyses from a gabbronorite of 2701 ± 8 Ma. A 100-pound (45 kilograms) sample of norite from the Basal Series yielded abundant zircon (Nunes and Tilton 1971) from which Nunes (1981) published a revised age of 2713 ± 3 Ma, an upper intercept $^{207}Pb/^{206}Pb$ age from highly discordant analyses of multi-grain zircon fractions (up to 1 mg) that is slightly older than the Sm-Nd isochron age. Premo et al. (1990) reported a range of U–Pb ages for zircon separated from footwall mafic sills and dikes (2711–2712 Ma) and a composite (7 multi-grain fractions) zircon-baddeleyite age (2705 ± 4 Ma) for two samples from the Lower Banded Series. The uncertainty of this age was reduced to 2704 ± 1 Ma ($^{207}Pb/^{206}Pb$ upper intercept ages) if the results of two fractions were omitted. Based on their associated uncertainties, these studies indicate a possible range of ~ 10–20 million years at ca. 2.7 Ga for crystallization of the Stillwater Complex.

U–Th–Pb-bearing Accessory Minerals in Stillwater Cumulates

To constrain the duration of magmatism in the Stillwater Complex, to test whether there were distinct magmatic episodes during emplacement of the intrusion, and to constrain the temporal relationship of PGE mineralization (J-M Reef) to the cumulates, we initiated a detailed U–Pb geochronological study of the Stillwater Complex from all the major units of the intrusion and from cross-cutting granophyres (Fig. 1.5). Samples were collected over the course of two field seasons. In summer

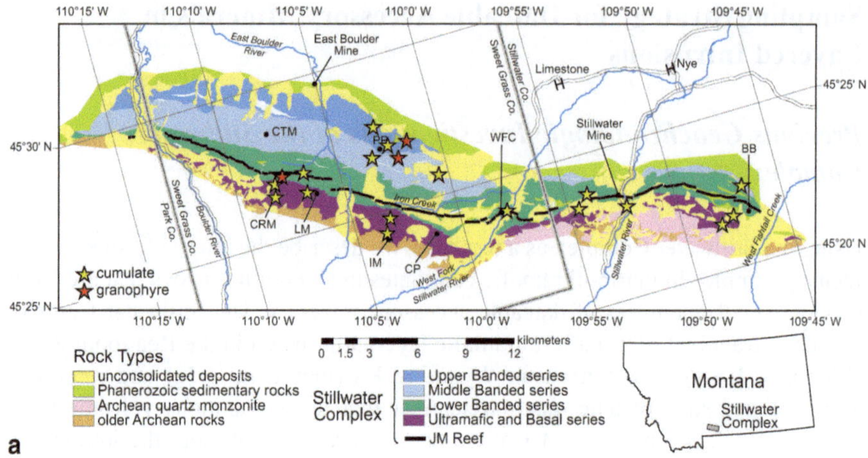

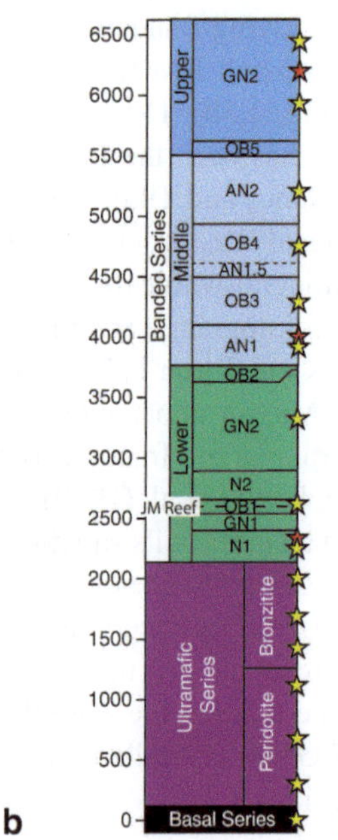

c

Zircon	Baddeleyite	Rutile	Apatite	Titanite	Weight (kg)
+	+	-	+	-	27.6
+	-	+	+	+	18.2
-	-	-	-	-	15.5
+	+	+	+	-	50.2
-	-	-	+	-	15.3
+	+	-	+	-	11.5
+	+	+	+	+	20.0
-	-	-	+	-	23.4
+	+	-	+	-	23.4
+	+	+	+	+	16.8
+	-	+	+	-	13.2
-	-	-	-	-	16.8
+	-	-	+	-	17.1
+	+	-	+	-	14.5
+	+	-	+	-	23.8
+	+	-	+	-	9.5
+	-	-	+	-	19.3

Fig. 1.5 Locations and mineral yields for U–Pb geochronological samples collected in the Stillwater Complex (Montana, USA). **a** Simplified geological map of the Stillwater Complex showing sample locations (adapted from Zientek et al. 2005); *stars* indicate sample locations. Also shown are the locations of the two PGE mines, Stillwater and East Boulder Mine. Abbreviations: *CTM* Contact Mountain; *CRM* Chrome Mountain, *IM* Iron Mountain; *PP* Picket Pin, *CP* Castle Point; *WFC* West Fork Creek, *BB* Benbow. **b** Stratigraphic section after McCallum (1996) showing the relative locations of samples—stratigraphic height in metres above the basal contact of the intrusion. **c** Table indicating range of U–Pb-bearing accessory minerals separated from each of the samples (+ = mineral recovered; − = mineral not recovered) and weight in kilograms of samples that were processed (range from 9.5 to 50.2 kg)

2005, the first set of samples ($n=8$) was collected from the Picket Pin, Chrome Mountain, and Contact Mountain areas of the high plateau in the west-central part of the Stillwater Complex and the sampling focused mainly on the gabbroic rocks of the Banded Series (Fig. 1.5a). Samples were selected based primarily on accessibility and with knowledge of existing whole rock trace element concentrations for samples from the two olivine-bearing zones (OB3 and OB4) in the Middle Banded Series (Meurer and Boudreau 1996; Meurer et al. 1999; WP Meurer, unpublished data). Samples with elevated incompatible trace element concentrations (e.g., Zr, U, Th, REE) were assumed to contain higher relative abundances of minerals that crystallized from fractionated interstitial melt and thus had an enhanced potential to contain zircon or baddeleyite. Based on these criteria, the success rate was approximately 50 % for the cumulates (four out of eight mafic-ultramafic samples yielded zircon ± baddeleyite) and 100 % for the four granophyre samples (all contained zircon ± rutile and titanite).

In summer 2011, sampling ($n=14$) was focused primarily on the Ultramafic Series as well as remaining gaps in the Banded Series, including samples from the uppermost units of the intrusion just below the unconformity with overlying Paleozoic and Mesozoic sedimentary rocks (Fig. 1.5a). Using knowledge gained from sampling mafic-ultramafic rocks in the intervening years (e.g., Bushveld Complex, Scoates and Friedman 2008; Muskox intrusion, Mackie et al. 2009; Thompson Nickel Belt, Scoates et al. 2010; Bird River Sill, Scoates and Scoates 2013), we targeted samples based primarily on their textural features (Fig. 1.6). Outcrops with heterogeneous structures (e.g., sulphide-bearing Basal Series norite, Fig. 1.6a), outcrops that contained coarse-grained (pegmatitic) patches, especially with coarse interstitial plagioclase (e.g., feldspathic pyroxenites and troctolites, Figs. 1.6b, c, and d), or plagioclase-rich rocks with relatively abundant and irregularly distributed poikilitic to sub-poikilitic pyroxene (ortho- and clinopyroxene) (Figs. 1.6e and f) were preferentially sampled. Outcrops that were avoided included those with no evidence for interstitial plagioclase (e.g., knobby peridotite with orthopyroxene oikocrysts, Fig 1.7a), with monomineralic rock types (e.g., orthopyroxenite pegmatite, Fig. 1.7b), consisting of one, two-, and three-phase cumulates with only minor interstitial material (i.e., "adcumulates" in the terminology of Wager et al. 1960) (e.g., bronzitite, Fig. 1.7c; gabbronorite, Figs. 1.7d and 1.7f), and plagioclase-rich rocks dominated by large (up to 5 cm) orthopyroxene oikocrysts (Fig. 1.7e). As a result, recovery of zircon (±baddeleyite) was over 90 % successful (13 out of the 14 samples processed) (Fig. 1.5c), including nearly all of the samples from the Ultramafic Series. Anorthosites, or other plagioclase-rich rocks, from the Banded Series are particularly good targets when multiple poikilitic to oikocrystic phases are present. Haskins and Salpas (1992) noted from their study of the geochemistry of ANII in the Middle Banded Series that samples with elevated incompatible trace element contents were richest in cumulus plagioclase and in intercumulus plagioclase and pyroxene. Preliminary results from the dated Stillwater samples indicate that the Banded Series crystallized at ca. 2709 Ma (Wall et al. 2010) and on-going results constrain the total duration of magmatism to a 3–4 million year window from 2709 to 2712 Ma (Wall et al. 2013).

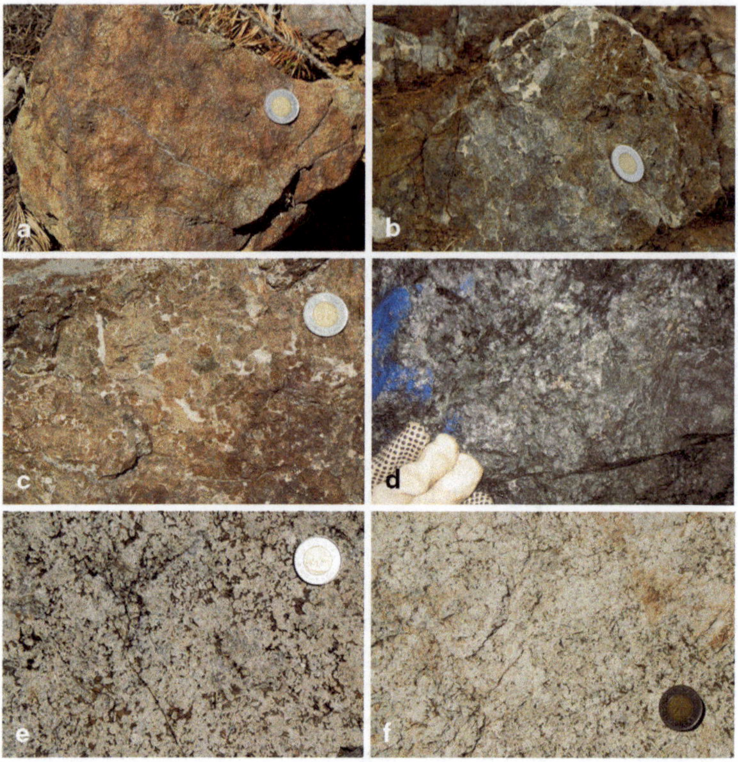

Fig. 1.6 Photographs of outcrops in the Stillwater Complex showing the macroscopic textures of successful sample targets that yielded zircon or baddeleyite for U–Pb dating (scale: *coin* is 2.75 cm in diameter). **a** Sulphide-bearing norite, Basal Series (Benbow). **b** Pegmatitic feldspathic orthopyroxenite *above* Chromitite J, Peridotite Zone (Benbow). **c** Pegmatitic feldspathic orthopyroxenite (note large *green* clinopyroxene), base of Bronzitite Zone (Iron Mountain). **d** Pegmatitic feldspathic olivine gabbronorite, J-M Reef (East Boulder Mine), Olivine-bearing zone I, Lower Banded Series; note coarse interstitial sulphide clots. Glove for scale (*blue* color is paint). **e** Leucogabbronorite with irregular poikilitic ortho- and clinopyroxene, Anorthosite-II zone, Middle Banded Series (Picket Pin). **f** Anorthosite with minor interstitial ortho- and clinopyroxene, uppermost Gabbronorite-III zone, Upper Banded Series (Picket Pin)

Dating the Merensky Reef in the Bushveld Complex

Background

The Paleoproterozoic Bushveld Complex in the northern Kaapvaal craton of South Africa (Fig. 1.8) (Cawthorn, Chap. 12, The Bushveld Complex) is host to some of the richest PGE, chromium, and vanadium deposits on Earth (Lee 1996) and currently produces the majority of the world's platinum (~80%), primarily from two stratiform sequences, the Merensky Reef and the Upper Group 2 (UG2) chromitite (Cawthorn et al. 2005). The age of the Bushveld Complex and of these major min-

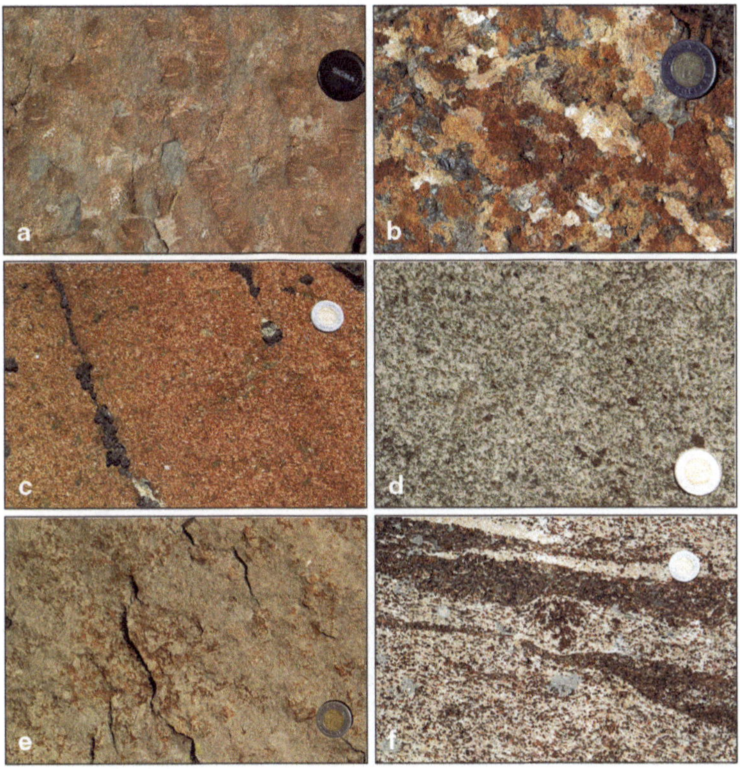

Fig. 1.7 Photographs of outcrops in the Stillwater Complex showing the macroscopic textures of sample targets that failed to yield zircon or baddeleyite or that are not considered prospective for U–Pb dating (scale: *coin* is 2.75 cm in diameter). **a** Peridotite with abundant orthopyroxene oikocrysts, Peridotite zone (Iron Mountain); lens cap is 7.25 cm in diameter. **b** Orthopyroxenite pegmatite, Peridotite zone (Chrome Mountain). **c** Clinopyroxene-bearing orthopyroxenite, Bronzitite zone (Chrome Mountain). **d** Gabbronorite, Gabbronorite-I (*below* J-M Reef), Lower Banded Series (West Fork); note that this is the sample locality of STL-100 from DePaolo and Wasserburg (1979). **e** Anorthosite with large orthopyroxene oikocrysts, Anorthosite-I, Middle Banded Series (Picket Pin). **f** Layered gabbronorite, Gabbronorite-III, Upper Banded Series (Picket Pin)

eralized horizons has long been of interest with comprehensive summaries provided in Walraven et al. (1990), Buick et al. (2001), and Nomade et al. (2004). Scoates and Friedman (2008) reported the first published precise age of crystallization for the Bushveld Complex by the CA-TIMS technique from a sample of the Merensky Reef (U–Pb zircon, 2054.4 ± 1.3 Ma, weighted mean $^{207}Pb/^{206}Pb$ age). In the section below, we provide an integrated U–Pb zircon/rutile/apatite and $^{40}Ar/^{39}Ar$ biotite geochronological approach from two widely separated (~310 km) samples of the Merensky Reef, including a revised value for the Scoates and Friedman (2008) age based on the significant changes in analytical protocols described in previous sections (Table 1.2).

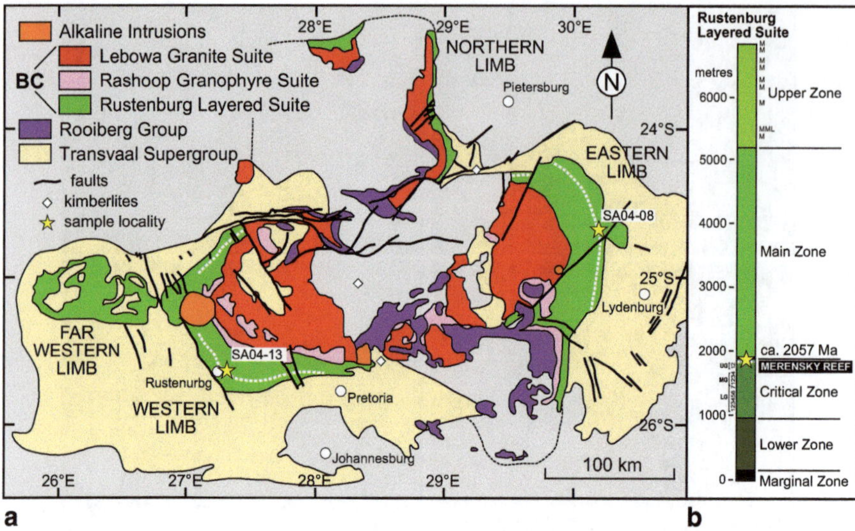

Fig. 1.8 Locations and geologic context of Merensky Reef samples in the Bushveld Complex. **a** Generalized geologic map of the Bushveld Complex (*BC*) showing the position of the Merensky Reef (*white dashed* line) and sample locations (*yellow* stars) (map modified from Kinnaird et al. 2005). Sample SA04-13 was collected underground from the Merensky Reef in West Mine (Rustenburg area). Sample SA04-08 was collected on surface from the Merensky Reef as exposed on the Farm Driekop. **b** Simplified stratigraphic section showing the major subdivisions of the layered mafic-ultramafic rocks (Rustenburg Layered Suite), the position of the platiniferous Merensky Reef near the *top* of the Critical zone, and the approximate positions of major chromitite horizons (*LG* Lower Group, *MG* Middle Group, *UG* Upper Group) in the Critical zone and major magnetite horizons in the Upper zone (*MML* main magnetite layer, *M* magnetite horizon)

Geologic Setting of the Merensky Reef in the Bushveld Complex

The Bushveld Complex consists of an 400 km-wide, 7- to 9-km-thick sequence of layered mafic-ultramafic rocks (referred to as the Rustenburg Layered Suite), which may be a shallowly dipping, interconnected sheet (Webb et al. 2004), that is associated with pre- and syn-layered suite marginal sills and intrusions (e.g., Eales and Cawthorn 1996; Kruger 2005) (Fig. 1.8a). The Bushveld Complex was emplaced at shallow levels in the crust (0.15–0.25 GPa: Pitra and de Waal 2001) into Paleoproterozoic rocks of the Transvaal Supergroup and the ca. 2061 Ma Rooiberg Group (Walraven 1997), a 6-km-thick sequence of precursor basaltic andesites to rhyolites (Buchanan et al. 2002; Mathez et al. 2013) that defines the upper age limit for the complex. The layered mafic-ultramafic rocks themselves are overlain and locally cut by a 2.5-km-thick sequence of ca. 2054 Ma granites (Rashoop Granophyre Suite and Lebowa Granite Suite: Walraven and Hattingh 1993).

Rocks of the Bushveld Complex are distributed in four discrete regions (Fig. 1.8a). The 200 km-long Eastern Limb contains the best surface exposures of the complex, including the Merensky Reef. The Western Limb, also ~200 km in length, generally has poor outcrop, although it has historically been the focus of much of the active mining for PGE due its higher average grades. The Northern Limb is partially covered beneath younger rocks and is notable for the presence of the PlatReef, a pyroxenitic-harzburgitic PGE–Cu–Ni-bearing package (up to 300 m-thick) that is situated directly on, or within a few tens of metres, of the contact with country rocks (e.g., Kinnaird et al. 2005; Mitchell and Scoon 2012). The least known of the regions, the Far Western Limb, is a poorly exposed eroded remnant of the Bushveld Complex. The Rustenburg Layered Suite, dominated by mafic-ultramafic rocks, is subdivided into the Marginal, Lower, Critical, Main, and Upper zones based on a variety of criteria (e.g., Eales and Cawthorn 1996), but most commonly on the appearance of a new cumulus mineral (Fig. 1.8b).

The PGE-rich Merensky Reef and UG-2 chromitite are both hosted in the Critical zone of the lower part of the Bushveld Complex. The Merensky Reef is a distinctive 40 to 120 cm-thick minable layer with a relatively constant grade of PGE (5–8 g/t or ppm) that occurs at the top of the Upper Critical zone and can be traced for several 100 km in the western and eastern limbs of the intrusion (white dashed line in Fig. 1.8a) (Cawthorn et al. 2005). The Reef is highly variable in thickness, structure, and texture and it can display strong vertical and lateral lithologic variations (Smith et al. 2003; Mitchell and Scoon 2007). The Merensky Reef in its highest or "normal" stratigraphic position in the Western limb consists of sulphide-bearing (2–4 vol %), coarse-grained (pegmatitic) feldspathic orthopyroxenite or melanorite bounded by thin chromitite stringers with the PGE concentrated near the chromitites (e.g., Barnes and Maier 2002; Smith et al. 2003; Cawthorn and Boerst 2006). The footwall to "normal" Reef facies varies from norite to anorthosite and the hanging wall is typically orthopyroxenite. The Merensky Reef in the Eastern limb shares some structural and compositional similarities with the classic localities described from the Western limb, although it is also characterized by some important differences. At Winnaarshoek, the 10–15 m-thick Reef package contains a 2–5 m-thick basal layer of feldspathic orthopyroxenite that includes the 1.8 m-thick Reef, which is bounded by thin chromitites (Mitchell and Scoon 2007). The Reef package here is underlain by layered norites and overlain by a sequence of plagioclase-rich rocks (norite-leuconorite, spotted-mottle anorthosite, mottled anorthosite). Pegmatitic feldspathic orthopyroxenites are known to occur in the Eastern limb, but unlike the Western limb they are not commonly mineralized.

Sample Locations and Characterization

Samples of pegmatitic feldspathic orthopyroxenite (SA04-11 and SA04-13) from the Western limb of the Rustenburg Layered Suite were collected on May 4, 2004 (erroneously given as April 4, 2004 in Scoates and Friedman 2008), from the mining face of the Merensky Reef (Fig. 1.9a) in the West mine, Townlands shaft (now known as the Khuselaka platinum mine, Khuseleka 1 shaft [25°37'29.1"S, 27°15'31.9"E]; level 22, ~850 m below surface), Rustenburg mining section. In the sampled area,

Fig. 1.9 Photographs showing sample localities of the Merensky Reef in the Bushveld Complex. **a** Photograph of the mining face of the Merensky Reef in the West Mine (Townlands shaft; level 22, ~850 m below surface), Rustenburg mining section, showing the coarse grain size and heterogeneous texture of the pegmatitic feldspathic orthopyroxenite. The *lower* bounding chromitite is a thin dense layer 2–3 cm thick, whereas the *upper* chromitite is an irregular diffuse layer ~2 cm thick. **b** Inset photograph focuses on the footwall contact of the Reef and underlying leuconorite; note 5 cm-thick anorthosite horizon immediately below lower chromitite and irregular anorthosite layers ~80 cm below lower chromitite. Photographs courtesy of Pat Hayman; hammer for scale. **c** Photograph of the Merensky Reef in outcrop at the Farm Driekop, Eastern Limb. Hand for scale

Reef thickness varies from ~5 to 40 cm (Figs. 1.9a and b). A sample (SA04-08) of the Merensky Reef exposed at surface from the Eastern Limb was collected on May 2, 2004, at Farm Driekop (24°31'11"S, 30°05'11.4"E) (Fig. 1.9c). The sample site is referenced in Bristow et al. (1993, Locality 3.4, p. 26) and the Reef was exposed in three old adits near the crest of a small hill at the time of sampling. The sample was collected from the northern edge of the adit from medium-grained equigranular pyroxenite beneath patchy pegmatitic feldspathic pyroxenite (Fig. 1.9c).

Sample SA04-13, a pegmatitic feldspathic orthopyroxenite, consists of large orthopyroxene crystals (0.5–3 cm diam), many displaying euhedral habit, enveloped by interstitial plagioclase that locally occupies pockets up to 5 cm across (Figs. 1.10a, b). Clots of biotite (0.5–1 cm), locally associated with zircon and interstitial quartz, and sulphide aggregates up to 0.5 cm diameter are heterogeneously distributed throughout the sample. Small (<0.4 mm) chromite grains occur in clusters and are concentrated in an ~5 mm thick stringer with plagioclase and orthopyroxene along the base of the pegmatitic feldspathic orthopyroxenite. Sample SA04-08, a variably equigranular and pegmatitic feldspathic pyroxenite, consists of small to large (0.1–2 cm diam) orthopyroxene crystals with both interstitial (poikilitic) clinopyroxene and plagioclase (Figs. 1.10c and 1.10d). Irregular rounded olivine grains (resorbed) occur within large orthopyroxene crystals. Minor interstitial biotite and quartz occur on the edges of the finer grained orthopyroxene and are locally intergrown with zircon. Only rare chromite grains (1–2 per thin section) are present in sample SA04-08 and there is no sulphide.

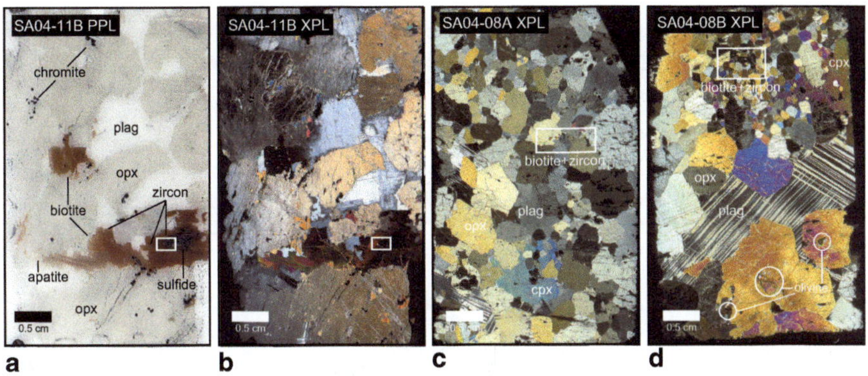

Fig. 1.10 Images of scanned thin sections for dated samples from the Merensky Reef (all sections are standard size, 45 × 25 mm). **a** Section SA04-11B (PPL) from the West Mine, Western Limb, showing the relative sizes and distribution of the main minerals (orthopyroxene, *opx*; plagioclase, *plag*; biotite, bio; *chromite*) and the locations of *zircon* and *apatite; white box* indicates location of zircon rim around *sulphide* shown in Fig. 1.11a. **b** Section SA04-11B (XPL) showing alteration along fractures in orthopyroxene and deformation in interstitial plagioclase (bent twins); *white box* indicates location of zircon rim around sulphide shown in Fig. 1.11a. **c** Section SA04-08A (XPL) from the Farm Driekop, Eastern Limb, showing equigranular orthopyroxene and clinopyroxene (*cpx*) with minor interstitial plagioclase; *white box* outlines area with minor late *interstitial biotite* and *zircon*. **d** SA04-08B (XPL) from the Farm Driekop, Eastern Limb, of feldspathic pyroxenite (both *opx* and *cpx*) with abundant large interstitial patches of plagioclase and no biotite; *white box* outlines area with minor late *interstitial biotite* and *zircon*

Accessory Minerals in the Merensky Reef

Both of the studied samples from the Merensky Reef, SA04-08 and SA04-13, are ideal targets for geochronology of layered intrusions as they are texturally heterogeneous with abundant interstitial plagioclase and local patches of biotite identifiable in hand sample (Figs. 1.9, 1.10). In thin section, both samples also reveal a wealth of potentially dateable accessory minerals that were identified and characterized prior to sample processing for mineral separates (Fig. 1.11).

Zircon, rutile, apatite, and biotite occur in the Merensky Reef (Figs. 1.11a, b, c, d, e, f, g, h, i and j), typically associated with interstitial quartz and alkali feldspar/plagioclase, locally as granophyric or myrmekitic intergrowths (Figs. 1.11k, and l). Zircon (~20–500 microns) is most commonly intergrown with interstitial biotite and, in one example, observed to rim sulphide and is in turn enveloped by biotite (Figs. 1.11a, b, c and d). In most cases, zircon is anhedral or irregularly shaped in the Merensky Reef (Figs. 1.11a, b and c), although examples of euhedral morphology are present (Fig. 1.11d). Rutile was found throughout the Critical zone of the Bushveld Complex by Cameron (1979), but has rarely been reported in the more recent literature. We find rutile in most samples from the Bushveld Complex that we have examined. In sample SA04-13, rutile occurs primarily as large (150–250 microns) equant to irregular grains associated with chromite, included within, as

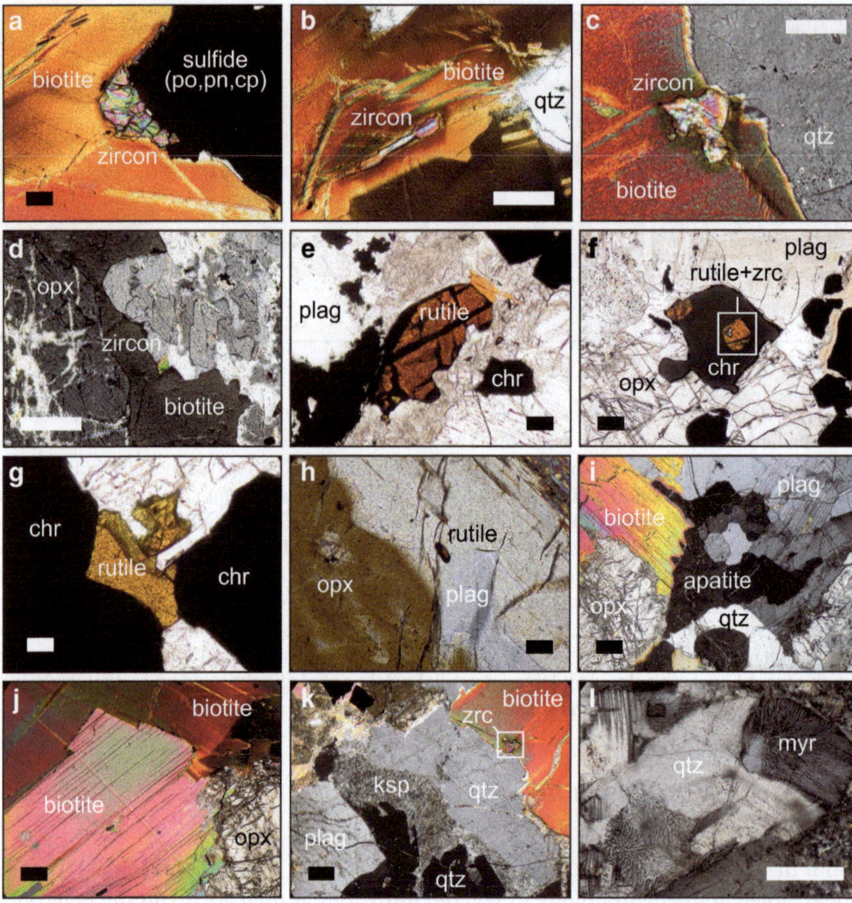

Fig. 1.11 Photomicrographs of accessory and other interstitial minerals in samples of the Merensky Reef (scale bar is 200 microns long in each panel). **a** Large grain of zircon on the margin of a composite sulphide (*po* pyrrhotite, *pn* pentlandite, *cp* chalcopyrite) encased in biotite (SA04-11B, XPL). **b** Zircon along cleavage plane in biotite with interstitial quartz (*qtz*) (SA04-11B, XPL). **c** Irregular zircon enclosed in the rim of biotite and in contact with interstitial quartz (SA04-11B, XPL). **d** Euhedral zircon on the margin of biotite in contact with orthopyroxene (*opx*) (SA04-11B, XPL). **e** Large single grain of rutile (brown-colored, high-relief) enclosed by orthopyroxene from near the base of the pegmatitic feldspathic orthopyroxenite (plagioclase, plag; chromite, *chr*) (SA04-13, PPL). **f** Rutile associated with chromite from the lower chromitite stringer; rutile apparently included in chromite contains an inclusion of zircon (zrc) (SA04-13, PPL). **g** Rutile on the edges of several chromite grains in contact with orthopyroxene (SA04-13, PPL). **h** Small euhedral crystal of rutile at the contact between orthopyroxene and plagioclase (SA04-08, XPL). **i** Large composite apatite cluster (at extinction) with interstitial biotite, quartz and plagioclase (SA04-11B, XPL). **j** Intergrown biotite grains (note euhedral outline to highly birefringent grain) showing deformation (*bent cleavage planes in lower centre*) and incipient alteration along the grain margin in contact with orthopyroxene (SA04-11B, XPL). **k** Interstitial biotite, quartz, and granophyre patch (quartz is at extinction, K-feldspar [*ksp*] is partially altered to sericite); note zircon in biotite rim (same as panel C) (SA04-11B, XPL). **l** Interstitial quartz and myrmekite (myr; vermicular intergrowth of quartz and plagioclase) (SA04-11B, XPL)

overgrowths on, and as discrete grains adjacent to chromite in the chromitite stringers (Figs. 1.11e, f and g). Smaller (20–100 microns), acicular grains of rutile are also observed enclosed in biotite, plagioclase, and orthopyroxene within the pegmatitic feldspathic orthopyroxenite of SA04-13 (Fig. 1.11h), and fine skeletal rutile-quartz intergrowths are found in SA04-08. Apatite is present as irregular interstitial aggregates intergrown with biotite and quartz (Fig. 1.11i). Biotite is present in all thin sections (~ 10) prepared from the Merensky Reef samples, ranging in size from large (1–2 cm) macroscopic sheaths (Figs. 1.10a, 1.11a and j) to fine (0.1–0.5 mm) platelets with quartz in interstitial patches between orthopyroxene crystals.

Microscopic quartz and quartz-alkali feldspar intergrowths are heterogeneously distributed throughout the Merensky Reef samples and are commonly spatially associated with zircon, biotite, apatite, and more rarely, rutile (Figs. 1.11b, c, i, k and l). Although rarely noted in the literature, Wagner (1929, p. 106) identified quartz and quartz-feldspar intergrowths in samples of the Merensky Reef from the Eastern Limb (Lydenburg district). This "granitic" association of minerals (quartz, alkali feldspar, biotite, zircon, apatite, rutile) indicates that much of the zircon and the other U–Th–Pb accessory minerals crystallized from fractionated interstitial melt likely at temperatures approaching the solidus of the Merensky Reef samples.

Zircon Textures and Zoning in Cathodoluminescence Images

The external morphology and internal textures of zircon reflect the geological history of the mineral (e.g., Pupin 1980; Vavra 1993; Corfu et al. 2003). Zircon imaging is accomplished initially using a binocular microscope during grain selection for geochronology based on a variety of features such as color, degree of transparency, shape and external morphology, and the absence of inclusions (mineral and fluid) and fractures. Backscattered electron (BSE) and cathodoluminescence (CL) imaging provide the highest resolution of internal structures (i.e., zoning, xenocryst cores, subsolidus modifications) of zircon and CL imaging is now considered standard practice in the final selection of grains for geochronology. Corfu et al. (2003) provide an excellent overview of zircon selection and imaging techniques for examining morphology and textures. Cathodoluminescence emission in zircon is complex with the possibility of quenching by optical absorption processes, however, it is dominated by narrow emission bands generally related to Dy^{3+}impurities or other rare earth elements such as Sm^{3+}, Eu^{2+}, Tb^{3+}(Nasdala et al. 2003).

For detailed imaging of zircon from the Merensky Reef samples in this study, all selected zircons were mounted in epoxy in 2.5 cm diameter circular grain mounts and polished until the grains were sectioned in half. The polished epoxy sections were carbon coated and prepared in the Electron Microbeam/X-Ray Diffraction Facility at the University of British Columbia, Vancouver. Back-scattered electron (BSE) and cathodoluminescence (CL) imaging were carried out on a Philips XL-30 scanning electron microscope (SEM) equipped with a Bruker Quanta 200 energy-dispersion X-ray microanalysis system and a Robinson cathodoluminescence detector. An operating voltage of 15 kV was used, with a spot diameter of 6 µm, and peak count time of 30 s.

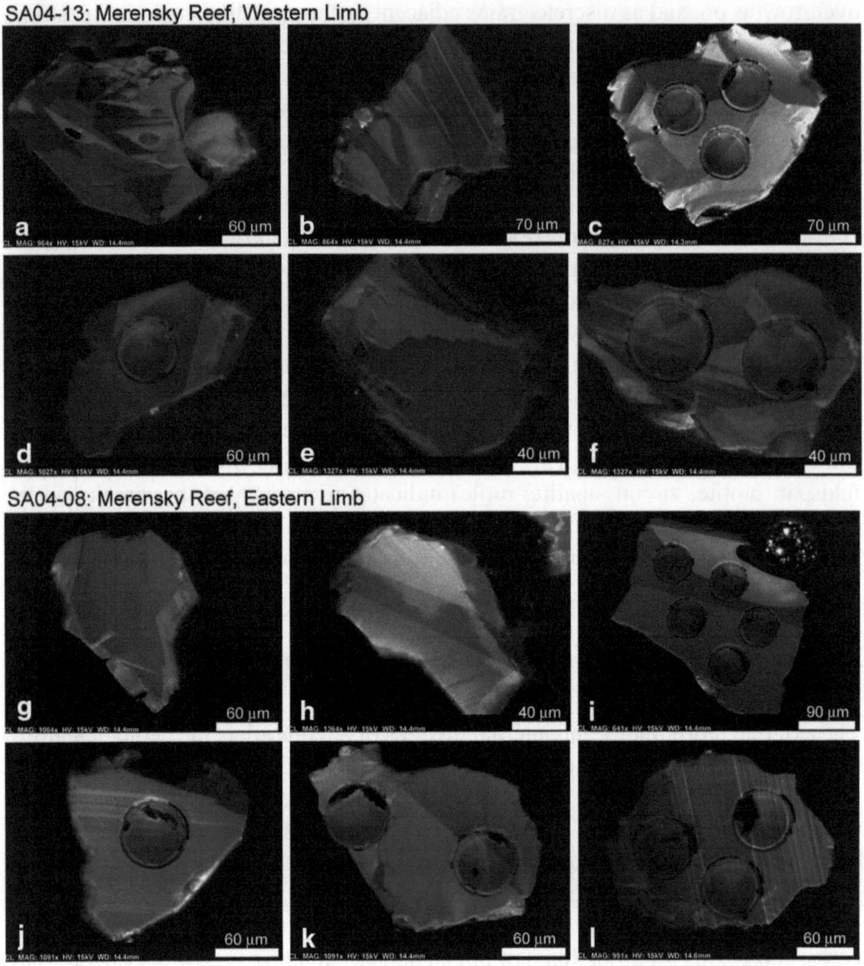

Fig. 1.12 Cathodoluminescence (*CL*) images of zircon from the Merensky Reef. Scale indicated with scale bars on each image. **a–f** Zircon grains from sample SA04-13, Western Limb, showing the typical irregular, anhedral morphology of interstitial zircon and simple to complex sector zoning. **g–l** Zircon grains from sample SA04-08, Eastern Limb again showing the typical irregular, anhedral morphology of interstitial zircon and presence of oscillatory zoning in addition to broad sector zones. *Circular* pits indicate locations of laser ablation-ICP-MS spot analyses (60 µ diameter) for trace element concentrations in zircon (note that the orientation of some zone contacts can be observed within the pits)

The BSE images of the Merensky zircon grains (not shown) reveal little internal structure with only slight changes in brightness, reflecting changes in average atomic number, present along select grain margins. In contrast, CL images provide clear details of the range of zoning present in zircon (Fig. 1.12), even though the CL emission is relatively low compared with zircon grains from other layered intrusions (e.g., Stillwater, work in progress). There is no evidence of inheritance or xenocrystic cores in any of the analyzed Merensky zircons. The external morphology is typically irregular with only local development of rational crystal faces (e.g., Figs. 1.12a, d, g and h) as is typical of zircon from gabbroic rocks that crystallized from fractionated interstitial melt late in the consolidation history of a cumulate (e.g., Scoates and Chamberlain, 1995; Scoates and Friedman 2008). Sector zoning is predominant in sample SA04-13 (Figs. 1.12a, b, c, d, e and f) and fine to coarse oscillatory zoning is characteristic of zircon from sample SA04-08 (Figs. 1.12g, h, i, j, k and l), likely reflecting different growth rates for zircon from these two different, and widely separated, samples of Merensky Reef (e.g., Hanchar and Miller 1993; Watson and Liang 1995; Corfu et al. 2003). The CL images also show that there are no prominent truncation surfaces indicative of resorption and there are few inclusions (Fig. 1.12a, black elongate feature in centre-left; Fig. 1.12k, black spot at base on rightmost laser ablation pit).

Methods

Treatment and Dissolution of Zircon, Rutile and Apatite

Mineral treatment protocols were carried out at the Pacific Centre for Isotopic and Geochemical Research (PCIGR) at the University of British Columbia, Vancouver (Canada). Zircon grains were separated from samples using conventional crushing, grinding, and Wilfley table techniques, followed by final concentration using heavy liquids and magnetic separation. Fractions of zircon, rutile, and apatite for analysis were hand-picked in methanol under magnification using a binocular microscope and selected based on grain morphology, quality, size and low magnetic susceptibility. Biotite grains were also handpicked under magnification using a binocular microscope and selected based on size and lack of inclusions and alteration.

All handpicked zircon grains were anhedral, clear, colorless to light pink, devoid of mineral and fluid inclusions, and ~50 μm in diameter. Prior to dissolution, the zircon grains were annealed and chemically abraded following the procedures outlined in Mattinson (2005). The zircon grains selected for CA-TIMS analysis were annealed in quartz glass crucibles in a muffle furnace at 900°C for 60 h. They were transferred into 10 mL Pyrex® beakers, agitated in ultrapure 3 N HNO_3 in an ultrasonic cleaner for 15 min, warmed to 60 ± 10°C for 30 min, and rinsed with ultrapure acetone and water. The zircon grains were then transferred to 3.5 mL PFA screwtop beakers, ultrapure HF (50%, 500 μL) and HNO_3 (14 N, 50 μL) were added, and caps were closed finger-tight. Following this, the beakers were placed in 125 mL PTFE Teflon® liners (up to four per liner) and about 2 mL HF and 0.2 mL concentrated HNO_3 was added to the liners. The liners were slid into stainless steel jackets of Parr® high-pressure dissolution devices, which were sealed and brought up to 200 °C for 16 h. After cooling, the beakers were removed from the liners and the remaining zircon grains were separated from the leachate.

The annealed, leached, and rinsed grains were then weighed, each transferred into 300 μL PFA microcapsules, and 50 μL of 50% HF and 5 μL of 14 N HNO_3

were added. Single zircon grains were transferred into individual microcapsules and each was spiked with EARTHTIME ^{205}Pb–^{233}U–^{235}U isotope tracer (verified by comparison with three mixed U–Pb gravimetric reference solutions from Massachusetts Institute of Technology, the University of California, Santa Barbara, and the National Environment Research Council Isotope Geosciences Laboratory, United Kingdom, made available through EARTHTIME), and each was subsequently capped and placed in a Parr® liner (8–13 microcapsules per liner). HF and nitric acids in a 10:1 ratio, respectively, were added to the liner, which was then placed in the jacket, sealed, and brought up to 240°C for 40 h for complete dissolution. The resulting solutions were dried on a hotplate at 130 °C. The fluorides were dissolved in 3.1 N HCl in the high-pressure devices for 12 h at 210 °C. Following dissolution, samples were transferred into clean 7 mL PFA beakers and dried after adding 2 μL of 0.5 N H_3PO_4. No chemical separation of U and Pb was done for zircon in this study.

Rutile was most abundant in the M2 and M5 magnetic fractions of both samples, whereas apatite was most abundant in the N2 magnetic fractions. Hand-picked rutile ranged in color from amber to chocolate brown, in shape from euhedral (rare) to subhedral wafers and irregular fragments, and in size from 100 to 200 μm across (5–10 grains per fraction). Handpicked apatite grains were colorless, subhedral and irregular fragments, and ranged in size from 50 to 100 μm across (25–85 per fraction). Rutile fractions were transferred into 10 mL Pyrex® beakers. About 2 mL of 1 M HNO_3 was added and grains were ultrasonicated for 5 min and warmed to 60 ± 10 °C for 10 min, rinsed with > 18 MΩ•cm water followed by sub-boiled acetone, and then air-dried. Fractions were weighed and transferred to 3.5 mL screwtop PFA Teflon® beakers. One mL of a 10:1 mixture of sub-boiled 50 % HF and 14 M HNO_3 was added followed by a weighed drop of EARTHTIME ^{205}Pb –^{233}U –^{235}U isotope tracer. The beakers were capped tightly and placed in a Parr® liner (four beakers per liner). HF and nitric acids in a 10:1 ratio, respectively, were added to the liner, which was then placed in the jacket, sealed, and brought up to 240°C for 40 h for complete dissolution. The resulting solutions were dried on a hotplate at 130°C. The fluorides, along with the weighed apatite fractions, were dissolved in 3.1 N HCl in the high-pressure devices for 12 h at 210 °C. Uranium was eluted into a beaker with 0.5 N HBr followed by elution of Pb into a separate beaker with 6.2 N HCl. The U was taken nearly to dryness at 110 °C, and 1 mL of sub-boiled 6.2 N HCl was added. The resin in the columns was washed with > 18 MΩ•cm water and conditioned with 6.2 N HCl, then the U solution was reintroduced into each column and washed with 8M HNO_3 to remove iron. Uranium was then eluted with water into the same beaker into which Pb was eluted. Samples were dried after addition of 2 μL of 0.5 N H_3PO_4.

U–Pb TIMS Isotopic Analysis

All zircon, rutile, and apatite samples were loaded onto single zone-refined Re filaments with 2 μL of silicic acid activator (Gerstenberger and Haase 1997). Both Pb and U isotopes were analyzed in the peak-hopping mode using a modified thermal ionization mass spectrometry (TIMS) on a VG354S equipped with Sector 54 electronics and a Daly photomultiplier housed at PCIGR (Table 1.2). All major Pb and U isotopes were collected in an analogue Daly detector. Isotope ratios of Pb were corrected for a fractionation of 0.25 %/amu ± 0.02 % (2s) based on replicate analysis of the NBS-982 Pb reference material throughout the course of this study. Due to the presence of two U isotopes in the spike, U is self-corrected in each run. All analytical errors were numerically propagated through the entire age calculation using the technique of Roddick (1987). Standard concordia diagrams were constructed and U–Pb concordia ages calculated with U–Pb_Redux (McLean et al. 2011); all errors are quoted at the 2s level. Additional U–Pb analytical results and details of the analytical procedure can be found in the footnote to Table 1.4.

Biotite and $^{40}Ar/^{39}Ar$ Isotopic Analysis

Biotite separates were hand-picked, washed in acetone, dried, wrapped in aluminum foil and stacked in an irradiation capsule with similar-aged samples and neutron flux monitors (Fish Canyon Tuff sanidine). The samples were irradiated on October 29 through November 4, 2005 at the McMaster Nuclear Reactor in Hamilton, Ontario, for 225 MWH, with a neutron flux of approximately 3×10^{16} neutrons/cm^2. Analyses (n = 36) of 12 neutron flux monitor positions produced errors of < 0.5 % in the J value. The samples were analyzed on January 3 and 4, 2006, at PCIGR. The mineral separates were step-heated at incrementally higher powers in the defocused beam of a 10W CO_2 laser (New Wave Research MIR10) until fused. The gas evolved from each step was analyzed by a VG5400 mass spectrometer equipped with an ion-counting electron multiplier. All measurements were corrected for total system blank, mass spectrometer sensitivity, mass discrimination, radioactive decay during and subsequent to irradiation, as well as interfering Ar from atmospheric contamination and the irradiation of Ca, Cl and K (isotope production ratios: $(^{40}Ar/^{39}Ar)K = 0.0302 \pm 0.00006$, $(^{37}Ar/^{39}Ar)Ca = 1416.4 \pm 0.5$, $(^{36}Ar/^{39}Ar)Ca = 0.3952 \pm 0.0004$, $Ca/K = 1.83 \pm 0.01(^{37}ArCa/^{39}ArK)$).

The results of the $^{40}Ar/^{39}Ar$ analyses, including plateau (spectrum) and inverse correlation ages, are shown in Table 1.5. The plateau and correlation ages were calculated using Isoplot ver. 3.09 (Ludwig 2003). Errors are quoted at the 2s (95 % confidence) level and are propagated from all sources, except for mass spectrometer sensitivity and age of the flux monitor. The best statistically justified plateau and plateau age were picked based on the following criteria: (1) three or more contiguous steps comprising more than 50 % of the ^{39}Ar; (2) probability of fit of the weighted mean age greater than 5 %; (3) slope of the error-weighted line through the plateau ages equals zero at 5 % confidence; (4) ages of the two outermost steps on a plateau are not significantly different from the weighted-mean plateau age (at 1.8s, six or more steps only); (5) outermost two steps on either side of a plateau must not have non-zero slopes with the same sign (at 1.8s, nine or more steps only).

Table 1.4 U–Th–Pb results for zircon, rutile and apatite from the Merensky Reef, Bushveld Complex

Composition							Isotopic Ratios					
Fraction[a]	Mass[b] (μg)	U[c] (ppm)	Th/U[d]	Pb*[c] (ppm)	Pb_c[e] (pg)	Pb*/Pb_c[e]	$^{206}Pb/^{204}Pb$[f]	$^{206}Pb/^{238}U$[g]	±2σ %[h]	$^{207}Pb/^{235}U$[g]	±2σ %[h]	$^{207}Pb/^{206}Pb$[g]
SA04-13 Zircon (Western Limb)												
CA1	24	7	3.68	5	0.92	135	4558	0.37507	0.11287	6.56936	0.16953	0.12703
CA2	21	18	4.00	13	0.91	304	9857	0.37566	0.10223	6.57788	0.15167	0.12700
CA3	14	44	2.80	27	1.05	364	13913	0.37548	0.10336	6.57508	0.15255	0.12700
CA4	11	103	1.28	50	0.91	607	29947	0.37562	0.09453	6.57833	0.14305	0.12702
CA5	12	31	1.59	16	1.70	110	5130	0.37394	0.08802	6.54894	0.15185	0.12702
CA6	12	22	4.30	17	1.68	118	3688	0.37491	0.09299	6.56664	0.16005	0.12703
SA04-08 Zircon (Eastern Limb)												
CA1	15	57	0.781	25	0.16	2277	124250	0.37428	0.31004	6.55404	0.32845	0.12700
CA2	12	28	1.37	14	1.63	103	4992	0.37548	0.30770	6.57611	0.32809	0.12702
CA3	14	23	0.747	10	0.42	324	17835	0.37435	0.32152	6.55595	0.34111	0.12702
CA4	11	11	1.17	5	4.34	14	700	0.37421	0.28923	6.56826	0.30842	0.12730
CA5	30	13	1.40	6	2.75	69	3318	0.37514	0.09252	6.56900	0.17149	0.12700
CA6	15	97	0.933	44	0.29	2279	120432	0.37527	0.15043	6.57124	0.18561	0.12700
CA7	15	28	1.62	14	0.70	310	14361	0.37571	0.10594	6.57931	0.15223	0.12701
CA8	13	33	1.21	16	0.75	276	13787	0.37528	0.08851	6.57157	0.14331	0.12700
CA9	15	65	1.41	32	1.29	376	18119	0.37534	0.09457	6.57102	0.14143	0.12697
CA10	20	155	1.19	74	0.82	1803	90453	0.37553	0.13116	6.57681	0.17235	0.12702
SA04-13 Rutile (Western Limb)												
R1	185	2	0.027	1.2	79.2	1.81	121	0.37488	0.257774	6.55513	0.43809	0.12682
R2	147	2	0.028	1.2	67.4	1.57	107	0.37604	0.288944	6.56994	0.46378	0.12671
R3	195	1	0.056	0.6	42.5	1.89	126	0.37597	0.260431	6.57130	0.44901	0.12676
R4	113	4	0.032	1.9	31.9	5.57	343	0.37498	0.182064	6.54762	0.32536	0.12664
R5	138	1	0.087	0.7	41.6	1.32	91.9	0.37755	0.324335	6.59989	0.43663	0.12678
SA04-08 Rutile (Eastern Limb)												
R1	207	37	0.008	14	33.0	84	5037	0.37464	0.13880	6.54559	0.17613	0.12672
R2	128	63	0.005	23	17.7	165	9878	0.37504	0.16505	6.55339	0.19694	0.12673
R3	111	62	0.005	23	18.2	136	8164	0.37275	0.37654	6.51103	0.39637	0.12669
R4	122	64	0.015	23	23.3	121	7217	0.37436	0.10208	6.54140	0.14781	0.12673
SA04-13 Apatite (Western Limb)												
A2	99	1	1.75	2.9	233	0.23	25.0	0.35892	4.11762	6.93679	7.32729	0.14017
A3	190	0.2	5.34	1.1	185	0.18	20.0	0.37108	10.72227	8.07141	17.46022	0.15776
SA04-08 Apatite (Eastern Limb)												
A1	140	2	3.11	10	1268	0.17	22.8	0.48566	3.10164	8.77285	4.16761	0.13101
A2	125	3	2.30	12	1309	0.17	23.7	0.45811	2.77341	7.97486	3.87527	0.12626
A3	64	3	1.37	18	1065	0.08	21.0	0.54384	4.09778	10.06558	5.52211	0.13424
A4	141	1	5.25	12	1472	0.12	19.9	0.60975	4.99752	11.45443	6.60221	0.13624
A5	144	1	4.76	14	1865	0.10	19.2	0.67948	5.80171	12.88478	7.33831	0.13753

[a] CA1, etc.: chemically abraded single zircon grains or fragments; chemical abrasion after Mattinson (2005) and Scoates and Friedman (2008); *R* rutile, *A* apatite

[b] Fraction weights of zircon determined from volume estimates just prior to final dissolution; apatite and rutile weights done by weighing on a microbalance

[c] U and Pb concentrations subject to uncertainty in weight of grains

[d] Th contents calculated from radiogenic ^{208}Pb and the $^{207}Pb/^{206}Pb$ date of the sample, assuming concordance between U–Th and Pb systems

[e] Pb* and Pb_c represent radiogenic and common Pb, respectively; mol% $^{206}Pb^*$ with respect to radiogenic, blank, and initial common Pb

[f] Measured ratio corrected for spike and fractionation only; mass discrimination of 0.25%/amu ± 0.03% (2σ abs) based on analysis of NBS-982 during the duration of this study; all Daly analyses

[g] Corrected for fractionation, spike, and common Pb; up to 5 pg (zircon), 5–10 pg (rutile), 20–25 pg (apatite) of common Pb was assumed to be procedural blank: $^{206}Pb/^{204}Pb = 18.50 \pm 1.0\%$; $^{207}Pb/^{204}Pb = 15.50 \pm 1.0\%$; $^{208}Pb/^{204}Pb = 38.40 \pm 1.0\%$ (all uncertainties 1-sigma); excess over blank was assigned to intial common Pb determined by leached feldspar analyses: $^{206}Pb/^{204}Pb = 15.174 \pm 0.006$, $^{207}Pb/^{204}Pb = 15.241 \pm 0.006$, and $^{208}Pb/^{204}Pb = 34.896 \pm 0.007$; 2σ errors

		Dates (Ma)						
±2σ %[h]	rho[i]	$^{206}Pb/^{238}U$[j]	±2σ abs[h]	$^{207}Pb/^{235}U$[j]	±2σ abs[h]	$^{207}Pb/^{206}Pb$[j]	±2σ abs[h]	% disc[k]
0.08786	0.877	2053.21	1.98	2055.25	1.49	2057.29	1.57	0.20
0.07032	0.913	2055.97	1.80	2056.39	1.34	2056.80	1.27	0.04
0.06605	0.932	2055.13	1.82	2056.01	1.34	2056.89	1.19	0.09
0.06251	0.935	2055.81	1.66	2056.45	1.26	2057.08	1.13	0.06
0.08130	0.897	2047.89	1.54	2052.50	1.34	2057.14	1.46	0.45
0.09225	0.858	2052.48	1.63	2054.88	1.41	2057.29	1.65	0.23
0.06376	0.981	2049.51	5.44	2053.19	2.89	2056.88	1.15	0.36
0.09010	0.961	2055.13	5.41	2056.15	2.89	2057.17	1.61	0.10
0.07382	0.976	2049.84	5.65	2053.44	3.01	2057.07	1.33	0.35
0.06828	0.975	2049.19	5.08	2055.10	2.72	2061.02	1.23	0.40
0.09623	0.898	2053.55	1.63	2055.20	1.51	2056.85	1.72	0.16
0.06296	0.947	2054.14	2.65	2055.50	1.64	2056.86	1.14	0.13
0.06954	0.910	2056.22	1.87	2056.58	1.34	2056.94	1.25	0.03
0.07392	0.895	2054.19	1.56	2055.54	1.26	2056.90	1.33	0.13
0.06722	0.905	2054.50	1.66	2055.47	1.25	2056.44	1.21	0.09
0.06412	0.942	2055.37	2.31	2056.24	1.52	2057.12	1.16	0.09
0.42184	0.356	2053.33	3.86	2052.30	4.53	2054.38	7.45	0.05
0.43732	0.400	2055.32	4.09	2057.76	5.09	2052.88	7.72	−0.12
0.41142	0.428	2055.51	3.96	2057.44	4.59	2053.56	7.26	−0.09
0.25028	0.645	2052.33	2.87	2052.79	3.20	2051.86	4.42	−0.02
0.33315	0.653	2059.33	3.85	2064.82	5.73	2053.85	5.88	−0.27
0.06443	0.944	2051.21	2.44	2052.05	1.55	2052.90	1.14	0.08
0.06524	0.950	2053.09	2.90	2053.10	1.73	2053.11	1.15	0.00
0.09517	0.971	2042.31	6.59	2047.39	3.49	2052.51	1.68	0.50
0.06293	0.938	2049.89	1.79	2051.49	1.30	2053.10	1.11	0.16
7.4621	0.248	1977.1	70.1	2103.4	65.0	2229.3	129.2	11
17.9313	0.263	2034.5	187.1	2239.1	157.7	2431.8	303.9	16
4.9958	0.079	2551.9	65.4	2314.7	38.0	2111.5	87.6	−21
4.5814	0.080	2431.2	56.2	2228.2	35.0	2046.5	80.9	−19
6.5841	0.087	2799.5	93.1	2440.8	51.0	2154.1	114.9	−30
7.9616	0.078	3069.0	122.0	2560.9	61.7	2180.0	138.6	−41
9.0480	0.066	3342.4	151.3	2671.3	69.2	2196.3	157.2	−52

[h] Errors are 2-sigma, propogated using the algorithms of McLean et al. (2011).

[i] R = correlation coefficient

[j] Isotopic dates are calculated using the decay constants $\lambda^{238} = 1.55125E^{-10}$ and $\lambda^{235} = 9.8485E^{-10}$ (Jaffey et al. 1971); $^{206}Pb/^{238}U$ and $^{207}Pb/^{206}Pb$ ages corrected for initial disequilibrium in $^{230}Th/^{238}U$ using Th/U [magma] = 3

[k] % discordance = 100 − (100*($^{206}Pb/^{238}U$ date)/($^{207}Pb/^{206}Pb$ date))

Table 1.5 $^{40}Ar/^{39}Ar$ results for biotite from the Merensky Reef, Bushveld Complex

Laser Power (%)	$\%^{39}Ar$	$^{40}Ar/^{39}Ar$	±1σ	$^{38}Ar/^{39}Ar$	±1σ	$^{37}Ar/^{39}Ar$	±1σ	$^{36}Ar/^{39}Ar$	±1σ	$\%^{40}Ar$ atm	$^{40}Ar^*/^{39}ArK$	±1σ	Age (Ma)	±1σ
SA04-13, J = 0.030017 ± 0.000054, Biotite (Western Limb)														
2.0	0.76	374.16	0.009	0.688	0.041	0.047	0.129	1.120	0.020	87.7	46.12	6.099	1568	139
2.1	1.27	71.442	0.006	0.417	0.037	0.054	0.097	0.096	0.030	39.1	43.54	0.880	1508	21
2.3	2.26	71.688	0.007	0.347	0.033	0.030	0.075	0.051	0.040	20.7	56.89	0.708	1797	14
2.5	5.61	74.785	0.005	0.283	0.020	0.010	0.171	0.034	0.028	13.3	64.93	0.445	1951	8
2.7	4.28	72.669	0.006	0.274	0.019	0.003	0.345	0.019	0.038	7.5	67.33	0.473	1995	8
2.8	4.84	69.384	0.005	0.267	0.022	0.002	0.351	0.006	0.065	2.7	67.61	0.355	2000	6
2.9	4.36	69.339	0.006	0.272	0.030	0.001	1.806	0.004	0.103	1.8	68.16	0.435	2009	8
3.0	5.71	68.500	0.005	0.264	0.022	0.002	0.340	0.003	0.141	1.2	67.74	0.382	2002	7
3.1	4.41	69.110	0.006	0.269	0.016	0.001	1.415	0.004	0.232	1.7	68.06	0.494	2008	9
3.2	6.20	68.152	0.005	0.271	0.024	0.001	0.410	0.002	0.177	1.0	67.55	0.377	1999	7
3.3	4.58	68.597	0.005	0.267	0.021	0.002	0.552	0.003	0.223	1.1	67.92	0.414	2005	7
3.4	4.34	68.723	0.005	0.271	0.023	0.001	0.631	0.004	0.141	1.7	67.64	0.404	2000	7
3.5	3.82	69.053	0.007	0.272	0.022	0.002	0.443	0.003	0.179	1.4	68.15	0.502	2009	9
3.6	5.93	67.991	0.005	0.268	0.014	0.001	0.674	0.002	0.203	0.9	67.49	0.355	1997	6
3.7	4.51	68.539	0.006	0.263	0.022	0.001	1.752	0.002	0.195	0.8	68.09	0.414	2008	7
3.8	4.97	68.358	0.005	0.268	0.028	0.001	0.632	0.002	0.233	0.9	67.84	0.382	2004	7
3.9	4.60	68.465	0.005	0.264	0.025	0.001	0.524	0.002	0.333	0.9	67.93	0.401	2005	7
4.0	5.42	68.006	0.005	0.258	0.014	0.002	0.339	0.002	0.154	0.9	67.51	0.327	1998	6
4.1	5.50	67.928	0.005	0.252	0.016	0.001	0.643	0.002	0.283	0.8	67.45	0.368	1997	7
4.2	2.81	68.823	0.006	0.258	0.035	0.002	0.884	0.003	0.357	1.3	68.03	0.508	2007	9
4.3	2.67	69.356	0.005	0.257	0.022	0.003	0.571	0.004	0.211	1.6	68.37	0.438	2013	8
4.4	3.57	68.938	0.006	0.249	0.029	0.002	0.609	0.003	0.135	1.4	68.09	0.422	2008	8
4.5	3.16	69.193	0.005	0.245	0.020	0.003	0.258	0.003	0.251	1.1	68.52	0.404	2016	7
4.7	1.90	69.592	0.006	0.245	0.034	0.002	0.626	0.005	0.207	2.0	68.26	0.498	2011	9
4.9	2.53	69.403	0.007	0.229	0.029	0.000	5.805	0.002	0.456	0.9	68.87	0.559	2022	10

Table 1.5 (continued)

Laser Power (%)	%^{39}Ar	^{40}Ar/^{39}Ar	±1σ	^{38}Ar/^{39}Ar	±1σ	^{37}Ar/^{39}Ar	±1σ	^{36}Ar/^{39}Ar	±1σ	%^{40}Ar atm	^{40}Ar*/^{39}ArK	±1σ	Age (Ma)	±1σ
SA04-11A, J = 0.030012 ± 0.000056, Biotite (Western Limb)														
2.0	0.20	73.253	0.019	0.266	0.106	0.152	0.125	0.161	0.054	64.5	26.02	2.449	1041	74
2.5	1.27	61.879	0.006	0.211	0.037	0.076	0.075	0.035	0.057	16.5	51.77	0.678	1691	14
2.8	4.73	68.465	0.005	0.197	0.025	0.008	0.135	0.003	0.157	1.2	67.71	0.354	2001	6
2.9	3.50	68.596	0.006	0.189	0.061	0.009	0.196	0.004	0.215	1.8	67.43	0.507	1996	9
3.0	4.14	68.996	0.006	0.194	0.021	0.009	0.117	0.004	0.208	1.8	67.81	0.513	2003	9
3.1	3.67	69.007	0.005	0.194	0.036	0.008	0.190	0.003	0.245	1.2	68.27	0.392	2011	7
3.2	4.87	68.915	0.005	0.200	0.020	0.010	0.106	0.004	0.129	1.8	67.74	0.381	2002	7
3.3	6.19	67.783	0.005	0.198	0.021	0.022	0.048	0.002	0.309	0.7	67.41	0.373	1996	7
3.4	3.66	68.613	0.005	0.199	0.032	0.021	0.083	0.003	0.302	1.1	67.97	0.402	2006	7
3.5	5.57	67.955	0.005	0.198	0.021	0.008	0.177	0.001	0.300	0.6	67.62	0.389	2000	7
3.6	5.06	68.480	0.005	0.190	0.022	0.006	0.155	0.003	0.166	1.2	67.76	0.392	2002	7
3.7	2.49	68.772	0.007	0.192	0.028	0.005	0.271	0.003	0.394	1.3	67.94	0.612	2005	11
3.8	5.14	67.697	0.005	0.197	0.018	0.004	0.099	0.001	0.286	0.6	67.36	0.378	1995	7
3.9	6.27	67.549	0.005	0.188	0.013	0.006	0.185	0.001	0.205	0.5	67.27	0.320	1993	6
4.0	6.13	67.649	0.005	0.193	0.025	0.007	0.148	0.001	0.400	0.6	67.35	0.383	1995	7
4.1	4.12	68.508	0.005	0.198	0.030	0.014	0.141	0.003	0.386	1.1	67.84	0.469	2004	8
4.2	5.58	67.923	0.005	0.191	0.037	0.009	0.126	0.001	0.743	0.4	67.73	0.416	2002	7
4.3	4.43	67.842	0.007	0.190	0.038	0.015	0.135	0.000	1.196	0.2	67.81	0.533	2003	10
4.4	3.99	67.932	0.008	0.193	0.025	0.076	0.049	0.001	1.355	0.2	67.88	0.600	2004	11
4.5	4.62	68.426	0.004	0.189	0.015	0.078	0.033	0.002	0.178	1.0	67.85	0.334	2004	6
4.7	4.59	67.173	0.005	0.193	0.024	0.092	0.031	−0.001	1.204	−0.1	67.30	0.436	1994	8
4.9	5.63	67.630	0.007	0.193	0.013	0.245	0.021	0.002	0.494	0.6	67.38	0.502	1995	9
5.3	4.15	68.499	0.005	0.183	0.021	0.069	0.057	0.002	0.354	0.9	67.97	0.402	2006	7

J, flux; 1σ, error quoted at 68 % confidence

Results

U–Pb Zircon

U–Pb data for all 10 chemically abraded zircon grains from the Eastern Limb sample of the Merensky Reef (SA04-08) are only slightly discordant (0.03–0.57 %), yield overlapping $^{207}Pb/^{206}Pb$ dates ranging from 2056.44 to 2061.02 Ma (Table 1.4), and give a weighted $^{207}Pb/^{206}Pb$ average age of 2056.88 ± 0.41/0.41/6.0 Ma (2s, MSWD = 0.12) (Fig. 1.13a). The U–Pb errors are reported in the ± X/Y/Z format of Schoene et al. (2006) with internal error in the absence of all systematic errors (± X), tracer calibration error (± Y), and uncertainty due to decay-constant errors (± Z). The MSWD, mean square of the weighted deviates is a useful parameter for evaluating the probability that a weighted-mean population is statistically equivalent (Wendt and Carl 1991). For the Western Limb sample (SA04-13) that was previously dated by Scoates and Friedman (2008), the U–Pb data for the six new analyses of chemically abraded zircon grains are slightly discordant (0.04–0.45 %), yield overlapping $^{207}Pb/^{206}Pb$ dates ranging from 2056.80 to 2057.29 Ma (Table 1.4), and give a weighted average of 2057.04 ± 0.55/0.55/6.0 Ma (2s, MSWD = 0.10) (Fig. 1.13b). This age is older than the previously published age of 2054.4 ± 1.3 Ma by several million years and has significantly better precision. These differences are attributed to the combination of changes in analytical protocols at PCIGR since 2008 (Table 1.2), including the systematic use of EARTHTIME reference synthetic solutions, reduced blanks, and a revised internal value for the NBS-982 Pb standard. The U–Pb zircon ages for these samples of the Merensky Reef supersede the value published by Scoates and Friedman (2008) for SA04-13 as well as the preliminary results reported in abstract form for both samples in Scoates et al. (2011) and Scoates et al. (2012). These two U–Pb zircon ages for the Merensky Reef, 2056.88 ± 0.41 Ma and 2057.04 ± 0.55 Ma, are identical within uncertainty and are interpreted as the age of crystallization of this stratabound horizon. These results for two samples separated by ~ 310 km indicate that the Merensky Reef formed concurrently across the entire length of the Bushveld Complex.

U–Pb Rutile

Data from four concordant rutile fractions (5–10 grains each) from SA04-08 yield a weighted mean $^{207}Pb/^{206}Pb$ age of 2052.96 ± 0.61 Ma (2s, MSWD = 0.14) (Fig. 1.13c) and data from five concordant multigrain fractions from SA04-13 yield a weighted mean $^{207}Pb/^{206}Pb$ age of 2053.0 ± 2.74 Ma (2s, MSWD = 0.13) (Fig. 1.13d). Both of the U–Pb rutile ages are younger than the U–Pb zircon ages from their respective samples of Merensky Reef, most notably in the high-precision result of SA04-08. These slightly younger ages are interpreted as cooling ages resulting from Pb diffusion in rutile. For relatively fast cooling rates between 100 and 1000 °C/million years, estimated closure temperatures range from ~ 720–800 °C based on the formulation of Dodson (1973) and experimental Pb diffusion data for rutile from Cherniak

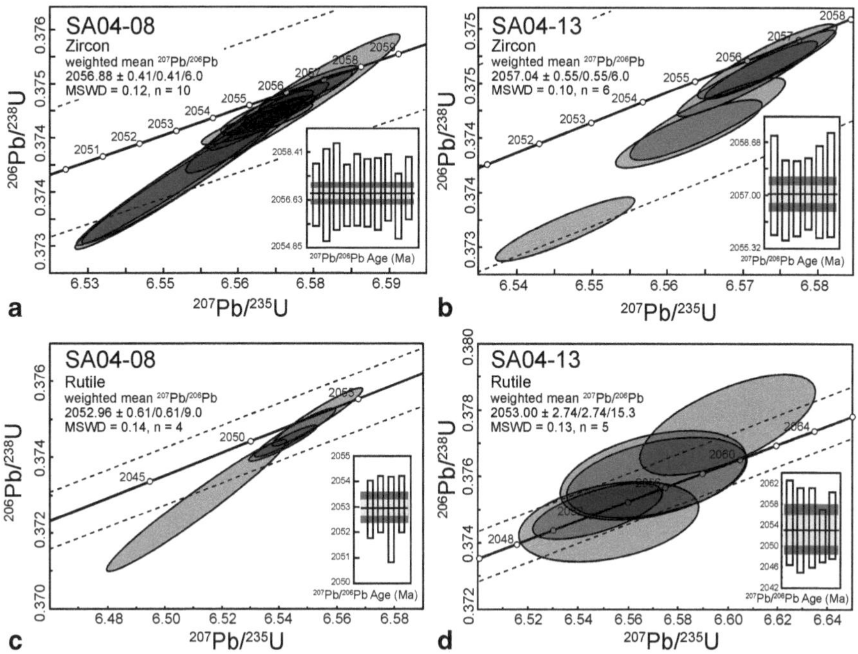

Fig. 1.13 Concordia diagrams showing U–Pb geochronological results for rutile and zircon from the Merensky Reef. **a** SA04-08 zircon. **b** SA04-13 zircon. **c** SA04-08 rutile. **d** SA04-13 rutile. Calculated weighted mean $^{207}Pb/^{206}Pb$ ages are indicated (uncertainty reported as 2s, including tracer calibration errors; decay-constant errors ignored). Each *shaded ellipse* indicates the analysis of a single zircon grain or 2–3 rutile grains (2s uncertainty). *Dashed* lines show the error bounds of the concordia curve. Inset in *lower right* of each panel show bar diagrams where each bar represents the analysis of a single fraction; the *horizontal black* line indicates the weighted mean age, the *light grey* band reflects the external reproducibility, and the *darker grey* band represents the total uncertainty. All dates Th-corrected for initial disequilibrium in $^{230}Th/^{238}U$ using Th/U [magma] = 3

(2000) (assuming cylindrical diffusion geometry and an average diffusion dimension of 125 μm). However, field-based studies consistently show that the closure temperature for rutile must be in the range of 400–450 °C (Schmitz and Bowring 2003). In this study, we assume that the ca. 2053 Ma ages from the U–Pb systematics of rutile indicate closure to Pb diffusion through a temperature of 450 °C following crystallization at ca. 2057 Ma.

U–Pb Apatite

Apatite from the two Merensky Reef samples has very low U concentrations (0.2–3 ppm) and large amounts of common Pb (< 23 % of the Pb in apatite is radiogenic) with correspondingly low $^{206}Pb/^{204}Pb$ (19.2–25.0) (Fig. 1.14a, Table 1.4). Age corrections for the initial common Pb composition were made using the results from co-existing leached plagioclase (Tam 2005; Scoates et al. 2006). The resulting U/Pb

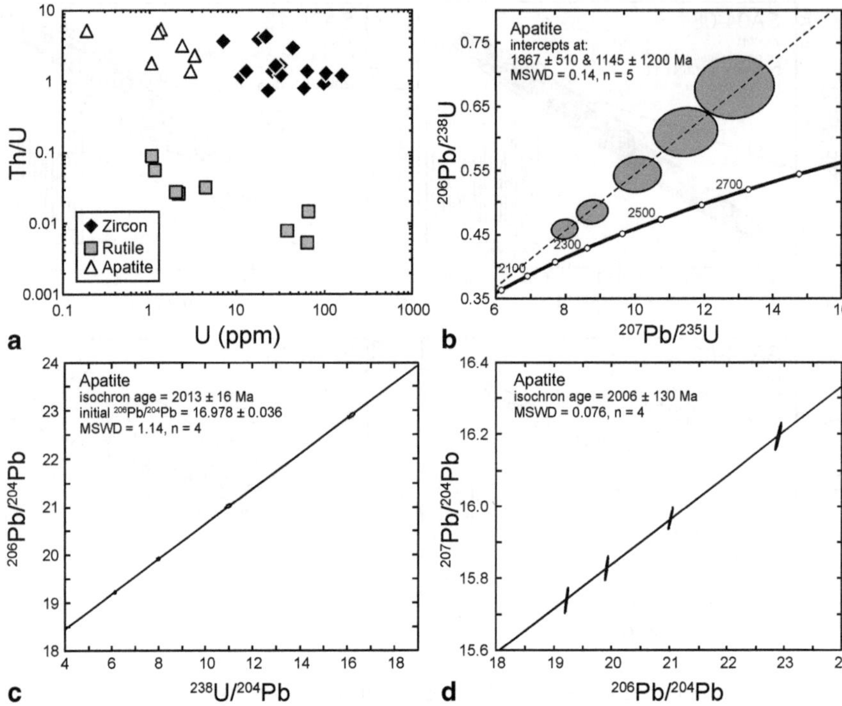

Fig. 1.14 Diagrams showing trace element and U–Pb geochronological results for apatite from the Merensky Reef. **a** Th/U vs. U (ppm) for all analyzed zircon, rutile, and apatite from the two samples of Merensky Reef in this study. Note that the plot axes are logarithmic to more effectively show the low U concentrations of apatite and rutile, and the very low Th/U of rutile. **b** Concordia diagram showing U–Pb geochronological results for apatite from SA04-08. **c** $^{206}Pb/^{204}Pb$ vs. $^{238}U/^{204}Pb$ plot showing results for apatite from SA04-08. **d** $^{207}Pb/^{204}Pb$ vs. $^{206}Pb/^{204}Pb$ plot showing results for apatite from SA04-08. Ellipses in **b–d** indicate 2s uncertainty. All plots produced using Isoplot (Ludwig 2003)

ages are extremely reversely discordant (up to 28 %; Fig. 1.14b), older than the U–Pb zircon crystallization age of the Merensky Reef (apatite $^{207}Pb/^{206}Pb$ dates = 2162–2467 Ma), and have large uncertainties. Reverse discordance is likely the result of U-loss at some time subsequent to crystallization (e.g., Kruger et al. 1998). The intercepts with concordia are highly imprecise with an upper intercept of ca. 1.9 Ga, close to the U–Pb zircon age, and a lower intercept of ca. 1.1 Ga. Use of a 1.1 Ga common Pb correction (Stacey and Kramers 1975) rather than the composition of the co-existing plagioclase yields concordant U/Pb ages. As they do not require a choice of initial Pb isotopic compositions, isochron plots (U–Pb, Pb–Pb) can also be used to investigate potential age relationships (Chamberlain and Bowring 2000). In a U–Pb isochron plot ($^{238}U/^{204}Pb$ vs. $^{206}Pb/^{204}Pb$), the resultant isochron age is 2013 ± 16 Ma (Fig. 1.14c). In a Pb–Pb isochron plot ($^{206}Pb/^{204}Pb$ vs. $^{207}Pb/^{204}Pb$), the isochron age is similar at 2006 ± 130 Ma (Fig. 1.14d), but with significantly larger uncertainty. Although these ages are likely to be inaccurate given the discordant

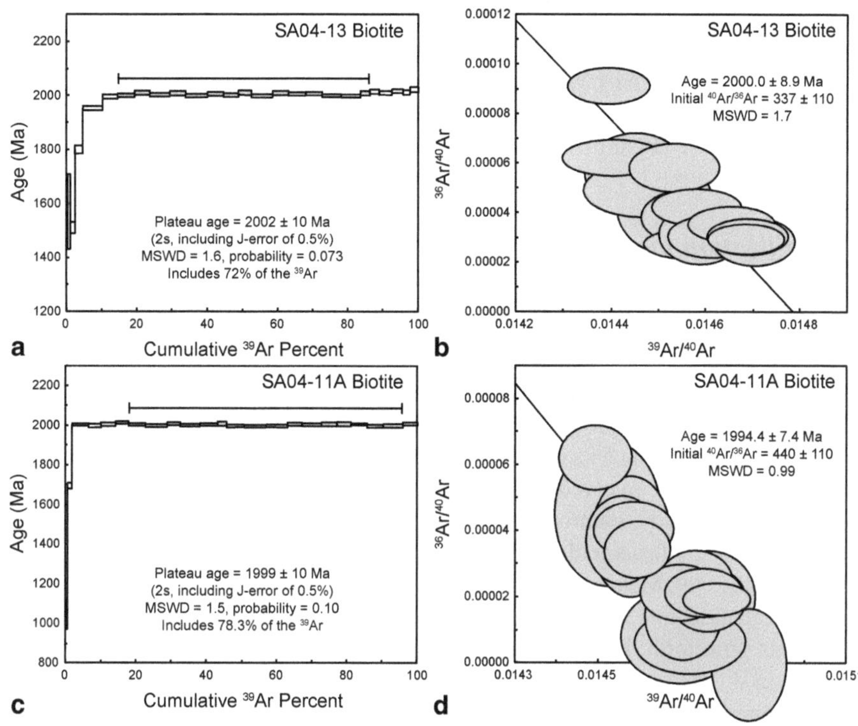

Fig. 1.15 Diagrams showing $^{40}Ar/^{39}Ar$ geochronological results for biotite from the Merensky Reef. **a** Apparent age spectrum for biotite from SA04-13, Eastern Limb. Grey-shaded boxes are those included in the plateau age calculation (range indicated by the *solid* line above the spectrum) and rejected steps are open boxes; 2s uncertainty level, errors include analytical uncertainty in J value. **b** Inverse isochron plot for biotite from SA04-13 (data-point error ellipses are 2s). **c** Apparent age spectrum for biotite from SA04-11A, Western Limb. **d** Inverse isochron plot for biotite from SA04-11A

nature of the U–Pb data and evidence for open-system behavior (Chamberlain and Bowring 2000), they may record hydrothermal resetting of the U–Pb systematics of apatite in the Merensky Reef at ca. 2.0 Ga, an event more strongly supported from the $^{40}Ar/^{39}Ar$ biotite results to follow.

Age spectra and inverse isochron diagrams for the two analyzed biotite samples (SA04-11A, SA04-13) from the Merensky Reef are shown in Fig. 1.15 with analytical data reported in Table 1.5. The plateau ages are 1999 ± 10 Ma (includes 78.3 % of the total ^{39}Ar released) and 2002 ± 10 Ma (includes 72 % of the total ^{39}Ar released), respectively, similar to the apatite U–Pb isochron age documented above. Both spectra have low apparent ages in the earliest stages of gas released (1–2 % for SA04-11A, 14 % for SA04-13). This is consistent with the early release of ^{39}Ar at the lowest extraction temperatures from minor chlorite interlayers in biotite that are artificially enriched in ^{39}Ar during irradiation and transfer from the surrounding biotite (Nomade et al. 2004). The inverse isochron ages, 1994.4 ± 7.4 Ma and 2000.0 ± 8.9 Ma, are indistinguishable from their respective plateau ages (Fig. 1.15).

Both isochrons have large uncertainties in initial $^{40}Ar/^{36}Ar$ with values that either overlap (sample SA04-13) with that determined for the precise estimate of atmospheric argon ($^{40}Ar/^{36}Ar = 298.56 \pm 0.31$; Lee et al. 2006; Mark et al. 2011) or that are higher (sample SA04-11A) indicative of excess ^{40}Ar introduced into the biotite at some time since crystallization (e.g., Kuiper 2002).

Crystallization and Cooling of the Merensky Reef and Bushveld Complex

The U–Pb zircon ages for the Merensky Reef, 2056.88 ± 0.41 Ma (Eastern Limb) and 2057.04 ± 0.55 Ma (Western Limb), are indistinguishable from each other and signify that the Reef is an intrusion-wide time marker in the crystallization history of the Bushveld Complex. Since its discovery in 1924 on the Farm Maandagshoek (Eastern Limb), the processes involved in the formation of the Merensky Reef have been the focus of intense study, controversy, and debate (e.g., Wagner 1929; Cousins 1969; Vermaak 1976; Ballhaus and Stumpfl 1986; Mathez 1995; Barnes and Maier 2002; Naldrett et al. 2009; Maier et al. 2013). From the perspective of chronology, the single most important feature of the Merensky Reef is that it is a regionally stratabound, but locally not stratiform, horizon with rocks of different composition comprising the footwall in different parts of the intrusion (e.g., mottled anorthosite in the Western Limb compared to layered norite from the Eastern Limb; Mitchell and Scoon 2007). This discontinuity ranges from kilometres at regional scale down to centimetres to metres at local scale (Eales and Cawthorn 1996), including the remarkable and persistent "potholes" (e.g., Ballhaus 1988; Carr et al. 1994; Smith and Basson 2006). Based on the results of this study, the absolute timing of the "Merensky discontinuity" is restricted to a relatively narrow interval (± 500,000 years or 0.025 %) at ca. 2057 Ma.

The duration of magmatism and cooling of the Bushveld Complex and associated mafic-ultramafic rocks across the northern Kaapvaal craton remains a fundamental question in the origin of the world's largest layered intrusion (Cawthorn and Walraven 1998; Cawthorn and Webb 2013). Whether the duration is 1, 5, or 10 Ma or longer, directly impacts estimates for rates of mantle melting, melt extraction, magma transport, magma flux, and cooling, and even potential environmental impacts (i.e., degassing associated with eruptive equivalents). Recent compilations of available geochronological data, combining in some cases published (analytical technique and data tables available) and unpublished (ages reported in abstracts with no associated data), suggest that most of the products of Bushveld-related magmatism may have crystallized in the interval of 2054–2061 Ma (Mapeo et al. 2004; Scoates and Friedman 2008; Yudovskaya et al. 2013; Rajesh et al. 2013). Comparison of recently published ages filtered for apparent uncertainties of less than ± 5–10 Ma, however, is problematic. The compiled ages include a wide range of minerals analyzed (e.g., zircon, baddeleyite, titanite, monazite, biotite), analytical techniques (e.g., TIMS [+ thermal evaporation], CA-TIMS, SHRIMP, LA-ICP-MS), and age interpretations (e.g., upper intercept $^{207}Pb/^{206}Pb$ ages from discordant results, weighted average $^{207}Pb/^{206}Pb$ ages for concordant results, Pb–Pb evaporation ages, $^{40}Ar/^{39}Ar$ plateau or inverse isochron ages). Few studies involve common reference materials, reference values, and data reduction techniques. In this study alone, the ca. 2057 Ma ages for the Merensky Reef samples are ~2.5 million years older than our previously published result from the same materials, an age range that may encompass the crystallization and cooling of the entire thickness of the intrusion (Cawthorn and Webb 2013). Additionally, in the absence of a common technique referenced to the same standards, a true comparison of the published geochronological results for the Bushveld Complex and related rocks requires use of all external errors, including uncertainty in decay constants for the different isotopic systems (e.g., Min et al. 2000; Renne et al. 2010), an exercise that will inevitably lead to significantly larger and overlapping uncertainty. Determination of the absolute and precise duration of Bushveld-related magmatism will require careful application of the single-grain CA-ID-TIMS U–Pb zircon method, referenced to EARTHTIME synthetic reference solutions and common data reduction protocols, to select samples of previously dated samples and intrusions and to newly targeted samples.

There remain, as of yet, few published datasets for samples that directly date the timing of crystallization of mafic-ultramafic rocks from the Rustenburg Layered Suite. The CA-TIMS U-Pb zircon ages for the Merensky Reef, 2056.88±0.41 and 2057.04±0.55 Ma, establish the synchronous crystallization of this horizon within the uppermost Upper Critical Zone across the Bushveld Complex (see Fig. 1.8a). Olsson et al. (2010) report a U–Pb age of 2057.7±1.6/6.4 Ma (±X/Z) for baddeleyite separated from a coarse-grained norite collected from the Marginal Zone (Eastern Limb). The date is an upper intercept $^{207}Pb/^{206}Pb$ age regressed through the slightly discordant (0.8–2.7%) U–Pb results of four multi-grain (n=10–23) baddeleyite fractions. This age is indistinguishable from the Merensky Reef ages within 2s uncertainty (internal errors only) and it is permissible that the lowermost part of the intrusion, from the Marginal Zone through the Upper Critical Zone, crystallized at ca. 2057 Ma. Direct comparison between the two datasets, however, is not possible at this time as Olsson et al. (2010) use an "in-house program" for data reduction and do not report use of EARTHTIME reference synthetic solutions to allow for interlaboratory comparison and calibration. The Merensky Reef ages are now closer to that of Buick et al. (2001) for titanite that formed in a retrogressed xenolith in the Upper Zone (Eastern Limb). Their U–Pb age of 2058.9±0.8 Ma (2s, weighted $^{207}Pb/^{206}Pb$ age, internal errors) from three large multi-grain titanite fractions (5–12 mg) provide an indirect age of crystallization for the uppermost part of the Bushveld Complex that appears to be older than the Merensky Reef ages by at least 500,000 years assuming that there is no interlaboratory bias. Whether or not titanite growth in the calc-silicate protolith occurred during hydrothermal alteration related to Rooiberg magmatism with subsequent partial resetting during incorporation of the xenolith into the Bushveld magmas (Scoates and Friedman 2008) or that titanite growth indeed records the age of Bushveld magmatism remains to be tested.

This study demonstrates that it is possible to extract a wide range of chronometric minerals from single samples of cumulates from layered intrusions (Fig. 1.16). The resultant geochronologic information can be used to constrain crystallization and cooling histories for individual samples and by extension to entire intrusions.

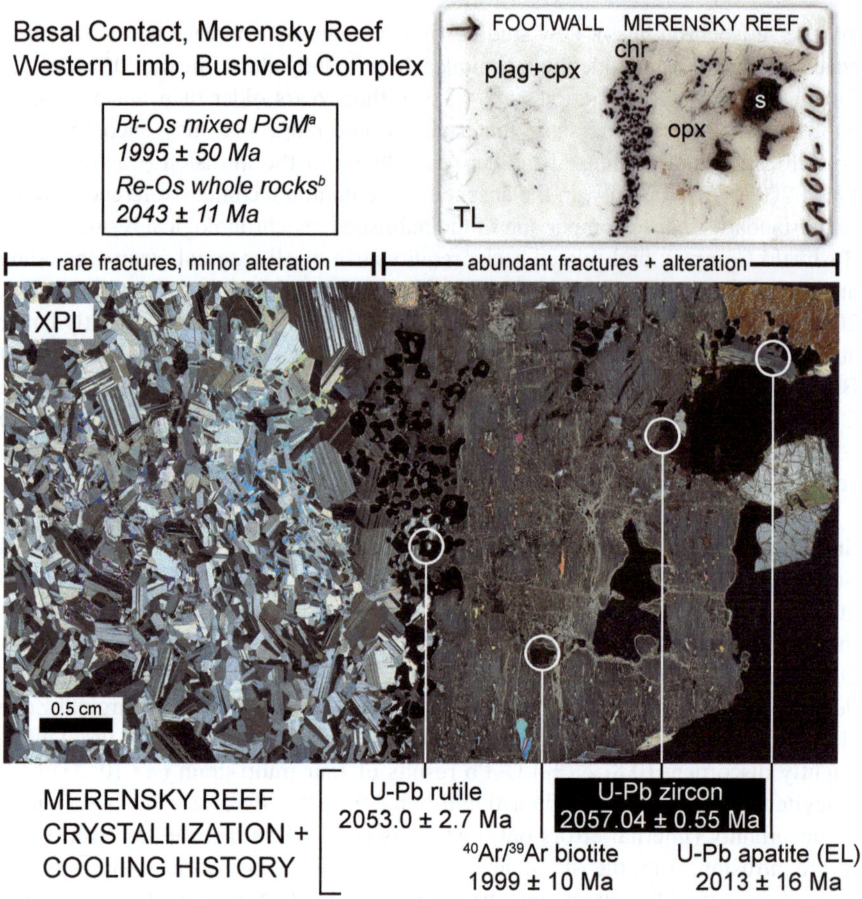

Fig. 1.16 Ages and petrologic setting of interstitial minerals dated from a sample of the Merensky Reef (SA04-13, Western Limb). A scanned thin section (45 × 25 mm) across the basal contact of the Reef is shown in the *upper right* of the figure in transmitted light (*TL*) for reference and in cross-polarized light (*XPL*) in the main part of the figure. The section displays the *uppermost* hanging wall leucogabbronorite to anorthosite (*left*), basal chromitite stringer (*centre*), and pegmatitic feldspathic orthopyroxenite (*right*). The *circled* areas indicate the approximate setting of interstitial minerals dated by the U–Pb method (zircon, rutile, apatite) and by the $^{40}Ar/^{39}Ar$ method (biotite); the ages determined in this study are shown below. Note that the apparent apatite age (U–Pb isochron) is from the Eastern Limb sample for comparison. This single sample captures the crystallization and cooling history of the Bushveld Complex from this locality of the Merensky Reef. Of note is the relatively pristine footwall plagioclase-rich rock with rare fractures and minor alteration compared to the Reef rock that is strongly fractured with abundant secondary alteration due to late hydrothermal alteration confined to the horizon of the Merensky Reef. Also indicated are previously published ages for the Merensky Reef package using different isotopic systems: (**a**) Merensky Reef (Western Limb), Pt–Os isochron age from mixed platinum group minerals (*PGM*), Coggon et al. (2012), (**b**) Bastard unit (Western Limb), whole rock Re–Os isochron for four poikilitic pyroxenites, Schoenberg et al. (1999). Abbreviations: *plag* plagioclase, *cpx* clinopyroxene, *chr* chromite, *opx* orthopyroxene, *s* sulphide

Most of the minerals analyzed in this study (zircon, apatite, biotite) demonstrably crystallized from late fractionated melt following crystallization of plagioclase-orthopyroxene-sulphide as they are found interstitial to these major rock-forming minerals (e.g., Figs. 1.2, 1.16) and they are observed in direct association with patches of intergrown quartz + alkali feldspar (Fig. 1.11). Rutile appears to have crystallized earlier, at higher temperatures, based on its presence as rims on chromite (Western Limb) that are poikilitically enclosed by plagioclase (Fig. 1.16) or as acicular crystals included in plagioclase, orthopyroxene and biotite (Eastern Limb). Following crystallization, the textures and mineralogy of the Merensky Reef sample in the western Bushveld Complex indicate that it was altered during fluid-rock interaction associated with subsequent hydrothermal activity. The footwall leucogabbronorite is unfractured with only trace patches of secondary minerals, whereas the feldspathic pegmatitic orthopyroxenite, particularly orthopyroxene, is very strongly fractured (sub-parallel to the basal contact defined by the chromitite stringer) with abundant secondary minerals (e.g., actinolite, epidote, talc, calcite) (Fig. 1.16). The margins of the coarse clots of base metal sulphides (pyrrhotite, pentlandite, chalcopyrite) have been replaced by secondary minerals, similar to the assemblages observed by Li et al. (2004) in their study of sulphide replacement from the UG2 and Merensky Reef (e.g., actinolite, epidote, calcite, magnetite). The presence of micro-fractures and evidence for interaction with an aqueous fluid (e.g., hydrous silicates, calcite) are particularly relevant to interpreting the chronologic information recorded by biotite. The relatively low closure temperature for Ar diffusion in biotite (~350–450 °C depending on cooling rate and diffusion dimension: Grove and Harrison 1996; Lee 1995; Nomade et al. 2004), combined with its weakly bound sheet structure, allow for the possibility of open-system behavior, argon loss, and resetting during fracture-controlled circulation of aqueous hydrothermal fluids. The anomalously young ca. 2000 Ma biotite $^{40}Ar/^{39}Ar$ ages for the Merensky Reef sample from Rustenburg (Western Limb), and the disturbed U–Pb systematics of apatite, are consistent with the effects of a post-magmatic hydrothermal event.

The thermal history of the Bushveld Complex, which is recorded in the mineral ages and their associated closure temperatures from the Merensky Reef, can be schematically divided into five stages (Fig. 1.17). Stage 1 is constrained by the high-precision U–Pb zircon results from both samples and dates the time of zircon crystallization (~2057±0.5 Ma) from highly fractionated interstitial melt at near-solidus temperatures (1000 °C or less). Stage 2 involved rapid cooling (~125 °C/Ma) down to temperatures of ~400–450 °C by 2053 Ma defined by the U–Pb rutile ages from the same samples. The geochronologic evidence for rapid cooling supports the results of thermal modeling of the Bushveld Complex indicating that the 8 km-thick layered intrusion was emplaced and cooled from magma liquidus temperatures down to the Curie temperature of magnetite (580 °C) in less than 700,000 years (Cawthorn and Walraven 1998; Cawthorn and Webb 2013). Assuming that their respective closure temperatures are similar, the $^{40}Ar/^{39}Ar$ results from Nomade et al. (2004) for biotite from the UG2 chromitite, Western Limb (2042.4±3.2 Ma), and from Cassata et al. (2009) for biotite from a gabbro near the MG-1 chromitite

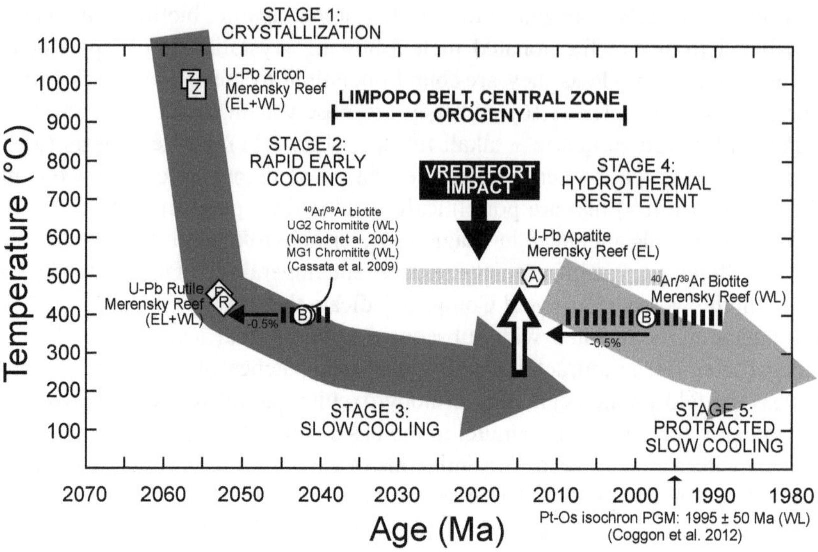

Fig. 1.17 Summary diagram showing a cooling curve for the Bushveld Complex as recorded by the different closure temperatures and ages for minerals from the Merensky Reef. The crystallization and cooling history of the Merensky Reef is divided into five stages that are constrained by U–Pb zircon results for crystallization of zircon at near-solidus temperatures (*Stage 1*), followed by rapid early cooling (*Stage 2*) defined by U–Pb rutile ages and from $^{40}Ar/^{39}Ar$ biotite results from Nomade et al. (2004) and Cassata et al. (2009) assuming a −0.5 % bias between the calibrations of the $^{40}Ar/^{39}Ar$ and U–Pb systems (indicated by the *arrow*), and initial slow cooling (*Stage 3*) below 300 °C. *Stage 4* is a postulated hydrothermal reset event of the Merensky Reef in the Western Limb of the Bushveld Complex recorded by the $^{40}Ar/^{39}Ar$ biotite ages (*arrow* indicates −0.5 % bias) and U–Pb isochron for apatite from this study. *Stage 5* involves protracted slow cooling from 300 to 150 °C over 600 Ma based on $^{40}Ar/^{39}Ar$ plagioclase results from Cassata et al. (2009). For reference, the timing of the Vredefort impact event to the south of the Bushveld Complex (ca. 2020 Ma: Kamo et al. 1996; Moser 1997) and orogeny in the Central Zone of the Limpopo Belt (Kramers and Mouri 2011) to the north are shown. Also shown is the Pt–Os isochron age from platinum group minerals from the Merensky Reef from Coggon et al. (2012). Analytical uncertainty on the U–Pb zircon and rutile ages is smaller than the symbol size. See text for discussion of specific closure temperatures. Abbreviations: *Z* zircon, *R* rutile, *A* apatite, *B* biotite, *WL* Western Limb, *EL* Eastern Limb

layer in the Critical Zone (2042±2 Ma), are both consistent with the U–Pb rutile ages when an −0.5 % bias between the calibrations of the $^{40}Ar/^{39}Ar$ and U–Pb systems is applied (e.g., Min et al. 2000; Nomade et al. 2004; Renne et al. 2010) (Fig. 1.17). If the generally accepted bias of 1 % is used instead (Renne et al. 2010), then the $^{40}Ar/^{39}Ar$ biotite ages from these samples that occur in rocks stratigraphically beneath the Merensky Reef, is 2062 Ma, which is older than the U–Pb zircon ages from the Merensky Reef. Stage 3 marks the onset of slow cooling of the Bushveld Complex below 300 °C based on high-precision $^{40}Ar/^{39}Ar$ data and Ar diffusion kinetics from plagioclase from a Critical Zone gabbro that indicate protracted slow cooling (Stage 5) through crustal temperatures of 150–200 °C over 600 Ma (Cassata et al. 2009). Stage 4 represents the hydrothermal event that locally affected the Merensky Reef in the sampled area of the Western Limb and that is supported both by the textural and mineralogical evidence for fracturing and secondary alteration of the reef and by the ca. 2000 Ma ages for biotite (Fig. 1.17). This hydrothermal event occurred in the interval from 2010 to 2020 Ma when considering a 0.5–1 % bias correction for comparison with the results from the U–Pb system. Mixed platinum group minerals (e.g., laurite, cooperate, sperrylite) yield a Pt–Os isochron age for the Merensky Reef of 1995 ± 50 Ma from mineral concentrates collected from the Frank Shaft (Coggon et al. 2012), ~9 km to the ESE of the Townlands Shaft (sample SA04-13 locality), and thus also appears to record the effects of this late, low-temperature hydrothermal event.

There are a number of potential heat sources that could have driven fluid flow and alteration in the Merensky Reef. Candidates include (1) the hydrothermal convection system of the Bushveld Complex itself, which is unlikely given that maintaining a hydrothermal system for 40–50 million years after initial emplacement even for the largest mafic layered intrusion on Earth vastly exceeds that now established even for giant gold deposits (e.g., Carlin, Nevada: Hickey et al. 2014) and porphyry systems where the duration of hydrothermal activity is constrained to be in the range of 10s of thousands of years up to 1–2 Ma (e.g., Cathles et al. 1997; Chiaradia et al. 2013), (2) hydrothermal activity that followed the Vredefort impact event at ca. 2020 Ma (Kamo et al. 1996; Moser 1997) (Fig. 1.17) when an asteroid struck the Kaapvaal craton 140 km to the south of the Bushveld Complex, ablating > 15 km of crust at the centre of the 70 km-diameter crater and producing a 300 km^2 thermal imprint, although impact-related metamorphism and hydrothermal circulation would likely have been limited to within the final crater itself (Turtle et al. 2003), and (3) hydrothermal circulation related to orogenesis at 2.02 ± 0.02 Ga (Fig. 1.17) in the Central Zone of the Limpopo Belt on the northern edge of the Kaapvaal craton that was associated with zones of transcurrent shearing at high-grade granulite-facies metamorphism (e.g., McCourt and Armstrong 1998; Buick et al. 2007; Kramers and Mouri 2011), which is potentially the most likely scenario given the presence of a regional heat source and stress field. Whether late hydrothermal activity is restricted only to the Merensky Reef in the Rustenburg area or whether it is more widespread will require detailed chronological studies of samples from across the Bushveld Complex (Western, Eastern, and Northern limbs) with multiple chronometric minerals.

Summary and Future Directions

Layered intrusions have long been considered to provide the clearest natural examples of the effects of fractional crystallization of magmas (Wager and Brown 1967). They formed throughout Earth history and they preserve rock records in such remarkable igneous bodies as the Bushveld Complex, Stillwater Complex, Muskox intrusion, and Skaergaard intrusion, and in innumerable smaller bodies exposed in crustal sections worldwide, that provide a template for how mafic magma differentiates in the Earth's crust. Yet, comprehensive age information involving integration of high-precision absolute ages, duration of magmatism, and cooling history is still lacking for most of them.

In this contribution, we have provided the field and petrologic context for identifying samples from layered intrusions with the potential to contain a wide array of U–Th–Pb-bearing accessory minerals, and K-bearing minerals, that are dateable by modern mass spectrometric techniques, especially zircon by the chemical abrasion-TIMS, or CA-TIMS, technique (Mattinson 2005). Prospective samples from mafic and ultramafic cumulates are those with heterogeneous textures, not equigranular or adcumulus textures, and with macroscopic evidence for interstitial minerals, typically poikilitic plagioclase, that crystallized from evolved interstitial melt. From the microscopic perspective, such samples may contain interstitial patches with minerals that crystallized from highly fractionated melt (e.g., quartz, alkali feldspar) at near-solidus temperatures. Dateable accessory minerals (e.g., zircon, baddeleyite, rutile, apatite), although typically present in minute quantities (a few tens to hundreds of grains per 10–20 kg of sample), are in sufficient quantity to provide high-precision geochronologic results where individual grains or grain fragments can be processed and analyzed. Single samples, separated by ~300 km, of the platiniferous Merensky Reef from the Bushveld Complex of South Africa can yield multiple chronometric minerals with different closure temperatures (e.g., Pb diffusion in zircon, rutile, and apatite; $^{40}Ar^{*}$ in biotite) that allow for the cooling histories of the samples, and by extrapolation that of the entire Bushveld Complex, to be established. And through the efforts of EARTHTIME, with isotopic tracers, U–Pb synthetic standards, and new data collection and reduction software available to the community to alleviate intra-laboratory bias, reductions in Pb blanks, and carefully characterized Pb fractionation of standard materials, it is possible to achieve internal uncertainties on the U–Pb analysis of single zircon grains to significantly less than 0.1 %. This now means that uncertainties of ± 500,000 years or less are attainable for Paleoproterozoic to Archean layered intrusions (e.g., Bushveld Complex, Stillwater Complex) and potentially ± 10,000–20,000 years for Phanerozoic intrusions, thus approaching the timescales of residence, magma storage, and differentiation of magmas in the crust (Turner and Costa 2007; Cooper and Kent 2014).

The discovery that zircon is more common in mafic-ultramafic rocks of layered intrusions than previously recognized provides the petrologic community with a wide range of new geochemical tools to decipher and quantify the processes that operate during their crystallization and consolidation. Careful petrographic studies involving the textural setting of zircon in cumulates and detailed SEM-BSE and SEM-CL imaging are required to fully document the timing of zircon saturation and crystal growth relative to other interstitial minerals (e.g., biotite, quartz, alkali feldspar, apatite, rutile). Because of the low abundance of zircon (<< 1 vol%) in mafic-ultramafic cumulates, the probability of intersecting crystals in thin sections is relatively small. An alternative approach to characterizing the distribution of zircon, and association with major cumulus minerals and pockets of trapped interstitial melt, is through the use of high-resolution X-ray computed tomography, a technique that has revolutionized our understanding of the spatial relationship of platinum group minerals to sulphide blebs and chromite grains in PGE-rich reefs from layered intrusions (e.g., Merensky Reef, Bushveld Complex; J-M Reef, Stillwater Complex; Platinova Reef, Skaergaard intrusion) (Godel et al. 2010; Godel et al. 2014).

In addition to U+Th, zircon can also incorporate a wide range of different elements in trace to wt% amounts, including Ti, Hf, and the rare earth elements (REE), through a variety of substitution mechanisms (e.g., Harley and Kelly 2007; Nardi et al. 2013). This will allow for future studies on the trace element and isotopic composition of zircon from layered intrusions that can be compared to information derived from the higher temperature cumulus minerals and thus potentially serve to map-out the complete geochemical evolution of cumulates from liquidus to solidus. The Ti content of zircon coexisting with rutile, which has a strong dependence on temperature (Watson et al. 2006; Ferry and Watson 2007; Fu et al. 2008), can be measured in situ by LA-ICP-MS or SHRIMP from zircon in thin sections or out-of-context zircon grains from mineral separates, so that the temperature at which zircon crystallized in cumulates can now be quantified. The REE contents of zircon from layered intrusions can be used to reconstruct the geochemical evolution of the fractionated melt from which it crystallized and help identify melt compositions and magma sources (e.g., Belousova et al. 2002; Claiborne et al. 2010; Nardi et al. 2013), which is a rapidly expanding field of research, especially for zircon in felsic rocks.

New methods have been developed to allow for the simultaneous acquisition of high-precision U–Pb geochronology and zircon trace element analysis (U–Pb TIMS-TEA, Schoene et al. 2010), a technique that can be expanded to include baddeleyite, titanite, apatite, rutile, monazite, and xenotime, if they are found to be present in cumulates under investigation. The radiogenic isotopic systematics of zircon (Lu–Hf, Sm–Nd) (e.g., Kinny and Maas 2003; Scherer et al. 2007; Barboni et al. (2013) and oxygen isotope signatures of zircon (Valley 2003; Valley et al. 2005) can provide information on variations in crustal contamination, crustal residency, and protolith and source compositions from the analysis (LA-MC-ICP-MS, SHRIMP) of single grains from cumulates sampled throughout the stratigraphic thicknesses of individual intrusions (e.g., Bushveld Complex, Zirakparvar et al. 2014). It is now also possible to conduct correlated microanalysis of zircon (trace elements, oxygen isotopes, Hf isotopes, U–Th–Pb geochronology) (Cavosie et al. 2006), which would be advantageous when zircon populations in layered intrusions are complex and spatial resolution is required (e.g., inherited cores, cracks, radiation-damaged zones), although with the trade-off of reduced precision. Depending on the particular layered intrusion and its specific emplacement and crystallization history, it may even be possible to identify the different types of zircon that are now recognized in arc plutons (e.g., autocrysts, antecrysts, xenocrysts, inherited zircon; Miller et al. 2007).

By combining these advances in technologies and applications with our current capabilities of determining high-precision crystallization and cooling ages from a range of rock types in layered intrusions, Archean to Cenozoic, the next 10–15 years of research will undoubtedly provide many important, and unexpected, discoveries based on the geochronology of layered intrusions with implications for the evolution of mafic magmatism in the Earth's crust. The geochronological tools are now available to address whether major layered intrusions are assembled in discrete magmatic episodes, whether all layered intrusions are simple stratigraphic sequences of cumulates with the oldest rocks at the base and the youngest at the top near the roof, whether discontinuities in cumulate sequences can be identified and time gaps measured, and whether their associated mineral deposits are directly related to the crystallization of their immediate host rocks or formed or were modified at some temporally resolvable time after crystallization. Answers to these questions will directly impact our understanding of the emplacement, crystallization, and cooling of these exceptional bodies of igneous rocks.

Acknowledgements

We thank Wolf Maier for leading a superb field trip on the Eastern Limb of the Bushveld Complex and to Anglo Platinum for allowing access and sampling in the West Mine, Rustenburg Section of the Western Limb of the Bushveld Complex. Our thanks to Bill Meurer for helping in initiating selection and collection of geochronology samples from the Stillwater Complex many summers ago. We are extremely grateful for the assistance, input, advice, and comments from Richard Friedman, who has been instrumental in maintaining the daily operation of the U–Pb geochronology stream at the PCIGR for so many years. We thank Hai Lin for help with sample processing (crushing, Wilfley Table, heavy liquids, and magnetic separation) and Yeena Lin for careful clean lab chemistry. Thanks to Mati Raudsepp for his expertise and management of the scanning electron microscope and associated cathodoluminescence detector at UBC. Dominique Weis has provided many useful ideas over the years on isotopic systematics and mass spectrometry that have significantly improved our U–Pb TIMS geochronology capabilities. Tom Ullrich conducted the $^{40}Ar/^{39}Ar$ analyses and data reduction. Many thanks to Jon Scoates for a timely and detailed review of the manuscript. We greatly appreciate the insightful review comments of Kevin Chamberlain, Wolf Maier, and an anonymous reviewer, in addition to the efficient editorial handling by Bernard Charlier. Funding for this study has been provided through a NSERC CGS-M and PGS-D awarded to Corey Wall and NSERC Discovery Grants awarded to Dr. James Scoates.

References

Alapieti T (1982) The Koillismaa layered igneous complex, Finland—its structure, mineralogy and geochemistry, with emphasis on the distribution of chromium. Geol Surv Finl Bull 319:116

Alapieti TT, Kujanpää J, Lahtinen JJ, Papunen H (1989) The Kemi stratiform chromitite deposit, northern Finland. Econ Geol 84:1057–1077 doi:10.2113/gsecongeo.84.5.1057

Amelin Y, Zaitsev AN (2002) Precise geochronology of phoscorites and carbonatites: the critical role of U-series disequilibrium in age interpretations. Geochim Cosmochim Acta 66:2399–2419. doi:10.1016/S0016-7037(02)00831-1

Amelin Y, Davis WJ (2006) Isotopic analysis of lead in sub-nanogram quantities by TIMS using a ^{202}Pb–^{205}Pb spike. J Anal At Spectrom 21:1053–1061. doi:10.1039/b606842a
Amelin YV, Heaman LM, Semenov VS (1995) U-Pb geochronology of layered mafic intrusions in the eastern Baltic Shield: implications for the timing and duration of Paleoproterozoic continental rifting. Precambrian Res 75:31–46. doi:10.1016/0301-9268(95)00015-W
Ballhaus CG (1988) Potholes of the Merensky Reef at Brakspruit Shaft, Rustenburg Platinum Mines: primary disturbances in the magmatic stratigraphy. Econ Geol 83:1140–1158. doi:10.2113/gsecongeo.83.6.1140
Ballhaus CG, Stumpfl EF (1986) Sulfide and platinum mineralization in the Merensky Reef: evidence from hydrous silicates and fluid inclusions. Contrib Mineral Petrol 94:193–204. doi:10.1007/BF00592936
Barboni M, Schoene B, Ovtcharova M, Bussy F, Schaltegger GA (2013) Timing of incremental pluton construction and magmatic activity in a back-arc setting revealed by ID-TIMS U/Pb and Hf isotopes on complex zircon grains. Chem Geol 340:76–93. doi:10.1016/j.chemgeo.2012.12.011
Barfod GH, Otero O, Albarède F (2003) Phosphate Lu-Hf geochronology. Chem Geol 200:241–253. doi:10.1016/S0009-2541(03)00202-X
Barkov AY, Savchenko YE, Men'shikov YP, Barkova LP (1996) Loveringite from the Last-Yavr mafic-ultramafic intrusion, Kola Peninsula; a second occurrence in Russia. Nor Geol Tidsskr 76:115–120
Barnes S-J, Maier WD (2002) Platinum-group elements and microstructures of normal Merensky Reef from Impala Platinum Mines, Bushveld Complex. J Petrol 43:103–128. doi:10.1093/petrology/43.1.103
Belousova EA, Griffin WL, O-Reilly SY, Fisher NI (2002) Igneous zircon: trace element composition as an indicator of source rock type. Contrib Mineral Petrol 143:602–622. doi:10.1007/s00410-002-0364-7
Boehnke P, Watson EB, Trial D, Harrison RM, Schmitt AK (2013) Zircon saturation re-revisited. Chem Geol 351:324–334 doi:10.1016/j.chemgeo.2013.05.02
Boudreau A (2011) The evolution of texture and layering in layered intrusions. Int Geol Rev 53:330–353. doi:10.1080/00206814.2010.496163
Boudreau AE, Hoatson DM (2004) Halogen variations in the Paleoproterozoic layered mafic-ultramafic intrusions of the East Kimberley, Western Australia: implications for platinum group element mineralization. Econ Geol 99:1015–1026. doi:10.2113/gsecongeo.99.5.1015
Boudreau AE, Mathez EA, McCallum IS (1986) Halogen geochemistry of the Stillwater and Bushveld complexes: evidence for transport of the platinum-group elements by Cl-rich fluids. J Petrol 27:967–986. doi:10.1093/petrology/27.4.967
Bowen NL (1928) The evolution of igneous rocks. Dover Publications, New York, NY p 332
Bowring JF, McLean NM, Bowring SA (2011) Engineering cyber infrastructure for U-Pb geochronology: Tripoli and U-Pb_Redux. Geochem Geophys Geosystems 12:(Q0AA19), p 19. doi:10.1029/2010GC003479
Bristow DG, Cawthorn RG, Harmer J, Lee CA, Tegner C, Vijoen MJ, Walraven F, Wilson JR (1993) Field excursion to the Bushveld Complex, 11–17th Sept 1993. Excursion guide—symposium on layering in igneous complexes, p 59
Buchanan PC, Reimold WU, Koeberl C, Kruger FJ (2002) Geochemistry of intermediate to siliceous volcanic rocks of the Rooiberg Group, Bushveld Magmatic Province, South Africa. Contrib Mineral Petrol 144:131–143. doi:10.1007/s00410-002-0386-1
Buick IS, Maas R, Gibson R (2001) Precise U-Pb titanite age constraints on the emplacement of the Bushveld Complex, South Africa. J Geol Soc Lond 158:3–6. doi:10.1144/jgs.158.1.3
Buick IS, Hermann J, Maas R, Gibson RL (2007) The timing of sub-solidus hydrothermal alteration in the Central Zone, Limpopo Belt (South Africa): constraints from titanite U–Pb geochronology and REE partitioning. Lithos 98:97–117. doi:10.1016/j.lithos.2007.02.002
Cabella R, Gazzotti M, Lucchetti G (1997) Loveringite and baddeleyite in layers of chromian spinel from the Bracco ophiolitic unit, Northern Apennines, Italy. Can Mineral 35:899–908
Cameron EN (1979) Titanium-bearing oxide minerals of the critical zone of the eastern Bushveld Complex. Am Mineral 64:140–150

Campbell IH (1987) Distribution of orthocumulate textures in the Jimberlana intrusion. J Geol 95:35–53

Campbell IH, Kelly PR (1978) The geochemistry of loveringite, a uranium–rare-earth-bearing accessory phase from the Jimberlana intrusion of Western Australia. Mineral Mag 42:187–193

Campbell IH, Naldrett AJ, Barnes SJ (1983) A model for the origin of the platinum-rich sulfide horizons in the Bushveld and Stillwater complexes. J Petrol 24:133–165. doi: 10.1093/petrology/24.2.133

Carr HW, Groves DI, Cawthorne RG (1994) The importance of synmagmatic deformation in the formation of Merensky Reef potholes in the Bushveld Complex. Econ Geol 89:1398–1410. doi:10.2113/gsecongeo.89.6.1398

Cassata WS, Renne PR, Shuster DL (2009) Argon diffusion in plagioclase and implications for thermochronology: a case study from the Bushveld Complex, South Africa. Geochim Cosmochim Acta 73:6600–6612. doi:10.1016/j.gca.2009.07.017

Cathles LM, Erendi AHJ, Barrie T (1997) How long can a hydrothermal system be sustained by a single intrusive event? Econ Geol 92:766–771. doi:10.2113/gsecongeo.88.8.1977

Cavosie AJ, Valley JW, Wilde SA, EIMF (2006) Correlated microanalysis of zircon: Trace element, $\delta^{18}O$, and U–Th–Pb isotopic constraints on the igneous origin of complex >3900 Ma detrital grains. Geochim Cosmochim Acta 70:5601–5616. doi:10.1016/j.gca.2006.08.011

Cawthorn RG (ed) (1996) Layered intrusions (Developments in Petrology 15). Elsevier, Amsterdam, The Netherlands p 531

Cawthorn RG, Boerst K (2006) Origin of the pegmatitic pyroxenite in the Merensky unit, Bushveld Complex, South Africa. J Petrol 47:1509–1530. doi:10.1093/petrology/egl017

Cawthorn RG, Walraven F (1998) Emplacement and crystallization time for the Bushveld Complex. J Petrol 39:1669–1687. doi:10.1093/petroj/39.9.1669

Cawthorn RG, Webb SJ (2013) Cooling of the Bushveld Complex, South Africa: implications for paleomagnetic reversals. Geology 41:687–690. doi:10.1130/G34033.1

Cawthorn RG, Barnes SJ, Ballhaus C, Malitch KN (2005) Platinum group element, chromium, and vanadium deposits in mafic and ultramafic rocks. In: Hedenquist JW et al. (eds) Economic Geology (One Hundredth Anniversary Volume), Littleton, CO pp 215–250 Society of Economic Geologists, Inc.

Chamberlain KR, Bowring SA (2000) Apatite–feldspar U–Pb thermochronometer: a reliable, mid-range (~450°C), diffusion-controlled system. Chem Geol 172:173–200. doi:10.1016/S0009-2541(00)00242-4

Chamberlain KR, Schmitt AK, Swapp SM, Harrison TM, Swoboda-Colberg N, Bleeker W, Peterson TD, Jefferson CW, Khudoley AK (2010) In situ U–Pb SIMS (IN-SIMS) micro-baddeleyite dating of mafic rocks: method with examples. Precambrian Res 183:379–387. doi:10.1016/j.precamres.2010.05.004

Chang Z, Vervoort JD, McClelland WC, Knaack C (2006) U-Pb dating of zircon by LA-ICP-MS. Geochem Geophys Geosystems 7:(Q05009), p 14. doi:10.1029/2005GC001100

Cherniak DJ (1993) Lead diffusion in titanite and preliminary results on the effects of radiation damage on Pb transport. Chem Geol 110:177–194. doi:10.1016/0009-2541(93)90253-F

Cherniak DJ (2000) Pb diffusion in rutile. Contrib Mineral Petrol 139:198–207

Cherniak DJ (2010) Diffusion in accessory minerals: zircon, titanite, apatite, monazite and xenotime. In: Zhang Y, Cherniak DJ (eds) Diffusion in minerals and melts, Reviews in Mineralogy and geochemistry, Chantilly, VA vol. 72. pp 827–869. doi:10.2138/rmg.2010.72.18 The Mineralogical Society of America

Cherniak DJ, Watson EB (2000) Pb diffusion in zircon. Chem Geol 172:5–24. doi:10.1007/PL00007671

Cherniak DJ, Watson EB (2003) Diffusion in zircon. In: Hanchar JM, Hoskin PWO (eds) Zircon, Reviews in Mineralogy and Geochemistry, Washington, DC vol. 53. pp 113–143. doi:10.2113/0530113 The Mineralogical Society of America

Cherniak DJ, Lanford WA, Ryerson FJ (1991) Lead diffusion in apatite and zircon using ion implantation and Rutherford backscattering techniques. Geochim Cosmochim Acta 55:1663–1673. doi:10.1016/0016-7037(91)90137-T

Chew DM, Petrus JA, Kamber BS (2014) U–Pb LA–ICPMS dating using accessory mineral standards with variable common Pb. Chem Geol 363:185–199. doi:10.1016/j.chemgeo.2013.11.006

Chiaradia M, Schaltegger U, Spiking R, Wotzlaw J-F, Ovtcharova M (2013) How accurately can we date the duration of magmatic-hydrothermal events in porphyry systems?—an invited paper. Econ Geol 108:565–584. doi:10.2113/econgeo.108.4.565

Chutas NI, Bates E, Prevec SA, Coleman DS, Boudreau AE (2012) Sr and Pb isotopic disequilibrium between coexisting plagioclase and orthopyroxene in the Bushveld Complex, South Africa: microdrilling and progressive leaching evidence for sub-liquidus contamination within a crystal mush. Contrib Mineral Petrol 163:653–668. doi:10.1007/s00410-011-0691-7

Claiborne LL, Miller CF, Wooden JL (2010) Trace element composition of igneous zircon: a thermal and compositional record of the accumulation and evolution of a large silicic batholith, Spirit Mountain, Nevada. Contrib Mineral Petrol 160:511–531. doi:10.1007/s00410-010-0491-5

Coggon JA, Nowell GM, Pearson DG, Oberthür T, Lorand J-P, Melcher F, Parman SW (2012) The ^{190}Pt–^{186}Os decay system applied to dating platinum-group element mineralization of the Bushveld Complex, South Africa. Chem Geol 302–303:48–60. doi:10.1016/j.chemgeo.2011.10.015

Compston W, Williams IS, Meyer C (1984) U–Pb geochronology of zircons from lunar breccia 73217 using a sensitive high mass-resolution ion microprobe. J Geophys Res Suppl 89:B525–B534

Condon D, Schoene B, Bowring S, Parrish R, Mclean N, Noble S, Crowley Q (2007) Earthtime: Isotopic tracers and optimized solutions for high-precision U-Pb ID-TIMS geochronology. American Geophysical Union, Fall Meeting 2007 [abstract #V41E-06].

Condon DJ, McKean N, Noble SR, Bowring SA (2010) Isotopic composition (^{238}U/^{235}U) of some commonly used uranium reference materials. Geochim Cosmochim Acta 74:7127–7143. doi:10.1016/j.gca.2010.09.019

Cooper KM, Kent AJR (2014) Rapid remobilization of magmatic crystals in cold storage. Nature 506:480–483. doi:10.1038/nature12991

Corfu F, Krogh TE, Ayres LD (1985) U–Pb zircon and sphene geochronology of a composite Archean granitoid batholith, Favorable Lake area, northwestern Ontario. Can J Earth Sci 22:1436–1451. doi:10.1139/e85-150

Corfu F, Hanchar JM, Hoskin PWO, Kinny P (2003) Atlas of zircon textures. In: Hanchar JM, Hoskin PWO (eds) Zircon, Reviews in Mineralogy and Geochemistry, Washington, DC vol. 53. pp 469–500. doi:10.2113/0530469 The Mineralogical Society of America

Cottle JM, Horstwood MSA, Parrish RR (2009) A new approach to single shot laser ablation analysis and its application to in situ Pb/U geochronology. J Anal At Spectrom 24:1355–1363. doi:10.1039/B821899D

Cottle JM, Kylander-Clark AR, Vrijmoed JC (2012) U-Th/Pb geochronology of detrital zircon and monazite by single shot laser ablation inductively coupled plasma mass spectrometry (SS-LA-ICPMS). Chem Geol 332–333:136–147. doi:10.1016/j.chemgeo.2012.09.035

Cousins CA (1969) The Merensky Reef of the Bushveld Igneous Complex. In: Wilson HDB (ed) Magmatic ore deposits: a symposium. Economic Geology Publishing Company, Lancaster, PA pp 239–251

Das A, Davis DW (2010) Response of Precambrian zircon to the chemical abrasion (CA-TIMS) method and implications for improvement of age determinations. Geochim Cosmochim Acta 74:5333–5348. doi:10.1016/j.gca.2010.06.029

Davis DW, Sutcliffe RH (1985) U-Pb ages from the Nipigon plate and northern Lake Superior. Geol Soc Am Bull 96:1572–1579. doi:10.1130/0016-7606(1985)96&#X1;1572:UAFTNP>2.0.CO;2

Davis DW, Williams IS, Krogh TE (2003) Historical developments of zircon geochronology. In: Hanchar JM, Hoskin PWO (eds) Zircon, Reviews in Mineralogy and Geochemistry, Washington, DC vol. 53. pp 145–181. doi:10.2113/0530145 The Mineralogical Society of America

DePaolo DJ (1985) Isotopic studies of processes in mafic magma chambers: I. The Kiglapait intrusion, Labrador. J Petrol 26:925–951. doi:10.1093/petrology/26.4.925

DePaolo DJ, Wasserburg GJ (1979) Sm-Nd age of the Stillwater Complex and the mantle evolution curve for neodymium. Geochim Cosmochim Acta 43: 999–1008. doi:10.1016/0016-7037(79)90089-9

Dickin AP (2005) Radiogenic isotope geology, 2nd edn. Cambridge University Press, Cambridge, UK p 512
Dodson MH (1973) Closure temperature in cooling geochronological and petrological systems. Contrib Mineral Petrol 40:259–274. doi:10.1007/BF00373790
Dodson MH (1978) A linear method for second-degree interpolation in cyclical data collection. J Phys E Sci Instrum 11:296
Eales HV, Cawthorn RG (1996) The Bushveld Complex. In: Cawthorn RG. (ed) Layered intrusions. Elsevier Science B.V., Amsterdam, pp 181–229. doi:10.1016/S0167-2894(96)80008-X
Ernst RE, Buchan KL (2001) Large mafic magmatic events through time and links to mantle-plume heads. Geol Soc Am Spec Pap 352:483–575. doi:10.1130/0-8137-2352-3.483
Ewing RC, Meldrum A, Wang L, Weber WJ, Corrales LR (2003) Radiation effects in zircon. In: Hanchar, JM, Hoskin PWO (eds) Zircon, Reviews in Mineralogy and Geochemistry, Washington, DC vol. 53. pp 387–425. doi:10.2113/0530387 The Mineralogical Society of America
Faure G, Mensing TM (2005) Isotopes: principles and applications, 3rd edn. Wiley, Hoboken, NJ p 928
Fenton MD, Faure G (1969) The age of the igneous rocks of the Stillwater Complex of Montana. Geol Soc Am Bull 80:1599–1604
Ferreira-Filho CF, Kamo SL, Fuck RA, Krogh TE, Naldrett JA (1994) Zircon and rutile U-Pb geochronology of the Niquelândia layered mafic and ultramafic intrusion, Brazil: constraints for the timing of magmatism and high grade metamorphism. Precambrian Res 68:241–255. doi:10.1016/0301-9268(94)90032-9
Ferry JM, Watson EM (2007) New thermodynamic models and revised calibrations for the Ti-in-zircon and Zr-in-rutile thermometers. Contrib Mineral Petrol 154:429–437. doi:10.1007/s00410-007-0201-0
Finch RJ, Hanchar JM (2003) Structure and chemistry of zircon and zircon-group minerals. In: Hanchar JM, Hoskin PWO (eds) Zircon, Reviews in Mineralogy and Geochemistry, Washington, DC vol. 53. pp 1–25. doi:10.2113/0530001 The Mineralogical Society of America
Fisher CM, Longerich HP, Jackson SE, Hanchar JM (2010) Data acquisition and calculation of U–Pb isotopic analyses using laser ablation (single collector) inductively coupled plasma mass spectrometry. J Anal Atomic Spectrom 25:1905–1920: doi:10.1039/c004955g
Francis D (2011) Columbia Hills—an exhumed layered igneous intrusion on Mars? Earth Planet Sci Lett 310:59–64. doi:10.1016/j.epsl.2011.08.003
French JE, Heaman LM, Chacko T (2002) Feasibility of chemical U–Th–total Pb baddeleyite dating by electron microprobe. Chem Geol 188:85–104. doi:10.1016/S0009-2541(02)00074-8
Frei D, Gerdes A (2009) Precise and accurate in situ U–Pb dating of zircon with high sample throughput by automated LA-SF-ICP-MS. Chem Geol 261:261–270. doi:10.1016/j.chemgeo.2008.07.025
Frost BR, Chamberlain KR, Schumacher JC (2000) Sphene (titanite): phase relations and role as a geochronometer. Chem Geol 172:131–148. doi:10.1016/S0009-2541(00)00240-0
Fu B, Page FZ, Cavosie AJ, Fournelle J, Kita NT, Lackey JS, Wilde SA, Valley JW (2008) Ti-in-zircon thermometry: applications and limitations. Contrib Mineral Petrol 156:197–215. doi:10.1007/s00410-008-0281-5
Fujimaki H (1986) Partition coefficients of Hf, Zr, and REE between zircon, apatite, and liquid. Contrib Mineral Petrol 94:42–45. doi:10.1007/BF00371224
Gatehouse BM, Grey IE, Campbell IH, Kelly P (1978) The crystal structure of loveringite—a new member of the crichtonite group. Am Mineral 63:28–36
Geisler T, Pidgeon RT, Kurtz R, Bronswijk WV, Schleicher H (2003) Experimental hydrothermal alteration of partially metamict zircon. Am Mineral 88:1496–1513
Geisler T, Schaltegger U, Tomsachek F (2007) Re-equilibration of zircon in aqueous fluids and melts. Elements 3:43–50. doi:10.2113/gselements.3.1.43
Gerstenberger H, Haase G (1997) A highly effective emitter substance for mass spectrometric Pb isotope ratio determinations. Chem Geol 136:309–312. doi:10.1016/S0009-2541(96)00033-2
Godel B, Barnes SJ, Barnes S-J, Maier WD (2010) Platinum ore in three dimensions: insights from high-resolution X-ray computed tomography. Geology 38:1127–1130. doi:10.1130/G31265.1
Godel B, Rudashevsky NS, Nielsen TFD, Barnes SJ, Rudashevsky VN (2014) New constraints on the origin of the Skaergaard intrusion Cu–Pd–Au mineralization: insights from high-resolution X-ray computed tomography. Lithos 190-191:27–36. doi:10.1016/j.lithos.2013.11.019

Grimes CB, John BE, Cheadle MJ, Mazdab FK, Wooden JL, Swapp S, Schwartz JJ (2009) On the occurrence, trace element geochemistry, and crystallization history of zircon from in situ ocean lithosphere. Contrib Mineral Petrol 158:757–783. doi:10.1007/s00410-009-0409-2

Grove M, Harrison TM (1996) ^{40}Ar diffusion in Fe-rich biotite. Am Mineral 81:940–951

Hamilton MA, Pearson DG, Thompson RN, Kelley SP, Emeleus CH (1998) Rapid eruption of Skye lavas inferred from precise U-Pb and Ar-Ar dating of the Rum and Cuillin plutonic complexes. Nature 394:260–263. doi:10.1038/28361

Hanchar JM, Miller CF (1993) Zircon zonation patterns as revealed by cathololuminescence and backscattered electron images: implications for interpretation of complex crustal histories. Chem Geol 110:1–13. doi:10.1016/0009-2541(93)90244-D

Harker A (1904) The Tertiary igneous rocks of Skye. (Memoir Geological Survey United Kingdom) Glasgow p 481 James Hedderwick and Sons

Harley SL, Kelly NM (2007) Zircon, tiny but timely. Elements 3:13–18. doi:10.2113/gselements.3.1.13

Harrison TM, Zeitler PK (2005) Fundamentals of noble gas thermochronometry. In: Reiner PW, Ehlers TA (eds) Low-temperature thermochronology: techniques, interpretations, and applications, Reviews in Mineralogy and Geochemistry, Chantilly, VA vol. 58, pp 123–149. doi:10.2138/rmg.2005.58.5 The Mineralogical Society of America

Haskin LA, Salpas PA (1992) Genesis of compositional characteristics of Stillwater AN-I and AN-II thick anorthosite units. Geochim Cosmochim 56:1187–212. doi:10.1016/0016-7037(92)90056-O

Heaman LM, LeCheminant AN (1993) Paragenesis and U-Pb systematics of baddeleyite (ZrO_2). Chem Geol 110:95–126. doi:10.1016/0009-2541(93)90249-I

Heaman LM, LeCheminant AN (2000) Anomalous U–Pb systematics in mantle-derived baddeleyite xenocrysts from Ile Bizard: evidence for high temperature radon diffusion? Chem Geol 172:77–93. doi:10.1016/S0009-2541(00)00237-0

Heaman L, Parrish R (1991) U-Pb geochronology of accessory minerals. In: Heaman L, Ludden JN (eds) Applications of radiogenic isotope systems to problems in geology. Mineralogical Association of Canada, short course handbook, Nepean, Canada vol 19. pp 59–102 Mineralogical Association of Canada

Heaman LM, Machado N, Krogh TE, Weber W (1986) Precise U-Pb zircon ages for the Molson dyke swarm and the Fox River sill: constraints for early Proterozoic crustal evolution in northeastern Manitoba, Canada. Contrib Mineral Petrol 94:82–89. doi:10.1007/BF00371229

Hemming SR, Rasbury ET (2000) Pb isotope measurements of sanidine monitor standards: implications for provenance analysis and tephrochronology. Chem Geol 165:331–337. doi:10.1016/S0009-2541(99)00174-6

Hess HH (1960) Stillwater igneous complex, Montana. Geol Soc Am Memoir 80:230

Hickey KA, Barker SLL, Dipple GM, Arehart GB, Donelick RA (2014) The brevity of hydrothermal fluid flow revealed by thermal haloes around giant gold deposits: implications for Carlin-type gold systems. Econ Geol 109:1461–1487

Hiess J, Condon, DJ, McLean N, Noble SR (2012) ^{238}U/^{235}U systematics in terrestrial uranium-bearing mineral. Nature 335:1610–1614. doi:10.1126/science.1215507

Higgins MD (2011) Textural coarsening in igneous rocks. Int Geol Rev 53:354–376. doi:10.1080/00206814.2010.496177

Higgins MD, van Breeman O (1998) The age of the Sept Iles layered mafic intrusion, Canada: implications for the Late Neoproterozoic/Cambrian history of Southeastern Canada. J Geol 106:421–432. doi:10.1086/516033

Hirschmann MM, Renne PR, McBirney AR (1997) ^{40}Ar/^{39}Ar dating of the Skaergaard intrusion. Earth Planet Sci Lett 146:645–658. doi:10.1016/S0012-821X(96)00250-6

Hoatson DM, Blake DH (eds) (2000) Geology and economic potential of the Palaeoproterozoic layered mafic-ultramafic intrusions in the East Kimberley, Western Australia. Aust Geol Surv Organ Bull 246:476

Holland HD, Gottfried D (1955) The effect of nuclear radiation on the structure of zircon. Acta Crystallogr 8:291–300. doi:10.1107/S0365110X55000947

Holness MB, Anderson AT, Mamrtin VM, MacLennan J, Passmore E, Schwindinger K (2007) Textures in partially solidified crystalline nodules: a window into the pore structure of slowly cooled mafic intrusions. J Petrol 48:1243–1264. doi:10.1093/petrology/egm016

Hulbert L (2005) Geology of the Muskox intrusion and associated Ni and Cu occurrences. Geological Survey of Canada, Open File 4881 (CD-ROM)

Hunter RH (1996) Texture development in cumulate rocks. In: Cawthorn RG (ed) Layered intrusions. Elsevier Science B.V., Amsterdam, pp 103–145. doi:10.1016/S0167-2894(96)80006-6

Ireland TR, Williams IS (2003) Considerations in zircon geochronology by SIMS. In: Hanchar JM, Hoskin PWO (eds) Zircon Reviews in mineralogy and geochemistry, Washington, DC vol. 53. pp 215–241.doi:10.2113/0530215 The Mineralogical Society of America

Irvine TN (1982) Terminology for layered intrusions. J Petrol 23:127–162. doi:10.1093/petrology/23.2.127-a

Irvine TN (1987) Appendix I. Glossary of terms for layered intrusions. In: Parsons I (ed) Origins of igneous layering. D. Reidel Publishing Company, Dordrecht, Holland pp 641–647. doi:10.1016/S0167-2894(96)80005-4

Ivanic TJ, Wingate MTD, Kirkland CL, Van Kranendonk MJV, Wyche S (2010) Age and significance of voluminous mafic-ultramafic magmatic events in the Murchison Domain, Yilgarn Craton. Aust J Earth Sci 57:597–614. doi:10.1080/08120099.2010.494765

Jackson ED (1961) Primary textures and mineral associations in the Ultramafic Zone of the Stillwater Complex, Montana. U. S Geol Surv Prof Pap 358:106

Jaffey AH, Flynn KF, Glendenin LE, Bentley WC, Essling AM (1971) Precision measurement of half-lives and specific activities of ^{235}U and ^{238}U. Phys Rev C 4(5):1889–1906. doi:http://dx.doi.org/10.1103/PhysRevC.4.1889

Jerram DA, Cheadle MJ, Philpotts AR (2003) Quantifying the building blocks of igneous rocks: are clustered crystal frameworks the foundation? J Petrol 44:2033–2051. doi:10.1093/petrology/egg069

Kamo SL, Reimold WU, Krogh TE, Colliston WP (1996) A 2.023 Ga age for the Vredefort impact event and a first report of shock metamorphosed zircons in pseudotachylitic breccias and granophyre. Earth Planet Sci Lett 144:369–387. doi:10.1016/S0012-821X(96)00180-X

Kinnaird JA, Hutchinson D, Schurmann L, Nex PAM, de Lange R (2005) Petrology and mineralization of the southern PlatReef: northern limb of the Bushveld Complex, South Africa. Miner Depos 40:576–597. doi:10.1007/s00126-005-0023-9

Kinny PD, Maas R (2003) Lu-Hf and Sm-Nd isotope systems in zircon. In: Hanchar JM, Hoskin PWO (eds) Zircon, Reviews in Mineralogy and Geochemistry, Washington, DC vol. 53. pp 327–341. doi:10.2113/0530327 The Mineralogical Society of America

Kosler J, Sylvester PJ (2003) Present trends and the future of zircon in geochronology: laser ablation ICPMS. In: Hanchar JM, Hoskin PWO (eds.) Zircon, Reviews in Mineralogy and Geochemistry, Washington, DC vol. 53. pp 243–275. doi:10.2113/0530243 The Mineralogical Society of America

Kramers J, Mouri H (2011) The geochronology of the Limpopo Complex: a controversy solved. Geol Soc Am Mem 207:85–106. doi:10.1130/2011.1207(06)

Krogh TE (1973) A low contamination method for the hydrothermal decomposition of zircon and extraction of U and Pb for isotopic age determinations. Geochim Cosmochim Acta 37:485–494. doi:10.1016/0016-7037(73)90213-5

Krogh TE (1982a) Improved accuracy of U-Pb zircon ages by the creation of more concordant systems using an air abrasion technique. Geochim Cosmochim Acta 46:637–649. doi:10.1016/0016-7037(82)90165-X

Krogh TE (1982b) Improved accuracy of U-Pb zircon dating by selection of more concordant fractions using a high gradient magnetic separation technique. Geochim Cosmochim Acta 46:631–635. doi:10.1016/0016-7037(82)90164-8

Krogh TE, Davis GL (1974) Alteration in zircons with discordant U-Pb ages. Carnegie Inst Wash Yearb 73:560–567

Kruger FJ (2005) Filling the Bushveld Complex magma chamber: lateral expansion, roof and floor interaction, magmatic unconformities, and the formation of giant chromitite, PGE and Ti-V-magnetitite deposits. Miner Depos 40:451–472. doi:10.1007/s00126-005-0016-8

Kruger FJ, Kamber BS, Harris PD (1998) Isotopic peculiarities of an Archaean pegmatite (Union Mine, Mica, South Africa): geochemical and geochronological implications. Precambrian Res 91:253–267. doi:10.1016/S0301-9268(98)00052-7

Krumrei TV, Villa IM, Marks MAW, Markl G (2006) A $^{40}Ar/^{39}Ar$ and U/Pb isotopic study of the Ilimaussaq complex, South Greenland: implications for the ^{40}K decay constant and for the duration of magmatic activity in a peralkaline complex. Chem Geol 227:258–273. doi:10.1016/j.chemgeo.2005.10.004

Kuiper YD (2002) The interpretation of inverse isochron diagrams in $^{40}Ar/^{39}Ar$ geochronology. Earth Planet Sci Lett 203:499–506. doi:10.1016/S0012-821X(02)00833-6

Kuiper KF, Deino A, Hilgen FJ, Krijgsman W, Renne PR, Wijbrans JR (2008) Synchronizing rock clocks of Earth History. Science 320:500–504. doi:10.1126/science.1154339

LeCheminant AN, Heaman LM (1989) Mackenzie igneous events, Canada: Middle Proterozoic hotspot magmatism associated with ocean opening. Earth Planet Sci Lett 96:38–48. doi:10.1016/0012-821X(89)90122-2

Lee JKW (1995) Multipath diffusion in geochronology. Contrib Mineral Petrol 120:60–82. doi:10.1007/BF00311008

Lee CA (1996) A review of mineralization in the Bushveld Complex and some other layered intrusions. In Cawthorn RG (ed) Layered intrusions. Elsevier Science B.V., Amsterdam, pp 103–145. doi:10.1016/S0167-2894(96)80006-6

Lee J-Y, Marti K, Severinghaus JP, Kawamura K, Yoo H-S, Lee JB, Kim JS (2006) A redetermination of the isotopic abundances of atmospheric Ar. Geochim Cosmochim Acta 70:4507–4512. doi:10.1016/j.gca.2006.06.1563

Li C, Ripley EM, Merino E, Maier WD (2004) Replacement of base metal sulfides by actinolite, epidote, calcite, and magnetite in the UG2 and Merensky Reef of the Bushveld Complex, South Africa. Econ Geol 99:173–184. doi:10.2113/gsecongeo.99.1.0173

Lindsley DH, Brown GM, Muir ID (1969) Conditions of the ferrowollastonite-ferrohedenbergite inversion in the Skaergaard intrusion. East Greenland. Mineral Soc Am Spec Pap 2:193–201

Ludwig KR (2003) Isoplot/Ex 3.00, a Geochronological Toolkit for Microsoft Excel: Berkeley Geochronology Center

Lumpkin GR (1999) Physical and chemical characteristics of baddeleyite (monoclinic zirconia) in natural environments: an overview and case study. J Nucl Mater 274:206–217

Luvizotto GL, Zack T, Meyer HP, Ludwig T, Triebold S, Kronz A, Münker C, Stockli DF, Prowatke S, Klemme S, Jacob DE, von Eynatten H (2009) Rutile crystals as potential trace element and isotope mineral standards for microanalysis. Chem Geol 261:346–369. doi:10.1016/j.chemgeo.2008.04.012

Mackie RA, Scoates JS, Weis D (2009) Age and Nd–Hf isotopic constraints on the origin of marginal rocks from the Muskox layered intrusion (Nunavut, Canada) and implications for the evolution of the 1.27 Ga Mackenzie large igneous province. Precambrian Res 172:46–66. doi:10.1016/j.precamres.2009.03.007

Maier WD, Barnes S-J, Gartz V, Andrews G (2003) Pt-Pd Reefs in magnetites of the Stella layered intrusion, South Africa: a world of new exploration opportunities for platinum group elements. Geology 31:885–888. doi:10.1130/G19746.1

Maier WD, Barnes S-J, Groves DI (2013) The Bushveld Complex, South Africa: formation of platinum-palladium, chrome- and vanadium-rich layers via hydrodynamic sorting of a mobilized cumulate slurry in a large, relatively slowly cooling, subsiding magma chamber. Mineral Depos 48:1–56. doi:10.1007/s00126-012-0436-1

Mapeo RBM, Kampunzu AB, Ramokate LV, Corfu F, Key RM (2004) Bushveld-age magmatism in southeastern Botswana: Evidence from U-Pb zircon and titanite geochronology of the Moshaneng Complex. S Afr J Geol 107:219–232. doi:10.2113/107.1-2.219

Mark DF, Stuart FM, de Podesta M (2011) New high-precision measurements of the isotopic composition of atmospheric argon. Geochim Cosmochim Acta 75:7494–7501. doi:10.1016/j.gca.2011.09.042.

Mathez EA (1995) Magmatic metasomatism and formation of the Merensky reef, Bushveld Complex. Contrib Mineral Petrol 119:277–286. doi:10.1007/BF00307287

Mathez EA, VanTongeren JA, Schweitzer J (2013) On the relationships between the Bushveld Complex and its felsic roof rocks, part 1: petrogenesis of Rooiberg and related felsites. Contrib Mineral Petrol 166:435–449. doi:10.1007/s00410-013-0884-3

Mattinson JM (2005) Zircon U-Pb chemical abrasion ("CA-TIMS") method: Combined annealing and multi-step partial dissolution analysis for improved precision and accuracy of zircon ages. Chem Geol 220:47–66. doi:10.1016/j.chemgeo.2005.03.011

Mattinson JM (2013) Revolution and evolution: 100 years of U-Pb geochronology. Elements 8, 53–57. doi: 10.2113/gselements.9.1.53

McBirney AR, Creaser RA (2003) The Skaergaard Layered Series, Part VII: Sr and Nd isotopes. J Petrol 44:757–771. doi:10.1093/petrology/44.4.757

McCallum IS (1996) The Stillwater Complex. In: Cawthorn RG (ed) Layered intrusions. Elsevier, Amsterdam, pp 441–484. doi:10.1016/S0167-2894(96)80015-7

McCourt S, Armstrong RA (1998) SHRIMP U–Pb zircon geochronology of granites from the Central Zone, Limpopo Belt, southern Africa: implications for the age of the Limpopo Orogeny. S Afr J Geol 101:329–338

McDougall I., Harrison T.M. (1999) Geochronology and thermochronology by the $^{40}Ar/^{39}Ar$ method. Oxford University Press, New York, NY pp 269

McLean NM, Bowring JF, S. A. Bowring (2011) An algorithm for U-Pb isotope dilution data reduction and uncertainty propagation. Geochem Geophys Geosyst 12(Q0AA18):1–26. doi:10.1029/2010GC003478.

McLelland JM, Chiarenzelli J (1990) Isotopic constraints on emplacement age of anorthositic rocks of the Marcy massif, Adirondack Mtns., New York. J Geol 98:19–41

Meinhold G (2010) Rutile and its applications in earth sciences. Earth-Sci Rev 102:1–28. doi:10.1016/j.earscirev.2010.06.001

Meurer WP, Boudreau AE (1996) Petrology and mineral compositions of the Middle Banded Series of the Stillwater Complex, Montana. J Petrol 37:583–607

Meurer WP, Boudreau AE (1998) Compaction of igneous cumulates. Part I. Whole-rock compositions as an indicator of the trapped liquid proportions in the Stillwater Complex, Montana. J Geol 106:281–292. doi:10.1086/516022

Meurer WP, Meurer MES (2006) Using apatite to dispel the "trapped liquid" concept and to understand the loss of interstitial liquid by compaction in mafic cumulates: an example from the Stillwater Complex, Montana. Contrib Mineral Petrol 151:187–201. doi:10.1007/s00410-005-0054-3

Meurer WP, Willmore CC, Boudreau AE (1999) Metal redistribution during fluid exsolution and migration in the Middle Banded series of the Stillwater Complex, Montana. Lithos 47:143–156. doi:10.1016/S0024-4937(99)00012-2

Mezger K, Hanson GN, Bohlen SR (1989) High-precision U-Pb ages of metamorphic rutile: application to the cooling history of high-grade terranes. Earth Planet Sci Lett 96:106–118. doi:10.1016/0012-821X(89)90126-X

Miller JS, Matzel JEP, Miller CF, Burgess SD, Miller RB (2007) Zircon growth and recycling during the assembly of large, composite arc plutons. J Volcanol Geotherm Res 167:282–299. doi:10.1016/j.jvolgeores.2007.04.019

Min K, Mundil R, Renne PR, Ludwig KR (2000) A test for systematic errors in $^{40}Ar/^{39}Ar$ geochronology through comparison with U/Pb analysis of a 1.1-Ga rhyolite. Geochim Cosmochim Acta 64:73–98. doi:10.1016/S0016-7037(99)00204-5

Minor DR, Mukasa SB (1997) Zircon U-Pb and hornblende $^{40}Ar-^{39}Ar$ ages for the Dufek layered mafic intrusion, Antarctica: implications for the age of the Ferrar large igneous province. Geochim Cosmochim Acta 61:2497–2504. doi:10.1016/S0016-7037(97)00098-7

Mitchell AA, Scoon RN (2007) The Merensky Reef at Winnaarshoek, Eastern Bushveld Complex: a primary magmatic hypothesis based on a wide reef facies. Econ Geol 102:971–1009. doi:10.2113/gsecongeo.102.5.971

Mitchell AA, Scoon RN (2012) The PlatReef of the Bushveld Complex, South Africa: a new hypothesis of multiple, non-sequential magma replenishment based on observations at the Akanani Project, north-west of Mokopane. S Afr J Geol 115:535–550. doi:10.2113/gssajg.115.4.535

Mondal SK, Ripley EM, Li C, Frei R (2006) The genesis of Archaean chromitites from the Nuasahi and Sukinda massifs in the Singhbhum Craton, India. Precambrian Research 148:45–66. doi:10.1016/j.precamres.2006.04.001

Morisset C-E, Scoates JS, Weis D, Friedman RM (2009) U-Pb and $^{40}Ar/^{39}Ar$ geochronology of the Saint-Urbain and Lac Allard (Havre-Saint-Pierre) anorthosites and their associated Fe–Ti oxide ores, Québec: Evidence for emplacement and slow cooling during the collisional Ottawan orogeny in the Grenville Province. Precambrian Res 174:95–116. doi:10.1016/j.precamres.2009.06.009

Morrison DA, Davis DW, Wooden JL, Gogard DD, Maczuga DE, Phinney WC, Ashwal LD (1985) Age of the Mulcahy Lake intrusion, northwest Ontario, and implications for the evolution of greenstone-granite terrains. Earth Planet Sci Lett 73:306–316. doi:10.1016/0012-821X(85)90079-2

Morse SA (1986) Convection in aid of adcumulus growth. J Petrol 27:1183–1214. doi:10.1093/petrology/27.5.1183

Morse SA (2008) Toward a thermal model for the Skaergaard liquidus. Am Mineral 93:248–251. doi:10.2138/am.2008.2792

Moser DE (1997) Dating the shock wave and thermal imprint of the giant Vredefort impact, South Africa. Geology 25:7–10. doi:10.1130/0091-7613(1997)0252.3.CO;2

Naldrett AJ, Wilson A, Kinnaird J, Chunnett G (2009) PGE tenor and metal ratios within and below the Merensky Reef, Bushveld Complex: implications for its genesis. J Petrol 50:625–659. doi:10.1093/petrology/egp015

Nardi LVS, Formoso MLL, Müller IF, Fontana E, Jarvis K, Lamarao C (2013) Zircon/rock partition coefficients of REEs, Y, Th, U, Nb, and Ta in granitic rocks: Uses for provenance and mineral exploration purposes. Chem Geol 335:1–7. doi:10.1016/j.chemgeo.2012.10.043

Nasdala L, Zhang M, Kempe U, Panczer G, Gaft M, Andrut M, Plötze M (2003) Spectroscopic methods applied to zircon. In: Hanchar JM, Hoskin PWO (eds) Zircon, Reviews in Mineralogy and Geochemistry, Washington, DC vol. 53. pp 427–467. doi:10.2113/0530427 The Mineralogical Society of America

Nasdala L, Hanchar JM, Kronz A, Whitehouse MJ (2005) Long-term stability of alpha particle damage in natural zircon. Chem Geol 220:83–103. doi:10.1016/j.chemgeo.2005.03.012

Nemchin AA, Horstwood MSA, Whitehouse MJ (2013) High-spatial-resolution geochronology. Elements 9:31–37. doi:10.2113/gselements.9.1.31

Nilsen O, Corfu F, Roberts D (2007) Silurian gabbro-diorite-trondhjemite plutons in the Trondheim Nappe Complex, Caledonides, Norway: petrology and U-Pb geochronology. Nor J Geol 87:329–342

Nomade S, Renne PR, Merkle RKW (2004) $^{40}Ar/^{39}Ar$ age constraints on ore deposition and cooling of the Bushveld Complex, South Africa. J Geol Soc Lond 161:411–420. doi:10.1144/0016764903-065

Norman MD, Nemchin AA (2014) A 4.2 billion year old impact basin on the Moon: U–Pb dating of zirconolite and apatite in lunar melt rock 67955. Earth Planet Sci Lett 388:387–398. doi:10.1016/j.epsl.2013.11.040

Nunes PD (1981) The age of the Stillwater complex—a comparison of U-Pb zircon and Sm-Nd isochron systematics. Geochim Cosmochim Acta 45:1961–1963. doi:10.1016/0016-7037(81)90028-4

Nunes PD, Tilton GR (1971) Uranium-lead ages of minerals from the Stillwater igneous complex and associated rocks, Montana. Geol Soc Am Bull 82:2231–2250

Nutman AP, McGregor VR, Friend CRL, Bennett VC, Kinny PD (1996) The Itsaq gneiss complex of southern West Greenland; the world's most extensive record of early crustal evolution (3900-3600 Ma). Precambrian Res 78:1–39

Oberthür T, Davis DW, Blenkinsop TG, Höhndorf A (2002) Precise U-Pb mineral ages, Rb-Sr and Sm-Nd systematics for the Great Dyke, Zimbabwe—constraints on late Archean events in the Zimbabwe craton and Limpopo belt. Precambrian Res 113:293–305. doi:10.1016/S0301-9268(01)00215-7

O'Neil J, Maurice C, Stevenson RK, Larocque J, Cloquet C, David J, Francis D (2007) The geology of the 3·8 Ga Nuvvuagittuk (Porpoise Cove) Greenstone Belt, northern Superior Province, Canada. In: Kranendonk MJ, Smithies RH, Bennett VC (eds) Earth's oldest rocks. Elsevier, Amsterdam, pp. 219–250

O'Neil, J, Francis D, Carlson RW (2011) Implications of the Nuvvuagittuq greenstone belt for the formation of Earth's early crust. J Petrol 52:985–1009. doi:10.1093/petrology/egr014

O'Neil, J, Carlson RW, Paquette J-L, Francis D (2012) Formation age and metamorphic history of the Nuvvuagittuq Greenstone Belt. Precambrian Res 220-221:23–44. doi:10.1016/j.precamres.2012.07.009

Olsson JR, Söderlund U, Klausen MB, Ernst RE (2010) U–Pb baddeleyite ages linking major Archean dyke swarms to volcanic-rift forming events in the Kaapvaal craton (South Africa), and a precise age for the Bushveld Complex. Precambrian Res 183:490–500. doi:10.1016/j.precamres.2010.07.009

Paces JB, Miller JD Jr (1993) Precise U-Pb ages of Duluth Complex and related mafic intrusions, northeastern Minnesota: geochronological insights to physical, Petrogenetic, paleomagnetic, and tectonomagmatic processes associated with the 1.1 Ga Midcontinent Rift System. J Geophys Res 98(B4):13,997–14,013

Page RW, Hoatson DM (2000) Geochronology of the mafic-ultramafic intrusions. In: Hoatson DM, Blake DH (eds) Geology and economic potential of the Paleoproterozoic layered mafic-ultramafic intrusions in the East Kimberley, Western Australia. AGSO Bulletin 246:163–172

Parrish RR (1987) An improved micro-capsule for zircon dissolution in U–Pb geochronology. Chem Geol Isot Geosci Sect 66:99–102. doi:10.1016/0168-9622(87)90032-7

Parrish RR, Noble SR (2003) Zircon U-Th-Pb geochronology by isotope dilution—thermal ionization mass spectrometry (ID-TIMS). In: Hanchar JM, Hoskin PWO (eds) Zircon, Reviews in Mineralogy and Geochemistry, Washington, DC vol. 53. pp 183–213. doi:10.2113/0530183 The Mineralogical Society of America

Parsons I (ed) (1987) Origins of igneous layering. NATO ASI series C 196. D. Reidel Publishing Company, Dordrecht, Holland p 666

Pasteels P, Demaiffe D, Michot J (1979) U-Pb and Rb-Sr geochronology of the eastern part of the south Rogaland igneous complex, southern Norway. Lithos 12:199–208. doi:10.1016/0024-4937(79)90004-5

Patchett PJ, Kouvo O, Hedge CE, Tatsumoto M (1981) Evolution of continental crust and mantle heterogeneity: evidence from Hf isotopes. Contrib Mineral Petrol 78:279–297. doi:10.1007/BF00398923

Pitra P, de Waal SA (2001) High-temperature, low-pressure metamorphism and development of prograde symplectites, Marble Hall Fragment, Bushveld Complex (South Africa). J Metamorph Petrol 19:311–325

Polat A, Frei R, Scherstén A, Appel PWU (2010) New age (ca. 2790 Ma), mantle source composition and geodynamic constraints on the Archean Fiskenaesset anorthosite complex, SW Greenland. Chem Geol 277:1–20. doi:10.1016/j.chemgeo.2010.06.016

Premo WR, Helz RT, Zientek ML, Langston RB (1990) U-Pb and Sm-Nd ages for the Stillwater Complex and its associated sills and dikes, Beartooth Mountains, Montana: Identification of a parent magma? Geology 18:1065–1068. doi:10.1130/0091-7613(1990)018 2.3.CO;2

Prendergast MD (2008) Archean komatiitic sill-hosted chromite deposits in the Zimbabwe craton. Econ Geol 103:981–1004. doi:10.2113/gsecongeo.103.5.981

Prevec SA, Ashwal LD, Mkaza MS (2005) Mineral disequilibrium in the Merensky Reef, western Bushveld Complex, South Africa: new Sm–Nd isotopic evidence. Contrib Mineral Petrol 149:306–315. doi:10.1007/s00410-005-0650-2

Pupin JP (1980) Zircon and granite petrology. Contrib Mineral and Petrol 73:207–220. doi:10.1007/BF00381441

Rajesh HM, Chisonga BC, Shindo K, Beukes NJ, Armstrong RA (2013) Petrographic, geochemical and SHRIMP U–Pb titanite age characterization of the Thabazimbi mafic sills: Extended time frame and a unifying petrogenetic model for the Bushveld Large Igneous Province. Precambrian Res 230:79–102. doi:10.1016/j.precamres.2013.02.002

Rasmussen B, Fletcher IR (2004) Zirconolite: a new U-Pb chronometer for mafic igneous rocks. Geology 32:785–788. doi:10.1130/G20658.1

Renne PR, Swisher CC, Deino AL, Karner DB, Owens T, DePaolo DJ (1998) Intercalibration of standards, absolute ages and uncertainties in $^{40}Ar/^{39}Ar$ dating. Chem Geol 145:117–152. doi:10.1016/S0009-2541(97)00159-9

Renne PR, Mundil R, Balco G, Min K, Ludwig KR (2010) Joint determination of ^{40}K decay constants and $^{40}Ar^*/^{40}K$ for the Fish Canyon sanidine standard, and improved accuracy for $^{40}Ar/^{39}Ar$ geochronology. Geochim Cosmochim Acta 74:5349–5367. doi: 10.1016/j.gca.2010.06.017.

Renne PR, Balco G, Ludwig KR, Mundil R, Mon K (2011) Response to the comment by W.H. Schwarz et al. on "Joint determination of ^{40}K decay constants and $^{40}Ar^*/^{40}K$ for the Fish Canyon sanidine standard, and improved accuracy for $^{40}Ar/^{39}Ar$ geochronology" by P.R. Renne et al. (2010). Geochim Cosmochim Acta 75:5097–5100. doi:10.1016/j.gca.2011.06.021

Rioux M, Bowring S, Dudas F, Hanson R (2010) Characterizing the U–Pb systematics of baddeleyite through chemical abrasion: application of multi-step digestion methods to baddeleyite geochronology. Contrib Mineral Petrol 160:777–801. doi:10.1007/s00410-010-0507-1

Rios S, Salje EKH, Zhang M, Ewing RC (2000) Amorphization in zircon: evidence for direct impact damage. J Phys Condens Matter 12:2401–2412. doi:10.1088/0953-8984/12/11/306

Roddick JC (1987) Generalized numerical error analysis with application to geochronology and thermodynamics. Geochim Cosmochim Acta 51:2129–2135. doi:10.1016/0016-7037(87)90261-4

Roelofse F, Ashwal LD (2012) The Lower Main Zone in the Northern Limb of the Bushveld Complex—a >1.3 km thick sequence of intruded and variably contaminated crystal mushes. J Petrol 53:1449–1476. doi:10.1093/petrology/egs022

Rollinson H, Appel PWU, Frei R (2002) A metamorphosed, Early Archean chromitite from West Greenland: Implications for the genesis of Archean anorthositic chromitites. J Petrol 43:2143–2170. doi:10.1093/petrology/43.11.2143

Saleeby JB (1992) Age and tectonic setting of the Duke Island ultramafic intrusion, southeast Alaska. Can J Earth Sci 29:506–522

Scherer EE, Münker C, Mezger K (2001) Calibration of the lutetium-hafnium clock. Science 293:683–687. doi:10.1126/science.1061372

Scherer EE, Whitehouse MJ, Münker C (2007) Zircon as a monitor of crustal growth. Elements 3:19–24. doi:10.2113/gselements.3.1.19

Schmitt AK, Chamberlain KR, Swapp SM, Harrison TM (2010) In situ U–Pb dating of micro-baddeleyite by secondary ion mass spectrometry. Chem Geol 269:386–395. doi:10.1016/j.chemgeo.2009.10.013

Schmitt AK, Perfit MR, Rubin KH, Stockli DF, Smith MC, Cotsonika LA, Zellmer GF, Ridley WI, Lovera OM (2011) Rapid cooling rates at an active mid-ocean ridge from zircon thermochronology. Earth Planet Sci Lett 302:349–358. doi:10.1016/j.epsl.2010.12.022

Schmitz MD, Bowring SA (2003) Constraints on the thermal evolution of continental lithosphere from U-Pb accessory mineral thermochronometry of lower crustal xenoliths, southern Africa. Contrib Mineral Petrol 144:592–618. doi:10.1007/s00410-002-0419-9

Schmitz MD, Kuiper KF (2013) High-precision geochronology. Elements 9:25–30. doi: 10.2113/gselements.9.1.25

Schmitz MD, Schoene B (2007) Derivation of isotope ratios, errors, and error correlations for U-Pb geochronology using ^{205}Pb-^{235}U-(^{233}U)-spiked isotope dilution thermal ionization mass spectrometry data. Geochem Geophys Geosyst 8(Q0800):1–20 doi:10.1029/2006GC001492

Schmitz MD, Bowring SA, Ireland TR (2003) Evaluation of Duluth Complex anorthositic series (AS3) zircon as a U-Pb geochronological standard: new high-precision isotope dilution thermal ionization mass spectrometry results. Geochim Cosmochim Acta 67:3665–3672. doi:10.1016/S0016-7037(03)00200-X

Schoenberg R, Kruger FJ, Nägler TF, Meisel T, Kramers JD (1999) PGE enrichment in chromitite layers and the Merensky Reef of the western Bushveld Complex; a Re–Os and Rb–Sr isotope study. Earth Planet Sci Lett 172:49–64. doi:10.1016/S0012-821X(99)00198-3

Schoene B (2014) U-Th-Pb geochronology. Treatise on geochemistry, 2nd edn. Amsterdam, The Netherlands pp 341–378. doi:10.1016/B978-0-08-095975-7.00310-7 Elsevier Ltd.

Schoene B, Crowley JL, Condon DJ, Schmitz MD, Bowring SA (2006) Reassessing the uranium decay constants for geochronology using ID-TIMS U–Pb data. Geochim Cosmochim Acta 70:426–445. doi:10.1016/j.gca.2005.09.007

Schoene B, Latkoczy C, Schaltegger U, Günther D (2010) A new method integrating high-precision U–Pb geochronology with zircon trace element analysis (U–Pb TIMS-TEA). Geochim Cosmochim Acta 74:7144–7159. doi:10.1016/j.gca.2010.09.016

Schwartz JJ, John BE, Cheadle MJ, Miranda EA, Grimes CB, Wooden JL, Dick HJB (2005) Dating the growth of oceanic crust at a slow-spreading ridge. Science 310:654–657. doi:10.1126/science.1116349

Scoates RFJ (1990) The Fox River Sill, Northeastern Manitoba—a major stratiform intrusion. Manitoba Energy and Mines, Geological Report GR82-3, p 192

Scoates JS, Chamberlain KA (1995) Baddeleyite (ZrO_2) and zircon ($ZrSiO_4$) from anorthositic rocks in the Laramie anorthosite complex, Wyoming: petrologic consequences and U-Pb ages. Am Mineral 80:1319–1329

Scoates JS, Chamberlain KA (2003) Geochronologic, geochemical and isotopic constraints on the origin of monzonitic and related rocks in the Laramie anorthosite complex, Wyoming, USA. Precambrian Res 124:269–304. doi:10.1016/S0301-9268(03)00089-5

Scoates JS, Friedman RM (2008) Precise age of the platiniferous Merensky Reef, Bushveld Complex, South Africa, by the U-Pb zircon chemical abrasion ID-TIMS technique. Econ Geol 103:465–471. doi:10.2113/gsecongeo.103.3.465

Scoates JS, Scoates RFJ (2013) Age of the Bird River Sill, Southeastern Manitoba, Canada, with implications for the secular variation of layered intrusion-hosted stratiform chromite mineralization. Econ Geol 108:895–907. doi:10.2113/econgeo.108.4.895

Scoates JS, Weis D, Williams GA, Henriques F, Tam L (2006) Initial lead isotopic compositions of plagioclase feldspar: leaching experiments, residue imaging, and applications. Geological Association of Canada-Mineralogical Association of Canada annual meeting, May 14-17, 2006, Montréal (abstract)

Scoates JS, Weis D, Franssens M, Mattielli N, Annell H, Frey FA, Nicolaysen K, Giret A (2007) The Val gabbro plutonic suite: a sub-volcanic intrusion emplaced at the end of flood basalt volcanism on the Kerguelen Archipelago. J Petrol 49:79–105. doi:10.1093/petrology/egm071

Scoates JS, Wall CJ, Friedman RM, Booth K, Scoates RFJ, Couëslan C, Macek J (2010) Recent progress in determining the precise age of ultramafic sills and mafic dikes associated with mineralization in the Thompson Nickel Belt, Manitoba, Canada. 11th international platinum symposium, Ontario Geological Survey, Miscellaneous Release Data 269:1–4

Scoates JS, Wall CJ, Friedman RM, Chamberlain KR (2011) Revisiting the age of the Merensky Reef, Bushveld Complex. Goldschmidt conference abstracts (Prague), Mineralogical Magazine 75 (3), p. 1831 (abstract)

Scoates JS, Wall CJ, Friedman RM, VanTongeren JA, Mathez EA (2012) Age of the Bushveld Complex. Goldschmidt conference abstracts (Montreal), Mineralogical Magazine 76, p. 2348 (abstract)

Silver LT, Deutsch S (1963) Uranium-lead isotopic variations in zircons: a case study. J Geol 71:721–758

Smith CH (1962) Notes on the Muskox intrusion, Coppermine River area, District of Mackenzie. Geol Surv of Can Pap 61–25, p 16

Smith SS, Basson IJ (2006) Shape and distribution analysis of Merensky Reef potholing, Northam Platinum Mine, western Bushveld Complex: implications for pothole formation and growth. Miner Depos 41:281–295. doi:10.1007/s00126-006-0059-5

Smith DS, Basson IJ, Reid DL (2003) Normal reef subfacies of the Merensky reef at Northam platinum mine, Zwartklip facies, Western Bushveld Complex, South Africa. Canadian Mineral 42:243–260. doi:10.2113/gscanmin.42.2.243

Smith ME, Chamberlain KR, Singer BS, Carroll AR (2010) Eocene clocks agree: coeval $^{40}Ar/^{39}Ar$, U-Pb, and astronomical ages from the Green River Formation. Geology 38:527–530. doi:10.1130/G30630.1

Söderlund U, Hofmann A, Klausen MB, Olsson JR, Ernst RE, Persson PO (2010) Towards a complete magmatic barcode for the Zimbabwe craton: Baddeleyite U–Pb dating of regional dolerite dyke swarms and sill complexes. Precambrian Res 183:388–398.

Stacey JS, Kramers JD (1975) Approximation of terrestrial lead isotope evolution by two-stage model. Earth Planet Sci Lett 26:207–221. doi:10.1016/0012-821X(75)90088-6

Stewart BW, DePaolo DJ (1990) Isotopic studies of processes in mafic magma chambers: II. The Skaergaard Intrusion, East Greenland. Contrib Mineral Petrol 104:125–141

Stewart BW, DePaolo DJ (1996) Isotopic studies of processes in mafic magma chambers: III. The Muskox intrusion, Northwest Territories, Canada. In: Basu A, Hart S (eds) Earth processes: reading the isotopic code, vol. 95. Geophysical Monograph, American Geophysical Union, Washington, DC pp 277–292

Storey M, Duncan RA, Swisher CC III (2007) Paleocene-Eocene Thermal Maximum and the opening of the Northeast Atlantic. Science 316:587–589. doi:10.1126/science.1135274

Tam LJ (2005) Pb and Sr isotopic compositions and trace element concentrations of plagioclase from the Merensky Reef in the Rustenburg sector of the Bushveld Complex, South Africa. Unpublished BSc thesis, University of British Columbia, p 86

Tarkian M, Mutanen T (1987) Loveringite from the Koitelainen layered intrusion, Northern Finland. Mineral Petrol 37:37–50

Tegner C, Duncan RA, Bernstein S, Brooks CK, Bird DK, Storey M (1998) ^{40}Ar–^{39}Ar geochronology of Tertiary mafic intrusions along the East Greenland rifted margin: relation to flood basalts and the Iceland hotspot track. Earth Planet Sci Lett 156:75–99. doi:10.1016/S0012-821X(97)00206-9

Tera F, Wasserburg GJ (1972) U–Th–Pb systematics in three Apollo 14 basalts and the problem of initial Pb in lunar rocks. Earth Planet Sci Lett 17:281–304. doi:10.1016/0012-821X(72)90128-8

Tilton GR, Davis GL, Wetherill GW, Aldrich LT (1957) Isotopic ages of zircon from granites and pegmatites. Trans Am Geophys Union 38:360–371

Turner S, Costa F (2007) Measuring timescales of magmatic evolution. Elements 3:267–272. doi:10.2113/gselements.3.4.267

Turtle EP, Pierazzo E, O'Brien DP (2003) Numerical modeling of impact heating and cooling of the Vredefort impact structure. Meteorit Planet Sci 38:293–303. doi:10.1111/j.1945-5100.2003.tb00265.x

Valley JW (2003) Oxygen isotopes in zircon. In: Hanchar JM, Hoskin PWO (eds) Zircon, Reviews in Mineralogy and Geochemistry, Washington, DC vol. 53, pp 343–385. doi:10.2113/0530343 The Mineralogical Society of America

Valley JW, Lackey JS, Cavosie AJ, Clechenko CC, Spicuzza MJ, Basei MAS, Bindeman IN, Ferreira VP, Sial AN, King EM, Peck WH, Sinha AK, Wei CS (2005) 4.4 billion years of crustal maturation: oxygen isotope ratios of magmatic zircon. Contrib Mineral Petrol 150:561–580. doi:10.1007/s00410-005-0025-8

Vavra G (1993) A guide to quantitative morphology of accessory zircon. Chem Geol 110:15–28. doi:10.1016/0009-2541(93)90245-E

Vermaak CF (1976) The Merensky Reef—thoughts on its environment and genesis. Econ Geol 71:1270–1298. doi:10.2113/gsecongeo.71.7.1270

Wagner PA (1929) The platinum deposits and mines of South Africa. Oliver and Boyd, Edinburgh, p 326

Wager LR, Brown GM (1967) Layered igneous rocks. Oliver & Boyd, Edinburgh, Great Britain p 588

Wager LR, Brown GM, Wadsworth WJ (1960) Types of igneous cumulates. J Petrol 1:73–85. doi:10.1093/petrology/1.1.73

Wall CJ (2009) Uranium-lead geochronology of granophyres from the Archean Stillwater Complex, Montana (USA): characterization of uranium-bearing accessory minerals (zircon, titanite, rutile) and preliminary dating results. Unpublished B. Sc. Honours Thesis, University of British Columbia, 60 pages

Wall CJ, Scoates JS, Friedman RM, Meurer WP (2010) Refining the precise age and duration of magmatism related to the Stillwater Complex. 11th international platinum symposium, Ontario Geological Survey, Miscellaneous Release–Data 269:1–4

Wall CJ, Scoates JS, Friedman RM, Meurer WP (2013) Identifying the timing of magma inputs and hiatuses during emplacement and crystallization of the Stillwater complex by high-precision U-Pb geochronology. William Smith Meeting of the Geological Society of London, London, UK, June 25–27, 2013 (abstract)

Walraven F (1997) Geochronology of the Rooiberg Group, Transvaal Supergroup, South Africa. Economic Geology Research Unit Information Circular, University of Witwatersrand, 316, p 21

Walraven F, Hattingh E (1993) Geochronology of the Nebo granite, Bushveld Complex. S Afr J Geol 96:31–41

Walraven F, Armstrong RA, Kruger FJ (1990) A chronostratigraphic framework for the north-central Kaapvaal craton, the Bushveld Complex and the Vredefort structure. Tectonophysics 171:23-48

Watson EB, Liang Y (1995) A simple model for sector zoning in slowly grown crystals: implications for growth rate and lattice diffusion, with emphasis on accessory minerals in crustal rocks. Am Mineral 80:1179-1187

Watson EB, Wark DA, Thomas JB (2006) Crystallization thermometers for zircon and rutile. Contrib Mineral Petrol 151: 413–433. doi:10.1007/s00410-006-0068-5

Webb SJ, Cawthorn RG, Nguuri T, James D (2004) Gravity modeling of Bushveld Complex connectivity supported by Southern African seismic experiment results. S Afr J Geology 107:207–218. doi:10.2113/107.1-2.207

Wendt I, Carl C (1991) The statistical distribution of the mean squared weighted deviation. Chem Geol 86:275–285. doi:10.1016/0168-9622(91)90010-T

Wetherill GW (1956) Discordant uranium–lead ages. Trans Am Geophys Union 37:320–326

White WM (2013) Geochemistry. Wiley-Blackwell, West Sussex, UK p 668.

Williams IS, Compston W, Collerson KD, Arriens PA, Lovering JF (1983) A reassessment of the age of the Windmill metamorphics, Casey area. In: Oliver RL, James PR, Jago JB (eds) Antarctic earth science. Australian Academy of Science, Canberra, pp 73–76

Wotzlaw J-F, Bindeman IN, Schaltegger U, Brooks CK, Naslund HR (2012) High resolution insights into episodes of crystallization, hydrothermal alteration and remelting in the Skaergaard intrusive complex. Earth Planet Sci Lett 355-356:199–212. doi:10.1016/j.epsl.2012.08.042

Yudovskaya M, Kinnaird J, Naldrett AJ, Rodionov N, Antonov A, Simakin S, Kuzmin D (2013) Trace-element study and age dating of zircon from chromitites of the Bushveld Complex (South Africa). Mineral Petrol 107:915–942. doi:10.1007/s00710-013-0269-3

Zaccarini F, Stumpfl EF, Garuti G (2004) Zirconolite and Zr-Th-U minerals in chromitites of the Finero Complex, Western Alps, Italy: evidence for carbonatite-type metasomatism in a subcontinental mantle plume. Can Mineral 42:1825–1845. doi:10.2113/gscanmin.42.6.1825

Zientek ML, Corson SR, West RD (2005) Geochemical surveys of soil and talus fines and the discovery of the J-M Reef, Stillwater Complex, Montana (Chapter 18). In: Mungall JE (ed) Exploration for platinum-group element deposits, vol. 5. Mineralogical Association of Canada, Short Course Series, pp 391–407

Zirakparvar NA, Mathez EA, Scoates JS, Wall CJ 2014 Zircon Hf isotope evidence for an enriched mantle source for the Bushveld Igneous Complex. Contrib Mineral Petrol 168:1050. doi:10.1007/s00410-014-1050-2

Zhong H, Zhu W-G (2006) Geochronology of layered mafic intrusions from the Pan-Xi area in the Emeishan large igneous province, SW China. Miner Depos 41:599–606. doi:10.1007/s00126-006-0081-7

Chapter 2
Igneous Layering in Basaltic Magma Chambers

O. Namur, Bénédicte Abily, Alan E. Boudreau, Francois Blanchette, John W. M. Bush, Georges Ceuleneer, B. Charlier, Colin H. Donaldson, Jean-Clair Duchesne, M. D. Higgins, D. Morata, Troels F. D. Nielsen, B. O'Driscoll, K. N. Pang, Thomas Peacock, Carl J. Spandler, Atsushi Toramaru and I.V. Veksler

Abstract Layering is a common feature in mafic and ultramafic layered intrusions and generally consists of a succession of layers characterized by contrasted mineral modes and/or mineral textures, including grain size and orientation and, locally, changing mineral compositions. The morphology of the layers is commonly planar, but more complicated shapes are observed in some layered intrusions. Layering displays various characteristics in terms of layer thickness, homogeneity, lateral continuity, stratigraphic cyclicity, and the sharpness of their contacts with surrounding layers. It also often has similarities with sedimentary structures such as cross-bedding, trough structures or layer termination. It is now accepted that basaltic

O. Namur (✉)
Institute of Mineralogy, University of Hannover, 30167 Hannover, Germany
e-mail: o.namur@mineralogie.uni-hannover.de

B. Abily · G. Ceuleneer
CNRS-UMR 5563, GET, OMP, University of Toulouse, 31400 Toulouse, France
e-mail: benedicte.abily@gmail.com

G. Ceuleneer
e-mail: georges.ceuleneer@get.obs-mip.fr

A. E. Boudreau
Division of Earth and Ocean Sciences, Nicholas School of the Environment and Earth Sciences, Duke University, Durham, NC 27708, USA
e-mail: boudreau@duke.edu

F. Blanchette
School of Natural Sciences, University of California Merced, Atwater, CA 95301, USA
e-mail: fblanchette@ucmerced.edu

J. W. M. Bush
Department of Mathematics, Massachusetts Institute of Technology, Cambridge, MA 02139, USA
e-mail: bush@math.mit.edu

B. Charlier · J.-C. Duchesne
Department of Geology, University of Liege, B 4000 Sart Tilman, Belgium
e-mail: b.charlier@ulg.ac.be

J.-C. Duchesne
e-mail: jc.duchesne@ulg.ac.be

B. Charlier et al. (eds.), *Layered Intrusions,* Springer Geology,
DOI 10.1007/978-94-017-9652-1_2

magma chambers mostly crystallize in situ in slightly undercooled boundary layers formed at the margins of the chamber. As a consequence, most known existing layering cannot be ascribed to a simple crystal settling process. Based on detailed field relationships, geochemical analyses as well as theoretical and experimental studies, other potential mechanisms have been proposed in the literature to explain the formation of layered igneous rocks. In this study, we review important mechanisms for the formation of layering, which we classify into dynamic and non-dynamic layer-forming processes.

C. H. Donaldson
School of Geography and Geosciences, University of St. Andrews, St. Andrews KY16 9AL, UK
e-mail: chd@st-andrews.ac.uk

M. D. Higgins
Sciences de la Terre, Université du Québec à Chicoutimi, 555 Blvd de l'université, Chicoutimi, Quebec G7H 2B1, Canada
e-mail: mhiggins@uqac.ca

D. Morata
Departamento de Geología & Andean Geothermal Center of Excellence (CEGA, Fondap-CONICYT), Facultad de Ciencias Físicas Matemáticas, Universidad de Chile, Santiago 13518, Casilla, Chile
e-mail: dmorata@ing.uchile.cl

T. F. D. Nielsen
Geological Survey of Denmark and Greenland, 1350 Copenhagen, Denmark
e-mail: tfn@geus.dk

B. O'Driscoll
School of Physical and Geographical Sciences, Keele University, Keele ST5 5BG, UK
e-mail: b.o'driscoll@keele.ac.uk

The University of Manchester School of Earth, Atmospheric and Environmental Science (SEAES) Williamson Building Oxford Road Manchester M13 9PL UK
email:brian.odriscoll@manchester.ac.uk

K. N. Pang
Department of Geosciences, National Taiwan University, Taipei, Taiwan, China
e-mail: knpang@hotmail.com

T. Peacock
Department of Mechanical Engineering, Massachusetts Institute of Technology, Cambridge, MA 02139, USA
e-mail: xray@mit.edu

C. J. Spandler
School of Earth and Environmental Sciences, James Cook University, Townsville, QLD 4811, Australia
e-mail: carl.spandler@jcu.edu.au

A. Toramaru
Department of Earth and Planetary Sciences, Kyushu University, Fukuoka 812 8581, Japan
e-mail: toramaru@geo.kyushu-u.ac.jp

I.V. Veksler
Helmholtz Centre Potsdam, GFZ German Research Centre for Geosciences, 14473 Potsdam, Germany
e-mail: ilya.veksler@gfz-potsdam.de

Dynamic processes occur during filling of the magma chamber or during its crystallization. They include differential settling or flotation of crystals with contrasted densities and/or grain sizes, flow segregation of crystal-laden magma and crystal segregation during convective liquid movement into the magma chamber. Double diffusive convection, which produces a stratified liquid column in the magma chamber, can also produce layering. Other dynamic processes include magma injection into the chamber, which results in magma stratification or magma mixing, and silicate liquid immiscibility either in the main magma chamber or within the solidifying crystal mush.

Non-dynamic layer-forming processes mainly include rapid changes in intensive conditions of crystallization (e.g. pressure, oxygen fugacity) that disrupt and change the stable liquidus assemblages, and transitory excursions about cotectic curves. Layering can also result from variation in nucleation rates and from mineral reorganization in a crystal mush through grain rotation, dissolution-precipitation due to initial heterogeneity in terms of grain size distribution, mineral modes or differential pressure. Many of these processes are driven by dissipation of energy and can be referred to as equilibration or self-organization processes.

Keywords Dynamic · Non-dynamic · Sedimentary features · Fluid dynamics · Dissipation of energy

Introduction

Igneous layering is a common feature of plutonic igneous bodies and occurs in most mafic-ultramafic layered intrusions (Wager and Brown 1968; Parsons 1987; Naslund and McBirney 1996; Irvine et al. 1998), in syenitic intrusions (Parsons 1979; Hodson 1998), in granites (Barbey 2009), in ophiolites (Nicolas et al. 1998; Jousselin et al. 2012) and at oceanic spreading ridges (Gillis et al. 2014). The origin of layering in mafic magma chambers was initially attributed to a crystal settling process combined with mineral sorting due to a density contrast between the different phases and the melt (Wager et al. 1960). However, experimental and theoretical models have shown that plagioclase is commonly slightly buoyant in basaltic melts (Campbell 1978; Campbell et al. 1978; Scoates 2000; Namur et al. 2011a) and should therefore not accumulate at the floor of the magma chamber. Plagioclase is a major component of basaltic layered intrusions which indicates that crystal settling cannot be responsible for the different kinds of layering in all known occurrences. It is therefore likely that layering formation involves other processes including in situ crystallization close to the margins of the magma chamber (Campbell 1978; McBirney and Noyes 1979; Naslund and McBirney 1996).

Given the wide range of layer types and layering features in terms of texture (including grain-size), mineral mode, composition, morphology, and the way layers are interstratified and repeat stratigraphically, it is unlikely that a single layer-forming process can explain all the known occurrences of igneous layering. Many processes have thus been proposed in the past 50 years to explain the origin of layering in individual localities (Wager and Brown 1968; McBirney and Noyes 1979; Parsons 1987; Naslund and McBirney 1996). Layering can form through processes occurring from the liquidus temperature down to subsolidus conditions (Naslund and McBirney 1996). Broadly, all of these processes can be subdivided into two groups: dynamic layer-forming processes and non-dynamic layer-forming processes (Irvine 1982; Boudreau and McBirney 1997; McBirney and Nicolas 1997).

In this chapter, we review the dynamic and non-dynamic processes that have been commonly proposed to explain the formation of igneous layering and describe examples of layering features resulting from these processes. We also take advantage of recent literature to show convincing examples for layer-forming mechanisms that we attribute to processes that have generally been not considered or not fully described in the past. Finally, we show examples of how numerical and experimental investigations might help to understand the formation of layering in real magma chambers. This chapter does not aim at describing all known occurrences of layering, but to describe important examples of layering features illustrating the most significant and best-documented layer-forming mechanisms occurring during the solidification of ultramafic-mafic magma chambers.

Layers and Layering

Layering is defined as an overall structure of cumulates which develops through the combination of individual layers, each forming sheet like cumulate units and each being a distinctive entity in its compositional (e.g. bulk rock and/or mineral compositions) and/or textural (e.g. grain size, mineral mode) features (Irvine 1982; Naslund and McBirney 1996). Individual layers therefore appear as a sheet-like discontinuity in the crystalline assemblage making up the sequence of rocks (Figs 2.1a–e).

Individual layers can be uniform or stratigraphically variable, therefore showing normal or reversed textural grading (e.g. morphology of the crystals), grain-size grading, modal grading (leucocratic vs melanocratic), or cryptic (chemical) grading (Table 2.1; Irvine 1982). They show a large range of forms (e.g. planar, lenticular, trough-shaped, synformal, antiformal, convoluted or basin-like) and can also be defined in terms of thickness (e.g. thin: 0–5 cm, medium-tick: 5–100 cm or thick: > 100 cm) and the nature of their contacts (sharp vs gradational). They can be laterally discontinuous in terms of mineral mode, texture or thickness (Fig. 2.1e). Individual layers may also show termination and locally merge together to form thicker layers (Fig. 2.1f)

Layering in igneous bodies is characterized by the succession of layers that may have variable characteristics in terms of grain-size (grain-size layering), mineral mode (modal layering), crystal shape (textural layering) or composition (cryptic layering; Irvine 1982; Table 2.1; Fig. 2.2). From one intrusion to another, or within a single intrusion, layering may be well defined (prominent layering) or poorly

Fig. 2.1 Examples of layers from various layered intrusions. **a** Alternating chromitite (*Cr*) and anorthosite (*An*) layers, UG1 footwall, Critical Zone of the Eastern Bushveld Complex, Dwars River, South Africa. **b** Troctolitic cumulates (*Tr*) with *thin*, sharply bounded, peridotite (*P*) layers. Younger Giant Dyke Complex, western Tugtutôq, Greenland. At larger scale, these layers show lateral discontinuities and bifurcation. Scale is 30 cm long. Courtesy of B. Upton. **c** Magnetitite layer (*Mt*) overlying leucogabbronorite, *Upper* Zone of the Eastern Bushveld Complex, Magnet Heights. Camera bag is 10 cm across. **d** Peridotite layers within the troctolite cumulates of the Younger Giant Dyke Complex, western Tugtutôq, Greenland. Note the perpendicular feldspar crystals growing up from the peridotite layers. Courtesy of B. Upton. **e** Peridotite layer within the troctolite cumulates of the Younger Giant Dyke Complex, western Tugtutôq, Greenland. Note that the thickness of the peridotite layer is increasing laterally. Courtesy of B. Upton. **f** Chromitite layers within anorthosite cumulates. UG1 footwall, Critical Zone of the Eastern Bushveld Complex, Dwars River, South Africa. Note that some chromitite layers are discontinuous laterally and locally merge in a single *thick* layer

defined, be planar or form other structures (e.g. trough, colloform, wavy, convoluted layering, cross stratified, inch-scale layering; Fig. 2.1f) and can be continuous (laterally and vertically) or discontinuous (or intermittent).

Layering deformation due to blocks falling on the still highly porous crystal mush has been described for the Sept Iles, Skaergaard (Figs 2.3a–b), Fongen-Hyllingen

Table 2.1 Important characteristics of the layers and common types of layering in layered intrusions

Characterization	Description	Important location	Lithology	Reference
Layer				
Thickness				
	Thin (0–5 cm)	Stillwater Complex, USA	Anorthosite—pyroxenite	Hess (1960)
	Medium-thick (5 cm–1 m)	Bushveld Complex, South Africa	Magnetitite	Cawthorn and Ashwal (2009)
	Thick (> 1 m)	Bushveld Complex, South Africa	Anorthosite	Cawthorn and Ashwal (2009)
Form				
	Planar	Skaergaard, Greenland	Leucogabbro—melanogabbro	Irvine et al. (1998)
	Lenticular	Kiglapait, Canada	Troctolite	Morse (1969)
	Trough-shaped	Skaergaard, Greenland	Anorthosite—gabbro	Irvine (1987)
	Synformal	Bjerkreim-Sokndal, Norway	Leuconorite—melanonorite	Bolle et al. (2000)
	Antiformal	Skaergaard, Greenland	Leucogabbro—melanogabbro	Irvine et al. (1998)
	Convoluted	Duke Island, USA	Dunite—peridotite	Irvine (1987)
	Colloform	Skaergaard, Greenland	Troctolite-gabbro	Namur et al. (2013)
Interal constitution				
	Homogeneous in mineralogy, grain size, texture	Bushveld Complex, South Africa	Chromite-anorthosite	Voordouw et al. (2009)
	Stratigraphically variable in mineralogy, grain size, texture	Panzhihua, China	Melanogabbro	Zhou et al. (2005)
	Laterally variable in mineralogy, grain size, texture	Duluth Complex	Fine-grained gabbro	Miller and Ripley (1996)
Composition				
	Leucocratic (high proportion of leucocratic phases)	Panzhihua, China	Plagioclase-rich gabbro	Zhou et al. (2005)
	Melanocratic (high proportion of mafic phases)	Panzhihua, China	Oxide-rich gabbro	Zhou et al. (2005)
	Monomineralic	Bushveld Complex, South Africa	Chromite-anorthosite-magnetitite	Eales and Cawthorn (1996)
Graded layers				
	Grain-size graded (variation in grain size of constituent minerals)	Duke Island, USA	Dunite—peridotite	Irvine (1974)

Table 2.1 (continued)

Characterization	Description	Important location	Lithology	Reference
	Modally graded (variation in mineral modes)	Panzhihua, China	Melanogabbro-Leucogabbro	Zhou et al. (2005)
	Cryptically graded (continuous variation of mineral compositions)	Skaergaard, Greenland	Gabbro	McBirney and Noyes (1979)
	Texturally graded (change in texture; poikilitic vs tabular grains)	Skaergaard, Greenland	Gabbro	McBirney and Noyes (1979)
Layer contact				
	Sharp	Sept Iles, Canada	Bottom of magnetitite layers	Namur et al. (2010)
	Gradational	Sept Iles, Canada	Top of magnetitite layers	Namur et al. (2010)
	Concordant	Skaergaard, Greenland	Gabbro	McBirney and Noyes (1979)
	Discordant	Duke Island, USA	Olivine pyroxenite	Irvine (1963)
Layering				
Lithologic variation				
	Modal layering	Ilímaussaq Complex, Greenland	Kakortokite	Sorensen and Larsen (1987)
	Grain size layering	Duke Island, USA	Olivine pyroxenite-peridotite	Irvine (1963)
	Textural layering	Rum Layered Suite, Scotland	Peridotite-allivalite	Brown (1956)
	Cryptic layering	Skaergaard, Greenland	Troctolite-gabbro	Wager and Brown (1968)
Distinctness				
	Prominent	Skaergaard, Greenland	Troctolite-gabbro	McBirney and Noyes (1979)
	Inconspicious	Sept Iles intrusion, Canada	Troctolite-gabbro	Namur et al. (2010)
Regularity				
	Macrorhytmic	Ilímaussaq Complex, Greenland	Kakortokite	Sorensen and Larsen (1987)
	Microrhythmic	Rum Layered Suite, Scotland	Allivalite	Brown (1956)
	Inch-scale (microrhythmic)	Stillwater Complex, USA	Orthopyroxenite-anorthosite	McCallum (1996)
Continuity				
	Laterally continuous	Bushveld Complex, South Africa	Chromitite	Mondal and Mathez (2007)

Table 2.1 (continued)

Characterization	Description	Important location	Lithology	Reference
	Laterally discontinuous	Rum Layered Suite, Scotland	Peridotite-allivalite	Bedard et al. (1988)
	Intermittent	Skaergaard, Greenland	Leucogabbro-melanogabbro	McBirney (1993)

and Sarqata Qaqa intrusions (Thompson and Patrick 1968; Wilson et al. 1981; Irvine et al. 1998; Higgins 2005; Namur et al. 2010; Morse 2011; see also Plates 2a and 3c in Irvine 1965). Layering often shows similarities with sedimentary features such as graded-bedded turbidity current deposits, loading, impact-structures, scour and fill, pinch and swell structures, angular discontinuities, slump structures and lateral grading (Figs 2.2d and 2.3d–f; Irvine 1963). Sedimentary-like features of layered cumulate rocks have been particularly well described for the ultramafic Duke Island intrusion (Irvine 1965) and the gabbroic Skaergaard and Fongen-Hyllingen intrusions (Thy and Wilson 1980; Wilson et al. 1981; Irvine et al. 1998). The vast majority of these structures have been interpreted as resulting from gravitational instability on sloping surfaces. Layering may also be deformed due to high-temperature faulting, generally resulting from currents of crystal-rich magmas (Figs. 2.4a, b). Additionally, layering can locally be recognized in the field due to contrasting degree of weathering of the constituting layers (Fig. 2.4c).

An important parameter of layering is the composition of constituent minerals. Mineral compositions within individual layers can be relatively constant or not showing any obvious stratigraphic evolution (Fig. 2.5a) or change (often continuously; Fig. 2.5b) due to liquid differentiation, or varying trapped liquid fraction (Barnes 1986). Between successive layers (or between layers and homogeneous rocks), mineral compositions can be relatively similar (Fig. 2.5c; see also Conrad and Naslund 1989) or may be different as a result of various processes such as contrasted interstitial liquid fractions (Barnes 1986; Conrad and Naslund 1989), injection of new liquid in the magma chamber (e.g. individual layers crystallized from different parent magmas; (Irvine 1975), interstitial liquid migration and chemical diffusion within the crystal mush (Holness et al. 2007; Namur et al. 2013; Leuthold et al. 2014), change in intensive conditions of crystallization (Pang et al. 2008; Cawthorn and Ashwal 2009), or simply melt evolution due to fractional crystallization. In some cases, some minerals may have similar compositions in successive layers (e.g. plagioclase in Fig. 2.5d), while other minerals may have different compositions (e.g. olivine in Fig. 2.5d) due to disequilibrium crystallization or contrasted effect of the trapped liquid shift (Conrad and Naslund 1989).

Various types of layering features have been described in the petrological literature and it is also worth noting that several of them can be observed in a single layered intrusion (see Fig. 2.6 for the Skaergaard intrusion). Most of these layering features have generally been attributed to two main types of layer-forming processes: dynamic processes and non-dynamic processes. However, as noted by Irvine et al. (1998), this subdivision may be too limited and it is obvious that some

Fig. 2.2 Examples of layering features in various layered intrusions. **a** Macro-rhythmic layering in the Skaergaard UZa, Greenland. Each sequence starts with a layer enriched in Fe–Ti oxides, followed by a clinopyroxene-rich gabbro and finally a plagioclase-rich layer. **b** Modally-graded layering in the Kiglapait intrusion, *Lower* Zone troctolite north of Slambang Bay, Canada; courtesy of S.A. Morse. **c** Micro-rhythmic layering in the Bjerkreim lobe of the Bjerkreim-Sokndal intrusion, Norway. This shows a succession of layers strongly enriched in orthopyroxene and oxide minerals followed by cm- to dm-thick layers dominated by plagioclase. **d** Very-fine scale layering in the Storgangen intrusion, Rogaland province, Norway. **e** Example of modally-graded layering in the magnetite layer 1 at Magnet Heights of the Bushveld Complex, South Africa; courtesy of R.G. Cawthorn. Note the progressive increase of plagioclase mode upwards. **f** Grain-size graded layering of olivine- and pyroxene-rich layers of the Duke Island ultramafic intrusion, Alaska; courtesy of E.M. Ripley

layer-forming processes may have both a dynamic component and non-dynamic component.

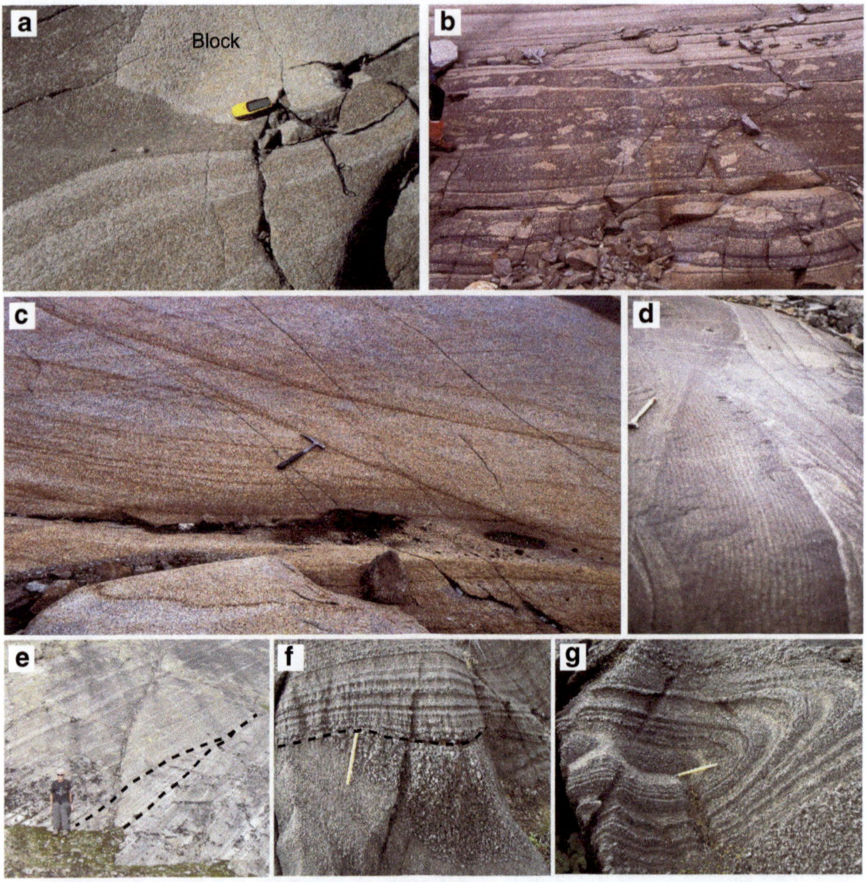

Fig. 2.3 **a–b** Example of deformation of igneous layering resulting from blocks falling; *Lower* Zone B of the Skaergaard intrusion on Kraemer Island (A) and *Middle* Zone on Kraemer Island, Greenland (B; courtesy of J. Koepke). **c** Modal layering in the cross-bedded belt of the Skaergaard intrusion, Uttental Plateau; courtesy of J. Koepke. **d** Sedimentary structures and inch-scale layering in Lower Series B gabbros of the Kap Edvard Holm Complex, Southern East Greenland. **e** Angular discontinuity (*dashed lines*) in layering of metagabbro from the Fongen-Hyllingen intrusion, eastern ridge of Fongen mountain. The rocks are two pyroxene gabbros that have been overprinted by amphibolite facies metamorphism but have retained their igneous texture. Many of the layers below the unconformity show modal grading, with relatively sharp, mafic-rich bases and gradually increasing proportion of plagioclase upwards; courtesy of R. Latypov. **f** Angular unconformity (*dashed line*) in the layered ultramafic sequence of the Duke Island intrusion, Alaska; courtesy of E.M. Ripley. **g** Trough feature in the layered ultramafic sequence of the Duke Island intrusion, Alaska; courtesy of E.M. Ripley

Dynamic Layer-Forming Processes

Dynamic layer-forming processes involve internal movement of melt, mush and crystals within the magma chamber and are generally most efficient near the margins of the magma chamber. Various types of dynamic processes have been pro-

Fig. 2.4 **a** High temperature faulting disrupting the layered sequence of troctolite and peridotite in the Younger Giant Dyke Complex, western Tugtutôq, Greenland. Scale is 50cm long. Courtesy of B. Upton. **b** High temperature faulting (*dashed line*) in alternating Fe–Ti oxide-rich gabbro (*FeTi*) and plagioclase- and pyroxene-rich layers (*P-Px*) at the contact between the Marginal Border Series and the Layered Series of the Skaergaard intrusion. Skaergaard Mini-Peninsula. Bar scale is 2 m. **c** Horizontal layering in the axial zone of the Younger Giant Dyke Complex, western Tugtutôq, Greenland. In-weathering peridotite layers (*P*) alternate with thin, more resistant, troctolite layers (*Tr*). See additional details in Upton et al. (1996). Courtesy of B. Upton

posed to explain the occurrence of some types of layering in various layered intrusions (Table 2.2). Many studies describe the layering associated with flow segregation, magma currents (McBirney and Nicolas 1997), continuous, intermittent or double diffusive convection (Wager and Brown 1968; McBirney and Noyes 1979), crystal settling (Naslund and McBirney 1996) and metasomatism (Boudreau 1982; McBirney 1985). Magma mixing within the main magma body or within the crystal mush is also often considered as an important process of layering formation, especially in the case of monomineralic layers (e.g. Irvine 1975; Holness et al. 2007; Cawthorn and Ashwal 2009).

Other processes such as silicate liquid immiscibility (McBirney and Nakamura 1974; Reynolds 1985) and crystal mush contraction (Petersen 1987) have been described in the literature but have generally not been considered as significant layer-forming mechanisms. In the following, we describe examples that demonstrate that these processes in fact are of major importance for the development of layering in magma chambers (Charlier et al. 2011; Namur et al. 2012a, 2013).

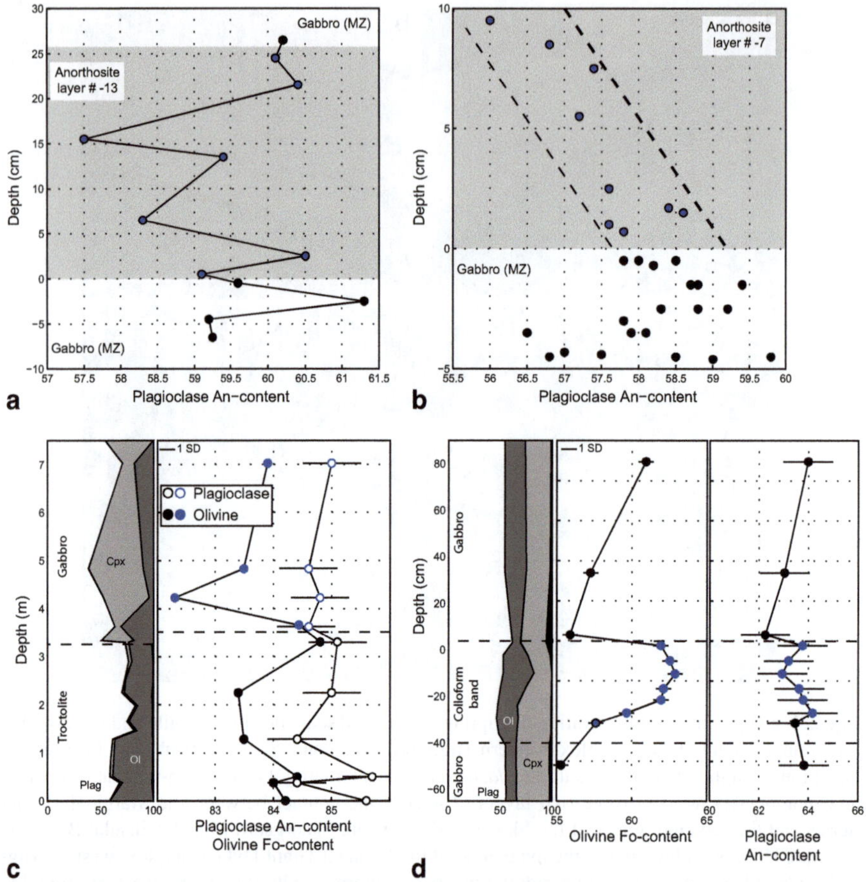

Fig. 2.5 Evolution of mineral compositions in layered sequences of cumulate rocks. **a** Plagioclase An-content in samples just below (*gabbro*) throughout the anorthosite layer −13 and above the anorthosite layer (*gabbro*) of the Bushveld Main Zone. Note the random distribution and the absence of well-defined evolution of An within the anorthosite layer. Note also the absence of significant change of plagioclase composition in the gabbro. Modified from Cawthorn and Ashwal (2009). **b** Plagioclase An-content in samples just below (*gabbro*) and throughout the anorthosite layer −7 of the Bushveld Main Zone. Note the relatively continuous decrease in An-content within the anorthosite layer. Modified from Cawthorn and Ashwal (2009). **c** Evolution of mineral modes and mineral compositions (*An* and *Fo*) in a troctolite-gabbro sequence of the Unit 8/9 allivalite of the Rum intrusion, Isle of Rum, Scotland. Note the change in olivine composition above the troctolite-gabbro interface and that plagioclase composition does not change significantly at this interface. Modified from Holness et al. (2007). **d** Evolution of mineral modes and mineral compositions (*An* and *Fo*) in a leucogabbro-gabbro (colloform band)-leucogabbro section of the Skaergaard Marginal Border Series. Note the change of olivine composition within the colloform band and that plagioclase composition is identical in the colloform and in the gabbro. (Modified from Namur et al. (2013))

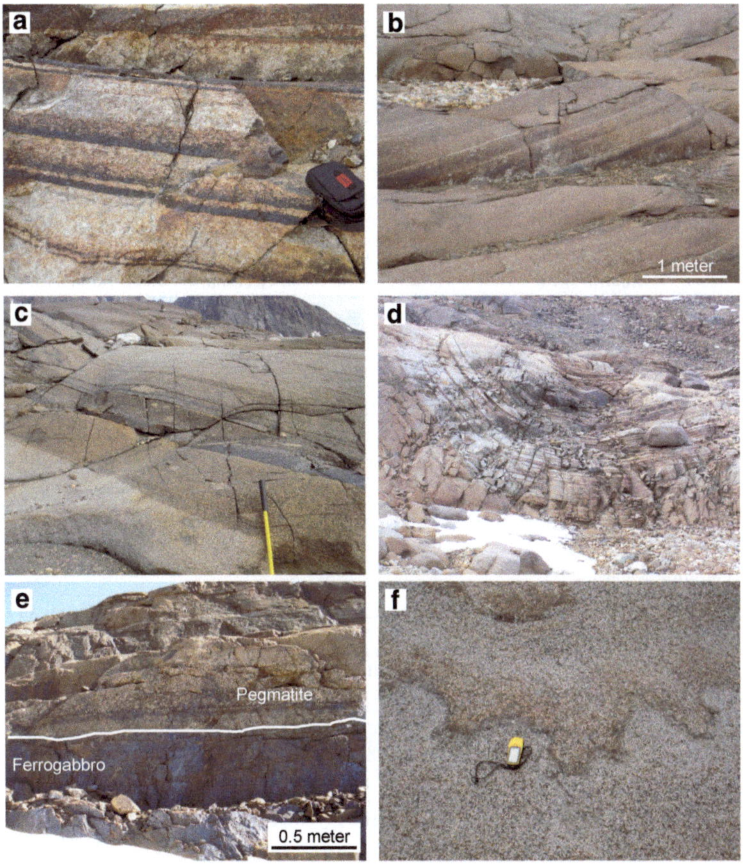

Fig. 2.6 Photographs from the Skaergaard intrusion (Greenland) illustrating that various types of layering can occur within a single layered intrusion. See additional layering features in Irvine et al. (1998). **a** Micro-rhythmic layering showing the alternation of cm-thick layers strongly enriched in Fe–Ti oxides and cm- to dm-thick layers dominated by clinopyroxene or plagioclase, UZa, mainland. **b** Rhythmic layering showing the succession dm-thick isomodal layers, UZa, mainland. **c** Macrorhytmic layering showing the succession of dm- to m-thick isomodal layers, LZc, Kraemer Island. **d** U-shaped spectacular layering defined as trough G; the dark layers are rich in olivine, pyroxene and Fe–Ti oxides, while the light are rich in plagioclase; the troughs were presumably channels that were swept by dense currents of crystal-laden magma; UZa, mainland. Trough band is ~30 m across; courtesy of J. Koepke. **e** Meter-thick pegmatitic layer within ferrogabbros from UZb, mainland. **f.** Well-defined colloform band showing contrasted colour and mineral modes with adjacent gabbros on either side of the colloform band, LZb*, Skaergaard Peninsula

Mechanical Sorting

Crystal Settling

Crystal settling is analogous to sedimentation. Minerals in a liquid with Newtonian properties (see McBirney and Noyes 1979 for the effect of non-Newtonian

Table 2.2 Important dynamic layer-forming processes

Process		Suggested layering	Important location	Reference
Mechanical sorting processes				
	Crystal settling	Size-graded cyclic layers	Imilik gabbro, Greenland	Brown and Farmer (1971)
	Crystal flotation	Anorthosite layer	Sept Iles, Canada	Namur et al. (2011a)
	Magma current and magma flow	Trough layers	Skaergaard, Greenland	Irvine (1987)
	Magma replenishment	Cyclic layering	Muskox, Canada	Irvine and Smith (1967)
Convection-related processes				
	Continuous convection	Modally-graded layers	Khibina, Russia	Kogarko and Khapaev (1987)
	Intermittent convection	Cyclic layering	Skaergaard, Greenland	Naslund et al. (1991)
	Double-diffusive convection	Cyclic and cryptic layering	Fongen-Hyllingen, Norway	Wilson et al. (1987)
Magma mixing processes				
	Magma replenishment	Magnetitite	Bushveld, South Africa	Harney et al. (1990)
	Magma contamination	Chromitite	Muskox, Canada	Irvine (1975)
Silicate liquid immiscibility		Nelsonite & Leucogabbro	Sept Iles, Canada	Charlier et al. (2011)
Processes within the crystal mush				
	Vertical liquid migration	Anorthosite layers	Stillwater, USA	Boudreau and McCallum (1992)
	Interstitial liquid migration	Colloform banding	Skaergaard, Greenland	Namur et al. (2013)
Deformation of crystal mush		Modally layered gabbro	Oman ophiolite	Jousselin et al. (2012)

properties) settle under the influence of gravity (g) with a settling velocity (V) defined by the Stokes law equation:

$$V = 2r^2 g(\rho_1 - \rho_2)/9\eta \tag{2.1}$$

which depends on the radius of the crystal (r; m), the density contrast between the crystal (ρ_1; kg/m^3) and the liquid (ρ_2; kg/m^3) and the viscosity of the liquid (η ; Ns/m^2). The net result is that minerals with different densities or grain sizes will settle at contrasting velocities (Fig. 2.7a), potentially resulting in the formation of layering with grain-size and/or mineral mode variations (Fig. 2.7b). This process is mostly efficient in magmatic systems where the magma has low viscosity and is close to Newtonian (low yield strength), and when density contrasts are significant. Layering formed by crystal settling according to the Stokes law should be graded in terms

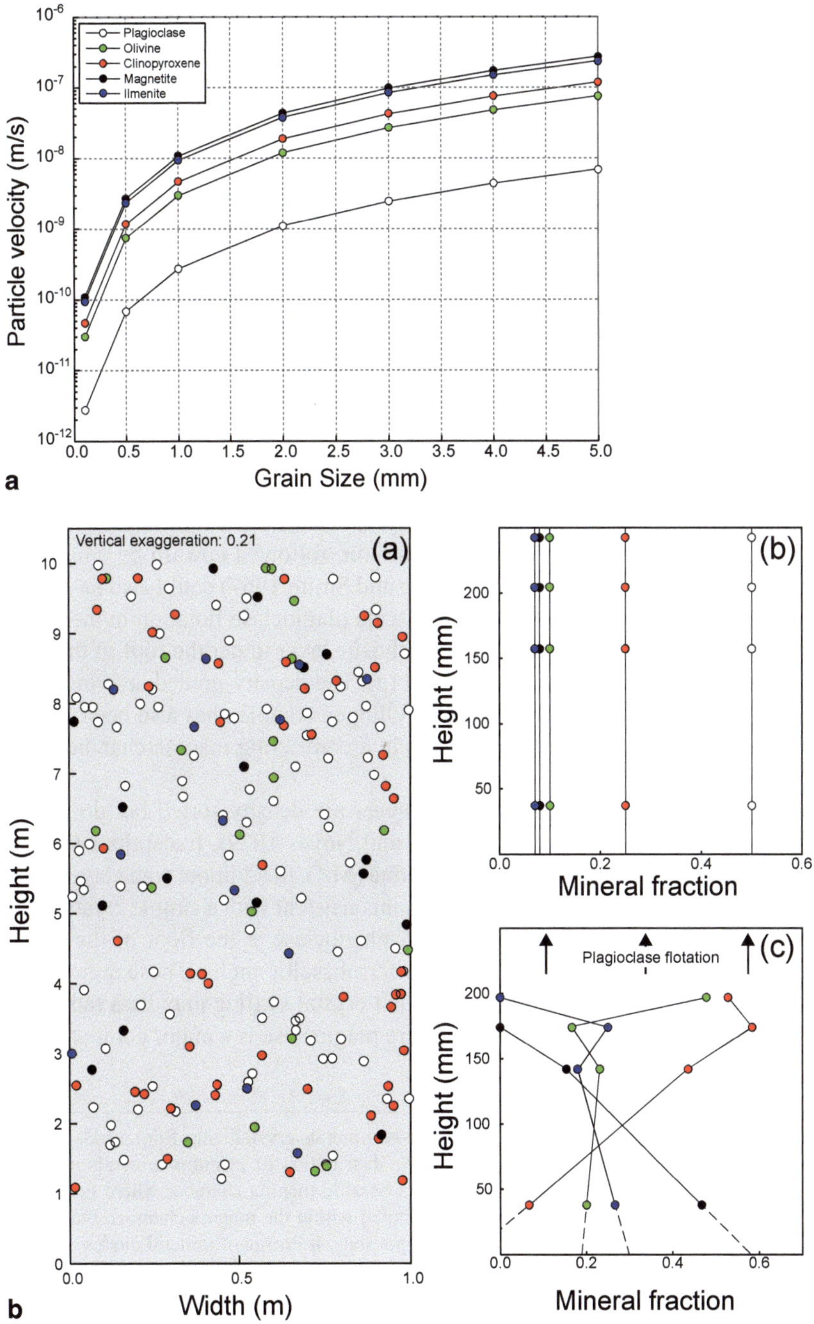

Fig. 2.7 Theoretical models showing the effect of mineral density and grain size on crystal settling in a basaltic magma chamber. **a** Evolution of particle velocity (settling or floating; m/s) as

of mineral modes, with the densest and coarsest minerals dominating the bottom of the layer and progressively grading upwards to minerals with lower densities and smaller grain sizes. In theory, a negative correlation between grain-size and mode should be observed for each mineral from base to top of a given layer (McBirney and Noyes 1979; Naslund and McBirney 1996). There should also be a difference in the crystal size distribution (Higgins 2002).

Mineral sorting occurs during silicate liquid crystallization, within the main magma body (Wager and Brown 1968) or during the filling of the magma chamber (initial filling or magma chamber replenishment; Irvine and Smith 1967; Jackson 1970; Marsh 1989) when the injected liquid contains a significant cargo of crystals. This may also take place during in situ crystallization if grains are not strongly attached to other grains or to the walls of the magma chamber (Namur et al. 2011a).

Layering formed by crystal settling has been suggested to have taken place in the classic Skaergaard intrusion (Wager and Brown 1968), the Imilik gabbro (Brown and Farmer 1971) and in the Vesturhorn eucritic gabbro (Roobol 1972). Well-defined grain-size sorting in the Duke Island peridotite (Fig. 2.2f) could also result from this process (Irvine 1974). The cyclic layered units in the Muskox intrusion with each cycle made up of a dunite basal unit, followed upward by a harzburgite and an uppermost orthopyroxenite (Irvine and Smith 1967) could also have formed through a crystal settling process. Large-scale plagioclase flotation in the Sept Iles intrusion formed a 100–500 m thick anorthosite layer under the roof of the magma chamber (Higgins 2005; Namur et al. 2011a). The density-graded layering (peridotite followed by gabbro) in the Fongen-Hyllingen complex has also been suggested to result from differential crystal settling in a convecting magma chamber (Nilsen 1973).

In the Skaergaard intrusion, graded layers are density-sorted but do not show significant grain-size sorting (McBirney and Noyes 1979). Exceptions to this are found in the Triple Group of the Middle Zone (MZ). In addition, some layers show a mafic top over a felsic base, which seems inconsistent with a simple crystal settling process. Another issue is the presence of plagioclase at the floor of the intrusion while plagioclase crystals should float in ferrobasaltic melts. These arguments led McBirney and Noyes (1979) to suggest that crystal settling may be a rather minor process in basaltic magma chambers where plagioclase is a major component. This

a function of grain size (radius; mm) for common minerals crystallizing from a basaltic melt. **b** Influence of contrasted particle velocities on the distribution of cumulus minerals in cumulate rocks. **a** Schematic illustration of a section of a basaltic magma chamber where homogeneous nucleation occurs. Minerals are randomly distributed within the magma chamber, but their relative proportion is in agreement with cotectic proportions; **b** Profile of mineral modes in cumulate rocks crystallizing from the liquid column in (**a**) if there is no mineral sorting due to deposition. A similar profile would develop if crystallization occurs in boundary layers (in situ crystallization). (**c**) Profile of mineral modes in cumulate rocks crystallizing from the liquid column in (**a**) when crystal sorting due to contrasted particle velocities is taken into account. All calculations were performed for a basaltic melt with a density of 2.75 g/cm^3 and a viscosity of 120 Pa s at 1 atm and 1100 °C. Mineral densities (g/cm^3) at 25 °C and thermal expansion coefficients are from Niu and Batiza (1991). Mineral grain sizes range from 1 to 5 mm seems to be the case in the Kilauea Iki lava lake and the Makaopuhi lava lake, which do not show evidence for density controlled sorting of crystals and which therefore cannot be explained by simple crystal settling in a homogeneous magma body (Moore and Evans 1967; McBirney and Noyes 1979; Vinet and Higgins 2011). However, as described below, many layering features of mafic layered intrusions can be explained by a crystal-settling process combined with current flows of crystal-laden magma.

Magma Currents

Currents of crystal-laden magmas lead to the concentration of crystals in parts of the flowing magma with minimum shear, locally forming nicely defined layering. Although this type of layering is mostly observed in dykes and sills, many layering features in basaltic layered intrusions may also result from such a process (Irvine et al. 1998).

Modally graded layers in various layered intrusions show characteristics identical to those observed in sedimentary rocks formed by turbidity currents (Irvine, 1965; see also Fig. 2.3). For many decades, this has led petrologists to assume that density currents were the main reason for layering (e.g. Wager and Brown 1968). A prime example is the Duke Island ultramafic intrusion that shows grain-size graded layering with scour-and-fill structures, slumps, angular unconformities and layer truncations as in cross-bedding (Irvine 1974; Naslund and McBirney 1996). The formation of the grain-size graded layering together with the abundant sedimentary features was explained by Irvine (1965) as resulting from density currents. Crystals first formed at the roof of the chamber and in the upper part of steeply inward-dipping walls, resulting in unstable accumulations that periodically slumped forming currents of crystal-laden magma at the floor of the magma chamber.

The origin of the olivine horizon (olivine-rich dolerite; 1–10 m thick) of the Palisades sill is still suggested by many to have formed by preferential olivine settling during magmatic flow into the sill. The initial olivine enrichment in the magma could result from a separate, olivine-rich, magma pulse or from an initially inhomogeneous magma in terms of crystal-content (Husch 1990; Gorring and Naslund 1995). In any case, the highly variable content of olivine in this horizon seems to indicate some preferential segregation of olivine.

In the Skaergaard intrusion, the Layered Series shows abundant cm- to dm-thick modally graded layers with olivine, magnetite and ilmenite concentrated at the bottom, pyroxene in the middle and plagioclase at the top (Fig. 2.2a) that extend for tens to hundreds of meters. Some layers are also associated with sedimentary-like features such as cross-bedding and layer termination. Modeling by Irvine et al. (1998) taking into account the density of the liquid, the densities of the different minerals, and their likely grain sizes, suggests that crystal deposition during density currents should result in layers with densest minerals at the bottom and the least dense minerals at the top. In addition, this process should also result in grain-size sorting with the largest grains of each mineral at the base of the layer. Comparison

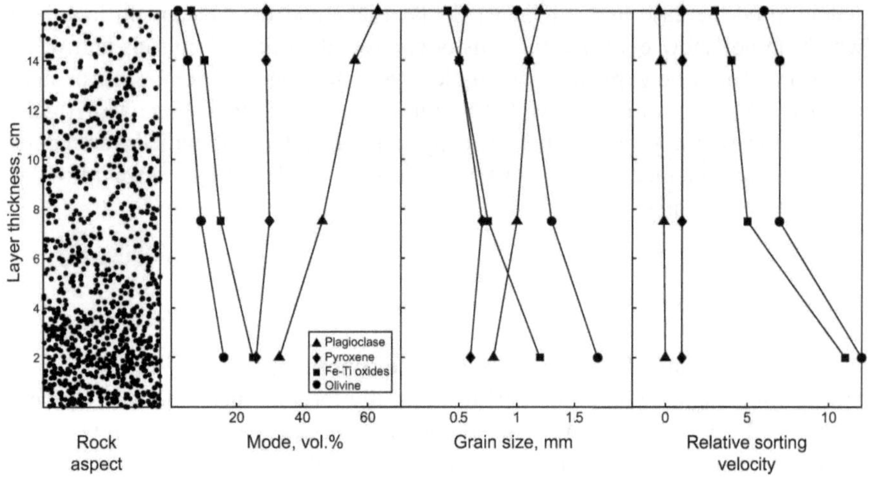

Fig. 2.8 Schematic model of layering development in the *Upper* Zone a of the Skaergaard intrusion. These layers were interpreted by Irvine et al. (1998) as forming during crystal-loaded magma currents at the *bottom* (floor) of the magma chamber. This diagram shows the rock aspect (relative proportion of mafic phases; *black dots* and plagioclase) and the upward evolution of mineral modes and grain sizes for the main minerals. It also shows the mineral sorting velocities relative to pyroxene. Modified from Irvine et al. (1998)

between theoretical models of crystal deposition by Irvine et al. (1998) and petrographic observations (mineral modes and grain-size distribution) by Conrad and Naslund (1989) reveals that the main modal and grain-size characteristics of the Skaergaard modally graded layers could be explained by a process of crystal deposition from crystal-rich density currents that broke away from the walls of the intrusion (Fig. 2.8). Current flows and density crystal-sorting in Skaergaard have been strongly criticized because plagioclase should float on dense ferrobasaltic melts and therefore not accumulate at the floor of the intrusion (see above). However, retention of plagioclase at the floor of the intrusion could be explained by plagioclase crystals being entangled with denser mafic minerals and because the plagioclase-enriched layer tops are immediately buried beneath layers of denser cumulates (Irvine et al. 1998). In addition, a significant contribution from a zone of in situ crystallization that would have been stirred by the flowing magma is also envisaged (Conrad and Naslund 1989).

At the top of the Skaergaard Upper Zone a (UZa), modally graded layering is joined by trough structures (Fig. 2.6d) made up of stacked synformal layers 10–50 m wide, up to 100 m high and 450 m in length (Irvine 1987). These rocks are strongly layered with some layers being almost monomineralic (e.g. anorthosite) and others strongly dominated by mafic phases (olivine, clinopyroxene and Fe-Ti oxides). The trough layers have been successively interpreted as intermittent and channelized density currents (Wager and Brown 1968), or as resulting from a combination of subparallel roller convection cells in which minerals are sorted and deposited by density currents (Irvine 1987). Trough structures also occur in the Fongen-Hyllingen intrusion where they show clear evidence for erosion at their bases (Wilson et al. 1981) and are therefore considered as an obvious example of dynamic layer-forming process.

Convection-Related Layering-Forming Processes

Convection in magma chamber can be either continuous or intermittent, each type potentially leading to the formation of layers. Convection generally results from density or temperature gradients in the magma chamber and details on its origin and the behavior of convective cells can be found in several papers including Turner and Campbell (1986), Martin et al. (1987) and Marsh (1989). Convection can keep the crystals in suspension in a magma chamber as long as settling velocities of crystals are small relative to turbulent fluid velocities (Marsh and Maxey 1985; Sparks et al. 1993). Settling will only start to be operative when a critical concentration of minerals in the liquid is reached. In a multi-saturated system, minerals may have contrasted settling velocities (Fig. 2.7) and also different critical concentrations, therefore leading to the formation of modally-graded layers (e.g. Khibina alkaline massif; Kogarko and Khapaev 1987; Ploumanac'h subalkaline granite; Barriere 1981). In contrast, continuous convection in a magma chamber can also result in the deposition of homogeneous massive gabbro without evidence for any layering feature (e.g. Skaergaard Layered Series; Wager and Brown 1968; Irvine 1987).

Convection can also be in the form of plumes (see also Section "Layering Formation During Crystallization Along Inclined Surface"), which break-off from the unstable boundary layers (thermal and compositional) at the roof of the magma chamber (where heat is predominantly lost; Morse 2008) and travel downwards to the bottom of the magma chamber (Brandeis and Jaupart 1986). The formation of these plumes is dependent on the temperature gradient within the boundary layer and the viscosity of the magma. In magma with low viscosity, crystals may initially form in the upper boundary layer and can be brought downwards when the plumes break-off (Brandeis and Jaupart 1986). In this case, layer formation proceeds both by bottom crystallization (i.e. in situ crystallization at the interface between the crystal mush and the main magma body) and by deposition from above. The crystals coming from above are deposited as discrete batches and not continuously as would happen during simple crystal settling from a large magma reservoir. Consequently, this process may interrupt the normal crystallization sequence (e.g. formation of homogeneous rocks) and result in the formation of well-defined intermittent layering such as the layering observed in Duke Island (Irvine 1974), Skaergaard (McBirney and Noyes 1979), Jimberlana Complex (Campbell 1978) and Stillwater (Hess 1960).

In a simpler way, convection can also be intermittent with episodes of convective overturn followed by periods of relative stagnation. Such a succession of convective and non-convective periods, together with crystal settling or flow along the magma chamber floor, was suggested as a potential mechanism for the production of cyclic layering in the Stillwater (Hess 1960) and Ilimaussaq intrusions (Engell 1973). In

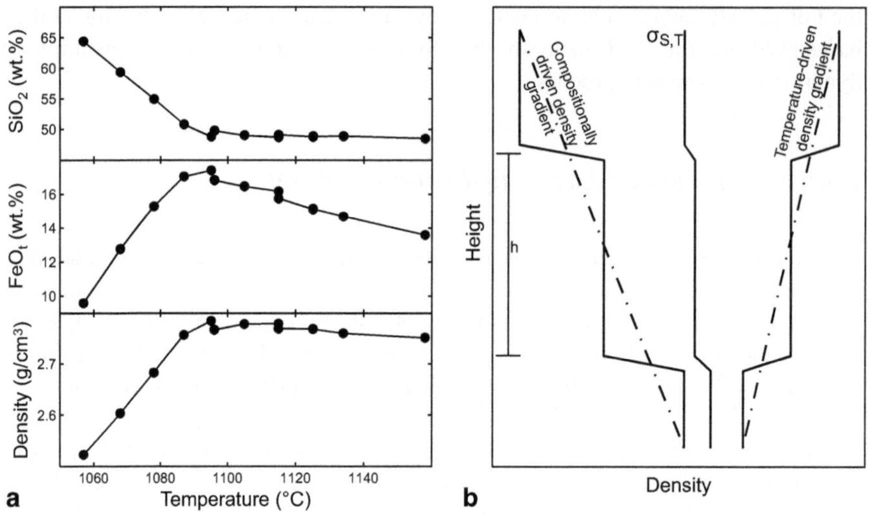

Fig. 2.9 **a** Evolution of melt composition (FeO_t and SiO_2) and density during differentiation of a ferrobasaltic magma chamber (decreasing temperature). Liquid compositions are based on experimental data from Toplis and Carroll (1995). See Namur et al. (2011a) for details on the procedure of calculation for liquid density. **b** Schematic diagram illustrating the initial distribution of density (*dashed lines*) resulting from gradients of composition and temperature in a ferrobasaltic magma chamber loosing heat at the roof and crystallizing at the floor. Double diffusive convection results in smooth gradients breaking down into convecting layers of height *h*, which are divided by step-like interfaces. Resulting density distribution due to combined effects of both factors is shown by $\sigma_{S,T}$

the Skaergaard intrusion, the alternation of m-thick plagioclase-rich layers and pyroxene-rich layers with sharp upper and lower boundaries in the Middle Zone was also proposed as resulting from the succession of convective episodes with periods of stagnation (Naslund et al. 1991).

Another type of convection inspired from fluid kinetics is double-diffusive convection and occurs in magma chambers when a density gradient driven by two parameters with different rates of diffusion develop. In basaltic magma chambers, it is assumed to be caused by variation in liquid composition (e.g. compositionally stratified liquid) and temperature, due to fractionation of the magma adjacent to a front of crystallization (in situ crystallization; McBirney and Noyes 1979; Chen and Turner 1980; Huppert and Sparks 1984; Naslund and McBirney 1996; Fig. 2.9). Double-diffusive convection leads to the formation of a stratified magma chamber with various liquid layers convecting as discrete units (Fig. 2.9). Crystallization of individual liquid layers could therefore result in igneous layering, with each layer showing contrasting mineral modes and potentially mineral compositions.

Compositional stratification in the magma can also result from magma chamber replenishment (see below for additional details), while diffusion of heat and chemical components across the interface between the magmas may lead to the formation of a vertical succession of double diffusive layers. Crystallization from contrasted magma layers along the sloping floor of the Honningsvåg intrusive suite and the Fongen-Hyllingen complex was suggested as explaining the evolution of mineral compositions and mineral proportions along the strike of individual layers (Robins et al. 1987; Wilson et al. 1987). A similar explanation was given for the lateral variation of pyroxene composition in the Great Dyke (Zimbabwe; Wilson 1982), while individual layers with contrasted mineralogy in the Stillwater complex and in Bjerkreim-Sokndal were suggested to have crystallized simultaneously from different liquid layers (Irvine et al. 1983; Jensen et al. 2003).

Magma Injection and Magma Mixing

Magma injection (e.g. magma recharge) in a crystallizing magma chamber is commonly envisaged as an efficient layer-forming process. It is now generally accepted that individual layers do not represent the crystallization product of individual magma batches because most layers of basaltic and ultramafic intrusions do not have a bulk-rock compositions identical to those of silicate melts at magmatic temperature. However, there are different ways magma injection may form layering: (1) injection of crystal-rich magmas can produce layering when the different batches of magma have contrasted crystal loads (Naslund and McBirney 1996). Modal (and grain-size) layering may also develop when differential crystal settling occurs after the injection of magma batches containing different types of crystals (Naslund and McBirney 1996); (2) injection of new batches of magma which can produce layering due to differentiation of each magma batch. The Muskox intrusion shows cyclic layered cumulate rocks in which each sequence is made up of dunite, followed by harzburgite and finally orthopyroxenite. It was suggested that each of these sequences resulted from the injection of a new batch of magma having olivine as the first liquidus phase followed by orthopyroxene (Irvine and Smith 1967). (3) Magma injection can also result in the formation of quench horizons of fine-grained material (e.g. gabbro) that produce layers within homogeneous cumulate rocks. This process was mainly described for the Kap Edvard Holm intrusion (Tegner et al. 1993) but may also have occurred in the Sept Iles intrusion (Namur et al. 2010; Namur et al. 2011b).

Magma chamber replenishment can also produce a stratified magma chamber when the new batch of magma has a density (and temperature) significantly different to that of the melt into the magma chamber (Fig. 2.10; Raedeke and McCallum 1984). Stratification of melt in a magma chamber can possibly lead to the formation of well-defined layering, as that observed in the cyclic units of the Rum Layered Suite (Huppert and Sparks 1980). When a basaltic magma chamber crystallizing plagioclase, olivine and pyroxene is replenished by a dense picritic magma with only olivine on the liquidus, this new batch of magma forms a basal layer at the bottom of the magma chamber. Crystallization of olivine from the basal layer produces a residual liquid that becomes progressively less dense. When its density becomes similar to that of the basaltic liquid, the two layers of liquid start to

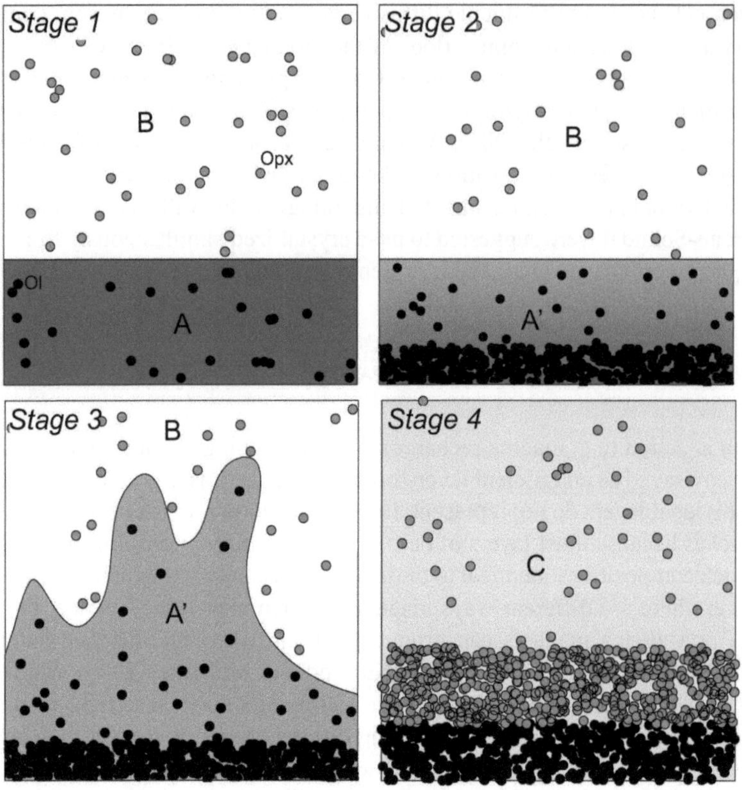

Fig. 2.10 Formation of the dunite-orthopyroxenite cycling layering in a subsection of the floor of an ultramafic magma chamber. The process starts by the injection of a primitive magma A (crystallizing olivine) into a magma chamber containing a liquid B with orthopyroxene as the sole liquidus phase (*Stage 1*). The density of olivine + A is higher than orthopyroxene + B, and liquid A (+ olivine) ponds at the floor of the magma chamber. $T_A > T_B$ and a non-turbulent interface develops between A and B, resulting in double diffusive convection with each liquid layer convecting as a discrete unit. As magma A cools down, it crystallizes olivine, accumulating at the floor, and orthopyroxene is resorbed due to latent heat release from olivine crystallization (*Stage 2*). When olivine crystallizes, the density (and composition) of A evolves to A' (see the color change in the lower layer at Stage 2) and when the density of A' + olivine becomes lower than B + orthopyroxene, the two layers start to mix (*Stage 3*). After mixing is complete, the relative proportion of A + olivine and B + orthopyroxene determines the composition of the hybrid magma (C) and the nature of the liquidus phase(s). In the case shown in *Stage 4*, orthopyroxene is the liquidus phase, forming an orthopyroxenite layer at the top of the dunite layer. If both olivine and orthopyroxene are stable phases, a harzburgite layer develops. *Ol* = olivine; *Opx* = orthopyroxene. (Modified from Raedeke and McCallum (1984))

mix convectively, producing a hybrid magma crystallizing plagioclase, olivine and pyroxene until interrupted by the influx of new batch of magma. A similar process has been invoked to explain the formation of the macro-rhythmic or cyclic units composed of dunite, harzburgite and orthopyroxenite in the Stillwater Complex (Raedeke and McCallum 1984; Fig. 2.10).

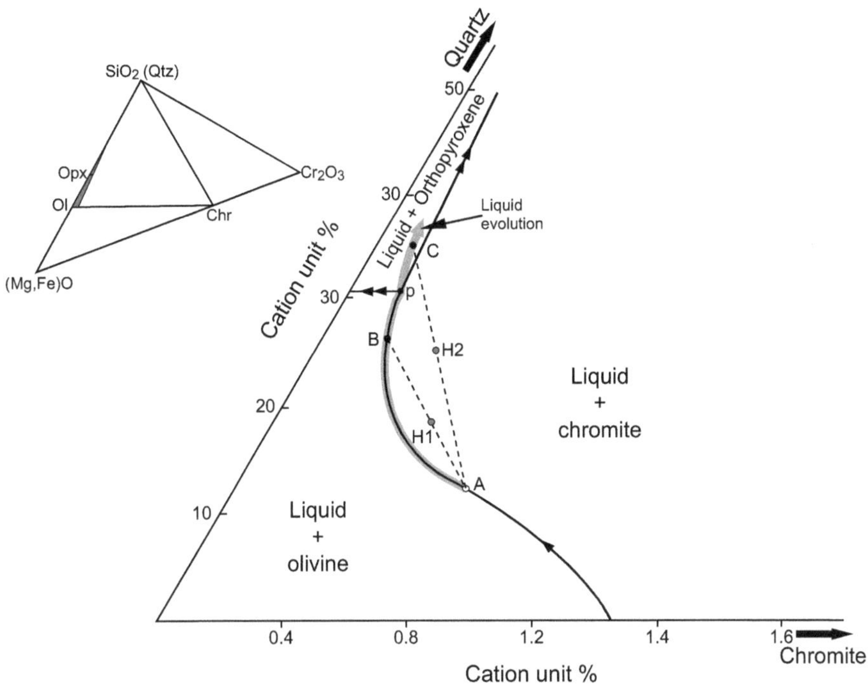

Fig. 2.11 Illustration of the influence of the magma mixing process in producing hybrid liquids saturated in only one phase, chromite in this case. This figure is a cation norm projection on the join olivine-chromite-quartz, which is part of the more extensive join (Mg, Fe)O–Cr_2O_3–SiO_2. Only a small part of the system olivine-chromite-quartz (*grey field* in the inset *triangle*) is represented and shows the stability fields of olivine, orthopyroxene and chromite. Boundary curves with double arrows are reaction boundaries, and cotectic curves are represented with a single arrow. A primitive magma of composition A evolves along the *A-B* curve and precipitates olivine and chromite. When the liquid reaches the point *p*, the combination of reaction boundaries requires that olivine and chromite stop forming as orthopyroxene starts to crystallize. The residual liquid evolves toward the composition symbolized by the point *C*. Mixing differentiated magma at *B* or *C* with primitive magma at *A* results in hybrid magmas *H1* or *H2* that will crystallize only chromite until they return on the olivine-chromite cotectic curve. (Modified from Irvine (1977))

Magma mixing, either due to magma chamber replenishment (with low density and viscosity contrast; Irvine 1977), contamination by country rocks (Irvine 1975) or interaction between magma in a basal boundary layer and the main magma body can also be a process responsible for the formation of layering (Wilson et al. 1996) in some cases with economic deposits as the result (Todd et al. 1982; Irvine et al. 1983; Naslund and McBirney 1996). The mechanism commonly proposed considers that mixing forces a liquid evolving along a convex saturation surface to fall within the stability of a single phase, therefore forming monomineralic layers, or at least layers with a very high proportion of one of the constituent phases (Fig. 2.11). This model has classically been invoked for the formation of chromitite layers in Muskox, Bushveld and Stillwater (Irvine 1975, 1977; Lipin 1993) but also to the formation of magnetitite layers in Bushveld (Harney et al. 1990). Finally, interstitial liquid migration and magma mixing, potentially accompanied by cumulate rock dissolution, has also been considered as a process for the formation for chromitite seams in the Rum Eastern Layered Intrusion (Young 1984; O'Driscoll et al. 2010).

Formation of Chromitite Layers

Chromitites are among the most distinctive and enigmatic geological features of layered mafic/ultramafic intrusions. Chromitites, here defined as a rock predominately composed of chromite or Cr-rich spinel, can be found in a diverse range of cumulate complexes, including alkaline mafic intrusions, Alaskan-type complexes and Archean anorthosites, but the most extensive, most economically significant, and most intensely studied examples are those from within the ultramafic sequences of Precambrian layered intrusions (e.g. Great Dyke, Stillwater Complex), or those of the (largely mafic) Critical Zone of the Bushveld Complex. Chromitites tend to form layers or seams that can be several mm to several meters in thickness and can be laterally continuous for many kilometers. In the extreme case of the Bushveld Complex, individual chromite horizons can be traced for 100s of kilometers (e.g. see Eales and Costin 2012). They often lie on, or close to, the boundary between cyclic units, and individual layered intrusions may host multiple chromite horizons. Chromitites often have sharp contacts with surrounding cumulate rocks (Fig. 2.12), although diffuse or gradational contacts, particularly along the upper contact, are also common (e.g. Campbell and Murck 1993). Despite their lateral continuity, individual layers can vary substantially in thickness and may bifurcate or anastomose

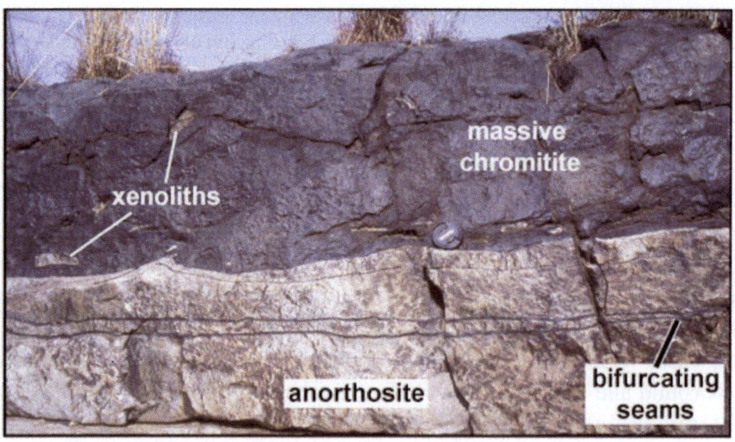

Fig. 2.12 UG1 chromitite seam of the Bushveld Complex, exposed at Dwars River. The massive chromitite seam contains abundant angular xenoliths of anorthosite, and has an irregular anastomosing contact with the underlying anorthosite. The two underlying thin chromitite seams bifurcate from a single layer at the *right* hand side of the photo

along strike (Fig. 2.1f; Voordouw et al. 2009; Lee 1981). Many chromitite seams preserve sedimentary-like features, such as graded bedding, braiding and slump structures, and can contain angular xenoliths of the surrounding cumulate rocks (Fig. 2.12; Voordouw et al. 2009). Despite modification due to subsolidus recrystallization, textural analysis reveals that chromite sometimes formed as an early cumulus phase, and in other cases occurs in the interstices of cumulus olivine (the latter often altered to serpentine), pyroxene or plagioclase.

Chromite is only a minor phase (<2 modal%) produced during closed-system cotectic crystallization of primitive mafic magma in the crust. Furthermore, mass balance calculations indicate that the volumes of magma required to form the chromitite layers of some intrusions (e.g. Bushveld, Stillwater) is much greater than their total cumulate rock volume (Eales and Costin 2012). It is therefore difficult to attribute the formation of thick chromitite seams found in layered intrusions to a simple model of fractional crystallization of primitive magma, followed by crystal accumulation. The apparent excess volume of chromite requires involvement of much larger magmatic systems than is usually considered, with influx of partially crystallized, chromite-bearing magmas from deeper staging chambers (e.g. Eales and Costin 2012) or removal of large volumes of residual liquids from the top of the chamber (Naldrett et al. 2012). However, neither of these scenarios alone can fully explain the formation of stratigraphically coherent cumulate layers consisting mostly of chromite (Cawthorn 2005; Cawthorn et al. 2005). It is therefore generally accepted that a mechanism (or mechanisms) is required to trigger extensive chromite crystallization, such as: (1) a pressure change in the magma chamber; (2) a change in oxygen fugacity of the magma; (3) mixing of primitive magma with fractionated residual magma; (4) assimilation of country rock by primitive magma (Cawthorn et al. 2005; Spandler et al. 2005); assimilation of plagioclase-rich footwall cumulate (O'Driscoll et al. 2009a, 2010). The magnitudes of pressure and oxygen fugacity changes required to produce the meter-thick chromitite layers found in many intrusions are not thought to be sustainable in upper crustal intrusive environments (Cawthorn et al. 2005).

There appears to be no universally applicable trends in the compositional variation through individual chromitite seams, or through a sequence of seams. For example, chromitites with the highest Cr/Fe include the lower group chromitites of the Bushveld Complex, the middle group of the Great Dyke and the thickest and most massive mid-upper units (G and H) of Stillwater Complex. In some cases chromitite units (e.g. Bushveld LG and UG2 seams, Stillwater G and H seams), have clear stratigraphic variation in metal contents (e.g. V, PGEs) and chromite major element composition (e.g. Cr/(Cr+Al) or $Mg/(Mg+Fe^{2+})$), which has been cited as support for in situ growth and settling of chromite from the magma chamber (e.g. Wilson 1982; Campbell and Murck 1993; Naldrett et al. 2012). In other cases, clear stratigraphic trends in the composition of chromite and the surrounding cumulate minerals are not observed (Mondal and Mathez 2007).

The most widely accepted model for the origin of chromitite layers and for their occurrence in various units of layered intrusions is mixing of primitive and fractionated magma in the chamber, which is consistent with the occurrence of chromitite at or near the base of cyclic units and conforms to the mineral chemistry trends and associated PGE and sulfide mineralization found in some seams (e.g. Irvine 1977; Campbell and Murck 1993). In some cases the expected trends in mineral chemistry are not observed (e.g. Mondal and Mathez 2007), which argues against significant magma mixing. Nevertheless, recent studies indicate that thick chromitite seams may be composed of multiple sublayers, each of which formed during discrete magmatic influx events (Junge et al. 2014). In the case of magma blending via repeated injections of relatively small melt volumes, variations in the major element composition of cumulate minerals above and below the chromitite seam may not be readily apparent. Enclosed within many chromite grains are multiphase inclusions that include unusual alkali phases not present within the surrounding cumulate rocks (Spandler et al. 2005). The most distinctive of these is aspidolite (Na-phlogopite), which is near to ubiquitous in these inclusions, but is almost exclusive to chromitites. These features are interpreted to be either remnants of contaminated silicate melts or hydrous fluids that were trapped during chromite growth, and hence provide clues to the origin of primary chromite formation. Radiogenic isotope studies of layered intrusions containing chromitite (e.g. Rb–Sr, Re–Os, Sm–Nd) also show that the strongest 'crustal' signatures are often associated with chromitite horizons (Kinnaird et al. 2002; Spandler et al. 2005; Day et al. 2008; O'Driscoll et al. 2009b). The isotopic data and occurrence of hydrous, alkali-rich inclusions in chromite would therefore argue in favour of chromite crystallization and the formation of almost mono-mineralic chromitite layers in response to crustal contamination of primitive magma (Spandler et al. 2005; Day et al. 2008).

Nevertheless, the variability of cumulate textures and sedimentary features preserved in chromitites attests to the complexity of crystal transport, accumulation and settling after chromite crystallization. In particular, the length-scales and mechanisms of chromite transport and remobilization prior to solidification remain debated (in-situ settling, with or without subsequent thickening, or long distance transport of chromite-laden slurries; e.g. Voordouw et al. 2009; Naldrett et al. 2012). All these suggest that chromitite formation needs not be restricted to a single mechanism; it is conceivable, and even likely, that a range of different processes are conducive to forming chromitite. Resolution and documentation of these issues are critical to the understanding of the genesis of chromitites and other monomineralic layers, and undoubtedly will be the focus of further research in the future.

Liquid Hybridization as the Origin of Cyclic Layering: An Example From the Oman Ophiolite

Hybridization of a silicate melt through magma mixing or magma hydration can change the sequence of crystallization and has the potential to produce igneous layering when this process is repeated. The layered cumulate unit of the Oman ophiolite (Fig. 2.13a) offers a good example of this process. This unit has an average thickness of 2–2.5 km and is characterized by a well-developed ubiquitous layering defined by the succession of mafic and ultramafic millimeter to meter-thick layers (Fig. 2.13b). Abily (2011) distinguished about fifty interlayered lithologies in this

Fig. 2.13 **a** Layered cumulate unit in the Oman ophiolite (Mahram area, Sumail massif). **b** Detail of layering in the Oman ophiolite showing the succession of mafic and ultramafic cm- to dm-scale isomodal layers

unit from modal and textural criteria. About 75 % of the layers are gabbroic with the rest being represented by peridotitic and pyroxenitic cumulates. The more common petrographic types are olivine gabbros and wehrlites.

Contacts between layers are usually marked by variations in mineral proportions (ratio layering; Jackson 1967), mineral assemblages (phase layering; Irvine 1982) and/or mineral habits (textural layering; Irvine 1982). Most cumulate layers are isomodal, but minor modally graded layers have also been reported in some areas of the ophiolite (Pallister and Hopson 1981; Juteau et al. 1988). Grain size-graded layers are rather scarce. In combination, this indicates that mechanical processes such as gravitational sorting were not dominant in Oman, as otherwise commonly proposed (Pallister and Hopson 1981).

The lower crustal section of the ophiolite is frequently cut by ultramafic bodies (mostly wehrlite) interpreted as late intrusions post-dating the crystallization of the layered cumulates. Consequently, many authors have proposed that most ultramafic layers were sills injected from ultramafic intrusions in pre-existing gabbroic layers (e.g. Juteau et al. 1988). This model calls for a two-stage evolution and it has been advocated that the second stage, i.e. ultramafic episode, would not be related to magma chamber activity but to the igneous activity occurring during the early stage of intra-oceanic thrusting (Ernewein et al. 1988). However, field observations (Clénet et al. 2010) indicate that peridotitic and pyroxenitic layers are not always present near ultramafic intrusions and do not show any systematic interlayering with gabbroic cumulates. Moreover, geochemical data indicate that minerals in the peridotitic/pyroxenitic and gabbroic layers globally have the same range of compositions in both major and trace elements (Fig. 2.14). This indicates that the melts parental to the ultramafic cumulates are not more primitive (in terms of Mg/Fe ratio) than those of gabbroic cumulates. Moreover, they are not systematically depleted

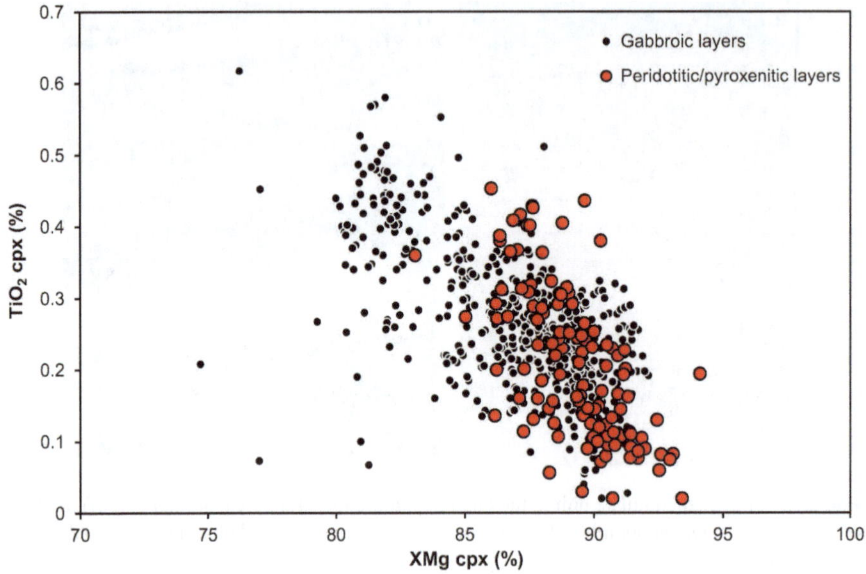

Fig. 2.14 Variation of TiO_2 (%) vs XMg in clinopyroxene (Cpx) for gabbroic and peridotitic/pyroxenitic layers in cumulate rocks of Oman ophiolite. XMg = 100 × Mg/(Mg + Fe_t)

in HFSE (e.g. TiO_2) as would be expected for a late magmatism originating from a depleted mantle source (Ernewein et al. 1988). The clear decoupling observed between the modal composition and the differentiation indexes (i.e. an ultramafic cumulate can be more chemically evolved than a mafic one) shows, however, that the apparent "crystallization trend" described in Fig. 2.14 by mafic and ultramafic layers cannot result from the fractional crystallization of a common parental melt.

Mineral composition profiles across the boundaries between peridotitic/pyroxenitic and gabbroic layers can be useful to constrain the origin of the layering (Abily 2011). An example is given in Fig. 2.15. In terms of modal composition, individual layers are quite homogeneous and the boundary between layers is clear-cut down to the grain scale. Despite the sharp nature of the lithological boundary, the chemical composition of minerals in both major and minor elements defines progressive evolutions toward the boundary. In the example of Fig. 2.15 these evolutions are observed in the lower layer only. The thickness of these "cryptic transition zones" varies according to the element and to the mineral considered. It ranges typically from a few mm (e.g. Cr_2O_3 in Cpx) to a few cm (e.g. Fo in Ol). These distances are too important to be consistent with solid-state intra-crystalline diffusion, but can be accounted for by diffusion in an interstitial liquid phase followed eventually by local re-equilibration between the solid and the interstitial liquid, although advection in the interstitial liquid is also likely. Both major and minor elements change toward the boundary (Fig. 2.15). This can potentially be explained by a mixing process between either the parental melts of each cumulate layer or intercumulus liquids percolating in the crystal mush. Whatever the details of the process, these observations

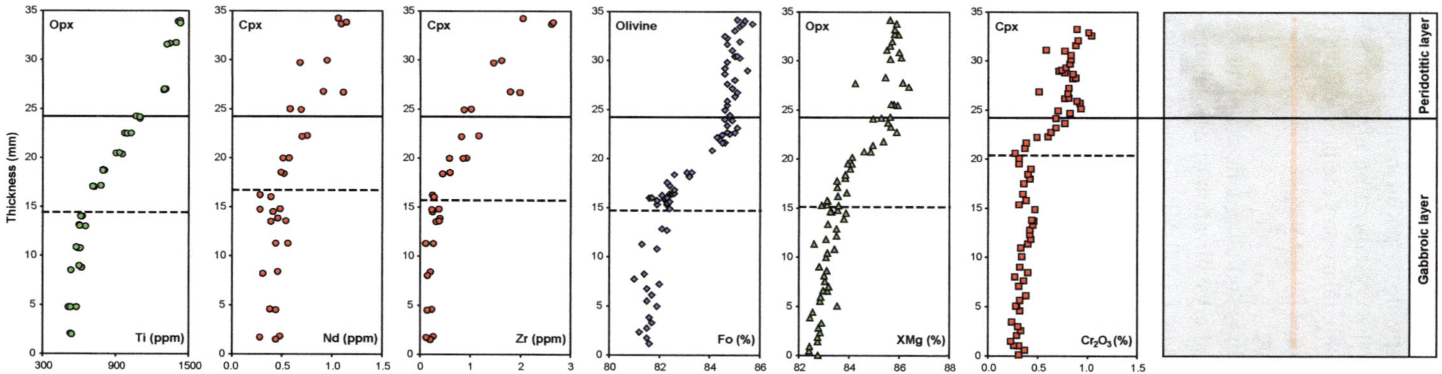

Fig. 2.15 Mineral compositions across the boundary between gabbroic and peridotitic layers in cumulate rocks of Oman ophiolite. *Bold line*—lithological boundary, *dotted line*—cryptic transition zone, *Cpx*=clinopyroxene, *Opx*=orthopyroxene, *Fo*=forsterite content of olivine ($100 \times Mg/[Mg+Fe^{2+}]$), $XMg = 100 \times Mg/[Mg+Fe_t]$

show that the crystallization of interlayered ultramafic and mafic cumulates are contemporary events, sufficiently close-spaced in time to maintain interstitial liquid in the gabbroic layer prior to the crystallization of ultramafic layer. The contrast between the sharp lithological boundary and the progressive chemical evolution in the lower layer could be explained by the sudden change of the cotectic assemblage after a certain amount of mixing.

In the Oman ophiolite, gabbros and ultramafic cumulates are largely considered as the products of the fractional crystallization of normal mid-ocean ridge basaltic (N-MORB) melts at low pressure (~200 MPa), owing to the predominance of the crystallization sequence olivine → plagioclase → clinopyroxene (Pallister and Hopson 1981). The study by Abily (2011) shows however that ultramafic layers characterized by the early crystallization of pyroxene are abundant in the cumulate unit (20 %) and that their crystallization is contemporary with that of gabbroic layers. The mafic melts parental to the ultramafic rocks are not systematically issued from an ultra-depleted source different from the MORB-like source of gabbroic cumulates. The different melts at the origin of the layered cumulates result from hybridization and hydration processes occurring in the magma chamber (Feig et al. 2006; Abily et al. 2011) and/or deeper in the lithosphere (Clénet et al. 2010). Hybridization and hydration can account for variations in both modal proportions and order of crystallization of variably evolved melts issued from a common mantle source.

Liquid and Fluid Migration in a Crystal Mush

Magma migration can occur in the crystal mush of solidifying layered intrusions and has the potential for solid-liquid disequilibrium and change in the stable liquidus assemblage. This process was shown to produce spectacular layering in the Skaergaard intrusion (Namur et al. 2013). Migration of interstitial liquid saturated in plagioclase into an olivine-bearing crystal mush was proposed as a mechanism for olivine dissolution and production of anorthosite layers as well as discordant bodies of troctolite and anorthosite in the olivine-bearing zones of the Stillwater complex (Boudreau and McCallum 1992; Meurer et al. 1997). In the same way, migration of a free fluid phase in a crystal mush can also produce partial melting of the cumulate rocks, a process which in some cases may lead to the formation of layering (Boudreau 1999; Leuthold et al. 2014).

Layering due to Magma Migration During Crystal Mush Contraction

The Marginal Border Series of the Skaergaard intrusion (East Greenland) is comprised of rocks having crystallized in situ along the vertical walls of the magma chamber. It is subdivided into an outer Unbanded Division and an inner Banded Division. The Banded division contains abundant cm- to dm-thick bands, here referred to as colloform bands, dominated by fine-grained mafic minerals and highly depleted in plagioclase (Fig. 2.16).

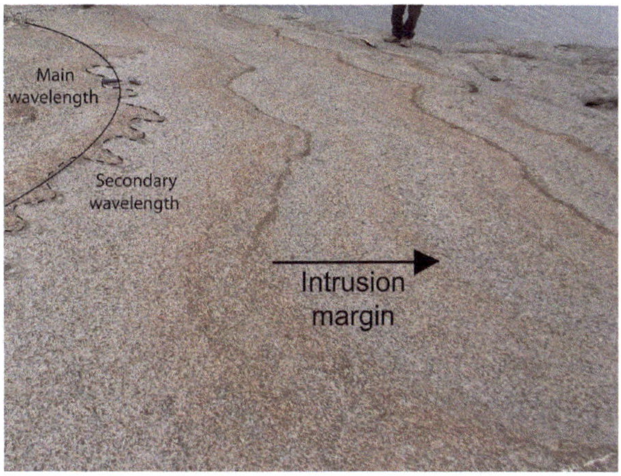

Fig. 2.16 Series of closely-spaced colloform bands showing two different wavelengths. Marginal Border Series of the Skaergaard intrusion, Skaergaard Mini-Peninsula, Greenland

The morphology of the colloform bands evolves significantly from the contact towards the center of the intrusion. The colloform layers nearest to the contact are almost planar in vertical cross-section, but are scalloped on horizontal rock surfaces. With increasing distance from the contact, the bands first lose their sub-planar length-sections and develop a cuspate shape in three-dimensions, then the cusps become increasingly elongate and finger-like and finally only the tips of the fingers remain, forming bowl-shaped features convex towards the intrusion margin (Namur et al. 2013). The colloform bands typically have a sharp outer margin, while the inner margin is sharp or gradational. The composition of plagioclase is similar in colloforms and surrounding gabbros, while olivine and pyroxenes are always more primitive (high Mg#) in colloforms compared to gabbros (Fig. 2.5d).

The cuspate morphology of the colloform bands, together with the fingers and bowls, is highly reminiscent of reaction fronts described in natural porous rocks caused by infiltration of liquid into a porous medium (Ortoleva et al. 1987). These similarities point towards an origin of the colloform banding by lateral melt migration within the crystal mush. The direction of the colloform cusps, the marked asymmetry of mineral modes, together with the convexity of the banding within the fingers, all indicate that silicate liquid flow was focused horizontally from the center of the intrusion towards the margin. This therefore suggests that the source of the infiltrating liquid was the main magma body, or at least, interstitial melt from younger parts of the crystal mush on the chamber walls. A possible mechanism for liquid migration in the Skaergaard crystal mush is volume contraction, or shrinkage, during solidification (Petersen 1987).

Infiltration of liquid from the inner part of the mush towards the margin results in significant thermo-chemical disequilibrium between solid and liquid, which results in reactive dissolution of pre-existing phases. Partial dissolution of minerals in the crystal mush explains the change in mush liquid composition to one capable of crystallizing the colloform bands with a high proportion of mafic minerals. Calculations based on thermodynamically- and empirically-derived equations indicate that such a liquid can be produced by preferential dissolution of mafic mineral phases, without significant contribution from plagioclase dissolution.

Experimental studies on thermally-induced dissolution of silicate phases in basaltic liquids suggest that plagioclase dissolves significantly faster than olivine and clinopyroxene (Edwards and Russell 1996). Thermal disequilibrium in the Skaergaard mush is therefore not the main process of mineral dissolution. Alternatively, it is suggested that the dissolution process is controlled by solid-liquid chemical disequilibrium in the crystal mush. Proposed liquid lines of descent for the Skaergaard intrusion (Toplis and Carroll 1996; Tegner and Cawthorn 2010) have been used by Namur et al. (2013) to calculate olivine and plagioclase equilibrium compositions (Toplis 2005; Namur et al. 2012b). Results indicate that for a given difference of liquid composition between the initial mush liquid and the new infiltrating liquid, the chemical disequilibrium between olivine (and presumably other mafic phases) and the new liquid will be much greater than for plagioclase. A significantly higher degree of olivine and clinopyroxene dissolution is therefore expected compared to plagioclase. The crystallization of the colloform bands is triggered by supersaturation of olivine- and pyroxene components in the infiltrating melt. As dissolution of evolved mafic phases in the mush proceeds, the melt becomes increasingly supersaturated in those phases. When some critical supersaturation state is reached, the melt abruptly starts to crystallize primitive mafic phases in a high-porosity region of the mush, producing a well-defined colloform band.

Hydrous-Fluid Fluxing Through Cumulate Rocks

Crystallization of silicate melt in the magma chamber, or within the crystal mush may lead to the exsolution of a free hydrous fluid-phase. Fluxing of fluid within the porous crystal mush may lead to dissolution of cumulate rocks with potential for the crystallization of rocks with other mineral assemblages and therefore to the development of some layering features (Boudreau 1999). The olivine-bearing zone I of the Stillwater complex is made up of a sequence with a troctolite subzone (with troctolite, dunite, gabbronorite), followed by an overlying anorthosite subzone. Following Boudreau (1999), the olivine-bearing rocks of the troctolite subzone are the product of partial melting of gabbronorite due to fluid fluxing of the cumulate rocks. Modal variations and texture of the grains (including grain size) in these rocks are consistent with progressive incongruent melt reactions (Boudreau 1999). A similar process of olivine-enrichment has been proposed for Paleoproterozoic layered mafic intrusions from East Kimberley, Australia (Boudreau and Hoatson 2004). The process starts with the production of olivine during incongruent melting of pyroxene, while plagioclase is incompletely melted (Boudreau 1988). At the same time, the chromium liberated during the melting of pyroxene may lead to the formation of chromite, also contributing to the formation of layering. With increasing fluid/rock ratio, other hydrous minerals may eventually form. Fluid fluxing and melting of cumulate rocks also result in the formation of hybrid melts due to mixing between the melting products and the resident liquid in the magma chamber. For the

Stillwater complex, it has been suggested that the hybrid melt was first saturated in plagioclase (followed by olivine or pyroxene) and that its crystallization produced the anorthosite subzone at the top of the olivine-bearing zone I (Boudreau 1999) as well as troctolitic/anorthositic bodies in olivine-bearing zone III and IV (Meurer et al. 1997). Where the fluid phase has a high sulphur component, it can also produce the crystallization of massive sulphides, also contributing to the formation of layers, locally with critical economic significance especially for Pt-group elements (Boudreau and Meurer 1999; Kanitpanyacharoen and Boudreau 2013) and other valuable elements (e.g. Au, Ag; Meurer et al. 1999).

Liquid Immiscibility

Silicate liquid immiscibility results in the separation of a single silicate melt into two or more conjugate immiscible liquids having contrasted compositions and physical properties (Fig. 2.17). It may occur during the differentiation of tholeiitic basalts and was shown to produce two contrasted melts: a ferrobasalt and a rhyolite (Philpotts 1982; Charlier et al. 2013). Similar immiscible melts have been observed in apatite-hosted melt inclusions from the Skaergaard intrusion (Jakobsen et al. 2005) and the Sept Iles intrusion (Charlier et al. 2011; Namur et al. 2012a). The compositional field for liquid immiscibility was recently experimentally investigated by Charlier and Grove (2012) who showed that the liquid before the onset of immiscibility may show a large range of composition. Immiscibility might therefore be considered as an important process of magma differentiation and was proposed as explaining the absence of intermediate compositions in volcanic tholeiitic provinces (Charlier et al. 2013). Important questions, and currently imperfectly solved, about immiscibility concern: (1) the efficiency of the two immiscible melts to segregate from each other at large scale and (2) where does immiscibility occur in the magma chamber? (e.g. in the whole magma body or in thin boundary layers at the interface between the crystal mush and the main magma body; Charlier et al. 2011; Jakobsen et al. 2011). Large-scale separation of the two immiscible melts may have significant implications for the formation of layering and ore deposits (e.g. layers enriched in Fe-Ti oxides and apatite) in layered intrusions (Scoon and Mitchell 1994; Charlier et al. 2011; Jakobsen et al. 2011).

Segregation of Immiscible Melts in the Main Magma Chamber

Silicate liquid immiscibility between rhyolitic and ferrobasaltic melts has been recognized to occur during the late-stage evolution of the Skaergaard and the Sept Iles layered intrusions (McBirney 1975; Jakobsen et al. 2005; Charlier et al. 2011). This process is an efficient mechanism to produce igneous layering if the paired immiscible melts segregate from each other. This type of igneous layering is observed in the Critical Zone of the Sept Iles intrusion (Charlier et al. 2011; Namur

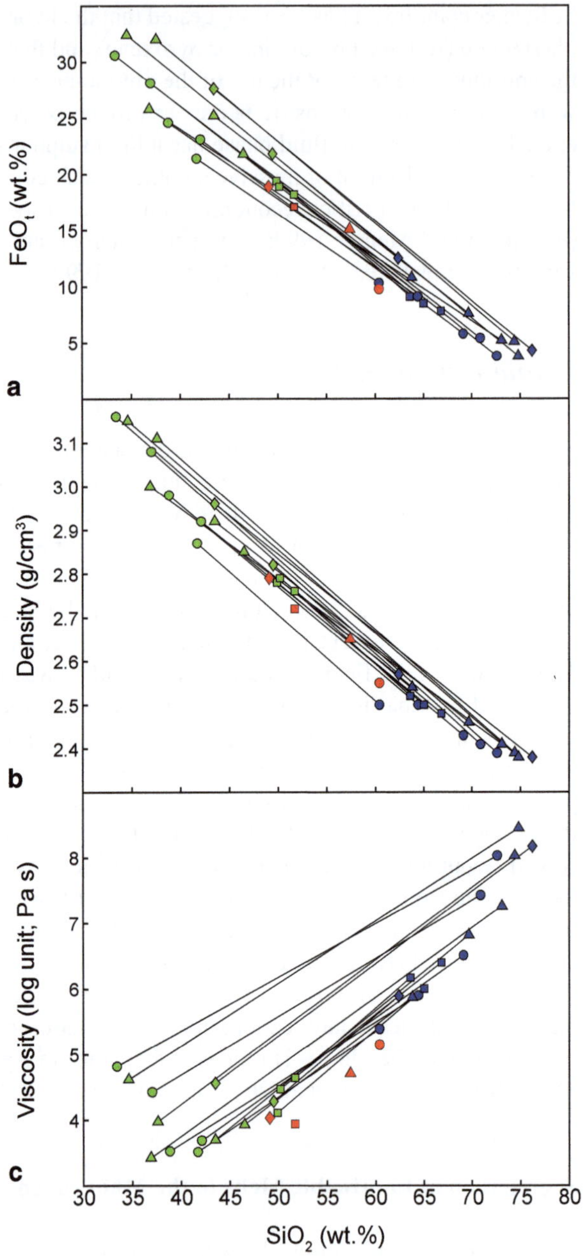

Fig. 2.17 Composition and physical properties of immiscible melts experimentally produced by Charlier and Grove (2012). **a** FeOt vs SiO_2 diagram. **b** Density of the melts. **c** Viscosity of the melts. *Blue symbols* = Si-rich immiscible melts; *Green symbols* = Fe-rich immiscible melts; *Red symbols* = homogenous melts before the onset of immiscibility. *Circles* = Sept Iles compositions; *diamonds* = Iceland (Thingmuli); *squares* = Snake River Plain; *triangles* = Mull

et al. 2012a). This horizon contains the most evolved cumulates of the megacyclic units II. They are made up of plagioclase (An_{55-34}), olivine (Fo_{66-21}), clinopyroxene ($Mg\#_{75-55}$), ilmenite, magnetite, apatite, ±pigeonite and show a bimodal distribution in terms of bulk-rock composition and density (Fig. 2.18). One group,

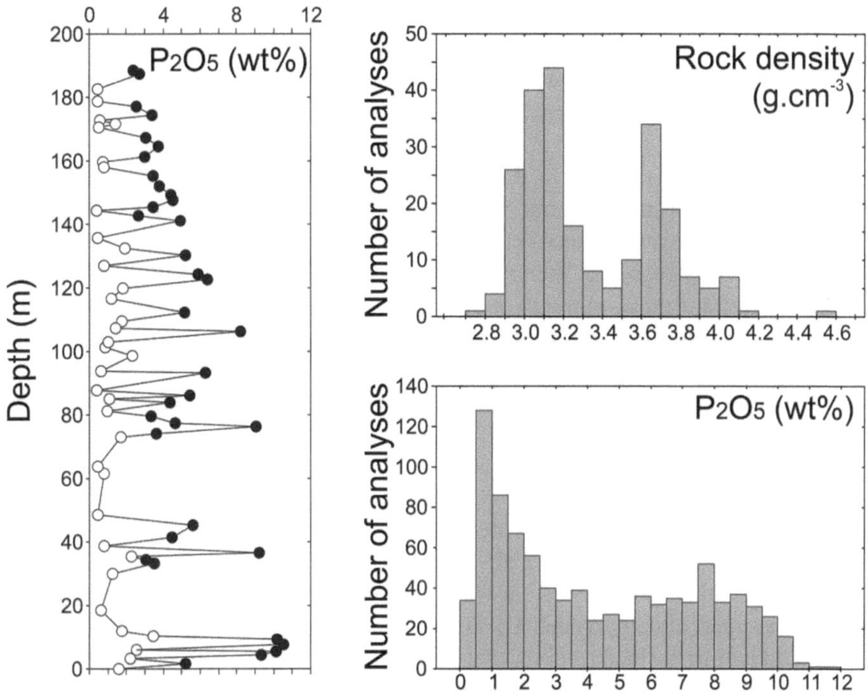

Fig. 2.18 Igneous layering in gabbros from the Critical Zone of the Sept Iles layered intrusion illustrated by stratigraphical variations of the P_2O_5 content and histograms for density (n=228) and P_2O_5 content (n=895) of bulk rocks. *Black* symbols are for dense P_2O_5-rich gabbros and *white* symbols are for light P_2O_5-poor gabbros

enriched in plagioclase and representing two thirds of the samples, has an average density of 3.0–3.2 g/cm^3 and contains 0–3.5 wt% P_2O_5. The other group, one third of the samples, has an average density of 3.6–3.8 g/cm^3 with 3–12 wt% P_2O_5. The standard interpretations of layer-forming processes in mafic intrusions by crystal settling and sorting or in situ oscillatory nucleation can produce the bimodal distribution of rock densities and bulk P2O5 contents with the same mineral compositions in the contrasted layers. However, the existence of two populations of melts inclusions trapped in apatite (Si-rich vs. Fe-rich; Charlier et al. 2011) support that igneous layering in the Critical Zone of the Sept Iles intrusion is produced by the large-scale segregation of the two immiscible melts in the magma chamber. This observation is crucial because the two equilibrium immiscible melts crystallize the same liquidus phases. Consequently, cumulates produced by the crystallization of two immiscible melts is hardly distinguishable from the crystallization products of a homogeneous melt followed by crystal sorting.

Immiscibility in the Crystal Mush

Silicate liquid immiscibility was recognized to occur in the crystal mush of mafic layered intrusions where it produced reactive microstructures (e.g. symplectites; Holness et al. 2011; Namur et al. 2012a) and contrasted types of plagioclase zoning profiles (Humphreys 2011). In the Skaergaard intrusion, only the Fe-rich immiscible melts is kept in the crystal mush, while most of the Si-rich conjugate is lost due to its low density (Holness et al. 2011). Crystallization of the Fe-rich melt in the high-porosity region of the crystal mush may therefore produce rocks significantly enriched in mafic minerals and Fe-Ti oxides compared to theoretical cotectic proportions. Immiscibility in the crystal mush has never been considered as a major layer-forming process but it may contribute to enhance some existing layering features, especially where initial crystal mush porosity is heterogeneous (Namur et al. 2014).

An example of this process may be given by the Triple Group (TG) of the Skaergaard Layered Series, which is a very noticeable succession of leucogabbro-melanogabbro-mesogabbro. TG is divided in 7–8 cycles with an average stratigraphic height of ca. 14 m (Andersen et al. 1998). The first order layering at the liquidus of the bulk liquid of the intrusion was in the TG governed by the distribution of plagioclase and clinopyroxene. Fe-Ti oxides-rich layers (dominated by ilmenite) are found within the leucogabbro, melanogabbro and mesogabbro units (Bollinberg 1995). Fe-Ti oxides are by far the most dense phases in the gabbros, followed by pyroxenes and plagioclase. A density-controlled accumulation in MZ should therefore lead to the common upward deposition Fe-Ti oxide-rich rocks, followed by pyroxene-rich rocks and plagioclase-rocks. This is not observed. The presence of Fe–Ti oxide-rich layers within various types of rocks in the TG could therefore be explained by crystallization of abundant Fe–Ti oxide minerals from an immisible Fe-rich melt in the high-porosity region of the crystal mush. This process would also explain the reactive microstructures observed at the interface between Fe–Ti oxides and silicate phases in most rocks from the TG (Holness et al. 2011).

Non-Dynamic Layer-Forming Processes

Non-dynamic layer-forming processes do not involve any large-scale movement of liquid or crystals but occur during crystallization as a result of changing the conditions of crystallization or by self-organization of the mineral assemblage in a crystal mush. There is increasing evidence that rock texture and igneous layering evolve together to produce striking examples of modal igneous layering. In this respect, igneous layering has many parallels with modal banding that develops in high-grade metamorphic rocks in what were either homogeneous protoliths (e.g. banded amphibolites from basalt protoliths) or banding that develops at an angle to relic sedimentary bedding. It is apparent that the rock texture and the mineral segregation that characterize modal layering in igneous and metamorphic rocks both evolve over time such that the final rock may have little resemblance to how the rock initially started. The background to this has been summarized by Boudreau (2011). Mineral segregation can be driven by a number of processes, such as dissipation of energy.

The models performed by Boudreau (2011) illustrate that textural and modal irregularities initially induced during nucleation or by other mechanisms continue to develop and sharpen with time. For example, a layer with poorly defined upper and lower contacts with surrounding rocks or with a weakly defined interior structure can develop much more strongly defined modal layering as the rock texture coarsens. In short, igneous layering may be self-defining in that modal inhomogeneities will become enhanced with time. Initial heterogeneities may be due to changing intensive parameters of crystallization (e.g. oxygen fugacity, pressure) and any dynamic stratification, but most of them presumably result from a fluctuation of nucleation rates in multisaturated melts.

Changes of Intensive Parameters of Crystallization

Formation of layering in cumulate rocks may result from transitory excursions about the cotectic proportions of precipitating minerals. As described above, the excursions can originate from convective overturn, injection of new magma or contamination by country rock material but can also result from changing the intensive parameters of crystallization such as pressure, nucleation rate, temperature and oxygen fugacity.

Nucleation and Crystal Growth Rate Fluctuations

Oscillatory nucleation and crystal growth have frequently been considered as a likely process to form layering not only in basaltic magma chambers (e.g. Wager and Brown 1968; McBirney and Noyes 1979; Bloomer et al. 1991) but also in granites (Barbey 2009). Magma must be supersaturated and undercooled in order to nucleate and grow crystals. The growth rate of a crystal depends on the volume free energy change associated with the transfer of chemical components from the melt to the forming crystals, while nucleation depends on both the volume free energy and the surface free energy. The volume free energy terms increases with supersaturation while the surface free energy terms remains relatively constant. The ratio of nucleation rate vs growth rate will therefore change significantly with the degree of supersaturation. The degree of undercooling at which crystallization occurs also significantly influences the shape and size of the crystals (Kirkpatrick 1975; Faure et al. 2006, 2007; Welsh et al. 2012).

The difference between growth rate and nucleation rate was considered as a layer-forming process in several studies. Wager and Brown (1968) attributed the formation of crescumulate layers in Skaergaard to this process, while Wager (1959) suggested that the cyclic layering in Bushveld might result from contrasted nucleation rates of minerals with different crystal structures (chromite, orthopyroxene and plagioclase). Oscillatory nucleation and growth at the magma chamber floor has also been proposed as a mechanism for the origin of fine-scale modally-graded layering with Fe-Ti oxides, olivine, pyroxenes and plagioclase and for the origin of non-cumulate layered fine-grained granular amphibole-rich gabbro in the Fongen-Hyllingen intrusion (Thy and Esbensen 1982; Thy et al. 1988). For the Ilimaussuaq intrusion, it has also been suggested that increasing vapour pressure caused an increase in the nucleation rate of feldspar and nepheline relative to pyroxene, producing normally graded layers (Sorensen and Larsen 1987).

In a supersaturated system, oscillatory nucleation of various phases may also occur. Theoretical models for this process have been proposed by McBirney and Noyes (1979) and Hort et al. (1993) and the latter authors indicated that this process is strongly dependent on the magma layer thickness, its viscosity and the growth rate of the equilibrium crystals. The nucleation and growth of a phase (e.g. A) depletes the surrounding region by diffusion of chemical species to the growing crystals. McBirney and Noyes (1979) have shown that the width of this zone will advance at a rate proportional to $t^{1/2}$ (with t = time), while the thermal gradient in the crystal mush advances at a rate proportional to t. Initially, the zone of depletion initially advances faster than the slowly moving thermal gradient due to the latent heat released by growing phase A. However, when increasing distance from the growing crystal, the thermal gradient eventually overtakes the boundary of the depletion zone. At this stage, the magma is sufficiently cold (undercooled) to initiate a new burst of nucleation of phase A. Between two episodes of nucleation of phase A, other phases (e.g. B) will be supersaturated in the depleted melt and crystallize. A combination of these processes therefore produces an alternation of rocks dominated by A and rocks dominated by B (Fig. 2.19).

Cyclic layering in the Bushveld complex is characterized by graded units with chromite-rich gabbro, followed upwards by orthopyroxenite and finally by plagioclase-rich rocks and could result from a process of diffusion-controlled oscillatory nucleation and crystal growth. In a similar way, this process could explain the layering with basal layers rich in olivine and pyroxene followed by plagioclase-rich layers in the Freetown complex, Sierra Leone (Hawkes 1967). This process was also suggested to explain the macro-rhythmic layering and modally-graded layering in the Layered Series of the Skaergaard intrusion which show an alternation of layers dominated by mafic phases and layers dominated by plagioclase (Wager 1959; Maaloe 1978; McBirney and Noyes 1979).

Layering by Oscillatory Nucleation in Bjerkreim Sokndal

The Bjerkreim-Sokndal layered intrusion (BKSK; Rogaland anorthosite province, Norway) intruded at a depth of ca. 15 km, corresponding to ca. 5 kbar, in a high temperature environment (ca. 1000 °C) due to the proximity of anorthosite diapirs

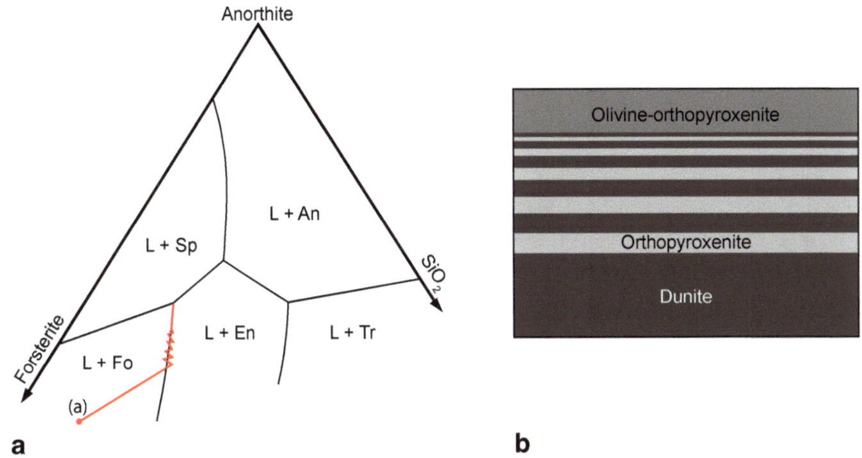

Fig. 2.19 Schematic models showing the effect of oscillatory nucleation during fractional crystallization on the development of layering in cumulate rocks. **a** Sub-section of the Forsterite-Anorthite-Quartz phase diagram with mineral stability fields and boundary curves from Longhi (1987). The liquid (**a**) starts by crystallizing forsterite. When it approaches the forsterite (*Fo*)—enstatite (*En*) peritectic curve, it becomes oversaturated in enstatite. Instead of laying on the curve, the liquid ends up into the enstatite stability field and crystallizes enstatite. After some enstatite has crystallized, the liquids get oversaturated in forsterite and starts crystallizing olivine again. This cycle may repeat until the liquid finally falls on the forsterite-enstatite peritectic curve. **b** Example of oscillatory nucleation described in *A* on the development of dunite-orthopyroxenite layering

(Drüppel et al. 2013). The consequence is that the BKSK intrusion cooled down and solidified very slowly. It comprises cumulate rocks dominated by anorthosite (p-C), ilmenite leuconorite (phi-C), magnetite leuconorite and norite (phim-C), and apatite-bearing gabbronorite (phimca-C) with p as plagioclase, h as Ca-poor pyroxene (hypersthene or pigeonite), i as ilmenite or hemo-ilmenite, m as magnetite, c as Ca-rich pyroxene, a as apatite, and –C as cumulate (Wilson et al. 1996). The parental magma was a ferrobasalt (jotunite/hypersthene ferrodiorite) with density values ($\rho = 2.63$–$2.66\ g/cm^3$) similar to that of the plagioclase density (An_{45-50}, $\rho = 2.62$–$2.63\ g/cm^3$; Vander Auwera and Longhi 1994). Plagioclase has thus little tendency to sink or float in the magma.

Classic characteristics of layered rocks are met in BKSK: macrocyclic units, isomodal layers (m- to tens of m-thick), small scale modally graded layers, inch-scale layering, igneous lamination and progressive and regressive cryptic layering. The shape of the magma chamber where layering was the most developed was sill-like with a high aspect ratio (up to 13) and a gently-sloping floor (Wilson et al. 1996).

An intriguing characteristic of the BKSK intrusion is the bulk composition of its cumulate rocks. For the most abundant rock types, phi-C, phim-C and phimca-C, Duchesne and Charlier (2005) showed that the composition of these rocks corresponds to a mixture between a plagioclase pole and a mafic pole (Fig. 2.20). In each type of rock, the relative proportion of the different phases in the mafic pole

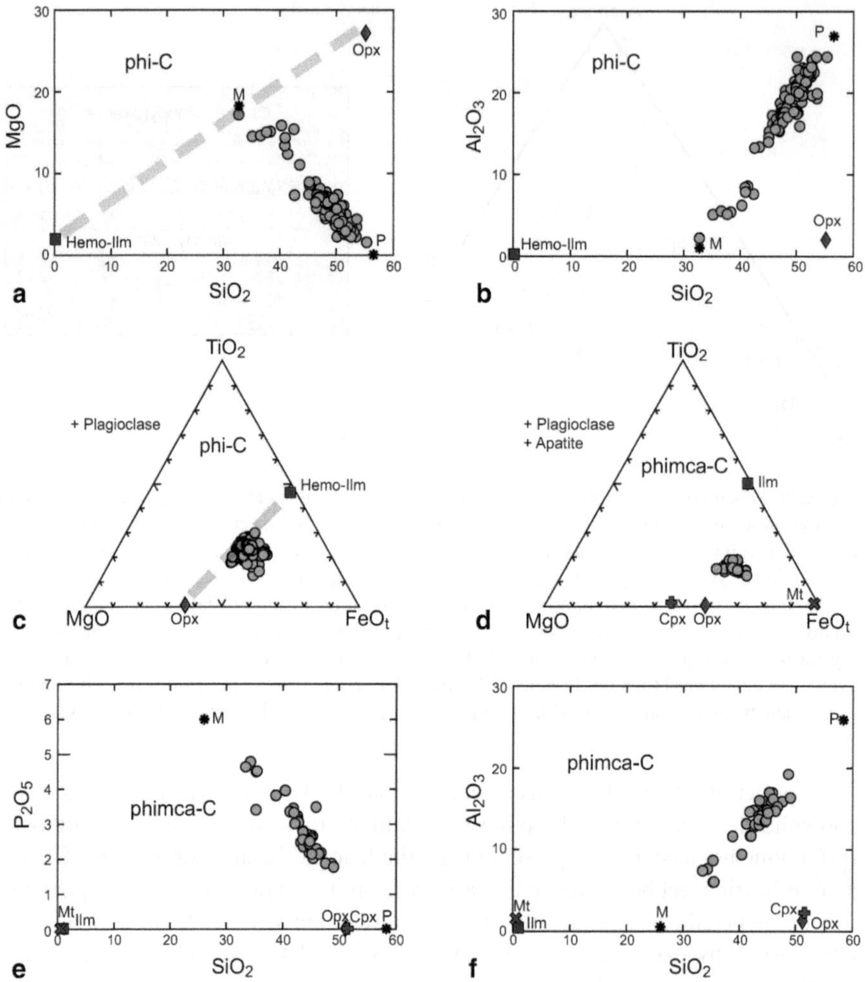

Fig. 2.20 Characteristic features of 2-pole cumulates in the Bjerkreim-Sokndal layered intrusion. **a** and **b** Harker diagrams in which 124 samples of phi-C compositions plot on a straight line joining a plagioclase (*P*) and a mafic pole (*M*) made up of Ca-poor pyroxene (*Opx*) and hemo-ilmenite (Hemo-Ilm). **c** Same samples projected in the MgO–TiO_2–FeO_t diagram. **d** 41 phimca-C compositions in the MgO–TiO_2–FeO_t diagram. **e** 41 phimca-C compositions plotted in P_2O_5–SiO_2 Harker diagram showing the role of apatite in the mafic pole (*M*). **f** Same samples in Al_2O_3–SiO_2 diagram showing the mixing relationship between the plagioclase pole (*P*) and the mafic pole (*M*). *Cpx* = Ca-rich pyroxene, *Ilm* = ilmenite, *Mt* = magnetite. Data from Duchesne and Charlier (2005) and Robins and Chiodoni (2007)

is constant. The major factor responsible for the modal layering in BKSK therefore seems to be the variation in modal ratio of plagioclase to the sum of the mafic phases (Duchesne and Charlier 2005).

The concept of mafic pole implies that there is no mechanical sorting of the heavy minerals although their densities vary from 3.2 for apatite to 5.2 for magnetite. The Stokes's law would predict a deposition speed approximately 5 times larger for magnetite than for apatite. Gravity controlled processes can thus be ruled out. It is classically considered that nucleation begins at the roof in a supercooled sheet of magma that is gravitationally unstable and descends through the magma chamber to finally wash the floor (Morse 2008). In the BKSK intrusion, this view can be questioned and different objections such as the evidence for double diffusive stratification in the chamber, the presence of a melted roof zone during solidification, and the absence of sorting of mafic minerals point to nucleation at the (inclined) floor of the magma chamber. A lack of sorting of mafic minerals and density similarities of plagioclase and magma have led Duchesne and Charlier (2005) to propose in situ growth by heterogeneous nucleation at the topmost layer of cumulus grains in direct contact with the magma (Sparks et al. 1985). In this model, variation in the plagioclase mode is due to variation in nucleation rate and degree of supersaturation. What causes these variations remains partly enigmatic. In isomodal layers a constant degree of supercooling could result from a balance between conductive heat loss and heat production (latent heat of crystallization) and could depend on the thickness of the double-diffusive layer. In modally graded layers, continuous increase in supersaturation due to faster heat loss than heat production resulting from thinner liquid layers could explain the upwards increase in plagioclase content.

Formation of Harrisitic Peridotite

Magma undercooling and the degree of supersaturation significantly influences the growth rate and shape of the forming crystals (Kirkpatrick 1975, Faure et al. 2006). Alternating periods of crystallization with contrasted degrees of undercooling and supersaturation may therefore produce well-defined layering features in basaltic magma chambers. An example of this process is given by the peridotite of the Rum Layered Suite.

The Rum Layered Suite, NW Scotland is divided into the Western (older) and Eastern Layered Intrusions (WLI and ELI) and the Central Intrusion (Emeleus 1997). The ELI has 16 macro-rhythmic units, each comprising a peridotite base and troctolite top; the peridotite is considered to represent the olivine cumulate product of new influxes of relatively primitive magma that subsequently fractionated to produce the troctolite (olivine + plagioclase cumulate; Brown 1956). The relatively poorly studied WLI exposes the lowest structural levels of the Rum intrusion and is sub-divided into the Harris Bay, Transitional and Ard Mheall members (Wadsworth 1961; Emeleus 1997).

The name 'harrisite' was originally assigned by Harker (1908) to a spectacular variety of coarse-grained feldspathic peridotite and olivine gabbro that is commonest in the WLI. The term has textural connotations and harrisite exhibits equant skeletal to extremely elongate skeletal or dendritic olivine crystal morphologies that can exceed 1 m in length (Fig. 2.21a). Elongate harrisitic olivines are commonly oriented perpendicular to magmatic layering and referred to as comb-layered. They

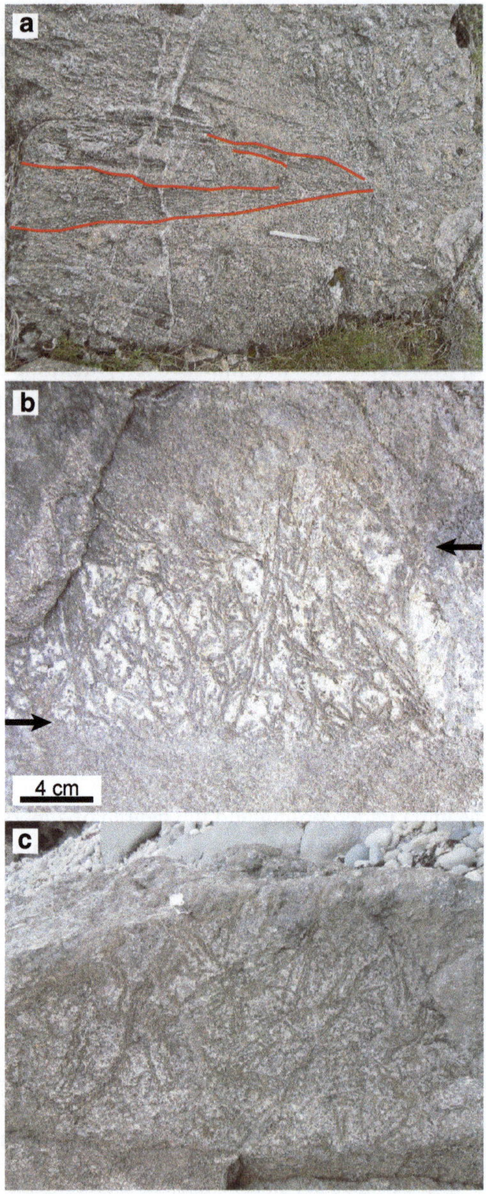

Fig. 2.21 Field photographs of harrisitic peridotite in the Rum Layered Suite. **a** Horizontally-oriented skeletal olivine crystals at Ard Mheall (pen in *lower* image *centre* is 13.5 cm long). The edges of single harrisitic crystal are highlighted in *red*. **b** Olivine dendrites in a layer of harrisite oriented approximately perpendicular to the plane of magmatic layering in Harris Bay. The top and base of the harrisite layer are arrowed. C Stellate olivine dendrites at Harris Bay. The hand lens close to top *center* of image gives scale

were initially called 'harrisitic cumulates' by Wager et al. (1960) in their formulation of cumulus theory, but later included in the more general group of rocks called 'crescumulates' by Wager (1963). Wager and Brown (1951), Brown (1956), Wager et al. (1960) and Wadsworth (1961) envisaged harrisitic olivines growing upward from a previously deposited substrate of cumulus olivines. Those crystals with the a-axis steeply inclined to the plane of the magmatic layering outcompeted other crystals in further growth (Fig. 2.21b). However, examples of harrisite in which elongated olivines are parallel to layering or completely random in orientation are also observed. Stellate growths of olivine, resembling spinifex texture in certain komatiite lava flows, are also common (Fig. 2.21c). In addition to olivine, harrisitic peridotite contains 40–60 vol.% of coarse-grained intercumulus plagioclase crystals and clinopyroxene may also be an intercumulus constituent of the rock. Harrisite is abundant throughout the Harris Bay and Ard Mheall members of the WLI, typically interlayered on the cm-to-m scale with granular-textured peridotite and olivine gabbro. This led Wadsworth (1961) to suggest that the conditions responsible for the formation of harrisite alternated with periods of relatively 'normal' crystal settling processes to produce equigranular peridotite.

Donaldson (1976, 1977) combined field observations in the WLI with laboratory experiments that reproduced some of the complexity of olivine crystal shapes preserved in harrisite. A gradational change of olivine morphology was revealed, dependent on the degree of undercooling, in which small (~ 1 mm) equant olivine crystals are overlain by progressively larger and more deeply indented skeletal olivines. Granular olivine crystals transition into tabular-shaped crystals. Irregularly shaped 'hopper' skeletal olivines form an intermediate stage between tabular crystals and highly elongate dendritic forms. Donaldson (1977) suggested that harrisitic olivine grew rapidly and attributed its growth to in situ crystallization from olivine-supersaturated magma. He also showed that some harrisite layers are intrusive into the cumulate above and below them.

O'Driscoll et al. (2007) applied crystal size distribution (CSD) analyses to a range of harrisitic olivine crystal shapes from the WLI and calculated, using crystal growth rates for disequilibrium growth forms from the literature, that harrisite crystals may have grown on the order of hours to days. They linked the formation of harrisite to the hypothetized 'leaky' open-system nature of the Rum Layered Suite (Emeleus et al. 1996) and suggested that the required supersaturation of the magma in olivine arose through deep undercooling of thin picrite sheets emplaced at the base of the Rum magma chamber, beneath cooler resident magma. They also suggest that the newly arriving picrite sheets were largely free of suspended olivine crystals, meaning that undercooling of the magma triggered homogeneous nucleation of olivine. These authors also highlighted field observations that suggest current or flow-induced erosion and re-deposition of harrisitic olivine crystals and concomitant deposition of granular-textured peridotite. These observations were interpreted as representing magma chamber processes that operated between replenishment events, compatible with the ideas of Wadsworth (1961).

Certain features of the WLI harrisites remain unexplained. One concerns the close association of chromitite (>60 vol.% Cr-spinel) with harrisite (Wadsworth 1961). Elsewhere on Rum, e.g. the ELI, chromitites have been explained by models invoking the contamination of picrite parental liquids with troctolitic cumulate (O'Driscoll et al. 2010). However, no such felsic contaminant obviously exists in the WLI. The chromitite seams in the WLI are more abundant than in the ELI (Figs 2.22a–c). Holistic models for chromitite formation on Rum (and associated PGE enrichment; O'Driscoll et al. 2009b) thus require adaptation by incorpora-

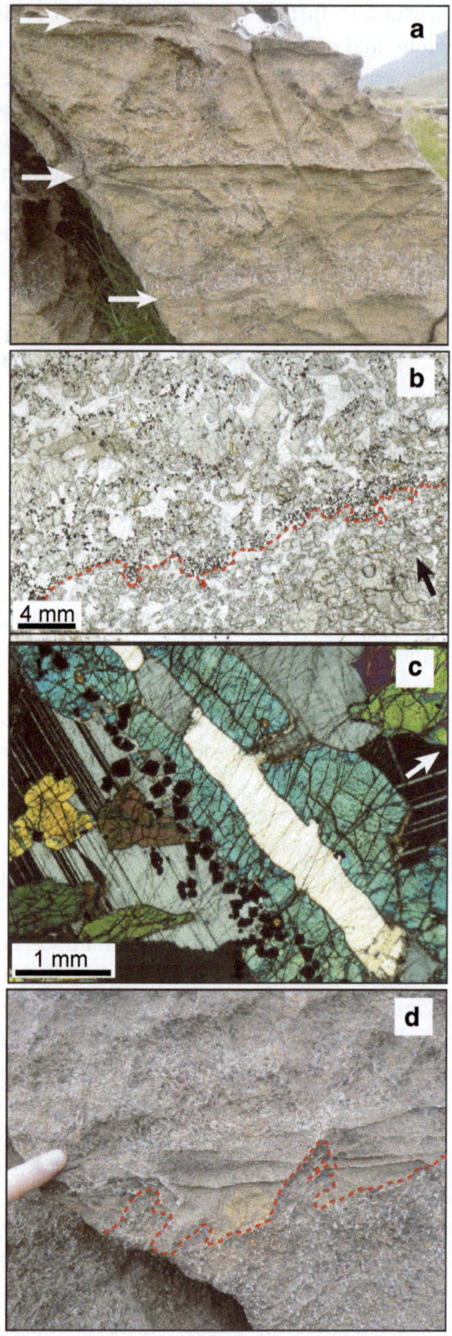

Fig. 2.22 Field photographs and microphotographs of harrisitic peridotite in the Rum Layered Suite. **a** Layered harrisite and granular-textured peridotite in the Ard Mheall member of the WLI (western layered intrusion of the Rum Layered Suite). Rhythmic layering is observed, whereby a thin chromitite seam (*three arrowed* in the photograph) is overlain by coarse-grained harrisitic peridotite which is in turn overlain by granular-textured peridotite. Note that the boundaries between harrisite and granular-textured peridotite are diffuse, and elongate harrisitic crystals are present in the latter lithology oriented at a high angle to the plane of layering. The hand lens at the top of the image gives scale. **b** Thin section scan (plane-polarized light) of a chromitite-bearing harrisite with way-up marked by the *black arrow*. The peridotite *below* the base of the diffuse seam (highlighted in *red*) is characterized by a paucity of Cr-spinel compared to the peridotite above. Olivine crystals above the seam are subtly coarser-grained than those below. **c** Elongate harrisitic olivine crystal exhibiting concentrations of Cr-spinel crystals along its base (way up is given by the *white arrow*). **d** Contact between harrisite and granular-textured peridotite immediately above. Note the extremely irregular nature of the contact (highlighted in *red*); granular textured peridotite appears to infill depressions between upstanding harrisitic olivines. The grain size of the granular-textured peridotite increases up section. Index finger in the *left-centre* of the photograph gives scale

tion of the observations on chromitite occurrence in the WLI. A second noteworthy feature of the WLI harrisites is the very complex relationships they locally exhibit with granular-textured peridotites (Fig. 2.22d). An explanation for the relationships observed might be that the harrisitic olivine substrate formed an irregular floor to the accumulating granular-textured peridotite which was variably affected by processes that operated to erode and re-deposit harrisitic olivine clasts. However, the situation also occurs where harrisite and granular-textured peridotite apparently occur comprising the same layer, bounded on both sides by chromitite seams (e.g. Figure 2.22a).

Pressure Change

Changes in the total pressure in a crystallizing magma chambers can significantly change the solid-liquid equilibria and result in the formation of layering. As experimentally shown in the forsterite (Mg_2SiO_4) – anorthite ($CaAl_2Si_2O_8$) – quartz (SiO_2; Sen and Presnall 1984) and forsterite (Mg_2SiO_4) – hematite (Fe_2O_3) – anorthite ($CaAl_2Si_2O_8$) – quartz (SiO_2; Osborn 1978), increasing pressure (1 bar to 10 kbar) expands the stability fields of spinel and orthopyroxene at the expense of olivine and plagioclase fields. Drastic pressure change could thus theoretically lead to the formation of chromitite layers, magnetitite layers or dunite and anorthosite (Naslund and McBirney 1996). This process has been invoked to explain the succession of pyroxenite and anorthosite layers in basaltic layered intrusions (Fig. 2.23) as well as chromitite layers in the Stillwater complex and magnetitite layers in the Bushveld complex (Cawthorn and Ashwal 2009).

Oxygen Fugacity Fluctuation

Change in oxygen fugacity conditions also significantly changes the stable liquidus assemblages, especially for phases containing multivalent cations as major elements (e.g. Toplis and Carroll 1995). An excellent example of layering produced by changes in oxygen fugacity conditions is shown by the Panzhihua intrusion (SW China). The Middle zone a (MZa) of the intrusion consists of layered cumulates made up of clinopyroxene, plagioclase, titanomagnetite (now variably oxy-exsolved to magnetite and ilmenite) and minor olivine that crystallized from the base upwards. During crystallization of its lower portion, the magma chamber is suggested to have been opened to recharge of more primitive magmas, which is reflected in the upward change of plagioclase composition from $\sim An_{60}$ to $\sim An_{70}$ (Pang et al. 2009). Over this stratigraphic interval of ~ 100 m, abundant cm- to dm-thick layers of Fe–Ti oxides are found intercalated with gabbroic rocks having lower abundance of oxide minerals (Fig. 2.24a).

The oxide ore layers are sheet-like bodies having planar surfaces against rocks above and below them. They are readily identifiable in the field because of their high density and low plagioclase contents. Both sharp and gradational contacts are

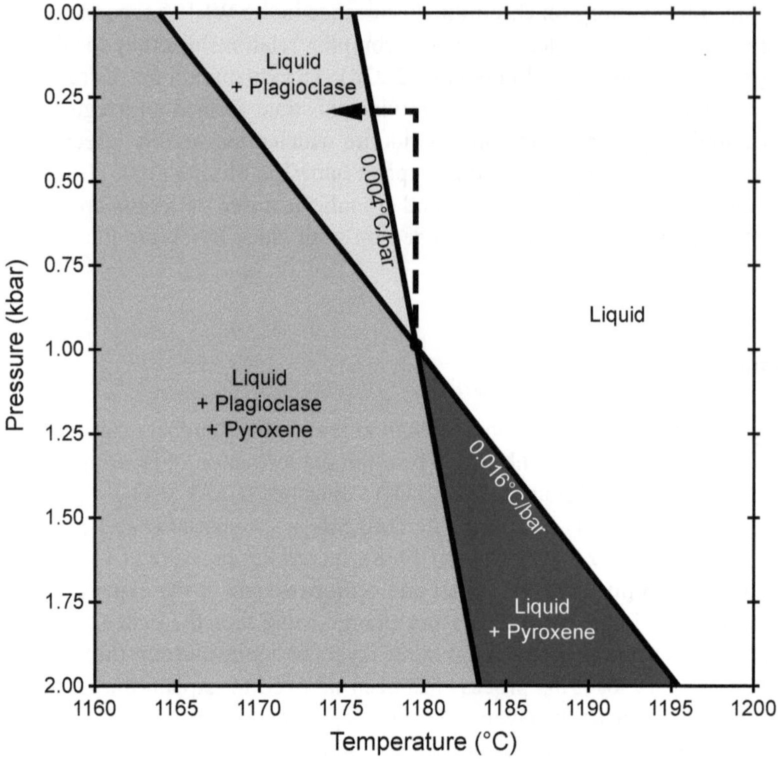

Fig. 2.23 Effect of pressure on the relative stabilities of plagioclase and pyroxene crystallizing from a basaltic melt. Note that the *dP/dT* gradient is less for plagioclase (0.04°C/MPa) than for pyroxene (0.16°C/MPa) and so changing pressure changes the stability of the two phases. Modified from Cawthorn and Ashwal (2009)

observed between the ore layers and the adjacent layers. The thickness of a given ore layer is largely uniform and the layers show no signs of termination, at least at the scale of the outcrops. They are either virtually monomineralic with respect to Fe-Ti oxides or semi-massive containing dispersed silicate grains; the oxide content is greater than 70 % in most cases. The silicate grains are mineralogically similar to the gabbroic layers (i.e. clinopyroxene, plagioclase, hornblende ± olivine), and in places they show corroded boundaries in contact with oxides. Some of these boundaries are rimmed by thin stripes of brown hornblende and/or olivine. The modal abundance of Fe–Ti oxides in the gabbros ranges from ~5 to 30 %; these minerals are interpreted to be of cumulus origin despite showing amoeboid shapes presumably due to subsolidus grain boundary readjustment (Duchesne 1999; Pang et al. 2008). Planar lamination is developed in the gabbroic layers in which plagioclase grains (or their aggregates) show a preferred orientation sub-parallel to the layering (Fig. 2.24a). Interestingly, elongate grains of plagioclase are observed in some oxide ore layers aligned sub-parallel to the layering (Fig. 2.24b), reminiscent

Fig. 2.24 Features of Fe–Ti oxide ore layers in the MZa of the Panzhihua intrusion. **a** Outcrop showing alternating layers of Fe–Ti oxide ores (*FeTi*) and gabbros (*gb*) in the Jianshan open-pit mine. **b** Sample from the ore layers showing preferred alignment of silicate grains and aggregates of plagioclase crystals

of trachytic texture in volcanic rocks. This rock texture is also observed in other layered intrusions such as Sept Iles (Namur et al. 2010) and the Bushveld Complex (Tegner et al. 2006).

The magmatic origin of the Fe-Ti oxide ore layers is of little doubt. The similar mineralogy (albeit with large difference in modal abundance) and orientation between the ore and gabbroic layers indicate that both are integral parts of the layered sequence. No large and systematic mineral compositional reversals are noted across individual layers, despite the upward increase in the An of plagioclase noted above. This indicates that while the ore layers might have formed during a period of magma addition, individual layers did not record replenishment events. Pang et al. (2008) advocated gravitational settling and sorting of titanomagnetite to explain occurrences of thick (~20–60 m), conformable lenses of Fe-Ti oxides in the LZ of the Panzhihua intrusion. However, occurrences of thin and multiple oxide ore layers in the MZa cannot be reconciled exclusively with the same mechanism and thus require an additional explanation.

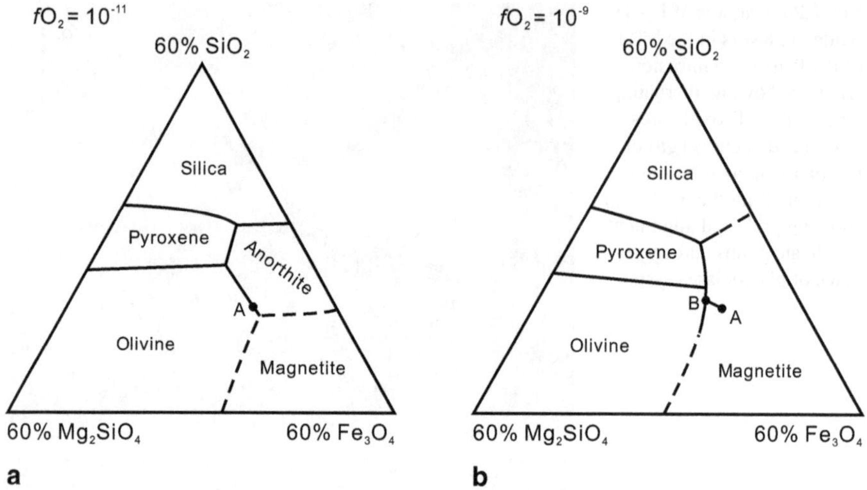

Fig. 2.25 Phase relations on the 40 % anorthite join in the system $MgSi_2O_4$–FeO–Fe_2O_3–$CaAl_2Si_2O_8$–SiO_2 at different oxygen fugacity conditions (fO_2), modified from Ulmer (1969) and Naslund and McBirney (1996). **a** $fO_2 = 10^{-11}$. **b**. $fO_2 = 10^{-9}$. Oxidation of magma with composition **A** on the olivine-anorthite join at low fO_2 results in one saturated with magnetite only at higher fO_2

Changes in intensive parameters, such as oxygen fugacity and pressure, provide alternative mechanisms for the formation of the oxide ore layers. Melting experiments in ferrobasaltic systems demonstrate that the stability of Fe–Ti oxides is strongly dependent on oxygen fugacity, which controls Fe_2O_3/FeO of the melt (Toplis and Carroll 1995). An increase in oxygen fugacity is not only able to induce magnetite saturation at higher temperatures (see the magnetite-in boundary in Fig. 2.20 from Toplis and Carroll 1995) but also to drive a crystallizing magma from the cotectic to the magnetite only phase field (point A, Figs 2.25a–b; Ulmer 1969). In theory, this causes magnetite to crystallize alone and its fractionation drives the magma composition back to the cotectic along which an oxide gabbroic assemblage crystallizes (point B, Fig. 2.25b). At Panzhihua, this mechanism deserves special attention because evidence for the interaction between magma and carbonate wall-rocks, a process capable of releasing large amount of oxidizing CO_2-rich fluid, has been found (Ganino et al. 2008).

A similar mechanism involving shifting of phase boundaries, but caused by a pressure increase, has also been proposed to explain excessive magnetite crystallization (Cawthorn and McCarthy 1980; Cawthorn and Ashwal 2009). Pressure variations might be triggered by vapour exsolution, magma addition, eruptions from the chamber and external tectonic events (Naslund and McBirney 1996 and references therein). This mechanism has the advantage of instantaneous effect over great horizontal distance in the magma chamber. In response to changing oxygen fugacity and/or pressure, a sequence of alternating oxide ore and gabbroic layers can be formed. It is noteworthy that changes in intensive parameters do not preclude subsequent settling and sorting of minerals and could occur together with gravitational accumulation.

Processes of Equilibration in the Crystal Mush

Crystal Mush Compaction

As experimentally shown by Coates (1936) and Boudreau and McBirney (1997), crystals of contrasted sizes and densities tend to sort themselves in rough layers as they compact under the influence of gravity. This is generally accompanied by the development of planar fabrics (mineral lamination) due to crystal re-orientation (Meurer and Boudreau 1998; Boorman et al. 2004; Williams et al. 2006). Based on the detailed analysis of laminated anorthosite rocks from the Sept Iles intrusion together with theoretical modeling, Higgins (1991) however indicated that this process is not sufficient to produce strong fabrics and that well-defined lamination can only form where there is a considerable component of simple shear. This is in agreement with observations from McBirney and Hunter (1995) indicating that there is no obvious relationship between the intensity of lamination and deformation in plagioclase-rich rocks of the Skaergaard intrusion.

The development of the planar fabrics can be significantly enhanced by mechanisms of mineral dissolution-precipitation (see below; Boudreau and McBirney 1997). The effectiveness of mineral segregation in a compacting mush may depend on the density contrast between the solid phases, their shapes, their relative proportions, the proportion of interstitial melt, its physical properties (density and viscosity), and its velocity (if moving). Boudreau and McBirney (1997) suggest that plagioclase-rich layers in the Yorkhaven sill and the alternation of cm-thick plagioclase-rich and pyroxene-rich layers in the Ferrar Dolerites of Antarctica could result from compaction of a crystal mush at the floor of the sills. Compaction of the crystal mush is also considered as an important process of formation of modal layering in oceanic magma chambers at slow-spreading ridges (Thy and Dilek 2000) but may be of marginal importance for layered intrusions due to slow accumulation of the crystals and extensive recrystallization (Boudreau and McBirney 1997).

Where a component of simple shear is present during crystal mush compaction, the intensity of deformation is a function of the crystal content of the mush (Thayer 1963; Benn and Allard 1989; Nicolas 1992; Jousselin et al. 2012). Although no obvious evidence for layering due to simple shear has been reported in layered intrusions, this process may be responsible for major layering-features in lower crustal sections (Thayer 1963; Benn and Allard 1989; Jousselin et al. 2012). It could also explain the formation of lenses and schlieren in felsic plutons (Barbey 2009).

Layering Formation Through Dissolution-Precipitation

In a compacting crystal mush, points where stress is concentrated tend to dissolve while points under smaller stress grow. The material dissolved in areas of high stress may then be deposited in regions of lower stress or can be expelled from the crystal mush. In a mixture of two minerals (± liquid), the more pressure sensitive phase has a greater free energy when its proportion is large than when it is small, and the stress is taken up by grains of the more resistant mineral. This might result in mineral mode reorganization through dissolution-precipitation and a rock with an initially weak inhomogeneity can develop into layers that are increasingly more monomineralic (Boudreau and McBirney 1997; Boudreau 2011). Layering in ultramafic rocks with olivine and pyroxene from ophiolites is thought to have formed by such a process (Dick and Sinton 1979). Because olivine dissolves more easily than pyroxene, crystals of pyroxene in olivine-rich rocks are subject to a greater stress than pyroxene grains in olivine-depleted rocks. They thus dissolve in regions where they are less abundant and reprecipitate where they are more abundant. This may well explain the formation of pyroxenite and dunite layers.

Initial inhomogeneity in grain size can also lead to the formation of layering because the system is energetically unstable (Boudreau 1994, 1995; Williams et al. 2006). The free energy of small crystals and their solubility are higher than those of larger crystals. This is the driving force of Ostwald ripening that leads to a general coarsening of grain size as larger grains grow at the expense of the smaller ones. This process can occur under isothermal and isochemical conditions because the heat absorbed and components released during dissolution of the small grains is balanced by crystal growth on large grains. The chemical potential at which small grains dissolve exceeds that at which large grain grow, which helps in the transfer of components between grains. Any zone or layer where grains are initially larger than in the surroundings will be energetically favored and will grow by diffusion of components from the regions where grains are dissolving (Boudreau 1987; Boudreau and McBirney 1997). In a rock with plagioclase and some large pyroxene crystals, this process might lead to a situation where the growth of a coarse-grained pyroxene-rich layer has depleted the surrounding rock of pyroxene, creating a pyroxene-free anorthosite layer. This process may ultimately lead to the formation of a series of regularly spaced layers, with the exact spacing between them depending on the growth (and dissolution) rate and the diffusion rate. This process is thought to be responsible for the formation of the inch-scale layering with pyroxenite layers and anorthosite layers in the Stillwater (Fig. 2.26). This is evidenced by the positive correlation between the pyroxene grain size and the layer spacing (Fig. 2.26; Boudreau 1987). A similar process is thought to have produced concentric layering in a rhyolitic dike near the eastern margin of the Skaergaard intrusion (McBirney et al. 1990) and minor layering in the Skaergaard Layered Series (Boudreau and McBirney 1997). It also contributed significantly to the texture and grain size distribution in various units of the Bushveld complex (Boorman et al. 2004; Williams et al. 2006). Modeling has also shown that a number of sedimentary-like features can also be developed by the same process, including the development of sharp modal contacts and size-graded layers (Fig. 2.27). The long characteristic times for mineral segregation to occur by these processes are consistent with the observation that well-developed modal layering is relatively rare in thin or small intrusions. It occurs only in the central parts of mafic intrusions the size of the Palisades Sill or the Skaergaard intrusion (or larger) where the mush zone may persist long enough for the reorganization process to be fully effective.

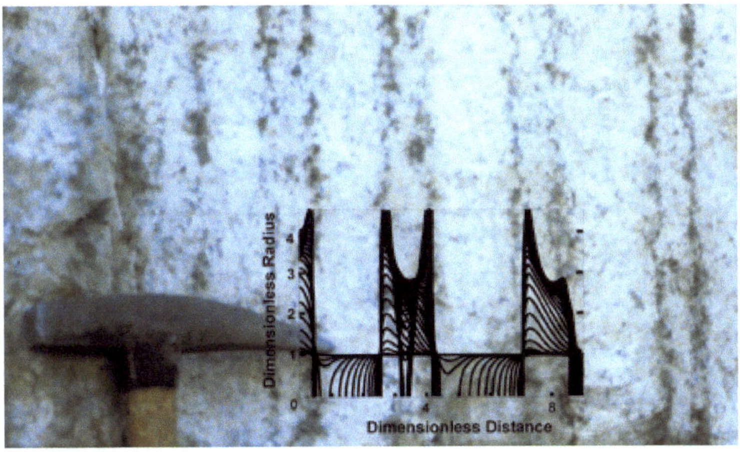

Fig. 2.26 Inch-scale doublets as exposed on the Mt. View Mine road in the Stillwater complex. Superimposed is the doublet pattern produced by numerical simulation of layer development during crystal aging at different times. (From Boudreau (1995))

The surface free energy of a crystal is also dependent on the nature of the surrounding crystals. In a system with two minerals (e.g. plagioclase—pyroxene, ± liquid), any initial heterogeneity in the distribution of the crystals is unstable and will lead to a process of reorganization through dissolution and reprecipitation (Fig. 2.27; Boudreau and McBirney 1997). In any region of the crystal mush where plagioclase-plagioclase contacts are more abundant than in the surrounding rock, plagioclase crystals have lower solubility and will grow at the expense of pyroxene crystals in this area and plagioclase crystals in the surrounding rock. The result is that the number of plagioclase-plagioclase contacts will progressively increase whereas the number of pyroxene-pyroxene contacts will progressively decrease. The opposite will happen in the surrounding rock, where the initial proportion of pyroxene is higher than in the average rock. This process of dissolution followed by precipitation is therefore self-propagating and the regions dominated by plagioclase-plagioclase contact cause the surrounding regions to be dominated by pyroxene-pyroxene contacts. Such a process can ultimately lead to the formation of cumulate rock layers being increasingly monomineralic and similar to those shown in Fig. 2.26.

As discussed above, dissolution-reprecipitation may occur in rocks with initial heterogeneity in terms of differential pressure, grain size and mineral mode and lead to the formation of increasingly monomineralic layers. If liquid in the crystal mush does not move, mass transport of the dissolved components result from diffusion only and the scale of these combined effects is on the order of centimeters (Boudreau and McBirney 1997). If liquid is moving through the mush (Fig. 2.27), this helps in the transport of dissolved components and surmounts the limitations of diffusion, with the overall effect of intensifying the layering and its vertical extent.

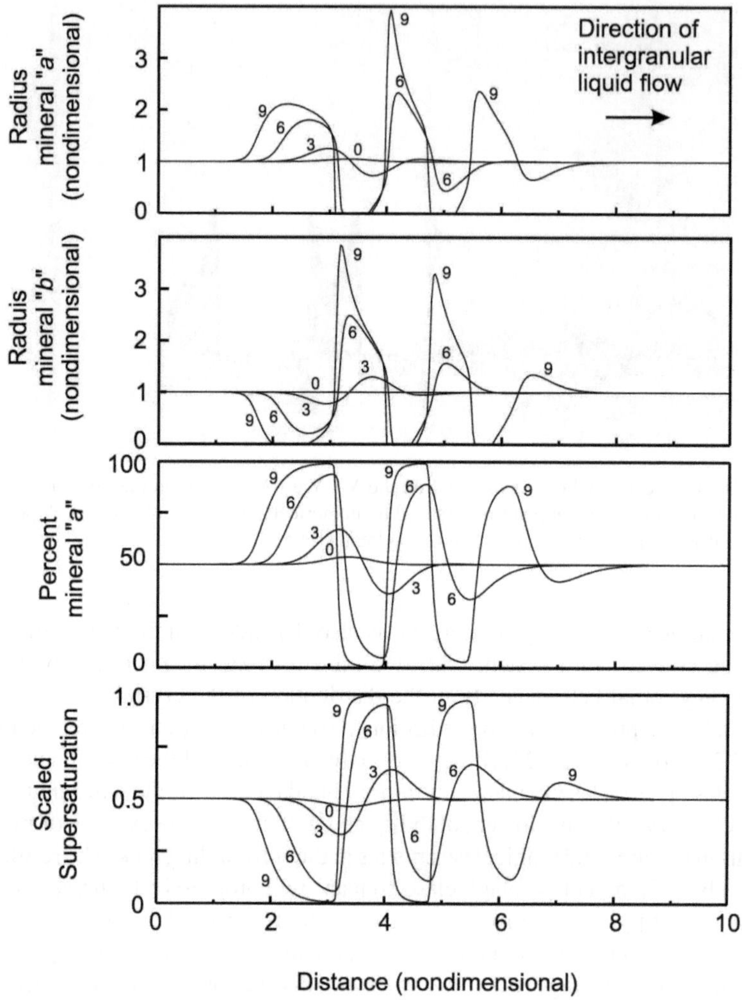

Fig. 2.27 Example of layer formation in a two crystal + liquid system in which the local equilibrium liquid concentration is a function of texture, in this case texture is defined solely as the mineral modes. All units are dimensionless, and system evolution is shown at the dimensionless times of 0, 3, 6, and 9. Advective liquid is moving from left to right through a crystal assemblage that has a small modal variation at three distance units from the origin. Note the development of size-graded layers with sharply define bases. (From Boudreau and McBirney 1997)

Layering by Grain Coarsening and Changing Liquid Composition

The La Cordadera gabbro intrusion is a minor component of the Illapel Plutonic Complex (IPC) batholith of central Chile, which ranges from medium-grained gabbros to trondhjemites, with hornblende and biotite ± clinopyroxene bearing tonalites and granodiorites as the most abundant lithologies. From the petrological and geochemical perspective, the IPC rocks define a typical subalkaline trend, with low

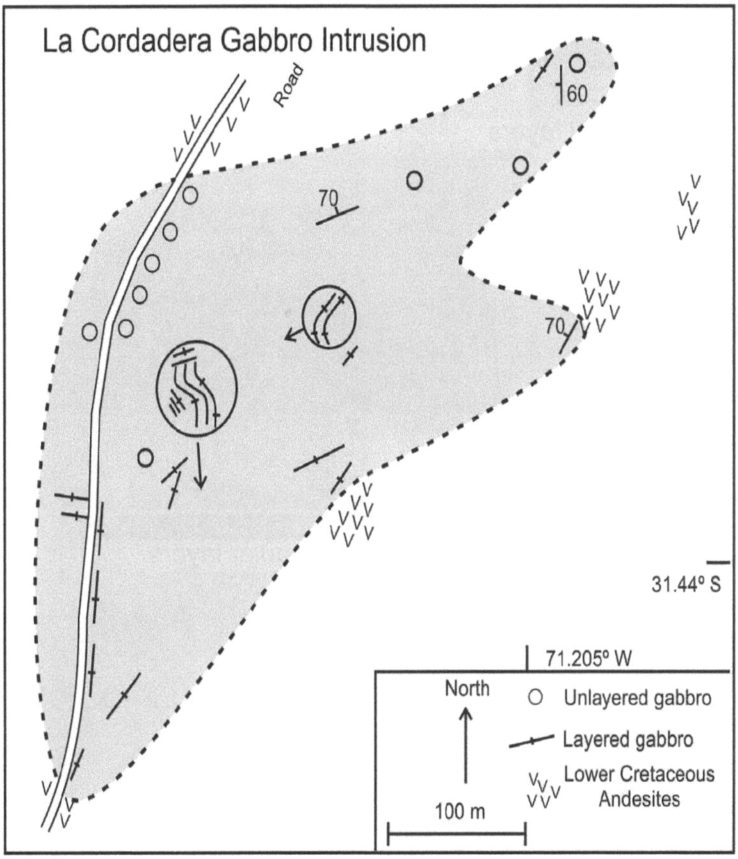

Fig. 2.28 Geological sketch map of the La Cordadera gabbro intrusion from the Illapel Plutonic Complex, Coastal Range of central Chile. Exposure is almost perfect along the road, but sparse elsewhere. Field measurements indicate that layering is mostly vertical to sub-vertical

$(^{87}Sr/^{86}Sr)_I$ ratios (0.7034–0.7039) and high epsilon Nd (up to +5) values (Parada et al. 1999).

The La Cordadera intrusion is an ovoid body measuring ~400 by ~500 m outcropping in the northern edge of the IPC. The western part of the intrusion is very well exposed along an N–S road cut, but elsewhere outcrops are limited (Fig. 2.28). The intrusion is unmetamorphosed, undeformed and does not appear to be cut by major faults. The northern edge of the unit is dominated by massive, dark, fine to medium grained gabbro, but the rest has well-developed layering (Figs 2.28 and 2.29). The layering is commonly defined by variations in the proportions of plagioclase and amphibole (Fig. 2.29). Mafic-rich layers are dominated by hornblende oikocrysts, whereas more felsic layers are mostly composed of plagioclase. At some locations mineral layering is accompanied by changes in grain-size. Although lamination or lineation of mineral phases was not observed in the field, it is visible in

Fig. 2.29 Layering and pegmatites in the La Cordadera gabbro intrusion. **a** The simplest example of layering occurs in the discovery outcrop along the road cut. Layers are parallel and spaced from 8 to 20 mm apart. **b** In many locations layers are not laterally continuous, but branch and terminate. **c** Some outcrops have very irregular layering. **d** Erosional contacts between sets of layers are common. **e** Pegmatites are especially well developed along the western edge of the intrusion. Here a small patch of pegmatite cuts planar layering and clearly post-dates it

thin section. Field measurements show that layering is generally vertical to subvertical with a strike that varies from north to east (Fig. 2.28). At several places almost orthogonal vertical layers are present in different parts of an outcrop, or adjacent outcrops (Fig. 2.29). This is significant because such outcrops cannot be tilted in any way to make the layering horizontal and therefore parts of the layering were originally very steep.

In some places layers are almost perfectly planar, with regular repeated cycles of plagioclase-rich rock alternating with amphibole-rich rock. The best example of this facies is in the road cut at the discovery outcrop (Figs. 2.28 and 2.29a) and also on the top of a small hill above the road, allowing limited three-dimensional observations which permit a clearer understanding of the different field relationships. Over a 3-m long block the E-W vertical layers are parallel and rhythmic. In the southern part of the block the layers are about 8 mm thick, increasing steadily to the north to about 20 mm. Layers are continuous, showing no branches or terminations. A similar outcrop occurs just across the road, but the vertical layering is N–S here (Fig. 2.28).

Elsewhere in the intrusion layering is more irregular. Layers can have spacings of 5–20 mm, but with branching and terminating layers (Fig. 2.29b). Layers can be arrhythmic with variable composition, width and spacing (Fig. 2.29c). Some areas have thin (~5 mm) layers of almost pure plagioclase set in poorly layered or unlayered amphibole-plagioclase rock. Some layers terminate abruptly against other planar layers (Fig. 2.29d). There is no evidence of mechanical displacement along such contacts and it appears that they were produced by erosion of older layered rocks. In some cases strongly curved layers that resemble channels were cut into older layered rocks. In some layers crystal size changes so abruptly that the contact appears to have formed by the chilling of a new magma pulse against existing layered rocks. In some places the layering is interrupted by patches of pegmatite, which appear to have replaced the earlier layered rocks (Fig. 2.29e). Very similar layering, both rhythmic and isolated, has also been observed in a similar-sized gabbroic intrusion associated with a granitic batholith in Sardinia (Hauser and Bussy 2014).

Although some of the structures observed in outcrop resemble channel flow, the subvertical orientation renders this interpretation impossible (Boudreau 2011). It is proposed that layering developed by dissipation of crystal surface energy during equilibration (Higgins 2011), the same mechanism that is credited with creating the 'inch scale' layering at Stillwater (Boudreau 1987). It is generally thought that such a mechanism needs the temperature to be maintained close to the liquidus for a long period of time, and hence that it rarely occurs in small intrusions. The presence of similar phase layering in some skarns (e.g. diopside-wollastonite layers) suggests another option (Thompson 1975). For the La Cordadera intrusion it is proposed that the system was directed towards equilibrium during cycles of replacement of clinopyroxene by amphibole in the semi-consolidated magma. Each cycle is envisaged as starting with the emplacement of relatively dry gabbroic magma, with some crystals, in a dyke that was fed from the base of the IPC and continued upwards, possibly to the surface. Crystallization of plagioclase, mafic minerals and oxides on the walls produced a semi-rigid network of crystals. Meanwhile, the composition of the magma flowing in the dyke became richer in water. The water diffused from the magma into the crystal network, transforming some of the clinopyroxene into amphibole. It is also possible that pyroxene was converted to amphibole at constant PH_2O at a peritectic. At the same time the system equilibrated, coarsening plagioclase and amphibole, and producing the layering. A new cycle would then start by the injection of a fresh batch of magma into the dyke, which eroded the earlier-formed layered rocks before it stalled and began to crystallise.

Modal and Cryptic Layering due to Chemical Diffusion in the Crystal Mush

The Upper Critical Zone (UCZ) of the Bushveld Igneous Complex displays spectacular layering with cyclic units comprising a basal chromitite layer overlain by a sequence of silicate cumulates in the order, from bottom to top, pyroxenite-norite-anorthosite (Naldrett et al. 1986; Eales and Cawthorn 1996). Four to five major chromitite layers reaching 1 m in thickness are recognized within the UCZ. In addition, there are numerous thin chromite seams and stringers of only a few centimeters or even millimeters in thickness. Extensive modal layering in cyclic units is accompanied by cryptic layering. The most pronounced and systematic are the changes of the orthopyroxene Mg# (the molar $Mg/(Mg+Fe^{2+})$ ratio), which tends to decrease in vertical sections from pyroxenitic cumulates to overlying leucocratic rocks.

Chromite compositions change abruptly at contacts of chromitite layers and thin chromite seams with adjacent anorthosites and pyroxenites (Fig. 2.30). High and constant $Mg/(Mg+Fe^{2+})$ and Fe^{2+}/Fe^{3+} ratios of spinel crystals in chromite cumulates change to variable and generally lower values in accessory chromite disseminated in silicate layers. Trace element composition of Cr-spinel also varies with lithology. For example, disseminated chromite from pyroxene-rich cumulates are higher in Zn and Ga but lower in Sc than chromite from anorthosites.

Several studies noted that Na_2O concentrations in plagioclase from individual UCZ cyclic units are fairly constant in contrast to strong variation trends of Mg# in orthopyroxene and spinel from the same units (Naldrett et al. 1986; Eales et al. 1990; Eales and Cawthorn 1996). Other anomalous features of the UCZ plagioclase composition include occurrences of reverse zoning near contacts with chromite seams (Fig. 2.30) and strong depletion of intercumulus plagioclase within chromitite layers in Li, K and Rb. The depletion of intercumulus plagioclase in K (Fig. 2.30) is even more puzzling in view of the relative enrichment of the same poikilitic crystals in La and other light REE, trace elements with similar plagioclase-melt distribution coefficients.

There might be a genetic link between the abrupt changes of chromite compositions at contacts of chromite and silicate cumulates, and compositional anomalies in cumulus and inter-cumulus plagioclase. Transient surges of chromite crystallization due to some temporal fluctuations of crystallization conditions (e.g. new magma

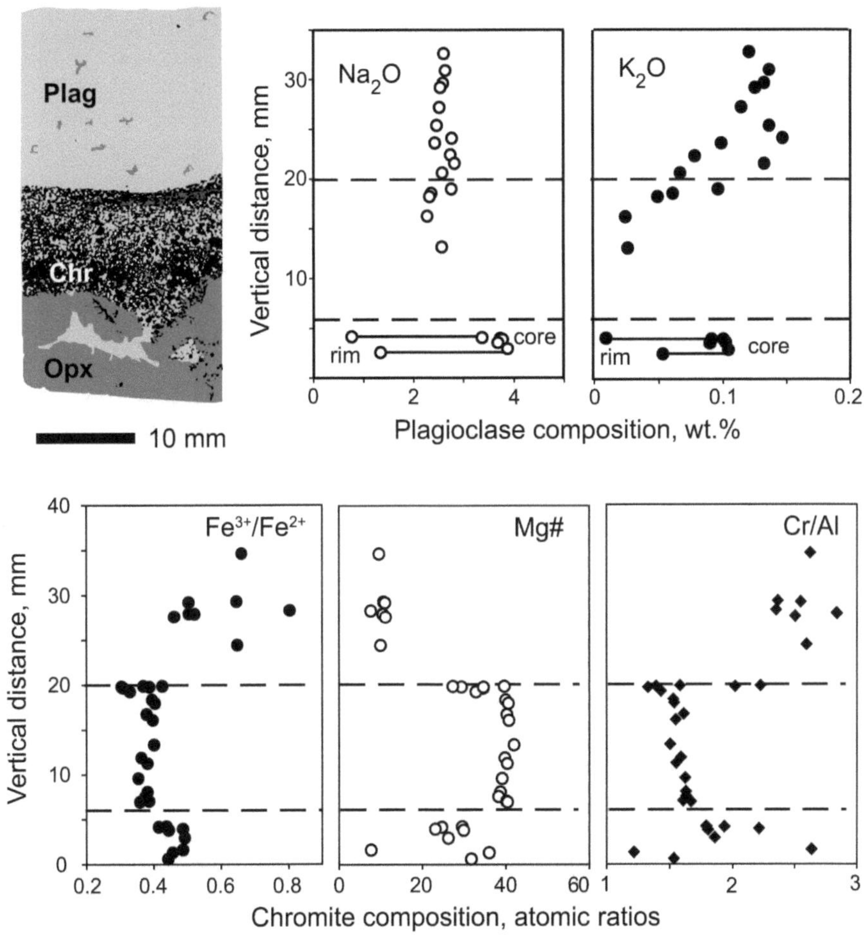

Fig. 2.30 Variations of chromite and plagioclase compositions across a thin chromite seam between anorthosite and footwall feldspathic pyroxenite in the Bushveld Upper Critical Zone. (Data from Veksler et al. (2013))

injections or periodic pressure fluctuations; see above) can produce cumulus layers with anomalously high chromite mode and peritectic reactions between chromite, silicates and inter-cumulus melt generate redox gradients at contacts of the chromite-rich layers with adjacent silicate cumulates. The gradients may result in Na diffusion in interstitial. Na diffusion is very efficient due to (1) the high electrical conductivity of magmas and the fact that magmatic silicate melts are strong ionic electrolytes (Waff and Weill 1975); (2) the predominant role of alkali and especially Na diffusion in the charge transfer (Waff and Weill 1975; Pommier et al. 2008) due to high diffusivity of Na^+ compared to other cations (Zhang et al. 2010); (3) the high basicity of the Na_2O component. This process may also produce changes in the mineralogy of the crystal mush and form well defined layering features.

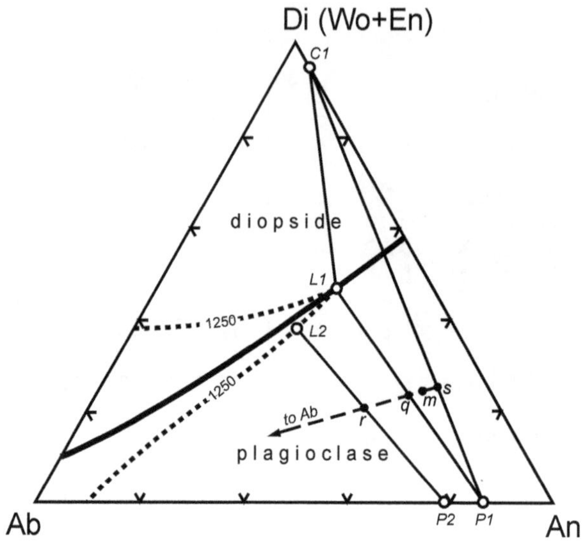

Fig. 2.31 Phase diagram of the system albite (*Ab*)-anorthite (*An*)—diopside (*Di*) showing effects of Na migration (addition and removal) on a cumulus mush m composed of diopsidic pyroxene *C1*, plagioclase *P1* and liquid *L1* equilibrated at 1250 °C. *Dashed line* and the points *r*, *q* and *s* show changes of the bulk composition *m* with Na addition or removal. See text for further discussion

The principles of isothermal partial melting due to Na migration are illustrated in Fig. 2.31 using the haplobasaltic system albite (Ab)-anorthite (An)—diopside (Di). At 1250 °C, liquid *L1* is in equilibrium with pyroxene and plagioclase solid solutions *C1* and *P1* (Kushiro 1973). The modal composition of a plagioclase-rich, "leucogabbroic" cumulus mush indicated by point *m* is 61 % plagioclase, 14 % pyroxene and 25 % liquid (by mass). Sodium addition to the mush will move the bulk composition towards the Ab apex in the direction indicated by the arrow. At the same time, the compositions of coexisting phases should remain constant until pyroxene is completely dissolved at point *q*, where the mush becomes a mixture of approximately equal amounts of liquid *L1* and plagioclase *P1*. With further Na addition, plagioclase should change its composition along the Ab-An binary towards Ab, while the equilibrium liquid composition should follow the 1250 °C isotherm in the same direction (e.g. the compositions *L2* and *P2*). In the same manner, Na addition to a cumulate with excess pyroxene (a mixture within the *L1*-*C1*-*P1* triangle but above the cotectic curve, not shown in Fig. 2.31) will cause complete dissolution of plagioclase and formation of pyroxenitic cumulate. It also follows from the diagram that the opposite process of isothermal Na removal from the original composition *m* will cause crystallization, so that at the point *s* the mush is transformed into a solid assemblage of 27 % pyroxene *C1* and 63 % plagioclase *P1*.

The simplified haplobasaltic system demonstrates two important points. First, a significant buffering capacity of the cumulus mush (e.g. constancy of phase compositions at a constant temperature until one of the phases disappears completely). Second, the strong effect of Na migration on phase proportions in the mush. Notably, less than 20 % of the original Na content in composition *m* is enough to re-melt all of the pyroxene (in case of Na addition) or to completely crystallize the mush (Na removal). The post-cumulus stage in the crystal-liquid mush can be viewed as adcumulus crystallization in a system open for Na and other alkalis, and closed for all the other components. The scenario is similar to that of metasomatism with mobile components, which produces sharply-divided monomineral zones, as developed theoretically by Korzhinskii (1970). This has therefore the potential to change significantly the mineral modes in the mush, and to produce layering with individual layers being almost monomineralic.

Theoretical and Experimental Models for the Origin of Layering

Theoretical and numerical modeling, together with low- and high-temperature experiments, have commonly been used to constrain the origin of igneous layering in mafic magma chambers. McBirney and Noyes (1979) used numerical simulations to suggest that intermittent layering might result from gravitational stratification of the liquid column, while cyclic layering might result from an oscillatory process of nucleation and crystal growth. Boudreau (1994, 1995, 2011) and Boudreau and McBirney (1997) used numerical simulations to show how crystal aging can enhance initial heterogeneity in mineral modes and result in strong layering in gabbroic rocks (see above). Sparks et al. (1993) experimentally investigated the behavior of convecting, particle-laden fluids cooled from above and suggested that modal and rhythmic layering results from deposition of minerals with various settling velocities. The settling process would only start when a critical concentration of minerals in the fluid is reached, explaining the cyclic distribution of the layers. High temperature experiments have also shown that diffusion and reaction in a thermal gradient can also change mineral modes of gabbroic rocks and result in well-defined igneous layering (Lesher and Walker 1988; Lundstrom et al. 2005).

Development of Cyclic Layering

Cyclic layering (Fig. 2.32a) is distinguished from normal rhythmic layering by the remarkably regular spacing of the layers, although the spacing generally increases with distance from the contact, roughly according to a geometric progression. The layers are parallel to the contact, even in the case of originally inclined or curved contacts and individual layers show spatial variation in the abundance of vesicles or crystals. The waveform of the eroded surface exhibits great variety and complexity in 2D and 3D spaces, indicating nonlinear oscillatory profiles of textural variables such as number density, size and volume fractions. Although some attempts to bring constraints on the origin of cyclic layering have been made in the past (McBirney

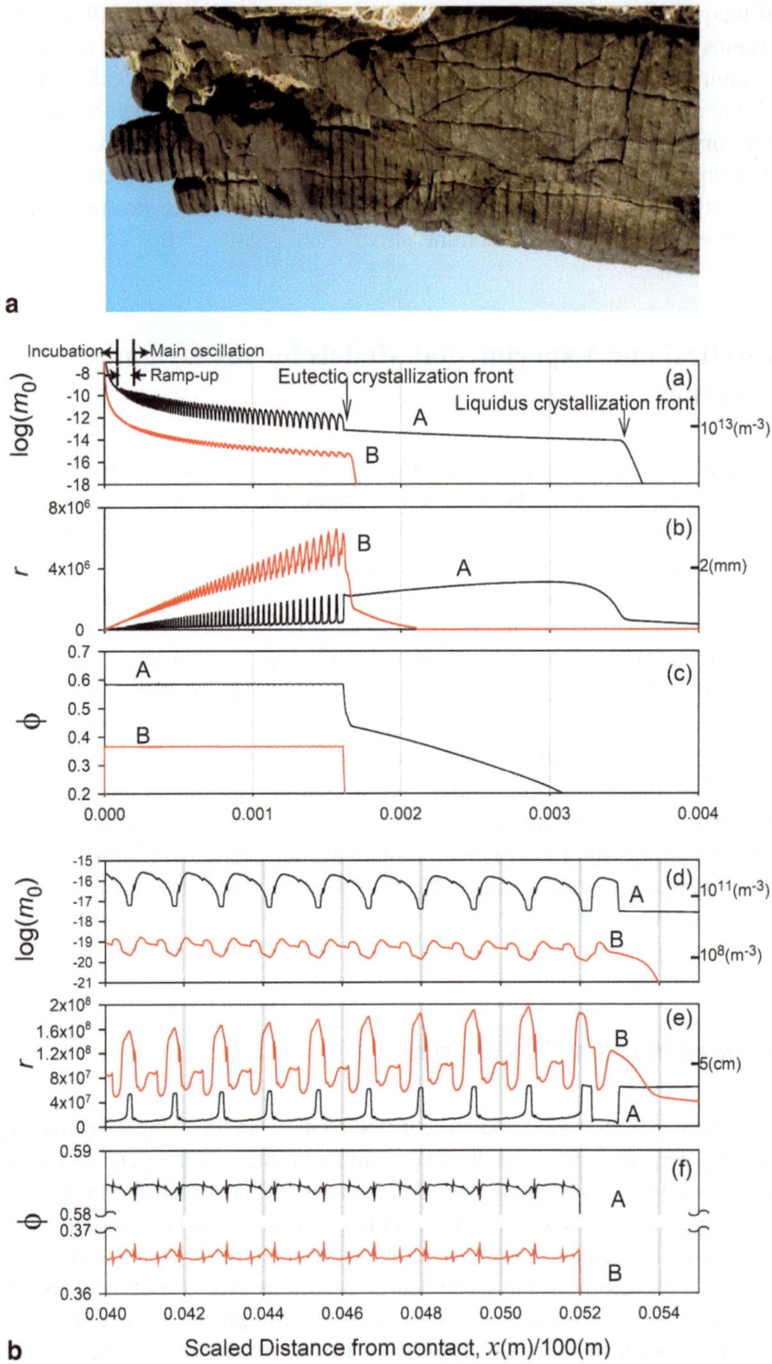

Fig. 2.32 **a** Example of cyclic layering from Ogi, Japan. The photograph has been rotated by 90° and the *left* hand side was originally at the top. **b** Example of numerical simulation of L-type

and Noyes 1979; Boudreau 1987; Brandeis and Jaupart 1987; Toramaru et al. 1996; Rogerson and Cardoso 2000), no comprehensive understanding of the formation mechanism has been reached.

Toramaru and Matsumoto (2012) carried out numerical simulations in a binary eutectic system simulating the Di-An system by taking into account the thermal and compositional diffusion as well as crystallization kinetics (homogeneous nucleation and diffusion-limited growth; see details in Toramaru and Matsumoto 2012). They used governing equations with two dimensionless controlling parameters: the Stefan number (St) representing the effects of latent heat release and the Lewis number (Le) defined as the ratio of thermal diffusivity to compositional diffusivity. For realistic values of Stefan number (e.g. 0.2; Fig. 2.32b) and Lewis number (e.g. 10^5), a two phase (A and B) layering develops through eutectic oscillation (compositional and thermal oscillation below the eutectic point) with the exception of the marginal region where no layering develops. Phases A and B have a similar polarity of oscillation in crystal density and crystal radius but the peaks are higher for phase B than for phase A. The wavy profiles are highly nonlinear and the spacing of layers follows a geometric progression.

By changing the values for St and Le, Toramaru and Matsumoto (2012) determined the conditions under which cyclic layering develops (Fig. 2.33a). Based on the morphology of the oscillatory profiles, they defined three types of layering (S = short-scale layering; L: long-scale layering; M: multiple-scale layering). Two important conclusions can be summarized as follows: (1) St, namely the release of latent heat, plays the most significant role in development of layering; (2) The layering type (S, L or M) and textural properties (e.g. number density, crystal size) depend essentially on Le. As St plays an important role on whether layering will develop or not, it is suggested that the release of latent heat by the first crystallizing phase just below the eutectic point controls the layering development. In addition, a multicomponent system making a eutectic system is also important because nonlinear crystallization rate coupled with textural feedback occur and contribute to the development of cyclic layering. This contrasts with the thermal oscillation in a single component reaction-limited system investigated by Brandeis and Jaupart (1987). Finally, the coupling between thermal and mass diffusions together with disequilibrium kinetics (Toramaru et al. 1996; Rogerson and Cardoso 2000) is shown to play a minor role on layering development at realistic values of Le and St.

The most important parameter characterizing the layering structure is a common ratio (a) in the geometric progression which describes the spacing between layers; $\Delta x_n = \Delta x_1 a^{n-1}$, where Δx_n is the spacing of n-th band. The numerical simulation shows that the common ratio converges to a unique value (a: 1.04–1.05, *log*a:

---

layering, which shows the spatial variation of dimensionless crystal number density (m_o), dimensionless mean radius (r) and volume fraction of crystals (ϕ) for mineral phases *A* and *B*. Lewis number (Le) = 10^3 and Stefan number (St) = 0.2 are assumed. Subfigures *a–c* represent the textural profiles from the contact to the liquidus crystallization front of A (3.6×10^4 s) and subfigures d-f represent the textural profiles in the vicinity of the eutectic crystallization front (1.2 yr). Representative values of number density and mean radius are indicated along the *right* sides of the graphs

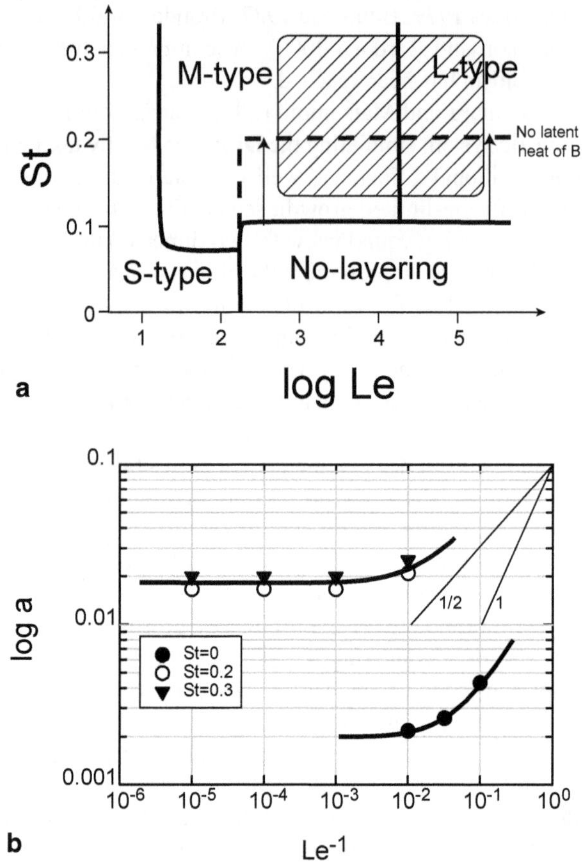

Fig. 2.33 **a** Development conditions of cyclic layering in terms of Stefan number (*St*) and Lewis number (*Le*). Hatched area corresponds to realistic values in magmatic systems. The case where the second phase releases no latent heat is also depicted to simulate the silicate-water systems. *S-type* = short-scale layering, *L-type*: long-scale layering, *M-type*: multiple-scale layering. **b** Common ratio (*a*) as a function of Lewis number and Stefan number

0.017–0.02) with increasing Le (Fig. 2.33b). However the value of common ratio is not exactly consistent with natural data, which diverse from 1.007 to 1.6. Further work has thus to be done to fully understand the formation process of cyclic layering as the simplest case and must include the vesiculation in crystallizing magmas and pressure changes caused by the mechanical interaction with the surrounding crust.

Cyclic vs Comb Layering

Comb layering is observed in mafic and felsic intrusions such as Skaergaard (Lofgren and Donaldson 1975), Rum (Donaldson 1977; O'Driscoll et al. 2007) and the Sierra Nevada batolith (Moore and Lockwood 1973), in alkaline plutons (Petersen 1985) and also in carbonatite dykes (Katz and Keller 1981). In Skaergaard, comb layers have been named perpendicular feldspar rocks, while they are named harrisite or crescumulate in Rum (Wager and Brown 1968). Compared to typical igneous layers, comb layers have some very specific characteristics: (1) constituent crystals are elongated, relatively large and preferentially oriented, and locally branching (Lofgren and Donaldson 1975; Donaldson 1977); (2) the direction of the crystals is nearly perpendicular to cumulus layers, to contact walls, or to the surface of exotic blocks (Moore and Lockwood 1973); (3) comb layer can exist as a single unit with limited lateral extent or can repeat several times with alternating fabrics and domains and therefore defining rhythmic layering.

The formation of comb layers (and constitutive crystals) has traditionally been interpreted as resulting from a sudden increase in the degree of supercooling (see above; Lofgren and Donaldson 1975; Petersen 1985; Faure et al. 2006), possibly with the contribution of some gravitational processes such as crystal removal, migration of interstitial liquid or liquid convection (Donaldson 1982; O'Driscoll et al. 2007). The formation of layering (repetition of comb layers) is thought to originate from oscillatory chemical processes, which might be caused by repeated nucleation and competitive growth in near-cotectic compositions (McBirney and Noyes 1979; Petersen 1985) or magma chamber replenishment. The contrasted diffusivities of heat and chemical components would therefore result in alternating periods with high and low nucleation and growth rates.

Some genetic characteristics of comb layering are similar to those of cyclic layering observed in shallow magmatic intrusions. In some intrusions, complete gradation between comb-textured feldspar dendrites (comb layering) to contact parallel laminar-textured feldspar phenocrysts (cyclic layering) is observed in a single outcrop (Petersen 1985), therefore rising an important question about the processes and factors controlling whether comb layering or cyclic layering would develop during silicate melt crystallization.

Crystallization experiments in gels have been performed to understand the role of chemical diffusion, crystal growth and nucleation rates on the origin of igneous layering (Henisch 1988). Two types of precipitation, regular banding of Liesegang bands and tree-like crystal aggregates have been reported but no clear transition between them has yet been documented. Toramaru et al. (2003) performed experiments based on inorganic reaction (PbI_2 precipitation) with solutions of KI and $Pb(NO_3)_2$ in gel. They used a diffusion medium of agarose (Sigma Company) to prevent hydrodynamic disturbance, convection and crystal settling. Experiments were designed with a gel containing iodide anion (I^-) placed above a gel containing lead cation (Pb^{2+}) in a test tube. Diffusion of I^- into Pb^{2+} gel results in monomer formation (PbI_2). When the concentration of monomer PbI_2 exceeds some critical value, the nucleation of PbI_2 crystals takes place. The reaction proceeds periodically in space and time forming Liesegang bands or other types of precipitations. Experiments were run with various concentrations of KI, $Pb(NO_3)_2$, and gel (see details in Toramaru et al. 2003). In analogy to natural magmatic systems, the concentration of $Pb(NO_3)_2$ represents the crystal content of the magma, diffusion in the gel is analogous to thermal and mass diffusion in magma, and the I/Pb ratio can be considered as a proxy for the temperature contrast between the magma and country rocks. Mass diffusion of I^-corresponds to temperature decrease in the magma by thermal diffusion, which induces supercooling for crystallizing phases in a similar way to the supersaturation of PbI_2 in the analog system.

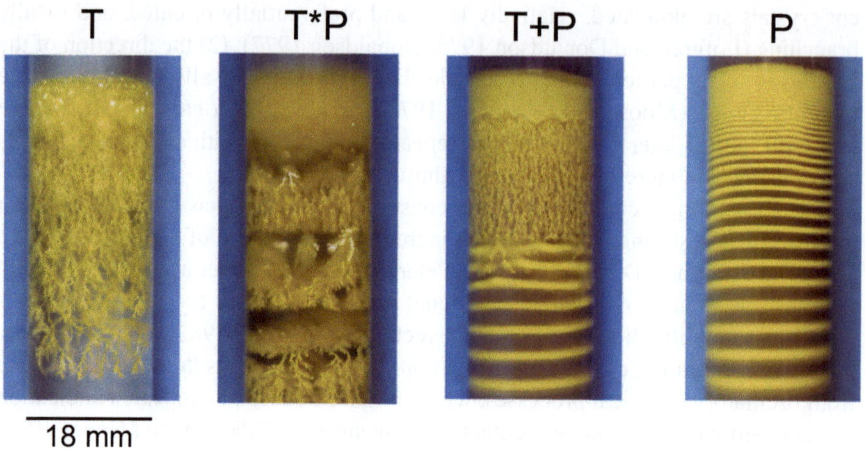

Fig. 2.34 Typical precipitation patterns in experiments of the diffusion-crystallization system. T-type; gel concentration is 0.25 wt.%, $[Pb(NO_3)_2] = 0.004$ mol/l and $[KI] = 16x[Pb(NO_3)_2]$; T*P-type; gel concentration is 0.5 wt.%, $[Pb(NO_3)_2] = 0.002$ mol/l and $[KI] = 32x[Pb(NO_3)_2]$; T+P-type; gel concentration is 2.0 wt.%, $[Pb(NO_3)_2] = 0.004$ mol/l and $[KI] = 32x[Pb(NO_3)_2]$; P-type; gel concentration is 2.5 wt.%, $[Pb(NO_3)_2] = 0.004$ mol/l and $[KI] = 16x[Pb(NO_3)_2]$. Yellow bands in P-type are microscopically suspensions of tiny PbI_2 grains of micro-meter scale in diameter

Toramaru et al. (2003) described two extreme (P and T) and two intermediate (T+P and T*P) precipitation patterns (Fig. 2.34) depending on the gel concentration. For high gel concentration (>3 wt.%), periodic patterns (P; Liesegang bands) develop. Band-width increases with distance from the contact. For low gel concentration (<1.4 wt.%) tree-like patterns (T) develop. For intermediate gel concentration (1.5–2.5 wt.%) a succession of tree-like aggregates and periodic precipitation patterns (T+P) develops. The tree-like crystal aggregates appear after a periodic precipitation pattern, or are sometimes sandwiched between periodic patterns. Where tree-like crystal aggregates develop, their continuity is dependent on the $Pb(NO_3)_2$ content. Where too low (<0.004 mol/l), periodically discontinuous tree-like crystal aggregates form, developing a hybrid (T*P) between T-type and P-type patterns.

Toramaru et al. (2003) considered that P-type banding is an analogy of cyclic layering in magma chambers, while T-type banding corresponds to comb layering. They proposed morphological phase diagrams for layering development as a function of crystal content, bulk nucleation rate (J_B), surface nucleation rate (J_S) and crystal growth rate (G; Figs 2.35a–b). Increasing the gel concentration for a given level of supersaturation results in an increase in the complexity of microscopic network of the gel structure and therefore in the amount of nucleation sites (Toramaru and Iochi, 2000). This results in homogeneous nucleation and the formation of Liesegang bands. On the other hand, decreasing the gel concentration dramatically reduces the number of microscopic nucleation sites of precipitation, resulting in the small nucleation rate and relative increase of growth rate. Consequently the growth of a crystal dominates the bulk nucleation of new crystals in the surrounding media,

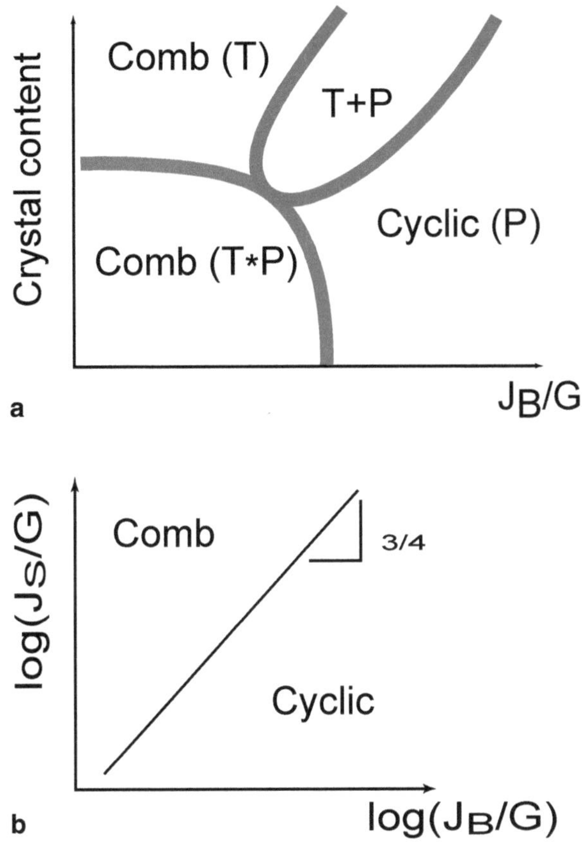

Fig. 2.35 **a** Experimentally integrated phase diagram showing the occurrence of comb layering and cyclic layering in terms of crystal content and agar concentration or J_B/G. J_B = bulk nucleation rate; G = crystal growth rate. **b** Diagram showing the kinetic conditions of occurrence of comb and cyclic layering. J_S = surface nucleation rate

forming fewer number of tree-like crystal aggregates due to surface nucleation on growing crystals.

In natural systems, olivine, pyroxene and feldspar are common minerals to form comb and cyclic layering. Cyclic layering is also characterized by the presence of vesicles induced by crystallization (second boiling). From the experiments by Toramaru et al. (2003), it can be suggested that the physical processes included in comb layering and cyclic layering are basically identical but that the transition from one to the other is controlled by nucleation (and the amount of nucleation sites) during crystallization.

Layering Formation During Crystallization Along Inclined Surface

Development of layering during crystallization along inclined surfaces is known to occur in mafic magma chambers. It was described in the Fongen-Hyllingen intrusion (Wilson et al. 1987) and the Panzhihua intrusion (Pêcher et al. 2013) and was experimentally reproduced in aqueous solutions (Huppert et al. 1986). This process was re-investigated by Blanchette et al. (2004) and explained in terms of Boycott effect (Boycott 1920).

When heavy particles are left to settle in a tilted container, the total settling time is less than it would be in a vertical container, as the average distance particles travel before reaching a boundary is reduced by the tilt. This acceleration of the settling process is called the Boycott effect, and it is simply understood from a geometric argument (Boycott 1920). The strong flow generated along the inclined wall, however, is more surprising.

When particles settle on an inclined roof, they leave behind a layer of clear fluid. The density of this clear fluid is less than that of the suspension away from the inclined roof, and the particle-free fluid is thus buoyant (Huppert et al. 1986; Blanchette et al. 2004). The clear fluid layer therefore begins to rise, hugging the inclined boundary, until it reaches the top of the suspension. The thickness of the layer of clear fluid grows with distance from the bottom corner, as more clear fluid gets released as particles settle away from the inclined roof. The parameters governing this growth are the distance along the inclined roof, x, the particle radius, a, the volume fraction of particles, ϕ_0, and the settling speed of individual particles, U_s. The thickness of the clear fluid layer, δ, and the velocity at which it flows upward, v, have been found to scale as (Acrivos and Herbolzheimer 1979)

$$\delta \sim \frac{a^{2/3}x^{1/3}}{\phi_0^{1/3}} \tag{2.1}$$

$$\text{and } \upsilon \sim \frac{x^{2/3}\phi_0^{1/3}}{a^{2/3}}U_s \tag{2.2}$$

Most noticeably, this process yields a flow that is much more vigorous than one would expect from settling particles. Note that the same process can occur in reverse when bubbles rise above an inclined floor and the bubble-free layer sinks along the inclined floor.

Consider now a system where the fluid in which particles are suspended is density stratified, either because of temperature or compositional variations (Huppert et al. 1986). For simplicity, we focus here on a constant density gradient $\Gamma = -\frac{1}{\rho 0}\frac{\partial \rho}{\partial Z}$, with Z the vertical coordinate and $\rho 0$ a reference density, but a similar effect would be observed in any stable stratification (Fig. 2.36). The presence of stratification has no noticeable effect on the early stages of the Boycott effect, and the clear fluid layer still forms beneath the inclined roof and begins to rise. However, as it rises, the density difference between the clear fluid layer and the suspension at that same height will decrease, as the fluid in the suspension becomes lighter with increasing height. If the ambient stratification is sufficiently pronounced, the clear fluid layer will eventually have the same density as the suspension at a level Z which satisfies:

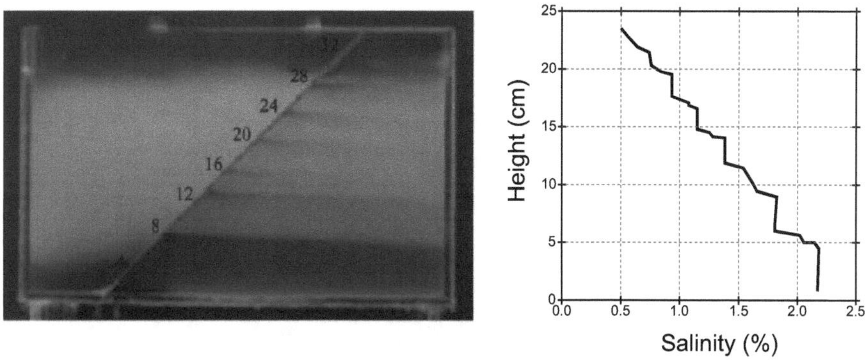

Fig. 2.36 Photograph of layering arising from strong salinity gradient and corresponding salinity profile taken at a distance of approximately 8 cm from the right-hand wall. Numbers represent the height (cm) in the tank. Note the larger primary layer at the base of the tank and the smaller overlying secondary layers. The salinity profile shown to remain once all the particles had settled

$$\phi 0(\rho_P - \rho_0) = \rho_0 Z\Gamma \quad (2.3)$$

where ρ_P is the particle density. At that point, the fluid in the particle-free layer is no longer buoyant and therefore it will no longer rise. Below the level Z, fluid in the clear layer is still buoyant and keeps rising, which forces the clear fluid to intrude into the bulk of the suspension (Fig. 2.37).

Experiments with glass particles settling in a salt-stratified fluid have confirmed that such intrusions do take place (Peacock et al. 2005). Near the bottom of the container, experiments showed an intrusion at or very near the level Z, while a series of smaller intrusions later appeared above the first intrusion. The entire bottom layer below the first intrusion then resembled a convection cell and became well-mixed. The secondary layers formed above also behaved as convection cells, but they had more complicated dynamics. Their spacing was observed to scale as $\Gamma^{-0.38}$ and to increase, though somewhat weakly, with particle concentration. Over time, these layers were also observed experimentally to merge with each other, or to split. Particle plumes were also seen between layers, revealing complicated non-linear behavior as the formation of lower layers affected the dynamics of higher ones. However, once the particles had settled out, the remaining stratification was clearly layered and those layers remained stable thereafter.

In magma chambers, two constituents may trigger the Boycott effect: settling crystals and rising bubbles. In the presence of vertical gradients of silicon dioxide (SiO_2; Spera et al. 1986), intrusions may form and generate meter-size layers (Blanchette et al. 2004). These convecting layers would eventually result in a step-like density profile. As the chamber cools and viscosity increases, the Boycott effect would be progressively slowed down, but remain present. The resulting layer formation could therefore survive the crystallization of the entire chamber and yield

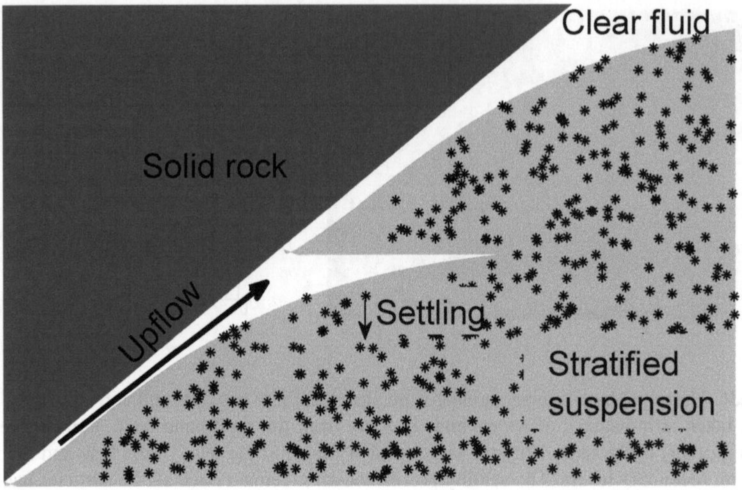

Fig. 2.37 Schematics of the Boycott effect in a stratified ambient. As particles settle beneath an inclined wall, they leave behind a layer of clear fluid (*white*). This fluid is buoyant and rises along the wall. Upon reaching its level of neutral buoyancy, it intrudes horizontally into the suspension

observable, layered rock formations. Such formation would exhibit parallel layers, and presumably be bounded by a thinner inclined structure corresponding to the solidified clear fluid.

Conclusions

Layering has been described in many ultramafic-mafic layered intrusions worldwide. It also occurs in oceanic crust sections and in silicic plutons. Many mechanisms may have contributed to the formation of the known occurrences of layering in magma chambers. These processes can generally be subdivided into two main types: dynamic processes, which mostly occur during the filling of the magma chamber or during the early stages of crystallization (high liquid fraction) and non-dynamic processes, which very broadly may occur at high liquid fraction (e.g. oscillatory nucleation, changes in intensive parameters of crystallization), or by reorganization/recrystallization in a mineral-dominated crystal mush. Other processes such as tectonic activity, plastic deformation in solid (or almost solid) rocks or seismic shocks have not been intensively considered in this review chapter but may locally contribute to the formation of some layering features (Naslund and McBirney 1996; Jousselin et al. 2012). As such, it is thus very unlikely that a single process of layering formation can explain all of the layering features observed in layered mafic-ultramafic intrusions and it is also very likely that a combination of several mechanisms may occur to form some layered cumulates.

In this review chapter, we have described and explained several dynamic and non-dynamic processes of layering formation and presented new examples and occurrences illustrating these processes. We have also shown examples of numerical and experimental simulations that we used to put constraints on some layer-forming processes and under which conditions they develop. The work that has been done in the past to understand the formation of layering helped us understanding the crystallization processes of basaltic magma and the physical properties of crystal mushes developing in basaltic magma chambers. Future work will presumably be done and will help further constraining several critical processes such as the formation of monomineralic layers and ore deposits, the mobility of interstitial melt within crystal mushes and ultimately the fundamental processes of magma differentiation in magma chambers.

Acknowledgments ON was supported by a Junior Research Fellowship from Magdalene College (University of Cambridge) and by a von Humboldt Fellowship (University of Hannover). This work has benefited from discussions with Madeleine Humphreys, Marian Holness, Jacqueline Vander Auwera, Richard Wilson and Brian Robins. The authors would like to thank J. Koepke, C. Li, S.A. Morse, R. Latypov, E. Ripley and B. Upton who kindly accepted to share field photographs.

Refrences

Abily B (2011) Caractéristiques pétrographique, géochimique et structurale de la section crustale profonde de l'ophiolite d'Oman : implications pour la genèse des magmas et le fonctionnement des chambres magmatiques à l'aplomb d'un centre d'expansion océanique. Paul Sabatier University, Toulouse

Abily B, Ceuleneer G, Launeau P (2011) Synmagmatic normal faulting in the lower oceanic crust: evidence from the Oman ophiolite. Geology 39:391–394

Acrivos A, Herbolzheimer E (1979) Enhanced sedimentation in settling tanks with inclined walls. J Fluid Mech 92:435–457

Andersen JCØ, Rasmussen H, Nielsen TFD, Ronsbo JC (1998) The Triple Group and the Platinova gold and palladium reefs in the Skaergaard Intrusion: stratigraphic and petrographic relations. Econ Geol 93:488–509

Barbey P (2009) Layering and schlieren in granitoids: a record of interactions between magma emplacement, crystallization and deformation in growing plutons. Geol Belg 12:109–133

Barnes SJ (1986) The effect of trapped liquid crystallization on cumulus mineral compositions in layered intrusions. Contrib Mineral Petrol 93:524–531

Barriere M (1981) On curved laminae, graded layers, convection currents and dynamic crystal sorting in the Ploumanac'h (Brittany) subalkaline granite. Contrib Mineral Petrol 77:214–224

Bédard JH, Sparks RSJ, Renner R, Cheadle MJ, Hallworth MA (1988) Peridotite sills and metasomatic gabbros in the Eastern Layered Series of the Rhum Complex. J Geol Soc Lond 145:207–224

Benn K, Allard N (1989) Preferred mineral orientations related to magmatic flow in ophiolite layered gabbros. J Petrol 30:925–946

Blanchette F, Peacock T, Bush JWM (2004) The Boycott effect in magma chambers. Geophys Res Lett 31:L05611

Bloomer SH, Meyer PS, Dick HJB, Ozawa K, Natland JH (1991) Textural and mineralogical variations in gabbroic rocks from Hole 735B. In: Von Herzen RP Robinson PT (eds) Proceed-

ings of the ocean drilling program, scientific results, vol 118. Ocean Drilling Program, Collge Station, Texas, pp 21–39
Bolle O, Diot H, Duchesne J-C (2000) Magnetic fabric and deformation in charnockitic igneous rocks of the Bjerkreim-Sokndal layered intrusion (Rogaland, Southwest Norway). J Struct Geol 22:647–667
Bollinberg K (1995) Textural and chemical evolution of the FeTi -minerals during the late- and post-magmatic cooling of the Skaergaard intrusion, PhD thesis, 300p, East Greenland. University of Copenhagen
Boorman S, Boudreau A, Kruger FJ (2004) The lower zone-critical zone transition of the bushveld complex: a quantitative textural study. J Petrol 45:1209–1235
Boudreau A (1982) The main platinum zone, Stillwater complex, MT: Evidence for bimetasomatism and a secondary origin for olivine. In: Walker D, McCallum IS (eds) Workshop on magmatic processes of early planetary crusts: magma oceans and stratiform layered intrusions, vol 82-01. Lunar and planetary institute technical report Houston, pp 59–61
Boudreau A (1987) Pattern formation during crystallization and the formation of fine-scale layering. In: Parsons D (ed) Origins of igneous layering. Reidel, Dodrecht, pp 453–471
Boudreau A (1988) Investigations of the Stillwater complex. IV. The role of volatiles in the petrogenesis of the J-M reef, Minneapolis adit section. Can Miner 26:193–208
Boudreau A (1994) Crystal aging in two component, two crystal systems. S Afr J Geol 97:473–485
Boudreau A (1995) Crystal aging and the formation of fine-scale layering. Mineral Petrol 54:55–69
Boudreau A (1999) Fluid fluxing of cumulates: the J-M reef and associated rocks of the Stillwater complex, Montana. J Petrol 40:755–772
Boudreau A (2011) The evolution of texture and layering in layered intrusions. Int Geol Rev 53:330–353
Boudreau A, Hoatson DM (2004) Halogen variations in the Paleoproterozoic layered mafic-ultramafic intrusions of East Kimberley, Western Australia: implications for platinum group element mineralization. Econ Geol 90:1015–1025
Boudreau A, McBirney AR (1997) The Skaergaard Layered Series. Part III. Non-dynamic layering. J Petrol 38:1003–1020
Boudreau A, McCallum IS (1992) Infiltration metasomatism in layered intrusions: an example from the Stillwater Complex, Montana. J Volcanol Geotherm Res 52:171–183
Boudreau A, Meurer AE (1999) Chromatographic separation of the platinum-group elements, gold, base metals and sulfur during degassing of a compacting and solidifying igneous crystal pile. Contrib Mineral Petrol 134:174–185
Boycott AE (1920) Sedimentation of blood corpuscles. Nature 104:532
Brandeis G, Jaupart C (1986) On the interaction between convection and crystallization in cooling magma chambers. Earth Planet Sci Lett 77:345–361
Brandeis G, Jaupart C (1987) The kinetics of nucleation and crystal growth and scaling laws for magmatic crystallization. Contrib Mineral Petrol 96:24–34
Brown GM (1956) The layered ultrabasic rocks of Rhum (Inner Hebrides). Phil Trans Royal Soc Lond B 240:1–53
Brown PE, Farmer DG (1971) Size-graded layering in the Imilik gabbro, East Greenland. Geol Mag 108:465–476
Campbell IH (1978) Some problems with the cumulus theory. Lithos 11:311–323
Campbell IH, Murck BW (1993) Petrology of the G and H chromitite zones in the Mountain view area of the Stillwater Complex, Montana. J Petrol 34:291–316
Campbell IH, Roeder PL, Dixon JM (1978) Plagioclase buoyancy in basaltic liquids as determined with a centrifuge furnace. Contrib Mineral Petrol 67:369–377
Cawthorn RG (2005) Contrasting sulphide contents of the Bushveld and Sudbury igneous complexes. Miner Deposita 40:1–12
Cawthorn RG, Ashwal LD (2009) Origin of anorthosite and magnetitite layers in the Bushveld Complex, constrained by major element compositions of plagioclase. J Petrol 50:1607–1637
Cawthorn RG, McCarthy TS (1980) Variations in Cr content of magnetite from the upper zone of the Bushveld Complex—evidence for heterogeneity and convection currents in magma chambers. Earth Planet Sci Lett 46:335–343

Cawthorn RG, Barnes SJ, Ballhaus C, Malitch KN (2005) Platinum group elements, chromium, and vanadium deposits in mafic and ultramafic rocks. In: Hedenquist JW, Thompson JFH, Goldfarb RJ, Richards JP (eds) Econ geol, one hundredth anniversary, vol 1905–2005. Econ, Geol, pp 215–249

Charlier B, Grove TL (2012) Experiments on liquid immiscibility along tholeiitic liquid lines of descent. Contrib Mineral Petrol 164:27–44

Charlier B, Namur O, Toplis MJ, Schiano P, Cluzel N, Higgins MD, Vander Auwera J (2011) Large-scale silicate liquid immiscibility during differentiation of tholeiitic basalt to granite and the origin of the Daly gap. Geology 39 :907–910

Charlier B, Namur O, Grove TL (2013) Compositional and kinetic controls on liquid immiscibility in ferrobasalt-rhyolite volcanic and plutonic series. Geochim Cosmochim Acta 113:79–93

Chen CF, Turner JS (1980) Crystallisation in a double-diffusive system. J Geophys Res 85:2573–2593

Clénet H, Ceuleneer G, Pinet P, Abily B, Daydou Y, Harris E, Amri I, Dantas C (2010) Thick sections of layered ultramafic cumulates in the Oman ophiolite revealed by an airborne hyperspectral survey: petrogenesis and relationship to mantle diapirism. Lithos 114:265–281

Coates RR (1936) Primary banding in basic plutonic rocks. J Geol 44:407–419

Conrad ME, Naslund HR (1989) Modally-graded rhythmic layering in the Skaergaard intrusion. J Petrol 30:251–269

Day JMD, Pearson DG, Hulbert LJ (2008) Rhenium-osmium isotope and platinum-group element constraints on the origin and evolution of the 1.27 Ga Muskox layered intrusion. J Petrol 49:1255–1295

Dick HJB, Sinton JM (1979) Compositional layering in alpine peridotites: evidence for pressure solution creep in the mantle. J Geol 87:403–416

Donaldson CH (1976) An experimental investigation of olivine morphology. Contrib Mineral Petrol 57:187–213

Donaldson CH (1977) Laboratory duplication of comb layering in the Rhum pluton. Mineral Mag 41:323–336

Donaldson CH (1982) Origin of some of the Rhum harrisite by segregation of intercumulus liquid. Mineral Mag 45:201–209

Drüppel K, Elsäber L, Brandt S, Gerdes A (2013) Sveconorwegian mid-crustal ultrahigh-temperature metamorphism in Rogaland, Norway: U-Pb LA-ICP-MS geochronology and pseudosections of sapphirine granulites and associated paragneisses. J Petrol 54:305–350

Duchesne JC (1999) Fe-Ti deposits in Rogaland anorthosites (South Norway): geochemical characteristics and problems of interpretation. Miner Deposita 34:182–198

Duchesne JC, Charlier B (2005) Geochemistry of cumulates from the Bjerkreim-Sokndal layered intrusion (S. Norway). Part I: Constraints from major elements on the mechanism of cumulate formation and on the jotunite liquid line of descent. Lithos 83:229–254

Eales HV, Cawthorn RG (1996) The bushveld complex. In: Cawthorn RG (ed) Layered intrusions. Elsevier, Amsterdam, pp 181–229

Eales HV, Costin G (2012) Crustally contaminated komatiite: primary source of the chromitites and Marginal, Lower and critical zone magmas in a staging chamber beneath the Bushveld complex. Econ Geol 107:645–665

Eales HV, de Klerk WJ, Teigler B (1990) Evidence for magma mixing processes within the critical and lower zones of the northwestern Bushveld complex, South Africa. Chem Geol 88:261–278

Edwards BR, Russell JK (1996) A review and analysis of silicate mineral dissolution experiments in natural silicate melts. Chem Geol 130:233–245

Emeleus CH (1997) Geology of Rum and The Adjacent Islands. Memoir for 1:50000 Geological Sheet 60, Scotland

Emeleus CH, Cheadle MJ, Hunter RH, Upton BGJ, Wadsworth WJ (1996) The rum layered suite. In: Cawthorn RG (ed) Layered intrusions. Elsevier, Amsterdam, pp 403–440

Engell J (1973) A closed system crystal fractionation model for the agpaitic Ilimaussaq intrusion, South Greenland, with special reference to lujavrites. Bull Geol Soc Den 22:334–362

Ernewein M, Pflumio C, Whitechurch H (1988) The death of an accretion zone as evidenced by the magmatic history of the Sumail ophiolite (Oman). Tectonophysics 151:247–274

Faure F, Arndt N, Libourel G (2006) Formation of spinifex texture in komatiites: an experimental study. J Petrol 47:1591–1610
Faure F, Schiano P, Trolliard G, Nicollet C, Soulestin B (2007) Textural evolution of polyhedral olivine experiencing rapid cooling rates. Contrib Mineral Petrol 153:405–416
Feig S, Koepke J, Snow J (2006) Effect of water on tholeiitic basalt phase equilibria: an experimental study under oxidizing conditions. Contrib Mineral Petrol 152:611–638
Ganino C, Arndt N, Zhou M-F, Gaillard F, Chauvel C (2008) Interaction of magma with sedimentary wall rock and magnetite ore genesis in the Panzhihua mafic intrusion, SW China. Miner Depos 43:677–694
Gillis KM, Snow JE, Klaus A, Abe N, Adriao AB, Akizawa N, Ceuleneer G, Cheadle MJ, Faak K, Falloon TJ, Friedman SA, Godard M, Guerin G, Harigane Y, Horst AJ, Hoshide T, Ildefonse B, Jean MM, John BE, Koepke J, Machi S, Maeda J, Marks NE, McCaig AM, Meyer R, Morris A, Nozaka T, Python M, Saha A, Wintsch RP (2014) Primitive layered gabbros from fast-spreading lower oceanic crust. Nature 505:204–207
Gorring M, Naslund HR (1995) Geochemical reversals within the lower 100 m of the Palisades sill, New Jersey. Contrib Mineral Petrol 119:263–276
Harker A (1908) The geology of the small isles of inverness-shire. Mem Geol Survey Scotland. Sheet 60
Harney DM, Merkle RK, Von Gruenewaldt G (1990) Platinum-group element behaviour in the lower part of the upper zone, eastern Bushveld complex-implications for the formation of the main magnetite layer. Econ Geol 85:1777–1789
Hauser AC, Bussy F (2014) Architecture and construction mechanisms of an incrementally built, vertically layered pluton (Punta Falcone, Sardinia). Paper presented at European Geophysical Union (EGU meeting), EGU, Geophysical Research Abstracts 16, EGU2014-10794
Hawkes DD (1967) Order of abundant crystal nucleation in a natural magma. Geol Mag 104:473–486
Henisch HK (1988) Crystals in gels and liesegang rings. Cambridge University Press, Cambridge
Hess HH (1960) Stillwater igneous complex. Geol Soc Am Mem 80:1–230
Higgins MD (1991) The origin of laminated and massive anorthosite, Sept Iles layered intrusion, Québec, Canada. Contrib Mineral Petrol 106:340–354
Higgins M (2002) A crystal size-distribution study of the Kiglapait layered mafic intrusion, Labrador, Canada: evidence for textural coarsening. Contrib Mineral Petrol 144:314–330
Higgins MD (2005) A new interpretation of the structure of the Sept Iles Intrusive suite, Canada. Lithos 83:199–213
Higgins MD (2011) Textural coarsening in igneous rocks. Int Geol Rev 53:354–376
Hodson ME (1998) The origin of igneous layering in the Nunarssuit Syenite, South Greenland. Mineral Mag 62:9–27
Holness MB, Hallworth MA, Woods A, Sides RE (2007) Infiltration metasomatism of cumulates by intrusive magma replenishment: the Wavy Horizon, Isle of Rum, Scotland. J Petrol 48:563–587
Holness MB, Stripp G, Humphreys MCS, Veksler IV, Nielsen TFD, Tegner C (2011) Silicate liquid immiscibility within the crystal mush: late-stage magmatic microstructures in the Skaergaard intrusion, East Greenland. J Petrol 52:175–222
Hort M, Marsh B, Spohn T (1993) Igneous layering through oscillatory nucleation and crystal settling in well-mixed magmas. Contrib Mineral Petrol 114:425–440
Humphreys MCS (2011) Silicate liquid immiscibility within the crystal mush: evidence from Ti in plagioclase from the Skaergaard intrusion. J Petrol 52:147–174
Huppert H, Sparks RS (1980) The fluid dynamics of a basaltic magma chamber replenished by influx of hot, dense ultrabasic magma. Contrib Mineral Petrol 75:279–289
Huppert HE, Sparks RSJ (1984) Double-diffusive convection due to crystallization in magmas. An Rev Earth Planet Sci 12:11–37
Huppert HE, Sparks RSJ, Wilson JR, Hallworth MA (1986) Cooling and crystallization at an inclined plane. Earth Planet Sci Lett 57:345–357
Husch JM (1990) Palisades sill: origin of the olivine zone by separate magmatic injection rather than gravity settling. Geology 18:699–702

Irvine TN (1963) Mineral Petrol of the ultramafic complex at Duke Island, Southeast Alaska. Min Soc Am Spec Pap 1
Irvine TN (1965) Sedimentary structures in igneous intrusions with particular reference to the Duke Island ultramafic complex. In: Middleton GV (ed) Primary sedimentary structures and their hydrodynamic interpretation, vol 12. Spec Pub Soc Econ Paleont Mineral, pp 220–232
Irvine TN (1974) Petrology of the Duke Island ultramafic complex southeastern Alaska. Geol Soc Am Mem 138:1–240
Irvine TN (1975) Crystallization sequences in the Muskox intrusion and other layered intrusions—II. Origin of chromitite layers and similar deposits of other magmatic ores. Geochim Cosmochim Acta 39:991–1020
Irvine TN (1977) Origin of chromitite layers in the Muskox intrusion and other stratiform intrusions: a new interpretation. Geology 5:273–277
Irvine TN (1982) Terminology for layered intrusions. J Petrol 23:127–162
Irvine TN (1987) Layering and related structures in the Duke Island and Skaergaard intrusions: similarities, differences and origins. In: Parsons D (ed) Origins of igneous layering. Reidel, Dodrecht, pp 185–245
Irvine TN, Smith CH (1967) The ultramafic rocks of the Muskox intrusion, Northwest Territories, Canada. In: Wyllie PJ (ed) Ultramafic and related rocks. Wiley, New York, pp 38–49
Irvine TN, Keith DW, Todd SG (1983) The J-M platinum-palladium reef of the stillwater complex, Montana. II. Origin of double-diffusive convective magma mixing and implications for the Bushveld Complex. Econ Geol 78:1287–1334
Irvine TN, Andersen JCO, Brooks CK (1998) Included blocks (and blocks within blocks) in the Skaergaard intrusion: geologic relations and the origins of rhythmic modally graded layers. Geol Soc Am Bull 110:1398–1447
Jackson ED (1967) Ultramafic cumulates in the Stillwater, Great Dyke, and Bushveld intrusions. In: Wyllie PJ (ed) Ultramafic and related rocks. Wiley, New York, pp 20–38
Jackson ED (1970) The cyclic unit in layered intrusions: a comparison of a repetitive stratigraphy in the ultramafic parts of the Stillwater, Muskox, Great Dyke and Bushveld Complexes. Spec Pub Geol Soc S Afr 1:391–424
Jakobsen JK, Veksler IV, Tegner C, Brooks CK (2005) Immiscible iron- and silica-rich melts in basalt petrogenesis documented in the Skaergaard intrusion. Geology 33:885–888
Jakobsen JK, Veksler IV, Tegner C, Brooks CK (2011) Crystallization of the Skaergaard intrusion from an emulsion of immiscible iron- and silica-rich liquids: evidence from melt inclusions in plagioclase. J Petrol 52:345–373
Jensen KK, Wilson JR, Robins B, Chiodoni F (2003) A sulphide-bearing orthopyroxenite layer in the Bjerkreim-Sokndal Intrusion, Norway: implications for processes during magma-chamber replenishment. Lithos 67:15–37
Jousselin D, Morales LF, Stephant A (2012) Gabbro layering induced by simple shear in the Oman ophiolite Moho transition zone. Earth Planet Sci Lett 331:55–66
Junge M, Oberthür T, Melcher F (2014) Cryptic variation of chromite chemistry, platinum group element and platinum group mineral distribution in the ug-2 chromitite: an example from the Karee Mine, western Bushveld Complex, South Africa. Econ Geol 109:795–810
Juteau T, Ernewein M, Reuber I, Whitechurch H, Dahl R (1988) Duality of magmatism in the plutonic sequence of the Sumail Nappe, Oman. Tectonophysics 151:107–135
Kanitpanyacharoen W, Boudreau AE (2013) Sulfide-associated mineral assemblages in the Bushveld Complex, South Africa: platinum-group element enrichment by vapor refining by chloride-carbonate fluids. Miner Deposita 48:193–210
Katz K, Keller J (1981) Comb-layering in carbonatite dykes. Nature 294:350–352
Kinnaird JA, Kruger FJ, Nex PAM, Cawthorn RG (2002) Chromitite formation—a key to understanding processes of platinum enrichment. Trans Institut Mining Metal 111:23–35
Kirkpatrick RJ (1975) Crystal growth from the melt: a review. Am Mineral 60:798–814
Kogarko LN, Khapaev VV (1987) The modeling of formation of apatite deposits of the Khibina massif (Kola Peninsula). In: Parsons D (ed) Origins of igneous layering. D. Reidel, Dodrecht, pp 589–611

Korzhinskii DS (1970) Theory of metasomatic zoning. Clarendon, Oxford
Kushiro I (1973) The system diopside-anorthite-albite: determination of compositions of co-existing phases. Yearb Carnegie Inst Wash 72:502–507
Lee CA (1981) Post-deposition structures in the Bushveld Complex mafic sequence. J Geol Soc London 145:327–341
Lesher CE, Walker D (1988) Cumulate maturation and melt migration in a temperature gradient. J Geophys Res 93:10295–10311
Leuthold J, Blundy JD, Holness MB, Sides R (2014) Successive episodes of reactive liquid flow through a layered intrusion (Unit 9, Rum Eastern Layered Intrusion, Scotland). Contrib Mineral Petrol 168:1–27
Lipin BR (1993) Pressure increases, the formation of chromitite seams, and the development of the Ultramafic series in the Stillwater Complex, Montana. J Petrol 34:955–976
Lofgren G, Donaldson C (1975) Curved branching crystals and differentiation in comb-layered rocks. Contrib Mineral Petrol 49:309–319
Longhi J (1987) Liquidus equilibria and solid solution in the system $CaAl_2Si_2O_8$—Mg_2SiO_4—$CaSiO_3$—SiO_2 at low pressure. Am J Sci 287:265–331
Lundstrom C, Boudreau A, Pertermann M (2005) Diffusion-reaction in a thermal gradient: Implications for the genesis of anorthitic plagioclase, high alumina basalt and igneous mineral layering. Earth Planet Sci Lett 237:829–854
Maaloe S (1978) The origin of rhythmic layering. Mineral Mag 42:337–345
Marsh BD (1989) On convective style and vigor in sheet-like magma chambers. J Petrol 30:479–530
Marsh BD, Maxey MR (1985) On the distribution and separation of crystals in convecting magma. J Volcanol Geotherm Res 24:95–150
Martin D, Griffiths R, Campbell I (1987) Compositional and thermal convection in magma chambers. Contrib Mineral Petrol 96:465–475
McBirney AR (1975) Differentiation of the Skaergaard intrusion. Nature 253:691–694
McBirney AR (1985) Further considerations of double-diffusive stratification and layering in the skaergaard intrusion. J Petrol 26:993–1001
McBirney AR (1993) Layered igneous rocks 25 years after Wager and Brown. Paper presented at the Symposium on Layering in Igneous Complexes, Wager and Brown 25th Anniversary Commemorative Meeting, Johannesburg, South Africa
McBirney AR, Hunter RH (1995) The cumulus paradigm reconsidered. J Geol 103:114–122
McBirney AR, Nakamura N (1974) Immiscibility in late-stages magmas of the Skaergaard intrusion. Yearb Carnegie Inst Wash 73:348–352
McBirney AR, Nicolas A (1997) The skaergaard layered series. Part II. Magmatic flow and dynamic layering. J Petrol 38:569–580
McBirney AR, Noyes RM (1979) Crystallization and layering of the Skaergaard intrusion. J Petrol 20:487–554
McBirney AR, White CM, Boudreau AE (1990) Spontaneous development of concentric layering in a solidified siliceous dike, East Greenland. Earth Sci Rev 29:321–330
McCallum IS (1996) The stillwater complex. In: Cawthorn RG (ed) Layered Intrusions. Elsevier, Amsterdam, pp 441–484
Meurer WP, Boudreau AE (1998) Compaction of igneous cumulates. Part II: compaction and the development of igneous foliation. J Geol 106:293–304
Meurer WP, Klaber S, Boudreau AE (1997) Discordant bodies from olivine-bearing zones III and IV of the Stillwater Complex, Montana—evidence for postcumulus fluid migration and reaction in layered intrusions. Contrib Mineral Petrol 130:81–92
Meurer WP, Willmore CC, Boudreau AE (1999) Metal redistribution during fluid exsolution and migration in the Middle Banded series of the Stillwater Complex, Montana. Lithos 47:143–156
Miller JD, Ripley EM (1996) Layered intrusions of the Duluth Complex, USA. In: Cawthorn RG (ed) Layered intrusions. Elsevier, Amsterdam, pp 257–301
Mondal SK, Mathez EA (2007) Origin of the UG2 chromitite layer, Bushveld complex. J Petrol 48:495–510

Moore JG, Evans BW (1967) The role of olivine in the crystallization of the prehistoric Makaopuhi tholeiitic lava lake, Hawaii. Contrib Mineral Petrol 15:202–223

Moore JG, Lockwood JP (1973) Origin of Comb Layering and Orbicular Structure, Sierra Nevada Batholith, California. Geol Soc Am Bull 84:1–20

Morse SA (1969) The Kiglapait layered intrusion, Labrador. Geol Soc Am Mems 112:1–204

Morse SA (2008) Compositional convection trumps silicate liquid immisciblity in layered intrusions: a Discussion of 'Liquid immiscibility and the evolution of basaltic magma' by Veksler et al. J Petrol 48:2187–2210, 49:2157–2168

Morse SA (2011) The fractional latent heat of crystallizing magmas. Am Mineral 96:682–689

Naldrett AJ, Gasparrini EC, Barnes SJ, Von Gruenewaldt G, Sharpe MR (1986) The upper critical zone of the Bushveld Complex and the origin of merensky-type ores. Econ Geol 81:1105–1117

Naldrett AJ, Wilson A, Kinnaird J, Yudovskaya M, Chunnett G (2012) The origin of chromitites and related PGE mineralization in the Bushveld complex: new mineralogical and petrological constraints. Miner Deposita 47:209–232

Namur O, Charlier B, Toplis MJ, Higgins MD, Liégeois J-P, Vander Auwera J (2010) Crystallization sequence and magma chamber processes in the ferrobasaltic Sept Iles layered intrusion, Canada. J Petrol 51:1203–1236

Namur O, Charlier B, Pirard C, Hermann J, Liégeois J-P, Auwera JV (2011a) Anorthosite formation by plagioclase flotation in ferrobasalt and implications for the lunar crust. Geochim Cosmochim Acta 75:4998–5018

Namur O, Charlier B, Toplis MJ, Higgins MD, Hounsell V, Liégeois J-P, Vander Auwera J (2011b) Differentiation of tholeiitic basalt to A-type granite in the Sept Iles layered intrusion, Canada. J Petrol 52:487–539

Namur O, Charlier B, Holness MB (2012a) Dual origin of Fe–Ti–P gabbros by immiscibility and fractional crystallization of evolved tholeiitic basalts in the Sept Iles layered intrusion. Lithos 154:100–114

Namur O, Charlier B, Toplis M, Vander Auwera J (2012b) Prediction of plagioclase-melt equilibria in anhydrous silicate melts at 1-atm. Contrib Mineral Petrol 163:133–150

Namur O, Humphreys MCS, Holness MB (2013) Lateral reactive infiltration in a vertical gabbroic crystal mush, Skaergaard intrusion, East Greenland. J Petrol 54:985–1016

Namur O, Humphreys MCS, Holness MB (2014) Crystallization of interstitial liquid and latent heat buffering in solidifying gabbros: skaergaard Intrusion, Greenland. J Petrol 55:1389–1427

Naslund HR, McBirney AR (1996) Mechanisms of formation of igneous layering. In: Cawthorn RG (ed) Layered intrusions. Elsevier, Amsterdam, pp 1–43

Naslund HR, Turner PA, Keith DW (1991) Crystallization and layer formation in the middle zone of the Skaergaard intrusion. Bull Geol Soc Den 38:165–171

Nicolas A (1992) Kinematics in magmatic rocks, with special reference to gabbro. J Petrol 33:891–915

Nicolas A, Reuber K, Benn K (1998) A new magma chamber model based on structural studies in the Oman ophiolite. Tectonophysics 151:87–105

Nilsen O (1973) Petrology of the Hyllingen gabbro complex, Sor-Trondelag, Norway. Nor Geol Tidsskr 1:213–231

Niu YL, Batiza R (1991) In situ densities of MORB melts and residual mantle - Implications for buoyancy forces beneath midocean ridges. J Geol 99:767–775

O'Driscoll B, Donaldson CH, Troll VR, Jerram DA, Emeleus CH (2007) An origin for harrisitic and granular olivine in the rum layered suite, NW Scotland: a crystal size distribution study. J Petrol 48:253–270

O'Driscoll B, Donaldson CH, Daly JS, Emeleus CH (2009a) The roles of melt infiltration and cumulate assimilation in the formation of anorthosite and a Cr-spinel seam in the Rum Eastern Layered Intrusion. Lithos 111:6–20

O'Driscoll B, Day JMD, Daly JS, Walker RJ, McDonough WF (2009b) Rhenium-osmium isotopes and platinum-group elements in the Rum Layered Suite, Scotland: implications for Cr-spinel seam formation and the composition of the Iceland mantle anomaly. Earth Planet Sci Lett 286:41–51

O'Driscoll B, Emeleus CH, Donaldson CH, Daly JS (2010) Cr-spinel seam petrogenesis in the rum layered suite, NW Scotland: cumulate assimilation and in situ crystallization in a deforming crystal mush. J Petrol 51:1171–1201
Ortoleva P, Merino E, Moore C, Chadam J (1987) Geochemical self-organization I; reaction-transport feedbacks and modeling approach. Am J Sci 287:979–1007
Osborn EF (1978) Changes in phase relations in response to change in pressure from 1 atm to 10 kbar for the system Mg_2SiO_4-iron oxide-$CaAl_2Si_2O_8$–SiO_2. Yearb Carnegie Inst Wash 77:784–790
Pallister JS, Hopson CA (1981) Samail ophiolite plutonic suite: field relations, phase variation, cryptic variation and layering, and a model of a spreading ridge magma chamber. J Geophys Res 86:2593–2644
Pang K-N, Zhou M-F, Lindsley D, Zhao D, Malpas J (2008) Origin of Fe-Ti oxide ores in mafic intrusions: evidence from the Panzhihua intrusion, SW China. J Petrol 49:295–313
Pang K-N, Li C, Zhou M-F, Ripley EM (2009) Mineral compositional constraints on petrogenesis and oxide ore genesis of the late Permian Panzhihua layered gabbroic intrusion, SW China. Lithos 110:199–214
Parada MA, Nyström JO, Levi B (1999) Multiple sources for the Coastal Batholith of central Chile (31–34°S): geochemical and Sr-Nd isotopic evidence and tectonic implications. Lithos 46:505–521
Parsons I (1979) The Klokken gabbro-syenite complex, South Greenland: cryptic variation and origin of inversely-graded layering. J Petrol 20:653–694
Parsons I (1987) Origins of Igneous Layering. D. Reidel, Dodrecht
Peacock T, Blanchette F, Bush JMW (2005) The stratified boycott effect. J Fluid Mech 529:33–49
Pêcher A, Arndt N, Jean A, Bauville A, Ganino C, Athurion C (2013) Structure of the Panzhihua intrusion and its Fe–Ti–V deposit, China. Geosci Front 4:571–581
Petersen JS (1985) Columnar-dendritic feldspars in the Lardalite intrusion, Oslo region, Norway: 1. Implications for unilateral solidification of a stagnant boundary layer. J Petrol 26:223–252
Petersen JS (1987) Solidification contraction: another approach to cumulus processes and the origin of igneous layering. In: Parsons I (ed) Origins of igneous layering. D. Reidel, Dodrecht, pp 505–526
Philpotts AR (1982) Compositions of immiscible liquids in volcanic rocks. Contrib Mineral Petrol 80:201–218
Pommier A, Gaillard F, Pichavant M, Scaillet B (2008) Laboratory measurements of electrical conductivities of hydrous and dry Mount Vesuvius melts under pressure. J Geophys Res: Solid Earth 113:B05205
Raedeke LD, McCallum IS (1984) Investigations in the stillwater complex: Part II. Petrology and petrogenesis of the ultramafic series. J Petrol 25:395–420
Reynolds I (1985) The nature and origin of titaniferous magnetite-rich layers in the upper zone of the Bushveld complex: a review and synthesis. Econ Geol 80:1089–1108
Robins B, Chiodoni F (2007) Poles apart: a discussion of "Geochemistry of cumulates from the Bjerkreim-Sokndal Intrusion (S. Norway). Part I: constraints from major elements on the mechanism of cumulate formation and on the jotunite liquid line of descent" by J.C. Duchesne and B. Charlier (Lithos 83, 2005, 229–254). Lithos 98:335–338
Robins B, Haukvik L, Jansen S (1987) The organization and internal structure of cyclic units in the Honningsvag intrusive suite, North Norway: Implications for intrusive mechanisms, double-diffusive convection and pore-magma infiltration. In: Parsons I (ed) Origins of igneous layering. D. Reidel, Dodrecht, pp 287–312
Rogerson MA, Cardoso SS (2000) Patterns of bubble desorption during the solidification of a multicomponent melt. J Fluid Mech 419:263–282
Roobol MJ (1972) Size-graded igneous layering in an Icelandic intrusion. Geol Mag 109:393–403
Scoates JS (2000) The plagioclase-magma density paradox re-examined and the crystallization of Proterozoic anorthosites. J Petrol 41:627–649
Scoon RN, Mitchell AA (1994) Discordant iron-rich ultramafic pegmatites in the Bushveld complex and their relationship to iron-rich intercumulus and residual liquids. J Petrol 35:881–917

Sen G, Presnall DC (1984) Liquidus phase relationships on the join anorthite-forsterite-quartz at 10 kbar with applications to basalt petrogenesis. Contrib Mineral Petrol 85:404–408

Sorensen H, Larsen LM (1987) Layering in the Ilimaussaq alkaline intrusion. In: Parsons I (ed) Origins of igneous layering. D. Reidel, Dodrecht, pp 1–28

Spandler C, Mavrogenes J, Arculus R (2005) Origin of chromitites in layered intrusions: evidence from chromite-hosted melt inclusions from the Stillwater Complex. Geology 33:893–896

Sparks RSJ, Huppert HE, Kerr RC, McKenzie DP, Tait SR (1985) Postcumulus processes in layered intrusions. Geol Mag 122:555–568

Sparks RSJ, Huppert HE, Koyaguchi T, Hallworth MA (1993) Origin of bimodal and rhythmic igneous layering by sedimentation in a convecting magma chamber. Nature 361:246–249

Spera FJ, Yuen DA, Clark S, Hong HJ (1986) Double-diffusive convection in magma chambers: single or multiple layers? Geophys Res Lett 13:153–156

Tegner C, Cawthorn R (2010) Iron in plagioclase in the Bushveld and Skaergaard intrusions: implications for iron contents in evolving basic magmas. Contrib Mineral Petrol 159:719–730

Tegner C, Wilson JR, Brooks CK (1993) Intraplutonic Quench Zones in the Kap Edvard Holm Layered Gabbro complex, East Greenland. J Petrol 34:681–710

Tegner C, Cawthorn RG, Kruger FJ (2006) Cyclicity in the Main and Upper Zones of the Bushveld complex, South Africa: crystallization from a zoned magma sheet. J Petrol 47:2257–2279

Thayer TP (1963) Flow-layering in alpine peridotite-gabbro complexes. Spec Pap Mineral Soc Am 1:55–62

Thompson AB (1975) Calc-silicate diffusion zones between marble and pelitic schist. J Petrol 16:314–346

Thompson RN, Patrick DJ (1968) Folding and slumping in a layered gabbro. Geol J 6:139–146

Thy P, Dilek Y (2000) Magmatic and tectonic controls on the evolution of oceanic magma chambers at slow-spreading ridges: perspectives from ophiolitic and continental layered intrusions. In: Dilek Y, Elthon EM, Nicolas A (eds) Ophiolites and oceaninc crust: new insights from field studies and the ocean drilling program, vol 349. Special paper of the geological society of America, pp 87–104

Thy P, Esbensen KH (1982) Origin of certain types of small-scale igneous layering from the Fongen-Hyllingen basic complex, Norway. Geologiska Foren i Stockh Forh 104:33–42

Thy P, Wilson JR (1980) Primary igneous load-cast deformation structures in the fongen-hyllingen layered basic intrusion, Trondheim Region, Norway. Geol Mag 117:363–371

Thy P, Jakobsen NN, Wilson JR (1988) Fine-scale graded layers in the Fongen-Hyllingen gabbroic complex, Norway. Can Miner 26:235–243

Todd SG, Keith DW, Le Roy LW, Schissel DJ, Mann EL, Irvine TN (1982) The J-M platinum-palladium reef of the Stillwater Complex, Montana; I, Stratigraphy and petrology. Econ Geol 77:1454–1480

Toplis MJ (2005) The thermodynamics of iron and magnesium partitioning between olivine and liquid: criteria for assessing and predicting equilibrium in natural and experimental systems. Contrib Mineral Petrol 149:22–39

Toplis MJ, Carroll MR (1995) An experimental study of the influence of oxygen fugacity on Fe–Ti oxide stability, phase relations, and mineral-melt equilibria in ferro-basaltic systems. J Petrol 36:1137–1170

Toplis MJ, Carroll MR (1996) Differentiation of ferro-basaltic magmas under conditions closed and open to oxygen: implications for the Skaergaard intrusion and other natural systems. J Petrol 37:837–858

Toramaru A, Iochi A (2000) Transition between periodic precipitation and tree-like crystal aggregates: a detail experimental study. Forma 15:365–376

Toramaru A, Matsumoto M (2012) Numerical experiment of cyclic layering in a solidified binary eutectic melt. J Geophys Res 117:B02209

Toramaru A, Ishiwatari A, Matsuzawa M, Nakamura M, Arai S (1996) Vesicle layering in solidified intrusive magma bodies: a newly recognized type of igneous structure. Bull Volcanol 58:393–400

Toramaru A, Harada T, Okamura T (2003) Experimental pattern transitions in a Liesegang system. Phys D 183:133–140

Turner JS, Campbell IH (1986) Convection and mixing in magma chambers. Earth Sci Rev 23:255–352

Ulmer GC (1969) Experimental investigations of chromite spinels. Econ Geol Monograph 4:114–131

Upton GGJ, Parsons I, Emeleus CH, Hodson ME (1996) Layered alkaline igneous rocks of the Gradar Province, South Greenland. In: Cawthorn RG (ed) Layered Intrusions. Elsevier, Amsterdam, pp 331–363

Vander Auwera J, Longhi J (1994) Experimental study of a jotunite (hypersthene monzodiorite): constraints on the parent magma composition and crystallization conditions (P, T, fO_2) of the Bjerkreim-Sokndal layered intrusion. Contrib Mineral Petrol 118:60–78

Veksler I, Reid DL, Dulski P, Keiding JK, Trumbull RB (2013) Electrochemical effects of magmatic crystallization: Cyclic units of the Bushveld Igneous Complex. Paper presented at the EGU General Assembly, Vienna

Vinet N, Higgins MD (2011) What can crystal size distributions and olivine compositions tell us about magma solidification processes inside Kilauea Iki lava lake, Hawaii? J Volcanol Geoth Res 208:136–162

Voordouw R, Gutzmer J, Beukes NJ (2009) Intrusive origin for the Upper Group (UG1, UG2) stratiform chromitite seams in the Dwars River area, Bushveld Complex, South Africa. Mineral Petrol 97:75–94

Wadsworth WJ (1961) The layered ultrabasic rocks of south-west Rhum, Inner Hebrides. Phil Trans Royal Soc Lond B 244:21–64

Waff HS, Weill DF (1975) Electrical conductivity of magmatic liquids: effects of temperature, oxygen fugacity and composition. Earth Planet Sci Lett 28:254–260

Wager LR (1959) Differing powers of crystal nucleation as a factor producing diversity in layered igneous intrusions. Geol Mag 96:75–80

Wager LR (1963) The mechanism of adcumulus growth in the layered series of the Skaergaard intrusion. Spec Pap Mineral Soc Am 1:1–9

Wager LR, Brown GM (1951) A note on rhythmic layering in the ultrabasic rocks of Rhum. Geol Mag 88:166–168

Wager LR, Brown GM (1968) Layered Igneous Rocks. Oliver & Boyd, Edinburgh, p. 588

Wager LR, Brown GM, Wadsworth WJ (1960) Types of igneous cumulates. J Petrol 1:73–85

Welsch B, Faure F, Famin V, Baronnet A, Bachèlery P (2012) Dendritic crystallization: a single process for all the textures of olivine in basalts? J Petrol 54:539–574

Williams E, Boudreau AE, Boorman S, Kruger FJ (2006) Textures of orthopyroxenites from the Burgersfort bulge of the eastern Bushveld Complex, Republic of South Africa. Contrib Mineral Petrol 151:480–492

Wilson AH (1982) The geology of the "Great Dyke", Zimbabwe: the ultramafic rocks. J Petrol 23:240–292

Wilson JR, Esbensen KH, Thy P (1981) Igneous petrology of the Synorogenic Fongen-Hyllingen layered basic complex, South-Central Scandinavian Caledonides. J Petrol 22:584–627

Wilson JR, Menuge JF, Pedersen S, Engell-Sorensen O (1987) The southern part of the Fongen-Hyllingen layered mafic complex. In: Parsons I (ed) Origins of igneous layering. D. Reidel, Dodrecht, pp 145–184

Wilson JR, Robins B, Nielsen F, Duchesne JC, Vander Auwera J (1996) The Bjerkreim-Sokndal layered intrusion, Southwest Norway. In: Cawthorn RG (ed) Layered Intrusions. Elsevier, Amsterdam, pp 231–256

Young IM (1984) Mixing of supernatant and interstitial fluids in the Rhum layered intrusion. Mineral Mag 48:345–350

Zhang Y, Ni H, Chen Y (2010) Diffusion data in silicate melts. Rev Mineral Geochem 72:311–408

Zhou MF, Robinson PT, Lesher CM, Keays RR, Zhang C-J, Malpas J (2005) Geochemistry, petrogenesis and metallogenesis of the Panzhihua gabbroic layered intrusion and associated Fe–Ti–V oxide deposits, Sichuan Province, SW China. J Petrol 46:2253–2280

Chapter 3
Quantitative Textural Analysis of Rocks in Layered Mafic Intrusions

Michael D. Higgins

Abstract Quantification of rock textures (= microstructures), in the form of crystal sizes, shapes and orientations, complements chemical and isotopic approaches to the determination of the petrogenesis of rocks in layered mafic intrusions. Textures develop initially by nucleation and kinetic growth, but this process is commonly obscured in layered mafic intrusions by later textural modifications. Foliation in layered mafic intrusions is common but lineation of non-equant crystals is generally weak or absent. This makes it difficult to determine if simple shear (flow) or pure shear (compaction) is dominant. However, if pure shear is important then it must be accompanied by significant pressure-solution to produce the strength of the observed foliations. Correlation of fabric strength and overall crystal shape suggest that growth and shear were concurrent. But by far the most important textural process is equilibration (coarsening), testified by crystal size distributions. Almost all rocks in layered mafic intrusions have equilibrated, but the degree of progress towards equilibrium is very variable, even within layered mafic intrusions. It appears that the process is controlled by local variations in liquid content, mineral mixture and temperature.

Keywords Crystal size distribution · Igneous rock · Textural quantification · Petrography

Introduction

Petrology developed as a separate discipline from geology in the nineteenth century largely due to the development of the petrographic microscope (Kile 2003). This enabled qualitative determinations of the structure of rocks, properly termed petrography. Although such studies continue to the present day, development of chemical and isotopic analysis of whole rocks, and later point analysis of mineral grains, have revolutionised the subject to the extent that petrology is now viewed by many as

M. D. Higgins (✉)
Sciences de la Terre, Université du Québec à Chicoutimi, 555 Blvd de l'université, Chicoutimi, Québec G7H 2B1, Canada
e-mail: mhiggins@uqac.ca

B. Charlier et al. (eds.), *Layered Intrusions,* Springer Geology,
DOI 10.1007/978-94-017-9652-1_3

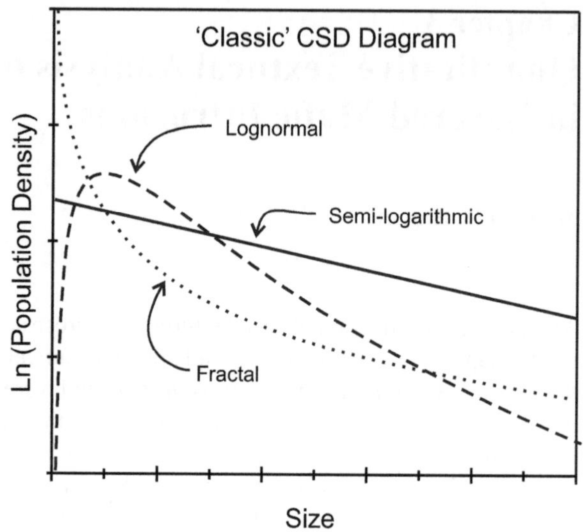

Fig. 3.1 Stereotypic crystal size distributions, plotted on the most commonly used diagram. Actual CSDs never completely follow these forms, but may approach them. Other CSD diagrams are more appropriate for verifying lognormal or fractal (power-law) distributions (Higgins 2006b)

an extension of geochemistry. One of the reasons is that such methods are quantitative, and hence have the power to constrain petrologic models of rock development numerically (Merriam 2004). However, it is also possible to quantify aspects of the structure of igneous rocks, and hence unleash the power of numerical methods for study of crystal growth and dissolution, amongst other processes (Higgins 2006a). In this paper I will use the term texture (= microstructure) for such structures. Texture can be quantified using measurements of crystal size, shape and orientation. Dihedral angles, an aspect of crystal shape, are discussed in another chapter (The influence of interfacial energies on igneous microstructures). Other textural parameters can be imagined, but are not usually measured for plutonic igneous rocks.

The need for quantification of textures was realised early. The first quantitative study of crystal size was published in 1898 on plagioclase megacrysts in the Isle Royale basalts (Lane 1898). However, the first study of relevance to the subject of this book was that of Jackson (1961) on the Stillwater complex in Montana. He measured the length of olivine, orthopyroxene and chromite crystals and found distributions of crystal intersection diameters close to lognormal. Crystal size distribution (CSD) continues to be the most important and widely determined quantitative textural parameter. CSD can have many different forms (Fig. 3.1): semilogarithmic ('classic straight CSD'); lognormal (inverted 'U'; hump-shaped); fractal (scale-independent; power-law); or combinations of these distributions (Higgins 2006b). Therefore, it is important to clearly describe and illustrate actual CSD and only those studies in which CSD are available in the original publications will be discussed in this survey.

Quantitative textural studies started in earnest when methods developed by engineers for materials processing were applied to rocks (Marsh 1988; Randolph and Larson 1971). Although early studies did not examine plutonic rocks, they did establish a consistent and meaningful way of displaying CSD data, by plotting ln(crystal population density) versus crystal size. These studies, and almost all subsequent

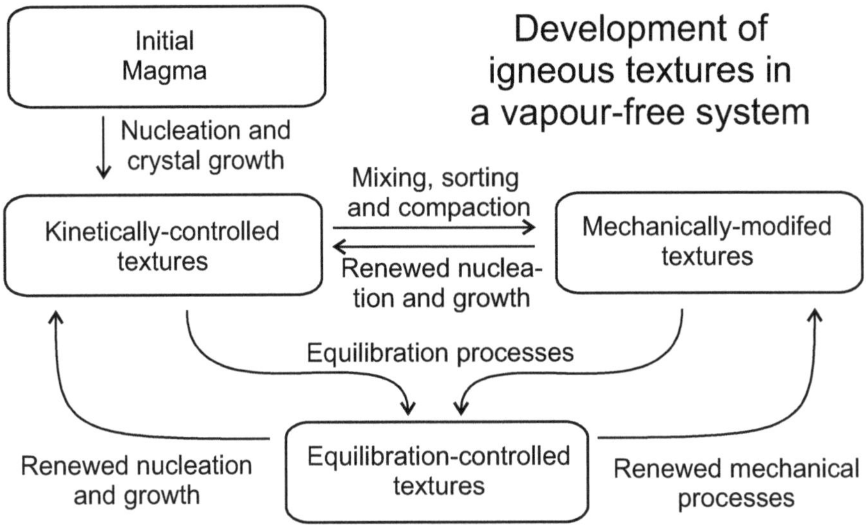

Fig. 3.2 Main vapour-absent petrologic processes that produce and modify textures in igneous rocks (Higgins 2011)

ones, were based on two dimensional measurements of thin sections which need to be converted to true three dimensional parameters such as CSD. The importance of these conversions and the correct way to carry them out were not appreciated immediately by all workers. However, this changed with the development of a more rigorous stereological conversion algorithm and distribution of a simple computer program to do these corrections (Higgins 2000). Nevertheless, it should not be forgotten that stereological conversion needs an estimate of the crystal shape and that it introduces errors which accumulate as the algorithm sequentially processes progressively smaller intersection bins. This is an unavoidable mathematical fact.

Direct measurement of CSD by tomography or serial sectioning is possible and avoids problems with stereology (Baker et al. 2012). However, the method has its limitations and no CSD studies of major phases in mafic plutonic rocks have been published so far. The most important constraint is the problem of distinguishing touching crystals of the same phase—X-ray absorption is isotropic even in optically anisotropic minerals like plagioclase. Clearly, this becomes more important as the phase abundance increases.

Development of Textures in Igneous Rocks

The texture of a igneous rock is the final product of all the petrologic processes since the inception of solidification, but some processes may have been overwritten and hence invisible in the final texture. In some situations textures representing intermediate phases of development are preserved and these can be used to define better the exact path of textural development (e.g. Higgins 1998). Figure 3.2 is a flow diagram of some of the more important vapour-absent petrologic processes. If we start with a crystal-poor magma, nucleation of new phases and their subsequent growth will produce kinetically-controlled textures. Texture has developed only by nucleation and growth of crystals. Volcanic rocks commonly sample these textures before the magma is completely solidified. Mechanical processes, such as mixing, sorting and compaction, can modify these kinetic textures while liquid is still present in the system. Such textures record the mechanical effects of magmatic deformation. If crystallisation continues during magmatic deformation then there will be a textural response, which is also kinetic. Textures produced by kinetic and mechanical processes have relatively high ratios of surface area to volume and hence high total surface energies. If the system is maintained at a temperature close to the mineral liquidus then the surface energy can be minimised by solution of small crystals and growth of larger ones (Higgins 2011). Solid-state equilibration may also occur at lower temperatures. This equilibration process never goes to completion as the energy loss gets progressively smaller as the crystals grow. Equilibrated materials with melt present can be rejuvenated by an increase in undercooling and will take on aspects of kinetic textures. Similarly, equilibrated textures can also be mechanically modified.

In this survey I will start by examining natural examples of kinetic, mechanical and equilibration textures, as much as they can be identified and separated. I will then discuss quantitative textural studies of layered mafic intrusions (LMI), and related intrusions, and try to interpret their textures in terms of these broad processes.

Kinetic Textures

Textures controlled only by nucleation and growth are called kinetic textures. Clearly, they are most apparent in volcanic rocks, where the crystallinity is generally low and the crystals have space to grow. Early work on CSD of rocks was based on this premise and involved a steady-state model in which magma is continuously fed and drained from a magma chamber, in which nucleation and growth occur (Marsh 1988; Randolph and Larson 1971). Such a model produces a straight CSD on a graph of ln(crystal population density) versus crystal size. No geological system corresponds exactly to this model, but many volcanic rocks do have straight CSD (e.g. Higgins and Roberge 2007). As a rock solidifies crystals grow until they meet their neighbours, giving an impingement texture, and the volume of liquid in which crystals can nucleate diminishes, modifying the controls on crystal size distribution. Work on closed-system (batch) crystallisation shows a straight CSD, except for a turn-down for small crystals towards the end of solidification (Marsh 1998).

The first CSD of a gabbro was measured on core samples from the Makaopuhi lava lake, Hawai'i (Cashman and Marsh 1988). Although this is clearly not an LMI, it is instructive to examine the textural characteristics of a gabbro, or more strictly a microgabbro, under these conditions. Samples were taken from different depths in a core that was drilled before the lake was completely solid. Consecutive samples represent different degrees of solidification, from 20 to 83 % (Fig. 3.3). All CSD

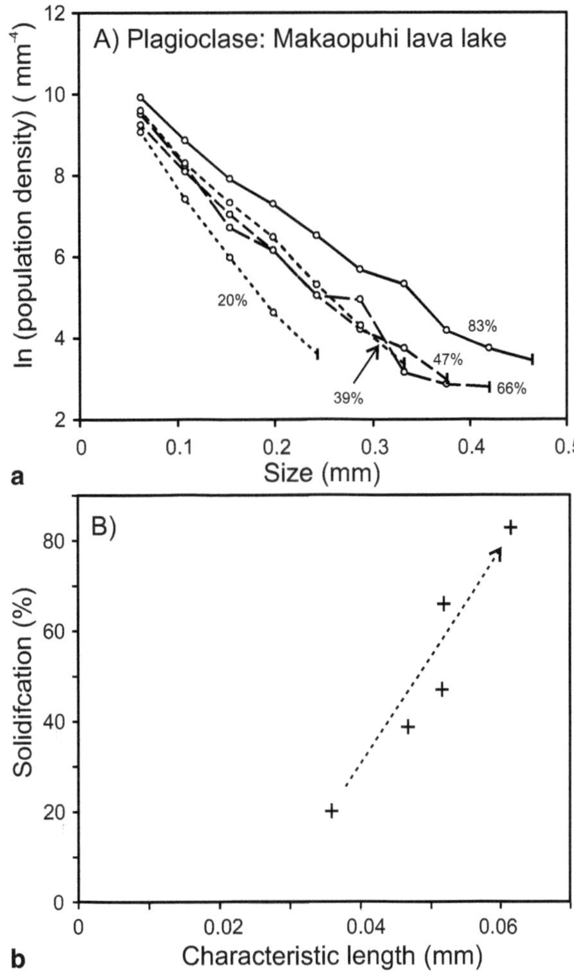

Fig. 3.3 **a** Plagioclase CSD of gabbros from the Makaopuhi lava lake (Cashman and Marsh 1988) recalculated from the original data using modern stereological methods (Higgins 2006a). The percent solidification is indicated for each sample. CSD for sizes less than 0.07 mm are undefined. **b** Textural parameters for Makaopuhi samples. The characteristic length (− 1/slope) increases almost linearly with solidification

were almost straight down to 0.07 mm—below this size the CSD is undefined. The slope of CSD decreased with increasing solidification. If the CSD are plotted on a graph of plagioclase abundance versus characteristic length (− 1/slope) they define a straight line. Cashman and Marsh (1988) interpreted this behaviour as a kinetic solidification sequence, but other interpretations are possible (Higgins 2011).

Crystallisation conditions can affect kinetic growth of the different faces of a crystal, changing the overall shape of the crystal. One well-known example is that microlites tend to be more acicular than larger crystals (Higgins 2006a). A more relevant example is the commonly observed difference in shape between crystals in massive and foliated rocks of the same composition.

The upper parts of the Sept Iles layered mafic intrusion are dominated by anorthosite, thought to have formed by floatation of plagioclase during the crystallisation

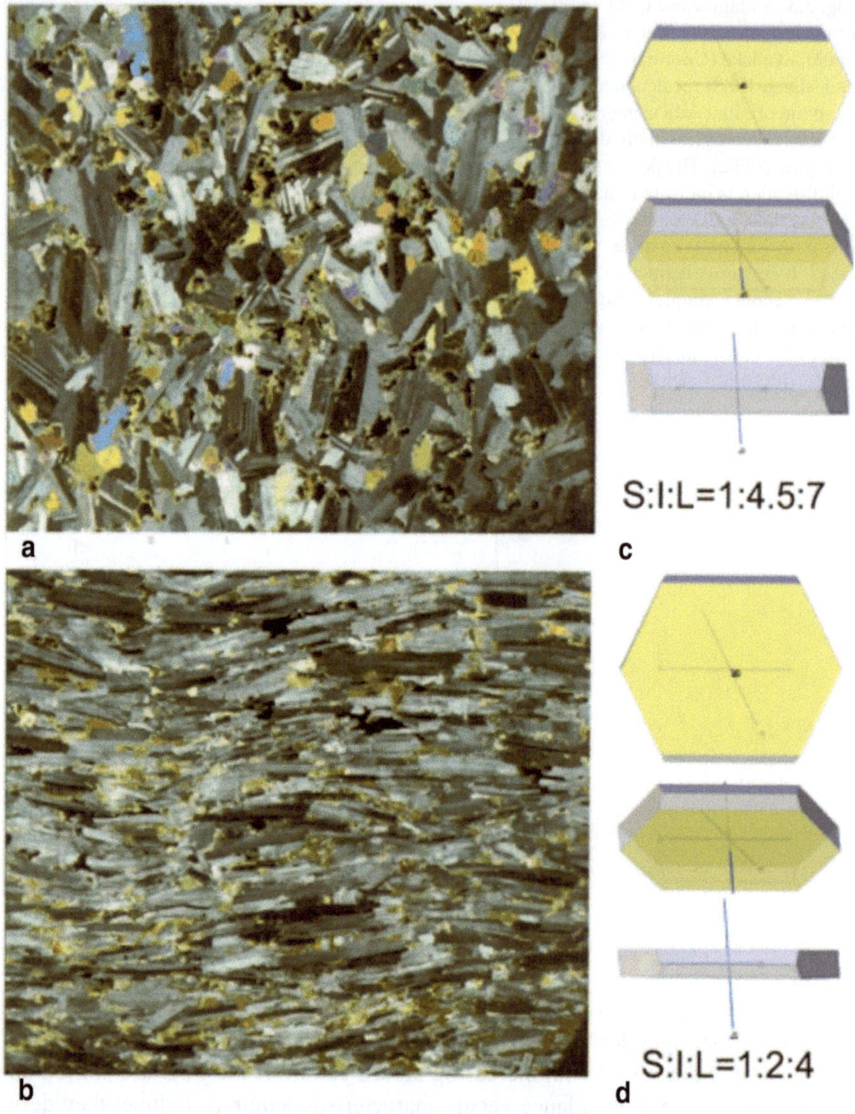

Fig. 3.4 Thin sections of anorthosite from the Sept Iles mafic intrusion in linearly *cross-polarised* light (Higgins 1991) **a** Massive anorthosite. **b** Well-foliated anorthosite sectioned orthogonal to foliation. **c, d** Model of typical plagioclase crystal shapes from the *massive* and *laminated* samples with overall aspect ratios, drawn using WinXMorph (Kaminsky 2005)

of the main layered series (Higgins 2005; Namur et al. 2011). Samples from a well-laminated facies and an adjacent massive facies of the anorthosite have identical compositions and CSD, but their crystal shape is different (Fig. 3.4) (Higgins 1991). This shape variation is attributed to variations during crystallisation in the Peclet number, which is the ratio of mass transport by advection to diffusion (Higgins and

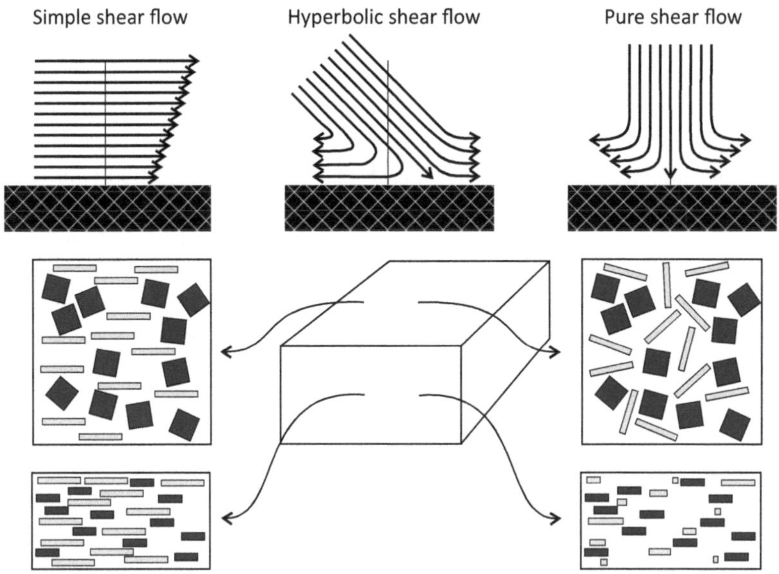

Fig. 3.5 Shear regimes for magma adjacent to a solid surface and idealised fabrics produced for tabular (*dark*) and prismatic (*pale*) crystals in sections *orthogonal* and *parallel* to the foliation. *Intermediate* flow regimes or *crystal* shapes may produce foliated and lineated fabrics. Foliations of tabular crystals cannot be used to distinguish between pure and simple shear. Misalignment of a section that is supposed to be parallel to fabric orientation produces preferential orientations of crystal outlines which are sometimes misinterpreted as a lineation

Chandrasekharam 2007). A high Peclet number will lead to a high chemical potential gradient around the growing crystal and hence a greater dispersion of growth rates for the different faces of the crystal.

Mechanically Modified Textures

Different flow regimes of magma containing non-equant crystals will produce different fabrics which may be preserved in the final rock (Fig. 3.5). Simple shear will rotate tabular crystals to produce a foliation (lamination) and prismatic crystals to produce a lineation. Tabular-prismatic (brick-shaped) crystals will produce a foliated and lineated fabric. Pure shear (coaxial shear; compaction) will produce a foliation for all crystal shapes. Hyperbolic shear flow will produce a foliation and lineation for all non-tabular crystals. Such fabrics can be quantified from measurements of individual crystals in sections orthogonal and parallel to the fabric directions. However, if a section of a foliated rock is not cut completely parallel to the foliation a preferred orientation of the crystal outlines may be interpreted as a lineation. This problem can be avoided if the fabric is quantified using a 3D technique, such as X-ray tomography or anisotropy of magnetic susceptibility (AMS). The latter is treated in another chapter of this book (O'Driscoll, et al.). Measurements of crystal orientations can be used to quantify fabric strength using such parameters as the 'alignment factor' (AF), which varies from zero for massive rocks to one for strong fabrics (Launeau and Cruden 1998; Meurer and Boudreau 1998). For instance, the massive anorthosite in Fig. 3.4A has an AF of 0.20 and the foliated anorthosite in Fig. 3.3B has an AF of 0.87.

Many quantitative studies of fabric in igneous rocks have been determined from the texture of plagioclase, which commonly has a tabular form in plutonic rocks (Higgins 2006a). Hence lineations cannot be recorded by plagioclase, although they can be determined from the fabric commonly prismatic crystals such as pyroxene. AMS can record lineations as well as foliations (O'Driscoll, Ferré, Stevenson and Magee).

Expulsion of liquid and compaction of crystals is an important process. It is generally driven by density differences, but can also be produced by flowage differentiation. Compaction (pure shear) will migrate CSD to higher population densities without changing the overall shape (Higgins 1998). It may also produce a weak foliation orthogonal to the compaction direction, but it is doubtful if this could be detected in most rocks (Higgins 1991). Compaction cannot generally deliver an adcumulate as close packing of the crystals will lead to the retention of an important proportion of trapped liquid, making an orthocumulate. If however compaction is accompanied by pressure-solution, then the crystal shapes will be changed and porosity reduced (Meurer and Boudreau 1998). It is not easy to model this process and it is perhaps best to consider it as akin to compaction with equilibration (see below).

An important process for changing the composition of magmas is fractional crystallisation, which can occur by separation of crystals from liquid by gravity. Crystals will settle or float in their surrounding liquid if (1) there are significant density differences (2) the liquid has Newtonian behaviour or the imposed stress exceeds the yield strength of the liquid (3) The advection velocity of the liquid is less that the separation velocity. The simplest approach to model textural changes during settling or floatation uses Stokes' equation to calculate grain velocity for a Newtonian fluid (Fig. 3.6) (Higgins 2002). This model predicts specific behaviour for crystals with densities greater or less than the fluid for different positions in the layer. Experimental studies of dense suspensions (30%) of olivine in basalt show that descent velocities are much less than those predicted by the Stokes' equation, which is for unhindered crystals in dilute suspensions (Schmidt et al. 2012). Crystal sizes were also measured in these experiments, but unfortunately the data were not treated correctly and it is difficult to draw any firm conclusions as to the effects of settling on CSDs.

Equilibration Textures

Reduction of the total surface energy of the crystals in an igneous rock is called equilibration or relaxation (Higgins 2011). The process starts at the grain boundar-

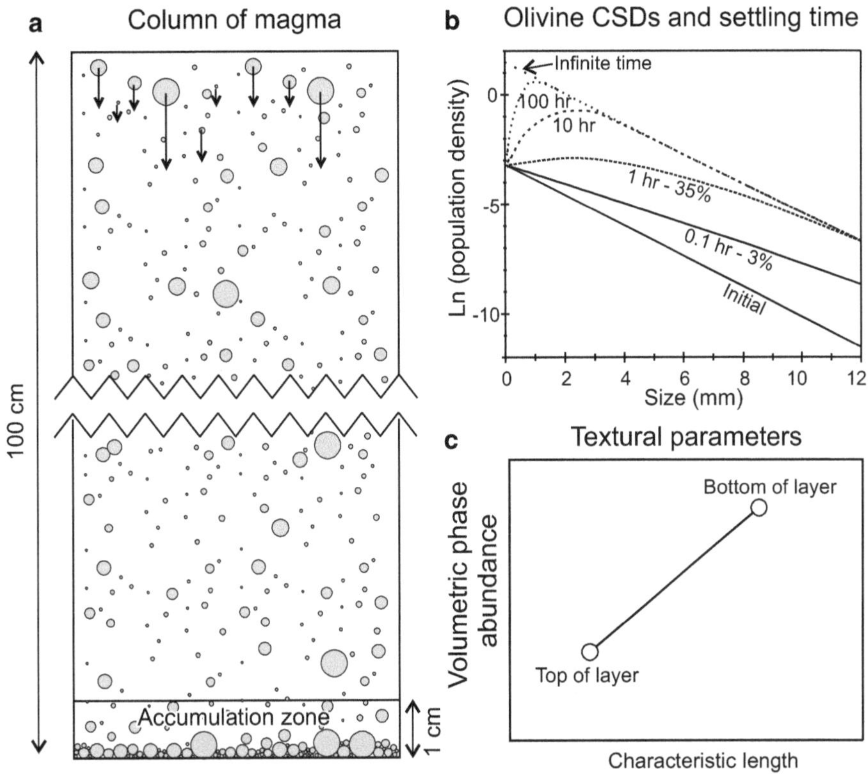

Fig. 3.6 Textural variations produced by settling of grains (Higgins 2002). **a** A population of *grains* with a straight *CSD* is distributed in a column and *grains* are allowed to descend into an accumulation zone at the base. **b** For *grains* denser that the fluid, the *CSD* steepens with time. **c** The characteristic length of the *CSD* varies systematically with height in the layer for crystals denser than the liquid (*olivine*)

ies where dihedral angles adjust to equilibrium values (Fig. 3.7) (The influence of interfacial energies on igneous). The process then continues along the grain boundaries, evening out the curvature. Finally, the greatest amount of energy is generally liberated by coarsening (= annealing, Ostwald ripening): the dissolution of small crystals and the concomitant growth of large crystals. This process occurs close to the mineral liquidus and the speed of the process is considerably enhanced by cycling of the temperature of the magma, even by only a few degrees (Mills and Glazner 2013; Simakin and Bindeman 2008). There have been many formulations of the coarsening process but few observations of the progression of the process in geological systems (Higgins 2011). One study from a layered troctolite body examined plagioclase textures preserved in an olivine oikocryst (Fig. 3.8) (Higgins 1998). A clear pattern of textural readjustment shows loss of small crystals and 'rotation' of the CSD during equilibration.

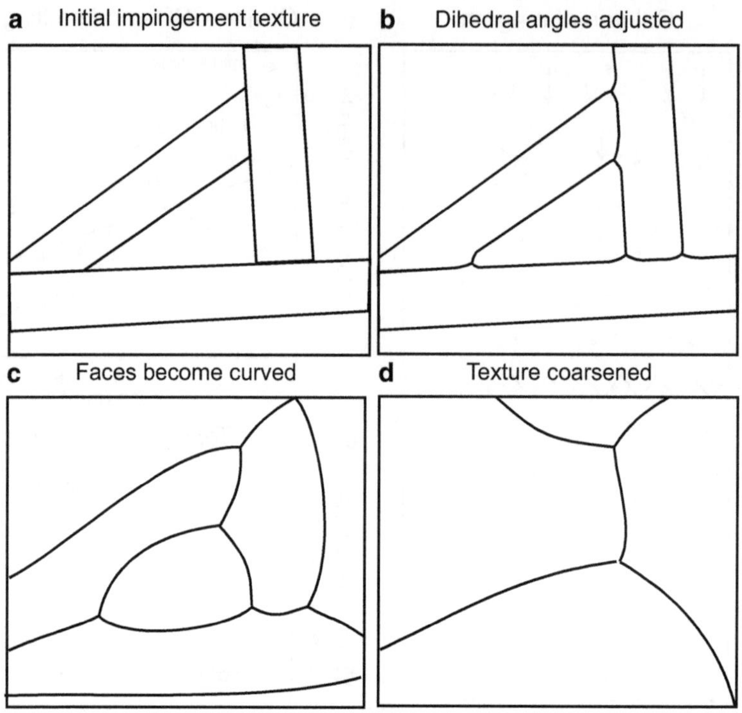

Fig. 3.7 Textural adjustments during equilibration (Higgins 2011). **a** Original kinetic *texture*. **b** *Dihedral* angles adjusted. **c** *Curvature* of faces. **d** *Coarsening*

Equilibration will also affect the shape of crystals which will tend to be transformed to more equant forms so that the surface area per unit volume and hence total surface energy may be minimised (Higgins 1998; Higgins 2011). In the example discussed above the overall aspect ratio of less-equilibrated plagioclase in the oikocryst is 1:3:3, whereas the shape of equilibrated plagioclase is 1:1.5:1.5 (Fig. 3.8). The overall effect of size and shape changes reduced the surface area per unit volume, and hence the surface energy, by 60 % (Higgins 2011).

During equilibration crystals are free to rotate, so that they can accommodate changes in size of their neighbours. Such rotation is random, so that the strength of mineral fabric is reduced. This was quantified for plagioclase crystals in troctolite (Higgins 1998; Fig. 3.8) where the alignment factor changes progressively from 0.73 in the 40 % crystallised material through 0.70 (50 %) and 0.63 (80 %) to 0.28 in the matrix (100 %). %). It has also been described qualitatively for rocks in the Stillwater Intrusion by Meurer et al. (1997).

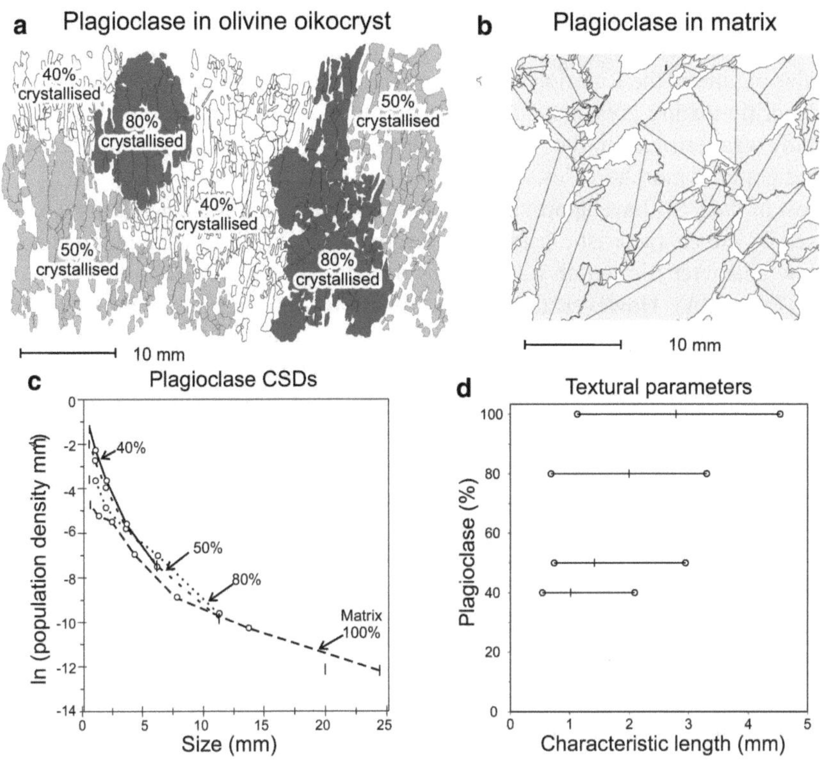

Fig. 3.8 Equilibration of plagioclase in troctolite (Higgins 1998). **a** Plagioclase textures preserved in an *olivine* oikocryst. Division into zones of different plagioclase abundance shows progressive textural development. **b** *Plagioclase* outside the oikocryst is much closer to equilibration. **c** *Plagioclase* CSDs show a progressive development with loss of small crystals and growth of larger crystals. All crystals were quantified; hence the *left* termination is the actual size limit. **d** Segmentation of the *curved* CSD into the characteristic lengths of the *steepest* and *shallowest parts* gives a clearer idea of the textural characteristics of equilibration

Interpretation of Rock Textures in LMI

I will now discuss various quantitative studies of textures in LMI in terms of kinetic, mechanical and equilibration textures. However, the nature of the rocks, and the parameters measured in each study, means that such processes cannot always be separated.

Dendritic Textures

Purely kinetic textures commonly occur in volcanic rocks, but are somewhat rare in a plutonic setting (Higgins 2006a). One possible example is the occurrence of dendritic crystals (crescumulate texture), which are a very minor component of many LMI. O'Driscoll et al. (2007) studied dendritic (= 'harrisitic'), hopper and granular olivine in the Rum Layered Suite (Fig. 3.9). Quantification of the branching dendritic texture presents a difficult stereological problem and for this reason the study was restricted to the simpler dendritic crystals ('skeletal'), as well as hopper and granular morphologies.

Most of the CSD have approximately straight right sides and hence can be paramatised by the characteristic length. The granular CSD tend to have hump shapes and low characteristic length (CL) values that do not correlate with phase abundance (Fig. 3.9A). However, there is a correlation between the maximum crystal size and CL (Fig. 3.9B). The dendritic olivines do not tend to have significant turn-downs for small crystals, and have higher CL and maximum crystal sizes.

O'Driscoll et al. (2007) interpreted the dendritic texture as a primary magmatic texture unmodified by coarsening or compaction. It was produced rapidly (~8 days) following emplacement of thin sheets of picritic magma into the base of the magma chamber. The granular layers formed by the accumulation of olivine, partly derived from broken dendritic crystals. The textural characteristics of the granular samples suggest that coarsening may have involved cycles of equilibration and nucleation/growth. It is difficult to estimate development times for these layers, but they must have accumulated much slower than the dendritic layers.

Lamination in Two Small LMI

Many LMI show significant lamination, but this has been rarely quantified or correlated with other textural parameters. O'Driscoll et al. (2008) have examined samples from two small intrusions using AMS and crystal texture measurements. In both cases AMS and shape-preferred orientations (SPO) are concordant, despite the fact that AMS quantifies magnetite texture and SPO is here based on plagioclase, and to a lesser extent clinopyroxene textures.

In the Ardnamurchan Centre 3 layered intrusion (Scotland) there are significant foliations and lineations that point towards the centre of the intrusion. O'Driscoll et al. (2008) proposed that fabric was largely developed by mechanical movement of unconsolidated crystal mush towards the centre of the intrusion and that the lineation was produced by stretching. If the foliation was produced at the same time then it must have been caused by a component of compaction. There is evidence for equilibration, but in most samples it does not seem to have affected the intensity of the fabric strongly (Fig. 3.10A). The lack of correlation between CL and phase abundance shows that any equilibration that did occur was not accompanied by overall crystal growth (Fig. 3.10B).

Although compositionally similar, the textures in the Druim Hain intrusion (Scotland) are very different from those in the Ardnamurchan Centre 3 layered intrusion. Unmodified impingement textures indicate that late equilibration was not important. AMS and textural measurements reveal that foliations are generally

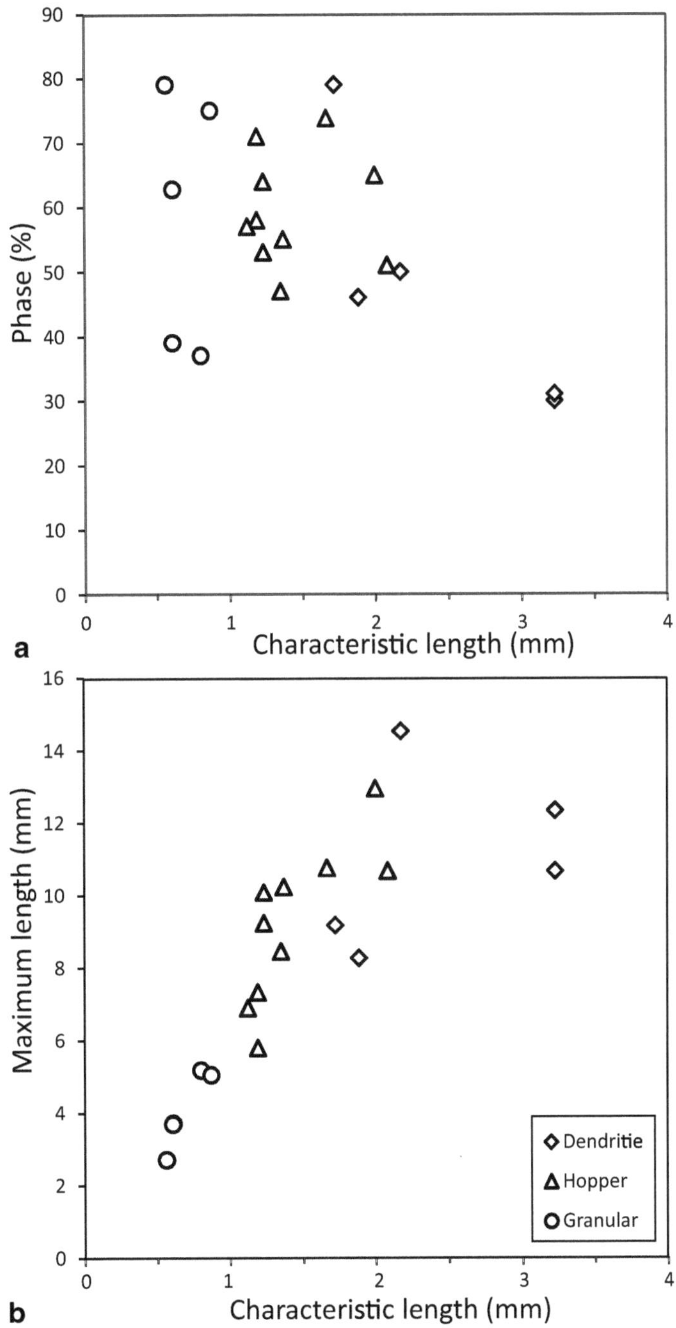

Fig. 3.9 Textural parameters of *dendritic* ('harrisitic'), *hopper* and *granular* olivine from the Rum layered intrusion (O'Driscoll et al. (2007). **a** Phase abundance and characteristic length are not correlated. **b** There is a strong correlation between characteristic length (a measure of the slope) and the size of the largest crystal

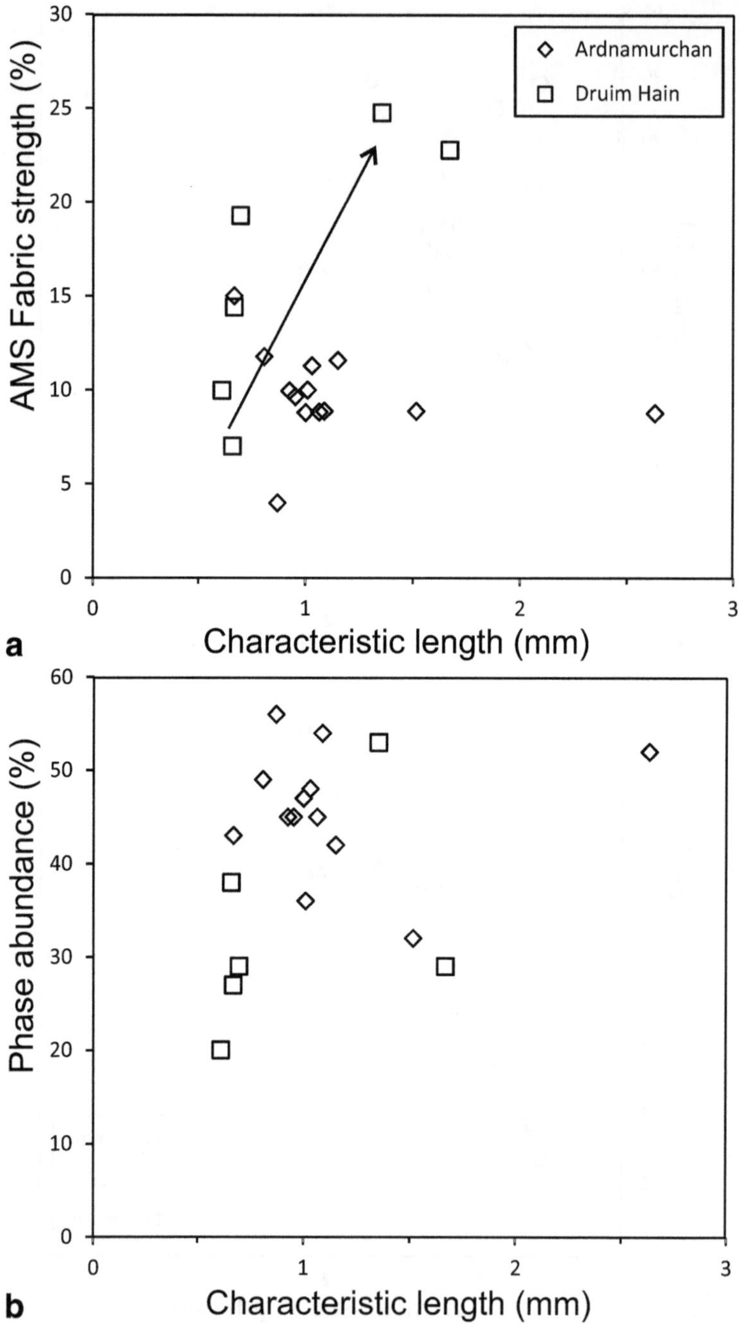

Fig. 3.10 Textural data from the *Ardnamurchan* centre 3 and Druim Hain intrusions. **a** AMS fabric strength versus characteristic length for plagioclase and pyroxene (O'Driscoll et al. 2008). With one exception, all of the *Ardnamurchan* samples, there is no correlation between fabric strength and characteristic length. For *Druim Hain*, the samples with the strongest fabric have the greatest characteristic length. **b** Phase abundance versus characteristic length of plagioclase (O'Driscoll et al. 2008)

stronger, but that lineations are essentially weak or absent. Fabric strength is correlated with CL and possibly with phase abundance (Fig. 3.10). O'Driscoll et al. (2008) considered that the foliation was not produced by mechanical processes, but instead by oriented growth of crystals parallel to planes of equal chemical potential, following the ideas of McBirney and Noyes (1979).

Variations in CL simply reflect the amount of growth in such a model and hence perhaps the rate of advance of the solidification front. Presence of deformed crystals in the coarsest rocks suggested that compaction may have also played a role in the development of the foliation. It will be necessary to examine other, similar intrusions before this novel process can be verified.

Bushveld Intrusion

The Bushveld (South Africa) is the largest LMI in the world (Cawthorn 2015), but there have been very few textural studies. Two papers have examined orthopyroxene textures in the lower zone dominated by orthopyroxenites and in the overlying critical zone marked by the entry of plagioclase: Boorman et al. (2004) looked at a thicker section near Jagdlust and Williams et al. (2006) examined similar rocks in a thinner section near Bergersfort.

In both sections, orthopyroxene CSDs have a hump shape, considered characteristic of coarsening and the CL of the right side can be used as a measure of the degree of coarsening. In the lower zone, CSDs from the thicker section have almost double the CL of those from the thinner section (Fig. 3.11A). This suggests that the degree of coarsening is partly controlled by the length of time available: the thicker section would have cooled more slowly. CSDs in samples from the critical zone are more diverse. Only those from the Jagdlust section are sufficiently numerous for discussion. Those samples with high phase abundances (pyroxenites) resemble the LZ CSD, whereas those with lower phase abundances (gabbros) have lower CL values. This may reflect the effects of a second phase in immobilising (pinning) grain boundaries (Berger and Herwegh 2004).

Quantification of fabric showed that most samples have a well-defined foliation, but that lineation in the plane of the foliation is weak or absent. Both studies reveal a broad correlation between the mean shape of the crystals, as expressed by the ratio of breadth to length, and the quality of foliation (Fig. 3.11B). This is similar to what was observed for plagioclase (Higgins 1991; Higgins and Chandrasekharam 2007) and probably again reflects the role of the Peclet ratio during crystal growth. Hence, some growth appears to have occurred during deformation of the magma. Pyroxenes are more equant in the thicker section (Fig. 3.11B), which may reflect the greater degree of coarsening of these rocks, as shown by their greater vales of CL.

The weak lineation of pyroxene, which presumably has a prismatic form, was used to suggest that the fabric was mostly formed by pure shear (Boorman et al. 2004; Williams et al. 2006). However, it is also possible that coarsening weakened any lineation and that a hyperbolic shear regime is more appropriate.

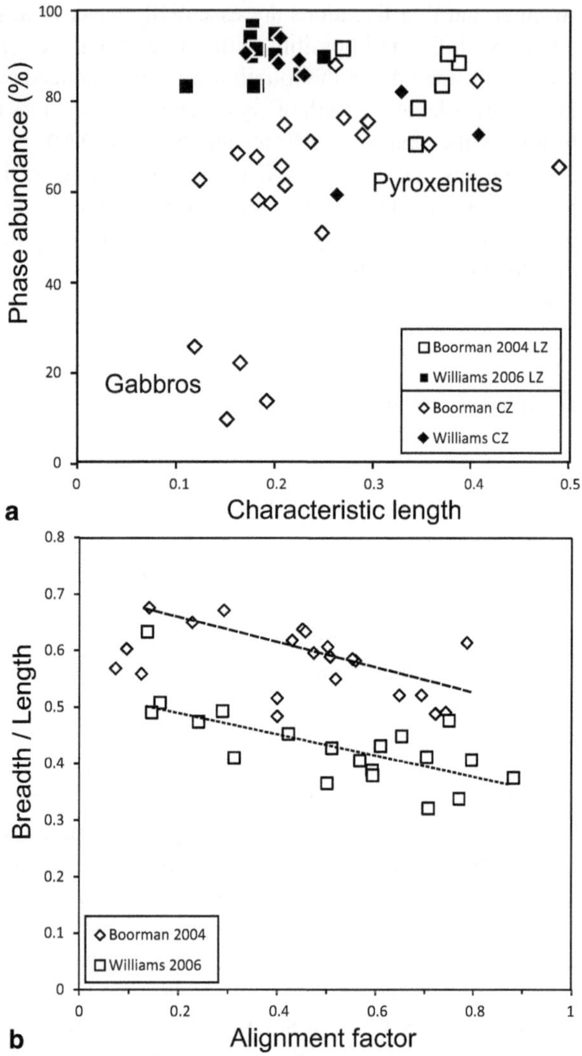

Fig. 3.11 Textural parameters for orthopyroxene from the lower and critical zones of the Bushveld intrusion. Data were compiled from a *thicker* section near Jagdlust (Boorman et al. 2004) and a *thinner* section near Bergersfort (Williams et al. 2006). **a** Phase abundance versus characteristic length. **b** Intersection breadth/length, a characteristic of the overall crystal aspect ratio, versus alignment factor, a measure of the quality of foliation

Kiglapait Intrusion

The Kiglapait intrusion (Labrador) is a medium sized body with a very regular bowl shape (Morse 1969; Morse 2015). Most of the magma is thought to have been injected in a single event, giving a somewhat simpler history than other LMI intrusions. The textures of all major phases have been determined for a complete section of the intrusion (Higgins 2002). CSD of plagioclase, olivine and pyroxene all have a turn-down for small crystals characteristic of coarsening. The right sides are sufficiently straight that the characteristic length could be calculated (Fig. 3.12). Plagioclase characteristic lengths in the layered rocks at the base are small, but just above

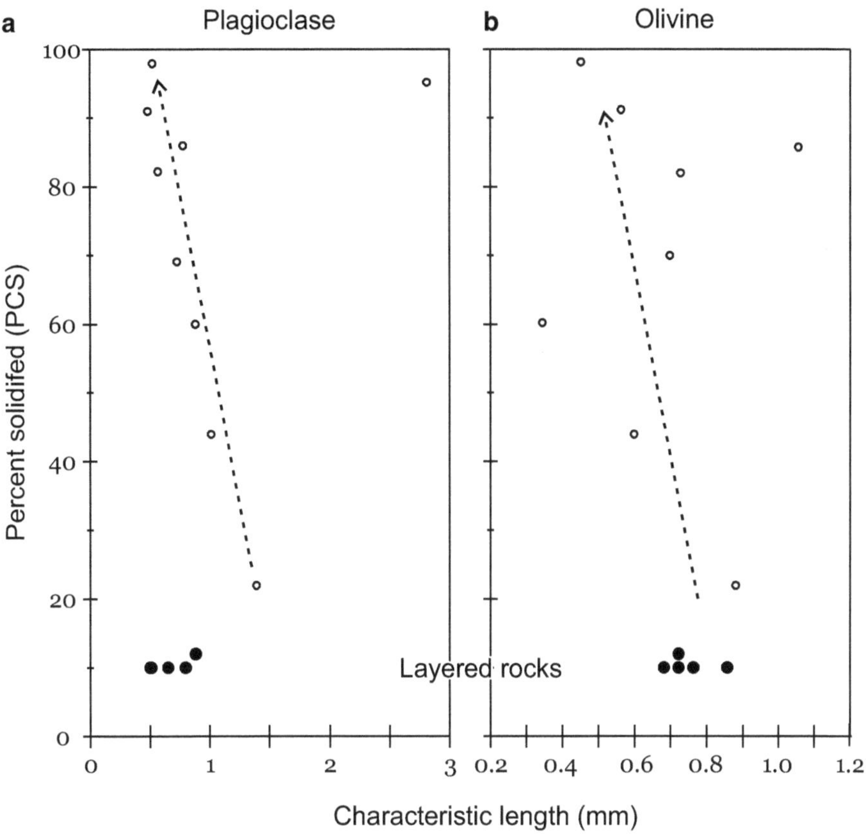

Fig. 3.12 Characteristic length of CSD of **a** *plagioclase*, and **b** *olivine* versus solidification in the Kiglapait intrusion (Higgins 2002)

they jump to the highest values, which then decrease regularly with solidification, except for one sample near the top. The pattern for olivine characteristic lengths is much less regular, but seems to display a decrease with solidification.

If CL values are interpreted in terms of coarsening, then the layered rocks appear to be less coarsened than the overlying massive rocks. However, examination of a single layer discussed below does suggest that the layered rocks are indeed coarsened. The high CL just above the layered rocks suggests that coarsening was at a maximum here, because cooling was slowest at this time. As the body solidified, the lower volume of the magma enabled faster cooling, and hence less time for coarsening. Although plagioclase and olivine appear to have nucleated and initially grown simultaneously, the regular variation in CL of the plagioclase suggests that this mineral controlled textural development during coarsening. Olivine crystals just accommodated themselves around the growing plagioclase crystals.

Textural variations were investigated within a single layer at the base of the Kiglapait intrusion (Higgins 2002). Olivine and plagioclase crystal outlines were

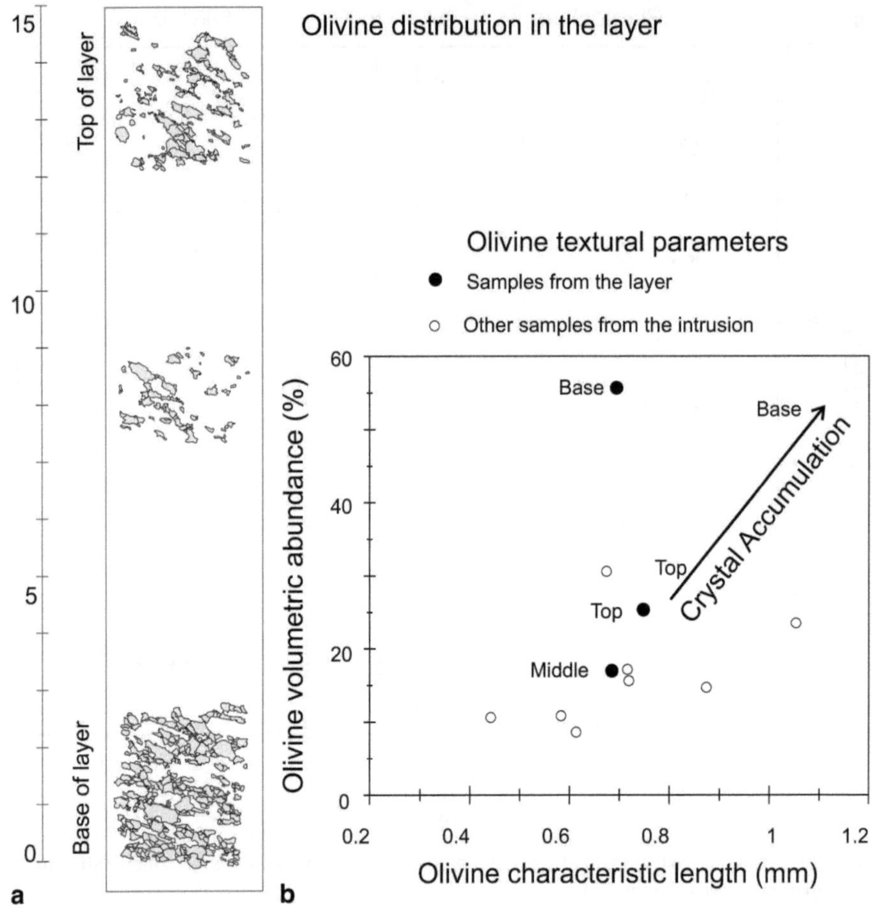

Fig. 3.13 Troctolite layer from the Kiglapait intrusion (Higgins 2002). **a** Olivine crystal outlines. **b** Olivine textural parameters

digitised in an 18 cm section starting at the olivine-rich base of the layer and continuing up into the base of the overlying layer (Fig. 3.13a). This layer is thought to have formed from a crystal-charged density current that swept down the walls and onto the floor of the magma chamber. There, olivine sank to the bottom whilst plagioclase was carried upwards by flow that balanced the descending olivine. These processes are partly controlled by crystal size and hence the CSD should also change. Simple modelling of textural changes during such a process (Fig. 3.6) suggests that at the base of the layer olivine should be abundant and have a shallow CSD (high CL). The abundance of olivine should decrease with increasing height and the CSD should become steeper (low CL). Although olivine abundances broadly fit a simple 'sedimentation' model, their CSD do not (Fig. 3.13b). Olivine subsamples from the base, middle and top of the layer all have very similar CSD. Hence if the layer was

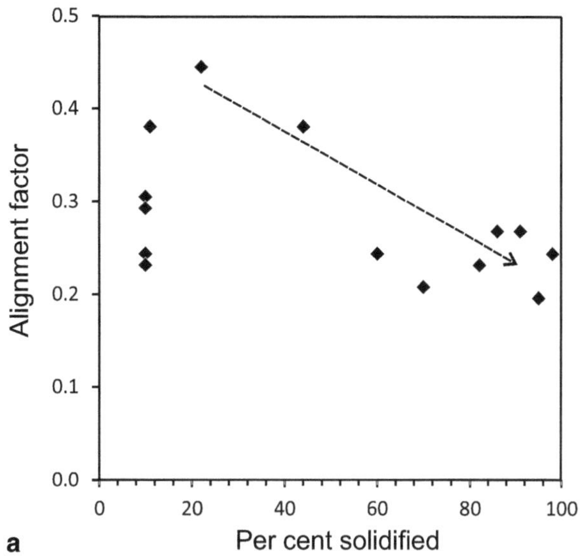

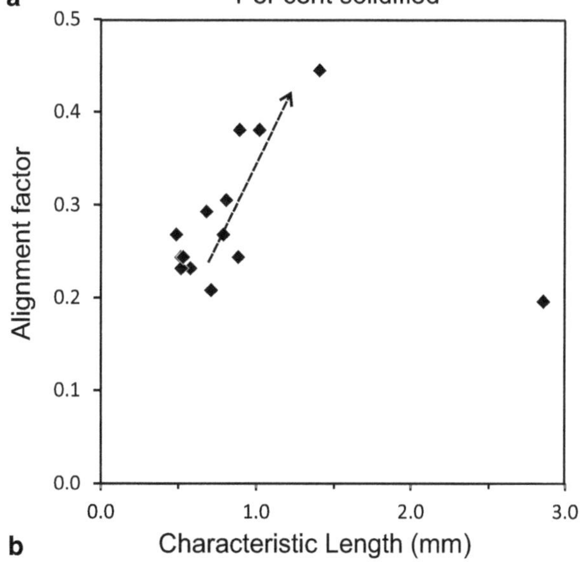

Fig. 3.14 Quantification of plagioclase textures in the Kiglapait intrusion. **a** Alignment factor (quality of foliation) versus percent solidified. **b** Alignment factor versus characteristic length of the CSD (Higgins 2002)

produced by crystal settling, then the CSD have been modified after the creation of the layer. The most likely responsible process is coarsening.

All samples from the Kiglapait intrusion have moderately strong foliations of plagioclase crystals (Fig. 3.14a). The layered section at the base seems to have some of the weakest foliations. The foliation is strongest just above this zone, then diminishing linearly with height in the intrusion. There is a correlation between foliation strength and degree of coarsening, as expressed by the characteristic length, except for one anomalously coarsened sample (Fig. 3.14b).

Variations in foliation strength can be interpreted in terms of convection vigour and other processes. The lower layered section would have been produced by sporadic currents that could have produced a strong foliation. However, descent of olivine crystals and displacement of plagioclase upwards may have reduced the quality of the foliation. Up section these would have been replaced by stronger, continuous currents, which slowly diminished in strength as the intrusion solidified. Coarsening does not seem to have reduced foliation strength, as was observed in the Lac-St-Jean anorthosite (Higgins 1998). It is possible that compaction was accompanied by pressure-solution (Meurer and Boudreau 1998), which would have maintained or even enhanced fabric strength.

Lilloise Layered Intrusion

The Lilloise mafic layered intrusion (Greenland) closely resembles the Kiglapait intrusion at a smaller scale (Magee et al. 2010). It is thought to represent a single filling event and the only complexity appears to be a period of subsidence and deformation during the solidification of the middle zone. The textures of all major minerals have been quantified in eight samples from a 2.8 km section (Fig. 3.15).

All samples lack crystals smaller than 0.1 mm and many have turn-downs for crystals slightly larger than this limit. This suggests that all samples and all minerals have been coarsened, but the degree and timing seems to be quite variable. Plagioclase CSD are curved so the overall CL values are not very informative. The left sides of the CSD are all similar, but the right sides flatten as they extend to greater sizes. A similar behaviour for plagioclase was observed in andesites from Montserrat, where it was interpreted to form in response to cycles of kinetic growth and equilibration (Higgins and Roberge 2003). The sequence of textural development is preserved by plagioclase chadocrysts in amphibole, as was also observed at Montserrat. Three of the four olivine samples have similar CSD, but the fourth is displaced upwards (Fig. 3.15a). The similarity of the right-side slope of all olivine samples suggests that coarsening may have been quite early and a subsequent period of kinetic growth of all crystals in one sample may have increased the overall olivine content, but not changed the shape of the CSD. Most clinopyroxene samples display a more common correlation of CL and phase abundance. However, the most clinopyroxene-rich sample lies above, but parallel to the most coarsened sample, suggesting again a period of kinetic growth of all crystals following coarsening. Only two samples of amphibole were quantified, but they have a curved CSD and again cycles of kinetic growth and equilibration may be involved.

The CSD parameters were also plotted against height in the intrusion (Fig. 3.15b). There is no clear and simple variation with height for any of the minerals, as was observed in the Kiglapait intrusion (Higgins 2002). This may reflect the low number of samples, but it may also be due to the perturbation of the intrusion during solidification (Magee et al. 2010). This may have opened up fluid circulation channels that would have led to focused coarsening.

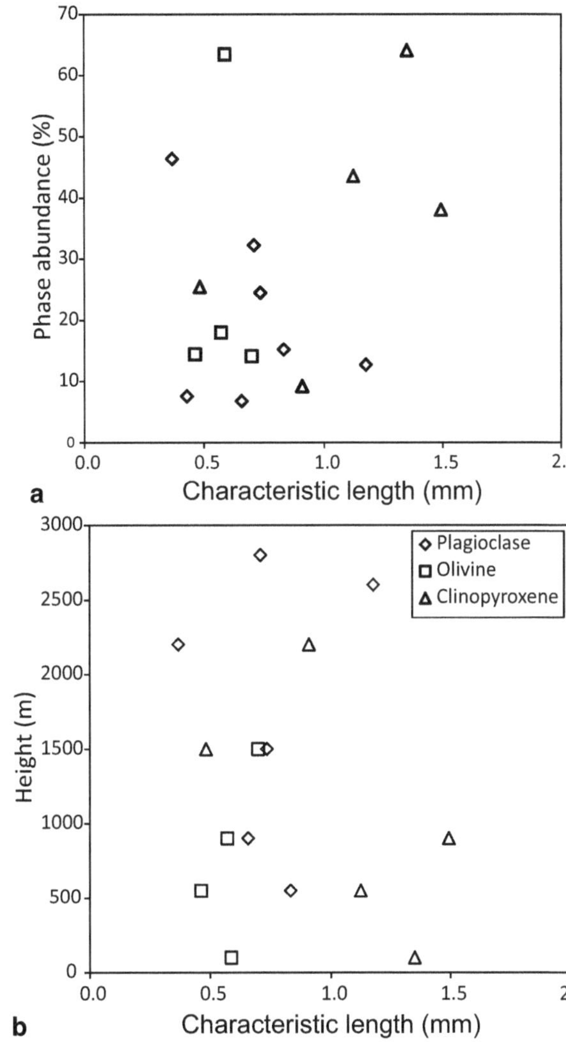

Fig. 3.15 Textural characteristics of major minerals of the Lilloise layered intrusion (Magee et al. 2010). Plagioclase CSD are *curved* and the overall CL parameter used here is probably not representative

Oman Ophiolite

The Oman ophiolite is one of the best preserved and least deformed sections of oceanic crust available on land. The section starts in the mantle and crosses the Moho into the layered (lower) gabbros which make up the lowest section of the oceanic crust. Layering is parallel to the Moho, which suggests a link with LMI. This transitions upwards into foliated and lineated gabbros (upper and middle gabbros), which have structures orthogonal to the Moho and parallel to the sheeted dyke complex. These are overlain by texturally variable 'High-level' gabbros, which transition abruptly into the sheeted dyke complex.

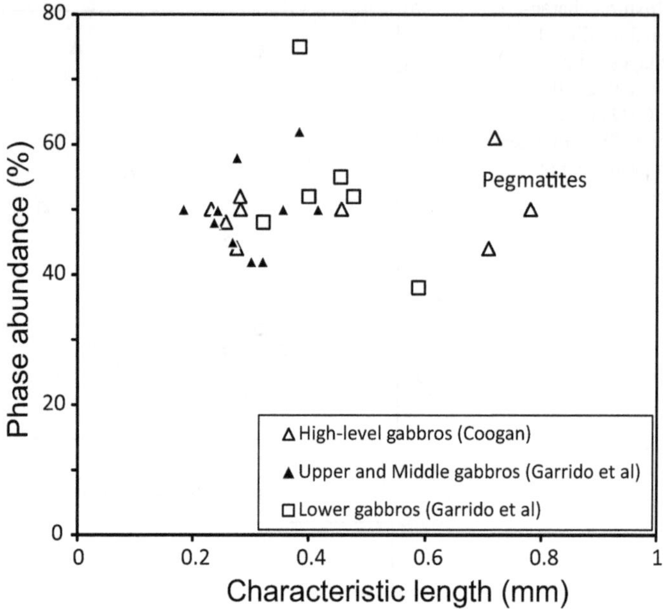

Fig. 3.16 Textural parameters of plagioclase in gabbros from the Oman ophiolite (Coogan et al. 2002; Garrido et al. 2001)

Two studies have examined plagioclase textures of gabbros in this body: Garrido et al. (2001) determined CSD from the whole section of gabbros and Coogan et al. (2002) only examined textures in the high level gabbros (Fig. 3.16). It is not very easy to combine these two studies as the methodology was slightly different, but they tell the same broad story. All samples had concave down shapes, with straight right-hand limbs, which are characteristic of textures produced by coarsening. Garrido et al. (2001) found that the lowest 2 km of the section was more coarsened than the upper 2.5 km (Fig. 3.16). Coogan et al. (2002) found broadly similar results for the upper gabbros, except for three 'pegmatite' samples, which had greater characteristic lengths. The overall story seems to be that slower cooling at the base of the crust enabled coarsening of the gabbros, while a small component of melt was still present. The upper gabbros were generally less coarsened except in some places, probably in response to focussed heat or silicate liquid flow that maintained the rocks at high temperatures. Such focussing of late liquids has been proposed as an origin for coarsened zones in megacrystic granites (Higgins 1999).

Chromite Layers in LMI

The origin of chromite seams in LMI is of both economic and petrogenetic importance. Textural quantification of chromite is relatively easy and the strong contrast

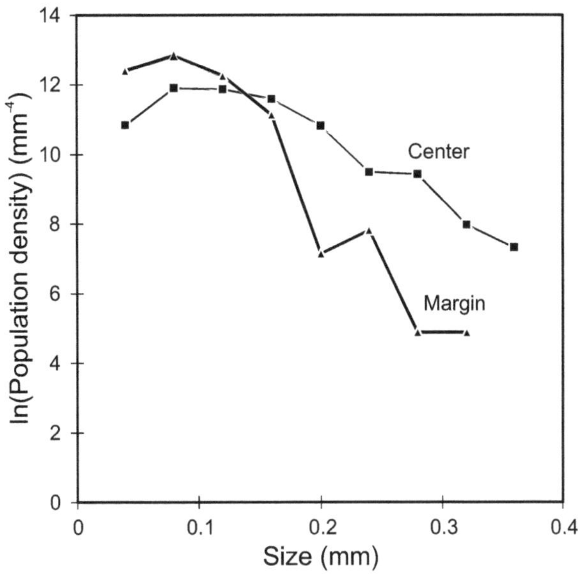

Fig. 3.17 CSD of chromite in a 1-cm-thick layer from the Stillwater intrusion. (Recalculated from Waters and Boudreau 1996)

between the density of chromite and mafic magmas means that the textural consequences of crystal settling should be very evident.

One of the earliest studies of crystal sizes was by Jackson (1961) of chromite and olivine in the ultramafic zone of the Stillwater intrusion (Montana). Unfortunately, the areas that were measured were not recorded and hence the data cannot be converted to more modern ways of displaying crystal size distributions. However, mean crystal sizes can be calculated from his data and these were used by Boudreau (1994) to show that crystal size varied between adjacent layers in a way consistent with layer formation by aging (equilibration).

Waters and Boudreau (1996) examined in detail a 1-cm-thick layer (Fig. 3.17). The layer was divided into central and marginal sections. The CSD of both parts had a down-turn for small crystals that is characteristic of coarsening, but the centre was much more coarsened than the margins. This was also observed in the Dawros intrusion, Ireland (Hunt et al. 2012). This is consistent with a geometrically controlled coarsening model (Dehoff 1991), as the crystals at the centre are much closer to each other. There was no textural evidence of settling, which may have been effaced by coarsening as was seen at Kiglapait (Fig. 3.13)

Chromite occurs in the Rum Layered Suite (Scotland) and has been studied in detail by O'Driscoll et al. (2009, 2010) (Fig. 3.18). The chromite occurs close to the boundaries of major cyclic units, which are interpreted to represent repeated injections of mafic and ultramafic magmas. Chromite occurs in several different contexts, all of which were investigated texturally for two different cyclic boundaries. The 'main seams' are 2-4 mm thick and lie along the boundary of the cyclic unit. The CSD have turn-downs for small crystals and straight right sides. There is little difference between the CSD of the top, middle and bottom of the seams, unlike

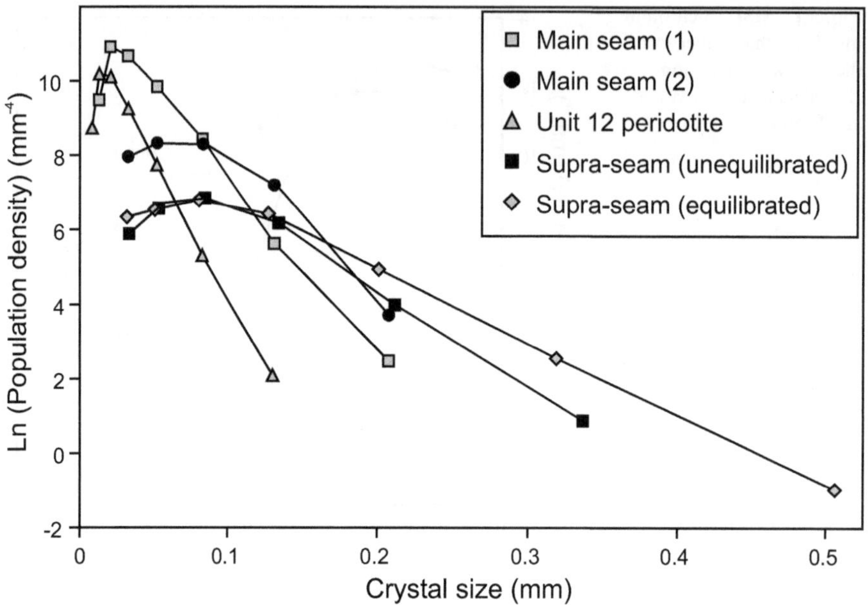

Fig. 3.18 CSD of chromite in the Rum Layered Suite at the cycle 11-12 boundary. (O'Driscoll et al. 2010)

the findings of Waters and Boudreau (1996) and Hunt et al. (2012). Different main seams and different parts of the same main seam have CSD with different maxima and right slopes. These textural characteristics exclude any evidence of crystal settling and indicate that the seam was coarsened to variable degrees.

Other chromite seams occur a few cm above and below the main seam in the overlying peridotite. A supra-seam has CSD that indicates a higher, but variable, degree of coarsening than the main seams. The most coarsened sample also has compositionally uniform grains, showing a link between chemical equilibration and coarsening. A subsidiary seam is less coarsened than the overlying main seam (O'Driscoll et al. 2009). A sample of minor, dispersed chromite in the overlying peridotite has the least coarsened CSD (O'Driscoll et al. 2010), but another similar sample is more coarsened than the adjacent seam (O'Driscoll et al. 2009).

O'Driscoll et al. (2010) interpret these textures as showing that chromite in the main seam crystallised *in situ*, as a result of the dissolution of existing cumulates by an influx of new hotter magma. Some of the chromite was injected into the peridotite to form a supra-seam or mixed to give the dispersed chromites. It is possible that the latter preserved the early texture of the chromite as the low concentration of this mineral in the peridotite retarded coarsening. In the seams, the greater concentration of chromite meant that the crystals were closer together and hence could coarsen more rapidly. The more extreme coarsening seen in some parts of the supra-seam may reflect temperature cycling produced by the influx of magma or possibly the greater abundance of olivine, which can exchange with chromite to relatively low temperatures. The dispersed chromite around the cycle 7-8 boundary is more coarsened than the one in the subsidiary seam, but less coarsened than the main seam.

All these variations in CSD reflect the variable degree of coarsening following crystallisation of chromite. The concentration of chromite is certainly one factor, with high concentrations favouring coarsening, as was also found by Waters and Boudreau (1996), but there are clearly other factors. O'Driscoll et al. (2010) suggest that pressure-solution compaction may also be a factor (Meurer and Boudreau 1998): This will depend on local conditions of strain, which could vary rapidly with position.

Summary

There is quantitative textural evidence for kinetic, mechanical and equilibration processes: Their extent and timing give the range of textures observed in LMI. Although all rocks must have passed through a stage of kinetically-controlled textures, it is not generally visible as it has been commonly effaced by equilibration. However, some rocks in small intrusions still retain impingement textures, which are evidence of pure kinetic control. Dendritic textures (crescumulates) are another example of pure kinetic textures, in this case represented by extreme growth.

Some mechanical processes are preserved in LMI rocks. Compositionally layered rocks preserve evidence for crystal settling, but it is not reflected by the CSD. This is probably due to subsequent equilibration of the textures. Foliations were produced while a liquid component was still present by simple shear (flow) or pure shear (compaction) accompanied by pressure-solution. Growth of crystals during deformation is recorded by a correlation between crystal aspect ratios and foliation strength.

The most important textural process that has affected almost all rocks in LMI is equilibration, although its extent and timing are vary variable. The role of this process is clearly documented by hump-shaped CSD; its progression can be monitored by decreasing slopes (increasing characteristic lengths) of the right-side of the CSD.

Many of the textural processes active in LMI have only been observed in single studies, hence it is difficult to verify their significance. Clearly this field would benefit from further studies, particularly of outcrop-scale structures like mineral layering. Recent technological advances in quantification in both two and three dimensions have considerably sped up data acquisition and should enable more textural studies to be completed (Table 3.1).

Table 3.1 Compilation of quantitative textural studies of layered mafic intrusions and related rocks

Unit name	Composition	Minerals quantified	Reference	Notes
Stillwater, USA	Ultramafic-mafic	Ol, Opx, Chromite	(Jackson 1961)	No areas or shapes recorded—cannot reconvert to CSD
Stillwater, USA	Ultramafic-mafic	Chromite	(Waters and Boudreau 1996)	
Dawros, Ireland	Ultramafic-mafic	Chromite	(Hunt et al. 2012)	
Kiglapait, Canada	Gabbro	Plag, Ol, Cpx	(Higgins 2002)	
Lilloise, Greenland	Gabbro	Plag, Ol, Cpx, Amph	(Magee et al. 2010)	
Ardnamurchan, UK	Gabbro	Plag, Cpx	(O'Driscoll et al. 2008)	Laminated gabbro
Druim Hain, UK	Gabbro	Plag, Cpx	(O'Driscoll et al. 2008)	
Rum, UK	Gabbro	Chromite	(O'Driscoll et al. 2010)	
Rum, UK	Gabbro	Ol	(O'Driscoll et al. 2007)	
Bushveld, South Africa	Ultramafic-mafic	Plag	(Maier 1994)	Upper Critical Zone. Only the largest 30 grains measured—cannot calculate CSD
Bushveld, South Africa	Ultramafic-mafic	Plag, Opx	(Boorman et al. 2004)	Lower Zone—Critical Zone Transition—thickest section
Bushveld, South Africa	Orthopyroxenite	Opx	(Williams et al. 2006)	Lower Zone—Critical Zone Transition—thinner section
Oman ophiolite, Oman	Gabbro	Plag	(Coogan et al. 2002)	High-level gabbros
Oman ophiolite, Oman	Gabbro	Plag	(Garrido et al. 2001)	Complete gabbro section; Incorrect plag shape used for corrections
Sept Iles, Canada	Anorthosite	Plag	(Higgins 1991)	One massive and one laminated sample. Data reconverted in Higgins (2006a)
Lac-St-Jean anorthosite suite, Canada	Troctolite	Plag	(Higgins 1998)	Alma sector. Data reconverted in Higgins (2006a)

Table 3.1 (continued)

Unit name	Composition	Minerals quantified	Reference	Notes
Sudbury, Canada	Norite	Plag	(Zieg and Marsh 2002)	Not a true LMI: No CSD graphs in paper
Ferrar Dolerites, Antarctica	Gabbro	Plag	(Zieg and Marsh 2002)	Not a true LMI: No CSD graphs in paper
Great Dyke, Zimbabwe	Ultramafic	Chromite	(Blenkinsop and Fernandes 2000)	Igneous texture modified by cataclasis
Makaopuhi lava lake	Gabbro	Plag, …	(Cashman and Marsh 1988)	Not a true LMI. Data reconverted in Higgins (2006a)

Acknowledgements I would like to thank NSERC (Canada) for supporting my research over the last 25 years. Lawrence Coogan and Alan Boudreau kindly supplied original data. Judit Ozoray, Alan Boudreau, Brian O'Driscoll and Christian Tegner offered helpful comments.

References

Baker DR, Mancini L, Polacci M, Higgins MD, Gualda GAR, Hill RJ, Rivers ML (2012) An introduction to the application of X-ray microtomography to the three-dimensional study of igneous rocks. Lithos 148:262–276

Berger A, Herwegh M (2004) Grain coarsening in contact metamorphic carbonates: effects of second-phase particles, fluid flow and thermal perturbations. J Metamorph Geol 22(5):459–474

Blenkinsop TG, Fernandes TRC (2000) Fractal characterization of particle size distributions in chromitites from the Great Dyke, Zimbabwe. Pure Appl Geophys 157(4):505–521

Boorman S, Boudreau A, Kruger FJ (2004) The lower zone-critical zone transition of the Bushveld Complex: a quantitative textural study. J Petrol 45(6):1209–1235

Boudreau AE (1994) Crystal aging in two component, two crystal systems. S Afr J Geol 97:473–485

Cashman KV, Marsh BD (1988) Crystal size distribution (CSD) in rocks and the kinetics and dynamics of crystallisation II. Makaopuhi lava lake. Contrib Mineral Petrol 99:292–305

Cawthorn RG (2015) The Bushveld Complex, South Africa. In: Charlier et al (eds) Layered intrusions. Sprinegr Geology. doi:10.1007/978-94-017-9652-1_12

Coogan LA, Thompson G, MacLeod CJ (2002) A textural and geochemical investigation of high level gabbros from the Oman ophiolite: implications for the role of the axial magma chamber at fast-spreading ridges. Lithos 63:67–82

Dehoff RT (1991) A geometrically general theory of diffusion controlled coarsening. Acta Metall Mater 39(10):2349–2360

Garrido CJ, Kelemen PB, Hirth G (2001) Variation of cooling rate with depth in lower crust formed at an oceanic spreading ridge; plagioclase crystal size distributions in gabbros from the Oman ophiolite. Geochem Geophys Geosyst 2:2000GC000136

Higgins MD (1991) The origin of laminated and massive anorthosite, Sept Iles intrusion, Quebec, Canada. Contrib Mineral Petrol 106:340–354

Higgins MD (1998) Origin of anorthosite by textural coarsening: quantitative measurements of a natural sequence of textural development. J Petrol 39:1307–1325

Higgins MD (1999) Origin of megacrysts in granitoids by textural coarsening: a crystal size distribution (CSD) study of microcline in the Cathedral Peak granodiorite, Sierra Nevada, California. In: Fernandez C, Castro A (eds) Understanding granites: integrating modern and classical techniques special publication (Vol. 158). Geological Society of London, London, pp. 207–219
Higgins MD (2000) Measurement of crystal size distributions. Am Mineral 85:1105–1116
Higgins MD (2002) A crystal size-distribution study of the Kiglapait layered mafic intrusion, Labrador, Canada: evidence for textural coarsening. Contrib Mineral Petrol 144:314–330
Higgins MD (2005) A new model for the structure of the Sept Iles Intrusive suite, Canada. Lithos 83:199–213
Higgins MD (2006a) Quantitative textural measurements in igneous and metamorphic petrology. Cambridge University Press, Cambridge
Higgins MD (2006b) Use of appropriate diagrams to determine if crystal size distributions (CSD) are dominantly semi-logarithmic, lognormal or fractal (scale invariant). J Volcanol Geotherm Res 154:8–16
Higgins MD (2011) Textural coarsening in igneous rocks. Int Geol Rev 53(3):354–376
Higgins MD, Chandrasekharam D (2007) Nature of sub-volcanic magma chambers, Deccan province, India: evidence from quantitative textural analysis of plagioclase megacrysts in the Giant Plagioclase Basalts. J Petrol 48:885–900
Higgins MD, Roberge J (2003) Crystal size distribution (CSD) of plagioclase and amphibole from Soufriere Hills volcano, Montserrat: evidence for dynamic crystallisation/textural coarsening cycles. J Petrol 44:1401–1411
Higgins MD, Roberge J (2007) Three magmatic components in the 1973 eruption of Eldfell volcano, Iceland: evidence from plagioclase crystal size distribution (CSD) and geochemistry. J Volcanol Geotherm Res 161:247–260
Hunt E, O'Driscoll B, Daly JS (2012) Parental magma composition of the syntectonic Dawros Peridotite chromitites, NW Connemara, Ireland. Geol Mag 149(4):590
Jackson ED (1961) Primary textures and mineral associations in the ultramafic zone of the Stillwater complex, Montana. U.S. Geol Surv Prof Pap 358:106
Kaminsky W (2005) WinXMorph: a computer program to draw crystal morphology, growth sectors and cross sections with export files in VRML V2.0 utf8-virtual reality format. J Appl Crystallogr 38:566–567
Kile DE (2003) The petrographic microscope: evolution of a mineralogical research instrument. Mineralogical Record, Inc., Tucson
Lane AC (1898) Geological report on Isle Royale, Michigan. Michigan Geological Survey, Lansing
Launeau P, Cruden AR (1998) Magmatic fabric acquisition mechanisms in a syenite: results of a combined AMS and image analysis study. J Geophys Res 103:5067–5089
Magee C, O'Driscoll B, Chambers AD (2010) Crystallization and textural evolution of a closed-system magma chamber: insights from a crystal size distribution study of the Lilloise layered intrusion, East Greenland. Geol Mag 147(3):363–379
Maier WD (1994) Grain-size variations of Cumulus plagioclase in the upper critical zone of the Bushveld Complex. Mineral Petrol 50(4):245–258
Marsh BD (1988) Crystal size distribution (CSD) in rocks and the kinetics and dynamics of crystallization I. Theory. Contrib Mineral Petrol 99:277–291
Marsh BD (1998) On the interpretation of crystal size distributions in magmatic systems. J Petrol 39:553–600
McBirney AR, Noyes RM (1979) Crystallisation and layering of the Skaergaard intrusion. J Petrol 20:487–554
Merriam DF (2004) The quantification of geology: from abacus to Pentium: a chronicle of people, places, and phenomena. Earth-Sci Rev 67(1–2):55–89
Meurer WP, Boudreau AE (1998) Compaction of igneous cumulates; Part II, Compaction and the development of igneous foliations. J Geol 106(3):293–304

Meurer WP, Klaber S, Boudreau AE (1997) Discordant bodies from Olivine-bearing zones III and IV of the Stillwater complex, Montana—evidence for post-cumulus fluid migration and reaction in layered intrusions. Contrib Mineral Petrol 130:81–92
Mills RD, Glazner AF (2013) Experimental study on the effects of temperature cycling on coarsening of plagioclase and olivine in an alkali basalt. Contrib Mineral Petrol 166(1):97–111
Morse SA (1969) The Kiglapait layered intrusion, Labrador. GSA Memoir 112:204
Morse SA (2015) Kiglapait intrusion, Labrador. In: Charlier et al (eds) Layered Intrusions. Springer
Namur O, Charlier B, Pirard C, Hermann J, Liégeois J-P, Auwera JV (2011) Anorthosite formation by plagioclase flotation in ferrobasalt and implications for the lunar crust. Geochim Cosmochim Acta 75(17):4998–5018
O'Driscoll B, Donaldson CH, Troll VR, Jerram DA, Emeleus CH (2007) An origin for harrisitic and granular olivine in the Rum Layered Suite, NW Scotland: a crystal size distribution study. J Petrol 48(2):253–270
O'Driscoll B, Stevenson CE, Troll VR (2008) Mineral lamination development in layered gabbros of the British palaeogene igneous province: a combined anisotropy of magnetic susceptibility, quantitative textural and mineral chemistry study. J Petrol 49(6):1187–1221
O'Driscoll B, Donaldson CH, Daly JS, Emeleus CH (2009) The roles of melt infiltration and cumulate assimilation in the formation of anorthosite and a Cr-spinel seam in the Rum Eastern layered intrusion, NW Scotland. Lithos 111(1–2):6–20
O'Driscoll B, Emeleus CH, Donaldson CH, Daly JS (2010) Cr-spinel Seam Petrogenesis in the Rum Layered Suite, NW Scotland: cumulate assimilation and in situ crystallization in a deforming crystal mush. J Petrol 51(6):1171–1201
O'Driscoll B, Ferré EC, Stevenson CTE, Magee C (2015) The significance of magnetic fabrics in layered mafic-ultramafic intrusions. In: Charlier et al (eds) Layered intrusions and their ore deposits. Springer Geology
Randolph AD, Larson MA (1971) Theory of particulate processes. Academic, New York
Schmidt MW, Forien M, Solferino G, Bagdassarov N (2012) Settling and compaction of olivine in basaltic magmas: an experimental study on the time scales of cumulate formation. Contrib Mineral Petrol 164(6):959–976
Simakin AG, Bindeman IN (2008) Evolution of crystal sizes in the series of dissolution and precipitation events in open magma systems. J Volcanol Geotherm Res 177(4):997–1010
Waters C, Boudreau AE (1996) A reevaluation of crystal size distribution in chromite cumulates. Am Mineral 81:1452–1459
Williams E, Boudreau A, Boorman S, Kruger F (2006) Textures of orthopyroxenites from the Burgersfort bulge of the eastern Bushveld Complex, Republic of South Africa. Contrib Mineral Petrol 151(4):480–492
Zieg MJ, Marsh BD (2002) Crystal size distributions and scaling laws in the quantification of igneous textures. J Petrol 43(1):85–101

Chapter 4
The Influence of Interfacial Energies on Igneous Microstructures

Marian B. Holness and Ron H. Vernon

Abstract The thermal history of igneous rocks is encoded in their microstructures via the control by cooling rate on crystal growth. If temperatures remain close to the liquidus, with small undercoolings, growth rates are low and microstructures are strongly affected by constraints imposed by interfacial energies. The minimisation of interfacial energy results in the melt topology becoming a function of dihedral angle and porosity. Further effects include the loss of the smallest crystals (Ostwald ripening), though this process is restricted to very small grainsizes. For large crystals, interfacial energy controls the microstructure by determining the shapes, but not the size or number of crystals. In mafic-ultramafic magmas, minerals with a larger structural anisotropy (e.g., plagioclase) commonly form crystals with low-energy faces, with or without some rounding of corners, whereas minerals with lower structural anisotropy (e.g., olivine) form crystals with rounded to locally planar boundaries.

Microstructures in partially solidified material from lava lakes and entrained glassy enclaves suggest that interfacial energies play only a minor role in the supersolidus of rapidly-cooled or coarse-grained rocks. They are most important for slowly-cooled bodies of hot magma crystallizing as crystallographically simple minerals, especially adcumulates in layered complexes.

Interfacial energies may also play a role in the sub-solidus, leading to the development of a granular microstructure in some adcumulates. This is, however, only important for very fine-grained rocks such as chill zones, or for monomineralic rocks. Elsewhere there is little or no evidence for significant grain growth in the sub-solidus, demonstrating that the end-stage of microstructural equilibration generally is not reached in crustal layered intrusions.

Keywords Interfacial energy · Microstructure · Layered intrusion · Solidification

M. B. Holness (✉)
Department of Earth Sciences, University of Cambridge, Downing Street,
Cambridge CB2 3EQ, UK
e-mail: marian@esc.cam.ac.uk

R. H. Vernon
Department of Earth and Planetary Sciences and National Key Centre for GEMOC, Macquarie University, Sydney NSW 2109, Australia
e-mail: rvernon@els.mq.edu.au

B. Charlier et al. (eds.), *Layered Intrusions,* Springer Geology,
DOI 10.1007/978-94-017-9652-1_4

Introduction

The microstructure of igneous rocks has been the subject of study since the pioneering work of H.C. Sorby in the nineteenth century. The distribution, size and shape of mineral grains provide a record of the chemical and thermal evolution of the magma as it reaches thermodynamic equilibrium with its surroundings. The history of this thermal equilibration can be deduced from the resultant microstructure because the rate of cooling controls both the primary microstructure formed during solidification (via the rate of crystal nucleation and growth (Cashman 1993), crystal shape (e.g. Lofgren 1974, 1983; Donaldson 1976; Hammer et al. 1999; Faure et al. 2003) and the distribution of crystal sizes (Cashman 1993; Burkhard 2001; Chap. 3)), and the extent to which this primary microstructure may be modified subsequently in the sub-solidus.

In rapidly cooled intrusions, such as small sills and dykes, primary microstructures formed during solidification are strongly affected by the kinetics of crystal growth. However, for the lower cooling rates experienced by large intrusions (or by small intrusions with an initial magmatic temperature close to that of their surroundings), the departure from equilibrium is smaller, with an associated reduction in the driving force for reaction. Under such circumstances another driving force comes into play, that of interfacial energy reduction. Interfacial energies are small (of the order 0.1 J m^{-2}, Sutton and Baluffi 1996; Hiraga et al. 2002) and consequently can only affect microstructures under conditions in which either the rate of crystallisation is low (for primary microstructures), or if the intrusion is neither deformed nor metamorphosed in the sub-solidus (for sub-solidus microstructural modification).

Early publications on the effect of interfacial energy on rock microstructure include Voll (1960), Weedon (1965) and Vernon (1970). However, a major step forward in our understanding of igneous rocks was made by R.H. Hunter, in two landmark contributions (Hunter 1987, 1996). In the first of these, written shortly after the igneous petrology community had recognised the significance of interfacial energies in controlling melt distribution, he provided a comprehensive description of how the minimisation of interfacial energies controls grain shape, drives grain coalescence and the expulsion of interstitial liquid, and controls the final microstructure of fully solidified cumulates. The second of Hunter's contributions placed textural equilibration in a temporal and spatial context, showing how the relative timings of nucleation and growth of different solid phases, together with compaction and deformation, resulted in a wide range of microstructures in layered intrusions.

Since Hunter's premature death in 1998, significant progress has been made in our understanding of microstructural development during and after solidification, much of it prompted by the insights that Hunter brought to the subject. Importantly, recent work suggests that the close approach to textural equilibrium suggested by Hunter (1987) is rare in crustal plutonic igneous environments; kinetic factors play a significant role in determining microstructures even in plutons as large as the Sept Iles and Bushveld intrusions (Holness et al. 2013). This contribution, which is intended as a continuation of Hunter's work, is a general review what is known about the extent to which interfacial energies play a role in controlling microstructures in igneous rocks, providing the essential background required to understand microstructural evolution in layered intrusions.

Caveats About Interpretation of Microstructure

The history of microstructural analysis of rocks is littered with argument and dissent, different authors interpreting the same structures in different ways. The reason behind many of these disagreements is that fully-solidified igneous rocks have microstructures that are the end-point of what might, or might not, have been a complex evolution, many stages of which may, or may not, have been obliterated by succeeding stages. An additional complication is that experimental studies in which detailed visual records were made of microstructural evolution have shown that some microstructures that, at first glance, could be interpreted in one way, formed in a completely different way (Means and Park 1994; Schiavi et al. 2009, 2010).

To minimize the potential for misinterpretation, petrographers should examine microstructures with a wide range of thermal histories, and be familiar with as wide a range of rock types as possible. Interpretations should be based on experimental observations and theoretical understanding. Fruitful ways of approaching the problem are to study suites of samples retrieved by repeated drilling through the crust of a lava lake (e.g. Helz 2009), or rocks that preserve a succession of developmental stages such as those recorded by poikilitic microstructures (e.g. Higgins 1998; Higgins and Roberge 2003) or compositional zoning patterns in individual mineral grains.

An example of this requirement to examine a broad range of rocks is Hunter's (1987) conclusion that equilibrium dihedral angles for melt-free silicate rocks may be lower than 30°. Work on the same minerals in high-grade metamorphic rocks demonstrates that equilibrium dihedral angles are much higher, with major implications for interpretation of rock history (e.g. Vernon 1968). The corollary of this more accurate and broader understanding of microstructural variation is that, though three-grain junctions in some cumulates are indeed close to equilibrium (Vernon 1970), the observations of Hunter (1987) demonstrate that microstructures in many cumulates are far from textural equilibrium. Subsequent work has shown that a wealth of information is preserved by this disequilibrium state (e.g. Holness et al. 2012a). In this contribution we discuss microstructures found in a wide range of igneous bodies such as sills, lava lakes and flows. By understanding these we are able to correctly decode the microstructural record in layered intrusions.

What is Interfacial Energy?

All interfaces (either between two solid phases, solid and liquid, solid and vapour, liquid and vapour, or liquid and liquid) have an excess free energy caused by the transition from one phase to the other and the associated compositional and structural inhomogeneities. This excess free energy is modified by both the composition (Gibbs 1961) and the curvature (Thomson 1871; Harker and Parker 1945; Bulau et al. 1979) of the interface. Interfacial energies tend to vary as a function of crystallographic orientation, with low energies for surfaces parallel to closely-spaced lattice planes.

The structural component of the interfacial energy, which is primarily entropic, means that there is a finite interfacial energy between a pure solid and its melt, as well as between a solid and a chemically different liquid. The energy of a fluid-solid interface may be reduced if chemical species from the liquid adsorb onto the solid surface, either by van der Waals forces (physical adsorption) or by chemical bonding (chemical adsorption). Chemically adsorbed molecules usually form a monolayer, with the adsorbed molecules bonded only at specific sites on the surface. A chemically adsorbed layer makes the interfacial energy highly sensitive to variations in temperature and pressure (Holness 1993, 1997a). The extent of chemical segregation at surfaces in metallic systems varies with their crystallographic orientation (Johnson et al. 1978; Sens et al. 1989). Wanamaker and Kohlstedt (1991) demonstrated that this is also the case in the olivine-basalt system.

The importance of absorption in controlling the energy of fluid-solid interfaces means that the magnitude of the interfacial energy correlates with departures from ideal mixing of the fluid and solid phases. The greater the positive deviation from ideal mixing (i.e. the greater the repulsion between the components, and the steeper the compositional and structural gradient in the transition zone between the two phases (Glicksman 2011)) the greater is the solid-fluid interfacial energy (Passerone et al. 1979). As the affinity between fluid and solid increases, there is an increase of adsorption at the interface and a consequent decrease in the fluid-solid interfacial energy. This means that in any given liquid the more soluble solids generally have lower interfacial energies (Nielsen and Söhnel 1971; Holness 1997a; Takei and Shimizu 2003).

When Does Interfacial Energy Matter?

Interfacial energies are small, and so any driving force they might exert for microstructural change is also small. For this reason, interfacial energies are important in situations in which the temperature changes slowly, tracking close to chemical equilibrium. For crystallising igneous rocks this means a low rate of increase of the solid fraction. Interfacial energies thus provide a force to be reckoned with when the extent of under-cooling, or supersaturation, is small. Because the contribution of interfacial energy to the total energy budget is dependent on the amount of surface area in the system, the significance of interfacial energies increases as the grain size decreases.

The shape and size of mafic magma bodies range widely, from sub-metre scale sills and dykes to 100 km-scale bodies such as the Bushveld Intrusion. There is therefore a correspondingly wide range of cooling and solidification time-scales between and within individual bodies that is reflected in the microstructure. For example, pockets of residual silica-rich liquid in small basaltic dykes and sills are commonly glassy. As the size of the body increases (or as we move into the centres of smaller bodies) the late-stage pockets are filled with aggregates of acicular and/or dendritic crystals. Only in more slowly cooled basaltic rocks are late-stage pockets filled with large single crystals or granophyric intergrowths of quartz and feldspar (Philpotts and Dickson 2000, 2002). Given the importance of a small under-cooling in determining whether interfacial energies significantly affect microstructure, we might expect interfacial energy to play a relatively important role in larger intrusions or in the centres of small intrusions. However, this may be offset by the effect of grain size; fine-grained rocks are more likely than coarse-grained rocks to experience microstructural evolution driven by interfacial energy reduction both during crystallisation and in the sub-solidus, because of the greater influence of the large interfacial area and because of the shorter diffusion distances. The important effect of interfacial energies on the microstructures of the chill zones of large intrusions will be addressed later.

Magma solidification is typified by a progression from a crystal-free liquid, through a stage where crystals and liquids coexist, to a completely solid rock. In this contribution we will discuss the effect of interfacial energies on microstructure in the context of this progression.

Crystal-Poor Systems: Isolated Grains

Our understanding of what happens in the earliest stages of crystallisation, when the system is dominated by liquid with few isolated mineral grains, is necessarily based on inference. This is especially true when we attempt to decode the microstructural evolution of solidified intrusions. We can perhaps circumvent this difficulty by setting up experiments to observe the earliest stages of crystallisation, but these are hampered by scaling issues, the short duration of experiments (and the niceties of experimental design) necessitating a fine grain size. This means that interfaces can assume a dominance that might be unrepresentative of naturally solidifying magma bodies. We can also study the phenocryst load of erupted magma to infer what is happening at depth, but this may be giving us only a partial view of solidification in the source.

Experiment and theory indicate that the nucleation rate and number density of viable nuclei are controlled by the degree of undercooling of the magma, with higher nucleation rates in systems with large departures from equilibrium (generally those which are cooled rapidly, or have undergone a rapid reduction in pressure). The nucleation process is unknown. The predominance of single, isolated phenocrysts in many volcanic rocks suggests homogeneous nucleation (e.g. Cashman 1988), though some petrologists have suggested that submicroscopic crystalline material is always present in magmas, causing heterogeneous nucleation to be the norm (e.g. Berkebile and Dowty 1982; Lofgren 1983; Marsh 1998). However, the nucleation mechanism does not concern us here.

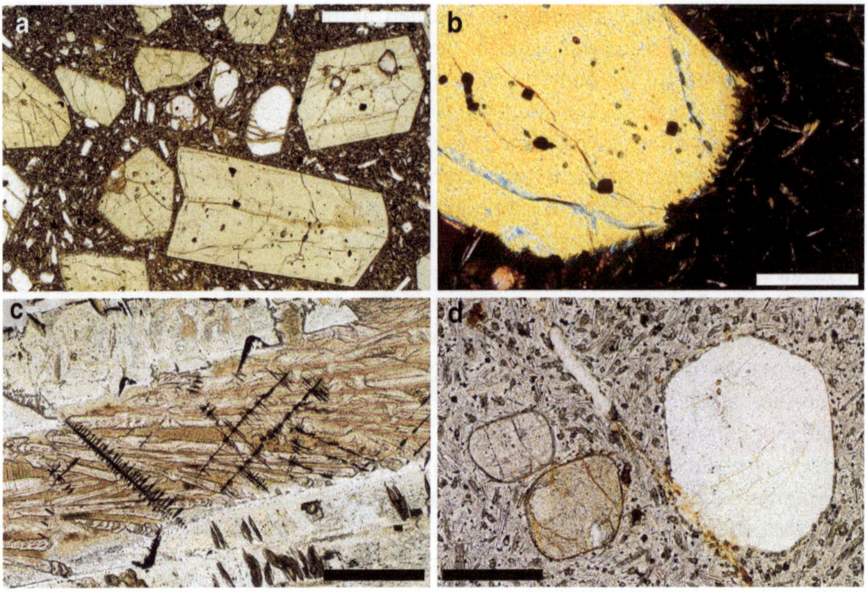

Fig. 4.1 **a** An ankaramite from Fuji, with abundant euhedral grains of clinopyroxene (*grey*) and rounded smaller grains of olivine (clear with curving cracks) set in a fine-grained groundmass. The pyroxene grains have shapes governed by crystal symmetry and must therefore have grown an isolated, free-floating grains. Plane-polarised light. Scale bar is 1 mm long (**b**) An olivine phenocryst in a mid-ocean ridge basalt. Note the cellular morphology of one growth face. This is caused by rapid growth occasioned by the large undercoolings associated with eruption. Plane-polarised light. Scale bar is 200 μm long. **c** Dendrites of magnetite in a spinifex-textured komatiite. The dendrites grow between blades of olivine (now completely serpentinised) and are set within a matrix of elongate clinopyroxene grains with a ladder-style morphology. Note the cubic symmetry of the magnetite dendrites. Plane-polarised light. Scale bar is 1 mm long. **d** Inninmorite from the Isle of Mull (Harker Collection number 13485), containing phenocrysts of plagioclase and pigeonite. The pyroxene and plagioclase grains are bounded by some plane faces that are joined by curved surfaces—these shapes are probably close to the minimum energy shape for these minerals. Plane-polarised light. Scale bar is 200 μm long

Crystal Shape

Numerous examples of isolated phenocrysts in rapidly cooled erupted magma (e.g. Fig. 4.1a) demonstrate that a stage of independent crystal growth follows nucleation (Vernon 2004; McDonnell et al. 2004). The shapes of these crystals are controlled by a combination of energetic and kinetic factors. Under conditions of relatively rapid growth, the shape is dominated by the slowest-growing faces. As these tend to be those with a high lattice density and closely-spaced lattice planes, the shape may therefore be dominated by faces with low interfacial energy. The relative areal importance of the different low energy faces depends on the details of the growth mechanisms, such as the orientation of screw dislocations and steps on the interface (Kirkpatrick 1975).

If growth is sufficiently rapid that diffusion within the liquid becomes rate-controlling, planar interfaces become unstable (Fig. 4.1b). Greater degrees of undercooling cause crystals to grow as dendrites with a high ratio of surface area to volume. The morphology of dendrites is generally crystallographically controlled (Fig. 4.1c) and, although much of the surface is formed of low energy faces, the total interfacial energy of each crystal is very high. These non-compact morphologies are therefore highly unstable and evolve towards lower energy shapes if the opportunity arises.

The minimisation of interfacial energy results in predictable shapes for the equilibrium form of both isolated crystals suspended in a liquid and of inclusions of liquid wholly enclosed by a single mineral grain. For a completely isotropic material this shape is a sphere, as this has the lowest possible ratio of surface area to volume. For materials in which the interfacial energy varies with crystallographic orientation, the lowest energy shape is that which maximises the area of low energy surfaces, and thus can have some similarities with shapes created during rapid growth. All minerals of geological relevance are anisotropic to some extent (e.g. Kretz 1966; Vernon 1968; Waff and Faul 1992; Laporte and Watson 1995; Kruhl 2001; Schafer and Foley 2002; Kruhl and Peternell 2002; Leibl et al. 2007), so the equilibrium shape of both isolated grains and of melt inclusions is generally non-spherical. The equilibrium shape may vary, depending on the medium surrounding isolated grains or the contents of melt inclusions, via the action of orientation-specific adsorption at the fluid-solid interface (Camel et al. 1981; Balluffi 1979). It comprises some planar faces that are joined by curved surfaces (Fig. 4.1d). Macroscopically these surfaces look smooth, but in detail they consist of planar segments, with faceting down to the nanometer scale (Kruhl 2001; Kruhl and Peternell 2002). Apparently, minimisation of the proportion of high-energy atoms is not achieved by the formation of smoothly curved boundaries, but by the formation of many local low-energy facets (interspersed with local high-energy adjustment segments) that macroscopically resemble curved boundaries. The facets reflect local attempts to minimize the proportion of high-energy atoms by arranging them in low-energy arrays.

Little work has been done to determine the equilibrium shape for common geological minerals. Detailed information is available only for olivine (Cabane et al. 2005) and quartz (Manley 1996; Cabane et al. 2001; Asada et al. 2002; Saiki et al. 2003).

Structurally simple minerals, such as quartz and olivine, attain their equilibrium shapes rapidly under isothermal experimental conditions (Cabane et al. 2001, 2005), with slower rates for more complex minerals such as plagioclase (Cabane et al. 2005). Quartz crystals have rounded to partly faceted shapes, more rounded shapes being developed in experiments at 900 °C and 1 GPa, whereas hexagonal shapes with rounded edges are developed in experiments at 800 °C and 0.2 GPa, suggesting that the equilibrium shape can vary with temperature and/or pressure (Cabane et al. 2001). The minimum energy shape for olivine crystals includes planar faces joined by rounded edges, whereas that for plagioclase crystals has planar faces and sharp to slightly rounded edges (e.g., Cabane et al. 2005, and Fig. 4.1d). Textural equilibrium is reached fastest in those systems in which the equilibrium melt-solid-solid dihedral angle (and hence the fluid-solid interfacial energy) is relatively low (Laporte 1994; Holness 1995). The link between fluid-solid interfacial energy and the rate of equilibration is most probably because for low dihedral angle phases the solubility is higher, promoting rapid textural equilibration predominantly via mass movement through the liquid. However, phenocrysts with smoothly rounded shapes consistent with textural equilibrium (Holness 2006) are rarely observed in erupted lavas (Fig. 4.1d), more strongly facetted, presumably growth-dominated, forms being the norm.

Ostwald Ripening

Because an atom on the surface of a small crystal suspended in a solution has a higher chemical potential compared to an atom of the same material on a planar surface (*i.e.* on the surface of an infinitely large, spherical, crystal) in contact with the same solution, it is more likely to leave the crystal (*i.e.* the small crystal will dissolve). To maintain chemical equilibrium, the small crystal must be in contact with a stronger solution than that needed to stabilise a larger crystal. The concentration of solute, C, in equilibrium with a spherical crystal of radius r is given by:

$$\frac{2\gamma}{r} = \frac{RT}{V}\ln\left(\frac{C}{C_0}\right)$$

(this is the Gibbs–Thomson, or Ostwald–Freundlich, equation) where γ is the energy of the solid-liquid interface (assumed constant regardless of crystal size), R the gas constant, T the temperature, V the molar volume of the solid phase, and C_0 is the solubility of a liquid in equilibrium with an infinitely large spherical crystal or a planar interface (Cahn 1980; Adamson 1990). The size effect on solubility equates to an effective superheating, ΔT_{eff}, of the smaller crystals, via the Gibbs-Thomson effect:

$$\Delta T_{eff} = \frac{2\gamma T_m}{r\rho\Delta H_m}$$

where ρ is the density of the crystal, ΔH_m is the enthalpy of melting, and T_m is the melting temperature (e.g. Cabane et al. 2005; Simakin and Bindeman 2008). The consequence of this relationship for a system containing suspended crystals of varying grain size within a medium, such as magma, is that the small crystals dissolve and transfer their mass to larger crystals (Lifshitz and Slyozov 1961). This process is Ostwald ripening.

How Relevant are These Processes in Igneous Rocks?

For Ostwald ripening to be the prime control on the evolution of the grain size distribution the system must be very close to equilibrium (equivalent to undercoolings of < 1°C, Cabane et al. 2005) otherwise crystallisation will swamp the interfacial energy driving force, particularly for large grain sizes. The importance of Ostwald ripening increases with increasing interfacial energy and diffusivities, and decreases as the grain size increases.

Because the driving force for ripening is greatly reduced as the grain size increases, a significant role for Ostwald ripening requires long times, although Ostwald ripening operating for short times can have an effect on the final grain size by eliminating the smallest crystals at the immediate post-nucleation stage. On timescales relevant to laboratory experiments and materials manufacturing, Ostwald ripening is important only for small grains (< 10 μm for crystals in aqueous solution (Boistelle and Astier 1988), < 1 μm for precipitates in a metallic alloy (Martin and Doherty 1976), or less than a few μm for mineral grains in a silicate liquid (Park and Hanson 1999; Cabane et al. 2005)).

For any given crystal size distribution there is a critical radius at which grains remain the same size, with smaller ones dissolving and larger ones growing. Park and Hanson (1999) introduced the concept of the *effective* critical radius below which all grains dissolve before the magma solidifies. The effective critical radius increases with the solidification time (and hence pluton size). Cabane et al. (2001) concluded that Ostwald ripening of quartz grains in silicic magma is sluggish and, even over 1 Ma, is unlikely to affect grains larger than 1 mm across. They suggest Ostwald ripening is only likely to be important for sizes < 10 μm. Similar results were obtained for plagioclase in andesitic magmas by Cabane et al. (2005), who suggested that Ostwald ripening is only efficient for grain sizes <20 μm, even over time scales of 100,000 years (i.e. commensurate with solidification times for km-scale intrusions). In contrast, Ostwald ripening may be important for olivine grains in basaltic liquid over these timescales (Park and Hanson 1999; Cabane et al. 2005).

The crystal size distributions (CSDs) of many igneous rocks are characterized by a deficit of small grains compared to the predicted logarithmic-linear correlation (Higgins and Roberge 2003). This absence of small grains has been attributed to Ostwald ripening (e.g. Higgins 1998; Park and Hanson 1999; Higgins and Roberge 2003; Johnson and Glazner 2010). However, the grain size over which Ostwald ripening is postulated to have occurred is somewhat at odds with predictions from experiments. Whereas a drop-off in number density occurs for grains smaller than 7 μm for pyroxene microlites in obsidian (Castro et al. 2003), in other igneous rocks the downturn on CSD curves generally occurs for grains several orders of magnitude larger than this (Waters and Boudreau 1996; Holness 1997b; Higgins 1998).

At least part of this apparent discrepancy may be a consequence of the isothermal nature of many of the experiments. Simakin and Bindeman (2008) demonstrated that fluctuations in the extent of supersaturation can result in ripening, as small crystals dissolve more readily than large crystals. Suitable thermal fluctuations may be common in magmatic systems, as a result of convective circulation, periodic degassing, episodic chamber replenishment, or as a natural consequence of nucleation kinetics (e.g. Brandeis et al. 1984). The end result depends on the number of fluctuations and their magnitude, larger individual fluctuations having more effect than the smaller. Johnson and Glazner (2010) used this idea to suggest that megacrysts of K-feldspar in granodiorites grow late in the magmatic history during thermal cycling, though the suggestion of late-stage growth of the megacrysts has been strongly challenged (Vernon 1986, 2004, 2010; Moore and Sisson 2008; Paterson et al. 2005; Vernon and Paterson 2008). This process (as with Ostwald ripening) is most effective for minerals with a high solubility and with high diffusivities in the liquid.

The importance of Ostwald ripening may be enhanced sporadically in layered intrusions. A major nucleation event, as suggested by Campbell (1987), might accompany the arrival of a new liquidus phase in a fractionating magma body. This would result in a large number of nuclei and a reduction of the supersaturation, thus permitting a time window in which ripening might occur (Cabane et al. 2005).

Another possibility is that the absence of large numbers of small crystals may simply be due to a reduction in nucleation. Magmas close to the liquidus, especially those that are multiply-saturated and thus thermally buffered during crystallisation, have very low undercoolings and these might not be great enough to drive nucleation. Instead solidification occurs by overgrowth on pre-existing grains (Bindeman 2003).

Crystal-Poor Systems: Grain Aggregations in the Early Stages of Solidification

The next step of the solidification process is when isolated grains cluster together, or aggregate, perhaps by heterogeneous nucleation (Kirkpatrick 1977) or by synneusis. Synneusis ("swimming together") occurs when previously isolated grains touch and remain in contact because the energy of the resultant solid-solid interface is less than that of the area of liquid-solid interface (e.g. Ikeda et al. 2002). Synneusis is perhaps most likely to occur if the system contains sufficient melt for the crystals to move relative to each other (Schwindinger and Anderson 1989) and is typically manifest in volcanic and plutonic rocks by grain clusters (e.g. Schwindinger 1999; Jerram et al. 2003; Vinet and Higgins 2010). Another possible mechanism for creating aggregates is the disintegration of a mushy layer forming within the magma chamber (Passmore et al. 2012).

Each member of a glomeroporphyritic aggregate typically shows evidence of independent growth, in the form of oscillatory zoning, before aggregation (in the case of fortuitous impingement) (Fig. 4.2) or after aggregation (in the case of heterogeneous nucleation). Generally, the grains in a glomeroporphyritic aggregate are of similar size, suggesting independent nucleation and growth, rather than coarsening of finer-grained aggregates.

Welsch et al. (2013) inferred that olivine aggregates and phenocrysts in basalts had an earlier history involving a period of dendritic crystallization at high degrees of undercooling (>60 °C) that preceded aggregation. However, such strong undercooling is unlikely to be a general feature of the solidification history of coarse-grained mafic and ultramafic rocks.

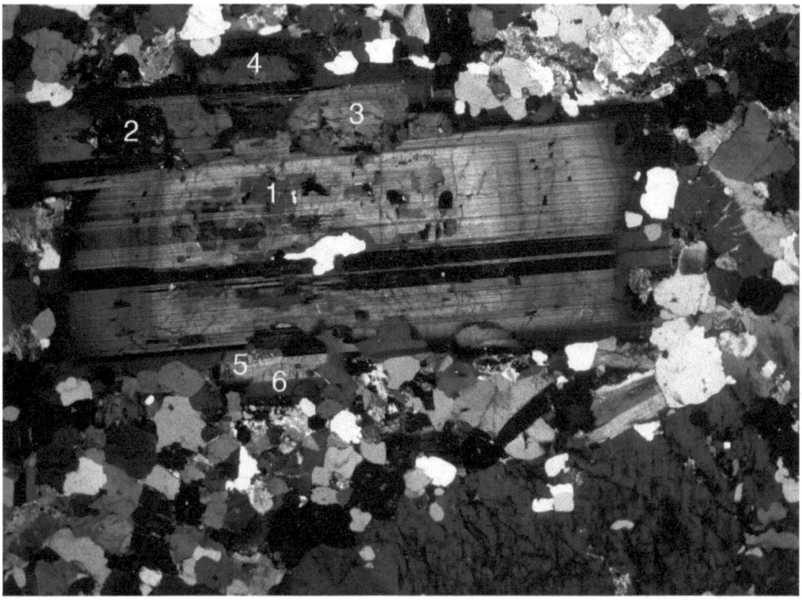

Fig. 4.2 Six zoned plagioclase phenocrysts (numbered) in parallel, apparent "synneusis" orientation, set in a much finer-grained groundmass of anhedral quartz, K-feldspar, and plagioclase. Crossed polars. The base of photo is 7 mm long. (Reproduced from Vernon and Collins (2011) by permission of University of Chicago Press)

Grain Coalescence

Schiavi et al. (2009) described the coalescence of plagioclase crystals during crystallization of a synthetic manganiferous high-K andesitic basalt, whereby adjacent crystals with similar or slightly misaligned orientation joined to form a single crystal of uniform crystallographic orientation. Thus coalescence therefore sometimes involves some grain rotation. This process is a form of Ostwald ripening, whereby the smaller of two touching grains is rapidly dissolved and incorporated into the larger (Davies et al. 1980). Smaller grains were reported to have been overgrown by migrating grain boundaries, though this process was not illustrated. The grains undergoing coalescence were predominantly in the 5–15 μm range, consistent with the main driving force being the reduction of interfacial energy. It is not known how important such a process would be for larger grains, but it is unlikely to be significant because the surface area to volume ratio would be correspondingly much lower.

Texturally Equilibrated Microstructures

The chemical potential of a system with curved interfaces (i.e. with rounded solid particles) is higher than that in a system with planar interfaces (Thomson 1871;

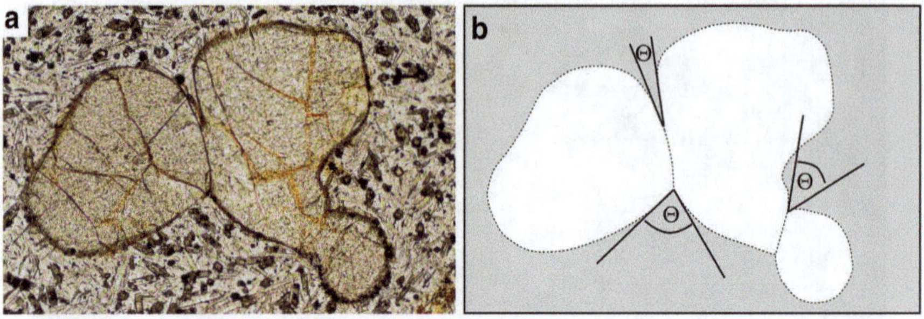

Fig. 4.3 **a** A cluster of three pigeonite phenocrysts from the inninmorite illustrated in Fig. 1d (Harker Collection number 13485). **b** A line drawing of (**a**) showing the tangents to the melt-pyroxene interfaces at the interface. The dihedral angle, Θ, is the angle between the two tangents at each junction. The variation in Θ is due partly to the variable orientation of the three-grain junction relative to the plane of the thin section (necessitating measurement with a universal stage), and partly to crystalline anisotropy

Bulau et al. 1979). This difference drives textural equilibration as all interfaces modify their shape to achieve the same mean radius of curvature. Thus, at near-liquidus conditions, when two crystals come into contact they develop curved interfaces (Fig. 4.3a) that meet at a characteristic angle (Fig. 4.3b) known as the melt-solid-solid equilibrium dihedral angle (Θ):

$$\Theta = 2\cos^{-1}\left(\frac{\gamma_{gb}}{2\gamma_{sl}}\right)$$

where γ_{gb} and γ_{sl} are the interfacial energies of the solid-solid grain boundary and the solid-liquid interface respectively (Smith 1948). For almost all silicate systems this dihedral angle is <60°, and is commonly in the region of 30° (reviewed by Holness 2010), consistent with the similarity of composition and structure of silicate liquids and silicate minerals. Anisotropy of both grain boundary energy and the energy of the fluid-solid interface causes the population of dihedral angles in any sample to vary about a median value, with a standard deviation of the order of 15° (Holness 2006). The composition of the silicate liquid appears to be unimportant. Melt-olivine-olivine dihedral angles vary little as the liquid composition changes from carbonatite through picrite to basalt and andesite (Holness 2006). This suggests that the integrated surface activity of complex natural silicate liquids is likely to lead to the same or similar interfacial energies.

A detailed study of a simple basaltic analogue system demonstrated a measurable temperature effect on melt-solid-solid dihedral angles, most probably a consequence of varying liquid composition (Ikeda et al. 2002). In the experiments of Ikeda et al. (2002) the measured melt-solid-solid dihedral angle displayed hysteresis during cooling and heating cycles, with values oscillating above and below 60°, most probably due to the formation of a compositional boundary layer at the melt-solid interface. Ikeda et al. (2002) described how the microstructure varies with

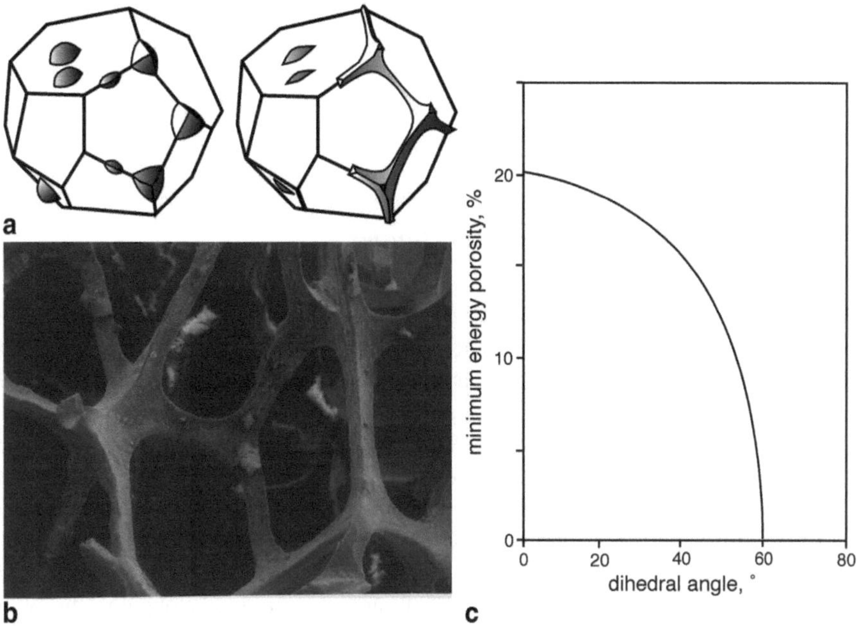

Fig. 4.4 **a** cartoons showing the effect of melt-solid-solid dihedral angle, Θ, on the equilibrium topology of the melt in an isotropic solid aggregate. LH—$\Theta > 60°$ and the melt forms isolated pockets at four-, three- and two-grain junctions. RH—$\Theta > 60°$ and the melt forms a stable fully interconnected network of channels on all three-grain junctions. After Watson and Brenan (1987). **b** A mould of the fluid-filled pore network on three-grain junctions, visible after the dissolution of the solid. Photograph by R.M. German. **c** The minimum energy porosity as a function of the equilibrium melt-solid-solid dihedral angle (Cheadle 1989)

the dihedral angle: grain clustering occurs when $\Theta > 60°$, whereas grain clusters become disaggregated when $\Theta < 60°$. The underlying cause for the varied extent of crystal clustering is the reduction of interfacial energy.

In texturally equilibrated melt-bearing polycrystals, melt topology is a function of Θ, with a stable interconnected network of fluid-filled channels on all three-grain boundaries if $\Theta < 60°$ (Figs. 4.4a, b. Smith 1948, 1964; Beere 1975; Bulau et al. 1979). This is the case for virtually all silicate systems of interest to igneous petrologists. For systems with little or no surface chemical affinity between liquid and solid (such as, for example, Fe-S liquids in olivine matrices under certain conditions of oxygen fugacity) $\Theta > 60°$ and melt is confined to isolated pockets on three- and four-grain junctions (Fig. 4.4a): see Holness (2010) for a review. Furthermore, the total interfacial energy of a melt-bearing polycrystal is a function of Θ and the volumetric fraction of melt (Beere 1975; Park and Yoon 1985; von Bargen and Waff 1986; Cheadle 1989) and, if possible, liquid migrates either into and out of a polycrystalline aggregate in order to attain the volumetric fraction associated with the lowest energy state (Watson 1982). For $\Theta > 60°$ this is achieved only when the polycrystal is melt-free, leading to the formation of melt-poor crystal clusters in the experiments of Ikeda et al. (2002). Conversely, if $\Theta < 60°$ the minimum energy state in one in which crystal aggregates contain interconnected melt in amounts that depend on Θ. This minimum energy porosity varies from zero if $\Theta = 60°$, to ~20 vol.% if $\Theta = 0°$ (Fig. 4.4c) (Park and Yoon 1985; Cheadle 1989), and leads to the break-up of crystal clusters observed in the experiments of Ikeda et al. (2002).

Observations of Natural Magma Bodies

Direct observation of the melt-rich stage of solidification at conditions of strong undercooling is possible from sequential drilling of actively solidifying bodies of magma such as lava lakes. The Kilauea Iki and Makaopuhi lava lakes have provided a wealth of information about this stage of the solidification history (Wright and Okamura 1977; Helz and Thornber 1987; Helz 2009). By pumping water down the drill hole, it is possible to extract coherent core containing up to 35 vol.% quenched glass, which is essentially the limit for mechanical coherence of a crystal framework.

The melt-rich parts of the crust of these basaltic lava lakes contain olivine phenocrysts and some blades of ilmenite, but are otherwise dominated by plagioclase and augite. The plagioclase forms loose chains of randomly oriented elongate laths, with more equant (commonly facetted) grains of augite forming a separate and less well-defined clustered framework interlocked with that of the plagioclase (Fig. 4.5a). Chains of plagioclase are thought to be a major feature of the intermediate stages of solidification of mafic magma, and can form a rigid and coherent network even in systems with porosities as high as 75% (Philpotts et al. 1998, 1999). Clustering of other minerals has also been suggested, olivine acting as a framework during the earliest stages of solidification of more mafic liquids such as komatiites (Jerram et al. 2003).

A critical observation is that the minerals of these rapidly-cooled melt-rich systems form grains with shapes apparently dominated by growth kinetics: both plagioclase and augite form facetted grains, with little or none of the partial curvature of melt-solid interfaces indicative of an approach to textural equilibrium (compare Fig. 4.2d with Fig. 4.5a). Instead the population of melt-plagioclase-plagioclase dihedral angles is controlled by the random impingement of plane-sided grains (Elliott et al. 1997), rather than by the post-impingement minimisation of interfacial energies (Holness et al. 2012a, 2012b).

The dominance of growth-controlled microstructures in Kilauea Iki is undoubtedly primarily a consequence of the relatively rapid cooling rate, but the nature of the minerals involved is also important. This is because structurally complex minerals, such as plagioclase, are slow to attain their minimum energy shape (Cabane et al. 2005). Growth-controlled plagioclase morphology is common in glassy, crystal-rich enclaves entrained in lavas (Holness et al. 2007a). These enclaves are thought to form by the break-up of crystal mushy layers in the magma plumbing system and are therefore likely to have cooled at rates similar to those of layered intrusions. The plagioclase crystals in these enclaves are predominantly plane-sided, with little to no curvature developed at two-grain junctions (Fig. 4.5b). It is possible

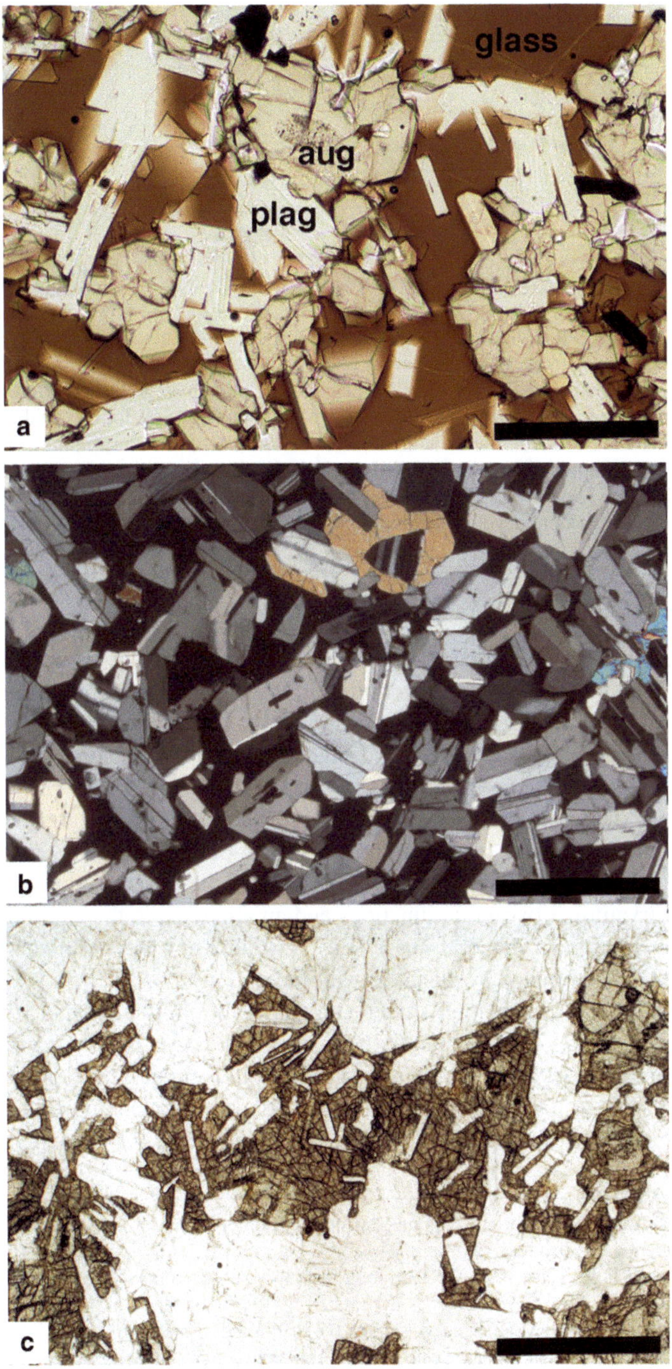

Fig. 4.5 **a** Sample KI76–145.3 from the crust of the Kilauea Iki lava lake. Note the tabular shape of the plagioclase and the equant and faceted shape of the augite. The plagioclase grains form

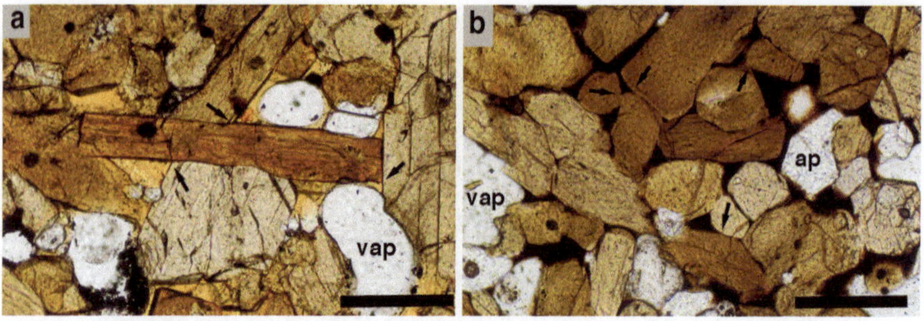

Fig. 4.6 Amphibole-dominated glassy crystal-rich enclaves from alkali basalt flows of the Kula Volcanic Province, Turkey, photographed in plane-polarised light. See Holness et al. (2005) for further details. **a** Sample K96–006 in which plane-sided amphibole grains are set in brown glass containing vapour bubbles (*vap*). The angles at two-grain junctions are formed by the intersection of plane-sided melt-solid interfaces, with no curvature associated with the minimisation of interfacial energies (examples are *arrowed*). This is an impingement microstructure (Elliott et al. 1997). Scale bar is 1 mm long. **b** Sample K96–021, in which a limited approach to textural equilibrium is evident in the curvature developed at two-grain junctions (examples are *arrowed*). This sample contains primocrysts of apatite (*ap*) in addition to amphibole, glass and vapour bubbles. Scale bar is 1 mm long

that crystal growth rates in large, deep-seated intrusions can be reduced to be commensurate with the rate of textural equilibration for plagioclase, but the microstructures in fully solidified gabbros are commonly dominated by euhedral plane-sided plagioclase grains (Fig. 4.5c), again suggestive of growth-controlled plagioclase morphology during the melt-rich phase of solidification.

Evidence of at least partial textural equilibration of melt-rich aggregates of structurally complex minerals is seen in the amphibole-dominated enclaves of the Kula Volcanic Province (Holness et al. 2005). Many of these display limited development of curvature at two-grain junctions and an approach to the equilibrium melt-amphibole-amphibole dihedral angle (Fig. 4.6) although the grain shapes on the whole are still dominated by planar growth faces.

The closest approach to melt-solid textural equilibrium is most likely to occur for olivine, because it is structurally relatively simple, it crystallises in hot magma and also because phenocryst populations in picritic liquids suggest the dominant growth form of olivine at low undercooling is polyhedral, with rounding of a significant proportion of the surface area (Figs. 4.7a, b). Juxtaposition of two rounded grains is

a framework of randomly oriented grains, while the augite grains form irregular chains. There is little evidence of formation of two-phase, three-grain junctions at this stage of solidification. Plane polarised light. Scale bar is 200 μm long. **b** Plagioclase-rich glassy enclave from Brandur, Iceland. The plagioclase morphology is dominated by the (010) faces, although the ends of the tabular crystals are somewhat rounded, suggestive of limited approach to textural equilibrium in the super-solidus. Crossed polars. Scale bar is 1 mm long. **c** Olivine-plagioclase orthocumulate with interstitial pyroxene from LZa of the Skaergaard Layered Series (sample 118698 from the Harker Collection). The plagioclase morphology is similar to that shown in (**b**). Plane polarised light. Scale bar is 1 mm long likely to create a two-grain junction with a geometry that is already close to the equilibrium melt-solid-solid dihedral angle (Elliott et al. 1997), making it more likely that the equilibrium geometry can be attained at any given cooling rate (Fig. 4.7c).

Microstructures, and hence melt topology, in melt-rich parts of gabbroic (i.e. plagioclase-dominated) mushy layers in layered intrusions are therefore likely to be far from textural equilibrium, whereas those of olivine-dominated (peridotitic) cumulates are most likely to be controlled by interfacial energies. This suggests that the clustering behaviour observed by Ikeda et al. (2002) may not play a significant role in gabbros but may occur in olivine-dominated regions. Similarly, the suggestion that expulsion of interstitial liquid is driven by the minimisation of interfacial energies as the porosity in the mushy layer is driven towards the minimum energy porosity (Campbell 1987; Hunter 1987) is unlikely to hold true for many gabbroic plutons.

The extent to which coalescence of the type described by Means and Park (1994) and Schiavi et al. (2009) occurs in igneous rocks is not clear. The experiments in which coalescence was observed were of short duration (hours) and might be relevant to natural rapidly cooled rocks in which strong undercooling induces abundant nucleation. The numerous nuclei rapidly impinge on (or nucleate heterogeneously on) neighbouring small crystals and thereby provide the potential for grain boundary migration through adjacent crystals. However, their application to coarse-grained rocks is uncertain, because they cannot provide direct and unequivocal evidence that coarse-grained igneous rocks can develop from abundant small crystals or aggregates. A particular difficulty is that grain coalescence/annexation and solid-state grain-boundary migration cannot account for the commonly observed concentric oscillatory zoning in igneous minerals. Moreover, experimental crystallization does not necessarily produce aggregates, but instead can produce many more individual crystals than aggregates (e.g. Shea et al. 2009).

Crystal-Rich Mush: The Final Stages of Solidification

The final stages of solidification begin when the solid components of the mushy layer form a fully connected, rigid mass in which the remaining liquid occupies pores of variable size and connectivity. Access to this stage of the solidification history is via crystal-rich glassy enclaves entrained in erupted lavas (Holness et al. 2007a), or via the crystal load of volcanic rocks (e.g. Vinet and Higgins 2010), although these sources provide no spatial information. Rapidly-cooled versions of this stage are also available from the drilled crust of lava lakes. Other sources of information are provided by fully solidified intrusions. However, coarse-grained rocks may have followed one of several possible pathways to arrive at their present microstructure, and the impossibility of direct observation of partially solidified slowly cooled magma bodies means that we have to make inferences about this microstructural evolution.

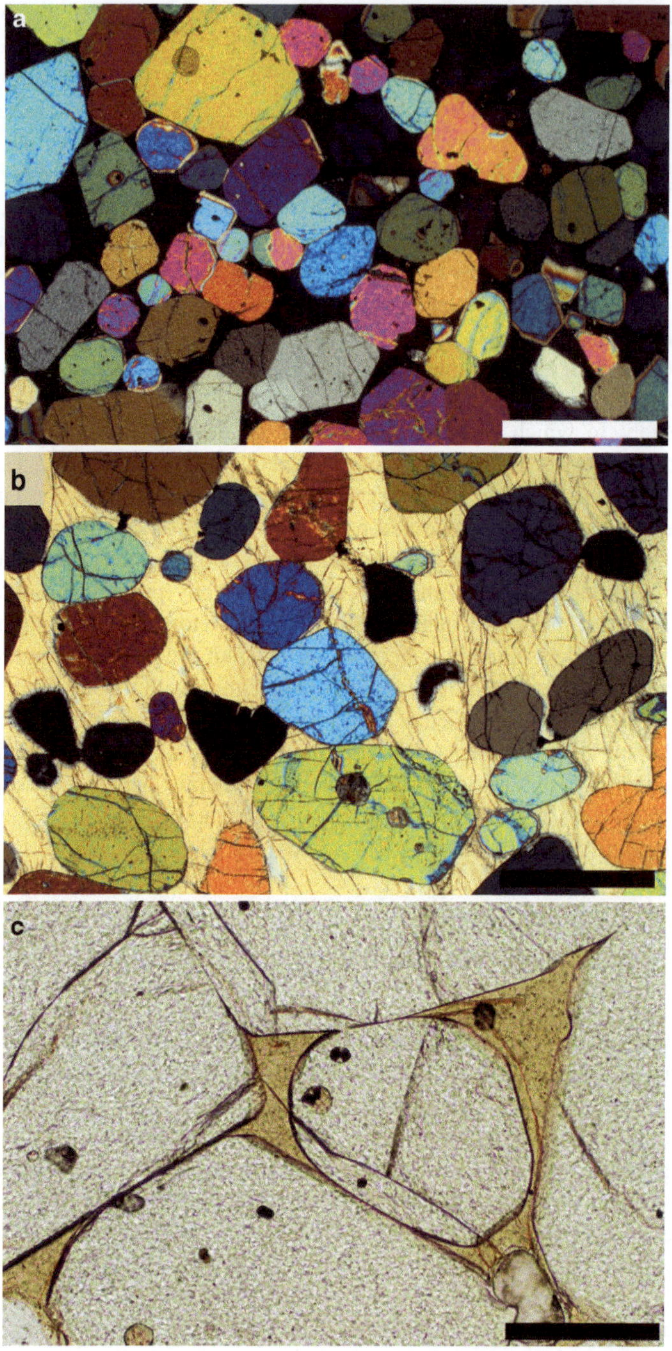

Fig. 4.7 **a** A glassy olivine-rich nodule from Damon's Cone, Mauna Loa, Hawaii (see Holness et al. 2007a for further details). Note the equant grain shape, with much of the surface formed of curved grains. Crossed polars. Scale bar is 1 mm long. **b** Peridotite cumulate from the Eastern

What did Coarse-Grained Rocks Used to Look Like?

A key difference between intrusions of widely differing size, and hence cooling rate, is the grain size. Larger, more slowly cooled, intrusions are generally coarser-grained than their smaller equivalents, except for pegmatites, which owe their exceptionally large grain size to retention of volatile components in the magma (Nabelek et al. 2010). Experimental work and inferences from natural examples (such as chill zones) suggests that this grain size difference is a consequence of the difference in nucleation rate; higher nucleation rates in the more rapidly cooled intrusions lead to a higher number of (small) crystals (e.g. Lofgren 1983; Cashman 1988; Wilhelm and Wörner 1996; Hammer et al. 1999). The corollary of this is that the slower cooling rates that aren't accessible via experimental studies lead to a low nucleation rate and therefore a smaller number of large crystals. The commonly held view is that the grain shapes observed in coarse-grained rocks result from free growth in liquid, followed by impingement and filling of the resulting interstices (e.g. Bryon et al. 1994, 1995). That these large crystals underwent a significant period of euhedral growth is commonly revealed by oscillatory zoning (e.g. Vernon 2010; du Bray et al. 2011), even though the external boundaries of zoned crystals may be irregular (e.g. du Bray et al. 2011). These independent large crystals eventually impinge on one another, after which the grain size is essentially fixed (see later section on grain growth in the sub-solidus).

However, it has recently been suggested that coarse-grained plutonic rocks such as gabbros began as fine-grained rocks with a basaltic or doleritic microstructure (e.g. Marsh 1998, 2007). According to this model, gabbroic microstructures develop by coarsening of small isolated grains and fine-grained aggregates, the process involving consumption (annexation) of some grains by others with which they come in contact during crystallization (e.g., Mock and Jerram 2005) or high-temperature grain growth, in an attempt to minimize interfacial energy (Marsh 2007; Hersum and Marsh 2007). It is inferred that the gradual attachment of crystals to each other and coalescence by grain boundary migration led to the formation of large crystals, reducing the overall number of grains and increasing the average grain size (Marsh 1998; Park and Hanson 1999; Hersum and Marsh 2007). Hersum and Marsh (2007) suggested that the average grain size can continue increasing by the process of annexation, even when there is little melt remaining. The evidence supporting this model is sparse. Marsh (1998) presented crystal size distribution data from the Makaopuhi lava lake that is suggestive of loss of small grains and a consequent

---

Layered Intrusion of Rum. Rounded olivine grains are enclosed by a single grain of clinopyroxene. Note the similarity of olivine morphology with that in the glassy Hawaiian example in (**a**). Cross polars. Scale bar is 1 mm long. **c** Detail of the Hawaiian enclave shown in (**a**), with olivine grains set in brown glass. Note the curvature at two-grain junctions, with some junctions formed by the intersection of two curved interfaces, while others are formed by the intersection of one curved and one planar interface. The presence of planar interfaces is likely to result from a combination of growth-controlled morphology and an approach to textural equilibrium. Plane polarised light. Scale bar is 200 μm long growth of the larger grains. Hersum and Marsh (2007) presented thin section images that purport to demonstrate the process in action, but in fact only show that a range of grain sizes is present in the partially solidified material.

The main argument against grain annexation and coalescence being the primary processes resulting in coarsely grained plutonic rocks is based on the likely extent of undercooling of large intrusions. A fine-grained starting material with abundant crystallites requires significant departure from equilibrium. Though large undercoolings are readily generated in small bodies such as lava lakes, thin sills and dykes, in a slowly cooled pluton a large undercooling necessitates inhibition of nucleation until the temperature is well below the liquidus. It isn't clear why or how such inhibition should occur. It is also difficult to envisage a situation whereby this mechanism can result in coarse microstructures during progressive solidification from the margins to the centre of an intrusion. Each successive layer would require nucleation inhibition in a continuous rolling cycle, despite an abundance of sites for heterogeneous nucleation. Following the abundant nucleation event, the coarsening of the initially fine-grained aggregate would then require slow cooling with near-liquidus conditions, with rates specific to each mineral concerned. Experimental work suggests that olivine would coarsen more readily than plagioclase, for example (Cabane et al. 2005). This continuous, repeated, and spatially localised sequence of high undercooling followed by low undercooling is unlikely to occur in deep-seated plutons.

A further argument against this new model of microstructural development for deep-seated magmas is that the grain size of crystal clots in lavas is generally large. These clots are believed to have been derived from deep-seated crystal-rich material (e.g. Vinet and Higgins 2010) and suggest that plutonic rocks are likely to have crystallised under conditions of small undercoolings, with low nucleation rates.

To What Extent Does Ostwald Ripening and Recrystallisation Occur in Melt-Poor Crystal Mushy Layers?

Ostwald ripening has been invoked as a driver for pattern formation in such systems via the amplification of an initially small perturbation in grain size in an otherwise unstructured or monotonically changing crystal population (Feinn et al. 1978; Feeney et al. 1983; Holness 1997b). Growth of the larger crystals at the expense of the smaller generates a macroscopic pattern in the distribution of the originally scattered particles, concentrating them into bands or some other pattern depending on the geometry of the physical or chemical gradients imposed on the system.

Boudreau (1994, 1995) and Boudreau and McBirney (1997) have proposed Ostwald ripening as the process responsible for cm-scale modal layers in cumulates. It was assumed that patterning occurred in the crystal mush, with diffusive mass transport within the remaining liquid. In such a system, the crystals form a touching framework and thus the rate of coarsening cannot be described by the classical Ostwald ripening theory originally developed by Lifshitz and Slyozov (1961).

Instead one must take into account the effects of high particle density and touching (non-isolated) particles. Both result in increased coarsening rates as a consequence of reduced diffusion paths (Ardell 1972) and grain coalescence (Davies et al. 1980) although it has not been established whether this increase is sufficient in the absence of thermal cycling (e.g. Simakin and Bindeman 2008).

Ostwald ripening has also been cited as the reason for the commonly observed reduction in the number of small crystals in plutonic rocks compared to those expected for a log-linear crystal size distribution. Higgins (1998) argued that the absence of small plagioclase grains in an anorthosite was caused by Ostwald ripening after ~40 % crystallisation (i.e. once the plagioclase had formed a coherent framework or mushy layer). He suggested that Ostwald ripening, rather than nucleation and crystallisation, was able to significantly influence the crystal size distribution because the mushy layer became thermally buffered by the release of latent heat of crystallisation that reduced the supersaturation. However, Ostwald ripening is not the only possible reason for a lower than expected population of small grains. Marsh (1988) suggested that loss of small crystals might be caused by compaction expelling the liquid that contains the youngest nuclei. Gravitational stress leading to compaction might facilitate growth of plagioclase and hence, perhaps, drive loss of the smaller grains (Boorman et al. 2004; Magee et al. 2010). The absence of small crystals may also be caused by a reduction of nucleation rate caused by a reduction in diffusivity as the temperature falls, by nucleation inhibition, both because the pore size decreases (see later) and because it is easier to overgrow existing grains, or by the reduction in driving force for nucleation as chemical and thermal equilibrium is approached.

The extent to which recrystallisation occurs in the mushy layer is likely to be small, although Philpotts and Dickson (2002) and Philpotts and Philpotts (2005) have suggested that significant recrystallisation occurs near the base of large flood basalt flows. The upper, rapidly-cooled, entablature of such flows has ophitic intergrowths of elongate plagioclase and pyroxene (Fig. 4.8a), whereas pyroxene in the more slowly cooled underlying colonnade forms granular aggregates surrounded by chains of plagioclase (Fig. 4.8b). The pyroxene grains in the entablature have compositions that bear no relationship to the pyroxene solvus, whereas those in the colonnade divide clearly into augite and pigeonite according to the expected compositions (Philpotts and Dickson 2000, 2002). These differences are interpreted as resulting from recrystallisation of pyroxene grains originating from the roof zone after they have sunk to the floor (Philpotts and Philpotts 2005), driven by a reduction in free energy as the pyroxene compositions approach chemical equilibrium. Philpotts and Philpotts (2005) suggested that during this recrystallisation most of the associated plagioclase laths were expelled from the pyroxene patches and recrystallised into shorter crystals, although it is not clear why they did not recrystallise to form granular aggregates as is commonly the case in other recrystallised mafic rocks. Note that this model involving extensive recrystallisation is based on the inference that large volumes of roof-nucleated crystals sank to the floor, and the assumption that the microstructure in the entablature is representative of that in the

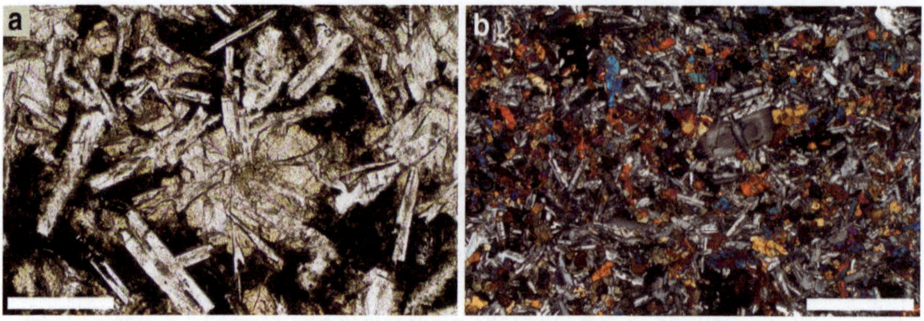

Fig. 4.8 Microstructures in the Holyoke flood basalt, sampled at Tariffville. **a** The entablature is characterised by ophitic pyroxene containing radiating bundles of elongate plagioclase. Plane polarised light. Scale bar is 200 μm long. **b** The pyroxene in the colonnade is granular, and the plagioclase does not form radiating clusters. Cross polars. Scale bar is 1 mm long

earliest history of the colonnade (i.e. the microstructure in Fig. 4.8a is the precursor of that in Fig. 4.8b).

Melt Distribution at Low Porosities

Critical to understanding the development of cumulates is the topology and distribution of melt in the mushy layer. The interconnectivity of this liquid determines the extent to which mass can be moved within the mush, with important implications for the formation of adcumulates (e.g. Campbell 1987; Hunter 1996). The extent to which interfacial energies control this topology is currently not well known because we don't yet know the kinetics of the various processes involved.

In texturally equilibrated melt-bearing rocks, melt generally forms a continuous network along three-grain junctions between the solid grains (Figs. 4.4a, b). This network is fully interconnected, even at vanishingly low porosities, if the interfacial energies are isotropic. In practice, the melt network is likely to become disconnected at porosities below some critical percolation threshold, due to anisotropy of interfacial energies (Laporte and Watson 1995; Laporte and Provost 2000). This threshold is of the order a few vol. % (e.g. Minarik and Watson 1995). Interestingly, if a rock crystallises far from textural equilibrium, with the pore geometry and connectivity controlled entirely by random impingement of grains growing with planar faces, the percolation threshold is likely to be in the range 10 vol.% (Cheadle et al. 2004; Hersum and Marsh 2006), very similar to that of a typically anisotropic, texturally equilibrated, partially molten rock.

Observation of glassy crystalline enclaves demonstrates that melt topology is generally far from this equilibrium ideal (Holness et al. 2007a). A very striking feature of these enclaves is the presence of thin, almost parallel-sided melt films on grain boundaries (Fig. 4.9a). These melt films locally contain small mineral grains (Fig. 4.9b), showing that they are a primary feature of the enclaves and not a conse-

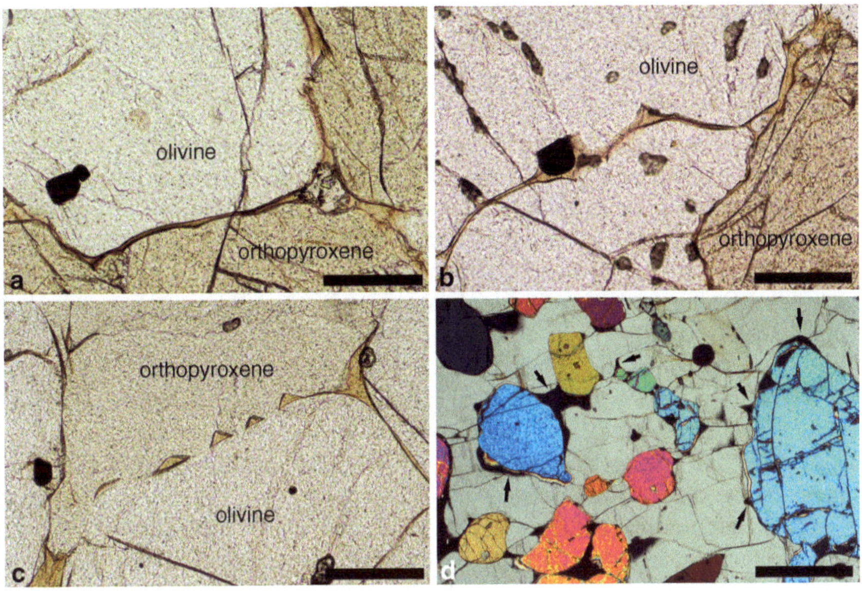

Fig. 4.9 Microstructures in glassy crystal-rich enclaves entrained in erupted lavas at Damon's Cone, Mauna Loa, Hawaii (for more details see Holness et al. 2007a) **a** Brown glass forms a film of irregular *thickness* along grain boundaries. Plane polarised light. Scale bar is 200 µm long. **b** The melt film in this image contains a primary grain of Fe-Ti oxide, demonstrating the primary nature of this film. Plane polarised light. Scale bar is 200 µm long. **c** Apparently isolated pockets of *brown* glass on the grain boundary between olivine and orthopyroxene. The asymmetric shapes of these pockets suggest they are not controlled by interfacial energies but are a growth feature. Plane polarised light. Scale bar is 200 µm long. **d** Orthopyroxene oikocryst with abundant included grains of olivine. Irregular pockets of melt (some examples of which are arrowed) form at the boundaries of the included grains. Crossed polars. Scale bar is 1 mm long

quence of depressurisation during entrainment. Their presence cannot be explained in terms of textural equilibrium: although melt films are a stable topology if the equilibrium melt-solid-solid dihedral angle is zero, all silicate systems examined to date have non-zero equilibrium angles (Holness 2010). Melt also forms irregular, texturally unequilibrated, pockets along grain boundaries (Fig. 4.9c) and apparently isolated pools associated with included grains in oikocrysts (Fig. 4.9d).

Further information about the melt distribution during the last stages of solidification is provided by augite-plagioclase-plagioclase junctions in dolerites. Progressive snapshots of the process of formation of these three-grain junctions are obtainable from different depths through the crust of the Kilauea Iki lava lake. Three-grain junctions between two grains of plagioclase and one of pyroxene do not commonly form during the earliest stages of solidification. At the bottom of the drillable crust, where the core contains ~35 vol.% melt (glass), only a few junctions are visible in a single thin-section. With progressive solidification, more junctions form, the wider junctions between two plagioclase grains being filled before lower-angle junctions (Holness et al. 2012a, 2012b). Formation of all three-grain junctions is complete when ~10 vol.% liquid remains. The three-grain junctions in rapidly cooled dolerites and basalts such as Kilauea Iki are formed by the direct replacement of melt by augite, with little discernible growth of plagioclase (Fig. 4.10a). The three-grain junctions are thus formed by the intersection of straight augite-plagioclase grain boundaries, with no relationship to that expected for textural equilibrium. The median *melt*-plagioclase-plagioclase dihedral angle in the Kilauea Iki crust is 60°, the expected value for randomly impinging tabular grains. However, the median angle of the *augite*-plagioclase-plagioclase junctions is considerably higher than this, at 78°, because augite tends to occupy the widest junctions, with the narrower junctions filled instead with silica-rich late-stage liquid that has quenched to glass (Fig. 4.10a).

Both the thick melt films in the glassy enclaves (Figs. 4.9a, b) and the preference for augite to grow into wide, rather than narrow, pores point to important effects of pore size on crystallisation. The reasons for this behaviour are complex and are likely to be due both to kinetic effects and to a fundamental control of interfacial energy on the growth of crystals in confined spaces.

Abundant empirical evidence demonstrates that crystal nucleation is inhibited as the volume of liquid is decreased. This is observed for the freezing of water droplets (Bigg 1953), the nucleation of salts in aqueous solution (Melia and Moffitt 1964; Kubota et al. 1988; Putnis and Mauthe 2001), and the crystallisation of melt inclusions in mineral grains (Cesare et al. 2009; Holness et al. 2012a). This is most probably a consequence of the stochastic nature of nucleation and a reduction in sites for heterogeneous nucleation in small, compared to larger, liquid volumes (Kubota et al. 1988). A further reason why a confined fluid can become more supersaturated, compared to an unconfined fluid of the same composition, is described by the Gibbs-Thomson (Ostwald-Freundlich) equation given earlier in the context of Ostwald ripening.

The undercooling required for a crystal to grow into a confined space depends on the shape of the crystal (i.e. whether or not it wets the sides of the pore) and the pore radius (Scherer 1999). For the geologically most likely situation of a crystal with a convex interface growing into a pore, the narrower the pore and the more convex the crystal, the greater the supercooling required for the crystal to enter and fill the pore. This interfacial energy effect means that augite does not grow into the narrower pores formed by two plagioclase grains impinging at low angle, because this requires such a great supersaturation that the glass transition temperature is encountered before it can happen.

There may also be a kinetic effect related to the reduction of mass transport into the narrower pores that suppresses further crystal growth. As the porosity decreases, the permeability decreases also. This means that the rate at which material can be brought to the growing crystal surfaces is decreased and this, together with the silica-rich composition of evolved late-stage liquids, makes it more likely that the remaining melt cools through the glass transition.

Evidence that pore size becomes important during the last stages of solidification in geological environments is provided by the increasing albite content of plagioclase grains growing into progressively narrower pores in a migmatite (Holness and

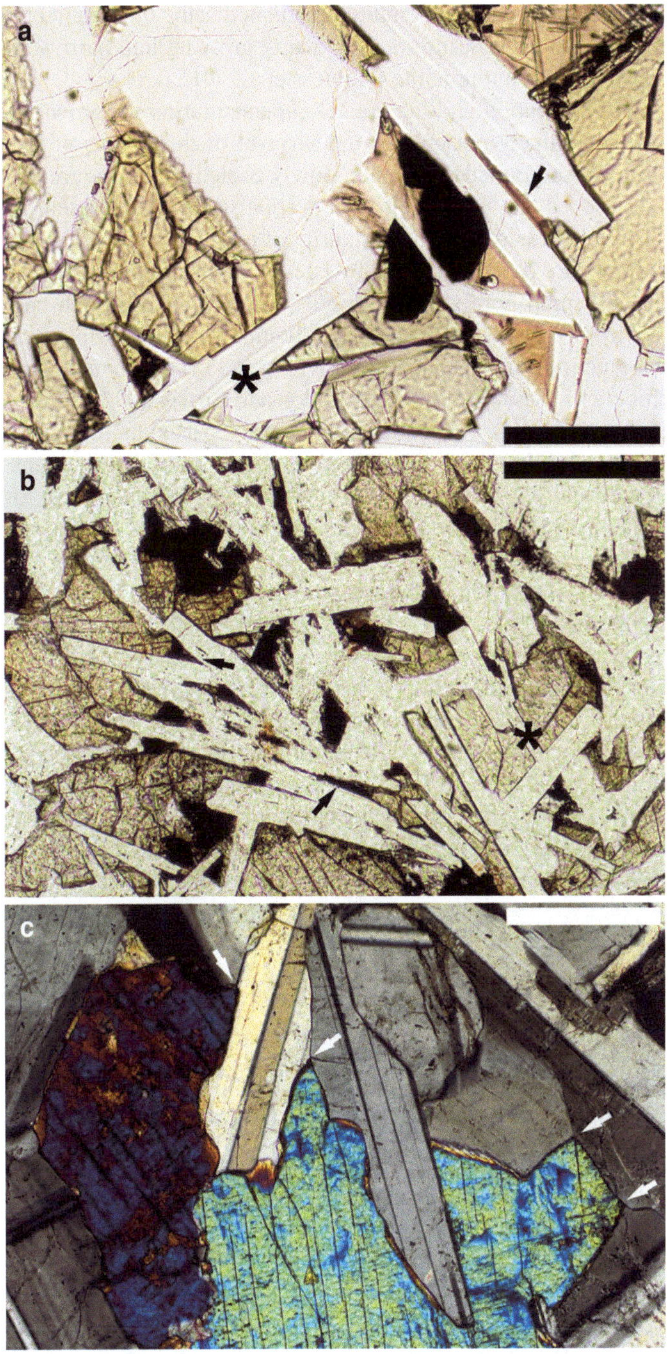

Fig. 4.10 **a** Sample KI76–143.3 from the crust of the Kilauea Iki lava lake, showing elongate plagioclase grains with augite infilling the spaces between. The melt is quenched to *brown* glass, and some opaque ilmenite grains are present. Note how the three-grain junctions are formed by

Sawyer 2008) and by lower temperatures calculated using the TitaniQ thermometer of Wark and Watson (2006), inferred for quartz growing into narrow spaces compared to larger pores in a migmatite (Holness et al. 2012a).

Inhibition of nucleation and the greater supersaturation for growth into narrow pores are the most likely reasons for the survival of thick films of melt on grain boundaries in coarse-grained (and hence slowly cooled) crystal aggregates. Holness et al. (2007a) argued that these melt films were also formerly present in now fully solidified layered intrusions, citing similarity of microstructures to suggest that the last drops of melt are pseudomorphed by late-crystallising minerals.

The effect of small pore size on crystal nucleation and growth has been suggested as a way of enhancing textural equilibration in melt-bearing rocks. Cuspate grains at the junctions of three larger grains (Fig. 4.11a) are common in high-grade metamorphic rocks (Sawyer 1999; Marchildon and Brown 2002; Holness and Sawyer 2008) and are used as a diagnostic indicator of partial melting. These cuspate grains are inferred to have solidified from a silicate melt, filling a pore geometry that was in melt-present textural equilibrium. The cuspate shape and low dihedral angles are inherited. Holness and Sawyer (2008) showed that the cuspate morphology is partly a consequence of nucleation inhibition in small pores, a delay in the full crystallization of the melt leading to greater opportunities for textural equilibration of the pore walls.

These cuspate grains are not generally seen in mafic intrusive rocks, perhaps because their development requires much slower cooling rates and a smaller undercooling than are commonly experienced by mafic intrusions in the shallow crust. Though similar microstructures could be present in more slowly-cooled gabbro bodies emplaced in the deep crust, exhumation is likely to involve orogeny with associated regional metamorphism and a consequent overprinting of early microstructure. However, cuspate augite grains are locally present in several layered intrusions and are believed to have formed during melt infiltration (Holness et al. 2012a).

Spontaneous Infiltration

An important consequence of the minimum energy state associated with equilibrium dihedral angles below 60° (Fig. 4.4c) is that, as long as the temperature is above the solidus, melt will spontaneously infiltrate a previously melt-free rock (Watson

intersection of planar augite-plagioclase grain boundaries (an example is denoted by the *asterisk*). Augite has only incompletely filled the very narrow pore marked by the arrow. $\Theta_{cpp}=78°$ in this sample. Plane polarised light. The scale bar is 200 μm long. **b** Dolerite from the 3.5 m *thick* Traigh Bhàn na Sgùrra sill, Isle of Mull. The same planar augite-plagioclase boundaries as those in (**a**) are present here (examples are marked with the *asterisk*). Augite has only incompletely filled several of the narrower pores (marked by *arrows*). $\Theta_{cpp}=78°$ in this sample. Plane polarised light. Scale bar is 200 μm long. **c** Dolerite from Karoo, Namibia. The three-grain junctions in this sample are formed of curved augite-plagioclase junctions (examples are *arrowed*), suggesting coupled growth of both minerals during the formation of the junction and leading to $\Theta_{cpp}>78°$. (sample 55448 from the Harker Collection). Crossed polars. Scale bar is 200 μm long

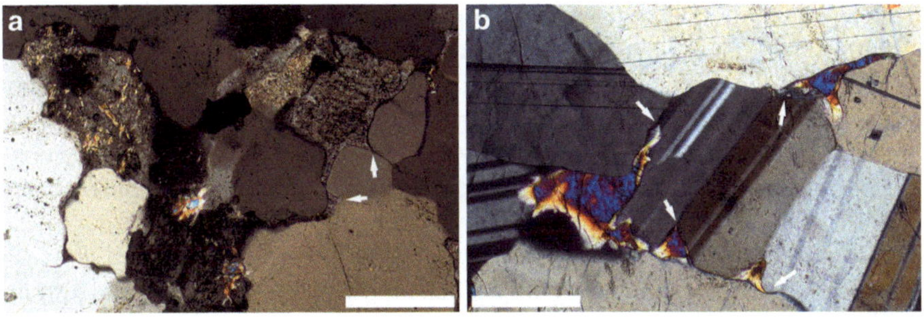

Fig. 4.11 **a** Quartzo-feldspathic migmatite from the aureole of the Ballachulish Igneous Complex, Scotland. Elongate and cuspate apophyses of feldspar extend along quartz-quartz grain boundaries (examples are *arrowed*). The feldspar-quartz-quartz dihedral angle population is close to that expected for texturally equilibrated melt-quartz-quartz, suggesting that the feldspar pseudomorphed an equilibrated melt topology. Crossed polars. Scale bar is 200 μm long. **b** Cuspate grains of augite in troctolitic cumulate of the Unit 10 allivalite of the Rum Eastern Layered Intrusion (sample 146199 from the Harker Collection). $\Theta_{cpp}=67.5°$ (Holness 2007), lower than expected for a Rum troctolite (Holness et al. 2007c), but consistent with the pseudomorphing of an equilibrated melt-filled porosity by augite. Crossed polars. Scale bar is 200 μm long

1982; Stevenson 1986). This occurs because a reduction in free energy is associated with the presence of a fully interconnected network of fluid-filled channels on three-grain junctions. Infiltration therefore occurs until the porosity reaches a value with the minimum energy which, for melt-solid-solid dihedral angles of ~30°, typical for basaltic systems (Holness 2006) is 20 vol.% (Fig. 4.4c).

Evidence that this process may have affected mafic cumulates is found in the Rum Eastern Layered Series, in which the bases of several of the plagioclase-bearing cumulate horizons (known by the local name of allivalite) are characterised by distinctive cuspate grains of interstitial augite (Fig. 4.11b) and anomalously low values of the augite-plagioclase-plagioclase dihedral angle, Θ_{cpp} (Holness et al. 2012a). These cuspate grains have been interpreted as the solidified remains of upwardly migrating late-stage melt derived from the underlying peridotite. This liquid infiltrated a previously melt-free rock, driven along plagioclase three-grain junctions by the consequent reduction in internal energy (Watson 1982; Stevenson 1986; Hammouda and Laporte 2000). The low angles in the Unit 10 allivalite were therefore inherited from a texturally equilibrated melt topology via a pseudomorphing process analogous to that seen in migmatites.

For infiltration to occur by this mechanism it is essential that the solidus of the infiltrating liquid is lower than that of the solidified rock being infiltrated, and that there is sufficient time to establish the equilibrium melt topology. The Unit 10 allivalite must therefore have been essentially completely solid but still relatively hot, at the time the cool, late-stage liquids were expelled from the underlying peridotite.

The similarity between three-grain junctions formed during the infiltration process and those seen in migmatites from regional metamorphic terranes is a consequence of both a small pore size (low volume of liquid) and the low growth rates of the pore walls. In migmatites these low growth rates are a consequence of the slow cooling rates typical of regional metamorphism and to nucleation inhibition in small pores. In the Rum example, the low growth rates were probably a consequence of low undercooling, meaning there was sufficient time to approach textural equilibrium. The final solidification history of the infiltrating liquid is directly analogous to the migmatite situation, with overgrowth of plagioclase in the pore walls by the plagioclase component of the late liquid and final crystallization of augite (and rare olivine) in the remaining space.

The features observed in the Unit 10 allivalite appear to have been a consequence of multiple injections of compositionally very different liquids into a stratified magma chamber and hence are unlikely to be common in layered intrusions formed from magma batches of similar composition. However, infiltration driven by interfacial energies might be a relatively common process in the mantle (e.g. Hammouda and Laporte 2000), although the record of this happening might be obliterated in by subsequent high-temperature recrystallisation.

Three-Grain Junction Geometries

Since Hunter (1987) wrote the first comprehensive description of microstructures in cumulates from the standpoint of textural equilibration we have realised that dihedral angles preserved in layered intrusions only very rarely reflect the influence of interfacial energies in either the precursor melt-bearing system or the completely solidified system. Instead the disequilibrium geometry of most three-grain junctions records a wealth of information about cooling histories and liquidus phase assemblages (e.g. Holness et al. 2012a). Here we will discuss the preservation of information about super-solidus processes. In the later section on sub-solidus microstructural evolution we will specifically address the extent to which interfacial energies control three-grain junction morphology in cumulates.

Three-grain junctions are formed during solidification by the expulsion or replacement of the last pockets of liquid. The exact process by which this occurs is not well known, but must involve mass migration through low permeability pore networks. Recent studies have shown that the geometry of three-grain junctions, specifically those between two grains of plagioclase and one of clinopyroxene (augite), varies systematically in individual intrusions (Holness et al. 2007b, 2012b, 2013). Although the mechanism by which this variation is achieved is not well understood, it is possible to draw some empirical conclusions from what has become a substantial body of data, quantified by the median dihedral angle, Θ_{cpp}.

Augite-plagioclase-plagioclase three-grain junctions in rapidly cooled dolerites and lava lakes are characterised by the intersection of straight augite-plagioclase grain boundaries and a median dihedral angle, denoted Θ_{cpp}, of 78° (Figs. 4.10a, b, Holness et al. 2012a). This value is far from that expected for sub-solidus equilibrium (109±4°, Holness et al. 2012a). In more slowly cooled dolerites (crystallising over times >10 years) the grain boundaries are curved in the immediate vicinity of the three-grain junction, leading to a higher value of Θ_{cpp} (Fig. 4.10c). The difference is caused by a change in the relative rate of growth of the plagioclase and

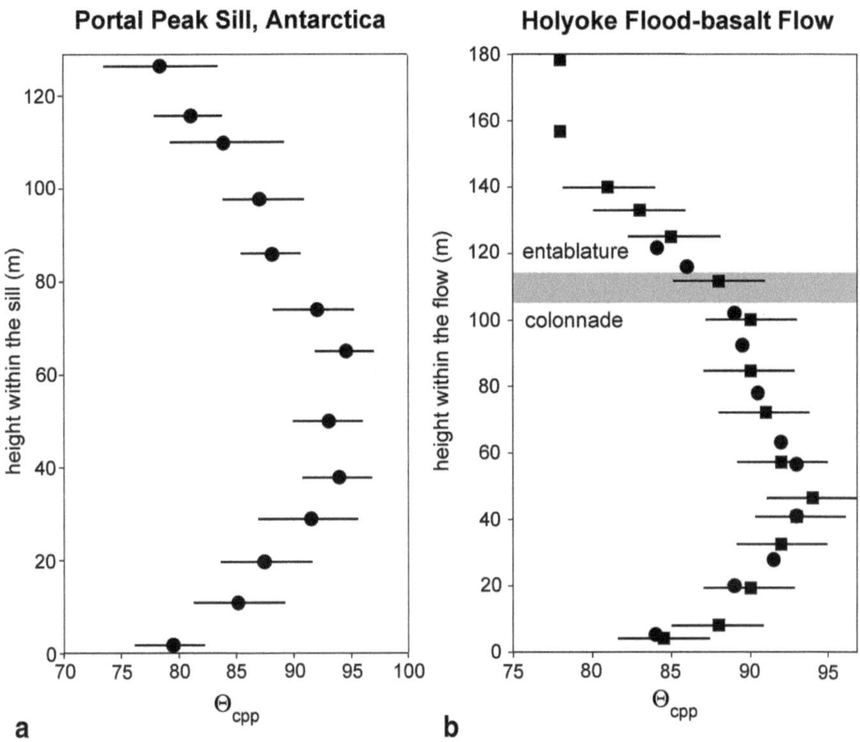

Fig. 4.12 The variation of Θ_{cpp} as a function of height in **a** the Portal Peak sill, Antarctica and **b** dolerite sill (**b**) the Holyoke flood basalt flow. The value of Θ_{cpp} for each sample is based on a population of 100 individual measurements, and the 2σ uncertainty is calculated according to the method of Stickels and Hücke (1964). The data for Holyoke was collected from two separate traverses. The squares show data for North Branford, where the flow is 187 m *thick*, and the circles show data collected at Tariffville, where the flow is 174 m *thick*. For clarity, uncertainties are shown only for the North Branford data. The approximate position of the boundary between the entablature and the colonnade is shown as a thick grey line. Note the symmetric variation of Θ_{cpp} with height in the Portal Peak sill, whereas the maximum value of Θ_{cpp} is found close to the base of the Holyoke flow, consistent with asymmetric heat loss from the latter

augite, most probably reflecting changes in the undercooling (Holness et al. 2012a). Θ_{cpp} varies symmetrically across dolerite sills, with the highest values in the centre (Fig. 4.12a. Holness et al. 2012b). Θ_{cpp} is a function of the time taken to crystallise each sample, assuming symmetrical heat loss from the roof and floor of the sills. In contrast, Θ_{cpp} varies asymmetrically through thick flood basalt flows, with the highest values close to the floor (Fig. 4.12b), consistent with the same control by cooling rates that is seen for sills.

Whereas metre-scale sills display symmetric grain size variation (e.g. Holness and Humphreys 2003), similar to metre-scale dykes (Cashman 1993), in thicker sills the horizon with maximum grain size always occurs within the upper half (e.g. Wilhelm and Wörner 1996). Maximum grain size is found in the middle of a lava lake or flow (e.g. Moore and Evans 1967; Philpotts et al. 1996). The coarse-grained zones are generally more silicic than the rest of the body. The position of the coarse grained material above the horizon with maximum Θ_{cpp} (i.e. that horizon that took longest to cool), demonstrates that the grain size variations are related to mass migration during solidification. Marsh (1995) suggested that the coarse-grained region is caused by gravitational tearing of an unstable solidification front on the roof, with crystallisation of locally-derived late-stage evolved liquid within the tear. However Wilhelm and Wörner (1996) pointed out that the crystal size distribution in the coarse-grained region is not consistent with a single-stage late crystal growth event. Instead they suggested that the coarse-grained regions are those where the last, evolved, volatile-rich residual liquid was concentrated, in which nucleation rates were low and growth rates high. This is similar to the explanation of Philpotts et al. (1996), who inferred that the late-stage silicic liquid is sourced from the lower compacting layers in the body.

Layered intrusions display a different spatial variation of Θ_{cpp}. Early work demonstrated that Θ_{cpp} in the plagioclase-rich cumulates of Rum varies, depending on whether the cumulates are troctolites (in which $\Theta_{cpp} \sim 80°$) or gabbros (in which $\Theta_{cpp} \sim 90°$) (Holness et al. 2007c). In the Skaergaard intrusion, step-wise increases in Θ_{cpp} occur when augite, followed by oxides and finally by apatite, are added to the initial troctolitic cumulus assemblage (Fig. 4.13a. Holness et al. 2007b). In the Upper Zone of the Bushveld intrusion, step-wise increases are observed on the addition of oxides and apatite to the liquidus assemblage (Fig. 4.13b. Holness et al. 2013). No change in Θ_{cpp} is associated with the replacement of olivine by orthopyroxene as the silica activity increases in the bulk magma in either Skaergaard or Bushveld. The sequence of liquidus assemblages in the lower part of the Sept Iles layered intrusion of Quebec is defined by the progressive addition of oxides, augite and apatite, each of which is associated with a step-wise increase in Θ_{cpp} (Fig. 4.13c. Holness et al. 2013). In a similar way to Bushveld and Skaergaard, increasing fractionation of the Sept Iles bulk magma results in the loss of olivine, but it is not replaced by orthopyroxene: Θ_{cpp} drops as a result (Fig. 4.13c). Θ_{cpp} therefore changes in a step-wise fashion to reflect the number (but not the nature) of the minerals on the liquidus.

Comparison of microstructures of plutonic rocks with those of high-grade metamorphic rocks (see below) demonstrates that the values of Θ_{cpp} in the gabbroic rocks are not those of textural equilibrium. Given that equilibrium values in metamorphic rocks do not change measurably as a result of bulk rock composition, the variations in Θ_{cpp} in plutonic rocks cannot be changing in response to minor compositional differences (i.e. there is no influence by interfacial energies). Instead, the main control on Θ_{cpp} appears to be the kinetics of crystal growth in confined spaces. This primary control must be thermal, possibly via the changes in the enthalpy budget associated with changes in the liquidus assemblage (Holness et al. 2013). These ideas still need to be tested and refined.

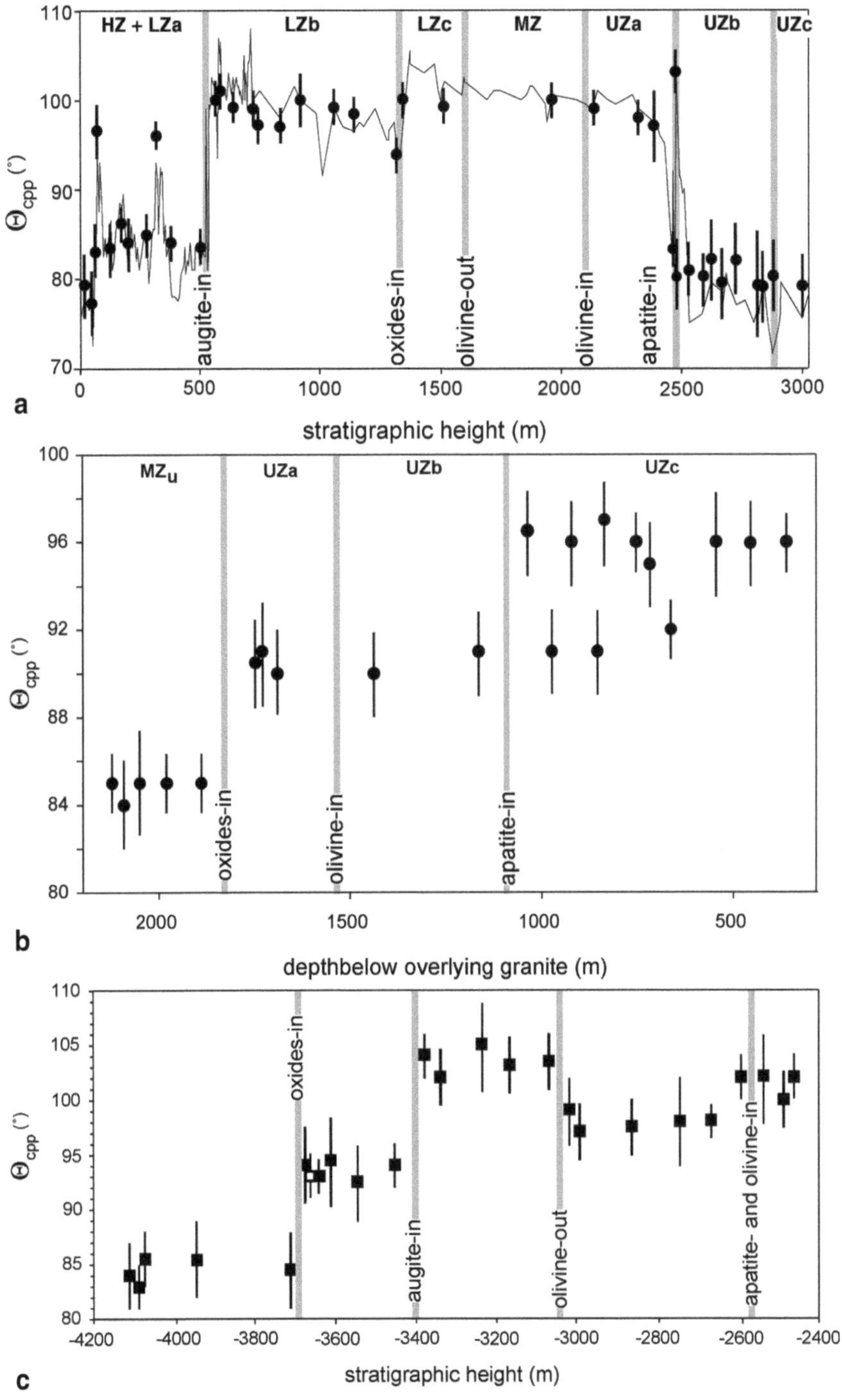

Fig. 4.13 The variation of Θ_{cpp} through the floor cumulates in **a** the Skaergaard intrusion, East Greenland, **b** the Upper Zone of the Bushveld Intrusion, South Africa and **c** the Mega-cyclic unit

Microstructural Evolution in the Sub-Solidus

As we have documented earlier, the primary igneous microstructure in rapidly-cooled rocks is dominated by the kinetics of crystal growth. Early-crystallising minerals (such as liquidus minerals in cumulate rocks) are commonly euhedral, with grain boundary orientations controlled by grain impingement during growth. Any curvature of grain boundaries will generally reflect the shape of early-crystallising phases that grew unimpeded in abundant liquid. Three-grain junction geometries are far from equilibrium. These growth-dominated microstructures are commonly entirely unmodified during the sub-solidus history of dolerites (in the absence of subsequent metamorphic heating). However, the slower cooling of layered intrusions leads to a significant temperature-time window in the sub-solidus during which primary igneous microstructures can evolve, driven by the minimisation of interfacial energies as they approach melt-absent textural equilibrium. This evolution is essentially metamorphic, with grains changing shape and size by diffusion and grain boundary migration. Sub-solidus microstructural evolution can lead to difficulty in interpreting microstructures in layered intrusions, necessitating the distinction between features that are primary (produced during solidification) and those formed in the sub-solidus (Hunter 1987; Means and Park 1994). Rounding of grains during textural equilibration, for example, can make interpretation of the cumulus assemblage difficult as the grain shape becomes ambiguous (Hunter 1987).

Sub-solidus textural equilibrium is characterised by a distinctive granular microstructure, with smoothly curved interfaces and equilibrium dihedral angles at all three-grain junctions (Fig. 4.14). Because solid-solid interfacial energies are very similar, equilibrium dihedral angles involving two silicate phases are in the range 100–160° (reviewed by Holness 2010). The anisotropy inherent in all geological materials (e.g. Kretz 1966; Schafer and Foley 2002) means that there is no single value of the equilibrium dihedral angle (Herring 1951; Laporte and Watson 1995; Laporte and Provost 2000; Kruhl 2001): the angle varies according to the relative orientation of the grains at the junction with a greater spread of possible angles for the more anisotropic minerals (e.g. Vernon 1968; Laporte and Provost 2000). This effect may be exacerbated by the presence of stable, compositionally distinct, nanometre-scale films on grain boundaries with particular orientations (Baram et al. 2011). Anisotropy also means that the minimum energy microstructure may include planar as well as curved interfaces (Herring 1951; Vernon 1968; Waff and Faul 1992; Kruhl 2001; Kruhl and Peternell 2002; Leibl et al. 2007).

---

1 of the Sept Iles Intrusion of Quebec. The position of the arrival or disappearance of cumulus phases in marked by *grey lines*. The continuous line in (**a**) shows the published data from Holness et al. (2007b) for which each data point was based on populations of 30 individual measurements of dihedral angle. The dots show the medians of population of 100 individual measurements. All data in both (**b**) and (**c**) is based on populations of 100 measurements. The 2σ uncertainty is calculated according to the method of Stickels and Hücke (1964). The bimodal Θ_{cpp} in UZc of the Bushveld (**b**) is caused by the sporadic appearance and disappearance of cumulus apatite (Tegner et al. 2006)

Fig. 4.14 A well-equilibrated microstructure seen in a bytownite-edenite gneiss from Sittampundi, Madras (ample 87975 from the Harker Collection). Note the generally smoothly curved grain boundaries and the ~120° dihedral angls developed at all three-grain junctions. The amphibole grain marked with an asterisk shows two faces that are parallel to the {110] cleavages, suggesting some anisotropy of interfacial energies. Crossed polars. Scale bar is 1 mm long. The grain marked A is bounded by faces that are concave outwards, whereas that marked B is bounded by faces that are convex outwards. This means the grain A will grow at the expense of grain B during normal grain growth

The Process of Textural Equilibration

The mechanisms by which microstructures evolve towards textural equilibrium are poorly known and we don't yet have firm kinetic constraints on the processes involved. Our current understanding is based on observations from suites of related samples that have each experienced slightly different thermal histories.

The glassy, crystal-rich, enclaves in alkali basaltic lavas from the Kula Volcanic Province in Western Turkey illustrated in Fig. 4.6 display little evidence of an overall rounding of melt-amphibole interfaces which instead are generally defined by the smooth low-index crystal faces expected to have formed during growth. However, the morphology of two-grain junctions varies from those characterised by the meeting of planar melt-amphibole interfaces (Fig. 4.6a) to those with well-defined curvature (Fig. 4.6b). This variation results in a smooth variation of the median melt-amphibole-amphibole dihedral angles between different enclaves in the suite (Fig. 4.15), from values expected for random impingement of planar-sided grains (~60°) to values consistent with melt-present textural equilibrium (~25°) (Holness et al. 2005).

These observations demonstrate two important features of textural equilibration. First, the continuity within the suite of enclaves of the range of median angles from impingement to equilibrium shows that equilibration of two-grain junctions is achieved by the gradual rotation of relatively large areas of the melt-solid interface, with a progressive change in the median and skew of the population. This is in contrast to the commonly held view that equilibrium angles are established instanta-

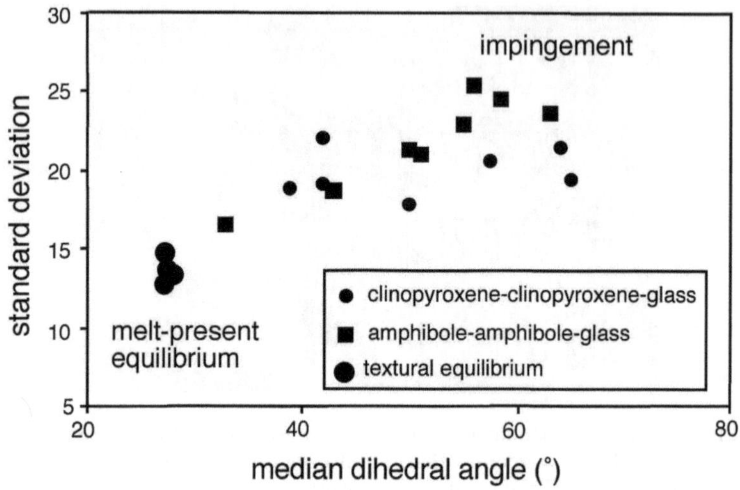

Fig. 4.15 Variation of melt-solid-solid dihedral angle populations in a suite of glassy crystalline enclaves from the Kula Volcanic Province, Turkey. There is a continuous sequence from populations typical of impingement towards melt-solid-solid textural equilibrium, demonstrating that angle populations change by rotation of relatively large areas of the melt-solid interfaces. Data from Holness et al. (2005)

neously in the immediate vicinity of the junction, with progressive outwards propagation of the equilibrium interfacial curvature (e.g. Mullins 1957). Second, textural equilibration begins with this gradual establishment of the equilibrium morphology in the vicinity of the two-grain junction: the development of smoothly curved grains at the expense of the planar, growth-controlled melt-crystal interfaces takes longer and may not be achieved before the equilibrium dihedral angle is attained.

Similar observations of a gradual approach to the equilibrium dihedral angle are seen in fully solidified rocks. The clearest examples are provided by the chill zones of large (>hundreds of metres) mafic bodies. The rapid cooling experienced by chill zones in smaller bodies such as thin sills and dykes results in a typical doleritic microstructure, with tabular plagioclase and planar plagioclase-augite grain boundaries (Fig. 4.16a). It is likely that all rapidly crystallised mafic rocks have this microstructure, and thus that the chill zones of the larger bodies also looked like this immediately after solidification. However, the observed microstructure in the chill zones of large bodies is characterised by rounded grains of augite and smoothly curved plagioclase-augite grain boundaries (Fig. 4.16b). As the size of the body increases, the microstructures of the chill zone becomes progressively more granular (Fig. 4.16c). The change from a fine-grained doleritic microstructure to an equally fine-grained granular one is consistent with significant sub-solidus microstructural adjustment driven by the minimisation of interfacial energies.

The microstructural evolution can be quantified using Θ_{cpp}. The initial rapidly-cooled microstructure has Θ_{cpp} ~78°, but the chill zone illustrated in Fig. 4.16c has a value of Θ_{cpp} within error of sub-solidus textural equilibrium. The influence of

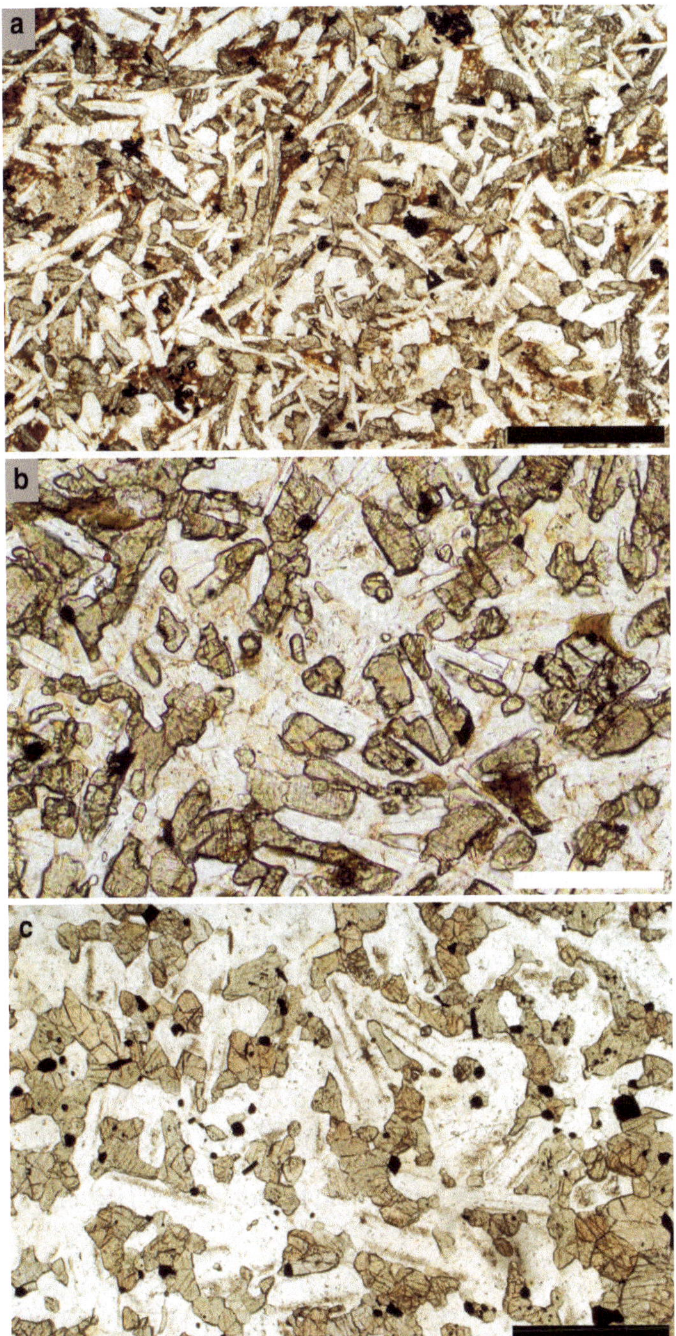

Fig. 4.16 **a** Microstructure in the 3.5 m *thick* Traigh Bhàn na Sgùrra dolerite sill, Isle of Mull. Plagioclase forms elongate laths with planar sides and augite (*pale brown*) and glass (*dark brown*) fill in the spaces between. Θ_{cpp} in this sample is 78 ± 3° (Holness et al. 2012a). Plane polarised light.

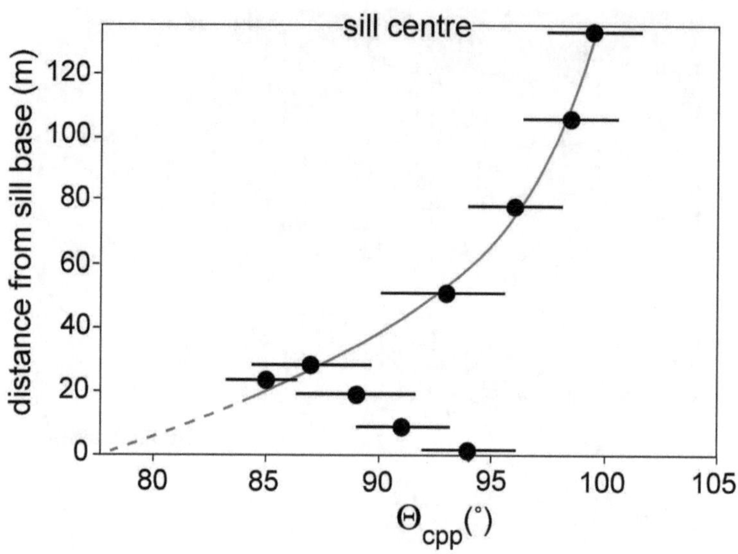

Fig. 4.17 Median augite-plagioclase-plagioclase dihedral angle (Θ_{cpp}) as a function of position within the lower half of the Basement Sill of Antarctica. The *dashed line* shows the expected trajectory of the spatial variation in Θ_{cpp}. The observed values depart significantly from the expected values due to sub-solidus textural equilibration of the primary igneous microstructure. Data from Holness et al. (2012b)

sub-solidus textural modification in large sills results in significant deviations of Θ_{cpp} from values consistent with calculated crystallisation times at the margins (Fig. 4.17). These higher than expected values are associated with the development of the granular microstructure, demonstrating a smooth increase of dihedral angle via the rotation of large areas of grain boundary. This rotation has a commensurately greater effect on the grain shape of the fine-grained rocks of the chill zone than it does on the much coarser-grained enclaves from Kula. In chill zones, therefore, the increase in dihedral angle occurs at the same time as the overall rounding of the grain shape.

Similar effects are observed in intra-plutonic chill zones, such as those formed in the Kap Edvard Holm gabbros by the rapid crystallisation of incoming batches of magma. The chill zones are marked by a fine grain size that increases upwards within each unit (Tegner et al. 1993). Θ_{cpp} is highest at the base of each chill zone, and decreases upwards, in an analogous manner to the sills (Fig. 4.17. Holness et al. 2012a, b). However, the microstructure in the chill zones is recognisably gabbroic

---

Scale bar is 200 μm long. **b** The chill zone of the 266 m *thick* Basement Sill of Antarctica (sample A-38). The augite grains are rounder than those illustrated in (**a**), with three-grain junctions formed by the intersection of curved grain boundaries (not clearly visible at this magnification). Θ_{cpp} in this sample is 96.5±2°. Plane polarised light. Scale bar is 200 μm long. **c** The chill zone of the Bushveld intrusion of South Africa (sample 114425 from the Harker Collection). The original tabular morphology of the plagioclase is still visible but the augite grains are now granular. Θ_{cpp} in this sample is 107±2.5° (Holness et al. 2012b). Plane polarised light. Scale bar is 1 mm long throughout. That they are not granular attests to the importance of grain size in controlling the extent of textural equilibration. Rotation of grain boundaries towards the high dihedral angles associated with sub-solidus textural equilibrium was not on a sufficiently large scale to have much effect on the grain shape.

The last stage of textural equilibration has been called textural coarsening (Higgins 2010). The simplest case to consider is that of a monomineralic aggregate, such as a plagioclase-rich domain in a troctolite or an olivine-rich domain in a peridotitic adcumulate. In such an aggregate, the primary influence on microstructure is grain growth (Hunter 1987). Normal grain growth is the term for sub-solidus grain growth driven by the differences in chemical potential between interfaces of different curvature (Harker and Parker 1945; Smith 1948; Martin and Doherty 1976; Vernon 2004). In a fully texturally equilibrated aggregate, in which dihedral angles are of the order 120°, large grains (those bounded by more than six other grains, such as the grain marked A in Fig. 4.14) are bounded by interfaces that are concave outwards, whereas small grains (those bounded by fewer than six other grains) have interfaces that are convex outwards. Because boundaries generally migrate towards their centres of curvature, large grains grow bigger at the expense of smaller grains. This process results in an increase of average grain size and, in the limit, a unimodal grain size.

Critically, normal grain growth is driven by differences in chemical potential created by interfacial curvature, and curvature is a function of grain size (rather than of growth-controlled morphology) only when both the equilibrium dihedral angle has been established and the curvature associated with this angle has been fully propagated along the entire length of grain boundaries. Therefore normal grain growth (in the strict sense) cannot begin until the microstructure has closely approached textural equilibrium. (Note that grain growth driven by grain boundary curvature can occur in the super-solidus if the mushy layer contains (texturally equilibrated) melt-free poly-crystalline aggregates. It should not be confused with Ostwald ripening which affects dispersed crystals in a uniform matrix).

How Common is Sub-Solidus Textural Equilibrium in Igneous Rocks?

The extent to which sub-solidus textural equilibrium is approached depends on thermal history, the spatial distribution of different minerals, the complexity of mineral crystal structure, and grain size.

While we don't yet have any constraints on the kinetics of the processes involved, observation shows that the rate of textural equilibration in monomineralic rocks (and in monomineralic domains of polymineralic rocks) is much faster than that in polymineralic rocks (Figs. 4.18a, b, c), because the former requires only movement of atoms across a grain boundary, whereas the latter requires mass migration along grain boundaries (Hunter 1987). Thus the extent of sub-solidus equilibration may vary on the scale of a thin section, depending on the spatial distribution of the

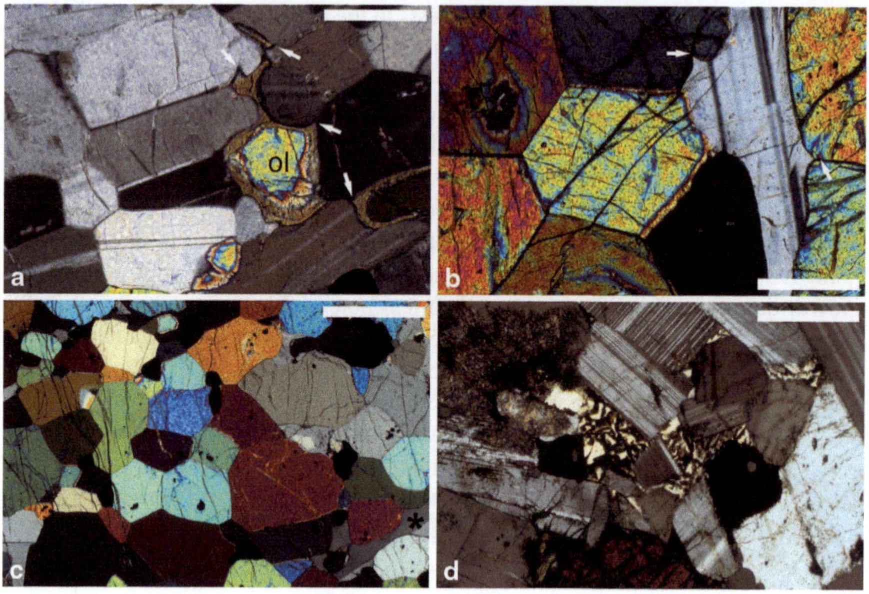

Fig. 4.18 **a** Troctolitic cumulate from the Rum Eastern Layered Intrusion. Cumulus grains of olivine are surrounded by rims of augite, which also extend down plagioclase-plagioclase grain boundaries. Θ_{cpp} is ~80°, denoting a significant departure from sub-solidus equilibrium (109°). Examples of low dihedral angles are *arrowed*. In contrast the microstructure of the neighbouring plagioclase-rich region is equilibrated, with smoothly curved grain boundaries, and dihedral angles of ~120°. Crossed polars. Scale bar is 200 μm long. **b** Olivine norite from the Colony Gabbro Complex, Sierra Leone (sample 80387 from the Harker Collection). To the *left* of the image, the olivine grains form a texturally equilibrated microstructure whereas the geometry of the olivine-plagioclase grain boundaries on the *right* are not in equilibrium. The two *arrowed* junctions have disequilibrium, low values of dihedral angle. Crossed polars. Scale bar is 200 μm long. **c** Olivine gabbro from the Freetown Complex of Sierra Leone (sample 143468 from the Harker Collection). This image shows olivine, with some grains of Fe-Ti oxide and a grain of plagioclase at the bottom right (marked by an *asterisk*). While the olivine-oxide region is well equilibrated, the shape of the plagioclase grain retains a significant inheritance from the original igneous morphology. This suggests that the rate of sub-solidus equilibration is dependent on the complexity of mineral structure and on the extent to which the original grains deviated from equant shapes. Crossed polars. Scale bar is 1 mm long. **d** Pocket of late-stage silicic liquid in gabbro from the Upper Zone of the Skaergaard Layered Series (sample 90–22 170). The pocket is bounded by plane-sided plagioclase grains, with no evidence for either super- or sub-solidus modification of the original growth-controlled morphology. Crossed polars. Scale bar is 200 μm long

different minerals. In layered intrusions, sub-solidus textural equilibrium is most commonly developed in monomineralic adcumulates (Hunter 1987), such as the olivine or clinopyroxene adcumulates of the Lower Zone of the Lilloise Intrusion (Magee et al. 2010), the plagioclase adcumulates of the Rum intrusion and the olivine cumulates of the Sgurr Dubh intrusion, Skye (Vernon 1970).

In contrast to mafic rocks, the lower solidus temperatures of granitoids make it difficult to modify their microstructures in the sub-solidus; granitoids therefore tend to retain their primary igneous microstructures (Vernon 2004; Vernon and Paterson

2008). The influence of solidus temperature is clearly visible in gabbroic rocks, in which the shapes of pockets of late-stage silicic liquid that crystallised as quartzo-feldspathic intergrowths display no sign of subsolidus modifications (Fig. 4.18d), even though adjacent higher-temperature mafic regions can be partially texturally equilibrated.

The grain size influences the extent of textural equilibration in several ways. First, the grain shape in a fine-grained rock is more affected by the changing curvature of grain boundaries as the three-grain junctions rotate towards the equilibrium dihedral angle. Second, the requirement for mass transport along grain boundaries to achieve textural equilibrium in polyphase rocks means that fine-grained rocks approach equilibrium more rapidly than coarse-grained equivalents. This is evident in the granular microstructure of chill zones of large intrusions, whereas the central coarser-grained parts of the same intrusion can display essentially unmodified gabbroic (i.e. primary igneous) microstructures. The onset of grain growth in such granular rocks (the last stage of textural equilibration) is rarely observed in crustal intrusions, confirming that grain-boundary adjustment under igneous cooling conditions does not lead to coarsening.

Conclusions

Microstructures in igneous rocks form by the nucleation and growth of minerals during cooling or depressurisation. The size, shape and distribution of these grains are functions of the extent to which the system is undercooled. If temperatures remain close to the liquidus, with very small undercooling, interfacial energy constraints can act to modify microstructures. Under these conditions Ostwald ripening results in the loss of the smallest crystals, and microstructures approach textural equilibrium: crystals assume the shape with the lowest interfacial energy, two- and three-grain junction geometries have the equilibrium dihedral angle, grain boundaries are smoothly curved, and liquid is either expelled or infiltrated in order to attain the minimum energy porosity. Once solidification is complete, and if temperatures remain sufficiently high to permit significant mass transport by diffusion, the minimization of the interfacial energy budget drives microstructural evolution towards the granular microstructure indicative of subsolidus textural equilibrium.

In contrast to the conclusions of earlier work on microstructures in layered intrusions (Hunter 1987), subsequent work has shown that interfacial energies are not generally the most significant control on grain shape, melt topology or porosity in crustal mafic-ultramafic intrusions. Instead, igneous microstructures, particularly in plagioclase-dominated rocks and in large intrusions with a coarse grain size, record the influence of the kinetics of grain growth, with abundant faceted grains and evidence for grain boundary geometries controlled by impingement during solidification. Ostwald ripening in such rocks is unlikely to play an important role. In small- to medium-sized crustal mafic intrusions there is little evidence of a tendency to adopt lower-energy crystal shapes during super-solidus cooling and microstructures formed during crystallisation are generally unmodified in the sub-solidus. Significant textural equilibration towards granular microstructures is evident only in some adcumulates (particularly those containing few minerals) and originally fine-grained rocks such as chill zones. Coarsening of fine-grained minerals does not contribute to the production of coarse-grained igneous microstructures.

Acknowledgements We are grateful to Fred Anderson, Roz Helz, Tony Philpotts, Bruce Marsh, Jane Hergt, Gordon Lister, John Maclennan and Ed Sawyer who have contributed to our developing understanding of microstructures in layered intrusions by loaning samples, and to Steve Laurie and Dan Pemberton who facilitated access to the Harker Collection. The previous two compilations of reviews on layered intrusions (edited by Ian Parsons and Grant Cawthorn) each contained a chapter on the microstructures of plutonic igneous rocks, written by Bob Hunter. Bob's groundbreaking work began within a Cambridge-based group that included Dan McKenzie and Mike Cheadle. It was the work of this group, and of Bob in particular, that triggered MBH's interest in rock microstructure.

References

Adamson AW (1990) Physical chemistry of surfaces, 5th edn. Wiley, New York

Ardell AJ (1972) The effect of volume fraction on particle coarsening: theoretical considerations. Acta Metall 20:61–71

Asada R, Shimobayashi N, Kitamura M (2002) Equilibrium form of negative crystals in igneous quartz. J Mineral Petrol Sci 97:59–69

Balluffi RW (1979) Grain boundary structure and segregation. in Interfacial Segregation (eds. Johnson WC, Blakely JM) Am Soc Metals, Ohio. pp 193–238

Baram M, Chatain D, Kaplan WD (2011) Nanometer-thick equilibrium films: the interface between thermodynamics and atomistics. Science 332:206–209

Beere W (1975) A unifying theory of the stability of penetrating liquid phases and sintering pores. Acta Metall 23:131–138

Berkebile CA, Dowty E (1982) Nucleation in laboratory charges of basaltic composition. Am Mineral 67:886–899

Bigg EK (1953) The supercooling of water. Proc Phys Soc (Lond), 66B:688–694

Bindeman IN (2003) Crystal sizes in evolving silicic magma chambers. Geology 31:367–370

Boistelle R, Astier JP (1988) Crystallisation mechanisms in solution. J Crystal Growth 90:14–30

Boorman S, Boudreau A, Kruger FJ (2004) The lower zone—critical zone transition of the Bushveld Complex: a quantitative textural study. J Pet 45:1209–1235

Boudreau AE (1994) Crystal aging in two component, two crystal systems. S Afr J Geol 97:473–485

Boudreau AE (1995) Crystal aging and the formation of fine-scale layering. Mineral Petrol 54:55–69

Boudreau AE, McBirney AR (1997) The Skeargaard Layered Series. Part III. Non-dynamic layering. J Petrol 38:1003–1020

Brandeis G, Jaupart C, Allegre CJ (1984) Nucleation, crystal growth and the thermal regime of cooling magmas. J Geophys Res 89:10161–10177

Bryon DN, Atherton MP, Hunter RH (1994) The description of the primary textures of "Cordilleran" granitic rocks. Contrib Mineral Petrol 117:66–75

Bryon DN, Atherton MP, Hunter RH (1995) The interpretation of granitici textures from serial thin sectioing, image analysis and 3-dimensional reconstruction. Min Mag 59:203–211

Bulau JR, Waff HS, Tyburczy JA (1979) Mechanical and thermodynamic constraints on fluid distribution in partial melts. J Geophys Res 84:6102–6108

Burkhard DJM (2001) Kinetics of crystallization: example of micro-crystallisation in basalt lava. Contrib Mineral Petrol 142:724–737
Cabane H, Laporte D, Provost A (2001) Experimental investigation of the kinetics of Ostwald ripening of quartz in silicic melts. Contrib Mineral Petrol 142:361–373
Cabane H, Laporte D, Provost A (2005) An experimental study of Ostwald ripening of olivine and plagioclase in silicate melts: implications for the growth and size of crystals in magmas. Contrib Mineral Petrol 150:37–53
Cahn JW (1980) Surface stress and the chemical equilibrium of small crystals: I. The case of the isotropic surface. Acta Metall 28:1333–1338
Camel D, Lesoult G, Eustathopoulos N (1981) Metastable equilibrium states of solid-liquid interfaces in metallic binary alloys. J Crystal Growth 53:327–336
Campbell IH (1987) Distribution of orthcumulate textures in the Jimberlana Intrusion. J Geology 95:35–53
Cashman KV (1988) Crystallisation of Mount St. Helens 1980-1986 dacite: a quantitative textural approach. Bull Volcanol 50:194–209
Cashman KV (1993) Relationship between plagioclase crystallization and cooling rate in basaltic melts. Contrib Mineral Petrol 113:126–142
Castro JM, Cashman KV, Manga M (2003) A technique for measuring 3D crystal-size distributions of prismatic microlites in obsidian. Am Mineral 88:1230–1240
Cesare B, Ferrero S, Salvioli-Mariani E, Pedron D, Cavallo A (2009) "Nanogranite" and glassy inclusions: the anatectic melt in migmatites and granulites. Geology 37:627–630
Cheadle MJ (1989) Properties of texturally equilibrated two-phase aggregates [Ph.D. thesis]: Cambridge, University of Cambridge, 163 p
Cheadle MJ, Elliott MT, McKenzie D (2004) Percolation threshold and permeability of crystallizing igneous rocks. Geology 32:757–760
Davies CKL, Nash P, Stevens RN (1980) The effect of volume fraction of precipitate on Ostwald ripening. Acta Metall 28:179–189
Donaldson CH (1976) An experimental investigation of olivine morphology. Contrib Mineral Petrol 57:187–213
du Bray EA, Bacon CR, John DA, Wooden JL, Mazbad FK (2011) Episodic intrusion, internal differentiation, and hydrothermal alteration of the Miocene Tatoosh intrusive suite south of Mount Rainier, Washington. Geol Soc Am Bull 123:534–561
Elliott MT, Cheadle MJ, Jerram DA (1997) On the identification of textural equilibrium in rocks using dihedral angle measurements. Geology 25:355–358
Faure F, Trolliard G, Nicollet C, Montel J-M (2003) A developmental model of olivine morphology as a function of the cooling rate and the degree of undercooling. Contrib Mineral Petrol 145:251–263
Feeney R, Schmidt SL, Strickholm P, Chadam J, Ortoleva P (1983) Periodic preciptation and coarsening waves: applications of the competitive particle growth model. J Chem Phys 78:1293–1311
Feinn D, Ortoleva P, Scalf W, Schmidt S, Wolff M (1978) Spontaneous pattern formation in precipitating systems. J Chem Physc 69:27–39
Gibbs JW (1961) The scientific papers of J. Willard Gibbs, 1. Dover Publications Inc, New York, p 219
Glicksman ME (2011) Principles of Solidification: an introduction to modern casting and crystal growth concepts. Springer, New York. ISBN: 978-1-4419-7343-6
Hammer JE, Cashman KV, Hoblitt RP, Newman S (1999) Degassing and microlite crystallization during pre-climactic events of the 1991 eruption of Mt. Pinatubo, Philippines. Bull Volcanol 60:355–380
Hammouda T, Laporte D (2000) Ultrafast mantle impregnation by carbonatite melts. Geology 28:283–285
Harker D, Parker ER (1945) Grain shape and grain growth. Trans Am Soc Metals 34:156–195
Helz RT (2009) Proceses active in mafic magma chambers: the example of Kilauea Iki Lava Lake, Hawaii. Lithos 111:37–46

Helz RT, Thornber CR (1987) Geothermometry of Kilauea Iki lava lake, Hawaii. Bull Colcanol 49:651–668
Herring C (1951) Some theorems on the free energies of crystal surfaces. Phys Rev 82:87–93
Hersum TG, Marsh BD (2006) Igneous microstructures from kinetic models of crystallsiation. J Volcanol Geotherm Res 154:34–47
Hersum TG, Marsh BD (2007) Igneous textures: on the kinetics behind the words. Elements 3:247–253
Higgins MD (1998) Origin of anorthosite by textural coarsening: quantitative measures of a natural sequence of textural development. J Petrol 39:1307–1323
Higgins MD, Roberge J (2003) Crystal size distribution of plagioclase and amphibole from Soufriere Hills Volcano, Montserrat: evidence for dynamic crystallization—textural coarsening cycles. J Petrol 44:1401–1411
Hiraga T, Nishikawa O, Nagase T, Akizuki M, Kohlstedt DL (2002) Interfacial energies for quartz and albite in pelitic schist. Contrib Mineral Petrol 143:664–672
Holness MB (1993) Temperature and pressure dependence of quartz-aqueous fluid dihedral angles: the control of adsorbed H_2O on the permeability of quartzites. Earth Planet Sci Lett 117:363–377
Holness MB (1995) The effect of feldspar on quartz-H_2O-CO_2 dihedral angles at 4kbar, with consequences for the behaviour of aqueous fluids in migmatites. Contrib Mineral Petrol 118:356–364
Holness MB (1997a) Surface chemical controls on pore-fluid connectivity in texturally equilibrated materials. In: Jamtveit B, Yardely B (eds) Fluid flow and transport in rocks. Chapman and Hall, London, pp 149–169
Holness MB (1997b) Geochemical self-organisation of olivine-grade contact metamorphosed chert nodules in dolomite marble, Kilchrist, Skye. J Met Geol 15:765–775
Holness MB (2006) Melt-solid dihedral angles of common minerals in natural rocks. J Petrol 47:791–800
Holness MB (2007) Textural immaturity of cumulates as an indicator of magma chamber processes: infiltration and crystal accumulation in the Rum Eastern Layered Intrusion. J Geol Soc Lond 164:529–539
Holness MB (2010) Decoding dihedral angles in melt-bearing and solidified rocks. In: Forster MA, Fitz Gerald JD, Lister GS (eds). J Virt Expl 35: paper 3, doi:10.3809/jvirtex.2010.00265
Holness MB, Anderson AT, Martin VM, Maclennan J, Passmore E, Schwindinger K (2007a) Textures in partially solidified crystalline nodules: a window into the pore structure of slowly cooled mafic intrusions. J Petrol 48:1243–1264
Holness MB, Humphreys MCS (2003) The Traigh Bhàn na Sgùrra sill, Isle of Mull: flow localization in a major magma conduit. J Petrol 44:1961–1976
Holness MB, Sawyer EW (2008) On the pseudomorphing of melt-filled pores during the crystallization of migmatites. J Petrol 49:1343–1363
Holness MB, Cheadle MJ, McKenzie D (2005) On the use of changes in dihedral angle to decode late-stage textural evolution in cumulates. J Petrol 46:1565–1583
Holness MB, Tegner C, Nielsen TFD, Stripp G, Morse SA (2007b) A textural record of solidification and cooling in the Skaergaard intrusion, East Greenland. J Petrol 48:2359–2377
Holness MB, Nielsen TFD, Tegner C (2007c) Textural maturity of cumulates: a record of chamber filling, liquidus assemblage, cooling rate and large-scale convection in mafic layered intrusions. J Petrol 48:141–157
Holness MB, Humphreys MCS, Sides R, Helz RT, Tegner C (2012a) Toward an understanding of disequilibrium dihedral angles in mafic rocks. J Geophy Res 117, B06207: doi:10.1029/2011JB008902
Holness MB, Richardson C, Helz RT (2012b) Disequilibrium dihedral angles in dolerite sills: a new proxy for cooling rate. Geology 40:795–798
Holness MB, Namur O, Cawthorn RG (2013) Disequilibrium dihedral angles in layered intrusions: a microstructural record of fractionation. Journal of Petrology 54:2067–2093

Hunter RH (1987) Textural equilibrium in layered igneous rocks. In: Parsons I (ed) Origins of igneous layering. Dordrecht: D. Reidel. pp 473–503.

Hunter RH (1996) Texture development in cumulate rocks. In: Cawthorn RG (ed) Layered Intrusions. Developments in petrology. Amsterdam: Elsevier Science B.V, pp 77–101 (15)

Ikeda S, Toriumi M, Yoshida H, Shimizu I (2002) Experimental study of the textural development of igneous rocks in the late stage of crystallization: the importance of interfacial energies under non-equilibrium conditions. Contrib Mineral Petrol 142:397–415

Jerram DA, Cheadle MJ, Philpotts AR (2003) Quantifying the building blocks of igneous rocks: are clustered crystal frameworks the foundation? J Petrol 44:2033–2051

Johnson BR, Glazner AF (2010) Formation of K-feldspar megacrysts in granodioritic plutons by thermal cycling and late-stage textural coarsening. Contrib Mineral Petrol 159:599–619

Johnson WC, Chavka NG, Ku R, Bomback JL, Wynblatt PP (1978) Orientation dependence of surface segregation in a dilute Ni-Au alloy. J Vacuum Sci Tech 15:467–469

Kirkpatrick RJ (1975) Crystal growth from the melt: a review. Am Mineral 60:798–814

Kirkpatrick RJ (1977) Nucleation and growth of plagioclase, Makaopuhi and Alae lava lakes, Kilauea volcano, Hawaii. Geol Soc Amer Bull 88:78–84

Kretz R (1966) Interpretation of the shape of mineral grains in metamorphic rocks. J Petrol 7:68–94

Kruhl JH (2001) Crystallographic control on the development of foam textures in quartz, plagioclase and analogue material. Int J Earth Sci 90:104–117

Kruhl JH, Peternell M (2002) The equilibration of high-angle grain boundaries in dynamically recrystallized quartz: the effect of crystallography and temperature. J Struct Geol 24:1125–1137

Kubota N, Fujisawa Y, Tadaki T (1988) Effect of volume on the supercooling temperature for primary nucleation of potassium nitrate from aqueous solution. J Crystal Growth 89:545–552

Laporte D (1994) Wetting behaviour of partial melts during crustal anatexis: the distribution of hydrous silicic melts in polycrystalline aggregates of quartz. Contrib Mineral Petrol 116:486–499

Laporte D, Provost A (2000) Equilibrium geometry of a fluid phase in a polycrystalline aggregate with anisotropic surface energies: dry grain boundaries. J Geophys Res 105:25937–25953

Laporte D, Watson EB (1995) Experimental and theoretical constraints on melt distribution in crustal sources: the effect of crystalline anisotropy on melt interconnectivity. Chem Geol 124:161–184

Leibl C, Kuntcheva B, Kruhl JH, Kunze K (2007) Crystallographic orientations of quartz grain-boundary segments formed during dynamic recrystallisation and subsequent annealing. Eur J Mineral 19:735–744

Lifshitz IM, Slyozov VV (1961) The kinetics of precipitation from supersaturated solid solutions. J Phys Chem Solds 19:35–50

Lofgren G (1974) An experimental study of plagioclase crystal morphology: isothermal crystallization. Am J Sci 274:243–273

Lofgren G (1983) Dynamc crytallisation of chondrule melts of porphyritic olivine composition: textures experimental and natural. Geochim Cosmochim Acta 53:461–470

Magee C, O'Driscoll B, Chambers AD (2010) Crystallisation and textural evolution of a closed-system magma chamber: insights from a crystal size distribution study of the Lilloise layered intrusion, East Greenland. Geol Mag 147:363–379

Manley CR (1996) Morphology and maturation of melt inclusions in quartz phenocrysts from the Badlands rhyolite lava flow, southwestern Idaho. Am Mineral 81:158–168

Marchildon N, Brown M (2002) Grain-scale melt distribution in two contact aureole rocks: implications for controls on melt localization and deformation. J Met Geol 20:381–396

Marsh BD (1988) Crystal size distribution (CSD) in rocks and the kinetics and dynamics of crystallization I: theory. Contrib Mineral Petrol 99:277–291

Marsh BD (1995) Solidification fronts and magmatic evolution. Mineral Mag 60:5–40

Marsh BD (1998) On the interpretation of crystal size distributions in magmatic systems. J Petrol 39:553–599

Marsh BD (2007) Crystallisation of silicate magmas deciphered using crystal size distributions. J Am Ceram Soc 90:746–757

Martin JW, Doherty RD (1976) Stability of microstructures in metallic systems. Cambridge University Press, pp 298

McDonnell S, Troll VR, Emeleus CH, Meighan IG, Brock D, Gould RJ (2004) Intrusive history of the Slieve gullion ring dyke, Ireland: implications for the internal structure of silicic subcaldera magma chambers. Min Mag 68:725–738

Means WD, Park Y (1994) New experimental approach to understanding igneous texture. Geology 22:323–326

Melia TP, Moffitt WP (1964) Crystallisation from aqueous solution. J Colloid Sci 19:433–447

Minarik WG, Watson EB (1995) Interconnectivity of carbonate melt at low melt fraction. Earth Planet Sci Lett 133:423–437

Mock A, Jerram DA (2005) Crystal size distributions (CSD) in three dimensions: insights from the 3D reconstruction of a highly porphyritic rhyolite. J Petrol 46:1525–1541

Moore JG, Evans BW (1967) The role of olivine in the crystallization of the prehistoric Makaopuhi lava lake, Hawaii. Contrib Mineral Petrol 15:202–223

Moore JG, Sisson TW (2008) Igneous phenocrystic origin of K-feldspar megacrysts in granitic rocks from the Sierra Nevada batholith. Geosphere 4:387–400

Mullins MW (1957) Theory of thermal grooving. J Appl Phys 28:333–339

Nabelek PI, Whittington AG, Sirbescu MLC (2010) The role of H_2O in rapid emplacement and crystallisation of granite pegmatites: resolving the paradox of large crystals in highly undercooled melts. Contrib Mineral Petrol 160:313–325

Nielsen AE, Söhnel O (1971) Interfacial tensions, electrolyte crystal-aqueous solution, from nucleation data. J Crystal Growth 11:233–242

Park H-H, Yoon DN (1985) Effect of dihedral angle on the morphology of grains in a matrix phase. Metall Trans 16A:923–928

Park Y, Hanson B (1999) Experimental investigation of Ostwald-ripening rates of forsterite in the haplobasaltic system. J Volcanol Geotherm Res 90:103–113

Passerone A, Sangiorgi R, Eustathopoulos N, Desre P (1979) Microstructure and interfacial tensions in Zn-In and Zn-Bi alloys. Metal Sci 13:359–365

Passmore E, Maclennan J, Fitton G, Thordarson T (2012) Mush disaggregation in basaltic magma chambers: evidence from the AD 1793 Laki eruption. J Petrol 53:2593–2623

Paterson SR, Vernon RH, Zák J (2005) Mechanical instabilities and accumulations of K-feldspar megacrysts in granitic magma, Tuolumne Batholith, California, USA. In: Kšhn, D. (Ed.), General Contributions 2005. J Virtual Expl Electronic Edition 18, Paper 11441–8142.

Philpotts AR, Dickson LD (2000) The formation of plagioclase chains during convective transfer in basaltic magma. Nature 406:59–61

Philpotts AR, Dickson LD (2002) Millimeter-scale modal layering and the nature of the upper solidification zone in thick flood-basalt flows and other sheets of magma. J Struct Geol 24:1171–1177

Philpotts AR, Philpotts DE (2005) Crystal-mush compaction in the Cohassett flood-basalt flow, Hanford, Washington. J Volcanol Geotherm Res 145:192–206

Philpotts AR, Carroll M, Hill JM (1996) Crystal-mush compaction and the origin of pegmatitic segregation sheets in a thick flood-basalt flow in the Mesozoic Hartford Basin, Connecticut. J Petrol 37:811–836

Philpotts AR, Shi J, Brustman C (1998) Role of plagioclase crystal chains in the differentiation of partly crystallised basaltic magma. Nature 395:343–346

Philpotts AR, Brustman C, Shi J, Carlson WD, Denison C (1999) Plagioclase-chain networks in slowly cooled basaltic magma. Am Min 84:1819–1829

Putnis A, Mauthe G (2001) The effect of pore size on cementation in porous rocks. Geofluids 1:37–41

Saiki K, Laporte D, Vielzeuf D, Nakashima S, Boivin P (2003) Morphological analysis of olivine grains annealed in an iron-nickel matrix: experimental constraints on the origin of pallasites and on the thermal history of their parent bodies. Meteorit Planet Sci 38:427–444

Sawyer EW (1999) Criteria for the recognition of partial melting. Phys Chem Earth. A24:269–279

Schafer FN, Foley SF (2002) The effect of crystal orientation on the wetting behaviour of silicate melts on the surfaces of spinel peridotite minerals. Contrib Mineral Petrol 143:254–261
Scherer GW (1999) Crystallisation in pores. Cement Concrete Res 29:1347–1358
Schiavi F, Walte N, Keppler H (2009) First in situ observation of crystallization processes in a basaltic-andesitic melt with the moissanite cell. Geology 37:963–966
Schiavi F, Walte N, Konschak A, Keppler H (2010) A moissanite cell apparatus for optical in situ observation of crystallizing melts at high temperature. Am Mineral 95:1069–1079
Schwindinger KR (1999) Particle dynamics and aggregation ofcrystals in a magma chamber with application to Kilauea Iki olivines. J Volcanol Geothermal Res 88:209–238
Schwindinger KR, Anderson AT (1989) Synneusis of Kilauea Iki olivines. Contrib Mineral Petrol 103:187–198
Sens H, Eustathopoulos N, Camel D (1989) Anisotropy of solid Si-liquid (Al, Si) interfacial tension in the binary and Sr-doped Al-Si eutectic. J Crystal Growth 98:751–758
Shea T, Larsen JF, Gurioli L, Hammer JE, Houghton BF, Cioni R (2009) Leucite crystals: surviving witnesses of magmatic processes preceding the 79AD eruption of Vesuvius, Italy. Earth Planet Sci Lett 281:88–98
Simakin AG, Bindeman LN (2008) Evolution of crystal sizes in the series of dissolution and precipitation events in open magma systems. J Volcanol Geotherm Res 177:997–1010
Smith CS (1948) Grains, phases and interfaces: an interpretation of microstructure. AIME Trans 175:15–51
Smith CS (1964) Some elementary principles of polycrystalline microstructure. Metall Rev 9:1–48
Stevenson DJ (1986) On the role of surface tension in the migration of melts and fluids. Geophys Res Lett 13:1149–1152
Stickels CA, Hücke EE (1964) Measurement of dihedral angles. Trans Metall Soc AIME 230:795–801
Sutton AP, Baluffi RW (1996) Interfaces in crystalline materials. (Monograph Phys Chem Materials 51). Clarendon, Oxford
Takei Y, Shimizu I (2003) The effects of liquid composition, temperature and pressure on the equilibrium dihedral angles of binary solid-liquid systems inferred from a lattice-like model. Phys Earth Planet Int 139:225–242
Tegner C, Wilson JR, Brooks CK (1993) Intraplutonic quench zones in the Kap Edvard Holm Layered Gabbro Complex, East Greenland. J Petrol 34:681–710
Tegner C, Cawthorn RG, Kruger FJ (2006) Cyclicity in the main and upper zones of the Bushveld complex, South Africa: crystallization from a zoned magma sheet. J Petrol 47:2257–2279
Thomson W (1871) On the equilibrium of vapour at a curved surface of liquid. Philos Mag 42:448–452
Vernon RH (1968) Microstructures of high-grade metamorphic rocks at Broken Hill, Australia. J Petrol 9:1–22
Vernon RH (1986) K-feldspar megacrysts in granites—phenocrysts, not porphyroblasts. Earth Sci Rev 23:1–63
Vernon RH (1970) Comparative grain-boundary studies in some basic and ultrabasic granulites, nodules and cumulates. Scot J Geol 6:337–351
Vernon RH (2004) A practical guide to rock microstructure. Cambridge University, Cambridge
Vernon RH (2010) Granites really are magmatic: using structural evidence to refute some obstinate hypotheses. In: Forster MA, Fitz Gerald JD, Lister GS (eds). J Virt Expl 35: paper 1, doi: 10.3809/jvirtex.2010.00264
Vernon RH, Collins WJ (2011) Structural criteria for identifying granitic cumulates. J Geol 119:127–142
Vernon RH, Paterson SR (2008) How late are K-feldspar megacrysts in granites? Lithos 104:327–336
Vinet N, Higgins MD (2010) Magma solidification processes beneath Kilauea Volcano, Hawaii: a quantitative textural and geochemical study of the 1969–1975 Mauna Ulu lavas. J Petrol 51:1297–1332
Voll G (1960) New work on petrofabrics. Liverpool Manchester Geol J 2:503–567

Von Bargen N, Waff H (1986) Permeabilities, interfacial areas and curvature of partially molten systems: results of numerical computations of equilibrium microstructures. J Geophys Res 91:9261–9276

Waff HS, Faul UH (1992) Effects of crystalline anisotropy on fluid distribution in ultramafic partial melts. J Geophys Res 97:9003–9014

Wanamaker BJ, Kohlstedt DL (1991) The effect of melt composition on the wetting angle between silicate melts and olivine. Phys Chem Minerals 18:26–36

Wark DA, Watson EB (2006) TitaniQ: a titanium-in-quartz geothermometer. Contrib Mineral Petrol 152:743–754

Waters C, Boudreau AE (1996) A re-evaluation of crystal-size distributions in chromite cumulates. Am Mineral 81:1452–1459

Watson EB (1982) Melt infiltration and magma evolution. Geology 10:236–240

Watson EB, Brenan JM (1987) Fluids in the lithosphere. 1. Experimentally-determined wetting characteristics of CO_2-H_2O fluids and their implications for fluid transport, hostrock, physical properties, and fluid inclusion formation. Earth Planet Sci Lett 85:497–515

Weedon DS (1965) The layered ultrabasic rocks of Sgurr Dubh, Isle of Skye. Scot J Geol 1:41–68

Welsch B, Faure F, Famin V, Baronnet A, Bachèlery P (2013) Dendritic crystallisation: a single process for all the textures of olivine in basalts? J Petrol 54:539–574

Wilhelm S, Wörner G (1996) Crystal size distribution in Jurassic Ferrar flows and sills (Victoria Land, Antarctica): evidence for proceses of cooling, nucleation and crystallization. Contrib Mineral Petrol 125:1–15

Wright TL, Okamura RT (1977) Cooling and crystallization of tholeiitic basalt, 1965 Makaopuhi Lava lake, Hawaii. USGS Prof paper 1004. pp 83

Chapter 5
Silicate Liquid Immiscibility in Layered Intrusions

Ilya V. Veksler and Bernard Charlier

Abstract More and more evidence for the development of silicate liquid immiscibility during cooling of magmas in layered intrusions have been presented. Here, we review some theoretical principles with a focus on the separation of two silicate melts, i.e. silica-rich vs. iron-rich. We discuss the role of melt structure and present phase equilibria relevant to stable and metastable immiscibility. The understanding of immiscibility in magmas has strongly benefited from recent progress in experimental approaches. Kinetics studies evidence the importance of nucleation barriers in producing unmixing, coarsening and potential separation of equilibrium melts. Improvement of analytical tools has also enabled detailed study of major and trace element partitioning. The study of immiscible emulsion in volcanic rocks also brings important information on the evolution of plutonic systems and on the potential formation of compositional gap along liquid lines of descent. We then present the most recent evidence for immiscibility in some major layered intrusions, i.e. the Skaergaard, Sept Iles, intrusions of the Emeishan province, and the Bushveld complex. Paired melts are identified as contrasted melt inclusions trapped in apatite and their segregation can be responsible for the formation of Fe–Ti–P-rich rocks. We finally discuss more broadly the potential effect of immiscibility in interstitial melt and the implications on the evolution of the crystal mush.

Keywords Unmixing · Binodal · Skaergaard · Sept Iles · Bushveld

I. V. Veksler (✉)
Department of Mineralogy and Petrology, Technical University Berlin,
Ackerstrasse 74, 13355 Berlin, Germany
e-mail: veksler@gfz-potsdam.de

Department of Geology, Perm State University, Bukireva 15, Perm, Russia, 614990

B. Charlier
Department of Earth, Atmospheric and Planetary Sciences, Massachusetts Institute of Technology, Cambridge, MA 02139, USA
e-mail: b.charlier@ulg.ac.be

Department of Geology, University of Liège, 4000 Sart Tilman, Belgium

B. Charlier et al. (eds.), *Layered Intrusions*, Springer Geology,
DOI 10.1007/978-94-017-9652-1_5

Introduction

Popular views on the significance and petrogenetic role of silicate liquid immiscibility have changed several times since the start of modern igneous petrology, from early enthusiasm (Daly 1914) to scepticism (Greig 1927; Bowen 1928) and to revived interest (De 1974; McBirney and Nakamura 1974; Philpotts 1979, 1982; Dixon and Rutherford 1979; Roedder 1979). Igneous petrologists have stumbled upon traces of liquid immiscibility in lavas and plutonic rocks sometimes by serendipity as, for example, described by Roedder (1978) who discovered glassy immiscibility products in samples of lunar basalts while trying to find hydrous fluid inclusions in olivine and plagioclase phenocrysts. Referring to another fairy tale, one could call liquid immiscibility Cinderella of igneous petrology as it is not getting proper recognition despite numerous glass and crystal "slippers" left behind in magmatic rocks. Recent years have seen a new wave of interest and significant progress in experimental and petrographic studies of silicate liquid immiscibility and its magmatic products. Hopefully, this will be the last major change of fortunes and immiscibility will hold a proper place among important magma differentiation processes.

It is believed that immiscibility between silicate melts and molten Fe–Ni alloys was responsible for the global, first-order planetary differentiation, which resulted in the formation of metal core and silicate shell during accretion of the Earth and other Earth-like planets. Immiscibility between salt, sulfide and silicate melts probably leads to formation of some types of carbonatites and magmatic sulfide deposits. Immiscibility between the drastically different chemical types of magmatic liquids has been reviewed elsewhere (e.g. Veksler 2004; Thompson et al. 2007) and is not discussed in this chapter. Here we consider only immiscibility between a pair of silicate melts. This restriction also excludes from our consideration hypothetical, essentially silica-free apatite-magnetite-ilmenite and apatite-magnetite-sulfide immiscible melts, which may have been responsible for the formation of nelsonites and Kiruna-type ore deposits in massif anorthosites and some layered intrusions (e.g., Philpotts 1967; Kolker 1982; Duchesne 1999; Hou et al. 2011).

In this chapter we briefly cover the progress since the last major reviews of the late 1970s and early 1980s (Roedder 1978, 1979; Philpotts 1979, 1982). The emphasis of this review is on layered intrusions because they are the topic of this book, and also because plutonic environment is especially favorable for full development of immiscibility and liquid-liquid fractionation. We first review some theoretical backgrounds on melt structure, phase equilibria, kinetics, and element partitioning. Separate sections deal with intrusions where the studies of immiscibility have been most extensive. We finally discuss more broadly the potential effect of immiscibility in the crystal mush.

Theoretical Principles and Definitions

Liquid Immiscibility at the Atomic Level

Liquid phases do not have long-range ordering on the atomic and molecular scale and therefore have much greater structural tolerance than crystalline solid solutions. Immiscibility in the liquid state usually occurs when the conjugate phases fundamentally differ in the type of chemical bonds. Therefore, immiscible fluid phases can be classified by their dominant types of bonding: ionic, covalent, metallic or intermolecular. Silicate melts are characterized by interplay of covalent and ionic bonds between components and they are correspondingly poorly miscible with liquids having metallic bonding (molten metals and sulphide melts) and fully dissociated ionic and molecular liquids (molten salts and supercritical C–O–H fluids).

Silicate liquid immiscibility is rooted in fundamental properties of silicate melts and highlights the distinction between two types of components, network formers and network modifiers. Network formers are small, highly-charged cations having, because of the small size, coordination numbers no greater than 4 in melts, glasses and crystalline silicates. Four-fold coordinated Si^{4+} is the most abundant and important network former of natural magmas. The group of network-formers in natural and technological melts also includes P^{5+}, B^{3+} and some other cations, which in nature are usually present only in trace amounts (e.g. Ge^{4+}, Sb^{3+} and As^{3+}).

Network-modifying cations have ionic radii greater than 87.2 pm and coordination numbers equal to, or higher than 5 (Hudon and Baker 2002). The group comprises alkali and alkaline earth cations excluding Li^{+}, Be^{2+} and Mg^{2+}, REE cations, Th^{4+} and U^{4+}. Cations with ionic radii greater than Si^{4+} (26 pm) but smaller than 87.2 pm belong to the amphoteric group. According to spectroscopic studies, amphoteric cations have at least two coordination numbers. One coordination number is four and another one is usually five or six (see Hudon and Baker 2002 for details). The most important amphoteric cation of natural magmatic melts is Al^{3+} and the group also includes Li^{+}, Be^{2+}, Mg^{2+}, Mn^{2+}, Zn^{2+}, Sc^{3+}, Ga^{3+}, Ti^{4+}, Zr^{4+}, Hf^{4+}, Nb^{5+}, Ta^{5+} and some other high field strength cations.

It has been noticed since long ago (see review of earlier works by Hess 1995) that the size of miscibility gaps in binary metal oxide—silica systems MO–SiO_2 correlate with Coulombic properties of the metal cations, such as ionic potential. Ionic potential is the ratio of Z/r, where Z is the nominal charge and r is the atomic radius of the cation. The correlation is especially evident in the binaries of alkali and alkaline earth cations (Fig. 5.1). To rationalize the correlation, Hess (1995, 1996) proposed that the forces responsible for melt unmixing at the atomic level are Coulombic and repulsive in nature. He pointed out that network-modifying cations are surrounded in silicate melts by both bridging and non-bridging oxyanions. Oxygen ions in bridging (Si–O–Si) and non-bridging positions (Si–O—M, where M is a network-modifying cation) isolate the network-modifier cations from each other by providing screens that mask the positive charges. However, modifier cations that are partly or wholly coordinated by bridging oxygen are poorly shielded from

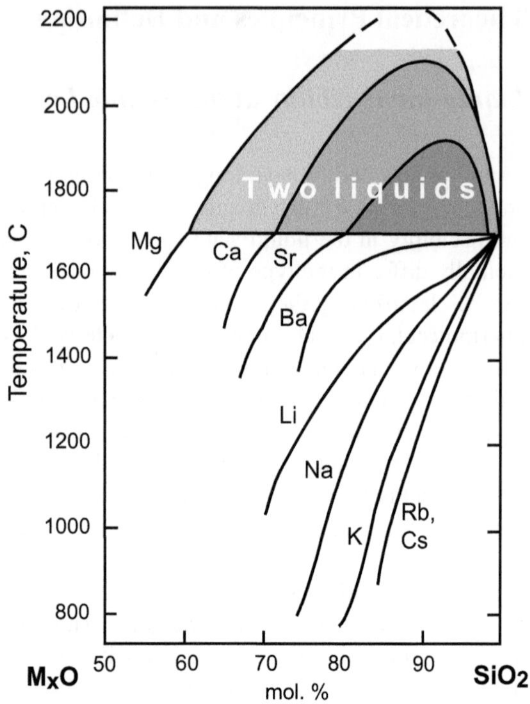

Fig. 5.1 Silica-rich portions of alkali and alkaline earth oxides binaries with two liquid fields (after Levin et al. 1964)

each other because the bridging oxygen ions are bonded to Si^{4+} cations by strong covalent bonds. Consequently, significant Coulombic repulsions build up between network-modifying cations, and they may eventually result in melt unmixing and phase separation. The higher the ionic potential of modifier cations, the greater are the Coulombic repulsions between them, and the larger is the width of the resulting miscibility gap.

In a detailed analysis of binary systems, Hudon and Baker (2002) noticed that the relationships between the ionic potential of network-modifying cations and the size of miscibility gaps are not linear. Furthermore, each group of homovalent cations (cations with the same Z) forms its own trend. For each group, the compositional width of miscibility gap and the unmixing temperature increase with Z/r, reach a maximum and then decrease. The maximum is reached at $r = 87.2$ pm, which, as mentioned above, defines the transition from true network modifiers to amphoteric cations. The authors proposed the existence of structural control on immiscibility in which the cation radius appears to have an important role. To rationalize the convex trends, they pointed out the increasing degree of covalency of the M–O bonds in homovalent groups of elements with decreasing r. Therefore, amphoteric cations capable of 4-fold coordination seem to undergo weak or no Coulombic repulsion and tend to produce little or no phase separation.

Liquid immiscibility in lunar and terrestrial magmas has been mostly observed in Fe-rich compositions and the link between immiscibility and high Fe contents has been interpreted as a consequence of specific properties of Fe cations. Cations of transition elements from the first row in the periodic table, such as Fe^{2+}, Co^{2+}, Ni^{2+}, Cu^{2+}, V^{3+}, and Cr^{3+}, have five d-electron orbitals, which are known to poorly shield the atomic nucleus (Hudon and Baker 2002). Notably, miscibility gaps associated with the divalent cations in corresponding oxide-silica binaries are larger than expected from the general *Z/r* trend. Immiscibility gaps are even greater in binaries of trivalent Cr^{3+}, V^{3+}, and Fe^{3+} (Hudon and Baker 2002).

Binodal and Spinodal

In thermodynamic terms, liquid immiscibility is linked with variations of the Gibbs free energy of mixing *G* with composition *X* at constant pressure and temperature (Fig. 5.2a). Detailed discussions of free energy in relation to immiscibility can be found in numerous reviews and textbooks (e.g. James 1975; Doremus 1994; Shelby 2005). The simplest case is that of a binary liquid where free energy varies smoothly with composition in the whole range between pure components A and B (Fig. 5.2a). Here the *G-X* curve is characterized in the region of unmixing by a local maximum. Tangent points *a* and *b* of the common tangent line define the compositions of the equilibrium immiscible liquids L_a and L_b. In the compositional interval between L_a and L_b a single homogenous liquid has a higher free energy than a mixture of the two immiscible liquids. Thus the two-liquid phase assemblage is the thermodynamically stable one, whereas the *a-b* segment of the *G-X* curve is metastable. The compositions L_a and L_b depend on temperature, and the loci of points *a* and *b* define a *T-X* curve that is termed binodal (Fig. 5.2b). All dry immiscible silicate systems known to date are characterized by convex binodals with upper consolute (or critical solution) points, at which the distinction between coexisting liquid phases vanishes. Some researchers (Navrotsky 1992; Hess 1996) have proposed that concave binodals and lower consolute points may also exist. At the consolute points liquids undergo second order (continuous) phase transition. As they approach this transition with falling temperature they pass through a region of critical opalescence (Fig. 5.2b). This region is the site of anomalously large compositional fluctuations and abnormal variations of viscosity, density, compressibility, and other physical properties (Simmons et al. 1970; Simmons and Macedo 1971; Mazurin and Porai-Koshits 1984). Similarly, the loci of inflexion points s_1 and s_2 in the *G-X* curves (Fig. 5.2a) where the second derivative $(\delta^2G/\delta x^2)_{T,P}=0$ defines the spinodal (Fig. 5.2b). The importance of both the spinodal and the phenomenon of critical opalescence in the understanding of the mechanism of liquid unmixing will be considered below.

Stable and Metastable Immiscibility

With regards to relationships between liquid immiscibility and crystallization, two topological cases are possible. In the first case, a stable two-liquid field emerges in

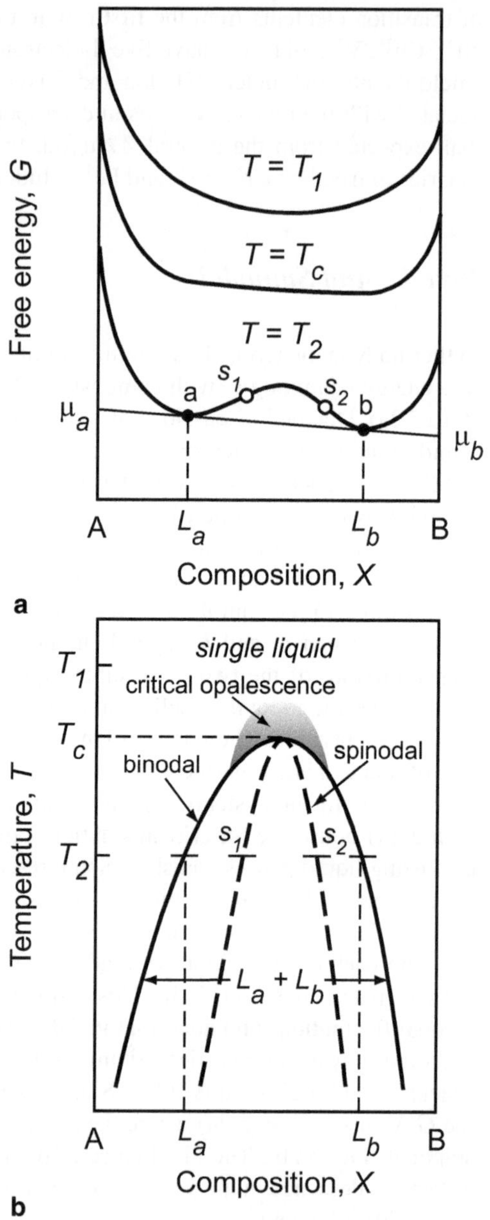

Fig. 5.2 Binodal and spinodal in a schematic binary *A–B*. **a** Gibbs free energy (G) versus composition (X) curves for temperature T_1 above the consolute point, the consolute temperature T_c, and temperature T_2 below the consolute point. Immiscible liquids *a* and *b* lie on a common tangent line to the energy curve. X_a and X_b are equilibrium compositions; μ_a and μ_b are chemical potentials of the components *A* and *B*. Inflexion points s_1 and s_2 define the spinodal. **b** Two-liquid region $a + b$, binodal and spinodal in *T-X* coordinates. T_1, T_c and T_2 are the same as in plot a. *Shaded area* shows a region of critical opalescence

a region where segments of the binodal dome overlie the liquidus surface. One or more crystalline phases may coexist with two immiscible liquids at the intersections between the binodal and the liquidus. In the second case, the binodal may be completely submerged (i.e. subliquidus), and in that case liquid immiscibility is metastable. These two types of relationships are schematically illustrated for a binary system in Fig. 5.3. The systems $CaO–SiO_2$ and $BaO–SiO_2$ (Fig. 5.1) serve as examples of the stable and metastable cases, respectively (Greig 1927; Hudon and Baker 2002). In *T-x* diagrams, the liquidus is flattened above the metastable binodal,

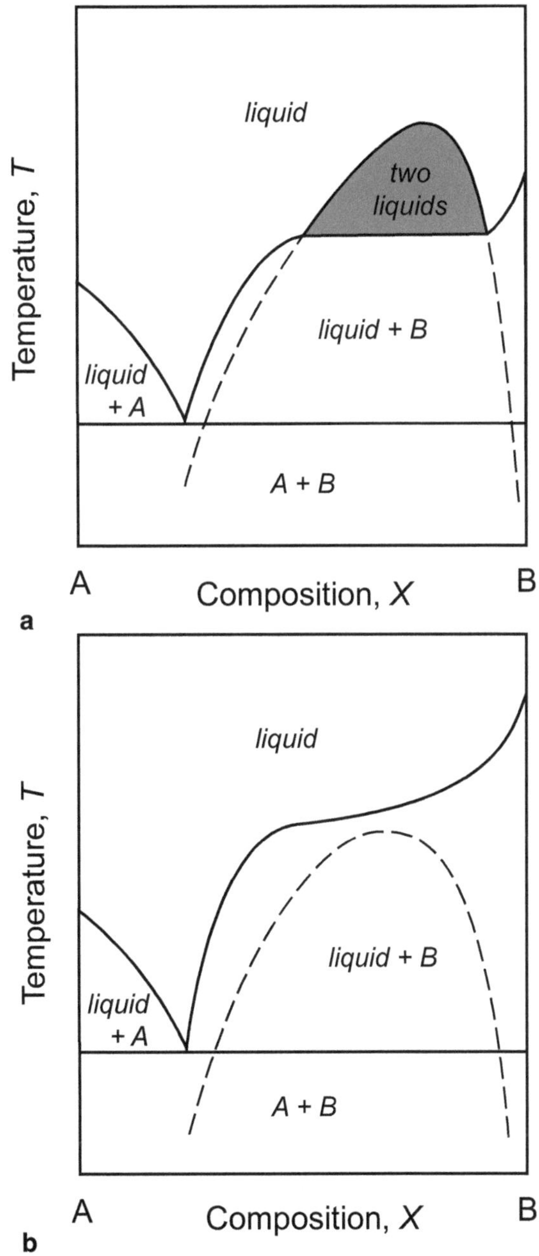

Fig. 5.3 Schematic binary phase diagrams showing **a** stable two-liquid field (*shaded area*) with its metastable extention (*dashed curves*) and **b** entirely sub-liquidus metastable immiscibility

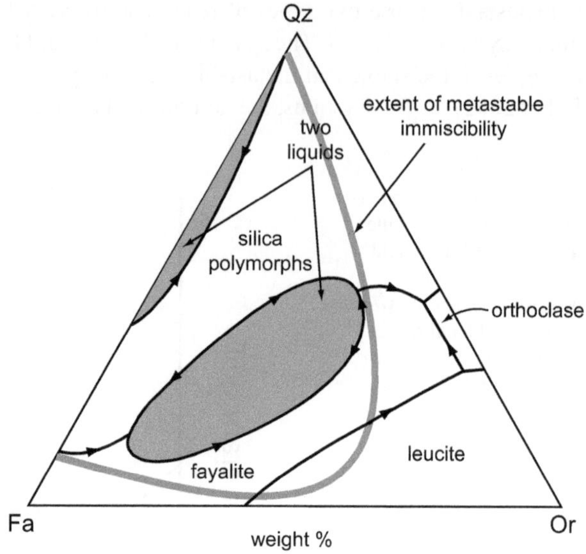

Fig 5.4 Liquid immiscibility in the fayalite *Fa*—orthoclase *Or*—silica *Qz* pseudo-ternary join after Roedder (1951, 1978). *Solid grey curve* shows the extent of sub-liquidus immiscibility according to Visser and Koster van Groos (1979)

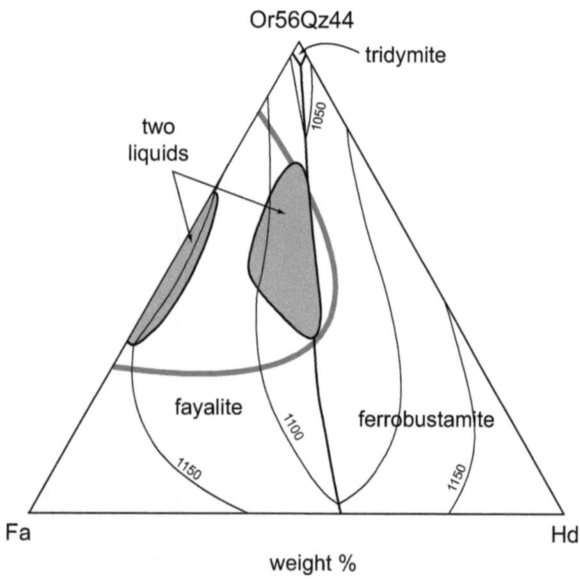

Fig. 5.5 Phase diagram of the join fayalite (*Fa*)—hedenbergite (*Hd*)—orthoclase (*Or*)—silica (*Qz*) after Hoover and Irvine (1978) showing fields of liquid immiscibility (*shaded areas*), cotectics and isotherms 1050, 1100 and 1150 °C. *Solid grey curve* shows the extent of two-liquid field in compositions containing 1.7–1.8 P_2O_5 according to Bogaerts and Schmidt (2006)

exhibiting a characteristic sigmoidal shape (Fig. 5.3b). Similar relationships are known in many ternaries and pseudo-ternaries, and some examples are presented in Figs. 5.4 and 5.5. It follows from Figs. 5.4 and 5.5 that seemingly separate stable miscibility gaps in multicomponent systems may actually represent fragments of a single binodal surface.

Nucleation Barriers

In the framework of currently accepted kinetic models (James 1975; Mazurin and Porai-Koshits 1984; Shelby 2005 and references therein), liquid immiscibility in compositional regions between the binodal and spinodal proceeds via nucleation and growth. It thus follows that they may be hampered by the thermodynamics and kinetics associated with nucleation and growth. The origins of thermodynamic barriers can be explained using the same *G-X* curve as introduced in Fig. 5.2a, and the relationships are illustrated in Fig. 5.6 following James (1975). For the bulk composition *x* lying between points *a* (binodal), and s_1 (spinodal), the overall free energy decrease due to unmixing to equilibrium liquids *a* and *b* (excluding interfacial effects) graphically corresponds to the vertical segment ΔG between the free energy curve at composition *b* and the tangent drawn to the curve at the composition *x*. This energy drop is the driving force for separation of the equilibrium phases. It should be noted however that when phase separation develops in a nucleation and growth regime, small sub-critical compositional fluctuations (represented for example, by composition *f*) will produce an increase in free energy (the segment Δg). This increase represents a thermodynamic barrier for unmixing. Clearly a fluctuation must develop beyond the composition *e* before the free energy starts to decrease, and the barrier is overcome. Thus, in nucleation region, the system is metastable with respect to infinitesimal compositional fluctuations. In contrast, there is no thermodynamic barrier for phase separation within the spinodal region between the inflection points s_1 and s_2 as the free energy change is negative for an infinitesimal fluctuation. Nonetheless, kinetic limitations on phase separation are universal. These limitations exist regardless of the exact mechanism of phase separation or the position of initial

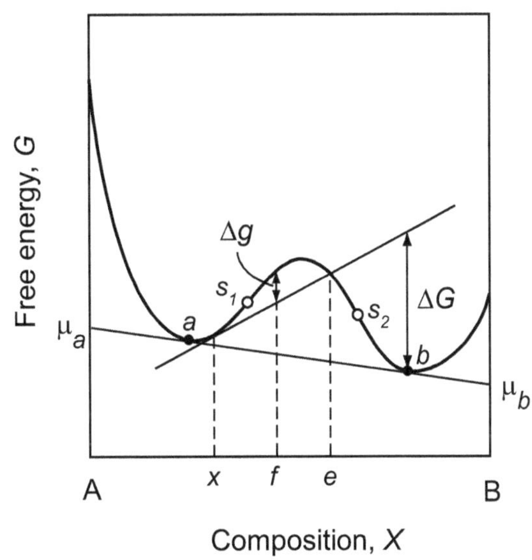

Fig. 5.6 Free energy (*G*) *vs*. composition (*X*) diagram showing graphical method of determining the thermodynamic driving force and the barrier for liquid-liquid phase separation. See text for details

composition x, because they arise from material transport and dynamic parameters, such as diffusion rates and viscosity.

Recent Progress in Experimental Studies

Magma Evolution Paths and Immiscibility Region

With the exception of rare cases of strongly overheated impact melts (e.g. the parental magma of the Sudbury complex right after the impact) terrestrial magmas are produced by partial melting of mantle or crustal rocks and therefore are crystal-saturated. The liquid line of descent and thus the potential for silicate melts to reach a two-liquid field is controlled by solid phases crystallization with decreasing temperature and their effect on the composition of residual melts. In dry systems, extreme iron enrichment (>18–19 wt.% FeO) was usually considered as necessary for the onset of unmixing (Dixon and Rutherford 1979; Philpotts and Doyle 1983). This is easily accomplished in alkali-poor systems such as lunar rocks or eucrites (Rutherford et al. 1974; Powell et al. 1980), but alkali-rich liquids evolve ultimately to iron depletion and silica enrichment (Veksler 2009). However, as illustrated by the experimental study of Charlier and Grove (2012), immiscibility can also develop along the liquid line of descent during the iron depletion and silica enrichment that occurs after Fe–Ti oxides appear in the crystallizing assemblage. The liquid line of descent must reach the appropriate compositional field to unmix, but the other condition is that the liquidus temperature crosses a super-liquidus binodal during differentiation (Fig. 5.7). Two classical study of the evolution of tholeiitic basalts by Juster et al. (1989) for the Galapagos spreading center and by Toplis and Carroll (1995) for the parental magma of the Skaergaard layered intrusion did not satisfy this condition.

In ternary systems and systems with greater number of components, stable immiscibility domes can sit either above cotectics (like, for example in the Roedder system illustrated in Fig. 5.4) or a dome may be positioned above a thermal divide (Fig. 5.8). In the first case, mechanical separation of immiscible melts does not change the liquid line of descent and the general direction of system evolution. In the Roedder system, both melts eventually follow the fayalite-silica, or (at higher fO_2) magnetite-silica cotectics and end up in the same eutectic with orthoclase. In some ternaries (e.g. $CaO–B_2O_3–SiO_2$) immiscibilty domes sit on top of thermal divides and mechanically separated immiscible melts evolve by further crystallization in different directions. The topology of natural multicomponent ferrobasaltic-andesitic systems is very complex but it seems that at low pressure, when plagioclase is stable and forms a thermal divide, the situation corresponds to the second case. That is, mechanically separated Fe-rich melt would probably further evolve along the "Fenner" Fe-enrichment trend ending in a Fe-rich eutectic, whereas the silica-rich liquid follows a more familiar Bowen trend of further silica enrichment, all the way down to the granitic eutectic.

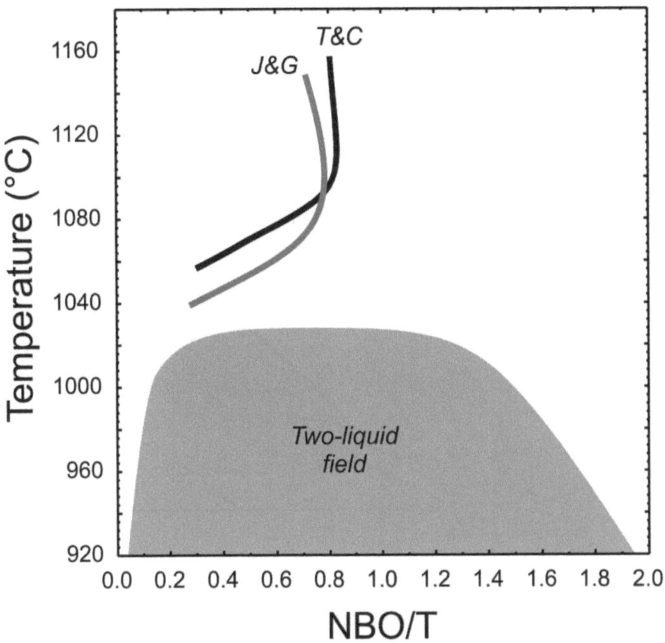

Fig. 5.7 Temperature as a function of the degree of polymerization (NBO/T) of experimental melts. The *grey* area represents the maximum compositional space of the two-liquid field from Charlier and Grove (2012). Liquid lines of descent of Toplis and Carroll (1995) (*T&C*; *black*) and Juster et al. (1989) (*J&G*; *grey*) are plotted

Fig. 5.8 Two topological types of liquid immiscibility regions in relationship to liquidus surface. **a** On top of a cotectic. **b** On top of a thermal barrier

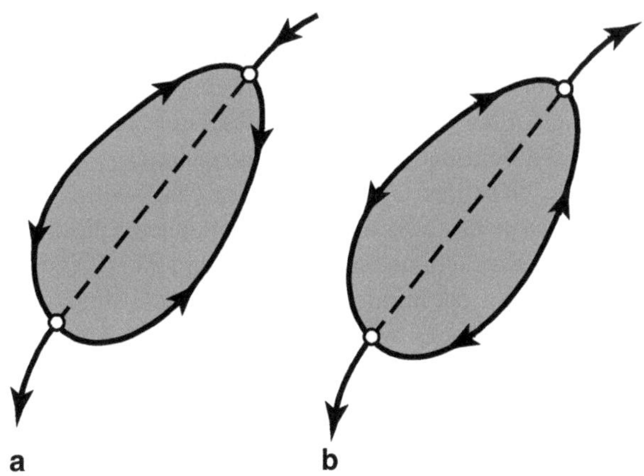

Silicate liquid immiscibility is difficult to identify also because it can be a furtive agent of fractionation (Longhi 1998). Indeed, a melt entering a two-liquid field upon cooling can leave this field because of crystallization and compositional evolution of the bulk residual liquid. This is schematically illustrated on Fig. 5.9 in

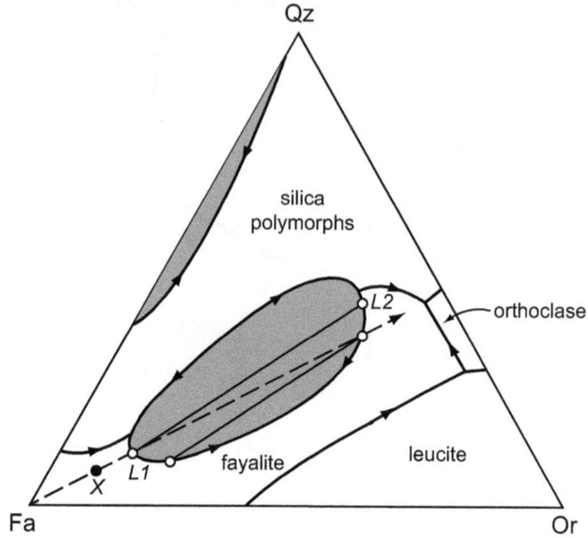

Fig. 5.9 Illustration of a crystallization path for liquid composition X in the fayalite (*Fa*)—orthoclase (*Or*)—silica (*Qz*) pseudo-ternary join after Roedder (1951, 1978), showing the furtive nature of liquid immiscibility upon cooling. *Grey* areas are fields of stable immiscibility. See text for discussion

the fayalite (Fa)—orthoclase (Or)—silica (Qz) pseudo-ternary with the equilibrium crystallization trend for liquid *X*. Liquid *X* located in the stability field of fayalite crystallizes this phase until it reaches the two liquid field, marking the separation of a high-silica liquid *L2*. The paired low-silica liquid *L1* reacts with crystallizing fayalite to produce the high-silica liquid. As the reaction proceeds upon cooling, compositions of the two liquids move closer together, and when a line from the high-silica liquid to fayalite passes through the bulk composition (*X*), the last of the low-silica liquid is consumed and the high-silica liquid moves directly away from fayalite in the one-liquid stability field (Longhi 1998). Thus, after complete solidification of the system, there is no remaining trace of silicate liquid immiscibility. The high-silica liquid produced ultimately has only recorded the fractionation of fayalite, K-feldspar, and silica at the eutectic.

In the Sept Iles layered intrusion (see Section "Sept Iles"), immiscibility has been interpreted to be a passing phenomenon limited to a narrow temperature interval, evaluated to be between 1015 and 985 °C (Charlier et al. 2011; Namur et al. 2012, 2015). Both immiscible melts evolved along the binodal upon cooling, forming cumulate rocks with identical mineral compositions but contrasted mineral modes. The bulk liquid composition was evolving towards silica-enrichment and iron-depletion by fractional crystallization to eventually leave the two-liquid field. Immiscibility had thus no influence on the liquid line of descent, except for the absence of intermediate melts.

Liquid-Liquid Element and Isotope Partitioning

Major and Trace Elements

Our knowledge of element partitioning between immiscible silicate melts has greatly improved in recent years owing to the introduction of new experimental and analytical methods such as high-temperature centrifuge separation (Schmidt et al. 2006; Veksler et al. 2006) and laser ablation ICP-MS analyses. It is now possible to analyze dozens of elements at concentrations varying over several orders of magnitude in the same sample and even at the same spot. Therefore, comparisons between individual elements and/or different chemical groups can be done with greater confidence.

The general picture that emerges from recent studies implies close links between ionic properties (e.g. the nominal charge *Z* and ionic radius *r* combined in the form of ionic potential *Z/r*) and liquid-liquid distribution of an element, which is usually expressed in the form of Nernst distribution coefficient $D_i = C_i^{Lfe}/C_i^{Lsi}$ where *C*'s are weight concentrations, subscripts denote the element and superscripts denote phases (*Lfe*—Fe-rich melt; *Lsi*—silica-rich melt). The connection is hardly surprising in view that Coulombic interactions appear to be the principal driving force for silicate liquid immiscibility at the atomic level (see Section "Liquid Immiscibility at the Atomic Level"). Liquid-liquid *D* values closely correlate with the widths and consolute temperatures of miscibility gaps in corresponding metal oxide-silica binaries (Hudon and Baker 2002; Schmidt et al. 2006; Veksler et al. 2006). Elements, with broader miscibility gaps in the binaries with silica tend to have higher *D* values and stronger concentrate in the silica-poor, Fe-rich conjugate liquid. When plotted against ionic potential Z/r, liquid-liquid *D* values of isovalent elements tend to form convex trends similar to those described by Hudon and Baker (2002) for the consolute temperatures.

A typical plot of the liquid-liquid *D* values against *Z/r* is presented in Fig. 5.10. Only three trace elements out of 37 shown on the plot, together with the major components K, Al and Si, concentrate in the silica-rich liquid. The elements include network modifiers with the lowest ionic potentials (Cs and Rb) and a network former Sb. The highest *D* values are reached by REE and trivalent transition elements Cr and V. Absolute *D* values vary with temperature, redox conditions and melt chemistry but the general pattern of element partitioning and proportionality between the *D* values of individual elements hold. On the basis of experimental data, Schmidt et al. (2006) proposed a modified regular solution model, which allows to calculate D_M of a trace element M in multicomponent compositions from the width of miscibility gap expressed as X_{FeO}^{Lfe}—X_{FeO}^{Lsi} and the consolute temperature of the binodal in the MO–SiO_2 binary.

Oxygen Isotopes

Kyser et al. (1998) and Lester et al. (2013b) studied oxygen isotope partitioning between Fe- and Si-rich immiscible liquids in the system K_2O–FeO–Fe_2O_3–Al_2O_3–SiO_2 at the atmospheric pressure and 1180 °C, and in similar compositions

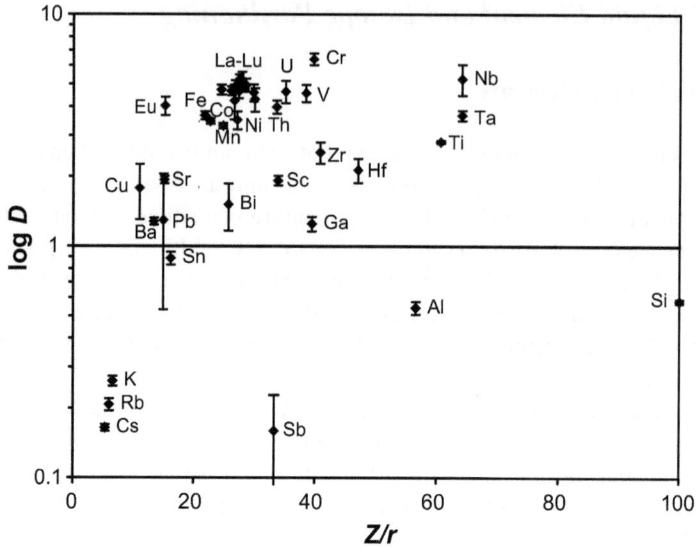

Fig. 5.10 Liquid-liquid Nernst partition coefficients *D* plotted against ionic potential *Z/r*. average values with standard deviations (shown as *vertical bars*) for a series of experiments on synthetic compositions in the system K_2O–FeO–Fe_2O_3–Al_2O_3–SiO_2 at 1200 °C and log fO_2 between −8.3 and −8.7 (Veksler unpublished data)

doped with H_2O, P and S at 1100–1200 °C and 200 MPa. In both studies Si-rich liquids were found to have $\delta^{18}O$ values higher than those of the Fe-rich liquids by 0.4–1 ‰. The observed preferential partitioning of ^{18}O into the Si-rich liquid is consistent with crystal-melt oxygen isotope equilibria and oxygen isotope fractionation in Soret (thermal) diffusion experiments. In all the cases, crystallochemical principles seem to work in the same way and phases or liquid fractions with more polymerized structure tend to be enriched in ^{18}O.

The Kinetics of Liquid Immiscibility

It has been recognized that silicate melts with compositions in vicinity of miscibility gaps often quench to opalescent glasses showing heterogeneity at sub-micron scale. Correct interpretation of sub-micron exsolutions in the glasses is crucial for distinguishing between stable and metastable immiscibility (see Section "Stable and Metastable Immiscibility") and requires good knowledge of the kinetics of liquid immiscibility. Kinetic issues were at the core of the discussion about the actual extent of the stable miscibility gap in the system K_2O–FeO–Al_2O_3–SiO_2 (Visser and Koster van Groos 1976, 1977; Freestone and Hamilton 1977; Roedder 1978) and in multicomponent compositions used in centrifuge experiments (Veksler et al. 2007,

2008a, b; Philpotts 2008). Two alternative views have been formulated. First is that the fine heterogeneity in glasses is due to instant melt unmixing during quenching. In other words, melt was homogenous before quenching and unmixed when temperature dropped below the metastable segment of the binodal (Fig. 5.3) but was still above the glass transition temperature. Second interpretation of the turbid, opalescent glasses is that they form from melts, which were heterogeneous before quench and represented thermodynamically stable but slowly growing sub-micron emulsions.

The mechanisms for melt unmixing and kinetics of droplet nucleation and growth have been studied extensively in technological melts and glasses (James 1975; Mazurin and Porai-Koshits 1984). In a typical case, initial nucleation of nanometer-sized immiscible droplets is believed to be very fast and further growth is dominated by process of coarsening, which is termed Ostwald ripening. Ripening is preferential growth of larger droplets at the expense of smaller ones driven by overall decrease in interfacial free energy γ. Theoretical models of the coarsening stage predict that the average droplet radius r should increase proportionately either to the square root or the cubic root of time (e.g. see review by James 1975). The exact relationship depends on the degree of super-saturation, and is also determined by the process that controls the growth (long range diffusion *versus* surface reaction rate). In any case, coarsening is faster when the value of γ is high.

First experimental measurements of interfacial tension γ have been recently carried out for immiscible Fe-rich and silica-rich melts in the system K_2O–FeO–Fe_2O_3–Al_2O_3–SiO_2 (Veksler et al. 2010). The measurements were performed in air for the most oxidized and chemically contrasting melt compositions. The measured γ values in the range of 8–16 mN/m are very low. For comparison with more familiar examples, γ values between water and immiscible organic liquids, such as mineral oil, benzene, or purified vegetable oils vary at room temperature between 30 and 50 mN/m. Veksler et al. (2010) pointed out that γ values for natural, less compositionally contrasting ferrobasaltic and rhyolitic melts should be even lower by a factor of 2 or 3. In the context of kinetic theories of liquid immiscibility (James 1975), low interfacial tension implies easier nucleation, higher nucleation density but, on the other hand, slow coarsening of the droplets by Ostwald ripening and protracted stability of fine emulsions.

The kinetics of silicate liquid immiscibility in natural magmas has been studied only briefly (Martin and Kushiro 1991) and its experimental coverage remains insufficient. In view of recent kinetic studies by Veksler et al. (2008a, 2010), silicate liquid immiscibility in natural ferrobasaltic, andesitic or rhyolitic magmas is likely to produce very fine, possibly sub-micrometer emulsions that may be stable over considerable time, and have significant effects on magma dynamics. Slow coarsening of the emulsions should hamper gravitational separation of immiscible melts, and may be among the reasons why traces of immiscibility are usually hard to detect in partly or fully crystallized igneous rocks.

The effects of H_2O and Other Volatile Components on Liquid Immiscibility

Lester et al. (2013a) studied the effects of H_2O, S, P, Cl and F additions on liquid immiscibility in the system $K_2O–FeO–Fe_2O_3–Al_2O_3–SiO_2$. The additions of H_2O at 200 MPa were shown to dramatically suppress liquidus surface temperatures, expand the stability field of magnetite and decrease that of silica minerals, thereby increasing the *T–X* range of the miscibility gap. Thus, the composition range of the miscibility gap at the fO_2 level of the magnetite-hematite (MH) buffer is increased relative to the range in equivalent anhydrous melts. At fO_2 equivalent to the Ni–NiO buffer H_2O addition displaced the miscibility gap to more silica-rich compositions and presumably extends the two-liquid field. Sulfur and phosphorus were shown to further expand the miscibility gap. Three-liquid immiscibility between Fe-rich silicate liquid, a Si-rich silicate liquid and a Fe-sulfide liquid was observed in some hydrated compositions at fO_2 = MH.

Silicate Liquid Immiscibility in Magmatic Systems

Volcanic Systems and the Daly Gap

Most mantle-derived lavas and pyroclastic materials erupt at the surface after extensive differentiation in magma chambers at depth. Volcanic products, which can be viewed as spills and overflows from intrusive systems, carry important information about magma evolution in the process of magma transformation into a layered intrusion. Therefore, microscopic and macroscopic evidence of liquid immiscibility in volcanic rocks is relevant for the present discussion.

Immiscible droplets of glasses trapped in the mesostasis of basalts have been recognized worldwide in both tholeiitic and alkaline series (e.g. De 1974; Sato 1978; Philpotts 1982; Kontak et al. 2002; Sensarma and Palme 2013). These immiscible interstitial textures consist of brown spheres, usually smaller than 10 μm, enclosed in a silica-rich glass. Dark globules are commonly finely crystallized to Fe–Ti oxides, iron-rich pyroxene and other fine-grained material whereas the glassy texture of the silica-rich component is better preserved. Examples of immiscibility textures in volcanic rocks are presented in Fig. 5.11.

Evidence for unmixing provided by these small-scale textures is usually not considered as sufficient to support the potential role of immiscibility as a large-scale differentiation process along the basalt-rhyolite evolution trend. However, an implication of silicate liquid immiscibility is that compositions between unmixed pairs cannot exist as homogeneous melt. Indeed, when a liquid line of descent of homogeneous melts reach a binodal surface, two contrasting liquids are produced. This is responsible for a gap of intermediate homogeneous melts during further cooling and differentiation. This absence of intermediate compositions is a major characteristic

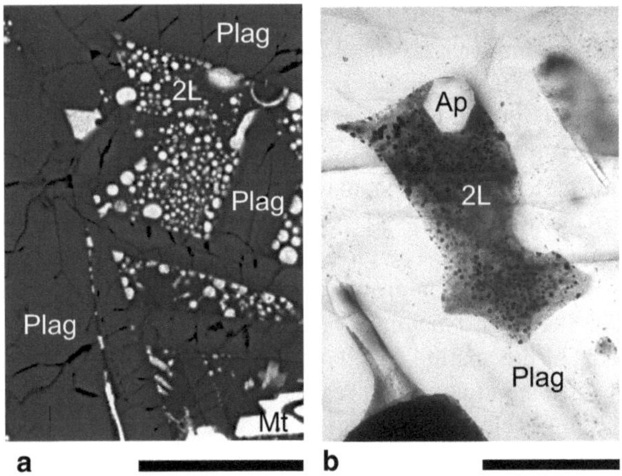

Fig. 5.11 Immiscibility textures in volcanic rocks. **a** Fe-rich droplets in silicic glass in mesostasis of tholeitic picrite from Koolau, Hawaii. Back scattered electron image. **b** Immiscible melt inclusions (*black* Fe-rich droplets in transparent silicic glass) in plagioclase phenocrysts in basaltic andesite from Mull, Scotland. Apatite crystal was trapped with the melt inclusion. Optical transmitted *light* image. Scale bars correspond to 50 microns. Abbreviations for phases: *Ap* apatite, *Plag* plagioclase, *Mt* magnetite, *2L* immiscible glasses

of tholeiitic series (Chayes 1963; Thompson 1972; Charlier et al. 2013). In the Sept Iles intrusive suite (see Section "Sept Iles"), silicate liquid immiscibility has been invoked to explain the dearth of intermediate (monzodioritic) melts along the evolution from tholeiitic basalt to A-type granite (Charlier et al. 2011; Namur et al. 2011).

Immiscibility in Mafic Layered Intrusions

Skaergaard

Since the pioneering study by Wager and Deer (1939), the Skaergaard intrusion in East Greenland became a classical example of a mafic layered intrusion. General information about the intrusion can be found in numerous reviews (e.g. McBirney 1996). First evidence for silicate liquid immiscibility in the Skaergaard intrusion came from experiments on synthetic compositions modeling evolved liquid fractions of the Skaergaard magma (McBirney and Nakamura 1974). Immiscibility was observed in mixtures of Fe-rich liquids and silica-rich granophyre precipitating mineral assemblages of the Upper Zone at temperatures around 1010–1015 °C, which is no more than 10 °C above the liquidus temperature. The composition of the Fe-rich liquid was derived from partial melting experiments on natural rocks from the top of the Upper Zone (UZc) and contained 30.5 wt.% FeO_t (McBirney and Naslund 1990). Direct confirmation that immiscibility had indeed taken place in the Skaergaard intrusion was obtained in a study of melt inclusions in cumulus apatite from the Upper Zone of the Layered Series (Jakobsen et al. 2005). Many cumulus apatite crystals in apatite-rich layers of the Upper Zone were found to contain melt inclusions of two distinct types easily recognizable under optical microscope as dark-coloured and light-coloured. Electron microprobe analyses of the inclusions showed that the compositions of the two types compared well with the Fe-rich and silica rich immiscible liquids previously produced in experiments (McBirney and Nakamura 1974) and reported as glasses in mesostasis of natural basaltic and andesitic lavas (Philpotts 1982). Therefore, the inclusions were interpreted as immiscible liquid droplets trapped by growing apatite crystals during crystallization and unmixing of the Skaergaard magma.

The discovery of the apatite-hosted melt inclusions posed further questions: (1) when exactly was magma unmixing reached at Skaergaard and (2) how much impact, if any, did immiscibility and liquid-liquid fractionation have on magma evolution? But regardless of the timing and material transport capacity, new proof of liquid immiscibility revived an old debate about the general direction of magma evolution at Skaergaard. Some researchers (Hunter and Sparks 1987, 1990; Toplis and Carroll 1995) argued in the past that the onset of magnetite and ilmenite crystallization must have turned liquid evolution in the Skaergaard magma towards silica enrichment, while others (e.g. McBirney and Naslund 1990; Thy et al. 2006) argued that Fe enrichment continued despite crystallization of Fe–Ti oxides and reached 30 wt.% FeOt. Numerous intermediate trends were proposed between the extremes (e.g. Brooks and Nielsen 1990; Morse 1990; Nielsen 2004). On the other hand, if silicate liquid immiscibility and separation of immiscible liquids by gravity played a significant role in magma differentiation not only in the Upper Zone but also earlier in the Middle Zone and at the top of the Lower Zone, liquid-liquid fractionation would account for the apparent diversity of the liquid evolution trends and finally resolve the long-lasting debate.

Veksler et al. (2007) tried to establish the exact timing of immiscibility at Skaergaard and in other examples of tholeiitic magma in a series of high-temperature centrifuge and static reverse experiments. The study revealed significant kinetic problems hampering experimental reproduction of liquid immiscibility in multicomponent system analogous to Skaergaard magma. Unmixing of a synthetic analogue of the Skaergaard parental liquid at 1110–1120 °C corresponding to the formation of the Middle Zone did not develop beyond sub-micron, nanoscale emulsions. Nevertheless, the emulsions showed signs of macroscopic separation in centrifuge experiments and were interpreted as stable, super-liquidus immiscibility. Such interpretation and geological implications of the study were challenged (McBirney 2008; Morse 2008; Philpotts 2008). In reply, Veksler et al. (2008b) admitted that experimental evidence for early magma unmixing at Skaergaard was inconclusive but stressed the importance of new petrographic observations confirming that immiscibility had played an important role in magma differentiation at Skaergaard.

Mineralogical and petrographic traces of silicate liquid immiscibility were revealed in the Skaergaard Layered Series from the top of the Lower Zone upwards by detailed studies of plagioclase zoning (Humphreys 2011), plagioclase-hosted melt inclusions (Jakobsen et al. 2011) and late-stage intercumulus reactive microstructures including serrated grain boundaries between pyroxene and plagioclase primocrysts, olivine rims around magnetite and ilmenite crystals, fish-hook pyroxenes and mafic symplectites (Holness et al. 2011). The reactive symplectites are characterized by extensive reverse zoning of plagioclase and replacement of orthopyroxene by fayalitic olivine and ferroaugite. The reactive microstructures document significant material transport in the cumulus mush of the Layered Series, which resulted in the depletion of late-stage intercumulus liquid in alkalis and silica, and enrichment in Ca and Fe. In view of the absence of reactive symplectites from the Marginal Border Series, where effective separation of immiscible liquids by gravity was impossible, the reactive microstructures in the Layered Series were interpreted as products of gravitational liquid-liquid separation leading to late-stage replacement reactions between residual Fe-rich liquid and cumulus primocrysts.

In summary, there is very strong evidence that silicate liquid immiscibility took place in the Skaergaard intrusion and magma unmixing resulted in fractionation of macroscopic volumes of Fe-rich and silicic immiscible products between the Upper Zone of the Layered Series and the Upper Border Series. There is also little doubt that starting from the top of the Lower Zone and through the Middle Zone of the Layered Series, intercumulus crystallization at the bottom of the magma chamber differed from crystallization in the main overlying magma and at the roof of the intrusion. Strong enrichment of intercumulus liquid at the bottom in Fe oxides was accompanied by depletion in alkalis and silica, which is inconsistent with experimental phase equilibria constraints and mass balance in a closed system (Veksler 2009). Intercumulus liquid immiscibility and gravitational liquid-liquid fractionation appear to be the most viable explanation of the observed macroscopic trends and the origin of reactive microstructures in the Layered Series.

Sept Iles

The Sept Iles layered intrusion is part of an intrusive suite on the north shore of the Saint-Lawrence River (Quebec; Higgins 2005; Namur et al. 2010, 2015). The intrusion crystallized from a ferrobasaltic parent magma and is subdivided into a Layered Series of troctolite and gabbro, an anorthositic Upper Border Series and a granitic Upper Series. The liquid line of descent followed a tholeiitic trend, and the cooling of the chamber was interrupted by two major and a series of minor influxes of magma (Namur et al. 2010, 2011).

Silicate liquid immiscibility and segregation of iron-rich and silica-rich paired melts have been recognized as a major differentiation process in the Sept Iles layered intrusion (Charlier et al. 2011; Namur et al. 2012). Evidence is provided in rocks from the Critical Zone of the intrusion. This horizon contains the most evolved cumulates of the megacyclic units II. They are made up of plagioclase (An_{55-34}), olivine (Fo_{66-21}), clinopyroxene (Mg# 75–55), ilmenite, magnetite, apatite, ±pigeonite. Two independent characteristics of these rocks support they have crystallized from an emulsion of two immiscible melts: (1) contrasting melts inclusions trapped in apatite and, (2) bimodal bulk-rock compositions of ferrogabbros.

Polycrystalline melt inclusions are ubiquitously trapped in apatite of ferrogabbros from the Critical Zone (Charlier et al. 2011). Two types of inclusions are clearly distinguished and can occur in the same apatite grain: dark inclusions which contain chlorite, amphibole, plagioclase, Fe–Ti oxides and fine-grained material; and light-coloured inclusions with albite, orthoclase, biotite, amphibole, fine-grained material and rarely quartz. These inclusions have homogenisation temperature between 1100 and 1060 °C and clearly display bimodal compositions: granitic Si-rich and ferrobasaltic Fe-rich. These inclusions clearly represent snapshots of trapped conjugate immiscible liquids.

Large-scale separation of these two immiscible melts is supported by the bimodal character of ferrogabbros. One group, enriched in plagioclase, has an average density of 3.0–3.2 g/cm^3 and contains 0–3.5 wt.% P_2O_5. The other group has an average density of 3.6–3.8 g/cm^3 with 3–12 wt.% P_2O_5. The two types of rocks have variable thicknesses and alternate at a scale of 5–20 m. The bimodality is interpreted to result from the segregation of the two immiscible melts, followed by crystallization and formation of two types of cumulates.

Layered Intrusions of the Emeishan Province

The 260 Ma Emeishan large igneous province in SW China host some major layered intrusions that are spatially and temporally associated with flood basalts and granitoids in the province (Zhou et al. 2008; Pang et al. 2010). They occur in the Panzhihua-Xichang (Panxi) region, and four of them host world-class resources of Fe, Ti and V: Panzhihua, Hongge, Baima and Xinjie (Pang et al. 2010).

Parental magmas to these intrusions are Fe–Ti-enriched basalts. Although Fe–Ti oxides saturate early in these basalts (Pang et al. 2008a), silicate liquid immiscibility has been invoked to explain the origin of the huge amounts of Fe–Ti-ores (Zhou et al. 2005). In this model, the separation of an immiscible oxide melts would be responsible for the formation of Fe–Ti–V oxide ore bodies in the lower part of the intrusion. Unmixing would have been triggered by the addition of fluids from upper crustal wall-rocks (Zhou et al. 2005). This model is supported by textural and mineralogical constraints from Hongge Fe–Ti–V oxide deposit showing the presence of two populations of Cr-rich and Cr-poor titanomagnetite (Wang and Zhou 2013). Cr-rich titanomagnetite is interpreted as an early liquidus phase whereas clusters of Cr-poor titanomagnetite, ilmenite, and apatite are thought to have formed from an Fe–Ti–(P)-rich immiscible melt. Ultimately, large-scale silicate liquid immiscibility in the region would be responsible for the formation of contrasting suites of rocks, with Si-rich immiscible melts forming syenites and Fe-rich immiscible melts crystallizing layered gabbros and massive oxides (Zhou et al. 2008).

Alternative models invoking continuous fractional crystallization have also been proposed. In this scenario, formation of huge amounts of Fe–Ti oxide is a simple result of gravitational sorting and settling (Pang et al. 2008b; Zhang et al. 2012). Frequent Fe–Ti-rich magma replenishment (Song et al. 2013) is coupled to sorting and settling to explain the formation of unusually thick stratiform Fe–Ti oxide layers.

Bushveld

The Paleoproterozoic (2054.4 ± 1.3 Ma; Scoates and Friedman 2008) Bushveld Complex in South Africa is a gigantic plutonic system comprising an inwardly dipping lopolith of mafic-ultramafic cumulate rocks, the Rustenburg Layered Suite (RLS), overlain in the centre by a large mass of contemporaneous granites. In terms of the total volume of basaltic magma, the Bushveld Complex by itself constitutes a large igneous province, with the intrusive equivalent of continental flood basalts. Present-day outcrops of the complex cover a surface area of 66,000 km^2 and the total thickness of igneous rocks is about 10 km or more. The mafic-ultramafic layered intrusion is sub-divided into stratigraphic zones comprising from the bottom to the top the Lower Zone (pyroxenites and harzburgites), Critical Zone (chromitites, harzburgites, pyroxenites, norites and minor anorthosites), Main Zone (norites and gabbronorites), and Upper Zone (gabbronorites and augite diorites). There is abundant evidence that the cumulates formed in an open system with multiple magma injections, volcanic eruptions and significant assimilation of continental crust. The open-system nature of magma evolution is evidenced in the RLS by trace element variations, breaks in initial Sr isotope ratios, and numerous reversals in compositional trends of cumulate minerals (Eales and Cawthorn 1996; Maier et al. 2013 and references therein). Studies of the chilled margins (Wilson 2012) and comagmatic sills associated with the RLS (Barnes et al. 2010) have revealed at least three distinct geochemical types of parental magmas. Recent studies of the Bushveld Complex are reviewed by Cawthorn (2015).

Silicate liquid immiscibility in the Bushveld Igneous Complex has been discussed in the context of magma differentiation in the Upper Zone including the origin of magnetite and nelsonite layers (e.g., Reynolds 1985; von Grunewaldt 1993) and especially in connection with the origin of iron-rich ultramafic pegmatites (IRUP). The latter are coarse-grained rocks forming distinct discordant bodies, ranging in size from a few meters to a few kilometers and randomly cutting through the layered rocks of the normal stratigraphy in the vertical interval from the Upper Zone down to the upper Critical Zone (Scoon and Mitchell 1994; Reid and Basson 2002). The bodies have irregular shapes resembling discs, tubes, sheets, dykes and complex branching forms (Fig. 5.12). Contacts with the host rocks are usually sharp and imply magmatic replacement reactions. Typical mineral assemblages of the IRUP comprise ferro-augite ($Wo_{34-46}En_{21-40}Fs_{18-40}$), olivine ($Fo_{30-50}$), magnetite, ilmenite and little or no plagioclase. Sulfides, Fe-amphibole and apatite may be locally abundant.

A genetic link between IRUP and silicate liquid immiscibility has been proposed primarily because typical whole-rock compositions of IRUP are high in FeOt, low in Al_2O_3 and resemble the compositions of Fe-rich immiscible melts (Scoon and Mitchell 1994; Reid and Basson 2002). Unmixing has been proposed to take place in parental magma of the Upper Zone because the initial $^{87}Sr/^{86}Sr$ isotope ratios of the IRUP cutting through the Upper Critical and the Main Zones are significantly different from those of the country rocks and similar to the rocks of the Upper Zone (Scoon and Mitchell 1994; Reid and Basson 2002). An origin of the IRUP-forming

Fig. 5.12 Exposures of iron-rich ultramafic pegmatites (IRUP; *dark*) replacing host leuconorite (*light*) in the U2 open pit of the Lonmin Marikana mine, Bushveld Complex (photo by DL Reid)

liquids in the parental magma of the Upper Zone is further supported by the Nd isotope data, Mg# of the Fe–Mg silicates, and the presence of liquidus magnetite (Reid and Basson 2002). In the Upper Critical Zone, the IRUP preferentially replace anorthosites and norites but in fully developed cases pyroxenite and even chromitite layers are also affected. The replacement reactions result in (1) reverse zoning of plagioclase in the IRUP in contact with norite or anorthosite, and complete disappearance of plagioclase from the inner, central parts of the IRUP bodies; (2) the replacement of low-Ca pyroxene by ferro-augite and olivine; (3) the replacement of chromite by magnetite. The nature and the origin of the IRUP replacement bodies defy a simple explanation, and remain contentious. Immiscibility hypothesis has been favored by some researches; however, it does not give all the answers to the problems posed by IRUP. For example, if the source of IRUP-forming liquids was indeed in the Upper Zone, one has to explain how the melts penetrated downwards through the 3 km thick cumulate pile of the Main Zone without changing their radiogenic isotope ratios, and how the melts became under-saturated in plagioclase.

Tegner and Cawthorn (2010) argued that the tholeiitic parental magma of the Upper Zone, in contrast with the parental magmas of the Skaergaard and Sept Iles intrusions, contained less FeOt from the start and did not follow the trend of strong Fe enrichment. Nevertheless, they did not rule out immiscibility and immiscible origin for prominent magnetite and nelsonite (apatite-magnetite) layers, which mark the Upper Zone stratigraphy. Recently VanTongeren and Mathez (2012) interpreted a sharp, 3-fold increase of REE concentrations in cumulus apatite at the top of the Upper Zone as a result of large-scale silicate liquid immiscibility and argued that the uppermost 325 m of the cumulates crystallized from gravitationally separated layer of the silica-rich immiscible liquid. Cawthorn (2013) challenged that interpretation and put forward an alternative model. He proposed that higher REE concentrations in the uppermost apatite were due to post-cumulus re-equilibration of apatite primocrysts with a greater amount of inter-cumulus liquid. According to his calculations, a change of the apatite-liquid mass ratio in cumulus mush from 0.5 to 0.09 would account for the 3-fold increase of the final REE concentrations in re-equilibrated apatite. In conclusion, evidence for complete gravitational separation of Fe-rich and silica-rich immiscible liquids in two major, macroscopic layers may be missing at Bushveld, as well as from other layered intrusions. However, as discussed below in Section "Intercumulus Crystallization and Liquid Immiscibility", liquid immiscibility in mafic layered intrusions does not need to form major, fully separated liquid layers dozens of hundreds meters in thickness in order to play a significant role in post-cumulus crystallization and material transport.

Other Layered Intrusions

In the Duluth layered intrusion (Minnesota, USA), Ripley et al. (1998) proposed that nelsonites and possibly oxide- and apatite-rich rocks were formed by liquid immiscibilty together with a sulfide melt. The authors suggest an important role of evolved ferrogabbroic melt produced during crystallization of tholeiitic magmas.

In the Bjerkreim-Sokndal layered intrusion, Norway, it has been suggested that immiscibility might have occured in the Transition Zone, between the Layered Series and the overlying silicic evolved rocks (Wilson and Overgaard 2005). Experiments have shown that mixtures of jotunite and quartz mangerite, the most common melt compositions associated with anorthosites, are indeed immiscible (Philpotts 1981). Alternatively, the Transition Zone cumulates have been interpreted as resulting from crystallization of a single magma (Duchesne et al. 1987; Duchesne and Wilmart 1997).

Evidence for coexisting Fe-rich and Si-rich melts have also been found in the Sudbury igneous complex in Canada. This intrusion crystallized from a melt sheet produced by a meteorite impact (Grieve et al. 1991). Bimodal compositions have been interpreted to result from emulsion of superheated norite and granophyre assemblages (Zieg and Marsh 2005). Melt inclusions trapped in apatite have preserved the composition of the two melts (Watts 2014), considered to be produced by immiscibility during cooling of the superheated melt sheet.

Loferski and Arculus (1993) described multiphase crystallized melt inclusions in cumulus plagioclase crystals from anorthosite layers and other rocks in the Middle Banded series of the Stillwater Complex (Monatana, USA) and interpreted the inclusions as traces of immiscible Fe-rich liquid concentrating Ti, P, Zr and REE. The authors proposed that silicate liquid immiscibility in Stillwater could either have occurred locally in compositional boundary layers around crystallizing plagioclase grains, or it could have occurred pervasively throughout the parent melt. Large masses of monomineral anorthosites may have crystallized from the main, feldspathic, silica-rich immiscible liquid.

Intercumulus Crystallization and Liquid Immiscibility

Despite disagreements about the exact timing of silicate liquid immiscibility in some specific cases, the broadly accepted view is that immiscibility in mafic layered intrusions starts at a late stage of magma evolution after extensive magma differentiation by fractional crystallization. The lowest temperatures and the most evolved melt fractions are likely to be found in partly crystallized marginal zones of a magma chamber, at the bottom, sides and at the roof. Therefore, unmixing is likely to start at first in intercumulus melt inside partly solidified crystal mush. If mush permeability is low and immiscible melts cannot separate by gravity, immiscibility will leave no traces in fully crystallized rocks. Both melts should crystallize identical mineral assemblages and crystallization products of a melt mixture would be indistinguishable from those of a single homogenous melt of the same bulk composition.

However, if the crystal mush is permeable enough for liquid-liquid fractionation by gravity, immiscibility would provide effective means of material transport in the form of phase convection and that may significantly affect intercumulus crystallization. Consider, for example, a section of gabbroic cumulus mush at the bottom of a magma chamber (Fig. 5.13). Suppose a vertical temperature gradient across the mush with the temperature increasing upwards. Suppose the porosity and permeability of the mush increase in the same direction. Part of the mush below the im-

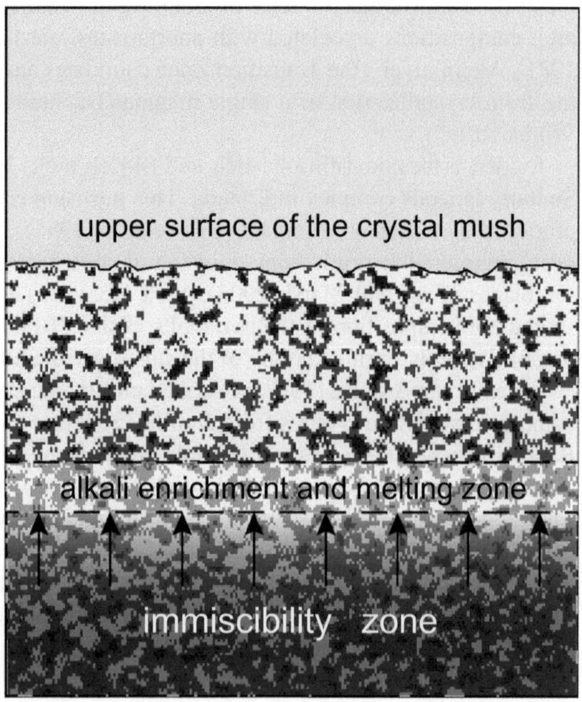

Fig. 5.13 Immiscibility in a cumulus crystal mush. See text for discussion

miscibility isotherm is labeled in Fig. 5.13 as immiscibility zone. Within the zone, immiscible melts are stable and separate by gravity. The lighter silica-rich melt moves upwards where it should eventually cross immiscibility isotherm, above which immiscibility is unstable. At this point, the immiscible silica-rich melt must re-dissolve in the resident intercumulus liquid. The dissolving melt brings extra silica and alkalis in the form of albite and orthoclase aluminosilicate components. The additions of such components to the multiply-saturated intercumulus melt are expected to trigger a series of crystal-melt reactions. For example, additional alkalis would offset the Ca/Na balance and the plagioclase-melt equilibrium. This should result in partial melting of cumulus plagioclase right above immiscibility zone. Pyroxenes and other minerals should also melt to keep the intercumulus liquid at equilibrium cotectic composition. Partial melting of earlier formed cumulates may have a positive feedback on the melt migration because of the increasing porosity and permeability in the melting zone. The extent of melting would depend on the balance between the cooling rate of the magma chamber and the rate of melt migration. Other factors, such as variations in melt viscosity or the size of immiscible melt droplets may also play significant roles. In principle, the process may lead to complete dissolution of some cumulus minerals and formation of monomineral layers. In any case, the consequences of intercumulus immiscibility appear to be complex and the effects of the resulting material transport may be far-reaching.

Conclusions and the Outlook

More and more evidence for immiscibility between two silicate melts in layered intrusions have been provided. Contrasted melt inclusions trapped in cumulus phases provide the most obvious record for the coexistence of equilibrium paired melts, as described in the Skaergaard and Sept Iles intrusions. The process affects the liquid line of descent during its late-stage evolution although the exact onset of unmixing is debated. Even if the two melts have contrasting physical properties (density, viscosity), their ability to segregate on large scale has still to be proven. Kinetics issues hamper nucleation and growth of immiscible droplets, so that small-scale segregation in the crystal mush might be more common than large-scale separation of the paired melts. The process has nevertheless implications for the formation of Fe–Ti–P ore deposits in layered intrusions, and by analogy, could also be important for the origin of Kiruna-type iron ore deposit. Further work might focus on the understanding of nucleation, growth, and potential segregation of immiscible melts.

Acknowledgments This paper has been reviewed by Jean Clair Duchesne and Jakob Klove Keiding. Olivier Namur is thanked for the editorial handling. BC acknowledges support by a Marie Curie International Outgoing Fellowship within the 7th European Community Framework Programme. IVV has been supported by the German Science Foundation (DFG) grants FR 557/23-2 and VE 619/2-1, and the Russian Science Foundation (RNF) grant No. 14-17-00200.

References

Barnes S-J, Maier WD, Curl EA (2010) Composition of the marginal rocks and sills of the Rustenburg Layered Suite, Bushveld Complex, South Africa: implications for the formation of the platimum-group element deposits. Econ Geol 105:1491–1511

Bogaerts M, Schmidt MW (2006) Experiments on silicate melt immiscibility in the system Fe_2SiO_4–$KAlSi_3O_8$–SiO_2–CaO–MgO–TiO_2–P_2O_5 and implications for natural magmas. Contrib Mineral Petrol 152: 257–274

Bowen NL (1928) The evolution of the igneous rocks. Princeton University Press, Princeton, p 332

Brooks CK, Nielsen TFD (1990) The differentiation of the Skaergaard intrusion. A discussion of Hunter and Sparks (Contrib Mineral Petrol 95:451–461). Contrib Mineral Petrol 104:244–247

Cawthorn RG (2013) Rare earth element abundances in apatite in the Bushveld Complex—A consequence of the trapped liquid shift effect. Geology 41:603–606

Cawthorn RG (2015) The Bushveld Complex, South Africa. In: Charlier B et al (eds) Layered Intrusions. Springer, Heidelberg

Charlier B, Grove TL (2012) Experiments on liquid immiscibility along tholeiitic liquid lines of descent. Contrib Mineral Petrol 164:27–44

Charlier B, Namur O, Toplis MJ, Schiano P, Cluzel N, Higgins MD, Vander Auwera J (2011) Large-scale silicate liquid immiscibility during differentiation of tholeiitic basalt to granite and the origin of the Daly gap. Geology 39:907–910

Charlier B, Namur O, Grove TL (2013) Compositional and kinetic controls on liquid immiscibility in ferrobasalt-rhyolite volcanic and plutonic series. Geochim Cosmochim Acta 113(0):79–93

Chayes F (1963) Relative abundance of intermediate members of the oceanic basalt-trachyte association. J Geophys Res 68(5):1519–1534

Daly RA (1914) Igneous rocks and their origin. McGraw-Hill, NewYork, p 563

De A (1974) Silicate liquid immiscibility in the Deccan traps and its petrogenetic significance. Geol Soc Am Bull 85:471–474

Dixon S, Rutherford MJ (1979) Plagiogranites as late-stage immiscible liquids in ophiolite and mid-ocean ridge suites: an experimental study. Earth Planet Sci Lett 45(1):45–60

Doremus RH (1994) Glass science. Wiley and, New York

Duchesne JC (1999) Fe–Ti deposits in Rogland anorthosites (South Norway): geochemical characteristics and problems of interpretation. Miner Depos 34:182–198

Duchesne JC, Wilmart E (1997) Igneous charnokites and related rocks from the Bjerkreim-Sokndal layered intrusion (Southwest Norway): a jotunite (hyperstene monzodiorite)-derived A-type granitoide suite. J Petrol 38:337–369

Duchesne JC, Denoiseux B, Hertogen J (1987) The norite-mangerite relationships in the Bjerkreim-Sokndal layered lopolith (SW Norway). Lithos 20:1–17

Eales HV, Cawthorn RG (1996) The Bushveld Complex. In: Cawthorn RG (ed) Layered intrusions. Elsevier, Amsterdam, pp 181–229

Freestone IC Hamilton DL (1977) Liquid immiscibility in K_2O–FeO–Al_2O_3–SiO_2: discussion. Nature 267:559

Greig JW (1927) Immiscibility in silicate melts. Am J Sci 13:133–154

Grieve RAF, Stöffler D, Deutsch A (1991) The Sudbury structure: Controversial or misunderstood? J Geophys Res Planets 96(E5):22753–22764

Hess PC (1995) Thermodynamic mixing properties and the structure of silicate melts. In: Stebbins JF, McMillan PF, Dingwell DB (eds) Structure, dynamics and properties of silicate melts. Mineral Soc America, Washington, DC Rev Miner 32:145–190

Hess PC (1996) Upper and lower critical points: Thermodynamic constraints on the solution properties of silicate melts. Geochim Cosmochim Acta 60:2365–2377

Higgins MD (2005) A new interpretation of the structure of the Sept Iles Intrusive suite, Canada. Lithos 83(3–4):199–213

Holness MB, Stripp G, Humphreys MCS, Veksler IV, Nielsen TFD (2011) Silicate liquid immiscibility within the crystal mush: Late-stage magmatic microstructures in the Skaergaard Intrusion, East Greenland. J Petrol 52:175–222

Hoover JD, Irvine TN (1978) Liquidus relations and Mg–Fe partitioning on part of the system Mg_2SiO_4–Fe_2SiO_4–$CaMgSi_2O_6$–$CaFeSi_2O_6$–$KAlSi_3O_8$–SiO_2. Carnegie Inst Wash Yearb 77, 774–784

Hou T, Zang Z, Kusky T (2011) Gushan magnetite-apatite deposit in the Ningwu basin, lower Yangtze River Valley, SE China: hydrothermal or Kiruna-type? Ore Geol Rev 33:333–346

Hudon P, Baker DR (2002) The nature of phase separation in binary oxide melts and glasses. I. silicate systems. J Non-Cryst Solids 303:299–345

Humphreys MCS (2011) Silicate liquid immiscibility within the crystal mush: evidence from Ti in plagioclase from the Skaergaard intrusion. J Petrol 52:147–174

Hunter RH, Sparks RSJ (1987) The differentiation of the Skaergaard intrusion. Contrib Mineral Petrol 95:451–461

Hunter RH, Sparks RSJ (1990) The differentiation of the Skaergaard intrusion. A reply. Contrib Mineral Petrol 104:248–254

Jakobsen JK, Veksler IV, Tegner C, Brooks CK (2005) Immiscible iron- and silica-rich melts in basalt petrogenesis documented in the Skaergaard intrusion. Geology 33:885–888

Jakobsen JK, Veksler IV, Tegner C, Brooks CK (2011) Crystallization of the Skaergaard intrusion from an emulsion of immiscible iron- and silica-rich liquids: evidence from melt inclusions in plagioclase. J Petrol 52: 345–373

James PF (1975) Liquid-phase separation in glass-forming systems. J Mater Sci 10:1802–1825

Juster TC, Grove TL, Perfit MR (1989) Experimental constraints on the generation of FeTi basalts, andesites, and rhyodacites at the Galapagos spreading center, 85°W and 95°W. J Geophys Res 94(B7):9251–9274

Kolker A (1982) Mineralogy and geochemistry of Fe–Ti oxides and apatite (nelsonite) deposits and evaluation of the liquid immiscibility hypothesis. Econom Geol 77:1146–1158

Kontak DJ, De Wolfe De Young MY, Dostal J (2002) Late-stage crystallization history of the Jurassic North Mountain Basalt, Nova Scotia, Canada. I. Textural and chemical evidence for pervasive development of silicate-liquid immiscibility. Can Miner 40:1287–1311

Kyser TK, Lesher CE, Walker D (1998) The effects of liquid immiscibility and thermal diffusion on oxygen isotopes in silicate liquids. Contrib Mineral Petrol 133:373–381

Lester GW, Clark AH, Kyser TK, Naslund HR (2013a) Experiments on liquid immiscibility in silicate melts with H_2O, P, S, F, and Cl: implications for natural magmas. Contrib Mineral Petrol 166:329–349

Lester GW, Kyser TK, Clark AH (2013b) Oxygen isotope partitioning between immiscible silicate melts with H_2O, P and S. Geochim Cosmochim Acta 109:306–311

Levin EM, Robbins CR, McMurdie HF (1964) Phase Equilibria Diagrams, vol 1. American Ceramic Society, Westerville

Loferski PJ, Arculus RJ (1993) Multiphase inclusions in plagioclase from anorthosites in the Stillwater Complex, Montana: implications for the origin of the anorthosites. Contrib Mineral Petrol 114:63–78

Longhi J (1998) Silicate liquid immiscibility: a furtive agent of fractionation. 29th Lunar and Planetary Science Conference, abstract No 1903

Maier WD, Barnes S-J, Groves DI (2013) The Bushveld Complex, South Africa: formation of platinum-palladium, chrome- and vanadium-rich layers via hydrodynamic sorting of a mobilized cumulate slurry in a large, relatively slowly cooling, subsiding magma chamber. Miner Depos 48(1):1–56

Martin B, Kushiro Y (1991) Immiscibility synthesis as an indicator of cooling rates in basalts. J Volcanol Geotherm Res 45:289–310

Mazurin OV, Porai-Koshits EA (1984) Phase separation in glass. Amsterdam, North-Holland, p 382

McBirney AR (1996) The Skaergaard intrusion. In: Cawthorn RG (ed) Layered intrusions. Developments in Petrology, 15. Elsevier, Amsterdam, pp 147–180

McBirney AR (2008) Comments on: 'liquid Immiscibility and the Evolution of Basaltic Magma'. J Petrol 48:2187–2210, 49:2169–2170

McBirney AR, Nakamura Y (1974) Immiscibility in late-stage magmas of the Skaergaard intrusion. Carnegie Inst Wash Yearb 73:348–352
McBirney AR, Naslund HR (1990) The differentiation of the Skaergaard intrusion. A discussion of Hunter and Sparks (Contrib Mineral Petrol 95:451–461). Contrib Mineral Petrol 104:235–240
Morse SA (1990) The differentiation of the Skaergaard intrusion. A discussion of Hunter and Sparks (Contrib Mineral Petrol 95:451–461). Contrib Mineral Petrol 104:240–244
Morse SA (2008) Compositional convection trumps silicate liquid immiscibilty in layered intrusions: a discussion of 'Liquid immiscibility and the evolution of basaltic magma' by Veksler et al. J Petrol 48:2187–2210, 49:2157–2168
Namur O, Charlier B, Toplis MJ, Higgins MD, Liégeois J-P, Vander Auwera J (2010) Crystallization sequence and magma chamber processes in the ferrobasaltic Sept Iles layered intrusion, Canada. J Petrol 51:1203–1236
Namur O, Charlier B, Toplis MJ, Higgins MD, Hounsell V, Liégeois JP, Vander Auwera J (2011) Differentiation of tholeiitic basalt to A-type granite in the Sept Iles layered intrusion, Canada. J Petrol 52:487–539
Namur O, Charlier B, Holness MB (2012) Dual origin of Fe–Ti–P gabbros by immiscibility and fractional crystallization of evolved tholeiitic basalts in the Sept Iles layered intrusion. Lithos 154(0):100–114
Namur O, Higgins MD, Vander Auwera J (2015) The Sept Iles intrusive suite, Quebec, Canada. In Charlier B et al (eds) Layered intrusions. Springer, Heidelberg
Navrotsky A (1992) Unmixing of hot inorganic melts. Nature 360:306
Nielsen TFD (2004) The shape and volume of the Skaergaard intrusion, Greenland: implications for mass balance and bulk composition. J Petrol 45:507–530
Pang K-N, Li C, Zhou M-F, Ripley EM (2008a) Abundant Fe–Ti oxide inclusions in olivine from the Panzhihua and Hongge layered intrusions, SW China: evidence for early saturation of Fe–Ti oxides in ferrobasaltic magma. Contrib Mineral Petrol 156:307–321
Pang KN, Zhou MF, Lindsley D, Zhao D, Malpas J (2008b) Origin of Fe–Ti oxide ores in mafic intrusions: Evidence from the Panzhihua intrusion, SW China. J Petrol 49:295–313
Pang KN, Zhou MF, Qi L, Shellnutt G, Wang CY, Zhao D (2010) Flood basalt-related Fe–Ti oxide deposits in the Emeishan large igneous province, SW China. Lithos 119(1–2):123–136
Philpotts AR (1967) Origin of certain iron-titanium oxide and apatite rocks. Econ Geol 62:303–315
Philpotts AR (1979) Silicate liquid immiscibility in tholeiitic basalts. J Petrol 20:99–118
Philpotts AR (1981) A model for the generation of massif-type anorthosites. Can Miner 19:233–253
Philpotts AR (1982) Compositions of immiscible liquids in volcanic rocks. Contrib Mineral Petrol 80:201–218
Philpotts AR (2008) Comments on: Liquid immiscibility and the evolution of basaltic magma. J Petrol 49:2171–2175
Philpotts AR, Doyle CD (1983) Effect of magma oxidation state on the extent of silicate liquid immiscibility in a tholeiitic basalt. Am J Sci 283(9):967–986
Powell MA, Walker D, Hays JF (1980) Controlled cooling and crystallization of a eucrite: Microprobe studies. Proc Lunar Planet Sci Conf 11th 2:1153–1168
Reid DL, Basson IJ (2002) Iron-rich ultramafic pegmatite replacement bodies within the upper critical zone, Rustenburg layered suite, Northam Platinum Mine, South Africa. Miner Mag 66:895–914
Reynolds IM (1985) The nature and origin of titaniferous magnetite-rich layers in the upper zone of the Bushveld Complex: a review and synthesis. Econ Geol 80:1089–1108
Ripley EM, Severson MJ, Hauck SA (1998) Evidence for sulfide and Fe–Ti–P-rich liquid immiscibility in the Duluth Complex, Minnesota. Econ Geol 93(7):1052–1062
Roedder E (1951) Low-temperature liquid immiscibility in the system K_2O–FeO–Al_2O_3–SiO_2. Am Miner 36:282–286
Roedder E (1978) Silicate liquid immiscibility in magmas and in the system K_2O–FeO–Al_2O_3–SiO_2: an example of serendipity. Geochim Cosmochim Acta 43:1597–1617

Roedder E (1979) Silicate liquid immiscibility in magmas. In: Yoder HS Jr (ed) The evolution of igneous rocks. Fiftieth Anniversary perspectives. Princeton University Press, Princeton, pp 15–58
Rutherford MJ, Hess PC, Daniel GH (1974) Experimental liquid line of descent and liquid immiscibility for basalt 70017. Proc Proceedings of the 5th Lunar Science Conference, pp 569–583
Sato H (1978) Segregation vesicles and immiscible liquid droplets in ocean-floor basalt of Hole 396B, IPOD/DSDP Leg 46. In: Dimitriev L, Heirtzler J (eds) Initial reports of the deep sea drilling project v 46, U.S. Gov Printing Office, Washington, pp 283–291
Schmidt MW Connolly JAD, Günther D, Bogaerts M (2006) Element partitioning—the role of melt structure and composition. Science 312:1646–1650
Scoates JS, Friedman RM (2008) Precise age of the platiniferous Merensky Reef, Bushveld Complex, South Africa, by the U–Pb zircon chemical ID-TIMS abrasion technique. Econ Geol 103:465–471
Scoon RN, Mitchell AA (1994) Discordant iron-rich ultramafic pegmatites in the Bushveld Complex and their relationship to iron-rich intercumulus and residual liquids. J Petrol 35:881–917
Sensarma S, Palme H (2013) Silicate liquid immiscibility in the ~2.5 Ga Fe-rich andesite at the top of the Dongargarh large igneous province (India). Lithos 170–171:239–251
Shelby JE (2005) Introduction to glass science and technology. The Royal Society of Chemistry, Cambridge
Simmons JH, Macedo PB (1971) Analysis of viscous relaxation in critical oxide mixtures. J Chem Phys 54:1325–1331
Simmons JH, Napolitano A, Macedo PB (1970) Supercritical viscosity anomaly in oxide mixtures. J Chem Phys 53:1165–1170
Song XY, Qi HW, Hu RZ, Chen LM, Yu SY, Zhang JF (2013) Formation of thick stratiform Fe–Ti oxide layers in layered intrusion and frequent replenishment of fractionated mafic magma: evidence from the Panzhihua intrusion, SW China. Geochem Geophys Geosyst 14(3):712–732
Tegner C, Cawthorn RG (2010) Iron in plagioclase in the Bushveld and Skaergaard intrusions: implications for iron contents in evolving basic magmas. Contrib Mineral Petrol 159:719–730
Thompson RN (1972) Evidence for a chemical discontinuity near the basalt-andesite transition in many anorogenic volcanic suites. Nature 236(5342):106–110
Thompson AB, Aerts M, Hack AC (2007) Liquid immiscibility in silicate melts and related systems. Rev Mineral Geochem 65:99–127 (In: Liebscher A, Heinrich CA (eds) Fluid-fluid interactions)
Thy P, Lesher CE, Nielsen TFD, Brooks CK (2006) Experimental constraints on the Skaergaard liquid line of descent. Lithos 92:154–180
Toplis MJ, Carroll MR (1995) An experimental study of the influence of oxygen fugacity on Fe–Ti oxide stability, phase relations, and mineral-melt equilibria in ferro-basaltic systems. J Petrol 36:1137–1171
VanTongeren JA, Mathez EA (2012) Large-scale liquid immiscibility at the top of the Bushveld Complex, South Africa. Geology 40:491–494
Veksler IV (2004) Liquid immiscibility and its role at the magmatic-hydrothermal transition: a summary of experimental studies. Chem Geol 210:7–31
Veksler IV (2009) Extreme iron enrichment and liquid immiscibility in mafic intrusions: experimental evidence revisited. Lithos 111:72–82
Veksler IV, Dorfman AM, Danyushevsky LV, Jakobsen JK, Dingwell DB (2006) Immiscible silicate liquid partition coefficients: Implications for crystal-melt element partitioning and basalt petrogenesis. Contrib Mineral Petrol 152:685–702
Veksler IV, Dorfman AM, Borisov AA, Wirth R, Dingwell DB (2007) Liquid immiscibility and evolution of basaltic magma. J Petrol 48:2187–2210
Veksler IV, Dorfman AM, Rhede D, Wirth R., Borisov AA, Dingwell DB (2008a) Kinetics and the extent of liquid immiscibility in the system K_2O–CaO–FeO–Al_2O_3–SiO_2. Chem Geol 256:119–130

Veksler IV, Dorfman AM, Borisov AA, Wirth R, Dingwell DB (2008b) Liquid immiscibility and evolution of basaltic magma. Reply to SA Morse, AR McBirney and AR Philpotts. J Petrol 49:2177–2186

Veksler IV, Kähn J, Franz G, Dingwell DB (2010) Interfacial tension between immiscible liquids in the system K_2O–FeO–Fe_2O_3–Al_2O_3–SiO_2 and implications for the kinetics of silicate melt unmixing. Am Miner 95:1679–1685

Visser W, Koster van Groos AF (1976) Liquid immiscibility in K_2O–FeO–Al_2O_3–SiO_2. Nature 264:426–427

Visser W, Koster van Groos AF (1977) Liquid immiscibility in K_2O–FeO–Al_2O_3–SiO_2: reply to Roedder, Freestone and Hamilton. Nature 267:560

Visser W, Koster van Groos AF (1979) Phase relations in the system K_2O–FeO–Al_2O_3–SiO_2 at 1 atmosphere with special emphasis on low temperature liquid immiscibility. Am J Sci 279:70–91

von Gruenewaldt G (1993) Ilmenite-apatite enrichments in the upper zone of the Bushveld Complex: a major titanium-rock phosphate resource. Int Geol Rev 35: 987–1000

Wager LR, Deer WA (1939) Geological investigation in East Greenland, Part III, The petrology of the Skaergaard intrusion, Kangerdluqssuaq, East Greenland. Meddelser om Grønland 105(4)283–296

Wang C, Zhou MF (2013) New textural and mineralogical constraints on the origin of the Hongge Fe–Ti–V oxide deposit, SW China. Miner Depos 48:787–798

Watts K (2014) A melt inclusion study of the Sudbury Igneous Complex (Ontario, Canada): evidence for two-liquid immiscibility and constraints on trace element distribution. Master Thesis, Saint Mary's University, Halifax, p 129

Wilson AH (2012) A chill sequence to the Bushveld Complex: insight into the first stage of magma emplacement and implications for parental magmas. J Petrol 53:1123–1168

Wilson JR, Overgaard G (2005) Relationship between the layered series and the overlying evolved rocks in the Bjerkreim-Sokndal intrusion, Southern Norway. Lithos 83(3–4):277–298

Zhang XQ, Song XY, Chen LM, Xie W, Yu SY, Zheng WQ, Deng YF, Zhang JF, Gui SG (2012) Fractional crystallization and the formation of thick Fe–Ti–V oxide layers in the Baima layered intrusion, SW China. Ore Geol Rev 49:96–108

Zhou M-F, Robinson PT, Lesher CM, Keays RR, Zhang C-J, Malpas J (2005) Geochemistry, petrogenesis and metallogenesis of the Panzhihua gabbroic layered intrusion and associated Fe–Ti–V oxide deposits, Sichuan Province, SW China. J Petrol 46:2253–2280

Zhou M-F, Arndt NT, Malpas J, Wang CY, Kennedy AK (2008) Two magma series and associated ore deposit types in the Permian Emeishan large igneous province, SW China. Lithos 103:352–368

Zieg MJ, Marsh BD (2005) The Sudbury Igneous Complex: viscous emulsion differentiation of a superheated impact melt sheet. Geol Soc Am Bull 117(11–12):1427–1450

Chapter 6
Basal Reversals in Mafic Sills and Layered Intrusions

Rais Latypov

Abstract Fractional crystallization of ponded basaltic magma results in rocks that become increasingly more evolved in composition inwards. However, almost all mafic sills and layered intrusions show basal reversals tens to hundreds of meters thick in which minerals and rocks become compositionally more primitive upwards. Two main types of basal reversals can be distinguished with different relationships with the overlying Layered Series. Most common are fully-developed reversals that gradually pass into the overlying Layered Series through a crossover horizon with the most primitive rock and mineral compositions. Above the crossover, the Layered Series becomes progressively more evolved. This type of reversal presents a mirror image of the overlying Layered Series. In contrast, aborted reversals are separated from a more primitive or evolved Layered Series by a sharp textural and compositional break. Both types of reversal are attributed to three principal processes: (1) inflowing magma becomes progressively more primitive, (2) magma undercooling gradually decreases and (3) adcumulus growth becomes increasingly more effective. The major difference between these types is that fully-developed reversals result from a continuous evolution and therefore show a gradual transition into the overlying Layered Series. In contrast, aborted reversals form when the evolution is abruptly interrupted by a pulse of primitive magma that resets the crystallization history in the chamber. This generally results in basal reversals in sharp contact with a more primitive Layered Series. In some lopolithic layered intrusions, aborted basal reversals are discordant with the Layered Series and therefore can be in contact with more evolved rocks along the upper sections of the inward-dipping floor.

Keywords Basal reversals · Layered intrusions · Mafic sills · Magma chamber processes · Magma emplacement · *In situ* growth · Chilled margins · Parental magmas · Adcumulates

---

R. Latypov (✉)
School of Geosciences, University of the Witwatersrand,
Private Bag 3, 2050 Wits, South Africa
e-mail: rais.latypov@wits.ac.za

B. Charlier et al. (eds.), *Layered Intrusions*, Springer Geology,
DOI 10.1007/978-94-017-9652-1_6

Introduction

Basal reversals or regressions are a common feature of mafic-ultramafic sills (e.g., Gunn 1966; Simkin 1967; Lightfoot and Naldrett 1984; Frenkel' et al. 1988; Froelich and Gottfried 1988; Gorring and Naslund 1995; López-Moro et al. 2007; Galerne et al. 2010) and layered intrusions (e.g. Campbell 1978; Pattison 1979; Irvine 1980; Alapieti 1982; Raedeke and McCallum 1984; Wilson and Engell-Sørensen 1986; Jaupart and Tait 1995; Miller and Ripley 1996; Ariskin et al. 2003; Egorova and Latypov 2012a, b). These are zones along the margins of igneous intrusions in which rocks become compositionally more primitive upwards. In particular, they are commonly characterized by a decrease in the number of crystallizing (liquidus) phases in successively formed rocks (e.g. *Ol* + *Pl* + *Cpx, Ol* + *Pl, Ol*) and minerals that become gradually more primitive (e.g. plagioclase An_{30}–An_{80}) inwards. Paradoxically, their crystallization sequences and compositional trends are opposite to that resulting from closed-system fractional crystallization of basaltic magma. Igneous petrologists over the last century have proposed several different explanations for basal reversals including contamination, magma undercooling, compositionally-zoned magma, crystal settling, prolonged magma emplacement, flow differentiation, reaction with varying amounts of intercumulus melt and shrinkage-induced magma inflow (see Latypov 2003a; Latypov et al. 2007; Chistyakova and Latypov 2009). None of these mechanisms provide an entirely satisfactory explanation and new insights into the origin of basal reversals are required. Alternative mechanisms that have been suggested recently are Soret magma differentiation (Latypov 2003a, b), compositional reconstitution of solidifying cumulates in response to a thermal gradient (Lundstrom 2009; Huang et al. 2009) and postcumulus redistribution of interstitial melt caused by the development of zones of low pressure in solidifying rocks (Aarnes et al. 2008; Galerne et al. 2010).

In this contribution I will review some examples of typical basal reversals and then illustrate how they can be explained in the light of the "three-factor model" (Latypov et al. 2011; Latypov and Egorova 2012). The examples include two major types of basal reversals in sills and layered intrusions: "fully-developed" and "aborted" reversals (Latypov et al. 2011). A fully-developed reversal shows a crystallization sequence and whole-rock compositional trend that are essentially the opposite of those in the overlying Layered Series into which it passes via an inflexion or crossover with the most primitive rock compositions. Two varieties of fully-developed reversals can be distinguished: those in which both whole-rock and mineral compositions show reverse trends (Fig. 6.1a) and those in which only whole-rock composition show reverse trends whereas minerals exhibit normal trends (Fig. 6.1b). Examples of fully-developed basal reversals are abundant in the literature (e.g. Krivenko et al. 1980; Alapieti 1982; Lightfoot and Naldrett 1984; Wilson and Engell-Sørensen 1986; Frenkel' et al. 1988, 1989; Froelich and Gottfried 1988; Koptev-Dvornikov et al. 2001; Gibb and Henderson 1992, 2006; Galerne et al. 2010; Ubide et al. 2012). An aborted reversal shows reverse whole-rock and mineral compositional trends that are separated by a sharp textural and

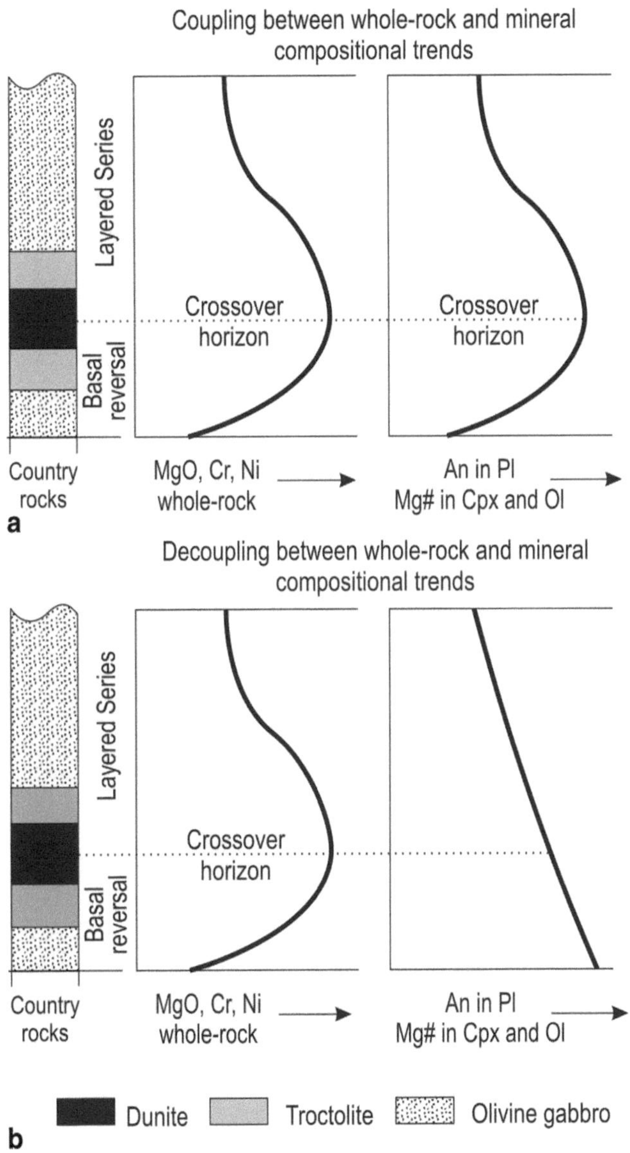

Fig. 6.1 Sections illustrating two varieties of fully-developed basal reversals in mafic sills and layered intrusions. Both varieties show crystallization sequences and whole-rock compositional trends that are essentially the opposite of those in the overlying Layered Series into which they pass via a crossover horizon with the most primitive rock composition. The main difference between these varieties is that in **a** both whole-rock and minerals show reverse trends whereas in **b** only the whole rock composition exhibits a reverse trend while the minerals have normal trends. Hereafter in figures and text: *Pl* plagioclase, *Ol* olivine, *Cpx* Ca-rich pyroxene, *Opx* orthopyroxene, *Pig* pigeonite, *Am* amphibole, *Qtz* quartz, *Kfs* alkali feldspar, *Chr* chromite. An = 100*An/(An+Ab); Mg# = 100*Mg/(Mg + Fe)

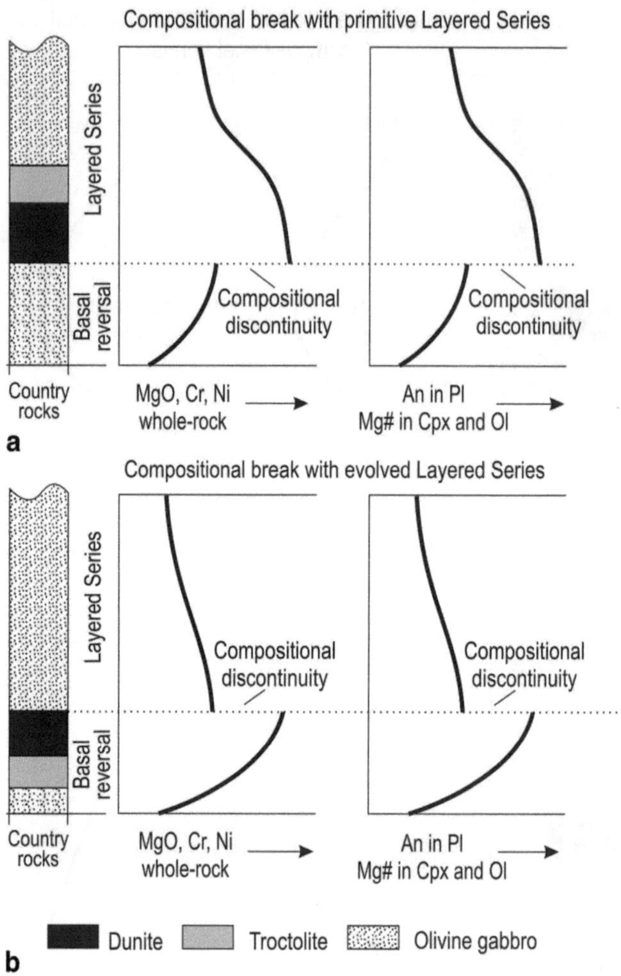

Fig. 6.2 Sections illustrating two varieties of aborted basal reversals in mafic sills and layered intrusions. Both varieties show reverse trends in whole-rock and mineral compositions that are terminated by a sharp compositional break with the overlying Layered Series. The main difference between these varieties is that in **a** the reversal is overlain by a more primitive Layered Series while in **b** the Layered Series is more evolved

compositional break from the overlying Layered Series. Two varieties of aborted reversals can also be identified: those in which basal reversals are overlain by a more primitive Layered Series (Fig. 6.2a) and those in which they are succeeded by a more evolved Layered Series (Fig. 6.2b). The latter has been described earlier from lopolithic layered intrusions (Irvine and Smith 1967; Alapieti and Lahtinen 2002) whereas the former has been identified in sheet-like layered intrusions only quite recently (Latypov et al. 2011; Egorova and Latypov 2012a, b).

Examples of Typical Fully-Developed Reversals

The Kuz'movsky Sill

The Kuz'movsky sill is exposed in the valley of the Podkamennaya Tunguska River in SE Siberia, Russia. It was emplaced during the igneous event that led to generation of the Siberian flood basalts close to the Permian-Triassic boundary. The sill has a thickness of about 85 m and is exposed from base to top (Frenkel' et al. 1988) (Fig. 6.3). It is intruded into marls and preserves chilled margins at both the floor and roof. The sill is composed of ophitic dolerite (pl-ol cumulate with oikocrysts of cpx) at the base and top and subophitic dolerite (pl-ol-cpx cumulate with subhedral, prismatic crystals of cpx) in the interior. The transition from ophitic to subophitic dolerite thus marks the appearance of cumulus clinopyroxene. In terms of petrography and whole-rock chemistry, the sill is subdivided into floor and roof sequences that meet at a Sandwich Horizon. The floor sequence of the sill comprises a basal reversal and a Layered Series while the roof sequence consists only of an Upper Border Series. The boundary between the reversal and the Layered Series is characterized by the most primitive rocks (crossover horizon) in the sill. The basal reversal is characterized by an upward increase in MgO (from ~8 wt% to ~12 wt%) and a decrease in TiO_2 (from ~1.3 to ~0.9 wt%). Above the crossover horizon, the rocks

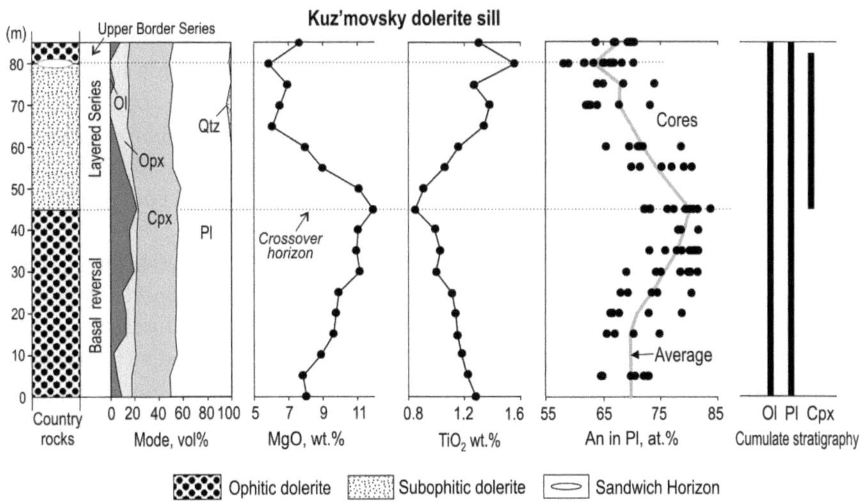

Fig. 6.3 Simplified section through the Kuz'movsky dolerite sill, Siberia, showing its stratigraphic subdivision, modal composition, whole-rock geochemistry and mineral composition. Note the thick sequence with reverse trends in whole-rock MgO and TiO_2 and plagioclase compositions that passes through a crossover horizon into a Layered Series showing normal cryptic layering. The reversal can be classified as fully-developed with coupled whole-rock and mineral compositions (Fig. 6.1a). The *thick vertical lines* give the distribution of cumulus (liquidus) minerals. (Modified from Egorova and Latypov 2013)

become increasingly more evolved in terms of these indices up to the Sandwich Horizon. The compositional profile of the sill can thus be described as S-shaped (Fig. 6.3). The composition of the cores and rims of plagioclase crystals in the Kuz'movsky sill also define an S-shaped profile. Plagioclase cores show a systematic upward increase in An-content from 72 to 84 at.% up to the crossover horizon and then a decrease through the Layered Series, with the most evolved plagioclase cores (An 58–59) occurring in the Sandwich Horizon. Plagioclase is normally zoned with compositional differences between cores and rims of up to 10–20% An. Since chemical diffusivity in plagioclase is sluggish, its original composition is commonly preserved (e.g. Morse 1984). This suggests that the observed compositional trends are primary and not the result of subsolidus re-equilibration. Basal reversals with similar whole-rock and mineral compositional trends are observed in several other mafic sills and layered intrusions. In particular, such trends have been documented in the basal zones of the Duluth (Chalokwu and Grant 1990) and Hasvik (Tegner et al. 1999) layered intrusions. Such trends are also evident in the margins of some Antarctic dolerite sills (Gunn 1962; 1966), Greenland dolerite sills (Gisselø 2001), the Golden Valley Sill, South Africa (Aarnes et al. 2008; Galerne et al. 2010) in two other Siberian dolerite sills (Egorova and Latypov 2013).

The Sudbury Igneous Complex

The 1.8 Ga Sudbury Igneous Complex (SIC) is a 1–5 km thick sheet of crystallized impact melt of andesitic composition in the south-eastern part of the Canadian Shield (Deutsch et al. 1995). A section through the SIC at Creighton in the South Range (Fig. 6.4) consists of three major rock units (from base to top): (1) norite composed of cumulus orthopyroxene and plagioclase (2) quartz gabbro lacking orthopyroxene and containing cumulus clinopyroxene and amphibole; (3) granophyre consisting of cumulus quartz and K-feldspar. The systematic and continuous geochemical and mineralogical variation from norite through quartz gabbro to granophyre have been interpreted as strong evidence for the crystallization of Sudbury from a single melt system (Therriault et al. 2002). This has been supported by recent numerical modelling that reproduced almost all geochemical features of the SIC by fractional crystallization of a single sheet of granodioritic magma (Lavrenchuk et al. 2010). A notable feature of the Creighton section is a well-developed basal reversal occupying a thickness of about 1200 m (Fig. 6.4; Lightfoot and Zotov 2005). The basal reversal is especially well evident from an upward increase in MgO (from ~4 wt% to ~6.5 wt%) and a decrease in La (from ~55 ppm to ~10 ppm). However, the most An-rich cores of plagioclase crystals show a steady normal trend with a systematic upward decrease in An-content from ~70 to ~60 at.% (Fig. 6.4). The lack of any reverse trend in plagioclase compositions has also been demonstrated by Naldrett et al. (1970) in two other traverses through the South Range. Since plagioclase crystals exhibit normal zoning of up to 20–30 at.% An throughout the entire noritic part of the complex, the observed compositional trends are primary.

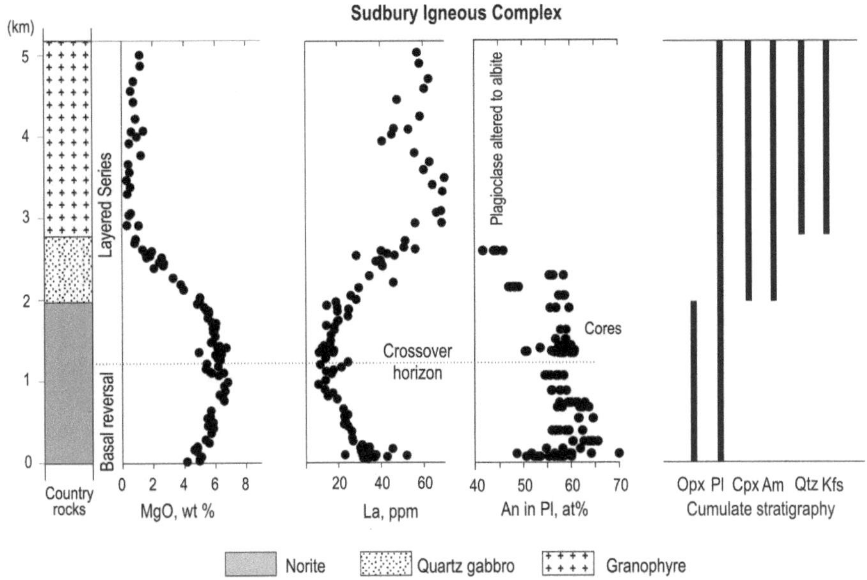

Fig. 6.4 Simplified section through the Sudbury Igneous Complex, Canada, along the Creighton traverse of the South Range showing its stratigraphic subdivision and whole-rock geochemistry. Note the basal reversal with reverse trends in whole-rock MgO and La that passes through a cross-over horizon into a Layered Series with normal cryptic layering. Cumulus plagioclase, however, has a normal trend (an upward decrease in An) across the basal reversal. The reversal can be classified as fully-developed with decoupled whole-rock and mineral compositions (Fig. 6.1b). The *thick vertical lines* give the distribution of cumulus (liquidus) minerals. Whole-rock data are mainly from Lightfoot and Zotov (2005). Plagioclase compositions are data not published earlier

This decoupling of whole-rock and mineral compositions contrasts with the more common situation exemplified by the Kuz'movsky sill (Fig. 6.3). At present, however, the SIC shows the only documented example of a basal reversal with this kind of decoupling. Further studies are required to show whether or not similar reversals are present in other intrusive bodies.

Examples of Typical Aborted Reversals

The Koitelainen Layered Intrusion

The Koitelainen layered intrusion in northern Finland was emplaced about 2.44 Ga ago as a sheet-like body at the boundary between the Archaean basement and its cover of Paleoproterozoic supracrustals (Mutanen and Huhma 2001). The intrusion consists of a ~100–200 m basal reversal overlain by ~3 km thick Layered Series (Mutanen 1997; Hanski et al. 2001; Latypov et al. 2011). In the section studied (Fig. 6.5) the basal reversal is within pigeonite gabbro (pl-cpx-pig cumulate)

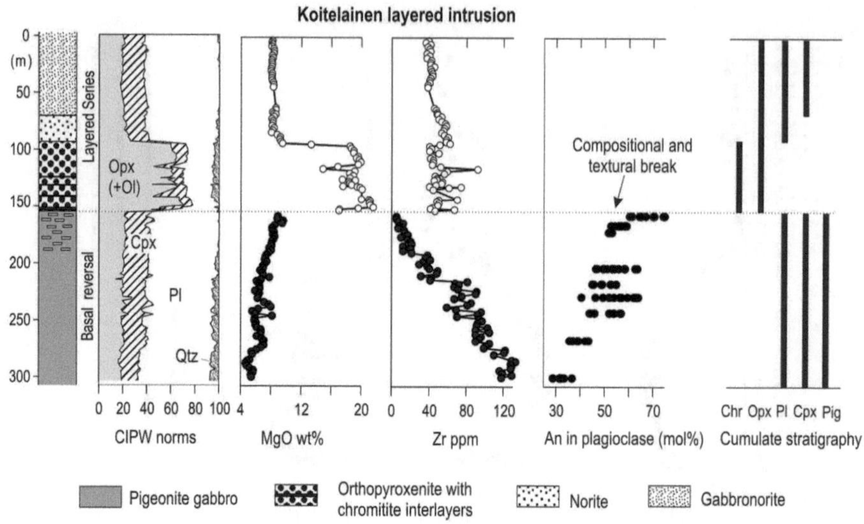

Fig. 6.5 Simplified section through the basal part of the Koitelainen Layered Intrusion, Finland, showing its stratigraphic subdivision, CIPW norms, whole-rock geochemistry and mineral compositions. Note the basal reversal (*filled circles*) with reverse trends in whole-rock MgO, Zr and An-content in plagioclase thatterminate abruptly at the contact with the overlying Layered Series (*open circles*). The reversal can be classified as an aborted one that shows a compositional break with a more primitive Layered Series (Fig. 6.2a). *Rectangles* indicate intervals of fine-grained pigeonite gabbro with a strong plagioclase lamination. Stratigraphic '0 meter' is defined by the top of borehole R-352. The *thick vertical lines* give the distribution of cumulus (liquidus) minerals. (Modified from Latypov et al. 2011)

whereas the Layered Series consists of orthopyroxenites (opx cumulate) with layers of massive chromitite passing through norite (pl-opx cumulate) to gabbronorite (pl-opx-cpx cumulate). Orthopyroxenite that overlies the reversal is generally very coarse-grained. There is no sign of chilling of the orthopyroxenite against the underlying pigeonite gabbro. The pigeonite gabbro becomes, however, finer-grained and acquires strong plagioclase lamination toward the contact with the orthopyroxenite. The pigeonite gabbro shows consistently reverse compositional trend in terms of whole-rock major and incompatible trace element composition from the base to top (Fig. 6.5). This is particularly well exemplified by the upward increase in MgO (from ~5 wt.% to 10 wt.%) and the decrease in Zr (from ~130 ppm to 3 ppm). The uppermost fine-grained pigeonite gabbro is also strongly depleted in other incompatible components (Latypov et al. 2011). The pigeonite gabbro has a sharp contact with orthopyroxenite with much higher MgO and Zr belonging to the overlying Layered Series. Plagioclase cores in the pigeonite gabbro also show a systematic upward increase in An-content from ~30 to ~70 at.% at the contact with the Layered Series. This trend must be primary in origin since plagioclase is normally zoned. Thus in this case whole-rock and mineral compositional trends are coupled; they are both reversed throughout the whole of the pigeonite gabbro. The dramatic changes in liquidus assemblages and geochemistry at the boundary between the basal reversal and the Layered Series suggest that the latter must have

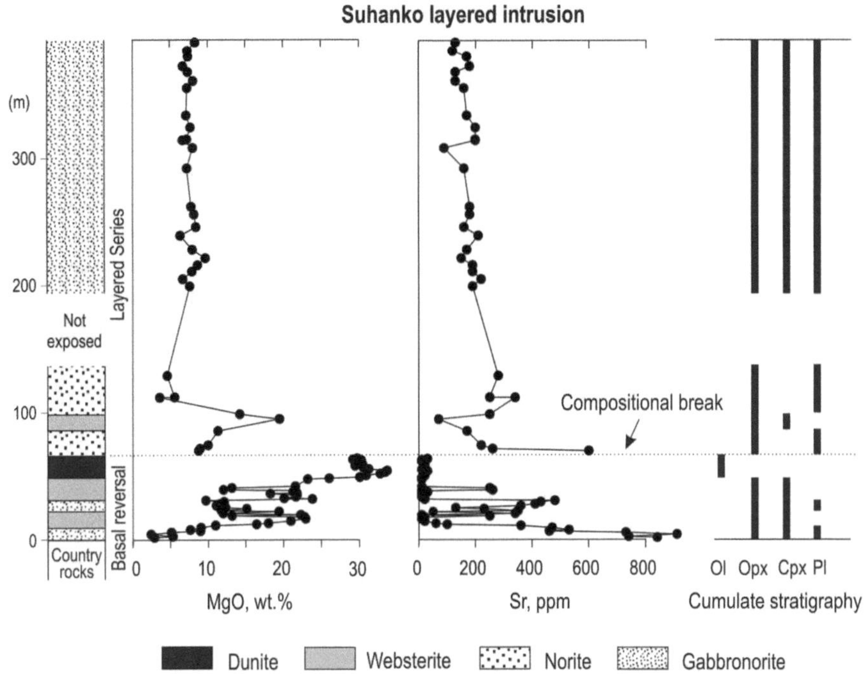

Fig. 6.6 Simplified section through the Suhanko Layered Intrusion, Finland, showing its stratigraphic subdivision and whole-rock geochemistry. Note the basal reversal with reverse trends in whole-rock MgO and Sr that terminate abruptly at the contact with the overlying Layered Series. The reversal can be classified as an aborted one that shows a compositional break with a more evolved Layered Series (Fig. 6.2b). The *thick continuous vertical lines* show the distribution of cumulus (liquidus) minerals. (Modified from Iljina et al. (1989) and Iljina (1994))

crystallized from much more primitive magma. Similar aborted reversals below more primitive Layered Series have been documented in the Imandra (Egorova and Latypov 2012a), Fongen-Hyllingen (Wilson and Engell-Sørensen 1986; Wilson and Larsen 1985; Egorova and Latypov 2012a) and Akanvaara (Mutanen 1997) layered intrusions.

The Suhanko Layered Intrusion

The 2.4 Ga old Suhanko layered intrusion, northern Finland (Alapieti and Lahtinen 2002), is underlain by Archaean granitoids and overlain by either similar granitoids or by a younger sequence of volcanic and sedimentary rocks (Iljina et al. 1989; Iljina 1994). The intrusion underwent amphibolite facies metamorphism and almost all its primary magmatic minerals have been altered, although cumulus textures are well preserved. It can be subdivided into a ~50 m thick basal reversal which comprises mainly ultramafic cumulates and an up to 500 m thick Layered Series of noritic to gabbronoritic cumulates (Fig. 6.6). The basal reversal comprises a sequence grading from gabbronorite (pl-cpx-opx cumulate) at the base through websterite (opx-cpx cumulate) to dunite (ol cumulate) at the top, thus constituting a remarkable reverse crystallization sequence. The dunite is included in the reversal because its lower contact with olivine-bearing websterite is gradational. The Layered Series has norite (pl-opx cumulate) at its base that is followed by rather monotonous sequence of gabbronorite (pl-cpx-opx cumulate). The norite in the lower part of the Layered Series contains a single layer of websterite that appears to give rise to a second basal reversal. The basal reversal shows distinctly reverse trends in whole-rock major and trace element composition. In particular, the reversal is marked by an overall upward increase in MgO (from ~5 to ~35 wt%) and a decrease in Sr (from ~850 ppm to ~10 ppm). As in the example above, the trends are truncated by the sharp compositional break at the contact with the Layered Series. However, in this case the norites of the Layered Series are much more evolved than the underlying dunite of the basal reversal. Such significant changes in crystallization sequence and rock geochemistry at the boundary between basal reversal and the Layered Series suggest that the latter must have crystallized from a more evolved magma. Other examples of similar aborted reversals beneath a more evolved Layered Series are the Muskox intrusion (Irvine and Smith 1967) and the Koillismaa layered intrusion (Alapieti and Lahtinen 2002; Karinen 2010).

Discussion

Current Models for the Origin of Basal Reversals

Existing explanations of basal reversals in sills and layered intrusions fall into two major groups, as depicted in Fig. 6.7. The first group (a) suggests that reversals are produced from crystal-free melt or crystal-poor magma that becomes more primitive in composition upwards. These can be generated by: (1) *in situ* contamination of the parental magma by assimilation of wall rocks (e.g. Tyson and Chang 1984; Severson and Hauck 1990; Gauert et al. 1995); (2) prolonged or multiple injections of magma that becomes more primitive with time (e.g. Morse 1979, 1981; Grant and Molling 1981; Henderson and Gibb 1987; Henderson et al. 2000; Gibb and Henderson 2006; Latypov et al. 2011); (3) gradual elevation of compositionally- and hence density-stratified magma so that increasingly primitive magma comes into contact with the sloping floor of the chamber (Wilson and Engell-Sørensen 1986); (4) Soret differentiation induced by a temperature gradient imposed by the cold country rock (Latypov 2003a; Latypov et al. 2007). All of these processes can be accompanied by (5) disequilibrium crystallization due to magma undercooling (Wager and Brown 1968; Miller and Ripley 1996; Chistyakova and Latypov 2012).

The hypotheses belonging to the second group (b) suggest that reversals are produced from crystal-melt mixtures that become more primitive in composition upwards. These can originated by: (6) a progressive increase in the amount of settled crystals (e.g. olivine) in response to a decrease in the solidification rate (e.g. Gray

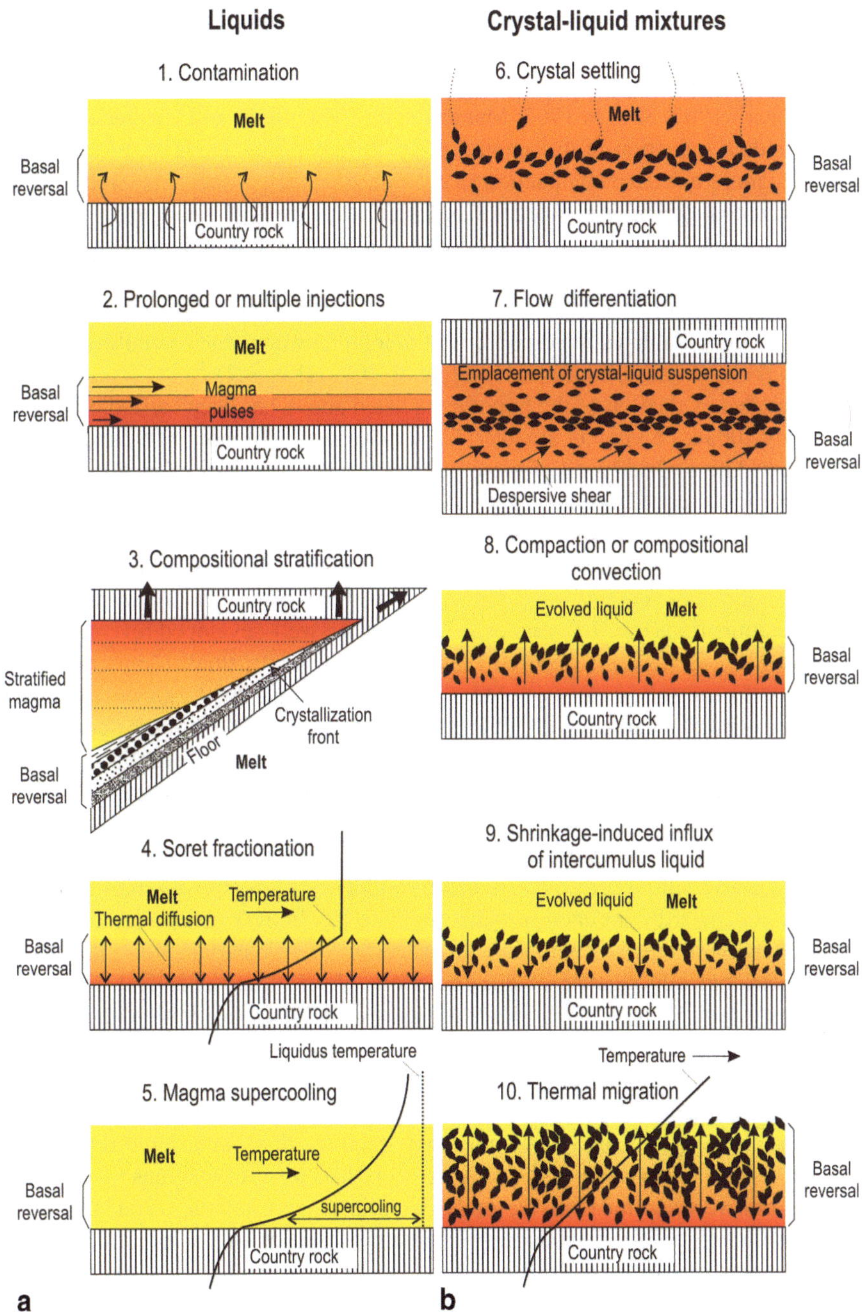

Fig. 6.7 Current hypotheses for the formation of basal reversals in mafic sills and layered intrusions subdivided into two major groups. The first group of hypotheses **a** appeals to crystallization of liquids of different composition whereas the second one **b** involves crystal-liquid mixtures of different composition. Also shown is a hypothesis invoking magma undercooling (5) that is assigned to group **a**. See text for more explanation. (Modified from Latypov 2003a)

and Crain 1969; Fujii 1974; Lightfoot and Naldrett 1984; Frenkel' et al. 1989; Marsh 1989; Helz et al. 1989; Ariskin et al. 2003); (7) hydrodynamic migration of phenocrysts away from contacts during magma flow (e.g. Bhattacharji and Smith 1964; Bhattacharji 1967; Simkin 1967; Mangan et al. 1993; Gibbs and Henderson 2005); (8) a gradual reduction in the amount of interstitial melt trapped in the rocks due either to more effective compaction (e.g. Boudreau and Philpotts 2002) or to compositional convection (e.g. Raedeke and McCallum 1984; Jaupart and Tait 1995; Tait and Jaupart 1996); (9) shrinkage-induced flow of melt from cumulates towards an intrusive contact (e.g. Cherepanov et al. 1982, 1983; Petersen 1987; Aarnes et al. 2008; Galerne et al. 2010; Namur et al. 2013); (10) the compositional reconstitution of solidifying cumulates in response to a thermal gradient due to cooling into the country rocks (Lundstrom 2009; Huang et al. 2009) (Fig. 6.7).

A detailed evaluation of the hypotheses listed above has been carried out earlier (Latypov 2003a; Latypov et al. 2007; Chistyakova and Latypov 2009) and will not be repeated here. Briefly, all of these petrological processes are plausible and may under particular conditions contribute in one way or another to the final composition of basal reversals of sills and layered intrusions. However, some of these processes are either of limited importance for physico-chemical reasons or are inadequate as explanations of some of the characteristic features of reversals. For instance, *in situ* contamination (Fig. 7.1) cannot account for the well-developed reversals in the Insizwa sill (Lightfoot and Naldrett 1984), the Noril'sk intrusions (Czamanske et al. 1994, 1995) and the Shiant Isles Main Sill (Foland et al. 2000), all of which lack evidence of crustal contamination including isotopic systems. Similarly, crystal settling (Fig. 7.6) cannot explain reversals that develop along steeply inclined margins and the walls of feeder dykes (e.g. Muskox intrusion, Irvine and Smith 1967) or those against the roof of a magma chamber (e.g. Noril'sk-type intrusions, Latypov 2003a). Similarly, stratified magma (Fig. 7.3) fails to account for reversals at the bases of mafic sills where there is neither evidence of compositionally stratified magma nor inwardly inclined margins (e.g. the Kuz'movsky Sill, Fig. 6.3). Soret differentiation in magma chambers, a hypothesis that I favoured earlier, is unlikely to be petrologically significant due to serious physical and chemical limitations (e.g. Lesher and Walker 1991; Gibb and Henderson 2006). In addition, there is one general problem of most hypotheses: they attempt to explain reversals in a *closed system* which is incompatible with primary reverse trends in plagioclase composition (Figs. 6.3 and 6.5; Latypov et al. 2011; Latypov and Egorova 2012a, 2012b; Egorova and Latypov 2013). The only two processes that can account for this important feature of reversals is the prolonged emplacement of magmas that become more primitive with time (Fig. 7.2) and disequilibrium crystallization due to magma undercooling (Fig. 7.4). For this reason, these processes as well as inward decreases in the amount of trapped melt (Fig. 7.8), are included as the principal factors in our generalised model as presented below.

A generalised "Three-Factor Model" for Reverse Compositional Trends

A critical review of the ten different processes suggested over the last century to explain reverse compositional trends (Latypov 2003a; Latypov et al. 2007; Chistyakova and Latypov 2009b) has prompted us to develop a generalised model that combines the most prominent features of existing concepts (Latypov et al. 2011). The model involves three principal petrological processes: (1) the inflowing magma becomes progressively more primitive in composition; (2) minerals crystallize from less undercooled magma with distance from the cold country rocks; and (3) evolved melt is removed from growing crystals more effectively with distance from intrusive contact. The combination of these factors is referred to as a "three-factor model" (Fig. 6.8). A compositional reversal that can result in the context of this model from parental magma X lying in the olivine field of the pseudoternary system *Ol-Pl-Cpx* is shown in Fig. 6.9.

The three processes are inferred to operate during the initial stages of magma chamber development. The temporal change in the composition of inflowing magmas that become more primitive is the main reason for reverse trends in whole-rock and mineral compositions (Fig. 6.8a). In the literature this is commonly regarded as involving "leading-edge fractionated magma", quite an old concept in petrology that makes perfect sense in an igneous system involving melting of a mantle source and magma staging. It implies the initial emplacement of batches of evolved magma produced by three principle mechanisms: (a) the early stages of fractional melting without complete mixing in the mantle source; (b) fractional crystallization and, possibly, magma stratification in a subjacent chamber; (c) fractional crystallization of magma flowing through conduits (Morse 1979, 1981, 2008; Henderson and Gibb 1987; Henderson et al. 2000; Gibb and Henderson 2006). In this paper I focus mainly on the latter option in which magmas producing a hypothetical basal reversal (Fig. 6.9a) represent the leading evolved fractionates of the parental magma X (somewhere along the path E → Y → X in Fig. 6.9b) that crystallized partially in a conduit during its ascent (e.g. Fig. 6.10a, b). However, this is unlikely to be the only process involved because crystallization in a conduit can hardly produce evolved liquids in an amount adequate for the formation of the large volume of evolved rocks observed in basal reversals. This is especially true for intrusive bodies in which basal reversals may reach thicknesses of up to several kilometres (e.g. 2.5 km in the Kiglapait intrusion, Morse, pers. comm., 2013; 7 km in the Kunene complex, Maier et al. 2013). It seems that some other processes, such as evacuation of a compositionally-stratified magma chamber may also be involved in the generation of such thick basal reversals.

A factor of lesser importance for the generation of reverse compositional trends is probably a gradual decrease in the degree of magma undercooling. This may have contributed to a temporal increase in the An-content of plagioclase cores and the Mg# of mafic minerals (Fig. 6.8b, insert), a possibility that was first mentioned by Wager and Brown (1968). This concept can be illustrated using the binary phase

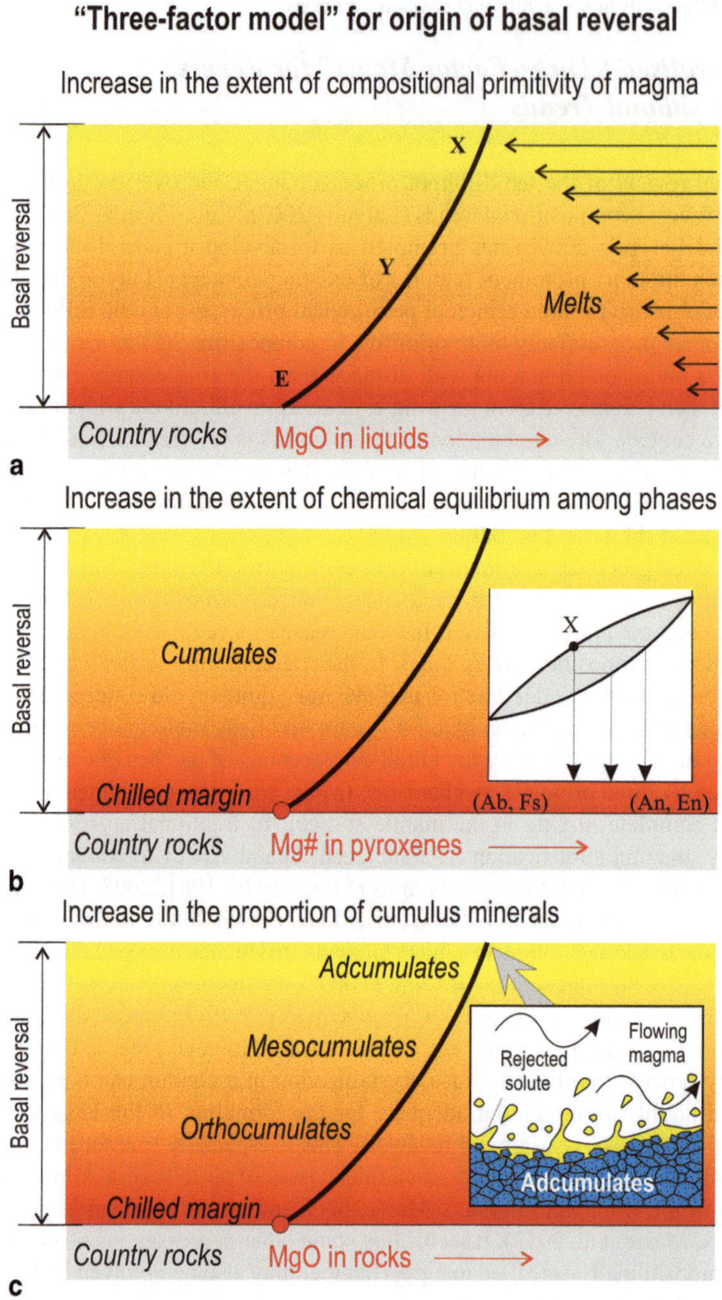

Fig. 6.8 Cartoon illustrating the three factors involved in the generation of reverse compositional trends in basal reversals: **a** Increasingly more primitive magma; **b** A progressive approach to chemical equilibrium; **c** An increasing proportion of cumulus minerals in the rocks. The simplified binary phase diagram (Ab, Fs)-(An, En) in **b** illustrates how magma, as it becomes gradually less undercooled, can crystallize minerals that are progressively enriched in the high temperature end member (e.g. An and En). The inset in **c** shows the formation of adcumulates at the top of a cumulate pile due to the effective removal of evolved melts (rejected solute) from crystals growing *in situ* by flowing magma. See text for further discussion. (Slightly modified from Latypov et al. 2011)

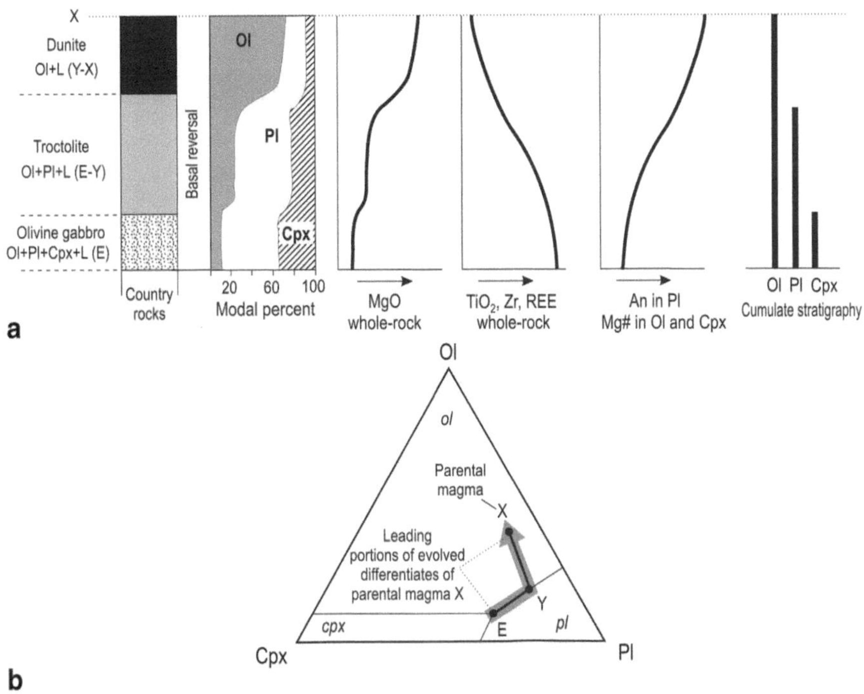

Fig. 6.9 The anatomy of a basal reversal **a** produced from the batches of evolved magma, following the path E → Y → X in **b**, derived from parental magma X lying in the olivine field in the pseudoternary system *Old-Pl-Cpx*. The evolved differentiates are supposed to have resulted from fractional crystallization of the parental magma during its ascent through feeder channels (Fig. 6.10). The basal reversal shows a reverse crystallization sequence *Ol+Pl+Cpx* → *Ol+Pl* → *Ol* with approximately cotectic proportions of the crystallizing phases, reverse trends in whole-rock MgO, TiO_2, Zr, and REE and reverse trends in An in plagioclase and Mg# of olivine and clinopyroxene. (Modified from Latypov 2003a)

diagram for plagioclase (Fig. 6.8b). The process is an inevitable consequence of the cooling that any magma experiences at the margins of plutonic bodies. Furthermore, it appears to be the only process that can result in the crystallization of minerals that become progressively more primitive from magma of uniform composition. The most recent experiments involving crystallization of natural andesitic melts at varying cooling rates indicate that the difference between equilibrium and disequilibrium plagioclase compositions can be as much as 30 mol.% An (Iezzi et al. 2011). Plagioclase compositions in basaltic lava flows also vary by ~30 mol.% An (e.g. Hoffer 1966). This gives some idea about the range of An in plagioclase that can be produced by decreasing undercooling.

A further contributory factor is the tendency for rocks to become more adcumulate-like (Fig. 6.8c). An increase in the proportion of cumulus minerals has a strong effect on the concentration of highly incompatible elements in rocks. This is commonly attributed to a return of interstitial melt into the main magma, by either compositional convection or compaction of cumulates that may become more effec-

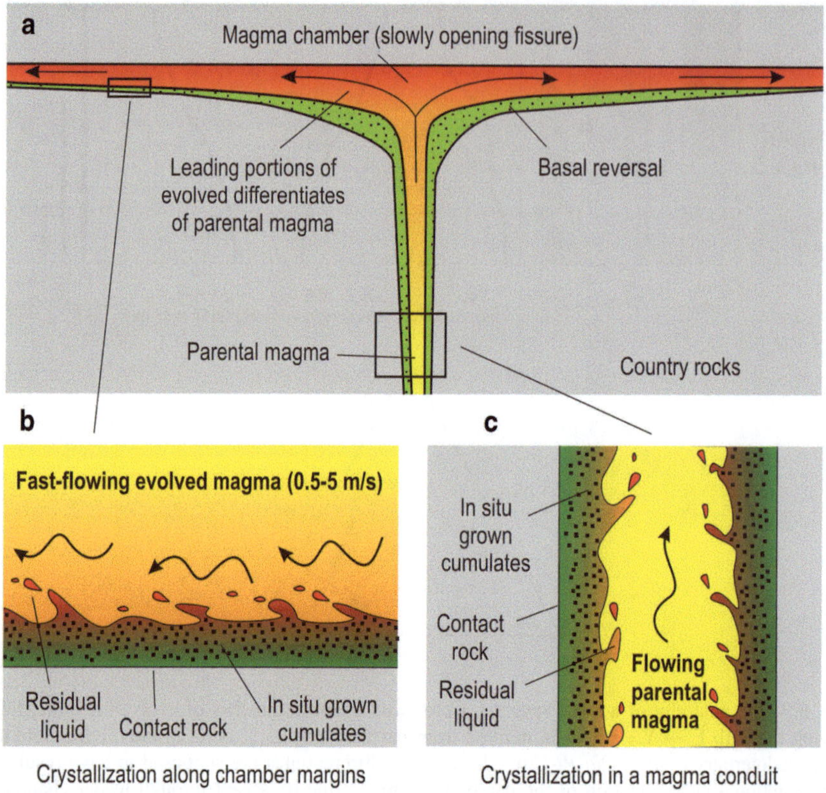

Fig. 6.10 A section through an intrusion illustrating the hypothesis that magma chambers start filling with evolved differentiates of the parental magma **a** due to its fractional crystallization in deeper parts of the magma conduit **b** The inflowing magma becomes more primitive in composition with time as a result of the decreasing degree of fractional crystallization as the walls of the conduit are heated. The rocks crystallizing from the inflowing magma acquires the compositional features of a typical basal reversal, with rocks and minerals becoming more primitive in composition upwards. Note that rock of the basal reversal tend to be meso- to adcumulates because magma flowing quickly along the base of the intrusion removes evolved liquid from growing crystals effectively. (Slightly modified from Egorova and Latypov 2012a)

tive with distance from an intrusive contact (e.g. Raedeke and McCallum 1984; Tait and Jaupart 1996). Alternatively, this may be accomplished by removal of evolved melt from attached crystals growing *in situ* by magma continuously flowing by (e.g. Hill et al. 1995) (Fig. 6.8c, insert; Fig. 6.10c). This process may become more effective with time since a decrease in the rate of magma crystallization allows more effective stripping of excluded evolved liquid from the crystallizing minerals. Morse (1986) was the first to realize that magma flow in a chamber must aid adcumulus growth. In his analysis he suggested thermal convection as a mechanism to strip a compositional boundary layer from growing crystals. The velocity of inflowing magma during emplacement is, however, many millions times higher than that resulting from thermal convection (0.5–5 m/s, Turcotte 1987 versus 0.5–5 km/year, Morse 1986). This means that inflowing magma must be much more effective in removal of a liquid boundary layer than thermally convecting magma in the chamber. In some cases compaction may also contribute to an increase in the proportion of cumulus minerals in basal reversals (e.g. Egorova and Latypov 2012a).

The basal reversal that can emerge as a result of the co-operation of these three processes (Fig. 6.9) will show: (a) an inverse crystallization sequence *Ol + + Pl + Cpx (olivine gabbro)* → *Ol + Pl(troctolite)* → *Ol (dunite)* with roughly cotectic proportions of the crystallizing phases; (b) reverse trends in whole-rock geochemistry (e.g. upward increase in MgO and reduction in incompatible TiO_2, Zr, and the REE); (c) reverse trends in mineral composition (e.g. increase in Mg# in mafic minerals and An in plagioclase upwards). It should be noted that the reverse trends arise as a result of all three of the individual processes. It is therefore important to clearly understand the specific contribution of each process in the reverse trends. The change in composition of the inflowing magma is almost certainly the most important process because it causes reverse trends in both mineral and whole-rock compositions. Disequilibrium crystallization due to magma undercooling is able to cause reverse trends in mineral compositions but not in whole-rock compositions. In contrast, increasing adcumulus growth is only able to produce whole-rock reverse trends. It has no effect on the primary composition of minerals and is not therefore able to produce reverse trends in mineral composition.

The Origin of Fully-Developed Reversals

The basic principles of the "three-factor model" can now be applied to explain the origin of both varieties of fully-developed basal reversals (Fig. 6.1). I start with those reversals in which both whole-rock and minerals show reverse compositional trends.

Basal Reversals with Coupled Whole-Rock and Mineral Compositional Trends

Mafic sills and layered intrusions with such reversals are not formed instantaneously from a single pulse of parental magma of uniform composition. Rather, they pass through an initial stage where the inflowing magma becomes increasingly more primitive with time (Fig. 6.11). The emplacement of leading pulses of evolved magma can be a consequence of either fractional crystallization of excess mafic minerals in magma flowing through the conduit, or fractional crystallization and magma stratification in a subjacent chamber. As a result, minerals crystallizing on the floor of a chamber from the inflowing magma become more primitive upwards. In particular, plagioclase and olivine show upward increases in An-content and Mg#, respectively. An additional factor that may have contributed to the formation of reverse compositional trends in minerals is a gradual decrease in the degree

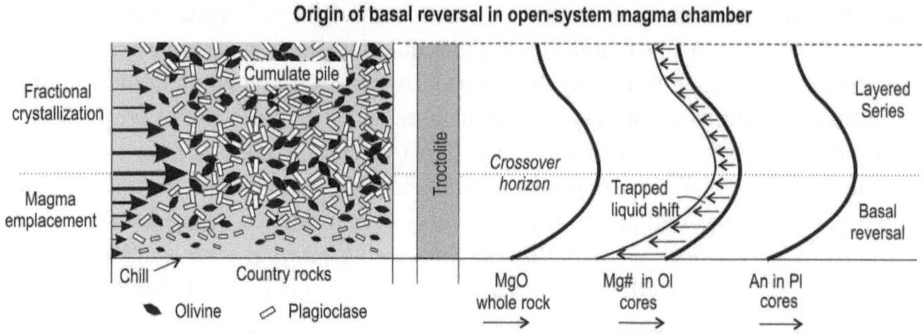

Fig. 6.11 Cartoon illustrating the formation of a fully-developed basal reversal with coupled whole-rock and mineral compositions (Fig. 6.1a) due to crystallization from flowing magma. The model assumes that crystals grow essentially *in situ* from increasingly more primitive magmas flowing into the chamber. The crossover horizon marks the time when magma emplacement ended and the body started to crystallize as a closed system to form increasingly more evolved minerals and rocks. Subsequent reaction of olivine with interstitial melt makes it more evolved. The length of the arrows on the left-hand side of the cartoon indicates how primitive the magma was during the evolution of the intrusion. (Modified from Latypov and Egorova 2012)

of magma undercooling. Also, the proportion of cumulus minerals in successively-formed rocks can increase away from intrusive contacts to give rise to basal reversals in whole-rock composition. Thus, the basal reversal essentially fossilizes a systematic change in composition of the inflowing magma. This is because inflation takes place slowly enough for the crystallization front to advance fast enough to freeze the varying influx. The previous fractionation history of the magma emplaced become thus preserved in reversals. This explanation is valid for mafic sills and relatively small intrusions, but is unlikely for large intrusions in which basal reversals can be up to several km thick (e.g. the Kiglapait intrusion and the Kunene complex), which implies that the cooling regime in the interior of these intrusions will likely be not fast enough to preserve the variable influx. In these intrusions the basal reversals may possibly form in response to mixing of resident, relatively evolved magma in the chamber with increasing amounts of the inflowing, relatively primitive magma. One can assume that the batches of primitive magma were injected over some time and that the rates of emplacement and mixing with resident magma were so slow that the hybrid magma in the main reservoir was able to cool and partially crystallize, producing a basal reversal. A similar hybridisation scenario has recently been postulated to explain the origin of a basal reversal associated with the Pyroxenite Marker of the Bushveld Complex (VanTongeren and Mathez 2013). Finally, when the filling of an intrusion ends, closed-system fractional crystallization takes over as the predominant mechanism of differentiation (Fig. 6.11). The onset of this stage is marked by a crossover horizon displaying the most primitive mineral and rock compositions. Subsequently, olivine and plagioclase become more evolved in composition upwards through the Layered Series. Postcumulus reaction of olivine with trapped melt commonly takes place in the course of several hours to days (Chakraborty 1997) and will change its original composition, enhancing its initial reverse trend. The ultimate result emerging from this scenario is the gen-

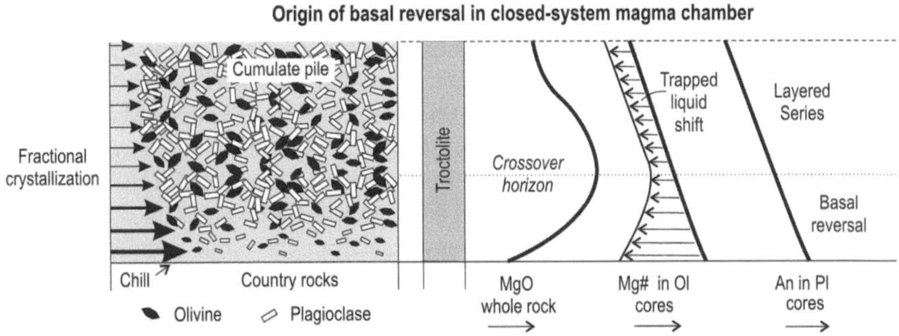

Fig. 6.12 Cartoon illustrating the formation of a fully-developed basal reversal with decoupled whole-rock and mineral compositions (Fig. 6.1b) by fractional crystallization under closed-system conditions. The model assumes that reverse trends in whole-rock MgO is a consequence of a decreasing solidification rate that results in a reduction in the amount of interstitial melt trapped in the rocks. Note that the primary compositions of the minerals show no reverse trends because closed-system fractional crystallization results in a unidirectional evolution from the base upwards. The final composition of olivine may acquire a reverse trend in response to reaction with interstitial liquid that decreases in abundance upwards. The length of the arrows on the left-side of the figure indicates how primitive the magma was during the evolution of the intrusion. (Modified from Latypov and Egorova 2012)

eration of an intrusive body, such as the Kuz'movsky sill (Fig. 6.3), with a fully-developed basal reversal in both mineral and whole-rock compositions.

Basal Reversals with Decoupled Whole-Rock and Mineral Compositional Trends

It is widely accepted that the SIC, that shows the only currently known example of such a basal reversal (Fig. 6.4), crystallised from a sheet of impact melt under closed-system conditions (e.g. Therriault et al. 2002). This implies that emplacement of magma that became more primitive with time cannot have been responsible for the origin of the reversal. In addition, closed-system fractional crystallisation is supported by a normal trend in plagioclase compositions (Fig. 6.4). The normal trend also indicates that disequilibrium crystallization due to magma undercooling cannot be an important factor in the formation of the basal reversal. It thus appears that only one of the factors in the "three-factor model", an inward increase in the proportion of cumulus minerals, is the most probable cause of the reversal (Fig. 6.12). Thus the whole-rock reverse trend is a result of an upward decrease in the amount of trapped melt in the rocks. This hypothesis can be tested by estimation of the amount of trapped liquid in the cumulates of the studied section using the bulk concentration of a highly incompatible element (e.g. REE) in rock samples and its concentration in the parental melt (e.g. Wager 1963). Because the SIC crystallised under closed-system conditions, the concentration in successive melts can be determined for each individual sample from the bulk composition of overlying rock sequence. In our calculations we have used La that behaved as an incompatible element during the crystallization of the mafic part of the SIC. As expected, the lowest estimated amount of trapped melt is at the crossover horizon (about 30–40 %), below which the amount systematically increases to 80–90 % in the lowermost norites. Preliminary modelling indicates that the upward decrease in the estimated amount of trapped melt is sufficient to explain the whole-rock reverse trends observed in the SIC (Latypov et al. in prep.). Naldrett et al. (1970) reached essentially the same conclusion based on the decreasing amount of interstitial phases (mainly quartz and alkali feldspar) upwards from the basal contact. One can also expect that the abundance of interstitial melt will be reflected in the degree of postcumulus re-equilibration of mafic minerals (e.g. olivine and pyroxenes) so that their final compositions have acquired reverse trends (Fig. 6.12). Reverse trends in pyroxene compositions are in fact observed in all sections through the South Range of the SIC (Naldrett et al. 1970) and can be attributed to the reaction of minerals with trapped melt.

The origin of Aborted Basal Reversals

The underlying tenets of the "three-factor model" are also valid for the interpretation of both varieties of aborted reversals (Fig. 6.2). However, the compositional discontinuity between the reversals and the overlying Layered Series suggests that some complicating additional factor was involved. Let us start with those reversals that are overlain by compositionally more primitive Layered Series.

Basal Reversals Beneath More Primitive Layered Series

As in the Koitelainen Layered Intrusion (Fig. 6.5), such basal reversals commonly show reverse whole-rock and mineral trends which can be interpreted in using the "three-factor model" (Fig. 6.8). The particular issue here is why these reversals do not pass gradually to the overlying Layered Series through a crossover horizon? Why are the trends instead so abruptly truncated at the base of the Layered Series? I suggest that at some stage the development of the basal reversals was interrupted by a new pulse of hot, dense, primitive magma that was parental to the overlying Layered Series (Fig. 6.13). The magma was emplaced along the temporary floor of the chamber, above the top of the basal reversal. During this process the magma chamber may have inflated both vertically and laterally. Since rocks of the basal reversal were still hot, there was no chilling of this new magma against them. In some cases the new magma caused some mechanical erosion of the underlying reversals, for instance in the Imandra Intrusion (Egorova and Latypov 2012a). In other cases, the new magma caused intensive textural and chemical reconstitution of the uppermost rocks of the basal reversal, as in the Koitelainen Intrusion, (Latypov et al. 2011). The emplacement of new magma can result in partial melting, compaction, and recrystallization of the pre-existing rocks. Partial melting and the transfer of partial

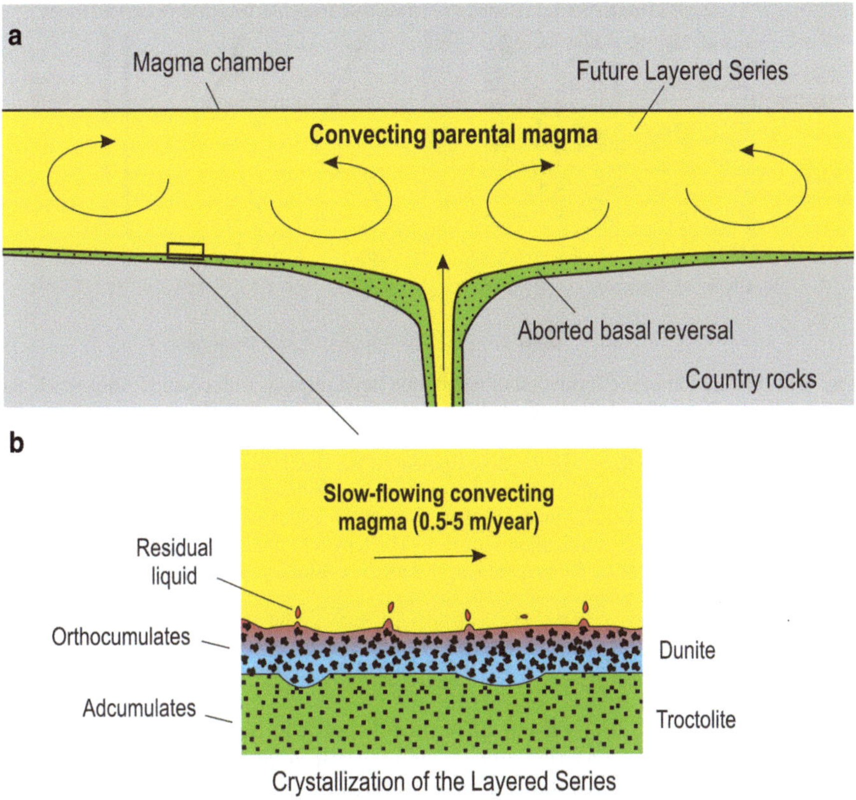

Fig. 6.13 Illustration of the hypothesis that a new pulse of parental magma **a** terminates the development of a basal reversal, causes mechanical and thermal erosion of its surface and results in crystallization of the more primitive mineral assemblages of the Layered Series. Note that dunite of the Layered Series tend to have orthocumulate composition **b** because slowly-convecting magma in the chamber does not effectively flush away evolved melt from growing crystals. (Slightly modified from Egorova and Latypov 2012a)

melts into the overlying magma may cause strong depletion in highly incompatible elements, hence the very low concentrations of Zr at the top of the basal reversal in the Koitelainen Intrusion (Fig. 6.5). Melting, compaction and related recrystallization of originally medium-grained and massive rocks may result in their transformation into finer-grained rocks with a marked plagioclase lamination (Latypov et al. 2011). After the textural and chemical reconstitution of the uppermost rocks of the basal reversal, renewed crystallization of the magma in the chamber led to the formation of primitive cumulates (e.g. orthopyroxenite). Slow cooling and crystallization gave rise to coarse-grained cumulates which, together with their different composition, results in the sharp textural and compositional break between the basal reversal and the Layered Series (Fig. 6.14).

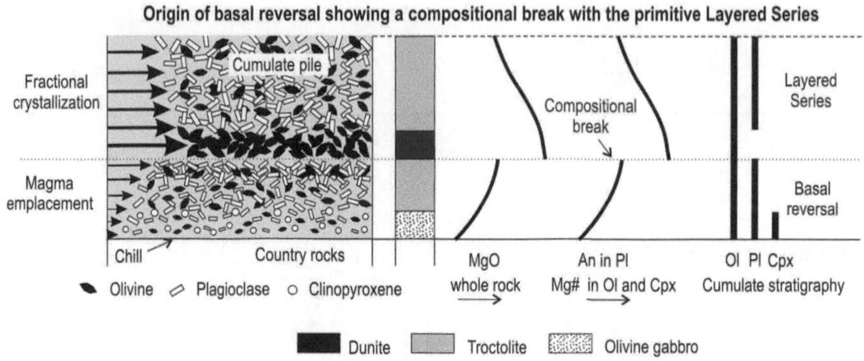

Fig. 6.14 Illustration of the formation of an aborted basal reversal with coupled whole-rock and mineral compositions (Fig. 6.2a) by crystallization from flowing magma. The hypothesis proposes that crystals grow essentially *in situ* from increasingly more primitive magmas flowing into the chamber. At some particular stage the development of the basal reversal was interrupted by the arrival of a new batch of primitive magma that led to the crystallization of the Layered Series. This event resulted in a sharp boundary between these two stratigraphic units of the intrusion in terms of texture, composition and crystallization sequence. The length of the arrows on the left-side of the figure indicates how primitive the magma was during the evolution of the intrusion

Basal Reversals Beneath More Evolved Layered Series

The reverse trends in this variety of basal reversal can be explained in a similar way to the case above. At first glance, it seems logical to interpret a compositional break with the overlying Layered Series as resulting from a new pulse of magma. The only difference is that in this case the magma emplaced must have been more evolved than that resident in the chamber. Although this scenario seems quite plausible, there is another possible explanation. It follows from the anatomy of lopolithic or V-shaped layered intrusions (e.g. the Muskox Intrusion) in which basal reversals and Layered Series are discordant to each other (Fig. 6.15). As a result, moving up the inwardly-inclined floor of the intrusion, the basal reversal comes into contact with increasingly more evolved rocks of the Layered Series. Any section from the middle and upper parts of the inclined sidewalls would then show a compositional break between a basal reversal and the Layered Series (Fig. 6.16). This appears to be the simplest possible explanation for the origin of such aborted reversal as observed in the Suhanko Layered Intrusion (Fig. 6.6). The saw-toothed compositional trends within the reversal itself are likely due to multiple emplacement episodes that culminated in the crystallisation of the dunite. The structure of V-shaped layered intrusions is, however, still poorly understood. In particular, it is not yet clear how the unconformity between a basal reversal and Layered Series in these intrusions actually forms. One possible scenario is that a basal reversal formed during an early stage when the intrusion was a sheet-like body. This was followed by the emplacement of a larger pulse of magma (or many heaves of magma) that caused the sagging of the central part of the intrusion to form a lopolithic body. The subsequent

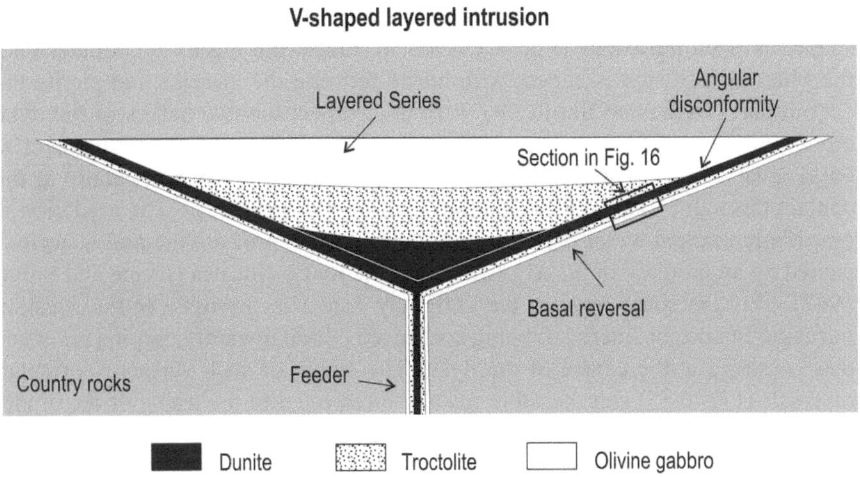

Fig. 6.15 A generalized section through a hypothetical V-shaped layered intrusion (e.g. the Muskox Intrusion) showing basal reversals developed along the inclined floor of the chamber and walls of the feeder dyke. In such intrusions the basal reversals and the Layered Series are discordant. As a result, moving up along the inclined floor of the intrusion, the basal reversal comes into contact with increasingly more evolved rocks in the Layered Series. Note that sections through the *middle* and *upper* parts of the inclined floor (see Fig. 6.16) show a compositional break between the basal reversal and the more evolved Layered Series

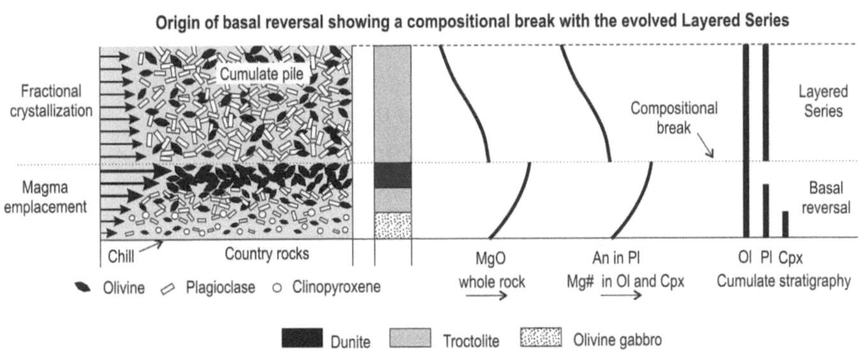

Fig. 6.16 Cartoon illustrating the formation of an aborted basal reversal with coupled whole-rock and mineral compositions (see Fig. 6.2b) from flowing magma. The hypothesis suggests that crystals grow essentially *in situ* and become more primitive upwards because of their formation from increasingly more primitive magmas flowing into the chamber. The compositional break between the basal reversal and the more evolved Layered Series is interpreted as a consequence of the discordance between these two structural units in a V-shaped layered intrusion (see Fig. 6.15). The length of the *arrows* on the *left-side* of the figure indicates how primitive the magma was during the evolution of the intrusion

in situ crystallization of magma in the chamber from the base upwards or settling and accumulation of minerals predominantly on the floor of the chamber will produce a Layered Series showing an angular disconformity with the basal reversal. Of special interest here is the origin of reversals in feeder dykes of V-shaped layered intrusions (Fig. 6.15). For instance, the feeder dyke underlying the Muskox Intrusion is zoned, with norite forming the margins and picrite the central part (Irvine and Smith 1967). In this respect the two halves of the dyke are reminiscent of the basal reversal on the inclined floor of the intrusion. The basal reversal generally shows a reverse gradation from bronzite gabbro at the contact through picrite and feldspathic peridotite to peridotite. The gradation is essentially marked by an increase in the concentration of olivine and is accompanied by an increase in the Mg/Fe ratio of the mafic silicates (Irvine and Smith 1967). Thus, in some respect, the vertically zoned feeder dyke of the Muskox intrusion can be considered as being composed of two inwardly dipping reversals that converge at the centre of the dyke. The origin of such vertically-oriented reversals (Fig. 6.15) can therefore also be interpreted in the frame of the model presented above (Fig. 6.8).

Implications

Chilled Margins are Not Always Representative of Parental Magmas

The model for origin of basal reversals as presented here has serious implications for the common practice of equating chilled margins with parental magmas. I suggest that this practice is not justified, for at least two reasons. Firstly, it follows from the analysis above, that mafic sills and layered intrusions usually do not form instantaneously from a single pulse of magma of uniform composition. Rather, they fill gradually with magma that becomes progressively more primitive with time. This being so, marginal chilled rocks may represent samples of evolved fractionates of the parental magma and not of this magma itself. This may explain some perplexing features of layered intrusions, such as: (a) the disparity between the compositions of chilled margins and the bulk of the intrusion, the former being commonly more evolved than the latter; (b) the occurrence of compositionally different chilled margins in the same intrusion; (c) the tendency for evolved chilled margins to have compositions close to low temperature cotectics; (d) the compositional similarities of chilled margins of intrusions with different parental magma compositions (Latypov et al. 2007). The latter feature is illustrated in Fig. 6.17. The figure shows that the leading evolved differentiates of parental magmas with any liquid line of descent in the pseudoternary system *Ol-Pl-Cpx*, such as A, B, C, D, E and F, can result in the crystallisation of basal reversals with similar gabbroic chilled margins.

Secondly, it is unlikely that chilled margins are true quenched liquids; they are rather fine-grained cumulates. Fine-grained cumulates may arise if evolved melt that surrounds growing crystals is continuously removed by flowing magma (Figs. 6.8c and 6.10c). In other words, cumulates at cooling margins may originate when the

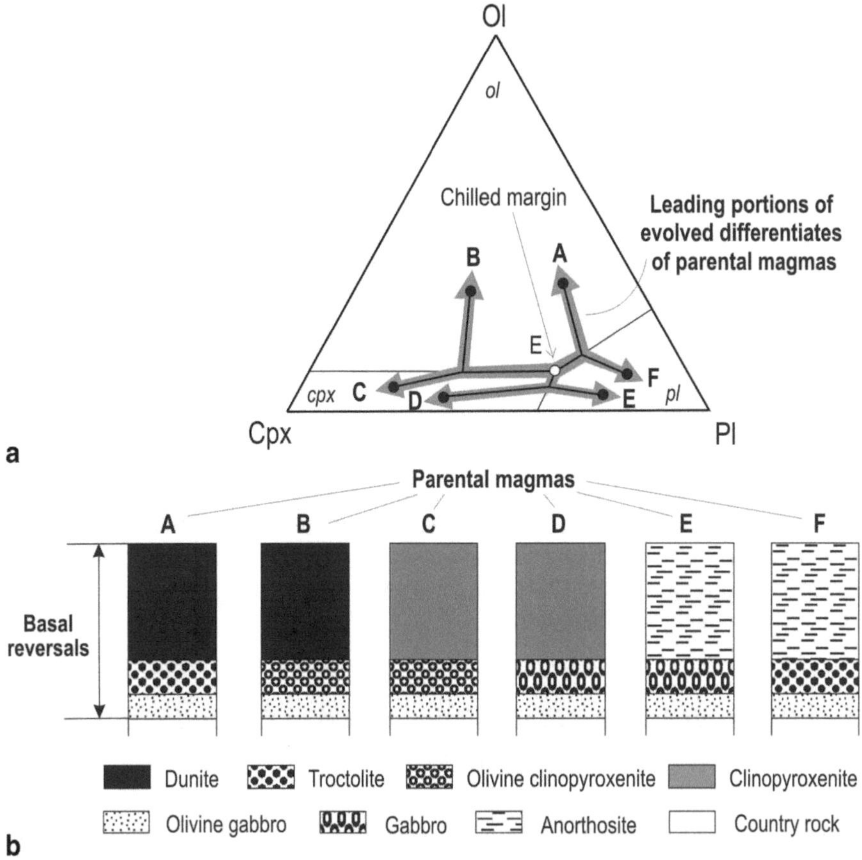

Fig. 6.17 The pseudoternary phase diagram Old-Pl-Cpx **a** showing that leading portions of evolved differentiates of compositionally different parental magmas *A*, *B*, *C*, *D*, *E* and *F* can result in basal reversals with similar marginal olivine gabbro **b** *Olivine gabbro* crystallizes from the most evolved magmas, corresponding to the low temperature eutectic *E*. (Modified from Latypov et al. 2007)

cooling effect of country rocks is overwhelmed by the effect of removal of evolved liquid by flowing magma, favouring the formation of cumulates. Such fine-grained marginal rocks are thus very poor approximations of injected magmas. This is probably why magma compositions estimated from marginal rocks are often too high in MgO (Berg 1980; Tegner et al. 1993). Evidence of the cumulate origin of marginal rocks comes from our recent study of compositional zoning in narrow mafic dykes (Chistyakova and Latypov 2010, 2011, 2012). The study clearly demonstrated that even very fine grained rocks in narrow dykes (10–20 cm thick) are cumulates, not quenched liquids. The margin-to-center increase in whole-rock MgO and normative An in these dykes results from an inward increase in the proportion of cumulus minerals. The compositional trends in dykes probably reflects a tendency of rocks to become cumulates with time owing to *in situ* crystallization on dyke walls while flowing magma flushes away residual liquid. If the margins of narrow dykes do not preserve liquid compositions then there is little possibility that the margins of large intrusions do. These bodies crystallized more slowly and there is therefore much more time for flowing magma to strip off melt from crystals growing *in situ* to form basal cumulates. If so, what we have commonly interpreted as chilled margins are probably fine-grained cumulates. To avoid confusion I would suggest therefore using the descriptive term “contact rocks”, rather than “chilled margins”.

Doubts about the reliability of chilled margins as indicators of parental magma composition are far from novel (e.g. Wager and Brown 1968; Cox et al. 1979; Gibb and Henderson 2006; Latypov et al. 2007; Morse 2008), but it is worth emphasizing in light of the current use of chilled margins in petrological modelling. A pertinent question to ask is: What method can be used to define parental magma composition if contact rocks are not reliable? It appears that there is no universal method that can be applied to all intrusive bodies to obtain a reliable estimate of their parental magma composition. However, despite many shortcomings (e.g. incomplete sections due to erosion, multiple magma pulses and the possible presence of phenocrysts) the best approach is to use the bulk composition of intrusions and the liquid line of descent as indicated by the crystallization sequence and mineral compositions (e.g. Irvine 1970; Frenkel’ et al. 1988; Vander Auwera and Longhi 1994; Latypov et al. 2001; Ariskin and Barmina 2004).

Two Different Magma Flow Regimes During Crystallization of Basal Reversals and Layered Series

The fine-grained pigeonite gabbros of the aborted reversals in the Koitelainen (Fig. 6.5; Latypov et al. 2011) and Imandra (Egorova and Latypov 2012a) layered intrusions show an approach to meso- and adcumulate textures and compositions. It is surprising that these fine-grained gabbroic rocks contain considerably less of incompatible elements (e.g. Rb, Ba, Ti, P, Zr, Y, REE) than the overlying coarse-grained orthopyroxenites of the Layered Series (e.g. Fig. 6.5). The rocks of the basal reversals apparently crystallized quite rapidly and should therefore be compositionally more like orthocumulates. In contrast, the rocks of the Layered Series crystallized slowly, there was more time to expel evolved interstitial liquid and they should therefore be adcumulates. The opposite is, however, the case. Pigeonite gabbros of the basal reversals are fine-grained meso- to adcumulates whereas orthopyroxenites of the Layered Series are generally orthocumulates containing up to 40–45 % of interstitial phases. One possible explanation of this paradox is fractional melting of the uppermost part of the pigeonite gabbro by new hot magma (Latypov et al. 2011). However, this cannot be the sole reason since the pigeonite gabbro of the Imandra layered intrusion shows no textural evidence for fractional melting (Egorova and Latypov 2012a). I suggest therefore that the principal cause of this contrast lies in different regimes of magma flow that operated during the formation of these two stratigraphic units.

Our hypothesis for the origin of basal reversals emphasizes inflowing magma (Fig. 6.10). In contrast, the crystallisation of a Layered Series after filling of the chamber took place from convecting magma (Fig. 6.13). Since magma flow in a chamber aids adcumulate growth (Morse 1986), one can expect that the process of adcumulate formation in marginal zones and Layered Series of mafic-ultramafic intrusions will be significantly different. In the former, the production of adcumulates may be aided by removal of evolved liquid by inflowing magma, whereas in the latter it is removed by convecting magma (Latypov et al. 2011). As pointed out above, the velocity of inflowing magma during emplacement is much higher than that of convecting magma and therefore the inflowing magma is much more effective in stripping a liquid boundary layer from growing crystals than thermally convecting magma in the chamber. Therefore, taking into account the effect of inflowing magma, marginal rocks may be more adcumulate than those of the Layered Series. This may explain why rapidly crystallizing pigeonite gabbros in the basal reversal are more adcumulate than the more slowly crystallizing orthopyroxenites of the Layered Series. The fine-grained pigeonite gabbros can thus be classified as primary (versus secondary) meso- to adcumulates (Campbell 1987). A good example of primary adcumulates, although not fine-grained, are olivine adcumulates in komatiite-related dunites that are believed to form by *in situ* growth of olivine crystals from flowing komatiitic lavas (e.g. Hill et al. 1995).

Basal reversals as Proxies of a Particular Regime of Cooling and Crystallization

Apart from the cases with extremely thick (several km) basal reversals, the cooling imposed by country rocks is likely the most important factor determining whether a basal reversal forms or not. In the absence of strong cooling into country rocks all processes involved in the "three-factor model" are unable to produce reverse compositional trends: (1) slow crystallization at the margins will give time sufficient for effective mixing of inflowing magmas so that crystallization will ultimately take place from a homogeneous magma body; (2) a lack of magma undercooling will make disequilibrium crystallization impossible; (3) slow crystallization will not promote trapping of interstitial melts in the cumulates. The latter factor is especially sensitive to the cooling regime because it is the thermal gradient between cold country rocks and hot magma in the chamber that governs the degree of trapping of interstitial melt in marginal rocks: A steep thermal gradient along an intrusive contact results in marginal rocks forming rapidly and therefore trapping a large amount of interstitial liquid. With increasing distance from the contact the thermal gradient becomes shallower and the rocks have more time to crystallize and, as a result, they contain decreasing amounts of trapped interstitial liquid.

The importance of a thermal gradient is supported by the fact that basal reversals are absent from intrusions emplaced into hot rocks. A well-known example is the 660 m thick Basistoppen Sill emplaced as a single pulse of magma into the upper part of the Skaergaard intrusion shortly after its solidification (Naslund 1989).

Slow crystallization in a hot environment resulted in one of the most extensive differentiation sequences known. The compositional ranges of olivine, plagioclase, and pyroxene from the Basistoppen sill are comparable to those in the much larger Skaergaard and Bushveld intrusions. The sill exhibits, however, no basal reversal (Naslund 1989). Another impressive example is the 100 m thick Nadezhda gabbronorite, a sill-like body in the Lukkulaisvaara Layered Intrusion, Russia. The body is interpreted as having formed from a portion of magma that was emplaced into the evolving chamber and became trapped within still hot, partly consolidated norite-gabbronoritic cumulates (Latypov et al. 2008). Detailed geochemical sampling has revealed no marginal reversals at either contacts of the body. Gabbronorites with the highest MgO and Cr occur at the margins of the intrusion and the cumulates become more evolved inwards (Latypov et al. 2008). Basal reversals in these two bodies failed to develop because their parental magmas were in contact with hot rocks and the thermal gradient was insufficient to either record any changes in the composition of the inflowing magmas or trap interstitial melt in the cumulates.

Thus a basic relationship is clear: no temperature gradient—no basal reversals. This is a significant inference because basal reversals can be reliable proxies of a particular regime of cooling and crystallization. In particular, the absence of any basal reversal would indicate that magma was intruded into hot country rocks and therefore cooled and crystallized slowly from the very beginning. In contrast, the occurrence of a pronounced basal reversal testifies to magma being intruded into relatively cold rocks and it therefore cooled and crystallized rapidly initially. An empirical rule is thus the larger temperature contrast between magma in the chamber and surrounding country rocks, the thicker basal reversals should be.

In this respect the extremely thick reversal at the base of the SIC (Fig. 6.4) is quite puzzling. This complex is currently interpreted as formed from a sheet of impact melt that was highly superheated (Grieve 1994; Ivanov and Deutsch 1999; Naldrett 2004). The melt sheet is thought to have cooled for 100,000–250,000 years from about 1700 °C to a liquidus temperature of around 1200 °C (Ivanov and Deutsch 1999; Naldrett 2004). Such a long cooling history from an extremely high initial temperature implies that the country rocks around the crater must have been strongly heated and partially to totally melted (Prevec and Cawthorn 2002). This means that by the onset of crystallization, the magma and country rocks must have reached thermal equilibrium. As a result the SIC must have been crystallized extraordinarily slowly, unlike any other terrestrial intrusive body. Under these conditions no basal reversal can be expected in the SIC. However, the opposite is true (Fig. 6.4) suggesting that our current understanding of the cooling history of this curious layered intrusion is likely incomplete (Latypov et al. 2010).

Conclusions

The majority of mafic sills and layered intrusions show basal reversals that are characterized by crystallization sequences, whole-rock and mineral compositional trends that are opposite to those expected from the fractional crystallization of basaltic magma under closed-system conditions. Some previous studies (e.g. Gibb and Henderson 2006; Morse 2008) and our current research (e.g. Latypov et al. 2011; Latypov and Egorova 2012) have shown that this enigma of basal reversals can be best understood in the context of open-system evolution of magma chambers during the initial stage. We have developed a generalised concept (the "three-factor model") that attributes the reverse compositional trends to three principle factors: (1) a change in the composition of the intruding magma; (2) an gradual reduction in undercooling and an increasing approach to equilibrium crystallisation; (3) an increase in the proportion of cumulus minerals in rocks related to effective removal of evolved melts from the surfaces of growing crystals by flowing magma. These processes appear to operate most effectively along cooling margins during the initial stages of magma chamber evolution. If cooling of the inflowing magma was slow, then basal reversals could not develop, as supported by their absence from intrusions that crystallized from magmas emplaced into pre-heated country rocks. This suggests that a thermal gradient plays a decisive role in the formation of reversals in intrusive bodies: no temperature gradient—no basal reversals. It is conceivable that the underlying principles of the "three-factor model" may be an appropriate explanation for the origin of most basal reversals, including even those that formed under closed-system conditions (e.g. Sudbury Igneous Complex).

Two different types of basal reversals, fully-developed and aborted, occur in mafic-ultramafic layered intrusions. Fully-developed basal reversals show a crystallization sequence and mineral compositional trends that are essentially the opposite of those in the overlying Layered Series to which they gradually pass through a crossover horizon. This type exhibits a continuous evolution during the formation of the reversal and is the most common in intrusive bodies. Aborted reversals are rarer and develop when the evolution is interrupted by a pulse of more primitive magma that results in the compositionally contrasting Layered Series. As a result, the basal reversal has a truncated crystallization sequence and mineral compositional trends, as well as a sharp compositional contact with the overlying Layered Series. I suggest that the basal reversals probably crystallise from evolved liquids derived by fractional crystallization of parental magma during its ascent along feeder conduits. In contrast, Layered Series appears to record the fractional crystallisation of little-modified parental magma. The hypothesis implies that chilled rocks in intrusive bodies are not representative of parental magma compositions. Rather, they crystallised from evolved precursors.

Detailed studies of the mineralogical and isotopic composition of basal reversals, as well as numerical simulations of their crystallization histories are required in order to gain a deeper insight into the origin of basal reversals. At present such studies are rare and not sufficient to constrain the formation of basal reversals. Most of the current petrological and numerical hypotheses are mainly based on *whole-rock* compositions; mineral compositions have not been used to further refine these models. However, primary mineral compositions have proven to be very useful for discriminating between alternative models for origin of basal reversals (e.g. Latypov and Egorova 2012). Much more work needs to be done along this and other related lines to ensure further progress in our understanding of the initial stages in the development of magma chambers.

Acknowledgements The study was inspired by and has benefited from numerous discussions of various aspects of magma chambers over many years with my colleagues and collaborators: Alexey Ariskin, Nikolay Balabonin, Victor Balagansky, Sveta Bolihosvkaya, Yana Bychkova, Sofya Chistyakova, Michail Dubrovsky, Vera Egorova, Valery Fedorenko, Georg Fedotov, Eduard Konnikov, Evgeny Koptev-Dvornikov, Andrey Izokh, Andrey Lavrenchuk, Egor Nikolaev, Victor Melezhik, Jury Podladchikov, Evgeny Pushkarev, Evgeny Sharkov, Vladimir Skiba, Ilya Veksler, Alexey Yaroshevsky, Tuomo Alapieti, Jean Bédard, Alan Boudreau, Ian Campbell, Grant Cawthorn, Bernard Charlier, Mark Ghiorso, Fergus Gibb, Mike O'Hara, Eero Hanski, Neil Irvine, Richard James, Chip Lesher, Bruce Marsh, Jim Miller, Alex McBirney, Tony Morse, Richard Naslund, Tony Philpotts, Steve Prevec, Brian Robins, Frank Spera, Steve Tait, Christian Tegner, Dave Walker, Bruce Watson, Marjorie Wilson, and Richard Wilson. I am also grateful to Brian Robins for his thorough editing that significantly improved the manuscript. I would also like to thank Markku Iljina for providing the original data for Fig. 6.6. Richard Naslund and Olivier Namur are thanked for their thoughtful reviews of the paper. Editorial handling by Bernard Charlier is highly appreciated. The research was supported by Fellow Research Grant from the Finnish Academy of Science and by the National Research Foundation (NRF) of South Africa for the Grants 87677, 90834 and 91812. Any opinion, finding and conclusion or recommendation expressed in this material is that of the author and the NRF does not accept any liability in this regard.

References

Aarnes I, Podladchikov YuY, Neumann ER (2008) Post-emplacement melts flow induced by thermal stresses: implications for differentiation in sills. Earth Planet Sci Lett 276:152–166

Alapieti TT (1982) The Koillismaa layered igneous complex, Finland—its structure, mineralogy and geochemistry, with emphasis on the distribution of chromium. Espoo Geol Surv Finl Bull 319:116

Alapieti TT, Lahtinen JJ (2002) Platinum-group element mineralization in layered intrusions of Northern Finland and the Kola Peninsular, Russia. In: Cabri LJ (ed) The geology, geochemistry, mineralogy and mineral beneficiation of platinum-group elements, Special Volume 54, Canadian Institute of Mining, Metallurgy and Petroleum, pp 507–546

Ariskin AA, Barmina GS (2004) COMAGMAT: development of a magma crystallization model and its petrologic applications. Geochem Intern 42(Supplement 1):S1–157

Ariskin AA, Konnikov EG, Kislov EV (2003) Modelling of the equilibrium crystallization of ultramafic rocks with application to the problems of formation of phase layering in the Dovyren pluton, Northern Baikal region, Russia. Geochem Int 41:107–129

Berg JH (1980) Snowflake troctolite in the Hettasch intrusion, Labrador: evidence for magma-mixing and supercooling in a plutonic environment. Contrib Mineral Petrol 72:339–351

Bhattacharji S (1967) Scale model experiments on flowage differentiation in sills. In: Wyllie PJ (ed) Ultramafic and related rocks. Wiley, New York, pp 69–70

Bhattacharji S, Smith CH (1964) Flowage differentiation. Science 145:150–153

Boudreau AE, Philpotts AJ (2002) Quantitative modeling of compaction in the Holyoke flood basalt flow, Hartford Basin, Connecticut. Contrib Mineral Petrol 144:176–184

Campbell IH (1978) Some problems with the cumulus theory. Lithos 11:311–323

Campbell IH (1987) Distribution of orthocumulate textures in the Jimberlana intrusion. J Geol 95:35–54

Chakraborty S (1997) Rate and mechanisms of Fe–Mg interdiffusion in olivine at 980 °C–1300 °C. J Geophys Res 102:12317–12331

Chalokwu CI, Grant NK (1990) Petrology of the Partridge River intrusion, Duluth Complex, Minnesota: I. Relationships between mineral compositions, density and trapped liquid abundance. J Petrol 31:265–293

Cherepanov AN, Sharapov VN, Krivenko AP (1982) A possible mechanism for antidromic differentiation in basic plutons. Dokladi Academii Nauk USSR 267:1204–1207 (in Russian)

Cherepanov AN, Sharapov VN, Krivenko AP (1983) The dynamic model of disequilibrium crystallization differentiation of magmas in basic plutons. Russ Geol Geofiz 3:28–36

Chistyakova SY, Latypov RM (2009) Two independent processes responsible for compositional zonation in mafic dykes of the Åland-Åboland Dyke Swarm, Kestiö Island, SW Finland. Lithos 112:382–396

Chistyakova SY, Latypov RM (2010) On the development of internal chemical zonation in small mafic dykes. Geol Mag 147:1–12

Chistyakova SY, Latypov RM (2011) Small dacite dyke, Southern Urals, Russia: rapidly quenched liquid or fine-grained cumulate? 6th International Dyke Conference volume "Dyke swarms: keys for geodynamic interpretation". In: Srivastava RK (ed) Dyke Swarms: keys for geodynamic interpretation. Springer-Verlag, Berlin, 583–601

Chistyakova SY, Latypov RM (2012) Magma differentiation and crystallization in basaltic conduits by two competing petrogenetic processes. Lithos 148:142–161

Cox KG, Bell JD, Pankhurst RJ (1979) The interpretation of igneous rocks. Unwin Hyman Ltd, London, 450

Czamanske GK, Wooden JL, Zientek ML, Fedorenko VA, Zen'ko TE, Kent J, King B-SW, Knight RJ, Siems DF (1994) Geochemical and isotopic constraints on the petrogenesis of the Noril'sk-Talnakh ore-forming system. Proceedings of the Sudbury-Noril'sk symposium. Ont Geol Surv Spec Publ 5:313–342

Czamanske GK, Zen'ko TE, Fedorenko VA, Calk LC, Budahn JR, Bullock JH(Jr), Fries TL, King B-SW, Siems DF (1995) Petrographic and geochemical characterization of ore-bearing intrusions of the Noril'sk type, Siberia; with discussion of their origin. Res Geol Special Issue 18:1–48

Deutsch A, Grieve RAF, Avermann M, Bischoff L, Brockmeyer P, Buhl D, Lakomy R, Muller-Mohl V, Ostermann M, Stoffler D (1995) The Sudbury sructure (Ontario Canada): a tectonically deformed multi-ring impact basin. Geol Rundsch 84:697–709

Egorova V, Latypov RM (2012a) Processes operating during the initial stage of magma chamber evolution: Insights from marginal reversal of the Imandra layered intrusion, Russia. J Petrol 53:3–26

Egorova V, Latypov RM (2012b) Prolonged magma emplacement as a mechanism for the origin of marginal reversal of the Fongen-Hyllingen layered intrusion, Norway. Geol Mag 53:1–18

Egorova V, Latypov RM (2013) Mafic-ultramafic sills: new insights from M- and S-shaped mineral and whole-rock compositional profiles. J Petrol 54:2155–2191

Foland KA, Gibb FGF, Henderson CMB (2000) Pattern of Nd and Sr isotopic ratios produced by magmatic and postmagmatic processes in the Shiant Isles Main Sill, Scotland. Contrib Mineral Petrol 139:655–671

Frenkel' MYa, Yaroshevsky AA, Ariskin AA, Barmina GS, Koptev-Dvornikov EV, Kireev BS (1988) Dynamics of in situ differentiation of mafic magmas. Nauka, Moscow, p 216

Frenkel' MYa, Yaroshevsky AA, Ariskin AA, Barmina GS, Koptev-Dvornikov EV, Kireev BS (1989) Convective-cumulative model simulating the formation process of stratified intrusions. In: Bonin N, Didier J, Le Fort P, Propach G, Puga E, Vistelius AB (eds) Magma-crust interactions and evolution. Athens: Theophrastus, pp 3–88

Froelich AJ, Gottfried D (1988) An overview of early Mezozoic intrusive rocks in the Culpeper basin, Virginia and Mariland. Geol Soc Am Publ 1776:151–164.

Fujii T (1974) Crystal settling in sills. Lithos 7:133–137

Galerne CY, Neumann ER, Aarnes I, Planke S (2010) Magmatic differentiation processes in saucer-shaped sills: evidence from the Golden Valley Sill in the Karoo Basin, South Africa. Geosphere 6:163–188

Gauert CDK, de Waal SA, Wallmach T (1995) Geology of the ultrabasic to basic Uitkomst complex, eastern Transvaal, South Africa: an overview. J Afr Earth Sci 21:553–570

Gibb FGF, Henderson CMB (1992) Convection and crystal settling in sills. Contrib Mineral Petrol 109:538–545

Gibb FGF, Henderson CMB (2005) Chemistry of the Shiant Isles Main Sill, NW Scotland, and wider application for petrogenesis of mafic sills. J Petrol 47:191–230
Gibb FGF, Henderson CMB (2006) Chemistry of the Shiant Isles Main Sill, NW Scotland, and wider application for petrogenesis of mafic sills. J Petrol 47:191–230
Gisselø PG (2001) Sorgenfri Gletscher Sill complex, East Greenland: solidification mechanisms of sheet-like bodies and the role of sill complexes in large igneous provinces. Ph.D. thesis, University of Aarhus, Denmark
Gorring ML, Naslund HR (1995) Geochemical reversals within the lower 100 m of the Palisades sill, New Jersey. Contrib Mineral Petrol 119:263–276
Grant NK, Molling PA (1981) A strontium isotope and trace element profile through the Partridge River troctolite, Duluth Complex, Minnesota. Contrib Mineral Petrol 77:296–305
Gray NH, Crain IK (1969) Crystal settling in sills: a model for suspension settling. Canad J Earth Sci 6:1211–1216
Grieve RA (1994) An impact model for the Sudbury structure. In: Lightfoot PC, Naldrett AJ (eds) Proceedings of the Sudbury-Noril'sk Symposium: Ontario Geological Survey Special Volume 5, p 119–132
Gunn BM (1962) Differentiation in Ferrar Dolerites, Antarctica. New Zealand. J Geol Geoph 5:820–863
Gunn BM (1966) Modal and element variation in Antarctic tholeiites. Geochim Cosmochim Acta 30:881–920
Hanski E, Walker RJ, Huhma H, Suominen I (2001) The Os and Nd isotopic systematics of c. 2.44 Ga Akanvaara and Koitelainen mafic layered intrusions in northern Finland. Precamb Res 109:73–102
Helz RT, Kirschenbaum H, Marinenko JW (1989) Diapiric transfer of melt in Kilauea Iki lava lake, Hawaii: a quick, efficient process of igneous differentiation. Geol Soc Am Bull 101:578–594
Henderson CMB, Gibb FGF (1987) The petrology of the Lugar sill, SW Scotland. Trans R Soc Edinb (Earth Sci) 77:325–347
Henderson CMB, Gibb FGF, Foland KA (2000) Mineral fractionation and pre- and post-emplacement processes in the uppermost part of the Shiant Isles Main Sill. Mineral Mag 64:779–790
Hill RET, Barnes SJ, Gole MJ, Dowling SE (1995) The volcanology of komatiites as deduced from field relationships in the Norseman-Wiluna greenstone belt, Western Australia. Lithos 34:159–188
Hoffer JM (1966) Compositional variations of plagioclase feldspar from a basaltic lava flow. Am Mineral 51:807–813
Huang F, Lundstrom CC, Glessner J, Ianno A, Boudreau A, Li J, Ferré EC, Marshak S, DeFrates J (2009) Chemical and isotopic fractionation of wet andesite in a temperature gradient: experiments and models suggesting a new mechanism of magma differentiation. Geochim Cosmochim Acta 73:729–749
Iezzi G, Mollo S, Torresi G, Ventura G, Cavallo A, Scarlato P (2011) Experimental solidification of an andesitic melt by cooling. Chem Geol 283:261–273
Iljina MJ (1994). The Portimo layered igneous complex with emphasis on diverse sulphide and platinum-group element deposits. Acta Universitatis Ouluensis. Series A, Scientiae Rerum Naturalium 258:158
Iljina MJ, Alapieti TT, Lahtinen JJ, Lerssi JM (1989) The Suhanko-Konttijärvi intrusion and related sulphide and PGE mineralization. Guide to the post-symposium field trip to 5th International Platinum symposium. Geological Survey of Finland. Guide 29:163–187
Irvine TN (1970) Crystallization sequences in the Muskox intrusion and other layered intrusions, I. Olivine-pyroxene-plagioclase relations. Spec Publ Geol Soc S Afr 1:441–476
Irvine TN (1980) Magmatic infiltration metasomatism, double diffusive fractional crystallization and adcumulus growth in the Muskox Intrusion and other layered intrusions. In: Hargraves RB (ed) Physics of magmatic processes. Princenton University press, Princenton, pp. 325–383
Irvine TN, Smith CH (1967) The ultrabasic rocks of the Muskox intrusion, Northwest Territories, Canada. In: Wyllie PJ (ed) Ultramafic and related rocks. Wiley, New York, pp 38–49
Ivanov BA, Deutsch A (1999) Sudbury impact event: cratering mechanics and thermal history. In: Dressler BO, Sharpton L (eds) Large meteorite impacts and planetary evolution II: Geological Society America Special Paper 339, pp 389–398

Jaupart C, Tait S (1995) Dynamic of differentiation in magma reservoirs. J Geophys Res 100:17617–17636
Karinen T (2010) The Koillismaa intrusion, northeastern Finland: evidence for PGE Reef forming processes in the layered series. Doctoral thesis. Geological Survey of Finland, Bulletin of Geological Survey of Finland 404, 176 p
Koptev-Dvornikov EV, Kireev BS, Pchelintseva NF, Khvorov DM (2001) Distribution of cumulative mineral assemblages, major and trace elements over the vertical section of the Kivakka intrusion, Olanga group of intrusions, Northern Karelia. Petrology 9:1–24
Krivenko AP, Balikin PA, Polyakov GV (1980) Distribution of alkalis in basic plutons. Russian Geologiya i Geofizica 243:144–149
Latypov RM (2003a) The origin of marginal compositional reversals in basic-ultrabasic sills and layered intrusions by Soret fractionation. J Petrol 44:1579–1618
Latypov RM (2003b) The origin of basic-ultrabasic sills with S-, D- and I-shaped compositional profiles by in situ crystallization of a single input of phenocryst-poor parental magma. J Petrol 44:1619–1656
Latypov RM, Egorova V (2012) Plagioclase compositions give evidence for in situ crystallization under horizontal flow conditions in mafic sills. Geology 40:883–886
Latypov RM, Smolkin VF, Alapieti TT (2001) Trend of differentiation and parental magma of the Pechenga Ni-bearing gabbro-wehrlite intrusions, Kola Peninsula, Russia. Petrology 9:381–397
Latypov RM, Chistyakova SY, Alapieti TT (2007) Revisiting problem of chilled margins associated with marginal reversals in mafic-ultramafic intrusive bodies. Lithos 99:178–206
Latypov RM, Chistyakova SY, Alapieti TT (2008) Fine-grained mafic bodies as preserved portions of magma replenishing layered intrusions: the Nadezhda gabbronorite body, Lukkulaisvaara intrusion, Fennoscandian shield, Russia. Mineral Petrol 92:165–209
Latypov RM, Lavrenchuk A, Lightfoot PC (2010) Was the parental magma of the Sudbury igneous complex really superheated? Conference proceeding of the 11th International Platinum Symposium, Sudbury, Canada
Latypov RM, Hanski E, Lavrenchuk A, Huhma H, Havela T (2011) A 'three-increase model' for origin of marginal reversal in the Koitelainen layered intrusion, Finland. J Petrol 52:733–764
Lavrenchuk A, Latypov RM, Lightfoot PC (2010) The Sudbury igneous complex, Canada: numerical modeling confirms fractionation of a single parental magma. Conference proceeding of the 11th International Platinum Symposium, Sudbury, Canada
Lesher CE, Walker D (1991) Thermal diffusion in petrology. In: Ganguly J (ed) Diffusion, atomic ordering, and mass transport: selected topics in geochemistry. Advances in physical geochemistry 8. Springer, New York, pp 396–451
Lightfoot PC, Naldrett AJ (1984) Chemical variation in the Insizwa complex, Transkei, and the nature of the parent magma. Can Mineral 22:111–123
Lightfoot PC, Zotov IA (2005) Geology and geochemistry of the Sudbury igneous complex, Ontario, Canada: origin of nickel sulfide mineralization associated with an impact-generated melt sheet. Geol Ore Deposits 47:349–381
López-Moro FJ, Murciego A, López-Plaza M (2007) Silurian/Ordovician asymmetrical sill-like bodies from La Codosera syncline, W Spain: a case of tholeiitic partial melts emplaced in a single magma pulse and derived from a metasomatized mantle source. Lithos 96:567–590
Lundstrom CC (2009) Hypothesis for the origin of convergent margin granitoids and Earth's contintental crust by thermal migration zone refining. Geochim Cosmochim Acta 73:5709–5729
Maier WD, Rasmussen B, Fletcher IR, Li C, Barnes S-J, Huhma H (2013) The Kunene anorthosite complex, Namibia, and its satellite intrusions: geochemistry, geochronology, and economic potential. Econ Geol 108:953–986
Mangan TM, Marsh BD, Froelich AJ, Gottfried D (1993) Emplacement and differentiation of the York Haven diabase sheet, Pennsylvania. J Petrol 34:1271–1302
Marsh BD (1989) On convective style and vigor in sheet-like magma chambers. J Petrol 30:479–530
Miller JD, Ripley EM (1996) Layered intrusions of the Duluth Complex, Minnesota, USA. In: Cawthorn RG (ed) Layered intrusions. Developments in petrology 15. Elsevier Science, B. V., pp 257–301

Morse SA (1979) Kiglapait geochemistry II: petrography. J Petrol 20:591–624
Morse SA (1981) Kiglapait geochemistry IV: the major elements. Geochim Cosmochim Acta 45:461–479
Morse SA (1984) Cation diffusion in plagioclase feldspar. Science 225:504–505
Morse SA (1986) Convection in aid of adcumulus growth. J Petrol 27:1183–1215
Morse SA (2008) The internal magma reservoir of large intrusions revealed by multiphase Rayleigh fractionation. J Petrol 49:2081–2098
Mutanen T (1997). Geology and ore petrology of the Akanvaara and Koitelainen mafic layered intrusions and the Keivitsa-Satovaara layered complex, northern Finland. Bull Geol Surv Finl 395:233
Mutanen T, Huhma H (2001) U-Pb geochronology of the Koitelainen, Akanvaara and Keivitsa layered mafic intrusions in central Finnish Lapland. In: Vaasjoki M (ed) Radiometric age determinations from Finnish Lapland and their bearing on the timing of Precambrian volcano-sedimentary sequences. Geological Survey of Finland, (Special Paper) 33:229–246
Naldrett AJ (2004) Magmatic sulfide deposits: geology, geochemistry, and exploration. Springer, Berlin, New York, 727
Naldrett AJ, Bray JG, Gasparrini EL, Podolsky T, Rucklidge JD (1970) Cryptic variation and the petrology of the Sudbury nickel irruptive. Econ Geol 65:122–155
Namur O, Humphreys MCS, Holness MB (2013) Lateral reactive infiltration in a vertical gabbroic crystal mush, Skaergaard intrusion, East Greenland. J Petrol 54:985–1016
Naslund HR (1989) Petrology of the Basistoppen Sill, East Greenland: a calculated magma differentiation trend. J Petrol 30:299–319
Pattison EF (1979) The Sudbury layer. Can Mineral 17:257–274
Petersen JS (1987) Solidification contraction: another approach to cumulus processeses and the origin of igneous layering. In: Parson I (ed) Origin of igneous layering. Reidel Publishing Company, Dordrecht, Holland, 505–526
Prevec SA, Cawthorn RG (2002) Thermal evolution and interaction between impact melt sheet and footwall: a genetic model for the contact sublayer of the Sudbury igneous complex, Canada. J Geophys Res 107(B8):ECV 5-1–ECV 5-14
Raedeke LD, McCallum IS (1984) Investigations in the stillwater complex: part II. Petrology and petrogenesis of the ultramafic series. J Petrol 25:395–420
Severson MJ, Hauck SA (1990) Geology, geochemistry, and stratigraphy of a portion of the Partridge River intrusion: Duluth, Minnesota, Natural Resources Research Institute, Technical Report 89-11, 236 p
Simkin T (1967) Flow differentiation in the picritic sills of north skye. In: Wyllie PJ (ed) Ultramafic and related rocks. Wiley, New York, pp 64–69
Tait S, Jaupart C (1996) The producing of chemically stratified and adcumulate plutonic igneous rocks. Mineral Mag 60:99–114
Tegner C, Wilson JR, Brooks CK (1993) Intraplutonic quench zones in the Kap Edvard Holm layered gabbro complex, East Greenland. J Petrol 34:681–710
Tegner C, Robins B, Reginiussen H, Grundvig S (1999) Assimilation of crustal xenoliths in a basaltic magma chamber: Sr and Nd isotopic constraints from the Hasvik layered intrusion, Norway. J Petrol 40:363–380
Therriault AM, Fowler AD, Grieve AF (2002) The Sudbury igneous complex: a differentiated impact melt sheet. Econ Geol 97:1521–1540
Turcotte DL (1987) Physics of magma segregation processes, In Mysen BO (ed) Magmatic processes: physicochemical principles. Geochemical Society Special Publication 1:69–74
Tyson RM, Chang LLY (1984) The petrology and sulfide mineralization of the Partridge River troctolite, Duluth Complex, Minnesota. Can Mineral 22:23–38
Ubide T, Arranza E, Lagoa M, Galéa C, Larreaa P (2012) The influence of crystal settling on the compositional zoning of a thin lamprophyre sill: a multi-method approach. Lithos 132–133:37–49
Vander Auwera J, Longhi J (1994) Experimental study of a jotunite (hypersthene monzodiorite): constraints on the parent magma composition and crystallization conditions (P, T, fO_2) of the Bjerkreim-Sokndal layered intrusion. Contrib Mineral Petrol 118:60–78

VanTongeren JA, Mathez EA (2013) Incoming magma composition and style of recharge below the pyroxenite marker, eastern bushveld complex, South Africa. J Petrol 54:1585–1605
Wager LR (1963) The mechanism of adcumulus growth in the layered series of the Skaergaard intrusion. Mineral Soc Am Spec Pap 37:837–858
Wager LR, Brown GM (1968) Layered igneous rocks. Oliver and Boyd, Edinburgh, p 588
Wilson JR, Engell-Sørensen O (1986) Marginal reversals in layered intrusions are evidence for emplacement of compositionally stratified magma. Nature 323:616–618
Wilson JR, Larsen SB (1985) Two-dimensional study of a layered intrusion; the hyllingen series, Norway. Geol Mag 122:97–124

Chapter 7
The Significance of Magnetic Fabric in Layered Mafic-Ultramafic Intrusions

Brian O'Driscoll, Eric C. Ferré, Carl T. E. Stevenson and Craig Magee

Abstract Anisotropy of magnetic susceptibility (AMS) has been recognised as a well-established fabric analysis tool for intrusive igneous rocks since the 1990s. The AMS technique provides directional information for magnetic foliation and magnetic lineation fabric components of the AMS ellipsoid, potentially coupled with a quantification of the overall fabric strength and geometry. The magnetic susceptibility (and therefore the AMS) of igneous rocks is often dominated by ferromagnetic mineral phases such as magnetite or low-Ti titanomagnetite, even where present in very minor amounts (e.g., ~0.1 vol.%). Fe-bearing silicates exhibit subordinate paramagnetic behaviour but are volumetrically much more important constituents of igneous rocks than Fe-Ti oxides, so may also contribute considerably to the AMS.

A significant application of AMS is in the characterisation, constraint and quantification of very weak or subtle mineral fabrics related to flow or tectonic deformation. In particular, studies of magnetic fabrics in sheet intrusions and in granite plutons have enormously enhanced our understanding of the magma flow regimes and emplacement kinematics in these settings. Studies of AMS in layered mafic-ultramafic intrusions have been comparatively sparse. This is despite the fact that

B. O'Driscoll (✉)
School of Physical and Geographical Sciences, Keele University, Keele ST5 5BG, UK
e-mail: brian.odriscoll@manchester.ac.uk

Now at School of Earth, Atmospheric and Environmental Sciences, The University of Manchester, Williamson Building, Oxford Road, Manchester M13 9PL, UK

E. C. Ferré
Department of Geology, Southern Illinois University at Carbondale, Carbondale, IL, 62901-4324, USA
e-mail: eferre@geo.siu.edu

C. T. E. Stevenson
School of Geography, Earth and Environmental Sciences, University of Birmingham, Birmingham B15 2TT, UK
e-mail: c.t.stevenson@bham.ac.uk

C. Magee
Department of Earth Science and Engineering, Imperial College, London SW7 2BP, UK
e-mail: c.magee@imperial.ac.uk

B. Charlier et al. (eds.), *Layered Intrusions,* Springer Geology,
DOI 10.1007/978-94-017-9652-1_7

magnetic fabrics from layered cumulates may provide information on a range of magma chamber processes, from initial magma emplacement to cumulate textural evolution and solidification, if accompanied by careful petrographic documentation and an understanding of the 'magnetic mineralogy' of the rock. A wide array of rock magnetic and complementary quantitative fabric analysis techniques can be employed to support an AMS dataset in this regard. With studies of layered mafic-ultramafic intrusions currently proceeding at unprecedented (micro-)scales of textural and geochemical detail, AMS offers petrologists a unique approach to investigating the microstructure of cumulates and the textural complexity they exhibit.

Keywords Anisotropy of Magnetic Susceptibility · Cumulate textures · Mineral lamination · Magnetic mineralogy · Magma chamber processes

Introduction: Magnetic Fabrics and Their Application to Mafic-Ultramafic Plutonic Rocks

Research on the magnetic fabrics of igneous rocks, pioneered by Balsley and Buddington (1960) and Khan (1962), has greatly contributed to our understanding of magma dynamics in lava flows, sheet intrusions and plutons over the past five decades (cf. Tarling and Hrouda 1993; Borradaile and Henry 1997; Martín-Hernández et al. 2004; Michel et al. 2008). In particular, the anisotropy of magnetic susceptibility (AMS) technique has proved to be a powerful tool in the systematic quantification and mapping of structural (magnetic) fabrics in granite intrusions (reviewed in Bouchez 1997), playing a major role in revisiting the diapir paradigm for the emplacement of many granite intrusions in the latter part of the twentieth century (e.g., Tarling and Hrouda 1993; Borradaile and Henry 1997). Through the application of AMS, many granitic plutons have been shown to have emplaced as sub-horizontal sheets, with an internal structural architecture typically controlled by the ambient tectonic regime (e.g., Améglio and Vigneresse 1999; Wilson et al. 2000; Stevenson et al. 2007a, b; Michel et al. 2008; Petronis et al. 2012). Our understanding of the rheology of granitic magmas (e.g., Petford 2009) indicates that the AMS of porphyritic granites often records the final increments of magmatic (e.g., emplacement-related) deformation. In sub-volcanic systems and lava flows, the AMS fabric can provide valuable information on the magma transport direction (e.g., Airoldi et al. 2012).

Considerably fewer magnetic fabric studies have been dedicated to large mafic-ultramafic intrusions, and in particular to layered mafic-ultramafic 'lopolithic' intrusions (e.g., Wagner et al. 1981; Mushayandebvu et al. 1995; MacLeod et al. 1996; Richter et al. 1996; Hargraves et al. 1999; Selkin et al. 2000; Ferré et al. 2002, 2009; O'Driscoll 2006; O'Driscoll et al. 2006, 2007a, 2008; Maes et al. 2007, 2008), despite the fact that such rocks may preserve a large range of small-scale kinematic structures potentially related to important dynamic magma chamber processes (Nicolas 1992; Benn and Allard 1989). This may be partly due to the fact that mafic-ultramafic cumulates commonly exhibit visible planar fabrics (mineral lamination), as well as compositional layering, in contrast to the frequent absence of such features in granite bodies or fine-grained mafic lava flows. Indeed, debates in the 1970's and 1980's on the development of layering and mineral fabrics in mafic-ultramafic intrusions, focused around the crystal settling versus *in situ* crystallisation paradigms, are classic in the subject of igneous petrology (McBirney and Noyes 1979; Cawthorn 1996). Central to these debates is that a wide range of magma chamber processes are envisaged in layered mafic-ultramafic intrusions that are not frequently considered to occur in their relatively viscous granitic counterparts; in essence, the latter have historically been viewed as much more likely to 'freeze-in' a primary magma flow fabric whilst mafic-ultramafic intrusions are subjected to a more protracted solidification history. This wide array of potential initial sources for layering and mineral fabrics in layered mafic-ultramafic intrusions, together with the possible modification of fabrics at the postcumulus stage, demands a cautious application of any fabric analysis and presents a problem well-suited to interrogation by AMS techniques.

The purpose of this review article is to introduce the principles of the AMS method and to provide specific context on the application of this technique to layered mafic-ultramafic intrusions. A broad overview of research on the magnetic fabrics of layered intrusions is provided, via a discussion of topics including magma emplacement models and cumulate textural evolution; two important stages of layered intrusion formation that can potentially be evaluated using AMS. The obstacles and issues surrounding the interpretation of magnetic fabrics are discussed, and the integration of the AMS technique with several important complementary tools is also evaluated with reference to several well-studied layered intrusions. Finally, we offer our perspectives on future applications and directions for the measurement and interpretation of magnetic fabrics in layered intrusions.

Principles of Rock Magnetism and the Origin of AMS

The magnetic behaviour of igneous rocks can be considered to reflect the variation of magnetisation (M) with that of an externally applied field (H). The susceptibility of an igneous rock to an induced magnetisation is controlled by its mineralogy and exhibits one of three forms of behaviour: diamagnetism, paramagnetism or ferromagnetism (Fig. 7.1). Diamagnetism is characterised by a slightly negative and non-permanent response in M to increasing H, whereas paramagnetic grains exhibit a positive (non-permanent) relationship between M and H. In contrast, ferromagnetic behaviour (*sensu lato*) is related to Fe^{3+}-bearing spinel phases and is characterised by a strongly positive relationship between M and H (Fig. 7.1), with a maximum threshold at the point of saturation magnetisation (M_s). Ferrimagnetism is a style of ferromagnetism (Fig. 7.1) that describes the specific behaviour of magnetite (Fe_3O_4). Removal of the external field leaves ferromagnetic (*s.l.*) grains with a remanent magnetisation for a variable period of time depending on grainsize (Dunlop and Özdemir 1997).

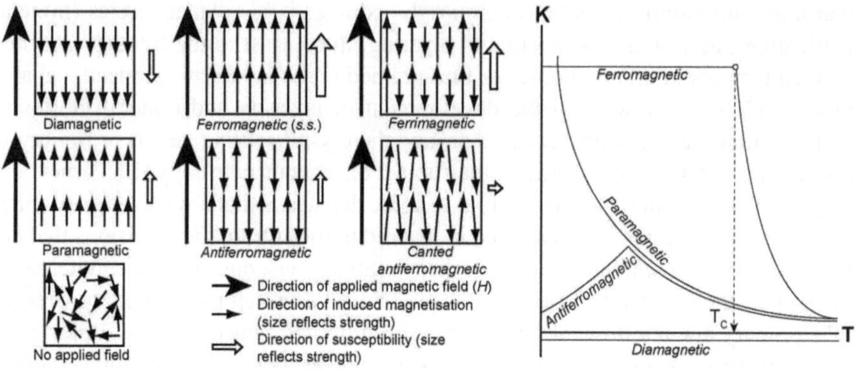

Fig. 7.1 Schematic diagram depicting the different types of rock magnetism (diamagnetism, paramagnetism and ferromagnetism), with associated applied magnetic fields (*H*) and net magnetisations (adapted from Tarling and Hrouda 1993). A sketch of the idealised relationship between K and temperature (*T*), illustrating the derivation of T_C via the inflection method is also shown

In igneous rocks, paramagnetic behaviour is predominantly controlled by silicate minerals that contain Fe in their crystal lattice, especially ferromagnesian minerals. Paramagnetic grains show strong temperature dependence whereby their susceptibility decreases with increasing temperature, according to the Curie law (Fig. 7.1; Dunlop and Özdemir 1997). However, ferromagnetic grains dominate the susceptibility signal when they are present, even in trace amounts in rocks (Martín-Hernández and Ferré 2007). Titanomagnetite, part of the magnetite-ulvöspinel solid solution, is a common accessory phase in gabbroic rocks and therefore typically determines the magnetic response of mafic rocks to *H*. The close coordination of the Fe^{2+} and Fe^{3+} cations in such spinels inhibits the thermal disruption observed in paramagnetic grains, up to the Curie temperature (T_C; Fig. 7.1) above which the magnetic susceptibility in ferromagnets decreases significantly and paramagnetic behaviour ensues according to the Curie-Weiss law (Dunlop and Özdemir 1997). For certain minerals, T_C is known (e.g., ~580 °C for pure magnetite) and is therefore used in many AMS and rock magnetic studies of igneous rocks as a diagnostic indicator of the mineral phases responsible for controlling the magnetic fabrics. Remanent magnetisations of ferromagnetic (*s.l.*) minerals are often thermally locked in between the T_C and the Blocking Temperature (T_B) (Dunlop and Özdemir 1997). Note that a remanent magnetisation does not affect the measurement of the *anisotropy* of magnetic susceptibility.

The ratio of *M* to *H* defines the magnetic susceptibility (*K*) – $K = M/H$. When magnetisations and magnetic fields are measured in Am^{-1}, *K* is expressed in International System [SI] units and is dimensionless. In polymineralic rocks, the bulk magnetic susceptibility is the sum of contributions from all minerals (i.e., diamagnetic, paramagnetic and ferromagnetic). The magnetic susceptibility of a rock generally varies with the direction along which it is measured and such variations are idealised by a symmetric second rank tensor. This tensor can be represented by an ellipsoid with three mutually perpendicular axes that correspond to the maximum

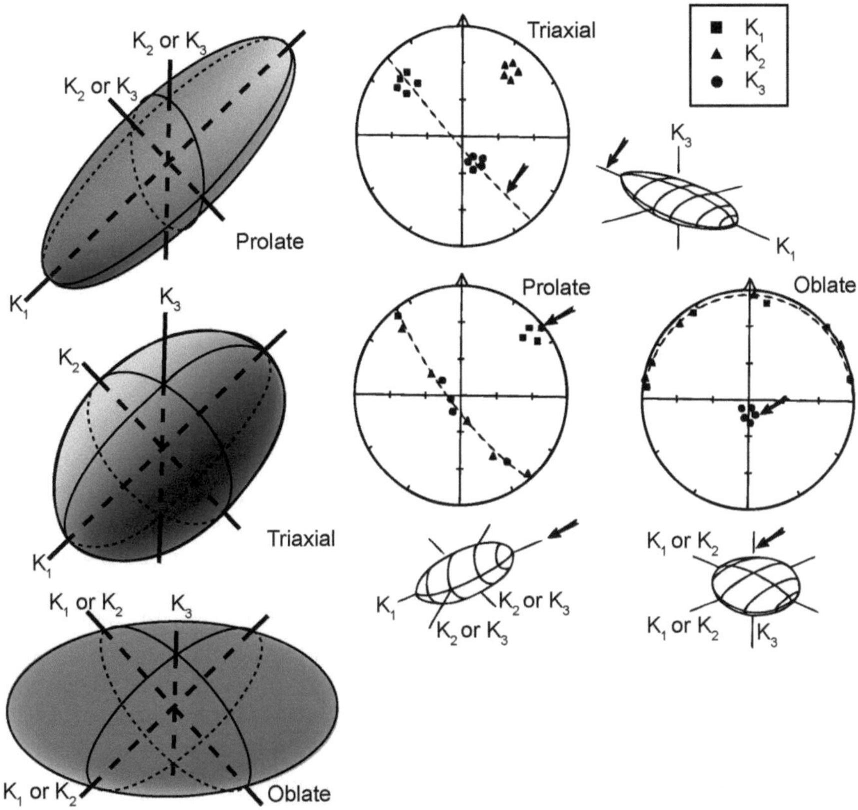

Fig. 7.2 Cartoons showing idealised prolate, triaxial and oblate susceptibility ellipsoids, with the magnitude of the three orthogonal axes (i.e., K_1, K_2 and K_3) also illustrated. Also depicted are equal-area projections for each type of ellipsoid geometry, adapted from Tarling and Hrouda (1993; see text for details)

[K1], intermediate [K2] and minimum [K3] susceptibility values (Fig. 7.2). The origin of the AMS is a combination of magnetocrystalline (i.e., lattice-preferred orientation [LPO]), magnetostatic (i.e., shape-preferred orientation [SPO]) and distribution anisotropy. Distribution anisotropy occurs when ferromagnetic grains are close to each other and uniformly arranged (Hargraves et al. 1991; Stephenson 1994; Gaillot et al. 2006). AMS therefore provides valuable information not only on fabric orientation but also on fabric strength and symmetry (Fig. 7.2). The corrected degree (or fabric strength) of anisotropy of magnetic susceptibility, which generally increases with fabric strength, is commonly reported as P'. A value of P' = 1 describes an isotropic fabric whilst a P' value of 1.15 describes a sample with 15 % anisotropy, and so on:

$$P' = \exp\sqrt{\left[2\Sigma\left(\ln K_i - \ln K_{mean}\right)^2\right]} \quad \left(\begin{array}{l}i = 1 \text{ to } 3,\ K_{mean} \text{ is the arithmetic mean} \\ \text{magnetic susceptibility}\end{array}\right)$$

The shape parameter T is a quantitative measure of the shape of the magnetic susceptibility ellipsoid, ranging from +1 to −1 (Jelínek 1981).

$$T = \frac{2\ln K_2 - \ln K_1 - \ln K_3}{\ln K_1 - \ln K_3} \quad (\text{oblate ellipsoid for } 0 < \mathrm{T} \leq 1, \text{ prolate for} -1 \leq \mathrm{T} < 0).$$

Owens (1974, 2000) developed an alternative set of AMS parameters:

$$\mathrm{H} = \mathrm{L} + \mathrm{F} = (K_1 - K_3)/K_{mean} \text{ with } \mathrm{L} = (K_1 - K_2)/K_{mean} \text{ and}$$
$$\mathrm{F} = (K_2 - K_3)/K_{mean} \;\; \mathrm{m} = \mathrm{L/F}, \quad = \tan^{-1}\mathrm{m}$$

H (quoted as a percentage, and not to be confused with *H*, or magnetic field) describes the strength of the magnetic fabric whilst the shape of the AMS ellipsoid is described by μ, where μ = 0–45° describes an oblate AMS ellipsoid and μ = 45–90° describes a prolate AMS ellipsoid. It has been suggested that Jelínek's (1981) parameters, which are based on the natural logarithms of the ratios of all three principal susceptibility axes, are best suited for rocks in which the anisotropy is large (> 15 %; O'Driscoll et al. 2008). Owens' (1974, 2000) parameters are based on the differences between the principal susceptibility axes and are arguably more useful in reporting relatively weak anisotropies. In either case, once the magnitude values of the three principal susceptibility axes and K_{mean} are known (measured), calculation of either set of parameters or translation between one set and the other is easily possible.

Many magnetic fabric studies of igneous rocks have focused on the emplacement of lava flows (e.g., Cañón-Tapia 2004), sheet intrusions or granites (e.g., Borradaile and Henry 1997; Bouchez 1997). Principally, it is understood that the long axes of minerals crystallising in magma become aligned within the magma flow regime. These studies broadly consider that K_1, the long axis of the AMS ellipsoid, corresponds to the magma flow direction while K_3, the short axis, is perpendicular to the flow plane (e.g., Rochette et al. 1992). The principal axes of the AMS fabric generally correspond with the principal axes of the mineral shape-preferred orientation (SPO) determined by image analysis (e.g., Archanjo et al. 1994; Grégoire et al. 1998; Launeau and Cruden 1998), so that K_1 defines the magnetic lineation and K_3 is the pole to the magnetic foliation. Hence, AMS has been determined to be a good proxy for the magma flow fabric in some igneous rocks. The determination of flow fabrics and AMS fabrics in analogue models (e.g., Arbaret et al. 1996, 1997, 2013) confirms the broad correlation between AMS and flow fabrics, at least in the simple shear regime.

However, in layered mafic-ultramafic intrusions the flow of magma (or crystal slurries) is often not the dominant texture (fabric)-forming process. For example, primary magmatic processes (e.g., crystal settling and *in situ* crystallisation), as well as secondary processes (e.g., postcumulus compaction) may result in macroscopic mineral lamination in mafic-ultramafic cumulates (Wager and Brown 1968; McBirney and Noyes 1979; Meurer and Boudreau 1998). In such rocks, AMS

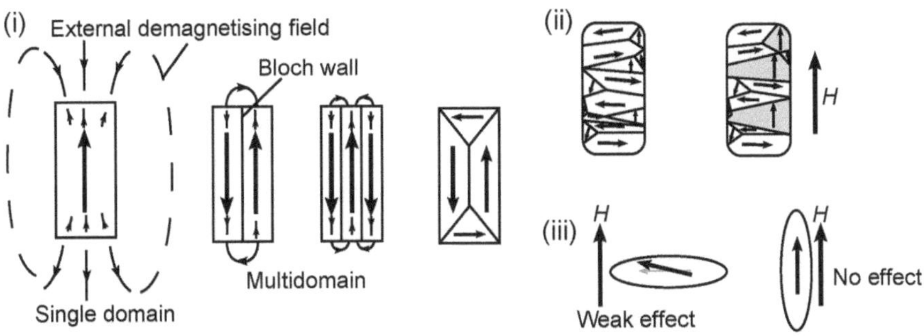

Fig. 7.3 Domain state and Bloch walls in ferromagnetic grains. (*i*) *SD* and *MD* configurations serve to minimise the external demagnetising field. (*ii*) Movement of the Bloch wall in *MD* grains gives rise to a magnetisation when an external field is applied. (*iii*) Inverse susceptibility anisotropy of a *SD* grain. The application of a weak magnetic field (*H*) parallel to the elongation direction produces no net magnetisation, whereas the application of a weak magnetic field perpendicular to the elongation direction produces a weak induced magnetisation parallel to the field. This figure is adapted from Appendix 1 of O'Driscoll et al. (2008)

studies have focused on using magnetic fabrics as a means to interpret the textural evolution of the cumulate, instead of concentrating on magma flow dynamics, the evidence for which may no longer exist in the rocks (e.g., O'Driscoll et al. 2007a, 2008). From a petrological standpoint, the importance of the order of crystallisation of different silicates versus oxides to AMS fabric interpretation cannot be overestimated. Careful characterisation of the ferromagnetic (and to a lesser extent the paramagnetic phases) as either cumulus or intercumulus is crucial. This is partly due to the fact that magnetic fabrics controlled by paramagnetic grains are generally crystallographic in origin, whilst those controlled by ferromagnetic grains are grain shape-dependent. The internal structure of ferromagnetic grains is defined by sub-grain magnetic domains (Fig. 7.3). Within each domain, the electron spins are parallel and give rise to an internal magnetic field that has a corresponding external demagnetising field oriented in the opposite direction. As large external demagnetising fields are energetically undesirable, the size of individual sub-grain domains is ~0.1 μm. In grains with multiple domains (>0.1 μm; MD), Bloch walls separate each single domain of parallel electron spins (Fig. 7.3). Grains significantly smaller in size than this are single domain (SD). Pseudo-single domain (PSD) behaviour may arise in grains approximately 0.1 μm in size and show behaviour intermediate between that of SD and MD grains. Mixtures of single- and multidomain titanomagnetites may yield intermediate fabrics, where two of the principal magnetic susceptibility axes are switched (Rochette et al. 1999; Ferré 2002; Fig. 7.3). In mafic-ultramafic intrusive igneous rocks the AMS fabric is typically dominated by (titano-) magnetite, which may occur as an accessory phase disseminated throughout the rock texture or as inclusions in ferromagnesian silicates (e.g., O'Driscoll et al. 2007a). The influence of diamagnetic minerals, including plagioclase and quartz, is so weak that just 5 wt.% of paramagnetic minerals (e.g., biotite, olivine, clinopyroxene) or an even smaller amount of ferromagnetic grains (0.1–1 wt.%) will dominate the susceptibility (Rochette 1987; Tarling and Hrouda 1993; Bouchez 1997). Secondary ferromagnetic minerals (e.g. formed during low-temperature alteration such as serpentinisation or exsolution of inclusions) may also affect the AMS, either detrimentally to the fabric interpretation or beneficially, by enhancement of an existing fabric (e.g., Stevenson et al. 2007b). In cumulates that comprise layered intrusions, paramagnetic and ferromagnetic grains can both be significant, either together or individually. Although paramagnetic and ferromagnetic fabric axes commonly coincide, it is preferable to analyse these fabrics separately (e.g., Archanjo et al. 1994; O'Driscoll et al. 2007a; Stevenson et al. 2007b; Martín-Hernández and Ferré 2007). The contribution of magmatic sulphides (e.g., pyrrhotite) to magnetic fabrics is negligible compared to that of titanomagnetite, with the exception of massive sulphide deposits (Ferré et al. 2002).

Aids to Understanding Magnetic Fabrics in Layered Intrusions

Understanding the Origin of AMS

The interpretation of magnetic fabrics in layered intrusions rests upon the identification of the dominant carrier and the determination of the origin of the magnetic fabric. The dominant carrier must be a mineral species that has an intrinsic magnetic anisotropy sufficiently high to cause the observed rock anisotropy. For example, a population of pyroxene crystals with intrinsic anisotropies of $P'_{pyr} = 1.50$ would be insufficient to explain whole rock P' values greater than 1.50, even if all crystals were oriented perfectly parallel to one another. The cause of the LPO, in the case of magnetocrystalline anisotropy, or SPO, in the case of magnetostatic anisotropy, needs to be evaluated in the light of complementary microstructural and rock magnetic studies. For example, the magnetic hysteresis properties of a coarse-grained titanomagnetite-bearing mafic rock may reveal that its ferromagnetic grains are SD in size, in turn suggesting that owing to their significantly smaller size these grains most likely grew under subsolidus conditions.

Stepwise heating/cooling of a powdered sample (e.g., from 40 °C to 680 °C to 40 °C), using a furnace attachment on a magnetic susceptometer, can also be used to evaluate the composition of a ferromagnetic mineral assemblage. Ferromagnetic (*s.l.*) minerals are highly sensitive to temperature fluctuations as heating causes super-exchange coupling forces to breakdown at the T_C (Dunlop and Özdemir 1997). For example, as the Ti-content of titanomagnetite increases, its T_C exhibits a corresponding decrease. In comparison, paramagnetic minerals always display an inverse slope with susceptibility proportional to 1/T (Fig. 7.1). The susceptibility of diamagnetic grains is T independent (Dunlop and Özdemir 1997). The overall morphologies of the heating-cooling curves therefore allow these three magnetic behaviours to be distinguished. The T_C estimates, which represent the transition from ferromagnetic (*s.l.*) to paramagnetic behaviour, are inferred using the inflection point or Hopkinson Peak method (Tauxe 1998; Petrovský and Kapička 2006). Fabian et al. (2013) recently proposed a more sophisticated method based on the measurement of magnetic hysteresis quarter loops. It is important to note that although most ferromagnetic (*s.l.*) minerals described in the literature are attributed a single unique T_C, this strongly depends on mineral composition. For example, titanomagnetites with variable Ti content have a range of T_C from 150–578 °C (Tarling and Hrouda 1993). Curie Points are often preceded by an increase in susceptibility (the Hopkinson effect; Hopkinson 1889), either represented by a shallow positive slope and/or by the development of a convex-upwards 'bump' in susceptibility, termed a Hopkinson Peak (Liss et al. 2002). The morphology of the Hopkinson Peak is dictated by the size and shape of the respective mineral grains.

Graphical Representation of AMS Data

AMS results are reported for block averages of specimen AMS tensors, measured on a spinning magnetometer (e.g., an AGICO Kappabridge instrument), which measures specimen susceptibility axes to within ± 2° (95 % confidence limits). AMS data is normalised by the specimen mean susceptibility, on the assumption that the specimens from a sample or sampling station are representative of a homogeneous multi-normal population. Magnetic fabric data are most often presented as lower hemisphere equal area plots of the mean principal susceptibility axes for each block sample (e.g., Fig. 7.4). This style of data presentation gives a good estimate of both the shape and the orientation of the mean susceptibility ellipsoid for each sample. The mean shape of the ellipsoid is represented by the shape and overlap of the 95 % confidence ellipses around each susceptibility axis. This is because the size of the ellipse semi-axis in each principal plane varies inversely as the difference between the two principal susceptibilities in that plane (Jelínek 1978; Owens 2000). Therefore, an oblate fabric will have a small circular confidence ellipse around the minimum susceptibility axis, with the other two confidence ellipses elongated in the foliation plane, whilst a triaxial fabric will have confidence ellipses around the maximum and minimum axes that are both elongated toward the intermediate axis (Fig. 7.4). The confidence ellipse area around the intermediate axis in the case of the latter will be larger and more circular than the other two. The overall size of the confidence ellipses for the mean principal directions for a sample, if based on a constant number of specimens, gives a relative indication of the strength or scatter of specimen anisotropies within a block; i.e., the larger the ellipse, the greater the scatter and the weaker the fabric (Owens 2000). AMS ellipsoid shapes may also be estimated from an L/F plot (Fig. 7.4), which plots the mean linear element of the magnetic fabric of a block against the mean planar element. The data presented in Fig. 7.4 illustrate standard plots for AMS data and represent two foliated cumulate samples from Unit 9 of the Rum Eastern Layered Intrusion (ELI; Rum Layered Suite). The data in Fig. 7.4a are from an olivine-gabbro at the topmost portion of

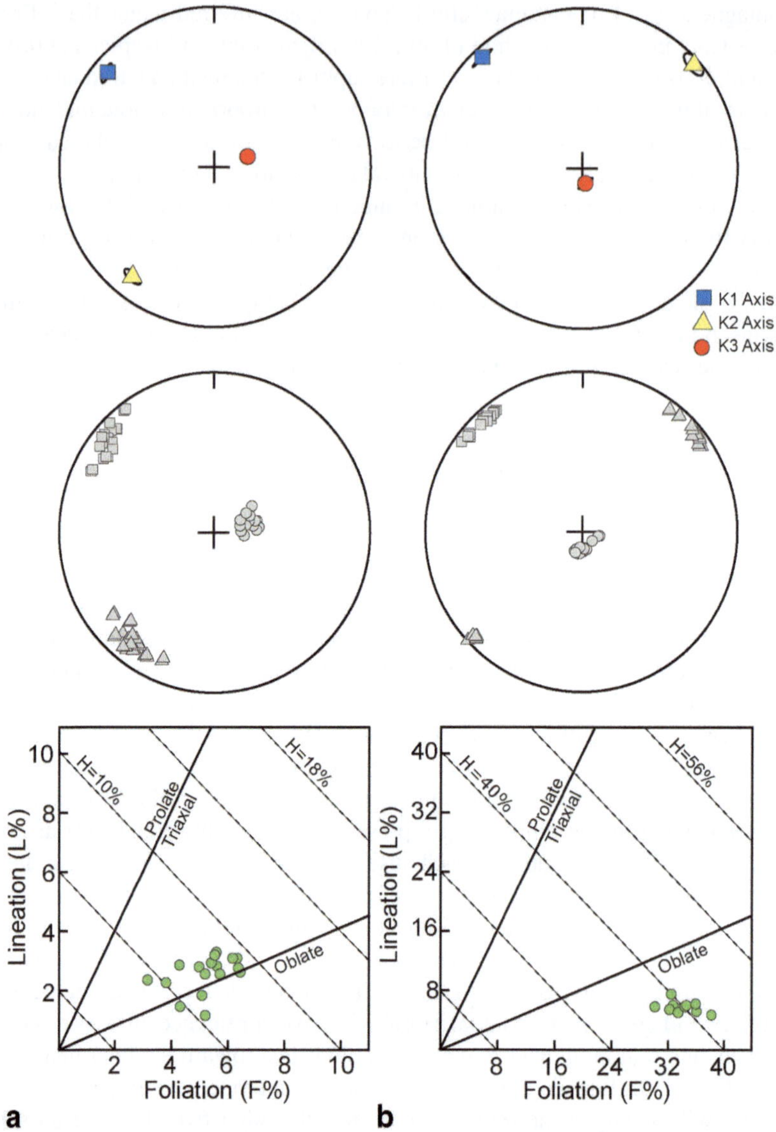

Fig. 7.4 The (**a**) and (**b**) columns illustrate the AMS fabric data for two Rum Eastern Layered Intrusion cumulate samples. The first row contains the normalised fabric data, with 95% confidence ellipses (note that these may be smaller than the actual susceptibility axis symbol). The second row contains the unnormalised data from all of the sample specimens for comparison. The final row shows how the corresponding AMS magnitude parameters for each sample (*green* filled *circles*; *L*, *F*, *H*) manifest on an *L-F* plot. Note the x-axis is differently scaled in each instance. The (**a**) column shows block sample data for a Unit 9 olivine gabbro (from 17 specimens), whilst the (**b**) column represents a Unit 9 troctolite (from 12 specimens), as further detailed in the text

Unit 9 on the north slopes of Hallival, whilst the data in Fig. 7.4b are from a laminated troctolite immediately (several m) underneath this. In both cases the AMS foliations show a good match to the macroscopic mineral lamination; the latter predominantly dips very gently to the WSW in this area, though local variations occur (Fig. 7.4b). Magnetic lineations are well defined in both cases. These are oriented sub-parallel to the strike of magnetic foliation (and therefore mineral lamination), a conclusion reached by O'Driscoll et al. (2007a) as well. Values for K_{mean} range between $0.97-2.5 \times 10^{-3}$, indicating a ferrimagnetic control on the AMS fabrics. This observation is supported by previously published data that suggests the magnetic susceptibility of the Rum cumulates is controlled by olivine-hosted magnetite dendrites (Housden et al. 1996). A final feature of these datasets worth highlighting is the relative differences in H quantified by the two AMS fabrics. Note that the magnetic anisotropy strength is considerably greater in the Unit 10 troctolite (~40%), this sample is also more oblate, than in the olivine-gabbro, where H=4–10%.

Image Analysis and Quantitative Textural Techniques

Where an obvious preferred orientation (foliation or lineation) of crystals in a rock occurs, there are a number of techniques that may support AMS in quantifying the strength and type of fabric preserved. Such techniques can be applied to silicate (or oxide) mineral fabrics and provide useful checks on the veracity and the meaning of an AMS dataset. For example, the collection of grain orientation data from three orthogonally cut thin-sections was utilised by Launeau and Robin (1996) and Launeau and Cruden (1998) for their Intercept method. The approach involves identifying the number and orientation of intersections between the boundaries of the mineral phases to produce a 'rose of intercept counts, which is a polar histogram of the number of counts of a given phase versus the angle of analysis' (Launeau and Robin 1996). Inherently, the minimum count direction corresponds to the elongation of the phase boundaries. The intercept method examines only crystal boundaries, thus it is not necessary for single crystals to be separated from each other for the 2D shape preferred orientation (SPO) of a given mineral phase in a sample to be calculated. Launeau and Cruden (1998) and O'Driscoll et al. (2008) provide examples of an integration of the AMS and Intercept method. Measurement of crystals from differently oriented thin sections was also used by Gee et al. (2004) to calculate a cumulative distribution function for the orientation of mineral phases in a sample. This technique is time-consuming, but produces second-rank tensor data with a precision approaching that of the AMS technique.

Electron back-scatter diffraction (EBSD) is a technique that can measure the lattice-preferred orientation of a mineral fabric. Measurements are carried out within a scanning electron microscope (SEM) and data can be collected from three orthogonal sections of a sample to assess lineation and foliation fabric components. For example, EBSD was combined with AMS by Feinberg et al. (2006) to show that a plagioclase lamination from gabbronorites of the Bushveld Complex controlled the distribution of oxide minerals within the lamination plane. Cheadle et al. (2007) and Gee et al. (2009) also report a combination of AMS and EBSD data from > 800 cores of the lowermost 500 m of the Jurassic Dufek layered intrusion (Antarctica), in order to link the magnetic fabric to the silicate mineral texture. Utilisation of EBSD data allowed these workers to determine separate silicate mineral lineation data for plagioclase, clinopyroxene and orthopyroxene. The orthopyroxene lineation in particular showed an unambiguous correlation with the magnetic lineation, whilst the plagioclase and clinopyroxene lineations could be linked to the AMS, though less convincingly (Gee et al. 2009).

A number of useful quantitative textural techniques that are not directly related to the measurement of mineral alignment also exist. For example, the crystal size distribution (CSD) method (see also the example of combining AMS and CSD data for the Stillwater cumulates below) provides a quantitative measure of the number of crystals of a given mineral phase, per unit volume within a series of defined size intervals (review in Higgins 2006 and also the chapter by M.D. Higgins in this volume). Kinking or curvature in a CSD profile is usually attributed to different processes that modify or add to the original crystal population (Marsh 1998; Higgins 2006). Specific examples that are relevant to cumulate formation include crystal accumulation (Marsh 1998), crystal mush compaction (Boorman et al. 2004), and post-nucleation crystal coarsening caused by annealing or Ostwald ripening (Boorman et al. 2004; Higgins 2006). O'Driscoll et al. (2008) combined CSDs of cumulus plagioclase crystals from laminated cumulates of the Ardnamurchan and Skye layered intrusions (NW Scotland) with AMS data from the same rocks. The resulting data illustrated that postcumulus textural coarsening may have enhanced the degree of foliation in each case, as progressively shallower CSD slopes were observed with increasing magnetic fabric (H) strength. The measurement of crystal spatial distributions (Big R values; Jerram et al. 2003) has also been used to assess textural coarsening and compaction in cumulates, though this approach has not yet been integrated with an AMS fabric study.

Magma Emplacement in Mafic Sheet Intrusions Versus Lopoliths

Intrusion of mafic magma occurs through a wide spectrum of emplacement styles, from localised sub-volcanic sheet intrusion in igneous central complexes to the intrusion of swarms of saucer-shaped sills in well-layered sedimentary basins (Hutton 2009). Such bodies may develop magmatic layering to a certain degree, depending on a host of factors including sheet thickness and magma flow regime (Marsh 2004). However, the world's classic anorogenic layered mafic-ultramafic intrusions (e.g., the Bushveld and Stillwater Complexes, the Skaergaard Intrusion and the Rum Layered Suite) have been interpreted as lopolithic in geometry (Wager and Brown 1968). The nature and significance of AMS fabrics from examples of both types of intrusion are considered in more detail below.

Magma Flow in Mafic Sheet Intrusions and Saucer-Shaped Sills: An AMS Perspective

It is not the aim of this contribution to provide an extensive review of sheet intrusion emplacement, but as some excellent examples of layered intrusions are known to be sill-like in morphology, it is worth briefly considering examples of published AMS data from some of these bodies. The majority of mafic sheets are intruded at the 1–10's of metres thickness scale. Within sheet intrusions, AMS fabrics, particularly the magnetic lineation, are predominantly employed to evaluate magma flow patterns. For example, Glen et al. (1997) carried out a detailed AMS study of 28 lava flows and three associated sills from the ~132 Ma Paraná-Etendeka igneous province and interpreted the results as reflecting a crystal shape-preferred magma flow fabric carried by phenocrysts in the fine-grained mafic rocks. Interestingly, consistent magma flow directions were observed between extrusive and intrusive rocks, an observation that was attributed to the control exerted on flood volcanism by large-scale rift-related crustal structures. Where macroscopic flow indicators can be observed to correlate with magnetic fabrics, magma flow can be inferred with a greater degree of confidence. Liss et al. (2002) compared ropy flow structures at the margins of the Whin Sill (UK) with AMS fabric data and found a close correlation between their orientations. The latter workers also drew attention to an apparent dichotomy in flow regimes between the sill margins and the interior, as reflected in the AMS dataset. In the central portion of the intrusion, AMS lineations traced a consistent but different flow regime to that preserved in the rocks and ropy flow structures of the contact regions. Magmatic (mineral or grainsize) layering is typically only sporadically developed in fine and medium-grained sheet intrusions and usually parallels host-rock contacts. In the ~60 Ma Ben Hiant Dolerite intrusion (Ardnamurchan, NW Scotland), mineral layering is observed in the coarser-grained deeper levels of the body and parallels the magnetic foliation (Magee et al. 2012). However, there are also numerous examples of sheet intrusions where AMS fabrics do not bear an obvious relationship to layering and warrant careful interpretation. Archanjo et al. (2012) utilised AMS fabrics to study the Neoproterozoic Apiaí gabbronorite in south-eastern Brazil. The central question that these workers addressed was whether intrusion occurred before regional tectonic folding of the host rocks or not. The magnetic fabric data were integrated with silicate SPO fabric determinations by using the Intercept method. Archanjo et al. (2012) concluded from their study that the Apiaí body was a horizontal sheet-like intrusion that was probably tilted by the regional deformation.

Saucer-shaped sills occur in sedimentary basins and may be precursors for lopolith intrusion geometries, typically forming from complex networks of closely-spaced sills (Chevallier and Woodford 1999; Malthe-Sørenssen et al. 2004; Hastie et al. 2011). They may also be variably well-layered. Polteau et al. (2008) reported AMS fabric data from the well-known Golden Valley saucer-shaped sill (Karoo Basin, South Africa) and identified magma flow-related macroscopic structures that they correlated with the magnetic fabrics. Magnetic foliations consistently dip away from the centre of the sill, and are linked with magma 'channels' and ropy flow structures that imbricate the magnetic foliation. Polteau et al. (2008) proposed multiple inflation/deflation cycles for the emplacement model of the Golden Valley sill, controlled to some degree by interaction with the surrounding wall-rock (Schofield et al. 2010). The magnetic lineation data in the Golden Valley intrusion is complex, with several apparent controls on lineation development. However, it is large sill intrusions, such as the Insizwa sill (South Africa), that display some of the best layering observed in sheet intrusions. The Insizwa sill is a 25 km diameter, >1 km thick layered mafic intrusion (Ferré et al. 2002; Fig. 7.5) that was emplaced as part of the Karoo Igneous Province. It exhibits layering at multiple scales and a sporadically developed macroscopic magmatic lineation, carried by tabular plagioclase crystals. A gross compositional layering, comprising peridotitic rocks at the sill base and monzodiorites and granodiorites at the sill top is observed, superimposed upon which is a modal layering at the cm-scale. Layering at all scales is reported to dip inward toward the intrusion centre, becoming steeper at the outer margins suggesting that the sill has a saucer-shaped geometry. Ferré et al. (2002) reported that the magnetic lineations matched the plagioclase lineations recorded, and that there was remarkable consistency in AMS lineation throughout all samples analysed (Fig. 7.5). This was interpreted by these workers as the magma flow direction, and aided in the narrowing down of a single magma source toward the SE of the intrusion. However, a disparity of 90° between modal layering and magnetic foliations was measured, attributed to the presence of intermediate magnetic fabrics (i.e., the effective switching of K_2 and K_3 axes) arising from the presence of both SD and MD magnetic grainsizes in the rocks. This important caveat to understanding the AMS of fine-medium-grained mafic rocks was also highlighted by Aubourg et al. (2008) in their study of the Okavango giant dyke swarm (Botswana). Despite this, the latter authors were able to discriminate between lateral and vertical flow in multiple dykes.

AMS Fabrics in Lopoliths: Primary Magma Inflow or Central Subsidence?

Lopolith Geometries and Petrofabrics in the Skaergaard Intrusion

Lopoliths are traditionally viewed as 'funnel-shaped' intrusions, emplaced broadly concordantly to their country rock host with a pronounced concave downward upper surface (roof-zone). Models for lopolithic styles of magma emplacement emphasise the need for intrusion inflation by floor depression to create the space for magma intrusion (Cruden 1998; Bolle et al. 2002), as opposed to the laccolithic-style roof-doming that is more typical of granites. Cruden (1998) proposed cantilever-type syn-to-post emplacement inward tilting of lopolith floors, possibly facilitated by normal faulting close to the site of magma ascent, to explain how lopoliths fill with magma. The important implication of this 'central subsidence' for such intru-

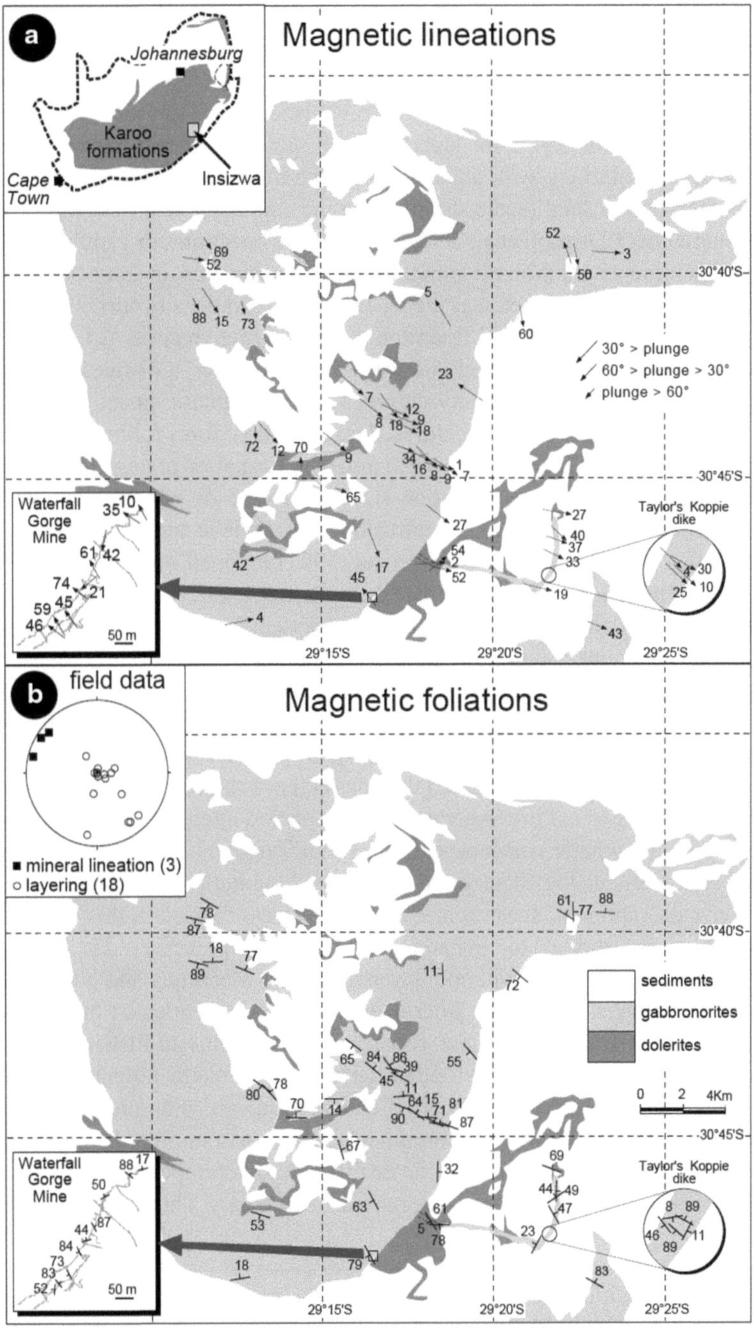

Fig. 7.5 Maps of magnetic fabrics in the Insizwa layered sill, Karoo, South Africa (adapted from Ferré et al. 2002). *Box*: Details of magnetic fabrics at the Waterfall Gorge Mine. *Circle*: Details of magnetic fabric at Taylor's Koppie dike. **a** Magnetic lineations ($K1$ axes). *Inset*: Location of the

sions is that primary magma flow patterns related to initial emplacement may be difficult to discern as they are overprinted by subsidence-related fabrics, although exceptions apparently do exist (see below). Many classic layered mafic-ultramafic lopoliths preserve such evidence of 'central subsidence'. For example, a common feature of such intrusions is for layering at all scales to steepen inwards toward a central focal point along with an increase in the intensity of syn-magmatic deformation, recorded in small-scale structures in the cumulates (e.g., the Rum Layered Suite, Scotland; O'Driscoll et al. 2007a). Magmatic lineations, or visible alignments of cumulus minerals on mineral lamination planes, also exist in some layered intrusions. These may be interpreted as the result of localised density current erosion and deposition at the side-wall and floor zones of a magma chamber (cf. Irvine 1980; Irvine et al. 1998; McBirney and Nicolas 1997). However, a distinction should be made, if possible, between density current-related magmatic lineations and lineations produced as postcumulus features. The latter may form following large-scale central subsidence and cumulate (crystal mush) deformation acting along an overall vertically-oriented extension direction (i.e., solid-state or syn-magmatic-to-solid-state deformation) in many layered intrusions (e.g., Wiebe and Collins 1998; Bolle et al. 2000, 2002). McBirney and Nicolas (1997) mapped out centrally plunging macroscopic lineations throughout the Skaergaard intrusion carried predominantly by cumulus plagioclase crystals and to a lesser extent by pyroxene or aggregates of Fe-Ti oxides contained within the plane of lamination. Lineations are particularly conspicuous in the Lower and Middle Zones of the outer western and northern portions of the intrusion. Elsewhere, magmatic lineations are typically only present in layers that have been locally deformed by drop-stones or trough-shaped layers. McBirney and Nicolas (1997) noted the convergence in their lineation data toward the south-central parts of the Skaergaard Intrusion and interpreted this to signify the interaction of magmatic currents with the solidifying crystal mush on the western steeper side walls of the Skaergaard Intrusion. Magmatic layering and lineations (together with schlieren, shear zones and S/C fabrics) formed in this way were attributed by these authors to 'dynamic' layer-forming processes. These processes were contrasted with 'non-dynamic' layering, which McBirney and Nicolas (1997) argued had occurred toward the interior of the body and reflected the final stages of crystallisation when steep inner contacts and long run-out distances no longer existed for the magmatic crystal-laden currents. It is worth noting that although McBirney and Nicolas (1997) reported abundant inward plunging lineations, the magnitude of plunge of these important structures was not recorded. This may be due to the fact that lineation plunge can be difficult, if not impossible, to measure on flat glacial slabs and 3D exposures are normally necessary. This situation underlines

---

Insizwa sill in South Africa. **b** Magnetic foliations (Planes to *K*3 axes). *Inset*: Macroscopic structures measured in the field. The magnetic fabric data shown here illustrate how AMS ellipsoids can be simplified (reduced) to a magnetic lineation and magnetic foliation component, for the purposes of magnetic fabric interpretation. The style of presentation of the AMS data in this figure should be compared with that in Fig. 7.3

the exciting potential for an AMS fabric study of the classic Skaergaard Intrusion which, provided the magnetic lineations could be linked to the overall rock fabric (note McBirney and Nicolas reported clusters of Fe-Ti oxides carrying the lineation in some instances), would likely resolve these issues.

Insights From Intrusion-Wide AMS Fabric Patterns in Lopoliths

Overall, magnetic fabric analyses have proved very powerful in elucidating magma flow and central subsidence processes in intrusions where layering and magmatic lineations are scarce or absent, or are not easily interpretable alone. An example of a layered mafic-ultramafic intrusion in which primary magma inflow has been inferred through the application of AMS is the ~1096 Ma Sonju Lake layered intrusion (Minnesota, USA; Maes et al. 2007). This study revealed quite oblate AMS ellipsoids throughout the layered sequence, for which the magnetic foliation plane broadly paralleled magmatic layering (and a locally well-developed mineral lamination). Associated magnetic lineations plunge definitively 'inward', and there is an apparent lack of cumulate deformation toward the magnetic lineation focal point. Maes et al. (2007) interpreted these lineations as reflecting a primary magma inflow direction, inferring a feeder zone located toward the southwest of the intrusion. In contrast, other studies have coupled magnetic lineations with chaotic slumping and loss of coherency of layered cumulate toward purported ascent sites in open system magma chambers and argued for subsidence related to periodic fluctuations in the magmastatic pressure (e.g., O'Driscoll et al. 2006, 2007a). For example, the lower parts of the 60 Ma Great Eucrite intrusion (Ardnamurchan, Scotland) comprise a coarse-grained orthocumulate gabbro in which layering is rare and mineral fabrics are absent. In this intrusion, considered for decades to be an archetypal mafic ring-dyke complex related to central (cauldron)-style collapse (Richey and Thomas 1930), AMS lineations generally plunge inward (O'Driscoll et al. 2006). Combined with localised inward dipping magmatic layering and rare small-scale structures in the layered rocks consistent with cumulate deformation on inward-dipping slopes, the intrusion was re-interpreted as a broadly lopolithic body in which central 'sagging' had played an important role. This interpretation was re-emphasised by O'Driscoll et al. (2008), who carried out an AMS study of strongly laminated (fluxion) gabbros at the topmost part of the Great Eucrite. The magnetic foliations measured were ubiquitously oriented parallel to the mineral lamination, the latter carried predominantly by cumulus plagioclase crystals (Fig. 7.6). Although no visible lineation was observed on the lamination planes, magnetic lineations occurred in the plane of lamination and plunged down-dip in almost all cases, mapping out a confocal arrangement that was sympathetic to the magnetic lineations reported by O'Driscoll et al. (2006) for the lower part of the intrusion. In a similar study, the ~60 Ma Carlingford Later Gabbros (NE Ireland), also originally believed to be a mafic ring-dyke (Le Bas 1960), were re-interpreted by O'Driscoll (2006) as a lopolith. The Later Gabbros comprise coarse-grained orthocumulate gabbro that only very sparsely exhibits any internal structure, although a large-scale macro-rhythmic

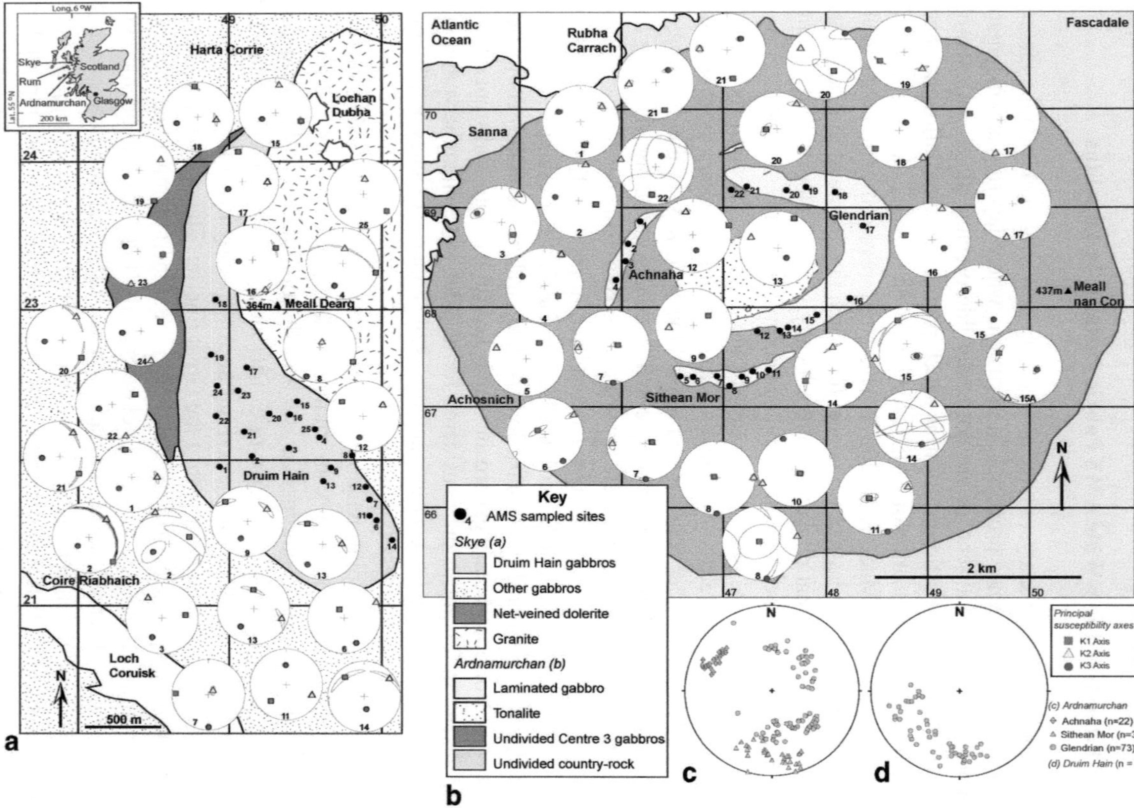

Fig. 7.6 AMS data for (**a**) the Druim Hain layered gabbros and (**b**) the Ardnamurchan Centre 3 laminated gabbros. Part (**a**) contains an inset map of Scotland, showing the location of the Skye, Rum and Ardnamurchan Palaeogene igneous centres. Data are plotted on equal-area projections and parts (**a**) and (**b**) are

cryptic layering was identified by Le Bas (1960). In the northern portion of the intrusion where folded and slumped magmatic layering occurs locally, magnetic lineations plunge inward. For the bulk of the Carlingford intrusion, O'Driscoll (2006) did not recognise consistent AMS fabric patterns in the dataset. However, it is possible (in light of more recent studies of other felsic and mafic igneous complexes in the region; Stevenson et al. 2007a; Stevenson et al. 2008; Stevenson and Bennett 2011) that a broadly NE-directed magnetic lineation represents a regional filling direction for not just the Carlingford intrusion but other Palaeogene igneous centres in NE Ireland too (e.g., Slieve Gullion and the Mourne Mountains; Fig. 7.7). This intriguing model is an example of an interpretation of AMS data that has linked multiple (coeval) igneous centres through lateral magma flow at a regional scale.

A very interesting study was presented by Cowan (1999) on the magnetic fabrics of the Sudbury Igneous Complex, in which the question of whether the intrusion was emplaced as a lopolith or as a flat sheet with subsequent (tectonic) folding was addressed. Sections of undeformed cumulate were sampled for AMS and it was revealed that magnetic foliation and lineation correlated well with a visible mineral lamination and associated magmatic (silicate) lineation. Together with a lack of evidence for solid-state strain, the effects of regional deformation were deduced to have been minimal, and Cowan (1999) concluded that Sudbury represented an originally lopolithic-shaped intrusion, rather than a (horizontal) melt sheet.

Cumulate Stratigraphy and AMS Fabrics in Layered Intrusions

Finally, it is worth making note of a corollary, albeit no less powerful, application of the AMS technique to sequences of cumulate in layered mafic-ultramafic intrusions. A number of workers have attempted to correlate AMS magnitude parameters with cumulate (lithology) stratigraphy (e.g., Ashwal et al. 2005; Maes et al. 2008; Ferré et al. 2009). Maes et al. (2008) and Ferré et al. (2009) reported AMS data from boreholes through the Insizwa sill, the Great Dyke (Zimbabwe) and the Bushveld Complex (South Africa). In all cases, distinct variations in AMS parameters were observed to correlate with layering. Titanomagnetite is the dominant control on magnetic susceptibility in each intrusion. As might be expected, magnetic susceptibility is observed to be principally a function of composition, but alone is not sufficient to recognise new batches of magma input in open system intrusions. However, variations in K_{mean} together with other AMS parameters may be indicative of new magma replenishment in such bodies. Maes et al. (2008) further showed the

adapted from Figs. 7.7 and 7.8 of O'Driscoll et al. (2008). Note the orientation of magnetic lineation and foliation data. It is useful to contrast the data presentation style here with that in Figs. 7.4 and 7.5. In addition to illustrating magnetic fabric (AMS ellipsoid) orientation data, the equal area projections in this figure allow the shape of the ellipsoid to be deduced, by examination of the three susceptibility axes and their associated confidence ellipses (see also Fig. 7.4). **c** and **d** Equal-area projections of poles to mineral lamination planes from the Ardnamurchan and Druim Hain gabbros, respectively, with (n) the number of measurements presented. Note the similarity of the data in each case to the magnetic foliation orientations for each of the intrusions

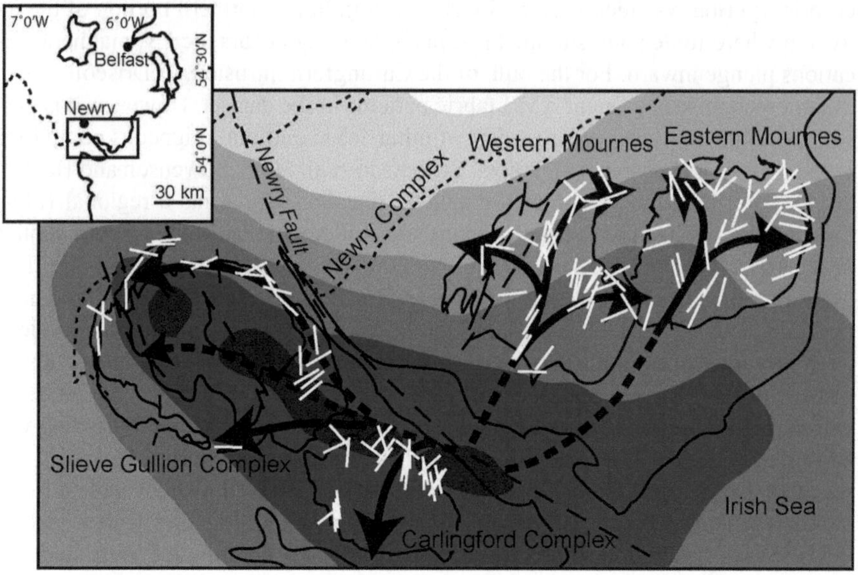

Fig. 7.7 Sketch map showing the outline geology of the four main Palaeogene igneous centres in NE Ireland (inset for location). The Caledonian Newry Complex is outlined with a *dotted line*. Main faults are shown with *dashed lines*. Depicted here are linear data from various AMS studies represented with a *white* bar highlighting the trend of each lineation from each study (plunge is not indicated). All lineation data, where a lineation may confidently be derived, have been used. This is not an exhaustive fabric map and detailed analysis of fabric data may be found in the relevant sources. What this diagram shows is the suggested relationship between the magma flow into each complex and a common source. Bouguer gravity anomaly data from Reay (2004) has been redrawn in greyscale with contours in increments of 10 mGal from 0–50, 50 being darkest. Gravity data highlights the confluence of our suggested common source and a gravity high which may represent a deeply penetrating feeder zone. The gravity high ridge extends westward beneath Slieve Gullion suggesting large volumes of mafic igneous rocks extend in this direction. Outline geology from Cooper and Johnson (2004). AMS data are taken from O'Driscoll (2006) for the Carlingford Complex, Stevenson et al. (2007a) for the Eastern Mourne pluton, Stevenson et al. (2008) for the Slieve Gullion Complex, Stevenson and Bennett (2011) for the Western Mourne pluton

potential of magnetic borehole stratigraphy in the Insizwa sill via the correlation between laterally discontinuous compositional units. At a more fundamental level, the close stratigraphic correlations (cumulate composition versus AMS) observed by Ferré et al. (2009) indicated that the bulk of the Fe-Ti oxides controlling the AMS is magmatic in origin. Both Ashwal et al. (2005) and Ferré et al. (2009) highlighted the way in which magnetic susceptibility dramatically increases at the Main Zone-Upper Zone junction in the Bushveld Complex. Below this layer, norites and gabbronorites dominate with relatively consistent (low) susceptibilities of the order of ≤ 0.05 SI, whereas above this level, susceptibilities vary considerably (0.1–5 SI) in anorthosites, ferrogabbros and magnetitites. The sympathetic variation of magnetic susceptibility and lithology was recorded to a vertical resolution of <5–10 cm by Ashwal et al. (2005). Interestingly, the latter authors found that anorthosite adcumulates commonly had higher susceptibilities than orthocumulates above and below, due to the presence of micro-inclusions of magnetite in the former.

Insights From Magnetic Fabrics Into Cumulate Textural Evolution in Layered Intrusions

Previous Work

Layer-parallel mineral lamination (foliation), referred to as 'igneous lamination' by Wager and Brown (1968) is a prominent texture in some of the world's best known layered mafic-ultramafic intrusions (e.g., the Bushveld and Stillwater Complexes, the Skaergaard Intrusion and the Rum Layered Suite). Such mineral lamination is usually carried by tabular or platy cumulus minerals. The processes responsible for the development of these fabrics, however, are still not well understood. In some cumulate sequences, the association of sedimentary or syn-magmatic deformation structures with layering and mineral lamination suggests dynamic processes were involved in lamination formation. The existence of a magmatic lineation within lamination planes is generally diagnostic of an origin through the motion of magma (± crystals). Still, even when present, such lineations remain difficult to measure accurately in the field and few studies have systematically mapped lineation patterns in cumulates over the scale of a whole layered intrusion (e.g., McBirney and Nicolas 1997). Magmatic lineation and foliation are more commonly observed and measured in distinctly limited sequences of cumulate in a given layered intrusion. For example, Quadling and Cawthorn (1994) reported detailed microstructural analyses of a layered sequence of laminated melagabbronorite and leucogabbronorite (10–20 m thick) in the Main Zone of the eastern limb of the Bushveld Complex. These authors noted a very interesting configuration of magmatic lineation through this sequence of layers. The leucocratic layers displayed a lineation, carried by plagioclase laths, which dominantly plunged down-dip on lamination planes. However, in the mafic layers, plagioclase lineation is parallel to strike, an effect that is particularly noted at the bases of the mafic layers. Quadling and Cawthorn (1994) attributed these microstructures to a complex fluid dynamic environment in which plagioclase in the leucocratic layers was deposited directly from flowing crystal laden currents, whilst the plagioclase in mafic layers was not only subject to deposition but was 'rolled' along the floor to some degree as well, in order to explain the strike-parallel lineation. Magmatic lineations of olivine, parallel to dip direction on lamination planes, were optically quantified in cumulates of the Lower Zone of the Kiglapait Intrusion by Belkin (1983) and were also attributed by him to crystal settling from a flowing magma.

The AMS technique has been applied as a textural tool to the study of layered and foliated gabbros in ophiolites (e.g., Abelson et al. 2001, 2002; Kawamura et al. 2005), which may have undergone larger shear strains than rocks from layered intrusions (Nicolas 1992). In oceanic gabbros, the AMS is typically interpreted as a proxy for shear strain at high temperature. The application of the AMS technique to detailed study of the microstructure of cumulate in layered intrusions, whether a visible fabric is present or not, can reveal fascinating insight into texture development in such rocks. The state-of-the-art in quantitative textural studies of layered intrusions (see this volume) emphasises the need to understand the supra-to-subsolidus cooling and solidification of the crystal mush. Intimately linked to this problem are issues of quantifying crystal mush rheology and the effect increasingly smaller volumes of intercumulus melt have on the cumulus framework as the solidus is approached, as well as the thicknesses which the crystal mush pile reaches. Concerning the latter, mush thicknesses from 10's of metres up to 100's of metres are currently suggested for the Skaergaard Intrusion in the literature (Tegner et al. 2009; McKenzie 2011; Namur et al. 2013), highlighting how poorly this issue is resolved. AMS is a very sensitive indicator of fabric, and combined with careful petrographic observations would be a very useful tool to investigate these issues, via high spatial resolution correlations of AMS fabric shape and intensity variations with cumulate texture throughout stratigraphic height.

A fundamental question regarding planar and linear alignments in layered cumulate involves the extent to which they are present when not visible. In such cases, what are the processes that inhibit fabric formation? Another issue that might be raised surrounds the relative importance of primary versus secondary cumulus processes in developing mineral lamination in the absence of a magmatic lineation. In an attempt to address these matters and to assess what information could be ascertained on textural development from the AMS method as applied to laminated cumulates, O'Driscoll et al. (2008) measured the AMS of two sequences of layered gabbro from British Palaeogene Igneous Province (BPIP; NW Scotland). The first locality was the Ardnamurchan 'fluxion' gabbros and the second was the Druim Hain gabbros, from the Isle of Skye. Both sequences show good development of mineral lamination, and in the former, Fortey (1980) quantitatively measured a magmatic lineation in one sample which he recorded as plunging down-dip on the plane of lamination. In the Ardnamurchan and Druim Hain gabbros, the AMS dataset revealed magnetic foliations that showed an excellent correlation with the silicate lamination plane (Figs. 7.6a and b). For the Ardnamurchan gabbros, the observation of Fortey (1980) was supported by the magnetic lineations, which consistently plunged down-dip on lamination planes defining a confocal arrangement around the intrusion. The AMS ellipsoids are predominantly triaxial in shape, indicating that the planar and linear elements of the fabric were for the most part equal in strength (Fig. 7.6b). By contrast, the Druim Hain gabbros exhibit oblate AMS ellipsoids that do not reveal a discernible intrusion-scale pattern in magnetic lineations at the scale of sampling (Fig. 7.6a), although the magnetic lineations were locally well developed. O'Driscoll et al. (2008) noted that whilst magnetite undoubtedly carried the magnetic fabrics in both cases, its intercumulus textural status in the Druim Hain rocks might be the reason why, compared to the Ardnamurchan fluxion gabbros, no lineation was observed. In particular, it was suggested that subtle postcumulus textural readjustment either above or below the solidus might have overprinted weak magnetic lineations, especially if the recrystallisation strengthened the planar component of the fabric. In both cases, quantitative shape-preferred orientation analyses (intercept counting; Launeau and Robin 1996) were carried out on thin sections of the gabbros and confirmed that the cumulus silicate framework controls the AMS fabrics. In a similar study of laminated troctolite from the Eastern Layered Intrusion (ELI) of the Rum Layered Suite, O'Driscoll et al. (2007a) recorded AMS foliations that were a good match for the plagioclase lamination that occurs throughout the upper portions of the Units 8, 9 and 10 cumulates (see also Fig. 7.4). Magnetic lineations, though weak, are fairly consistently oriented parallel to the strike of foliation planes. This result matches an earlier quantitative petrofabric study by Brothers (1964), which reported plagioclase lineation oriented perpendicular to dip direction in a number of Rum troctolite samples, although such lineation was almost never macroscopically obvious. Given the pervasive structural evidence throughout the upper part of the ELI, these lineations were attributed to intrusion-wide slumping of the cumulate back down into the magma feeder conduit, the Long Loch Fault.

Application of the AMS Technique to Unusual Cumulate Textures

Two further examples of the application of AMS as a petrofabric tool in studying some classic cumulate textures are detailed below. Both examples deal with cumulates in which mineral lamination is not the primary texture of interest and each serves to highlight some of the strengths and weaknesses of the application of AMS to cumulates. The first example is from the classic 'crescumulate' harrisite of the Western Layered Intrusion (WLI) of the Rum Layered Suite (Fig. 7.8) and the second is the 'inch-scale' layering of the Lower Banded Series in the Stillwater Complex (USA; Fig. 7.9).

The Rum Layered Suite Harrisitic Peridotites (Crescumulates)

Harrisite is a coarse-grained mafic or ultramafic plutonic rock comprising $\geq$25 % olivine crystals that have a dendritic or skeletal habit (e.g., Fig. 7.8a). Elongated olivines are frequently oriented at high angles to layer planes, an observation that gave rise to the terms 'crescumulate' and 'comb-layering' (Wager et al. 1960; Taubeneck and Poldervaart 1960; Donaldson 1977; O'Driscoll et al. 2007b; see also Ch. 2 of this volume for a fuller description on the morphology and mode of formation of the Rum Layered Suite harrisites). Thus, although harristic textures may be characterised by a pronounced magmatic lineation, this lineation is often oriented orthogonal to the plane of layering. AMS fabrics are presented for two samples of harrisite and a sample of fine-grained granular-textured peridotite from a ~60 cm thick sequence, at the Ard Mheall locality of the WLI. In the granular-textured peridotite, coarse harrisitic olivine 'clasts' are imbricated and possess a lineation that plunges gently (parallel to the dip of the layer plane) toward the centre of the Rum intrusion (Fig. 7.8b). No macroscopic foliation is visible in the granular-textured

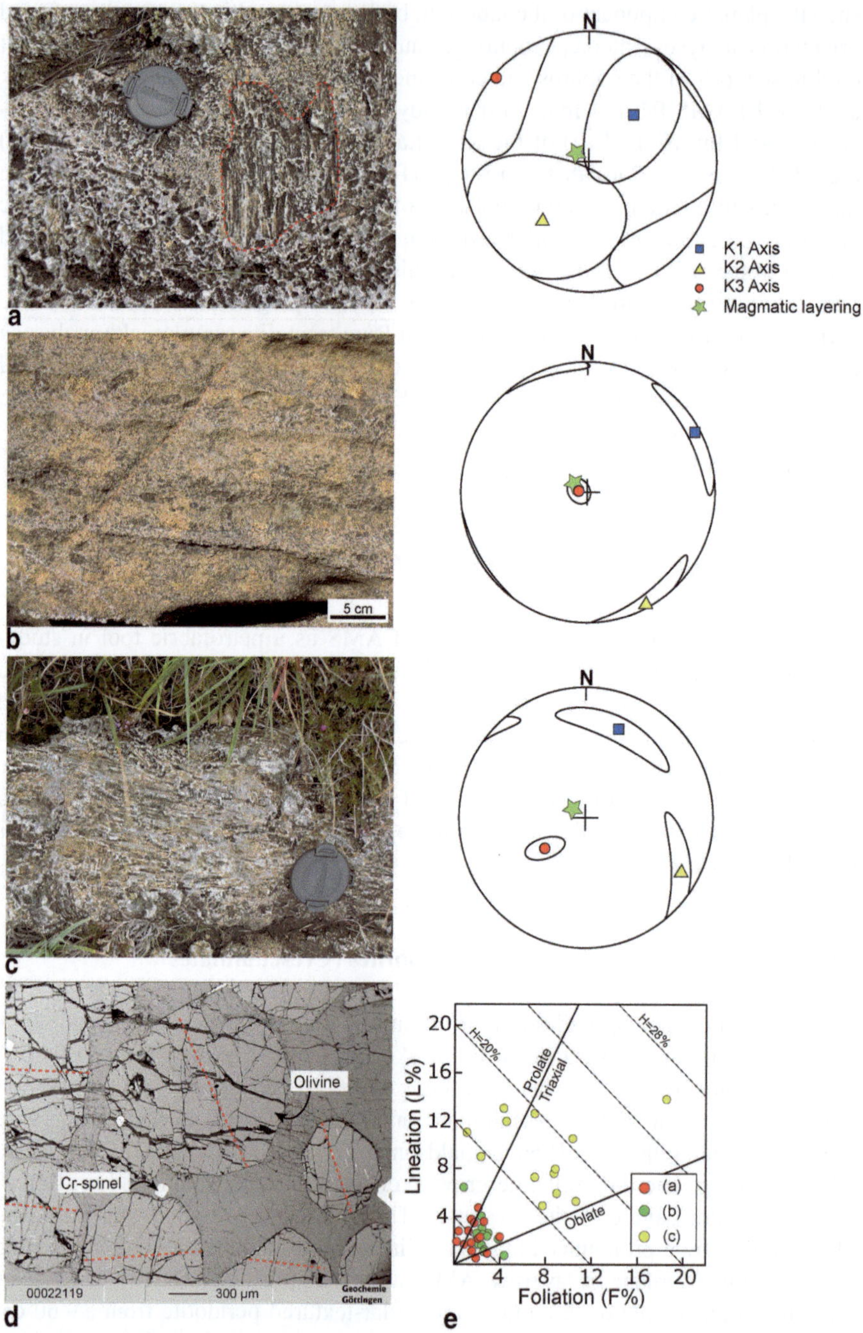

Fig. 7.8 AMS fabrics in harrisitic and granular-textured peridotite from the Western Layered Intrusion, Isle of Rum, Scotland. See inset in Fig. 7.6 for the regional location of Rum. **a**, **b** and **c** illustrate field photographs of the cumulate texture at each outcrop, with adjacent normalised AMS fabrics for each sample discussed, illustrated by equal-area projection. The data in the equal

peridotite. The principal fabric in the harrisite is controlled by the growth attitude of the skeletal olivine crystals, but is not consistent at the outcrop scale (Fig. 7.8a), or from one layer to the next. Of the harrisite samples, one shows a fairly well-defined AMS foliation with corresponding lineation (Fig. 7.8c), whilst the other is quite poorly developed (Fig. 7.8a) with overlapping confidence ellipses. In the case of the former, no relationship is apparent between the magnetic foliation measured and the attitude of the layering of the surrounding rocks, which strikes NW-SE and dips gently to the SE. The sample itself is very coarse-grained, and likely predominantly composed of a single ($\geq$20 cm long) crystal (Fig. 7.8c). The AMS fabric measured is probably therefore the crystallographic magnetic fabric of a single olivine crystal (see below; Fig. 7.8d). In the case of the fabric illustrated in Fig. 7.8a, numerous skeletal and tabular-shaped olivine crystals between 0.5 and 2 cm long occur in various orientations throughout the sample, explaining why a single well-developed AMS fabric is not present. The AMS fabric for the granular-textured peridotite reveals that the magnetic lineation is oriented approximately down-dip on the magnetic foliation, broadly paralleling the flow direction suggested by the imbricated harrisitic olivine fragments (Fig. 7.8b). The magnetic foliation parallels that of the magmatic layering, strengthening the case for the existence of a mineral lamination. The L-F plot reveals that the AMS fabric of the single harrisitic olivine crystal has variable H, ranging between 8 and 32 % (Fig. 7.8e). The shape of the magnetic fabric from the sample in Fig. 7.8c is triaxial to strongly prolate, perhaps signifying a control on the AMS by a single, highly elongate olivine crystal. By contrast, the granular-textured peridotite contains a much weaker (H $\approx$4 %) but more consistent fabric. The bulk magnetic susceptibilities (K_{mean} = 4–14 $\times$ 10^{-3} SI) reveal that magnetite is likely the dominant mineral in controlling the magnetic fabrics; olivine in all three samples contains abundant crystallographically-constrained magnetite dendrites (e.g., Fig. 7.8d). The AMS fabrics from the granular-textured peridotite therefore reflect a subtle olivine crystallographic fabric, the planar element of which is probably parallel to intracrystalline 010 planes. A similar suggestion was made

area plots show block average data for 10–14 specimens normalised by specimen mean susceptibility, on the assumption that the specimens from a block represent a homogenous multi-normal population. The specimen tensor-averaging process allows within-block variability to be characterised through calculation of 95 % confidence limits (cf. Jelínek 1978; Owens 2000). The *green star* on each equal area plot gives an indication of the orientation (pole to the plane) of local magmatic layering. The centre of the Rum Layered Suite is roughly to the ENE on each of the equal area plots. **a** and **c** are harrisite samples, whilst **b** is a granular- textured peridotite. A single dendritic olivine crystal is highlighted in *red* in **a**. Note the reworked and imbricated harrisitic olivine clasts in **b**, offering an independent opportunity to measure *magmatic* lineation in the sample, which corresponds well with (parallels) the AMS lineation. **d** Back-scattered electron micrograph showing some of the finer-grained rounded olivine crystals that comprise the harrisite sample from **a** *above*. Note the presence of abundant bright crystallographically-constrained magnetite dendrites within the olivine crystals. The larger bright equant crystals adjacent to the olivines are Cr-spinels and the *light-grey* intercumulus mineral is plagioclase. The *dashed red* lines in the olivine crystals highlight the orientation of the magnetite dendrites for each: note the apparently random orientation even at this scale of observation. **e** L-F plot illustrating the strength and shape of the AMS ellipsoid for each of the three samples illustrated *above* (inset at *bottom-right* of plot gives key) for Rum peridotites in the ELI by O'Driscoll et al. (2007a) and for mantle olivine AMS fabrics from the Twin Sisters (Washington State, USA) dunites (Ferré et al. 2005). Given the evidence for imbrication of elongate harrisite fragments in the granular-textured peridotite, it is probable that its mineral lamination is a primary depositional fabric related to sedimentary reworking of the crystal mush. The intra-crystalline magnetite dendrites are late-stage, formed as the solidified crystal mush cools (e.g., Putnis 1979; Housden et al. 1996) during which time they inherit the pre-existing olivine fabric. Tabular-shaped olivine crystals in peridotites throughout the Rum Layered Suite have been reported by several workers to define a visible localised planar fabric (Brothers 1964; Worrell 2002). The AMS data presented for the Rum ELI cumulates by O'Driscoll et al. (2007a) as well as those illustrated here highlight that subtle fabrics carried by relatively equant olivines may be more wide-spread than previously considered and highlight the potential of the AMS technique in mapping these out.

The Stillwater Complex 'Inch-Scale' Layering

The inch-scale 'doublets' exposed on the Stillwater Complex Mountain View Mine Road (southern Montana, USA) are one of the classic examples of fine-scale rhythmic layering, traditionally attributed to *in situ* crystallisation (cf. Boudreau 1987, 1995, 2011). A sequence of anorthosite in the Lower Banded Zone, several 10's of metres thick, contains multiple pairs of pyroxenite (dominantly orthopyroxene) layers that become more widely spaced with stratigraphic height through the section (Fig. 7.9a), until the doublets become indistinct and give way to a mottled bronzite-plagioclase cumulate. Boudreau (1987) developed numerical models that explained the development of the inch-scale layering in terms of solid-state textural adjustment driven by crystal aging. In summary, Boudreau (1987, 1995) explained the pattern formation as a style of Liesegang banding, where nucleation of the constituent crystals had little or no influence on the eventual layering; instead, layering was imposed on a relatively structureless crystal pile at a postcumulus stage. Hence, crystal aging, or the continued growth of large crystals and consequent dissolution of smaller crystals, controlled by vertically oriented thermal or compositional gradients, drove layer formation. Boudreau (1987) noted that within the plane of layering, pyroxene crystals are not randomly oriented, but are arranged in polygonal patterns. It was also noted that the spaces between pyroxenite layers and the mean plagioclase grainsize both increase as the overall modal abundance of pyroxene decreases, and a positive relationship is also observed between the spacing separating any two layers forming a doublet (α) and the space between sets of doublets (β) (Boudreau 1987, 2011). Figures 7.9b, c and d illustrate the AMS data for three samples of inch-scale layered anorthosite collected along the Mountain View Mine Road. These are block samples that were subsequently drilled in the laboratory to produce cores for AMS analysis. Drilling was conducted perpendicular to the plane of layering as deduced from block surfaces, i.e., on the AMS equal area plots, layering is oriented approximately horizontally and therefore denoted by the outer

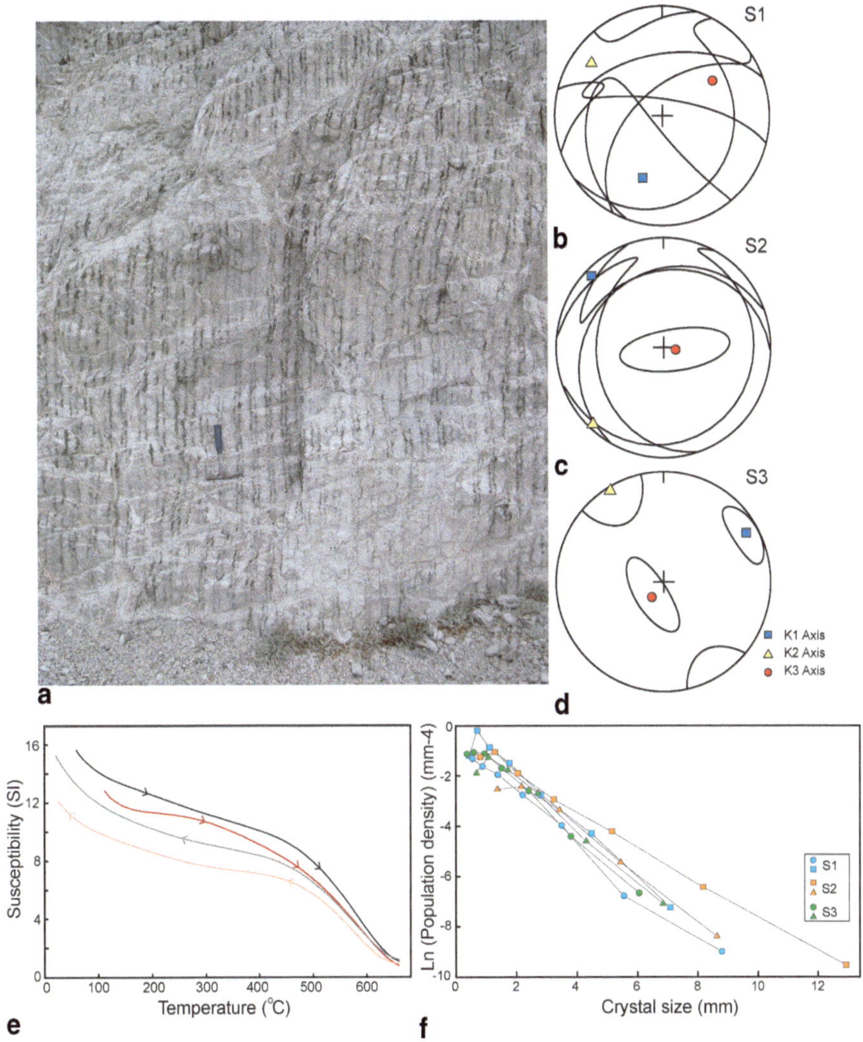

Fig. 7.9 **a** Field photograph of an exposure of the Stillwater Complex inch-scale layering outcrops exposed on the Mountain View Mine Road (geological hammer *left* of centre gives scale). Way up is to the *left* of the image. **b**, **c** and **d** are normalised AMS fabrics for each of the samples S1, S2 and S3, respectively, as described in the text. Between 5 and 9 specimens were averaged to calculate the confidence ellipses illustrated. **e** Temperature dependence of susceptibility (heating and cooling paths *arrowed*) for two samples of Stillwater cumulate. **f** Plagioclase crystal size distribution curves for samples S1, S2 and S3 illustrated *above*. Two CSDs (measured from two different sections cut perpendicular to layering) are presented for each of the three samples. CSDs from the same hand-specimen show broad agreement, with the possible exception of S2, one of whose CSDs extends to shallower slopes. The hand specimen from which this CSD comes is the coarsest-grained of the samples and also exhibits the finest development of layering (see text for details) but does not exhibit a well-developed AMS fabric. The S2 CSD with the shallower slope has been digitised from plagioclase crystals closest to the layering

limit of the equal area projection. The plots labelled S1 and S2 illustrate magnetic fabric data for two samples collected adjacent to one another (at the locality shown in Fig. 7.9a), whilst that labelled S3 is collected ~10 m away (up sequence), where the doublets become more diffuse and irregular. The data are intriguing. Bulk susceptibilities range between 86 and 168×10^{-5} SI for the 3 samples. The sample in Fig. 7.9b reveals no discernible magnetic fabric. That in Fig. 7.9c shows a weakly developed magnetic foliation, oriented parallel to layering, with no corresponding lineation. However, Fig. 7.9d, illustrating the AMS fabric for S3 shows the best developed fabric, with a magnetic foliation that approximately parallels the orientation of the layering and a well-defined magnetic lineation. This is interesting as S3 is the sample with the most poorly-developed layering. Petrographic observation reveals a marked paucity in oxide mineral phases that might dominate the susceptibility, so several T_C experiments were carried out to determine the temperature dependence of susceptibility for S3 (Fig. 7.9e). The data provide support for the bulk susceptibilities quoted above, in that the measurements also indicate low susceptibilities. Both heating curves exhibit broadly negative slopes, suggesting that paramagnetic behaviour is dominant. This implies that the pyroxene, which is intercumulus, probably carries the AMS (plagioclase is known to be diamagnetic; Tarling and Hrouda 1993). In the samples with well-defined layering, some specimens comprise only anorthosite, whilst others contain pyroxenite and anorthosite. However, in S3, where the inch-scale layering becomes quite diffuse, orthopyroxene is distributed much more evenly throughout the rock although its distribution still defines a crude layering. Thus, all of the individual AMS cores contain some of the paramagnetic AMS carrier, whereas in samples S1 and S2, some specimens contain only massive anorthosite. No macroscopic planar or linear alignment of plagioclase crystals was observed in any of the three samples. As described above, measurements of CSDs can be a useful aid to interpreting AMS fabrics in cumulates. The CSDs for S1, S2 and S3 are also illustrated in Fig. 7.9f and add to the observations outlined above, lending support to the crystal aging model of Boudreau (1987, 1995), in that the sample with the best developed doublets (S2) exhibits the coarsest plagioclase grainsize (i.e., shallowest CSD slope).

Outlook and Future Directions

AMS has been regarded as a well-established fabric analysis tool since the early 1990s (Tarling and Hrouda 1993; Bouchez 1997) and has provided invaluable insights into the magmatic flow regime and emplacement history of intrusive igneous rocks from a variety of crustal settings and magma compositional backgrounds. Magnetic fabrics potentially hold key information on static and dynamic fabric-forming processes operating in the magma chambers that produce layered intrusions. A key observation that can be drawn from the discussion in this paper is that static magmatic processes (e.g., in situ crystallisation) should yield dominantly oblate fabrics, whereas dynamic processes should yield prolate fabrics as a consequence of viscous shear of the magma (+crystals). Exceptions do exist, however, where prolate AMS fabrics may occur in directionally oriented crystals that have grown in situ as a response to thermal and/or compositional gradients between one magma and another in a chamber, or between magma and crystal mush (e.g, the example of harrisitic olivine discussed above; see also Huang et al. 2009 and Lundstrom et al. 2011).

We stand to learn much about the magma emplacement dynamics and the geometries of layered mafic-ultramafic intrusions from AMS. As has been indirectly alluded to in this review, much of the current volume of study of layered intrusions (including geochemical, isotopic, quantitative textural, numerical approaches) broadly focus on the formation and evolution of the cumulate during solidification, rather than detailed consideration of how the magma was emplaced in the first instance (e.g., in terms of space constraints and supply rates). AMS may have the potential to further our understanding of lopolithic geometries of igneous intrusions, in much the same way as it has for laccolith development in granite intrusions (cf. Bouchez 1997). The best approach here may be the selection of intrusions that do not appear to have undergone widespread postcumulus textural evolution. Relatively rapidly solidified orthocumulates, such as the example of the Carlingford Later Gabbros referred to above, perhaps represent the best 'target' layered intrusions for this sort of study.

It is also worth noting that layered mafic-ultramafic intrusions are amongst the most enriched geological environments in the world in terms of some varieties of base metal and precious metal mineralisation. Notable examples here include enrichment in some layered intrusions of Cr in chromitite (≥Cr-spinel) seams, Ni and the platinum-group elements (PGE; Os, Ir, Rh, Ru, Pd, Pt). The Bushveld Complex, the largest known layered intrusion on Earth, produces ~75 % of the world's exploited Pt, as well as the majority of the other accessible PGE in addition to ~16 % of global Ni reserves (Naldrett 2004). In the Bushveld Complex, as well as other layered intrusions (e.g., the Rum Layered Suite), the PGE are frequently concentrated in the chromitite seams, indicating that these horizons represent extraordinary magma chamber processes. Studies of the AMS and broader rock magnetic properties of such horizons, as well as the cumulate sequences stratigraphically above and below, may offer significant insight into these processes. Specifically, if coupled with an appropriate understanding of the rock magnetic properties of the different mineral phases (metals, oxides, sulphides and silicates), a combination of AMS and rock magnetic experiments could potentially allow detailed information on properties such as particle grainsize, distribution and modal abundance to be extracted from the rocks. Cr-spinel in layered mafic-ultramafic intrusion chromitites is not typically ferromagnetic (O'Driscoll; Unpublished Data), and it may be Fe-sulphide that is the most important constituent magnetic mineral. Indeed, experimental data exists for the behaviour of pyrrhotite [$Fe_{(1-x)}S$ (x=0–0.17)] that suggests it behaves ferrimagnetically (Dunlop and Özdemir 1997).

The most exciting future direction for AMS in layered intrusions may be its continued application to well-layered and laminated cumulates. Rocks in which planar (and linear) silicate mineral alignments are present offer immense potential for understanding AMS in layered rocks, if accompanied by careful microstructural work. The integration of the EBSD technique with AMS in this regard may turn out to be key; indeed, this is apparently being borne out by ongoing studies of the Antarctic Dufek layered intrusion cumulates (Cheadle et al. 2007; Gee et al. 2009). As also noted earlier, previous reports of incompletely measured (trend without plunge data), though widespread, macroscopic mineral alignments in parts of the Skaergaard Intrusion stratigraphy suggest that AMS measurements would provide real insight into the dynamic evolution of the crystal mush there. Small-scale syn-magmatic linear structures are also abundant in the Skaergaard rocks, so directional information on the kinematic processes involved could be gleaned from multiple sources. More than 80 years after its discovery, research on the Skaergaard intrusion continues to inform and inspire studies on layered cumulates everywhere else, and is at the forefront of our efforts to understand the magma chamber processes that give rise to igneous layering. Recent work on the Skaergaard cumulates by Holness et al. (2011) and Tegner et al. (2009) has begun to quantitatively piece together the postcumulus history of the crystal mush from the standpoint of the evolution and migration of the intercumulus liquid at an intrusion-wide scale. Given the high levels of sensitivity of AMS to such effects, i.e., its ability to quantify mineral fabrics in terms of directionality as well as strength and shape, the technique offers real scope to assist in these specific problems at high spatial resolutions of stratigraphic (vertical and lateral) sampling.

Acknowledgements BO'D acknowledges numerous discussions with John Reavy, Mike Petronis, Henry Emeleus, Colin Donaldson, Mike Cheadle, Dougal Jerram and Michael Higgins on the origin of mineral alignments in cumulate rocks. The authors are grateful to editors Bernard Charlier, Rais Latypov, Olivier Namur and Christian Tegner for their time spent organising this book and for soliciting this review. Bernard Charlier provided excellent editorial support. Olivier Bolle and Bernard Henry provided thoughtful reviews that improved an early manuscript version.

References

Abelson M, Baer G, Agnon A (2001) Evidence from gabbro of the Troodos ophiolite for lateral magma transport along a slow-spreading mid-ocean ridge. Nature 409:72–75

Abelson M, Baer G, Agnon A (2002) Fossil ridge-transform intersection in the Troodos ophiolite: new perspectives from rock magnetism in the gabbro suite and fracture mechanism analysis. Geochem Geophys Geosyst 3(8):1–17. 10.129/2001GC 000245, 002002

Airoldi G, Muirhead JD, Zanella E, White JDL (2012) Emplacement process of Ferrar Dolerite sheets at Allan Hills (South Victoria Land, Antarctica) inferred from magnetic fabric. Geophys J Int 188(3):1046–1060

Améglio L, Vigneresse J.L (1999) Geophysical imaging of the shape of granitic intrusions at depth: a review. Geol Soc Lond Spec Pub 168(1):39–54

Arbaret L, Diot H, Bouchez J-L (1996) Shape fabrics of particles in low concentration suspensions: 2D analogue experiments and application to tiling in magma. J Struct Geol 18(7):941–950

Arbaret L, Diot H, Bouchez J-L, Saint-Blanquat M, Lespinasse P (1997) Analogue 3D simple shear experiments of magmatic biotite subfabric. In: Bouchez J-L, Hutton DHW, Stephens

WE (eds) Granites: from seggregation of melt to emplacement fabrics. Kluwer, Dordrecht, pp 129–143

Arbaret L, Launeau P, Diot H, Sizaret S (2013) Magnetic and shape fabrics of magnetite in simple shear flows. J Volc Geotherm Res 249:25–38

Archanjo CJ, Bouchez J-L, Corsini M, Vauchez A (1994) The Pombal granite pluton: magnetic fabric, emplacement and relationships with the Brasiliano strike-slip setting of NE Brazil (Paraiba State). J Struct Geol 16(3):323–335

Archanjo CJ, Campanha GAC, Salazar CA, Launeau P (2012) Using AMS combined with mineral shape preferred orientation analysis to understand the emplacement fabrics of the Apiaí gabbro-norite (Ribeira Belt, SE Brazil). J Earth Sci (Geol Rundsch) 101:731–745

Ashwal LD, Webb SJ, Knoper MW (2005) Magmatic stratigraphy in the Bushveld Northern Lobe: continuous geophysical and mineralogical data from the 2950 m Bellevue drillcore. S Afr J Geol 108:199–232

Aubourg C, Tshoso G, le Gall B, Bertrand H, Tiercelin J-J, Kampunzu AB, Dyment J, Modisi M (2008) Magma flow revealed by magnetic fabric in the Okavango giant dyke swarm, Karoo igneous province, Northern Botswana. J Volc Geotherm Res 170:247–261

Balsley JR, Buddington AF (1960) Magnetic susceptibility anisotropy and fabric of some Adirondack granites and orthogneisses. Am J Sci A 258:6–20

Belkin HE (1983) Petrofabric analyses of selected rocks from the Kiglapait layered intrusion, Labrador. GSA Abs Prog Ann Mtg:20077

Benn K, Allard B (1989) Preferred mineral orientations related to magmatic flow in ophiolite layered gabbros. J Petrol 30:925–946

Bolle O, Diot H, Duchesne JC (2000) Magnetic fabric and deformation in charnockitic igneous rocks of the Bjerkreim-Sokndal layered intrusion (Rogaland, Southwest Norway). J Struct Geol 22:647–667

Bolle O, Trindade RIF, Bouchez J-L, Duchesne JC (2002) Imaging downward granitic magma transport in the Rogaland Igneous Complex, SW Norway. Terr Nova 14:87–92

Boorman S, Boudreau A, Kruger FJ (2004) The lower zone-critical zone transition of the Bushveld Complex: a quantitative textural study. J Petrol 45:1209–1235

Borradaile GJ, Henry B (1997) Tectonic applications of the magnetic susceptibility and its anisotropy. Earth-Sci Rev 42(1–2):49–93

Bouchez J-L (1997) Granite is never isotropic: an introduction to AMS studies of granitic rocks. In: Bouchez J-L, Hutton DHW, Stephens WE (eds) Granite: from segregation of melt to emplacement fabrics, vol 8. Kluwer, Dordrecht, pp 95–112

Boudreau AE (1987) Pattern formation during crystallization and the formation of fine-scale layering. In: Parsons I (ed) Origins of igneous layering. D. Reidel, Boston, pp 453–471

Boudreau AE (1995) Crystal aging and the formation of fine-scale layering. Miner Petrol 54:55–69

Boudreau AE (2011) The evolution of texture and layering in layered intrusions. Int Geol Rev 53(3–4):330–353

Brothers RN (1964) Petrofabric analysis of Rhum and Skaergaard layered rocks. J Petrol 5:255–274

Cañón-Tapia E (2004) Flow direction and magnetic mineralogy of the lava flows from the central parts of the Peninsula of Baja California, Mexico. Bull Volc 66:431–442

Cawthorn RG (1996) Layered intrusions. Elsevier, Amsterdam, 531 p

Cheadle MJ, Meurer WP, Grimes CB, Gee JS, McCullough BC (2007) Understanding the magmatic construction of the Dufek Complex, Antarctica. EOS Trans AGU 88(52):Fall Meet Suppl Abstract V53D-02

Chevallier L, Woodford A (1999) Morpho-tectonics and mechanism of emplacement of the dolerite rings and sills of the western Karoo, South Africa. S Afr J Geol 102(1):43–54

Cooper MR, Johnston (2004) Palaeogene intrusive igneous rocks (Chapter 15). In: Mitchell WI (ed) The geology of Northern Ireland. Geological Survey of Northern Ireland, Belfast, pp 179–198

Cowan EJ (1999) Magnetic fabric constraints on the initial geometry of the Sudbury igneous complex: a folded sheet or basin shaped igneous body? Tectonophysics 307:135–162

Cruden AR (1998) On the emplacement of tabular granites. J Geol Soc Lon 155:853–862
Donaldson CH (1977) Laboratory duplication of comb layering in the Rhum pluton. Mineral Mag 41:323–336
Dunlop D, Özdemir Ö (1997). Rock magnetism. Cambridge University Press, Cambridge, pp 1–573
Fabian K, Shcherbakov VP, McEnroe SA (2013) Measuring the Curie temperature. Geochem Geophys Geosyst 14(4):947–961
Feinberg JM, Wenk H-W, Scott GR, Renne PR (2006) Preferred orientation and anisotropy of seismic and magnetic properties in gabbronorites from the Bushveld layered intrusion. Tectonophysics 420:345–356
Ferré EC (2002) Theoretical models of intermediate and inverse AMS fabrics. Geophys Res Lett 29(7):1–31. 10.1029/2001GL014367
Ferré E.C, Améglio L (2000) Preserved magnetic fabrics vs annealed microstructures in the syntectonic recrystallised George granite, South Africa. J Struct Geol 22:1199–1219
Ferré EC, Bordarier C, Marsh JS (2002) Magma flow inferred from AMS fabrics in a layered mafic sill, Insizwa, South Africa. Tectonophysics 354:1–23
Ferré EC, Tikoff B, Jackson M (2005) The magnetic anisotropy of mantle peridotites: example from the Twin Sisters dunite, Washington. Tectonophysics 398(3–4):141–166
Ferré EC, Maes SM, Butak KC (2009) The magnetic stratification of layered mafic intrusions: Natural examples and numerical models. Lithos 111(1–2):83–94
Fortey NJ (1980) Petrofabrics of laminated gabbros from the centre 3 igneous complex, Ardnamurchan, Scotland. Mineral Mag 43:989–994
Gaillot P, de Saint-Blanquat M, Bouchez J-L (2006) Effects of magnetic interactions in anisotropy of magnetic susceptibility: models, experiments and implications for igneous rock fabrics quantification. Tectonophysics 418(1–2):3–19
Gee JS, Meurer WP, Selkin PA, Cheadle MJ (2004) Quantifying three-dimensional silicate fabrics in cumulates using cumulative distribution functions. J Petrol 45(10):1983–2009
Gee JS, Lusk MW, Cheadle MJ, Grimes CB, Meurer WP (2009) Origin and significance of magnetic anisotropy in the Dufek layered intrusion. EOS Trans AGU 90(52):Fall Meet Suppl Abstract GP42A-07
Glen JMG, Renne PR, Milner SC, Coe RS (1997) Magma flow inferred from anisotropy of magnetic susceptibility in the coastal Paraná-Etendeka igneous province: evidence for rifting before flood volcanism. Geology 25(12):1131–1134
Grégoire V, Darrozes J, Gaillot P, Nédélec A (1998) Magnetite grain shape fabric and distribution anisotropy vs. rock magnetic fabric: a three-dimensional case study. J Struct Geol 20:937–944
Hargraves RB, Chan CY, Johnson D (1991) Distribution anisotropy: the cause of AMS in igneous rocks? Geophys Res Lett 18:2193–2196
Hargraves RB, Briden JC, Daniels BA (1999) Palaeomagnetism and magnetic fabric in the Freetown Complex, Sierra Leone. Geophys J Int I 36:705–713
Hastie WW, Watkeys MK, Aubourg C (2011) Significance of magnetic and petrofabric in Karoo-feeder dykes, Northern Lebombo. Tectonophysics 513(1–4):96–111
Higgins MD (2006) Quantitative textural measurements in igneous and metamorphic petrology, 1st edn. Cambridge University Press, Cambridge, p 270
Holness MB, Stripp G, Humphreys MCS, Veksler IV, Nielsen TFD, Tegner C (2011) Silicate liquid immiscibility within the crystal mush: Late-stage magmatic microstructures in the Skaergaard intrusion, East Greenland. J Petrol 52:175–222
Hopkinson J (1889) Magnetic properties of alloys of nickel and iron. J Proc R Soc Lon 47:23–24
Housden J, O'Reilly W, Day SJ (1996) Variations in magnetic properties of Unit 10, Eastern Layered Intrusion, Isle of Rum, Scotland: implications for patterns of high temperature hydrothermal alteration. Trans R Soc Edinbur (Earth Sci) 86:91–112
Huang F, Lundstrom CC, Glessner J, Ianno A, Boudreau A, Li J, Ferré EC, Marshak S, DeFrates J (2009) Chemical and isotopic fractionation of wet andesite in a temperature gradient: Experiments and models suggesting a new mechanism of magma differentiation. Geochim Cosmochim Acta 73:729–749

Hutton DHW (2009) Insights into magmatism in volcanic margins: bridge structures and a new mechanism of basic sill emplacement—Theron Mountains, Antarctica. Petrol Geosci 15:269–278

Irvine TN (1980) Magmatic density currents and cumulus processes. Am J Sci 280A:1–58

Irvine TN, Anderson JC, Brooks CK (1998) Included blocks (and blocks within blocks) in the Skaergaard intrusion: geologic relations and the origins of rhythmic modally graded layers. Geol Soc Am Bull 110:1398–1447

Jelínek V (1978) Statistical processing of anisotropy of magnetic susceptibility measured on groups of specimens. Stud Geophys et Geodaet 22:50–62

Jelínek V (1981) Characterization of the magnetic fabric of rocks. Tectonophysics 79:63–67

Jerram DA, Cheadle MJ, Philpotts AR (2003) Quantifying the building blocks of igneous rocks: are clustered crystal frameworks the foundation? J Petrol 44:2033–2051

Kawamura K, Hosono T, Allawati HM, Gogawa Y, Taniguchi H (2005) Origin of layering in cumulate gabbros in the Oman ophiolite: insights from magnetic susceptibility measurements in the Wadi Sadm area. Isl Arc 14:564–570

Khan MA (1962) The anisotropy of magnetic susceptibility of some igneous and metamorphic rocks. J Geophys Res 67:2873–2885

Launeau P, Cruden AR (1998) Magmatic fabric acquisition mechanisms in a syenite: results of a combined anisotropy of magnetic susceptibility and image analysis study. J Geophys Res 103:5067–5089

Launeau P, Robin P-YF (1996) Fabric analysis using the intercept method. Tectonophysics 267:91–119

Le Bas MJ (1960) The petrology of the layered basic rocks of the Carlingford Complex, County Louth. Trans R Soc Edinb 64:169–200

Liss D, Hutton DHW, Owens WH (2002) Ropy flow structures: a neglected indicator of magma-flow in sills and dikes. Geology 30:715–718

Lundstrom CC, Marshak S, Defrates J, Mabon J (2011) Alternative processes for developing fabric and mineral compositional zoning in intrusive rocks. Int Geol Rev 53(3–4):377–405

MacLeod CJ, Célérier B, Früh-Green GL, Manning CE (1996) Tectonics of Hess Deep: a synthesis of drilling results from Leg 147. In: Mével C, Gillis KM, Allan JF, Meyer PS (eds) Proc ODP, Sci Results 147: College Station, TX (Ocean Drilling Program), pp 461–475

Maes SM, Tikoff B, Ferré EC, Brown PE, Miller JDJ (2007) The Sonju Lake layered intrusion, northeast Minnesota: internal structure and emplacement history inferred from magnetic fabrics. Precam Res 157:269–288

Maes SM, Ferré EC, Tikoff B, Brown PE, Marsh JS (2008) Magneto-stratigraphy of a mafic layered sill: a key to the Karoo volcanics plumbing system. J Volc Geotherm Res 172:75–92

Magee C, Stevenson CTE, O'Driscoll B, Petronis MS (2012) Local and regional controls on the lateral emplacement of the Ben Hiant Dolerite intrusion, Ardnamurchan (NW Scotland). J Struct Geol 39:66–82

Malthe-Sørenssen A, Planke S, Svensen H, Jamtveit B (2004) Formation of saucer-shaped sills. In: Breitkreuz C, Petford N (eds) Physical geology of high-level magmatic systems, vol 234. Geological Society, London, Bath, UK, Special Publications, pp 215–227

Marsh BD (1998) On the interpretation of crystal size distributions in magmatic systems. J Petrol 39:553–599

Marsh BD (2004) A magmatic mush column Rosetta Stone: the McMurdo Dry Valleys of Antarctica. EOS (Trans Am Geophys Union) 85:497–502

Martín-Hernández F, Ferré EC (2007) Separation of paramagnetic and ferrimagnetic anisotropies: a review. J Geophys Res Sol Earth 112:1–16. doi:10.1029/2006JB004340

Martín-Hernández F, Luneburg CM, Aubourg C, Jackson M (2004) Magnetic fabric: methods and applications-an introduction. In: Martín-Hernández F, Luneburg CM, Aubourg C, Jackson M (eds) Magnetic fabric: methods and applications, vol 238. Geological Society, London, Bath, UK, Special Publications, pp 1–7

McBirney AR, Nicolas A (1997) The Skaergaard layered series: part II. Dynamic layering. J Petrol 38:569–580

McBirney AR, Noyes RM (1979) Crystallization and layering in the Skaergaard intrusion. J Petrol 38:487–554
McKenzie D (2011) Compaction and crystallization in Magma chambers: towards a model of the Skaergaard intrusion. J Petrol 52(5):905–930
Meurer WP, Boudreau AE (1998) Compaction of igneous cumulates; Part II, Compaction and the development of igneous foliations. J Geol 106:293–304
Michel J, Baumgartner L, Pultiz B, Schaltegger U, Ovtcharova M (2008) Incremental growth of the Patagonian Torres del Paine laccolith over 90 k.y. Geology 36(6):459–462
Mushayandebvu, MF, Bates, MP, Jones, DL (1995) Anisotropy of magnetic susceptibility results from the Mashonaland dolerite sills and dykes of Northeast Zimbabwe. In: Baer G, Heimann A (eds) Physics and chemistry of dykes. Balkema, Rotterdam, pp 151–161
Naldrett AJ, (2004) Magmatic sulfide deposits: geology, geochemistry and exploration. Springer, Berlin, pp 1–728
Namur O, Humphreys M, Holness MB (2013) Lateral reactive infiltration in a vertical gabbroic crystal mush, Skaergaard intrusion, East Greenland. J Petrol 54(5):985–1016
Nicolas A (1992) Kinematics in magmatic rocks with special reference to gabbros. J Petrol 33:891–915
O'Driscoll B (2006) Magmatic layering and magnetic fabrics in the Carlingford 'Later Gabbros', Co. Louth, Ireland. Ir J Earth Sci 24:37–50
O'Driscoll B, Troll VR, Reavy RJ, Turner P (2006) The Great Eucrite intrusion of Ardnamurchan, Scotland: re-evaluating the ring-dyke concept. Geology 34(3):189–192
O'Driscoll B, Hargraves RB, Emeleus CH, Troll VR, Donaldson CH, Reavy RJ (2007a) Magmatic lineations inferred from anisotropy of magnetic susceptibility fabrics in Units 8, 9, and 10 of the Rum Eastern Layered Series, Scotland. Lithos 98:27–44
O'Driscoll B, Donaldson CH, Troll VR, Jerram DA, Emeleus CH (2007b) An origin for harrisitic and granular olivine from the Rum Layered Suite, NW Scotland: a crystal size distribution study. J Petrol 2:253–270
O'Driscoll B, Stevenson CTE, Troll VR (2008) Mineral lamination development in layered gabbros of the British Palaeogene Igneous Province: A combined anisotropy of magnetic susceptibility, textural and mineral chemistry study. J Petrol 49(6):1187–1221
Owens WH (1974) Mathematical model studies on factors affecting the magnetic anisotropy of deformed rocks. Tectonophysics 24:115–131
Owens WH (2000) Statistical applications to second-rank tensors in magnetic fabric analysis. Geophys J Int 142:527–538
Petford, N (2009) Which effective viscosity? Mineral Mag 73(2):167–191
Petronis MS, O'Driscoll B, Stevenson CTE, Reavy RJ (2012) Controls on emplacement of the Caledonian Ross of Mull Granite, NW Scotland: anisotropy of magnetic susceptibility, magmatic, and regional structures. Geol Soc Am Bull 124(5–6):906–927
Petrovský E, Kapička A (2006) On determination of the Curie point from thermomagnetic curves. J Geophys Res: Solid Earth and Planets 111 (B12S27). doi:10.1029/2006JB004507
Polteau S, Ferré EC, Planke S, Neumann E-R, Chevallier L (2008) How are saucer-shaped sills emplaced? Constraints from the Golden Valley Sill, South Africa. J Geophys Res: Solid Earth 113:B12104
Putnis A, (1979) Electron petrography of high-temperature oxidation in olivine from the Rhum layered intrusion. Mineral Mag 43:293–297
Quadling K, Cawthorn RG (1994) The layered Gabbronorite Sequence, Main Zone, eastern Bushveld Complex. S Afr J Geol 97(4):442–454
Reay D (2004) Geophysics and concealed geology (Chapter 19). In Mitchell WI (ed) The geology of Northern Ireland. Geological Survey of Northern Ireland, Belfast, pp 227–248
Richey JE, Thomas HH (1930) The geology of Ardnamurchan, North-West Mull and Coll Mem Geol Surv Scot, Edinburgh. HMSO
Richter C, Kelso P, MacLeod CJ (1996) Magnetic fabrics and sources of magnetic susceptibility in lower crustal and upper mantle rocks from Hess Deep. Proc ODP Sci Res 147:393–403. College Station TX (Ocean Drilling Program)

Rochette P (1987) Magnetic-susceptibility of the rock matrix related to magnetic fabric studies. J Struct Geol 9(8):1015–1020
Rochette P, Jackson M, Aubourg C (1992) Rock magnetism and the interpretation of anisotropy of magnetic-susceptibility. Rev Geophys 30(3):209–226
Rochette P, Aubourg C, Perrin M (1999) Is this magnetic fabric normal? A review and case studies in volcanic formations. Tectonophysics 307 (1–2):219–234
Schofield N, Stevenson CTE, Reston T (2010) Magma fingers and host rock fluidization in the emplacement of sills. Geology 38:63–66
Selkin PA, Gee JS, Tauxe L, Meurer WP, Newell AJ (2000) The effect of remanence anisotropy on paleointensity estimates: a case study from the Archean Stillwater Complex. Ear Plan Sci Lett 183:403–416
Stephenson A (1994) Distribution anisotropy: two simple models for magnetic lineation and foliation. Phys Earth Planet Int 82:49–53
Stevenson CTE, Bennett N (2011) The emplacement of the Palaeogene Mourne Granite Centres, Northern Ireland: new results from the Western Mourne Centre. J Geol Soc Lond 168:831–836
Stevenson CTE, Owens WH, Hutton DHW, Hood DN, Meighan IG (2007a) Laccolithic, as opposed to cauldron subsidence, emplacement of the Eastern Mournes pluton, N. Ireland: evidence from anisotropy of magnetic susceptibility. J Geol Soc Lond 164:99–110
Stevenson CTE, Owens WH, Hutton DHW (2007b) Flow lobes in granite: the determination of magma flow direction in the Trawenagh Bay Granite, N.W. Ireland, using anisotropy of magnetic susceptibility. Geol Soc Am Bull 119:1368–1386
Stevenson CTE, O'Driscoll B, Holohan EP, Couchman R, Reavy RJ (2008) The structure, fabrics and AMS of the Slieve Gullion ring-complex, N. Ireland: testing the ring-dyke emplacement model. In: Thomson K, Petford N (eds) Structure and emplacement of high-level magmatic systems, vol 302. Geological Society, London, Bath, UK, Special Publications, pp 159–184
Tarling DH, Hrouda F (1993) The magnetic anisotropy of rocks. Chapman and Hall, New York, pp 1–232
Taubeneck WH, Poldervaart A (1960) Geology of the Elkhorn mountains, Northeastern Oregon: part 2. Willow Lake intrusion. Geol Soc Am Bull 71:1295–1322
Tauxe L (1998) Paleomagnetic principles and practice. In: Of modern approaches in geophysics, vol 17. Kluwer, Dordrecht, 299 p
Tegner C, Thy P, Holness MB, Jakobsen JK, Lesher CE (2009) Differentiation and compaction in the Skaergaard Intrusion. J Petrol 50(5):813–840
Wager LR, Brown GM (1968) Layered igneous rocks. Oliver and Boyd, London, pp 1–588
Wager LR, Brown GM, Wadsworth WJ (1960) Types of igneous cumulates. J Petrol 1:73–85
Wagner J, Hedley IG, Steen D, Tinkler C, Vuagnat M (1981) Magnetic anisotropy and fabric of some progressively deformed ophiolitic gabbros. J Geophys Res 86(B1):307–315. doi:10.1029/0JGREA0000860000B1000307000001. Issn: 0148-0227
Wiebe RA, Collins WJ (1998) Depositional features and stratigraphic sections in granite plutons: Implications for the emplacement and crystallization of granite magma. J Struct Geol 20:1273–1289
Wilson J, Ferré EC, Lespinasse P (2000) Repeated tabular injection of high-level alkaline granites in the Eastern Bushveld, South Africa. J Geol Soc Lond 17:1077–1088
Worrell L (2002) The origin of igneous cumulates: integrated studies of peridotites from the Western Layered Series of the Rum layered intrusion. Unpublished PhD thesis, Department of Earth and Ocean Sciences. Liverpool, University of Liverpool

Chapter 8
Physical Controls of Nucleation, Growth and Migration of Vapor Bubbles in Partially Molten Cumulates

James E. Mungall

Abstract The interstitial melt in partially molten cumulate piles in layered intrusions must at some point reach saturation with a volatile phase such as water vapor or hydrosaline melt. A number of models have been proposed in which orthomagmatic fluids migrate through partially solidified cumulates and participate in the formation of ore deposits. Here I examine the topology of the crystal—melt—vapor system in a cumulate and relate this to the role of capillary forces in governing the size and mobility of individual vapor bubbles. Capillary forces will play a dominant role in setting the number density and sizes of bubbles. In any cumulate rock with crystals smaller than several cm in diameter, bubbles of the postcumulus aqueous phase will be unable to migrate away from their sites of nucleation and growth. Although bubble stranding does not preclude the eventual loss of volatile constituents from the cumulate after complete solidification, it does prevent such constituents from migrating while the system remains partially molten and therefore prohibits them from participation in magmatic-hydrothermal interactions within the magma chamber.

Keywords Fluid · Immiscibility · Compaction · Trapped liquid · PGE

Introduction

The nucleation, growth, and migration of vapor bubbles in partially molten rocks are processes that may play key roles in the evolution of cumulate sequences in layered intrusions. The transfer of ore metals in orthomagmatic fluids is a central process in the formation of ore deposits in convergent margins (Richards 2011). Although little is known directly about the distribution of fluid during the maturation of cumulate rocks, the migration of fluid has been invoked as a major contributor to the development of ore deposits such as the Merensky Reef and UG2 chromitite of

J. E. Mungall (✉)
Department of Earth Sciences, University of Toronto,
22 Russell St, Toronto, ON M5S 3B1, Canada
e-mail: mungall@es.utoronto.ca

B. Charlier et al. (eds.), *Layered Intrusions*, Springer Geology,
DOI 10.1007/978-94-017-9652-1_8

the Bushveld Complex in South Africa, or the J-M Reef of the Stillwater Complex in Montana, USA (Boudreau and McCallum 1992; Meurer et al. 1997, 1999; Meurer and Claeson 2002; Boorman et al. 2002). It is generally assumed by both advocates and opponents of such models that once a fluid phase has formed, the cumulate pile acts as a kind of Maxwell's demon, allowing vapor bubbles to move unimpeded while preventing any other intercumulus phase except melt from moving at all (Boudreau 1999, 2004). This presumption is founded on the assumption that because fluids are deformable, there can be no impediment to their motion even in confined spaces.

There can be no doubt that immiscible aqueous fluids do exist in some layered cumulates at some times. A number of studies have described CO_2–H_2O fluid inclusions co-entrapped with silicate melt within plutonic xenoliths erupted with basalts (Huraiova et al. 1996; Bureau et al. 1998a, b; Sachs and Hansteen 2000; Upton 2000; Schiano et al. 2004; Krumrei et al. 2007). Mathez and Webster (2005) showed that late crystallized apatite crystals in Bushveld and Stillwater were equilibrated with brine. Hanley et al. (2008) documented the presence of brine as secondary fluid inclusions in trails crossing igneous olivine in the Stillwater Complex without any associated hydrous mineral alteration. The question at issue here is whether or not the aqueous fluid is capable of migration while the cumulate still contains interstitial silicate melt.

Because fluid phases are mobile along microfracture networks after complete solidification of the rock, and because the solidus of many mafic assemblages is above the temperature of many important dehydration reactions that would otherwise allow the formation of hydrous minerals from deuteric fluids, little of the volatile constituents originally present will remain in the rocks after cooling is complete, regardless of when they may have been mobile. As a result, any discussion of the possible existence and physical behaviour of fluids under conditions between the liquidus and the solidus must depend on indirect evidence and on an understanding of the chemistry and physics controlling the formation and motions of vapor bubbles in partially molten rocks.

Several studies have explored similar systems comprising two fluid phases in a solid granular matrix. Gibert et al. (1998) did an experimental study of fluid immiscibility and selective CO_2 entrapment. Minarik (1998) did experiments on a system comprising basaltic melt, carbonatite liquid, and olivine. The morphologies of systems comprising sulfide melt, silicate melt, and solids including olivine and chromite were studied by Holzheid et al. (2000), Barnes et al. (2008), and Chung and Mungall (2009). Mungall and Su (2005) addressed the upward motion of sulfide drops in compacting mantle rock and concluded that the capillary forces tending to restore sulfide droplets to spherical shapes would dominate over the hydraulic gradient driving upward melt extraction pores smaller than the droplets, leaving sulfide droplets stranded in the rock while silicate melt escaped. In most cases these investigators found that the immisicible phase with relatively low volume, such as sulfide melt, vapor, or carbonatite melt, tended not to wet the solid framework of silicate crystals, leading to stranding and trapping of the immiscible phase by capillary forces unless it could persist to temperatures below the solidus of the silicates.

In this article I examine some physical constraints on bubble formation and migration in cumulates. The problem can be broken down into several distinct topics: the texture of the solid framework of crystals that define the topology of the network of voids occupied by melt and vapor; the constraints on motion of bubbles as functions of the body and surface forces exerted on the bubbles as they pass through constrictions in the melt network; and the size distribution of vapor bubbles within the cumulate. I show here that by considering all of these issues together one can conclude that it is highly unlikely that vapor is able to migrate significant distances through partially molten cumulates. Even if a solid framework of crystals fractures before the interstitial melt is completely solidified, the interstitial melt will invade the newly formed fractures, leaving the vapor bubbles once again in a melt-filled channel. Instead, vapor bubbles probably remain trapped in small pockets of interstitial melt until the entire system has reached its solidus, upon which they are free to migrate along fractures.

Initial Considerations

To set the stage for this discussion it is convenient to establish some terms that will be used without further qualification from here on (Fig. 8.1; notation in Table 8.1). In the detailed discussions following this section the relevant concepts will be elaborated using quantitative descriptions.

In the present context the word *cumulate* refers to a partially molten assembly of crystals at rest in grain-grain contact. The mode of formation of the cumulate is of secondary importance to what follows. The crystals are bathed in an *interstitial melt* which is at equilibrium with the crystals. Where crystals have accumulated to some depth the lithostatic pressure exerted by the pile of relatively dense crystals will significantly exceed the hydrostatic pressure exerted by the interstitial melt, leading to *compaction* of the cumulate. Although mechanical compaction (e.g., grain boundary sliding) will be important initially as the crystals come to rest, the dominant mode of compaction will be by pressure solution or grain boundary diffusion creep, which is accomplished by diffusion along grain-grain contacts, dissolution of solute into the melt, and reprecipitation of solute on portions of crystals not already in contact with other grains.

As a result of the compaction process, the grain contacts evolve from point contacts in an orthocumulate texture in which all crystal surfaces are *wetted* by the melt, to *dry* sub-planar grain-grain boundaries coexisting with wetted surfaces facing into *pore chambers* between four or more grains and *pore throats* between three grains in a mesocumulate texture, and the degree of *contiguity* of grains thus evolves from zero in orthocumulates toward a maximum value of 1 if all interstitial melt has been expelled to form an adcumulate texture. In mesocumulates the distribution of melt between pore chambers and pore throats is highly dependent on the *dihedral angle* between adjacent grains in contact with melt.

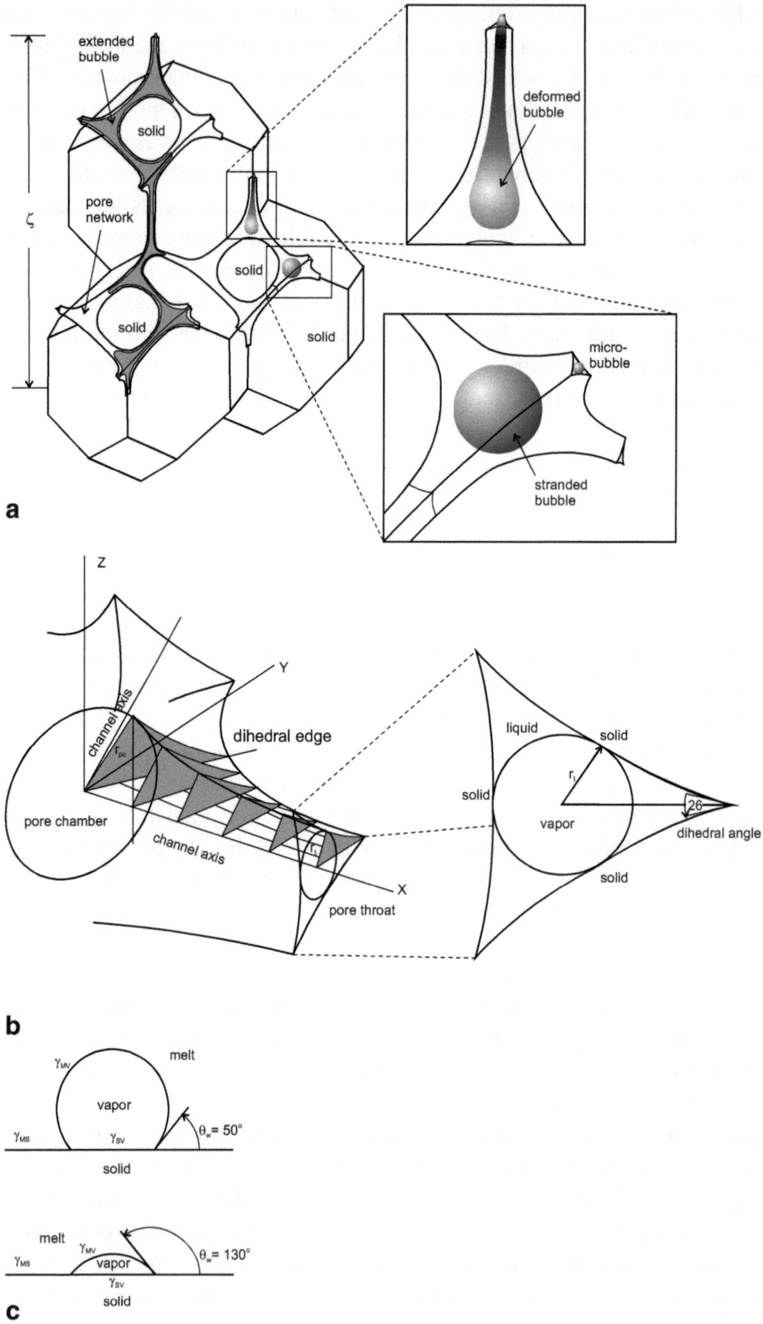

Fig. 8.1 **a** Distribution of melt and vapor in a well-compacted crystal aggregate. The microbubble, stranded bubble, and extended pore network regimes of bubble migration are illustrated. The examples shown here and in subsequent figures are all for the case where $\gamma_{SV} > \gamma_{SM} + \gamma_{MV}$, i.e., where vapor does not wet crystals at all. **b** Detail of a section of the pore network containing a pore

Table 8.1 Notation

Variable	Meaning		Value used	Dimensions
A	Cross-sectional area of the narrowest portion of a pore throat			m^2
a	Radius of a solid grain in the cumulate			m
C^*_{CO2}	Equilibrium concentration (i.e., solubility) of CO_2 in the silicate melt			$\mu g g^{-1}$
C^*_{H2O}	Equilibrium concentration (i.e., solubility) of H_2O in the silicate melt			wt%
C_{CO2}	Concentration of CO_2 in the silicate melt			$\mu g g^{-1}$
C_{H2O}	Concentration of H_2O in the silicate melt			wt%
C_i^{cum}	Concentration of an element in the cumulate rock			wt%
C_i^0	Concentration of an element in the silicate melt when accumulation occurred			wt%
d_g	Average distance between water molecules in silicate melt			m
D_{H2O}	Diffusivity of H_2O in silicate melt			$m^2 s^{-1}$
$D_{Cl}^{fluid/melt}$	Partition coefficient of Cl			None
F_{melt}	Volume fraction of melt in the cumulate (PALLADIUM model)			None
g	Acceleration due to gravity		9.83	ms^{-2}
h	Height			m
I	Bubble nucleation rate			$m^{-3}s^{-1}$
J^*	Rate constant for bubble nucleation			$m^{-3}s^{-1}$
k_B	Boltzmann's constant		1.38×10^{-23}	JK^{-1}
N	Number density			m^{-3}
n_g	Number of water molecules per unit volume of melt			m^{-3}
P	Pressure		200	MPa
P_{cap}	Capillary pressure inside a spherical bubble			Pa
ΔP_{cap}	Change in capillary pressure within a bubble due to its deformation			Pa

Table 8.1 (continued)

Variable	Meaning		Value used	Dimensions
q	Permeability of the cumulate pore network			m^2
R	Ideal gas constant		8.314	$Jmol^{-1}K^{-1}$
r_v	Bubble radius			m
r_b	Mean radius of curvature of the lower meniscus on a bubble in a hydrostatic gradient			m
R_b	Radius of curvature of a bubble enclosed within a pore chamber			m
r_{crit}	Critical bubble nucleus size allowing spontaneous growth			m
r_{pc}	radius of the largest sphere that can fit into a pore chamber			m
r_t	Radius of the largest sphere that can pass through a pore throat			m
r_{tc}	Critical value of r_t allowing rise of a deformed bubble for a given pore chamber size			m
T	Temperature			K
t	Time			s
$\mathbf{t}_i$	A vector in the plane of a two-phase interface i which is perpendicular to a dihedral edge			None
V_{melt}	Upward velocity of melt during compaction of the cumulate (PALLADIUM model)			None
v_g	Volume of a water molecule in silicate melt			m^3
v_t	Terminal velocity of a bubble			ms^{-1}
w_0	Relative velocity of melt and solids at the top of the cumulate pile			ms^{-1}
x	Distance			m
X_{cum}	Thickness of cumulate pile			m
ΔG_N	Extensive free energy of formation of a single bubble, dependent on bubble radius			J
$\Delta G_{N,crit}$	Extensive free energy of the critical bubble nucleus			J
ΔG_R	molar free energy of the reaction:	$H_2O_{(melt)} = H_2O_{(vapor)}$		$Jmol^{-1}$

Table 8.1 (continued)

Variable	Meaning		Value used	Dimensions
Γ	Angle subtended by a conical cavity in a solid substrate			Degrees
α	Accumulation rate constant			ms^{-1}
δ	Compaction length			m
ϕ	Porosity of cumulate			None
$\gamma_{MV,}\ \gamma_{SM,}\ \gamma_{SV}$	Surface tensions between melt M, solid S, and vapor V			Nm^{-1}
η_M	Viscosity of the water-saturated felsic melt		2750	PaS
η_S	Viscosity of the solid matrix in the cumulate			PaS
θ	Dihedral angle between two crystals at the corner of a melt-filled pore			Degrees
θ_w	Dihedral angle between vapor and solid at the contact with melt (wetting angle)			Degrees
ψ	Supplement of θ_w			Degrees
ρ_M	Density of silicate melt			kgm^{-3}
ρ_S	Density of solid phases in the cumulate			kgm^{-3}
ρ_V	Density of vapor			kgm^{-3}
τ_d	Diffusion time scale			s
τ_h	Compaction time scale			s
τ_0	Characteristic time of compaction			s
ξ_S	Bulk viscosity of the solid matrix in the cumulate			PaS
ζ	Height of connected vapor column			m

If the melt is initially vapor-undersaturated but evolves to higher volatile concentration with cooling and precipitation of added material to the crystal surfaces then it will eventually attain a state of equilibrium vapor saturation at which vapor would be a thermodynamically stable phase and *bubbles* should form. Here the word bubble applies whether the exsolved fluid is a low density vapour, an aqueous or carbonic fluid, or a hydrosaline melt. Alternatively, the melt may be at equilibrium with a vapor phase which is quantitatively extracted by flotation during crystal accumulation, leaving a cumulate pile containing vapor-saturated melt but not containing any bubbles. The existence of a surface tension between melt and vapor prevents *homogeneous bubble nucleation* at the exact concentration corresponding to equilibrium vapor saturation because the existence of a bounding surface adds to the total free energy of a bubble. This phenomenon of inhibited nucleation of bubbles may force the melt to evolve into a state of supersaturation before stable bubbles can form unless suitable sites exist where the effects of surface tension are overcome by geometric effects involving the surfaces of a phase other than melt and vapor, so *heterogeneous nucleation* is possible. The number of bubbles existing in a given volume of cumulate is the *number density N*. *Nucleation rate* is the number of new bubbles that nucleate per unit time in a given melt volume and depends on the degree of volatile supersaturation and on the melt-vapor surface tension; growth rate depends on the diffusivity of volatile constituents through the melt and the degree of volatile supersaturation.

Because the total internal energy of the system is minimized by reducing the total area of surfaces within it, there is a thermodynamic driving force for the growth of larger bubbles at the expense of smaller ones, the process of *Ostwald ripening*. The rate of Ostwald ripening depends on surface tension and diffusivity of the components of the phases in question through the melt.

The existence of a surface tension generally denoted as γ, acting parallel to an interface between two phases, exerts a pressure called *capillary pressure*, P_{cap}, on the phase inside the concavity of a curved interface, which varies inversely with the radius of curvature of the interface. Capillary pressure forces bubbles into spherical shapes when they are suspended in liquid or into forms having uniform mean curvature if they have some free surfaces but others are constrained by other phases. Similarly, the surface tension on a portion of a crystal-melt interface which is sharply concave toward the crystal causes the crystal to have a higher internal energy than it would have it the interface had everywhere the same radius of curvature; this causes crystals to assume regular shapes and dictates the magnitude of the dihedral angle and the topology of the pore network at textural equilibrium. Anisotropy of the crystal-melt surface tension causes the formation of faceted rational faces on crystals at equilibrium with the melt. Depending on the relative magnitudes of the surface tensions in various orientations these rational faces may meet at sharp corners or may be separated by smoothly curved segments of interface.

chamber and pore throat. After von Bargen and Waff (1986, Figs. 8.3 and 8.4). **c** Illustration of the wetting angle. When the bubble is in contact with a planar solid interface it takes the shape of a truncated sphere facing into the melt with the contact angle as shown, depending on the relative magnitudes of the surface tensions of the three types of interface (Eq. 8.1)

If $\gamma_{SV} < \gamma_{SM} + \gamma_{MV}$, where the subscripts refer to solid S, vapor V, and melt M, then bubbles will tend to attach themselves to solids. The resultant texture for the simple case of a planar crystal face which does not react with the vapor comprises a bubble with the shape of a truncated sphere (i.e., a spherical cap) sitting on the crystal, with a three-phase contact line forming a circle where they meet. The angle measured in the melt phase between the crystal face and the plane tangent to the melt-vapor interface is the wetting angle θ_w as shown in Fig. 8.1c and defined by:

$$\cos\theta_w = \frac{(\gamma_{SV} - \gamma_{SM})}{\gamma_{MV}} \tag{8.1}$$

If the vapor and the crystal are able to react and if the anisotropy of the crystal is low enough, then the form of the vapor-solid interface may become embayed, and the geometry is less simple. However the concept of a wetting angle is not affected by this complication.

Few measurements exist for vapor-melt-crystal wetting angles. From observations of the Bishop Tuff, Gualda and Ghiorso (2007) report θ_w of 5–25° for felsic silicates, and 45–50° for magnetite. On the other hand, the extremely low oversaturations required for bubble nucleation in experimental studies(Hurwitz and Navon 1994; Navon et al. 1998; Gardner and Denis 2004) seem to require that the wetting angle for biotite is much larger than these values (see discussion of heterogeneous nucleation, below). Indeed it is very difficult to infer a wetting angle from an explosively degassed and then quenched rhyolite glass. Cluzel et al. (2008) have inferred wetting angles by inverting the classical heterogeneous nucleation theory for several experimental determinations of oversaturation at the onset of heterogeneous bubble nucleation and they conclude that $\theta_w \approx 160°$ for magnetite, while they concur with previous workers in estimating that θ_w is probably $\leq 20°$ for feldspar. The wetting angles for olivine, enstatite, and diopside in CO_2 vapor were determined by Schafer and Foley (2002) to be strongly dependent on crystal orientation but all to be in range between 2 and 24°. Their measured wetting angles for Mg–Al spinel were all between 35 and 42°.

The bubbles shown in Fig. 8.1 do not wet the crystals at all, for simplicity of presentation. The following discussion of the role of capillary forces in controlling the movement of bubbles is based on the assumption that the bubbles are non-wetting against the principal silicate minerals composing the solid framework of the pore network. However the presence of a non-zero wetting angle will serve to ease the passage of a bubble through a pore throat because the spherical cap on the tip of the bubble where it occupies the pore throat has a radius equal to $r_t/\cos(\theta_w)$. When the enclosing solid is plagioclase with a wetting angle of 20°, the effective pore throat radius is increased by a factor of 1.06, which is a trivial difference from the case of a non-wetting bubble.

A further issue to consider for bubbles that slightly wet the solids is the additional energy required to detach a migrating bubble from the surfaces of crystals at the bottom of a pore chamber to allow it to move up into the next pore throat.

Gualda and Ghiorso (2007) presented a detailed analysis of the energy balance for attachment and detachment of bubbles on mineral surfaces, concluding that for small values of θ_w less than 45° the effect is negligible. However they also find that detachment of bubbles from magnetite crystals is strongly discouraged because of its much greater wetting angle. Although detachment may be less of an issue in monomineralic cumulates, the occurrence of detachment forces in polymineralic cumulates will tend to diminish mobility; hence the ideal non-wetting cases dealt with in this chapter offer the greatest possible mobility.

Models

Geometry of the Pore Network

The physical framework within which bubbles will nucleate, grow, and migrate is dictated by the geometry of the pore network within which the solid cumulus phases are bathed. Melt distribution at textural equilibrium depends on the solid-solid-melt dihedral angle and the porosity.

The dihedral angle depends on the relative magnitudes of solid-solid and solid-melt interfacial energy or surface tension, denoted γ_{SS} and γ_{SM}, respectively, which has dimensions of force/length = energy/area. Surface tension acts due to the existence between two phases of an interface where the components of each phase are not able to arrange themselves in their lowest energy configurations because they are not within a homogeneous phase (Herring 1951). The internal energy of a system is minimized when the surface areas of interfaces are arranged so as to minimize the total surface energy of the aggregate. The force balance due to the surface tensions of the two coexisting crystals and the melt at a point along the dihedral edge (Fig. 8.1) is given by

$$\sum_{i=1}^{3}\left(\gamma_i \boldsymbol{t}_i + \frac{\partial \gamma_i}{\partial \boldsymbol{t}_i}\right) = 0 \tag{8.2}$$

where $\boldsymbol{t}_i$ is the unit vector perpendicular to the dihedral edge lying in the plane of interface i and $\frac{\partial \gamma_i}{\partial \boldsymbol{t}_i}$ expresses the change in surface tension with change in crystallographic orientation of the interface i, where i might be melt-solid, solid-solid, or melt-vapor in the present context (Herring 1951; Hoffman and Cahn 1972). If the force balance is not equal to zero then there is traction on the dihedral edge that tends to drive recrystallization of the solids so as to shift the orientations of the wetted surfaces relative to the solid-solid surface. If the crystals are isotropic then the surface tension does not depend on the orientation of the unit vector $\boldsymbol{t}_i$ so $\frac{\partial \gamma_i}{\partial \boldsymbol{t}_i} = 0$. If the two solids are crystals of the same isotropic phase, the equilibrium dihedral angle θ is given by

$$2\cos\left(\frac{\theta}{2}\right)=\frac{\gamma_{SS}}{\gamma_{SM}} \tag{8.3}$$

where γ_{SS} and γ_{SM} are the surface tensions of the solid-solid and solid-liquid interfaces (e.g., Laporte and Provost 2000).

If the solid-melt surface tension is smaller than half of the solid-solid surface tension and the melt fraction is high then most crystal surfaces will be wetted and the solid-phase contiguity will be low even if the forms of the crystals are tightly packed with apparently planar grain-grain boundaries (A. Provost, pers. comm. 2013); if the solid-melt surface tension predominates then contiguity will be high and the melt will be confined to grain edges or isolated pore chambers. Deformation rates during compaction of cumulates are very sensitive to contiguity and therefore to interfacial tension (Takei and Holtzman 2009).

If the dihedral angle is less than 60° then the melt will form an interconnected network of pore chambers where four grains meet and pore throats or channels along three-grain contacts regardless of the porosity, as illustrated in Fig. 8.1, after von Bargen and Waff (1986). Dihedral angles have been measured in intersertal-textured rocks by Holness (2006), who report that all important rock-forming minerals have dihedral angles with melt of less than 60°. The dihedral angles for plagioclase and pyroxene are 29 and 37°, respectively. The persistence of the pore network down to arbitrarily low porosity theoretically should permit the total extraction of melt from a cumulate undergoing compaction if the texture remains perfectly equilibrated (e.g., von Bargen and Waff 1986; Bulau et al. 1979), however kinetic constraints described below will hinder this.

The existence of crystal anisotropy causes pore network textures to depart from the ideal forms depicted in Fig. 8.1. The dihedral angles depend on the relative orientation of neighboring grains even of the same mineral, and the existence of planes with lower surface energies causes the formation of some planar interfaces between crystals and melt (e.g., von Bargen and Waff 1988; Laporte and Provost 2000). Anisotropy introduces a significant dispersion of equilibrium dihedral angles (Holness et al. 2006). The more complex textures developed in anisotropic crystals are beyond the scope of this chapter.

In the context of bubble migration through any one particular pore, we are interested in the minimum radius of the most constricted portions of the pore throat between pore chambers. Figure 8.2a shows the ratio of the cross-sectional area A of the pore throat to the square of the particle radius a as a function of porosity, after von Bargen and Waff (1986; their Fig. 8.9). To find the radius of the largest bubble that can pass through the pore throat I approximate the cross-section of the pore throat to an equilateral triangle. The resulting estimate of the pore throat radius normalized to the radius a of the solid grains is shown in Fig. 8.2b. At porosity extrapolated to 0.2 the narrowest part of the pore throat is $0.1a$ for dihedral angles in the range from 20° to 50°. These dihedral angles span the range of values for common rock-forming minerals. In Fig. 8.1b the ratio of the radius of the pore chamber to that of the pore throat (i.e., r_{pc}/r_t) is about 2.2 for that example, which illustrates a porosity less than 0.05 (van Bargen and Waff 1986).

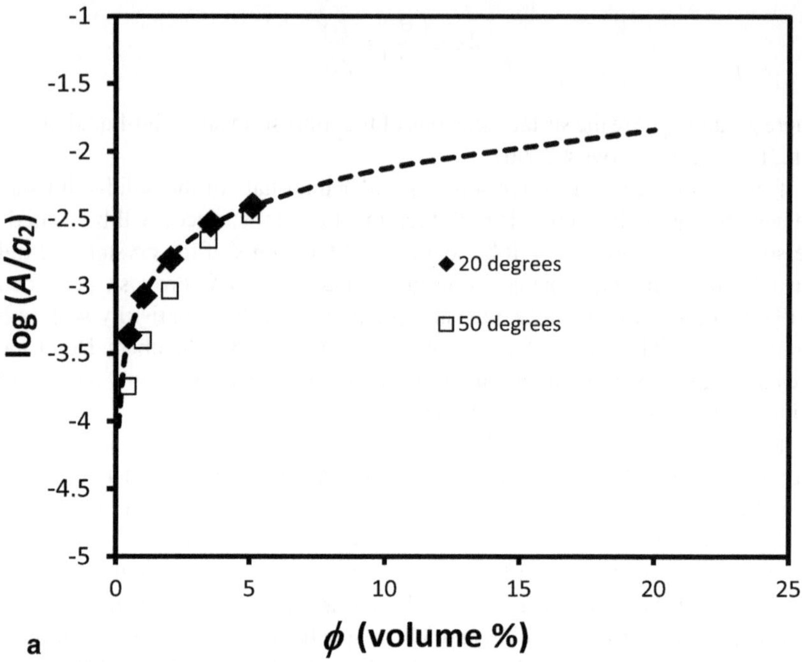

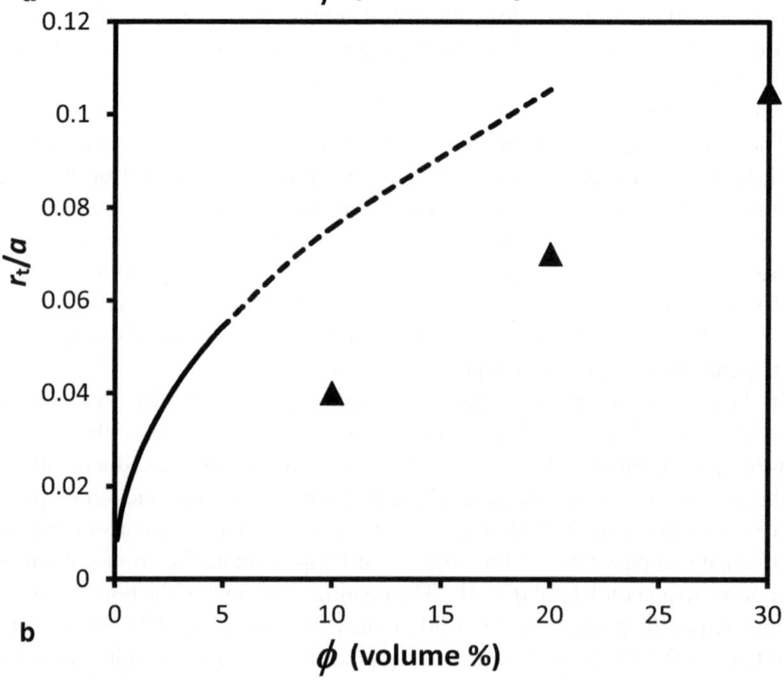

Fig. 8.2 **a** Ratio of cross-sectional area of pore throat to *square* of particle radius. Symbols represent values calculated by von Bargen and Waff (1986). The *dashed line* is an extrapolation of a fit to the data for 20° dihedral angles ($\log(A/a^2) = 0.954 \log \phi$—3.08). **b** Pore throat radius normalized to grain size. *Solid line*: calculated from van Bargen and Waff (1986) results; *dashed line* extrapolated from the fit shown in **a**; *triangles*: calculated values from Sweeney and Martin (2003) for

The extrapolation in Fig. 8.2b is not entirely satisfactory so it is worth examining this from another viewpoint. Sweeney and Martin (2003) modeled the distribution of pore space in a compacting assembly of ductile spheres without any consideration of the dihedral angles. In their model matter is transferred from the overlapping volumes of impinging monodisperse spheres so as to minimize the distance travelled and form a perfectly conical overgrowth within the remaining pore space. As a result, their dihedral angles change from zero at the beginning of the compaction process to values greater than 60° as compaction continues. They found that pore throats and pore chamber dimensions vary approximately linearly with the degree of compaction. Their estimate of r_{pc}/r_t varies from around 1.35 in a random packing with with 35 % porosity before compaction to 1.5 at 20 % porosity and 2.2 at 10 % porosity. They did not explicitly address the occlusion of residual pore space at low porosity.

It is easily shown that the ratio of pore chamber radius to pore throat radius for the tetrahedrally coordinated pore chamber between four spherical crystals on a close-packed lattice is 1.45, and that the ratio of the pore chamber radius to that of the crystals is 0.225.

I conclude that in cumulates with dihedral angle less than the critical value of 60° for occlusion of pores, the ratio r_{pc}/r_t will increase from about 1.45 prior to compaction to values greater than 2.2 at texturally equilibrated porosities below 10 %. The key observation here is that there is no reason to anticipate that r_{pc}/r_t will ever be less than about 1.35 because compaction always causes it to increase from this value regardless of the dihedral angle. This conclusion will be important in the section on the migration of bubbles through pore throats.

In experimental studies with grain sizes on the order of 100 μm the textures are completely equilibrated in the sense that dihedral angles are at local equilibrium after times of 100 h (e.g., Waff and Faul 1992; Laporte and Watson 1995). The timescale of diffusion-controlled textural re-equilibration will vary as a^2. Increasing the size of crystals from 100 to 1000 μm would therefore increase the time required by a factor of 100. In other words, textural equilibrium should still be attained in compacting cumulates on a timescale measured in years at most, and the assumption of continuous textural equilibrium at the scale of individual pores is probably safe.

Permeability of partially molten rocks has been widely investigated (Frank 1968; Maaløe and Scheie 1982; von Bargen and Waff 1986; Wark and Watson 1998) and a variety of functional forms have been proposed to relate permeability q to porosity ϕ raised to some exponent and grain size squared. Here I adopt the equation of Wark and Watson (1998),

$$q = \frac{\phi^3 a^2}{50} \tag{8.4}$$

which matches both experimental data and idealized theoretical pore networks (von Bargen and Waff 1986) very adequately. It should be noted that a in this case is the radius of the grains in the aggregate, hence the smaller value of the constant in the numerator than in the equation as it was presented by Wark and Watson (1998).

plastic spheres without textural equilibration. The extrapolated value of $r_t/a=0.1$ at $\phi=20\%$ for calculations in the text is a maximum plausible pore throat radius

Cheadle et al. (2004) addressed a somewhat different problem more akin to the static growth of diabasic or orthocumulate texture. They modeled the growth of crystals during cooling either within a loosely packed cumulate or in an initially crystal-free volume within which no compaction occurs. They show that if the rock fails to reach textural equilibrium, some of the pore space will be occluded by crystals that grow so as to close off the network of pore throats and maintain dihedral angles larger than their equilibrium values. They estimate that the limiting porosity, or percolation threshold, is between 8 and 10%.

Accumulation and Compaction

Although it might seem to be a major digression from the principal subject of this chapter, a moderately thorough discussion of the process and timescale of accumulation and compaction of cumulates is necessary to provide a framework for the timescales of bubble migration.

Accumulation rates are difficult to constrain directly but minimum rates can be established based on anticipated rates of heat loss from intrusions and the consequent rate of formation of new solids (e.g., Irvine 1970) if the intrusion is assumed to have crystallized as a closed system. His heat flow model can be reduced to the simple expression

$$X_{\text{cum}} = \alpha t^{1/2} \tag{8.5}$$

where X_{cum} is the thickness of the cumulate pile, t is time, and α is a constant depending on thermal conductivity, heat capacity, the temperature difference between magma and country rock, and the initial porosity of the cumulates. The accumulation rate is therefore

$$\frac{dX_{\text{cum}}}{dt} = \frac{\alpha}{2t^{1/2}} \tag{8.6}$$

Estimates of α range from 42.5 m $yr^{-1/2}$ for the Muskox intrusion (Tharp et al. 1998) to 19.6 m $yr^{-1/2}$ for the Bushveld Complex (Irvine 1970). Accumulation rates are very high initially but for these values of α, after 10,000 years of closed system behaviour, they slow to 0.1–0.2 m $yr^{-1}$and after the 200,000 year solidification time estimated by Irvine (1970) for the Bushveld Complex they are as low as 0.02 m yr^{-1}. Since the Bushveld Complex is the largest known layered intrusion on Earth, this sets an approximate lower limit of 0.02 m yr^{-1} to accumulation rates for any layered intrusion assumed to have evolved as a closed system.

However the Bushveld Complex has probably functioned as an open system onto the base of which crystals were deposited from influxes of crystal-charged magma which subsequently exited to make room for more fresh magma (e.g., Cawthorn and Walraven 1998). In this case accumulation rate is limited only by the rate at which fresh magma is assumed to have passed through the system. A rough upper bound on the accumulation rate can be estimated by dividing the approximately 7.5 km thickness of the layered rocks of the Lower, Critical, and Main Zones by the 75,000 year cooling time estimated by Cawthorn and Walraven (1998), to give an average accumulation rate of 0.1 m yr^{-1}.

The significance of accumulation rate to the issue of bubble migration is that the accumulation rate sets an absolute mimimum value for the upward ascent rate of bubbles if they are to be able to escape from the cumulate as more crystals continue to be added to the top of the pile.

McKenzie (1984) laid a theoretical groundwork for the treatment of compaction of partially molten rocks. He described a hypothetical layer comprising crystals and interstitial melt with a constant porosity ϕ. Melt distribution and dihedral angles are assumed to be always at textural equilibrium, sitting above an impermeable base. Restricting his treatment to a look at the rate of compaction at time $t=0$, where analytical solutions were tractable, he established the concept of a compaction length

$$\delta_c = \left[\frac{q\left(\xi_S + \frac{4\eta_S}{3}\right)}{\eta_M} \right]^{1/2} \tag{8.7}$$

which is the distance above the base of the layer over which the compaction rate (i.e., rate of shortening due to compaction) decreases by a factor of e (i.e., 2.718). The compaction length depends on the relative magnitudes of the bulk and shear viscosity of the crystalline matrix, ξ_S and η_S, the viscosity of the interstitial melt, η_M, and the permeability q which is a function of porosity ϕ as discussed above. The compaction length is a property of the system depending on viscosity and permeability and is unaffected by gravitational acceleration or the relative densities of the liquid and solid phases. Although the equilibrium distribution of melt in a system with dihedral angles <60 theoretically permits complete removal of interstitial melt, the compaction length increases exponentially as porosity is reduced, due to the exponential decrease in permeability. As a result, cumulates rarely attain a true adcumulate composition but always retain some melt even if none of the melt is truly occluded in closed pores.

The sum of viscosity terms in Eq. (8.7) is poorly known and it is customary to select a single effective viscosity in their place. Chosen values have ranged from 10^{18} Pa s for upper mantle peridotite (McKenzie 1984), 10^{15} Pa s for Skaergaard gabbroic cumulates (Tegner et al. 2009) to as low as 5×10^{11} Pa s for plagioclase networks undergoing grain-boundary sliding compaction (Boudreau and Philpotts 2002). Takei and Holzman (2009) have shown that the effective viscosity of a partially molten aggregate undergoing grain boundary diffusion creep is a sensitive function not only of porosity but also of the contiguity, which in turn is a function of dihedral angle. For minerals with dihedral angles similar to 20°, the effect of diminishing contiguity is to diminish effective bulk viscosity by two orders of magnitude as porosity increases from 0 to 0.1.

Evidently compaction length is a difficult parameter to pin down to precise values, but qualitatively it is clear that compaction lengths are large for high porosities and diminish rapidly as porosity is eliminated by compaction. Choosing appropriate values of the parameters, Tegner et al. (2009) have estimated a compaction length of 847 m for the Skaergaard cumulates. Using a matrix viscosity of 10^{15} Pa s, melt viscosity of 1 Pa s, porosity of 50 % and grain diameter of 1 mm I find a similar value for δ_c of 774 m.

The relative velocity of melt and matrix w_0 at the top of the compacting layer, where no compaction is occurring, is

$$w_0 = \frac{k_\phi}{\mu}\frac{(1-\phi)}{\phi}(\rho_S - \rho_M)g \tag{8.8}$$

where ρ_S and ρ_M are the densities of solids and melt, respectively (McKenzie, 1984). Sparks et al. (1985) showed that for a given compaction length the characteristic time τ_0 for porosity to be reduced by a factor of e is given as

$$\tau_0 = \frac{\delta_c}{w_0(1-\phi)} \tag{8.9}$$

The compaction timescale for the layer of thickness h is given by McKenzie (1984) as

$$\tau_h = \tau_0\left(\frac{h}{\delta_c} + \frac{\delta_c}{h}\right) \tag{8.10}$$

Tegner et al. (2009) have used Eq. (8.10) to estimate compaction timescales of a few 100 years for olivine gabbro layers with thicknesses of 50–847 m. That is, it will take less than 1000 years for a layer to compact from its initial state to one in which porosity has been reduced to approximately 22 % if it begins its existence with porosity of 60 %. At accumulation rates on the order of 0.1 m yr^{-1}, a 50 m layer will therefore be compacted in a time similar to the time taken for it to be deposited, with the result that compaction occurs near the top of the cumulate pile and is complete by the time a layer has been buried under a few tens of metres of subsequent cumulates.

When compaction has reached its limiting value of decreased permeability and increased bulk viscosity, a small amount of melt will remain in the cumulate and can be referred to as the trapped liquid. As a general observation, I note that the prevalent occurrence of essentially monomineralic and bimineralic adcumulate and mesocumulate rocks comprising almost nothing except the original accumulated grains attests to the fact that intercumulus mineral growth during compaction is very limited in extent. The growth of oikocrysts is the main exception to this rule, but even in poikilitic rocks it is common to see little other evidence of postcumulus mineral growth outside of very restricted volumes between compacted primary mineral grains. These observations indicate that compaction tends to occur in a nearly isothermal regime shortly after accumulation and before any significant amount of cooling, which would introduce new crystallizing phases at the liquidus, has occurred.

Going beyond the concepts of instantaneous compaction length and compaction time, Shirley (1986) implemented McKenzie's (1984) equations in a time-dependent numerical model. He examined the evolution of cumulate piles through time for a variety of physical parameters corresponding to a range of compaction lengths and times, and showed that within any cumulate there are essentially three compaction regimes that slowly grow upwards as material is added to the top of the pile and liquid is compacted upward out of it. In an upper compaction zone at the top of the cumulate pile and a lower compaction zone at the base of the pile, the amount of interstitial melt decreases downwards; between these two compacting zones is a zone throughout which the upward migration of melt being expelled from the lower compaction zone exerts a viscous force on the solid framework exactly sufficient to balance the negative buoyancy force driving compaction. This intermediate zone, called the fluidized zone, therefore experiences no compaction and the melt travels up through it at a constant rate. It is important to recognize that the cumulate within the fluidized zone is a solid rock with low porosity, and that the term "fluidized" applies only to the force balance between upward moving melt and downward-moving solids.

The amount of trapped liquid retained after compaction of real cumulate rocks can be estimated by considering the concentration of an incompatible trace element that is quantitatively rejected by growing crystals, if its concentration in the original melt is known. In the simplest case, where crystallization does not occur during compaction and the concentration of the element in the melt therefore does not change until after compaction has been completed, the final porosity ϕ_{TL} is given by

$$\phi_{TL} = \frac{C_i^{cum}}{C_i^0} \tag{8.11}$$

where the C_i^{cum} and C_i^0 are the concentrations of the incompatible element in the whole cumulate rock and the original melt, respectively (e.g., Cawthorn 1996).

Tegner et al. (2009) have estimated the amount of trapped liquid and have attempted to quantify the rate and extent of compaction in cumulates of the Skaergaard Intrusion. Trapped liquid estimation was done by first determining the trace element concentrations in the evolving melt that was at equilibrium with each successive layer in the cumulate pile assuming Rayleigh fractionation of the melt in the magma chamber and then applying Eq. (11) to each layer of cumulate. They conclude that outside of a thermal boundary layer in the lower zone where rapid cooling prevented effective compaction, trapped liquid proportions (i.e., final porosities after compaction) are consistently around 10 % in olivine gabbros and closer to 5 % in gabbros containing cumulus oxide minerals.

In light of their estimates of the timescale of compaction, Tegner et al. (2009) conclude that most compaction occurred within the uppermost few tens of metres of the cumulate pile. In other words, a continuous process of lithification turned the crystal pile into a solid as it formed. This rapid induration of the cumulate pile into a porous solid with a few percent remaining liquid is consistent with the observation of widely varied final porosities from layer to layer and suggests that despite the continued accumulation of hundreds of metres of crystals above such hardened layers there was no further compaction after these low values of trapped liquid had been reached. The model indicates that the entire compaction process was complete before accessory postcumulus minerals began to grow, providing a line of evidence independent of the application of Eq. (11) to estimate the amount of trapped liquid.

Mathez et al. (1997) studied the Atok section of the Merensky Reef in the Bushveld Complex. Observing that the apparent amount of trapped liquid within the Reef pyroxenites is very high (c.f. Cawthorn 1996), they argued that interstitial melt from lower layers in the cumulate pile had migrated up to the level of the Reef, there to react with the Reef package and leading to complex effects on bulk composition, interstitial melt composition, and mineral assemblage—this effectively would nullify any arguments based on the simple-minded calculation of trapped liquid proportions using trace element systematics, and would be in direct contradiction to attempts to generalize the results of Tegner et al. (2009).

However recently obtained chronostratigraphic data for the Bushveld Complex has raised uncertainty regarding the relative ages of the Merensky Reef and its footwall, which might therefore have been completely solid and relatively cool at the time of emplacement of the Reef (Scoates et al. 2012). The obvious erosional unconformity below the Reef would tend to support such a view and certainly suggests that during magma recharge events in this dynamic open system the top of the cumulate was probably much cooler than the overlying melt. In this context, it is possible that the large amounts of trapped liquid inferred for the Reef package are a consequence of the same dynamics that led to the preservation of large trapped liquid proportions in the Lower Zone of the Skaergaard Complex; namely, rapid cooling through the bottom of the layer freezing in the form of the downward-increasing porosity profile in the lower compaction zone (Tegner et al. 2009).

Furthermore, Mathez et al. (1997) followed Irvine (1970) in setting the rates of accumulation and migration of isotherms during cooling by considering heat flux through the top of a magma chamber behaving as a thermodynamically closed system. Since there is abundance evidence for open system behaviour of the Bushveld Complex, whose cumulates are now thought to have been deposited from a succession of through-going magmas of varied compositions (e.g., Cawthorn and Walraven 1998), the very slow accumulation and cooling rates supposed by Mathez et al. (1997) are probably not appropriate choices.

I conclude from this discussion that the relatively well-constrained and self-consistent results obtained by Tegner et al. (2009) on the Skaergaard Intrusion are representative of accumulation and compaction rates in mafic layered intrusions.

In summary, cumulates that are not in chilled boundary layers appear to achieve a stable highly compacted state in a short time after they first accumulate. Postcumulus mineral growth generally occurs predominantly after compaction has finished; the growth of oikocrysts is the sole exception to this generalization. Within a span of several 100 years after its deposition the cumulate can be considered to be a rigid framework, however depending on the relative rates of compaction in the lower compaction zone and cooling through the base of the cumulate pile, the framework of solids in the fluidized zone may still experience upward flow of interstitial melt long after its initial phase of compaction in the upper compaction zone is complete.

Volatile Budget and Solubility

The separation of fluids and vapors from magmas is conceptually divisible into two distinct stages. If a magma rises and is emplaced in the crust then its primary volatile contents may exceed their solubilities, leading to immediate saturation of the melt with vapor. This is called first boiling, and since it occurs when the magma is in a largely liquid state, the resulting immiscible vapor can be expected immediately to be physically separated from the melt and removed from the system. Subsequent evolution of the melt, whether in the magma chamber or within the cumulate pile, will proceed by the crystallization of volatile-poor solid phases and the consequent enrichment of the residual melt in volatile constituents, causing immediate re-saturation of vapor if first boiling has occurred, or later causing the first occurrence of vapor saturation if first boiling has not occurred. This secondary passive mode of reaching vapor saturation is called second boiling. In the case of either first or second boiling, the cumulate may be expected not to contain any separate vapor phase as it accumulates and begins to compact, because any vapor bubbles produced during first boiling should easily have been separated from the denser solid cumulus phases at the time of deposition of the cumulate.

The behaviour of mixed H_2O–CO_2–NaCl fluids at equilibrium with silicate melts remains incompletely understood in some details, but some general points are well known. An important concept to grasp is that silicate melts containing some small amount of CO_2 do not become saturated with H_2O or NaCl at some fixed concentration. A CO_2-dominated mixed vapor phase forms at very low concentrations of H_2O and CO_2 in the magma, and then as the total volatile budget is increased during second boiling this vapor phase evolves to increasing H_2O contents. When the water content of the silicate melt has reached several wt% it reaches a plateau value at which it might be said to have reached saturation with water vapor, but this process has been a gradual one of increasing concentration of H_2O in the fluid rather than the sudden appearance of a water phase. Typical mantle-derived magmas containing trace amounts of CO_2 are therefore continuously saturated in a vapor throughout their evolution in the crust.

Mixed H_2O–CO_2 vapors are completely miscible at magmatic temperatures and pressures but the solubility of NaCl is very low in this phase except at the lowest CO_2 concentrations (Duan et al. 1995). If the amount of NaCl is increased, at some point the vapor may become saturated with a separate immiscible hydrosaline fluid (i.e., salt melt or brine; Duan et al. 1995; Webster 2004). The solvus between hydrosaline fluid and aqueous fluid opens upward with increasing temperature and is dramatically widened by the addition of small amounts of CO_2. Silicate melts may therefore be saturated with NaCl-poor CO_2–H_2O vapors, or with NaCl-rich hydrosaline fluid, or with both. If the system is rich enough in NaCl to be saturated with hydrosaline fluid then further increases to the NaCl content of the bulk system do not cause notable increases in the concentration of NaCl in the individual phases, but instead cause increasing amounts of hydrosaline fluid to form at equilibrium with a vapor and a silicate melt whose compositions remain essentially fixed. Measured fluid-melt partition coefficients are fairly constant in presence of a single aqueous vapor phase but once separation of the aqueous phase into hydrosaline fluid and vapor occurs the bulk 'fluid'/melt partition coefficient increases dramatically as Cl concentration in the combined hydrosaline and aqueous phases rises to several tens of wt% (Kravchuk and Keppler 1994; Webster 1997, 1999). The H_2O and NaCl concentrations in the melt at which both vapor and hydrosaline fluid are stable depend on melt composition (Webster 1999). At 200 MPa and temperatures > 1000 °C haplogranitic melt with about 4.5 wt% H_2O and 2800 ppm Cl (Cl/H_2O weight ratio of about 0.06) coexists with both immiscible H_2O vapor and hydrosaline fluid (Webster 1997, 2004). Although hydrosaline fluid saturation occurs at much higher Cl concentrations in mafic liquids, the behavior of felsic liquids is more germane to the problem at hand, where we are considering the final stages of crystallization of interstitial melt.

It is impossible to measure, directly or indirectly, the volatile content of the Bushveld magmas. Here I assume that the parental magma was essentially komatiitic (e.g., Barnes 1989; Barnes et al. 2010) and was derived from a depleted mantle source similar to the MORB source. Using canonical values of the ratios H_2O/Ce = 170 (Michael 1995), CO_2/Nb = 530 (Cartigny et al. 2008), K/Cl = 13 (Kendrick et al. 2012), and assuming that a typical Al-undepleted komatiite contains about 0.33 ppm Nb, 0.09 % K_2O, and 0.7 ppm Ce (e.g., compilation of Fiorentini et al. 2010), this implies that the primitive Bushveld magma may have contained approximately 175 ppm CO_2, 119 ppm H_2O, and 57 ppm Cl when it arrived in the crust. These values are very far below vapor saturation at 200 MPa at the ~ 1400 °C liquidus temperature of the putative parental Bushveld B1 magma as calculated using MELTS thermodynamic software (Ghiorso and Sack 1995; Iacono-Marziano et al. 2012). However it is thought that the primitive Bushveld magma assimilated approximately 40 % by weight of lower crustal rock (Barnes et al. 2010). The lower crust contains approximately 0.5 % CO_2, 1.5 % H_2O, and 200 ppm Cl (Gao et al. 1998) so mixing proportions of 0.4 lower crust with 0.6 komatiite would yield a melt I refer to henceforth as the parent magma containing 0.6 % H_2O, 0.2 % CO_2 and 114 ppm Cl.

The average Bushveld B1 marginal sill composition contains about 400 ppm Cl, providing an alternate estimate of the Cl concentration of the magma (Barnes et al. 2010). It is unknown how much Cl concentration in the marginal sills was changed from its original magmatic values by degassing during crystallization or by subsolidus or metamorphic growth of halogen-bearing hydrous minerals, so this figure gives a poor upper limit on the Cl budget of the magma. Here I use both 114 and 400 ppm Cl as bounds on the likely concentration in the parental melt.

According to a model of vapor solubility in mafic magmas (Iacono-Marziano et al. 2012) the average parent melt with 0.6 % H_2O would be vapor-saturated at 1035 ppm CO_2 at 200 MPa and 1400 °C. This implies that about half of the original

CO_2 would have boiled off upon emplacement of the parent magma at the crustal level of the Bushveld Complex. The mole fraction of CO_2 in the coexisting vapor would be 0.976. Although this event has a large effect on the amount of CO_2 in the melt, it represents the removal of only 0.1 % of the mass of the system and has a minimal effect on the H_2O or Cl budget of the magma.

Figure 8.3 illustrates the evolution of vapor and melt in a hypothetical Bushveld cumulate containing a residual felsic melt of granitic composition. The initial condition for the model is set by taking the estimated volatile budget for the parent magma (above) after removal of the first boiling vapor. The formation of a cumulate and its early stages of crystallization are represented by a single step in the calculation. In the model this is done by switching from the model of CO_2–H_2O fluid solubility in basalt (Iacono-Marziano et al. 2012) to one for felsic liquids (Liu et al. 2005) and doubling the total volatile budget to account for passive enrichment of the residual melt in volatile species during crystallization of 50 % of its mass as anhydrous minerals. The simulation shown in Fig. 8.3 thus begins at 200 MPa, 1000 °C and a felsic liquid containing 1.2 % H_2O and 2070 ppm CO_2. At the first step in the calculations this bulk composition is allowed to reach equilibrium and the vapor is removed from the system, leaving a melt containing 1.194 % H_2O and 1014 ppm CO_2.

The model simulates isothermal, isobaric crystallization of the haplogranitic residual melt close to its eutectic temperature leading to second boiling. Two conditions are addressed. In one, the vapor is assumed to remain trapped within the pore space of the cumulate and continuously re-equilibrates with the melt. In the other, at small intervals of crystallization of about 10 % or less, the evolved fluid is removed and the composition of the remaining melt is taken as a new bulk composition for the next stage of crystallization and vapor exsolution. Cl concentration in the fluid is treated with a constant Nernst partition coefficient $D_{Cl}^{fluid/melt}$ of 30, which is scaled linearly to the mole fraction of H_2O in the vapor phase. If the concentration of Cl in the silicate melt exceeds 2800 ppm then it is assumed that a hydrosaline fluid containing 50 wt% Cl has formed, and an amount of this phase sufficient to accommodate the excess melt Cl is formed.

Whether the system is open or closed to vapor transport, the vapor remains CO_2-dominated until about 70 % of the melt has crystallized. Recalling that the starting point of this simulation was a melt that had already undergone 50 % crystallization from the B1 parent magma, one sees that the first 85 % of the evolution of the mafic magma would have occurred in the presence of a carbonic fluid comprising less than 0.1 % of the mass of the entire magmatic system. A true aqueous phase does not appear until 80 % of the interstitial melt has crystallized and the remaining melt contains about 5 % H_2O.

Saturation of the fluid with a hydrosaline fluid does not occur at any time in either model if the initial concentration of Cl is 114 ppm. In either the open or closed system case the maximum Cl concentration achieved in the aqueous phase is less than 3 wt%. However, if the average B1 marginal chill Cl concentration of 400 ppm is used in the starting composition, then the system becomes saturated with an immiscible hydrosaline fluid when the melt contains about 2.5 % H_2O, even before the vapor phase has become H_2O -dominated (Fig. 8.4). In the closed system, the

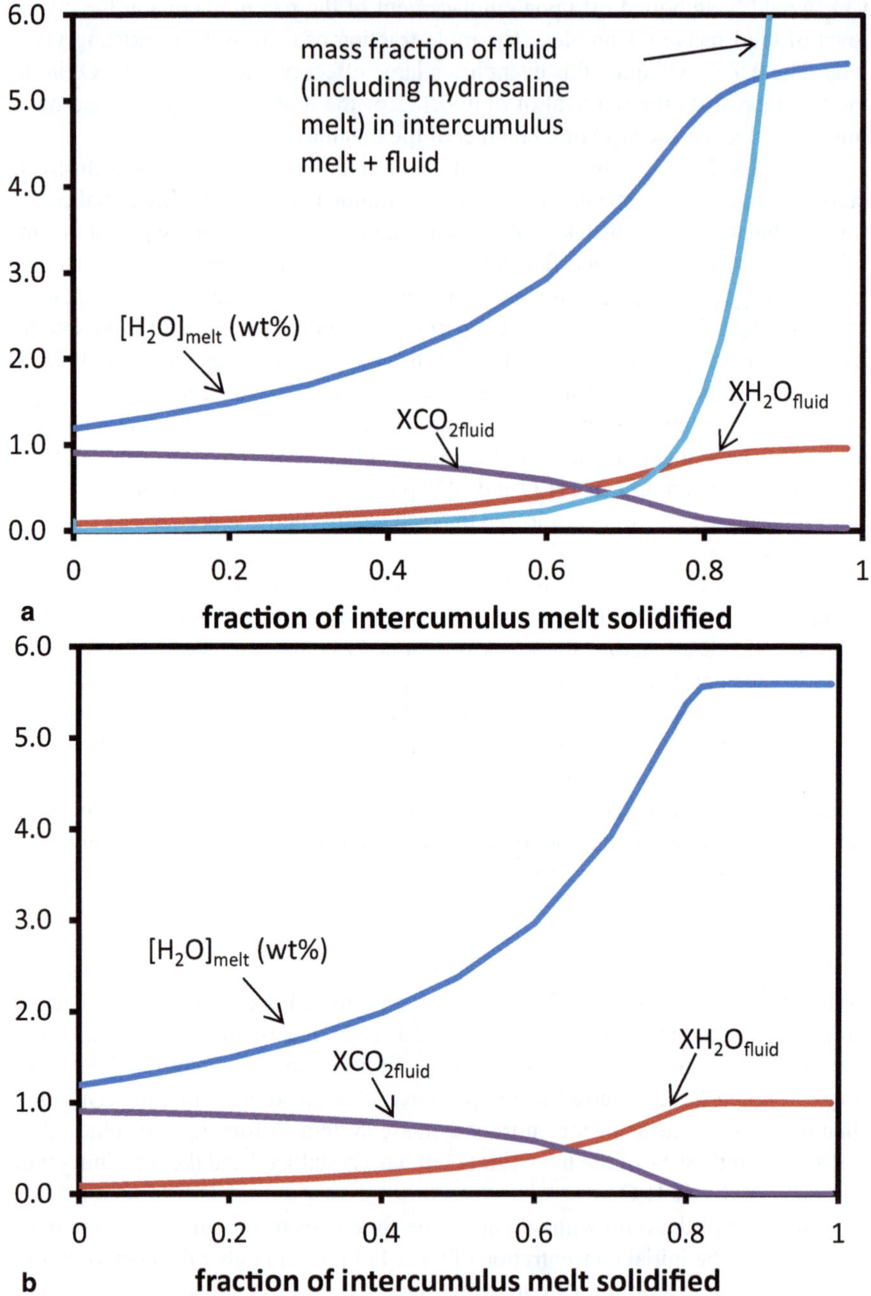

Fig. 8.3 Vapor solubility in fluid and melt. **a** closed system, **b** open system. Note that in the open system, fluid is removed as it forms so there its weight fraction in the intercumulus phases is always zero. In the closed system the fluid (including both vapor and hydrohaline fluid) rises to over 60 % of the mass of the residual melt + fluid when 98 % of the original intercumulus melt has solidified

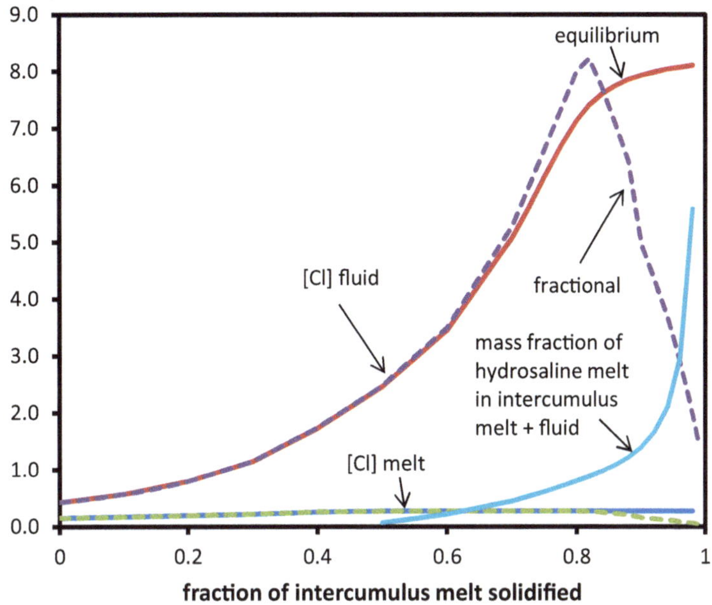

Fig. 8.4 Cl concentrations in melts and fluids. The weight fraction of hydrosaline fluid retained in the intercumulus phases (melt + vapor + hydrohaline fluid) in the closed system increases sharply at the end of the solidification process

weight fraction of hydrosaline fluid (relative to the sum of hydrosaline fluid + aqueous fluid + remaining silicate melt) reaches several wt% at the same time as the proportion of aqueous fluid rises to several tens of %.

If fractional removal of the evolved vapor occurs, then early-formed hydrosaline fluid strips the system of Cl until at later stages of crystallization the hydrosaline fluid is all gone and the concentrations of Cl in the vapor and melt decrease.

It is important to stress that the concentration of Cl in the starting liquid has a qualitatively decisive role in controlling the pathway of volatile exsolution in the interstitial melt. If Cl concentration is low, then a single volatile phase is present throughout the crystallization history of the magma, progressing from carbonic fluid to Cl-rich aqueous vapor in time. If Cl concentration is high, then in addition to the continuously evolved vapor phase, the system will reach saturation with an immiscible hydrosaline fluid about halfway through the crystallization of the interstitial melt. This second Cl-rich volatile phase may tend to concentrate ore metals.

Boudreau (2004) implemented a model of compaction and fluid saturation in cumulates based on Shirley's (1986) work in a software package called PALLADIUM, which he makes available for free download and which I have used here to test some plausible scenarios for cumulate compaction. The object of this exercise is to determine under which circumstances a vapor phase might be present while melt continues to migrate upward through the cumulates, potentially aiding the transport of vapor bubbles. The ability of the flowing melt to entrain bubbles is considered separately in a later section of this chapter.

Input parameters for PALLADIUM that are relevant to the current discussion were set as follows: $a=3\times10^{-4}$ m; permeability constant $K_0=1.5\times10^{-3}$; dimensionless deposition rate (Boudreau 2004) V* = either 0.1 or 0.01; liquid viscosity $\eta_M=1.0$ Pa s; magma column height = 1000 m. The porosity at which the cumulate was considered to have solidified completely was set at 0.1 %, and the limiting porosity below which further compaction was considered impossible due to occlusion of pores was set to 1 %. These two limiting porosities have both been set to unrealistically low values to ensure that in the resulting solutions the amounts of melt available for upward migration are always at a maximum. The cooling rate through the bottom of the cumulate pile (dT/dt) was set to one of three possible values of −1, −0.1, or −0.01 K yr^{-1}. The effective matrix viscosity η_S was set to one of three possible values of 10^{14}, 10^{15}, or 10^{16} Pa s. The initial concentration of H_2O in the melt $C_{H2O}=1.2$ wt% and the solubility of $C^*_{H2O}=5$ wt%. The permutations of parameter choices listed above led to eighteen different runs of the model, addressing a considerable range of plausible conditions during cumulate compaction. All runs were allowed to proceed until the height of the cumulate pile had reached 200 m. In all cases the mobility of the vapor phase was limited to be equal to that of the interstitial melt.

The results of the different trials of PALLADIUM are shown in Fig. 8.5, arranged in two grids of nine panels each with the solid matrix viscosity increasing to the right and the cooling rate increasing downwards. In part a the characteristic deposition rate is 0.1 and in part b it is 0.01. The mode of vapor in the system is expressed as a proportion of the volume of the porosity; where the system is vapor saturated and melt is absent this proportion rises to unity. The porosity is expressed as a volume fraction F_{melt}; the initial porosity is set to 0.6. The upward velocity of melt through the interstices of the cumulate V_{melt} is expressed in m yr^{-1} on the same scale as the melt and vapor fractions.

In most of the panels of Fig. 8.5 the upper compaction zone appears as a depth interval within which F_{melt} decreases rapidly and V_{melt} is high, reaching maximum values of slightly less than 0.5 m yr^{-1}; below this interval most panels show a fluidized zone in which both F_{melt} and V_{melt} are essentially constant at low values similar to 1 % and 0.01 m yr^{-1}, respectively. In the lower compaction zone the porosity falls toward zero. A wide range of behaviors is observed, with H_2O becoming saturated in the upper compaction zone, in the fluidized zone, in the lower compaction zone, or not at all, depending on the relative rates of cooling, deposition, and compaction. In many cases a separate vapor phase appears only below the lower compaction zone where melt migration has ceased at the specified minimum porosity. Depth intervals in which vapor is present within a portion of the system where upward flow of interstitial melt is occurring are indicated by grey background shading; these are the situations in which upward melt migration might be expected to propel vapor bubbles upward through pore throats, a subject that will be returned to in a later section.

In general, the occurrence of upward-migrating vapor-saturated melts is encouraged by combinations of high matrix viscosity, rapid cooling through the base of the system, and low rates of accumulation of crystals at the top of the pile.

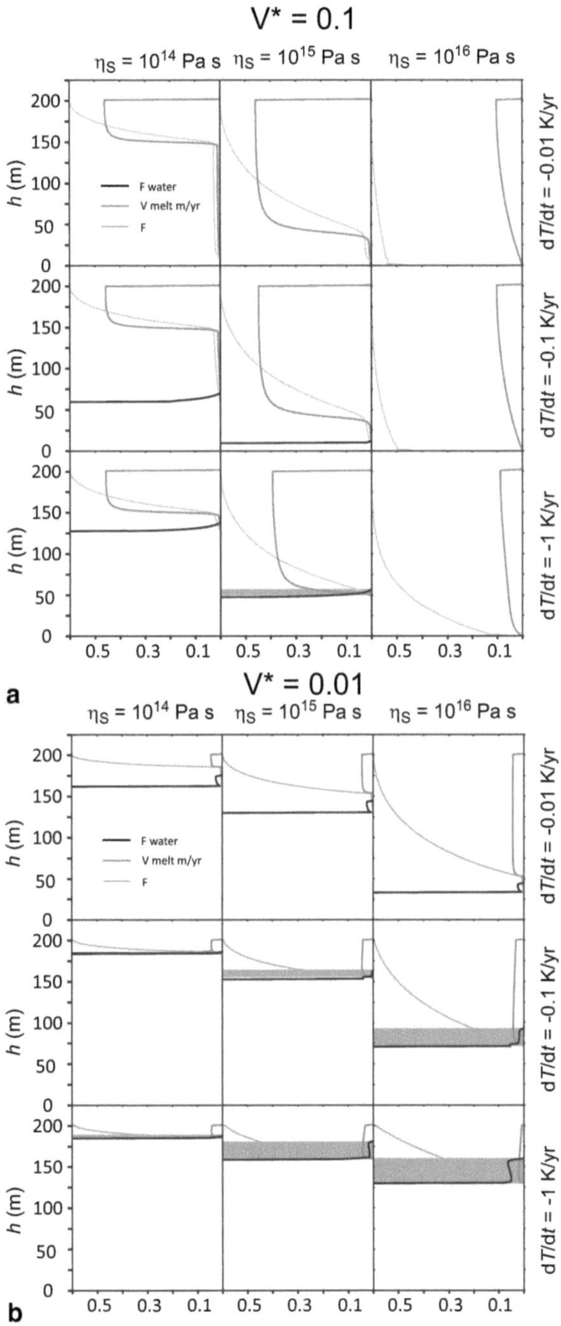

Fig. 8.5 Distribution of melt and vapor in compacting cumulates. All figures were calculated using PALLADIUM software (Boudreau 2004) with parameter selections as described in the main text. Black lines: volume fraction of vapor in intercumulus porosity; *heavy grey lines*: melt velocity upward through solid matrix; *light grey lines*: volume fraction of porosity. *Grey* shaded areas represent depth intervals in which vapor-saturated melt is migrating upwards

Mobility of Immiscible Volatile-Rich Phases

Dispersed droplets of immiscible liquid within a continuous liquid phase may have positive or negative buoyancy, causing them to tend to rise or fall through an enclosing liquid. Chung and Mungall (2009) described the physical constraints on mobility of dense immiscible sulfide liquid drops through crystal mush, taking into consideration the buoyant body forces tending to cause droplets to fall, the viscous drag forces causing their descent to slow, and the capillary forces that might stall their progress through restricted pore throats between pore chambers. The physics of the present problem are very similar but the direction of buoyancy body forces is opposite.

There are three different geometric states in which to consider bubble migration. In the first, bubbles are smaller than the most constricted portions of the pore throats joining pore chambers; this can be referred to as the microbubble regime. In the second, bubbles are larger than pore throats but similar in size to the pore chambers; this can be referred to as the stranded bubble regime. In the third, bubbles grow to sizes much larger than the individual pore chambers, occupying volumes of the pore network spanning several crystal diameters; this can be called the network or extended bubble regime.

In the microbubble regime, each bubble's experience can be likened to buoyant rise along a tortuous tubular path. Neglecting interactions with the walls of the pore throats and tortuosity, the bubble's ascent velocity can be described with a variant of Stokes' law for an inviscid liquid within a viscous medium (Hadamard 1911)

$$v_t = \frac{2g\Delta\rho g r_v^2}{3\eta} \tag{8.12}$$

where $\Delta\rho$ is the difference in density between melt and vapor, r_v is the radius of the bubble, and η is the viscosity of the melt. The viscosity of water-saturated rhyolite melt containing 5 wt% H_2O at 200 MPa and 1000 °C is 2750 Pa s (Hess and Dingwell 1996). The density of haplogranite melt containing 5 wt% H_2O is about 1900 kgm^{-3} (e.g., Lange and Carmichael 1990; Ochs and Lange 1999), whereas that of Cl-rich aqueous-carbonic vapor is about 600 kgm^{-3} at 1000 °C, 200 MPa (Bowers and Helgeson 1983; Duan et al. 1995). The density contrast $\Delta\rho$ is therefore 1300 kg m^{-3}.

If the bubble size approaches the size of the pore throat (presuming that it does not wet the crystals), then Eq. (8.12) must be modified (Francis 1933) to account for the need for the viscous melt to slip past the rising bubble in the constricted passageway:

$$v_t = \frac{2g\Delta\rho r_v^2}{3\eta}\left(1 - \frac{r_v}{r_t}\right)^{2.25} \tag{8.13}$$

where r_t is the radius of the largest sphere that can fit through the passageway undeformed. Equation 8.13 is not valid for bubbles with radii larger than those of the pore throats, where it gives negative results.

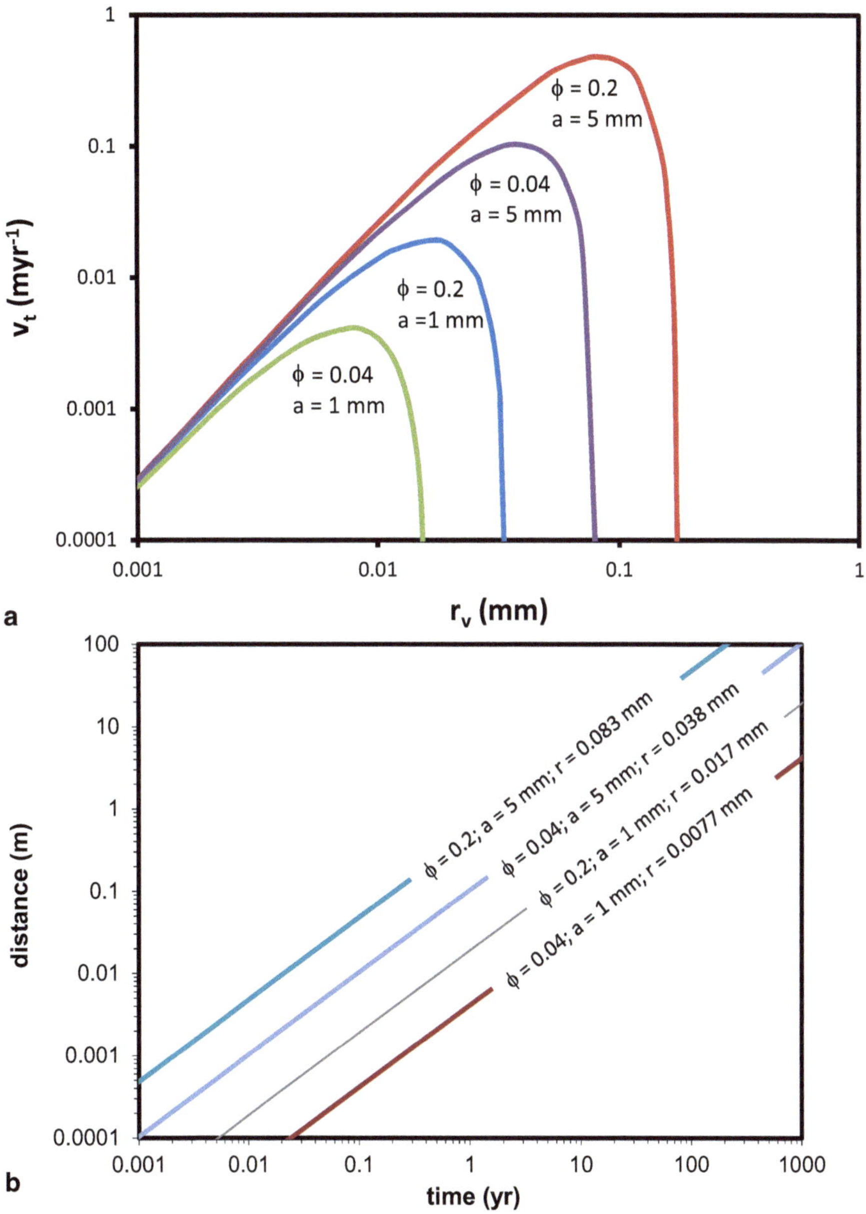

Fig. 8.6 **a** Ascent rates for bubbles smaller than the narrowest constrictions in the pore network in the microbubble regime. **b** Distance traveled by the fastest possible bubbles for each of the combinations of porosity and grain size shown in **a**

Figure 8.6 illustrates the upward velocities of bubbles and the maximum distances bubbles can travel as a function of time for cumulates with grain radii of 1 and 5 mm. For each grain size two stages of compaction are considered; one a relatively early stage of compaction, when the porosity is 0.2 and the vapor phase is a carbonic fluid, and another with porosity of 0.04 and a chloride-rich aqueous fluid or hydrosaline fluid. Each curve in Fig. 8.6b is calculated for the bubble size that has the highest velocity in Fig. 8.6a for the given porosity and grain size.

Comparison with the melt ascent velocities plotted in Fig. 8.5 shows that for cumulates with mm-scale crystals the buoyant rise of bubbles will be insignificant compared to the melt velocity as long as melt is still being extracted. For cm-scale crystals the opposite is true, and the rate of melt ascent during compaction will make a negligible contribution to the rate of bubble rise due to buoyancy forces alone.

If bubbles are larger than the minimum constriction of the pore throats, then their velocity as determined above will be zero. Another way of looking at this is to consider the force balance on a spherical bubble confined to the pore chamber as in Fig. 8.1. The pressure within a bubble is greater than the pressure outside by an amount called the capillary pressure P_{cap}, which is given by

$$P_{cap} = \frac{2\gamma_{MV}}{r_v} \tag{8.14}$$

If the bubble is to be moved into the pore throat, it will need to be deformed into a pear shape as shown, with a smaller radius of curvature in the pore throat than it has in the pore chamber. The difference in capillary pressure between the deformed state and the spherical state is given by

$$\Delta P_{cap} = 2\gamma_{MV}\left[\frac{1}{r_t} - \frac{1}{r_b}\right] \tag{8.15}$$

where r_t is the radius of the sphere that will just fit through the smallest constriction in the pore throat and r_b is the radius of the lower part of the pear-shaped bubble while it is squeezing into the pore throat. In order for the bubble to rise through the constriction, ΔP_{cap} must be balanced by the difference in buoyancy caused by the hydrostatic head resulting from the difference in density between the bubble and the melt $\Delta\rho$, over the distance h between the bottom and the top of the bubble:

$$2\gamma_{MV}\left[\frac{1}{r_t} - \frac{1}{r_b}\right] = \Delta\rho g h \tag{8.16}$$

Chung and Mungall (2009) explored this issue in some detail, deriving equations that must be satisfied simultaneously if the pore throat is simplified to a cylindrical shape emanating from a pore chamber which is a half-space:

$$\frac{1}{r_t} - \frac{1}{r_b} - \frac{\Delta\rho g}{2\gamma_{MV}}\left(r_b + r_t + \sqrt{r_b^2 - r_t^2}\right) = 0 \tag{8.17}$$

and

$$-2R_b^3 + r_b^3 + r_t^3 + \left(r_b^2 + \frac{1}{2}r_t^2\right)\sqrt{r_b^2 - r_t^2} = 0 \tag{8.18}$$

where R_b is the size of the spherical droplet prior to deformation. Setting the value of R_b, the critical value r_{tc} of r_t can be defined as the minimum pore throat radius below which the bubble is unable to rise. This critical value is an absolute upper limit which applies to pore throats with vertical axes. Any deviation from vertical orientation will increase critical pore throat radius. The larger the initial bubble size, the greater the buoyant force; therefore, r_{tc} is found by setting the initial bubble radius to the radius of the pore chamber r_{pc} and solving Eqs. (8.17, 8.18) simultaneously.

The melt-vapor surface tension of water-saturated rhyolite at 1000 °C has been measured by Bagdassarov et al. (2000). It varies strongly with dissolved water content, approximately obeying the relationship:

$$\gamma_{MV} = 236.6\exp\{-0.35P_{H2O}\} \tag{8.19}$$

where γ_{MV} is in mN m^{-1} and P is in kbar. Water-saturated felsic liquid at 200 MPa and 1000 °C has a surface tension of about 100 mN m^{-1}.

Figure 8.7 shows the pore throat radius r_t plotted against the initial radius of the bubble R_b for the surface tension and density contrast given. The curve representing the critical pore throat radius separates pore throats too small to allow bubble motion (below the r_{tc} line) from ones large enough for spontaneous rise of the bubbles (above the r_{tc} line). If the initial radius is smaller than the pore throat radius (line labeled spherical bubble/deformed bubble) then the bubble exists in the microbubble regime described above and is free to rise. If a bubble plots to the right of the 1:1

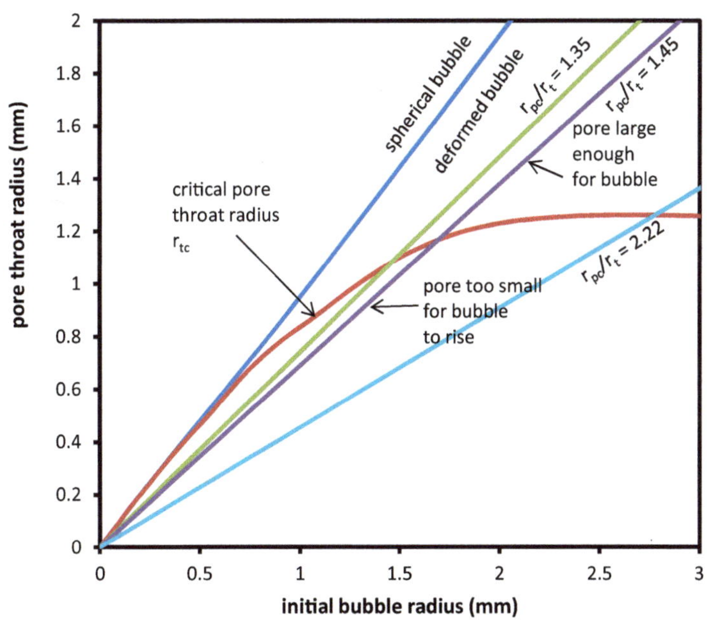

Fig. 8.7 Pore throat radius vs. initial radius of bubbles in the stranded bubble regime. See text for explanation

line then it is larger than the pore throat and can only rise if it is larger than the critical radius shown as r_{tc}.

In any given equilibrated texture there will be a constant ratio of r_{pc}/r_t, which was previously shown to fall between 1.35 and 2.2 for porosities greater than 0.1. Here we are treating the case of spherical bubbles, so the largest possible value of R_b is r_{pc}. A given r_{pc}/r_t ratio defines a line in Fig. 8.7, the two limiting values of which are shown. Coarse-grained orthocumulates fall at high values of r_t at the upper right end of this line, in the field of freely rising bubbles. If we consider progressively smaller grain sizes, the value of r_t diminishes until it crosses the line of r_{tc}. The intersection point defines the minimum pore throat radius, and hence the minimum grain size, below which bubbles are stranded by capillary forces. For orthocumulate textures with close-packed spherical crystals and ($r_{pc}/r_t = 1.45$; $r_t/a = 0.225$) this minimum pore throat size is approximately 1.2 mm, giving a minimum crystal radius of 1.2 mm/0.225 = 5.3 mm, i.e., 10.6 mm crystal diameters. Therefore in any orthocumulate texture, spherical bubbles cannot deform themselves to rise spontaneously through constrictions unless the grain size exceeds 1 cm.

Compaction reduces the sizes of both pore throats and pore chambers (Fig. 8.5), so the critical crystal radius allowing upward bubble migration will increase as porosity decreases. Using Sweeney and Martin's (2003) $r_{pc}/r_t = 1.5$ and $r_t/a = 0.1$ (Fig. 8.5) gives r_{tc} of 1.2 mm for porosity of 0.2. The minimum crystal radius for porosity of 0.2 is thus 10×1.2 mm = 12 mm, i.e., grain size of 2.4 cm. Finally, considering a porosity of 0.1, $r_{pc}/r_t = 2.2$, $r_t/a = 0.07$, and the minimum crystal grain size is 4.2 cm.

In compacted cumulates with porosity of 0.2, bubbles the size of pore chambers cannot force their way through the pore network unless the grain diameters are greater than 2.4 cm. If porosity is 0.1 then the minimum crystal diameter rises to 4.2 cm.

As mentioned in the Initial Considerations section vapor may have a small wetting angle against crystals that would increase the effective pore radius by several percent (e.g., 6 % in the case of plagioclase). As discussed previously, this would tend to be offset by the additional energy required to detach bubbles from crystals at the bottoms of pores before they could slide up along the other crystals to enter a pore throat higher up.

The third geometry of interest, the network regime, occurs when bubbles are larger than their enclosing pore chambers, extending over arbitrarily large heights within the pore network but not having spherical shapes (Fig. 8.1). The termination of such a bubble within each pore throat will resemble a spherical cap that has been squashed into a somewhat triangular capillary, and the capillary pressure within the bubble will be dictated by the mean radius of curvature of that termination. This situation can be called the network regime and is described by Eq. (8.16), with h allowed to increase to the value of ζ as shown in Fig. 8.1. When the bottom of the bubble is in a pore throat then the left hand size of Eq. (8.16) is small and the bubble can move upward. However at some point the bubble will encounter a situation in which its bottom arrives within a pore chamber while the top is in a pore throat and becomes stranded. The critical height for unimpeded rise of extended bubbles is

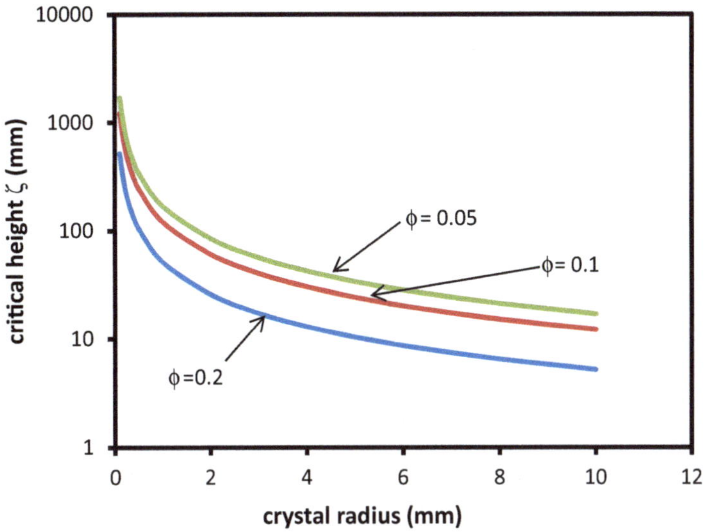

Fig. 8.8 Critical pore height vs. grain size. For porosities and crystal grain sizes indicated, extended vapor-filled networks must exceed the critical height before they can spontaneously rise through the cumulate

therefore determined by setting r_t to the minimum pore throat size and r_b to the pore chamber radius r_{pc}.

Using the relations already established to relate pore throat and chamber sizes to grain size and porosity, Fig. 8.8 has been drawn to show the critical height of an extended bubble as a function of grain size.

For crystals of radius 1 mm the critical height is 52 mm for $\phi = 0.2$ and 122 mm for $\phi = 0.1$. These geometries correspond to highly extended vapor networks 25 to 50 times larger than the diameters of the crystals and therefore hundreds of times longer than the diameter of the pore throats. Because of the relative viscosities of vapor and melt, the two-phase flow can be said to be destabilized and the network should resolve itself into long narrow fingers (e.g., Zhang et al. 2000). One can anticipate that such a long bubble (122 mm long, 0.07 mm radius) would tend to snap off to form a string of smaller unconnected bubbles (e.g., Unsal et al. 2009) but if it did not, it would be capable of moving upward. In very coarse-grained cumulates with 1 cm grain radius the critical height falls to 5.2 and 12.2 mm respectively, which are about the same size as the crystals themselves and would without exception be limited to individual narrow pores. The formation of such large bubbles would require that all of the vapor exsolved from a considerable volume of cumulate be collected into a single bubble. The volume of cumulate required to produce such large bubbles, and hence the number density of individual bubbles, can be estimated by considering the volume fraction of vapor exsolved at a given stage in the crystallization of the interstitial melt and the volume fraction of melt in the cumulate. For the example of a cumulate that had 20 % initial porosity after the initial stage of compaction was complete but prior to solidification of the remaining trapped melt, after 75 % solidification of the interstitial melt vapor makes up 0.3 % of the original volume of melt. The critical height is 171 mm for crystals of 1 mm radius, or 17 mm for 10 mm crystal radius. The volume of initial trapped melt that has given up its excess vapor to make a single bubble is 1/0.003 = 333 times the volume of the bubble, while porosity has diminished from 0.2 to 0.05. The volume of the cumulate from which this bubble was exsolved is equal to $333/\phi = 1665$ times the volume of the bubble.

Taking a horizontal slice through the cumulate at the equator of a single crystal one can envision it being bordered by six other crystals and six melt-filled pore channels. One vapor filled channel therefore occupies a cross-sectional area in the horizontal plane that corresponds to the area of a single crystal (e.g., Frank 1968). The volume of cumulate that contains a single vapor bubble occupying a single pore channel and its solid framework is that of a cylinder with cross sectional area of approximately πa^2 and height ζ; its volume is $\pi\zeta a^2$ which, in the present example, is 534 mm^3 if the crystal radius is 1 mm, or 5340 mm^3 if the crystal radius is 10 mm. The volume of the cumulate that has been degassed to produce the bubble of critical height is therefore $1665\pi\zeta a^2$ which is 8.9×10^5 mm^3 for the 1 mm crystal radius and 8.9×10^6 mm^3 for the 10 mm radius grain size. The cube roots of these volumes are 9.6 cm and 20.7 cm respectively, which are the edge lengths of nominal cubes of cumulate within which these bubbles could form.

Even if bubbles cannot spontaneously deform themselves to move up through pore throats, their ascent might be aided by the upward flow of the interstitial melt in a situation that might be likened to slug flow of a bubble within a pipe. The upward flow of interstitial melt is driven by the hydraulic gradient imposed by the contrasting densities and viscosities of the solid matrix and escaping melt, which can be described in terms of Hagen-Poiseuille flow (e.g., Durst 2008, p. 381). For a fluid with Newtonian viscosity flowing within a cylindrical pore throat, the melt velocity averaged over the entire cross-sectional area of the throat, taken here to be equivalent to V_{melt} in the PALLADIUM compaction model, can be related to the radius and length of the pipe r_t and l, the viscosity of the melt η_M, and the pressure difference between the top and bottom of the pore throat ΔP_{hy} as follows:

$$\Delta P_{hy} = \frac{8\eta_M l V_{av}}{r_t^2} \tag{8.20}$$

As also briefly discussed by Chung and Mungall (2009) the hydraulic gradient $\Delta P_{hy}/l$ driving melt migration arises from the different densities of solid matrix and interstitial melt, mediated by their viscosity contrast (McKenzie 1984), and it will tend to force bubbles upward through pores. If ΔP_{hy} exceeds the value of ΔP_{cap} imposed by forcing a bubble to deform itself and project into a narrow pore throat then sufficient force is exerted to overcome the surface tension, and the bubble can pass through the constriction. Since ΔP_{hy} varies inversely with pore radius, its maximum value will apply in the pore throat.

We can consider two limiting cases for the choice of pore length l; if the bubble is isolated in a system comprising many pores that are not occluded by bubbles, then the value of ΔP_{hy} can be taken over the length of the bubble itself; in the preceding discussion this value varies between the diameter of the pore chamber $2r_{pc}$ and the height of an extended bubble ζ. If all pores at a given horizon in the cumulate are occluded by bubbles then the effective distance over which the hydraulic gradient is integrated to press against the bottom of a single bubble is the vertical distance between bubbles, which in this case would be approximately equal to the grain diameter $2a$. Since our interest here is in those conditions under which hydraulic pressure might exceed the effects of bubble buoyancy alone, the absolute upper limit to this parameter is therefore about 100 mm for the case of the extended bubble capable of upward migration.

The PALLADIUM models found upward melt velocities of no more than 0.05 m yr^{-1} when a vapor phase was present, using a melt viscosity of 1 Pa s. Melt viscosity does not scale with temperature or composition in PALLADIUM, and since it is shown above that the viscosity of melt saturated with H_2O vapor in the coolest interstitial melt is probably much greater, this value of V_{melt} must therefore be an upper limit. However to consider the most favorable possible conditions for bubble migration I take the upper value of viscosity of 2768 Pa s and combine it with the length-scale l of 0.1 m and V_{melt} of 0.05 m yr^{-1}. With this unrealistically generous combination of parameter choices the hydraulic pressure exerted by melt flow is 351 Pa, which is sub-equal to the capillary pressure of 518 Pa resisting bubble migration. In all other cases the hydraulic pressure falls well below the critical value, being negligible in comparison with surface forces. For example, for isolated bubbles occupying individual pore chambers with $a=1$ mm, $r_t=0.1$ mm, $\eta_M=1$ Pa s, $V_{melt}=0.01$ m yr^{-1}, the hydraulic pressure is 5×10^{-5} Pa, seven orders of magnitude smaller than the capillary pressure. Slug flow of melt through pore throats is therefore unlikely to be a significant factor in the upward mobility of vapor bubbles in cumulates.

Homogeneous Bubble Nucleation

The reaction to form water vapor from water-saturated melt can be written:

H_2O (silicate melt) = H_2O (vapor)

If the melt is supersaturated with H_2O then the ΔG_R for this reaction is not zero because it is not at equilibrium; the irreversible reaction to form water vapor has a negative ΔG_R. The classic Gibbs treatment of nucleation barriers to growth from supersaturated solutions can be summarized by the expression

$$\Delta G_N = \frac{4}{3}\pi r^3 \frac{\Delta G_R}{V_m} + 4\pi r^2 \gamma_{MV} \tag{8.21}$$

where ΔG_N is the (extensive) free energy needed to form a nucleus of radius r out of supersaturated melt. ΔG_N is the sum of a volumetric free energy term and a surface energy term. ΔG_R is the molar free energy of the irreversible reaction to form the

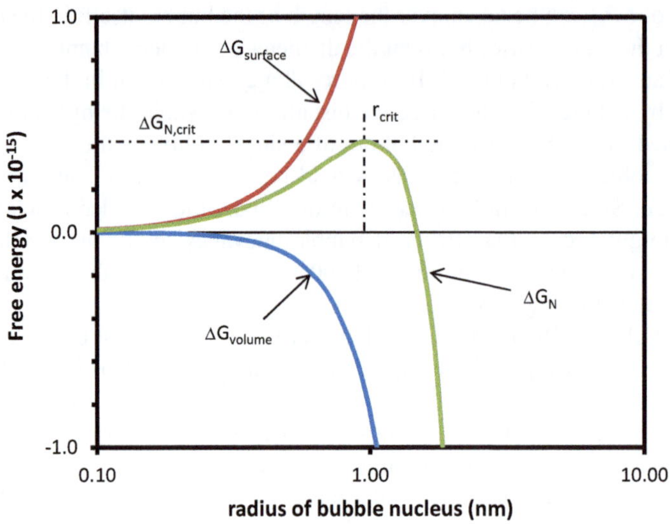

Fig. 8.9 Energy balance for the formation of a critical nucleus for a bubble. Bubble nuclei smaller than the critical radius spontaneously shrink and re-dissolve. Bubbles larger than the critical radius spontaneously grow. The free energy difference between the critical nucleus and the condition of stable equilibrium is the activation energy of bubble nucleation

stable vapor phase, V_m is the molar volume of the vapor and γ is the melt-vapor surface tension. Because the (negative) free energy change produced by growing the bubble depends on r^3 whereas the (positive) surface energy term depends on r^2, ΔG_N rises from zero at zero radius to a maximum and then falls as the net effect of the increasing volume of the bubble comes to outweigh the positive energy contribution of the surface. The energy balance involved in the formation of the bubble nucleus is illustrated in Fig. 8.9. The critical radius of the nucleus r_{crit}, below which it will spontaneously re-dissolve and above which it will spontaneously grow to become a stable bubble, is found by taking the derivative $d\Delta G_N/dr$ and solving for the maximum in ΔG_N where the derivative is equal to zero

$$r_{crit} = \frac{-2\gamma V_m}{\Delta G_R} \tag{8.22}$$

The magnitude of the energy barrier represented by the maximum in ΔG_N is found by substituting Eq. (8.22) into (8.21) to find

$$\Delta G_{N,crit} = \frac{16\pi\gamma^3 V_m^2}{3\Delta G_R^2} \tag{8.23}$$

Toramaru (1995) suggested a maximum value of 10^{-9} m for successful nucleation on the timescale of explosive vesiculation of rhyolite. Taking the radius of a single water molecule as 0.28 nm, giving a volume of 0.33 nm^{-3} per molecule (Fu et al. 2009), this critical nucleus would contain approximately 12 water molecules.

Figure 8.9 shows the variation of the volume and surface terms and the sum of the extensive energy required to form bubble as functions of the bubble radius with numerical parameters chosen so the critical radius is 1 nm. The value of ΔG_R required to produce this size of critical bubble is 6 kJ mol^{-1}, which can be calculated with MELTS (Ghiorso and Sack 1995) to correspond to a concentration of 10.46 wt% H_2O in the surrounding melt, double the equilibrium value at 200 MPa. The pressure at which this concentration of water would be at equilibrium with the melt is 470 MPa; the capillary pressure of water vapor inside a 1 nm bubble from Eq. (8.14) would be 200 MPa. Adding the capillary pressure to the system pressure of 200 MPa gives 400 MPa, which is similar in magnitude to the pressure at which melt containing 10.46 wt% H_2O would be saturated at equilibrium with water vapor.

Another way to estimate the degree of supersaturation required to stabilize a critical bubble is to assess the rate of formation I of supercritical nuclei per unit volume:

$$I = J^* exp\left\{\frac{-\Delta G_{N,crit}}{k_B T}\right\} \tag{8.24}$$

where the pre-exponential constant J^* is equal to

$$J^* = \frac{2n_g^2 v_g D_{H2O}}{d_g}\left[\frac{\gamma}{k_B T}\right]^{1/2} \tag{8.25}$$

k_B is Boltzmann's constant, n_g is the number of potential nucleation sites (taken to be equal to the number of water molecules) per unit volume of the melt, v_g is the volume of a water molecule in the melt, d_g is the average distance between water molecules in the melt, and D_{H2O} is the diffusion coefficient of water molecules (e.g., Hurwitz and Navon 1994). Equation (8.24) is a remarkable result insofar as it contains the product of the cube of surface tension and the square of the reaction affinity ΔG_R all within an exponential function. The free energy barrier to nucleation $\Delta G_{N,\ crit}$ can be seen to be the activation energy of nucleation, which is a thermally activated process. If the activation energy can be somehow reduced to zero, then the nucleation rate is equal to the pre-exponential factor, which in the present situation exceeds 10^{28} m^{-3} s^{-1}.

As shown in Fig. 8.10, homogeneous nucleation rates calculated using Eqs. (8.22–8.24) are effectively equal to zero at all H_2O contents <7.2 wt% (ΔG_R <3050 J mol^{-1}) but then rather suddenly increase to extremely high values at slightly higher H_2O contents. At 7.4 wt% H_2O in the melt the critical bubble radius is about 1.8 nm. The capillary pressure is about 100 MPa and the saturation pressure for 7.4 wt% H_2O is about 310 MPa; by either measure, $\Delta P_{nucleation}$ = about 100 MPa.

Thus by either arbitrarily setting a critical bubble nucleus size of 1 nm or by attempting to calculate nucleation rates one determines that supersaturations on the order of 100 MPa are necessary to cause homogeneous nucleation of bubbles in felsic liquids. These estimates are similar to values obtained during isothermal decompression experiments (Mangan and Sisson 2000; Gondé et al. 2011; Gardner 2012).

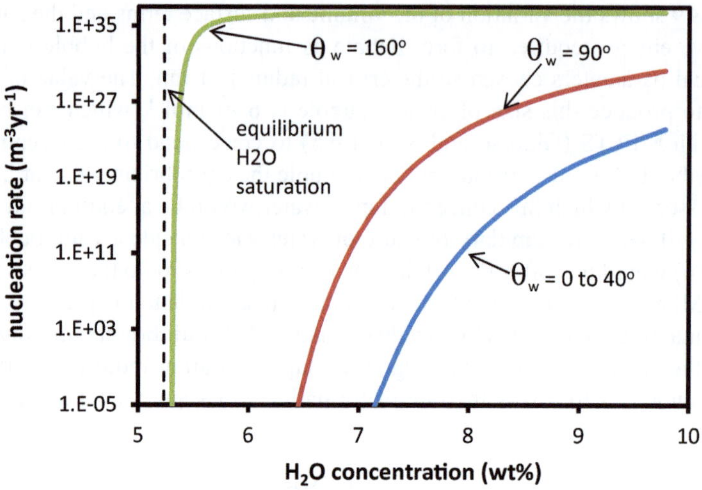

Fig. 8.10 Bubble nucleation rate as a function of H_2O concentration in rhyolite melt at 200 MPa pressure and 1000 °C. The equilibrium saturation concentration of H_2O is 5.24 wt% as shown with the *dashed line*. Homogeneous nucleation (*curve* labeled $\theta_w = 0$–$40°$) will not occur at observable rates unless H_2O concentration exceeds 7.3 %. The presence of silicate minerals with θ_w of < 40° has no discernible effect on nucleation rate, but phases with bubble wetting angles of 90° and 160° will markedly reduce the amount of supersaturation required to initiate bubble nucleation. In the case of magnetite, the required H_2O supersaturation is only 0.1 wt%

Vapor bubbles will not form in an initially bubble-free melt by isobaric homogeneous nucleation unless the melt evolves to extreme supersaturation. A melt nominally saturated with H_2O vapor containing 5.3 wt% H_2O would have to undergo about 30 % fractional crystallization before reaching 7.4 wt% H_2O and undergoing spontaneous homogeneous bubble nucleation.

This would not prevent bubble formation in the long term but would certainly inhibit the formation of more bubbles once one bubble finally did nucleate, causing the few bubbles that did nucleate to be able to grow to very large sizes. It is interesting to consider whether such widely spaced large bubbles could attain the sizes calculated above for mobility in the network migration regime. That is, can bubble nucleation be suppressed sufficiently that individual nucleation sites would be more than 20 cm apart in the cumulate?

After a bubble nucleates from a strongly supersaturated melt it will grow by diffusive mass transport through the melt. A diffusive boundary layer commonly referred to as a skin layer will form surrounding the bubble within which the concentration of H_2O falls below the critical saturation concentration. New bubbles can only nucleate outside this boundary layer. An approximate diffusive timescale τ_d in spherical geometry is

$$\tau_d = \frac{r^2}{4D_{H2O}} \tag{8.26}$$

where D_{H2O} is the diffusion coefficient of H_2O. In rhyolite melt at 1000 °C containing 5 % H_2O D_{H2O} is 6×10^{-11} m^2s^{-1} (Ni and Zhang 2008), giving $\tau_d = 1$–5 years for 10–20 cm diffusion lengths. Diffusion will thus erase concentration gradients between bubbles several tens of cm apart effectively instantaneously compared with the timescales of compaction, cooling, and bubble migration, and much faster than the anticipated rate of nucleation of new bubbles.

Inhibited homogeneous nucleation of bubbles at large vapor supersaturation in cumulates therefore appear to be one possible route to the formation of small numbers of very large vapor pockets within the intercumulus pore network that would be large enough to rise spontaneously through the cumulate on reasonable timescales.

Heterogeneous Nucleation

However there are other pathways to bubble nucleation. If the wetting angle between vapor and crystal is greater than zero then the free energy of nucleation is diminished by a factor Φ.

$$\Delta G_{\mathrm{N,crit}} = \frac{\Phi 16\pi\gamma^3 V_{\mathrm{m}}^2}{3\Delta G_{\mathrm{R}}^2} \tag{8.27}$$

and

$$\Phi = (2 - \cos\theta_{\mathrm{w}})(1 + \cos\theta_{\mathrm{w}})^2 / 4 \tag{8.28}$$

Volmer (1939), Turnbull (1950); note that Turnbull's θ_w is the supplement of θ_w as it is defined here). If θ_w is equal to zero then Φ is unity, the surface of the crystal is completely wetted by the silicate melt and there is no difference to $\Delta G_{N,\ crit}$; if $\theta_w = 180°$ then the vapor would form a continuous film on the crystals, the activation energy would fall to zero, and there would be no inhibition whatsoever of heterogeneous nucleation of bubbles. For a wetting angle of 40° Φ is 0.96, which has a negligible effect on the activation energy and rate of nucleation. However if θ_w exceeds 40° then the value of Φ begins to fall quasilinearly, which lowers the activation energy for heterogeneous nucleation (Fig. 8.9). The presence of magnetite, with θ_w of 160°, diminishes the required supersaturation to 0.1 wt% H_2O, corresponding to an excess vapor pressure in the bubble of about 5 MPa.

In vesiculation experiments containing microlites of iron oxides there is little or no nucleation lag, and supersaturations as low as 1–5 MPa are sufficient to initiate widespread and nearly instantaneous bubble nucleation in rhyolites (Gardner and Denis 2004; Hurwitz and Navon 1994). In these experiments the bubbles are observed to nucleate mostly on the tips of the microlites. I suggest here that the reason may have to do with the presence of surface roughness such as hopper-shaped or swallowtail terminations on the microcrystals. To illustrate the role of crystal morphology in heterogeneous nucleation, Fig. 8.11 illustrates a cross section through two hypothetical conical cavities in crystal surfaces or between crystals. At a critical value of the angle Γ subtended by the cone surface, the wetting angle produces a planar

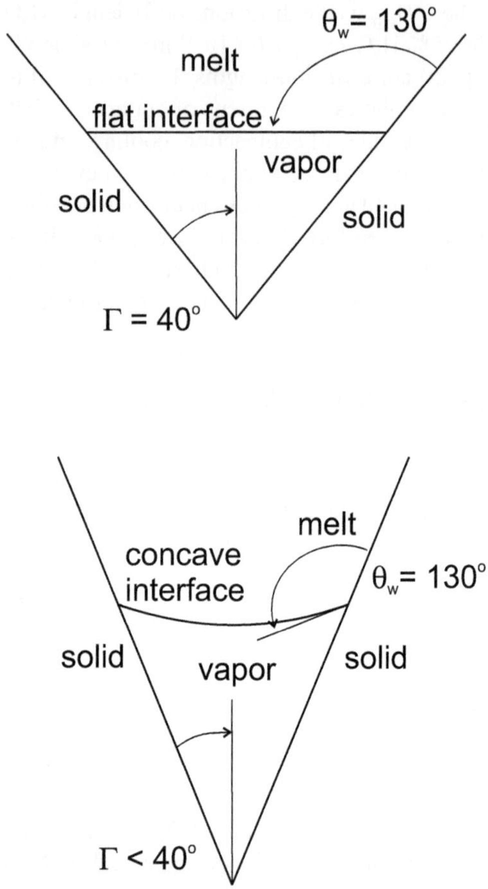

Fig. 8.11 Cross sectional sketch of heterogeneous nucleation in a conical cavity for a hypothetical wetting angle of 130°. If the half-angle Γ subtended by the surface of the cone is less than ($\theta_w - 90$) then the bubble interface is concave into the bubble and it will be stable even when the melt is mildly vapor-undersaturated

interface between vapor and melt inside the cavity. If the cone is any narrower, the vapor-melt interface will be convex rather than concave into the vapor bubble. In this case two the capillary pressure inside the bubble will be less than the pressure outside the bubble, and the calculated activation energy of nucleation will be negative. This situation was explored in detail by Turnbull (1950) and is reiterated by Kelton and Greer (2010) who showed that the nucleating phase (bubbles, in our case) can persist metastably even when the melt is undersaturated with the nucleating phase.

Turnbull gave equations that allow calculation of the free energy of formation of a bubble within a conical cavity:

$$\Delta G_{\mathrm{N}} = \frac{\pi r^3 \Delta G_{\mathrm{R}} \varphi}{v} + \pi r^2 \gamma (2v - \sin\Gamma \cos\psi) \tag{8.29}$$

where

$$\varphi = \frac{sin^2\Gamma cos\Gamma}{3} - \frac{2sin^3\Gamma}{3cos^3(\Gamma+\psi)} + \frac{sin^3\Gamma tan(\Gamma+\psi)}{cos^2(\Gamma+\psi)} - \frac{sin^3\Gamma tan^3(\Gamma+\psi)}{3} \tag{8.30}$$

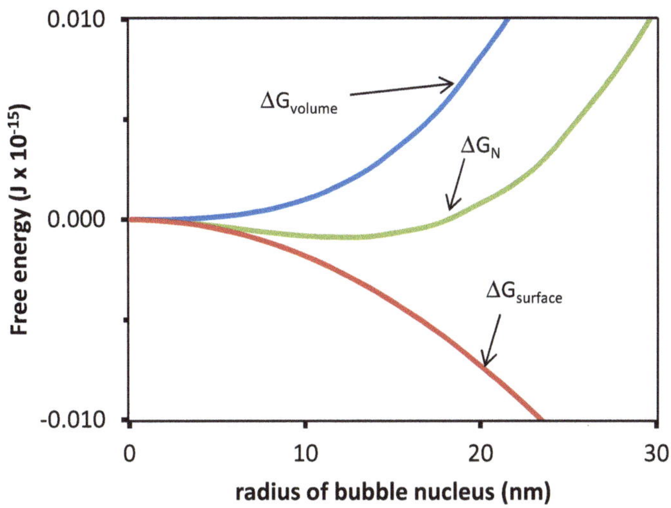

Fig. 8.12 Free energy of a bubble in a conical cavity with an internal angle $\Gamma = 60°$ in a mineral with a wetting angle of 160° (e.g., magnetite) for the case in which the melt contains 5.2 wt% H_2O at 1000 °C and 200 MPa. The melt is vapor-undersaturated by about 0.1 wt% H_2O, but the bubble in the cavity is stable with a radius of about 11 nm. If the H_2O concentration in the melt increases the bubble will grow; once H_2O reaches the saturation concentration the bubble will grow without limit and there will be no delay in bubble nucleation because the bubble already exists

and

$$v = \frac{\sin^2\Gamma}{\cos^2(\Gamma+\psi)} - \frac{\sin^2\Gamma\tan(\Gamma+\psi)}{\cos(\Gamma+\psi)} \tag{8.31}$$

and $\psi = 180 - \theta_w$.

The values of the surface, volume, and net free energies of bubbles in conical cavities determined with Eqs. (8.29–8.31) are plotted in Fig. 8.12 as functions of bubble radius. The key thing to observe in Fig. 8.12 is that there is a minimum in ΔG_N at a finite bubble radius even when the irreversible free energy change ΔG_R for bubble nucleation (i.e., the reaction affinity) is positive and the melt is vapor-undersaturated. If ΔG_R becomes negative (melt becomes vapor saturated) then there is no maximum in the ΔG_N curve, and bubbles can grow from infinitesimal sizes without any kinetic inhibition.

Although microlites with complex embayed geometries are unlikely to exist in texturally equilibrated cumulates, there are many possible instances of three-grain contacts with melt that satisfy the conditions for spontaneous heterogeneous nucleation under fluid undersaturated conditions. To understand this it is necessary to consider the effects of crystal anisotropy on the shapes of pores (e.g., Laporte and Watson 1995; Laporte and Provost 2000). When anistropic crystals form equilibrated textures with the melt pore network there are many orientations of the crystals that favor the formation of faceted solid-melt interfaces. If three faceted interfaces meet in the corner of a melt-filled pore, then the three-grain boundary that emerges from the apex of the pore will be free of melt even if the melt-crystal dihedral angle would normally be less than the limiting value of 60° that applies in isotropic crystal-melt systems (Price et al. 2006). Any pore bounded by at least three faceted crystals will therefore have the possibility of being subtended by a sufficiently small angle to stabilize metastable bubbles in marginally fluid-undersaturated melts. As the melt evolves into a state of equilibrium saturation and then supersaturation these metastable nuclei would grow without hindrance and there would be no delay as is seen in homogeneous nucleation. The dry three-grain contacts that emanate from faceted pores are very common in experimentally equilibrated aggregates of anisotropic crystals (e.g., van Bargen and Waff 1986; Laporte and Watson 1995; Laporte and Provost 2000; Ikeda et al. 2002). Even if only one pore in 1000 had the suitable geometry for heterogeneous bubble nucleation, a volume of cumulate 20 cm on each side containing 1 cm crystals would contain about 8 nucleation sites; if the crystals were about 1 mm in size then there would be thousands of sites.

It is therefore probable that vapor saturation occurs without any kinetic impediment in texturally equilibrated partially molten cumulates containing magnetite, at number densities much greater than those required for the isolated growth of very large vapor-filled pore networks capable of spontaneous upward migration.

Ostwald Ripening

As discussed above, the water content of melt at equilibrium with a curved melt-vapor interface is a decreasing function of the radius of curvature of the bubble wall. The concentration of H_2O in the melt next to small bubbles is therefore greater than that next to large bubbles, which sets up an activity gradient in a polydisperse emulsion. Diffusion works to erase the concentration gradient, causing small bubbles to redissolve and feed solute to the melt where it can diffuse to feed the growth of larger bubbles. The system will evolve through time until all of the vapor has been transferred to a single large bubble, through the process known as Ostwald ripening (e.g. Lifshitz and Slyozov 1961).

Because the timescale of Ostwald ripening is governed by diffusion between bubbles, it is similar to the diffusive timescale of bubble growth defined by Eq. 8.26, i.e., 1–10 years even for diffusion distances in the tens of cm. It is therefore likely that Ostwald ripening would affect the bubble distribution so as to favor the maturation of the cumulate from a high bubble number density to a low number density over periods of years. This process will tend to destroy bubbles moving in the microbubble regime, replacing them with smaller numbers of larger bubbles in the stranded bubble regime, and might conceivably also permit continued growth of stranded bubbles into the network regime where they are large enough to migrate.

However if the growth of a bubble within a pore chamber causes it to impinge on the all of the walls of the chamber, then any further growth will cause a decrease in the radius of curvature because it is henceforth required to protrude into progressively narrower cuspate terminations of the pore chamber where it enters pore throats; this is true whether or not the bubble partially wets the crystals as long as its wetting angle is less than 90° (see Fig. 8.1). In this case large bubbles only continue to grow larger than the inscribed sphere within their pore chambers until their internal capillary pressure has grown as large as that in neighboring smaller bubbles. Once this metastable state of equalized capillary pressure has been achieved, neither the large nor the small bubbles will spontaneously change their sizes. Since extended bubbles in the network bubble regime have many constricted terminations and must therefore exist in a state of high capillary pressure (Fig. 8.1), there is no driving force to cause them to grow at the expense of their neighbors, which will have the same capillary pressure even if they are much smaller. Increased global supply of water by continued crystallization of the interstitial melt will drive each bubble to grow incrementally, while maintaining a capillary pressure equivalent to that of its neighbors. Ostwald ripening may destroy microbubbles and prevent the vapor transport in the microbubble regime but once bubble sizes have ripened into the stranded bubble regime no further consolidation of the bubble size distribution is possible.

Discussion and Conclusions

Large degrees of melting of the mantle produces melts that contain rather low volatile contents but assimilation of continental crust containing hydrous or carbonate minerals would elevate them considerably. A plausible guess at the volatile content of the parental Bushveld B1 magma is 0.6 wt% H_2O, 0.2 wt% CO_2, and 100 to 400 ppm Cl. This composition would immediately be saturated with CO_2 vapor at the time of emplacement at a depth of several km in the crust. Subsequent crystallization would eventually drive a felsic intercumulus residual melt into saturation with a CO_2-poor aqueous fluid. If Cl concentration in the primitive melt is 100 ppm then the system will not evolve a separate hydrosaline fluid phase, but the more Cl-rich composition of the marginal B1 sills would. All of the possible fluid phases would be buoyant.

Current understanding of cumulate maturation suggests that the system achieves a stable topology at a relatively early stage, with an essentially static amount of trapped liquid and an interstitial melt-filled pore network that is locally in textural equilibrium. A typical mafic magma probably is saturated at mid- to upper-crustal conditions with a carbonic fluid containing very little H_2O. First boiling will most likely remove this early fluid from the system, so that cumulates may form initially with volatile concentrations such that a vapor phase would be at equilibrium with the melt if bubbles of it could nucleate.

Possible means of bubble migration can be divided into the microbubble, stranded, and extended or network regimes, depending on the relative sizes of bubbles and crystals. In the microbubble regime, bubbles are smaller than the narrowest constrictions in pore throats. For the 2–10 mm grain sizes typical of major layered intrusions (e.g., Bushveld Complex, Stillwater Complex, Skaergaard Intrusion), the mobility of individual bubbles through the pore network is extremely limited. In 2 mm-sized cumulates, the rate of ascent of bubbles is always less than the lowest plausible rate of accumulation of new cumulates. In coarser grain sizes, at early stages of compaction while carbonic fluids are being exsolved, particularly in the coarsest possible textures, the fluid may be able to escape from the cumulate at rates greater than the rate of addition of new cumulates on the top. At the stage in the evolution of the cumulate when chloride-rich aqueous fluids are present, it takes more than 1000 years for a bubble to escape from a 100 m layer even for the coarse-grained cumulate.

Comparing the 100's to 1000's of year timescale of microbubble ascent to the timescale of years for diffusive re-equilibration and Ostwald ripening it is clear that microbubble migration would be impossible for any significant distance. Small bubbles will grow to large enough sizes to become stranded by capillary forces in relatively short times whether number density is high (Ostwald ripening) or low (rapid growth from supersaturated melt).

In the stranded bubble regime the effect of capillary forces is to trap any bubble larger than the pore throat radius unless the grain size of the cumulate is greater than 2 cm for uncompacted orthocumulates, rising to 4 cm for cumulates with 10 % trapped liquid. Since the typical range of grain sizes in layered intrusions like the Bushveld Complex is on the order of 1–5 mm, I conclude that individual bubbles with sizes on the order of pore sizes will remain stranded. There may be rare examples of extremely coarse-grained plutons where bubbles can migrate but this will be the exception rather than the rule.

In cases where the interstitial melt is being driven upward through pore throats by compaction the resulting hydraulic pressure gradient is insufficient to overcome the capillary pressure tending to oppose deformation of the bubbles.

In the network or extended bubble regime, bubbles would only be capable of upward motion if a single bubble were to grow from each ~ 10 cm cube of cumulate for rocks similar to Bushveld cumulates with 2 mm diameter crystals, or for each ~20 cm cube of very coarse grained cumulates with 2 cm diameter crystals. Any bubbles forming at any higher number density would be stranded. These number densities correspond with the largest oikocrysts commonly observed in heteradcumulate textures and hence may be reasonable if one considers water vapor as a postcumulus phase like any other. However in order to grow to such large sizes bubbles must either nucleate extremely rarely, or must grow by devouring smaller neighboring bubbles through Ostwald ripening.

Nucleation theory indicates that although homogeneous nucleation of vapor bubbles is very strong discouraged by capillary effects, there are probably numerous locations in a maturing cumulate pile where heterogeneous nucleation is favored even under conditions of slight vapor undersaturation once magnetite has appeared as an intercumulus phase. Since magnetite is likely to appear long before saturation with an aqueous phase has occurred, the initial number density of vapor bubbles is therefore expected to be considerably higher than that required to promote the diffusive growth of individual bubbles large enough to rise in the network regime.

The driving force for Ostwald ripening is provided by the capillary pressure in the bubbles, which depends on the curvature of the melt-vapor interface. As soon as bubbles grow large enough to impinge on their pore walls their radii of curvature will begin to decrease, which removes the chemical potential gradients driving

Ostwald ripening. There is therefore no means for the system to diminish its total free energy by growing larger bubbles, and the bubbles must remain stranded in their pores.

The foregoing discussion of vapor mobility has presumed that the pore network exists in a solid framework comprising only the cumulus phases such as plagioclase, olivine, or pyroxenes. As long as the crystallization of interstitial melt proceeds by the addition of overgrowths to existing grains, such as growth of albitic rims on plagioclase primocrysts, the pore network should continue to show dihedral angles and topology that follow the assumptions laid out above. However by the stage at which a felsic interstitial melt remains and the vapor phase is water-rich, the process of second boiling is probably accompanied by the crystallization of accessory phases including biotite, magnetite, K-feldspar, quartz, rutile, etc. If late-forming crystals are not forming as epitaxial overgrowths then they will grow into the pores and impinge on the crystals on the opposite sides, occluding the pores. Modeling of this process by Cheadle et al. (2004) produced complete occlusion of the pore network at porosities on the order of 8–11 %; similar conclusions were reached by Bryon et al. (1996) from textural analysis of a natural granodiorite. The constraints on bubble migration in cumulates containing postcumulus mineral phases are therefore much more stringent than they would be for the simple geometries considered here.

I conclude that although the separation of an immiscible vapor or hydrosaline fluid is an inevitable consequence of cooling and solidification of interstitial melt in cumulate piles, it is difficult for any aqueous fluid bubble to move more than a few metres from its site of nucleation until the entire system has cooled to temperatures below its solidus.

References

Bagdassarov N, Dorfman A, Dingwell DB (2000) Effect of alkalis, phosphorus, and water on the surface tension of haplogranite melt. Am Mineral 85:33–40

Barnes SJ (1989) Are Bushveld U-type parent magmas boninites or contaminated komatiites? Contrib Mineral Petrol 101:447–457

Barnes SJ, Fiorentini ML, Austin P, Gessner K, Hough RM, Squelch AP (2008) Three-dimensional morphology of magmatic sulfides sheds light on ore formation and sulfide melt migration. Geology 36:655–658

Barnes S-J, Maier WD, Curl EA (2010) Composition of the marginal rocks and sills of the Rustenburg Layered Suite, Bushveld Complex, South Africa: implications for the formation of platinum-group element deposits. Econ Geol 105:1491–1511

Boorman SL, McGuire JB, Boudreau AE, Kruger FJ (2002) Fluid overpressure in layered intrusions: formation of a breccia pipe in the Eastern Bushveld Complex, Republic of South Africa. Mineral Dep 38:356–369

Boudreau AE (1999) Chromatographic separation of the platinum-group elements, gold, base metals and sulfur during degassing of a compacting and solidifying igneous crystal pile. Contrib Mineral Petrol 134:174–185

Boudreau AE (2004) Palladium, a program to model the chromatographic separation of the platinum-group elements, base metals and sulfur in a solidifying pile of igneous crystals. Can Mineral 42:393–403

Boudreau AE, McCallum IS (1992) Concentration of platinum-group elements by magmatic fluids in layered intrusions. Econ Geol 87:1830–1848

Boudreau AE, Philpotts AR (2002) Quantitative modeling of compaction in the Holyoke flood basalt flow, Hartford Basin, Connecticut. Contrib Mineral Petrol 144:176–184

Bowers TS, Helgeson HC (1983) Calculation of the thermodynamic and geochemical consequences of nonideal mixing in the system H2O–CO2–NaCl on phase relations in geologic systems: equation of state for H2O–CO2–NaCl fluids at high pressures and temperatures. Geochim Cosmochim Acta 47:1247–1275

Bulau JR, Waff HS, Tyburczy JA (1979) Mechanical and thermodynamic constraints on fluid distribution in partial melts. J Geophys Res 84(B11):6102–6108

Bureau H, Métrich N, Pineau F, Semet MP (1998a) Magma-conduit interaction at Piton de la Fournaise volcano (Réunion Island): a melt and fluid-inclusion study. J Volcan Geoth Res 84:39–60

Bureau H, Pineau F, Métrich N, Semet MP, Javoy M (1998b) A melt and fluid inclusion study of the gas phase at Piton de la Fournaise volcano (Réunion Island). Chem Geol 147:115–130

Bryon DN, Atherton MP, Cheadle MJ, Hunter RH (1996) Melt movement and the occlusion of porosity in crystallizing granitic systems. Mineral Mag 60:163–171.

Cartigny P, Pineau F, Aubaud C, Javoy M (2008) Towards a consistent mantle carbon flux estimate: insights from volatile systematics (H2O/Ce, dd, CO2/Nb) in the North Atlantic mantle (14° N and 34° N). Earth Planet Sci Lett 265:672–685

Cawthorn RG (1996) Re-evaluation of magma compositions and processes in the uppermost Critical Zone of the Bushveld Complex. Mineral Mag 60:131–148

Cawthorn RG, Walraven F (1998) Emplacement and crystallization time for the Bushveld Complex. J Petrol 39:1669–1687

Cheadle MJ, Elliott MT, McKenzie D (2004) Percolation threshold and permeability of crystallizing igneous rocks: the importance of textural equilibrium. Geology 32:757–760

Chung H-Y, Mungall JE (2009) Physical controls on the migration of immiscible fluids through partially molten silicates, with special reference to magmatic sulfide ores. Earth Planet Sci Lett 286:14–22

Cluzel N, Laporte D, Provost A, Kannewischer I (2008) Kinetics of heterogeneous bubble nucleation in rhyolitic melts: implications for the number density of bubbles in volcanic conduits and for pumice textures. Contrib Mineral Petrol 156:745–763

Duan Z, Moller N, Weare JH (1995) Equation of state for the NaCl–H2O–CO2 system: prediction of phase equilibria and volumetric properties. Geochim Cosmochim Acta 59:2869–2882

Durst F (2008) Fluid mechanics. Springer, Berlin, pp 727

Fiorentini ML, Barnes SJ, Maier WD, Burnham OM, Heggie G (2010) Global variability in the platinum-group element contents of komatiites. J Petrol 52:83–112

Francis AW (1933) Wall effect in falling ball method for viscosity. Journal of Applied Physics 4:403–406.

Frank FC (1968) Two-component flow model for convection in the Earth's upper mantle. Nature 220:350–352

Fu L, Bienenstock A, Brennan S (2009) X-ray study of the structure of liquid water. J Chem Phys 131 doi:10.1063/1.3273874

Gao S, Luo T-C, Zhang B-R, Zhang H-F, Han Y-W, Zhao Z-D, Hu Y-K (1998) Chemical composition of the continental crust as revealed by studies in East China. Geochim Cosmochim Acta 62:1959–1975

Gardner JE (2012) Surface tension and bubble nucleation in phonolite magmas. Geochim Cosmochim Acta 76:93–102

Gardner JE, Denis M-H (2004) Heterogeneous bubble nucleation on Fe–Ti oxide crystals in high-silica rhyolitic melts. Geochim Cosmochim Acta 68:3587–3597

Ghiorso MS, Sack RO (1995) Chemical mass transfer in magmatic processes IV. A revised and internally consistent thermodynamic model for the interpolation and extrapolation of liquid-solid equilibria in magmatic systems at elevated pressures and temperatures. Contrib Mineral Petrol 119:197–212

Gibert F, Guillaume D, Laporte D (1998) Importance of fluid immiscibility in the H2O–NaCl–CO2 system and selective CO2 entrapment in granulites: experimental phase diagram at 5–7 kbar, 900 °C and wetting textures. Eur J Mineral 10:1109–1123
Gondé C, Martel C, Pichavant M, Bureau H (2011) In situ bubble vesiculation in silicic magmas. Am Mineral 96:111–124
Gualda GAR, Ghiorso MS (2007) Magnetite scavenging and the buoyancy of bubbles in magma. Part 2: energetics of crystal-bubble attachment in magmas. Contrib Mineral Petrol 154:479–490
Hadamard J (1911) Mouvement perament lent d'une sphere liquid et visqueuse dans un liquide visqueux. Compt Rend Acad Paris 152:1735–1738
Hanley JJ, Mungall JE, Pettke T, Spooner ETC, Bray CJ (2008) Fluid and halide melt inclusions of magmatic origin in the ultramafic and lower banded series, Stillwater Complex, Montana, USA. J Petrol 49:1133–1160
Herring C (1951) Some theorems on the free energies of crystal surfaces. Phys Rev 82:87–93
Hess K-U, Dingwell DB (1996) Viscosities of hydrous leucogranitic melts: a non-Arrhenian model. Am Mineral 81:1297–1300
Hoffman DW, Cahn JW (1972) A vector thermodynamics for anisotropic surfaces I. Fundamentals and application to plane surface junctions. Surf Sci 31:368–388
Holness MB (2006) Melt-solid dihedral angles of common minerals in natural rocks. J Petrol 47:791–800
Holzheid A, Schmitz MD, Grove TL (2000) Textural equilibria of iron sulfide liquids in partially molten silicate aggregates and their relevance to core formation scenarios. J Geophys Res B 105:13555–13567
Holness MB (2006) Melt-solid dihedral angles of common minerals in natural rocks. Jour Petrol 47:791–800.
Huraiova M, Konecny P, Konecny V, Simon K, Hurai V (1996) Mafic and salic xenoliths in late tertiary alkaline basalts: fluid inclusion and mineralogical evidence for a deep-crustal magmatic reservoir in the Western Carpathians. Eur J Mineral 8:901–916
Hurwitz S, Navon O (1994) Bubble nucleation in rhyolitic melts: experiments at high pressure, temperature, and water content. Earth Planet Sci Lett 122:267–280
Iacono-Marziano G, Morizet Y, Le Trong E, Gaillard F (2012) New experimental data and semi-empirical parameterization of H2O–CO2 solubility in mafic melts. Geochim Cosmochim Acta 97:1–23
Ikeda S, Toriumi M, Yoshida H, Shimizu I (2002) Experimental study of the textural development of igneous rocks in the late stages of crystallization: the importance of interfacial energies under non-equilibrium conditions. Contrib Mineral Petrol 142:397–415
Irvine TN (1970) Heat transfer during solidification of layered intrusions. I. Sheets and sills. Can J Earth Sci. 7:1031–1061.
Kelton KF, Greer AL (2010) Nucleation in condensed matter applications in materials and biology. Pergamon materials series 15. 743 pp. Elsevier, New York
Kendrick MA, Kamenetsky VS, Phillips D, Honda M (2012) Halogen systematics (Cl, Br, I) in mid-ocean ridge basalts: a Macquarrie Island case study. Geochim Cosmochim Acta 81:82–93
Kravchuk IF, Keppler H (1994) Distribution of chlorite between aqueous fluids and felsic melts at 2 kbar and 800 °C. Eur J Mineral 6:913–923
Krumrei TV, Pernicka E, Kalimoda M, Markl G (2007) Volatiles in a peralkaline system: abiogenic hydrocarbons and F–Cl–Br systematics in the naujaite of the Ilimaussaq intrusion, South Greenland. Lithos 95:298–314
Lange RL, Carmichael ISE (1990) Thermodynamic properties of silicate liquids with emphasis on density, thermal expansion, and compressibility. Rev Mineral 24:25–64
Laporte D, Provost A (2000) Equilibrium geometry of a fluid phase in a polycrystalline aggregate with anisotropic surface energies: dry grain boundaries. J Geophys Res 105(B11):25937–25953
Laporte D, Watson EB (1995) Experimental and theoretical constraints on melt distribution in crustal sources: the effect of crystalline anisotropy on melt interconnectivity. Chem Geol 124:161–184

Lifshitz IM, Slyozov VV (1961) The kinetics of precipitation from supersaturated solid solutions. J Phys Chem Solids 19:25–50
Liu Y, Zhang Y, Behrens H (2005) Solubility of H2O in rhyolitic melts at low pressures and a new empirical model for mixed H2O–CO2 solubility in rhyolitic melts. J Volcanol Geotherm Res 143:219–235
Maaløe S, Scheie A (1982) The permeability controlled accumulation of primary magma. Contrib Mineral Petrol 81:350–357
Mangan M, Sisson T (2000) Delayed, disequilibrium degassing in rhyolite magma: decompression experiments and implications for explosive volcanism. Earth Planet Sci Lett 183:441–455
Mathez EA, Webster JD (2005) Partitioning behavior of chlorine and fluorine in the system apatite-silicate melt-fluid. Geochim Cosmochim Acta 69:1275–1286
Mathez EA, Hunter RH, Kinzler R (1997) Petrologic evolution of a partially molten cumulate: the Atok section of the Bushveld Complex. Contrib Mineral Petrol 129:20–34
McKenzie D (1984) The generation and compaction of partially molten rock. J Petrol 25:713–765
Meurer WP, Claeson DT (2002) Evolution of crystallizing interstitial liquid in an arc-related cumulate determined by LA ICP-MS mapping of a large amphibole oikocryst. J Petrol 43:607–629
Meurer WP, Klaber S, Boudreau AE (1997) Discordant bodies from olivine-bearing zones III and IV of the Stillwater Complex, Montana—evidence for postcumulus fluid migration and reaction in layered intrusions. Contrib Mineral Petrol 130:81–92
Meurer WP, Willmore CC, Boudreau AE (1999) Metal redistribution during fluid exsolution and migration in the middle banded series of the Stillwater Complex, Montana. Lithos 47:143–156
Michael P (1995) Regionally distinctive sources of depleted MORB: evidence from trace elements and H2O. Earth Planet Sci Lett 131:301–320
Minarik WG (1998) Complications to carbonate melt mobility due to the presence of an immiscible silicate melt. J Petrol 39:1965–1973
Mungall JE, Su Shanguo (2005) Interfacial tension between magmatic sulfide and silicate liquids: constraints on kinetics of sulfide liquation and sulfide migration through silicate rocks. Earth Planet Sci Lett 234:135–149.
Navon O, Chekmir A, Lyakhovsky V (1998) Bubble growth in highly viscous melts: theory, experiments, and autoexplosivity of dome lavas. Earth Planet Sci Lett 160:763–776
Ni H, Zhang Y (2008) H2O diffusion models in rhyolitic melt with new high pressure data. Chem Geol 250:68–78
Ochs FA, Lange RA (1999) The density of hydrous magmatic liquids. Science 283:1314–1317
Price JD, Wark DA, Watson EB, Smith AM (2006) Grain-scale permeabilities of faceted polycrystalline aggregates. Geofluids 6:302–318
Richards JP (2011) Magmatic to hydrothermal metal fluxes in convergent and collided margins. Ore Geol Rev 40:1–26
Sachs PM, Hansteen TH (2000) Pleistocene underplating and metasomatism of the lower continental crust: a xenolith study. J Petrol 41:331–356
Schafer FN, Foley SF (2002) The effect of crystal orientation on the wetting behaviour of silicate melts on the surfaces of spinel peridotite minerals. Contrib Mineral Petrol 143:254–261
Schiano P, Clocchiatti R, Boivin P, Medard E (2004) The nature of melt inclusions inside minerals in an ultramafic cumulate from Adak volcanic center, aleutian arc: implications for the origin of high-Al basalts. Chem Geol 203:169–179
Shirley DN (1986) Compaction of igneous cumulates. J Petrol 94:795–809
Scoates JS, Wall CJ, Friedman RM, VanTongeren JA, Mathez EA (2012) Age of the Bushveld Complex. Goldschmidt 2012 Conference Abstracts
Sparks RSJ, Huppert HE, Kerr RC, McKenzie DP, Tait SR (1985) Postcumulus processes in layered intrusions. Geol Mag 122:555–568
Sweeney SM, Martin CL (2003) Pore size distributions calculated from 3-D images of DEM-simulated powder compacts. Acta Material 51:3635–3649
Takei Y, Holtzman BK (2009) Viscous constitutive relations of solid-solid composites in terms of grain boundary contiguity: 1. Grain boundary diffusion control model. J Geophys Res 114:B06025. doi:10.1029/2008JB005850

Tegner C, Thy P, Holness MB, Jakobsen JK, Lesher CE (2009) Differentiation and compaction in the Skaergaard intrusion. J Petrol 50:813–840
Tharp TM, Loucks RR, Sack RO (1998) Modeling compaction of olivine cumulates in the Muskox intrusion. Am J Sci 298:758–790
Toramaru A (1995) Numerical study of nucleation and growth of bubbles in viscous magmas. J Geophys Res 100(B2):1913–1931
Turnbull D (1950) Kinetics of heterogeneous nucleation. J Chem Phys 18:198–203
Unsal E, Mason G, Morrow NR, Ruth DW (2009) Bubble snap-off and capillary-back pressure during counter-current spontaneous imbibition into model pores. Langmuir 25:3387–3395
Upton BGJ, Semet MP, Joron J-L (2000) Cumulate clasts in the Bellecombe Ash Member, Piton de la Fournaise, Réunion Island, and their bearing on cumulative processes in the petrogenesis of the Réunion lavas. J Volc Geoth Res 104:297–318.
Volmer M (1939) Kinetik der Phasenbildung (Steinkopf, Leipzig), pp 438
von Bargen N, Waff HS (1986) Permeabilities, interfacial areas and curvatures of partially molten systems: results of numerical computations of equilibrium microstructures. J Geophys Res 91(B9):9261–9276
von Bargen N, Waff HS (1988) Wetting of enstatite by basaltic melt at 1350 °C and 1.0- to 2.5-GPa pressure. J Geophys Res 93(B2):1153–1158
Waff HS, Faul UH (1992) Effects of crystalline anisotropy on fluid distribution in ultramafic partial melts. J Geophys Res 97(B6):9003–9014
Wark DA, Watson EB (1998) Grain-scale permeabilities of texturally equilibrated, monomineralic rocks. Earth Planet Sci Lett 164:591–605
Webster JD (1997) Exsolution of magmatic volatile phases from Cl-enriched mineralizing granitic magmas and implications for ore metal transport. Geochim Cosmochim Acta 61:1017–1029
Webster JD (1999) Chloride and water solubility in basalt and andesite melts and implications for magmatic degassing. Geochim Cosmochim Acta 63:729–738
Webster JD (2004) The exsolution of magmatic hydrosaline chloride fluids. Chem Geol 210:33–48
Zhang Y, Shariati M, Yortsos YC (2000) The spreading of immiscible fluids in porous media under the influence of gravity. Transp Porous Media 38:117–140

Chapter 9
Platinum-Group Element Deposits in Layered Intrusions: Recent Advances in the Understanding of the Ore Forming Processes

Bélinda Godel

Abstract The major deposits of platinum-group elements (PGE) of the Earth are associated with ultramafic and mafic igneous rocks. The bulk of current PGE production is extracted from narrow stratiform horizons referred to as reefs located in the lower to central portions of large layered intrusions and is dominated by the Bushveld Complex in South Africa. The PGE-mineralized horizons occur in most case as laterally continuous and uniform layers that can extend over hundreds of kilometres along strike. The origins of these extensive PGE-rich layers remain controversial and subject to debate. Over the past ten years, technological developments have allowed the acquisition of multidisciplinary dataset at spatial resolution, detection limits and precisions that were impossible to achieve in the past. The results obtained provide additional insights into the complexity of the ores and the variability both within and between different ore deposits and highlight the necessity of adapting (for each deposit) models of formation based on all of the information available. Models of formation of PGE-reefs need to consider and integrate the superimposition of both physical and geochemical processes, over a range of scales and over a wide range of temperatures.

Keywords PGE · Cumulates · Bushveld · Reef · Ultramafic

Introduction

The platinum-group elements (PGE: Pt, Pd, Ir, Os, Rh and Ru) are highly-siderophile elements used in a range of applications including the automobile industry (e.g. autocatalysts), jewellery, medicine (e.g. pacemakers, dental alloys, cancer treatments) and the high technology sector (e.g. electric contacts and production of glass fibres). Over the past ten years, the demand for PGE has continued to increase, resulting in a boom in mineral exploration to find new resources. Currently, most of the

B. Godel (✉)
1CSIRO Mineral Resources Flagship, 26 Dick Perry Avenue,
Kensington, WA 6151, Australia
e-mail: belinda.godel@csiro.au

B. Charlier et al. (eds.), *Layered Intrusions*, Springer Geology,
DOI 10.1007/978-94-017-9652-1_9

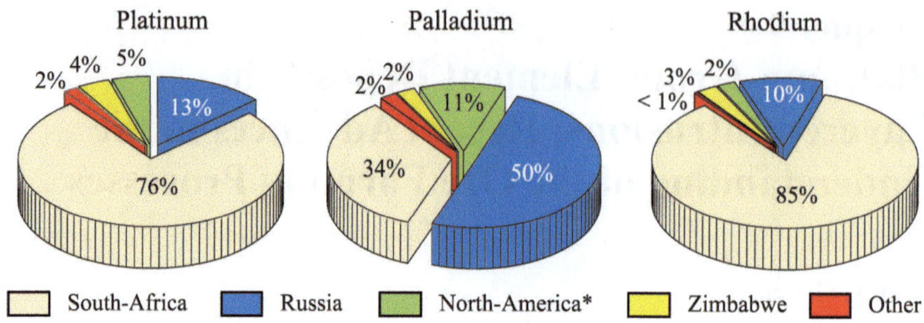

Fig. 9.1 Average world's platinum-group elements supply by country for the period 2007–2011. Data are extracted from Johnson–Mattheys (2011) and presented by commodity

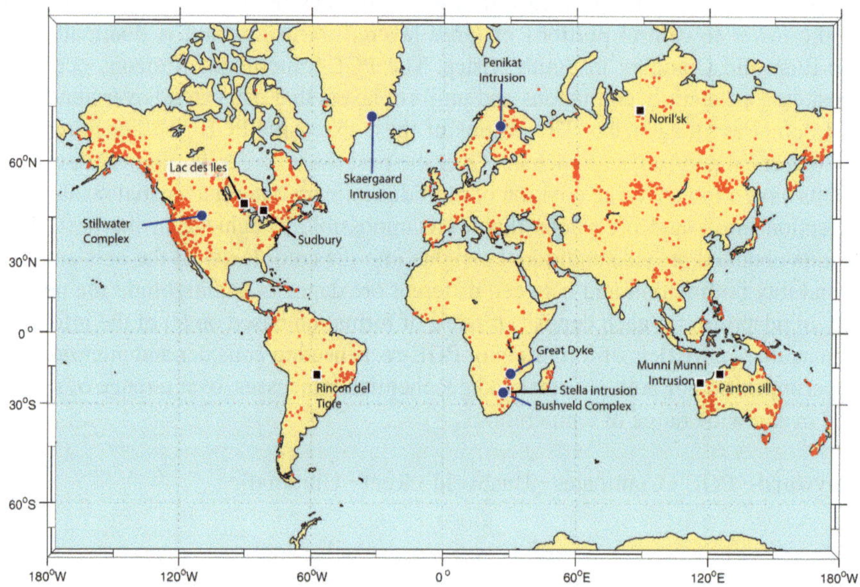

Fig. 9.2 World's map showing the main economic PGE-deposits and PGE occurrences in mafic and ultramafic rocks. The locations of PGE deposits and occurrences are extracted from the USGS PGE–Ni–Cr deposit and occurrence bibliographic database (Causey et al. 2009)

world's supply of PGE is provided by only few countries (Johnson-Matthey 2011): South Africa, Zimbabwe, Russia, Canada and the U.S.A (Fig. 9.1). There are many occurrences of PGE in the world (Fig. 9.2) with three main groups of mineralization (Cawthorn et al. 2005; Maier 2005; Naldrett 2004) occurring as: (i) stratiform PGE mineralization within mafic-ultramafic intrusions (referred to as "reefs"); (ii) a by-product of major Ni–Cu sulfide deposits and; (iii) placers.

The major deposits of PGE are associated with ultramafic and mafic igneous rocks (e.g. layered and unlayered intrusions and zoned ultramafic complexes) that usually occur in tectonically stable areas (Eales and Cawthorn 1996; Maier and Groves 2011; Wager and Brown 1968). Platinum-group element mineralization has been identified in various rock types and stratigraphic levels in layered intrusions of any age, size and origin (Maier 2005). However, most of the PGE resources on Earth are currently mined from narrow stratiform reefs located in the lower to central portions of large layered intrusions. These intrusions are Archean to early Proterozoic in age and have parental magmas with magnesian compositions (e.g. the Bushveld Complex in South Africa (Chap. 12), the Great Dyke in Zimbabwe (Oberthür 2002; Wilson and Tredoux 1990) and the Stillwater Complex in the U.S.A (McCallum et al. 1980; Raedeke and McCallum 1984, Zientek, 2002). Other PGE-rich zones have been observed in the upper portions of layered intrusions and are associated with sulfur-poor and Fe–Ti-rich rocks (e.g. Skaergaard intrusion in West Greenland (Andersen et al. 1998; Nielsen 2001), the Stella intrusion in South-Africa (Maier et al. 2003), the Rio Jacaré intrusion in Brazil (Sa et al. 2005), the Rincon del Tigre in Bolivia (Prendergast 2000), the Sonju Lake intrusion in the Duluth Complex in the U.S.A (Miller et al. 2002) or the Kotelainen intrusion in Finland (Mutanen 1997)). Platinum-group element mineralization observed in the upper part of the intrusion is in most cases uneconomic with only a few exceptions (e.g. the Stella and the Skaergaard intrusions). Significant PGE mineralization is also observed in poorly-layered (e.g. Noril'sk–Talnakh in Russia, (Arndt et al. 2005)) or unlayered intrusions (e.g. the Lac des Iles intrusion in Canada (Lavigne and Michaud 2001)).

Several schools of thoughts for the origin of PGE mineralization in layered intrusions have emerged over the years, with some authors suggesting that: (i) the PGE crystallize directly from the magma in the form of platinum-group minerals (PGM) that accumulate on the crystal pile (Hiemstra 1979); (ii) the PGE were collected by a sulfide liquid that segregated from the magma and accumulated on the cumulate pile (Campbell and Naldrett 1979; Campbell et al. 1983; Naldrett et al. 1986); or (iii) the PGE were collected by a fluid migrating upwards through the cumulate pile during compaction (Boudreau 1999; Boudreau 2008; Boudreau and Mc Callum 1992b; Willmore et al. 2000). Although research on the PGE deposits has increased in the past twenty years, the processes that controlled the PGE enrichment in layered intrusions remain controversial and vigorously debated as none of the models named above have successfully explained the structure, petrology, geochemistry and mineralogy of the PGE-rich horizons (e.g. Cawthorn et al. 2005; Mungall and Naldrett 2008).

Recent technological developments (e.g. laser ablation inductively coupled mass spectrometry (LA-ICP-MS), high-resolution X-ray computed tomography, isotope-ratio mass spectrometers, or automated scanning-electron microscope) have allowed the acquisition of multi-disciplinary datasets at spatial resolution, detection limits and precisions that were impossible to achieve in the past. These new results have provided additional insight into the genesis of the PGE deposits. This contribution presents short descriptions of selected layered intrusions hosting significant PGE mineralization and reviews the main type of PGE deposits observed. This is followed by a discussion on the residency of PGE in the rocks and the processes that may occur during ore genesis, from the partial melting of the Earth's mantle to the post-cumulus redistribution of the PGE.

Geology of Selected Layered Intrusions

Current production and future reserves of Pt, Pd and Rh are overwhelmingly dominated by deposits associated with the Bushveld Complex in South Africa (Fig. 9.1). Consequently, a large proportion of the research effort in the last ten years has been undertaken to provide additional insights into the genesis of economic PGE-reefs focussed on the Bushveld Complex. In comparison, a relatively limited number of studies published in the literature have focussed on the genesis of other significant PGE-deposits hosted within layered intrusions (e.g. the Great Dyke, the Stillwater Complex, the Penikat or the Skaergaard intrusions). The geology and the petrography of these intrusions are summarized below to provide a basis for the descriptions of mineralization types and characteristics of PGE reefs occurring in these layered intrusions.

Bushveld Complex

The Bushveld Complex in South-Africa (Eales and Cawthorn 1996; Chap. 12; Wager and Brown 1968) was intruded at 2057. 7 ± 1.6 Ma (Olsson et al. 2010) into the Kaapvaal craton subparallel to the sedimentary layering of the Transvaal Supergroup (Fig. 9.3). It is by far the largest layered intrusion in the world (Eales and Cawthorn 1996) and extends over 450 km east-west and 350 km north-south. The ultramafic and mafic rocks of the Bushveld Complex are referred as the Rustenburg Layered Suite (South African Committee For Stratigraphy 1980). The 6.5–8.7 km thick Rustenburg Layered Suite has undergone little deformation or metamorphism after its solidification (Eales et al. 1993). It comprises four outcropping areas referred to as limbs (Fig. 9.3): the western limb, eastern limb, the far western limb, the northern limb. A fifth limb (referred to as "the Bethal limb", in the south crops out as a small inlier of granophyre and comprises only Upper Zone stratigraphy that is covered by Karoo sediments (Buchanan 1979; Kruger 2005). Seismic and gravity data suggested that the Eastern and Western Limbs originally formed a continuous intrusion (Cawthorn and Webb 2001; Cole et al. 2014; Webb et al. 2004). The northern limb is separated from the other limbs by the Thabazimbi–Murchison regional fault zone that formed around 2.9 Ga and is a terrane boundary that separates the southern Archean crust and lithospheric mantle of the Kaapvaal Shield from the Pietersburg Block to the north (McDonald and Holwell 2011 and references therein). The ultramafic and mafic rocks of the Rustenburg Layered Suite have been sub-divided on the basis of cumulus mineral assemblages into five major zones (South African Committee for Stratigraphy, 1980): the Marginal zone mainly formed of fine-grained norite; the Lower Zone consisting of cyclic units of orthopyroxenite, minor harzburgite, and minor chromitite); the Critical Zone formed by pyroxenite, norite, anorthosite and chromitite; the Main Zone consisting mainly of gabbronorite and; the Upper Zone formed by anorthosite, diorite and magnetitite (Fig. 9.3 and Fig. 9.4). Although there are a number of layers enriched in PGE within the intrusion (Barnes and Maier 2002a; Naldrett 2004) only three are sufficiently enriched in PGE, and continuous enough, to be called reefs: the Upper

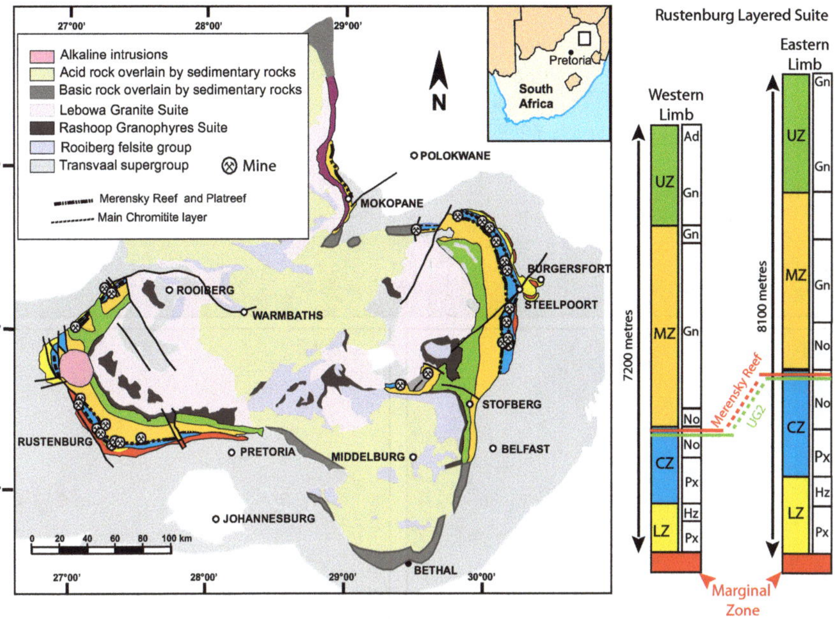

Fig. 9.3 Simplified geological map of the Bushveld Complex (South Africa) and stratigraphy of the Rustenburg Layered Suite. The geological map is adapted from Von Gruenewaldt 1979 and the stratigraphy for the Eastern and Western Limbs are modified after Cawthorn et al. ((2005). *LZ* Lower Zone; *CZ* Critical Zone, *MZ* Main Zone; *UZ* Upper Critical Zone; *Px* pyroxenite, *Hz* harzburgite; *No* Norite; *Gn* gabbronorite; *AD* Apatite Diorite

Group Chromitite 2 (UG-2), the Merensky Reef, and the Platreef (Fig. 9.3). These reefs represent the world's three largest PGE horizons (Cawthorn 1999; Naldrett 2011; Naldrett et al. 2011a). The Platreef is situated at the base of the intrusion in the Northern Limb. In contrast, the UG-2 chromitite and the Merensky Reef are located in the upper part of the Critical Zone some 2 km above the base of the intrusion and crop out both in the Western and Eastern Limbs (Fig. 9.3).

Stillwater Complex

The 2705 +/− 4 Ma old (Premo et al. 1990) Stillwater Complex is a layered ultramafic to mafic intrusion that is located in southwestern Montana in the U.S.A. (Fig. 9.5). The base of the Stillwater complex is in contact with middle to late Archean metasedimentary rocks, whereas the uppermost part of the complex is unconformably overlain by Phanerozoic sedimentary rocks. The original size of the complex is unknown as only a part (~ 44 km along its WNW strike length and ~ 6 km in thickness) of the complex is exposed along the Beartooth uplift (Page and Zientek 1985). Detailed descriptions of the petrography and stratigraphy of the complex were provided by Zientek et al. (2002 and references therein). To summarise, the layered rocks of the complex have been divided into five series (Fig. 9.5), from

Fig. 9.4 Photographs of field exposures showing different lithologies from the Bushveld Complex in South Africa, modified after Godel (2007). **a** View of lithological variations from the floor rocks to the Upper Critical Zone in the Olifants River Trough in the Eastern Limb of the Bushveld Complex. This section is referred to as the "Cameron section". **b** Field exposure of the UG1 chromitite layers and interlayered plagioclase-rich rocks (Upper Critical Zone) at Dwars River, Eastern Limb of the Bushveld Complex. **c** Field exposure showing the contact between the Lower and the Upper Critical Zones ("Cameron Section") with pyroxenite overlain by a thin layer of undulating chromitite overlain by anorthosite. **d** and **e** Examples of field outcrop of Merensky Reef in the Eastern Limb of the Bushveld Complex

bottom to top (Barnes and Naldrett 1986; McCallum et al. 1980; Page and Zientek 1985; Zientek et al. 1985): the Basal series consists of norites and bronzitites which contain minor amounts of base-metal sulfide minerals; the Ultramafic series is formed essentially of harzburgite and bronzitite, with minor chromite seams; the

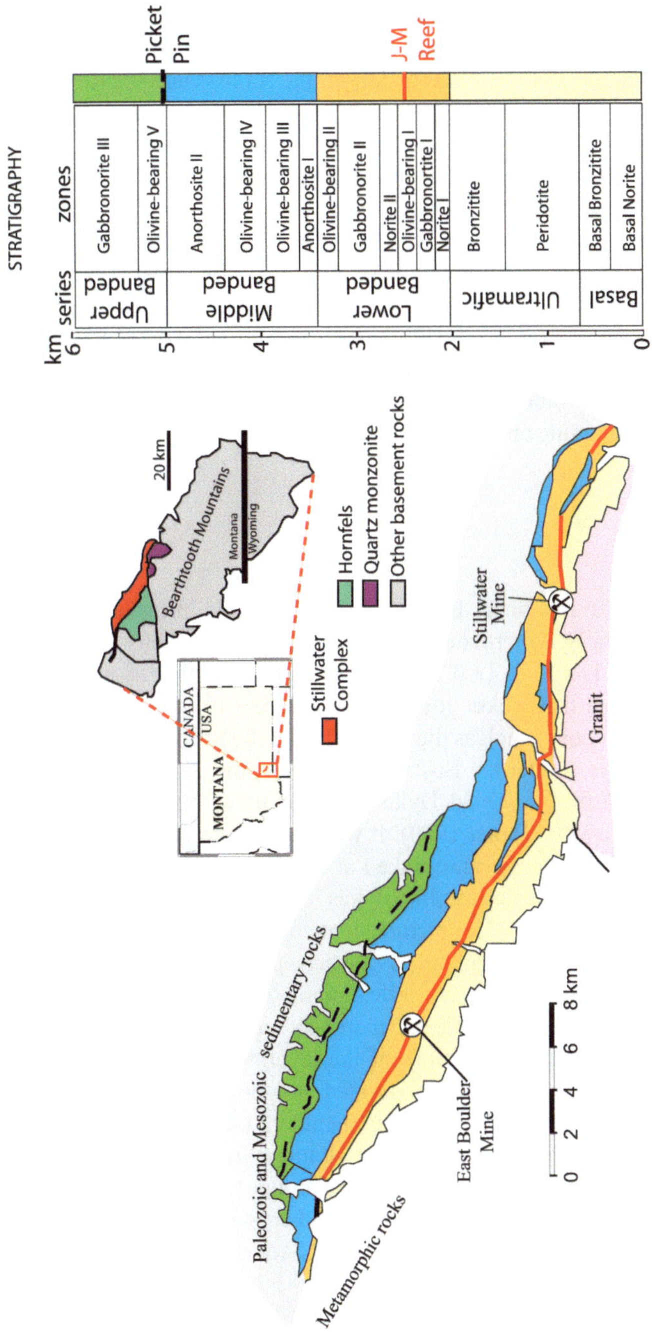

Fig. 9.5 Simplified geological map and stratigraphy of the Stillwater Complex (Montana). The geological map and the stratigraphy are modified after Zientek et al. (2002, 1985)

Lower, Middle, and Upper Banded series contain mainly norite, gabbronorite, olivine gabbronorite, troctolite and anorthosite. Several layers enriched in PGE are hosted within the complex, amongst them the Johns-Manville Reef (J-M Reef) and the Picket Pin deposit (Fig. 9.5). In contrast to the Bushveld Complex, the complex has been affected by several metamorphic and/or hydrothermal processes which have triggered alteration of the silicate and the sulfide minerals (Czamanske and Loferski 1996; Godel and Barnes 2008a; Godel and Barnes 2008b; Lechler et al. 2002; Page 1976; Polovina et al. 2004). The alteration commonly observed in the Stillwater Complex may notably be related to: (i) the initial crystallisation or cooling of the complex; (ii) the intrusion of quartz monzonite near the base of the complex (Nunes and Tilton 1971); (iii) the intrusion of mafic dikes and sills that crosscut the complex (Czamanske and Zientek 1985); (iv) greenschist facies metamorphism at 1.7 to 1.6 Ga (Nunes and Tilton 1971; Page 1977) or (v) the Creteaceous Rocky Mountains Laramide uplift (Polovina et al. 2004).

Great Dyke

The Great Dyke is a linear layered mafic-ultramafic intrusion that cuts across the Archean and early Proterozoic granites and greenstone belts of the Zimbabwe craton (Fig. 9.6). The 2575.4 ± 0.7 Ma Great Dyke (Oberthür et al. 2002) has a highly elongated shape that makes it distinctly different to other large layered intrusion hosting PGE deposits such as the Bushveld and the Stillwater Complexes (Fig. 9.6). The Great Dyke is after the Bushveld Complex, the second largest single resource of PGE on Earth. The Great Dyke aligns along ~550 km in the NNE direction and is ~11 km wide (Worst 1960). It was emplaced as a series of initially isolated magma chambers that became linked at their highest levels during later stages of intrusion. The presence of satellite dykes located parallel to the intrusion are interpreted to reflect craton wide fractures that controlled the emplacement of the Great Dyke (Wilson and Prendergast 1989). The aspect ratio of the intrusion profoundly affected the rock types, layering style, the mineral composition and the mineralized ore bodies (Wilson 1996). Magma chambers and sub-chambers have been defined based on the style and continuity of the layering and the structure and comprise (Fig. 9.6) the North Chamber (subdivided into the Musengeiz, Darwendale and Sebakwe Subchambers) and the South Chambers (subdivided into the Selukwe and Wedza Subchambers). The Great Dyke stratigraphy (Fig. 9.6) is formally subdivided into a lower Ultramafic Sequence and an upper Mafic Sequence. The dynamic interplay between magma emplacement and crystallization gave rise to the succession of cyclic units within the Ultramafic Sequence (Wilson 1982). These cyclic units (Fig. 9.6) consist in the top of the sequence of a lower dunite or harzburgite layer overlain by a pyroxenite layer whereas, in the lower part the cyclic units are defined by chromitite layers intercalated within dunite (Wilson 1982). The PGE-mineralized zones (Oberthür 2011) are associated with the uppermost portion of the Ultramafic Sequence and those associated with sulfides represent by far the most important PGE concentrations (e.g. the Main Sulfide Zone).

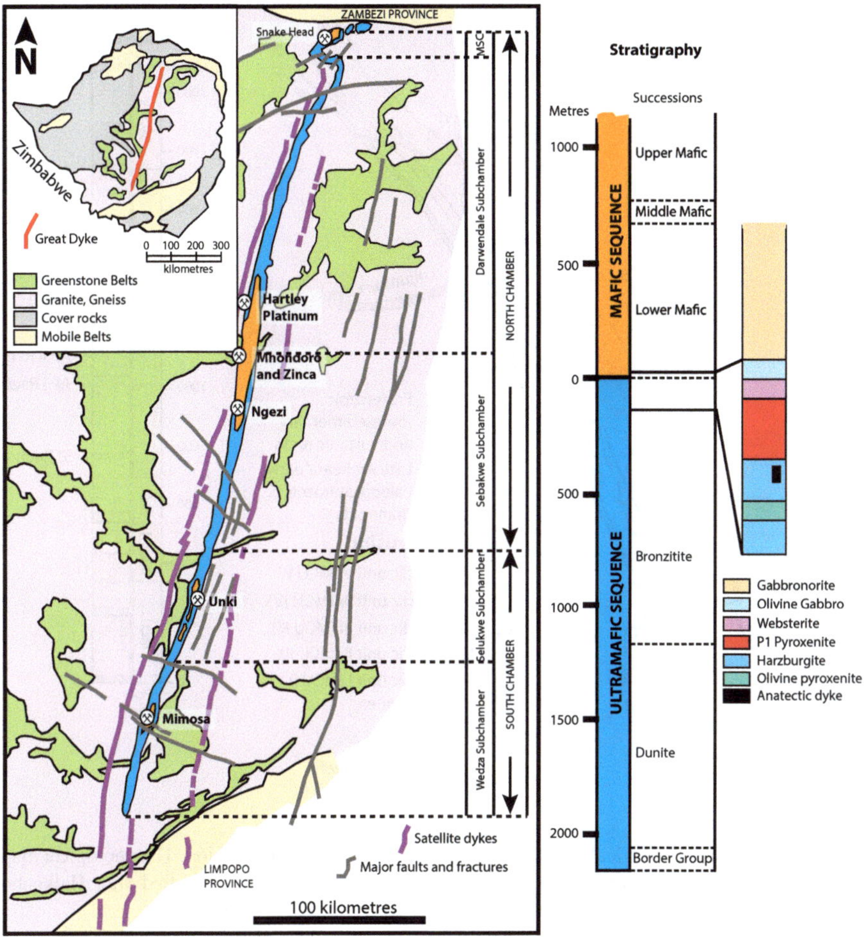

Fig. 9.6 Simplified geological map and stratigraphy of the Great Dyke of Zimbabwe. The geological map and the stratigraphy are modified after Wilson (1996) and Wilson and Prendergast (1989)

Penikat Intrusion

The 2440 Ma old (Alapieti and Lahtinen 1986) Penikat intrusion belongs to a group of intrusions forming the 300 km long Tornio–Näränkävaara Belt that extends from east Sweden to south-west Finland. The Penikat intrusion is located ~70 km south of the Arctic Circle on the northern side of the Gulf of Bothnia (Fig. 9.7). The original shape of the intrusion is uncertain but its present surface is ~23 km long and 1.5 to 3.5 km wide. The original intrusion has been fragmented into five blocks (referred to as Sompujärvi, Kilka, Yli-Penikka, Keski-Penikka, and Ala-Penikka blocks, Fig. 9.7) during the Svecokarelidic orogeny. The mafic Loljunmaa Dyke located east of the Kilka block is co-magmatic to the intrusion and may represent a part of its feeder channel. The intrusion has been divided into three main

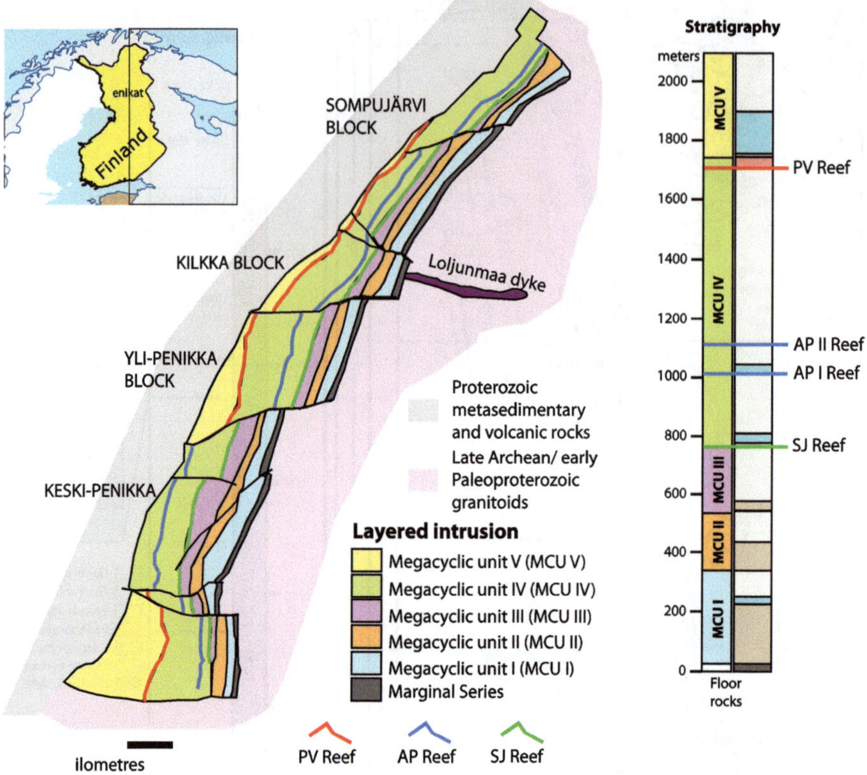

Fig. 9.7 Simplified geological map and stratigraphy of the Penikat intrusion. The geological map is modified after Alapieti and Lahtinen (1986) and the stratigraphy is modified after Halkoaho et al. (1990)

units: the Marginal series, the Layered series and the granophyres (Alapieti and Lahtinen 1986). The Layered series consists of cyclic units of ultramafic, gabbroic, and anorthositic rocks which have further been divided, based on the variation in occurrence of cumulus minerals, into five megacyclic units referred to as Megacyclic Unit I to V (Fig. 9.7), from base to top. The basal units (Megacyclic units I to III) consist mainly of ultramafic rocks with minor mafic rocks where the upper units (Megacyclic units IV and V) are mainly formed by mafic rocks with minor ultramafic rocks (Alapieti and Lahtinen 2002). The PGE-rich mineralization is found in three distinct layers (SJ, AP and PV reefs) within the Megacyclic unit IV consisting, from bottom to top, of a 10 to 20 m thick layer of ultramafic cumulates overlain by a thick sequence of gabbroic and anorthositic cumulates (Alapieti and Lahtinen 1986). The Penikat Intrusion underwent complex deformation and metamorphism during the Svecokarelian Orogeny resulting in pervasive alteration of the cumulate rocks (Alapieti and Halkoaho 1995). The majority of the magmatic minerals have been altered to varying extents however augite grains are relatively well preserved (Alapieti and Halkoaho 1995).

Skaergaard Intrusion

The 55.65 ± 0.3 Ma old (Hirschmann et al. 1997) Skaergaard Intrusion in East Greenland formed during the continental break up in the North Atlantic and was emplaced within the Tertiary East Greenland plateau basalts (Nielsen 1975). The intrusion is in a plan view ~11 km long and ~8 km wide (Fig. 9.8). The intrusion was originally interpreted to have formed by single event of magma injection into a magma chamber that crystallized and fractionated under generally closed-system conditions (McBirney 1996; Wager and Brown 1968). Holness et al. (2007) showed that several pulses of tholeitic magmas were involved in the formation of the Skaergaard Intrusions and that intrusion cooled in a closed system after the magmas had homogenised into the chamber. The internal structure of the intrusion is divided into three series: the Layered Series (LS), the Marginal Border Series (MBS) and the Upper Border Series (UBS). Details of the stratigraphy, petrography and geochemistry of the intrusion are detailed in Tegner et al. (this volume) and are not repeated here. Palladium and gold mineralization are observed in the uppermost past (~100 m) of the Middle Zone (Andersen et al. 1998; Holwell and Keays 2014; Nielsen 2001). This stratiform group of layers rich in Pd and Au is referred to as the Platinova Reef (Bird et al. 1991).

Characteristics of the Different Deposit Types

Platinum-group element mineralization classification have mainly been proposed based on the types of parental magmas from which the rocks crystallized (Naldrett 2004) or based on the stratigraphic setting of the mineralization and the rock types they are associated with (Maier 2005; Maier et al. 2013). Based on a simplified version of Maier's (2005) classification, the PGE-deposits associated with layered intrusions can mainly be classified into: (i) "Contact-type" PGE mineralization; (ii) PGE-mineralization associated with large chromitite layers at or near the base of macrorhythmic cycles; (iii) PGE-mineralization located in the lower- to middle part of layered intrusions and; (iv) PGE-mineralization located in the upper part of layered intrusions and associated with Fe–Ti oxide-rich rocks (Fig. 9.9).

"Contact-Type" PGE-Mineralization

The "contact-type" PGE mineralization owes its name to its location at or near the base or the sidewall of layered intrusions (Fig. 9.9). The mineralized zones are characterised by the presence of disseminated to massive base metal sulfides. Deposits of this type comprise the Platreef of the Bushveld Complex in South Africa (McDonald and Holwell 2011 and references therein) layered intrusions in northern Finland such as the Portimo complex (Alapieti and Lahtinen 2002), and the Feodorov-Panski instrusion in Russia (Dubrovsky and Rundqvist 2009). Typically,

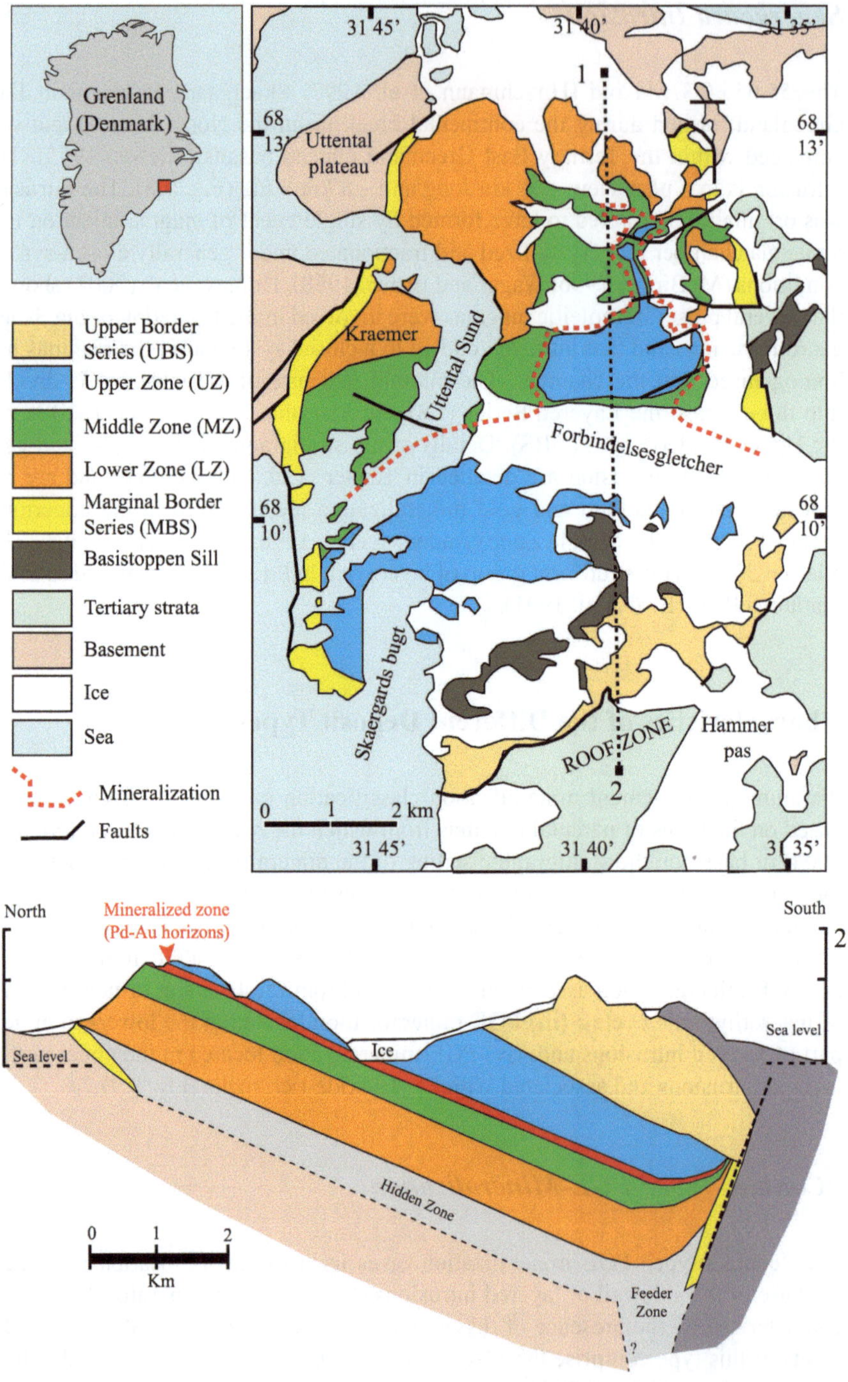

Fig. 9.8 Simplified geological map and stratigraphy of the Skaergaard intrusion (East Grenland). The geological map (**a**) and the cross-section are modified after Wager and Brown (1968) and Nielsen (2004)

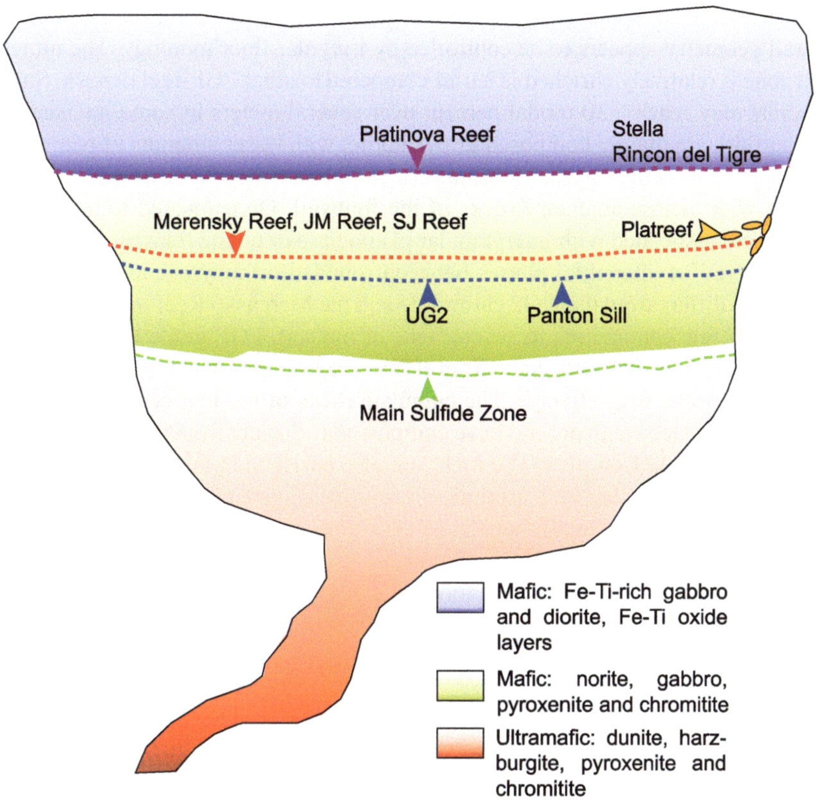

Fig. 9.9 Schematic diagram showing the location of selected PGE-horizons in ultramafic-mafic layered intrusions. The figure is modified after Maier (2005)

the sulfide abundances are higher than other PGE reef types with proportions commonly averaging 5 to 10 volume percent. The sulfides are in most cases associated with hetereogeneous mafic rocks and less commonly with ultramafic rocks except for the Platreef. The textures of the host rocks vary widely in terms of mineral abundances and crystal size and are referred to as "taxitic" or "variably textured" rocks. Xenoliths of the country rocks and veins of felsic rocks are commonly observed. Although the base metal sulfides are more abundant, the platinum-group elements tenors are usually lower (typically in the order of 10's ppm combined PGE) than those observed in reefs located in the middle portion of the intrusions.

The Platreef of the Bushveld Complex (due to its large size and attractive metal tenors) is currently the only contact-type PGE mineralization mined. The Platreef is defined as "a lithologically variable unit dominated by pyroxenite, which is irregularly mineralized with PGE, Cu and Ni, between the Transvaal metasedimentary footwall or the Archean basement and the overlying Main Zone gabbronorite" (Kinnaird et al. 2005; Manyeruke et al. 2005; McDonald and Holwell 2011; Yudovskaya et al. 2012). The mineralization is hosted within vari-textured pyroxenites, norites and gabbros with rare peridotites and chromitites. The reef thickness (10 to 400 m thick) varies from south to north with largest thickness observed in the south. Its overall geometry appears to be controlled by irregular floor topology. The mineralized zone is relatively enriched in sulfur compared to other PGE-reef deposit. Sulfide minerals may reach >30 modal percent over several meters in some intersections. The sulfides are mainly composed of pyrrhotite, with lesser amounts of pentlandite, chalcopyrite and minor pyrite. The massive sulfides are mostly localized towards the contact with metasedimentary rocks of the footwall. Disseminated to net-textured sulfides are associated with intergranular plagioclase or quartz feldspar symplectites and observed as interstitial phases between rounded orthopyroxenes. The Platreef (Fig. 9.10) differs from the UG2 chromitite and the Merensky Reef in terms of thickness (from few centimetres to several tens of meters), PGE grade (lower, ~3 ppm Pt + Pd + Rh), Pt/Pd ratios vary from 0.2 to 1 and higher Cu and Ni (0.1–0.25% and 0.15–0.36%, respectively). The cumulate rocks of the Platreef are enriched in incompatible trace elements and have compositions distinct from the cumulate rocks from the Bushveld Complex. The rocks are also enriched in initial Sr, εNd and γOs isotopes and exhibit heavy S isotopic signatures suggesting crustal contamination (Barnes and Maier 2002a; Holwell et al. 2007; Penniston-Dorland et al. 2008). Recent, multi-sulfur isotope analysis (Sharman et al. 2013) indicated that a restricted stratigraphic horizon in the Platreef footwall (the Duitschland formation) contributed to the crustal sulfur isotopic signature of the ore. The addition of crustal sulfur lead to an apparent dilution of pre-existing PGE-rich sulfide melt (Sharman et al. 2013).

PGE-Mineralization Associated With Large Chromitite Layers

The enrichment of chromitites in Ir, Os and Ru (referred to as IPGE) is almost universal whereas enrichment in the other PGE, Pt, Pd and Rh (referred to as PPGE) are rarer. Chromitite layers observed in layered intrusions are usually enriched in all PGE compared to their silicate hosts (Hiemstra 1979; Lee and Parry 1988; Naldrett et al. 2011b; Teigler and Eales 1993; Von Gruenewaldt 1986). Another example is the sub-economic Panton Sill in Western Australia (Fig. 9.2). All chromitite layers of the Bushveld Complex are enriched in PGE with concentrations varying from 0.1 to 40 ppm. However, to date only one layer is currently of grade sufficient to be mined, the UG-2 chromitite. It is the second stratiform chromitite layer within the upper group of chromitite layers (Upper Critical Zone). The UG-2 sequence (60 to 120 cm thick) consists of thick chromitite overlying a pegmatoidal feldspathic pyroxenite and, more rarely anorthosite that extends over several hundred kilometres of strike length (Davey 1992). The chromite content varies between 60 to 90 modal percent with Cr/Fe ratios of the chromites ranging from 1.26 to 1.4 (Lee 1996 and references therein; Mondal and Mathez 2007; Naldrett et al. 2011b). The UG-2 contains major PGE-mineralization that represents the largest PGE resource in the world (Cawthorn et al. 2005; Lee 1996; Mungall and Naldrett 2008). The PGE concentrations average 5 ppm with zones hosting up to 10 ppm (Gain 1985; Hiemstra 1986; Maier and Barnes 2008; McLaren and De Villiers 1982; Naldrett et al. 2011b; Von Gruenewaldt 1986). The PGE (Figs. 9.10 and 9.11) tend to be more

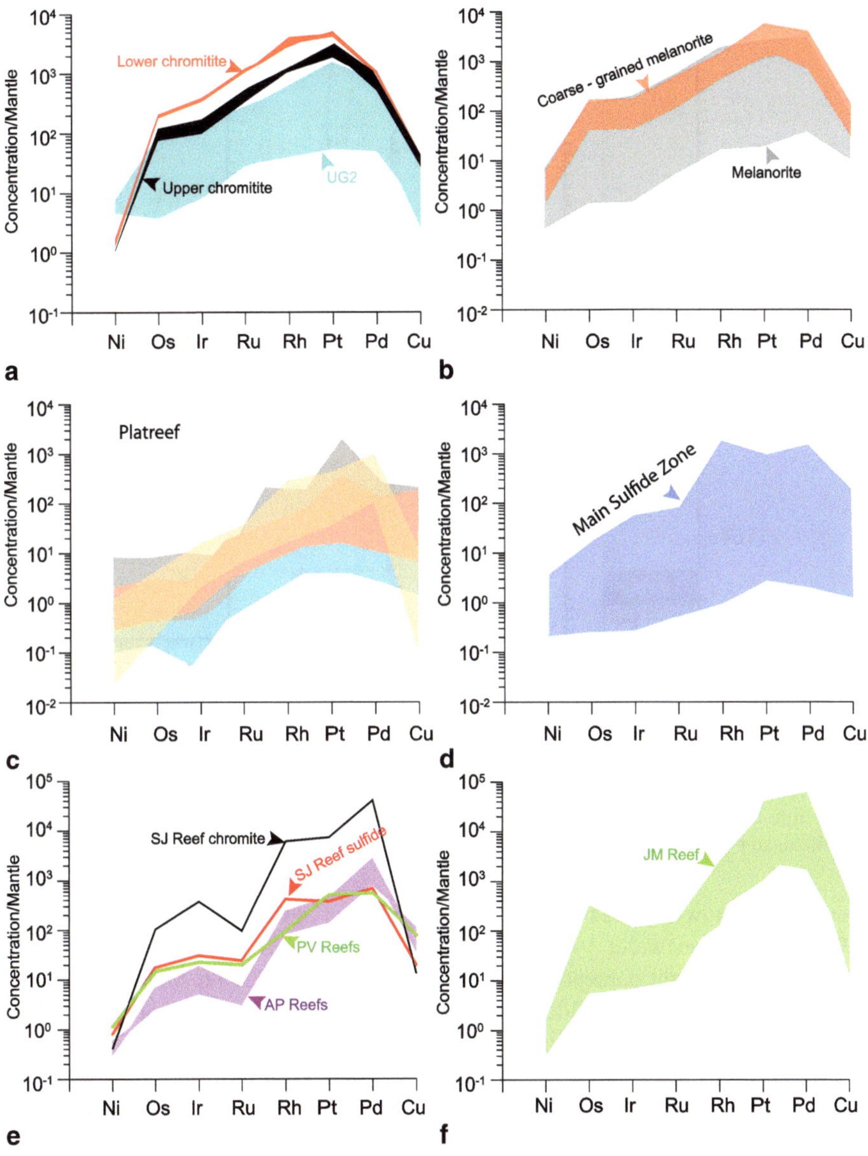

Fig. 9.10 Whole-rock base-metal and platinum-group element concentrations from various PGE deposits represented as mantle-normalized spectras **a** Base-metal and platinum-group element concentrations of chromitite layers in the Merensky Reef (Lower and Upper chromitites) and the UG2 chromitite. The data are extracted from Godel et al. (2007), Maier and Barnes (2008), Barnes and Maier (2002b). **b** Base-metal and platinum-group element concentrations of coarse-grained (i.e pegmatoidal) melanorite and melanorite from the Merensky Reef. The data are extracted from Godel et al. (2007), Barnes and Maier (2002b), Naldrett et al. (2009b) and Osbahr et al. (2013). **c** Base-metal and platinum-group element concentrations of pyroxenite from various localities within the Platreef. The data are extracted from Holwell and McDonald (2006, 2007) and Manyeruke et al. (2005). **d** Base-metal and platinum-group element concentrations across the

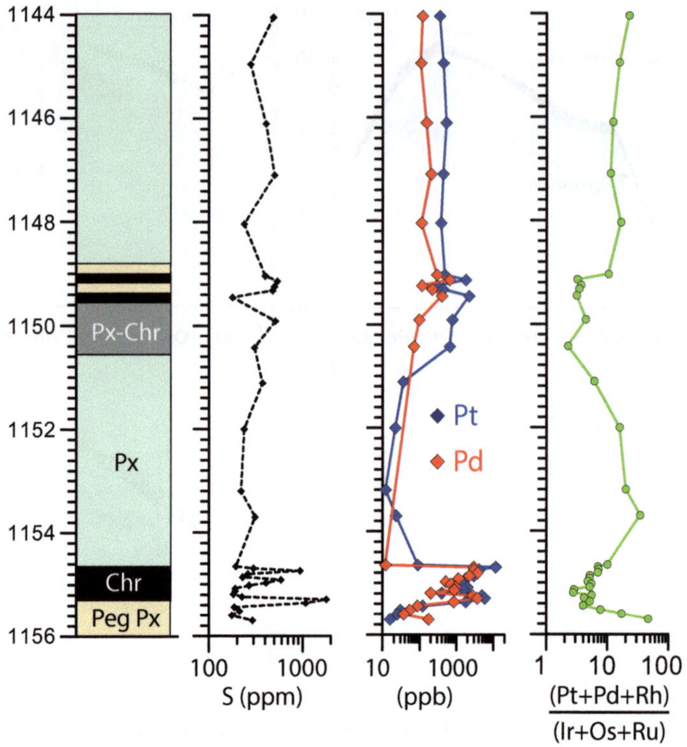

Fig. 9.11 Stratigraphic variation of sulfur, PGE and base-metal concentration across the UG2 chromitite layer. The data are extracted from Maier and Barnes (2008). *Chr* chromitite; *Peg Px* Pegmatoidal pyroxenite and *Px* pyroxenite

concentrated towards the base and the upper contact of the layer (Maier and Barnes 2008; Von Gruenewaldt 1986). The sulfur concentration within the chromitite is less than 0.1 weight percent in most cases (Barnes et al. 2009; Maier and Barnes 2008) with exceptional sulfide abundance of up to 2.5 volume percent observed the Messina Mine (Von Gruenewaldt et al. 1990). Cawthorn et al. (2005) reported that the Pt/Pd ratios are significantly different from the Eastern to the Western Limb. However, no systematic variations in Pt/Pd ratios have been observed between the Northwest and Eastern Limbs (Naldrett et al. 2011b).

Main Sulfide Zone of the Great Dyke. The data are extracted from Wilson and Prendergast (2001). **e** Base-metal and platinum-group element concentrations in mineralized horizons from the Penikat intrusion in Finland. The data are extracted from Alapieti and Lahtinen (2002). **f** Base-metal and platinum-group element concentrations in the JM-Reef of the Stillwater Complex. The data are extracted from Godel and Barnes (2008b) and Barnes and Naldrett (1986). The coloured areas represent the range of concentrations observed. The concentrations have been normalized to the primitive mantle values from McDonough and Sun (McDonough and Sun 1995)

PGE-Reef Types Located in the Lower to Middle Parts of Layered Intrusions

Platinum-group element rich mineralization may occur in the lower part (<1000 m above the base) of layered intrusions. Deposits of this type are usually characterized by the presence of tens to hundreds of meters of disseminated base metal sulfides with sulfide abundances usually less than 5 volume percent. The base metal sulfides (in most case pyrrhotite, pentlandite and chalcopyrite) are hosted within peridotites and pyroxenites. Deposits of this type include the Santa Rita zone within the Mirabela intrusion (Barnes et al. 2011), the Kevitsa intrusion in Finland, and the Kapalagulu intrusion in Tanzania (Maier et al. 2008), and tend to have low PGE tenors. Platinum-group element mineralization observed higher (~1 to 2 km) in the stratigraphy are also characterized by the presence of disseminated base-metal sulfides (<5 volume percent). The PGE reefs usually occur in stratigraphic intervals where lithological and compositional changes occurred and where interlayering of ultramafic and mafic rocks is present (Barnes and Maier 2002a; Cawthorn et al. 2005). The host rocks may have various compositions from harzburgite, pyroxenite, troctolite, norite or anorthosite. Deposits of this type include the Merensky Reef of the Bushveld Complex, the JM Reef of the Stillwater Complex, the Main Sulfide Zone in the Great Dyke of Zimbabwe and the PGE reefs of the Penikat intrusion described below.

The Main Sulfide Zone of the Great Dyke

The Great Dyke of Zimbabwe hosts the world's second largest reserve of PGE (Oberthür 2011). The principal PGE resources of the Great Dyke are located in the Main Sulfide Zone (MSZ) and the Lower Sulfide Zone (LSZ) hosted within the P1 Pyroxenite of the uppermost Ultramafic Cyclic Unit (Fig. 9.6). The MSZ contains a higher abundance of sulfide and PGE than the LSZ and represent, by far the main PGE resource of the Great Dyke. The large majority of PGE resource (~75%) found in the MSZ is located in the Sebakwe and Darwendale Subchambers (Oberthür 2011). The PGE mineralization occurs in a narrow (1 to 2 m thick) zone towards the top of the ultramafic sequence. Disseminated base-metal sulfides composed of intergrowths of pyrrhotite, pentlandite, chalcopyrite and minor pyrite are hosted within mesocumulates composed mainly of orthopyroxene, augite, plagioclase, biotite, alkali feldspar and quartz (Wilson and Prendergast 2001). The base-metal sulfides represent 0.5 to 10 volume percent of the rock and occur as elongated blebs interstitial to silicate minerals. The MSZ is characterized by a particular distribution of the base-metal sulfide and the PGE referred to as "offset" reef in which the peaks both the Pt and Pd concentrations and the PGE and base metal sulfides are separated from each other stratigraphically (Fig. 9.12). The PGE mineralization has been defined into two main subzones with a lower Pd-rich (i.e Pd > Pt) and an upper Pt-rich (i.e. Pt > Pd) subzones. The average PGE concentrations from various localities are presented in Fig. 9.10. Recent studies have shown that a large

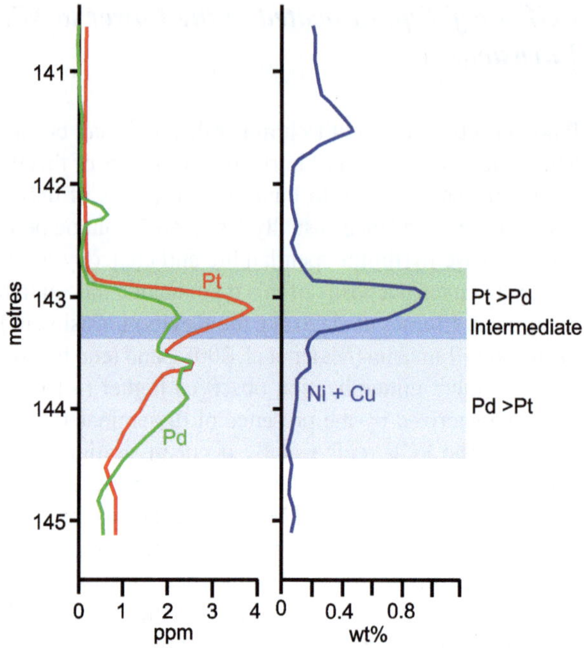

Fig. 9.12 Vertical distribution of PGE and base metals showing"Offset reef" of the Great Dyke of Zimbabwe. The data from a drillcore from the Mimosa mine and are extracted from Wilson and Prendgast (Wilson and Prendergast 2001) showing Pd and Pt-rich subzones

proportion of Pd and Rh are hosted in pentlandite whereas Pt is dominantly present in the form of platinum-group minerals (Oberthür 2011).

Merensky Reef of the Bushveld Complex

The Merensky Reef is a narrow PGE reef that constitutes, after the UG-2 chromitite, the second largest PGE resource in the world. It occurs 15 to 400 m above the UG-2 chromitite (Lee 1996). The Merensky Reef lies at the base of a specific and very laterally persistent cyclic unit that contains thin basal chromitite layers (Vermaak 1976). The base of the Merensky Reef was used by some workers to define the boundary between the Upper Critical Zone and the Main Zone (Kruger 2005). The PGE- mineralization and the vertical succession are variable depending on the localities considered but are consistent across the entire intrusion in comprising a package of thin chromite stringers and locally pegmatoidal orthocumulate orthopyroxenite overlain by norite (Cawthorn 2002; Naldrett et al. 2009b; Viljoen 1999). Details of all the different reef types are presented in several studies (Kinloch 1982; Naldrett et al. 2009b; Viljoen 1999) and are not detailed in this contribution. The mineralized zone has a thickness that rarely exceeds 1 m with average PGE concentrations varying from 5 to 10 ppm and Pd/Pt ratios of ~0.6 (Barnes and Maier 2002b; Cawthorn et al. 2005; Godel et al. 2007; Naldrett et al. 2009b; Wilson and Chunnett 2006). The so-called "normal" Merensky Reef (Fig. 9.13) consists, from the bottom upwards of (Barnes and Maier 2002b; Godel et al. 2006; Godel et al. 2008b; Naldrett et al. 2009b; Vukmanovic et al. 2013): (i) a footwall anor-

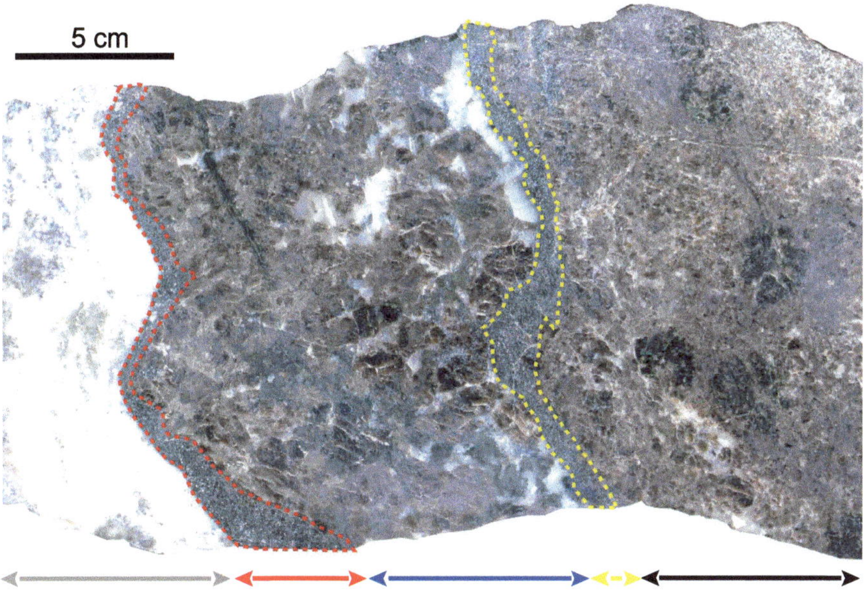

Fig. 9.13 Typical normal Merensky Reef, Rustenburg Platinum Mine, Western Limb of the Bushveld Complex. Modified after Godel (2007)

thosite, pyroxenite or norite; (ii) a lower layer of chromitite that undulates at the centimetre scale and ranges from 2 to >40 mm in thickness containing euhedral to amoeiboidal chromite crystals; (iii) a coarse-grained melanorite of varying thickness (from few centimetres to 10 m), with in some areas harzburgite and dunite; (iv) a thin (2 to 15 mm thick) upper chromitite layer containing euhedral chromite crystals overlain by (v) melanorite of thickness varying from 30 to 225 cm. Sulfide mineral proportions vary from 0.5 volume percent (in the chromitite layers) up to 10 volume percent in localized areas above the lower chromitite with average sulfide abundance of 3 volume percent over the reef (Barnes and Maier 2002b; Godel et al. 2007; Godel et al. 2008; Naldrett et al. 2009b). X-ray computed tomography and 3D image analysis reveal that in the pyroxenite, the sulfides are located at the boundaries of pyroxene and plagioclase crystals and form up to 10 cm interconnected networks. The sulfide networks fill paleovertical dilatancies formed during compaction, which facilitated the downward migration of sulfide liquid in the cumulate (Godel et al. 2006). These sulfide networks stop at the level of underlying chromitite layers where changes in sulfide wettability occur. The PGE concentrations are higher in the chromitite layers, vary at the centimetre scale and can reach locally 15 ppm total PGE (Figs. 9.14 and 9.10). Within both chromitite layers, high-resolution X-ray computed tomography combined with 3D quantitative image analysis and mineralogical analysis indicates that platinum group minerals occupy preferential sites at triple points between chromite, sulfide and silicate minerals at the edges of sulfide blebs (Godel et al. 2010). This observation implies that

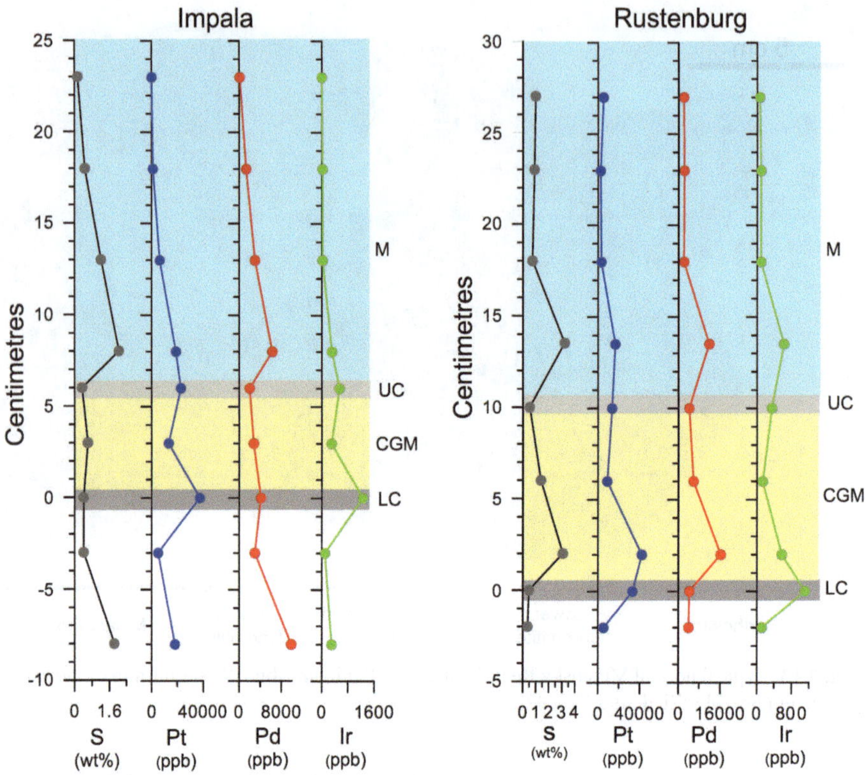

Fig. 9.14 Stratigraphic variation of sulfur, PGE and base-metal concentration across the Merensky at the Impala and Rustenburg Platinum Mines. The data from Impala Mine are extracted from Barnes and Maier (2002b) and the data from the Rustenburg Pt Mine are from Godel et al. (2007). *LC* lower chromitite; *CGM* coarse-grained melanorite; *UC* upper chromitite and *M* melanorite

platinum-group minerals nucleated in preferred sites subsequent to accumulation of the sulfide droplets (Godel et al. 2010).

J-M Reef of the Stillwater Complex

The Johns-Manville (J-M) Reef is the main PGE-rich layer within the complex and it is hosted within troctolite, olivine gabbronorite or anorthosite of the olivine-bearing cumulate zone I (OB-1) within the Lower Banded Series (Barnes and Naldrett 1986; Todd et al. 1982; Turner et al. 1985; Zientek et al. 2002 and references therein). Disseminated base metal sulfides (pyrrhotite, pentlandite, chalcopyrite and minor pyrite) locally represent 0.5 up to 5 % by volume of the rock and form sulfide networks interconnected over distances of few centimetres. Where the cumulate rocks have been altered, secondary silicates (mainly chlorite, tremolite and actinolite) are closely associated with the sulfides, forming alteration haloes around them. In the altered olivine gabbronorite, the sulfides are altered to secondary magnetite that form corona around the sulfides. Secondary chalcopyrite is also present and is associated with secondary silicates. The J-M Reef has the highest average Pt + Pd grade (~ 18 ppm with an average Pd Pt ratio of 3) of all known PGE ore deposits (Zientek et al. 2002). The whole-rock concentration of Pd and Pt can reach values of several hundred ppm (Fig. 9.10) at the hand sample scale (Barnes and Naldrett 1986; Godel and Barnes 2008a; Godel and Barnes 2008b; Keays et al. 2012; Zientek et al. 2002). The Pd/Pt ratios are independent of the host rock type.

The SJ, AP and PV Reefs of the Penikat Intrusion

The Sompujärvi Reef (referred to as SJ Reef) is located in the basal part of the Megacyclic Unit IV and is characterized by extremely low sulfide abundance and high PGE concentrations (up to tens of ppm PGE) associated with finely disseminated sulfides and chromite (Halkoaho et al. 1990). The Ala-Penikka Reefs (referred to as AP I and II Reefs) are observed ~ 200 to 400 m above the base of the Megacyclic Unit IV in the basal bronzitic portion and consist of narrow (0.4 to 1 m thick) PGE-rich mineralization associated with disseminated sulfides (4 volume percent). The AP I Reef is 2–70 cm thick and contains heterogeneous PGE concentrations associated with disseminated sulfides. The AP II Reef is found above the AP I Reef and exhibits similar characteristics. The Paasivaara Reef (referred to as PV Reef) is located at the transition between the Megacyclic Units IV and V (~ 700–1000 m above the AP Reefs) and is hosted within anorthosite (Alapieti and Lahtinen 1986; Huhtelin et al. 1990). Similarly to other reefs at Penikat, the PV PGE-mineralized zone is erratic in thickness and PGE concentration but contain the highest Pt/Pd ratios of all the reefs observed in the intrusion (Fig. 9.10). The PGE are associated with disseminated sulfides in the PV Reef (Alapieti and Lahtinen 1986; Huhtelin et al. 1990).

PGE Reef Types in the Upper Parts of Layered Intrusions

Occurrence of platinum-group element and gold mineralization have been described in the upper portion of layered intrusions (e.g. the Skaergaard Intrusion in west Greenland (Andersen et al. 1998), the Rio Jacaré complex in Brazil (Sa et al. 2005), the Sonju Lake and Birch Lakes intrusions in the Duluth complex in the U.S.A. (Miller et al. 2002), the Rincon del Tigre intrusion in Bolivia (Prendergast 2000), the Stella intrusion in South Africa (Maier et al. 2003), or the Kotalainen intrusion in Finland (Mutanen 1997)). In contrast to the other PGE-types, the mineralization is often hosted in gabbro-diorite and Fe–(Ti) oxide-rich layers. The deposits are relatively enriched in Cu, Pd, Au and Pt with considerably smaller Ru, Ir, Os concentrations. This is illustrated by the large Pd/Ir and Cu, Ni ratios of the rocks (e.g. average Pd/Ir ratios of ~ 150 and 250 for and Cu/Ni ratios of 2.45 and 3.122 for the Rio Jacaré and Stella intrusions, respectively. In most of the intrusions, the PGE grades are too low (typically 0.5 to 2 ppm) to be economic. The Skaergaard and Stella intrusions contain higher grades (ca. 5 to 15 ppm) over thin interval and feasibility studies are under way potentially to mine the PGE and Au.

The Platinova Reef is associated with tens-of-meter scale modal layers referred to as the Triple Group, formed by oxide-rich tholeitic gabbros (Wager and Brown 1968). The Reef occurs in repeated meter-thick layers of sulfide-poor gabbros located in the lower part of the Triple Group consisting of three conspicuous plagioclase-rich leuco-gabbro layers within plagioclase-pyroxene-oxide melanogabbros (Andersen et al. 1998; Nielsen et al. 2005). The Platinova Reef is inferred from compilation of drillcore data to comprise 8 continuous Pd-rich horizons, named from base to top: Pd5, Pd4b, Pd4a, Pd3b, Pd3a, Pd2b, Pd2a and Pd1 (Nielsen 2001). The PGE mineralization horizons are perfectly concordant with the layering in the host rocks and internal variations correlate throughout the intrusion (Nielsen et al. 2005). The number of Pd-enriched layers and the combined total PGE and Au contents increase toward the centre of the mineralization (Andersen et al. 1998; Holwell and Keays 2014; Nielsen et al. 2005). The main PGE reef (Pd5) of the Platinova Reef is a 4–5 m (at a cut off of 0.7 ppm) thick mineralized horizon with maximal grade reaching 5 ppm PGE (Nielsen et al. 2005). In contrast to other PGE reef types, the base-metal sulfides in the Pd5 layer are entirely dominated by bornite (Cu_5FeS_4) with exsolution lamellae of chalcocite (Cu_2S) and digenite (Cu_9S_5). These Cu-sulfides represent less than 0.05 modal percent of the samples. In 3D, the Cu-rich sulfides (bornite ± chalcocite assemblage) occur either as sub-spherical (sphericity >0.9) globules that are mainly observed entirely enclosed within Fe–Ti oxides or as more irregular Cu-rich sulfide blebs that are either located at the boundaries between silicates (pyroxene or olivine) and Fe–Ti oxides or located between silicates grains (Godel et al. 2014). More than 75 % (by volume of PGM) are associated with base-metal sulfides. Palladium, the main element that can potentially be mined, is almost entirely hosted by skaergaardite (PdCu) (Rudashevsky et al. 2009). Most of the Skaergaardite (PdCu) occurs (in 3D) as globular to elongated grains found attached to subspherical (sphericity >0.9) Cu-rich sulfide blebs enclosed within large ilmenite grains (Godel et al. 2014).

Hosts of the Platinum-Group Elements

As described above, all the PGE-reefs currently mined consist of narrow mineralized zones associated with based-metal sulfides disseminated within silicate and/or chromite crystal matrices or chromitites containing trace amounts of sulfides. Over the past ten years, multi-disciplinary studies (e.g. X-ray computed tomography, whole-rock geochemistry, in situ mineral analysis by laser ablation ICP-MS, and the detailed analysis of platinum-group minerals and their textural charasteristics) have been carried out on various reefs hosted within layered intrusions. The results of these studies have brought additional insights into the residency of the PGE in the mineralized rocks with important implications for ore genesis.

Oxides

Chromitites in mafic-ultramafic layered intrusions are often enriched in PGE relative to their silicate host lithologies (e.g. Great Dyke, Bushveld Complex, Rum intrusion). Two mechanisms have been proposed to explain the relationship between chromite and the PGE and the Ir-group PGE (referred to as IPGE and comprising Ir, Os and Ru) in particular: (i) the IPGE could partition into chromite crystals and be present in solid solution or ii) the IPGE could exhibit low solubility in mafic magma and crystallise in the form of IPGE-rich minerals or alloys during chromite crystallization. The first hypothesis is based on experimental studies that have shown that Ir, Ru, Rh and Pd could partition into spinel phases (Brenan et al. 2012; Capobianco and Drake 1990; Capobianco et al. 1994; Righter et al. 2004; Sattari et al. 2002). Recently, homogeneous Ru laser ablation ICP-MS spectra in chromite from komatiites have been interpreted as evidence of partitioning of the IPGE in solid solution within natural chromite crystals (Locmelis et al. 2011; Pagé et al. 2012). Several laser-ablation ICP-MS analyses have been carried out on chromite crystals from the Merensky Reef of the Bushveld Complex (Ballhaus and Sylvester 2000; Godel et al. 2008; Osbahr et al. 2013), and on chromite and secondary magnetite grains of the J-M Reef of the Stillwater Complex (Godel et al. 2008), in order to test whether the PGE partitioned into natural spinel phases in sulfide-rich PGE-reef environments. The results of all of these studies indicate that none of the PGE are present in detectable (i.e. in the range of 10s ppb) amounts in chromite (Fig. 9.15) or secondary magnetite observed within the reefs. In these minerals, the PGE are only found as discrete PGE-rich phases mainly Pt-rich in the Merensky Reef (Ballhaus and Sylvester 2000; Godel et al. 2008; Osbahr et al. 2013), Pd-rich in the J-M Reef (Godel and Barnes 2008a) with minor Ru, Os, Ir (probably laurite) in both cases. Although to date no data are available for other deposit, it seems improbable that chromite and spinel in general contribute significantly to the PGE mass balance in PGE reefs.

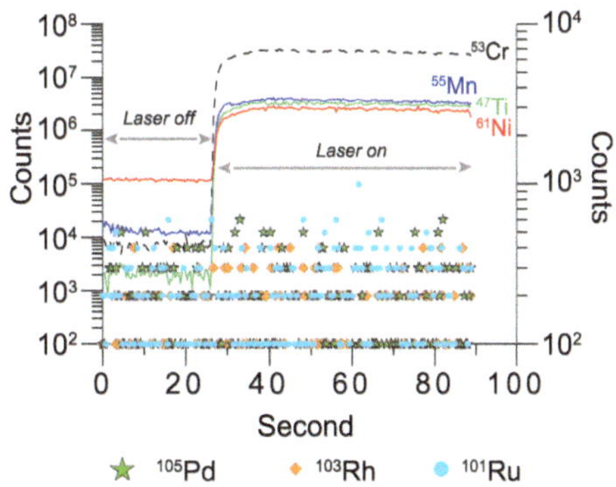

Fig. 9.15 Example of laser ablation spectra of chromite crystal from upper chromitite of the Merensky Reef of the Bushveld Complex showing the variation of selected platinum-group elements

Base-Metal Sulfides

Over the past ten years, the development of laser-ablation ICP-MS allowed the determination of trace element concentrations at ppb to ppm levels in base metal sulfides (pentlandite, pyrrhotite, chalcopyrite and pyrite) and provided a means for the evaluation of the contribution of base-metal sulfides to the overall PGE mass balance in PGE ore deposits. A large body of data now exists on the concentrations of PGE and semi-metals in sulfide minerals from different PGE-mineralization hosted within layered intrusions including the Merensky Reef (Ballhaus and Sylvester 2000; Godel et al. 2007; Osbahr et al. 2013); the Platreef (Holwell and McDonald 2006; Holwell and McDonald 2007; Hutchinson and McDonald 2008) and the UG2 (Osbahr 2012) of the Bushveld Complex; the JM-Reef of the Stillwater Complex (Godel and Barnes 2008b); the Main sulfide Zone of the Great Dyke (Barnes et al. 2008); and the AP and PV reefs of the Penikat intrusions (Barnes et al. 2008).

In the Bushveld (Merensky Reef, Platreef and UG2) and Stillwater (JM Reef) Complexes, pentlandite is the main carrier of PGE in solid solution followed by pyrrhotite (Fig. 9.16a–d). In contrast, in the Great Dyke (Main Sulfide Zone) and the Penikat Intrusion (AP and PV reefs) pentlandite and pyrrhotite contain similar amounts of PGE in solid solution (Figs. 9.16e and f). In all cases, chalcopyrite is almost entirely devoid of PGE in solid solution (Fig. 9.16). Overall, the shape of mantle normalized metal patterns of pentlandite (Fig. 9.17) and pyrrhotite (Fig. 9.18) are comparable from one deposit to the other. Palladium is the principal PGE hosted in the base-metal sulfides with concentrations in pentlandite from the JM Reef reaching several weight percent (Cabri et al. 1984; Godel and Barnes 2008b; Polovina et al. 2004) as illustrated in Fig. 9.17. The contribution of pentlandite on the Pd budget in the whole-rock varies between reefs, with values reaching average of 75 % in the Merensky Reef (Figs. 9.16a and b). Osmium, Ir, Ru and Rh concentrations in pentlandite (and to a lesser extent pyrrhotite) are broadly similar in the different reefs including the Merensky Reef chromitite layers and the UG2. The contribution of base-metal sulfide to the IPGE budget is smaller in the chromitite layers (e.g. UG2, Fig. 9.16c) than in sulfide-dominated PGE mineralization (Fig. 9.16). The similarity of IPGE patterns for all the deposits suggest that the processes that lead to the concentrations of these elements in the base-metal sulfides are comparable and would indicate that the concentrations have not significantly been influenced by post-magmatic processes such as alteration and metamorphism. Platinum concentrations in pentlandite and pyrrhotite are extremely low as illustrated by the strong negative anomalies observed on all the mantle-normalized metal patterns of pentlandite (Fig. 9.17) and pyrrhotite from the JM Reef (Fig. 9.18b). Only a small proportion (<5 %) of the whole-rock Pt budget is accounted by the base-metal sulfides. Based on the results described above, it is clear that although PGE in the whole-rocks are correlated to their sulfur contents in most cases, base-metal sulfides are not the main host of PGE in the rocks.

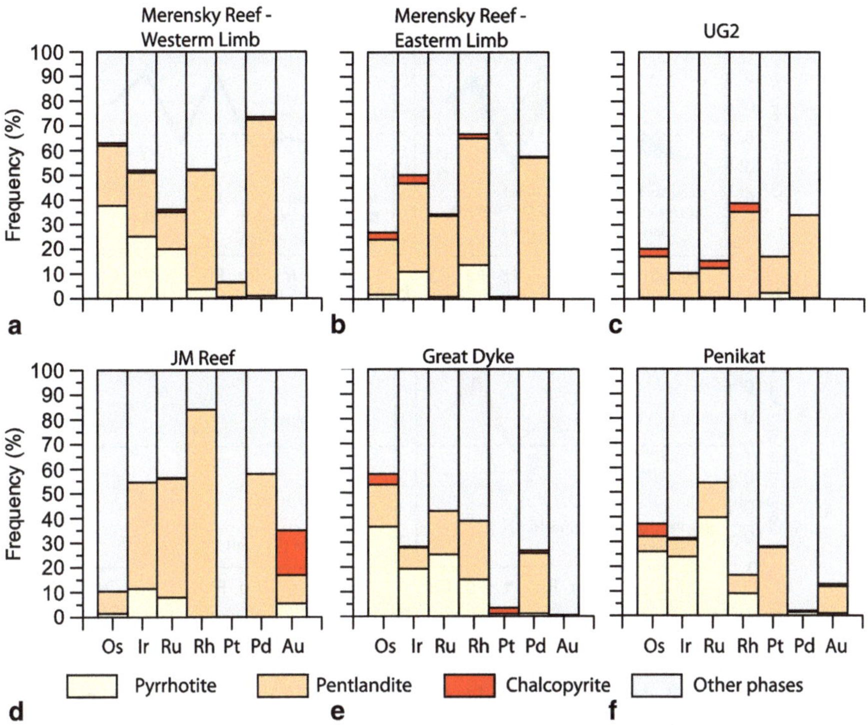

Fig. 9.16 Average mass balance of platinum-group elements between whole-rock, base-metal sulfides and other phases in various PGE reefs. **a** Merensky Reef (Western Limb of the Bushveld Complex). The data are extracted from Godel et al. (2007) and Osbahr et al. (2013). **b** Merensky Reef (Eastern Limb of the Bushveld Complex). The data are extracted from Osbahr et al. (2013). **c** UG2. The data are extracted from Osbahr (2012). **d** JM Reef of the Stillwater Complex. The data are extracted from Godel and Barnes (2008b). **e** Main sulfide Zone of the Great Dyke. The data are extracted from Barnes et al. (2008). **f** AP and PV Reefs of the Penikat intrusion. The data are extracted from Barnes et al. (2008)

Platinum-Group Minerals

The detailed LA-ICP-MS analysis and PGE-mass balances carried out on various deposits described above revealed that in a vast majority of cases, most of the PGE budget in the whole rock is controlled by platinum-group minerals (Fig. 9.16). The increased availability of scanning electron miscroscope (SEM), the development of SEM-based automated mineral liberation analyser (SEM-MLA), high-resolution X-ray computed tomography and image processing techniques have made it possible to constrain the location, composition, size and textural relationships in both 2D and 3D of the platinum-group minerals at the sample scales (Figs. 9.19 and 9.20). A large number of recent studies have reported platinum-group mineral (PGM) analyses for major PGE deposits (Barnes et al. 2008; Godel 2013; Godel and Barnes 2008a; Godel and Barnes 2008b; Godel et al. 2007; Godel et al. 2010; Holwell

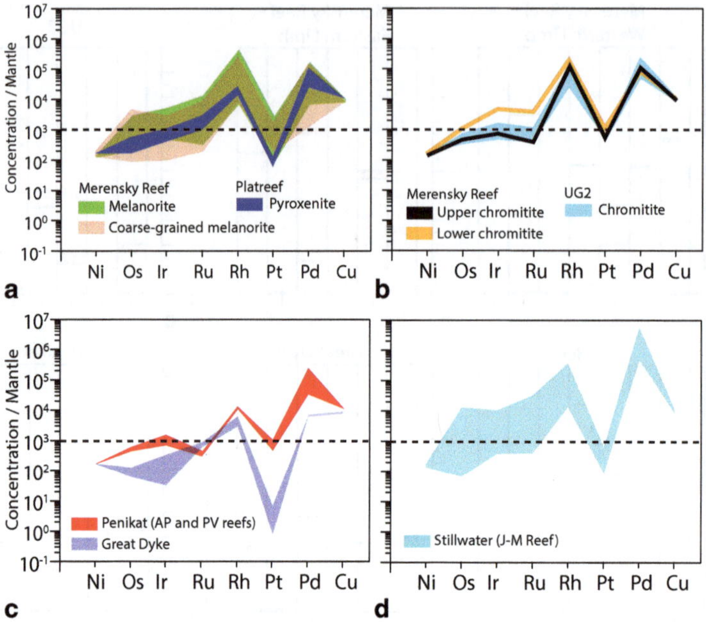

Fig. 9.17 Mantle-normalized base-metal and platinum-group element spectras of pentlandite analysed from various PGE deposits. **a** Pentlandite from the Merensky Reef coarse-grained melanorite and melanorite (Godel et al. 2007; Osbahr et al. 2013) and pentlandite from the Platreef pyroxenite (Holwell and McDonald 2007) **b** Pentlandite from the Lower and Upper chromitite layers from the Merensky Reef (Godel et al. 2007) and the UG2 chromitite (Osbahr 2012). **c** Pentlandite from the Ap and PV Reefs of the Penikat intrusion and from the Main Sulfide Zone of the Great Dyke (Barnes et al. 2008). **d** Pentlandite from the JM Reef of the Stillwater Complex (Godel and Barnes 2008b)

and McDonald 2007; Holwell et al. 2006; Locmelis et al. 2010; Oberthür et al. 2003; Osbahr 2012; Osbahr et al. 2013; Rudashevsky et al. 2009; Rudashevsky et al. 2004; van der Merwe et al. 2012; Yudovskaya et al. 2012). These studies complemented the extensive literature available (e.g. Alapieti and Lahtinen 1986; Cawthorn et al. 2002; Kingston and El-Dosuky 1982; Kinloch 1982; Kinloch and Peyerl 1990; Lee 1996; McLaren and De Villiers 1982; Prichard et al. 2004). A wide range of PGM has been reported in various deposits and localities around the world (see Cabri (2002) and references therein for details) and included notably PGE-rich sulfides, tellurides and bismuthotellurides, arsenides, antimonides or alloys. Overall, the PGM assemblages and their textural relationships with other phases vary from one deposit to the other and might vary from one location to the other within the same deposit (Figs. 9.21 and 9.22). It is out of the scope of the present contribution to discuss all the details of PGM mineralogy in various deposits. The PGM are predominantly, if not exclusively, formed at low temperatures after the PGE have reached their present sites of accumulation within homogeneous droplets of sulfide liquid. Hence their identities are of secondary importance in the discussion of petrogenesis of PGE mineralization. For further information the reader is invited to consult the references given.

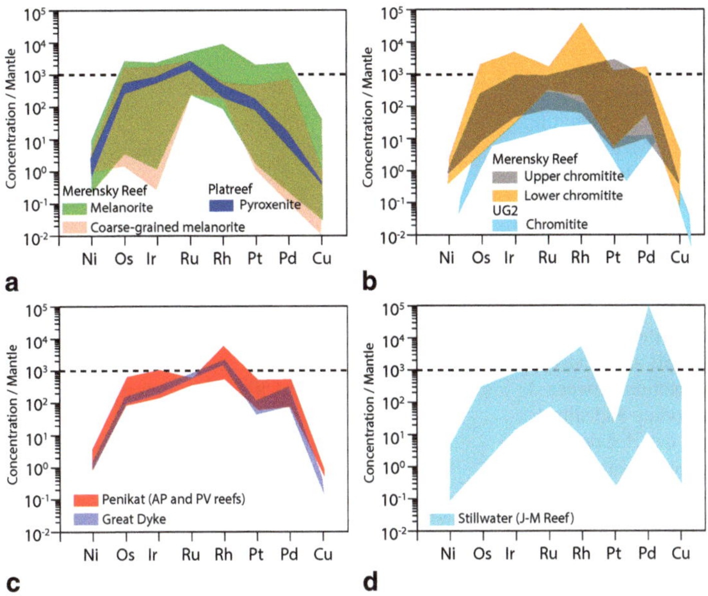

Fig. 9.18 Mantle-normalized base-metal and platinum-group element spectras of pyrrhotite analysed from various PGE deposits. **a** Pyrrhotite from the Merensky Reef coarse-grained melanorite and melanorite (Godel et al. 2007; Osbahr et al. 2013) and pyrrhotite from the Platreef pyroxenite (Holwell and McDonald 2007). **b** Pyrrhotite from the Lower and Upper chromitite layers from the Merensky Reef (Godel et al. 2007) and the UG2 chromitite (Osbahr 2012). **c** Pyrrhotite from the Ap and PV Reefs of the Penikat intrusion and from the Main Sulfide Zone of the Great Dyke (Barnes et al. 2008). **d** Pyrrhotite from the JM Reef of the Stillwater Complex (Godel and Barnes 2008b)

Ore Forming Processes

The formation of PGE ore deposits requires that the PGE be concentrated about three to four orders of magnitude relative to Earth's mantle and crust concentrations, and two to three orders of magnitude over concentrations in inferred parent magmas to reef-bearing intrusions. This is a similar degree of concentration to that involved in other magmatic ore layers such as chromite seams and vanadiferous magnetite layers.

The PGE concentrations within a given orebody are influenced by a combination of processes that may occur at various stages of its history during: (i) mantle melting; (ii) the magma ascent and its emplacement into the crust; (iii) the crystallization of the magma to form a layered intrusion and; (iv) late to post-magmatic events such as hydrothermal alteration and metamorphism. One of the main challenges is the evaluation of the influence of each of the above factors in the control of ore genesis. In most of the deposits, the processes are complex with superimposition of physical and geochemical processes over a large range of temperature. The main factors that have been considered as controlling the genesis of PGE deposits in layered intrusion are reviewed and discussed below.

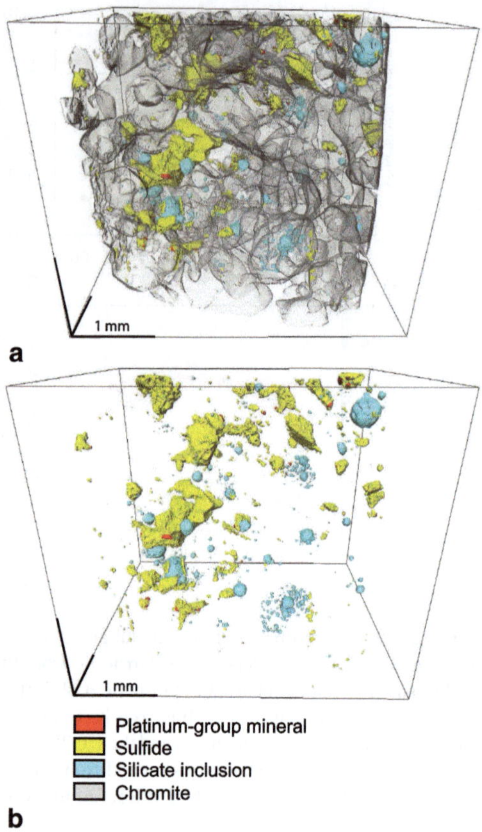

Fig. 9.19 3D distribution of chromite, base-metal sulfides, platinum group mineral and silicates inclusions in a sample of Lower Chromitite layer from the Merensky Reef. The 3D distribution of the minerals has been determine using high-resolution x-ray computed tomography and dedicated image analysis workflow. In most case, the platinum-group minerals are located at the triple junction between sulfide, chromite and silicate minerals. A detailed description of the platinum-group minerals textural relationship with sulfides and chromite is provided in Godel et al. (2010). The origin of the different chromite morphologies and silicate inclusions are discussed in Vukmanovic et al. (2013)

Mantle Processes and Parental Magmas

Platinum-Group Elements in the Earth's Mantle

Platinum-group elements concentrations in the Earth's primitive mantle vary from ~1 to 7 ppb (McDonough and Sun 1995). The PGE are typically hosted by micron-scale base-metal sulfides and alloys that represent an extremely minor volume proportion (typically less than 0.1%) of the mantle (Lorand et al. 2008b). Detailed mineralogical investigation coupled with in situ mineral geochemistry of peridotite samples provided information on the residency of PGE in the lithospheric mantle (Alard et al. 2000; Lorand and Alard 2001; Lorand et al. 2008a; Lorand et al. 2008b; Luguet et al. 2007). These studies have revealed that more than 90% of the PGE in mantle lherzolites are hosted by base-metal sulfides. Alard et al. ((2000) identified two different populations of sulfides: (i) Fe–Ni monosulfide solid solution (mss) observed as inclusions within olivine and (ii) Fe–Ni sulfide (pentlandite) and Cu–Fe sulfides (chalcopyrite and cubanite) observed at olivine contact boundaries.

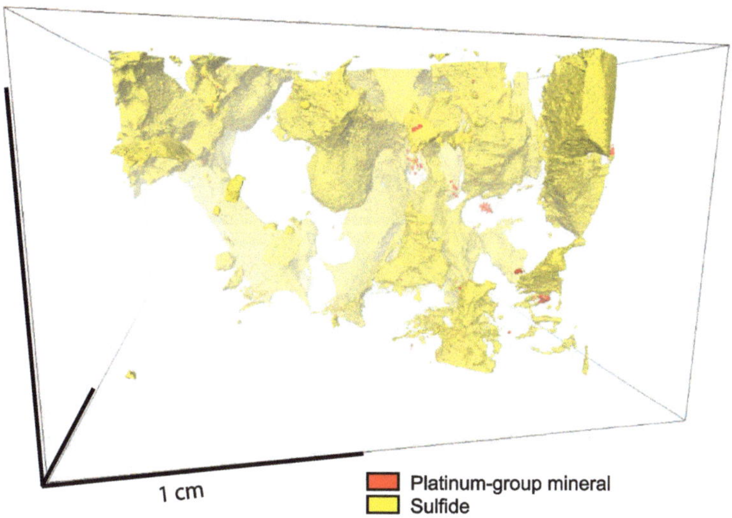

Fig. 9.20 3D distribution of base-metal sulfides and platinum-group minerals in a sample of olivine cumulate from the JM Reef Stillwater. In 3D, the platinum-group minerals are observed either at the contact between base-metal sulfides and silicates or in most cases in alteration haloes around the base-metal sulfides. In that cases, the platinum-group minerals are not touching the sulfide boundaries

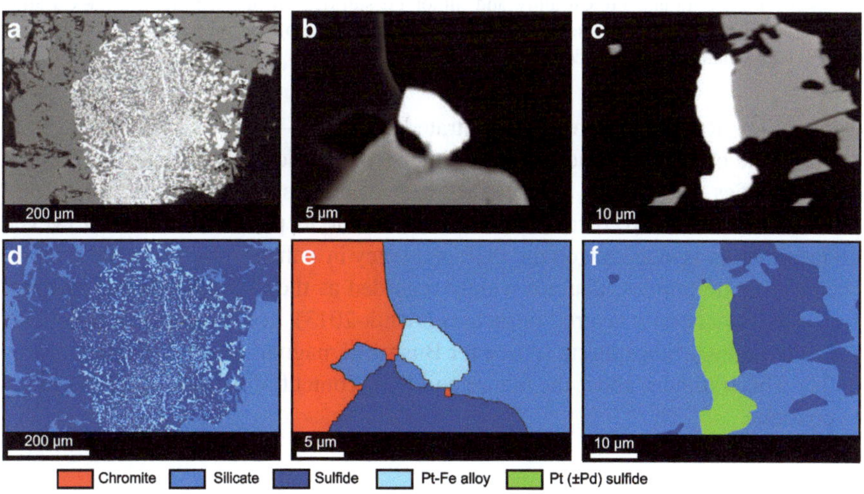

Fig. 9.21 Examples of platinum-group minerals observed within the Merensky Reef, Rustenburg Platinum Mine (Bushveld Complex). **a** Backscattered electron and analysed **d** images showing symplectites of Pt–Fe alloy and base-metal sulfide minerals (pyrrhotite) at the contact with orthopyroxene (silicate). The image is modified from Godel et al. (2007). **b** Backscattered electron and analysed **e** images of euhedral Pt–Fe alloy in contact with chromite, sulfide and silicate. **c** Backscattered electron and analysed **f** images of Pt(±Pd) sulfide located at the contact between sulfide and silicate

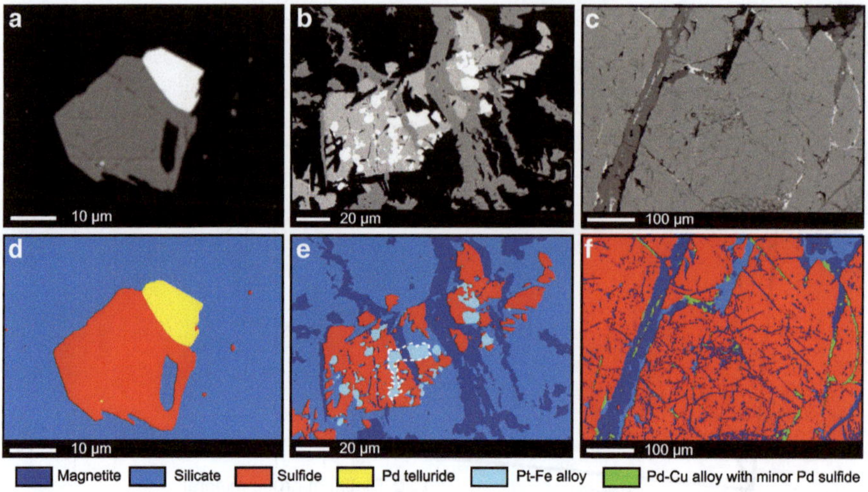

Fig. 9.22 Examples of platinum-group minerals observed within the JM Reef, East Boulder Mine (Stillwater Complex). **a** Backscattered electron and analysed **d** images showing the association between Pd telluride, base metal sulfides and silicates. **b** Backscattered electron and analysed **e** images showing the complex associated between base-metal sulfide, Pt–Fe alloy and secondary magnetite. The primary Pt–Fe alloy and base-metal sulfides are crosscut by secondary magnetite (e.g. white tie line on **e**. The figure is modified after Godel and Barnes (2007). **c** Backscattered electron and analysed **f** images showing the complex association between base-metal sulfides, secondary magnetite and Pd–Cu alloy and minor Pd sulfides. In this sample, the Pd–Cu alloy is observed crosscutting the base metal sulfide over a distance of several millimetres and is also observed associated with the secondary magnetite,

Osmium, Ir, Ru and Rh were concentrated in the Fe–Ni mss whereas Pd and to a lesser extent Pt was enriched in the interstitial sulfides. The particular textural characteristics, mineralogy and geochemistry of these sulfides influences the composition of melts derived from lithospheric mantle as presented below. However, no information is available on the PGE residency in asthenospheric mantle sources, such as mantle plumes, that are widely regarded as the likely sources of magmas for large intracratonic layered intrusions (Arndt 2013), notwithstanding arguments for a dominantly lithospheric source for Bushveld magmas (Richardson and Shirey 2008). The argument about the nature and PGE content of mantle sources is likely to continue.

Platinum-Group Elements Behaviour During Partial Melting of the Mantle

Partition coefficients of PGE between sulfide and silicate melts are in the order of 10^4–10^5(Barnes and Lightfoot 2005 and references therein), and possibly much higher (Fonseca et al. 2009, Mungall and Brenan, 2014). Hence, at equilibrium the concentration of a given PGE in a sulfide melt is typically at least four orders of magnitude larger than in coexisting silicate melt. Hence, the igneous geochemistry of sulfur, and in particular the mass balance of equilibrating silicate and sulfide liquid (Campbell et al. 1983) strongly controls the behaviour of the PGE. Primary mantle-derived magmas show PGE concentrations ranging from 0.2 to ~40 ppb (Crocket 2002) with concentrations mainly depending on the behaviour of base-metal sulfides during partial melting of the mantle (Barnes et al. 1985) and subsequent sulfide liquid saturation history of the silicate magma. The solubility of sulfur in mantle-derived melts mainly depends on pressure (i.e. depth) and temperature of melting (Mavrogenes and O'Neill 1999). Sulfur solubility in typical basalts broadly ranges from 500 to 1000 ppm with sulfur solubility of ~800 ppm at temperature of 1400 °C and a depth of 120 km. Low to moderate degrees of partial melting of typical mantle, thought to contain about 200–250 ppm sulfur, cannot led to the entire dissolution of the sulfides (Barnes et al. 1985; Hamlyn and Keays 1986). In that case, residual Fe–Ni sulfides sequestrating some PGE remains in the mantle (Bockrath et al. 2004a). As a result, magma derived from low to moderate degrees of partial melting (i.e $<20\,\%$) of the mantle (e.g. mid-ocean ridge basalts) have low PGE concentrations and exhibit poor potential for forming subsequently economic PGE deposits. In contrast, at high degree of partial melting (generating komatiites or picritic continental flood basalts) all the sulfur present in the mantle is dissolved and all the PGE are transferred to the silicate melt. The PGE concentration in the silicate magmas are dictated by the partition coefficient of host mineral and melt and fractionation of Pt and Pd relative to Ir, Os, Ru and Rh may be observed (Mungall and Naldrett 2008). All of the major known PGE-deposits on Earth originated from high-degree of partial melting of the mantle and are related to intraplate picritic, or possibly komatiitic magmatism (Wilson 2012). However at high oxygen fugacity (fO_2), sulfides in the mantle may (partially or entirely) convert to sulfate leading to a strong increase in sulfur solubility in the melt (Jugo 2009; Jugo et al. 2005; Mungall 2002b; Mungall et al. 2006). In that case, even a small degree of partial melting may result in the consumption of all the sulfur present in the mantle (Dale et al. 2012). Hence, magmas formed under these oxidizing-conditions may have potential to generate economic PGE deposits, provided that the dissolved sulfur can subsequently be reduced to sulfide. In addition, several events of melting can occur and complicate the predictions of PGE behaviours in the magma. To date, little is known about the form of the PGE in the melt and whether they are completely dissolved or form nano-to micro particles enriched in Ir, Os and Ru that may be transported by the melt upon its ascent into upper crustal level (Bockrath et al. 2004a).

Fractionation of Platinum-Group Elements and Parental Magmas

Fertile magmas form under sulfide-undersaturated conditions. Once a fertile silicate magma is formed it needs to ascend from the mantle to the upper level of the crust where it can potentially form a PGE deposit. During its ascent the magma becomes strongly undersaturated in sulfide (i.e. sulfide liquid are unlikely to form and collect PGE) as a result of decreasing pressure (Mavrogenes and O'Neill 1999; Wendlandt

1982). Fractionation (mostly resulting from the process of fractional crystallization of olivine) may potentially lead to a change in PGE concentrations. Experimental studies of the solubility and partitioning of PGE between olivine and silicate melt (Brenan et al. 2005; Brenan et al. 2003) suggested that olivine-silicate melt partition coefficients for Rh, Ru, and Ir increase with decreasing oxygen fugacity (fO_2) to reach value of ~2 at fO_2 relevant to mafic and ultramafic igneous systems. In contrast, Pt and Pd behave incompatibly during olivine fractionation. The difference in behaviour between the IPGE (Ir, Os, Ru,) and the PPGE (Pt, Rh, Pd) during olivine fractionation may account for some of the observed variation in IPGE/PPGE ratios variation in primitive high MgO, sulfur poor magmas (Puchtel and Humayun 2000). Overall the solubility of IPGE in silicate melt is very low suggesting that IPGE-rich alloys may form and be included with olivine and chromite (Brenan et al. 2005a). The incorporation of IPGE-rich alloys in chromite was demonstrated experimentally and justified with thermodynamic models by Finnigan et al. (2008). That hypotheses was confirmed Barnes and Fiorentini (2008), based on the analysis of a large body of data on komatiites. Many other studies have also documented the occurrence of PGE alloys in mantle peridotite (Kogiso et al. 2008; Lorand et al. 2010; Luguet et al. 2007). Recent work by Park et al. (2013) suggested the existence of Pt alloy in some arc magma sources.

Proposed parental magmas to most of the layered intrusions hosting PGE-deposits (e.g. Bushveld Complex, Stillwater Complex, and Great Dyke) are in equilibrium with high-magnesian olivine (~Fo90) and probably represent primary magmas that have not been significantly affected by fractionation during their ascent (Cawthorn et al. 2005). Hence, their PGE concentrations may be similar to those present in primary magma. The estimation of parental magma of a layered intrusion is relatively difficult as several issues arise in the estimation of the parental magma from the chilled margins or the fine-crystallized rocks spatially associated with an intrusion as they may not necessarily be representative of the parental magma of the intrusion that they are associated with (Godel et al. 2011 and references therein). Current estimates of the composition of the Bushveld Complex parental magmas (Barnes et al. 2010 and reference therein) indicate that Pt and Pd concentrations vary from 10 to 25 ppm ppb and 3 to 15 ppb, respectively (Fig. 9.23). Recently, Mungall and Brenan (2014) demonstrated, based on new experiments and modelling, that this unremarkable PGE concentrations in the parental magmas are enough to explain the occurrence of high grade ore occurences providing suitable conditions are reached.

Role of Base-Metal Sulfide Liquids

With the rare exception of some PGE-rich chromitites, most of the PGE deposits observed within layered intrusions presented above are associated with the presence of magmatic base- metal sulfides (typically forming intergrowths of pyrrhotite, pentlandite and chalcopyrite). These sulfides represent the accumulation of primary immiscible sulfide liquid that segregate from the silicate magma.

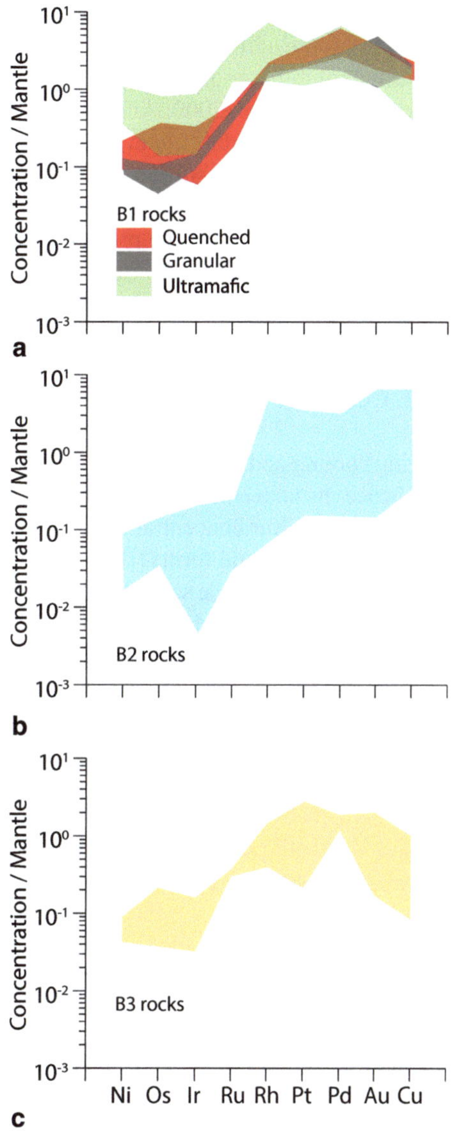

Fig. 9.23 Estimate of platinum-group element concentrations in the inferred parental magmas of the Bushveld complex. The concentrations for various magma types are extracted from Barnes et al. (2010) and presented are mantle-normalized base-metal and PGE patterns. The normalization values are from McDonough and Sun (1995)

Sulfur Solubility

The sulfur concentration in a silicate melt depends on parameters (total pressure, temperature, oxidation state and melt composition) that control the sulfur speciation and solution mechanisms (see detailed review in Mungall (2014) and Ariskin et al. (2013) for further information). The sulfur concentration at sulfide saturation (SCSS) depends mainly on the melt composition (particularly the Fe content), temperature, pressure and oxygen fugacity (fO_2). In addition to Fe, chalcophile elements such as

Ni and Cu (Ariskin et al. 2009; Ariskin et al. 2013; Ariskin et al. 2010; Evans et al. 2008) and some of the PGE (Laurenz et al. 2013) may also play a significant control on sulfur solubility in ultramafic and mafic magmas. Over the past few years, empirical (Ariskin et al. 2009; Ariskin et al. 2010; Li and Ripley 2009; Li and Ripley 2005) and experimental (Liu et al. 2007) models have been developed to predict the sulfur concentration at sulfide saturation. Typical basaltic magmas can dissolve or contain ~800–1000 ppm sulfur and are typically sulfide undersaturated in the crust.

Sulfide saturation may be provoked by external factors such as magma cooling and crystallization, mixing of magma with different compositions and the assimilation of crustal rocks by the primitive magmas as described below.

PGE Collection by Sulfide Liquid

During cooling and crystallization, sulfur is incompatible in the crystallising phases and hence, its concentration increases progressively as the magma fractionates. Ultimately, the sulfur concentration in the magma may reach the value at which immiscible sulfide liquid forms (i.e SCSS described above). In the absence of external factors favouring sulfide saturation, a large degree of crystallization may be required before that sulfide liquid form. Hence, the sulfide liquid may segregate from highly evolved silicate magma with ferrobasalt composition (e.g. after ~60–70% crystallization at Skaergaard or Sonju Lake intrusions). In general, the sulfide saturation will occur after Ni has been depleted as a result of olivine crystallization (Jenner et al. 2010; Ripley and Li 2013). Platinum, Pd and Cu are incompatible in the silicate and oxides crystallizing and hence their concentrations in the magma increase (Wohlgemuth-Ueberwasser et al. 2013). Lowering of Fe^{2+} activity as a result of extensive magnetite crystallization may lead to sulfide saturation. In that case, the sulfide that form are Cu–Fe and potentially PGE-rich (e.g. bornite-chalcocite- skaergaardite assemblage in the Platinova Reef of the Skaergaard intrusion). At low degrees of supersaturation, the immiscible sulfide droplets that form are expected to be widely spaced and will grow by diffusion processes (Mungall and Su 2005). Once the droplets have grown pass a certain diameter, they will sink toward the bottom of the magma chamber, a process similar to fractional crystallization of the silicate. The first immiscible sulfide to form will be highly enriched in the PGE with PGE concentrations depending mainly on their partition coefficients and their diffusion rate in growing sulfide droplets (Mungall 2002a). The variability in degree of supersaturation, diffusivity rates and to a lesser extent partition coefficients between different PGE may lead to an offset in sulfide abundance and concentrations of different PGE (Mungall 2002a; Mungall 2007) as observed for the Great Dyke (Wilson and Prendergast 2001) and the Munni Munni intrusion in Australia (Barnes 1993).

The most popular model of formation of PGE reef is a model whereby the PGE are collected by an immiscible sulfide liquid (Figs. 9.24a, b and c) that form as a result of magma recharge and mixing (Barnes and Maier 2002b; and Naldrett 1979a; Campbell et al. 1983; Godel et al. 2007; Maier and Barnes 1999; Mungall and Brenan 2014; Naldrett et al. 2009a). Li and Ripley (2009) tested that hypothesis and showed that mixing of two different magmas may result in the attainment of sulfide saturation in the hybrid magma. An important prerequisite is that each magma should have sulfur concentration sufficiently high to be close to their concentration at sulfur saturation. Small to moderate degree of saturation in the hybrid magma will lead to the formation of a small amount of sulfide liquid. In contrast, assimilation of crustal sulfur by the magma may enhance the sulfur saturation process and lead to the formation of large amounts of sulfides and a potentially to the dilution of preexisting PGE-rich sulfide melt (e.g. the Platreef (Hutchinson and McDonald 2008; Sharman et al. 2013). In all cases, once sulfide droplets have formed they will interact with their surrounding environment in a manner that mainly depends on physical properties and constraint. In a layered intrusion, the newly formed droplets are expected to grow and settle and accumulate to the bottom of the magma chamber when their size is large enough to overcome gravity constraint. Under particular conditions (Chung and Mungall 2009; Mungall and Su 2005 and Chap. 8), the sulfide droplets may coalesce to form 3D networks of interstitial sulfide liquid that may percolate downwards into the crystallizing cumulus mineral pile until it reaches an area where no further migration is possible (Godel et al. 2006).

The base-metal and PGE concentrations in the sulfide liquid (C_{Sul}) is mainly depending on the partition coefficient between sulfide and silicate melt and the bulk metal concentration and is best described, in a closed system, by Campbell and Naldrett (1979) equation:

$$C_{Sul} = C_0 \, D^{Sul/Sil} \frac{R+1}{R+D^{Sul/Sil}}$$

where Co is the concentration of a given metal in the original silicate liquid, $D^{Sul/Sil}$ is the partition coefficient (for each metal) between the sulfide and the silicate liquids, and R is the mass ratio of silicate to sulfide liquid. Hence, if small amounts of sulfide form in a magma chamber (i.e. high R-factor) their PGE concentration will be large, typically those found in PGE reef deposits (Campbell et al. 1983). In contrast, if the amount of sulfide that forms is significant relative to the host magma (i.e. low R-factor), the PGE concentration in the sulfide will be diluted and may not be sufficient to make economic PGE ore grades. However, if a large amount of sulfide is available, it may represent an interesting target for Ni and Cu deposit where PGE will be a byproduct. In rare case such as the Platreef, suitable parental magma combined with various amounts of assimilation of crustal sulfur may lead to the formation of an exceptional PGE-deposit (Sharman et al. 2013 and references therein).

The equation of Campbell and Naldrett (1979) has successfully and widely been applied to model metal concentrations observed in a variety of magmatic ore deposits (Naldrett 2004 and references therein). Notably, the application of this equation to the silicate rocks of the Merensky Reef (Barnes and Maier 2002b; Godel et al. 2007) indicates that the PGE present in solid-solution of the base-metal sulfide and observed as platinum-group minerals closely associated with the sulfide can be accommodated in a sulfide liquid. This sulfide liquid interacted with a large volume of magma with metal compositions similar to that suggested for the Bushveld

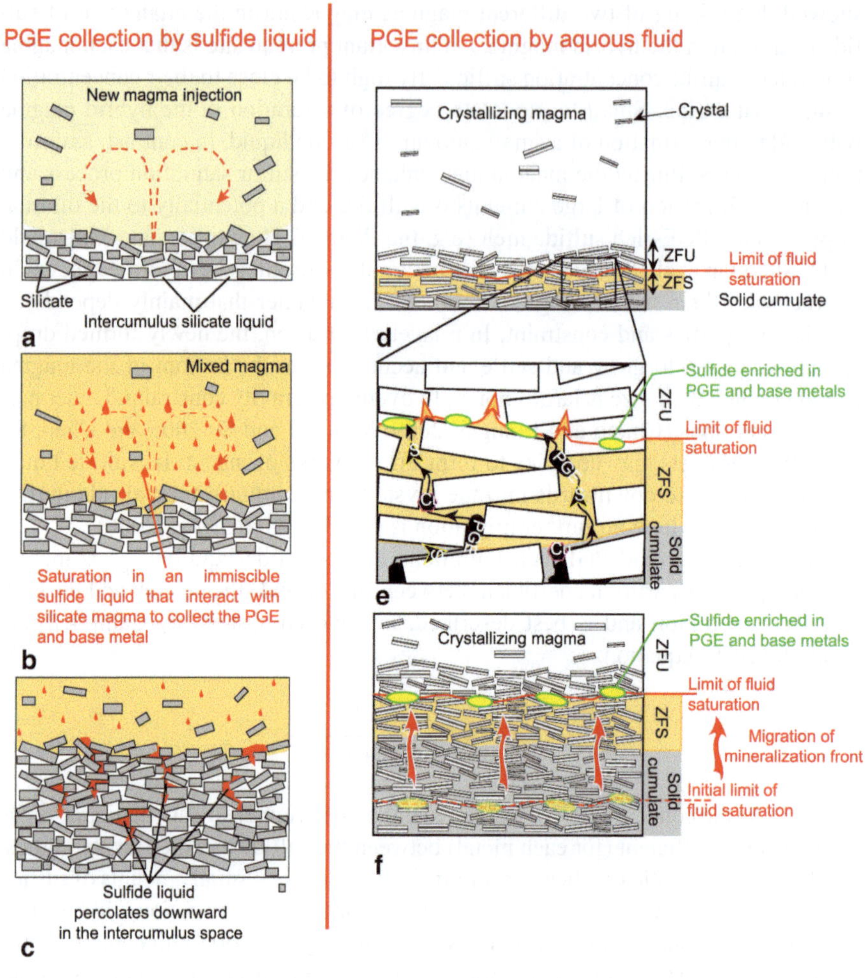

Fig. 9.24 Schematic diagram illustrating proposed model of formation of platinum-group element deposits in silicate rocks (modified after Godel, 2007). **a**, **b** and **c** Model of PGE collection by an immiscible magmatic sulfide liquid that segregate from the magma and then percolate downwards in the interstices between cumulus minerals. See text for further explanation and reference. **d**, **e**, and **f**) Model of PGE collection by an aqueous fluid that percolates upwards into the cumulate pile as it crystallises. *ZFU* zone of fluid undersaturation, *ZFS* zone of fluid saturation. See text for further explanation and reference

Complex (Barnes et al. 2010). During cooling, the PGE and Pt in particular may exsolved from the sulfide as platinum-group minerals or in some case partition into a semi-metal and sulfarsenide melt that subsequently crystallize as platinum-group minerals (Fig. 9.25). This combination of processes accounts for the range of geochemical and textural features observed in the silicate rocks of the Merensky Reef.

In contrast, the PGE (mainly Pt, Ir, Os and Ru) concentrations observed in PGE-rich (often sulfide poor) chromitite layers of the Merensky Reef (Barnes and

Maier 2002b; Godel et al. 2007) and large PGE-rich chromitite layers (Maier and Barnes 2008; Naldrett et al. 2011b; Naldrett 2004) are too high to be solely explained by a process of PGE collection by an immiscible sulfide liquid, unless partition coefficients are at the extreme limit of the range that has been determined on experimental evidence (Fonseca et al. 2009), and R-factor values are at the upper limit of plausibility. Cawthorn et al. (2005) have argued that this combination of circumstances is unlikely, although Kerr and Leitch (2005) have proposed a mechanism of multi-stage dissolution and re-enrichment of sulfide liquid that could account for extreme PGE enrichment. Application of this model to the Merensky Reef was presented in Godel et al. (2008a) and Naldrett et al. (2012).

PGE-Enrichment in Chromitite Layers

Several processes have been proposed to explain the particular PGE enrichment and metal ratios observed in chromitite layers and the predominance of platinum-group minerals on the PGE-budget. Naldrett and Lehmann (1988) suggested that primary PGE-rich magmatic sulfide trapped within the chromitite could react with chromite to produce a Fe-rich chromite and remove sulfur from the system according to the reaction:

$$4/3\ Fe_2O_3\ \left(Chromite\right) + 4/3\ FeS = Fe_3O_4\ \left(Chromite\right) + 1/6\ FeS$$

Consequently, this reaction lowers the sulfur fugacity (fS_2) and could have resulted in the partial desulfurization of the sulfide leading to the formation of residual monosulfide solid solution, Fe–Pt alloys and Cu–Pd rich sulfide melt (Peregoedova et al. 2004). This model can explain the fact that the sulfides in the chromitite layers of the Bushveld Complex exhibit high PGE/ sulfur ratios and are generally enriched in Cu and to a lesser extent Ni. By considering this hypothesis, the chromitite layers should have originally contained at five to ten times more sulfides than they do now and all the PGE were originally hosted in the sulfide liquid. The model does not explain the reason why the chromitite layers exhibit Pd concentrations lower than those of the other PGE. Two other possibilities have been envisaged to explain the particular features observed in chromitite layers and are related to the early formation of platinum-group minerals. Evidence of the direct crystallization of laurite (RuS_2) from a sulfide-undersaturated silicate melt has been provided experimentally and thermodynamically (Andrews and Brenan 2002; Bockrath et al. 2004b; Brenan and Andrews 2001). Based on PGE mass balance and geochemical modelling in the Merensky Reef chromitites, Barnes and Maier (2002b) and Godel et al. (2007) proposed that some of the PGE may crystallise as platinum-group minerals (mainly laurite and Pt alloy) before an immiscible sulfide liquid forms, and these PGMs are collected by crystallizing chromite. Subsequent, sulfide liquid formation collected the remaining PGE. Finnigan et al. (2008) carried out experiments involving the growth and reequilibration of chromite crystals in a silicate melt. In these experiments, platinum-group minerals (mainly laurite and IPGE alloys) formed at the in-

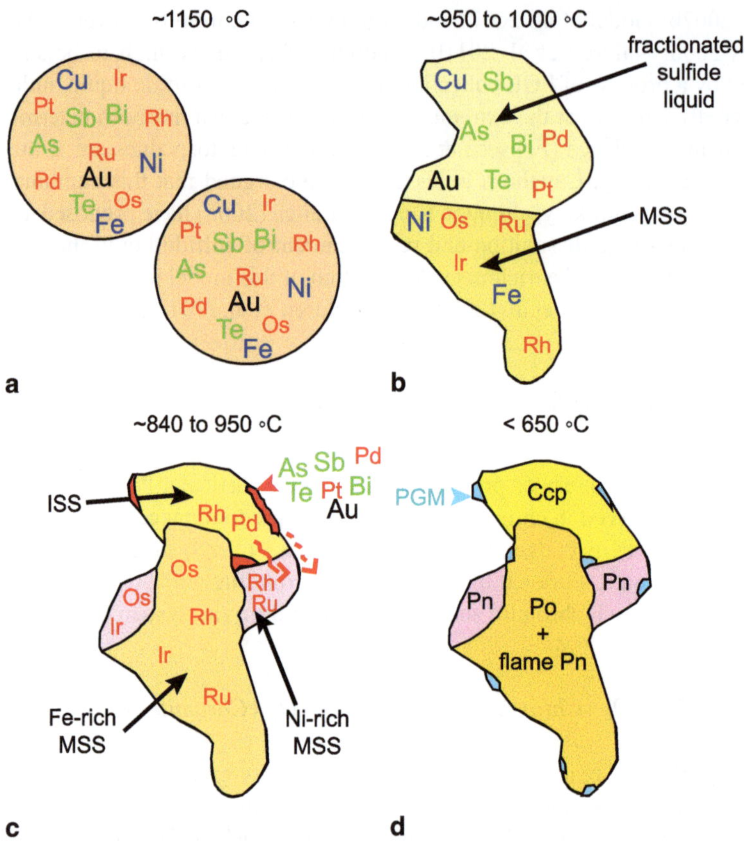

Fig. 9.25 Schematic diagram summarizing the behaviour of platinum-group elements, base metal and other chalcophile elements during the formation of a base metal sulfide liquid and its cooling (modified after Barnes et al. 2006). **a** Immiscible sulfide liquid droplets form and collect the PGE and other chalcophile elements. **b** On cooling, the sulfide liquid fractionates leading to the formation of monosulfide solid solution (*MSS*) and fractionated Cu-rich sulfide liquid. **c** Further cooling lead to the crystallization of intermediate sulfide solid-solution (*ISS*) and Ni-rich MSS from the fractionated sulfide liquid. In some cases, sulfosalt melts (i.e. As, Sb, Te or Bi bearing) may formed and collect some of the PGE. These melts may migrate along sulfide/silicate boundaries. **d** At temperature < At tC, pyrrhotite and pentlandite exsolve to form the *MSS* and chalcopyrite from the *ISS* and platinum-group minerals crystallized from the sulfosalt melt

terface between the chromite crystal and the silicate melt. Based on their results, the authors proposed that the platinum-group minerals may precipitate as a result of local reduction around chromite crystal, probably caused by the preferential uptake of Fe^{3+} and Cr^{3+} by the chromite crystal relative to the melt. During the crystallization of the chromite, the local reduction triggers a decrease (by few percent) of the metal solubility at the crystal/melt interface leading to the formation of the platinum-group minerals that are subsequently trapped by the growing chromite. To date no consensus exists on the origin of PGE enrichment in chromitite layers and it remains a controversial topic. An alternative model was proposed by Latypov et al. (2013 and references therein) to explain the lithological and textural relationships observed. In this model, the chromitites are interpreted to have formed by *in situ* growth directly at a crystal–liquid interface and the PGE are scavenged by sulfide droplets from fresh magma continuously brought towards the base by convection. The convection allows the sulfide to equilibrate with picritic magma of thousands to million times their own volume, resulting in sulfide high PGE tenor.

Role of Fluids on PGE Distribution and Redistribution

Experimental studies have shown that the base metals and the PGE may be transported by fluids of various compositions and within a large range of temperature (Baker et al. 2001; Hanley et al. 2005; Peregoedova et al. 2006; Wood 2002 and references therein). Details of the geochemistry of the PGE in low-to high temperature hydrothermal and magmatic fluids are presented in Hanley (2005) and a recent review by Barnes and Liu (2012). Evidence of PGE transport and reprecipitation has been observed at various scales and in different geological settings (Augé et al. 2005; Bursztyn and Olivo 2010; Cabral et al. 2002; El-Ghorfi et al. 2006; Molnar et al. 2001; Moroni et al. 2001; Olivo et al. 2001).

Platinum-Group Elements Collection by Deuteric Fluids

The common observation of pegmatoidal textures, the presence of fluid and hydrous melt inclusions and the presence of Cl-rich apatites (i.e. having high Cl/F ratios) associated with PGE mineralization has led some authors to propose that the PGE-rich horizons observed in layered intrusions resulted from the percolation of Cl-rich aqueous fluid that carried and redeposited the PGE (e.g. Ballhaus and Stumpfl 1986; Boudreau 2008; Boudreau and McCallum 1986; Boudreau and Mc Callum 1992a, b; Kanitpanyacharoen and Boudreau 2013; Vermaak and Hendriks 1976; Von Gruenewaldt 1979; Willmore et al. 2000). In this type of model (Figs. 9.24d, e and f), H_2O dissolved in silicate magmas and intercumulus fractionated silicate liquid becomes progressively saturated in aqueous Cl-rich fluid as the crystal pile solidifies. The Cl-rich aqueous fluid collects the base-metals and the PGE as it migrates upward through the compacting cumulate pile. The base-metal and PGE are transported in solution until it reaches the interface where the intercumulus silicate liquid is not saturated in Cl-rich aqueous fluid. At this interface, the fluid dissolves in the intercumulus silicate liquid and PGE-rich base-metal sulfides with a mineralization front moving upwards as the limit of fluid saturation in the intercumulus silicate liquid migrates, ultimately forming a base-metal and PGE-rich sulfide horizon (Boudreau 1999; Boudreau and Mc Callum 1992b). However, the application of this model to explain the entire complexity of the PGE-reefs remains controversial (Barnes and Maier 2002b; Cawthorn et al. 2005; Godel et al. 2007; Keays et al. 2012; Maier et al. 2013; Naldrett et al. 2009b; Osbahr 2012). The main arguments against this model are from the literature:

(i) No anomalously-poor PGE concentrations or PGE-depletion are observed in the rocks below the mineralized horizons of the Bushveld Complex (Barnes and Maier 2002a; Barnes and Maier 2002b; Naldrett et al. 2009b) and the Stillwater Complex (Keays et al. 2012);
(ii) PGE-reefs, in most of the cases, can be traced over large distances (100's km in the Bushveld Complex) and directly overly a magmatic unconformity. Upward migration of fluids would lead to discordant features that are not related to magmatic stratigraphy.
(iii) Iridium and Ru are considered to be insoluble in aqueous hydrothermal/magmatic fluids, hence it is difficult to reconcile this with the concentration observed in the reefs;
(iv) In the Bushveld Complex, the mantle-normalized base-metal and PGE patterns of the mineralized horizons (e.g. Merensky Reef, Fig. 9.10) show shapes identical to those of the B1 magma (Fig. 9.23), suggesting that the same process concentrated all the PGE at the level of the reef (Fonseca et al. 2009).
(v) The osmium isotopic composition of Bushveld Complex cumulates shows a sharp change towards more radiogenic compositions at the level of the Merensky reef (McCandless and Ruiz 1991). The Merensky Reef is as much a concentration of Os as it is of Pt and Pd, but the isotopic data imply that Os cannot have been derived from the underlying cumulates. In addition, there are places in the Bushveld Complex (e.g. the far southern limit of the eartern lobe) where the UG2 reef package sits on top of the footwall metasediments without underlying Rustenburg Layered Suite stratigraphy. Hence a model involving upward transport by fluids would require that concentrations of different PGEs occurred in the same rock by entirely different processes (Li and Ripley 2006; Li et al. 2004).

Hydrothermal Remobilization of the Platinum-Group Elements

Evidences of post-cumulus remobilization of PGE in PGE reefs hosted within layered intrusion have notably been described for the JM-Reef of the Stillwater Complex (Godel and Barnes 2008a; Godel and Barnes 2008b), the Platreef (Holwell and McDonald 2006; Hutchinson and Kinnaird 2005; Yudovskaya et al. 2012). In the JM-Reef of the Stillwater Complex, detailed mineralogical analysis (Godel and Barnes 2008a; Godel and Barnes 2008b) of altered samples revealed the presence of Pd–Cu alloys (with composition similar to Skaergaardite (Rudashevsky et al. 2004)) and to a lesser extent Pd sulfides observed either as interstitial phase between magnetite crystal or as small veinlets crosscutting the base-metal sulfide minerals (Fig. 9.22). This observation suggests that Pd was mobile after the base-metal sulfides solidified and possibly added to the system during alteration or metamorphic processes that lead to the oxidation of the base-metal sulfides to magnetite. The origin, the scale and the extent of the Pd remobilization remain poorly constrained.

Weathering and Remobilization of the Platinum-Group Elements

In the Main Sulfide Zone of the Great Dyke, primary magmatic PGE-rich sulfides have been oxidized during near-surface weathering and represent a challenge for geometallurgy (Locmelis et al. 2010; Oberthür and Melcher 2005). In this oxidized ore, the proportion of PGE found in solid solution of the base-metal sulfides is negligible but Pd concentration in relict pentlandite can reach 6500 ppm. Most of the Pt and Pd are hosted by relicts of primary and secondary platinum-group minerals. PGE released during the oxidation of the primary are redistributed and incorporated into Fe/Mn oxides/hydroxides and clay minerals leading to considerable local variation in the distribution of PGE.

The examples described above highlight the importance of considering hydrothermal processes over a range of temperatures (from low-temperature hydrothermal to high-temperature magmatic or hydrothermal) in the generation or modification of high grade PGE-ores.

Conclusions

Over the past 10 years, technological developments (e.g. LA-ICP-MS, high-resolution X-ray computed tomography, isotope-ratios mass spectrometers) have allowed the acquisition of multidisciplinary dataset at spatial resolution, detection limits and precisions that were impossible to achieve in the past, in both 2D and 3D. The results obtained provided additional insights into the complexity of the ores, the residency of PGE in the rocks and the variability within and between different ore deposits and highlighted the necessity of adapting (for each deposit) models of formation based on all of the information available. Each PGE deposit is different and the processes that form them are complex. Models of formation may range from the primary accumulation of immiscible base-metal sulfide liquid to the late-stage (post-cumulus alteration or metamorphism) remobilization of the PGE and base-metal sulfides. It is clear that any model of formation of PGE-reefs needs to consider the superimposition of both physical and geochemical processes, over a range of scale and over a wide range of temperature (Fig. 9.26) to successfully account for the observed petrographic and mineralogical complexities.

Acknowledgments James Mungall, Ian McDonald and Rais Latypov (associate editor) are thanked for their comments and suggestion that improve the quality of the manuscript. Steve Barnes, Patrick Nadoll and Ludovic Ricard are acknowledged for their suggestions during CSIRO internal review processes. This is a contribution from the CSIRO Mineral Down Under Flagship.

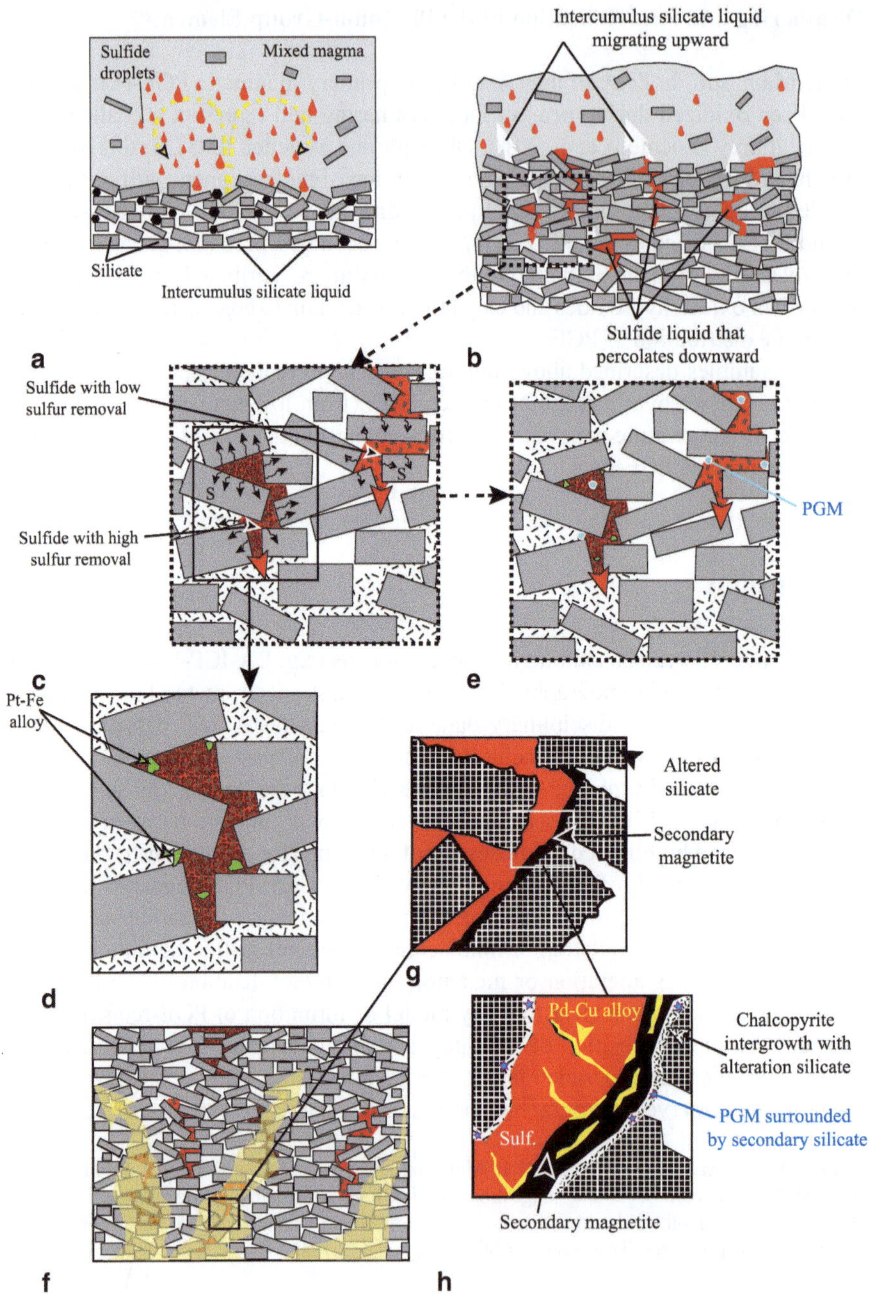

Fig. 9.26 Schematic diagram summarizing the processes that might be involved in the formation and modification of PGE- reef type deposits. **a** Immiscible magmatic sulfides form as a result of magma mixing, crustal contamination and/or crystal fractionation. **b** The sulfide droplets collect the base metal and PGE, settle in the magma chamber and potentially accumulate and coalesce while percolating downward into the cumulus mineral framework. **c** Intercumulus silicate melts or

References

Alapieti TT, Halkoaho TAA (1995) Cryptic variation of augite in the Penikat Layered Intrusion, Northern Finland, with reference to megacyclic units and PGE-enriched zones. Mineral Petrol 54(1–2):11–24

Alapieti TT, Lahtinen JJ (1986) Stratigraphy, petrology, and platinum-group element mineralization of the early Proterozoic Penikat layered intrusion, northern Finland. Econ Geol 81(5):1126–1136

Alapieti TT, Lahtinen JJ (2002) Platinum-group element mineralization in layered intrusions of Northern Finland and the Kola Peninsula, Russia. In: Cabri LJ (ed) Geology, Geochemistry, Mineralogy and Mineral benefication of Platinum Group Element, vol Special Volume 54. pp 507–546

Alard O, Griffin WL, Lorand JP, Jackson SE, O'Reilly SY (2000) Non-chondritic distribution of the highly siderophile elements in mantle sulfides. Nature 407(6806):891–894

Andersen JCO, Rasmussen H, Nielsen TFD, Ronsbo JG (1998) The Triple Group and the Platinova gold and palladium reefs in the Skaergaard Intrusion—stratigraphic and petrographic relations. Econ Geol 93:488–509

Andrews DRA, Brenan JM (2002) The solubility of ruthenium in sulfide liquid: implications for platinum group mineral stability and sulfide melt-silicate melt partitioning. Chem Geol 192:163–181

Ariskin AA, Barmina G, Bychkov K, Danyushevsky L (2009) Parental magmas of mafic layered intrusions: using an updated COMAGMAT model for calculations of sulfide-silicate cotectics during their crystallization. In: Proceedings of Xi-an International Ni–Cu(Pt) Deposit Symposium 2009, vol., Xi'an

Ariskin AA, Danyyushevsky LV, Bychkov K, Barmina GS (2010) Modeling sulfide solubility in MORB glasses and olivine cumulate piles: the opposite effect of NiO versus FeO in the melt. In: Survey OG (ed) 11th International Platinum Symposium, vol Abstract volume, Miscellaneous Release. Sudbury, Ontario, Canada, pp 1–4

Ariskin AA, Danyushevsky L, Bychkov KA, McNeill AW, Barmina G, Nikolaev GS (2013) Modelling solubility of Fe–Ni sulfides in basaltic magmas: the effect of Ni. Econ Geol 108(8):1983–2003

Arndt NT (2013) The lithospheric mantle plays no active role in the formation of magmatic ore deposits. Econ Geol 108(8):1953–1970

Arndt NT, Lesher CM, Czamanske GK (2005) Mantle derived magmas and magmatic Ni–Cu–PGE deposits. Econ Geol 100th Anniversary Volume:5–24

Augé T, Petrunov R, Bailly L (2005) On the origin of the PGE mineralization in the elatsite porphyry Cu–Au deposit, Bulgaria: comparison with the Baula-Nuasahi complex, India, and other alkaline PGE-rich porphyries. Can Mineral 43:1355–1372

Baker DR, Barnes S-J, Simon G, Bernier F (2001) Fluid transport of sulfur and metals between sulfide melt and basaltic melt. Can Mineral 39:537–546

---

magmatic fluids may percolate upward into the cumulate pile and potentially react with the base metal sulfide leading to area of localized desulfurisation **d** and the exsolution of some platinum-group minerals (e.g. Pt–Fe alloy). **e** On cooling, additional platinum-group minerals crystallise from the sulfides or from semi-metal or sulfosalt melts. The platinum-group minerals tend to form at the contact at the interface with the base-metal sulfides. **f** Post-cumulus hydrothermal or metamorphic events may lead to the remobilisation of PGE and to the crystallization of secondary PGM. In the example of the Stillwater Complex **g**, secondary magnetite is the host of secondary Pd–Cu alloy **h**. In addition, the base-metal sulfide may partially be replaced by secondary silicate. Insoluble platinum-group minerals are not affected and are observed as inclusion within secondary silicate without any visible textural relationship with sulfides

Ballhaus CG, Stumpfl EF (1986) Sulfide and platinum mineralization in the Merensky Reef: evidence from hydrous silicates and fluid inclusions. Contrib Mineral Petrol 94(2):193–204
Ballhaus C, Sylvester P (2000) Noble metal enrichment processes in the Merensky Reef, Bushveld Complex. J Petrol 41(4):545–561
Barnes SJ (1993) Partitioning of the platinum group elements and gold between silicate and sulfide magmas in the Munni-Munni Complex, Western-Australia. Geochim Cosmochim Acta 57(6):1277–1290
Barnes SJ, Fiorentini ML (2008) Iridium, ruthenium and rhodium in komatiites: Evidence for iridium alloy saturation. Chem Geol 257(1–2):44–58
Barnes S-J, Lightfoot PC (2005) Formation of magmatic nickel sulfide deposits and processes affecting their copper and platinum-group element contents. Econ Geol 100th Anniversary Volume:179–213
Barnes SJ, Liu W (2012) Pt and Pd mobility in hydrothermal fluids: Evidence from komatiites and from thermodynamic modelling. Ore Geol Rev 44:49–58
Barnes S-J, Maier WD (2002a) Platinum-group element distributions in the Rustenburg Layered Suite of the Bushveld Complex, South Africa. In: Cabri LJ (ed) Geology, Geochemistry, Mineralogy and Mineral benefication of Platinum Group Element, vol Special Volume 54. Canadian Institute of Mining, Metallurgy and Petroleum, pp 553–580
Barnes S-J, Maier WD (2002b) Platinum-group elements and microstructures of Normal Merensky Reef from Impala Platinum Mines, Bushveld Complex. J Petrol 43(1):103–128
Barnes SJ, Naldrett AJ (1986) Geochemistry of the J-M Reef of the Stillwater Complex, Minneapolis adit area II. Silicate mineral chemistry and petrogenesis. J Petrol 27(4):791–825
Barnes S-J, Naldrett AJ, Gorton MP (1985) The origin of the fractionation of platinum-group elements in terrestrial magmas. Chem Geol 53(3–4):303–323
Barnes S-J, Cox R, Zientek M (2006) Platinum-group element, Gold, Silver and Base Metal distribution in compositionally zoned sulfide droplets from the Medvezky Creek Mine, Noril'sk, Russia. Contrib Mineral Petrol 152(2):187–200
Barnes S-J, Prichard HM, Cox RA, Fisher PC, Godel B (2008) The location of the highly siderophile elements in platinum-group element rich ore deposits (a textural, whole rock geochemical and microbeam study): Implication for the formation of the deposits. Chem Geol 248:295–317
Barnes S-J, Savard D, Bédard LP, Maier W (2009) Selenium and sulfur concentrations in the Bushveld Complex of South Africa and implications for formation of the platinum-group element deposits. Miner Deposita 44(6):647–663
Barnes S-J, Maier WD, Curl EA (2010) Composition of the marginal rocks and sills of the Rustenburg Layered Suite, Bushveld Complex, South Africa: implications for the formation of the Platinum-Group Element Deposits. Econ Geol 105(8):1491–1511
Barnes SJ, Osborne G, Cook D, Barnes L, Maier WD, Godel B (2011) The Santa Rita nickel sulfide deposit in the Fazenda Mirabela Intrusion, Bahia, Brazil: geology, sulfide geochemistry and genesis. Econ Geol 106:1083–1110
Bird DK, Brooks CK, Gannicott RA, Turner PA (1991) A gold-bearing horizon in the Skaergaard Intrusion. Econ Geol 86(5):1083–1092
Bockrath C, Ballhaus C, Holzheid A (2004a) Fractionation of the platinum-group elements during mantle melting. Sciences 305:1951
Bockrath C, Ballhaus C, Holzheid A (2004b) Stabilities of laurite RuS2 and monosulfide liquid solution at magmatic temperature. Chem Geol 208(1–4):265–271
Boudreau A (1999) Fluid fluxing of cumulates: The J-M reef and associated rocks of the Stillwater complex, Montana. J Petrol 40(5):755–772
Boudreau A (2008) Modeling the Merensky Reef, Bushveld Complex, Republic of South Africa. Contrib Mineral Petrol 158:431–437
Boudreau A, McCallum IS (1986) Investigations of the Stillwater Complex: III. The Picket Pin Pt/Pd deposit. Econ Geol 81:1953–1975
Boudreau AE, Mc Callum IS (1992a) Concentration of platinum-group elements by magmatic fluids in layered intrusions. Econ Geol 87(7):1830–1848

Boudreau AE, Mc Callum IS (1992b) Infiltration metasomatism in layered intrusions—an example from the Stillwater Complex, Montana. J Volcanol Geotherm Res 52(1–3):171–183
Brenan JM, Andrews DRA (2001) High-temperature stability of laurite and Ru–Ir–Os alloy and their role on PGE fractionation in mafic magmas. Can Mineral 39:341–360
Brenan JM, McDonough WF, Dalpe C (2003) Experimental constraints on the partitioning of rhenium and some platinum-group elements between olivine and silicate melt. Earth Planet Sci Lett 212(1–2):135–150
Brenan JM, McDonough WF, Ash R (2005) An experimental study of the solubility and partitioning of iridium, osmium and gold between olivine and silicate melt. Earth Planet Sci Lett 237(3–4):855–872
Brenan JM, Finnigan CF, McDonough WF, Homolova V (2012) Experimental constraints on the partitioning of Ru, Rh, Ir, Pt and Pd between chromite and silicate melt: The importance of ferric iron. Chem Geol 302–303(0):16–32
Buchanan DL (1979) A combined transmission electron microscope and electron microprobe study of Bushveld pyroxenes from the Bethal area. J Petrol 20:327–354
Bursztyn NE, Olivo GR (2010) PGE-Rich Ni–Cu sulfide mineralization in the Flin Flon greenstone Belt, Manitoba, Canada: Implications for hydrothermal remobilization of platinum-group elements in basic-ultrabasic sequences. Econ Geol 105(8):1469–1490
Cabral AR, Lehmann B, Kwitko R, Costa CHC (2002) The Serra Pelada Au–Pd–Pt deposit, Carajas Mineral Province, Northern Brazil: reconnaissance mineralogy and chemistry of very high grade palladian gold Mineralization. Econ Geol 97(5):1127–1138
Cabri LJ (2002) The Platinum-Group Minerals. In: Cabri LJ (ed) Geology, Geochemistry, Mineralogy and Mineral benefication of Platinum Group Element, vol Special Volume 54. Canadian Institute of Mining, Metallurgy and Petroleum, pp 13–129
Cabri LJ, Blank H, El Goresy A, Laflamme JHG, Nobiling R, Sizgoric MB, Traxel K (1984) Quantitative trace-element analyses of sulfides from Sudbury and Stillwater by proton microprobe. Can Mineral 22:521–542
Campbell IH, Naldrett AJ (1979) The influence of silicate silicate: sulfide ratios on the geochemistry of magmatic sulfides. Econ Geol 74:1503–1505
Campbell IH, Naldrett AJ, Barnes SJ (1983) A model for the origin of the platinum-rich sulfide horizons in the Bushveld and Stillwater Complexes. J Petrol 24(2):133–165
Capobianco CJ, Drake MJ (1990) Partitioning of ruthenium, rhodium, and palladium between spinel and silicate melt and implications for platinum group element fractionation trends. Geoch Cosmochim Acta 54(3):869–874
Capobianco CJ, Hervig RL, Drake MJ (1994) Experiments on crystal/liquid partitioning of Ru, Rh and Pd for magnetite and hematite solid solutions crystallized from silicate melt. Chem Geol 113(1–2):23–43
Causey JD, Galloway JP, Zientek ML (2009) An index to PGE–Ni–Cr deposits and occurrences in selected mineral-occurrence databases. USGS Open-File Report 2009–1045
Cawthorn RG (1999) The platinum and palladium resources of the Bushveld Complex. S Afr J Sci 95(11/12):481
Cawthorn RG (2002) Platinum-group element deposits in the Bushveld Complex, South Africa. In: Cabri LJ (ed) Geology, Geochemistry, Mineralogy and Mineral benefication of Platinum Group Element, vol Special Volume 54. Canadian Institute of Mining, Metallurgy and Petroleum, pp 389–429
Cawthorn RG, Webb SJ (2001) Connectivity between the western and eastern limbs of the Bushveld Complex. Tectonophysics 330(3–4):195–209
Cawthorn RG, Lee CA, Schouwstra RP, Mellowship P (2002) Relationship between PGE and PGM in the Bushveld Complex. Can Mineral 40:311–328
Cawthorn RG, Barnes SJ, Ballhaus C, Malitch KN (2005) Platinum-group element, chromium, and vanadium deposits in mafic and ultramafic rocks. Eco Geol 100th Anniversary Volume:215–249

Chung H-Y, Mungall JE (2009) Physical constraints on the migration of immiscible fluids through partially molten silicates, with special reference to magmatic sulfide ores. Earth Planet Sci Lett 286(1–2):14–22

Cole J, Webb SJ, Finn CA (2014) Gravity models of the Bushveld Complex—have we come full circle? J Afr Earth Sci 92:97–118

Crocket JH (2002) Platinum-group element geochemistry of mafic and ultramafic rocks. In: Cabri LJ (ed) Geology, Geochemistry, Mineralogy and Mineral benefication of Platinum Group Element, vol Special Volume 54. Canadian Institute of Mining, Metallurgy and Petroleum, pp 553–580

Czamanske GK, Loferski PJ (1996) Cryptic trace-element alteration of anorthosite, Stillwater Complex, Montana. Can Mineral 34:559–579

Czamanske GK, Zientek GK (1985) The Stillwater Complex, Montana: geology and guide. Montana College of Mineral Science and Technology, Butte, Montana

Dale CW, Macpherson CG, Pearson DG, Hammond SJ, Arculus RJ (2012) Inter-element fractionation of highly siderophile elements in the Tonga Arc due to flux melting of a depleted source. Geochim Cosmochim Acta 89(0):202–225

Davey SR (1992) Lateral variations within the Upper Critical zone of the Bushveld complex on the farm Rooikoppies 297 JQ, Marikana, South Africa. S Afr J Geol 95:141–149

Dubrovsky MI, Rundqvist TV (2009) Petrology of the early Proterozoic platinum-bearing massif of Fedorov Tundra, Kola Peninsula. Geol Ore Deposits 51(7):577–587

Eales HV, Cawthorn RG (1996) The Bushveld Complex. In: Cawthorn RG (ed) Layered Intrusions. The Nederlands, Elsevier, Amsterdam, pp 181–230

Eales HV, Botha WJ, Hattingh PJ, de Klerk WJ, Maier WD, Odgers ATR (1993) The mafic rocks of the Bushveld complex: a review of emplacement and crystallization history, and mineralization, in the light of recent data. J Afr Earth Sci 16(1–2):121–142

El-Ghorfi M, Oberthür T, Melcher F, Lüders V, Boukhari A, Maacha L, Ziadi R, Baoutoul H (2006) Gold-palladium mineralization at Bleïda Far West, Bou Azzer-El Graara Inlier, Anti-Atlas, Morocco. Miner Deposita 41(6):549–564. doi:10.1007/s00126-006-0077-3

Evans KA, O'Neill HSC, Mavrogenes JA (2008) Sulfur solubility and sulfide immiscibility in silicate melts as a function of the concentration of manganese, nickel, tungsten and copper at 1 atm and 1400 °C. Chem Geol 255(1–2):236–249

Finnigan CS, Brenan JM, Mungall JE, McDonough WF (2008) Experiments and models bearing on the role of chromite as a collector of platinum group minerals by local reduction. J Petrology:egn041 doi:10.1093/petrology/egn041

Fonseca ROC, Campbell IH, O'Neill HSC, Allen CM (2009) Solubility of Pt in sulfide mattes: Implications for the genesis of PGE-rich horizons in layered intrusions. Geochim Cosmochim Acta 73(19):5764–5777

Gain SB (1985) The geologic setting of the platiniferous UG-2 chromitite layer on the farm Maandagshoek, eastern Bushveld Complex. Econ Geol 80(4):925–943

Godel B (2007) Rôles des liquides sulfurés dans la formation des minéralisations riches en éléments du groupe du platine: applications au complexe du Bushveld (Afrique du Sud) et au complexe de Stillwater (Etats-Unis) Ph. D. Université du Québec, 369 p

Godel B (2013) High resolution X-ray computed tomography and its application to ore deposits: case studies from Ni–Cu–PGE deposits. Econ Geol 108:2005–2019

Godel B, Barnes S-J (2008a) Image analysis and composition of platinum-group minerals in the J-M Reef, Stillwater Complex (U.S.A.). Econ Geol 103(3):637–651

Godel B, Barnes S-J (2008b) Platinum-group elements in sulfide minerals and the whole rocks of the J-M Reef (Stillwater Complex): implication for the formation of the reef. Chem Geol 248:272–294

Godel B, Barnes S-J, Maier WD (2006) 3-D distribution of sulfide minerals in the Merensky Reef (Bushveld Complex, South Africa) and the J-M Reef (Stillwater Complex, USA) and their relationship to microstructures using X-Ray Computed Tomography. J Petrol 47:1853–1872

Godel B, Barnes S-J, Maier WD (2007) Platinum-group elements in sulfide minerals, platinum-group minerals, and whole-rocks of the Merensky Reef (Bushveld Complex, South Africa): Implications for the formation of the reef. J Petrol 48(8):1569–1604

Godel B, Maier WD, Barnes S-J (2008) Platinum-group elements in the Merensky and J-M Reefs: a review of recent studies. J Geol Soc India 72:595–609

Godel B, Barnes SJ, Barnes S-J, Maier WD (2010) Platinum ore in 3D: Insights from high-resolution X-ray computed tomography. Geology 38(12):1127–1130

Godel B, Barnes S-J, Maier WD (2011) Parental magma composition inferred from trace element in cumulus and intercumulus silicate minerals: an example from the Lower and Lower Critical Zones of the Bushveld Complex, South Africa. Lithos 125:537–552

Godel B, Rudashevsky NS, Nielsen TFD, Barnes SJ, Rudashevsky VN (2014) Constraints on the origin of the Skaergaard intrusion mineralization: insights from high-resolution X-ray computed tomography. Lithos 190(27–36)

Halkoaho TAA, Alapieti TT, Lahtinen JJ (1990) The Sompujarvi PGE reef in the Penikat layered intrusion, northern Finland. Miner Petrol 42(1–4):39–55. doi:10.1007/bf01162683

Hamlyn PR, Keays RR (1986) Sulfur saturation and second stage melts: application to the Bushveld platinum metal deposits. Econ Geol 81:1431–1435

Hanley JJ (2005) The aqueous geochemistry of the platinum-group elements (PGE) in surficial, low-T hydrothermal and high-T magmatic-hydrothermal environments. In: Mungall JE (ed) Exploration for platinum-group element deposits, vol Short Course 35. Mineralogical Association of Canada, pp 35–56

Hanley JJ, Pettke T, Mungall JE, Spooner ETC (2005) The solubility of platinum and gold in NaCl brines at 1.5 kbar, 600 to 800[deg]C: A laser ablation ICP-MS pilot study of synthetic fluid inclusions. Geoch Cosmochim Acta 69(10):2593–2611

Hiemstra SA (1979) The role of collectors in the formation of the platinum deposits in the Bushveld Complex. Can Mineral 17:469–482

Hiemstra SA (1986) The distribution of chalcophile and platinum-group elements in the UG-2 chromitite layer of the Bushveld Complex. Econ Geol 81(5):1080–1086

Hirschmann MM, Renne PR, McBirney AR (1997) 40Ar39Ar dating of the Skaergaard intrusion. Earth Planet Sci Lett 146(3–4):645–658

Holness MB, Tegner C, Nielsen TFD, Stripp G, Morse SA (2007) A textural record of solidification and cooling in the Skaergaard Intrusion, East Greenland. J Petrology:egm064 doi:10.1093/petrology/egm064

Holwell DA, Keays RR (2014) The Formation of Low-Volume, High-Tenor Magmatic PGE-Au Sulfide Mineralization in Closed Systems: Evidence from Precious and Base Metal Geochemistry of the Platinova Reef, Skaergaard Intrusion, East Greenland. Econ Geol 109(2):387–406

Holwell D, McDonald I (2006) Petrology, geochemistry and the mechanisms determining the distribution of platinum-group element and base metal sulfide mineralisation in the Platreef at Overysel, northern Bushveld Complex, South Africa. Miner Deposita 41:575–598

Holwell DA, McDonald I (2007) Distribution of platinum-group elements in the Platreef at Overysel, northern Bushveld Complex: a combined PGM and LA-ICP-MS study. Contrib Mineral Petrol 154(2):171–190

Holwell DA, McDonald I, Armitage PEB (2006) Platinum-group mineral assemblages in the Platreef at the Sandsloot Mine, northern Bushveld Complex, South Africa. Mineral Mag 70(1):83–101

Holwell DA, Boyce AJ, McDonald I (2007) Sulfur isotope variations within the Platreef Ni–Cu–PGE Deposit: genetic implications for the origin of sulfide mineralization. Econ Geol 102(6):1091–1110

Huhtelin TA, Alapieti TT, Lahtinen JJ (1990) The Paasivaara PGE reef in the Penikat layered intrusion, northern Finland. Mineral Petrol 42(1–4):57–70

Hutchinson D, Kinnaird JA (2005) Complex multistage genesis for the Ni–Cu–PGE mineralisation in the southern region of the Platreef, Bushveld Complex, South Africa. Applied Earth Sc 114

Hutchinson D, McDonald I (2008) Laser ablation ICP-MS study of platinum-group elements in sulfides from the Platreef at Turfspruit, northern limb of the Bushveld Complex, South Africa. Miner Deposita 43(6):695–711

Jenner FE, O'Neill HSC, Arculus RJ, Mavrogenes JA (2010) The magnetite crisis in the evolution of arc-related magmas and the initial concentration of Au, Ag and Cu. J Petrol 51(12):2445–2464

Johnson-Matthey (2011) Platinum 2011 Interim Review. (40)

Jugo PJ (2009) Sulfur content at sulfide saturation in oxidized magmas. Geology 37(5):415–418

Jugo PJ, Luth RW, Richards JP (2005) An experimental study of the sulfur content in basaltic melts saturated with immiscible sulfide or sulfate liquids at 1300C and 10 GPa. J Petrol 46(4):783–798

Kanitpanyacharoen W, Boudreau AE (2013) Sulfide-associated mineral assemblages in the Bushveld Complex, South Africa: platinum-group element enrichment by vapor refining by chloride-carbonate fluids. Miner Deposita 48(2):193–210

Keays R, Lightfoot P, Hamlyn P (2012) Sulfide saturation history of the Stillwater Complex, Montana: chemostratigraphic variation in platinum group elements. Miner Deposita 47(1–2):151–173

Kerr A, Leitch AM (2005) Self-destructive sulfide segregation systems and the formation of high-grade magmatic deposits. Econ Geol 100:311–332

Kingston GA, El-Dosuky BT (1982) A contribution on the platinum-group mineralogy of the Merensky Reef at the Rustenburg Platinum Mine. Econ Geol 77(6):1367–1384

Kinloch ED (1982) Regional Trends in the Platinum-Group Mineralogy of the Critical Zone of the Bushveld Complex, South-Africa. Econ Geol 77(6):1328–1347

Kinloch ED, Peyerl W (1990) Platinum-group minerals in various rock types of the Merensky Reef: Genetic Implications. Econ Geol 85:537–555

Kinnaird J, Hutchinson D, Schurmann L, Nex P, de Lange R (2005) Petrology and mineralisation of the southern Platreef: northern limb of the Bushveld Complex, South Africa. Miner Deposita 40(5):576–597

Kogiso T, Suzuki K, Suzuki T, Shinotsuka K, Uesugi K, Takeuchi A, Suzuki Y (2008) Detecting micrometer-scale platinum-group minerals in mantle peridotite with microbeam synchrotron radiation X-ray fluorescence analysis. Geochem Geophys Geosyst 9(3):Q03018

Kruger F (2005) Filling the Bushveld Complex magma chamber: lateral expansion, roof and floor interaction, magmatic unconformities, and the formation of giant chromitite, PGE and Ti-V-magnetitite deposits. Miner Deposita 40(5):451–472

Latypov R, O'Driscoll B, Lavrenchuk A (2013) Towards a model for the in situ origin of PGE reefs in layered intrusions: insights from chromitite seams of the Rum Eastern Layered Intrusion, Scotland. Contrib Mineral Petrol 166(1):309–327

Laurenz V, Fonseca RIOC, Ballhaus C, Jochum KP, Heuser A, Sylvester PJ (2013) The solubility of palladium and ruthenium in picritic melts: 2. The effect of sulfur. Geochim Cosmochim Acta 108:172–183

Lavigne MJ, Michaud MJ (2001) Geology of the North American Palladium Ltd's Roby zone deposit, Lac des Iles. Exploration and Mining geology 10:1–17

Lechler PJ, Arehart GB, knight M (2002) Multielement and isotopic geochemistry of the J-M Reef, Stillwater Intrusion, Montana. In: 9th International Platinum symposium, vol Abstract Programm. Billings, pp 245–248

Lee CA (1996) A review of mineralization in the Bushveld Complex and some other layered intrusions. In: Cawthorn RG (ed) Layered Intrusions. Elsevier, Amsterdam, pp 103–145

Lee CA, Parry SJ (1988) Platinum-Group Element geochemistry of the lower and middle group chromitites of the eastern Bushveld Complex. Econ Geol 83(6):1127–1139

Li C, Ripley EM (2005) Empirical equations to predict the sulfur content of mafic magmas at sulfide saturation and applications to magmatic sulfide deposits. Miner Deposita:-46

Li C, Ripley EM (2006) Formation of Pt–Fe alloy by desulfurization of Pt–Pd sulfide in the J-M Reef of the Stillwater Complex, Montana. Can Mineral 44:895–903

Li C, Ripley EA (2009) Sulfur contents at sulfide-liquid or anhydrite saturation in silicate melts: empirical equations and example applications. Econ Geol 104:405–412

Li C, Ripley EM, Merino E, Maier WD (2004) Replacement of base metal sulfides by actinolite, epidote, calcite, and magnetite in the UG2 and Merensky Reef of the Bushveld Complex, South Africa. Econ Geol 99(1):173–184

Liu Y, Samaha N-T, Baker DR (2007) Sulfur concentration at sulfide saturation (SCSS) in magmatic silicate melts. Geochim Cosmochim Acta 71(7):1783–1799

Locmelis M, Melcher F, Oberthür T (2010) Platinum-group element distribution in the oxidized Main Sulfide Zone, Great Dyke, Zimbabwe. Miner Deposita 45(1):93–109. doi:10.1007/s00126-009-0258-y

Locmelis M, Pearson NJ, Barnes SJ, Fiorentini ML (2011) Ruthenium in komatiitic chromite. Geochim Cosmochim Acta 75(13):3645–3661

Lorand J-P, Alard O (2001) Platinum-group element abundances in the upper mantle: new constraints from in situ and whole-rock analyses of Massif Central xenoliths (France). Geoch Cosmochim Acta 65(16):2789–2806

Lorand J-P, Luguet A, Alard O, Bezos A, Meisel T (2008a) Abundance and distribution of platinum-group elements in orogenic lherzolites; a case study in a Fontete Rouge lherzolite (French Pyrenees). Chem Geol 248(3–4):174–194

Lorand JP, Luguet A, Alard O (2008b) Platinum-group elements: a new set of key tracers for the Earth's interior. Elements 4:247–252

Lorand J-P, Alard O, Luguet A (2010) Platinum-group element micronuggets and refertilization process in Lherz orogenic peridotite (northeastern Pyrenees, France). Earth Planet Sci Lett 289:298–310

Luguet A, Shirey SB, Lorand J-P, Horan MF, Carlson RW (2007) Residual platinum-group minerals from highly depleted harzburgites of the Lherz massif (France) and their role in HSE fractionation of the mantle. Geochim Cosmochim Acta 71:1553–1570

Maier WD (2005) Platinum-group element (PGE) deposits and occurrences: Mineralization styles, genetic concepts, and exploration criteria. J Afr Earth Sci 41:165–191

Maier WD, Barnes SJ (1999) Platinum-group elements in silicate rocks of the Lower, Critical and Main Zones at union section, western Bushveld Complex. J Petrol 40(11):1647–1671

Maier WD, Barnes S-J (2008) Platinum-group elements in the UG1 and UG2 chromitites, and the Bastard reef, at Impala platinum mine, western Bushveld Complex, South Africa: Evidence for late magmatic cumulate instability and reef constitution. S Afr J Geol 111:159–176

Maier W, Groves D (2011) Temporal and spatial controls on the formation of magmatic PGE and Ni–Cu deposits. Miner Deposita 46(8):841–857

Maier WD, Barnes SJ, Gartz V, Andrews G (2003) Pt–Pd reefs in magnetites of the Stella layered intrusion, South Africa: A world of new exploration opportunities for platinum group elements. Geology 31(10):885–888

Maier WD, Barnes SJ, Bandyayera D, Livesey T, Li C, Ripley E (2008) Early Kibaran rift-related mafic-ultramafic magmatism in western Tanzania and Burundi: Petrogenesis and ore potential of the Kapalagulu and Musongati layered intrusions. Lithos 101(1–2):24–53

Maier WD, Barnes SJ, Groves DI (2013) The Bushveld Complex, South Africa: formation of platinum, palladium, chrome- and vanadium-rich layers via hydrodynamic sorting of a mobilized cumulate slurry in a large, relatively slowly cooling, subsiding magma chamber. Miner Deposita 48(1):1–56

Manyeruke TD, Maier WD, Barnes S-J (2005) Major and trace element geochemistry of the Platreef on the farm Townlands, northern Bushveld Complex. S Afr J Geol 108(3):381–396

Mavrogenes JA, O'Neill HSC (1999) The relative effects of pressure, temperature and oxygen fugacity on the solubility of sulfide in mafic magmas. Geochim Cosmochim Acta 63(7–8):1173–1180

McBirney AR (1996) The Skaergaard intrusion. In: Cawthorn RG (ed) Layered intrusions. Elsevier, Amsterdam, pp 174–180

McCallum IS, Raedeke LD, Mathez EA (1980) Investigations of the Stillwater Complex: Part I. Stratigraphy and Structure of the Banded Zone. Am J Sci 280-A:59–87

McCandless TE, Ruiz J (1991) Osmium isotopes and crustal sources for platinum-group mineralization in the Bushveld Complex, South africa. Geology 19:1225–1228

McDonald I, Holwell DA (2011) Geology of the Northern Bushveld Complex and the Setting and Genesis of the Platreef Ni–Cu–PGE Deposit. Reviews in. Econ Geol 17:297–327

McDonough WF, Sun SS (1995) The composition of the Earth. Chem Geol 120:223–253

McLaren CH, De Villiers JPR (1982) The Platinum-Group Chemistry and Mineralogy of the UG-2 Chromitite Layer of the Bushveld Complex. Econ Geol 77:1348–1366

Miller JD, Green JC, Severson MJ, Chandler VW, Hauck SA, Peterson DM, Wahl TE (2002) Geology and mineral potential of the Duluth Complex and related rocks of northeastern Minnesota. Minnesota Geological survey Report of Investigations 58:207

Molnar F, Watkinson dH, Jones PC (2001) Multiple hydrothermal processes in footwall units of the North Range, sudbury Igneous Complex, Canada, and implications for the genesis of vein-type Cu–Ni–PGE deposits. Econ Geol 96:1645–1670

Mondal SK, Mathez EA (2007) Origin of the UG2 chromitite layer, Bushveld Complex. J Petrol 48(3):495–510

Moroni M, Girardi V, Ferrario A (2001) The Serra Pelada Au-PGE deposit, Serra dos Carajás (Pará State, Brazil): geological and geochemical indications for a composite mineralising process. Miner Deposita 36(8):768–785

Mungall JE (2002a) Kinetic controls on the partitioning of trace elements between silicate and sulfide liquids. J Petrol 43(5):749–768

Mungall JE (2002b) Roasting the mantle: slab melting and the genesis of major Au and Au-rich Cu deposits. Geology 30:915–918

Mungall JE (2007) Magmatic ore deposits. In: Holland HD, Turekian KK (eds) Treatise on Geochemistry, vol 3. Pergamon, Oxford, pp 1–33

Mungall JE (2014) Geochemistry of Magmatic ore deposits. In: Holland HD, Turekian KK (eds) Treatise on Geochemistry, vol, vol 13. Pergamon, Oxford, pp 195–218

Mungall JE, Brenan JM (2014) Partitioning of platinum-group elements and Au between sulfide liquid and basalt and the origins of mantle-crust fractionation of the chalcophile elements. Geochim Cosmochim Acta 125(0):265–289

Mungall JE, Naldrett A (2008) Ore deposits of the platinum-group elements. Elements 4:253–258

Mungall JE, Su S (2005) Interfacial tension between magmatic sulfide and silicate liquids: Constraints on kinetics of sulfide liquation and sulfide migration through silicate rocks. Earth Planet Sc Lett 234(1–2):135–149

Mungall JE, Hanley JJ, Arndt NT, Debecdelievre A (2006) Evidence from meimechites and other low-degree mantle melts for redox controls on mantle-crust fractionation of platinum-group elements. Proc Natl Acad Sci 103(34):12695–12700

Mutanen T (1997) Geology and petrology of the Akanvaara and Koitelainen mafic layered intrusion and Keivitsa-Satovaara layered Complex, northern Finland. Geological Survey of Finland Field Trip Guidebook, 10th International Platinum Symposium Bulletin 395:233

Naldrett AJ (2004) Magmatic sulfide deposits; geology, geochemistry and exploration. Berlin, Springer Verlag, Germany,

Naldrett A (2011) Fundamentals of magmatic sulfide deposits. Reviews in. Econ Geol 17:1–50

Naldrett AJ, Gasparrini EC, Barnes SJ, Von Gruenewaldt G, Sharpe MR (1986) The Upper Critical Zone of the Bushveld Complex and the origin of Merensky-type ores. Econ Geol 81(5):1105–1117

Naldrett AJ, Kinnaird J, Wilson A, Yudovskaya M, McQuade S, Chunnett G, Stanley C (2009a) Chromite composition and PGE content of Bushveld chromitites: Part 1 and 2; the Lower and Middle Groups. Appl Earth Sci 118(3–4):131–161

Naldrett AJ, Wilson A, Kinnaird J, Chunnett G (2009b) PGE tenor andmetal ratios within and below the Merensky Reef, Bushveld Complex: Implications for its Genesis. J Petrology:egp015 doi:10.1093/petrology/egp015

Naldrett A, Kinnaird J, Wilson A, Yudovskaya M, Chunnett G (2011a) Genesis of PGE-enriched Merensky Reef and chromitite seams of the Bushveld Complex. Reviews in. Econ Geol 17:235–296

Naldrett A, Wilson A, Kinnaird J, Yudovskaya M, Chunnett G (2011b) The origin of chromitites and related PGE mineralization in the Bushveld Complex: new mineralogical and petrological constraints. Miner Deposita:1–24

Naldrett A, Wilson A, Kinnaird J, Yudovskaya M, Chunett G (2012) The origin of chromitites and related PGR mineralization in the Bushveld Complex: new mineralogical and petrological constraints. Miner Deposita 47:209–232

Nielsen TFD (1975) Possible mechanism of continental break-up in the North Atlantic. Nature 253:182–184

Nielsen TFD (2001) The palladium potential of the skaergaard intrusion, South-East Greenland. Danmarks og Grønlands Geologiske Undersøgelse Rapport 2002/113:39

Nielsen TFD (2004) The Shape and Volume of the Skaergaard Intrusion, Greenland: Implications for Mass Balance and Bulk Composition. J Petrol 45(3):507–530
Nielsen TFD, Andersen JCO, Brook CK (2005) The Platinova Reef of the Skaergaard Intrusion. In: Mungall JE (ed) Exploration for Platinum-Group Element Deposits, vol 35. Mineralogical Association of Canada, pp 431–456
Nunes PD, Tilton GR (1971) Uranium-lead ages of minerals from the Stillwater Complex and associated rocks, Montana. Geol Soc Am Bull 82:2231–2250
Oberthür T (2002) Platinum-group element mineralization of the Great Dyke, Zimbabwe. Geology, Geochemistry, Mineralogy and Mineral benefication of Platinum Group Element Special Volume 54:483–506
Oberthür T (2011) Platinum-group element mineralization of the Main Sulphide Zone, Great Dyke, Zimbabwe. Reviews in. Econ Geol 17:329–350
Oberthür T, Melcher F (2005) Behaviour of platinum-group elements and platinum-group minerals in the supergene environment: a caase study of persistence and redistribution in the Main Sulfide Zone of the Great Dyke, Zimbabwe. In: Mungall JE (ed) Exploration for Platinum-Group Elements Deposits, vol Short Course Vol. 35. Mineralogical Association of Canada, pp 97–111
Oberthür T, Davis DW, Blenkinsop TG, Hohndorf A (2002) Precise U–Pb mineral ages, Rb–Sr and Sm–Nd systematics for the Great Dyke, Zimbabwe-constraints on late Archean events in the Zimbabwe craton and Limpopo belt. Precambrian Res 113(3):293–305
Oberthür T, Weiser T, Gast L (2003) Geochemistry and mineralogy of platinum-group elements at Hartley Platinum Mine, Zimbabwe. Mineral Deposita 344–355
Olivo GR, Gauthier M, Williams-Jones AE, Levesque M (2001) The Au–Pd Mineralization at the Conceicao Iron Mine, Itabira District, Southern Sao Francisco Craton, Brazil: an example of a Jacutinga-type deposit. Econ Geol 96(1):61–74
Olsson JR, Soderlund U, Klausen MB, Ernst RE (2010) U–Pb baddeleyite ages linking major Archean dyke swarms to volcanic-rift forming events in the Kaapvaal craton (South Africa), and a precise age for the Bushveld Complex. Precambrian Res 183(3):490–500
Osbahr I (2012) Platinum-group element distribution in base-metal sulfides of the Merensky Reef and UG2 from the eastern and western Bushveld Complex, South Africa. Ph. D. Friedrich-Alexander Universität Erlangen-Nürnberg
Osbahr I, Klemd R, Oberthür T, Brätz H, Schouwstra R (2013) Platinum-group element distribution in base-metal sulfides of the Merensky Reef from the eastern and western Bushveld Complex, South Africa. Miner Deposita 48(2):211–232
Page NJ (1976) Serpentinization and alteration in an olivine cumulate from the Stillwater Complex, southwestern Montana. Contrib Mineral Petrol 54:127–137
Page NJ (1977) Stillwater Complex, Montana: rock succesion, metamorphism and structure of the complex and adjacent rocks. US Geological Survey Professional paper, 999:79
Page NJ, Zientek ML (1985) Geologic and Structural setting of the Stillwater Complex. In: Czamanske GK, Zientek ML (eds) The Stillwater Complex, Montana: geology and guide, vol Special Publication, no. 92. Montana Bureau of Mines and Geology, Butte, Montana, pp 1–8
Pagé P, Barnes S-J, Bédard JH, Zientek ML (2012) In situ determination of Os, Ir, and Ru in chromites formed from komatiite, tholeiite and boninite magmas: Implications for chromite control of Os, Ir and Ru during partial melting and crystal fractionation. Chem Geol 302–303(0):3–15
Park J-W, Campbell IH, Arculus RJ (2013) Platinum-alloy and sulfur saturation in an arc-related basalt to rhyolite suite: Evidence from the Pual Ridge lavas, the Eastern Manus Basin. Geochim Cosmochim Acta 101(0):76–95
Penniston-Dorland SC, Wing BA, Nex PAM, Kinnaird JA, Farquhar J, Brown M, Sharman ER (2008) Multiple sulfur isotopes reveal a magmatic origin for the Platreef platinum group element deposit, Bushveld Complex, South Africa. Geology 36(12):979–982
Peregoedova A, Barnes S-J, Baker DR (2004) The formation of Pt–Ir alloys and Cu–Pd-rich sulfide melts by partial desulfurization of Fe–Ni–Cu sulfides: results of experiments and implications for natural systems. Chem Geol 208(1–4):247–264
Peregoedova A, Barnes S-J, Baker DR (2006) An experimental study of mass transfer of platinum-group elements, gold, nickel and copper in sulfur-dominated vapor at magmatic temperatures. Chem Geol 235(1–2):59–75

Polovina JS, Hudson DM, Jones RE (2004) Petrographic and geochemical characteristics of post-magmatic hydrothermal alteration and mineralization in the J-M Reef, Stillwater Complex, Montana. Can Mineral 42:261–277

Premo WR, Helz RT, Zientek ML, Langston RB (1990) U–Pb and Sm–Nd ages for the Stillwater Complex and its associated sills and dikes, Beartooth Mountains, Montana: Identification of a parent magma? Geology 18:1065–1068

Prendergast MD (2000) Layering and precious metals mineralization in the Rincon del Tigre Complex, Eastern Bolivia. Econ Geol 95:113–130

Prichard HM, Barnes S-J, Maier WD, Fisher PC (2004) Variations in the nature of the platinum-group minerals in a cross-section through the Merensky Reef at Impala Platinum: Implications for the mode of formation of the reef. Can Mineral 42:423–437

Puchtel I, Humayun M (2000) Platinum group elements in Kostomuksha komatiites and basalts: implications for oceanic crust recycling and core-mantle interaction. Geochim Cosmochim Acta 64(24):4227–4242

Raedeke LD, McCallum SI (1984) Investigations in the Stillwater Complex: Part II. Petrology and Petrogenesis of the Ultramafic Series. J Petrol 25:395–420

Richardson SH, Shirey SB (2008) Continental mantle signature of Bushveld magmas and coeval diamonds. Nature 453(7197):910–913

Righter K, Campbell AJ, Humayun M, Hervig RL (2004) Partitioning of Ru, Rh, Pd, Re, Ir, and Au between Cr-bearing spinel, olivine, pyroxene and silicate melts. Geochim Cosmochim Acta 68(4):867–880

Ripley E, Li C (2013) Sulfide saturation in mafic magmas: is external sulfur required for magmatic Ni–Cu–(PGE) ore genesis? Econ Geol 108:45–58

Rudashevsky NS, McDonald AM, Cabri LJ, Nielsen TFD, Stanley CJ, Kretzer YL, Rudashevsky VN (2004) Skaergaardite, PdCu, a new platinum-group intermetallic mineral from the Skaergaard instrusion, Greenland. Mineral Mag 68(4):615–632

Rudashevsky NS, Kretser YL, Rudashevsky VN, Nielsen TFD (2009) Gold, PGE ans sulfide phases of the precious mineralization of the Skaergaard intrusion, parts 6 and 7. Danmarks og Grønlands Geologiske Undersøgelse Rapport 2009 54 and 68:33 p

Sa JHS, Barnes S-J, Prichard HM, Fisher PC (2005) The distribution of base metals and platinum-group elements in magnetitite and its host rocks in the Rio Jacare Intrusion, northeastern Brazil. Econ Geol 100(2):333–348

Sattari P, Brenan JM, Horn I, McDonough WF (2002) Experimental Constraints on the Sulfide- and Chromite- Silicate Melt Partitioning Behavior of Rhenium and Platinum-Group Elements. Econ Geol 97(2):385–398

Sharman ER, Penniston-Dorland SC, Kinnaird JA, Nex PAM, Brown M, Wing BA (2013) Primary origin of marginal Ni–Cu–(PGE) mineralization in layered intrusions: Δ33S evidence from the Platreef, Bushveld, South Africa. Econ Geol 108(2):365–377

South African Committee For Stratigraphy (1980) Stratigraphy of Southern Africa. Part 1. Lithostratigraphy of South Africa, South West/ Namibia, and the Republics of the Boputhatswana, Trankei, and Venda. In: Geological Survey of South Africa Handbook, vol 8. p 690

Teigler B, Eales HV (1993) Correlation between chromite composition and PGE mineralization in the critical zone of the western Bushveld Complex. Miner Deposita 28(5):291–302

Todd SG, Keith DW, Le Roy LW, Schissel DJ, Mann EL, Irvine TN (1982) The J-M platinum-palladium reef of the Stillwater Complex, Montana: I. stratigraphy and petrology. Econ Geol 77:1454–1180

Turner AR, Wolfgram D, Barnes SJ (1985) Geology of the Stillwater County sector of the J-M Reef, including the Minneapolis Adit. In: Czamanske GK, Zientek ML (eds) The Stillwater Complex, Montana: geology and guide, vol Special Publication, no. 92. Montana Bureau of Mines and Geology, Butte, Montana, pp 210–230

van der Merwe F, Viljoen F, Knoper M (2012) The mineralogy and mineral associations of platinum group elements and gold in the Platreef at Zwartfontein, Akanani Project, Northern Bushveld Complex, South Africa. Mineral Petrol 106(1–2):25–38

Vermaak CF (1976) Merensky Reef—Thoughts on Its Environment and Genesis. Econ Geol 71(7):1270–1298

Vermaak CF, Hendriks LP (1976) A review of the mineralogy of the Merensky Reef, with specific reference to new data on the precious metal mineralogy. Econ Geol 71(7):1244–1269

Viljoen MJ (1999) The nature and origin of the Merensky Reef of the western Bushveld Complex based on geological facies and geophysical data. South Afr J Geol 102(3):221–239

Von Gruenewaldt G (1979) A review of some recent concepts of the Bushveld Complex, with particular reference to sulfide mineralization. Canadian mineralogist 17:233–256

Von Gruenewaldt G (1986) Platinum-group element-chromitite associations in the Bushveld Complex. Econ Geol 81:1067–1079

Von Gruenewaldt G, Horsch H, Dickst D, de Wet J (1990) PGE mineralization in the western sector of the eastern Bushveld Complex. Mineral Petrol 42(1–4):71–95

Vukmanovic Z, Barnes SJ, Reddy S, Godel B, Fiorentini ML (2013) Microstructure in chromite crystals of the Merensky Reef (Bushveld Complex, South Africa). Contrib Mineral Petrol 165(6):1031–1050

Wager LR, Brown GM (1968) Layered igneous rocks. Oliver and Boyd, Edinburgh

Webb SJ, Cawthorn KG, Nguuri T, James D (2004) Gravity modeling of Bushveld Complex connectivity supported by Southern African Seismic Experiment results. In: South Afr J Geol, vol 107. Geological Society of South Africa, pp 207–218

Wendlandt RF (1982) Sulfide saturation of basalt and andesite melts at high pressures and temperatures. Am Mineral 67(9–10):877–885

Willmore CC, Boudreau AE, Kruger FJ (2000) The halogen geochemistry of the Bushveld Complex, Republic of South Africa: Implications for chalcophile element distribution in the lower and critical zones. J Petrol 41(10):1517–1539

Wilson AH (1982) The Geology of the Great Dyke, Zimbabwe: The Ultramafic Rocks. J Petrol 23(2):240–292

Wilson A (1996) The great Dyke of Zimbabwe. In: Cawthorn RG (ed) Layered Intrusions, vol. Elsevier Science, Amsterdam, pp 365–402

Wilson AH (2012) A chill sequence to the Bushveld Complex: insight into the first stage of emplacement and implications for the parental magmas. J Petrol 53(6):1123–1168

Wilson AH, Chunnett G (2006) Trace element and platinum group element distributions antd the genesis of the Merensky Reef, Western Bushveld Complex, South Africa. J Petrol 47:2369–2403

Wilson AH, Prendergast MD (1989) The Great dyke of zimbabwe—I: tectonic setting, stratigraphy, petrology, structure, emplacement and crystallization. In: Prendergast MD, Jones MJ (eds) Magmatic sulfides—The Zimbabwe Volume, vol. The Institution of Mining and Metallurgy, London, pp 1–20

Wilson AH, Prendergast MD (2001) Platinum-group element mineralisation in the Great Dyke, Zimbabwe, and its relationship to magma evolution and magma chamber structure. South Afr J Geol 104:319–342

Wilson AH, Tredoux M (1990) Lateral and vertical distribution of platinum-group elements and petrogenetic controls on the sulfide mineralization in the P1 pyroxenite layer of the Darwendale Subchamber of the Great Dyke, Zimbabwe. Econ Geol 85:556–584

Wohlgemuth-Ueberwasser C, Fonseca R, Ballhaus C, Berndt J (2013) Sulfide oxidation as a process for the formation of copper-rich magmatic sulfides. Miner Deposita 48:115–127

Wood SA (2002) The aqueous Geochemistry of the platinum-group elements with applications to ore deposits. In: Cabri LJ (ed) Geology, Geochemistry, Mineralogy and Mineral benefication of Platinum Group Element, vol Special Volume 54. Canadian Institute of Mining, Metallurgy and Petroleum, pp 211–249

Worst BG (1960) The great Dyke of Southern Rhodesia. South Rhod Geol Surv Bull 47:234

Yudovskaya M, Kinnaird J, Naldrett AJ, Mokhov AV, McDonald I, Reinke C (2012) Facies variation in PGE mineralization in the central Platreef of the Bushveld Complex, South Africa. Can Mineral 49(6):1349–1384

Zientek ML, Czamanske GK, Irvine TN (1985) Stratigraphy and nomenclature for the Stillwater Complex. In: Czamanske GK, Zientek ML (eds) The Stillwater Complex, Montana: geology and guide, vol Special Publication, no. 92. Montana Bureau of Mines and Geology, Butte, Montana, pp 21–32

Zientek ML, Cooper RW, Corson SR, Geraghty EP (2002) Platinum-group element mineralization in the Stillwater Complex, Montana. In: Cabri LJ (ed) Geology, Geochemistry, Mineralogy and Mineral benefication of Platinum Group Element, vol Special Volume 54. Canadian Institute of Mining, Metallurgy and Petroleum, pp 459–481

Part II
Reviews of Major Layered Intrusions

Chapter 10
The Panzhihua Intrusion, SW China

Kwan-Nang Pang, J. Gregory Shellnutt and Mei-Fu Zhou

Abstract The ~260 Ma Panzhihua intrusion is an elongate layered gabbroic intrusion with a length of ~19 km and a thickness of ~2 km in the central part of the Emeishan large igneous province, SW China. It is made up from the base to top of microgabbros (Marginal zone, MGZ), gabbros, oxide gabbros and Fe-Ti oxide ores (Lower zone and Middle zone a, LZ and MZa), apatite-bearing leucogabbros (Middle zone b, MZb), and gabbros and oxide gabbros (Upper zone, UZ). Igneous layering is observed in rocks of the MZa, MZb and UZ. Textures and geochemical data indicate that, with the exception of the MGZ, rocks in the other zones represent cumulates. The most plausible parental magma, inferred from rock compositions from the MGZ, is a moderately fractionated ferrobasalt rich in FeO, TiO_2 and poor in SiO_2 relative to common tholeiitic liquids and such a composition, although rare, is known in the high-Ti basalts of the Emeishan large igneous province. Relatively low $(^{87}Sr/^{86}Sr)_{260\ Ma}$ (0.7039–0.7055) and high $\varepsilon Nd_{260\ Ma}$ (−2.1 to +4.3) indicate a juvenile mantle source and minimal contamination by old continental crust during crystallization. Differentiation of the intrusion was driven by fractionation of a magnetite gabbroic assemblage (clinopyroxene + plagioclase + titanomagnetite ± olivine), from which the Fe-Ti oxide ores formed primarily through gravitational accumulation in the LZ and MZa. Apatite and ilmenite joined the fractionation assemblage in the MZb. The early appearance of liquidus titanomagnetite relative to

Book chapter revised for *Layered intrusions (Springer)* 27th September 2014

K.-N. Pang (✉)
Institute of Earth Sciences, Academia Sinica, P.O. Box 1-55, 11529 Taipei, Taiwan, Republic of China
e-mail: knpang@earth.sinica.edu.tw

Department of Geosciences, National Taiwan University, P.O. Box 13-318, 106 Taipei, Taiwan, Republic of China

J. G. Shellnutt
Department of Earth Sciences, National Taiwan Normal University, 88 Tingzhou Road Section 4, 11677 Taipei, Taiwan, Republic of China
e-mail: jgshelln@ntnu.edu.tw

M.-F. Zhou
Department of Earth Sciences, The University of Hong Kong, Hong Kong, China
e-mail: mfzhou@hku.hk

B. Charlier et al. (eds.), *Layered Intrusions,* Springer Geology,
DOI 10.1007/978-94-017-9652-1_10

well-studied layered intrusions might be an effect of the unusual bulk composition of the parental magma, magma-wall rock interaction, the presence of volatiles, or any combination thereof. The differentiation trend was interrupted by at least two magma recharge events and some residual liquid escaped from the magma chamber. The A-type granitoids intimately associated with the intrusion have elemental and Sr-Nd isotopic features consistent with protracted differentiation of the ferrobasaltic parental magma, implying the process was driven towards silica enrichment.

Keywords Layered intrusion · Ferrobasalt · Fe-Ti oxides · Emeishan · Differentiation

Introduction

The Panzhihua intrusion is a type example of Fe-Ti oxide ore-bearing layered intrusions in the end-Guadalupian Emeishan large igneous province (ELIP), SW China (Ma et al. 2001; Zhou et al. 2005; Pang et al. 2010). The intrusion, together with similar intrusions in the province, account for a significant portion of Fe, Ti and V resource of the country and have attracted numerous studies over the past decade (Zhong et al. 2002; Zhou et al. 2005, 2008; Pang et al. 2008a, b, 2009, 2010; Zhang et al. 2009; Hou et al. 2011b, 2012a, b, 2013; Shellnutt and Pang 2012; Bai et al. 2012; Song et al. 2013; Howarth et al. 2013; Howarth and Prevec 2013). These intrusions represent key targets to study magma chamber processes associated with the intrusive facies of the ELIP. Earlier investigations of the Panzhihua intrusion and related magmatic rocks not only improved our understanding of its petrogenesis and ore genesis, but also revealed additional complexity and problems for the magmatic system. Consensus has not been reached regarding the magma source characteristics, the nature of the magmatic system, the role of magma-wallrock interaction and the key factors for ore genesis. This paper provides a review of the basic geological features and new information stemming from recent investigations of the Panzhihua intrusion and related magmatic rocks, aiming to stimulate further research on this enigmatic igneous body.

Geological Setting

The Panzhihua intrusion is one of the several Fe-Ti oxide ore-bearing layered gabbroic bodies emplaced in the western margin of the Yangtze Block, SW China during an intense magmatic episode associated with formation of the ELIP at ~260 Ma (Zhou et al. 2005; Pang et al. 2008a). The other oxide ore-bearing intrusions include Hongge (Zhong et al. 2002; Bai et al. 2012), Baima (Shellnutt et al. 2009), Taihe (Hou et al. 2012a) and Xinjie (Zhong et al. 2004, 2011; Wang et al. 2008) in the Panzhihua-Xichang region, and Anyi and Mianhuadi further to the south (Zhou et al. 2013). The volcanic succession of the ELIP, ranging in thickness from several hundred meters to several kilometers, is dominated by flood basalts with minor picritic and silicic lavas emplaced in a relatively rapid manner (<2 m.y.) as a result of mantle plume impingement (Chung and Jahn 1995; Xu et al. 2004; Xiao et al. 2004; Zhang et al. 2006). As in many other LIPs, the flood basalts have been divided into low-Ti and high-Ti series on the basis of compositions (Xu et al. 2001; Xiao et al. 2004), although a continuum appears to exist between them as more geochemical data become available (Shellnutt and Jahn 2011; Hou et al. 2011a). Nevertheless, the Panzhihua intrusion has been genetically linked to the high-Ti Emeishan basalts (Zhou et al. 2005, 2008; Pang et al. 2008a, 2010).

The Panzhihua intrusion was emplaced in Neoproterozoic dolomitic limestones (Fig. 10.1). Its lithology and cumulus mineralogy are summarized in Fig. 10.2. The intrusion has been widely interpreted as a differentiated sill (Ma et al. 2001, 2003; Zhou et al. 2005; Pang et al. 2009; Song et al. 2013), but a recent proposal is that its central portion might represent a discordant dyke linking two concordant segments (Pêcher et al. 2013). The wall rocks along the lower contact of the intrusion were transformed into marbles during contact metamorphism, forming an aureole up to ~300 m in thickness in places (Ganino et al. 2008). Intimately associated with the intrusion is an elongate A-type granitic pluton northwest of and sub-parallel to the strike of the intrusion (Shellnutt and Zhou 2007; Zhong et al. 2009; Shellnutt and Jahn 2010). Available geological maps show that the areal exposure between the A-type granitoids and the intrusion is comparable, and such a gabbro-granitoid association is also known for other ore-bearing intrusions (Shellnutt et al. 2011). The intrusion and the granitoids are unconformably overlain by Triassic clastic strata and the original upper contact of the intrusion is not preserved (Fig. 10.1). After solidification, the intrusion was cut into various fault-bounded blocks and was tilted to the northwest by ~25–65° as inferred from the dip of the igneous layering. Ultramafic dykes have been identified in the contact aureole and the lowermost portion of the intrusion and have been considered to be cogenetic with the intrusion itself (Hou et al. 2012b, 2013). In addition, the intrusion was cut by a swarm of thin felsic dykes but their relations to the main mass of the intrusion or the aforementioned granitic pluton remain uncertain. Despite the intense tectonic activity associated with the Cenozoic India-Eurasia collision, the intrusion remains largely intact, unmetamorphosed and undeformed.

Age

Precise radiometric ages have been obtained from the Panzhihua intrusion and related magmatic rocks (Table 10.1). The SHRIMP zircon U-Pb age of 263 ± 3 Ma for a leucogabbro has generally been considered the crystallization age of the intrusion (Zhou et al. 2005), contemporaneous with the main phase of Emeishan magmatism at ~260 Ma (Zhou et al. 2002). Zircon from picritic and anorthositic dykes cutting the intrusion also indicate similar ages within error (Hou et al. 2012b, 2013). Argon dating of plagioclase separated from a plagiogranite dyke cutting the intrusion

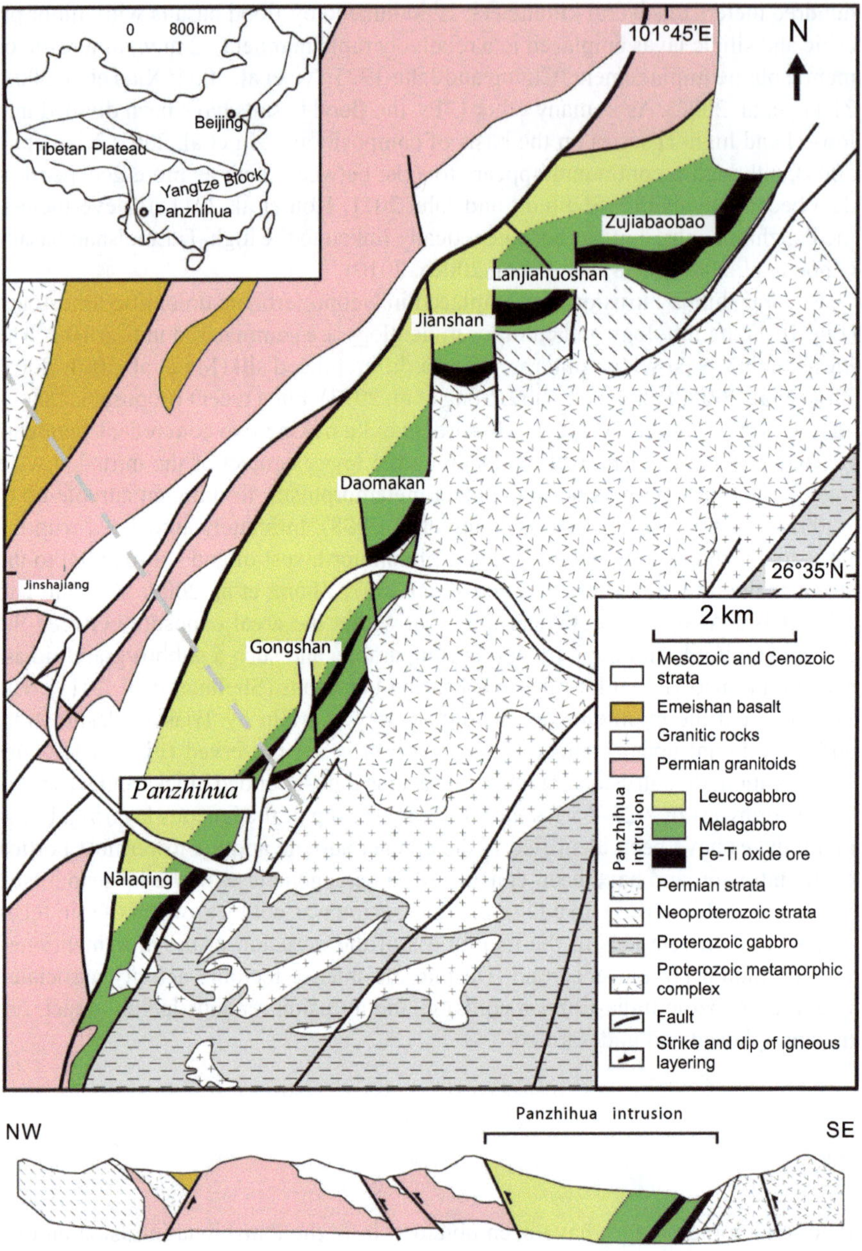

Fig. 10.1 Geological map of the Panzhihua intrusion (modified after Zhou et al. 2005). The cross section indicated by the *grey dotted line* on the map is after Ganino et al. (2008). Inset shows the geographic location in SW China

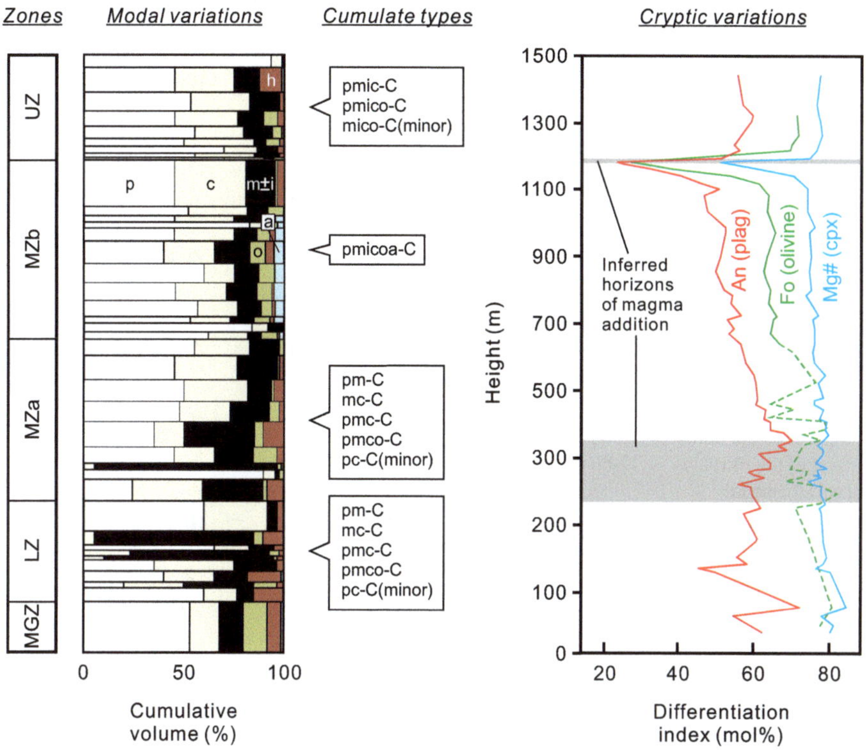

Fig. 10.2 A simplified stratigraphic column of the Panzhihua intrusion, showing the zonal division and the variations in cumulate types, cumulus minerals and their compositions. The cumulate nomenclature of Irvine (1982) was used by which rocks are named according to the first letter of the main minerals followed by "-C" meaning cumulus: plagioclase(p), magnetite(m), ilmenite(i), clinopyroxene(c), olivine(o), apatite(a) and hornblende(h). Data for mineral mode are after Zhou et al. (2005) and mineral compositions are after Pang et al. (2009). The *dotted line* means that olivine might or might not be cumulus according to its texture

yields an age of 246±4.1 Ma (Boven et al. 2002), which is younger than the above ages. Syenites from the A-type granitic pluton were dated at 254.6±1.3 Ma and 251.6±1.6 Ma by biotite $^{40}Ar/^{39}Ar$ geochronology (Lo et al. 2002) and at 253.1±1.9 Ma by SHRIMP zircon U-Pb geochronology (at Daheishan ~5 km north of the central part of the Panzhihua intrusion, Zhong et al. 2009). However, recent dating of a syenite from the same locality by high-precision CA-TIMS zircon U-Pb technique yields an age of 259.1±0.5 Ma (Shellnutt et al. 2012), suggesting at least that some A-type granitoids are related to the Panzhihua intrusion by the same magmatic event at ~260 Ma.

Table 10.1 Radiometric ages related to the Panzhihua intrusion

Sample	Method[a]	Age	Reference
Plagiogranite dyke cutting the lower part of the intrusion	Plagioclase $^{40}Ar/^{39}Ar$	246±4.1 Ma	Boven et al. (2002)
Syenites from the Panzhihua complex (location unknown)	Biotite $^{40}Ar/^{39}Ar$	254.6±1.3 Ma, 251.6±1.6 Ma	Lo et al. (2002)
Leucogabbro in the upper part of the intrusion	SHRIMP zircon U-Pb (n=13)[b]	263±3 Ma	Zhou et al. (2005)
Syenite to the north-west of the intrusion (at Daheishan)	SHRIMP zircon U-Pb (n=19)	253.1±1.9 Ma	Zhong et al. (2009)
Syenite to the north-west of the intrusion (at Daheishan)	CA-TIMS zircon U-Pb (n=5)	259.1±0.5 Ma	Shellnutt et al. (2012)
Anorthositic dyke cutting the intrusion	LA-ICP-MS zircon U-Pb (n=14)	259.77±0.79 Ma	Hou et al. (2012b)
Picritic dyke cutting the lower part of the intrusion	LA-ICP-MS zircon U-Pb (n=5)	261.4±4.6 Ma	Hou et al. (2013)

[a] *SHRIMP* sensitive high-resolution ion microprobe, *CA-TIMS* chemical-abrasion thermal ionization mass spectrometry, *LA-ICP-MS* laser ablation-inductively coupled plasma-mass spectrometer
[b] n = number of zircon analyses used to define the concordia age

Lithology and Stratigraphic Division

Based on internal structure and extent of Fe-Ti oxide mineralization, the Panzhihua intrusion was divided by mining geologists into four zones from the base upwards: the Marginal zone (MGZ), Lower zone (LZ), Middle zone (MZ) and Upper zone (UZ) (Figs. 10.2 and 10.3a) (Ma et al. 2001). This scheme is generally adopted by most subsequent studies (e.g. Zhou et al. 2005; Pang et al. 2009; Zhang et al. 2009; Song et al. 2013; Howarth et al. 2013). The MGZ is a 0–40 m-thick unit dominated by equigranular microgabbroic rocks (~0.2–0.5 mm) with some hornblende-rich lithologies, and is interpreted as the chilled base of the intrusion. The main minerals include clinopyroxene, plagioclase, hornblende, magnetite, ilmenite and apatite with or without olivine. The LZ is up to 110 m-thick and generally consists of gabbros, oxide gabbros and Fe-Ti oxide ores with increasing amount of oxide minerals. There appears to be a complete spectrum for the modal abundance of Fe-Ti oxides in the rocks, and both sharp and gradational contacts are observed between the ores and their host rocks (Figs. 10.3b and c). Minor anorthosites occur in the upper part of this zone (Ma et al. 2003). The LZ rocks display an isotropic, coarse-grained to pegmatitic texture that is typical of the gabbroic rocks, whereas the ores are coarse-grained and massive. The mineral paragenesis comprises dominantly clinopyroxene, plagioclase, titanomagnetite (oxy-exsolved to magnetite and ilmenite)

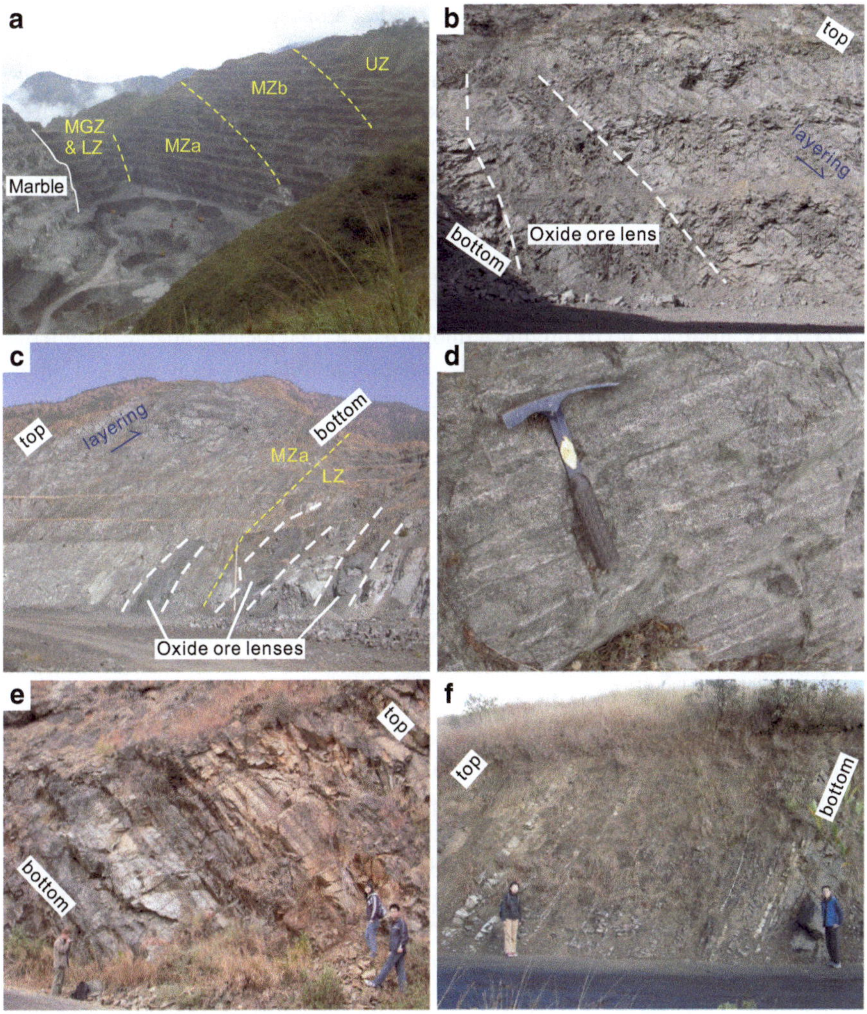

Fig. 10.3 **a** The Jianshan open-pit mine where a section from the carbonate wall rocks (now marble) to the overlying stratigraphic zones of the Panzhihua intrusion is exposed (after Pang et al. 2009). Each terrace has a height of ~ 15 m. **b** A stratiform lenticular mass of Fe-Ti oxides in contact with layered gabbros in the Jianshan mine (*lower* MZa). **c** Stratiform lenticular masses of Fe-Ti oxides in the Zujiabaobao mine (*upper* LZ and *lower* MZa). **d** Fine-scale modal layering at an outcrop scale marked by differences in the modal abundance of plagioclase (MZa). **e** Layering in leucogabbros along a roadcut section north of the Lanjiahuoshan mine (MZb). **f** Alternating layers of anorthosites, gabbros (*centre* of view) and olivine clinopyroxenites along a roadcut section north of the Lanjiahuoshan mine (UZ)

and minor olivine and hornblende. The MZ is up to 800 m-thick and is characterized by a medium grain size, igneous layering and planar lamination. The layering is marked by the relative abundance between plagioclase and mafic silicates + Fe-Ti oxides (modal layering), and can be observed at a distance from (Fig. 10.3b, c) or close to outcrops (Fig. 10.3d). The planar lamination is marked by the preferred orientation of plagioclase, and to a lesser extent clinopyroxene, sub-parallel to the modal layering. These features make rocks of the MZa readily distinguishable from the LZ. The MZ was sub-divided into a lower sub-zone MZa and an upper sub-zone MZb based on the appearance of cumulus apatite (Pang et al. 2009). Compared to the LZ, the MZa contains much less oxide ores in terms of relative volume, but the lithology and mineralogy between the two zones are similar. The MZb is made up of layered apatite-bearing leucogabbros and is free of oxide ores (Fig. 10.3e). The top of this zone is marked by a ~3 m-thick layer of anorthosite (Ma et al. 2003). The MZb rocks are made up of olivine, clinopyroxene, plagioclase, apatite, magnetite, ilmenite and minor hornblende. The UZ, ranging in thickness from 500 to 1500 m, is composed of layered gabbros and leucogabbros with minor olivine clinopyroxenites and anorthosites (Fig. 10.3f). No ore occurrences are known in this zone. The UZ rocks are made up of variable olivine, clinopyroxene, plagioclase, Fe-Ti oxides and minor hornblende.

Mineral Assemblages

Olivine

Olivine is unevenly distributed in the Panzhihua intrusion (Fig. 10.2). It is a ubiquitous cumulus phase in the MZb but is only found sporadically in the MGZ, LZ, MZa and UZ. Olivine in these zones occurs either as thin rims along silicate-Fe-Ti oxide interface (see Fig. 10.5d in Pang et al. 2009) or as granular crystals that can be reasonably interpreted as a cumulus phase. For the latter, no significant compositional zoning (>1 mol % in Fo content) is observed. The most magnesian cumulus olivine of the intrusion, found in the MGZ, has a composition of Fo_{78-81} (Pang et al. 2008b, 2009). Interestingly, Hou et al. (2012b) reported primitive olivine (up to Fo_{91}) from wehrlitic dykes cutting the contact aureole and the lower part of the intrusion, but their relations to the main mass of the intrusion remain uncertain (see later discussion). The most ferroan olivine of the intrusion, found in the uppermost MZb, has a composition of Fo_{28}. The Panzhihua olivine has low Ni contents (<250 ppm) except in the MGZ where olivine contains up to ~1300 ppm Ni.

Ca-rich Pyroxene

Clinopyroxene is common throughout the intrusion (Fig. 10.2) and has been studied by Pang et al. (2009). Its composition is titanaugitic with moderate Al_2O_3 (0.90–4.43 wt.%). Core and rim variations in Mg# [defined as $100x$Mg/

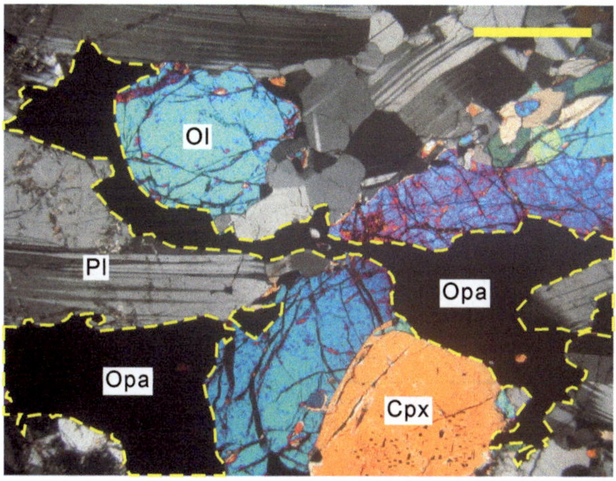

Fig. 10.4 The amoeboid texture of Fe-Ti oxides (highlighted by the *yellow dashed line*) in an oxide gabbro from the LZ (pmco-C). The *yellow bar* in the *upper right* denotes a length scale of 1 mm. Note that this type of rock is transitional to the Fe-Ti oxide ores in which the amoeboid Fe-Ti oxides form a connecting network enclosing the silicate grains

$(Mg+Fe^{2+}+Fe^{3+})$] are mostly within 1–3 mol % in a given clinopyroxene grain and 3–6 mol % in a given sample. The most magnesian clinopyroxene of the intrusion is found in the MGZ and has Mg# of 85. The most ferroan clinopyroxene is present at the uppermost MZb and has Mg# of 51. The Panzhihua clinopyroxene has TiO_2 ranging from 0.40 to 1.65 wt.% and V ranging from < 100 to ~ 520 ppm, and the lowest abundances of these elements are found in the uppermost MZb. The Cr content is generally low (< 100 ppm).

Plagioclase

Plagioclase is a ubiquitous mineral throughout the intrusion (Fig. 10.2) and has been studied by Pang et al. (2009). Core and rim compositions reveal no significant zoning (>2 mol % in An content), which is probably a result of recrystallization into fine aggregates that obscured the original zoning patterns. The variation of An content of plagioclase cores in a given sample is generally less than 5 mol %. The most calcic plagioclase of the intrusion is found in the MGZ and has a composition of An_{73}, whereas the most sodic plagioclase with An_{24} composition occurs in the uppermost MZb. The Panzhihua plagioclase has K_2O ranging from 0.06 to 0.25 wt.% and Fe ranging from ~ 1100 to ~2600 ppm.

Fe-Ti Oxides

Iron-titanium oxides occur in rocks throughout the intrusion and have been studied by Pang et al. (2008a). Magnetite is dominant over ilmenite in the LZ and MZa, where the amoeboid morphology exhibited by these minerals (Fig. 10.4) was

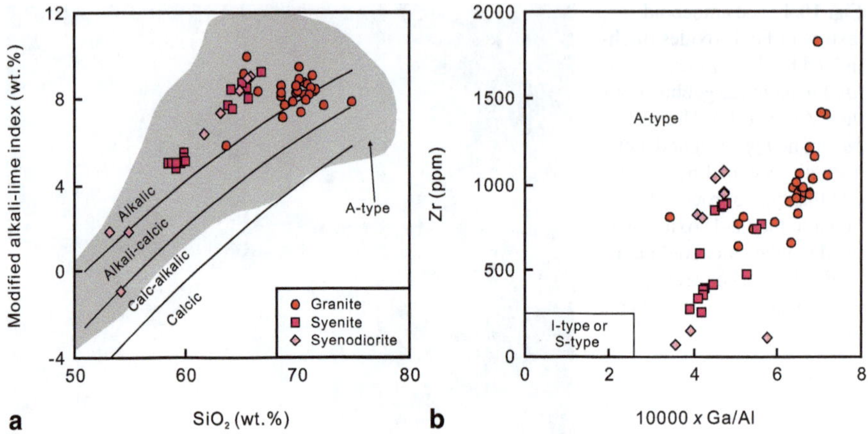

Fig. 10.5 Geochemical characteristics of the A-type granitoids associated with the Panzhihua intrusion. **a** Co-variation of modified alkali-lime index and SiO_2 (after Frost et al. 2001). **b** Co-variation of Zr and 10,000 *x* Ga/Al (after Whalen et al. 1987). The data are after Shellnutt and Zhou (2007), Zhong et al. (2009) and Shellnutt and Jahn (2010)

taken as indicative of crystallization from immiscible oxide melts at a late stage (Zhou et al. 2005). This interpretation is, however, questionable because euhedral titanomagnetite is found enclosed in clinopyroxene in the LZ (see Fig. 10.5d of Pang et al. 2008a), and its composition is indistinguishable from oxide grains showing amoeboid textures in a given rock (see Tables 10.1 and 10.2 of Pang et al. 2008a). In addition, the reliability of oxide textures in general has been questioned owing to the notoriously known effect of subsolidus coarsening and grain boundary readjustment (e.g. Duchesne 1996, 1999). Magnetite composition ranges from nearly pure Fe_3O_4 to titanomagnetite with ~40 mol % of ulvöspinel in solid solution. It contains Al_2O_3 and MgO at the level of a few weight percent and Mn, V and Cr in the parts per million level. These constituents, together with TiO_2, can be variable at a grain scale due to extensive subsolidus re-equilibration and exsolution. Ilmenite composition ranges from nearly pure $FeTiO_3$ to ferroan ilmenite with ~16 mol % hematite in solid solution with variable MgO at the level of a few weight percent, and Mn and V at the parts per million level. The primary oxide, reconstructed from compositions of titanomagnetite in the ores and exsolved phases, is an aluminous titanomagnetite (Usp_{40}) with 40 wt.% FeO, 34 wt.% Fe_2O_3, 16.5 wt.% TiO_2, 5.3 wt.% Al_2O_3, 3.5 wt.% MgO, 0.5 wt.% MnO and 4800 ppm V (Pang et al. 2008a). It was suggested that primary ilmenite is either volumetrically minor or absent in the LZ and MZa, and that the ilmenite in the oxide assemblage could be explained by external granule oxy-exsolution of ulvöspinel from the primary aluminous titanomagnetite. However, ilmenite is as abundant as or more abundant than magnetite in the MZb and possibly the UZ, implying a primary origin.

Table 10.2 Geochemical compositions of representative rocks from the Panzhihua intrusion and related magmatic rocks

	The Panzhihua intrusion							Ultramafic dykes		A-type granitoids		
Lithology	Micro-gabbro	Melano-gabbro	Oxide gabbro	Fe-Ti oxide ore	Anortho-site	Apatite-bearing gabbro	Gabbro	Picrite dyke	Wehrlitic dyke	Granite	Syeno-diorite	Syenite
Zone	MGZ	MGZ	LZ	LZ	MZa	MZb	UZ	MGZ	MGZ	–	–	–
Sample	LJ-03	PZH-1	LJ-09	LJ-13	LJ-17	LJ-30	LJ-45	PZH1004	PZH05	GS03-010	GS03-027	WB-0604
Major oxides (wt.%)												
SiO_2	44.70	39.63	29.02	3.21	49.73	38.54	45.75	42.02	39.63	71.45	63.13	59.98
TiO_2	2.78	4.31	7.66	15.05	0.38	4.83	1.98	1.51	0.90	0.39	0.79	1.28
Al_2O_3	15.27	9.94	9.79	5.16	27.40	13.03	16.16	6.46	6.87	11.62	14.34	13.73
Fe_2O_3[a]	13.58	18.02	35.67	72.10	1.60	19.10	10.56	12.02	9.77	5.09	6.31	11.19
MnO	0.19	0.24	0.28	0.37	0.02	0.23	0.14	0.16	0.14	0.14	0.19	0.40
MgO	7.16	11.63	5.78	5.24	0.57	7.24	7.01	20.96	31.91	0.06	0.58	0.65
CaO	10.22	12.67	9.42	0.47	11.90	12.53	12.98	12.35	4.91	0.59	2.07	3.56
Na_2O	2.57	1.51	1.36	0.07	4.43	2.06	2.53	1.01	1.36	5.18	5.87	6.64
K_2O	1.41	0.23	0.25	0.02	0.82	0.18	0.37	0.76	0.47	4.44	3.53	2.14
P_2O_5	0.38	0.27	0.03	0.00	0.08	2.43	0.07	0.20	0.15	0.03	0.19	0.31
Mg#	52.5	57.5	25.4	13.2	42.7	44.3	58.2	89.7	87.3	2.40	16.2	10.9
Trace elements (ppm)												
V	373	322	1350	4843	12.94	369	276	230	174	1.00	5.00	n.a.
Cr	198	839	43.1	777	1.92	15.5	225	1841	2150	9.00	6.00	4.70
Ni	55.7	312	75.8	273	11.16	0.17	97.0	891	1594	6.00	2.00	n.a.
Cu	85.5	n.a.[b]	148	376	76.8	13.8	216	273	169	5.00	8.00	n.a.
Zn	110	n.a.	215	466	10.3	112	44.3	n.a.	n.a.	267	171	196
Sc	29.5	n.a.	36.4	20.0	2.09	15.1	27.6	n.a.	18	14.0	16.0	14.7

Table 10.2 (continued)

	The Panzhihua intrusion							Ultramafic dykes		A-type granitoids		
Lithology	Micro-gabbro	Melano-gabbro	Oxide gabbro	Fe-Ti oxide ore	Anortho-site	Apatite-bearing gabbro	Gabbro	Picrite dyke	Wehrlitic dyke	Granite	Syeno-diorite	Syenite
Zone	MGZ	MGZ	LZ	LZ	MZa	MZb	UZ	MGZ	MGZ	–	–	–
Sample	LJ-03	PZH-1	LJ-09	LJ-13	LJ-17	LJ-30	LJ-45	PZH1004	PZH05	GS03-010	GS03-027	WB-0604
Co	47.5	75.0	125	254	10.4	56.9	44.1	69.1	124	0.00	3.00	n.a.
Rb	31.5	5.23	3.68	0.45	5.12	0.92	6.28	32.1	38.8	93.0	71.0	35.0
Sr	898	548	331	9.66	920	526	465	224	147	42.0	127	456
Y	27.3	22.9	6.68	0.94	2.39	26.5	5.52	10.0	9.15	76.0	60.0	50.4
Zr	138	123	22.0	13.21	11.6	17.8	8.61	60.1	59.5	1057	829	347
Nb	34.0	3.16	3.33	2.48	2.04	4.34	0.61	13.3	9.91	91.0	71.0	58.1
Ba	481	229	80.9	8.78	129	91.8	248	897	238	555	800	1413
La	26.3	11.4	1.88	0.61	3.01	15.4	1.70	7.18	7.68	107	79.6	54.2
Ce	60.0	37.5	5.11	1.36	6.13	41.5	4.73	16.7	17.7	277	162	112
Pr	8.26	6.92	0.89	0.20	0.80	6.84	0.86	2.26	2.28	26.5	20.2	14.9
Nd	34.8	33.9	4.93	0.94	3.41	36.1	5.02	10.6	9.64	103	76.7	62.8
Sm	7.53	8.47	1.54	0.23	0.70	8.94	1.54	2.40	2.3	20.6	15.4	13.8
Eu	2.60	2.40	0.86	0.09	0.72	3.39	1.14	0.76	0.73	4.20	3.90	6.68
Gd	7.25	7.53	1.65	0.22	0.71	9.23	1.67	2.29	2.52	18.2	13.9	10.2
Tb	1.02	1.02	0.27	0.04	0.09	1.21	0.24	0.41	0.35	2.80	2.10	1.90
Dy	5.59	5.26	1.53	0.22	0.50	6.03	1.28	2.10	1.91	15.7	12.4	11.1
Ho	1.03	0.94	0.28	0.04	0.09	0.98	0.22	0.47	0.37	3.00	2.50	2.12
Er	2.61	2.33	0.69	0.10	0.23	2.18	0.49	1.24	1.01	8.60	6.90	5.37
Tm	0.33	0.28	0.08	0.01	0.03	0.23	0.05	0.19	0.12	1.20	1.00	0.80

Table 10.2 (continued)

	The Panzhihua intrusion							Ultramafic dykes		A-type granitoids		
Lithology	Micro-gabbro	Melano-gabbro	Oxide gabbro	Fe-Ti oxide ore	Anortho-site	Apatite-bearing gabbro	Gabbro	Picrite dyke	Wehrlitic dyke	Granite	Syeno-diorite	Syenite
Zone	MGZ	MGZ	LZ	LZ	MZa	MZb	UZ	MGZ	MGZ	–	–	–
Sample	LJ-03	PZH-1	LJ-09	LJ-13	LJ-17	LJ-30	LJ-45	PZH1004	PZH05	GS03-010	GS03-027	WB-0604
Yb	1.95	1.66	0.48	0.07	0.16	1.09	0.29	1.08	0.7	7.40	6.10	4.56
Lu	0.30	0.23	0.07	0.01	0.02	0.15	0.04	0.17	0.12	1.00	0.90	0.63
Hf	3.71	4.03	0.84	0.49	0.32	0.63	0.35	2.20	1.81	20.1	16.8	9.40
Ta	1.74	0.16	0.25	0.22	0.14	0.35	0.05	n.a.	0.64	6.00	5.10	3.48
Pb	2.22	2.44	0.44	1.44	2.17	0.31	1.08	5.60	94.6	n.a.	n.a.	6.79
Th	0.34	0.10	0.06	0.06	0.21	0.29	0.03	2.30	1.05	15.5	13.2	5.65
U	0.12	0.08	0.02	0.02	0.05	0.07	0.01	2.90	0.26	2.70	3.20	1.29

Data sources: The Panzhihua intrusion (Zhou et al. 2005; Zhang et al. 2009), ultramafic dykes (Hou et al. 2012b, 2013), and A-type granitoids (Shellnutt and Zhou 2007; Zhong et al. 2009; Shellnutt and Jahn 2010)

[a] Total iron as Fe_2O_3

[b] n.a. = not analyzed

Apatite

Apatite occurs in the MGZ and MZb of the intrusion, but only that in the latter is considered as a cumulus phase according to texture and modal abundance. Apatite from the MGZ is fine-grained, acicular and is intergrown with fine crystals of silicate and oxide minerals, presumably formed by equilibrium crystallization as a result of rapid cooling of the intruded magma. Despite these differences, both types of apatite are rich in F (2.4–3.4 wt.%) and poor in Cl (mostly below detection limit to 0.15 wt.%) and H_2O (0.16–0.6 wt.% calculated from ideal mineral formula) (Pang et al. 2009). As shown by their data, apatite composition is largely uniform in the MZb irrespective of stratigraphic position.

Accessory Minerals

Hornblende is a widespread accessory mineral throughout the intrusion. It has pargasitic composition with relatively high TiO_2 (2.6–5.0 wt.%) but low F and Cl (<0.25 wt.% for each species) (Pang 2008). Biotite is less common than hornblende but no mineral chemical data have been reported yet. Minor amounts of hercynitic spinel, with average composition of ~75 mol % spinel *sensu stricto* and ~25 mol % hercynite, are associated with Fe-Ti oxides, consistent with formation by exsolution from titanomagnetite. Base-metal sulfides, including pyrrhotite, chalcopyrite and Co-pentlandite are common accessory minerals in the intrusion (Pang 2008).

Geochemistry

Major Elements

The bulk-rock major and trace element geochemistry of the Panzhihua intrusion have been determined by Zhou et al. (2005, 2008), Zhang et al. (2009), Pang et al. (2013) and Song et al. (2013). Data for representative rock types of the intrusion, and also for related magmatic rocks, are listed in Table 10.2. A full dataset of published geochemical analyses is available through written communication to the corresponding author. One common feature of the reported data is the wide range of SiO_2 (~2–50 wt.%) that correlate negatively with FeO and TiO_2 and positively with most other major elements. Iron-titanium oxide ores and anorthosites define the lowest and highest limits of the SiO_2 range respectively, with the gabbroic rocks intermediate between them. The above features indicate that the bulk Fe-Ti oxides exerts a dominant control on the major element compositions, that is, a dilution effect in all major elements other than FeO and TiO_2. Relatively high P_2O_5 (~1–3 wt.%) is characteristic of the MZb where cumulus apatite is present. There is no evidence for any rocks in the intrusion having SiO_2 significantly greater than

~50 wt.%, except the felsic dykes cutting the intrusion. Ultramafic dykes cutting the wall rocks and the lower portion of the intrusion are characterized by relatively low SiO_2 (~37–43 wt.%) and high MgO (~21–34 wt.%) (Hou et al. 2012b, 2013). The A-type granitoids associated with the intrusion are characterized by intermediate to acidic (~54–75 wt.% SiO_2) (Fig. 10.5a), peralkaline or metaluminous compositions, and A-type characteristics (Fig. 10.5b) (Shellnutt and Zhou 2007; Zhong et al. 2009; Shellnutt and Jahn 2010).

Trace Elements

With the exception of the MGZ, rocks of the intrusion have generally low concentrations of incompatible trace elements, consistent with a cumulate origin. For example, over 85 % of the published bulk-rock data show Zr concentrations of less than 50 ppm (Fig. 10.6). Based on the bulk-rock Zr budget, the amount of trapped liquid in the Panzhihua cumulates was estimated to be less than 20 % (Pang et al. 2009). The figure also shows that microgabbros from the MGZ have trace element abundances generally comparable to the high-Ti basalts.

The trace element abundances and patterns for different types of cumulates in the intrusion are variable as illustrated in Fig. 10.7a for the representative rock types shown in Table 10.2. The Fe-Ti oxide ore sample has the lowest abundances for most trace elements except Nb, Ta and Ti, elements that are preferred in Fe-Ti oxides (Zhang et al. 2009; Pang et al. 2013) and which show positive anomalies in

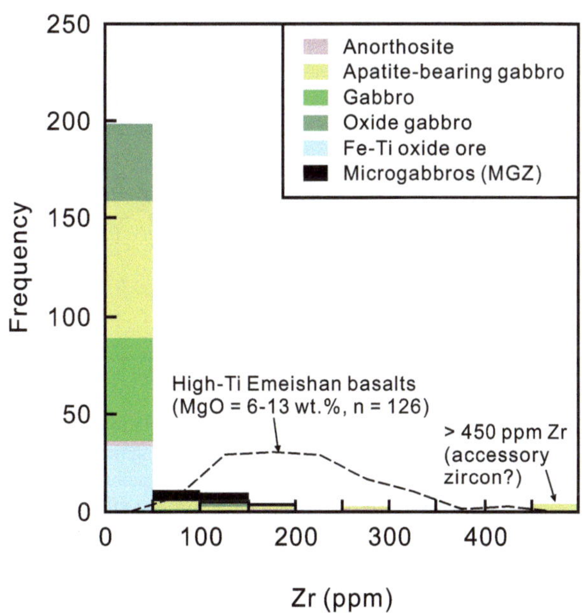

Fig. 10.6 Frequency distribution of bulk-rock Zr concentration for rocks of the Panzhihua intrusion. The data are after Zhou et al. (2005), Zhou et al. (2008), Zhang et al. (2009), Pang et al. (2013), Song et al. (2013) and unpublished data of the first author (KNP). Data for the Emeishan basalts are screened from an extensive database on the basis of MgO = 6–13 wt.% and Ti/Y > 500, the sources of which are available through written communication to the corresponding author

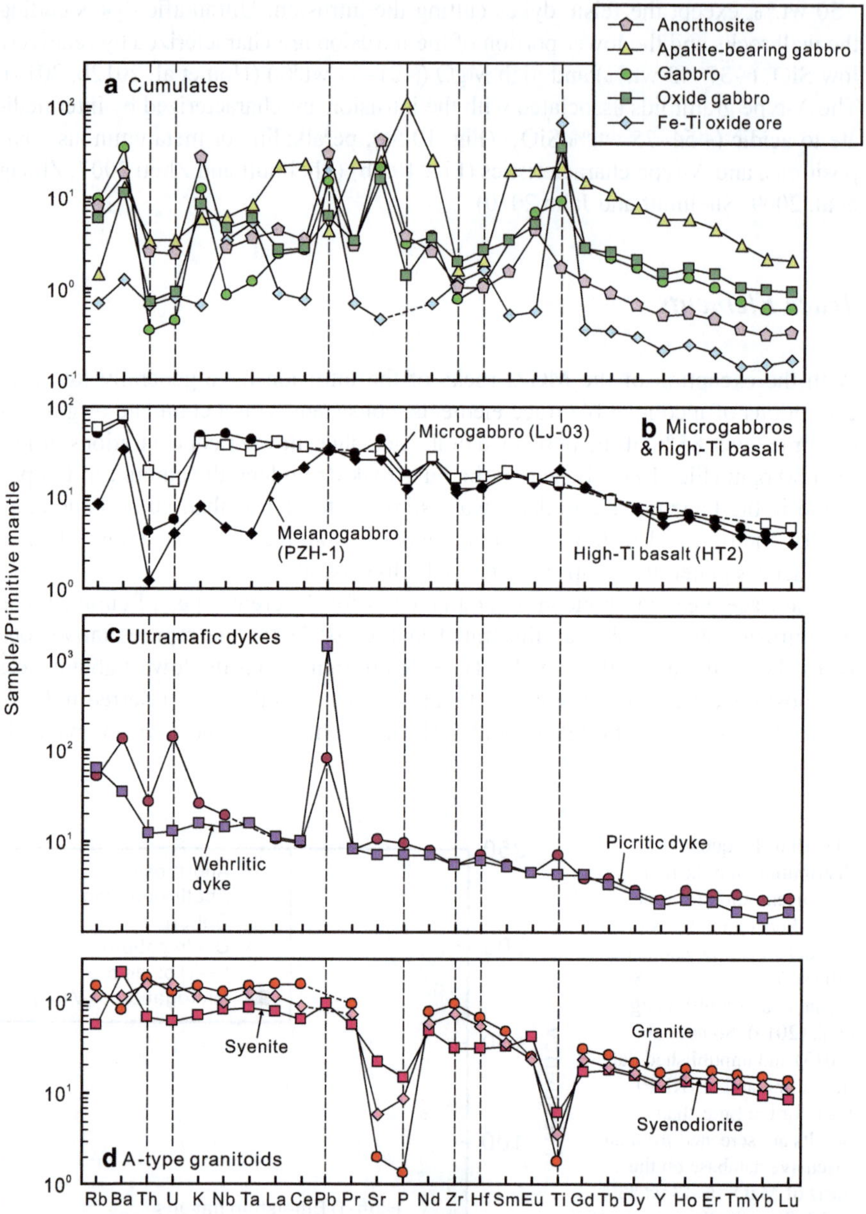

Fig. 10.7 Primitive mantle-normalized incompatible trace element patterns for representative rocks of the Panzhihua intrusion and related magmatic rocks. **a** Different types of cumulates in the intrusion (LZ, MZa, MZb and UZ). **b** Microgabbros (MGZ) and a high-Ti basalt sample (HT2 sub-group, Xu et al. 2001). **c** Ultramafic dykes cutting the *lower* part of the intrusion. **d** A-type granitoids associated with the intrusion.Normalizing values are after Sun and McDonough (1989). Key elements are highlighted using *dashed lines*. The data and references are given in Table 10.2

the mantle-normalized variation diagram. The positive Pb anomaly for the ores is due to the presence of sulfide minerals. The oxide gabbro sample and the gabbro sample have higher concentrations of trace elements but the above anomalies appear to be somewhat less pronounced. The apatite-bearing gabbro sample is marked by diagnostic positive P and negative Zr-Hf anomalies, presumably reflecting the preferential incorporation of rare-earth over high field strength elements by cumulus apatite (Prowatke and Klemme 2006). The anorthosite sample is characterized by positive Ba, Sr and Eu anomalies as would be expected for the presence of cumulus plagioclase. A diagnostic feature for the gabbroic rocks and the anorthosite is the prominent negative Th-U anomalies relative to Ba and K. Microgabbros from the MGZ contain high abundances of most trace elements and negative Th-U anomalies are variably developed (Fig. 10.7b). The ultramafic dykes are marked by lower trace element abundances than the microgabbros and positive Pb anomalies, which might be related to contamination by the carbonate country rocks (Fig. 10.7c). The A-type granitoids display negative anomalies in Sr, P and Ti relative to elements of similar incompatibilities, consistent with fractionation of plagioclase, apatite and Fe-Ti oxides, respectively (Fig. 10.7d).

Isotopic Compositions

The Sr-Nd isotopic compositions of the intrusion have been determined by Zhou et al. (2008), Pang et al. (2013) and Song et al. (2013). Initial Sr isotopic ratios and εNd values (calculated at 260 Ma) for bulk-rock samples vary from 0.7039 to 0.7055 and −0.1 to +4.3, respectively, consistent with the isotopic compositions obtained from clinopyroxene separates [$(^{87}Sr/^{86}Sr)_{260\ Ma}=0.7043–0.7051$ and $\varepsilon Nd_{260\ Ma}=-2.1$ to $+3.2$, Zhang et al. 2009]. The isotopic data plot near the high-$\varepsilon Nd_{260\ Ma}$, low-$(^{87}Sr/^{86}Sr)_{260\ Ma}$ end of the high-Ti Emeishan basalts (Fig. 10.8a). In a diagram of $^{143}Nd/^{144}Nd$ versus age, the data for the intrusion, the associated A-type granitoids and the picritic dykes cutting the lower part of the intrusion generally overlap between the depleted mantle and the CHUR curves (Fig. 10.8b). The Sr-Nd isotopic data indicate that magma contamination by old crust is insignificant in the petrogenesis of the intrusion.

The Pb isotopes of the intrusion have been studied by Zhang et al. (2009). The isotopic compositions of ten clinopyroxene separates vary in a limited way, i.e. $^{206}Pb/^{204}Pb=17.91–18.39$, $^{207}Pb/^{204}Pb=15.57–15.61$ and $^{208}Pb/^{204}Pb=38.29–38.56$, which fall in the fields of high-Ti Emeishan basalts and OIB (see Fig. 10.8 in Zhang et al. 2009). The O isotopes of the intrusion have been examined by Zhang et al. (2009) and Ganino et al. (2013a) but different results are obtained. The $\delta^{18}O$ values of clinopyroxene analyzed by Zhang et al. (2009) (4.2–6.8 ‰) is systematically lower than those of plagioclase analyzed by Ganino et al. (2013a) (7.4–8.9 ‰). The noble gas isotopes He and Ar of the intrusion have been studied by Hou et al. (2011b), who interpreted the low $^{3}He/^{4}He$ (0.45–2.32 Ra) and $^{40}Ar/^{36}Ar$ (472–2963) in olivine and clinopyroxene as indicative of addition of subduction-related fluids and melts stored in the lithospheric mantle.

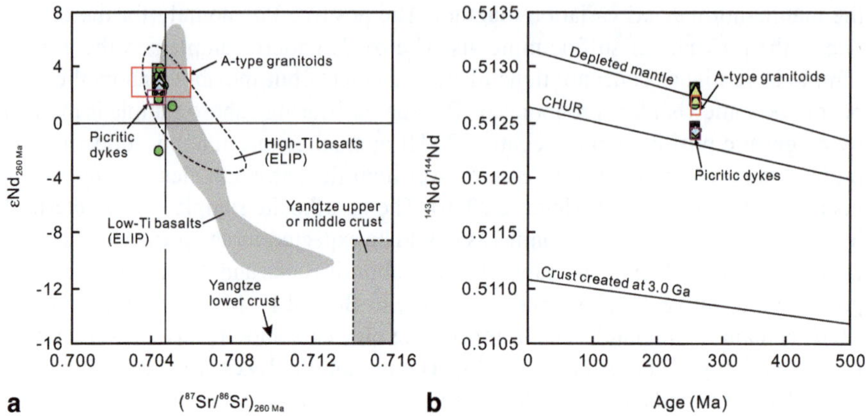

Fig. 10.8 Sr-Nd isotopic compositions of the Panzhihua intrusion. **a** Co-variation of εNd and $^{87}Sr/^{86}Sr$, calculated at 260 Ma, for bulk-rocks and clinopyroxene separates. Fields for the flood basalts of the ELIP (Anh et al. 2011 and references therein), the A-type granitoids associated with the intrusion (Shellnutt and Zhou 2007; Zhong et al. 2009; Shellnutt and Jahn 2010), the picritic dykes cutting the *lower* part of the intrusion (Hou et al. 2013), and the Yangtze *upper* or *middle* crust (Wang et al. 2007) are shown for comparison. **b** Co-variation of $^{143}Nd/^{144}Nd$ and age. The depleted mantle curve is computed using the average $^{143}Nd/^{144}Nd$ for MORB (0.51315) and the average Sm/Nd for depleted MORB mantle (0.411, Workman and Hart 2005). The crust curve is computed using the average Sm/Nd for the total crust (0.195, Rudnick and Gao 2003), assuming that it grew from CHUR at 3.0 Ga. Legend is the same as in Fig. 10.7

Discussion

Intensive Parameters

The temperature at which the Panzhihua intrusion crystallized has been estimated by mineral composition and comparison to high-Ti basalts of the ELIP. On the basis of olivine-liquid Fe-Mg exchange equilibrium, Pang et al. (2009) showed that the most magnesian olivine found in the intrusion (~Fo_{78-81}) was in equilibrium with a liquid having Mg# of ~51–56. Representative high-Ti basalts lying within or close to this compositional range were modeled by Pang et al. (2008b) and Shellnutt and Jahn (2010) using the petrological algorithm MELTS and by Ganino et al. (2008) using the algorithm PELE (the PC version of MELTS). The liquidi of the modeled compositions are all ~1200 °C on an anhydrous basis.

The pressure at which the intrusion crystallized has been inferred from mineral composition and geological relations. Pang et al. (2008a) noted that the primary titanomagnetite with ~5.3 wt.% Al_2O_3 in solid solution is similar to an Fe-Ti spinel synthesized in an experimental ferrodioritic melt at ~5 kbar (Lindsley, personal communication). Using the revised amphibole thermobarometer by Ridolfi and Renzulli (2012), the amphibole compositions reported by Pang (2008) are consistent with formation at ~3 kbar. Thus, a pressure range of ~3–5 kbar likely represents

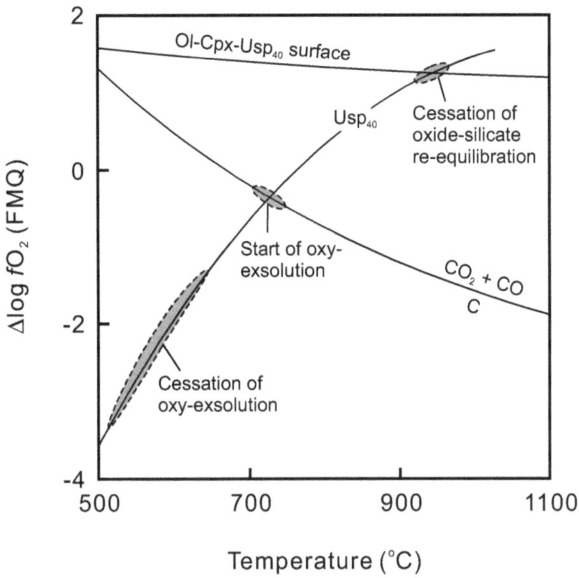

Fig. 10.9 Co-variation of $\Delta \log fO_2$ and temperature showing the inferred cooling trend for Fe-Ti oxides in the Panzhihua intrusion (Pang et al. 2008a). See text for further details

the pressure at which the intrusion crystallized. Shellnutt and Jahn (2010) noted an outcrop east of the city of Panzhihua where A-type granitoids associated with the intrusion intruded flood basalts of the ELIP. If the lithostatic pressure is solely contributed by the volcanic sequence, which has a maximum thickness of ~5 km (see earlier descriptions) and an average density of 3.0 g cm^{-3}, the granitoids and by inference the Panzhihua intrusion would have crystallized at ~1.5 kbar or below (Pang et al. 2010). One probable explanation for the discrepancy is that the intrusion crystallized at ~3–5 kbar and formed the residual silicic magma, which migrated upwards and crystallized in a chamber at the base of the flood basalt pile.

As mentioned above, titanomagnetite and ilmenite from the intrusion underwent extensive subsolidus re-equilibration and exsolution, hence cannot be used to estimate oxygen fugacity (fO_2) at magmatic temperatures. An additional complexity is the possibility that much ilmenite in the lower part of the intrusion where the Fe-Ti oxide ores accumulated formed from external granule exsolution of the primary aluminous titanomagnetite (~Usp_{40}) (Pang et al. 2008a). These authors, however, provided an indirect fO_2 estimate in the lower part of the intrusion. Using the QUILF program of Andersen et al. (1993), it was demonstrated that the primary aluminous titanomagnetite (~Usp_{40}) was last equilibrated with olivine and clinopyroxene at ~950 °C and fO_2 between 1 and 1.5 log units above the fayalite-magnetite-quartz (FMQ) buffer (Fig. 10.9). It is important that the calculated olivine-clinopyroxene-Usp_{40} surface lies above FMQ + 1 at higher temperatures up to ~1100 °C (Fig. 10.9). The value is comparable to the high end exhibited by most mantle-derived magmas (see Mallmann and O'Neill 2009), and in good agreement with the low V contents (~4800 ppm) of the primary titanomagnetite (Pang et al. 2010). The relatively oxidizing condition might be a result of magma-wallrock interaction (see later discussion) or an intrinsic character of the parental magma. In the upper part of the intrusion where primary ilmenite occurs together with magnetite (e.g. MZb), the fO_2 was presumably lower (Howarth et al. 2013).

Parental Magma and Source Characteristics

A genetic relation between the parental magma (defined herein as the magma emplaced in the intrusion at the present level of exposure, rather than the one that was last in equilibrium with the mantle) of the Panzhihua intrusion and the high-Ti basalts of the ELIP appears inescapable (Zhou et al. 2005, 2008; Pang et al. 2008a, 2010; Song et al. 2013; Howarth and Prevec 2013). There is no evidence for the emplacement of an ultramafic magma (e.g. ferropicrite) that differentiated in place to become the intrusion at the present level of exposure, although the parental magma might have been derived from a more primitive ferropicritic magma at depth (Hou et al. 2013). Two published bulk-rock analyses from the MGZ, which represents the chilled base of the intrusion, are available hitherto. Sample LJ-03 by Zhou et al. (2005) is a microgabbro having ~44.7 wt.% SiO_2, ~7.3 wt.% MgO and Mg# of ~53 (Table 10.2), which is consistent with the parental magma composition predicted by the most Mg-rich olivine in the intrusion using olivine-liquid equilibrium (see earlier discussion). Interestingly, the sample has a trace element signature in excellent agreement with the HT2 sub-group of the high-Ti Emeishan basalts characterized by marked relative Th-U depletions that are largely absent in other high-Ti basalts (Fig. 10.7b, also see Xu et al. 2001). Sample PZH-1 by Zhang et al. (2009) is a fine-grained melanogabbro having ~39.6 wt.% SiO_2, ~12 wt.% MgO and Mg# of ~75 (Table 10.2). The low SiO_2 and high MgO (both absolute and relative to FeO) of the sample indicate that it might have been influenced by accumulation of olivine + Fe-Ti oxides, in line with its depletion in highly incompatible trace elements compared with sample LJ-03 (Fig. 10.7b). It is important that both compositions are rich in FeO and TiO_2 and poor in SiO_2, but the extents of such features are stronger for sample PZH-1 than sample LJ-03.

The origin of the Fe- and Ti-rich character of the parental magma is open to question. Based on similar features exhibited by the high-Ti Emeishan basalts, Zhou et al. (2005) and Zhou et al. (2008) suggested that the enrichments were inherited from the plume source and were further enhanced by fractional crystallization. Pang et al. (2008a) noted, on the basis of melting experiments by Whitaker et al. (2007), that enrichments in FeO and TiO_2 with concomitant SiO_2 depletion might have been caused by high pressure fractionation (e.g. ~10 kbar). Zhang et al. (2009) and Hou et al. (2011b) emphasized the importance of an Fe- and Ti-rich component in the lithospheric mantle and proposed that its interaction with plume-derived magma contributed to the genesis of the Fe- and Ti-rich parental magma.

Magma-Wall Rock Interaction

The contact aureole variably developed along the lower contact of the intrusion has been studied by Ganino et al. (2008, 2013a, b). Ganino et al. (2008) demonstrated that the formation of brucite marble, the dominant lithology of the aureole, can be explained by decarbonation of dolostone and subsequent hydration of periclase during contact metamorphism. Ganino et al. (2013a) interpreted moderate REE enrichments and strong HFSE depletions in a subset of brucite marbles as indicative of partial melting of carbonate country rocks. If this is correct, the temperature condition in the aureole would have been heterogeneous and that carbonate assimilation might have locally occurred along the magma-wall rock contact. Ganino et al. (2013b) argued, on the basis of elevated $\delta^{18}O$ values of plagioclase in the intrusion, that the magma from which the intrusion crystallized might have assimilated ~8–14% carbonate wall rocks. This interpretation is, however, at odds with the low $\delta^{18}O$ values of clinopyroxene (~4.2–6.8 ‰) (Zhang et al. 2009) as would be expected for crystallization from uncontaminated mantle-derived magmas, because wall rock assimilation should affect the O-isotope composition of the contaminated magma and any minerals crystallized from it. Nevertheless, any of the above processes (fluid-driven decarbonation versus assimilation) has the potential of releasing abundant CO_2 from the wallrocks into the magmatic system and might explain the relatively high fO_2 and oxide ore occurrences in the lower part of the intrusion. If the intrusion represents a conduit-type magma plumbing system (see later discussion), the process might have contributed to the end-Guadalupian environmental crisis (Ganino and Arndt 2009).

Order of Crystallization and Differentiation Trend

Petrography and mineralogy of the MGZ and LZ indicate that the parental magma of the Panzhihua intrusion attained multiple saturation of clinopyroxene, plagioclase and titanomagnetite at the time of emplacement. As noted earlier, the amoeboid texture of Fe-Ti oxides does not necessarily imply that they crystallized late because of subsolidus grain boundary readjustment (Duchesne 1996, 1999; Pang et al. 2008a). Olivine is not ubiquitous in the MGZ and LZ; it is present in substantial amounts thus clearly a cumulus mineral in some horizons but occurs as occasional crystals or thin rims between clinopyroxene and Fe-Ti oxides in the majority of rocks in these zones. This indicates that the mineral could not have been far from the liquidus during the interval where these zones crystallized. This crystallization sequence continued upwards to the MZa despite the reversal in plagioclase composition in the lower part of this zone that presumably reflects magma addition (Pang et al. 2009) (Fig. 10.2). The appearance of apatite at the base of the MZb is followed closely by that of ilmenite and the ubiquitous presence of olivine. The disappearance of apatite in the UZ is possibly related to a major event of magma addition at the base of this zone.

The magma compositions at successive stages of differentiation have not been precisely constrained, partly because of the general lack of rock compositions that can be taken as representative of differentiated liquids. Pang et al. (2009) noted, on the basis of bulk-rock Zr budgets relative to the high-Ti basalts in the ELIP, that most rocks of the intrusions represent cumulates and the differentiated liquids would have escaped the magma chamber. The A-type granitoids associated with the intrusion are plausible candidates for this liquid, as evidenced by trace element and Sr-Nd isotopic data. The granitoids are characterized by depletions in Sr, P and Ti relative to elements of similar incompatibilities (Fig. 10.7d), consistent with fractionation of plagioclase, apatite and Fe-Ti oxides, i.e. the cumulus minerals in the intrusion (Shellnutt and Zhou 2007; Shellnutt and Jahn 2010). The Nd isotopic ranges exhibited by the intrusion and the granitoids overlap ($\varepsilon Nd_{260\ Ma}$ of +2 to +3.5) (Fig. 10.8). Further, the assemblage of aenigmatite + ilmenite in the granitoids points to a relatively reducing condition (Shellnutt and Iizuka 2011), presumably related to the abundant crystallization of magnetite in the intrusion. Thus, the granitoids most likely represent products of extreme differentiation of the parental magma, implying differentiation towards silica enrichment. This is in accordance with the onset of titanomagnetite crystallization and its fractionation at a relatively early stage (Pang et al. 2009).

Nature of the Magmatic System

A closed magmatic system involving a single influx of magma with little or no recharge, analogous to that represented by the Skaergaard and Kiglapait intrusions, does not satisfactorily account for the key features of the Panzhihua intrusion, including the abundance of Fe-Ti oxides (Song et al. 2013), the deficit in Zr (and other incompatible trace elements) (Pang et al. 2009), and the occurrence of reversals in mineral compositions (Pang et al. 2009; Song et al. 2013) and in bulk-rock compositions (Pang et al. 2013), and the inferred fO_2 variations and silicate disequilibrium textures (Howarth et al. 2013). The above features can be readily explained by a magmatic system involving periodic replenishment, tapping, or a combination of both.

Despite the affinity to an open system magma chamber, diverse opinions exist among existing petrologic models. Figure 10.10 illustrates, in a purely descriptive manner, the key elements of the proposed models for the development of the Panzhihua intrusion and associated Fe-Ti oxide ores. Interested readers should refer to the respective references for details. Pang et al. and Zhou et al. considered the intrusion as a differentiated sill and the A-type granitoids as part of the same magmatic system. In the Song et al. and Pêcher et al. models, however, the intrusion is taken to represent a magma plumbing system (magma flow through system or conduits) that presumably fed the high-Ti Emeishan basalts. Pêcher et al. further proposed that the intrusion consists of two concordant segments linked to each other by a discordant dyke in its central portion, rather than a differentiated sill. The Hou et al. model is

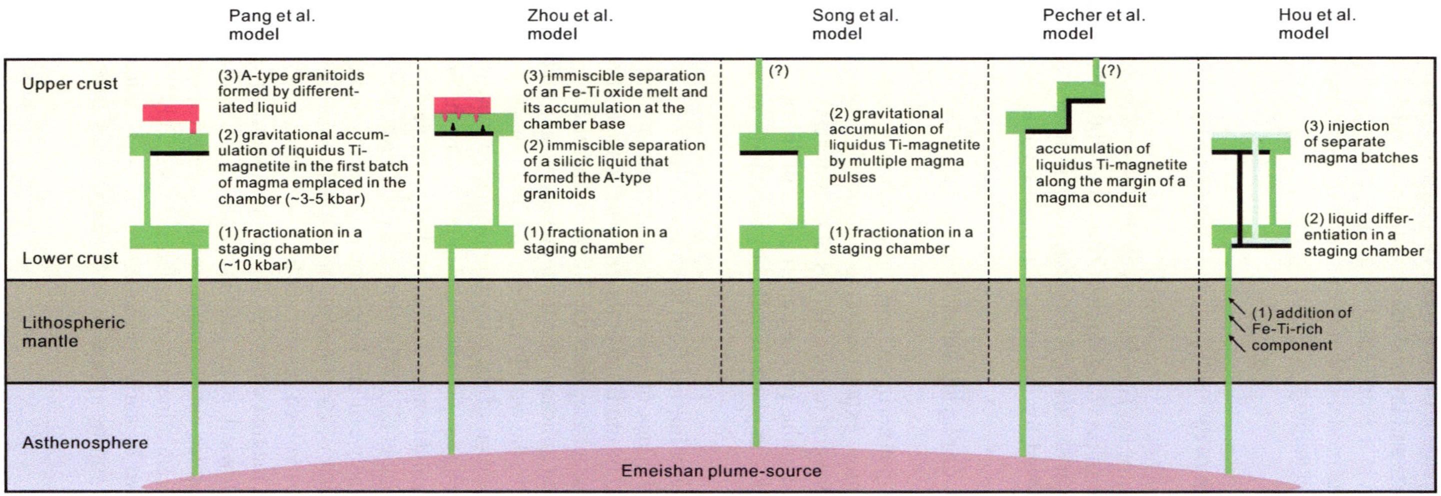

Fig. 10.10 A schematic diagram illustrating the proposed petrologic models for the formation of the Panzhihua intrusion and the associated Fe-Ti oxide ores. References are as follows: the Pang et al. model (Pang et al. 2008a, 2009), the Zhou et al. model (Zhou et al. 2005, 2013), the Song et al. model (Song et al. 2013), the Pêcher et al. model (Pêcher et al. 2013), and the Hou et al. model (Hou et al. 2012b)

complex, involving emplacement of five types of magma (wehrlite, massive ore, disseminated ore, anorthosite and layered gabbro) that were related to each other to a common parental magma in a deep-seated magma chamber by liquid differentiation. One implication for the model is that the intrusion was a blind system, involving magma addition without tapping.

Genesis of Fe-Ti oxide ores

The origin of thick, stratiform Fe-Ti oxide ores in the lower part of the Panzhihua intrusion has been the focus of previous studies (Zhou et al. 2005; Pang et al. 2008a; Song et al. 2013; Howarth et al. 2013). A magmatic origin for the ores is indisputable and any models proposed for ore formation must be able to account for the following features:

i. The ores are stratiform, displaying either sharp or gradational contacts against the gabbroic host rocks.
ii. The ores occur as multiple horizons, and those occupying lower stratigraphic positions are generally more extensive (up to ~60 m-thick) than those at higher stratigraphic positions (several to several tens of centimeters).
iii. The ores contain isolated grains (or their aggregates) of silicate minerals present in the gabbroic host rocks, in places mantled by amphibole or to a lesser extent olivine.

The hypothesis that the ores solidified from an immiscible Fe-Ti oxide melt is discredited at present due to the lack of experimental support and the strong reliance on oxide textures that could be modified at the subsolidus stage. If one accepts that titanomagnetite is a liquidus phase, then the silicate mineral compositions within and close to the ore horizons indicates that the mineral crystallized early compared to other layered intrusions in the world (Table 10.3). The probable reasons for this include the unusual composition of the parental magma, an increase in fO_2 during magma-wall rock interaction (e.g. Ganino et al. 2008), and/or the presence of volatiles (e.g. H_2O and F) that might cause the delayed appearance of silicate minerals relative to Fe-Ti oxides (Pang et al. 2008a, Howarth and Prevec 2013). Some workers proposed a causal link between ore formation and magma addition. Song et al. (2013) proposed, on the basis of Cr in magnetite and bulk-rock compositions, that ore formation presumably involved crystal settling in magma batches that replenished the lower part of the magma chamber due to their high density compared to the pre-existing magma. Howarth et al. (2013) proposed, on the basis of textures and MELTS modeling, that ore concentration might have involved preferred dissolution of pre-existing silicate minerals by replenished, variably hydrous magmas. In any case, the roles played by early crystallization of Fe-Ti oxides and their accumulation by settling and sorting appear to be essential in ore genesis.

Table 10.3 Compositions of olivine, Ca-rich pyroxene and plagioclase at the appearance of cumulus Fe-Ti oxides in the Panzhihua intrusion compared to major layered intrusions

	Fo% (olivine) (olivine)	Mg# (Ca-rich pyroxene)	An% (plagioclase)
Panzhihua	~71	~79	~69
Skaergaard[a] (the base of Lower Zone c)	~56	~65	~53
(the top of Middle Zone)	~40	~55	~46
Bushveld	–	~67	~61
(Upper Zone)			
Kiglapait	~56	~70	~45
Sept Iles (Megacyclic Unit I)	~66	~74	~61
Duluth	~58	~63–68	~48–56
Bjerkreim–Sokndal	–	~75	~50
Newark island	~55	~64	~47

Data sources: Pang et al. (2009), Namur et al. (2010) and references therein
[a] The base of the Lower Zone c is where abundant Fe-Ti oxides first occur from the base of the intrusion upwards, whereas the top of the Middle Zone is where an abrupt drop of V in pyroxene was identified by Jang et al. (2001) as indicative of the onset of magnetite crystallization

Summary and Conclusions

The ~260 Ma Panzhihua intrusion represents a smoking gun for the genesis of magmatic Fe-Ti oxide deposits and a petrologic enigma—how can rocks composed of essentially Fe-Ti oxides form in magma chambers. The intrusion records the evolution of a ferrobasaltic magma in a periodically replenished magma chamber in an intra-continental setting.

The most plausible parental magma is a moderately fractionated ferrobasalt rich in FeO and TiO_2 and poor in SiO_2, similar to some high-Ti Emeishan basalts. Generation of such an unusual magma composition might have involved mantle plume-lithosphere interaction and/or high-pressure fractionation. The magma was emplaced in Neoproterozoic dolomitic limestone at ~3–5 kbar, crystallized olivine, clinopyroxene, plagioclase and titanomagnetite, and also ilmenite and apatite at a later stage. Differentiation was interrupted by addition of more primitive magmas into the chamber and fractionation of magnetite gabbro drove the differentiated liquid towards silica enrichment, forming the A-type granitoids intimately associated with the intrusion. Early and abundant crystallization of titanomagnetite might be a result of the unusual parental magma composition, magma-wall rock interaction, presence of volatiles, or any combination thereof. Its accumulation by gravitational settling and sorting in one or more batches of magma formed the spectacular stratiform Fe-Ti oxide ores in the lower part of the intrusion.

Acknowledgments We thank Xiao Fu and Liang Qi for continuous analytical support for the Panzhihua intrusion and related magmatic rocks in the ELIP, Yuxiao Ma for fieldwork arrangement, and Xie-Yan Song for sharing mineral chemical and bulk-rock geochemical data for the intrusion. KNP is indebted to John Malpas, Don Lindsley, Chusi Li and Neil Irvine who gave detailed and constructive comments on his work. Sun-Lin Chung is gratefully acknowledged for giving to KNP the time to complete this manuscript. We acknowledge the constructive reviews by Jean-Clair Duchesne and Clément Ganino, and editorial handling by Bernard Charlier. This study is partially supported by the Research Grant Council of Hong Kong (HKU707012P) to MFZ.

References

Andersen DJ, Lindsley DH, Davidson PM (1993) QUILF: a Pascal program to assess equilibria among Fe-Mg-Mn-Ti oxides, pyroxene, olivine, and quartz. Comput Geosci 19:1333–1350

Anh TV, Pang K-N, Chung S-L, Lin H-M, Tran TH, Tran TA, Yang H-J (2011) The Song Da magmatic suite revisited: a petrologic, geochemical and Sr-Nd isotopic study on picrites, flood basalts and silicic volcanic rocks. J Asian Earth Sci 42:1341–1355

Bai Z-J, Zhong H, Naldrett AJ, Zhu W-G, Xu G-W (2012) Whole-rock and mineral composition constraints on the genesis of the giant Hongge Fe-Ti-V oxide deposit in the Emeishan large igneous province, Southwest China. Econ Geol 107:507–524

Boven A, Pasteels P, Punzalan LE, Liu J, Luo X, Zhang W, Guo Z, Hertogen J (2002) $^{40}Ar/^{39}Ar$ geochronological constraints on the age and evolution of the Permo-Triassic Emeishan Volcanic Province, Southwest China. J Asian Earth Sci 20:157–175

Chung S-L, Jahn B-M (1995) Plume-lithosphere interaction in generation of the Emeishan flood basalts at the Permian-Triassic boundary. Geology 23:889–892

Duchesne JC (1996) Liquid ilmenite or liquids ilmenite: a comment on the nature of ilmenite vein deposits. In: Demaiffe D (ed) Petrology and geochemistry of magnatic suites of rocks in the continental and oceanic crusts. Université Libre de Bruxelles, Tervuren, pp 73–82

Duchesne JC (1999) Fe-Ti deposits in Rogaland anorthosites (South Norway): geochemical characteristics and problems of interpretation. Miner Depos 34:182–198

Frost BR, Barnes CG, Collins WJ, Arculus RJ, Ellis DJ, Frost CD (2001) A geochemical classification for granitic rocks. J Petrol 42:2033–2048

Ganino C, Arndt NT (2009) Climate changes caused by degassing of sediments during the emplacement of large igneous provinces. Geology 37:323–326

Ganino C, Arndt NT, Zhou M-F, Gaillard F, Chauvel C (2008) Interaction of magma with sedimentary wall rock and magnetite ore genesis in the Panzhihua mafic intrusion, SW China. Miner Depos 43:677–694

Ganino C, Arndt NT, Chauvel C, Alexandre J, Athurion C (2013a) Melting of carbonate wall rocks and formation of the heterogeneous aureole of the Panzhihua intrusion, China. Geosci Front 4:535–546

Ganino C, Harris C, Arndt NT, Prevec SA, Howarth GH (2013b) Assimilation of carbonate country rock by the parent magma of the Panzhihua Fe-Ti-V deposit (SW China): evidence from stable isotopes. Geosci Front 4:547–554

Hou T, Zhang Z, Kusky T, Du Y, Liu J, Zhao Z (2011a) A reappraisal of the high-Ti and low-Ti classification of basalts and petrogenetic linkage between basalts and mafic-ultramafic intrusions in the Emeishan Large Igneous Province, SW China. Ore Geol Rev 41:133–143

Hou T, Zhang Z, Ye X, Encarnacion J, Reichow MK (2011b) Noble gas isotopic systematics of Fe-Ti-V oxide ore-related mafic-ultramafic layered intrusions in the Pan-Xi area, China: the role of recycled oceanic crust in their petrogenesis. Geochim Cosmochim Acta 75:6727–6741

Hou T, Zhang Z, Encarnacion J, Santosh M (2012a) Petrogenesis and metallogenesis of the Taihe gabbroic intrusion associated with Fe-Ti-oxide ores in the Panxi district, Emeishan Large Igneous Province, Southwest China. Ore Geol Rev 49:109–127

Hou T, Zhang Z, Pirajno F (2012b) A new metallogenic model of the Panzhihua giant V-Ti-iron oxide deposit (Emeishan Large Igneous Province) based on high-Mg olivine-bearing wehrlite and new field evidence. Int Geol Rev 54:1721–1745
Hou T, Zhang Z, Encarnacion J, Santosh M, Sun Y (2013) The role of recycled oceanic crust in magmatism and metallogeny: Os-Sr-Nd isotopes, U-Pb geochronology and geochemistry of picritic dykes in the Panzhihua giant Fe-Ti oxide deposit, central Emeishan large igneous province, SW China. Contrib Mineral Petrol 165:805–822
Howarth GH, Prevec SA (2013) Hydration vs. oxidation: modelling implications for Fe-Ti oxide crystallisation in mafic intrusions, with specific reference to the Panzhihua intrusion, SW China. Geosci Front 4:555–570
Howarth GH, Prevec SA, Zhou M-F (2013) Timing of Ti-magnetite crystallisation and silicate disequilibrium in the Panzhihua mafic layered intrusion: implications for ore-forming processes. Lithos 170–171:73–89
Irvine TN (1982) Terminology for layered intrusions. J Petrol 23:127–162
Jang YD, Naslund HR, McBirney AR (2001) The differentiation trend of the Skaergaard intrusion and the timing of magnetite crystallization: iron enrichment revisited. Earth Planet Sci Lett 189:189–196
Lo C-H, Chung S-L, Lee T-Y, Wu G (2002) Age of the Emeishan flood magmatism and relations to Permian-Triassic boundary events. Earth Planet Sci Lett 198:449–458
Ma Y, Liu JF, Wang HF, Mao YS, Ji XT, Wang DK, Yan ZZ (2001) Geology of the Panzhihua region. Sichuan science and technology press, Chengdu (in Chinese)
Ma Y, Ji XT, Li JC, Huang M, Kan ZZ (2003) Mineral Resources of the Panzhihua region. Sichuan Science and Technology Press, Chengdu (in Chinese)
Mallmann G, O'Neill H St C (2009) The crystal/melt partitioning of V during mantle melting as a function of oxygen fugacity compared with some other elements (Al, P, Ca, Sc, Ti, Cr, Fe, Ga, Y, Zr and Nb). J Petrol 50:1765–1794
Namur O, Charlier B, Toplis MJ, Higgins MD, Liégeois J-P, Vander Auwera J (2010) Crystallization sequence and magma chamber processes in the ferrobasaltic Sept Iles layered intrusion, Canada. J Petrol 51:1203–1236
Pang K-N (2008) Origin of the Permian Panzhihua layered gabbroic intrusion and the hosted Fe-Ti-V oxide deposits, Sichuan Province, SW China. Dissertation, The University of Hong Kong
Pang K-N, Li C, Zhou M-F, Ripley EM (2008a) Abundant Fe-Ti oxide inclusions in olivine from the Panzhihua and Hongge layered intrusions, SW China: Evidence for early saturation of Fe-Ti oxides in ferrobasaltic magma. Contrib Mineral Petrol 156:307–321
Pang K-N, Zhou M-F, Lindsley DH, Zhao D, Malpas J (2008b) Origin of Fe-Ti oxide ores in mafic intrusions: evidence from the Panzhihua intrusion. J Petrol 49:295–313
Pang K-N, Li C, Zhou M-F, Ripley EM (2009) Mineral compositional constraints on petrogenesis and ore genesis of the Panzhihua layered gabbroic intrusion, SW China. Lithos 110:199–214
Pang K-N, Zhou M-F, Qi L, Shellnutt G, Wang CY, Zhao D (2010) Flood basalt-related Fe-Ti oxide deposits in the Emeishan large igneous province, SW China. Lithos 119:123–136
Pang K-N, Zhou M-F, Qi L, Chung S-L, Chu C-H, Lee H-Y (2013) Petrology and geochemistry at the LZ-MZa transition of the Panzhihua intrusion, SW China: implications for differentiation and oxide ore genesis. Geosci Front 4:517–534
Pêcher A, Arndt N, Jean A, Bauville A, Ganino C, Athurion C (2013) Structure of the Panzhihua intrusion and its Fe-Ti-V deposit, China. Geosci Front 4:571–582
Prowatke S, Klemme S (2006) Trace element partitioning between apatite and silicate melts. Geochim Cosmochim Acta 70:4513–4527
Ridolfi F, Renzulli A (2012) Calcic amphiboles in calc-alkaline magmas: thermobarometric and chemometric empirical equations valid up to 1,130 °C and 2.2 GPa. Contrib Mineral Petrol 163:877–895
Rudnick RL, Gao S (2003) Composition of the continental crust. In: Rudnick RL (ed) The crust, treatise in geochemistry, vol 3. Elsevier-Pergamon, Oxford, pp 1–64

Shellnutt JG, Iizuka Y (2011) Mineralogy from three peralkaline granitic plutons of the Late Permian Emeishan large igneous province (SW China): evidence for contrasting magmatic conditions of A-type granitoids. Eur J Mineral 23:45–61

Shellnutt JG, Jahn B-M (2010) Formation of the Late Permian Panzhihua plutonic-hypabyssal-volcanic igneous complex: implications for the genesis of Fe-Ti oxide deposits and A-type granites of SW China. Earth Planet Sci Lett 289:509–519

Shellnutt JG, Jahn B-M (2011) Origin of Late Permian Emeishan basaltic rocks from the Panxi region (SW China): Implications for the Ti-classification and spatial-compositional distribution of the Emeishan flood basalts. J Volcano Geotherm Res 199:85–95

Shellnutt JG, Pang K-N (2012) Petrogenetic implications of mineral chemical data for the Permian Baima igneous complex, SW China. Mineral Petrol 106:75–88

Shellnutt JG, Zhou M-F (2007) Permian peralkaline, peraluminous and metaluminous A-type granites in the Pan-Xi district, SW China: their relationship to the Emeishan mantle plume. Chem Geol 243:286–316

Shellnutt JG, Zhou M-F, Zellmer G (2009) The role of Fe-Ti oxide crystallization in the formation of A-type granitoids with implications for the Daly gap: an example from the Permian Baima igneous complex, SW China. Chem Geol 259:204–217

Shellnutt JG, Wang K-L, Zellmer GF, Iizuka Y, Jahn B-M, Pang K-N, Qi L, Zhou M-F (2011) Three Fe-Ti oxide ore-bearing gabbro-granitoid complexes in the Panxi region of the Permian Emeishan large igneous province, SW China. Am J Sci 311:773–812

Shellnutt JG, Denyszyn SW, Mundil R (2012) Precise age determination of mafic and felsic intrusive rocks from the Permian Emeishan large igneous province (SW China). Gondwana Res 22:118–126

Song X-Y, Qi H-W, Hu R-Z, Chen L-M, Yu S-Y, Zhang J-F (2013) Formation of thick stratiform Fe-Ti oxide layers in layered intrusion and frequent replenishment of fractionated mafic magma: evidence from the Panzhihua intrusion, SW China. Geochem Geophys Geosyst 14:712–732

Sun S-S, McDonough WF (1989) Chemical and isotopic systematics in ocean basalt: implication for mantle composition and processes. In: Saunders AD, Norry MJ (eds) Magmatism in the Ocean Basins. Geological Society London Special Publication 42. Blackwell, Oxford, pp 313–345

Wang CY, Zhou M-F, Qi L (2007) Permian flood basalts and mafic intrusions in the Jinping (SW China)-Song Da (northern Vietnam) district: Mantle sources, crustal contamination and sulfide segregation. Chem Geol 243:317–343

Wang CY, Zhou M-F, Zhao D (2008) Fe-Ti-Cr oxides from the Permian Xinjie mafic-ultramafic layered intrusion in the Emeishan large igneous province, SW China: Crystallization from Fe- and Ti-rich basaltic magmas. Lithos 102:198–217

Whalen JB, Currie KL, Chappell BW (1987) A-type granites: geochemical characteristics, discrimination and petrogenesis. Contrib Mineral Petrol 95:407–419

Whitaker ML, Nekvasil H, Lindsley DH, Difrancesco NJ (2007) The role of pressure in producing compositional diversity in intraplate basaltic magmas. J Petrol 48:365–393

Workman RK, Hart SR (2005) Major and trace element composition of the depleted MORB mantle (DMM). Earth Planet Sci Lett 231:53–72

Xiao L, Xu YG, Mei HJ, Zheng YF, He B, Pirajno F (2004) Distinct mantle sources of low-Ti and high-Ti basalts from the western Emeishan large igneous province, SW China: implications for plume-lithosphere interaction. Earth Planet Sci Lett 228:525–546

Xu YG, Chung SL, Jahn BM, Wu GY (2001) Petrologic and geochemical constraints on the petrogenesis of Permian-Triassic Emeishan flood basalts in southwestern China. Lithos 58:145–168

Xu Y-G, He B, Chung S-L, Menzies MA, Frey FA (2004) Geologic, geochemical, and geophysical consequences of plume involvement in the Emeishan flood-basalt province. Geology 32:917–920

Zhang Z, Mahoney JJ, Mao J, Wang F (2006) Geochemistry of picritic and associated basalt flows of the western Emeishan flood basalt province, China. J Petrol 47:1997–2019

Zhang Z, Mao J, Saunders AD, Ai Y, Li Y, Zhao L (2009) Petrogenetic modeling of three mafic-ultramafic layered intrusions in the Emeishan large igneous province, SW China, based on isotopic and bulk chemical constraints. Lithos 113:369–292
Zhong H, Zhou XH, Zhou MF, Sun M, Liu BG (2002) Platinum-group element geochemistry of the Hongge Fe-V-Ti deposit in the Pan-Xi area, southwestern China. Miner Depos 37:226–239
Zhong H, Yao Y, Prevec SA, Wilson AH, Viljoen MJ, Viljoen RP, Liu B-G, Luo Y-N (2004) Trace-element and Sr-Nd isotope geochemistry of the PGE-bearing Xinjie layered intrusion in SW China. Chem Geol 203:237–252
Zhong H, Zhu W-G, Hu R-Z, Xie L-W, He D-F, Liu F, Chu Z-Y (2009) Zircon U-Pb age and Sr-Nd-Hf isotope geochemistry of the Panzhihua A-type syenitic intrusion in the Emeishan large igneous province, southwest China and implications for growth of juvenile crust. Lithos 110:109–128
Zhong H, Qi L, Hu R-Z, Zhou M-F, Gou T-Z, Zhu W-G, Liu B-G, Chu Z-Y (2011) Rhenium-osmium isotope and platinum-group elements in the Xinjie layered intrusion, SW China: Implications for source mantle composition, mantle evolution, PGE fractionation and mineralization. Geochim Cosmochim Acta 75:1621–1641
Zhou M-F, Malpas J, Song X-Y, Robinson PT, Sun M, Kennedy AK, Lesher CM, Keays RR (2002) A temporal link between the Emeishan large igneous province (SW China) and the end-Guadalupian mass extinction. Earth Planet Sci Lett 196:113–122
Zhou M-F, Robinson PT, Lesher CM, Keays RR, Zhang C-J, Malpas J (2005) Geochemistry, petrogenesis and metallogenesis of the Panzhihua gabbroic layered intrusion and associated Fe-Ti-V oxide deposits, Sichuan Province, SW China. J Petrol 46:2253–2280
Zhou M-F, Arndt NT, Malpas J, Wang CY, Kennedy A (2008) Two magma series and associated ore deposit types in the Permian Emeishan large igneous province, SW China. Lithos 103:352–368
Zhou M-F, Chen WT, Wang CY, Prevec SA, Liu P-P, Howarth GH (2013) Two stages of immiscible liquid separation in the formation of Panzhihua-type Fe-Ti-V oxide deposits, SW China. Geosci Front 4:481–502

Chapter 11
The Sept Iles Intrusive Suite, Quebec, Canada

Olivier Namur, Michael D. Higgins and Jacqueline Vander Auwera

Abstract The Sept Iles Intrusive Suite (Quebec, Canada) is made up of a large layered intrusion, late gabbro intrusions and a composite sill (Pointe du Criard Sill). The layered intrusion crystallized from a ferrobasaltic magma and is subdivided into a Layered Series of troctolite and gabbro, an anorthositic Upper Border Series and a granitic Upper Series. The formation of the Upper Border Series resulted from plagioclase flotation from the base to the roof of the magma chamber. Fractionation of troctolites and gabbros in the Layered Series resulted in SiO_2-enrichment and FeO_t-depletion of the residual melt, ultimately forming the granite of the Upper Series. The solidification history of the Layered Series was interrupted by two major and a series of minor influxes of ferrobasaltic melt, significantly enlarging the size of the initial magma chamber. As a consequence, the Layered Series can be subdivided into three megacyclic units (MCU I, II and III). Mixing between resident magma and undifferentiated melt during replenishments had an important influence on both mineral compositions and the liquid lines of descent during the crystallization of the various megacyclic units. It is shown that the liquid line of descent during crystallization of MCU II reached silicate liquid immiscibility. Immiscible melts crystallized two different types of apatite-bearing gabbros, one of which is a major P–Ti–Fe deposit. Cumulate rocks in the layered intrusion show a wide range of crystallized interstitial liquid content. Expelling of this liquid from the crystal mush during solidification is explained both by compaction and compositional convection, but the relative efficiency of these two processes is shown to change significantly with differentiation.

O. Namur (✉)
Institute of Mineralogy, University of Hannover, 30167 Hannover, Germany
e-mail: o.namur@mineralogie.uni-hannover.de

M. D. Higgins
Sciences de la Terre, Université du Québec à Chicoutimi, 555 Blvd de l'université, Chicoutimi, Quebec G7H 2B1, Canada
e-mail: mhiggins@uqac.ca

J. Vander Auwera
Department of Geology, University of Liege, Liege, Belgium
e-mail: jvdauwera@ulg.ac.be

B. Charlier et al. (eds.), *Layered Intrusions,* Springer Geology,
DOI 10.1007/978-94-017-9652-1_11

Keywords Ferrobasalt · Cumulate · Anorthosite · Granite · Magma chamber processes

Introduction

The Sept Iles intrusive suite (Quebec, Canada) is made up of a layered intrusion, some 80 km in diameter, a series of late gabbro intrusions and the Pointe du Criard composite sill. This magmatic event occurred during the later part of the Ediacarian period (564 Ma; Higgins and van Breemen 1998) and played an important role in the evolution of the Precambrian continental crust of Laurentia. The layered intrusion, estimated to be the third largest in the world (Namur et al. 2010), comprises a lower Layered Series with troctolite and gabbro, an anorthositic Upper Border Series and a granitic Upper Series. According to Higgins (2005), all parts of the layered intrusion are comagmatic and related by fractional crystallization and other physical processes that occurred during solidification of the magma chamber.

Rocks from the Layered Series of the layered intrusion are dominated by plagioclase, olivine, clinopyroxene, Fe–Ti oxide minerals and apatite. They resulted from the crystallization of a ferrobasaltic melt (Namur et al. 2010) and strongly resemble the ferrogabbros of the Skaergaard intrusion (McBirney 1989), and the Upper Zone of the Bushveld Complex (Tegner et al. 2006). This troctolite-gabbro succession contains more than 20 cm- to m-thick layers enriched in magnetite and ilmenite, similar to those observed in the Bushveld complex (Cawthorn and Ashwal 2009). It also contains many dm-m blocks of anorthosite, somewhat similar to those observed in the Layered Series of the Skaergaard intrusion (Irvine et al. 1998; Higgins 2005). Mineral compositions in the Layered Series evolve upwards but the main trend is interrupted by two large and several small regressions to more primitive compositions. The most evolved rocks of the Layered Series are made up of apatite-bearing olivine gabbros which represent an important Fe–Ti–P deposit. The uppermost part of the Layered Series does not crop out but is thought to be structurally overlain by a 200–500 m-thick layer of anorthosite, locally containing cm- to m-scale pockets of syenitic to granitic material. Continuous outcrops in the southern part of the Sept Iles archipelago show that the anorthosite is overlain by km-scale cupolas of syenite and granite of two types: the first one contains abundant magmatic mafic enclaves and the second one is free of any enclaves.

Here we present an overview of the present understanding of the origin and evolution of the Sept Iles Intrusive Suite. We describe the different units of the Intrusive Suite with major focus on the layered intrusion. We present the stratigraphic evolution of rocks and mineral compositions in the layered intrusion and use them to describe how the magma chamber was filled and replenished. We also present field relationships and compositional data for rocks of the Sept Iles chilled margin, dykes cutting the Layered Series and silicic rocks from the Upper

Series that we use to determine the liquid line of descent of the layered intrusion and potential implications for other ferrobasaltic intrusions. We then show how the liquid line of descent controls some specific magma chamber processes such as plagioclase flotation, resulting in the formation of anorthosite at the roof of the magma chamber, and the expelling and/or exchange of interstitial melt from the crystal matrix during solidification. We finally describe how a silicate-liquid immiscibility process exerted a control on the Sept Iles liquid line of descent and how it contributed to the formation of large-scale Fe–Ti–P ore deposits.

Geological Setting

Regional Geology

The Sept Iles intrusive Suite (SIIS) is located on the north shore of the St Lawrence River, about 500 km to the north-east of Quebec City, Canada. It is a non-deformed and non-metamorphosed magmatic body emplaced into high-grade gneisses (Higgins 2005) of the allochthonous polycyclic belt of the Grenville geological province (Rivers et al. 1989; Rivers 2008). The age of the SIIS was originally determined from Rb–Sr isochrons on granite and syenite that gave results ranging from 538 ± 17 to 555 ± 2 Ma, with an average of ca. 540 Ma (Higgins and Doig 1977, 1981). The SIIS was re-dated at 564 ± 4 Ma using the U–Pb method on zircons extracted from a granophyre pod in the gabbroic cumulates (Higgins and van Breemen 1998). This age coincides with a large magmatic event in Eastern Laurentia, dominated by volcanic rocks (basalts) and plutons (alkaline plutons, carbonatites and kimberlites; Doig and Barton 1968; Doig 1970; Kamo et al. 1989; Kumarapeli 1993; Higgins and van Breemen 1998). This magmatic event, including the SIIS, was attributed to the initiation of the St Lawrence rift system during the opening of the Iapetus Ocean (Kumarapeli and Saull 1966). Given the large volume of magma represented in the SIIS, it is suggested that it might be related to a mantle plume (Hill et al. 1992; Higgins and van Breemen 1998), although the location of the SIIS on a major lithospheric discontinuity suggests that it may also result from lithospheric re-organization (Namur et al. 2010). In any case, the Sr isotopic composition of bulk-rocks and mineral separates from the SIIS implies a mantle source (Higgins and Doig 1981; Namur et al. 2010).

Higgins (2005) proposed that magmatic activity in the Sept Iles region started with the eruption of an unknown volume of flood basalts and that the SIIS was subsequently emplaced at the unconformity between the lavas and the Grenville basement rocks. No flood basalts are preserved anywhere in the Sept Iles area, but flood basalts commonly erupt during continental rifting, such as the opening of the St Lawrence rift system. Basalts of similar age to the SIIS are present to the southwest of Sept Iles (the Tibbit Hill Formation) and elsewhere in the Appalachians (Higgins and van Breemen 1998).

The Sept Iles Intrusive Suite

The SIIS is dinner-plate shaped, with a diameter of about 80 km and a maximum thickness of 7 km. Only 10 % crops out on the mainland and on the islands of the Sept Iles archipelago (Fig. 11.1). The quality of outcrops is relatively poor on the mainland, with the exception of rare quarries and along a power line crossing the SIIS from west to east. In contrast, outstanding outcrops of anorthosite, syenite and granite occur on the Marconi Peninsula and on the islands of the archipelago. These outcrops were therefore the first to be investigated and the SIIS was first described as an anorthosite massif (Higgins and Doig 1977, 1981) similar to Proterozoic massif-type anorthosites. However, unpublished mapping by T. Ahmedali in the 1960s and T. Feininger in 1985–1990 followed by detailed mapping by Cimon (1998) and geophysical studies by Loncarevic et al. (1990) have revealed that the mainland outcrops are more complicated than previously thought and expose a sequence of massive troctolites and layered gabbros that are overlain by anorthosite, syenite and granite. In light of these new data, the SIIS was then described as a single layered intrusion (Higgins and Doig 1986; Higgins 1991). However, a recent reinvestigation by Higgins (2005) suggests that the igneous activity around Sept Iles was more complex and that the Sept Iles intrusive rocks are better described as an intrusive suite (Sept Iles Intrusive Suite) of three magmatic components: a layered intrusion, the Pointe du Criard composite sill and a series of late-stage gabbros.

The Sept Iles Layered Intrusion

General Overview

The Sept Iles layered intrusion is by far the most voluminous unit of the SIIS although the relative volumes of the different units, including the layered intrusion, are not precisely known. Its size is well defined by a gravity anomaly, some 80 km in diameter, with a maximum Bouguer anomaly of 80 mGal, which is the largest gravity anomaly in eastern North America (Loncarevic et al. 1990). The layered intrusion is also defined by an important magnetic anomaly (Fig. 11.2). There is no evidence that the roof of the intrusion is still preserved. As proposed for other layered intrusions (e.g. the Bushveld Complex; Cheney and Twist 1991) it has been suggested that the initial intrusion was a sill and that subsidence of the floor due to the weight of the cumulate rocks is responsible for its final lopolithic form (Higgins 2005). Interest in the Sept Iles layered intrusion increased at the end of the 1980s because the composition and size of the intrusion suggested that it might host significant Ni, Cu and PGE resources. Three cores (DC-84670, 450 m; DC-84698 (DC-8), 1920 m and DC-84699 (DC-9), 2550 m), currently kept at the Ministère des Ressources Naturelles et de la Faune du Québec, were drilled in

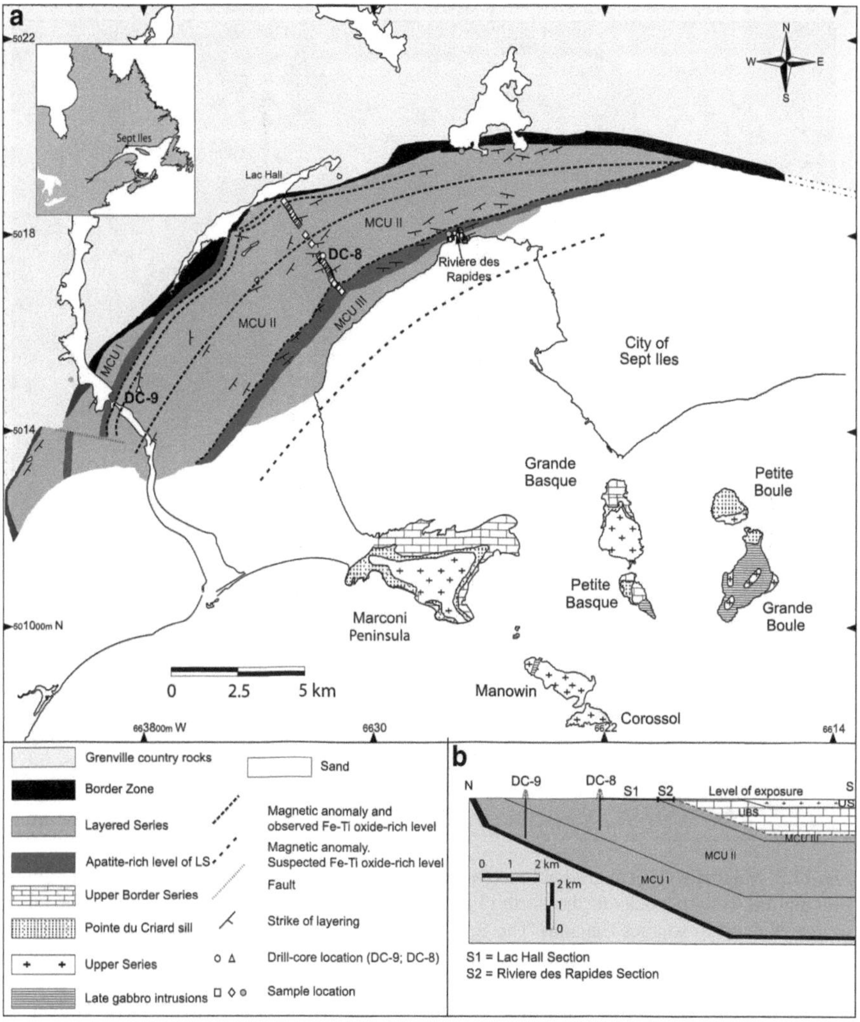

Fig. 11.1 **a** Geological map of the Sept Iles Intrusive Suite adapted from Higgins and Doig (1986) and Higgins (2005). Note the vertical and lateral subdivision of the Layered Series of the layered intrusion into three megacyclic units (*MCUI-III*). The inset map in the *upper left* corner shows the location of the map area in southeastern part of Canada. Locations of surface samples and drill-cores (*DC-8*; *DC-9*) from the Layered Series (Namur et al. 2010) are shown for reference. Detailed stratigraphic relationships between these samples can be seen in Figs 11.1b and 11.5. **b** Schematic cross-section of the Sept Iles layered intrusion. Note the vertical succession of the 3 MCUs of the Layered Series and the Upper Border Series and Upper Series. Note also the stratigraphic intervals intersected by the drill cores and where field samples were collected. *LS* Layered Series, *UBS* Upper Border Series, *US* Upper Series

the 1990s by Inco Inc (see location of DC-8 and DC-9 in Fig. 11.1) but no metallic ores were found. However, a 200 m thick sequence of apatite-bearing gabbro was discovered in DC-84699, while another interval, some 250 m in thickness,

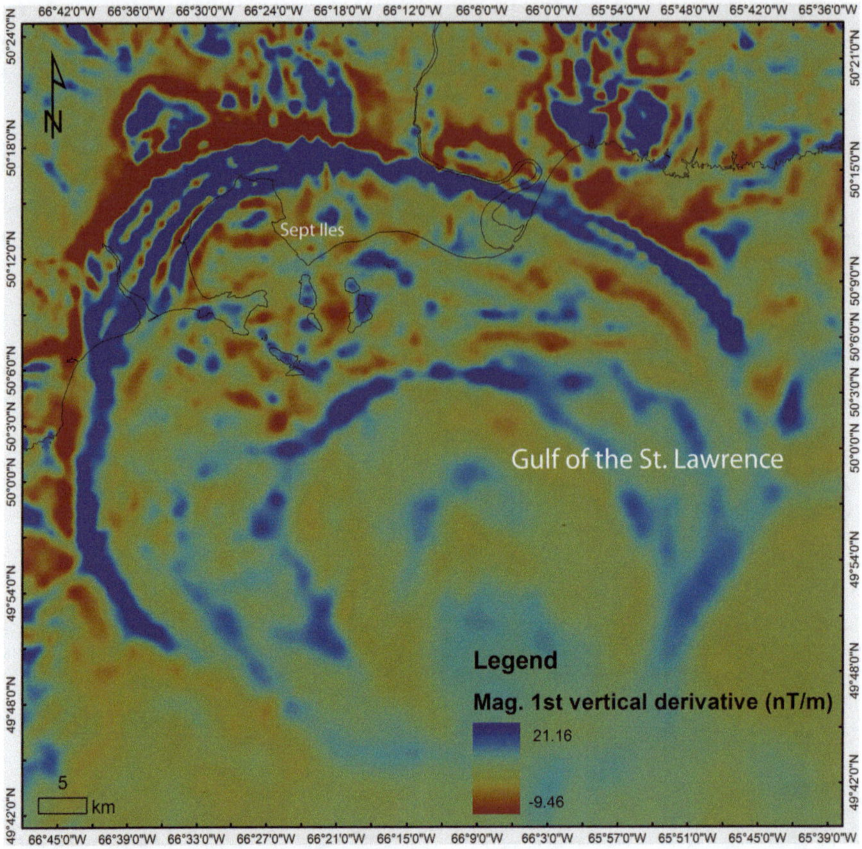

Fig. 11.2 Magnetic vertical gradient map of the Sept Iles area showing the Sept Iles layered intrusion and the country rocks to the north (1st vertical derivative in nT/m, resolution: 1 km) (data source: Natural Resources Canada). The Sept Iles layered intrusion is characterized by relatively high magnetic gradient compared to the country rocks. Curved east-west *blue bands* highlight the largest Fe–Ti oxide layers of the Sept Iles Layered Series. This map also highlights the size and circular shape of the layered intrusion. (Courtesy of Lajeunesse et al. (2013), Laval University, Quebec, Canada)

was known to occur on the mainland. The latter level, generally called the Critical Zone, was investigated at the end of the 1990s by the Soquem Inc. which drilled 30 cores 100–250 m long.

Using samples from the Inco drill-cores (see their stratigraphic distribution below) together with additional samples collected in the field, the Sept Iles layered intrusion was subdivided by Cimon (1998) into four series: a Lower Series dominated by massive troctolite, a Layered Series comprising layered gabbro, a Transitional Series made up of anorthosite and an Upper Series dominated by granite and minor syenite. However, more recent investigations by Higgins (2005) and Namur et al. (2010) along with comparisons with the Skaergaard intrusion have shown that the subdivision proposed by Cimon (1998) is inconsistent with the processes that occurred during crystallization (see below). A simpler subdivision into the gabbro-troctolite Layered Series, the anorthositic Upper Border Series and the granite-syenite Upper Series has therefore been adopted (Fig. 11.1). It should already be noted that in contrast to the Skaergaard intrusion, the Upper Border Series does not correspond to a mirror image of the Layered Series but to a 200–500 m-thick unit dominated by anorthosite. However, the anorthosite may have served as a nucleation site for minor *in situ* crystallization of troctolite and gabbro at the roof of the chamber.

No continuity in drill-cores or in the field is observed between the Layered Series and the Upper Border Series. However, the presence of blocks of anorthosite in the Layered Series which are petrographically and compositionally identical to the anorthosite observed in the Upper Border Series suggests a genetic link between these two series. This is also confirmed by the presence of abundant syenite-granite pods in the anorthosite of the Upper Border Series that resemble the granophyre pockets observed in the most evolved cumulates of the Layered Series (Higgins and Doig 1981; Namur et al. 2010, 2011a). Continuous outcrops on the Marconi Peninsula and several islands show a vertical relationship between the anorthosite of the Upper Border Series and the syenite-granite of the Upper Border Series. These field relationships all together strongly suggest that the Layered Series, Upper Border Series and Upper Series all belong to the same intrusive body and are genetically related by a single process of crystallization in the Sept Iles magma chamber.

To the north, the Sept Iles layered intrusion is surrounded by a thin zone of massive fine-grained dolerite called the Sept Iles Border Zone. Geophysical data are not sufficiently detailed to determine whether it continues beneath the sea or if it only occurs in the north. Based on bulk-rock and mineral compositions, Cimon (1998) interpreted this unit as a separate intrusive body, younger in age than the Sept Iles layered intrusion and having crystallized from a more evolved parental magma. However, by analogy with other layered intrusions, this zone might also correspond to a chilled margin representing rapidly cooled magma against cold country rocks (Wager and Brown 1968; Hoover 1989) or to a marginal reversal made up of fine-grained cumulate rocks becoming increasingly more primitive upwards in the stratigraphy (Egorova and Latypov 2012).

The Sept Iles Border Zone

The exposed part of the Sept Iles layered intrusion is surrounded by a Border Zone (at least 20 m thick) of massive fine-grained dolerite with minor amounts of medium-grained monzonite, representing an abrupt contrast with the coarse-grained gabbros of the Layered Series and the highly deformed host Grenvillian gneisses (Fig. 11.1). The most external part of this zone contains abundant pods and lenses of evolved material, probably resulting from the local partial melting of the country rocks. It also contains abundant cm- to m-scale angular blocks of Grenvillian gneiss

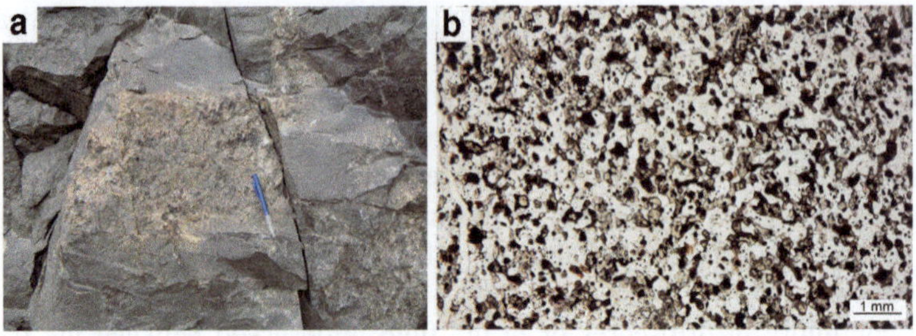

Fig. 11.3 **a** Photograph of fine-grained dolerite of the Sept Iles Border Zone. A dm-scale block of quartzite derived from the Grenvillian country rocks is included into the dolerite. Sept Iles mainland, looking north. **b** Photomicrograph of fine-grained dolerite from the Sept Iles Border Zone, dominated by plagioclase, clinopyroxene and Fe–Ti oxides. Transmitted light

(Fig. 11.3a). Dolerite from the Border Zone has granular texture and an average plagioclase size of 0.1–0.5 mm (Fig. 11.3b). It is dominated by lath-shaped, strongly zoned, locally antiperthitic, plagioclase, anhedral to subhedral clinopyroxene and anhedral Fe–Ti oxide patches dominated by magnetite (Fig. 11.3b). Anhedral to slightly rounded olivine and orthopyroxene are observed in some samples, together with minor needles of apatite (up to 300 μm long), and composite biotite-amphibole rims around clinopyroxene and Fe–Ti oxides. Traces of quartz, alkali-feldspar and zircon are observed in some samples. The monzonite is quite similar, except that it contains more quartz, amphibole and alkali-feldspar and less Fe–Ti oxides than the dolerite.

The Layered Series

Field Relationships, Modal Layering and Ore Deposits

From its contact with the Border Zone to the central part of the Layered Series, four main rock types have been recognized in the field: troctolite (Fig. 11.4a), Fe–Ti oxide-bearing troctolite, gabbro (Fig. 11.4b) and apatite-bearing gabbro. Troctolites and Fe–Ti oxide troctolites are coarse-grained massive rocks, while gabbros are finer-grained and exhibit mineral lamination. Igneous layering is commonly observed in gabbros (Fig. 11.4b), dipping 15–30 towards the centre of the intrusion, and showing a regular alternation of cm- to dm-thick mafic and more felsic layers. Layers commonly exhibit modal grading from olivine, pyroxene, ilmenite or magnetite-rich bases to plagioclase-rich tops. Modally-layered intervals locally with slump to trough structures and minor unconformities are especially evident in Fe–Ti oxide-rich gabbros. Most layers are ca. 10 cm-thick but may range from a few cm to more than 1 m. As described below, some Fe–Ti oxide-rich layers are however locally thicker than 15 m (Fig. 11.4c; Namur et al. 2010). Centimeter- to

Fig. 11.4 Photographs of field relationships in the Layered Series. **a** Coarse-grained troctolite from MCU III, Rivière des Rapides. **b** Centimeter to decimeter thick rhythmic layering in gabbro showing alternating plagioclase-rich leucocratic layers and mafic mineral-rich melanocratic layers. To the *right* there is an autolith of pale anorthosite. Hall lake section. **c** Representative outcrop of a meter-thick massive Fe–Ti oxide layer, Sept Iles mainland. **d** Anorthositic autolith block within gabbros of the Layered Series. Note the deformation of the layering at the base of the autolith, suggesting that the block has foundered into a partially solid crystal pile, Sept Iles mainland

meter-scale blocks of anorthosite are abundant in the gabbro and locally deform the layering of the underlying rocks (Fig. 11.4b, d). They are especially abundant in gabbros from the western side of the intrusion.

Fe–Ti oxide mineralization is an important feature of the Sept Iles layered intrusion and occurs either as tabular or lens-shaped massive bodies. Tabular ore bodies are isomodal layers, 5 cm- to 15 m-thick (Fig. 11.4c), dominated by magnetite and ilmenite, concordant with the general layering, which have sharp lower and diffuse upper contacts.

Two ca. 200 m-thick apatite-rich gabbro units occur in the lower and middle part of the Layered Series. The lower unit is comprised of homogeneous leucocratic gabbro with 2–5 vol.% apatite, while the upper unit shows an alternation of melanocratic layers dominated by Fe–Ti oxides and apatite and plagioclase-rich layers (with very minor Fe–Ti oxides) on a scale of 1–20 m (Charlier et al. 2011; Namur et al. 2012a).

Subdivision of the Layered Series and Lithological Stratigraphy

The presence of two units of apatite-bearing gabbro and the disappearance of cumulus apatite at certain levels suggest that the fractionation trend of the Sept Iles parent magma was temporarily interrupted. This allowed the Layered Series to be subdivided into three megacyclic units (MCUs; Irvine 1982) with repeated characteristic sequences of cumulates (Namur et al. 2010). MCU I (1785 m-thick) and MCU II (2533 m-thick) are topped by apatite-bearing gabbro units (Fig. 11.5). MCU III is at least 396 m-thick, but its upper part is not exposed. The Layered Series is at least 4800 m thick (from −4210 to 549 m, with the 0 m reference level corresponding to the appearance of cumulus apatite in MCU II; Namur et al. 2010).

MCU I is characterized by the following succession of rocks: troctolite (plagioclase, olivine; po-C; notation following Irvine 1982), Fe–Ti oxide-troctolite (+ magnetite and ilmenite; pomi-C), gabbro (+ clinopyroxene; pomic-C), olivine-free gabbro (- olivine; pmic-C) and apatite-bearing olivine gabbro (+ olivine and apatite; pomica-C; Fig. 11.5). These units show contrasting mineral (see below) and bulk-rock compositions (Tables 11.1, 11.2). Troctolite (Fig. 11.6a) consists of plagioclase (ca. 70 vol.%) and olivine (ca. 30 vol.%), with minor clinopyroxene, Fe–Ti oxides and apatite. It has been shown by Namur et al. (2011a) that some plagioclase is missing as compared to theoretical cotectic proportions and that olivine is therefore in excess. Fe–Ti oxide-troctolite (Fig. 11.6b) is identical to troctolite except that the proportion of Fe–Ti oxides is significantly higher (10–15 vol.%). The latter minerals form large anhedral grains (up to 10 mm), frequently polycrystalline and generally dominated by magnetite. Olivine gabbro, gabbro (Fig. 11.6c) and apatite-bearing olivine gabbro (or locally nelsonite; Fig. 11.6d) are medium grained (1–4 mm) with plagioclase and clinopyroxene generally defining a strong lamination.

MCU II consists of a complex succession of Fe–Ti oxide-troctolite (pomi-C), olivine gabbro (pomic-C) and gabbro (pmic-C), topped by a 200 m-thick unit of apatite-bearing olivine gabbro (pomica-C). Olivine is relatively abundant in the lower part of MCU II but tends to decrease upwards. It is very low (less than 10 vol.%) in two intervals (pmic-C) where olivine is interpreted as an intercumulus phase (Fig. 11.5). The mode of clinopyroxene is highly variable in the lowest 1300 m of MCU II, where this mineral alternates between cumulus and interstitial on a scale of ca. 100 m. The amount of clinopyroxene then becomes more uniform and higher (ca. 30 vol.%) towards the top of MCU II.

In MCU III, clinopyroxene and apatite are minor minerals and the rocks are therefore predominantly Fe–Ti oxide-troctolite (pomi-C) and troctolite (po-C). The upper part of MCU III is not exposed (Fig. 11.5).

Cryptic Layering

The base of MCU I displays a 60 m-thick basal reversal in the composition of plagioclase (An_{63-68}; An=[Ca/(Ca+Na]), olivine (Fo_{68-74}; Fo=[Mg/(Mg+Fe^{2+}]) and clinopyroxene ($Mg\#_{76-79}$; Mg#=[Mg/(Mg+Fe^{2+}]), followed by a continuous decrease of

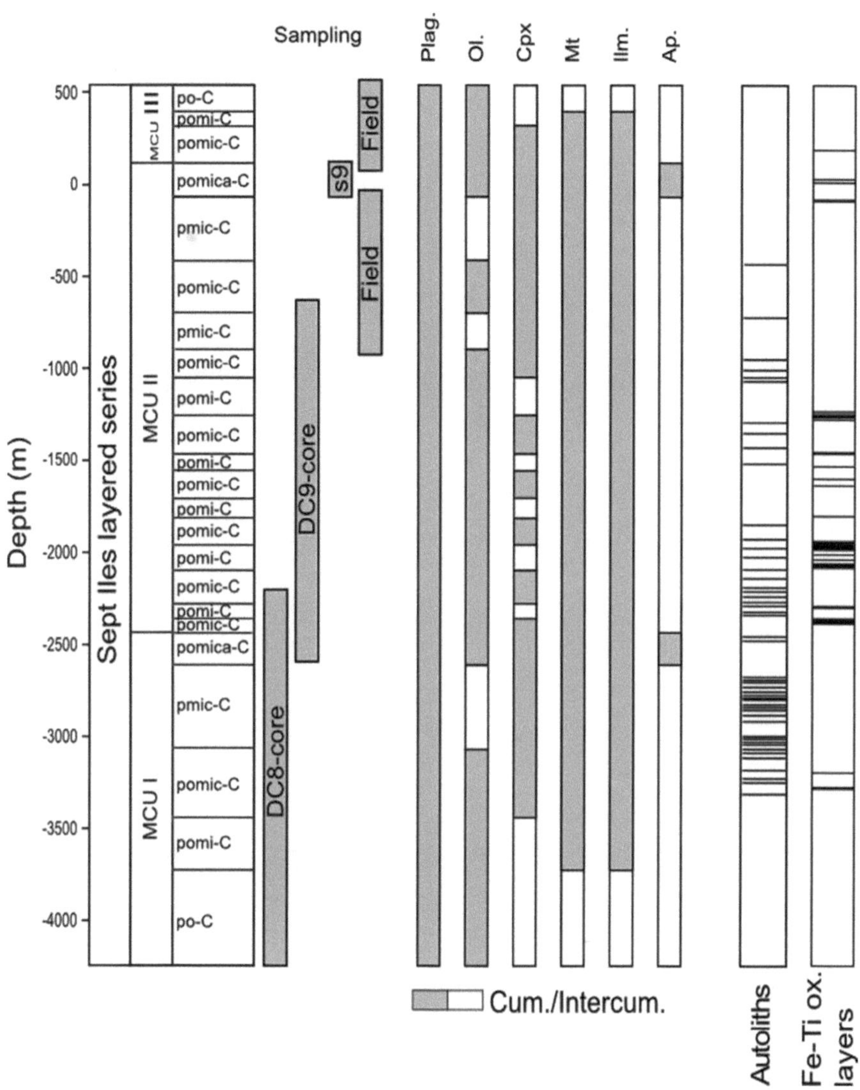

Fig. 11.5 Stratigraphy of the Sept Iles Layered Series as it crops out on the mainland and as intersected by the drill cores. The subdivision into three megacyclic units (*MCUI-III*) is shown. The stratigraphic intervals intersected by the drill cores are shown for reference together with the stratigraphic intervals where field samples were collected. The distribution of cumulus (*grey*) and intercumulus (*white*) assemblages is also shown and was determined based on geochemical constraints and mineral modes (see text for details). Classification of cumulus assemblages following Irvine (1982): *p* plagioclase, *o* olivine, *m* magnetite, *i* ilmenite, *c* clinopyroxene, *a* apatite, *-C* cumulate. The distribution of the main anorthosite blocks (autoliths; see details in the text for their origin) and Fe–Ti oxide layers is also shown. The '0-meter' reference level corresponds to the first appearance of cumulus apatite in MCU II

Table 11.1 Mineral assemblages and mineral compositions in the Sept Iles Layered Series. (Data from Namur et al. (2010); cumulate classification following Irvine (1982))

		Average mineral proportions (wt.%)						Range of silicate mineral compositions			Isotop. Comp.
Unit	Thickness (m)	Plagioclase	Olivine	Magnetite	Ilmenite	Cpx	Apatite	Plag (An%)	Olivine (Fo%)	Cpx (Mg#%)	$(^{87}Sr/^{86}Sr)_{564}$
MCU I											
po-C	522	67	33	–	–	–	–	68–61	73–66	–	0.70372–0.70392
pomi-C	340	53	25	13	9	–	–	61–60	66–66	–	0.70376
pomic-C	326	47	12	12	10	19	–	60–57	66–65	75–73	0.70382
pmic-C	432	47	–	13	12	28	–	57–51		74–71	0.70382
pomica-C	165	47	21	9	6	12	5	51–47	60–55	71–68	0.70393–0.70405
MCU II											
pomi-C	543	56	20	17	7	–	–	62–60	70–68	–	0.70381–0.70389
pomic-C	1116	47	14	13	9	17	–	60–56	68–64	76–73	0.70444
pmic-C	722	46	–	14	11	29		56–52		74–70	0.70396–0.70462
pomica-C	171	43	19	12	6	14	6	52–34	57–21	70–57	0.70493
MCU III											
po-C	132	68	32	–	–	–	–	69–62	71–65	–	0.70360
pomi-C	120	56	27	12	8	–	–	62–60	67–65	–	0.70382–0.70383
pomic-C	144	52	12	14	7	15	–	60–39	65–45	76–66	–

p plagioclase, *o* olivine, *m* magnetite, *i* ilmenite, *c* clinopyroxene, *a* apatite

Table 11.2 Average bulk rock compositions of cumulus units in the Layered Series (MCU I). (Data from Namur et al. (2010))

Unit	po-C	pomi-C	pomic-C	pmic-C	pomica-C
SiO_2 (wt.%)	45.23	40.03	41.08	40.04	39.90
TiO_2	0.58	3.87	4.70	6.75	4.45
Al_2O_3	20.87	18.24	16.05	14.01	14.69
FeO_t	8.26	18.32	17.47	18.95	17.22
MnO	0.11	0.16	0.17	0.20	0.26
MgO	8.71	7.73	7.50	6.44	6.50
CaO	9.75	8.47	10.07	10.97	10.89
Na_2O	2.58	2.57	2.58	2.48	3.01
K_2O	0.23	0.21	0.17	0.16	0.24
P_2O_5	0.06	0.04	0.04	0.05	2.39
LOI	0.15	0.28	0.15	0.04	0.20
Total	99.86	99.90	99.99	100.09	99.74

these parameters to the top of MCU I, where they reach An_{47}, Fo_{55} and cpx-$Mg\#_{68}$ (Fig. 11.7). The major element compositions of magnetite and ilmenite do not show any clear evolution throughout MCU I. In contrast, the Cr-content of magnetite first increases in po-C rocks (from the base of MCU I to −3600 m), where it reaches 10,000 ppm, before dropping quickly below the XRF detection limit (e.g. 5 ppm, Duchesne and Bologne 2009) after the appearance of cumulus Fe–Ti oxides. MCU II starts with a rapid, progressive, upwards increase in the An-content of plagioclase (An_{47} to An_{64}), Fo-content of olivine (Fo_{55} to Fo_{72}) and Mg# of clinopyroxene ($Mg\#_{68}$ to $Mg\#_{78}$). Mineral compositions (ca. An_{62}; Fo_{70}; cpx-$Mg\#_{75}$) are then relatively constant to − 590 m. The upper part of MCU II then shows a continuous evolution of minerals down to An_{34}, Fo_{21} and cpx-$Mg\#_{55}$ at the top of the megacyclic unit. The magnetite Cr-content drops rapidly below the XRF detection limit in this upper section. In MCU III, the compositions of silicate minerals show a significant, progressive regression to An_{70}, Fo_{75} and cpx-$Mg\#_{78}$. The Cr-content of magnetite also increases to 23,000 ppm at the top of the exposed MCU III.

Sr-isotope ratios $(^{87}Sr/^{86}Sr)_{564}$ in plagioclase separates from the Sept Iles Layered Series vary between 0.70360 and 0.70498. This ratio increases upwards in each MCU and its evolution is strongly anti-correlated with the evolution of plagioclase-An, olivine-Fo and clinopyroxene-Mg#. The highest $(^{87}Sr/^{86}Sr)_{564}$ values are observed at the top of MCU I (0.70406) and MCU II (0.70498; Namur et al. 2010).

Anorthosite Blocks

Abundant anorthosite blocks cm- to tens of meters across are enclosed within cumulates of the Layered Series, and have been interpreted to be autoliths (Higgins 2005; Namur et al. 2011a) derived from the Upper Border Series. They are

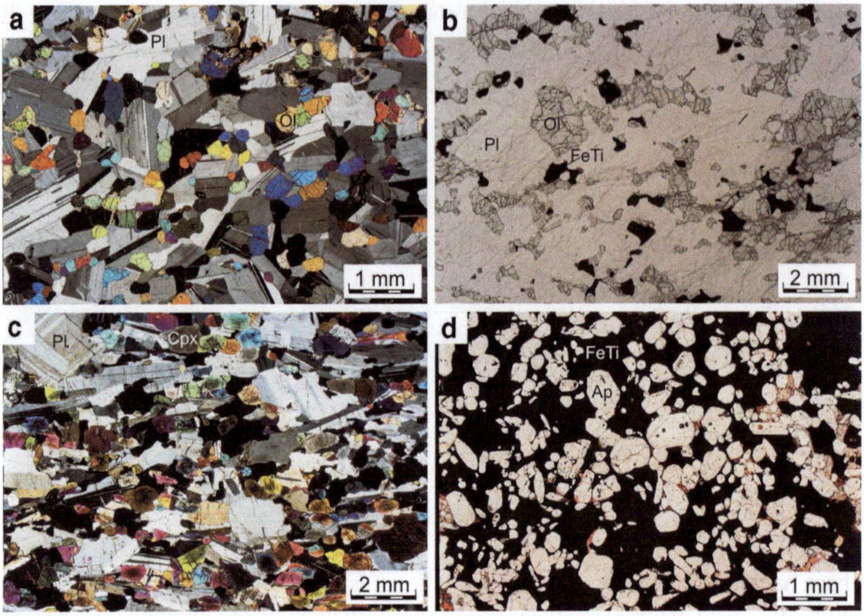

Fig. 11.6 Photomicrographs showing mineral textures and mineral assemblages in the Sept Iles Layered Series. **a** Troctolite (po-C) with randomly oriented, tabular plagioclase and millimeter-scale rounded olivine. DC9-2267.5; MCU I; cross-polarized transmitted light. This corresponds to the most primitive cumulus assemblage observed in the Sept Iles Layered Series. **b** Massive Fe–Ti oxide troctolite (pomi-C) containing large rounded grains of olivine, plagioclase and millimeter-scale anhedral patches of Fe–Ti oxides. DC 8-1309; MCU II; transmitted light. This photograph illustrates the appearance of large grains of cumulus Fe–Ti oxides (magnetite and ilmenite). **c** Gabbro (pmic-C) with long tabular plagioclase, prismatic clinopyroxene and millimeter-scale patches of Fe–Ti oxides. Sample ON05-35; MCU II; cross-polarized light. This photograph illustrates the appearance of cumulus clinopyroxene. Note the strong mineral lamination (especially for plagioclase and clinopyroxene), which is very common after the appearance of Fe–Ti oxides and clinopyroxene as cumulus phases. **d** Nelsonite sample from the apatite-bearing olivine gabbro unit (pomica-C) of MCU II. Note the presence of large crystallized melt inclusions (black spots) in apatite grains. Sample s9-218.8; MCU II; transmitted light. Other samples of apatite-bearing gabbro are more leucocratic than sample s9-218.8 and are dominated by plagioclase, olivine and clinopyroxene, with minor Fe–Ti oxides and apatite. *Pl* plagioclase; *Ol* olivine; *FeTi* Fe–Ti oxides; *Cpx* clinopyroxene; *Ap* apatite

mostly angular slabs and have sharp contacts with cumulates. Plagioclase ranges in colour from dark to light grey or white. Two types of autoliths are recognized: blocks of massive anorthosite (by far the most abundant) and blocks of laminated anorthosite. These blocks occur from the middle part of MCU I to the lower part of MCU III and are enveloped by gabbro (pomic-C; pmic-C; pomica-C) and Fe–Ti oxide troctolite (pomi-C). No autoliths have been observed in troctolite (po-C) in the lower part of MCU I and the upper part of MCU III (Fig. 11.5). Autoliths consist predominantly of plagioclase (84–96 vol.%), with minor olivine (0–6 vol.%), clinopyroxene (1–9 vol.%) and Fe–Ti oxides (0–4 vol.%). Low temperature alteration products are present in some of them.

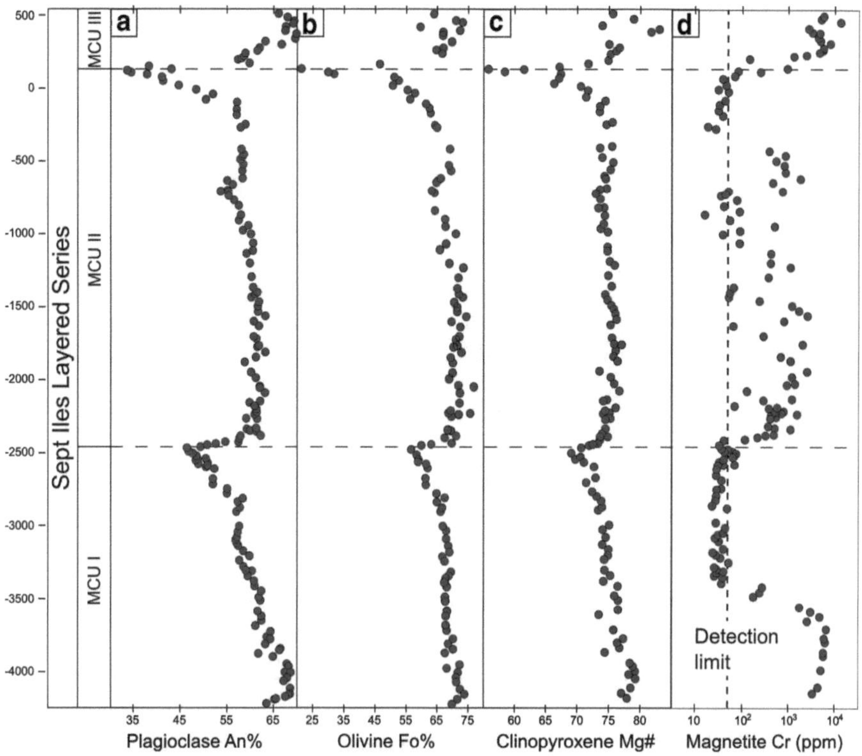

Fig. 11.7 Major- and trace-element variations in mineral compositions with stratigraphic position in the Layered Series. **a** Plagioclase (An%), **b** Olivine (Fo%), **c** Clinopyroxene (Mg#), **d** Magnetite Cr (ppm, log scale). *Dashed lines* represent boundaries between MCU. Note the evolution to lower An, Fo and cpx-Mg# in each MCU and the progressive reversals to more primitive compositions at the bottom of MCU II and MCU III. Also note the 60 m-thick marginal reversal at the base of MCU I. Data from Namur et al. (2010, 2012a). See Fig. 11.5. for details on where the samples were collected

Plagioclase shows a very restricted range of core compositions ($An_{60\text{-}68}$), with rims generally extending down to $An_{60\pm3}$. Olivine and clinopyroxene also show very restricted ranges of composition from Fo_{60} to Fo_{66} and cpx-$Mg\#_{68}$ to cpx-$Mg\#_{75}$, respectively. Sr-isotope ratios $(^{87}Sr/^{86}Sr)_{564}$ in plagioclase separates range from 0.70369 to 0.70386 and do not show any obvious trend when plotted against stratigraphic height in the Layered Series (Namur et al. 2011a).

The Upper Border Series

The Upper Border Series is composed of anorthosite, with minor leucotroctolite and leucogabbro (Higgins 1991). It is best exposed on the islands of the Sept Iles archipelago and the Marconi Peninsula. Two types of anorthosite with

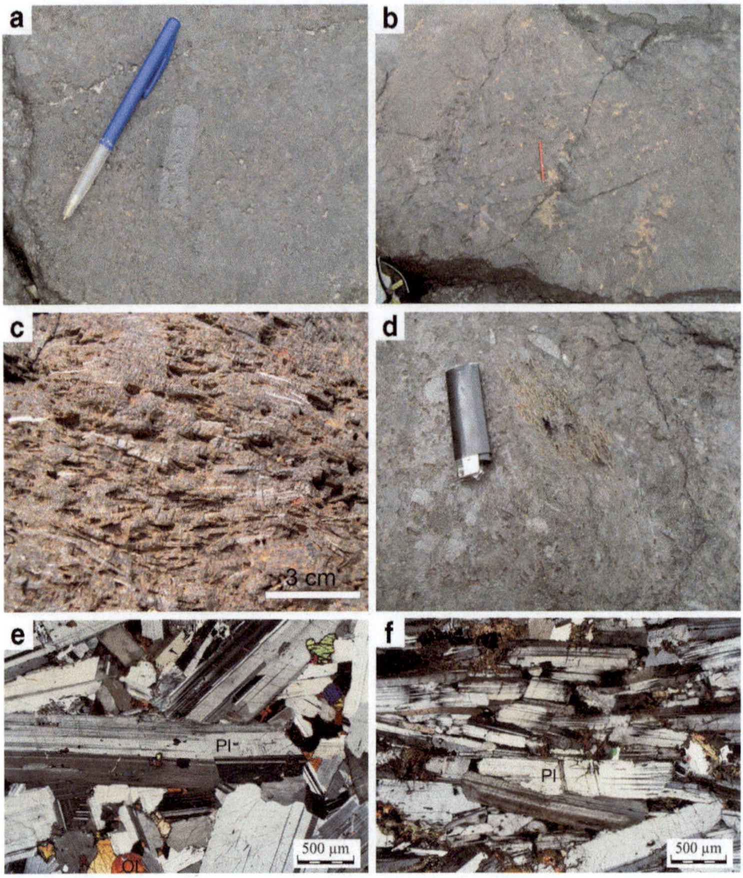

Fig. 11.8 Photographs (**a–d**) showing field relationships in the Upper Border Series and microphotographs (**e–f**) of representative samples of anorthosite. **a** Massive anorthosite of the Upper Border Series displaying a 5 cm-long plagioclase crystal set in a matrix of smaller plagioclase grains. **b** Massive anorthosite from the Upper Border Series showing interstitial syenitic to granitic material between plagioclase crystals. **c–d** Laminated anorthosite from the Upper Border Series. Altered clinopyroxene oikocrysts have been weathered out. **e** Massive anorthosite showing tabular, mm-scale, plagioclase grains and interstitial poikilitic olivine. M-07-55; cross-polarized transmitted light. **f** Strongly laminated anorthosite with large plagioclase grains, minor Fe–Ti oxide minerals and biotite in replacement of primary ferromagnesian minerals. GB-X1; cross-polarized transmitted light. *Pl* plagioclase; *Ol* olivine

very complex distribution are recognized (Higgins 1991): massive anorthosite (Figs 11.8a, b), which constitutes over 90 vol.% of the Upper Border Series, and laminated anorthosite (Figs 11.8c, d) which generally forms layers a few cm to tens of m-thick within the massive anorthosite. Where the layers can be traced, they generally terminate abruptly in the massive anorthosite within 1–20 m. The transition from massive to laminated anorthosite generally takes place over 1–5 cm. This lamination probably formed as a result of a simple shear due to strong convection currents at the upper border of the magma chamber (Higgins 1991). Convection of the anorthositic mush may have also taken place and could explain how blocks of laminated anorthosite were transported to other parts of the magma chamber and enclosed within the massive anorthosite. Within the laminated anorthosite, the orientation of the lamination is generally not parallel to the overall structure of the intrusion and there are areas where the lamination diverges and wraps around pods of massive anorthosite up to 1 m long (Higgins 1991). In some areas, the laminated anorthosite is only present as angular to rounded xenoliths, 10 cm to 1 m long, enclosed in the massive anorthosite. Massive and laminated anorthosites both locally contain interstitial syenitic to granitic material (Fig. 11.8b), forming cm- to dm-scale pods, resulting from the percolation of silicic melt through the partly solidified anorthosite mush (Higgins 2005; Namur et al. 2011a, b).

Massive anorthosites are comprised of randomly oriented subhedral to euhedral plagioclase, with minor olivine, poikilitc clinopyroxene (up to 10 cm across), Fe–Ti oxides and very minor apatite, quartz, alkali-feldspar, amphibole, sulfides and minerals resulting from low-temperature hydrothermal alteration (Fig. 11.8e). Plagioclase generally forms 0.1–40 mm long grains, but some granophyric segregations contain plagioclase grains up to 1 m across. The large size of the crystals probably results from crystal size maturation in a crystal mush with an initial high liquid fraction (Higgins and Chandrasekharam 2007; Higgins 2011). These segregations are more common in the massive anorthosite than in the laminated anorthosite (Higgins 1991). Laminated anorthosites show a similar mineralogy but plagioclase grains display a very strong lamination (Fig. 11.8f).

Plagioclase, olivine and clinopyroxene show very restricted ranges of composition from An_{62} to An_{68}, Fo_{59} to Fo_{65} and cpx-$Mg\#_{66}$ to cpx-$Mg\#_{75}$, respectively. Bulk-rock compositions are high in SiO_2, Al_2O_3, CaO and Na_2O, reflecting the plagioclase-rich nature of these rocks (Table 11.3). $(^{87}Sr/^{86}Sr)_{564}$ in plagioclase separates range from 0.70356 to 0.70379 (Namur et al. 2011a).

The Upper Series

The Sept Iles Upper Series is comprised of syenite, granite and minor monzonite, cropping out on Marconi peninsula and the islands. There are several components that have complex field relationships and internal structures, some of which are revealed by the distribution of magmatic mafic enclaves (MME; Barbarin 2005). The two main components are: silicic rocks free of MME and MME-bearing silicic rocks (Higgins 2005; Hounsell 2006; Fig. 11.9a, b). Some MME contain plagioclase megacrysts and petrographically resemble the dolerite dykes in the Layered and Upper Border Series and Late Gabbro intrusions. The MME are generally ovoid, 1–10 cm long. MME locally occur as aggregates, forming m-scale mafic masses, or dyke swarms, separated by silicic rocks. There is a gradation from isolated blob-like MME to homogeneous mafic dykes through disjointed

Table 11.3 Bulk rock compositions of representative samples from the Upper Border Series. (Data from Namur et al. (2011a))

Sample	M-07-29	M-07-24	M-07-23	M-07-22	M-07-17	M-07-14	GB-05-04	GB-X1
Type	Massive	Massive	Massive	Massive	Massive	Massive	Massive	Laminated
SiO_2 (wt.%)	51.03	49.80	50.22	49.89	51.57	51.66	50.27	50.23
TiO_2	1.61	1.13	0.91	1.51	1.24	0.90	0.37	1.33
Al_2O_3	26.54	23.65	24.72	23.30	23.65	23.71	26.53	23.54
FeO_t	1.71	4.67	3.96	4.52	3.36	4.30	1.83	5.09
MnO	0.15	0.11	0.09	0.12	0.14	0.07	0.07	0.09
MgO	0.23	0.83	2.08	2.03	2.60	2.14	0.79	2.27
CaO	11.57	12.82	11.05	12.25	11.61	11.08	12.35	12.11
Na_2O	3.12	3.38	3.65	3.53	2.46	3.12	3.79	3.01
K_2O	0.59	0.51	0.93	0.82	1.01	1.26	0.24	0.57
P_2O_5	0.13	0.12	0.25	0.18	0.14	0.16	0.05	0.13
LOI	1.89	0.99	2.62	1.88	2.04	1.68	0.53	2.35
Total	98.57	98.01	100.48	100.03	99.82	100.08	96.82	100.72

dykes made up of MME aggregates. MME-free granite generally forms dm- to m-scale 'pillows' enclosed in MME-rich material (Fig. 11.9c). The MME-rich units occur between the pillows. In vertical section of the Upper Series, the amount of pillows drastically decreases from base to top (Fig. 11.9d). These field relationships suggest that the MME-rich magma formed first, but was not fully solidified when the MME-free granite was intruded (Higgins 2005).

Silicic rocks of the Upper Series are dominated by alkali-feldspar, plagioclase, quartz and amphibole, with minor clinopyroxene, Fe–Ti oxides, olivine, biotite, apatite, zircon, titanite, chlorite and fluorite (Higgins and Doig 1981; Hounsell 2006; Namur et al. 2011b). Alkali-feldspar ($Ab_{31}An_1Or_{68}$-$Ab_8An_0Or_{92}$) forms anhedral to subhedral tabular crystals with inclusions of plagioclase, quartz and amphibole. Plagioclase ($An_{25\text{-}18}$) forms small, antiperthitic, zoned, anhedral to tabular grains. Quartz forms 0.2–3 mm rounded grains, while edenitic amphibole ($Mg\#_{54\text{-}13}$) is represented by anhedral to subhedral dark-green to brown grains. Clinopyroxene (from Mg#51; $Ens_{25}Fs_{24}Wo_{51}$ to $Mg\#_{29}$; $Ens_{18}Fs_{34}Wo_{48}$) is only found as cores within amphibole grains. Minor subhedral to euhedral reddish biotite, frequently associated with fluorite, is observed in some samples. Fe–Ti oxides (magnetite: $Usp_{17\text{-}3}$; Ilmenite: $Hem_{1\text{-}0}$) occur as small (<2 mm) anhedral patches dominated by magnetite. Zircon and apatite locally form small euhedral grains.

Silicic rocks in the Upper Series have geochemical affinities with A-type granitoids (Loiselle and Wones 1979; Clemens et al. 1986; Eby 1990; Bonin 2007; Table 11.4) and are best described as ferroan alkalic to alkali-calcic granitoids (Frost et al. 2001; see below). They have high REE contents and display highly fractionated REE patterns with large negative Eu anomalies (Eu/Eu*: 0.22–0.86). Sr-isotope

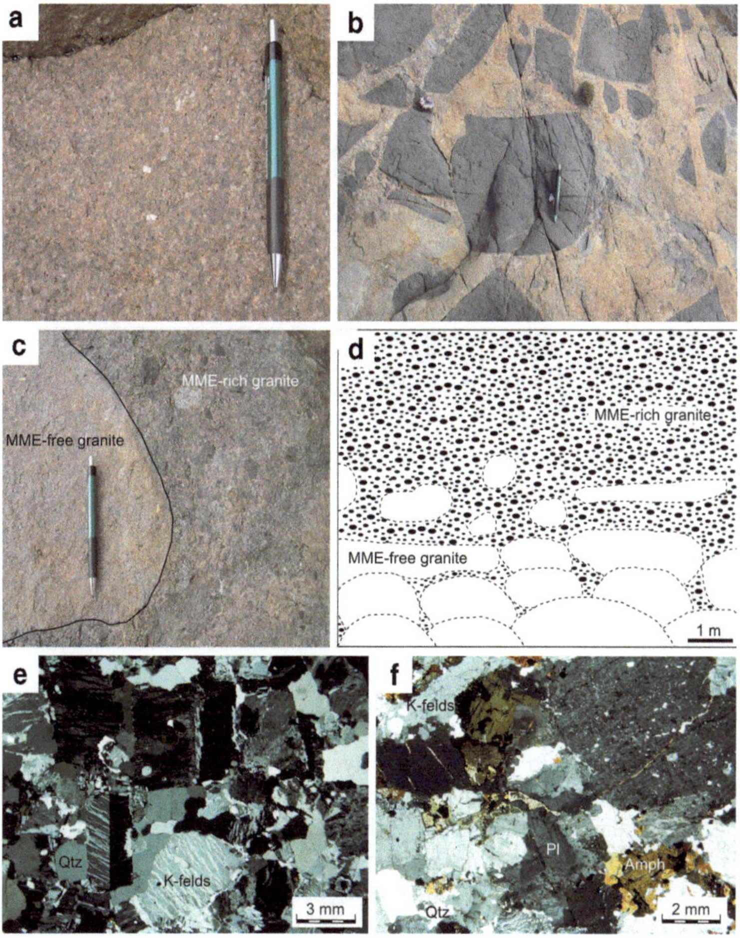

Fig. 11.9 Photographs (**a–c**) and sketch (**d**) showing field relationships in the Upper Series and microphotographs (**e–f**) of representative samples of granite. **a** Representative outcrop of *pink* magmatic mafic enclaves (*MME*)-free granite, Grande Basque Island. **b** Outcrop of *pink* granite containing abundant MME. The MME are mainly fine- to medium-grained dolerite similar to the dykes cutting the Upper Series, Marconi Peninsula. **c** Detail of the sharp contact between MME-rich granite facies and a 50 cm wide pillow of MME-free granite, Corossol Island. **d** Schematic relationships of MME-rich granite and MME-free granite as observed on Corossol Island (after Higgins 2005). Note that MME-free granite forms m-scale pillows surrounded by MME-rich material. MME-free granite pillows are abundant at the bottom of the section and their abundance decreases upwards. **e** Coarse-grained granite showing large alkali-feldspar and quartz grains. 03–05; cross-polarized transmitted light. **f** Coarse-grained granite with alkali-feldspar, plagioclase, quartz and amphibole grains. 03–42; cross-polarized transmitted light. *MME* magmatic mafic enclave; *K-felds* alkali-feldspar; *Pl* plagioclase; *Qtz* quartz; *Amph* amphibole

ratios have been measured in nine samples (from monzonite to granite) and were found to be slightly higher in MME-free rocks (($^{87}Sr/^{86}Sr)_{564}$: 0.70412–0.70548) than in MME-bearing rocks (($^{87}Sr/^{86}Sr)_{564}$: 0.70362–0.70369; Namur et al. 2011b).

Table 11.4 Bulk rock compositions of representative samples from the Upper Series and Pointe du Criard Sill. (Data from Namur et al. (2011b))

Sample	03-42	03-49	03-43	03-44	03-45	07-60	07-61
Unit	Monz.	Monz.	Syenite	Granite	Granite	Syenite	Granite
Type	US	US	US	US	US	PC	PC
SiO_2 (wt.%)	61.09	56.94	63.39	74.71	71.13	67.97	70.15
TiO_2	1.27	2.89	1.20	0.34	0.41	0.43	0.35
Al_2O_3	14.49	13.56	13.31	11.95	12.63	13.13	12.98
FeO_t	9.52	10.25	8.52	3.42	5.26	4.80	2.46
MnO	0.23	0.13	0.19	0.08	0.10	0.06	0.04
MgO	0.96	2.67	0.80	0.20	0.17	0.54	0.27
CaO	3.63	5.34	3.01	0.63	0.64	1.17	1.61
Na_2O	4.61	4.33	4.87	3.99	4.53	4.61	4.65
K_2O	3.40	2.24	3.69	4.68	4.32	4.87	4.92
P_2O_5	0.30	0.52	0.31	0.07	0.09	0.10	0.08
LOI	0.04	1.10	0.38	0.05	0.32	1.10	1.22
Total	99.51	99.94	99.64	100.09	99.57	98.75	98.70

US Upper Series, *PC* Pointe du Criard

Other Intrusive Components

Many dykes are found to crosscut the Layered Series and the Upper Border Series. They can be subdivided into two types: dolerite dykes and intermediate to silicic dykes. Dolerite dykes (0.1–1 m thick) are very fine-grained (0.1–0.5 mm) dominated by plagioclase, clinopyroxene and Fe–Ti oxides. They generally cut the layering at a high angle and dip steeply to the south. Intermediate and silicic dykes (and pods) are fine- to medium-grained (1–5 mm) and are generally found to cut apatite-bearing gabbros. They are dominated by subhedral, antiperthitic plagioclase, clinopyroxene, amphibole (primary grains or replacing clinopyroxene) and Fe–Ti oxides. Alkali-feldspar is a major phase in some samples. Minor quartz, biotite, apatite and zircon are observed in most samples. Dolerite and monzonite dykes in the Layered Series have bulk-rock compositions similar to the dolerite and monzonite in the Border Zone, with $(^{87}Sr/^{86}Sr)_{564}$ ratios ranging from 0.70353 to 0.70363.

The Pointe du Criard sill (Higgins 1990) is up to 50 m-thick and crops out over an area of 10 by 23 km, including all the islands and the Marconi peninsula. It was emplaced into the Upper Border Series and Upper Series of the Sept Iles layered intrusion. It is composed of three magmatic components: dolerite, leucogabbro and silicic rocks (Fig. 11.10a). The margins of the sill are made up of dolerite, which has fine-grained contacts with the host (anorthosite or granite). At the base of the sill, the dolerite unit is 0.1–6 m thick and locally contains plagioclase megacrysts that increase in proportion upwards. The dolerite grades into leucogabbro

Fig. 11.10 **a** Schematic sketch of the Pointe du Criard sill displaying the succession of the three magmatic components: dolerite, leucogabbro and felsic rocks (after Higgins 2005). Enclaves of dolerite and leucogabbro are abundantly found in the lower and upper part of the felsic component. **b** Typical outcrop of the Pointe du Criard sill showing the succession of two different magmatic components: leucogabbro and silicic rocks. It should be noted that the external part of the silicic rocks contains abundant MME and mafic xenoliths. Size and abundance of MME and xenoliths decrease progressively toward the central part of the sill. The first dolerite component of the Pointe du Criard sill is not observed on this outcrop. **c** Close-up of the leucogabbro component of the Pointe du Criard sill. It is made up of fine-grained dolerite containing abundant centimeter-scale plagioclase phenocrysts. Plagioclase crystals larger than 20 cm have been observed in some outcrops. **d** Photomicrograph of porphyritic granite from the Pointe du Criard sill showing large alkali-feldspar grains in a fine-grained matrix comprised of alkali-feldspar, quartz, plagioclase and amphibole. Sample 03–08; cross-polarized transmitted light. **e** Representative composite dyke of dolerite-granite as observed in the Upper Series. Marconi Peninsula. *K-felds* alkali-feldspar

over a few cm to dm. The latter is 0.3–10 m thick and is mostly composed of large plagioclase crystals (Figs 11.10b, c). The silicic rocks constitute the bulk of the sill. At the base, leucogabbro grades into the silicic rocks over a few cm. The lower 1–3 m of the silicic rocks contains abundant enclaves of dolerite and leucogabbro. Some enclaves are highly elongated, up to 2 m long and only a few cm wide. The dolerite is dominated by plagioclase, clinopyroxene and Fe–Ti oxides, with minor olivine and orthopyroxene. The most plagioclase-rich parts of the leucogabbro unit are almost indistinguishable from the anorthosite of the Upper Series. The silicic rocks of the sill (Table 11.4) are readily distinguishable from those of the Upper Series, by a distinctive porphyritic texture and a low content of quartz. Large, zoned, and rounded crystals (up to 3 cm; Fig. 11.10d) of alkali-feldspar ($Ab_{15}An_0Or_{85}$-$Ab_9An_0Or_{91}$) float in a fine-grained, equigranular matrix (<1 mm) dominated by perthitic alkali-feldspar, plagioclase (An_{1-5}), amphibole ($Mg\#_{18-22}$), Fe–Ti oxides (Usp_{3-14} and hem_1), ± quartz (Namur et al. 2011b). Relict cores of clinopyroxene ($Mg\#_{30-29}$) are locally observed in the central part of amphibole grains. Composite dolerite-granite dykes (Fig. 11.10e) may be part of the same event.

The Late Gabbro intrusions crop out on the southern part of Petite Basque Island and form the bulk of Grosse Boule Island (Higgins and Doig 1981). They consist of generally fine- to medium-grained gabbros, commonly with sub-horizontal layering. Angular blocks of anorthosite are locally present. They contain black plagioclase crystals up to a few centimeters long and resemble the anorthosite of the Upper Border Series. The petrography of these gabbros is little studied but they appear to be olivine gabbros. It is not clear if the abundant mafic dykes and locally composite dykes that cut the Upper Border Series and Upper Series are part of the same magmatic event.

Discussion

Composition and Origin of the Sept Iles Parental Magma

Petrography of dolerite samples from the Border Zone suggest that this unit might represent liquids chilled against cool country rocks during magma emplacement (Wager and Brown 1968; Hoover 1989; Namur et al. 2010). However, other examples of fine-grained rocks at the contact between country-rocks and coarse-grained gabbros have been interpreted as part of a marginal reversal made up of fine-grained cumulates (Egorova and Latypov 2012). In Sept Iles, the poor quality of outcrops and the absence of drill-cores in this unit makes impossible to evaluate the vertical and lateral mineralogical and geochemical evolution of this unit. However, Namur et al. (2010, 2011b) have shown that the most fine-grained dolerite samples (Fig. 11.3) have a ferrobasaltic composition (SiO_2: 45.9–52.4 wt.%; MgO: 6.29–3.61 wt.%; FeO_t: 13.2–17.1; Table 11.5) which is significantly different to

Table 11.5 Parent magma compositions of the Sept Iles Layered Series. (Data from Namur et al. (2010, 2011b)

Sample	05-45	07-01	07-48	07-180	Average
Location	Border zone	Dyke	Border zone	Border zone	
SiO_2 (wt.%)	48.56	48.51	48.65	48.53	48.56
TiO_2	2.91	2.88	2.78	2.82	2.85
Al_2O_3	14.62	14.13	14.59	14.67	14.50
FeO_t	15.10	14.82	15.06	14.50	14.87
MnO	0.29	0.23	0.21	0.21	0.24
MgO	5.53	5.56	5.14	5.65	5.47
CaO	9.97	9.61	9.47	9.79	9.71
Na_2O	2.90	2.76	2.81	2.63	2.78
K_2O	0.83	0.60	0.90	0.77	0.78
P_2O_5	0.36	0.72	0.36	0.82	0.57
LOI	0.00	0.23	0.23	0.03	–
Total	101.07	100.05	100.20	100.42	100.33
Rb (ppm)	8.97	16.2	18.1	8.03	12.8
Sr	472	531	517	523	511
Ba	458	509	421	515	476
Zr	90.8	126	264	126	152
Hf	2.81	3.52	–	–	3.17
Th	0.741	0.957	–	–	0.849
La	32.4	42.2	–	–	37.3
Ce	72.1	86.3	–	–	79.2
Pr	10.5	11.8	–	–	11.2
Nd	42.4	52.2	–	–	47.3
Sm	9.31	10.5	–	–	9.91
Eu	2.92	3.10	–	–	3.01
Gd	9.61	11.4	–	–	10.5
Tb	1.42	–	–	–	1.42
Dy	8.49	10.7	–	–	9.60
Ho	1.82	2.12	–	–	1.97
Er	4.96	6.00	–	–	5.48
Tm	0.699	–	–	–	0.699
Yb	5.58	5.59	–	–	5.59
Lu	0.581	0.876	–	–	0.729

the cumulates of the Layered Series. Moreover, most samples have slightly fractionated REE with a weak Eu anomaly (Eu/Eu*: 0.77–1.34) strongly suggesting that these samples have the composition of liquids having been chilled against the country rocks and that they do not represent cumulate rocks. Thermodynamic calculations of phase equilibria with the MELTS (Ghiorso and Sack 1995) and other algorithms (Toplis 2005) have furthermore shown that the most primitive dolerite (highest Mg#; Table 11.5) is an adequate parent to the most primitive cumulates of the Layered Series (MCU I: An_{68}, Fo_{72}; MCU III: An_{72}, Fo_{71}). The Sept Iles parent magma is therefore inferred to be a ferrobasaltic liquid, relatively low in SiO_2 (ca. 48 wt.%), high in FeO_t (ca. 15 wt.%) and TiO_2 (3 wt.%), and moderately enriched in trace elements compared to primary melts of the primitive mantle. The concentrations of FeO_t and TiO_2 are significantly higher than in other ferrobasaltic layered intrusions such as Skaergaard (Hoover 1989; Toplis and Carroll 1995) and the Main and Upper Zones of the Bushveld Complex (Tegner et al. 2006) which, as shown below, has important implications for the Sept Iles sequence of crystallization. Typical primitive mantle-derived basaltic magmas have FeO_t contents between 7 and 10 wt.% and TiO_2 contents between 0 and 1 wt.% (Falloon et al. 1999; Danyushevsky et al. 2003); however, fractional crystallization of silicate phases can result in FeO_t and TiO_2 enrichment of residual liquids (Juster et al. 1989; Snyder et al. 1993; Toplis and Carroll 1995). The Sept Iles magma most probably represents an evolved basalt resulting from a previous process of fractionation of a more primitive magma in a deep-seated magma chamber. High FeO_t and TiO_2 basalts have also been observed in the Galapagos Spreading Centre and were interpreted by Juster et al. (1989) as resulting from 50 % fractionation of a typical mid-ocean ridge basalt (MORB) composition.

Crystallization of the Layered Series

Sequence of Crystallization

Detailed determination of mineral modes (see summary in Table 11.1) together with petrographical relationships in the Sept Iles Layered Series and detailed study of plagioclase-plagioclase-clinopyroxene dihedral angles indicate the following sequence of cumulus mineral appearance: plagioclase and olivine, followed by magnetite and ilmenite, then clinopyroxene and finally apatite (Fig. 11.5; Namur et al. 2010; Holness and Vernon 2015). Olivine disappears temporarily, probably as a result of high SiO_2 activity in residual melts (Morse 1990). The relative timing of Fe–Ti oxides, clinopyroxene and apatite appearance is supported by changes in the evolution of trace element concentrations in minerals and in bulk-rock compositions; (1) the evolution of Cr in magnetite changes from a trend of increasing content when Fe–Ti oxides are interstitial phases (low mode; po-C rocks) to a trend of decreasing content when they become liquidus phases (high mode; pomi-C rocks; Fig. 11.11a). We interpret this as the result of fractionation of olivine and plagioclase (po-C cumulates), in which Cr is incompatible, from the main magma body. As a consequence, the Cr content of the melt continuously increases, as does the melt trapped in the crystal mush. The interstitial magnetite crystallizes earlier and from a melt progressively enriched in Cr as we go upwards in stratigraphy,

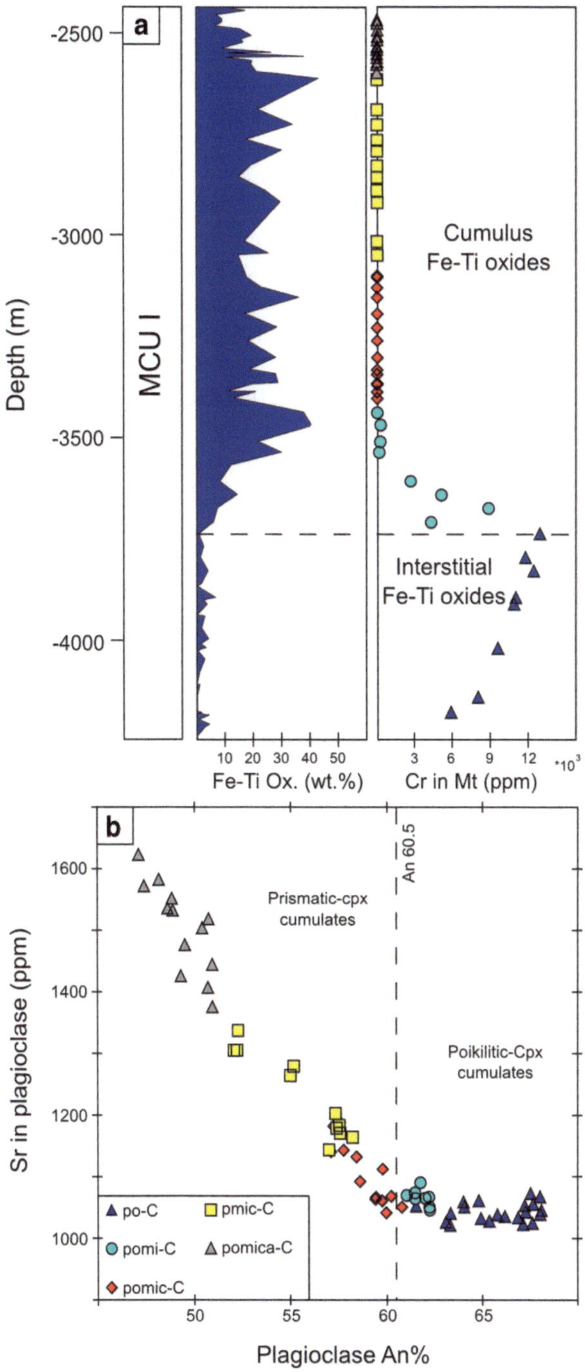

Fig. 11.11 **a** Stratigraphic variation of Fe–Ti (magnetite + ilmenite) oxide mode and Cr content in magnetite in MCU I. *Dashed line* represents the onset of cumulus Fe–Ti oxide crystallization which is marked by a change in the evolution of magnetite Cr-content. **b** Variation of An (%) vs Sr (ppm) in plagioclase separates from MCU I. Note the change in the slope at $An_{60.5}$, when clinopyroxene becomes prismatic and cumulus. *Ox* oxides; *Mt* magnetite; *Cpx* clinopyroxene

explaining the trend of increasing Cr in magnetite in po-C rocks. With the appearance of cumulus magnetite (and ilmenite; pomi-C rocks), Cr becomes highly compatible in the crystal matrix and the Cr-content of the liquid drops, explaining the upwards Cr-decrease observed in magnetite from the pomi-C unit; (2) the Sr-content in plagioclase starts to significantly increase at the appearance of cumulus clinopyroxene. This is because the saturation of this new phase changes the behaviour of Sr from a moderately compatible ($Kd^{Bulk} \sim 1$) element to an incompatible element ($Kd^{Bul} \sim 0.5$) in the cumulus matrix (Fig. 11.11b). The appearance of cumulus clinopyroxene is furthermore marked by a morphological change from poikilitic crystals to tabular crystals; (3) the saturation of apatite is recorded by a significant jump in the bulk-rock P_2O_5 content (Namur et al. 2010; Namur and Charlier 2012).

The most striking feature in the Sept Iles sequence of crystallization is the saturation of Fe–Ti oxides before clinopyroxene. This is in contrast with what is observed in many ferrobasaltic layered intrusions (McBirney 1989; Snyder et al. 1993; Wiebe and Snyder 1993) and other ferrobasaltic provinces (Carmichael 1964; Juster et al. 1989; Charreteur and Tegner 2013). Published experimental phase diagrams for ferrobasaltic liquids show that neither high water content nor high fO_2 can explain the sequence of crystallization observed in the Sept Iles Layered Series (Toplis and Carroll 1995; Botcharnikov et al. 2008). However, it has been shown that an increase in the FeO_t content of an anhydrous ferrobasalt from ca. 13 to ca. 16 wt.% results in the saturation of magnetite and ilmenite before clinopyroxene (Snyder et al. 1993; Toplis and Carroll 1995). It is therefore suggested that the early saturation of Fe–Ti oxides in Sept Iles is related to the high FeO_t content of the parent magma, and that differentiation took place with fO_2 close to the FMQ (Fayalite-Magnetite-Quartz) buffer.

Cryptic Layering and Filling of the Magma Chamber

The bottom of MCU I is characterized by the presence of a 60 m-thick marginal reversal of mineral compositions. By comparison with other intrusions, this may suggest that the initial filling of the magma chamber was gradual and involved a magma becoming increasingly primitive with time (Latypov et al. 2011). After the initial filling, fractional crystallization resulted in the stratigraphic succession of rocks with contrasting cumulus assemblages and an upward evolution towards lower An content of plagioclase, Fo content of olivine and Mg# clinopyroxene through MCU I and MCU II (Fig. 11.7). However, these trends are accompanied by gradual up-section increases in An%, Fo% and cpx-Mg# at the base of MCU II and MCU III. These changes in mineral compositions are accompanied by the disappearance of cumulus apatite and clinopyroxene. Cumulus Fe–Ti oxides also disappear at the base of MCU III. Large magma chamber replenishments are invoked to explain the cessation of magmatic differentiation in MCU I and MCU II as well as the observed shifts in mineral compositions and Sr isotopic ratios at the base of MCU II and MCU III. The retrograde change in mineral compositions in reversals indicates that extensive mixing must have occurred during replenishment. Similar progressive magma chamber replenishments have been observed in the Fongen–Hyllingen layered intrusion (Wilson and Larsen 1982) and the Bushveld Complex (Roelofse and Ashwal 2012).

Extensive mixing between the resident and the injected liquid in MCU II and MCU III can occur when there is a low density, low-viscosity, contrast or if there is a high injection rate (Campbell and Turner 1986, 1989; Wiebe and Snyder 1993), which in turns mostly depends on the diameter of the feeder dyke (Wiebe and Snyder 1993). Extensive fractionation in each MCU of the Layered Series is expected to change significantly the density of residual liquids, especially after the saturation of Fe–Ti oxides (Wiebe and Snyder 1993). Slow magma injection should therefore lead to the formation of a zoned magma chamber and to an abrupt regression of mineral compositions, in contrast with what is observed at the base of MCU II and MCU III. Although the size of the magma feeder beneath Sept Iles is unknown, it is suggested that it must be large enough to allow magma injection at high rate and thorough mixing of magmas with contrasting densities, and presumably viscosities. This injection may have been as a vigorous fountain such that the proportion of the new to residual magma in the hybrid gradually increased (Campbell and Turner 1986, 1989). Strong convection in the magma chamber also further contributed to mixing between the injected and resident magma (Brandeis and Jaupart 1986).

The rate at which the liquids that formed MCU II and MCU III were injected into the magma chamber is unknown. A progressive, rather than stepwise, evolution of mineral compositions over 200–300 m suggests that the rate of injection is more or less identical to the rate of advance of the crystallization front (e.g. top of the crystal mush) at the bottom of the magma chamber. The volumes of injected magma at the base of MCU II and MCU III are also unknown but some constraints can be obtained about the relative volumes of injected vs resident magmas. The basal reversal within MCU III reaches mineral compositions with An%, Fo% and cpx Mg# similar to those observed at the base of MCU I. In contrast, minerals from the basal reversal of MCU II do not reach such primitive compositions. This dissimilarity indicates that the hybrid magma that crystallized MCU III was more primitive than that forming MCU II. The most evolved minerals from the Sept Iles layered series are observed at the top of MCU II, suggesting that MCU II has probably experienced a longer differentiation interval than MCU I before magma replenishment. It can thus be inferred that the proportion of residual liquid left at the top of MCU II was low. Consequently, when replenishment occurred at the base of MCU III, the low proportion of resident magma residual from MCU II was mixed with the primitive replenishing magma and the hybrid magma crystallized minerals with high An%, Fo% and cpx Mg#. In contrast, the replenishing magma at the base of MCU II was mixed with a higher proportion of resident magma residual from MCU I and the hybrid liquid thus crystallized less primitive minerals at the base of MCU II.

Minor Magma Chamber Replenishments

After the main magma chamber replenishment that initiated the formation of MCU II, plagioclase, olivine and clinopyroxene show relatively constant compositions over a cumulate sequence of 1.5 km (Fig. 11.7). In detail, these minerals show compositional cycles at a scale of ca. 200–500 m in which they first evolve towards slightly lower An-, Fo- and Mg# and then show minor reversals (1–4 An%; 1–3 Fo%; 1–3 Mg# %) to more primitive compositions (Fig. 11.12). Regressive sequences are also accompanied by temporary disappearances of cumulus clinopyroxene. In contrast to the silicate minerals, magnetite displays very large compositional variation within these cycles, with the Cr-content increasing from less than 10 ppm to more than 1000 ppm over the successive sequences.

Namur et al. (2010) have shown that neither pressure changes in the magma chamber (Panjasawatwong et al. 1995; Toplis 2005), nor changes in fO_2 (Toplis and Carroll 1995; Namur et al. 2012b) or crystallization from a stratified magma chamber (Tegner et al. 2006) can produce these reversals. In contrast, they suggested that these reversals in the lower part of MCU II resulted from minor replenishments by undifferentiated magma that mixed with the resident magma. These events were used to subdivide MCU II into seven cyclic units, 100–500 m thick, each corresponding to the differentiation products of a new batch of hybrid magma.

Large sequences of cumulate stratigraphy with relatively constant mineral compositions are observed in other layered intrusions and have been described in detail in the Bushveld Complex. They have recently been considered as a potential evidence that layered intrusions may not form by fractional crystallization of large volumes of magma but may instead result from the stacking of crystal mushes, with constant mineral compositions, injected from deep-seated magma chambers (Roelofse and Ashwal 2012; see also details in Cawthorn 2012). Here, we show that this interpretation is largely biased by the use of silicate mineral compositions only; minerals that are not sensitive enough to record small-scale magma chamber replenishments by primitive magma. However, when other minerals such as magnetite are considered it is apparent that these large intervals of stratigraphy reflect cycles with progressive differentiation interrupted by many injections of primitive magma.

Expansion of the Magma Chamber

During the initial filling of the magma chamber that crystallized MCU I, the chamber expanded laterally until it reached an unknown initial diameter. After this initial filling, the two large magma chamber filling events that respectively formed MCU II and MCU III, as well as minor replenishments within MCU II, contributed to significantly enlarge the size of the Sept Iles magma chamber. Given the relative thickness each of MCU, the magma column was however never thicker than 2.5 km. The initial magma chamber was presumably in the

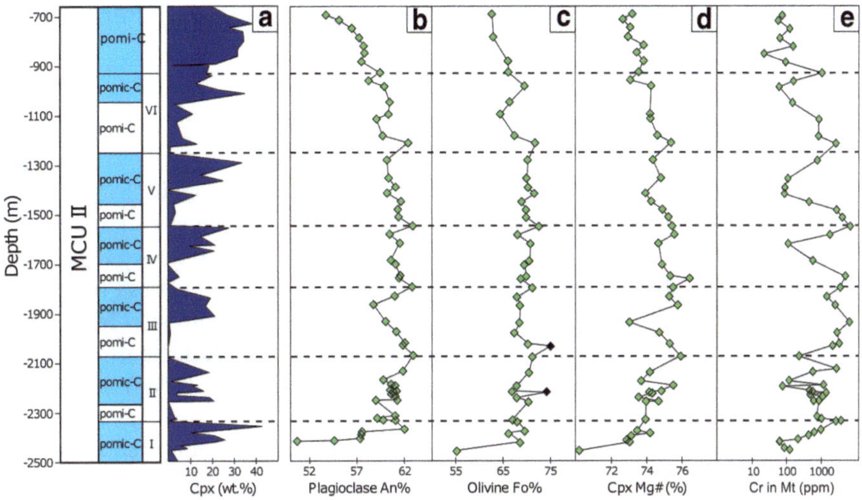

Fig. 11.12 Modal and compositional stratigraphic variations in the lower part of MCU II (from −2500 to −700 m). **a** Mineral mode of clinopyroxene; **b** An content of plagioclase; **c** Fo content of olivine; **d** Mg-number of clinopyroxene; **e** Cr content of magnetite. I–VII represent the seven cyclic units of MCU II. *Dashed lines* represent the boundaries between cyclic units. The two *black diamonds* in the olivine trend are samples that are interpreted to have undergone intense sub-solidus re-equilibration (see Namur et al. 2010). Note that at the bottom of each cycle minerals evolve to more evolved compositions (lower An, Fo, cpx-Mg-number, low magnetite Cr-content) while at the top of the cycles they evolve to more primitive compositions. The relatively low mode of clinopyroxene at the bottom of each cycle should also be noted

form of a sill that progressively deformed to a dinner plate shape by subsidence of the floor due to the weight of cumulate rocks, a process that occurred contemporaneously with crystallization (Higgins 2005). It probably started during the crystallization of MCU I and continued during crystallization of MCU II and MCU III. It is therefore likely that the liquids having formed MCU II and MCU III were injected into a dinner-plate shaped magma chamber, which may cause some angular inconformity in the layering (modal and/or cryptic) at the interface between rocks from two successive MCUs (Wilson and Larsen 1982). Similarly to other major layered intrusions such as Bjerkreim–Sokndal (Wilson et al. 1996) and the Bushveld Complex (Kruger 2005), it is likely that the magma chamber did not only expand vertically but also laterally during magma injections that formed MCU II and MCU III. The fact that MCU II is found to be in contact with the Border Zone in the north-eastern part of the Sept Iles layered intrusion may confirm this idea. This probably indicates that the Border Zone was formed incrementally by a series of magma injections that formed the major MCUs and that the magma chamber increases laterally after each of these injections. Unfortunately, the poor quality of outcrops on the Sept Iles mainland makes the relationship between the cumulate rocks of the Layered Series and the Border Zone poorly constrained.

The Sept Iles Liquid Line of Descent and the Origin of Ferroan Granites

Liquid Compositions and Formation of the Silicic Melts

A series of rocks from the Sept Iles Intrusive Suite have petrographic characteristics suggesting that they might represent liquid compositions. The most important of them are (1) rocks from the Sept Iles Border Zone (chilled margin; see representative samples in Table 11.5), (2) dykes cross cutting the Layered Series (Table 11.5) and (3) silicic rocks from the Upper Series and Pointe du Criard sill (Table 11.4). The major and trace element bulk-rock compositions of these samples as well as $(^{87}Sr/^{86}Sr)_{564}$ ratios have been presented in Higgins and Doig (1986) and Namur et al. (2011b).

Dolerite samples from the Border Zone, dykes cutting the Layered Series and silicic rocks from the Upper Series and the Pointe du Criard define a coherent compositional trend from basalt to rhyolite, with a small amount of intermediate trachy-andesite and trachyte compositions (see below), straddling the boundary between subalkaline and alkaline fields in a TAS diagram (Fig. 11.13a). All the samples are located within the tholeiitic field of the AFM diagram (Fig. 11.13b), while the most evolved compositions fall within the field of ferroan alkalic to alkali-calcic granitoids (Fig. 11.13c, d; Frost et al. 2001). Major elements from basalt to rhyolite display coherent trends when plotted as a function of a differentiation index such as SiO_2, with some elements continuously decreasing (e.g. CaO, MgO), continuously increasing (K_2O) or first increasing and then decreasing (FeO_t, TiO_2, Na_2O, P_2O_5). REE contents continuously increase from basalt to rhyolite, while a progressively deeper negative Eu anomaly develops. Other incompatible elements (e.g. Ba, Rb) increase from basalt to rhyolite, while elements such as P, Ti and Sr are progressively depleted, forming large negative anomalies in N-MORB normalized trace element diagrams.

Continuous major and trace element trends from basalt to rhyolite suggest that the fractionation of cumulus phases observed in the Layered Series from the Sept Iles parent magma might be responsible for the genesis of intermediate and silicic liquids in the Sept Iles layered intrusion. This was first modelled using mass-balance equations by Higgins and Doig (1986) and then in more detail by stepwise least-squares mass regressions by Namur et al. (2011b). Modelling indicates that the maximum enrichment in FeO_t (17 wt.%; saturation of Fe–Ti oxides) is reached when the liquid SiO_2 content is 47 wt.% and the proportion of residual liquid (F) is 0.86 (Fig. 11.14). The appearance of cumulus Fe–Ti oxides (Fig. 11.14) is responsible for a progressive SiO_2 enrichment of residual liquids. Clinopyroxene and apatite saturations are reached when $F = 0.72$ and 0.52, respectively. Further fractionation of ferrogabbros drives residual liquids towards further SiO_2 enrichment, with monzonitic and granitic compositions being produced when F is 0.32 and 0.08, respectively. Fractional crystallization has also been tested using trace elements,

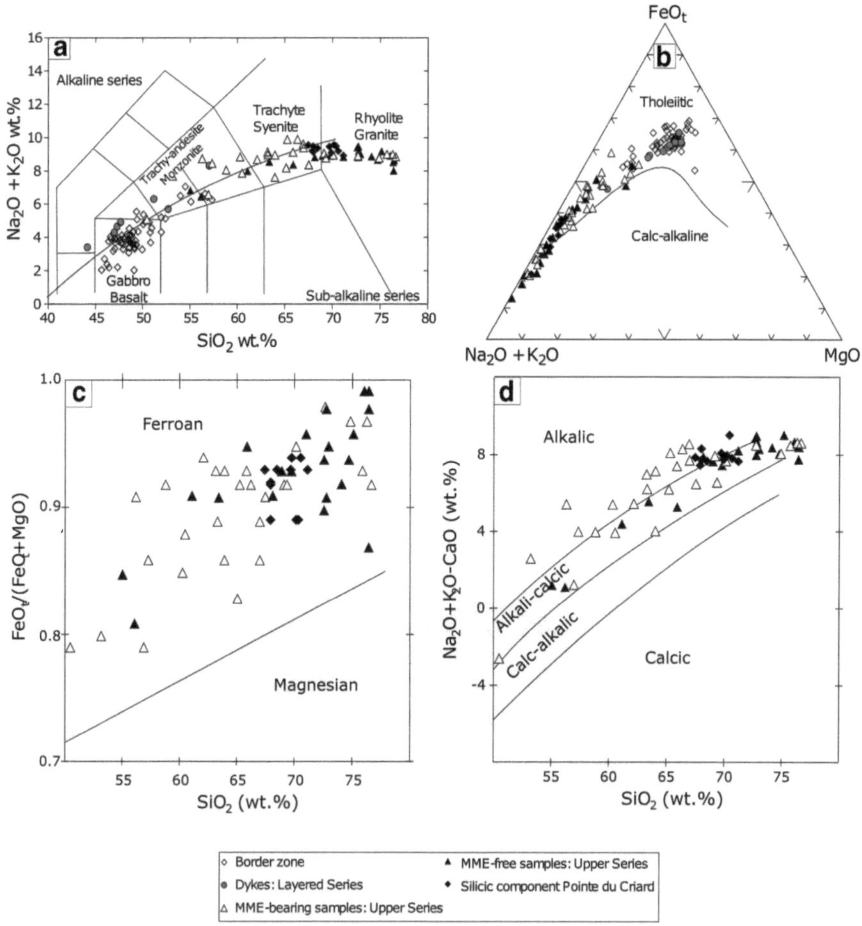

Fig. 11.13 The compositions of fine-grained dolerite samples (*Border Zone*), dykes (*Layered Series*) and silicic rocks (*Upper Series*) from the Sept Iles Intrusive Suite plotted in: **a** The total alkali-silica (*TAS*) diagram ($Na_2O + K_2O$) vs SiO_2. The limit between the alkaline and sub-alkaline series is shown by the *curved line* (Irvine and Baragar 1971). Observe that the whole set of samples forms a single consistent trend from basaltic to rhyolitic compositions at the boundary between the alkaline and sub-alkaline series. **b** The AFM diagram showing the limit between tholeiitic and calc-alkaline fields (Irvine and Baragar 1971). Sept Iles samples fall within the tholeiitic compositional field. **c** Major element geochemical classification of silicic samples in the $FeO_t/(FeO_t + MgO)$ vs SiO_2 diagram (Frost et al. 2001). **d** Major element geochemical classification of silicic samples in the $Na_2O + K_2O–CaO$ vs SiO_2 diagram (Frost et al. 2001)

with the Rayleigh equation, appropriate partition coefficients (see Namur et al. 2011b), together with mineral and liquid proportions obtained by mass-balance calculations for major elements. The trace element patterns of monzonitic, syenitic and granitic melts are well reproduced by fractional crystallization of the parent magma for F = 0.3, 0.2 and 0.1, respectively (Fig. 11.15).

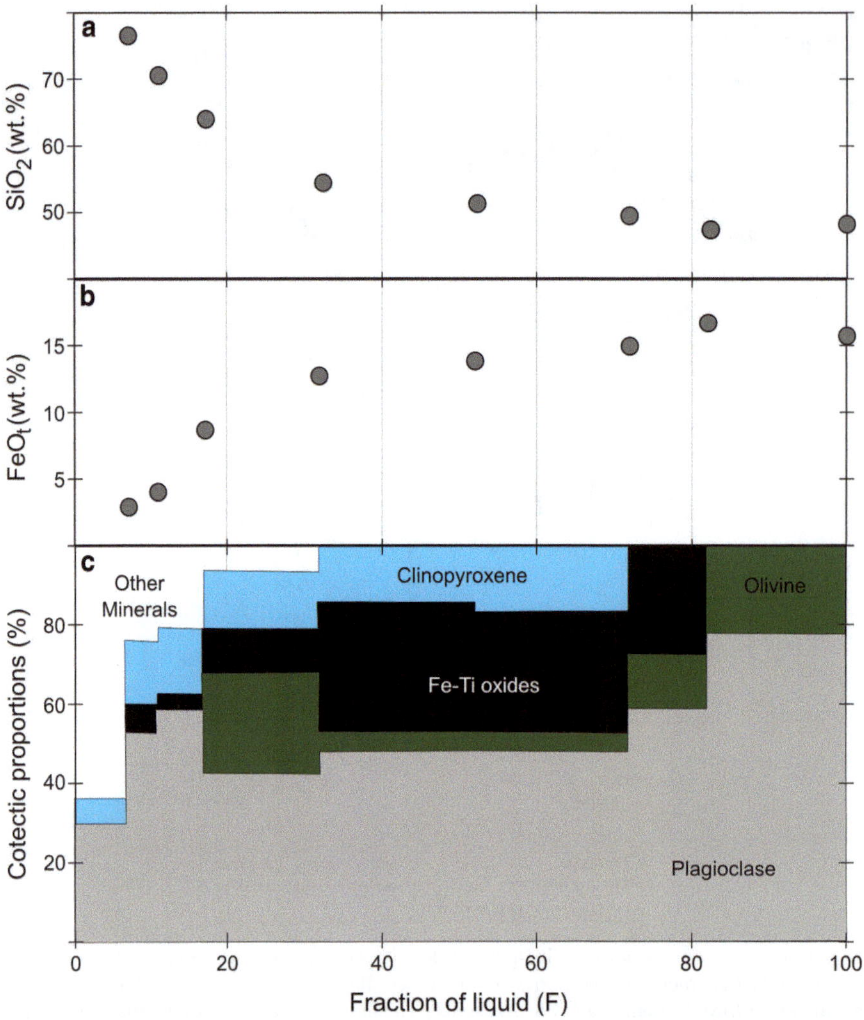

Fig. 11.14 Results of mass-balance calculations showing the best fit results for the Sept Iles liquid line of descent. **a** SiO_2 in residual melts vs F (fraction of remaining melt). **b** FeO_t vs F. **c** Calculated mineral cotectic proportions for each cumulus assemblages. Mineral cotectic proportions are relatively similar to mineral modes presented in Namur et al. (2010) and determined by point counting on polished thin sections. Other minerals = amphibole, biotite, alkali-feldspar and apatite. *Solid circles* in (**a**) and (**b**) represent the liquid compositions used in the calculations by Namur et al. (2011b). Calculations have been performed for successive liquids (*circles*; decreasing F) by removing minerals in the proportions shown in (**c**). The starting parental magma composition is presented in Table 11.5. Calculated mineral compositions can be found in Namur et al. (2011b). Note that the FeO content of residual melts decreases after the saturation of *Fe–Ti oxides*, a much discussed issue for the Skaergaard intrusion

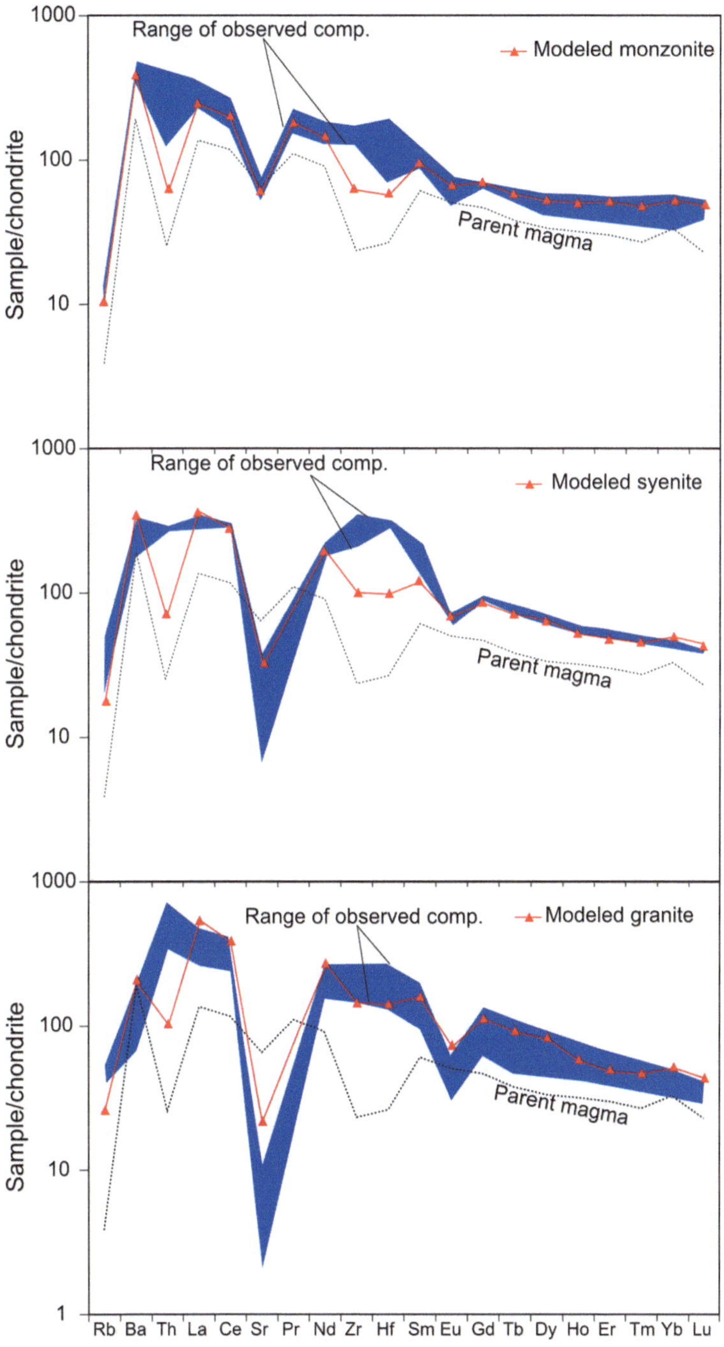

Fig. 11.15 Chondrite-normalized trace element variation diagrams showing the compositions of residual melts produced by fractional crystallization (Rayleigh fractionation) and comparison with the compositions of rocks (*blue fields*) observed in the Sept Iles layered intrusion. **a** monzonitic, **b** syenitic and **c** granitic liquids. Mineral proportions used in calculations are from major-element mass-balance calculations (see Fig. 11.14). Partition coefficients used in calculations can be found in Namur et al. (2011b)

Comparison of fractional crystallization models with the composition of Sept Iles rocks therefore suggests that protracted fractional crystallization is a realistic way to produce the silicic liquids observed in Sept Iles. Other processes such as partial melting of gneissic country rocks (Lightfoot et al. 1987; Patino Douce 1997) or a mafic source in the lower or upper crust (Beard and Lofgren 1991) can also produce A-type granites. However, theoretical models performed by Namur et al. (2011b) have shown that melting of none of these sources can produce melts with compositions similar to those of the granites observed in Sept Iles. It is therefore suggested that the continuous trend from basalt to rhyolite (Fig. 11.13) represents the actual Sept Iles liquid line of descent. This is also in agreement with field relationships convincingly showing that no trace of silicic melt is present in cumulate rocks more primitive than apatite-bearing gabbros. This unambiguously suggests that silicic magmas are residual after extensive fractionation of troctolites and gabbros in the Layered Series. In contrast to the Skaeergaard intrusion for which the liquid line of descent is highly debated (Hunter and Sparks 1987; McBirney and Naslund 1990; Toplis and Carroll 1996; Tegner 1997; Thy et al. 2009), the Sept Iles layered intrusion therefore shows an undisputable evidence of liquid line of descent evolving towards SiO_2-enrichment after Fe–Ti oxide saturation. This trend is moreover in perfect agreement with liquid compositions observed in volcanic ferrobasaltic provinces worldwide (Charlier et al. 2013).

Role of Country Rock Contamination

There is a remarkable covariation of the composition of cumulus minerals and their Sr isotopic ratios within each MCU in the Layered Series (Fig. 11.16a). This suggests that assimilation of country rocks was taking place continuously during the solidification history of Sept Iles (Tegner et al. 2005), and that this process, together with fractional crystallization, contributed to the geochemical composition of the cumulates and the residual liquids. Contamination probably resulted from the incorporation of blocks of country rocks into the magma before emplacement or possibly in the chamber itself. A contributing factor may have been strong tectonic activity, as major St Lawrence Rift system faults cut the intrusion. It is therefore of interest to determine whether or not country-rock assimilation contributed to the SiO_2 enrichment of the residual melts. The Sr content of the Sept Iles parent magma is estimated to be 511 ppm, while its initial Sr-isotopic composition was 0.70353 (Namur et al. 2010, 2011b). Three main types of country rocks occur in the Grenville province around Sept Iles: (1) metagabbro (Sr: 260 ppm, $(^{87}Sr/^{86}Sr)_{564}$: 0.7094); (2) gneissic monzonite (Sr: 230 ppm, $(^{87}Sr/^{86}Sr)_{564}$: 0.7189); (3) cordierite-bearing leucocratic gneiss (Sr: 97 ppm, $(^{87}Sr/^{86}Sr)_{564}$: 0.7746). The amount of country rock assimilated during the solidification of the Sept Iles layered intrusion was modelled using the assimilation-fractional crystallization (AFC) equations of DePaolo (1981) and the Sr

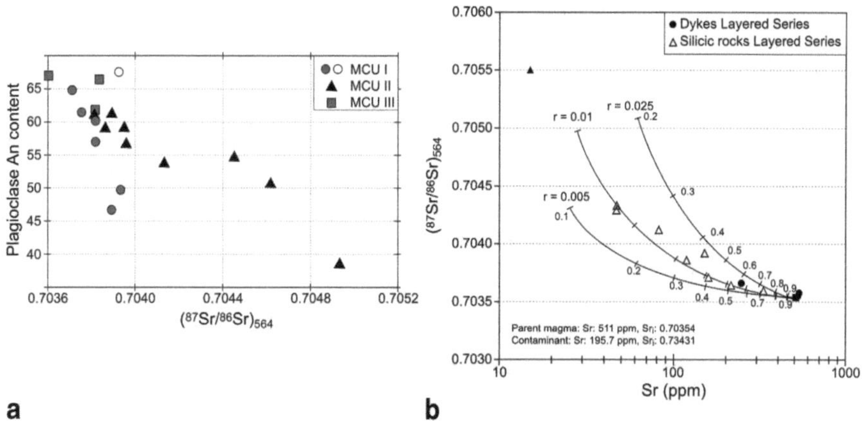

Fig. 11.16 Sr isotopic data in the Sept Iles layered intrusion. **a** Binary diagram showing the An-content and $(^{87}Sr/^{86}Sr)_{564}$ ratio of plagioclase separates from the Layered Series. Note how the Sr isotopic ratios are anticorrelated with the An-content of plagioclase, which is considered as an index of differentiation of the Sept Iles magma. This indicates that contamination from country-rock material was taking place during the solidification of the Sept Iles Layered Series. **b** $(^{87}Sr/^{86}Sr)_{564}$ vs Sr for Sept Iles liquids. The *curves* show AFC models calculated following DePaolo (1981) for various mass ratios (0.005–0.025) of assimilated to fractionated material. Numbers along AFC lines give the fractions of residual liquids. Sr and isotopic data for the Sept Iles parent magma and the contaminant are shown for reference. This diagram indicates that the evolution of Sr-isotopic ratios in the Sept Iles layered intrusion can be modelled with a very low amount of contamination from the country-rock material (e.g. 2.5 %; see text for details). Data from Namur et al. (2010, 2011b).

isotopic compositions of Sept Iles liquids (Namur et al. 2011b; Fig. 11.16b). The average composition of the three distinct types of country rocks was used as the contaminant component (Sr: 196 ppm, $(^{87}Sr/^{86}Sr)_{564}$: 0.7343). D_{Sr}^{Bulk} was set to 1.6 for the F range from 1.0 to 0.7, and to 2.1 for further differentiation. Results of AFC modelling indicate that the Sr isotopic composition of Sept Iles rocks can be reproduced with only very minor crustal assimilation. A ratio (r) between 0.01 and 0.025 of assimilated to crystallized material is sufficient to match adequately the observed Sr isotopic compositions of Sept Iles liquids (Fig. 11.16b). The amount of assimilation is even lower than 1 % when using the most radiogenic country rocks (cordierite-bearing gneiss) as the main contaminant. It is however slightly higher (5 %) when using the least-radiogenic country rocks (metagabbro). AFC models therefore suggest that country-rock assimilation played a very minor role in the geochemical evolution of Sept Iles melts.

Paucity of Intermediate Liquid Components

Most of intermediate compositions along the Sept Iles liquid line of descent contain abundant mafic enclaves (MME, Fig. 11.13). They therefore represent hybrid magma with intermediate bulk compositions, and create an artificial continuous differentiation trend between the most evolved basalts at 57 wt.% SiO_2 and the most primitive MME-free granitoids at 67 wt.% SiO_2. The very low proportion of true intermediate compositions observed along the Sept Iles liquid line of descent, generally referred to as a 'Daly gap' has been observed in other magmatic systems and has traditionally been interpreted as resulting from magma passing through all the intermediate compositions (intermediate liquid SiO_2 contents) for only a small drop of temperature, so only a small proportion of intermediate liquid is actually produced (Grove and Donnelly–Nolan 1986; Toplis and Carroll 1996; Peccerillo et al. 2003; Tegner et al. 2006). However, this model is inconsistent with mass-balance calculations showing that 15 % of fractionation is required to cover the whole range of intermediate compositions and that only a further 10 % of fractionation is required to drive residual liquids across the compositional range of highly silicic melts (67–77 wt.% SiO_2). Silicic rocks are more abundant than intermediate rocks in the Sept Iles layered intrusion (Fig. 11.13), suggesting that the low proportion of intermediate melts represented at Sept Iles cannot be related to a low volume of intermediate melt produced by fractional crystallization.

Minerals in apatite-bearing gabbros from the Layered Series MCU II, especially apatite, contain abundant, 10–120 μm, polycrystalline crystallized melt inclusions of two types: a granitic Si-rich composition and a Fe-rich ferrobasaltic composition. Single inclusions locally show evidence for unmixing between the two compositions, and Si-rich and Fe-rich compositions are therefore interpreted as immiscible melts (Charlier et al. 2011). Silicate liquid immiscibility in the Sept Iles magma is interpreted as being responsible for the paucity of intermediate liquid compositions along the liquid line of descent. Immiscibility started by the intersection of Sept Iles residual liquids with a subliquidus binodal (Fig. 11.17a). This process started when the residual liquid reached a monzonitic composition (57 wt.% SiO_2; 10 wt.% FeO_t), e.g. the start of the Daly gap, producing highly contrasting immiscible melts and no melt of intermediate composition (as opposed to fractional crystallization; Fig. 11.17a). Upon cooling, both immiscible melts kept solidifying and crystallized cumulate rocks with identical mineral compositions (but contrasted mineral modes), while the residual liquids evolved along the bimodal. During this process, the bulk liquid composition (i.e. the sum of the two immiscible melts) was continuously enriched in SiO_2 and depleted in FeO_t (Fig. 11.13). Ultimately, the bulk liquid compositions left the two liquid field (Fig. 11.17a), producing a single homogeneous liquid that further evolved to produce the residual granites of the Upper Series, which are satisfactorily modelled by a simple fractional crystallization process (Fig. 11.14). Immiscibility in Sept Iles was therefore a furtive event, with no influence on the composition of liquids along the liquid line of descent, except that this process was responsible for the absence of primary intermediate melts (Fig. 11.17b).

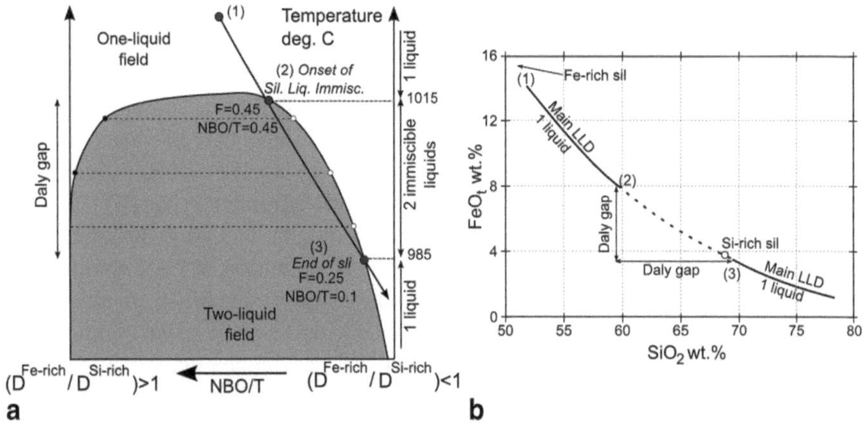

Fig. 11.17 Schematic illustration of the evolution of the Sept Iles liquid line of descent through a two-liquid field. The morphology of the binodal is based on Charlier and Grove (2012). The Sept Iles liquid reaches the binodal on the Si-rich side at ca. 1015 °C when F = 0.45 (Namur et al. 2012a) and splits into two conjugate immiscible melts (Fe-rich and Si-rich). Immiscible melts produce cumulate rocks (ferrogabbros) with identical minerals but in contrasting proportions. Fractional crystallization of ferrogabbros drives the bulk-liquid composition (sum of Si-rich and Fe-rich immiscible melts) towards further SiO_2-enrichment (see Fig. 11.14) and the bulk-liquid composition ultimately leaves the two-liquid field at ca. 985 °C (F = 0.25). This results in the formation of a single stable liquid after the episode of silicate liquid immiscibility. Further differentiation of a single liquid produces the granites from the Sept Iles Upper Series (see text for details and models in Fig. 11.14). *sli* silicate liquid immiscibility; $D^{Fe\text{-}rich}$ elements partitioning into the Fe-rich immiscible melt; $D^{Si\text{-}rich}$ elements partitioning into the Si-rich immiscible melt. **b** FeO_t vs SiO_2 diagram showing the result of the silicate liquid immiscibility episode on the Sept Iles liquid line of descent. The only evidence for the immiscibility episode is the absence of intermediate liquid compositions (Daly gap; Charlier et al. 2013). *sil* silicate immiscible melt

Implications of the Liquid Line of Descent for Magma Chamber Processes

Plagioclase Flotation in Ferrobasalts

Plagioclase in anorthosite from the Upper Border Series shows a very restricted range of composition ($An_{62\text{-}68}$), which is moreover similar to the plagioclase composition in anorthosite blocks (autoliths) in the Layered Series. It is also identical in composition to the plagioclase found in troctolites (po-c) and Fe–Ti oxide troctolites (pomi-C) in the Layered Series, but is significantly more primitive than the plagioclase from the gabbros (pomic-C, pmic-C and pomica-C; $An_{31\text{-}60}$; Fig. 11.7). A similar relationship is observed for trace elements and Sr-isotope ratios in plagioclase (Namur et al. 2011a). A genetic link between plagioclase in the Upper Border Series, in autoliths and in troctolites from the Layered Series seems therefore evident.

The origin of anorthosite in the Sept Iles layered intrusion has been investigated by Namur et al. (2011a) using density calculations for liquids and plagioclase (Bottinga and Weill 1970; Niu and Batiza 1991), oxygen fugacity along the FMQ buffer and a pressure of 3 kbar. Liquids used for density calculations correspond to the samples plotted in Fig. 11.13. A water content of 0.4 wt.% H_2O has been assumed for the parent magma. For each liquid sample, the composition of equilibrium plagioclase has been calculated using the equations of Namur et al. (2012b).

The calculated density of the Sept Iles parent magma is 2.70 g/cm^3, increasing due to fractionation of troctolites to 2.75 g/cm^3 at saturation in Fe–Ti oxides, and then decreasing to 2.16 g/cm^3 in granitic melts as the result of Fe–Ti oxide troctolite and gabbro fractionation (Fig. 11.18a). Calculated plagioclase compositions range from An_{68} to An_{34}, in perfect agreement with plagioclase compositions in the Layered Series, and plagioclase density decreases from 2.69 to 2.64 g/cm^3 with differentiation.

Density calculations indicate that plagioclase was buoyant in primitive melts, but that plagioclase became denser than the equilibrium melt soon after the saturation of Fe–Ti oxides, when the liquid reached 49 wt.% SiO_2. Primitive liquids in which plagioclase was buoyant were in equilibrium with a very restricted range of plagioclase composition ($An_{61\text{-}69}$), identical to that observed in Sept Iles anorthosite (Fig. 11.18a). In contrast, more evolved plagioclases ($An_{<60}$), were in equilibrium with evolved liquids (>49 wt.% SiO_2) and were no longer buoyant. Formation of the anorthosite therefore resulted from the flotation of primitive plagioclase ($An_{>60}$) and its accumulation at the roof of the magma chamber (Fig. 11.18b). The decrease of liquid density subsequent to the saturation of Fe–Ti oxides inhibited further flotation of plagioclase (Fig. 11.18b). Furthermore, the anorthosite from the Upper Border Series became denser than the main magma body. As a consequence, anorthositic blocks sank into the magma chamber, forming the autoliths blocks observed in the Layered Series.

As illustrated above, plagioclase ($An_{>60}$) that was buoyant is present in the Layered Series troctolites and the most primitive Fe–Ti oxide troctolites. This indicates that not all of the plagioclase grains floated and accumulated at the roof of the magma chamber. We interpret this as a consequence of *in situ* crystallization. This mechanism prevents flotation due to attachment of the crystals to the floor of the magma chamber. Formation of coherent three-dimensional chains of crystals (Philpotts and Carroll 1996; Philpotts et al. 1998) may also have contributed to the retention of plagioclase grains on the chamber floor. Comparison between the proportion of plagioclase in the troctolites of the Layered Series (67.5 % on average; Namur et al. 2010) and estimated plagioclase cotectic proportion in equilibrium melts (78 %; Namur et al. 2011b) can be used to put some constraints on the efficiency of *in situ* crystallization. These values confirm that plagioclase is missing in the Layered Series, and that more than 10 % of the liquidus plagioclase accumulated at the top of the magma chamber to form the Upper Border Series.

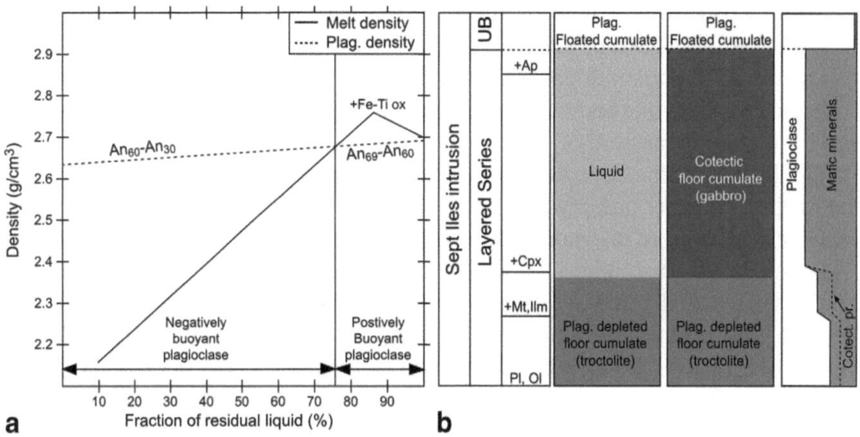

Fig. 11.18 Schematic models for the distribution of plagioclase in the Sept Iles layered intrusion (modified from Namur et al. 2011a). **a** Evolution of plagioclase and melt densities as a function of the proportion of residual liquid. Note the crossover between liquid and plagioclase density just after the saturation of Fe–Ti oxides (F = 0.76). At a higher residual melt fraction, plagioclase is positively buoyant in the basaltic melt whereas it becomes negatively buoyant when F < 0.76. **b** Simplified stratigraphic column of the Sept Iles layered intrusion showing the sequence of cumulus phases, the mechanisms of crystallization and the relative proportions of plagioclase and mafic minerals. Troctolite and Fe–Ti oxide-bearing troctolite crystallize *in situ* on the floor of the magma chamber. Plagioclase (An_{69-60}) is positively buoyant and some of the plagioclase crystals float to the top of the magma chamber where they accumulate to form a 100–500 m-thick anorthosite (Upper Border Series). After the saturation of Fe–Ti oxides (F = 0.76), the density of the magma becomes lower than that of equilibrium plagioclase (An_{60-34}), which is then no longer buoyant. Cotectic gabbro crystallizes *in situ* on the top of the previously formed troctolite crystal pile. At this stage, the anorthosite at the top of the magma chamber does not grow any further. Additional plagioclase flotation might have occurred during the crystallization of MCU III where the equilibrium liquid was locally not saturated in Fe–Ti oxides (see Fig. 11.5)

Development of Ferroan Granites

Evidence for silicic material is observed in many ferrobasaltic layered intrusions. In some cases, such as in Skaergaard, silicic rocks only form granophyre pods enclosed in the cumulates (McBirney 1989), while in other cases it forms masses of granite at the top of the intrusion (Wiebe and Snyder 1993). It has recently been suggested that the most evolved rocks of the Bushveld complex may also be of granitic composition (VanTongeren et al. 2010). Understanding the origin of these rocks is generally hampered by the lack of constraints on the ferrobasaltic liquid lines of descent in layered intrusions (Hunter and Sparks 1987; Toplis and Carroll 1996; Thy et al. 2009). The Sept Iles layered intrusion shows a clear example of ferrobasaltic layered intrusion with residual liquids evolving towards SiO_2 enrichment and FeO depletion after Fe-Ti oxide saturation. It is worth noting that the composition of Sept Iles granite is almost identical to the granophyre observed in Skaergaard and the rhyolite at the top of the Bushveld, which suggests that these two intrusions may have followed a differentiation trend identical to that of the Sept Iles layered intrusion (Tegner and Cawthorn 2010; VanTongeren et al. 2010; Namur et al. 2011b). It is therefore important to decipher how evolved, residual liquids can be extracted from the cumulate pile and collect at the top of the magma chamber to accumulate and form granitic masses.

The history of the Sept Iles layered intrusion started with the initial filling of the magma chamber and crystallization of MCU I in the Layered Series together with the formation of the anorthositic Upper Border Series by plagioclase flotation (Namur et al. 2011a, b; Fig. 11.19a). Fractionation of troctolite and gabbro in MCU I drove the residual liquid to a monzonitic composition (e.g. 55 wt.% SiO_2; 10 wt.% FeO_t) that lay between the top of the cumulates and the bottom of the Upper Border Series (Fig. 11.19b). Using expressions from Ghiorso and Sack (1995) and Bottinga and Weill (1970), Namur et al. (2011b) estimated that the residual liquid at this stage had a temperature of ca. 1050 °C and a density of <2.57 g/cm^3. This density is lower than that of typical anorthosite (ca. 2.70 g/cm^3; Cawthorn and Ashwal 2009; Namur et al. 2011a). Part of the residual liquid thus ascended through the partly molten anorthosite, probably helped by compaction, and accumulated at the top of the magma chamber where it formed cupolas and continued to differentiate (Fig. 11.19b). The passage of silicic liquid through the Upper Border Series anorthosite is recorded by a matrix of monzonite to syenite around plagioclase crystals and several pods and dykes of syenite within the anorthosite.

The Sept Iles layered intrusion presents two types of ferroan granite: one with magmatic mafic enclaves (MME-rich granite) and one free of enclaves. The origin of these two contrasted types of granite can be understood when magma chamber replenishments are taken into account. During the crystallization of the most evolved cumulates of MCU I (apatite-bearing gabbros), a major event of magma chamber replenishment occurred (Fig. 11.19c). Feeder dykes for this magma injection have not been clearly identified in the field, but they might be part of the same system that produced the fine- to medium-grained dykes of the Late Gabbro intrusions. Undifferentiated magma and monzonitic melts thoroughly mixed within the main magma chamber owing to vigorous convection and the highly turbulent injection. This process resulted in the formation of a very homogeneous hybrid melt (Fig. 11.19d). However, some basaltic melt reached the silicic cupolas at the top of the magma chamber (Upper Series). In these, the viscosity contrasts between basaltic and silicic melts and the absence of strong convection hampered significant mixing (Barbarin 2005). Injection of basaltic magma into the silicic magma was thus responsible for the formation of the mafic enclaves (MME) observed in one of the two granite facies form the Upper Series (Fig. 11.19d). Crystallization of MCU II also resulted in the formation of silicic melts that migrated through the anorthosite to collect at the bottom of the mostly solidified MME-bearing granite (Fig. 11.19e). Namur et al. (2011b) suggested that residual liquids from MCU II are responsible for the formation of the MME-free granite unit in the Upper Series (Fig. 11.19f). This is also in agreement with Sr isotopic ratios in cumulates from MCU II and MME-free granites.

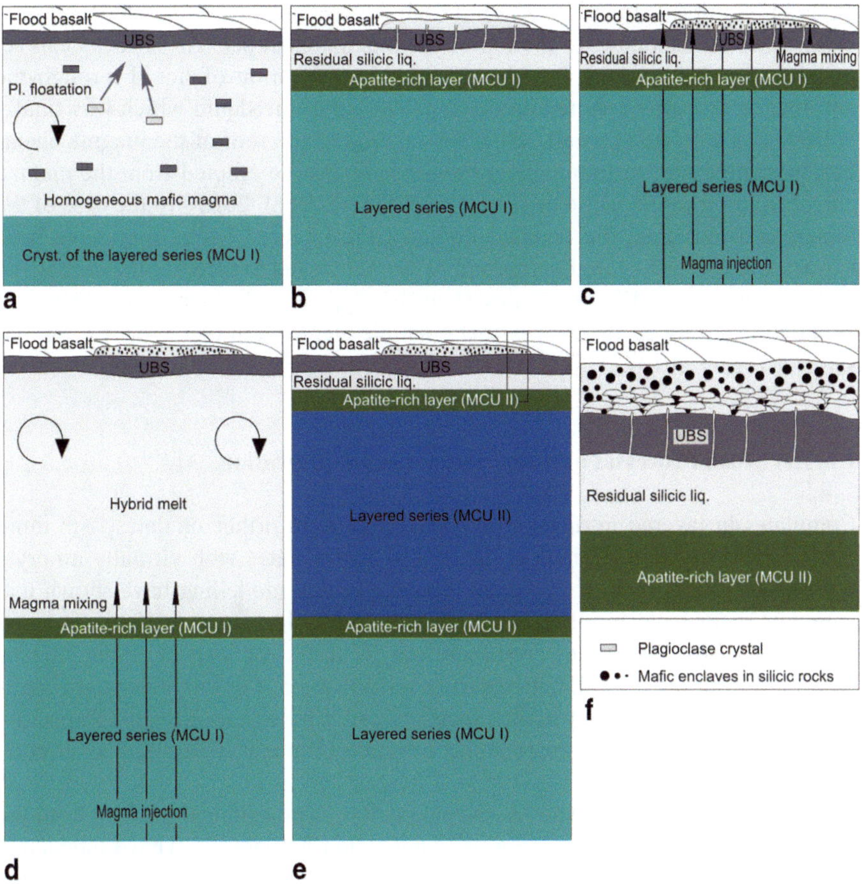

Fig. 11.19 Schematic model for the emplacement of silicic rocks in the Sept Iles layered intrusion (modified from Namur et al. 2011b). **a** Initial filling of the magma chamber and crystallization of MCU I together with simultaneous formation of the Upper Border Series (*UBS*) by plagioclase flotation (see Fig. 11.18). **b** The intermediate (monzonite) residual liquid of MCU I is less dense than the overlying anorthosite and partly ascends through the Upper Border Series to form cupolas of silicic melt at the top of the magma chamber where it further differentiates (Upper Series; *US*). **c** Magma chamber replenishment (injection of the liquid that will form MCU II) by undifferentiated magma. Complete mixing and hybridization with residual melt occurred in the main magma chamber. In contrast, mixing does not occur in silicic cupolas from the Upper Series and the injection of basalt is responsible for magma mingling and the formation of mafic enclaves (MME) in the silicic liquids of the Upper Series, forming the MME-rich granite facies (see text for details). **d** Crystallization of MCU II from a homogenous melt and differentiation of residual liquids towards SiO_2-enrichment. **e** Due to its low density, part of the residual intermediate liquids of MCU II (remaining after the crystallization of apatite-bearing olivine gabbros) ascends through the Upper Border Series. **f** These silicic liquids stay at the base of the MME-bearing silicic rocks and form the MME-free silicic rock unit of the Upper Series

Observations from the Sept Iles layered intrusion allow us to suggest that the formation of large volume of A-type granite at the top of a layered intrusion first requires a liquid line of descent evolving towards granitic composition but is helped by the presence of rocks with high permeability, such as a loose plagioclase flotation cumulate, at the top of the magma chamber. These rocks can act as pathways for residual, low-density liquids. Migration of liquid through the anorthosite probably resulted in a channelization of the liquid which was finally collected as a series of small, km-scale, cupolas at the roof of the magma chamber. In some cases, residual silicic liquids may also be erupted from the magma chamber and form rhyolitic deposits (VanTongeren et al. 2010). In the case of the Skaergaard intrusion, the residual liquid of the Layered Series may have been extracted from the Layered Series and have contributed to increase the proportion of intercumulus melt in cumulates from the Upper Border Series, which would explain the strong enrichment in incompatible elements in these rocks (Salmonsen and Tegner 2013).

Crystal Mush Processes and Interstitial Liquid Migration

Cumulates in layered intrusions span the range from orthocumulates with more than 30 % of crystallized interstitial melt to adcumulates with virtually no crystallized interstitial melt. Experiments and theoretical modelling have shown that cumulate rocks have an initial porosity of 40–60 % (Jerram et al. 1996; Philpotts et al. 1998). Some processes must therefore occur to expel part or all the interstitial liquid from the crystal matrix during solidification. For the Skaergaard intrusion, it was suggested that these processes could be gravity-driven compaction of the crystal mush or compositional convection (Toplis et al. 2008; Tegner et al. 2009; McKenzie 2011).

Using bulk-rock compositions and plagioclase compositional profiles from the Sept Iles MCU I of the Layered Series, Namur and Charlier (2012) have shown that these two processes are not mutually exclusive and can occur together within the same intrusion. They however observed that their relative efficiency changed with differentiation. In troctolites (po-C) with plagioclase and olivine, the fraction of interstitial liquid (e.g. 30 %) is high when calculated using elements compatible in clinopyroxene and Fe–Ti oxides (Zr, Cr and V) whereas it is much lower (0–10 %) when calculated using P which is only compatible in apatite (Fig. 11.20). These results were used to suggest that the porosity in troctolites was initially reduced to ca. 30 % by compaction. This process then stops presumably due to the low-density contrast between the crystal matrix and the equilibrium ferrobasalt and the thin crystal mush at the bottom of the intrusion (Holness et al. 2007). Further porosity reduction resulted from crystallization together with compositional convection. Saturation of intercumulus Fe–Ti oxides at the bottom of the troctolite crystal mush changed the density evolution of interstitial melt from a trend of increasing density to a trend of decreasing density. Crystallization of Fe–Ti oxides therefore results in an unstable density distribution between the Fe–Ti oxide saturated interstitial melt at the bottom of the mush and interstitial melt at the top of the mush which is not yet saturated in Fe–Ti oxides (Tait et al. 1984; Toplis et al. 2008). This initiated the onset of compositional convection. Compositional convection in troctolites is

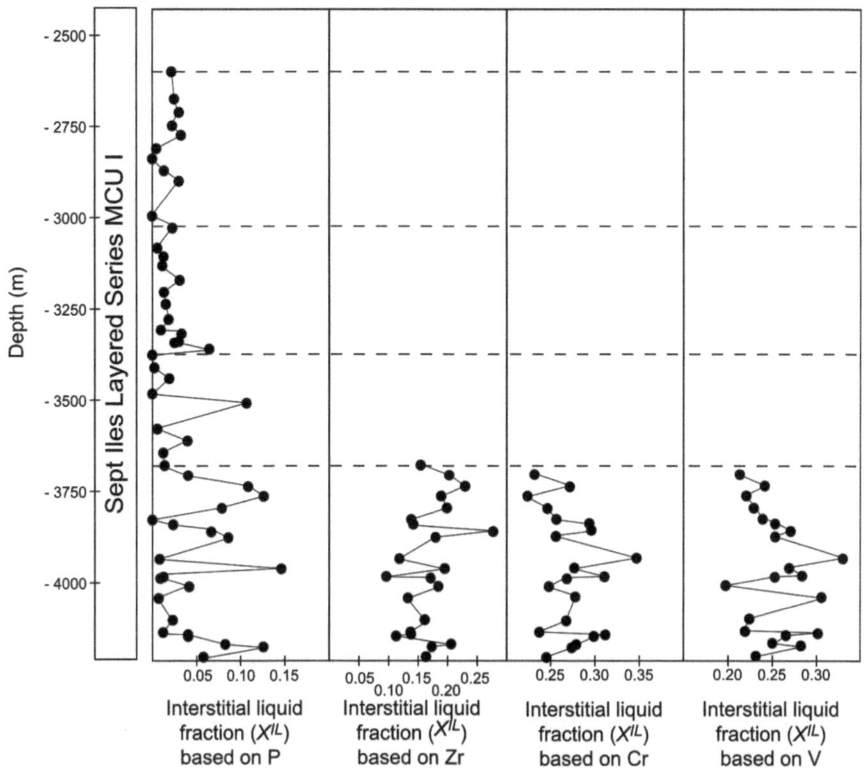

Fig. 11.20 Calculated fractions of crystallized interstitial liquid in troctolites and gabbros of the Layered Series MCU I (see Namur and Charlier 2012 for details of the calculation procedure). **a** Fraction of liquid calculated from the whole-rock and liquid P-contents. **b** Fraction of liquid calculated using Zr. **c** Fraction of liquid calculated using Cr. **d** Fraction of liquid calculated using V. Note that the liquid fractions in the po-C unit as calculated with elements of contrasting compatibilities in Sept Iles cumulus phases are quite different. Liquid fractions calculated with elements entering the structure of Fe–Ti oxides (Cr, V and Zr) are relatively high (0.15–0.40), while the liquid fractions calculated with P are very low (generally less than 0.10)

moreover revealed by plagioclase profiles with large external rims buffered to a composition of An_{61}, similar to those observed in Skaergaard and interpreted as forming during solidification of interstitial melt buffered at a constant composition due to liquid convection through the mush (Toplis et al. 2008).

In gabbros with cumulus Fe–Ti oxides, compositional convection and compaction are probably both efficient in expelling (or exchanging) melt from the crystal mush. The distribution of the last fraction of intercumulus melt in Sept Iles cumulates, as recorded by the whole-rock P content, is significantly different in troctolites (po-C) and Fe–Ti oxide-bearing rocks (pomi-C to pmic-C). In troctolites, the distribution of P is heterogeneous, and high and low values alternate at a scale of 30–250 m. In Fe–Ti oxide-bearing rocks, the distribution of the whole-rock P content is less variable, suggesting that the last fraction of intercumulus melt was homogeneously distributed within the mush. Namur and Charlier (2012) used this information to suggest that expulsion of interstitial liquid from Fe–Ti oxide-bearing cumulates is dominated by compaction, resulting in a porous flow, rather than compositional convection which may produce highly-channelized flow (Worster 1992) and which is therefore more likely to produce a heterogeneous distribution of the liquid within the crystal mush. The role of compaction in Fe–Ti oxide-bearing rocks is moreover illustrated by the high aspect-ratio of their plagioclase crystals and the development of well-defined mineral lamination (Namur and Charlier 2012).

Formation of Fe–Ti–P Ore Deposits

Two 200 m-thick sequences of apatite-bearing gabbro occur at the top of MCU I and MCU II. These sequences are significantly enriched in P, Fe and Ti. The uppermost level was drilled by the Soquem Inc in the 1990s and more than 30 cores, up to 250 m long, were retrieved along a 7 km long east-west traverse on the mainland. Apatite-bearing gabbros are the most evolved cumulates of MCU I and MCU II. In MCU I, they are homogeneous leucocratic gabbros with 3–4 vol.% apatite and 15–20 vol.% Fe–Ti oxides. In MCU II, they show an alternation on a 5–20 m scale between leucocratic gabbros (<5 vol.% apatite; <10 vol.% Fe–Ti oxides) and melanocratic gabbros (up to 25 vol.% apatite and 60 vol.% Fe–Ti oxides). Based on an incremental forward model of crystallization, it has been shown that the liquid lines of descent for MCU I and MCU II are slightly different for all the major elements (Namur et al. 2012a). This is related to the extent of magma mixing during the important replenishment event that initiated the crystallization of MCU II as well as minor replenishments that took place during its formation. Minor compositional differences can have important consequences for the physical and chemical evolution of the liquid as well as for the formation of cumulates. When liquid lines of descent for MCU I and MCU II are plotted in the ternary diagram Si–Al–Ca (Charlier and Grove 2012), it emerges that the bulk liquids of MCU II entered a two-liquid field and that immiscibility developed (Fig. 11.21). Immiscibility during crystallization of MCU II is shown by contrasting Fe-rich and Si-rich apatite-hosted melt inclusions (Charlier et al. 2011). In contrast, liquids formed during the crystallization of MCU I never encountered the two-liquid compositional field (Fig. 11.21). Crystallization from a homogeneous magma in MCU I resulted in the formation of homogeneous leucocratic apatite-bearing gabbros. In MCU II, Fe-rich and Si-rich immiscible liquids were segregated on a 5–20 m scale, although chemical equilibrium between both liquids was probably maintained. Whether the whole magma body reached immiscibility at the same time or immiscibility occurred repeatedly in basal layers at the top of the crystal pile is not known (Charlier et al. 2011). It is however suggested that crystallization from the two immiscible liquids resulted in the formation of alternating highly leucocratic gabbro and highly melanocratic gabbro (Charlier et al. 2011; Namur et al. 2012a).

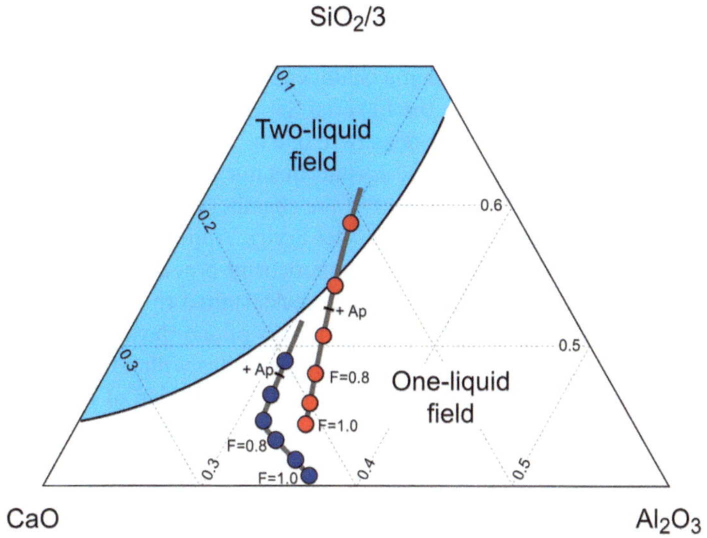

Fig. 11.21 Representation of liquid lines of descent for MCU I (*blue circles*) and MCU II (*red circles*) in a subsection of the SiO_2–CaO–Al_2O_3 compositional space. The positions of the one-liquid field (*white*), two-liquid field (*blue*) and the binodal are from Charlier and Grove (2012). *F* fraction of residual liquid; *+Ap* the initiation of apatite crystallization. Distance between each circle corresponds to 10 % of fractional crystallization. Incremental forward models stop at F = 0.45. Note that the residual liquids from MCU I do not encounter the two-liquid field unlike those from MCU II

Large-scale segregation of immiscible melts, as opposed to simple fractional crystallization of a homogeneous melt, is proposed as a very efficient process to concentrate economically important minerals such as apatite, magnetite and ilmenite in layered intrusions. Nevertheless, some more work is needed to understand how the immiscible liquids actually segregated (Chung and Mungall 2009) and were stratigraphically organized within the magma chamber.

Conclusions

The Sept Iles Intrusive Suite is made up of three igneous units, of which the layered intrusion is by far the most voluminous. The other units are late-stage gabbro intrusions and the Pointe du Criard composite sill. The cumulates in the layered intrusion crystallized from a ferrobasaltic parent magma and form a succession with the following sequence of crystallization: plagioclase and olivine, followed by Fe–Ti oxides, then clinopyroxene and finally apatite. This order of crystallization was interrupted by several replenishments of the magma chamber and the Layered Series can be subdivided into three megacyclic units (MCUs), each of them having crystallized similar sequences of cumulates. Magma chamber replenishments produced lateral and vertical extension of the magma chamber and therefore significantly enlarged its size compared to the initial magma chamber.

Saturation of Fe–Ti oxides in the Layered Series changed the evolution of melt density from a trend of increasing density to one of decreasing density. Before Fe–Ti oxide saturation, some plagioclase crystals that nucleated and grew at the floor of the magma chamber floated to the roof of the chamber, where they accumulated to produce the anorthosite of the Upper Border Series. Appearance of cumulus Fe–Ti oxides and the resulting decrease in magma density prevented plagioclase grains from further floating and cotectic cumulate rocks started to crystallize at the base of the magma chamber. Eventually, altered blocks of anorthosite fell from the roof to the top of the crystal pile and were buried as autoliths within the Layered Series. The evolution of liquid density associated with the saturation of Fe–Ti oxides has also implications for the evolution of the interstitial melt within the solidifying crystal mushes at the floor of the magma chamber. Appearance of liquidus Fe–Ti oxides is thought to have changed the main way the interstitial melt was expelled from compositional convection in troctolites to cumulate compaction in Fe–Ti oxide-bearing gabbros.

Fractionation of cumulates forming the Sept Iles Layered Series from the ferrobasaltic parent magma resulted in a liquid line of descent towards SiO_2 enrichment and FeO depletion after the saturation of Fe–Ti oxides. This liquid line of descent had a limited proportion of intermediate melts (e.g. 57–67 wt.% SiO_2). Minor country-rock assimilation took place concurrently with differentiation but did not play an important role in the geochemical evolution of Sept Iles liquids. SiO_2 enrichment during crystallization had important implications for the dynamic evolution of the magma chamber. Residual liquids produced during the crystallization of MCU II encountered a two-liquid immiscibility field. This produced contrasting Si-rich and Fe-rich immiscible melts and hence accounts for the absence of intermediate liquid compositions. On the other hand, residual liquids from MCU I never reached the two-liquid field. The limited proportion of intermediate liquids observed along the Sept Iles liquid line of descent is therefore related only to the differentiation of MCU I. The onset of immiscibility in MCU II also resulted in the crystallization of cumulate rocks from contrasting and spatially segregated immiscible melts. This produced alternating sequences of leucocratic and melanocratic apatite-bearing gabbros towards the top of MCU II. Immiscibility is proposed as an efficient process for the concentration of mafic minerals and apatite, and the production of Fe–Ti–P mineralization in layered intrusions (Namur et al. 2015; Veksler and Charlier 2015). In MCU I, fractional crystallization of a single homogeneous liquid also led to the crystallization of apatite-bearing gabbros, but these are comparatively leucocratic and are currently of no economic interest.

Acknowledgements Research on Sept Iles by the authors was funded by the Belgian National Fund for Scientific Research (FNRS), the Belgian Fund for Research in Industry and Agriculture (FRIA) and the Natural Science and Engineering Research Council of Canada, Discovery grants program. ON acknowledges the FRIA for a position as PhD student and the University of Liège for a teaching assistantship. ON also acknowledges support from the Natural Environment

Research Council (NERC; United Kingdom), Newton Trust and Magdalene College (University of Cambridge). The Ministère des Ressources Naturelles et de la Faune du Québec and Soquem Inc are thanked for access to drill-cores. We would to thank B. Charlier, J.C. Duchesne, M.J. Toplis, J.P. Liégeois, J. Hermann, C. Pirard and M.B. Holness for comments. Analytical support by G. Bologne, N. Delmelle, H.J. Bernhardt, C. Gilson, C. Allen and J.L. Devidal was also highly appreciated. Constructive reviews by B. Robins and J. VanTongeren as well as detailed comments and editorial handling by R. Latypov were highly appreciated and significantly improved the quality of the manuscript.

References

Barbarin B (2005) Mafic magmatic enclaves and mafic rocks associated with some granitoids of the central Sierra Nevada batholith, California: nature, origin, and relations with the hosts. Lithos 80:155–177

Beard JS, Lofgren GE (1991) Dehydration melting and water-saturated melting of basaltic and andesitic greenstones and amphibolites at 1, 3 and 6.9 kbar. J Petrol 32:365–401

Bonin B (2007) A-type granites and related rocks: evolution of a concept, problems and prospects. Lithos 97:1–29

Botcharnikov RE, Almeev RR, Koepke J, Holtz F (2008) Phase relations and liquid lines of descent in hydrous ferrobasalt: implications for the Skaergaard intrusion and Columbia river flood basalts. J Petrol 49:1687–1727

Bottinga Y, Weill DF (1970) Densities of liquid silicate systems calculated from partial molar volumes of oxide components. Am J Sci 269:169–182

Brandeis G, Jaupart C (1986) On the interaction between convection and crystallization in cooling magma chambers. Earth Planet Sci Lett 77:345–361

Campbell IH, Turner JS (1986) The influence of viscosity on fountains in magma chambers. J Petrol 27:1–30

Campbell IH, Turner JS (1989) Fountains in magma chambers. J Petrol 30:885–923

Carmichael IS (1964) The petrology of Thingmuli, a tertiary volcano in Eastern Iceland. J Petrol 5:435–460

Cawthorn RG (2012) Multiple sills or a layered intrusion: time to decide. S Afr J Geol 115:283–290

Cawthorn RG, Ashwal LD (2009) Origin of anorthosite and magnetitite layers in the Bushveld complex, constrained by major element compositions of plagioclase. J Petrol 50:1607–1637

Charlier B, Grove TL (2012) Experiments on liquid immiscibility along tholeiitic liquid lines of descent. Contrib Mineral Petrol 164:27–44

Charlier B, Namur O, Toplis MJ, Schiano P, Cluzel N, Higgins MD, Vander Auwera J (2011) Large-scale silicate liquid immiscibility during differentiation of tholeiitic basalt to granite and the origin of the daly gap. Geology 39:907–910

Charlier B, Namur O, Grove TL (2013) Compositional and kinetic controls on liquid immiscibility in ferrobasalt-rhyolite volcanic and plutonic series. Geochim Cosmochim Acta 113:79–93

Charreteur G, Tegner C (2013) Multiple ways of producing intermediate and silicic rocks within Thingmúli and other Icelandic volcanoes. Contrib Mineral Petrol 166:471–490

Cheney E S, Twist D (1991) The conformable emplacement of the Bushveld mafic rocks along a regional unconformity in the Transvaal succession of South Africa. Precambrian Res 52:115–132

Chung HY, Mungall JE (2009) Physical constraints on the migration of immiscible fluids through partially molten silicates, with special reference to magmatic sulfide ores. Earth Planet Sci Lett 286:14–22

Cimon J (1998) Le Complexe de Sept-Îles: I – L'Unité à apatite de Rivière des Rapides, Complexe de Sept-Îles; localisation stratigraphique et facteurs à l'origine de sa formation. ET 97–05, Ministère de l'Énergie et des Ressources du Québec, 1:1–33

Clemens JD, Holloway JR, White RJ (1986) Origin of an A-type granite: experimental constraints. Am Mineral 71:317–324

Danyushevsky L, Perfit M, Eggins S, Falloon T (2003) Crustal origin for coupled ultra-depleted and plagioclase signatures in MORB olivine-hosted melt inclusions: evidence from the Siqueiros Transform Fault, East Pacific Rise. Contrib Mineral Petrol 144:619–637

DePaolo DJ (1981) Trace element and isotopic effects of combined wallrock assimilation and fractional crystallization. Earth Planet Sci Lett 53:189–202

Doig R (1970) An alkaline rock province linking Europe and North America. Can J Earth Sci 7:22–28

Doig R, Barton JM (1968) Ages of carbonatites and other alkaline rocks in Quebec. Can J Earth Sci 5:1401–1407

Duchesne JC, Bologne G (2009). XRF major and trace element determination in Fe–Ti oxide minerals. Geol Belg 12:205–212

Eby GN (1990) The A-type granitoids: a review of their occurrence and chemical characteristics and speculations on their petrogenesis. Lithos 26:115–134

Egorova V, Latypov R (2012) Prolonged magma emplacement as a mechanism for the origin of the marginal reversal of the Fongen–Hyllingen layered intrusion, Norway. Geol Mag 149:909–926

Falloon T, Green D, Danyushevsky L, Faul U (1999) Peridotite melting at 1.0 and 1.5 GPa: An experimental evaluation of techniques using diamond aggregates and mineral mixes for determination of near-solidus melts. J Petrol 40:1343–1375

Frost BR, Barnes CG, Collins WJ, Arculus RJ, Ellis DJ, Frost CD (2001) A geochemical classification for granitic rocks. J Petrol 42:2033–2048

Ghiorso MS, Sack RO (1995) Chemical mass transfer in magmatic processes IV. A revised and internally consistent thermodynamic model for the interpolation and extrapolation of liquid-solid equilibria in magmatic systems at elevated temperatures and pressures. Contrib Mineral Petrol 119:197–212

Grove TL, Donnelly-Nolan JM (1986) The evolution of young silicic lavas at Medicine Lake Volcano, California: implications for the origin of compositional gaps in calc-alkaline series lavas. Contrib Mineral Petrol 92:281–302

Higgins MD (1990) A three-component composite dyke and its associated intrusion, the Pointe du Criard, Quebec, Canada. In: Parker A, Rickwood P, Tucker D (eds) Mafic dykes and emplacement mechanism. Balkema, Rotterdam, pp 63–68

Higgins MD (1991) The origin of laminated and massive anorthosite, Sept Iles layered intrusion, Quebec, Canada. Contrib Mineral Petrol 106:340–354

Higgins MD (2005) A new interpretation of the structure of the Sept Iles Intrusive Suite, Canada. Lithos 83:199–213

Higgins MD (2011) Textural coarsening in igneous rocks. Int Geol Rev 53:354–376

Higgins MD, Chandrasekharam D (2007) Nature of sub-volcanic magma chambers, Deccan Province, India: evidence from quantitative textural analysis of plagioclase megacrysts in the giant plagioclase Basalts. J Petrol 48:885–900

Higgins MD, Doig R (1977) 540-Myr-old anorthosite complex in the Grenville Province of Quebec, Canada. Nature 267:40–41

Higgins MD, Doig R (1981) The Sept Iles anorthosite complex: field relationships, geochronology, and petrology. Can J Earth Sci 18:561–573

Higgins MD, Doig R (1986) Geochemical constraints on the differentiation processes that were active in the Sept Iles complex. Can J Earth Sci 23:670–681

Higgins MD, van Breemen O (1998) The age of the Sept Iles layered mafic intrusion, Canada: implications for the late Neoproterozoic/Cambrian history of southeastern Canada. J Geol 106:421–432

Hill RI, Campbell IH, Davies G, Griffiths RW (1992) Mantle plumes and continental tectonics. Science 256:186–193

Holness MB, Vernon RH (2015) The influence of interfacial energies on igneous microstructures. In: Charlier et al. (eds) Layered Intrusions, Springer Geology

Holness MB, Tegner C, Nielsen TFD, Stripp G, Morse SA (2007) A textural record of solidification and cooling in the Skaergaard intrusion, East Greenland. J Petrol 48:2359–2377
Hoover JD (1989) The chilled marginal gabbro and other contact rocks of the Skaergaard intrusion. J Petrol 30:441–476
Hounsell V (2006) Origine des roches felsiques de la serie superieure de la suite intrusive de Sept-Iles, Quebec. Master's thesis, University of Quebec in Chicoutimi (UQAC)
Hunter RH, Sparks RS (1987) The differentiation of the Skaergaard intrusion. Contrib Mineral Petrol 95:451–461
Irvine TN (1982) Terminology for layered intrusions. J Petrol 23:127–162
Irvine TN, Baragar WR (1971) A guide to chemical classification of common volcanic rocks. Can J Earth Sci 8:315–341
Irvine TN, Andersen JC, Brooks CK (1998) Included blocks (and blocks within blocks) in the Skaergaard intrusion: geologic relations and the origins of rhythmic modally graded layers. Geol Soc Am Bull 110:1398–1447
Jerram D, Cheadle M, Hunter R, Elliott M. (1996) The spatial distribution of grains and crystals in rocks. Contrib Mineral Petrol 125:60–74
Juster T, Grove TL, Perfit M (1989) Experimental constraints on the generation of FeTi basalts, andesites, and rhyodacites at the Galapagos Spreading Center, 85 W and 95 W. J Geophys Res 94:9251–9274
Kamo SL, Gower CF, Krogh TE (1989) Birthdate for the Iapetus Ocean? A precise U–Pb zircon and baddeleyite age for the Lang Range Dykes, southeast Labrador. Geology 17:602–605
Kruger FJ (2005) Filling the Bushveld Complex magma chamber: lateral expansion, roof and floor interaction, magmatic unconformities, and the formation of giant chromitite, PGE and T-V-magnetitite deposits. Miner Deposita 40:451–472
Kumarapeli PS (1993) A plume-related segment in the rifted margin of Laurentia, southern Canadian Appalachians, seen through a completed Wilson cycle. Tectonophysics 219:47–55
Kumarapeli PS, Saull VA (1966) The St Lawrence valley system: a north American equivalent to the East African rift valley system. Can J Earth Sci 3:639–658
Lajeunesse, P. St-Onge, G. Locat, J., Duchesne, M.J., Higgins, M.D., Sanfacon, R., Ortiz, J. (2013). The Corossol structure: A possible crater on the seafloor of the northwestern gulf of St. Lawrence, Eastern Canada. Meteor Planet Sci 48:2542–2558
Latypov R, Hanski E, Lavrenchuk A, Huhma H, Havela T (2011) A 'three-increase model' for the origin of the marginal reversal of the Koitelainen layered intrusion, Finland. J Petrol 52:733–764
Lightfoot PC, Hawkesworth CJ, Sethna SF (1987) Petrogenesis of rhyolites and trachyes from the Deccan trap: Sr, Nd and Pb isotope and trace element evidence. Contrib Mineral Petrol 95:44–54
Loiselle MC, Wones D (1979) Characteristics and origin of anorogenic granites. Geol Soc Am, Abstracts with Program. 11, 468
Loncarevic BD, Feininger T, Lefebvre D (1990) The Sept-Iles layered mafic intrusion: geophysical expression. Can J Earth Sci 27:501–512
McBirney AR (1989) The Skaergaard Layered Series: I. structure and average compositions. J Petrol 30:363–397
McBirney AR, Naslund HR (1990) The differentiation of the Skaergaard intrusion: A discussion of Hunter and Sparks. Contrib Mineral Petrol 104:235–240
McKenzie D (2011) Compaction and crystallization in magma chambers: towards a model for the Skaergaard intrusion. J Petrol 52:905–930
Morse SA (1990) The differentiation of the Skaergaard intrusion: a discussion of Hunter and Sparks. Contrib Mineral Petrol 104:240–244
Namur O, Charlier B (2012) Efficiency of compaction and compositional convection during mafic crystal mush solidification: the Sept Iles layered intrusion, Canada. Contrib Mineral Petrol 163:1049–1068

Namur O, Charlier B, Toplis MJ, Higgins MD, Liegeois JP, Vander Auwera J (2010) Crystallization sequence and magma chamber processes in the ferrobasaltic Sept Iles Layered Intrusion, Canada. J Petrol 51:1203–1236
Namur O, Charlier B, Pirard C, Hermann J, Liegeois JP, Vander Auwera J (2011a) Anorthosite formation by plagioclase flotation in ferrobasalt and implications for the lunar crust. Geochim Cosmochim Acta 75:4998–5018
Namur O, Charlier B, Toplis MJ, Higgins MD, Hounsell V, Liegeois JP, Vander Auwera J (2011b) Differentiation of tholeiitic basalt to A-type granite in the Sept Iles Layered Intrusion, Canada. J Petrol 52:487–539
Namur O, Charlier B, Holness MB (2012a) Dual origin of Fe–Ti–P gabbros by immiscibility and fractional crystallization of evolved tholeiitic basalts in the Sept Iles layered intrusion. Lithos 154:100–114
Namur O, Charlier B, Toplis MJ, Vander Auwera J (2012b) Prediction of plagioclase-melt equilibria in anhydrous silicate melts at 1-atm. Contrib Mineral Petrol 163:133–150
Namur O, Abily B, Boudreau A, Blanchette F, Bush JWM, Ceuleneer G, Charlier B, Donaldson CH, Duchesne JC, Higgins MD, Morata D, Nielsen TFD, O'Driscoll B, Pang KW, Peacock T, Spandler C, Toramaru A, Veksler I (2015) Igneous layering in basaltic magma chambers. In: Charlier et al. (eds) Layered Intrusions, Springer Geology
Niu Y, Batiza R (1991) In situ densities of MORB melts and residual mantle: implications for buoyancy forces beneath mid-ocean ridges. J Geol 99:767–775
Panjasawatwong Y, Danyushevsky LV, Crawford AJ, Harris KL (1995) An experimental study of the effects of melt composition on plagioclase-melt equilibria at 5 and 10 kbar: implications for the origin of magmatic high-An plagioclase. Contrib Mineral Petrol 118:420–432
Patino Douce AE (1997) Generation of metaluminous A-type granites by low-pressure melting of calc-alkaline granitoids. Geology 25:743–746
Peccerillo A, Barberio MR, Yirgu G, Ayalew D, Barbieri M, Wu TW (2003) Relationships between Mafic and Peralkaline silicic magmatism in continental rift settings: a petrological, geochemical and isotopic study of the gedemsa volcano, central Ethiopian rift. J Petrol 44:2003–2032
Philpotts AR, Carroll M (1996) Physical properties of melted tholeiitic basalt. Geology 24:1029–1032
Philpotts AR, Shi J, Brustman C (1998) Role of plagioclase crystal chains in the differentiation of partly crystallized basaltic magma. Nature 395:343–346
Rivers T (2008) Assembly and preservation of lower, mid, and upper orogenic crust in the Grenville Province: Implications for the evolution of large hot long-duration orogens. Precambrian Res 167:237–259
Rivers T, Martignole J, Gower CF, Davidson A (1989) New tectonic divisions of the Grenville Province, southeastern Canadian shield. Tectonics 8:63–84
Roelofse F, Ashwal LD (2012) The lower and main zone in the northern limb of the bushveld complex: a 1.3 km thick sequence of intruded and variably contaminated crystal mushes. J Petrol 53:1449–1476
Salmonsen L, Tegner C (2013) Crystallization sequence of the Upper Border Series of the Skaergaard Intrusion: revised subdivision and implications for chamber-scale magma homogeneity. Contrib Mineral Petrol 165:1155–1171
Snyder D, Carmichael IS, Wiebe RA (1993) Experimental study of liquid evolution in an Fe-rich, layered mafic intrusion: constraints of Fe–Ti oxide precipitation on the T-fO2 and T-p paths of tholeiitic magmas. Contrib Mineral Petrol 113:73–86
Tait S, Huppert HE, Sparks RS (1984) The role of conpositional convection in the formation of adcumulate rocks. Lithos 17:139–146
Tegner C (1997) Iron in plagioclase as a monitor of the differentiation of the Skaergaard intrusion. Contrib Mineral Petrol 128:45–51
Tegner C, Cawthorn RG (2010) Iron in plagioclase in the Bushveld and Skaergaard intrusions: implications for iron contents in evolving basic magmas. Contrib Mineral Petrol 159:719–730
Tegner C, Wilson JR, Robins B (2005) Crustal assimilation in basalt and jotunite: constraints from layered intrusions. Lithos 83:299–316

Tegner C, Cawthorn RG, Kruger FJ (2006) Cyclicity in the main and upper zones of the Bushveld complex, South Africa: crystallization from a zoned magma sheet. J Petrol 47:2257–2279
Tegner C, Thy P, Holness MB, Jakobsen JK, Lesher CE (2009) Differentiation and compaction in the Skaergaard intrusion. J Petrol 50:813–840
Thy P, Lesher CE, Tegner C (2009) The Skaergaard liquid line of descent revisited. Contrib Mineral Petrol 157:735–747
Toplis MJ (2005) The thermodynamics of iron and magnesium partitioning between olivine and liquid: criteria for assessing and predicting equilibrium in natural and experimental systems. Contrib Mineral Petrol 149:22–39
Toplis MJ, Carroll M (1995) An experimental study of the influence of oxygen fugacity on Fe–Ti oxide stability, phase relations, and mineral-melt equilibria in ferro-basaltic systems. J Petrol 36:137–1170
Toplis MJ, Carroll MR (1996) Differentiation of ferro-basaltic magmas under conditions open and closed to oxygen: implications for the Skaergaard intrusion and other natural systems. J Petrol 37:837–858
Toplis MJ, Brown W, Pupier E (2008) Plagioclase in the Skaergaard intrusion. Part 1: core and rim compositions in the layered series. Contrib Mineral Petrol 155:329–340
VanTongeren J, Mathez E, Kelemen P (2010) A felsic end to Bushveld differentiation. J Petrol 51:1891–1912
Veksler IV, Charlier B (2015) Silicate liquid immiscibility in layered intrusions. In: Charlier et al. (eds) Layered Intrusions, Springer Geology
Wager LR, Brown GM (1968) Layered igneous rocks. Oliver & Boyd, Edinburgh
Wiebe RA, Snyder D (1993) Slow, dense replenishments of a basic magma chamber: the layered series of the Newark Island layered intrusion, Nain, Labrador. Contrib Mineral Petrol 113:59–72
Wilson JR, Larsen SB (1982) Discordant layering relations in the Fongen–Hyllingen basic intrusion. Nature 299:625–626
Wilson JR, Robins B, Nielsen FM, Duchesne JC, Vander Auwera J (1996) The Bjerkreim–Sokndal layered intrusion, Southwest Norway. In Cawthorn RG (ed) Layered intrusions. Elsevier, Amsterdam
Worster M (1992) Instabilities of the liquid and mushy regions during solidification of alloys. J Fluid Mech 237:649–669

Chapter 12
The Bushveld Complex, South Africa

R. Grant Cawthorn

Abstract The mafic rocks of the Bushveld Complex, South Africa, were emplaced into a stable cratonic shield some 2.06 b.y. ago, and have remained remarkably well preserved from deformation, metamorphism and low temperature alteration, at least partly by its isostatic impact on the entire crustal thickness. The generation and emplacement of possibly 1 million km^3 of magma within 65,000 years and the lateral continuity of layering (for up to 100 km) remain intriguing challenges to understanding the evolution of this igneous body. The intrusion is exposed as a three-lobed body, up to 7 km thick, with inward dipping layers that range from dunite to monzonite. Platinum-group element-rich orthopyroxenite, chromitite and vanadiferous magnetitite layers contain vast proportions of the World's deposits of these commodities. Modal layering, on scales from mm to tens of m, ranges from well-developed in some vertical sections to virtually absent in others. Distinctive layers (ranging from mm to many tens of m) can be identified in two or even three lobes testifying to their connectivity. Feeders to the intrusion cannot be identified, and the exact compositions (and numbers) of the parental magmas are still debated. Rates and effectiveness of their mixing also require resolution. Models of magma additions and extents of mixing lead to very conflicting interpretations in terms of rapidity and vertical extents of the sequence affected. As the largest known mafic intrusion it represents an end-member in terms of magmatic chamber processes.

I make the following variably contentious suggestions and conclusions. Many long vertical sections that show modal graded bedding (from orthopyroxenite to anorthosite) with no tendency for minerals to occur in their cotectic proportions may be best explained by grain settling and sorting due to gravity (applying Stokes' Law). In contrast, in the chromitite and magnetitite layers marked vertical changes in composition have been attributed to in situ growth. Magma additions may be inferred from upward reversals in the mg# in olivine and pyroxene in the Lower and Main Zones, and upward increase in the An content of plagioclase in the Main Zone. Such reversals are not abrupt, but are preserved through thicknesses of 100 s of metres. Chromitite layers usually have sharp basal contacts and show no change in the mg# of the mafic mineral above, compared to below, each layer. Models of

---

R. G. Cawthorn (✉)
School of Geosciences, University of the Witwatersrand,
PO Wits, Johannesburg 2050, South Africa
e-mail: Grant.Cawthorn@wits.ac.za

B. Charlier et al. (eds.), *Layered Intrusions,* Springer Geology,
DOI 10.1007/978-94-017-9652-1_12

magma addition and rapid mixing appear inconsistent with such observations and mineral data. Orthopyroxenites in the Upper Critical Zone have an mg# <82 that demonstrates that these minerals formed in equilibrium with plagioclase, but that plagioclase did not co-accumulate with pyroxene. Hence, these pyroxenites did not form from a magma saturated only in pyroxene. At the top of the Critical Zone there is Sr isotopic evidence for major addition of magma, but calculations using the Cr contents of pyroxene preclude any mixing between the magmas. The Cr content of magma needed for chromite saturation and the stability of olivine in the MELTS computer model do not produce crystallization sequences that match actual experimental observations on the MgO- and SiO_2-rich liquids that have been proposed for the parental magmas to the Bushveld Complex

As a consequence of the very slow accumulation rates of grains at the base of the chamber, combinations of Ostwald ripening and annealing caused consolidation into an essentially solid framework close to the grain-magma interface (no more than a few metres). Within most of the layered sequence, the small proportion of interstitial liquid that remained was effectively trapped. During its final solidification, this trapped liquid caused significant changes in mineral compositions, namely (i) decrease in mg# in pyroxenes where present in low abundance, and (ii) incompatible trace-element enrichment in all minerals relative to their original cumulus composition. These effects limit the ability to undertake geochemical modelling based on trace element abundances.

Models that envisage introduction of grain slurries into their present locations either from below or from the sides create more problems than they solve, primarily in terms of producing distinctive layers of near-constant thickness over enormous areas, again by application of Stokes' Law. Furthermore, the low Al_2O_3 content of orthopyroxene is inconsistent with their derivation from a deeper magma chamber.

This chapter summarizes some of the main features of the Bushveld Complex, and then examines some of the main debates and challenges to understanding its magmatic history. Aspects of platinum-group element mineralization are reviewed in a companion chapter. Possibly we (I) err in trying to identify a single (or dominant) process, whereas many mechanisms may have been working in tandem with each being more effective in different situations. Many enigmas remain to be resolved, and there is little agreement regarding almost all aspects of the genesis of this huge body of rock.

Keywords Layering · Cumulate rock · Platinum · Magnetitite · Trapped liquid

History

The first mention of igneous rocks and minerals in the area of South Africa to become famous as the Bushveld was to norite, chromite and magnetite on a map by a German explorer, Carl Mauch, in the late 1860s (Hargar 1934). The first use of the word Bushveld (which refers to savannah with smallish trees—often acacia thornbush—typical on the high plateau north of Pretoria), was the "Plutonic series of the

Bushveld" by Molengraaff (1901). Hall (1932) published a 532-page monograph that included 184 publications on the Bushveld, and by 1976 Molyneux et al. compiled an index of 541 papers and 113 theses, with an addendum (Knowles 1978) of a further 101 publications. It is impossible to speculate on the number of publications now about these rocks. Thus, numerous publications and authors must go unrecognised here, their efforts synthesised by Willemse (1964), Wager and Brown (1968) and Eales and Cawthorn (1996), and recently, primarily on the origin of its ores by Maier et al. (2012). Possibly the single most memorable event was the discovery of the platiniferous Merensky Reef in 1924 that hosts about 75% of the World's platinum-group element deposits, although platinum-bearing chromitites were known by 1908 (Cawthorn 1999).

Location

The Kaapvaal craton (Fig. 12.1a) became stabilized between 3 and 2.7 b.y., ago and has been intermittently submerged (and rocks in the volcano-sedimentary basins—Witwatersrand, Ventersdorp, Transvaal, Waterberg and Karoo—accumulated), but has not been subsequently significantly deformed or disrupted. Considerable time after the Transvaal sedimentary rocks had formed there were the following events:

- intrusion into the Transvaal sedimentary rocks of mafic sills (totalling at least 2 km in thickness) and eruption of mafic to felsic volcanic rocks (Dullstroom and Rooiberg Groups) up to 3 km thick, although relative ages of the sills and volcanic rocks are unresolved,
- intrusion of the layered ultramafic-monzonitic rocks (Bushveld Complex) up to 7 km thick, and
- intrusion of granites (Lebowa Granite Suite), possibly 2 km thick.The South African Committee on Stratigraphy (SACS 1980) proposed that the term "Bushveld Complex" be applied to all the above-mentioned igneous events occurring within a short time interval at about 2.06 b.y. They proposed the name Rustenburg Layered Series for the mafic rocks. For simplicity, and in keeping with many international publications, I will retain the terminology Bushveld Complex for the layered mafic rocks described here.

The preserved extent of the Bushveld Complex, a schematic section showing relations to country rocks, the Bouguer gravity anomaly map and an inferred cross-section are shown in Figs. 12.1b, c, d and e respectively.

Morphology

Floor and Roof

Almost everywhere the Bushveld Complex has intruded into the Transvaal Supergroup, more specifically the alternating quartzites and shales of the Pretoria Group.

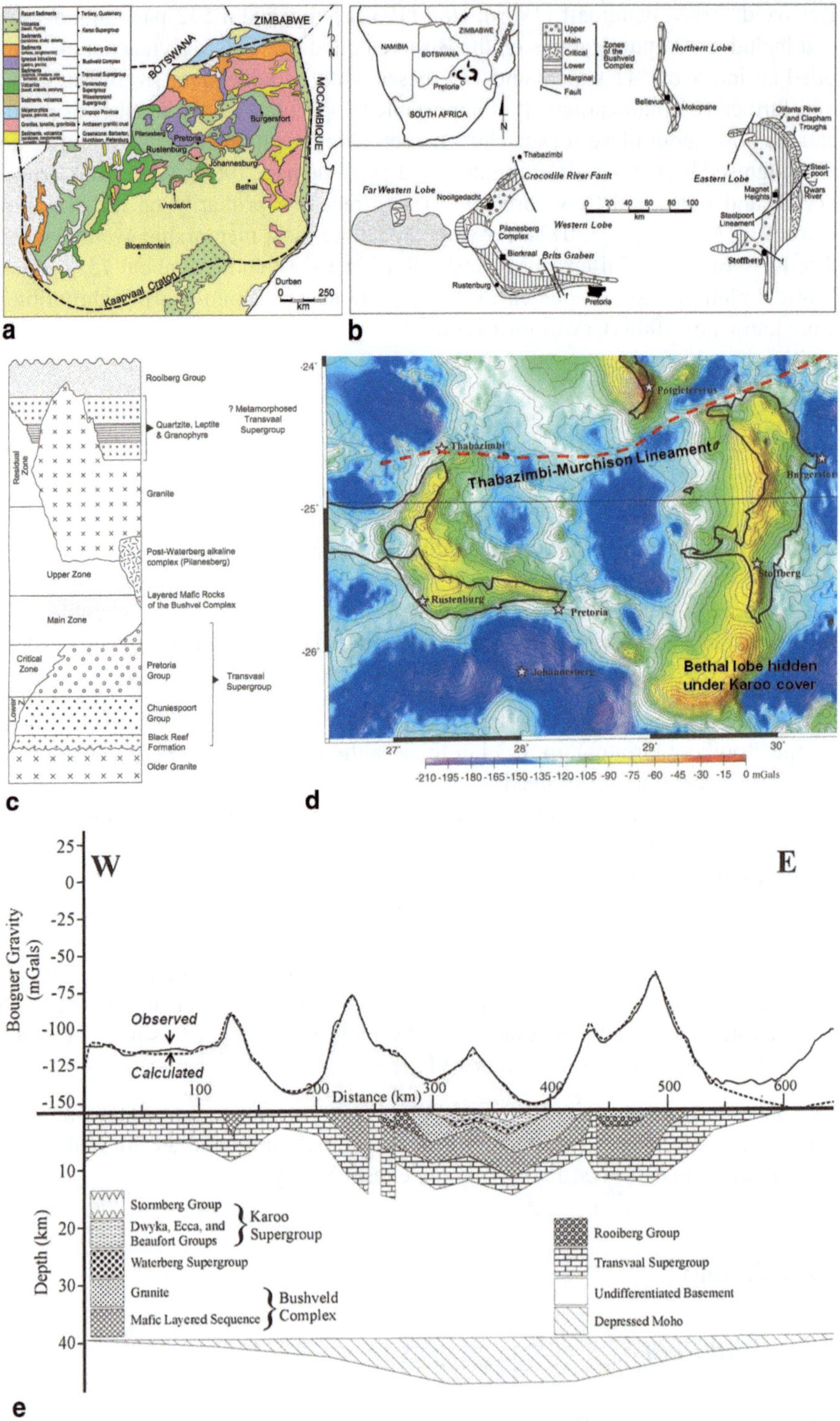

Fig. 12.1 Maps of Bushveld Complex. **a** Regional map showing the Archaean Kaapvaal craton and major sedimentary basins and igneous rocks. **b** Simplified map of the mafic rocks of the

In the west and northeast, emplacement was essentially concordant at the level of the Magaliesberg Quartzite, whereas from Steelpoort (Fig. 12.1b) southward the emplacement was discordant, being emplaced into higher stratigraphic units (Button 1976; Cawthorn 1998). Conversely, in the northern lobe the intrusion was into deeper levels, including the dolomitic Chuniespoort Group and the basement granitic gneisses. All floor rocks have been metamorphosed and variably partially melted (Johnson et al. 2003).

The nature of the roof represents a more enigmatic problem. It is impossible to envisage a mechanically rigid roof of dimensions of over 300 km north-south and east-west atop a magma chamber at up to 1300 °C. The observations of von Gruenewaldt (1972) might be (rather liberally) simplified here to suggest that the roof was composed of the Rooiberg felsic volcanic rocks that were dismembered into enormous blocks that floated on the mafic magma chamber (Cawthorn 2013a). They were continuously melted at their bases, the melt being injected upwards between the blocks, producing granophyres. Residual material was highly metamorphosed, destroying most primary volcanic features. The question of the nature of the original roof rocks is further complicated by the intrusion of the Bushveld Granites close to this boundary.

Vertical and Lateral Extents

The preserved, probably contiguous, sequence can be traced around an area of about 40,000 km^2 (exposed west, east and north lobes), but it is has been proven by drilling of gravity anomalies to occur under Karoo cover in the southeast (called the Bethal lobe; Fig. 12.1a) and immediately north of Thabazimbi (Fig. 12.1b, increasing its extent greatly. Furthermore, its metamorphic aureole can be traced much further outward in the west and east, and the original extent of mafic rocks must have been even greater (Cawthorn and Walraven 1998). They calculated that the volume of this body may be in the order of 1 million km^3.

A stratigraphic column is shown in Fig. 12.2. It implies a maximum thickness of 8 km. However, in many areas various sections, especially of the bottom part, are not developed, and so most profiles do not preserve the entire succession. It cannot be proven but it is now generally accepted that eastern and western lobes are connected through the centre at depth (Nguri et al. 2001), but whether the northern lobe was ever contiguous with the other two is unknown. Very simplistically, the combined eastern and western lobs may be considered to have the shape of a very wide, but irregular funnel with each zone, in places overlapping underlying zones (Kruger 2005). Furthermore, a combination of poor exposure and lack of layering to measure dip, especially in the Main Zone, makes estimates in any profile uncertain.

Bushveld Complex showing the major zones and lobes. At this scale, the outcrop of the Residual or Roof Zone near Stoffberg is too small to demarcate. **c** Schematic block diagram largely taken from Willemse (1964) of the relationships between the major rock formations. **d** Regional Buouger anomaly gravity map including the outcrop of the Bushveld Complex (outlined) as compiled by the Council for Geoscience. **e** West-east section, shown on Fig. 12.1d of gravity data and inferred structure (from Cawthorn et al. 1998). Dotted and solid lines are calculated and observed gravity respectively

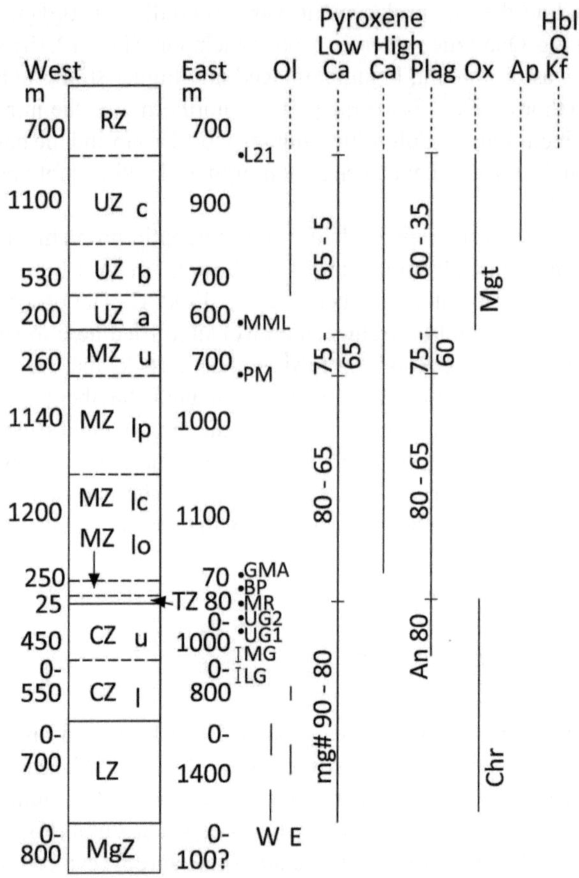

Fig. 12.2 Stratigraphic section through the Bushveld Complex (adapted and enlarged after Cawthorn 2013a). The maximum thicknesses are shown, and should not be summed to obtain actual total thickness anywhere. The solid lines denote cumulus mineral assemblages, dotted lines intercumulus status. Abbreviations of the zones: *MgZ* Marginal, *LZ* Lower, CZ_l Lower Critical, CZ_u Upper Critical, *TZ* Transition, MZ_{lo} Lower Main (primary orthopyroxene), MZ_{lc} Lower Main (cumulus clinopyroxene and orthopyroxene), MZ_{lp} Lower Main (primary pigeonite), MZ_u Upper Main, UZ_a Upper a, UZ_b Upper b, UZ_c Upper c, *RZ* Residual (or Roof). Mineral abbreviations: *Ol* olivine, Low Ca low-calcium pyroxene (orthopyroxene, inverted clinobronzite and inverted pigeonite), high Ca high-calcium clinopyroxene, *Plag* plagioclase, *Ox* oxide (*mgt* magnetite, *chr* chromite), *Ap* apatite, *Hbl* hornblende, *Q* quartz, *Kf* potassium feldspar. W and E in the olivine column refer to west and east lobes. Important marker horizons are indicated: *LG*, *MG UG* Lower, Middle and Upper Group Chromitites, *MR* Merensky Reef, *BP* Bastard Pyroxenite, *GMA* Giant Mottled Anorthosite, *PM* Pyroxenite Marker, *MML* and *L21* Main Magnetitite Layer and magnetitite layer 21. Generalized cumulus compositional trends are shown for low-calcium pyroxene (mg#) and plagioclase (anorthite content)

The vertical succession is divided into zones based mainly on mineral assemblages. The Lower Zone may reach 2 km thick, but is only present in a series of troughs. The most spectacular outcrops (Fig. 12.3a) are developed in the Olifants River section (Fig. 12.1b), described by Cameron (1978). Another detailed study

Fig. 12.3 Assorted photographs of aspects of the layering of the Bushveld Complex. **a** Dunite-pyroxenite layering on a scale of tens of m in the Lower Zone in the Olifants River trough. **b** and **c** Boundary between Lower Critical Zone and Upper Critical Zone on Jaglust (Cameron 1978). Feldspathic pyroxenite is overlain by a thin chromitite layer (MG2) and leuconorite. The chromitite layer displays undulations (like load structures) as displayed by the undersurface of that chromitite and anorthosite seen in an overhanging slab (**c**). Width of view 20 cm. **d** Thin chromitite layers with sharp upper and lower contacts interlayered with pyroxenite and pegmatitic pyroxenite. Note that the pegmatitic facies can occur directly above a chromitite and also terminate against a pyroxenite of normal grain size (Steelpoort chromite mine). **e** Sharp upper and lower

Fig. 12.3 (continued) contacts to thin chromitite layers in poikilitic anorthosite. Note the small lenses of anorthosite and birfurcations of the chromitite layers (Dwars River). **f** Rare outcrop of the Pyroxenite Marker, (on the farm Mooimeissiesfontein) eastern lobe, showing parallel cross bedding in the overlying gabbronorite. Width of view 1 m. The more massive and resistant nature of this rock is due to the inversion of many small grains of pigeonite into large poikiloblasts of orthopyroxene that make a powerful "cement". **g** Random spinifex textured pyroxenite about 10 m from the floor contact (near Burgersfort). Coin is 2.5 cm. **h** Thinly layered melagabbronorite—leucogabbronorite

Fig. 12.3 (continued) a short distance below the Pyroxenite Marker, Main Zone. All rocks contain cumulus plagioclase and two pyroxenes, but none are in their cotectic proportions (Quadling and Cawthorn 1994). Note the planar cross-bedded structure and termination of mafic layers. Photo from Dr TG Molyneux. **i** Disruption of thinly layered sequence shown in (**h**) and redeposition of angular fragments testifying to the force of horizontal currents, and the solid nature of the cumulate pile immediately below the mush-liquid interface. **j** Main Magnetitite layer with a sharp, planar contact to underlying anorthosite (Magnet Heights). Width of view 50 cm. **k** Rounded anorthosite fragment in thin chromitite layer. Note depression of the lower contact of the chromitite due to the impingement of the autolith (Dwars River). Width of view 25 cm. **l** Assorted fragments (autoliths) in disseminated MG2 chromitite overlying feldspathic pyroxenite. Two of the fragments are anorthosite, but there are allegedly no rocks with cumulus plagioclase below this level (2 km south of Jaglust). **m** Magnetitite layer 1 (first above the Main Layer) showing abrupt underlying contact with anorthosite and upper contact grading into anorthosite. Note the absence of pyroxene. **n** Thinly layered anorthosite—pyroxenite in footwall to UG2 at Impala Platinum Mines (Rustenburg). **o** and **p** two examples (analogous to the Dwars River locality) showing bifurcation of chromitite (**o**) and

Fig. 12.3 (continued) extremely thin lenses (**p**) of pyroxenite in LG6 Chromitite layer. Coin is 2.5 cm. **q** and **r** Disruption of thin chromitite layers **q** showing brittle fragmentation unlike the plastic deformation shown in **s**. The large oikocryst of orthopyroxene on left has grown across the boundary between the layered chromitite-anorthosite sequence and the disrupted vertical mass of anorthosite. Width of view 50 cm. Close-up (**r**) of edge of the chromitite layer at bottom left of (**q**). Note that where the anorthosite has re-intruded through the chromitite layer a trail of chromite grains has been smeared upward along the contact between the primary anorthosite and the remobilised anorthosite (Dwars River). **s** Extreme deformation emphasized by the thin chromitite layer. Note that the chromitite layer retains extremely sharp contacts despite deformation, and shows no tendency to disaggregate (Dwars River). **t** and **u** Plan view of deformed melanorite—anorthosite (**t**) and brecciation of anorthosite fragments (**u**) and redeposition in norite (below UG3 chromitite, Maandagshoek). Coin is 2.5 cm. **v** Vertical mine face showing "Flame Bed" of variably slumped norite and poikilitic anorthosite layers below the Merensky Reef on Impala Platinum Mines (Rustenburg). Width of view 50 cm. **w** Examples of the extreme thinness of chromitite layers, and even mere heavy disseminations in planar bedding in anorthosite. **x**, **y** and **z** Assorted features of disruptive iron-rich ultramafic bodies. Vertical dyke (**x**) turning into a horizontal structure. Width of view 5 m. Concordant body (**y**) with a perfectly planar basal surface but more irregular upper contact to anorthosite. Width of view 8 m. Highly irregular margins to two discordant bodies (**z**). Width of view 4 m. **x** and **y** are from Lonplats mines, and **z** is from Tweefontein (Tegner et al. 1994)

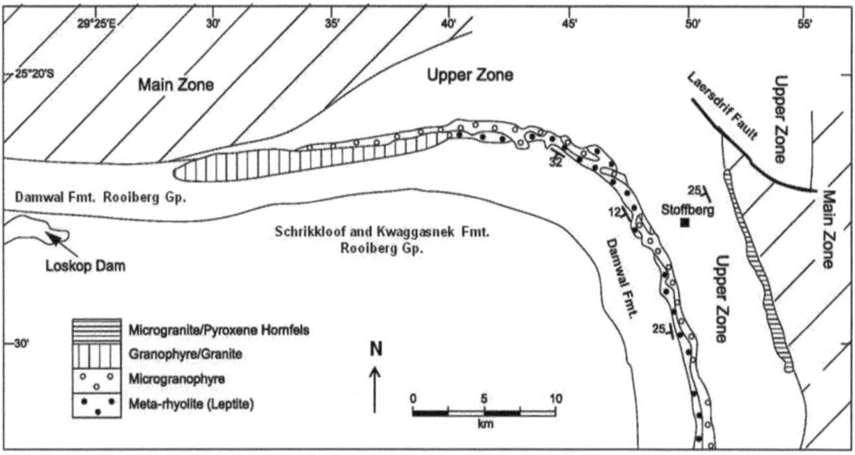

Fig. 12.4 Geological map of the Loskop Dam area in the southeastern Bushveld, showing how the mafic rocks at the upper contact with the roof rock felsites may range from Main Zone in the west to Upper Zone in the east. Mainly from the 1:250,000 maps of the Council for Geoscience, South Africa, and Groeneveld (1970)

comes from a borehole section at Nooitgedacht (Fig. 12.1b), Union Mine (Teigler and Eales 1996). The lateral extents of the Critical, Main and Upper Zones progressively increase (Kruger 2005) as the magma chamber expanded with sequential filling (Fig. 12.1b). This expansion is most obvious in the southeast. The extent to which Lower and Critical Zones may persist below the onlap of the Main Zone there is unknown.

The top contact is also discordant (Fig. 12.4), most obviously near Stoffberg (Fig. 12.1b). From northwest to southeast in this area, the topmost rocks range from Main Zone through Upper Zone to a poorly exposed sequence recorded by Hall (1932) as syenite and Groeneveld (1970) as fayalite diorite, for which the name Residual or Roof Zone (and rock type—monzonite) has recently been proposed (Cawthorn 2013a).

Continuity

Hall's (1932) cross-section from west to east showed the western and eastern lobes as connected in the form of a lopolith. Initial regional gravity interpretations suggested they were not connected, but reinterpretations of the regional gravity data from the Council for Geoscience, shown in Fig. 12.1d by Cawthorn et al. (1998) and Webb et al. (2011) reinstated the connectivity hypothesis (see modelled cross-section in Fig. 12.1e). Apart from demonstrating the isostatic effects of this thick, dense body, causing depression of the continental crust—mantle boundary, Cawthorn et al. (1998) noted the existence of numerous very distinct matching layers of rocks in eastern and western lobes. At the level of the Lower Zone discrete elongate finger-like bodies may have existed (Uken and Watkeys 1997a), but the occurrence in both lobes of the same number of distinctive chromitite layers of comparable compositions in the Lower Critical Zone suggests consanguinity from that level upward. The southern portion of the northern lobe appears lithologically different from the equivalent levels in the west and east. Lower Zone rocks are recognised, but debate continues as to the extent of equivalent Critical Zone rocks in the north (Kinnaird et al. 2005). The scarcity of chromitite layers suggests a discrete evolution of the northern chamber at this stage. Debate also continues as to whether the Platreef (a coarse-grained platiniferous orthopyroxenite) in the north is the equivalent of the Merensky Reef in the west and east. It is complicated by the fact that the Platreef has been extensively modified at the liquidus and subsolidus stages by interaction with the underlying floor rocks, especially the very reactive Chuniespoort dolomites (Kinnaird et al. 2005). Above this level (into the Main Zone), boreholes studied by Ashwal et al. (2005) and Roelofse and Ashwal (2012) in the northern lobe provide evidence for possible correlations with the east and west lobes, in that the mineral assemblages and compositions in the northern lobe have many similarities with the Main and Upper Zones elsewhere. A difference exists in the middle of the Main Zone where a chemically distinct Pyroxenite Marker in the west and east may have a lateral representation as a troctolite layer in the north. Again, above this layer the mineral compositions and distinctive magnetitite layers, especially in terms of thickness and vanadium content in the north find analogues in the west and east (Cawthorn and Molyneux 1986). Connectivity therefore probably also extended to the northern lobe at this level, but perfect lateral homogeneity may not have been achieved. The southeastern or Bethal lobe is only known from borehole information (Buchanan 1979), and consists of Upper Zone lithologies. It was probably connected with the eastern lobe over a structural arch that was possibly only overflowed by the transgression of Upper Zone magma (Kruger 2005). In the far western lobe, erosion has removed all except Marginal to Lower Critical Zones (Engelbrecht 1985), but the extensive thermal aureole (Engelbrecht 1990) suggests that a very thick and laterally extensive body originally existed here.

Feeders

The location(s) of vertical feeders that must have existed remain elusive. Dyke swarms have variably been proposed (Uken and Watkeys 1997b; Olssen et al. 2010), but none has the right age or distinctive composition (see "Parental Magmas") to indicate magma injection mechanisms (Maré and Fourie 2012). Eales (2002) proposed a feeder (of undefined shape) in the northwest, based on the gravity high. However, this "high" occupies an extent of more than 30 km north-northeast by several km perpendicularly. Such an enormous area (bigger than almost every other layered intrusion in the World) cannot represent a feeder, but indicates an area underlain by very thick ultramafic rocks. The east-northeast trending Thabazimbi-Murchison lineament (Fig. 12.1d) has also been proposed as a feeder direction, based on the fact that some of the best studied Lower Zone sections occur in the west, east and north close to this structure (Clarke et al. 2009). Apart from this spatial connection no other evidence exists, specifically, there are no gravity highs along that trend (Fig. 12.1d). Hatton and von Gruenewaldt (1987) documented fundamental differences in the thickness and spacing of most chromitite layers on either side of the Steelpoort lineament (a very straight 60 km-long valley, 2 km wide and devoid of outcrop, Fig. 12.1b). Such a dramatic change in detailed stratigraphic sequence is reminiscent of that observed in the Rum Intrusion where the eastern and western layered sequences cannot be correlated; they are separated by the Long Loch fault within which there is outcrop of chaotic melanges and breccias indicative of a feeder zone. By analogy, in the east (Steelpoort) and in the west (Crocodile and Rustenburg) conjugate faults defining the normal fault directions in a north-south compressive strain ellipsoid (du Plessis and Walraven 1990) could represent potential feeder directions. This issue remains unsolved.

Age

In the last 10 years many age determinations have been undertaken on different minerals from the Rooiberg Group lavas, the mafic layered rocks and the granites. Almost all suggest an age of 2.056–2.060 b.y. These data are reviewed by Scoates et al. (this volume). Intriguingly, they further suggest that there may a time break of 5 m.y. at the base of the Merensky Reef, top of the Critical Zone. Such an assertion creates problems. There is no chill at that level, as expected if there had been such a long time gap. The Sr isotopic evidence for a new magma occurs above the Merensky Reef (as discussed under "Parental Magmas") This disparity needs further investigation.

Stratigraphy

The gross structure of the Bushveld is relatively simple—gently radially inward dipping layered mafic rocks. However, exposure is very poor in the west and north and only generalizations of the stratigraphy are possible there. Exposure is intermittently better in the east. Fortunately, a number of deep boreholes have been drilled by exploration companies and the Council for Geoscience, and provide detailed vertical sections, but lateral variations are known to exist.

The names Marginal, Lower, Critical, Main and Upper Zones have long been used (Hall 1932), but proposed boundaries have changed (and still need reconsideration). The currently used boundaries and subdivisions were formalized by Wager and Brown (1968). Subsequently, the South African Committee on Stratigraphy (SACS 1980) proposed an alternative nomenclature, namely that the mafic layered rocks be termed the Rustenburg Layered Suite. It rejected the zonal nomenclature and proposed names that reflected local geographic features. Thus, the Main Zone was given the names Winnaarshoek Norite-Anorthosite, Leolo Mountain Gabbronorite and Mapochs Gabbronorite for the eastern lobe; Mathlagame Norite-

Table 12.1 Proposed parental magmas to different successions in the Bushveld Complex

	B1	B2	B3	Hendriksplaas Norite	Marginal Zone Rustenburg
SiO_2	5567	50.79	51.33	48.52	50.93
TiO_2	0.33	0.76	0.37	0.20	0.21
Al_2O_3	11.46	15.7	16.14	17.4	18.03
Fe_2O_3	9.96	12.54	10.45	10.19	7.70
MnO	0.17	0.19	0.18	0.23	0.16
MgO	13.36	6.91	7.69	8.53	8.81
CaO	6.32	10.7	11.25	11.92	9.85
Na_2O	1.59	1.94	1.91	1.95	2.28
K_2O	1.07	0.25	0.28	0.11	0.16
P_2O_5	0.06	0.16	0.03	0.03	0.01
LOI	−0.01	0.27	0.17		
(ppm)					
S		177(90)	97(36)		
Ba	293	192	139		91
Cr	1874	201	408		650
Cu	32	76(67)	46 (13)		26
Nb	4.5	3.9	1.4		
Ni	368	106	133		193
Rb	41	4.44	8.86	1	7
Sr	198	348	337	313	295
Y	11.4	19.4	9.92	10	9
Zr	71	54.3	22.9	10	13

Lower Zone (B1) magma—sample from locality described by Davies et al. (1980), with trace elements re-analysed by Barnes et al. (2010). Average B2 and B3 are from Barnes et al. (2010). B2 and B3 are remarkably similar despite the suggestion that they are parental to the Upper Critical and Main Zones respectively. Hendriksplaas norite is from the type locality of the Marginal Zone, eastern lobe (own unpublished data). Marginal Zone, Rustenburg is average from Davies (1982). Both of these two samples are from below Lower and Lower Critical Zones

Anorthosite and Pyramid Gabbronorite for the western lobe, and no proposed name for the same rocks in the northern lobe. I argue that such terminology destroys the important consanguinity of such rocks, and negates attempts to correlate, which I suggest we can do with a high degree of confidence.

The Marginal Zone (called the Hendriksplaas Norite in older literature) is the most enigmatic. Nearly everywhere that it is in contact with the Magaliesberg Quartzite floor rocks, the contact is covered by a thick quartzite scree. Its upper contact with the ultramafic rocks is also rarely exposed. It is a variably fine/medium grained norite showing poor layering, and can reach up to 800 m in the west (Vermaak 1976), based on borehole intersections, but may be totally absent (Wilson 2012). Such thicknesses and variable grain size, suggest multiple injections. However, its depletion in incompatible elements such as Zr (Table 12.1) suggests that it has partial cumulate characteristics. Its composition is not suitable as a parental magma to the Lower Zone (see "Parental Magmas") and may represent multiple, variably differentiated injections of magma, not directly related to the layered mafic rocks. However, no chilled facies can be found between these rocks and the Lower Zone to demonstrate an age difference.

The Lower Zone has a basal orthopyroxene-rich succession, followed by and interlayered with (thin) olivine-rich sequences (Fig. 12.2). (I know of only two clinopyroxenite layers in the entire intrusion, and so the term pyroxenite henceforth refers to orthopyroxenite.) In the Stillwater Intrusion Cooper (1997) has shown that specific layers within such alternations of dunite-harzburgite-orthopyroxenite packages cannot be traced laterally with confidence (based on many closely-spaced boreholes). The outcropping alternating sequence in the Olifants River Trough (Cameron 1978) provides the only opportunity to test for such similarity in the Bushveld. My attempted detailed mapping was not conclusive due to poor outcrop, but in general, olivine-rich layers could not be traced along strike for more than a very few hundred metres out of 6 km of the trough. (Magnesite—the product of surface breakdown of olivine by acid rain-water was usually the only tracer for such olivine-bearing layers.) In this trough, Cameron (1978) suggested that there was an upper pyroxenite within the Lower Zone above which an increase in interstitial plagioclase defined the top of the Lower Zone. However, in the west, the borehole study by Teigler and Eales (1996) showed no increase in interstitial plagioclase, and they argued for placing the top of the Lower Zone at the top of the uppermost olivine-rich layer. I would subscribe to that view. However, if one could totally redefine units, I would include all ultramafic rocks of the Lower Zone and Lower Critical Zone into a single zone, making it comparable to the Ultramafic Unit of the Stillwater Complex.

The Critical Zone was so named, according to Hall (1932), because of the remarkable layering inferred to result from some unspecified "critical" conditions operating in the magma chamber. Those critical conditions are still not understood. It is divided into a Lower and Upper Critical Zone based on the appearance of cumulus plagioclase. The Lower Critical Zone is dominated by pyroxenite with rare olivine-bearing units and chromitite layers. There is a package that contains up to 7 chromitite layers referred to as the Lower Group chromitites (LG1–LG7, numbered from the base upward), but only the LG6 is thick enough to be economic. The continuity of the lowest layers is not known and may be limited, which may indicate that the chamber was still not fully interconnected at these levels. (Layering in the Lower Critical Zone is not very pronounced, so perhaps it should not be designated as "Critical" Zone, but merely part of an Ultramafic Suite, as suggested above.) In the west there is a considerable gap between the LG chromitites and a Middle Group of chromitites (MG1–MG4), whereas in the east there is only a short vertical interval between the LG and MG. It is above the MG2 layer that cumulus plagioclase appears (Fig. 12.3b). The origin of the features analogous to load structures in sedimentary rocks (Fig. 12.3c) at the lower surface of the thin MG3 chromitite remains enigmatic. The relationship between overlying anorthosite (low density) and underlying pyroxenite (high density) negates the possibility that such features are density driven. Above this level the spectacular layering is seen in repetitive sequences typically of chromitite, (rarely harzburgite), pyroxenite, norite, and anorthosite. (They are called cycles, but should actually be called rhythmic units. A cycle ought to be of the kind ABCDCBA, whereas a rhythm is ABCD-ABCD, as discussed by Wager and Brown 1968, p. 545). It is thus only the Upper Critical Zone that results from these "critical" conditions of Hall (1932). Near the top of the Upper Critical Zone occur the Upper Group 1 and 2 chromitite layers (there is a third in the northeast). The uppermost chromitites occurs within the Merensky Unit (in which there may be one, two or three chromitite layers around the Bushveld), although a discontinuous chromitite may occur at the base of the uppermost rhythmic unit, known as the Bastard Unit (based on its remarkable similarity to the Merensky Unit in all aspects except being devoid of platinum-group elements).

Variation in thickness of chromitite layers exists within individual lobes, especially across the Steelpoort lineament (Hatton and von Gruenewaldt 1987). In the north of the eastern lobe the LG chromitites are traceable with constant thicknesses for 100 km. However, south of the Steelpoort lineament there are probably no LG chromitites or at most very trivially thin layers. In contrast, the MG chromitites are very robust south of Steelpoort compared to the north. Debate exists among chromite-mining companies as to whether it is the LG6 or MG1 that is the thick layer mined south of the lineament. I would suggest that it is the MG1 on the basis of whole-rock analyses that show the undisputed LG6 layer north of Steelpoort contains interstitial orthopyroxene only, whereas south of Steelpoort the mined layer has both orthopyroxene and plagioclase (Fig. 12.5), consistent with it having an affinity with the Upper Critical Zone. Faulting cannot change the thickness or vertical spacing of rock layers—in this case, chromitite layers, and so the role of this Steelpoort lineament in controlling the thickness of chromitite layers remains to be resolved. In the western lobe, the LG chromitites apparently do not occur to the east of the Spruitfontein structure about 20 km east of Rustenburg (Clarke et al. 2000), but detailed successions in this area have not been made available by the chromite-mining companies and exposure is lacking.

The sequences of pyroxenite-norite-anorthosite in each rhythmic unit are more complicated than the names imply (as discussed in the section "Origin of Layering"). Small-scale layering with modal variations of orthopyroxene and plagioclase exist, with both sharp and gradational contacts between units. Chromitite layers almost always have very sharp lower and upper contacts (Fig. 12.3d and e).

The boundary between the Critical Zone and the gabbronorite-dominated and largely unlayered Main Zone is not clearly defined. It has been taken within the Merensky Unit on the basis of a Sr isotopic break (see section "Parental Magmas"). An alternative boundary is at the top of the highest prominent rhythmic unit, known as the Bastard Unit. The top of this unit is marked by a poikilitic anorthosite with extremely large oikocrysts of pyroxene (the Giant Mottled Anorthosite). Above this layer are leuconorites which are difficult to distinguish from common similar rocks of the Critical Zone, but with a distinctly lower Cr/MgO whole-rock value (see below). Use of the principle espoused by Wager and Brown (1968) of using mineral

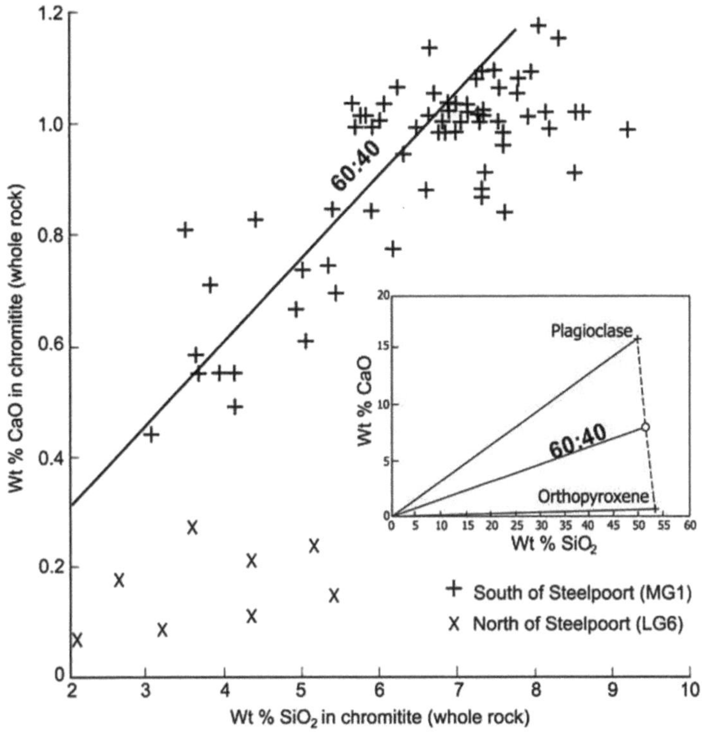

Fig. 12.5 Plot of whole-rock compositions (SiO_2 versus CaO) from the LG6 and MG1 chromitite layers. Note how the LG6 data show a trend toward orthopyroxene as the only interstitial phase, whereas the MG1 data show a trend toward orthopyroxene (40 %) plus plagioclase (60 %) as interstitial phases (as shown in the inset). The LG6 analyses are from north of the Steelpoort lineament where the designation as LG6 is undisputed. The MG1 is from south of the lineament where it has been proposed that this layer may be the same as the LG6 north of the lineament. This chemical difference demonstrates that the MG1 is genuinely different from the LG6. All data with less than 2 % SiO_2 have been excluded. (Data are own unpublished analyses)

(dis)appearance for stratigraphic subdivision would place a boundary some 300 m higher where cumulus clinopyroxene appears (Von Gruenewaldt 1973; Mitchell 1990). Thus, an entirely satisfactory definition for the boundary does not exist. An alternative scheme was proposed by Kruger (1990). He suggested the term Transition Unit for an interval from the Merensky unit up to the Giant Mottled Anorthosite. The reasons were elaborated by Seabrook et al. (2005) who showed that the initial Sr isotope ratio increased a short distance above, but not at, the Merensky chromitite, whereas the Cr content of pyroxene only decreased at the top of this succession. Thus, this interval contains geochemical features of both Critical and Main Zones, and hence the term "transition".

Somewhat higher the primary orthopyroxene is gradually replaced by pigeonite, now inverted to orthopyroxene, producing very large poikiloblasts. Thus, three cumulus pyroxenes appear to co-exist. Even higher, there occurs an important layer, known as the Pyroxenite Marker (Von Gruenewaldt 1973; Klemm et al. 1985a; VanTongeren and Mathez 2013). It has been traced for about 70 km in the east, but then laterally terminates (Von Gruenewaldt 1973), to be replaced by a magnetite gabbronorite. There is only one (borehole core) example of it in the west (Cawthorn et al. 1991), so its ubiquity must be considered doubtful. A correlatable stratigraphic unit in the northern lobe may be represented by a troctolite layer (Ashwal et al. 2005). In the east and west lobes this marker is overlain by a thick zone where primary orthopyroxene reappears. The mg# of pyroxene and the An value of plagioclase increase upward, and the initial Sr isotopic ratio decreases, all over an interval of about 300 m that straddles the Pyroxenite Marker (Von Gruenewaldt 1973; Klemm et al. 1985a; Cawthorn et al. 1991; VanTongeren and Mathez 2013). Locally developed below this marker the rocks may be quite strongly modally layered (Molyneux 1974; Quadling and Cawthorn 1994; Nex et al. 1998), and above it fine-scale modal layering is observed (Fig. 12.3f). Rare anorthosites occur throughout the Main Zone.

The base of the Upper Zone is defined by the appearance of cumulus titanomagnetite, a short vertical interval above which occur layers of almost monomineralic magnetite. 25 have been identified in the east, slightly fewer in the west and north. Two thick and distinctive layers appear to be continuous throughout all three lobes. Near the base is a 2 m-thick layer, mined for its 1 % vanadium. The uppermost layer is more disseminated and up to 6 m thick with only 0.2 % V. Of the many unresolved issues regarding these layers is the little mentioned fact that almost all layers are over- and underlain by anorthosites; cumulus mafic minerals are generally absent.

A three-fold subdivision of the Upper Zone; subzone (a) containing plagioclase, pyroxenes and magnetite; subzone (b) also with olivine and subzone (c) also with apatite was proposed by Wager and Brown (1968). I suggest this subdivision was strongly influenced by their subdivision of the Skaergaard Intrusion, and also the fact that there is good exposure of such a section near Magnet Heights in the east (Fig. 12.1b). Such a subdivision ignores the conclusions of Hall (1932), who reported rocks that he called syenite in all lobes at the top of the intrusion. They were described in more detail by Groeneveld (1970) from the southeast, but not included into any formal stratigraphy despite being 700 m thick according to him. I have proposed that their existence requires incorporation into the stratigraphy as the Residual or Roof Zone (Cawthorn 2013a). These rocks do not display an obvious cumulus texture, and contain increasing proportions of hornblende, quartz and potassium feldspar. It is relevant here to contrast the Bushveld and Skaergaard successions. The Skaergaard crystallized from bottom, top and sides (albeit different thicknesses). No sides are known for the Bushveld, and the crystallization is unidirectional with no accumulation from the top. This difference is almost certainly due to the nature of the roof rocks. In Skaergaard they are refractory basalt. In the Bushveld they are felsic volcanics that would have been intermittently melted by the underlying basaltic magma and hence, no crystals could have adhered to the upper surface.

Two important discordant features in the Bushveld need mention. The first occurs north of the Pilanesberg Complex (Fig. 12.1b) where the Upper Zone rocks cut through the Main and Critical Zone and abut the floor rocks. Faulting can be ruled out because the floor rocks are not displaced, and layers of magnetitite drape with centripetal dips into this large depression rather than dipping uniformly toward the centre of the complex. Wilson et al. (1994) presented a detailed study of the mineral compositions and initial Sr isotope ratio of samples from a borehole close to the edge of the northern transgression. They identified xenoliths of Main Zone rocks engulfed in the host Upper Zone and concluded that there had been major injection of the magma that produced the Upper Zone disrupting the original layered succession. The fate of the disrupted Main and Critical Zone rocks is unknown. (Each of these two transgressive Upper Zones lobes is the size of the Skaergaard intrusion.) The second feature is the presence of discordant (perpendicular to the layering) pegmatitic iron-rich ultramafic bodies (Scoon and Mitchell 1994). They are extremely irregularly distributed round the intrusion. They trend from olivine- and clinopyroxene-rich near the base to magnetite-rich near the top.

Mineral Compositions

Mineral compositions define a stratigraphy. Generalized trends of vertical changes in mineral compositions from many studies have been compiled previously, and summarized in Fig. 12.2. The overall fractionation trends of An value in plagioclase and mg# in mafic minerals belies a much more complicated pattern in detail. In strongly modally layered rocks mineral compositions (both major and trace elements) can be modified by reaction with the interstitial magma, especially for the mafic minerals, which greatly complicates interpretation (as discussed in the section "Trapped Liquid Shift Effect").

In the ultramafic rocks of the Lower and Critical Zones the minerals show a pattern of differentiation and reversals (Fig. 12.6), such that through a vertical interval of over 1500 m (at the extreme thickest) only minor net differentiation can be observed. Profiles from west and east are broadly similar, but it is not possible to suggest that any specific reversal can be traced that distance laterally. Recently, a section in the east containing even higher mg# phases ($Fo_{91.5}$) has been identified in borecore (Wilson 2012). In the north, comparable highly Mg-rich compositions have been identified (Hulbert and von Gruenewaldt 1985). The very minor interstitial plagioclase is quite strongly zoned and provides no petrological information.

With regard to Fig. 12.6, near the base of the Lower Zone there is an interval of rocks with cumulus plagioclase. It has been indicated in diagrams by Vermaak (1976) in the southwest, and Cameron (1978) in the northeast, although little information is available apart from its existence. A much more detailed study of these rocks in the northwest was presented by Teigler and Eales (1996), who reported that it was 60 m thick and contained an average of 20% plagioclase. Its location coincides with the most evolved orthopyroxene compositions (Fig. 12.6a and b), and they concluded that its occurrence was consistent with differentiation of the magma. Its existence at three such widely separated places might suggest that mag-

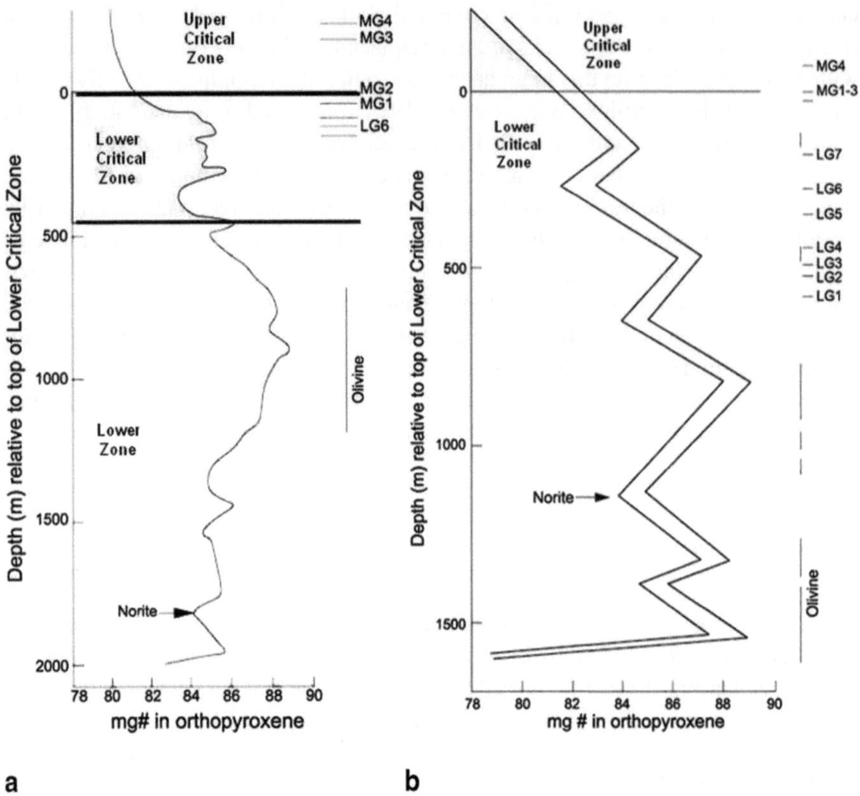

Fig. 12.6 Values of mg# in orthopyroxene in two sections of the Lower Zone and Lower Critical Zone, **a** in the east (Olifants River Trough) from Cameron (1978), and **b** in the west (Nooitgedacht) from Teigler and Eales (1996), showing their zigzag pattern. Localities shown in Fig. 12.1b. Note that there are sections that show differentiation (decreasing mg#), but equally sections with slowly increasing mg# due to slow, sustained recharge. There are no sudden increases in the mg#. In Fig. 12.6c is shown the mg# of orthopyroxene from the two samples immediately below and above each chromitite layer, showing that there is no observable increase in mg# across such layers and hence challenging the model for the origin of chromitite layers by addition of primitive magma

matic continuity existed over such an area early in the evolution of the chamber. Immediately above this unit there is a gradual increase in mg# of the orthopyroxene (over tens to hundreds of metres) from 84 to 88 (together with the appearance of olivine). Despite this convincing evidence of addition of primitive magma there are no chromitic intervals, suggesting that magma addition did not produce chromitite layers. Furthermore, there is no change in the mg# of orthopyroxene from below to above any of the chromitite layers (Fig. 12.6c), again suggesting that addition of primitive magma (and mixing) cannot be the cause of chromitite layer formation.

At the top of the Lower Critical Zone plagioclase becomes a cumulus phase with a composition of An_{80}, and changes little through the entire Upper Critical Zone. In contrast, the orthopyroxene composition oscillates with significant variation in

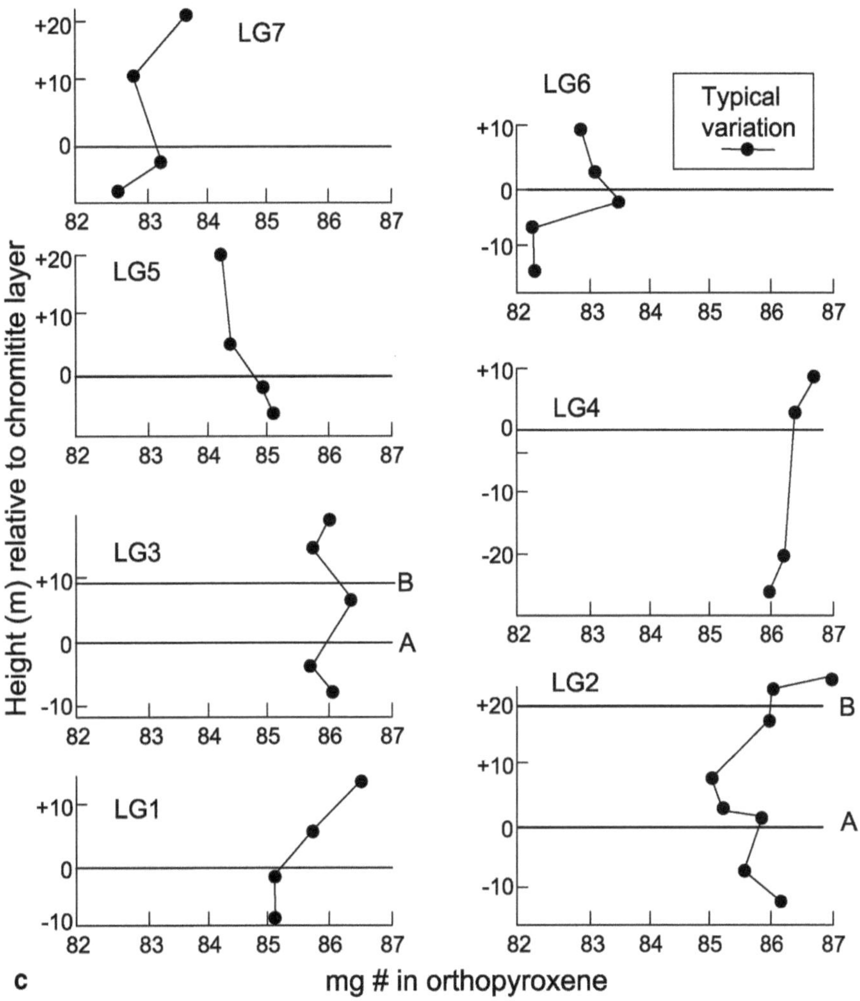

Fig. 12.6 (continued)

mg#. The Cr contents of the cumulus orthopyroxene remain remarkably constant over this interval (Fig. 12.7), despite this mineral being by far the most abundant cumulus phase and having a partition coefficient for Cr of over three and possibly as high as five (Barnes 1986a). Both mg# and Cr contents decrease significantly where the orthopyroxene is an intercumulus phase, especially close to the top of the Upper Critical Zone (Fig. 12.7).

Chromite compositions show a general systematic trend upward (Naldrett et al. 2012). From LG1 to MG2 the trend is one of decreasing mg# and cr#–Cr/(Cr + Al)–due to pyroxene fractionation, but above the MG2 the trend is toward higher cr# due to plagioclase (and pyroxene) fractionation.

Fig. 12.7 Plot of the general trends in wt % Cr_2O_3 contents in orthopyroxene from the Lower, Critical and Main Zones (mainly from a very large data base summarised by Eales 2000, but also Schürmann 1993, Maier and Eales 1997, and Mitchell and Manthree 2002). The first and last authors did not distinguish between cumulus and intercumulus pyroxene, but the general distinction has been made here based on their reported modal proportions

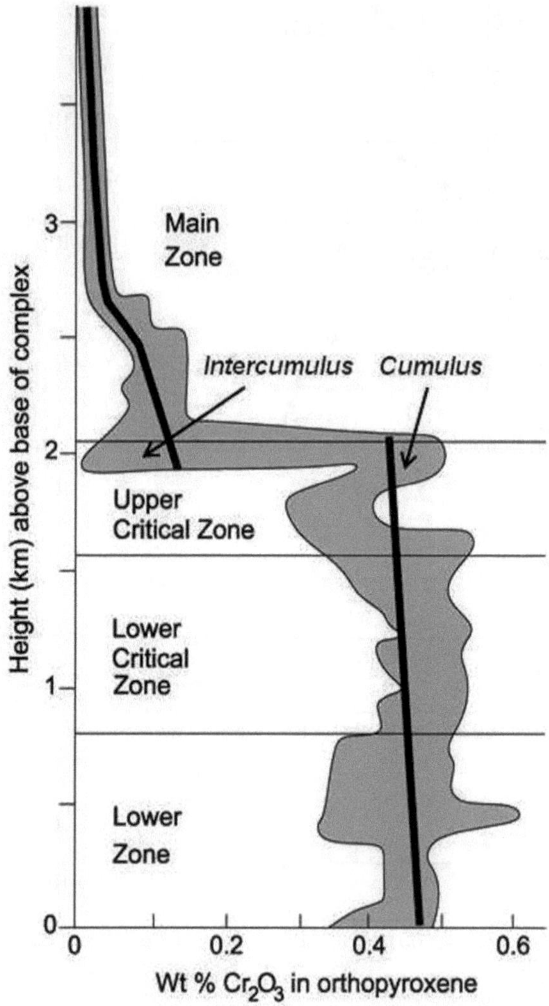

Across the Critical Zone—Main Zone boundary there is no significant change in An or mg# values of cumulus phases. However, immediately above the Giant Mottled Anorthosite in the Main Zone the Cr decreases (Fig. 12.8a) and the Ti increases markedly in the pyroxenes (Mitchell 1990), and a very detailed study shows the whole-rock Cr/MgO abruptly decreases above this level (Fig. 12.8b). Throughout the overlying 1000 m above the Giant Mottled Anorthosite the extent of fractionation in mineral compositions is very subdued, with many very minor oscillations (Mitchell 1990; Roelofse and Ashwal 2012). A major reversal in mineral compositions occurs gradually across the Pyroxenite Marker interval in the west and east (Fig. 12.9). In the southern part of the eastern limb, a reversal occurs even though there is no Pyroxenite Marker. Such lateral variations suggest inhomogeneity in the magma chamber. Sadly, the borehole cores available for the detailed mineralogical studies of Mitchell (1990) and Roelofse and Ashwal (2012) did not traverse this

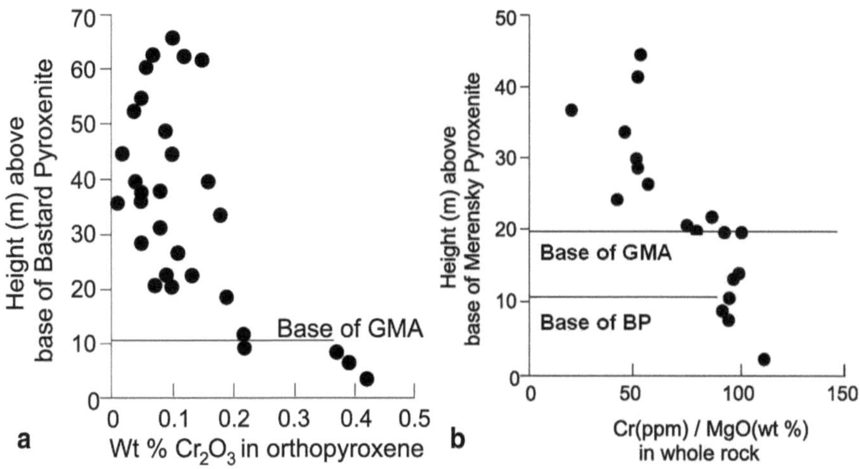

Fig. 12.8 **a** Plot of Cr_2O_3 in orthopyroxene in the section above the Bastard Pyroxenite, which occurs some 10 m above the Merensky Reef (data taken from Mitchell 1990), and **b** of whole-rock Cr/MgO (own unpublished data). GMA is Giant Mottled Anorthosite, the top of which is conventionally considered base of Main Zone

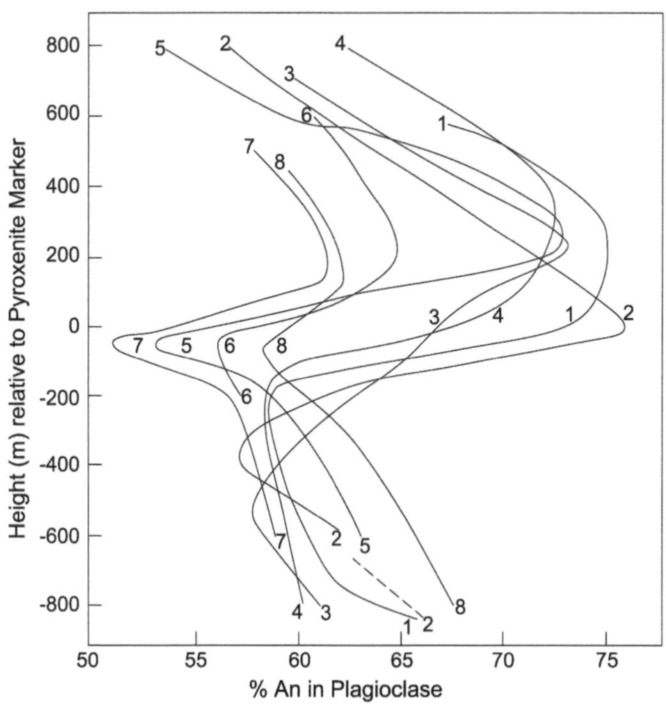

Fig. 12.9 Plot of An content in plagioclase versus height in the Main Zone, relative to the Pyroxenite Marker or top of magnetite gabbronorite, eastern lobe. The profiles are numbered from north to south over a distance of more than 100 km. Profiles 1, 4 and 8 from Lundgaard (2003); profile 2 from vanTongeren and Mathez (2013); profile 3 from von Gruenewaldt (1973); profiles 5, 6 and 7 from Klemm et al. (1985a)

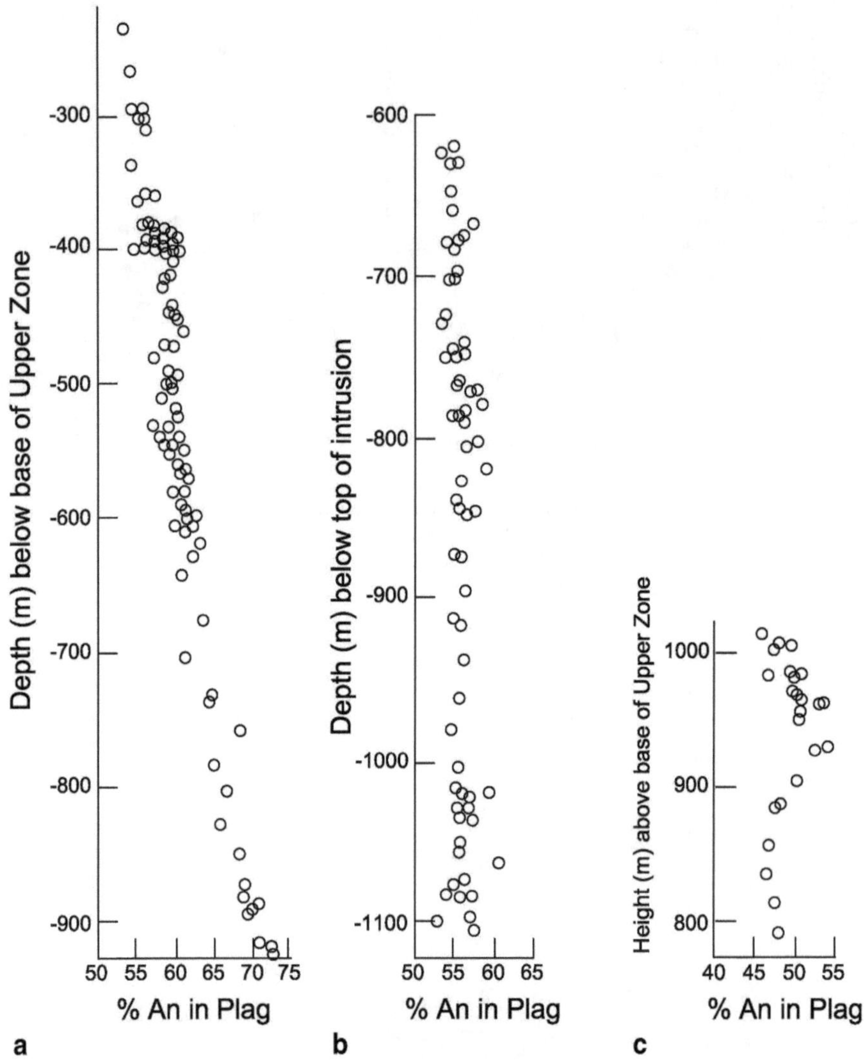

Fig. 12.10 Variations in plagioclase compositions from sections of the Upper Zone from Bellevue borehole (see Fig. 12.1b) using data from Ashwal et al. (2005). **a** shows a section with regular upward decrease in An content, **b** shows a long section with near-constant An content, and **c** shows an interval that includes a slow reversal in An content through about 90 m

interval, and so variability in the western and northern lobes across this interval remains to be investigated.

Oscillations in mineral composition are recognised in the Upper Main and Upper Zones (Ashwal et al. 2005; Tegner et al. 2006, Cawthorn and Ashwal 2009), although there is a distinct overall differentiation upward. The most evolved compositions reported are Fo_3 and An_{37}. But there are distinctly different patterns where the compositions of the plagioclase may show a systematic significant differentiation (Fig. 12.10a), or remain constant for considerable vertical intervals in another sequence (Fig. 12.10b), or show a slow reversal (Fig. 12.10c). The An content of the plagioclase remains constant across all the magnetitite and anorthosite layers (Cawthorn and Ashwal 2009). The mafic minerals show an even more irregular pattern, which does not coincide with the pattern for plagioclase. Such irregularity may result from re-equilibration with oxide phase and/or the trapped liquid shift effect.

In the magnetitite layers, the V content of magnetite decreases, albeit rather erratically, upward (Klemm et al. 1985b; Cawthorn and Molyneux 1986; Eales and Cawthorn 1996) from about 1% V at the base to 0.2% at the top. However, the analyses of disseminated magnetite between the layers show a more complex pattern, not consistent with differentiation of a single, homogeneous magma (Klemm et al. 1985b). Again, the extent of change of composition of the disseminated magnetite due to reaction with trapped liquid or subsolidus reaction with other silicate minerals has not been adequately investigated (cf. Duchesne 1972). Studies of the Cr content of magnetite in short vertical sections revealed dramatic upward decreases and abrupt reversals over very short intervals, even within single layers of magnetitite (Cawthorn and McCarthy 1980; Klemm et al. 1985b).

Textures

The types of cumulate textures stylised by Wager et al. (1960) are rarely seen in Bushveld rocks. Most rocks are chemically adcumulates, but the extent of overgrowth versus assorted maturation processes (mechanical compaction, annealing) and the depth in the cumulus pile at which those processes occurred are unresolved (Hunter 1996). Monomineralic pyroxenites, anorthosites and rare dunites show highly annealed textures (Figs. 12.11a, b and c), with triple junctions (120°) and planar surfaces (not crystallographic primary surfaces, but the result of annealing). Very small grains such as plagioclase and chromite (Fig. 12.11a) occur at triple point boundaries (Boorman et al. 2004). Preserved intercumulus grains with concave and irregular boundaries do occur but are of very minor modal proportions (Fig. 12.11d). However, evidence of considerable resorption can be identified in olivine-rich (replaced by orthopyroxene) and orthopyroxene-rich (replaced by clinopyroxene) rocks (Fig. 12.11e). In the Lower and Critical Zones the orthopyroxene shows multiple, extremely fine, twin lamellae (much thinner than typical albite twinning), which Wagner (1929) concluded were inverted clinobronzite, but they are generally referred to as orthopyroxene. The gabbronorites of the Main Zone do not display any classic cumulate textures; no mineral appears euhedral; all are variably anhedral and interlocking, suggesting significant post-cumulus modification (Fig. 12.11f). The inversion of pigeonite to orthopyroxene in the Main Zone results in very large poikiloblastic grains (Fig. 12.11g). Multiple different orientations of high-temperature, pre-inversion exsolution lamellae testify to the original (pigeonite) grains (von Gruenewaldt 1970). It is this recrystallization that produces a very powerful "cement", resulting in extremely strong (and unfractured) rocks that are used in the dimension stone industry from the well-exposed Pyramid Hills. The

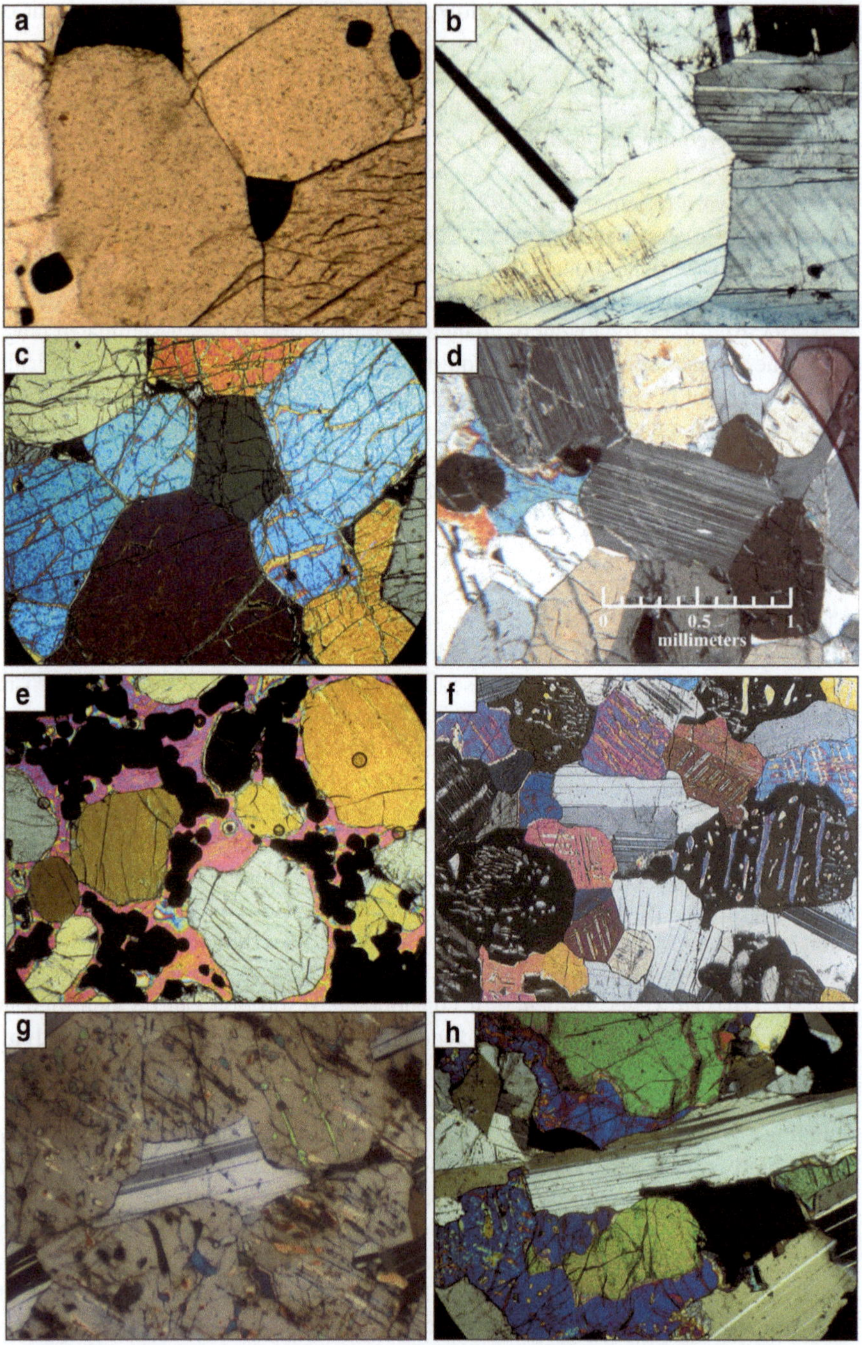

Fig. 12.11 Assorted photographs of textures within the Bushveld Complex. **a**, **b** and **c** Dunite, anorthosite and pyroxenite adcumulates with triple junctions and grain boundaries that are not natural crystal faces. **d** Cumulus orthopyroxene and minor chromite with interstitial plagioclase and clinopyroxene. **e** Cumulus orthopyroxene and chromite, the former showing marked resorption (gaps between orthopyroxene outline and chromite grains) and replacement by large poikilitic clinopyroxene. **f** Main Zone gabbronorite with plagioclase, clinopyroxene (with thin exsolution

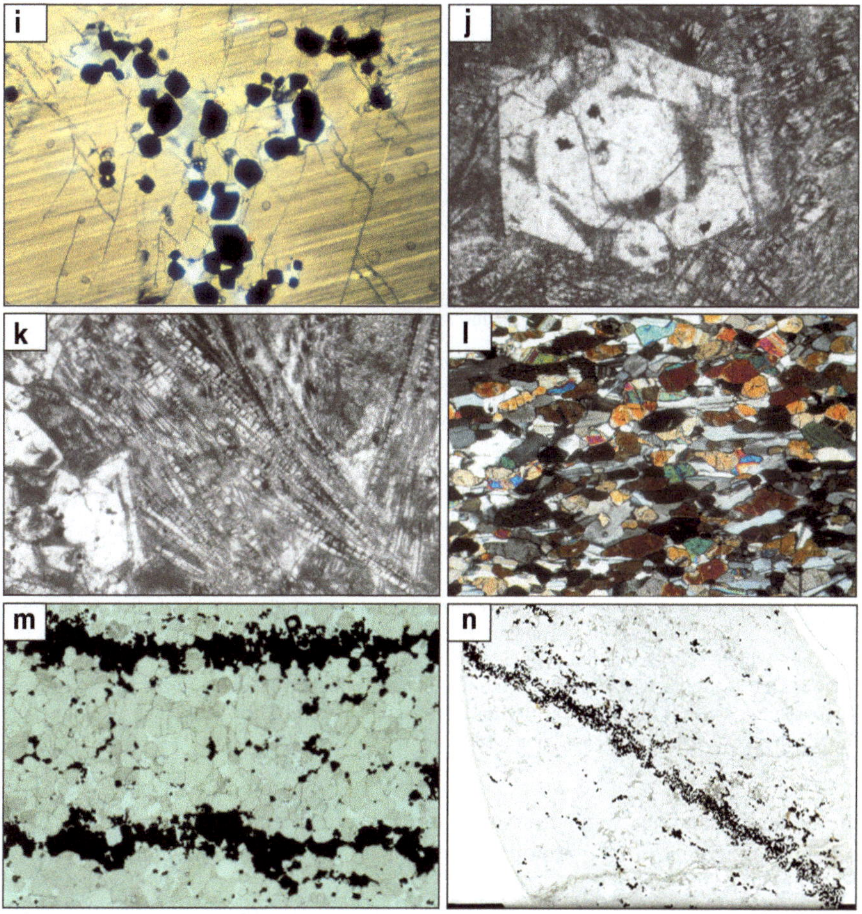

Fig. 12.11 (continued) lamellae) and inverted pigeonite (showing abundant and thick exsolution lamellae). Note that none of the minerals is convincingly euhedral in shape due to significant grain boundary readjustment. **g** A single large optically continuous orthopyroxene grain. However, it can be see that it was originally several grains of pigeonite. Each original grain shows different orientations of lamellae which exsolved prior to inversion to orthorhombic pyroxene. The plagioclase is interstitial to the original pigeonite grains and still preserves its original shape. (*Main* Zone below Pyroxenite Marker). **h** Olivine-magnetite gabbro from Upper Zone. Only plagioclase preserves a relatively euhedral shape. Olivine (three green grains) is irregular in shape and surrounded by anhedral clinopyroxene (blue). A large anhedral magnetite grains is present toward top left. **i** A single large grain of orthopyroxene in pegmatitic pyroxenite from the Merensky Reef. Note that it contains a tricuspate arrangement of chromite grains and anhedral plagioclase. They define the outlines of smaller original orthopyroxene grains that have recrystallized into the single, large grain (Cawthorn and Boerst 2006). **j** Hopper olivine grain in fine matrix from thin sill, the composition of which is considered to be representative of the earliest magma type in floor rocks east of Rustenburg (Cawthorn et al. 1981). **k** Spinifex pyroxene needles in parallel and radiating pattern, with triangular very fine grained matrix at bottom right, from same thin sill as in (j). **l** Microgabbronorite with granular texture of plagioclase and two pyroxenes, the plagioclase defining a weak parallel fabric. (Marginal Zone north of Steelpoort Chromite Mine) **m** Thin chromitic layers in pyroxenite. Note how the grains of chromite drape over and fill between the irregular surface defined by the larger pyroxene grains. **n** Weakly disseminated chromite grains in anorthosite above an extremely thin layer of chromitite

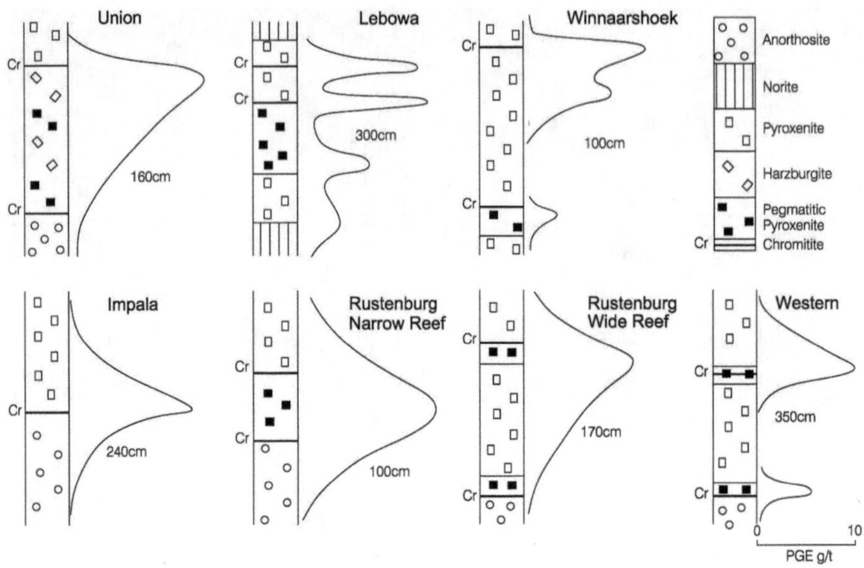

Fig. 12.12 Distribution (semi-quantitative) of platinum-group elements through the Merensky Reef from different mines (updated and modified but mainly from Cawthorn et al. 2002). Note that usually the highest grades do not coincide with pegmatitic pyroxenite and that the highest grade becomes more obviously top-loaded as the reef becomes thicker. The vertical thicknesses of each section in cm are indicated. For specific examples and values in thick reef which show the PGE distribution best, see Mitchell and Scoon (2007)

more massive and resistant nature of this rock is due to the inversion of many small grains of pigeonite into large poikiloblasts of orthopyroxene that make a powerful "cement".

All three lobes have this same thick unit just below the Pyroxenite Marker. In the Upper Zone the distinction between cumulus and intercumulus status deteriorates; only plagioclase has a vaguely euhedral shape; all other minerals are variably anhedral, especially magnetite (Fig. 12.11h). Apatite is euhedral, often very elongate, but always is in igneous rocks, so its texture is no proof of cumulus status.

The oxide minerals possibly show the greatest degree of textural re-equilibration because of their simple structure. Annealing of chromite has been described by Waters and Boudreau (1996). The magnetite also shows re-equilibration with almost monomineralic rocks resulting from massive coarsening of the grain size up to 2 cm (Reynolds 1985; Willemse 1969).

The coarse-grained pyroxenites present major problems in interpretation. They have been reported associated with chromitite layers, but can occur elsewhere. Often they occur below the chromitite where they have been interpreted as the result of trapping of evolved fluids (Mondal and Mathez 2007). However, they do occur within and above chromitites (Fig. 12.3d). They occur within the Merensky Unit (Fig. 12.12), usually above a thin chromitite. Here the texture has been attributed to the trapping of a fluid that caused almost complete hydration melting (Mathez et al.

1997). It is not clear to this author why such an almost totally melted layer did not penetrate the overlying cumulate package. The textures of these rocks pose further problems. The large orthopyroxene grains (up to 2 cm) contain many inclusions of tricuspate plagioclase often associated with trails of similarly clustered chromite grains (Fig. 12.11i) that have been interpreted as the result of textural annealing from many small grains close to the solidus (Cawthorn and Boerst 2006), aided by a temporary cessation in grain accumulation at the grain-liquid interface.

With reference to the genesis of the platinum mineralization it is often stated to be associated with the pegmatitic pyroxenite, an oversimplification, influenced by publications on the earliest mined sections at Rustenburg Platinum Mines (Vermaak 1976). Vermaak himself warned against assuming great lateral uniformity. Where the vertical succession through the Merensky is chromitite—pegmatitic pyroxenite—chromitite there is a convenient cut-off grade and width of 80–100 cm (Fig. 12.12). However, where the unit becomes thinner or thicker the separation between pegmatitic textures and mineralization becomes progressively clearer (Fig. 12.12). Sections that contain no pegmatitic pyroxenite may still contain good grade as shown for Impala mine, specifically its northern portion (Leeb-Du Toit 1986), and in thicker sections the mineralization and pegmatitic pyroxenite need not be correlated, as seen in the sections from Western/Lonplats (Davey 1992) and Winaarshoek/Marula (Scoon and Mitchell 2009) Platinum Mines.

The textures in the fine-grained rocks at the margins of the intrusion and as sills below, proposed as parental magmas deserve highlighting. Spinifex-like textures (random and parallel) of the pyroxenes were described by Davies et al. (1980), Sharpe (1981) and Wilson (2012) as shown in Figs. 12.3g and 12.11j and k. They have been termed the B1 compositions (Table 12.1).

Textures seen in the chilled sampled considered to represent the subsequent magmas, termed B2 or B3 compositions, are totally different. These rocks have microgranular textures with polygonal grain boundaries between the plagioclase and pyroxenes (Fig. 12.11l). They are reminiscent of the textures described by, for example, Latypov et al. (2011) as being the result of partial melting and/or annealing.

A final, cautionary note regarding textures needs to be mentioned. The literature (not just Bushveld) often contains phrases along the lines, "the pyroxenite contains 90 % cumulus pyroxene and 10 % intercumulus plagioclase", implying 10 % of intercumulus liquid. Texturally, it might be valid to say that there is 10 % anhedral or interstitial plagioclase, but the genetic implication is almost certainly wrong. Two alternative descriptions are more plausible. (1) This 10 % of plagioclase may have the same composition as the plagioclase in adjacent norite where it is demonstrably cumulus, and hence the rock should be termed heteradcumulate feldspathic pyroxenite (Wager et al. 1960). (2) The liquid forming the 10 % of plagioclase is almost certainly saturated in both plagioclase and pyroxene, and those minerals crystallize in approximately the proportions 60 % plagioclase and 40 % pyroxene (and maybe significant proportions of other minerals). Thus, if there is 10 % intercumulus plagioclase then there must also be 7 % of intercumulus pyroxene and variable proportions of other minerals. The correct restatement of the above quote would then become "the pyroxenite contains 83 % cumulus pyroxene and 10 and 7 % of intercumulus plagioclase and pyroxene respectively (and other minerals)." Both of these mechanisms are plausible (even contributing in the same rock), and both drastically change the implied proportions of cumulus and intercumulus minerals, and intercumulus liquid.

Parental Magmas

First Magma Type

Many dyke swarms are known on the Kaapvaal craton (Uken and Watkeys 1997b; Olssen et al. 2010), but none have the distinctive geochemical signature of the Bushveld. The enigmatic Marginal Zone varies along strike in thickness, in grain size (but generally medium grained) and composition, making rocks of the Marginal Zone unsuitable for predictions of the parental magma(s) compositions. Wager and Brown's (1968) proposed parental magma emphasized their perceived similarities between the Bushveld and Skaergaard, but whereas the Lower Zone at Skaergaard is dominated by troctolite cumulates, in Bushveld it is dominated by orthopyroxene (and less olivine). Thus, parental magmas cannot have been similar. Experimental work on Wager and Brown's proposed Bushveld parent magma demonstrated this fallacy (Tilley et al. 1968; Biggar 1974). Sills adjacent to the intrusion offered an alternative for estimating parental magma. Two types of sill were identified by Willemse (1969), one microgabbroic (and extensively metamorphosed) and one micronoritic. Further subtypes were distinguished by (Sharpe 1984). The latter includes a range of compositions characterised by relatively high SiO_2 and MgO. Actual analyses (Table 12.1) of fine-grained and quench textures members of this suite range from 10 % MgO (Maier et al. 2000) to 14 % MgO (Davies et al. 1980; Wilson 2012), although Wilson argued, on the basis of mg# in olivine, that a magma with 19 % MgO must have been parent to at least some of the lowest cumulates. No full analysis was given by him. If it accepted that the most voluminous ultramafic rocks contain olivine and orthopyroxene with mg# not greater than 91 then the parent magma ought to produce such compositions. Magma with 10 % MgO would produce less magnesian minerals. Magma with 19 % MgO would crystallize abundant olivine before orthopyroxene saturation, yielding dunite (which is very scarce) rather than harzburgite with cumulus orthopyroxene. There was perhaps no single magma composition that was repetitively injected into the chamber, but a suite of related compositions, all being characterised by high Cr, MgO and SiO_2. The orthopyroxene in the Lower and Lower Critical Zones oscillate in mg# up to 88 (Cameron 1978, 1980; Teigler and Eales 1996), which would have crystallized from a magma with about 13 % MgO and so that might be considered the dominant added magma composition, at least in the east and west.

Modelling the crystallization of chromite from parental magmas is another contentious issue. This author believes that the use of the MELTS crystallization programme to model these MgO- and SiO_2-rich magmas (as used by Naldrett et al.

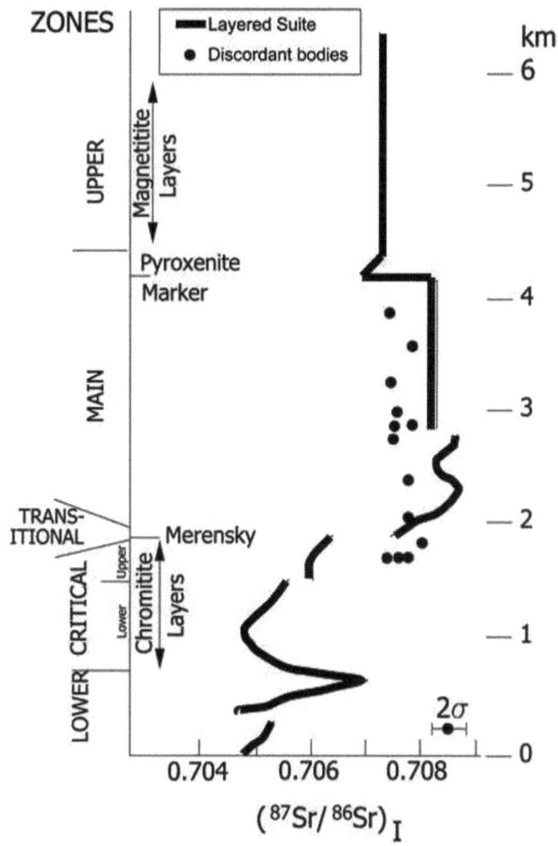

Fig. 12.13 Initial $^{87}Sr/^{86}Sr$ values from the entire succession as compiled by Kruger (1994). Also included are data from the discordant ultramafic bodies that cut the Critical and Main Zones compared to the ratio for the layered rocks (Data from Scoon and Mitchell 1994 and Cawthorn et al. 2000). The distinctive and constant values for these bodies shows that they are not derived from local material

2012; Eales and Costin 2012) to predict the crystallization of chromite is inappropriate. MELTS indicates that such magmas only become saturated in chromite at ±2000 ppm Cr, whereas the actual experimental studies of Barnes (1986a), Murck and Campbell (1986), Cawthorn and Biggar (1993) and Roeder et al. (2006) yielded chromite in equilibrium with magnesian liquid containing only 1000 ppm Cr at the QFM buffer.

Subsequent Magma Types

Wager and Brown (1968) discussed multiple magmatic injections filling the chamber, but implied that all had the same composition. The possibility that injected magmas changed composition fundamentally was proposed by Hamilton (1977) on the basis of variations in initial $^{87}Sr/^{86}Sr$ ratios in the cumulate succession, and refined by Kruger (1992; 1994) as shown in Fig. 12.13 The liquid compositions and the stratigraphic layers which they formed have been much debated. Note that each magma type may have been repeatedly injected; what is emphasised here is the injection of different compositions.

The Lower Zone and Lower Critical Zone are dominated by orthopyroxenite with minor olivine-rich layers, the mineral compositions and dominance of orthopyroxene over olivine suggest a magma with 13% MgO, using accepted K_D values for Mg and Fe. For magma compositions above the Lower Critical Zone even greater uncertainty exists. Harmer and Sharpe (1985) suggested that there was a different magma for the Upper Critical Zone. It was argued that the appearance of cumulus plagioclase required a different magma type. There was also a transitory increase in the initial Sr isotope ratio (Fig. 12.13). The idea was further developed by Naldrett et al. (2009), who suggested that the bulk rock Th/Nb ratio of the Upper Critical Zone indicated mixing with different magmas. With regard to the former issue it was shown that the mg# of orthopyroxene at this level is exactly that expected at the composition of magma that begins to crystallize plagioclase, based on the experimental studies reviewed by Cawthorn (2002). At this boundary are the Middle Group Chromitite layers. It has been shown that most chromitite layers are associated with increases in initial Sr ratio, but that the value decreases again above these layers (Schoenberg et al. 1999; Kinnaird et al. 2002). The Cr content of the proposed added magma is 200–400 ppm (Barnes et al. 2010), much lower than the underlying magma. Yet the orthopyroxene does not change in Cr content across this boundary (Fig. 12.7). The evidence for addition of a fundamentally different magma at this boundary therefore is very weak.

A fundamental change in initial Sr isotope ratio occurs close to the Merensky Reef. Hamilton (1977, p. 32) stated that the Critical Zone Sr isotopic signature (i.e. R_0 of 0.7065) could be traced to “3 m above the Merensky Reef” (based on a total of 11 samples). That the isotopic break occurs above the Merensky Reef is confirmed by all six subsequent detailed profiles in Fig. 12.14 that shows that all the samples of pegmatitic and normal grain sized pyroxenite in the Merensky Unit have R_0 values typical of the Critical Zone. Thus the inference that the magma addition is at the Merensky Reef is based on a rather-too-casual inspection of the data. The affinity of the Merensky (pegmatitic) pyroxenite is totally Critical Zone in its mineral composition and isotopic chemistry.

As well as this dramatic increase in the R_0 value into the Main Zone, there is also a major expansion of the eastern lobe toward the south at this level (Fig.12.1b), demonstrating the lateral extension of the Bushveld chamber (Kruger 2005). It should be noted that it is at the position of this expansion of Main Zone east of Roossenekal across the floor that Wager and Brown (1968) took their chilled sample of proposed parental magma. The same area was resampled in detail by Sharpe (1981), Maier et al. (2000) and Barnes et al. (2010), who proposed a magma composition similar to that of Wager and Brown (Table 12.1). All these samples from this area have a microgranular texture, with triple point junctions between plagioclase and pyroxenes (Fig. 12.11l). The samples analysed by Maier et al. (2000) have an initial Sr isotopic ratio of 0.706, and so are not suitable to explain the increase in this ratio in the Main Zone to 0.7085. These rocks were described by Sharpe (1981) and Maier et al. (2000) from the eastern Bushveld where the Main Zone abuts the floor. How-

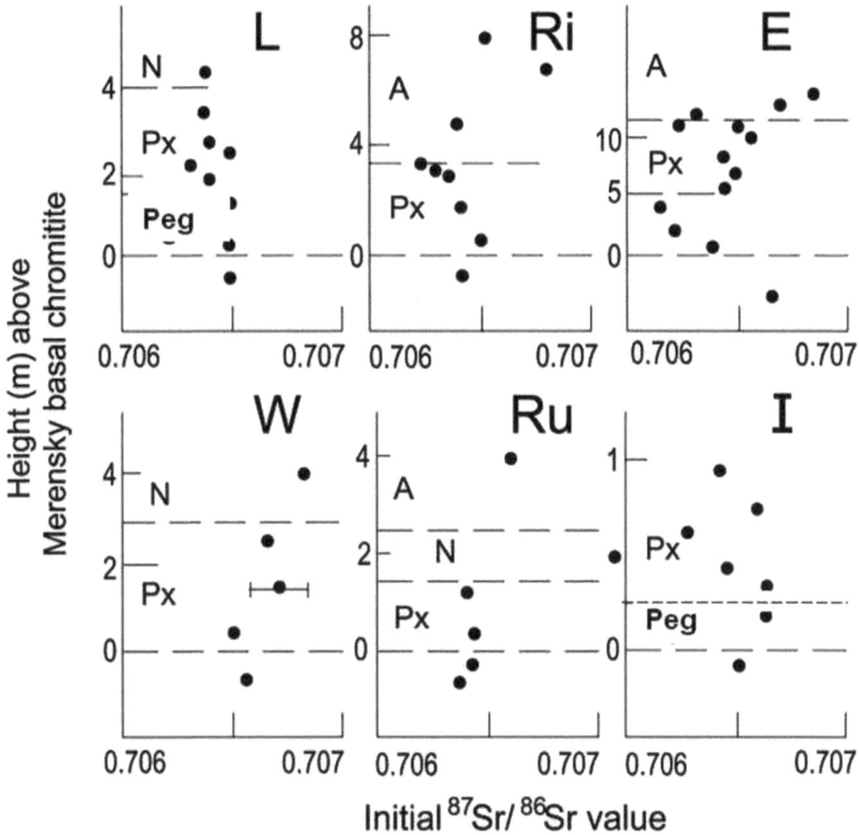

Fig. 12.14 Initial $^{87}Sr/^{86}Sr$ values through several sections of the Merensky Unit. Note that the break in ratio does not occur at the base of the unit as is commonly assumed. Abbreviations: Px and Peg refer to pyroxenite and pegmatitic pyroxenite. Mine names and data sources *L* Lebowa (Lee and Butcher 1990), *Ri* Richmond (Seabrook et al. 2005), *E* Eastern (Carr et al. 1999), *W* Western (Cawthorn 2011), *Ru* Rustenburg (Kruger and Marsh 1985), *I* Impala (Cawthorn and Boerst 2006)

ever exactly the same compositions (and textures) are found in rocks that underlie the Lower and Critical Zones in the east (to the north of Steelpoort) and in the west as described by Vermaak (1976). Analyses are given in Table 12.1. In these two settings they are not consistent with being parental to the Main Zone.

At the Pyroxenite Marker there is a major gradual increase in mg# in pyroxene and An in plagioclase, and a decrease in initial Sr isotope ratio, indicating addition of magma. The nature of this magma was excluded from the discussions of Sharpe (1981) and Cawthorn et al. (1981). The only samples identified as possible added magma was by Davies and Cawthorn (1984), based on complex intrusive relationships seen in the Upper Critical Zone. Sharpe (1981) argued for a very different genesis for this reversal at the Pyroxenite Marker, suggesting that these features do not represent a new magma but the continued crystallization of the stranded magma that originally overlay the Critical Zone/Merensky Reef, because mineral compositions and initial Sr ratio are similar. In this model the Lower Main Zone resulted from an injection of dense magma that did not mix with the residual magma. The limited range of differentiation in the Lower Main Zone as demonstrated by the decrease in An in plagioclase from 75 to 65 (and similar mg# of the pyroxenes) in more than 1.5 km of cumulates begs the question as to where and what happened to the remaining more evolved magma in this model.

Using initial Sr isotope ratios Kruger et al. (1987) suggested that there were no subsequent major influxes, but Tanner et al. (2014) suggested that there were two further additions based on Cr contents of clinopyroxene in the Upper Zone although other aspects of the chemistry of these magmas was not resolvable.

Mass Balance Models

Given that there are a number of different magmas with no consensus on their compositions or relative proportions, and that the original extent of the intrusion is undefined, attempts at mass balance modelling seem highly conjectural. However, two calculations warrant discussion, one at the base and one at the top. Eales (2002) and Eales and Costin (2012) discussed a mass balance up to the level of the top of the Critical Zone. They focused primarily on the Cr budget, and concluded that there was 8 times more Cr in this section than in the parental magma, and hence that there was an imbalance. This writer would contend that the one-dimensional view adopted in this model is too simplistic. The profile for which the calculation was done was through the section with the thickest Lowest Zone in the western lobe, and with well-developed chromitite layers (Teigler and Eales 1996). It is well-known that the ultramafic layers and the Lower Group chromitites thin away from this vertical section. Indeed, Maier and Eales (1994) suggested the terms proximal and distal to explain the lateral variation in thickness of these sequences. Hence, the validity of this calculation can be questioned. Another issue is important in this regard. At the top of the Critical Zone the plagioclase composition is about An_{78} and the cumulus orthopyroxene has mg# of about 78. These compositions are still relatively primitive (for example, more primitive than the initial Skaergaard magma). A typical basalt may have phenocrysts with those compositions. The extent of differentiation using a Bushveld primary magma (Table 12.1, column 1) and modelled using the MELTS, programme, suggests that there had only been 17 % fractionation (Li et al. 2001) up to this level. Hence, 5 times as much magma as the mass that has solidified still remained in the chamber. Thus, the mass balance anomaly is not just for Cr, but for the entire magmatic system. The fate of this volume of magma is still unknown, but may have been intruded laterally or extruded (Cawthorn and Walraven 1998; Naldrett et al. 2012). This enigma offers some support for the proposal of Sharpe (1985) mentioned above that this magma survived as an upper layer that produced the Upper Main Zone and Upper Zone.

A second mass balance calculation was presented for the end of differentiation. A calculation of the concentration of Zr and K_2O in the Upper Main and Upper Zone rocks yielded concentrations much less than found in the various proposed parental magmas (Cawthorn and Walraven 1998). They therefore proposed that a significant volume of evolved magma (with high Zr and K_2O) had escaped from the measured sections of the chamber, either vertically or laterally. Subsequent debates have challenged the vertical escape concept (Eales 2002), but VanTongeren et al. (2010) have resurrected this mechanism. This author claims that the lateral expulsion is a far more plausible model. The recognition of the existence of a Residual Zone with rocks originally called syenites by Hall (1932) may go part way to resolving this mass balance problem. There are few analyses of these Residual Zone rocks, but their Zr and K_2O contents are not high enough to totally account for the missing magma (Cawthorn 2013a). What rocks may have existed laterally, and have now been totally eroded, can never be more than topics of pure speculation.

Magma Mixing

The recognition of multiple pulses of different magma composition leads inevitably to discussion of the behaviour of such magmas once in the chamber, specifically mixing versus stratification. The assumption of magma mixing in Bushveld studies can be traced to two papers by Irvine (1973, 1975), which presented two ways of producing chromitite layers. However, whether magmas of different density (and temperature) would mix remains unresolved. In general, magmas of different density should not mix (as reviewed by Campbell 1996), but should remain stratified. Their subsequent cooling and crystallization, separately or with interaction, then becomes extremely difficult to predict. A review of both volcanic and intrusive examples shows that although stirring may produce extremely intimately convoluted (migmatite-like) textures, contacts between two different magmas remain remarkably sharp (Perugini and Poli 2012) and true homogenisation between mixed magmas is an extremely protracted process. A converse view exists, using observations from plagioclase compositions in calk-alkaline volcanic rocks, where rapid mixing of basic and silicic magmas has been claimed (Kent et al. 2010). Thus, fluid dynamic principles and mineral compositions yield two contrary views on whether magma mixing can occur or not. This writer suggests that resolving this impasse is one of the important challenges in understanding the mineralogical and geochemical evolution of layered complexes.

Three important features in the Bushveld can be re-examined in the light of these debates on magma addition and mixing.

1. Chromitite Layering. The chromitite layers were the issue addressed by Irvine in his magma mixing models. In all except one instance the lower contact of the chromitite layers is extremely sharp (Fig. 12.3d). Campbell and Murck (1993) presented calculations on the thicknesses of magmas that needed to be mixed

to produce such layers. Taking the thickest of the Bushveld layers as 1 m their Fig. 12.10 showed that many kilometres in total needed to be perfectly mixed. These estimates were the optimum for magmas mixed in equal proportions. Unequal proportions would demand even greater vertical thicknesses of magma. This writer would challenge whether several kilometres (vertically) of magma can be mixed as abruptly as would be required to produce the observed sharp contacts over an area of more than 30,000 km^2.

2. Critical Zone—Main Zone transition. The addition of isotopically different magma in the Merensky Unit interval triggered the concept of magma mixing and the formation of the platiniferous Reef itself (Campbell et al. 1983). Whereas magma addition is probably indisputable, the extent of mixing remains debatable. The Cr content of orthopyroxene can be used as a critical parameter for distinguishing Critical Zone versus Main Zone lineage. It was first shown by Mitchell (1990) that the Cr_2O_3 content of the orthopyroxene at the Merensky interval is 0.4 % and decreases abruptly to 0.1 % only 30 m higher in the Lower Main Zone (Fig. 12.8a). The major review by Eales (2002) reinforced this data set with additional profiles. A simple calculation for Cr using these mineral compositions shows that no mixing can have occurred in the approximately 30 m above the Merensky Unit (see Appendix 1).
   This author's view (adapted from the modelling of Cawthorn and Walraven 1998) is that the residual magma was pushed outward by the new injection(s) of magma. However, this non-mixing hypothesis would also be consistent with the previous suggestion of Sharpe (1985), as discussed above.
   In the interval between the Merensky Unit and some 30 m above there is an apparently anomalous sequence. The Sr isotopic ratios indicate that there is a fundamentally new magma at this level, but the Cr in orthopyroxene shows that this sequence has formed from typical Critical Zone liquid. This paradox was resolved by Seabrook et al. (2006) who suggested that the new Main Zone magma formed a layer below the residual Critical Zone magma with no mixing. Orthopyroxene and chromite continued to form and sank from the upper liquid layer whereas the interstitial plagioclase (that contained all the Sr) solidified from the lower liquid layer.
   Further challenging evidence against rapid mixing is based in initial Sr isotope ratios. The Critical Zone has a ratio of 0.7065, and the Main Zone a ratio of 0.7085. If the latter were the result of mixing of equal proportions of magma, then the added magma must have had a ratio of 0.7105. I suggest that is a highly improbable isotopic composition for an enormous volume of basic magma.
3. Pyroxenite Marker. One of the first and most convincing examples of addition of new magma in any intrusion is the Pyroxenite Marker in the Main Zone. It was identified by Lombaard (1934), but it was von Gruenewaldt (1970) who showed that the mg# of pyroxene and the An content of plagioclase increase by about 10 units upward, and there is now also known to be a decrease in the initial Sr isotopic ratio (Sharpe, 1985). A similar study in the western Bushveld (Cawthorn et al. 1991), reached similar conclusions, but they further emphasised that

the change in mineral composition was gradual and could be recognised over a vertical interval of more than 200 m, approximately equally disposed below and above the Pyroxenite Marker. Thus, this most convincing example of magma addition suggests that it was a very protracted process rather than abrupt (as also shown in the Lower Zone rocks discussed above).

Origin of Layering

Layering can range from very abrupt variations in modal proportions over short vertical intervals (as in the Critical and Upper Zones) to so diffuse (visually cryptic) over very many tens of metres that it requires detailed point counting of mode (Roelofse and Ashwal 2012) or detailed density measurements (Ashwal et al. 2005) to recognize.

Possibly the most important observation that must be satisfied by any model for modal layering in the Bushveld is the remarkable continuity of many distinctive layers (chromitites, magnetitites and Merensky Unit in particular) often in both western and eastern limbs, but also in some cases in the northern limb. It was the main argument in the suggestion that the west and east exposures are part of a single lopolith, and not discrete intrusions (Cawthorn et al. 1998).

Two opposing paradigms dominate views of solidification processes; as summarised in the Wager and Brown (1968) model of crystal settling and sorting, and the Marsh (2006) model of in situ crystallization. The analogies would be with clastic versus chemical sedimentary rocks. In the former model, heat would be lost through the roof causing an increase in magma density and/or crystallization at the top resulting in downward plunging packages, the continued crystallization in which would be promoted by the relatively greater steeper slope (δ temperature/δ pressure) of the liquidus than the adiabat with increasing pressure. The latter view would be that nucleation is heterogeneous not homogeneous, and will occur on a pre-existing substrate (floor, walls, roof—if solid) where heat loss occurs. Modelling which of these effects is more dominant is difficult.

Grain Settling

The Wager and Brown (1968) model is summarized on their pages 208–221, in which they envisaged grains nucleating in the magma chamber; this crystal-charged magma (proportion of grains unspecified) slowly circulated by convection, dropping out a steady rain of all cumulus minerals (in their approximate cotectic proportions) and producing the homogeneous successions. Intermittently, super-charged slurries descended the walls (or possibly elsewhere from the roof) of the (Skaergaard) intrusion; lost momentum as they travelled along the instantaneous floor;

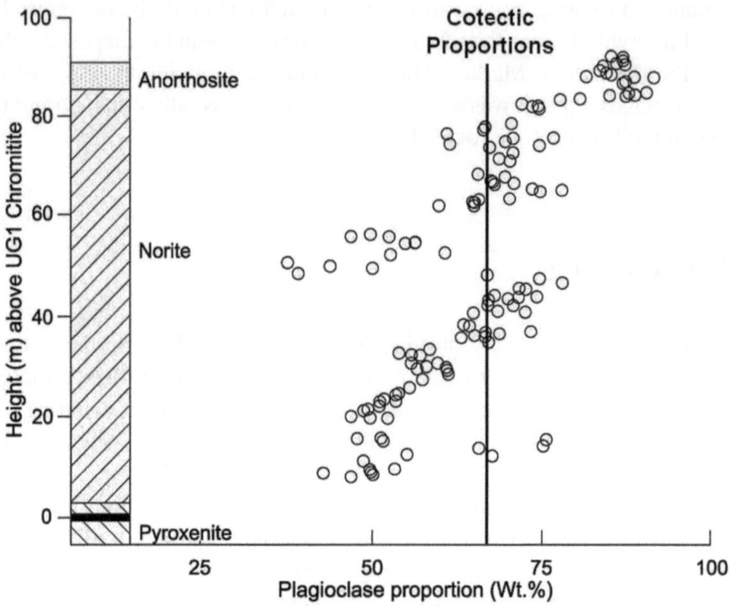

Fig. 12.15 Modal data through part of the Upper Critical Zone. Detailed modal proportions of plagioclase through the MG4 rhythmic unit, taken from a borehole in the western Bushveld Complex (own unpublished data), using the method described by Cawthorn and Spies (2003). The section that is called norite is not homogeneous but contains three sections that show a progressive upward increase in the plagioclase content from 35 to over 75 %. The probable cotectic proportion of plagioclase to orthopyroxene is about 65 %, and there is no tendency for rocks with that proportion to occur. The simplest explanation for the observed modal variations is by crystal settling and sorting

and gradually dropped their load of crystals in order of density (according to Stokes' Law). An intricately amplified development of these views was presented by Irvine et al. (1998; p. 1437). They emphasised the likely non-laminar nature of the horizontal flow of the magma close to the crystal-liquid boundary and suggested that shearing within the liquid permitted intermittent separation and reattachment of the magma close to this boundary that was capable of major redistribution and sorting of minerals of different density.

Can this first model be applied to the Bushveld? Two fundamental expressions of deposition by settling would be (i) a gradual decrease in mafic mineral proportion in the graded layers upward and, if horizontal turbidity currents were involved, in the direction of movement, and also (ii) erosive and possibly channel-like features displaying cross-bedding. Such features do occur in the Bushveld. Evidence for decreasing mafic mineral proportion upward on a large scale is shown in Fig. 12.15. Planar cross-bedding has been recorded (Fig. 12.3h) in the middle of the Main Zone (Quadling and Cawthorn 1994), and is strongly suggestive of horizontal flow of crystal-charged slurries. The disruption of such layering (Fig. 12.3i) demonstrates the erosive power of such currents. Quadling and Cawthorn (1994) also documented linear fabrics suggestive of horizontal flow. They reported alternating layers of gabbronorite with greater and less than the cotectic proportion of plagioclase (and none with the cotectic proportion). The plagioclase defined a linear fabric in the plane of the layering, but with very different orientations in the two layers. In the mafic-rich layer the lineation was parallel to strike, whereas in the felsic-rich layer it was down dip. These features were interpreted to be due to flow of grains in a current, the denser, mafic-rich portion being carried as a bed load, and the less dense, felsic portion being carried as a suspended load. Any plagioclase in the bed load would roll along the bottom with its long axis perpendicular to flow, whereas in the suspended load the long axis would be parallel to flow. The abundance of the lateral terminations of the mafic layers (Fig. 12.3h) suggests that these currents are of relatively limited extent horizontally. Many more studies would be needed to test if linear rather than planar fabrics were common in the Bushveld.

In contrast to the laterally discontinuous nature of the individual layers shown in Fig. 12.3h (although the entire package of layering is many metres thick and very continuous over tens of km), are the examples where modal layering is laterally very continuous. These features indicate a contrast with the Skaergaard layering (Irvine et al. 1998) where the laterally non-continuous layering is strongly suggestive of horizontal flow. Such a mechanism becomes questionable when a specific layer can be traced in the Bushveld for very great distances. Horizontal flow for many tens to possibly one hundred km that would be required by such observations is considered implausible. Such sequences are more consistent with the model of Sparks et al. (1993) where a convecting magma (of chamber-wide dimensions) held ever more grains in suspension until a critical concentration was reached, speed of convection decreased and settling and sorting of the grains from the suspension occurred. As the magma column was depleted of grains, the upper part of the column recommenced convecting, cooling and crystallization and another cycle resumed. A possible example of this effect is shown in Fig. 12.15 where there are three contiguous vertical sections that would all be classified as norite, but showing a regular upward increase in modal plagioclase content.

There is an important consequence of this model. Consider a magma that is crystallizing plagioclase and pyroxene. When settling did commence, pyroxene grains would concentrate at the floor. Such a pyroxene-rich layer has traditionally been inferred to have formed from a magma saturated only in pyroxene. The inference used in reaching such a conclusion is that whatever mineral assemblage crystallized from the magma was immediately and totally preserved by accumulation at the base. Given the concept of crystal settling this inference is invalid. It was shown by Cawthorn (2002) that pyroxenes in several pyroxenites had evolved mg# values (less than 82) that indicated that the pyroxene had crystallized together with plagioclase but that the plagioclase had not accumulated at the base (Fig. 12.16). This observation challenges the view that pyroxenite forms from magma saturated only in pyroxene.

Among the most dramatic exposures in the Bushveld Complex are the oxide-rich layers. They remain relatively constant in thickness for tens to even hundreds

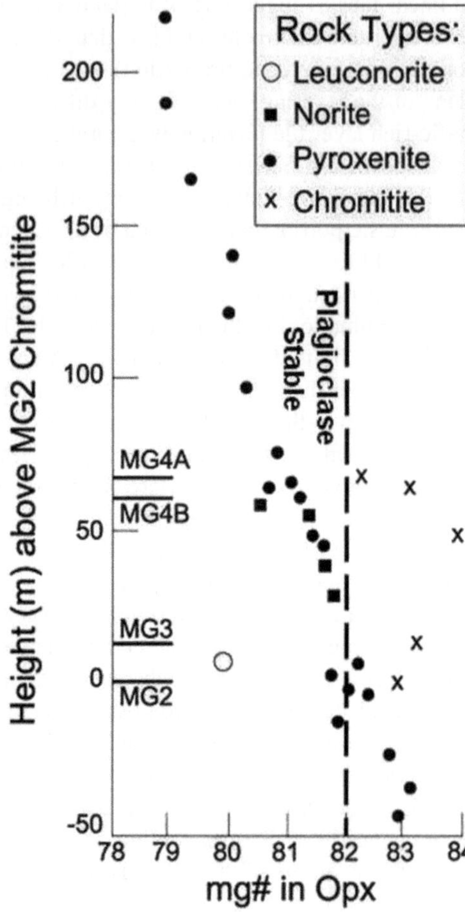

Fig. 12.16 mg# in orthopyroxene from a section above the MG2 (data from Teigler et al. 1992). Note that the mg# in orthopyroxene in pyroxenites is no greater than those in norites. All values in both norites and pyroxenites are less than 82 which is the composition reflecting the stage of crystallization when plagioclase becomes a cumulus phase. Thus, pyroxenites with mg# <82 are not the result of formation from a liquid crystallizing only pyroxene, but indicate that the crystallizing plagioclase did not accumulate on the floor

of km in the east and west. Some correlatable layers are even of similar thickness and composition in all three lobes, e.g.: the vanadium content of magnetitite layers (Cawthorn and Molyneux 1986). Their substantial thicknesses (especially for the chromitite layers) require formation from a very thick and laterally uniform volume of magma that was driven into the oxide stability everywhere abruptly. The abruptness of the event can be demonstrated by the extremely sharp basal contacts seen for chromitite (Fig. 12.3d) and magnetitite layers (Fig. 12.3j).

A challenge to crystal settling depends upon density considerations. This problem was first recognised by Darwin (1844) where he noted that a lava flow was enriched in albite at its base which he considered contrary to the inferred density of the host lava. When Bottinga et al. (1982) developed a rigorous magma density calculation scheme based on partial molar volumes of oxides they pointed out that basic magma may be denser than plagioclase, which would make an origin of anorthosites by grain settling impossible. With regard to anorthosites in the Upper

Critical Zone, the magma is probably quite siliceous and so could have had a lower density than plagioclase. However, in the Main Zone where the minerals become more evolved, but prior to magnetite formation, the density of the magma may have been greater than plagioclase. However, the original density calculations of Bottinga et al. (1982) and Campbell et al. (1978) and experiments assumed zero water. The subsequent inclusion of the effect of water into magma density calculations indicates a reduction in density of about 0.15 g cm^{-3} for each 1 % of water (Lange 1994; Ochs and Lange 1999). Hence, provided this 1 % water can be inferred for evolved Bushveld magma, plagioclase would probably have been dense enough to sink. 170 years after Darwin's observation this issue is still not satisfactorily resolved.

A significant feature encountered is the presence of autoliths that almost certainly have been transported before downward accumulation into the succession (Fig. 12.3k). A remarkable example is the presence of anorthosite autoliths in the MG2 chromitite (Fig. 12.3i). It is commonly stated that below the MG2 there is no cumulus plagioclase, yet thin fragments of anorthosite are found in this chromitite layer. Their origin is enigmatic. However, it was shown above that plagioclase may be a crystallizing but not an accumulating phase at levels where pyroxenite layers (with pyroxene having an mg# <82) were being preserved. Teigler and Eales (1993) also reported the presence of resorbed plagioclase inclusions in orthopyroxene grains below the MG2 chromitite, and reached a similar interpretation. Were the non-accumulating plagioclase gains forming rafts or mats that were capable of being deposited either laterally or at some higher level by currents?

In Situ Growth

In the other model for solidification, Marsh (2006) suggested that the most likely site for grain nucleation would be where the magma is coolest and where heterogeneous rather than homogeneous nucleation would occur. The margins of the intrusion are the most conducive to creation of these conditions. Hence, he proposed in situ growth of solid, with fractionation occurring by the remaining liquid migrating inward from the solidifying carapace. In the case of the Bushveld, once a moderate thickness of cumulates had formed at the floor, cooling would be mainly through the roof suggesting that would be the best site for grain formation. But no roof cumulates exist in the Bushveld unlike Skaergaard. This model can be tested against the repetitive sequence of pyroxenite—norite—anorthosite in the Bushveld Complex. In the simple binary pyroxene—plagioclase phase diagram a magma could begin to crystallize pyroxene. It would then reach the eutectic and produce pyroxene and plagioclase (norite), but it could not then produce only plagioclase (anorthosite). Considered another way, with in situ growth all multi-phase assemblages formed ought to be in cotectic proportions. However, the above-mentioned sequences often show perfectly gradational changes in modal proportions over significant vertical intervals that are not constrained by cotectic proportions (Fig. 12.15). A second

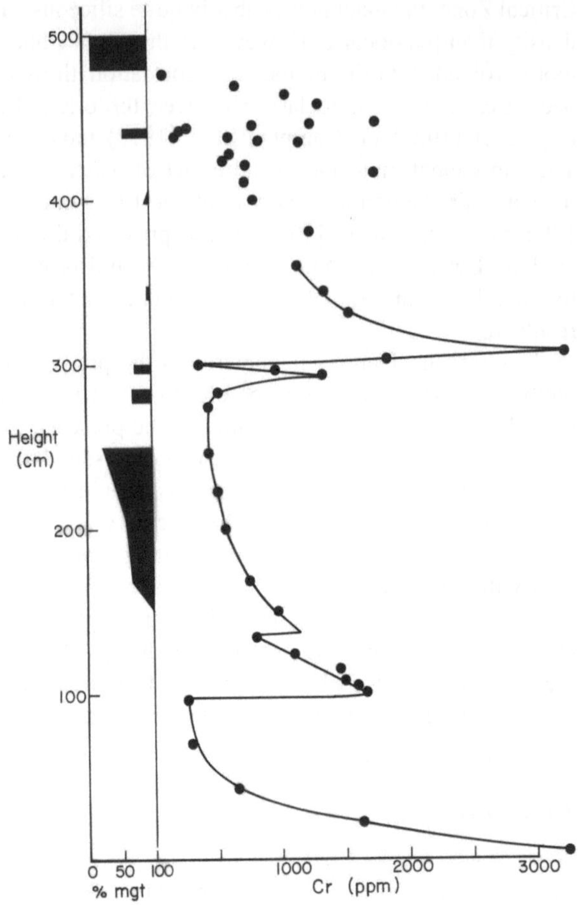

Fig. 12.17 Cr content of pure magnetite mineral separates taken from the Main Magnetitite Layer (from McCarthy and Cawthorn 1983). The proportion of plagioclase in this section is shown on left. The marked gradual decrease in Cr content demonstrates that the magnetite did not form from a large volume of magma or that settling occurred over a large distance. The reversals in Cr content demonstrate that these layers are not formed as one homogeneous layer, but are produced by several discrete layers superposed, in some cases recognisable by the presence of cumulus plagioclase (e.g. at 300 cm height)

example of such gradual changes in modal mineralogy from essentially one mineral to another is provided by the magnetitite layers, almost all of which show massive magnetite (only one cumulus phase) at the base increasing regularly upward in proportion of plagioclase to ultimately produce anorthosite (Fig. 12.3m). The abundance of layers with non-cotectic mineral proportions in the Critical and Upper Zones, especially, is a challenge to the model of in situ growth.

The Cr content of magnetite in magnetitite layers has been used to argue in favour of in situ growth (Cawthorn and McCarthy 1980). They showed a depletion of Cr from several thousand to less than one thousand ppm Cr with a number of abrupt reversals in less than 2 m of magnetitite (Fig. 12.17). Such massive depletion is difficult to model if grains were forming and settling from a large volume of magma. The chromitite and magnetitite provide an interesting contrast. A similar model for the formation of both might seem logical. However, the chromitite layers containing 300,000 ppm Cr do not appear to deplete the magma significantly in Cr based on the Cr content of overlying orthopyroxene (Teigler and Eales 1996). Junge et al. (2014) showed a number of rather subtle fractionation-reversal cycles in the UG2 chromitite layer qualitatively comparable to the trend identified by Cawthorn and McCarthy (1980) for magnetitite. The explanation for the lack of depletion in Cr in overlying orthopyroxene may be found in the fact that if Cr in magma that was producing chromite decreased slightly, then chromite crystallization would terminate (by reaction relation) or be swamped by cotectic silicate formation. However, if Cr in the magma that was producing magnetite decreased, magnetite could still continue crystallizing (because Cr is merely a trace element) and the magma could become further depleted in Cr.

Latypov et al. (2013) have argued that the thin chromitite layers (in Rum) at the boundaries between anorthosite and peridotite might have resulted from in situ growth. They showed that chromite grains and thin layers could be found on vertical or even overhanging surfaces within potholes, and argued that they could not have accumulated there by vertical settling. With regard to the potholes in the Merensky Reef, Schmidt (1952) presented petrofabric evidence that the pyroxene grains had been plastered with their long axes horizontal against the circular vertical walls of the pothole by swirling currents. Exactly the same process might have applied to the chromite grains. Thus, finding the definitive criteria to constrain mechanisms of grain accumulation remains elusive.

Other Mechanisms

A number of other factors and driving mechanisms have been discussed by Naslund and McBirney (1996). Ultimately, grains must sink or grow in place, but not necessarily exclusively one or the other. One of these processes proposed by Maaløe (1978) involves oscillatory grain growth, with the liquid becoming alternately supersaturated in one phase then another on either side of a cotectic. Possible examples of such features are noted (Fig. 12.3n), but are very rare compared to the gradual change in mode noted in Fig. 12.15. The sequence pyroxenite-norite-anorthosite observed in the Critical Zone is not consistent with oscillatory nucleation. As noted by Wager (1959) nucleation order may be controlled by simplicity of crystal structure, producing an order of oxide, olivine, pyroxene and plagioclase (as seen in the Bushveld rhythmic units). A refinement of that concept was presented by Duchesne and Charlier (2005) who suggested that variations in nucleation rate of plagioclase relative to other phases could produce a single graded unit in terms of increasing plagioclase proportion upward rather than multiple oscillations in modal proportions. Another model for layering involves pressure fluctuations, which, rather than changing the magma composition so it lies in a new mineral stability field, changes the position of phase boundaries relative to a constant magma composition. Naslund and McBirney (p. 27) stated "The effects of a pressure change would be felt nearly simultaneously over the entire magma chamber, and as a result, [is] attractive for explaining layers of great lateral extent". Closely similar to this model is one of seismic shocks that can trigger extreme rates of nucleation (tap the side of a glass containing a carbonated drink). Again, Naslund and McBirney (p. 19) concluded that "seismically induced layers should be laterally continuous over the entire chamber". These two quotes from Naslund and McBirney fit perfectly the descriptions of the chromitite and magnetitite layers.

Two further complexities to interpreting modal variations are provided by a consideration of what minerals might be expected to accumulate. In successions of pure anorthosite there is the implication that only plagioclase is accumulating (whether it be by settling or in situ growth). Such a process requires that no mafic minerals be forming anywhere in the magma chamber because they would definitely sink if plagioclase could sink. Hence, there must be periods when no significant nucleation and growth of mafic minerals is occurring anywhere in the chamber. Does such an observation imply that there were pulses of nucleation and grain growth followed by settling with no further nucleation? The second example comes from the Upper Zone that is mainly composed of magnetite gabbronorite. Above the Main Magnetitite Layer there are two closely spaced additional magnetitite layers. The intervals below and between the magnetitite layers consists of anorthosite (with minor oxide), but no cumulus pyroxene., which, being denser than plagioclase, would have sunk if grains of it were present. Hence, why is there no cumulus pyroxene present in this interval? A possible explanation could be that nucleation and growth of magnetite grains removed iron from the adjacent magma. Given the extremely low MgO content of the magma at this level of the intrusion, removal of iron could have destabilised pyroxene and driven the magma into the plagioclase stability field.

UG1 Chromitite at Dwars River

The thin bifurcating chromitite layers in the anorthositic footwall to the UG1 chromitite at the National Monument site, Dwars River, are the most photographed and possibly most enigmatic features in the Bushveld (Fig. 12.3e), and so are highlighted here. Similar bifurcations and lenses are seen wherever the footwall to the UG1 is exposed. However, these features are not unique to the UG1 footwall. They occur in all the chromitite layers, but usually with pyroxenite as the lenses. Two examples are shown in Fig. 12.3o and p. The difference is that the chromitite layers tend to be thicker and the silicate lenses thinner than the Dwars River anorthosite package. In the chromite mines the pyroxenite lenses are known as "internal waste" and "partings", and can range from a few grains to 30 cm thick and up to hundreds of metres long. I would suggest that a purely clastic sedimentological analogy the simplest interpretation, using the Dwars River locality for the explanation. It occurs at the top of the MG4 chromitite rhythmic unit, and only plagioclase was the accumulating phase. Any one of the assorted mechanisms for producing chromitites operated. (I prefer the possibility of tectonic instability triggering earthquakes with associated shock waves.) They initiated nucleation of chromite grains everywhere in the chamber, which accumulated as thin and, importantly, continuous layers. Repeated earthquakes caused the large number of thin layers, as implied by the compositional data for chromite reported by Junge et al. (2014). Between each layer the plagioclase continued to accumulate. Possibly as a result of variably horizontal flow rates in plagioclase-laden suspensions, the grains sedimented in elongate lenses analogous to braided stream deposition. Thus, the difference between discontinuous plagioclase lenses and perfectly continuous chromitite layers (when they first formed) resulted from the two mechanisms of accumulation, the former separating from horizontally flowing basal suspensions, and the latter from earthquake-triggered nucleation throughout the entire chamber. However, the chromitite layers are essentially devoid of cumulus silicate grains (although there is abundant intercumulus plagioclase and pyroxene). Thus, this model requires that there are no silicate grains accumulating during the period of chromite accumulation. The "sand volcano" model of Nex (2004) is slightly different. He emphasised the discordant upward emplacement of plagioclase suspensions disrupting the originally uniform chromitite layers (Fig. 12.3q and r), and suggested that the upward motion of a plagioclase suspension was promptly transferred into a sub-horizontal collapse as momentum was lost to produce gentle domes of accumulated plagioclase on the chromitite floor. My one reservation with this model is that the width of many of these plagioclase lenses is in the order of a few to many metres. However, the spacing between these (chromitite) disruptive structures is one to two orders of magnitude greater. It is therefore hard to conclude that there is a disruptive feature beneath each plagioclase dome. The search for definitive features to demonstrate any of the proposed models for Dwars River will doubtless continue to occupy geologists' imagination.

The many conjectural interpretations mentioned above demonstrate that there are major uncertainties in determining the origin(s) of layering. Searching for a single mechanism may be unrealistic.

Consolidation

The *in situ* crystallization model of Marsh (2006) envisages a vertical profile (of unspecified thickness) near the base of the magma chamber that ranges from 0% to close to 100% crystalline. Even the crystal settling model envisages a zone that contains variable interstitial liquid proportion. The thickness of that zone is difficult to demonstrate because it is so transitory (see Irvine et al. 1998). Also the fate of that liquid is variably interpreted. Wager and Brown (1968) considered that it might be trapped (and solidify to produce an orthocumulate rock with up to 30% post-cumulus minerals); or, close to the interface (thickness or depth of this zone unspecified), be capable of interaction with the overlying liquid to produce adcumulates (with overgrowth of the cumulus minerals with the cumulus composition). Alternatively, that interstitial liquid may have been driven out by textural annealing/physical compaction (Hunter 1996) as modelled by McKenzie (2011). Unravelling the relative contributions of all these processes is probably almost impossible and can vary within different sections of the same intrusion. Here, I document a number of field observations in the Bushveld that are relevant to this debate.

Compositional convection is a process in which the interstitial liquid changes its density and hence migrates vertically due to gravity, dependant on whether the crystallizing assemblage decreases or increases the density of the residual magma. As discussed by Naslund and McBirney (1996), metasomatism and zone refining are subtle consequences of this process. Any of these mechanisms could be channelized (and produce distinct discordant bodies), or pervasive (and hence be extremely cryptic). In the Upper Critical and Main Zones of the Bushveld the magma crystallized plagioclase and one or two pyroxenes, and the evolving magma would have become denser (McBirney 1985), and so gravity-driven upward migration would not have occurred. One specific example is also worthy of mention. Tait and Jaupart (1992) suggested that the discordant iron-rich bodies that occur in the Critical and Main Zones are examples of this compositional convection. Not only is the density change of the evolving liquid incorrect (Scoon 1994), but the initial Sr isotopic ratio (Fig. 12.13) of these rocks demonstrates that the iron-rich liquids are not derived from the adjacent cumulate succession (Scoon and Mitchell 1994; Cawthorn et al. 2000).

Most of the textures in the Lower, Critical and Main Zones show evidence of annealing, with triple point junctions and planar edges that are not major crystallographic directions (which tend to dominate crystal growth in a magma). Hence, they are not the product of just grain overgrowth as in the adcumulus model of Wager et al. (1960). Some combination of Ostwald ripening and other compaction processes has clearly been operative. The three most important factors are time, proportion of liquid and temperature gradient. These collectively result in a crystal pile with interstitial liquid, the thickness of which will depend upon the rate of cooling. Cooling of the magma produces grains that accumulate on the floor with quicker cooling producing a thicker mush zone. Being such a thick intrusion the Bushveld crystallized very slowly. Using the cooling curves of Cawthorn and Walraven (1998) the rate at which the crystal mush advanced decreased per zone in cm y^{-1}, from 40 (Lower) to 10 (Critical) to 4 (Main) to 2.5 (Upper). These rates are about ten times faster than calculated by Morse (1979; p. 578) for the Kiglapait Intrusion which is of comparable thickness to the Bushveld. The reason for this discrepancy is not obvious to me. Using our quicker cooling rates, the accumulation rate ranges from one grain (2 mm in size) per day to one grain per 16 days. The rate at which grains will anneal and so compact and exclude remaining liquid is not known, because it is the combination of a number of different processes. It will be quickest in monomineralic rocks because all touching grains are the same mineral, and slower when more cumulus minerals are present. Plagioclase would be the slowest mineral to recrystallize because of its tectosilicate structure and oxides the fastest. Hence, the relationship between depth in the crystal pile and proportion of liquid is difficult to predict. However, the rate of annealing toward a solid rock is probably significant compared to the rate of arrival of the grains at the floor (for Bushveld), and so only a thin zone of crystal mush is inferred to have existed from these considerations. Thus Ostwald ripening may be more significant than compaction forces.

Structures in the layered rocks provide some limiting examples. Below the UG1 chromitite at Dwars River is a zone of slumping (Fig. 12.3s) of alternating melanocratic and leucocratic norite. The thickness of this slumped zone is less than 3 m, and is then overlain by multiple thin chromitite layers devoid of this deformation.

A similar package of variably slumped and brecciated noritic rocks occurs between the UG2 and UG3 chromitites on Maandagshoek (Fig. 12.3t and u, and described in detail by Maier et al. 2012), underlain and overlain by UG2 and UG3 that are perfectly planar and undisturbed. An interpretation of these structures is that these very few m-thick zones contained sufficient liquid to be capable of soft-sediment slumping possibly due to floor instabilities driven by loading by this huge mass of magma. They may be interpreted as seismic-triggered instability structures (Carr et al. 1994) in a plastic mush. Deeper layers were unaffected. The slumping event pre-dated the formation of the overlying, planar chromitite layers. The thickness of this zone of soft-sediment therefore may be in the order of a few metres.

Below the Merensky Reef at Impala Platinum mine there is an area that has been significantly deformed. Locally on the mine it was called the "flame bed" (Fig. 12.3v). All the "flames" show evidence of slumping in the same direction, down the dip. These flames are much more regular than at the Dwars River UG1 locality and may indicate a greater degree of cohesion of the grains and less liquid (or a less powerful seismic trigger). Below and above that layer is an undeformed planar layered package limiting the slumped package to very few metres.

In the Main Zone there is a sequence of strongly planar layered rocks. There is one outcrop where they can be seen to have been brecciated into blocks tens of cm across and typically ten cm thick (Fig. 12.3i). The fragments are quite angular. The maximum thickness of this layered package is 10 m (Quadling and Cawthorn 1994). Hence, one infers that these crystal-charged slurries that produced this brecciation broke up layered rocks that were sufficiently rigid to produce angular blocks, and so contained very little liquid. This feature also shows that the layering was of a primary nature and not due to any post-depositional process. These four examples suggest that the rate at which loose uncompacted grains annealed into progressively more rigid bodies occurred within a few metres of the upper boundary with the magma (with most to least liquid suggested in the order presented).

In the Upper Zone the fractionating magma would have become less dense due to oxide segregation (McBirney 1985), and so compositional convection might be envisaged. Magnetite has a large partition coefficient for Cr and patterns of fractionation and importantly abrupt reversals in Cr content of magnetite within the magnetitite layers are observed (Fig. 12.17). Had there been significant upward migration of liquid through these layers these reversals would have been smoothed into more gentle upward increases or completely eliminated, using the interpretation of Irvine (1980) for upward displacement and smoothing of reversals in the Muskox Intrusion.

Another example from the Upper Zone provides a different test for these models. Below Magnetitite Layer number 1 (the first above the Main Magnetitite Layer) there is a xenolith of Dullstroom basalt that has been metamorphosed into a granular texture. Along the outcrop it is at least 20 m long. Its extent at right angles is obviously unknown. It is 40 cm thick. Such a large xenolith would be expected to have trapped upward percolating residual magma if such a process were operative. Six short vertical profiles were taken along the strike, three from below the xenolith and three where the normal succession of footwall anorthosite overlain by magnetite oc-

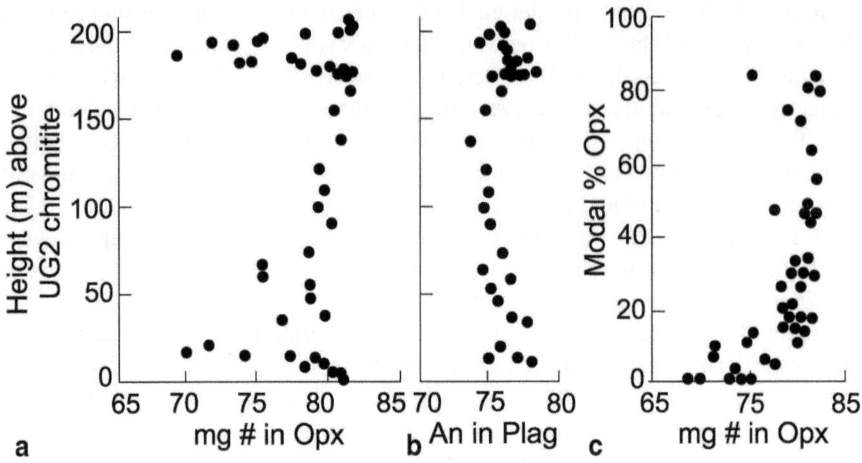

Fig. 12.18 **a** mg# in orthopyroxene and **b** An content in plagioclase versus height through the UG2 rhythmic unit. **c** mg# in orthopyroxene versus proportion of orthopyroxene showing that there is a pronounced decrease in mg# of the pyroxene where it is present at less than 20 % by mode, and hence is composed of a significant proportion of intercumulus material. These trends suggest that the orthopyroxene has been markedly affected by the trapped liquid shift effect where its abundance is low. (Data from Schürmann 1993)

curs. There is no evidence for accumulation of residual magma beneath the xenolith based on incompatible trace element abundances (Cawthorn and Street 1994).

It has been suggested that upward percolation of liquid could have caused the concentration of chromitite layers within anorthosite in the Upper Critical Zone (Boudreau 1994). At Dwars River there is another definitive outcrop (Fig. 12.3q and r). Thin layers of chromitite have been disrupted by intrusion of an anorthosite mush. Layers of chromitite beneath this feature are continuous and suggest that the anorthosite in the mobilized mush is very locally derived (Nex 2004). Along the near-vertical boundary between the layered anorthosite and the discordant anorthosite is a thin zone of chromite grains (Fig. 12.3r), smeared out from the layer by the intruding anorthosite mush. The two observations, that the chromitite layer is truncated by the anorthosite mush and the smearing of the chromite grains, show that the chromitite layers are syn-depositional features.

Models that envisage large-scale vertical melt migration and metasomatism or zone refining can be tested with detailed mineralogical trends in the uppermost Critical Zone. A section of 200 m of layered plagioclase and orthopyroxene-bearing rocks is shown in Fig. 12.18. The An content of the plagioclase is near-constant at An_{75-78}. However, the mg# of the orthopyroxene varies greatly, notably in two sections with very leucocratic rocks (Fig. 12.18). If there had been metasomatism that had equilibrated with the sequence as the liquid passed upward, all the mg# values should be the same. The extent of variation on quite small scales is more consistent with the trapping of liquid, as discussed below, than with zone refining.

The overall conclusions suggested by these assorted observations are that fine-scale layering is a primary feature, and that consolidation of a very slowly accumulating pile of grains, from a mush to a very coherent, essentially rigid body of rock, occurs within a few metres of the magma interface, and that pervasive interaction between grains and a migrating residual liquid is minimal.

The Trapped Liquid Shift Effect (TLSE)

Ideally, the compositions of cumulus minerals in layered intrusions ought to define processes such as fractionation and recharge of the magma. However, their interpretation is hampered by the need to consider the post-cumulus effect of the interstitial liquid. Cooling and crystallization of this liquid should produce zoned margins to grains. The rate of exchange between NaSi and CaAl in plagioclase is extremely slow (Morse 1984), even at magmatic temperatures and so any original zoning is preserved, and the cores of grains can be analysed to give a true reflection of the cumulus composition. However, Mg–Fe exchange in mafic minerals is fast, resulting in homogenisation of grains (at least, in the Bushveld), and variable distortion away from the original cumulus composition. Barnes (1986b) showed that the effect could be quantified provided that the liquid fraction was trapped. If pervasive migration of liquid occurred then reaction between cumulus grains and this liquid would occur but the magnitude of the effect would be completely random. Trace element concentrations can also be influenced by this effect, and examples for major oxides and compatible and incompatible elements in minerals are given below.

Major Oxides

Two examples are used here to demonstrate the TLSE for mg#. The variation in mg# in orthopyroxene in the uppermost Critical Zone is shown in Fig. 12.18. The near-constancy of the plagioclase composition shows that a very large volume of magma existed, such that 200 m of cumulates produced no change in plagioclase composition. However, the mg# of the orthopyroxene is apparently randomly variable. The value of the mg# in orthopyroxene can be compared with its modal proportion in those rocks (Fig. 12.18c). It can be seen that the mg# decrease correlates with the decrease in modal content below 20%. In samples with abundant orthopyroxene there is a relatively large proportion of cumulus mineral compared to that formed from the trapped liquid. As a result, the final analysed mineral has a composition still close to that of the true cumulus phase. However, where there is little orthopyroxene, the rock contains a much larger relative proportion of the intercumulus component and the final mg# of the pyroxene (and the whole-rock) composition decreases. A comprehensive study of the mineral compositions in the Main Zone in the northern lobe was presented by Roelofse and Ashwal (2012). It showed irregular variability vertically of the mg# of the mafic minerals (Fig. 12.19), but

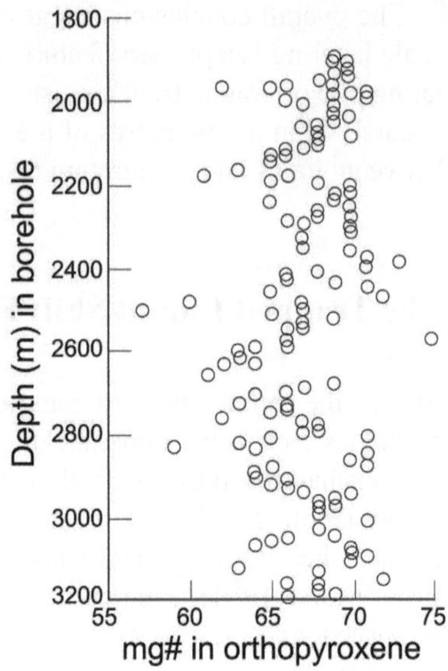

Fig. 12.19 Plot of mg# in orthopyroxene for a 1400 m thick section of the Main Zone. (Data taken from Roelofse and Ashwal 2012)

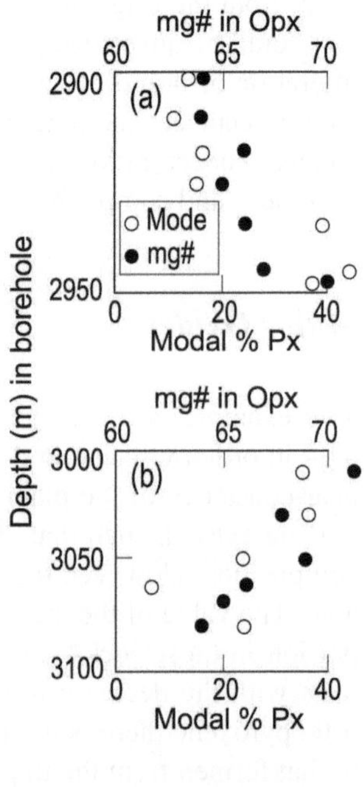

Fig. 12.20 Two short sections taken from Fig. 12.19 showing an upward decrease **a** and an upward increase **b** in mg# with height. Also plotted are the modal proportions of pyroxene in these sections. High mg# correspond to rocks with a high content of (cumulus) pyroxene, and the mg# is lower where there is less pyroxene and the trapped liquid contribution becomes more significant

little variation in the An content of plagioclase. Two detailed sections are shown in Fig. 12.20, illustrating this variability in mg#. In Fig. 12.20a there appears to be an upward decrease in mg# in orthopyroxene. Conversely, in Fig. 12.20b there is an upward increase. Such trends might be interpreted as due to fractionation and recharge respectively. However, when the modal proportion of pyroxene is considered in these plots it can be seen that there is a strong positive correlation between

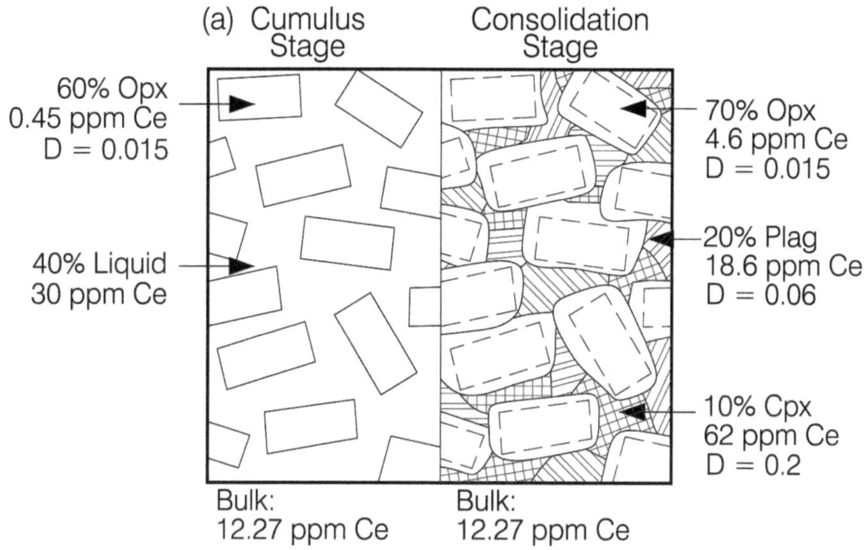

Fig. 12.21 Model showing the effect of the trapped liquid shift on the Ce abundance in mineral compositions. The cumulus orthopyroxene in equilibrium with liquid contains 0.45 ppm Ce, but after solidification of the trapped liquid its composition has increased to 4.6 ppm

mg# and modal pyroxene content, with the more evolved pyroxene compositions in pyroxene-poor rocks. Such a correlation is exactly that predicted by the trapped liquid shift model.

I conclude that using the mg# value of minerals is not a reliable method for modelling fractionation and recharge in a magma chamber. The An content of plagioclase may provide a better parameter for such modelling because the cores are not reset by reaction with interstitial liquid. However, inhomogeneity within the cores of plagioclase grains is typically ±2 % An (Cawthorn and Ashwal 2009) and so only sustained changes larger than that may be considered significant.

Trace Elements

The incompatible trace element contents in minerals behave in an analogous way to the mg# as a result of the TLSE. The trapped liquid contains very much higher concentrations of such elements than the cumulus minerals, and so contribute very disproportionately to the bulk rock concentration. On solidification of the trapped liquid these incompatible elements have to be partitioned between the cumulus and any new intercumulus minerals (according to their individual partition coefficients). Thus, the final minerals would contain a much higher concentration of such elements than the original cumulus phase. This effect is shown in Fig. 12.21 for Ce in orthopyroxene. The liquid may have contained 30 ppm Ce and so the cumulus pyroxene contained 0.45 ppm (D = 0.015). A model rock is considered, containing

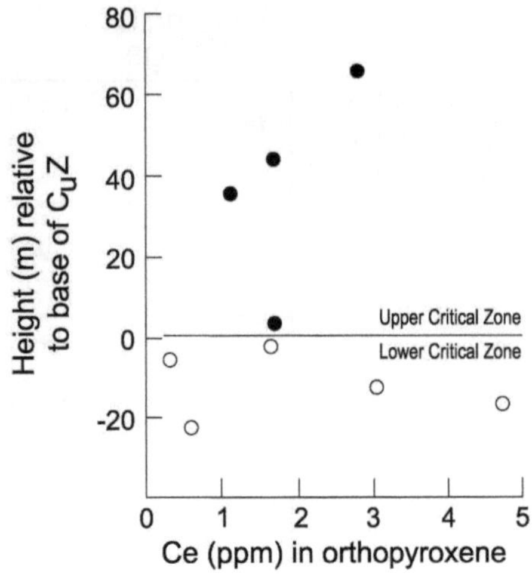

Fig. 12.22 Ce contents in orthopyroxene across the boundary between Lower (open symbols) and Upper (closed symbols) Critical Zone. (Own unpublished data)

60 % cumulus pyroxene and 40 % trapped liquid. The total Ce content is 12.27 ppm Ce. On solidification, the rock now contains 20 % plagioclase, 10 % clinopyroxene (both formed from the trapped liquid) and 70 % orthopyroxene. The 12.27 ppm Ce has distributed itself between the three minerals to yield Ce contents of 4.6, 62 and 18.6 ppm (for orthopyroxene, clinopyroxene and plagioclase, respectively, using the D values shown in Fig. 12.21). Thus the cumulus orthopyroxene with 0.45 ppm Ce has changed to contain 4.6 ppm, a ten-fold increase. (Smaller proportions of trapped liquid would yield smaller increases in final Ce concentration.) The precision of these calculations depends upon a knowledge of the D values and the proportion of trapped liquid (and the modal proportions produced by the trapped liquid).

This principle was used by Cawthorn (1996b) to show how the large variations observed in REE content of orthopyroxene in the Critical Zone could be explained. A similar demonstration of this effect in the Lower Zone was shown by Godel et al. (2011). Its effect is also shown in Fig. 12.22. The orthopyroxene and whole-rock concentrations from a section traversing the Lower Critical to Upper Critical Zones were determined. The Ce content of the orthopyroxene varies irregularly by a factor of 10 (0.3–5 ppm). If these values for Ce in orthopyroxene were considered the true cumulus compositions then they would demand that the liquid from which they formed also varied erratically by a factor of 10, which I suggest is implausible. The Zr whole-rock content varies considerably, and can be considered a proxy for the proportion of trapped liquid. A plot of the final Ce content in the pyroxene versus the whole-rock Zr content shows a strong positive correlation (Fig. 12.23). Hence, I suggest that the variation in Ce in pyroxene results from the variable proportions of and extent of re-equilibration with trapped liquid. By extrapolation to 8 ppm Zr (the likely content of Zr in pyroxene) the true cumulus pyroxene can be inferred to have less than 0.5 ppm Ce (as used in Fig. 12.21). If there had been infiltration and variable metasomatism during final compaction and solidification, as proposed by Mathez et al. 1997) for the Bushveld and McKenzie (2011) with specific reference to the Skaergaard Intrusion, such a correlation as seen in Fig. 12.23 would not result.

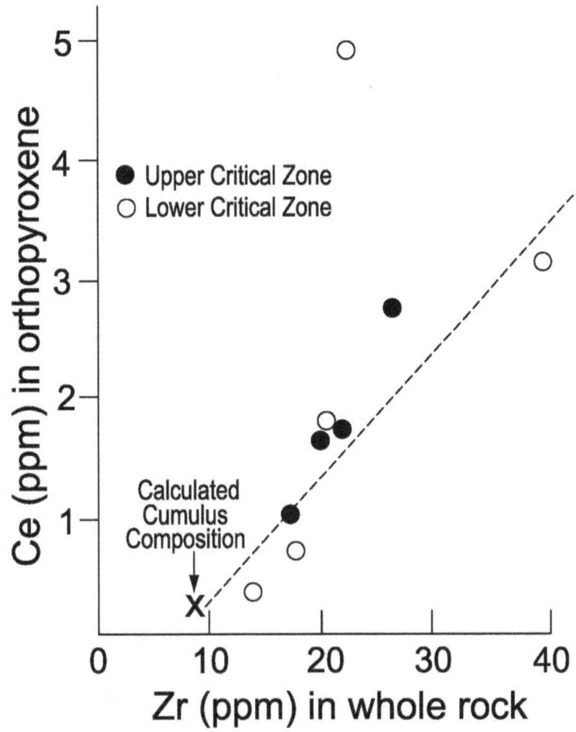

Fig. 12.23 Relationship between Ce content in orthopyroxene and Zr content in whole rock that proxies for the proportion of trapped liquid. Extrapolation to 8 ppm Zr yields the cumulus orthopyroxene composition. (Own unpublished data)

The same principle can be applied to the REE in apatite, even though they are highly compatible in apatite. The effect is shown in Fig. 12.24, which is analogous to Fig. 12.21. For simplicity of demonstration, it is assumed that the other minerals

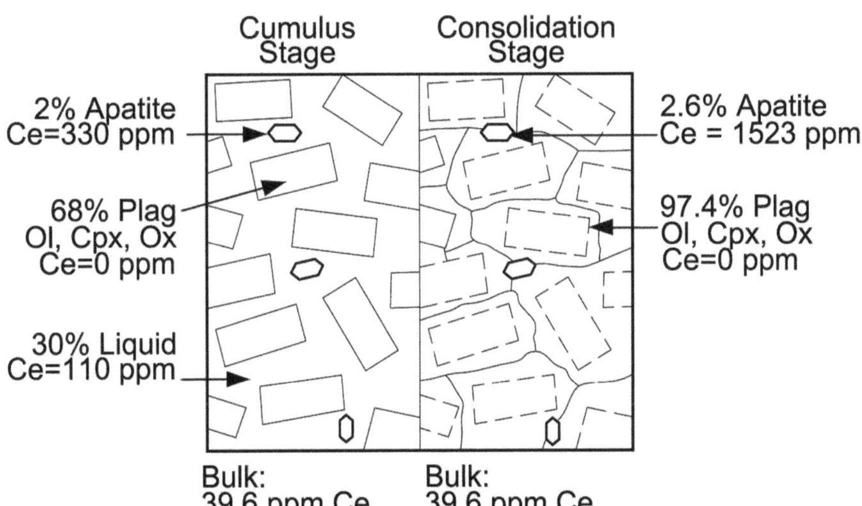

Fig. 12.24 Behaviour of Ce in apatite, analogous to that shown in Fig. 12.21 for orthopyroxene

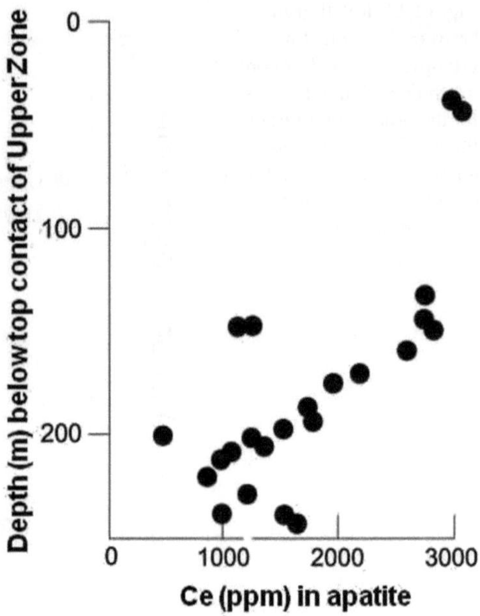

Fig. 12.25 Plot of Ce in apatite versus height below upper contact in Bierkraal borehole BK1 (Own unpublished whole-rock data, and calculated by the method described by Tollari et al. 2008)

contain no Ce. The Ce in apatite in a cumulate rock with 2 % cumulus apatite and 30 % trapped liquid will increase by a factor of 5 due to the TLSE. The concentration of Ce in apatite in a vertical section through the Upper Zone is shown in Fig. 12.25 and indicates a general upward increase, and a decrease in the normative proportion of apatite. This relationship is predicted by application of the TLS principle and has been used by Cawthorn (2013b) to explain the variations in REE in apatite in the Upper Zone, and is shown in Fig. 12.26. Data previously published was interpreted to have resulted from liquid immiscibility (VanTongeren and Mathez 2012). As can be seen, the proportions of apatite and the trapped liquid are crucial in determining the final apatite composition. The exact proportion of trapped liquid in these rocks is difficult to quantify. The inclusion into these calculations of the minor concentrations of Ce that enter the other minerals is of trivial significance to this principle.

Slurries

In the debate concerning the origin of cumulus grains the concept of injection of a slurry into the presently observable Bushveld is becoming increasingly promoted. The general principle of this model was presented by Marsh (2006). Whereas it is well-known that many lavas erupt carrying some crystals, in none of the models presented below do the authors state the proportion of grains in their proposed slurries. The density of such a slurry would increase linearly as the proportion of mafic phases increases, and the viscosity would increase exponentially (McKenzie 2011).

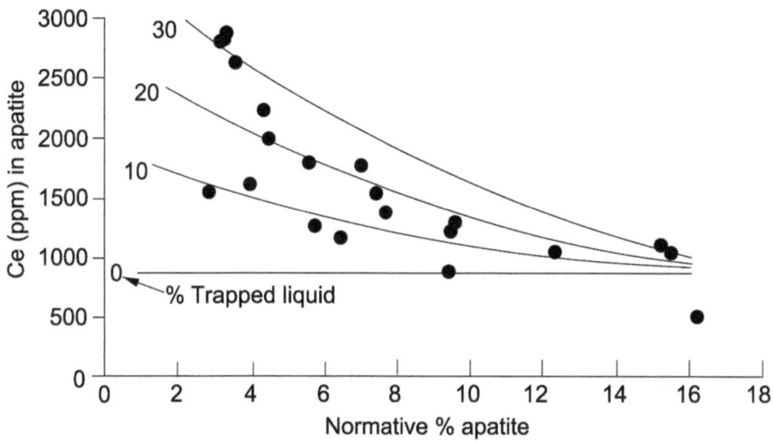

Fig. 12.26 Plot of normative apatite versus Ce content in apatite, showing the effect of changes in Ce content due to reaction with trapped liquid. (Own unpublished data and using the method described by Cawthorn 2013b). Where there is abundant cumulus apatite it buffers the effect of the trapped liquid shift effect and samples with high apatite content approximate to the true cumulus Ce content. However, where there is little cumulus apatite the trapped liquid shift effect becomes very pronounced

Hence, the ability of a slurry to be transported upward would become severely restricted. There are three variants to this model: (i) upward emplacement from a deeper chamber, (ii) lateral transport from the edges of an originally much wider Bushveld chamber, and (iii) injection into (not onto) the cumulate pile. They are discussed below.

Injection from a Deeper Chamber

Maier and Barnes (1998) compared the REE contents of Main Zone cumulates with the proposed parental magma to these rocks, and found them to be closely similar (Fig. 12.27). They proposed that the Main Zone must have been injected as a slurry of the pyroxene and plagioclase grains. The logic of this argument is not clear. If the proposed composition (referred to as B3 or B2) to this zone is a true liquid (i.e. not itself a crystal mush), then a rock that consists of an unspecified proportion of that liquid plus grains of plagioclase and pyroxene (which would be much lower in their REE contents than the liquid from which they formed) should be depleted in REE relative to the liquid. The similarity between the purported liquid (B2 or B3) and the Main Zone cumulates argues against this model. Furthermore, the proposed B2 and B3 compositions have a lower initial $^{87}Sr/^{86}Sr$ value (<0.706) than the Main Zone rocks (Maier et al. 2000).

Roelofse and Ashwal (2012) developed the idea of the injection of slurries. They proposed that the pyroxenes and the plagioclase came from two different magma

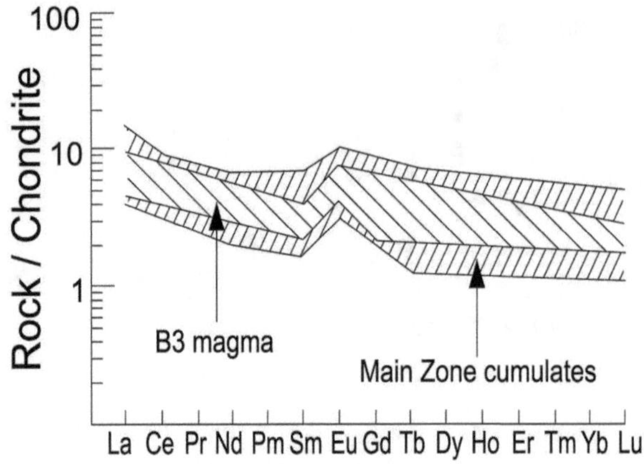

Fig. 12.27 Plot of REE abundance in typical Main Zone cumulates and of the purported parental magmas to the Main Zone (known as B2 or B3) taken from Maier and Barnes (1999). Their very similar abundances demonstrate that the one cannot be a cumulus-enriched assemblage and the other a true liquid. If there were about 10 % trapped liquid in a cumulate rock and if the D values for these REE are low, then the liquid should contain almost 10 times as much as any crystal mush or cumulus rock

chambers at depth and then mixed to produce the gabbronorites in the Bushveld chamber. It needs to be questioned why the dense pyroxenes did not settle (in the deeper magma chamber) to produce a solid layer of rock that would not be capable of being remobilized into a slurry. It must further be questioned how a dense slurry of pyroxene (pre-supposing it existed) could be so effectively and completely mixed with a less dense slurry of plagioclase, because clustering of the pyroxenes or plagioclase grains together has never been reported, as would be expected if the process had not been extremely efficient. The extent of liquids of different density mixing was discussed above. The proposed crystal mushes would have had much higher density contrast and much higher kinematic viscosities than between two liquids, and so I question whether they could mix efficiently. Finally, it is well-known that the Al content of pyroxene co-existing with an aluminous phase (and hence being saturated in Al) increases with increasing pressure (Charlier et al. 2010). Data from assorted experimental studies at crustal pressures are shown in Fig. 12.28. At the mid-crustal pressure envisaged by Roelofse and Ashwal (2012) the orthopyroxene ought to contain at least 3 % Al_2O_3. All Bushveld orthopyroxene grains contain less than 1.5 %, and most less than 1 % (Teigler and Eales 1996; Mitchell 1990; Schürmann 1993; Roelofse and Ashwal 2012).

One of the reasons for the proposed model by Roelofse and Ashwal (2012) was the fact that there was no mineral differentiation in terms of An content of plagioclase for their long vertical section of the Main Zone. They argued that differentiation should have occurred in the Bushveld chamber. They therefore proposed the existence of a deeper chamber providing grains of a constant composition. However, how this deeper magma chamber failed to differentiate, but provided minerals of

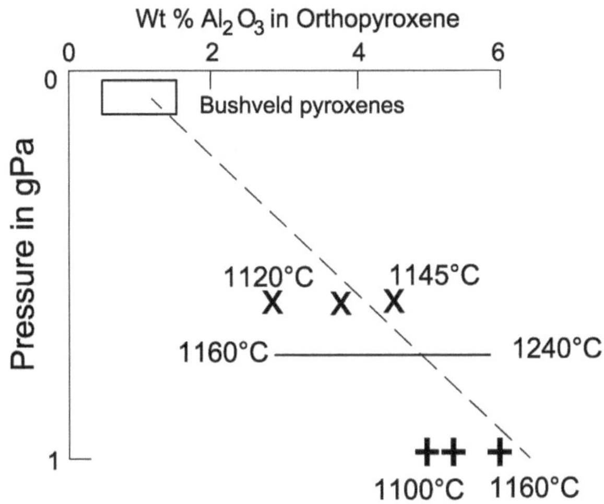

Fig. 12.28 Plot of weight % Al_2O_3 in orthopyroxene (in plagioclase-saturated assemblages) as a function of pressure, taken from assorted natural examples (the dashed line is a semi-quantitative trend from the data of Charlier et al. 2010 for anorthosites) and experimental studies (horizontal line is range of concentrations as a function of temperature from Chalott-Prat et al. 2010, and crosses and squares are individual data points from Vander Auwera et al. 1994 and Gudfinnsson and Presnall 2000)

constant composition was not discussed. The constancy of mineral composition can be explained by reference to lava compositions in large igneous provinces. In the Karoo and east Greenland, Marsh et al. (1997) and Tegner et al. (1998) showed that very many successive liquid compositions were erupted with a very constant composition. By analogy, for the Bushveld chamber, uniform liquids rather than grain mushes, could have been repeatedly injected producing minerals of near-constant composition.

Eales (2002) and Mondal and Mathez (2007) proposed that the chromite in the Critical Zone had been injected as a slurry. The validity of the summation undertaken by Eales has been discussed above. Given the high density of chromite relative to the magma, the possibility of it being so uniformly distributed over such large areas if injected at some point or line as a slurry seems implausible. In an expanded version of this concept, Eales and Costin (2012) suggested that olivine, orthopyroxene and chromite all accumulated in a deeper chamber to be injected as slurries. Given the high density of all three minerals it is not clear how this enormous mass of grains could have been held in suspension in this deeper chamber without coagulating on the floor. The model requires that the grains remain totally separated from each other because, upon injection, all the chromite grains were able to settle to the floor of the Bushveld chamber to form the chromitite layers. Paradoxically, the application of Stokes' Law clearly shows that the very small chromite grains, despite having higher density, would settle much more slowly than the much larger olivine and orthopyroxene. Hence, the order of accumulation proposed by them—chromite before the mafic silicates—contradicts the settling rates. Again, the Al content of orthopyroxene in the Critical Zone argues against this model (Cameron 1980; 1982; Teigler and Eales 1996; Schürmann 1993). Finally, a slurry existing at the bottom of a large funnel-shaped chamber would not be erupted until all the overlying magma had been pumped out (and presumably into the Bushveld magma chamber), and so this model does not avoid the problem of the mass of magma in the Bushveld chamber.

Lateral Injection

A variation of this slurry model has been presented by Maier et al. (2012). They suggested that there was inward slumping of a semi-consolidated (and mineralogically unsorted) grain mush from the periphery of the intrusion (beyond the present outcrop limit) into the centre (the observable Bushveld area) and that as the grains resettled they separated into the observed modal layers. This model raises a number of questions. If these grains were accumulating around the periphery of the intrusion what was accumulating in the centre of the intrusion (which is the area now preserved)? The known extent of the intrusion (300 km wide) does not contain any of the original unsorted material. Did the accumulated slurry form in a ring all round the margins of the original chamber and all slump inward (everywhere at the same time) from the outside to produce layers of constant thickness everywhere? Or did the mass slumping occur from one side only and hence had to traverse many hundreds of km right across the preserved Bushveld, again producing layers of constant thickness? How did the huge mass of grains remain in a semi-aggregated state such that it could be totally disaggregated into single grains that were then able to sort themselves based on Stokes' Law? A calculation of the mass involved is given in Appendix 2. I suggest that all the examples of deformation seen (Fig. 12.3s to v, and in Maier et al. 2012) show that a semi-consolidated mush existed in the known Bushveld (not some peripheral area) and that tectonic disturbance caused intense deformation (Carr et al. 1994). The horizontal distances moved during such disturbances and deformation were in the order of metres to many tens of metres, not many hundreds of kilometres. Most importantly, these semi-consolidated layers remained largely intact and coherent, and did not disaggregate despite the intensity of their deformation. As shown above, the thickness of the unconsolidated mush was probably in the order of metres, and so the total masses involved in such disturbances was trivial compared to that required to redistribute material to form thick layers over many tens of thousands of square kilometres. Also the model contradicts Stokes Law, again requiring that chromite sank more rapidly than the mafic silicates. Further, the absence of cumulus pyroxene associated with the magnetitite-anorthosite sequences (discussed above) is inconsistent with this model. Finally, both models of redeposition of a slurry fail to explain the observation that the interstitial silicate phases in chromitite layers have higher $^{87}Sr/^{86}Sr$ values than the immediately overlying silicate-rich layers (Kinnaird et al. 2002).

Both the chromitite and the magnetitite layers show evidence of systematic vertical variations in composition, which is not explicable if they formed by remobilisation of an unconsolidated mush either from below or the periphery.

Injection into the Cumulate Pile

Yet another set of views on slurries has evolved that relates primarily to chromitites. The bifurcating UG1 chromitite and its footwall stringers seen at Dwars River (Fig. 12.3e), but present everywhere that the UG1 is exposed (especially underground) have been taken as evidence that the chromitite was a slurry that was injected (discordantly) into an already formed sequence, not emplaced at the mush—magma interface as with the above models (Voordouw et al. 2009). This principle has been applied to the UG2 chromitite as well (Maier and Barnes 2008). I have to debate this model on the grounds that a dense crystal mush cannot be laterally distributed with uniform thickness over an area of 30,000 km^2 (for the UG1 and UG2, which have so many unique features that they are demonstrably the same layer in the east and west). A second issue relates to whether this mechanism only applies to the UG1 and UG2 or all chromitites. If the model applies only to these two layers are all other chromitites formed by accumulation at the mush-magma interface? If so, what criteria distinguish the two different mechanisms of formation? There are many chromitite layers, often associated with the thicker, named layers, that are only mm in thickness (Fig. 12.3w), and yet are traceable over great distances. I question if they could have been intrusive. When viewed in thin section it is clear that small grains of chromite have settled on top of grains of orthopyroxene, not as an intrusion through the middle of a pyroxenite package (Fig. 12.11m). Many of the thin chromitite stringers show a zone of disseminated chromite grains, usually above the chromitite layer (Fig. 12.11n), which cannot be explained by an intrusive origin. Further, the relations at Dwars River (Fig. 12.3q) clearly show that the chromitites can be cut by anorthosite bodies, demonstrating the primary accumulation of chromite.

Finally, the geometry observed in the well-known potholes in the UG2 chromitite, disproves the intrusive relationship. In a pothole, the footwall succession is variably removed and the subsequently accumulating chromitite layer drapes itself into the pothole. Directly above the pothole the overlying layers are thickened relative to the normal succession. These features are shown in Fig. 12.29. Figure 12.29a shows a slightly stylised normal succession of (upward) footwall pyroxenite (FP) and anorthosite (FA), overlain by the UG2 chromitite (chr), then hangingwall pyroxenite (HP), norite (HN) and anorthosite (HA). Figure 12.29b shows the relations observed in a pothole (Hahn and Ovendale, 1994) where FA and part of PF have been truncated, and HP and HN thickened. Figures 12.29c and d show the situation envisaged in models that propose later intrusion of the chromitite between FA and HP; Fig. 12.29c shows the situation before chromitite injection, and Fig. 12.29d after the postulated injection with slight discordances producing the pothole. Note

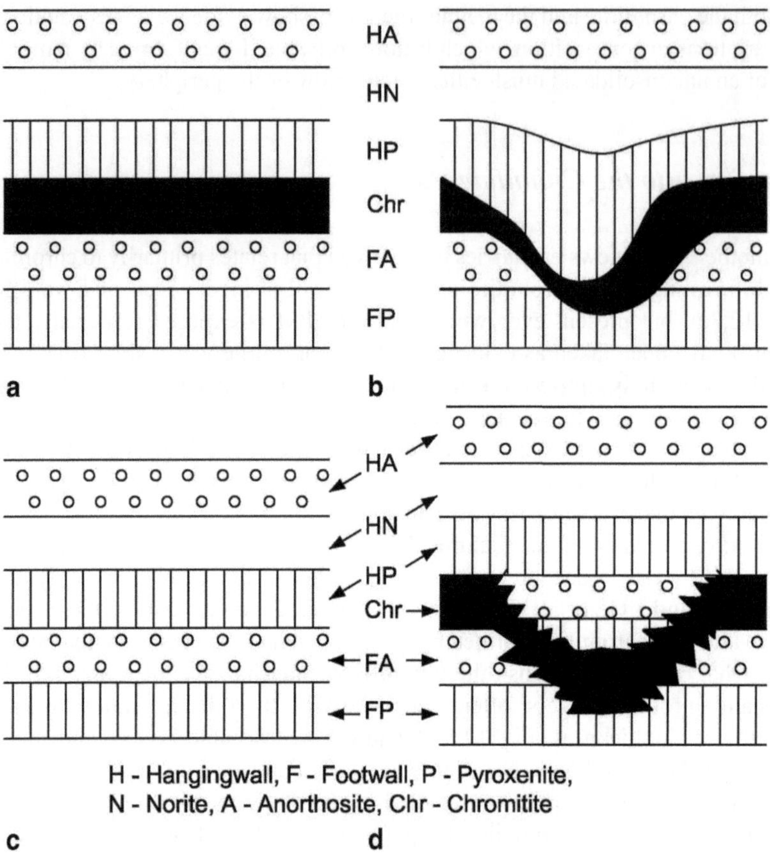

Fig. 12.29 Evidence against intrusive chromitites using observations from the UG2. **a** Illustration of typical layered succession. **b** Cross-section through a typical pothole at the level of the UG2 Chromitite layer (generalized from author's own observations and Hahn and Ovendale 1994). **c** Hypothetical succession pre-emplacement of intrusive chromitite. **d** Geometrical relations in a pothole demanded if chromitite is intrusive. Note that some typical footwall rocks(FP and FA) would be expected above the chromitite with this model. See text for explanation

that some of FP and FA are now displaced above the chromitite, and that HP and HN have not changed in thickness. The observed relations (Fig. 12.29b) show thickening of hangingwall layers and no preservation of footwall layers above the chromitite. All the observed features are very simply explained in a stratigraphic accumulation model, where the layers FA and FP were eroded by current activity. The chromitite was deposited on that eroded surface. The hanging wall layers infilled the depression and eventually obliterated the topographic effects of the pothole erosion event. Such (clastic) sedimentary processes to smooth temporary topographic irregularities were emphasised by Irvine et al. (1998) in the Skaergaard intrusion, although in their case there were autoliths forming positive features that were buried, with the layers becoming horizontal a very short vertical distance above the autoliths. For exactly the same reasons, the pothole geometry seen at the level of the Merensky Reef invalidates models (Mitchell and Scoon 2007) that propose an intrusive origin for the Merensky Reef.

Density Inversions

The association of anorthosite (density 2.75 g cm^{-3}) and overlying magnetitite (density >5), as shown in Fig. 12.3j, represents the biggest density contrast between any two layers of rock above the core-mantle boundary. Furthermore, they are gravitationally unstable (denser above less dense). If the two layers had existed as mushes with a considerable proportion of interstitial liquid it would be expected that diapiric structures ought to have developed. The remarkably planar boundary between such paired layers suggests that the underlying anorthosite was extremely rigid before any significant accumulation of magnetite occurred. This same observation applies to the MG3, MG4, UG1 and UG2 chromitites that are underlain by anorthosite have remarkably planar undeformed interfaces.

Discordant Bodies

The remarkably laterally extensive layering is interrupted by very irregularly distributed pegmatitic ultramafic bodies ranging from olivine- and clinopyroxene-rich to magnetite-dominated vertically. Vertical continuity of any single body cannot be established, except for the 300-m vertical exposure in the Kennedys Vale vanadiferous magnetitite plug, near Dwars River along the Steelpoort structure (Viljoen and Scoon 1985). Many are pipe-like, but many also are dyke-like (Fig. 12.3x) or are planar and concordant (Fig. 12.3y). Whereas most are iron-rich, those bodies in the Critical Zone are quite magnesian (Viljoen and Scoon 1985; Tegner et al. 1994). Exposures of contacts are limited to mining activities and road cuttings. They do not displace the surrounding rocks but clearly replace them. For example, opposing sides of bodies and not planar and do not match (Fig. 12.3z). They are replacive, and relic features such as remains of thin chromitites and platinum mineralisation at the level of the Merensky Reef have been recorded in them (Viljoen and Heiber 1986). Scoon and Mitchell (1994) interpreted them as either evolved magmas or iron-rich immiscible liquids derived from the Upper Zone. The initial Sr isotope ratio of these rocks is inconsistent with them having been derived from a local source (Fig. 12.13). If they had originated in the Upper Zone as immiscible liquids then they had to melt their way downward for distances of over 2 km through cumulates that would have a very high liquidus temperature, and that remelted material had to be expelled upward without mixing with the descending liquid. Where sill-like bodies are exposed, it is the upper contact that is irregular, not the lower contact, challenging the view that they were emplaced downward. Thus, the mechanism of emplacement must be considered unresolved.

Summary

The Bushveld Complex shows the most complete differentiation sequence of any layered intrusion with olivine Fo_{90} at the base, through iron-rich successions to monzonite with olivine Fo_1, and containing 65 % SiO_2 at the top. In preserved lateral extent (more than 60,000 km^2) the Bushveld Complex is by far the World's largest layered mafic intrusion. Many distinctive layers (chromitites, Merensky and Bastard Reef, and Pyroxenite Marker) can be traced in the east and west lobes; and the magnetitite layers traced also in the northern lobe. Identifying processes that explain the uniformity of such layers is probably the single biggest challenge to modelling this intrusion. Whether most of the accumulation of grains and modal layering is by settling and gravity sorting or by in situ growth is still totally unresolved, although great thicknesses showing gradual modal changes independent of cotectic proportions suggest the former. The rate of accumulation of grains on the floor of the intrusion was extremely slow, permitting textural re-equilibration close to the mush-liquid interface. Many features suggest that the vertical column in which there was considerable liquid was restricted to a few metres at maximum. The remaining ±10% of interstitial liquid became trapped and as it solidified it modified mg# in mafic minerals when that mineral was present as a minor cumulus phase, and increased the incompatible trace element contents of all minerals. Interpretations based on mg# to model fractionation processes, and of trace elements to model instantaneous liquid compositions are extremely unreliable. The remarkable state of preservation of this 2.06 b.y. old body permits studies (especially geochemical) that are impossible on many old igneous rocks, and provides an insight into the processes that have operated, but which still require conclusive resolution.

Acknowledgements I have been fortunate to have been based at the University of the Witwatersrand for 40 years, which has provided support for my research into the Bushveld Complex in many ways. Financial support has been generously provided by NRF (South Africa), Angloplatinum, Impala Platinum and Lonplats. Many mining company geologists have willingly given of their knowledge and their bore cores, without which many of my Bushveld studies would be impossible. Also academic geologists (many of whose names appear in the reference list below) have shared ideas and debates, variably enlightening and challenging in my attempts to grasp inklings into the genesis of the Bushveld. Richard Wilson, Jean-Clair Duchesne and James Roberts undertook very thorough reviews of earlier versions of this manuscript. Lyn Whitfield and Di du Toit drafted many diagrams. To all of them—very many thanks.

Appendix

1. Calculation to demonstrate that mixing cannot have occurred between resident and new magma at the base of the Main Zone.

Mitchell (1990) showed that the Cr_2O_3 content of the orthopyroxene below and in the Merensky interval is 0.4 % and decreases abruptly to 0.1 % 30 m higher in the

Lower Main Zone (Fig. 12.8a). Using partition coefficients that Barnes (1986a) showed to be temperature- and MgO-dependant, the Cr content of the magmas may be inferred. At the top of the Critical Zone the liquid contained about 7% MgO (Cawthorn 1996a), yielding a partition coefficient of 5. Thus the liquid contained 500–600 ppm Cr. The Main Zone liquid may have had similar or slightly less MgO. The partition coefficient may have been 6, yielding a liquid with 100 ppm Cr. (Uncertainties in the exact values of these numbers can be seen below to be irrelevant.) If it is assumed that the Main Zone is the product of mixing of residual Critical Zone and added Main Zone liquids an equation may be derived.

M (Cr in mixed magma in ppm) = R (Cr in residual Critical Zone magma)*F (fraction of residual magma involved) + A (Cr in added magma) * (1-F).

M = 100 (see above); R = 500 (see above); A = 100–200 ppm (Maier et al. 2000; Barnes et al. 2010). Hence, F < 0! There can be no mixing!

Note that the actual analyses of purported parental magma to the Main Zone (B2 or B3 in Table 12.1) contain 200 and 400 ppm Cr. These concentrations are implausible, based on the Cr content of the pyroxenes, even if no mixing had occurred.

2. Calculation of mass of material needed to make a 1 m thick layer of chromitite in the slumping model of Maier et al. (2012).

There would need to be a mass of chromite, pyroxene, (?) plagioclase and (?) olivine. One assumes that they are present in their cotectic proportions. Bulk density would be about 3 $t \cdot m^{-3}$. Suppose the cotectic proportion of chromite in that assemblage is 0.2%. To make 1 m of chromitite therefore requires 500 m of grain mush. The area over which the chromitite layers are preserved is at least 30,000 km^2. Hence, the mass of unconsolidated material would need to be $3 * 500 * 30000 * 10^6$ or $4.5 * 10^{13}$ t. All this material must remain in a completely disaggregatable state so that chromite grains could separate from silicate grains. This mass must have accumulated elsewhere for every 1 m-thick chromitite layer.

References

Ashwal LD, Webb SJ, Knoper MW (2005) Magmatic stratigraphy in the Bushveld northern lobe: continuous geophysical and mineralogical data from the 2950 m Bellevue drillcore. S Afr J Geol 108:199–232

Barnes SJ (1986a) The distribution of chromium among orthopyroxene, spinel and silicate liquid at atmospheric pressure. Geochim Cosmoschim Acta 50:1889–1909

Barnes SJ (1986b) The effect of trapped liquid crystallization on cumulus mineral compositions in layered intrusions. Contrib Miner Petrol 93:524–531

Barnes S-J, Maier WD, Curl E (2010) Composition of the marginal rocks and sills of the Rustenburg layered suite, Bushveld Complex, South Africa: implications for the formation of the PGE deposits. Econ Geol 105:1491–1511

Biggar GM (1974) Phase equilibrium studies of the chilled margins of some layered intrusions. Contrib Mineral Petrol 46:159–167

Boorman S, Boudreau AE, Kruger FJ (2004) The lower zone-critical zone transition of the Bushveld Complex—a quantitative textural study. J Petrol 45:1209–1235

Bottinga Y, Weill DF, Richet P (1982) Density calculations for silicate liquids: I. Revised method for aluminosilicate liquids. Geochim Cosmochim Acta 46:909–919

Boudreau AE (1994) Mineral segregation during crystal aging in two-crysal, two-component systems. S Afr J Geol 97:473–485

Buchanan DL (1979) A combined transmission electron microscope and electron microprobe study of bushveld pyroxenes from the Bethal area. J Petrol 20:327–354

Button A (1976) Stratigraphy and relations of the Bushveld floor in the eastern Transvaal. Trans Geol Soc S Afr 79:3–12

Cameron EN (1978) The lower zone of the eastern Bushveld Complex in the olifants river trough. J Petrol 19:437–462

Cameron EN (1980) Evolution of the lower critical zone, central sector, eastern Bushveld Complex. Econ Geol 75:845–871

Cameron EN (1982) The upper critical zone of the eastern Bushveld Complex—precursor of the merensky reef. Econ Geol 77:1307–1327

Campbell IH (1996) Fluid dynamic processes in basaltic magma chambers. In: Cawthorn RG (ed) Layered intrusions. Elsevier, Amsterdam, pp 45–76

Campbell IH, Murck BW (1993) Petrology of the G and H chromitite zones in the mountain view area of the stillwater complex, Montana. J Petrol 34:291–316

Campbell IH, Roedder PL, Dixon JM (1978) Plagioclase buoyancy in basaltic liquid as demonstrated with a centrifuge furnace. Contrib Miner Petrol 67:369–377

Campbell IH, Naldrett AJ, Barnes SJ (1983) A model for the origin of the platinum-rich sulfide horizons in the bushveld and stillwater complexes. J Petrol 24:133–165

Carr HW, Groves DI, Cawthorn RG (1994) The importance of syn-magmatic deformation in the formation of merensky reef potholes in the Bushveld Complex. Econ Geol 89:1398–1410

Carr HW, Groves DI, Kruger FJ, Cawthorn RG (1999) Petrogenesis of merensky reef potholes at the western platinum mines: Sr-isotopic evidence for synmagmatic deformation. Miner Deposita 34:335–347

Cawthorn RG (1996a) Re-evaluation of magma compositions and processes in the uppermost critical zone of the Bushveld Complex. Miner Mag 60:131–148

Cawthorn RG (1996b) Models for incompatible trace-element abundance and their application to plagioclase and pyroxene compositions in the Bushveld Complex. Contrib Miner Petrol 123:109–115

Cawthorn RG (1998) Geometrical relations between the Transvaal Supergroup, the Rooiberg Group and the mafic rocks of the Bushveld Complex. S Afr J Geol 101:275–280

Cawthorn RG (1999) The discovery of the platiniferous merensky reef in 1924. S Afr J Geol 102:178–183

Cawthorn RG (2002) Delayed accumulation of plagioclase in the Bushveld Complex. Miner Mag 66:881–893

Cawthorn RG (2011) Geological interpretations from the PGE distribution in the bushveld merensky and UG2 chromitite reefs. South Afr Inst Min Metall 11:67–79

Cawthorn RG (2013a) The residual or roof zone of the Bushveld Complex. J Petrol 54:1875–1900

Cawthorn RG (2013b) Rare earth element abundances in apatite in the Bushveld Complex—a consequence of the trapped liquid shift effect. Geology 41:603–606

Cawthorn RG, Ashwal LD (2009) Origin of anorthosite and magnetite layers in the Bushveld Complex, constrained by major element compositions of plagioclase. J Petrol 50:1607–1637

Cawthorn RG, Biggar GM (1993) Crystallization of titaniferous chromite, magnesian ilmenite and armalcolite in tholeiitic suites in the Karoo Province. Contrib Miner Petrol 114:221–235

Cawthorn RG, Boerst K (2006) Origin of pegmatitic pyroxenite in the merensky unit, Bushveld Complex, South Africa. J Petrol 47:1509–1530

Cawthorn RG, McCarthy TS (1980) Variations in Cr content of magnetite from the upper zone of the Bushveld Complex—evidence for heterogeneity and convection currents in magma chambers. Earth Planet Sci Lett 46:335–343

Cawthorn RG, Molyneux TG (1986) Vanadiferous magnetite deposits of the Bushveld Complex. In: Anhaeusser CR, Maske S (eds) Mineral deposits of Southern Africa. Geol Soc S Afr Johannesburg 2:1251–1266

Cawthorn RG, Spies L (2003) Plagioclase-rich cyclic units in the Bushveld Complex. Contrib Miner Petrol 145:47–60
Cawthorn RG, Street J (1994) Vertical migration of residual magma in the Bushveld Complex. Miner Petrol 51:345–354
Cawthorn RG, Walraven F (1998) Emplacement and crystallization time for the Bushveld Complex. J Petrol 39:1669–1687
Cawthorn RG, Davies G, Clubley-Armstrong A, McCarthy TS (1981) Sills associated with the Bushveld Complex, South Africa. Lithos 14:1–15
Cawthorn RG, Meyer PS, Kruger, FJ (1991) Major addition of magma at the Pyroxenite Marker in the western Bushveld Complex, South Africa. J petrol 32:739–763
Cawthorn RG, Cooper GRJ, Webb SJ (1998) Connectivity between the western and eastern limbs of the Bushveld Complex. S Afr J Geol 101:291–298
Cawthorn RG, Harris C, Kruger FJ (2000) Discordant ultramafic pegmatoidal pipes in the Bushveld Complex. Contrib Miner Petrol 140:119–133
Cawthorn RG, Merkle RKW, Viljoen MJ (2002) Platinum-group element deposits in the Bushveld Complex, South Africa. In: Cabri LJ (ed) Geology, geochemistry, mineralogy and mineral beneficiation of platinum-group elements. Can Inst Min Metal Petrol 54:389–430
Chalot-Prat F, Falloon TJ, Green DH, Hibberson WD (2010) An experimental study of liquid compositions in equilibrium with plagioclase-spinel lherzolite at low pressures (0.75 gPa). J Petrol 51:2349–2376
Charlier B, Duchesne JC, Vander Auwera J, Storme JY, Maquil R, Longhi J (2010) Polybaric fractional crystallization of high-alumina basalt parental magmas in the Egersund-Ogna massif-type anorthosite (Rogaland, SW Norway) constrained by plagioclase and high-alumina orthopyroxene megacrysts. J Petrol 51:2515–2546
Clarke BM, Uken R, Watkeys MK (2000) Intrusion mechanisms of the southwest Rustenburg Layered Suite as deduced from the Spruitfontein inlier. S Afr J Geol 103:120–127
Clarke BM, Uken R, Reinhardt J (2009) Structural and compositional constraints on the emplacement of the Bushveld Complex, South Africa. Lithos 111:21–36
Cooper RW (1997) Magmatic unconformities and stratigraphic relations in the Peridotite zone, Stillwater Complex, Montana. Can J Earth Sci 34:407–425
Darwin C (1844) Geological observations on the volcanic islands. Appleton, London
Davey SR (1992) Lateral variation within the upper critical zone of the Bushveld Complex on the farm Rooikoppies 297JQ, Marikana, South Africa. S Afr J Geol 95:141–149
Davies G (1982) The petrogenesis of the peripheral zone of the Rustenburg layered suite and associated sills between Hartbeestpoort and Buffelspoort dams, western Bushveld Complex. Unpubl PhD thesis Univ Witwatersrand, Johannesbg, p 444
Davies G, Cawthorn RG (1984) Mineralogical data on multiple intrusion in the Rustenburg layered suite of the Bushveld Complex. Miner Mag 48:469–480
Davies G, Cawthorn RG, Barton JM, Morton M (1980) Parental magma to the Bushveld Complex. Nature 287:33–35
Du Plessis CP Walraven F (1990) The tectonic setting of the Bushveld Complex in Southern Africa, part 1. Structural deformation and distribution. Tectonophysics 179:305–319
Duchesne JC (1972) Iron-titanium oxide minerals in the Bjerkreim-Sokndal massif, south-western Norway. J Petrol 13:57–81
Duchesne JC, Charlier B (2005) Geochemistry of cumulates from Bjerkreim-Sokndal layered intrusion (S. Norway). Part I: constraints from major elements on the mechanism of cumulate formation and on the jotunite liquid line of descent. Lithos 83:229–254
Eales HV (2000) Implications of the chromium budget of the western limb of the Bushveld Complex. S Afr J Geol 103:141–150
Eales HV (2002) Caveats in defining the magmas parental to the mafic rocks of the Bushveld Complex, and the manner of their emplacement: review and commentary. Miner Mag 66:815–832
Eales HV, Cawthorn RG (1996) The Bushveld Complex. In: Cawthorn RG (ed) Layered intrusions. Elsevier, Amsterdam, pp 81–229

Eales HV, Costin G (2012) Crustally contaminated komatiite: primary source of the chromitites and marginal, lower, and critical zone magmas in a staging chamber beneath the Bushveld Complex. Econ GeoI 107:645–665
Eales HV, Reynolds I (1986) Cryptic variations within chromitites of the upper critical zone, northwestern Bushveld Complex. Econ Geol 81:1056–1066
Engelbrecht JP (1985) The chromites of the Bushveld Complex in the Nietverdient area. Econ Geol 80:896–910
Engelbrecht JP (1990) Contact metamorphic processes related to the aureole of the Bushveld Complex in the Marico district, western Transvaal. S Afr J Geol 93: 339–349
Godel B, Barnes S-J, Maier WD (2011) Parental magma composition inferred from in situ trace elements in cumulus and intercumulus silicate minerals: example from the lower and lower critical zones of the Bushveld Complex (South-Africa). Lithos 125:537–552
Groeneveld D (1970) The structural features and the petrography of the Bushveld Complex in the vicinity of Stoffberg, eastern Transvaal. In: Visser DL, von Gruenewaldt G (eds) Symposium on the Bushveld Igneous Complex and other layered intrusions, Geol Soc South Africa, Johannesbg, pp 36–45
Gudfinnsson GH, Presnall DC (2000) Melting behaviour of model lherzolite in the system CaO–MgO–Al_2O_3–SiO_2–FeO at 0.7 to 2.8 gPa. J Petrol 41:1241–1269
Hahn UF, Ovendale B (1994) UG2 chromitite layer potholes at Wildebeestfontein North Mine, Impala Platinum Limited, XVth CMMI Congress, Johannesburg. S Afr Inst Min Metall 3:195–200
Hall AL (1932) The Bushveld Igneous Complex in the central Transvaal. Geol Surv S Afr Mem 28:544
Hamilton PJ (1977) Sr isotope and trace element studies of the great dyke and Bushveld mafic phase and their relation to early proterozoic magma genesis in Southern Africa. J Petrol 18:24–52
Hargar HS (1934) An early Transvaal geological map by Carl Mauch. Trans Geol Soc S Afr 37:1–4
Harmer RE, Sharpe MR (1985) Field relations and strontium isotope systematics of the marginal rocks of the eastern Bushveld Complex. Econ Geol 80:813–837
Hatton CJ, von Gruenewaldt G (1987) The geological setting and petrogenesis of the Bushveld chromitite layers. In Stowe CW (ed) Evolution of chromium ore fields. Van Nostrand, New York, pp 109–143
Hulbert LJ, von Gruenewaldt G (1985) Textural and compositional features of chromite in the lower and critical zones of the Bushveld Complex. Econ Geol 80:872–895
Hunter RH (1996) Texture development in cumulate rocks. In: Cawthorn RG (ed) Layered intrusions. Elsevier, Amsterdam, pp 77–102
Irvine TN (1975) Crystallization sequences in the Muskox intrusion and other layered intrusions: II origin of chromitite layers and similar deposits of other magmatic ores. Geochim Cosmochim Acta 39:991–1020
Irvine TN (1977) Origin of chromitite layers in the Muskox intrusion and other stratiform intrusions: a new interpretation. Geology 5:273–277
Irvine TN (1980) Magmatic infiltration metasomatism, double diffusive fractional crystallization and adcumulus growth in the muskox intrusion and other layered intrusions. In: Hargraves RB (ed) Physics of magmatic processes. Princeton University Press, NJ 325–384
Irvine TN, Andersen JCØ, Brooks CK (1998) Included blocks (and blocks within blocks) in the Skaergaard intrusion; geological relations and the origins of rhythmically modally graded layers. Geol Soc Am Bull 110:1398–1447
Johnson TE, Gibson RL, Brown M, Buick IS, Cartwright I (2003) Partial melting of metapelitic rocks beneath the Bushveld Complex, South Africa. J Petrol 44:789–813
Junge M, Oberthür T, Melcher F (2014) Cryptic variation of chromite chemistry, platinum group element and platinum group mineral distribution in the UG-2 chromitite: an example from the karee mine, western Bushveld Complex, South Africa. Econ Geol 109:795–810

Kent AJ, Darr C, Koleszar AM, Salisbury MJ, Cooper KM (2010) Preferential eruption of andesitic magmas through recharge filtering. Nat Geosci 3:631–636
Kinnaird JA, Kruger FJ, Nex, PAM, Cawthorn RG (2002) Chromite formation—a key to understanding processes of platinum enrichment. Trans Inst Miner Min Metal 111:B23–B35
Kinnaird JA, Hutchinson D, Schürmann L, Nex PAM, De Lange R (2005) Petrology and mineralization of the southern Platreef: northern limb of the Bushveld Complex. Miner Depos 40:576–597
Klemm DD, Ketterer S, Reichhardt F, Steindl J, Weber-Diefenbach K (1985a) Implication of vertical and lateral compositional variations across the Pyroxenite Marker and its associated rocks in the upper part of the main zone in the eastern Bushveld Complex. Econ Geol 80:1007–1005
Klemm DD, Henckel J, Dehm R, von Gruenewaldt G (1985b) The geochemistry of titanomagnetite in magnetite layers and their host rocks of the eastern Bushveld Complex. Econ Geol 80:1075–1088
Knowles RA (1978) A bibliography of the geology of the Bushveld Complex. Uni Pretoria Res Rep 12:1–20
Kruger FJ (1990) The stratigraphy of the Bushveld Complex: a reappraisal and relocation of the main zone boundaries. S Afr J Geol 93:376–381
Kruger FJ (1992) The origin of the merensky cyclic unit: Sr isotopic and mineralogical evidence for an alternative orthomagmatic model. Aust J Earth Sci 39:255–261
Kruger FJ (1994) The Sr-isotopic stratigraphy of the western Bushveld Complex. S Afr J Geol 97:393–398
Kruger FJ (2005) Filling the Bushveld Complex magma chamber: lateral expansion, roof and floor interaction, magmatic unconformities and the formation of giant chromitite, PGE and Ti-V magnetitite deposits. Miner Depos 40:451–472
Kruger FJ, Marsh JS (1985) The mineralogy, petrology and origin of the Merensky cyclic unit in the western Bushveld Complex. Econ Geol 80:958–974
Kruger FJ, Cawthorn RG, Walsh KL (1987) Strontium isotopic evidence against magma addition in the upper zone of the Bushveld Complex. Earth Planet Sci Lett 84:51–58
Lange RA (1994) The effect of H_2O, CO_2 and F on the density and viscosity of silicate melts. In: Carroll MR, Holloway Jr (eds) Volatiles in magmas. Rev Miner 30:331–369
Latypov R, Hanski E, Lavrenchuk A, Huhma H, Havela T (2011) A 'three-increase model' for the origin of the marginal reversal of the Koitelainen layered intrusion Finland. J Petrol 52:733–764
Latypov R, O'Driscoll B, Lavrenchuk A (2013) Towards a model for the in situ origin of PGE reefs in layered intrusions: insights from chromitite seams of the rum eastern layered intrusion, Scotland. Contrib Miner Petrol 166:309–327
Lee CA, Butcher AR (1990) Cyclicity in the Sr isotope stratigraphy through the merensky and bastard reefs, atok section, eastern Bushveld Complex. Econ Geol 85:877–883
Leeb-Du Toit A (1986) The Impala platinum mines. In: Anhaeusser CR, Maske S (eds) Mineral deposits of Southern Africa. Geol Soc S Afr Johannesburg 2:1091–1106
Li C, Maier WD, de Waal SA (2001) The role of magma mixing in the genesis of PGE mineralization in the Bushveld Complex: thermodynamic calculations and a new interpretation. Econ Geol 96:653–662
Lombaard BV (1934) On the differentiation and relationships of the rocks of the Bushveld Complex. Trans Geol Soc S Afr 37:5–52
Lundgaard KL (2003) Magma chamber processes studied in two layered intrusions. PhD thesis, Aarhus University, Denmark
Maaløe S (1978) The origin of rhythmic layering. Miner Mag 42:337–345
Maier WD, Barnes SJ (1998) Concentrations of rare earth elements in silicate rocks of the lower, critical and main zones of the Bushveld Complex. Chem Geol 150:85–103
Maier WD, Barnes SJ (2008) Platinum-group elements in the UG1 and UG2 chromitites, and the bastard reef, at impala platinum mine, western Bushveld Complex, South Africa: evidence of late magmatic cumulate instability and reef constitution. Geol Soc S Afr 111:159–176

Maier WD, Eales HV (1994) Facies model for interval between UG2 and merensky reef, western Bushveld Complex, South Africa. Trans Inst Min Metal 103:B22–B30
Maier WD, Eales HV (1997) Correlation within the UG2-Merensky reef interval of the western Bushveld Complex, based on geochemical, mineralogical and petrological data. Geol Surv S Afr Bull 120:56
Maier WD, Arndt NT, Curl EA (2000) Progressive crustal contamination of the Bushveld Complex: evidence from Nd isotopic analyses of the cumulate rocks. Contrib Miner Petrol 140:316–327
Maier WD, Barnes SJ, Groves DI (2012) The Bushveld Complex, South Africa: formation of platinum, palladium, chrome- and vanadium-rich layers via hydrodynamic sorting of a mobilized cumulate slurry in a large, relatively slowly cooling, subsiding magma chamber. Miner Depos 48:1–56
Maré LP, Fourie CJS (2012) New geochemical and palaeomagnetic results from neoarchaean dyke swarms in the Badplaas-Barberton area, South Africa. S Afr J Geol 115:145–170
Marsh BD (2006) Dynamics of magmatic systems. Elements 2:287–192
Marsh JS, Hooper PR, Rehacek J, Duncan RA, Duncan AR (1997) Stratigraphy and age of Karoo basalts of Lesotho and implications for correlations within the Karoo Igneous Province. Geophys Monogr 100:247–272
Mathez EA, Hunter RH, Kinzler R (1997) Petrologic evolution of partially molten cumulates: the atok section of the Bushveld Complex. Contrib Miner Petrol 129:20–34
McBirney AR (1985) Further considerations of double-diffusive stratification and layering in the skaergaard intrusion. J Petrol 26:993–1001
McCarthy TS, Cawthorn RG (1983) The geochemistry of vanadiferous magnetite in the Bushveld Complex: implications for crystallization mechanisms in layered complexes. Miner Depos 18:505–518
McKenzie D (2011) Compaction and crystallization in magma chambers: towards a model of the Skaergaard intrusion. J Petrol 52:905–930
Mitchell AA (1990) The stratigraphy, petrography and mineralogy of the main zone of the northwestern Bushveld. S Afr J Geol 93:818–831
Mitchell AA, Manthree R (2002) The giant mottled anorthosite: a transitional sequence at the top of the upper critical zone of the Bushveld Complex. S Afr J Geol 105:15–24
Mitchell AA, Scoon RN (2007) The merensky reef at winnarshoek, eastern Bushveld Complex: a primary magmatic hypothesis based on a wide reef facies. Econ Geol 102:971–1009
Molengraaff GAF (1901) Géologie de la République Sud Africaine du Transvaal. Bull Soc Geol France 4:13–92
Molyneux TG (1974) A geological investigation of the Bushveld Complex in Sekhukuneland and part of the Steelpoort valley. Trans Geol Soc S Afr 77:329–338
Molyneux TG, von Gruenewaldt G, Knowles RA (1976) A bibliography of the geology of the Bushveld Complex. Uni Pretoria Res Rep 2:1–63
Mondal SK, Mathez EA (2007) Origin of the UG2 chromitite layer, Bushveld Complex. J Petrol 48:495–510
Morse SA (1979) Kiglapait geochemistry I; systematics, sampling, and density. J Petrol 20:555–590
Morse SA (1984) Cation diffusion in plagioclase feldspar. Science 225:504–505
Murck BW, Campbell IH (1986) The effects of temperature, oxygen fugacity and melt composition on the behaviour of chromium in basic and ultrabasic melts. Geochim Cosmo Acta 50:1871–1888
Naldrett AJ, Wilson AH, Kinnaird J, Chunnett G (2009) PGE tenor and metal ratios within and below the Merensky Reef, Bushveld Complex: implications for genesis. J Petrol 50:625–659
Naldrett AJ, Wilson A, Kinnaird J, Yudovskaya M, Chunnett G (2012) The origin of chromitites and related PGE mineralization in the Bushveld Complex: new mineralogical and petrological constraints. Miner Depos 47:209–232
Naslund HR, McBirney AR (1996) Mechanisms of formation of igneous layering. In: Cawthorn RG (ed) Layered intrusions. Elsevier, Amsterdam, 1–44

Nex PA (2004) Formation of bifurcating chromitite layers of the UG1 in the bushveld igneous complex, an analogy with sand volcanoes. J Geol Soc Lond 161:903–909
Nex PA, Kinnaird JA, Ingle LJ, Van der Vyver BA, Cawthorn RG (1998) A new stratigraphy for the main zone of the Bushveld Complex, in the Rustenburg area. S Afr J Geol 101:215–223
Nguuri TK, Gore J, James DE, Webb SJ, Wright C, Zengeni TG, Gwavava O, Snoke JA (2001) Crustal signature beneath southern Africa and its implication for the formation and evolution of the Kaapvaal and Zimbabwe cratons. Geophys Res Lett 28:2501–2504
Ochs FA III, Lange RA (1999) The density of hydrous magmatic liquids. Science 283:1314–1317
Olsson JR, Söderlund U, Klausen MB, Ernst RE (2010) U-Pb baddelyite ages linking major Archean dyke swarms to volcanic rift forming events in the Kaapvaal craton (South Africa), and a precise age for the Bushveld Complex. Precamb Res 183:490–500
Perugini D, Poli G (2012) The mixing of magmas in plutonic and volcanic environments: analogies and differences. Lithos 153:261–277
Quadling K, Cawthorn RG (1994) The layered gabbronorite sequence, Main Zone, eastern Bushveld Complex. S Afr J Geol 97:442–454
Reynolds IM (1985) Contrasted mineralogy and textural relations in the uppermost Ti-magnetite layers of the Bushveld Complex in the Bierkraal area north of Rustenburg. Econ Geol 80:1027–1048
Roeder P, Goften E, Thornber C (2006) Cotectic proportions of olivine and spinel in olivine-tholeiitic basalt and evaluation of pre-eruptive processes. J Petrol 47:883–900
Roelofse F, Ashwal LD (2012) The lower main zone in the northern limb of the Bushveld Complex—a > 1.3 km thick sequence of intruded and variably contaminated crystal mushes. J Petrol 53:1449–1476
SACS (South African Committee for Stratigraphy) (1980) Kent LE (compiler) Stratigraphy of South Africa. Geol Surv S Afr Pretoria, Handbook, 8:690
Schmidt ER (1952) The structure and composition of the merensky reef and associated rocks on the Rustenburg platinum mine. Trans Geol Soc S Afr 55:233–279
Schoenberg, R, Kruger FJ, Nägler F, Meisel T, Kramers JD (1999) PGE enrichment in chromitite layers and the merensky reef of the Bushveld Complex, a Re–Os and Rb–Sr isotope study. Earth Planet Sci Lett 172:49–64
Schürmann LW (1993) The geochemistry and petrology of the upper critical zone in the boshoek section of the western Bushveld Complex. Geol Surv S Afr Bull 113:88
Scoon RN (1994) Comment on compositional convection in a reactive crystalline mush and melt differentiation. J Geophys Res 99:11913–11917
Scoon RN, Mitchell AA (1994) Discordant iron-rich ultramafic pegmatites in the Bushveld Complex and their relationship to iron-rich intercumulus and residual liquids. J Petrol 35:881–917
Scoon RN, Mitchell AA (2009) A multi-stage orthomagmatic and partial melting hypothesis for the driekop platiniferous dunite pipe, eastern limb of the Bushveld Complex, South Africa. S Afr J Geol 112:163–186
Seabrook CL, Cawthorn RG, Kruger FJ (2005) The Merensky reef, Bushveld Complex: mixing of minerals not mixing of magmas. Econ Geol 100:1191–1206
Seabrook CL, Cawthorn RG, Kruger FJ (2006) The Merensky Reef, Bushveld Complex: Mixing of minerals not mixing of magmas. Econ Geol 100:1191–1206
Sharpe MR (1981) The chronology of magma influxes to the eastern compartment of the Bushveld Complex, as exemplified by its marginal border group. J Geol Soc Lond 138:307–326
Sharpe MR (1984) Petrography, classification and chronology of mafic sill intrusions beneath the eastern Bushveld Complex. Geol Surv S Afr Bull 77:40
Sharpe MR (1985) Strontium isotope evidence for preserved density stratification in the main zone of the Bushveld Complex. Nature 316:119–126
Sparks RSJ, Huppert HE, Koyaguchi T, Hallworth MA (1993) Origin of modal and rhythmic igneous layering by sedimentation in a convecting magma chamber. Nature 361:246–249
Tait S, Jaupart C (1992) Compositional convection in a reactive crystalline mush and melt differentiation. J Geophys Res 98:6735–6756

Tanner D, Mavrogenes JA, Arculus RJ, Jenner FE (2014) Trace element stratigraphy of the belleview core, northern bushveld: multiple magma injections obscured by diffusion processes. J Petrol 55:859–882
Tegner C, Wilson JR, Cawthorn RG (1994) The dunite-clinopyroxenite pegmatoidal pipe, Tweefontein, eastern Bushveld Complex, South Africa. S Afr J Geol 97:415–430
Tegner C, Lesher CE, Larsen LM, Watt WS (1998) Evidence from rare-earth element record of mantle melting for cooling of the tertiary Iceland plume. Nature 395:591–594
Tegner C, Cawthorn RG, Kruger FJ (2006) Cyclicity in the main and upper zones of the Bushveld Complex, South Africa: crystallization from a zoned magma sheet. J Petrol 47:2257–2279
Teigler B, Eales HV (1993) Correlation between chromite composition and PGE mineralization in the critical zone of the western Bushveld Complex. Miner Depos 28:291–302
Teigler B, Eales HV (1996) The lower and critical zones of the western limb of the Bushveld Complex, as indicated by the nooitgedacht boreholes. Geol Surv S Afr Bull 111:126
Teigler B, Eales HV, Scoon RN (1992) The cumulate succession in the critical zone of the Rustenburg layered suite at brits, western Bushveld Complex. S Afr J Geol 95:17–28
Tilley CE, Yoder HS Jr, Schairer JF (1968) Melting relations of igneous rock series. Carnegie Inst Wash Yrbk 66:450–457
Tollari N, Barnes SJ, Cox RA, Nabil H (2008) Trace element concentrations in apatite from the sept-iles intrusive suite, Canada—implications for the genesis of nelsonites. Chem Geol 252:180–190
Uken R, Watkeys MK (1997a) Diapirism initiated by the Bushveld Complex. Geology 25:723–726
Uken R, Watkeys MK (1997b) An interpretation of mafic dyke swarms and their relationship with major mafic magmatic events on the Kaapvaal craton and limpopo belt. S Afr J Geol 100:341–348
Vander Auwera J, Longhi J (1994) Experimental study of a jotunite (hypersthenes monzodiorite): constraints on the parental magma composition and crystallization conditions (P, T, f_{O2}) of the Bjerkreim-Sokndal layered intrusion (Norway). Contrib Miner Petrol 118:60–78
VanTongeren J, Mathez EA (2012) Large-scale liquid immiscibility at the top of the Bushveld Complex, South Africa. Geology 40:491–494
VanTongeren JA, Mathez EA (2013) Incoming magma composition and style of recharge below the pyroxenite marker, eastern Bushveld Complex, South Africa. J Petrol. 54:1585–1605
VanTongeren JA, Mathez EA, Keleman PB (2010) A felsic end to Bushveld differentiation. J Petrol 51:1891–1912
Vermaak CF (1976) The Merensky reef—thoughts on its environment and genesis. Econ Geol 71:1270–1298
Viljoen MJ, Hieber R (1986) The Rustenburg section of Rustenburg Platinum Mines Limited, with reference to the merensky reef. In: Anhaeusser CR, Maske S (eds) Mineral deposits of Southern Africa. Geol Soc S Afr Johannesburg 2:1107–1134
Viljoen MJ, Scoon RN (1985) The distribution and main geological features of discordant pegmatoid in the Bushveld Complex. Econ Geol 80:1109–1128
Von Gruenewaldt G (1970) On the phase change orthopyroxene-pigeonite and resulting textures in the main and upper zones of the Bushveld Complex in the eastern Transvaal. In Visser DJL, von Gruenewaldt G (eds) Symposium on the bushveld igneous complex and other layered intrusions. Geol Soc S Afr Spec Publ 1:67–73
Von Gruenewaldt G (1972) The origin of the roof rocks of the Bushveld Complex between tauteshoogte and paardekop in the eastern Transvaal. Trans Geol Sof S Afr 75:121–134
Von Gruenewaldt G (1973) The Main and upper zones of the Bushveld Complex in the Roossenekal area, eastern Transvaal. Trans Geol Soc S Afr 76:207–227
Voordouw R, Gutzmer J, Beukes NJ (2009) Intrusive origin for Upper Group (UGI, UG2) stratiform chromitite seams in the Dwars River area, Bushveld Complex, South Africa. Miner Petrol 97:75–94
Wager LR (1959) Differing powers of crystal nucleation as a factor producing diversity in layered igneous intrusions. Geol Mag 96:75–80

Wager LR, Brown GM (1968) Layered igneous rocks. Oliver and Boyd, Edinburgh, p 588
Wager LR, Brown GM, Wadsworth WJ (1960) Types of igneous cumulate. J Petrol 1:73–85
Wagner P (1929) Platinum deposits and mines of South Africa. Oliver and Boyd, Edinburgh, p 326
Waters C, Boudreau AE (1996) A re-evalaution of crystal-size distributions in chromitite cumulates. Am Miner 81:1452–1459
Webb SJ, Ashwal LD, Cawthorn RG (2011) Continuity between eastern and western Bushveld Complex, confirmed by xenoliths from kimberlite. Contrib Miner Petrol 162:101–107
Willemse J (1964) A brief outline of the geology of the Bushveld Igneous Complex. In: Haughton SH (ed) The geology of some ore deposits in Southern Africa. Geol Soc S Afr Johannesburg 2:91–128
Willemse J (1969) The vanadiferous magnetic iron ore of the Bushveld Igneous Complex. Econ Geol Monogr 4:137–208
Wilson AH (2012) A chill sequence to the Bushveld Complex: insight into the first stage of emplacement and implications for the parental magmas. J Petrol 53:1123–1168
Wilson JR, Cawthorn RG, Kruger FJ, Grundvig S (1994) Intrusive origin for the unconformable upper zone in the northern gap area, western Bushveld Complex, South Africa. S Afr J Geol 97:462–472

Chapter 13
Kiglapait Intrusion, Labrador

Stearns A. Morse

Abstract The Kiglapait intrusion is the largest and youngest of a suite of troctolitic bodies in the Nain Plutonic Series of anorthosite and related rocks in Labrador. The intrusion is dated at 1307 Ma; the Manvers Granite cutting the intrusion is dated at 1300 Ma. The Kiglapait magma body was large and deep, with a 9.6 km thick roof and a depth of 8.4 km beneath an Upper Border Zone (UBZ) that records the entire history of fractional crystallization inverted, so that it runs upside down from troctolite to ferrosyenite. The magma cooled through the roof and the UBZ and only secondarily through the floor after the basal layers of cumulates were laid down. The crystallization time is estimated at 1 Myr. The lower 84 % of the intrusion is troctolite of the Lower Zone, succeeded by addition of augite, Fe-Ti oxide minerals, sulfide globules, and apatite in the Upper Zone. The slow arrivals of augite, oxide minerals and sulfide are succeeded by the sharp arrival of apatite. Apatite is dominated by fluorine, and the water content of the intrusion is estimated at 68 ppm. Plagioclase compositions vary from 68 to 10 %An and olivines from 72 %Fo at the base to 0 %Fo in the uppermost ferrosyenites. The An range and residual porosity decrease with time and stratigraphic height to 99 volume percent solidified (PCS) and then increase to the end of crystallization, suggesting the late influence of feldspar networks. Oxygen norms of whole-rock compositions are wildly scattered in a ternary augite-feldspar-olivine plot but cluster strongly near the ternary three-phase point. The UBZ has a higher range of residual porosity than the maximum values in the layered series. The partitioning behavior of nickel in olivine v. liquid shows a positive correlation between the partition ratio and the sulfur concentration in the melt; this result may be a clue to an early metal-sulfide saturation. Oxygen and sulfur isotopes are like those of the mantle. The low oxygen fugacity of the melt runs between the FMQ and WM oxygen buffers. Silica activity remained too low to crystallize low-Ca pyroxene and then rose sharply when Fe-Ti oxide minerals crystallized in excess. The cooling history of the magma from 0 to 99.99 PCS spans the range 1244–1023 °C referenced to 3 kbar and plots linearly with plagioclase composition except for three concave-up shelves related to the arrival of new phases. The magma source region has been explored to 15 kbar and the results suggest

S. A. Morse (✉)
Department of Geosciences, University of Massachusetts,
611 North Pleasant Street, Amherst, MA 01003-9297, USA
e-mail: tm@geo.umass.edu

B. Charlier et al. (eds.), *Layered Intrusions,* Springer Geology,
DOI 10.1007/978-94-017-9652-1_13

the melting of harzburgite with the assemblage Al-cpx, Al-opx, garnet, spinel, and olivine at > 1,400°C. The decompression of this liquid to the emplacement pressure of ~3 kbar would involve the removal of Mg-olivine to the extent of >25 % and consequent evolution of the ascending magma. Thermal analysis of roof and floor regions provides useful insights on the adiabatic constraints of crystal and liquid gradients and the removal of latent heat.

Keywords Heat loss · Trace element partitioning · Sulfide saturation · Source region · Emplacement

Introduction

The Kiglapait Layered Intrusion is visible from space (Fig. 13.1) and from commercial airliners on the northern route from Europe, revealed by its outward-dipping onion-skin jointing. With an area of 560 km^2 and an original volume of ~3500 km^3 (Morse 1969) it is the eighth largest single layered mafic intrusion in area on Earth (Namur et al. 2010). At an age of 1.307 Ga (see Morse 1996) it is also about fourth youngest. It is about 8 times larger in area and 13 times larger in volume than the Skaergaard Intrusion (Nielsen 2004), with which it shares a location on an east coast

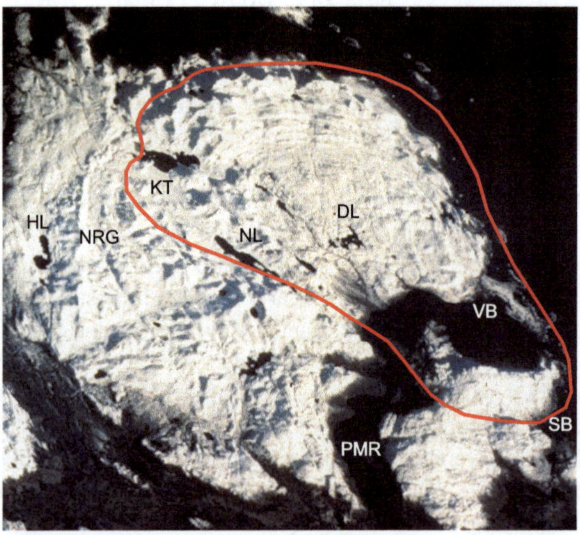

Fig. 13.1 Image from Space Shuttle showing the Kiglapait Intrusion with no sea ice (*black* is water). The outline of the intrusion is shown in *red*. North is to the *upper left*; length of intrusion is 32 km. The curved grain of the intrusion is caused by the outward-dipping expansion jointing, roughly orthogonal to the layering. Abbreviations of landmarks: *HL* Hettasch Lake in the Hettasch Intrusion (Berg 1980); *NRG* North Ridge Gabbro outside Kiglapait Intrusion; *KT* Kiglapait Tessialua (lake); *NL* North Lake; *DL* David Lake; *VB* Village Bay base camp; *SB* Slambang Bay; *PMR* Port Manvers Run. Hasselblad photo courtesy L. D. Ashwal, Lunar & Planetary Institute

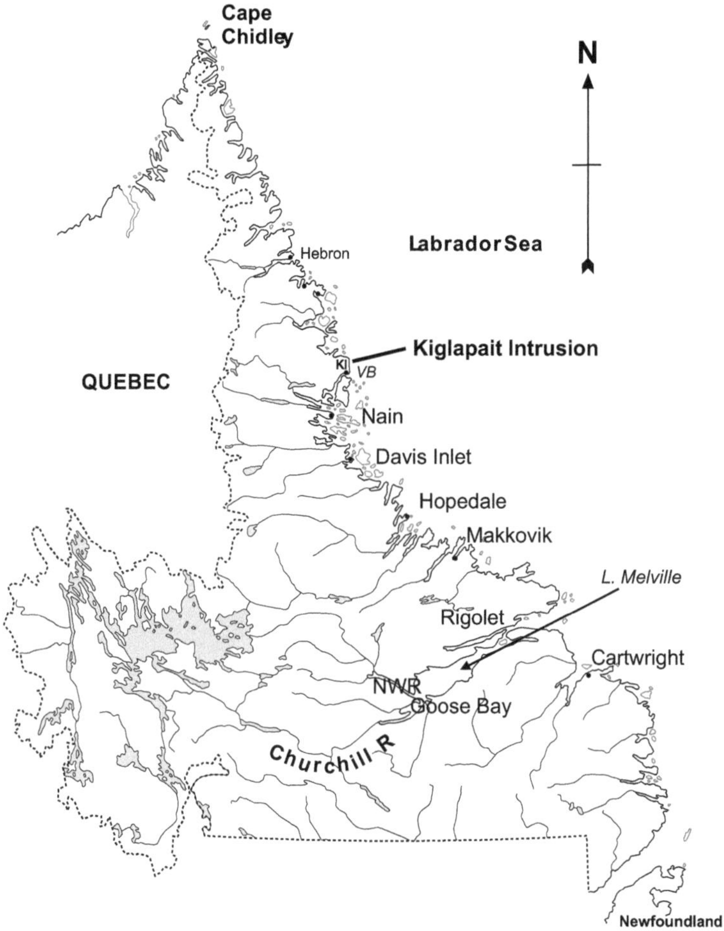

Fig. 13.2 Outline map of Labrador, NL, Canada, showing the location of the Kiglapait intrusion and major towns. Symbols: *VB* Village Bay; *NWR* North West River

(Fig. 13.2) in the thermal Arctic and a great deal of compositional similarity. Its calculated magma depth from roof to base of 8.4 km is among the greatest anywhere known. If filled at the rate of an oceanic hotspot it might have taken 70,000 years to fill (White et al. 2006); if at the rate of a flood basalt 3500 years (idem), or at the rate of the Bishop Tuff, 35 days (Hildreth and Wilson 2007). Take your pick.

Kiglapait is one of a baker's dozen troctolitic bodies among the anorthosites of the Nain Plutonic Series (Ryan 1990); Fig. 13.3. They tend to be the youngest of the plutonic bodies after the noritic anorthosites. With its extreme fractionation to ferrosyenite, Kiglapait looks not unlike a deeper and older counterpart to the Gardar Province in west Greenland, with which it has been tentatively linked (e.g., Myers et al. 2008; Upton 2013) but with lower K_2O in the Kiglapait.

When I began studying the Kiglapait Intrusion in 1957 under the auspices of British Newfoundland Exploration Ltd. (BRINEX) we only knew that layering had

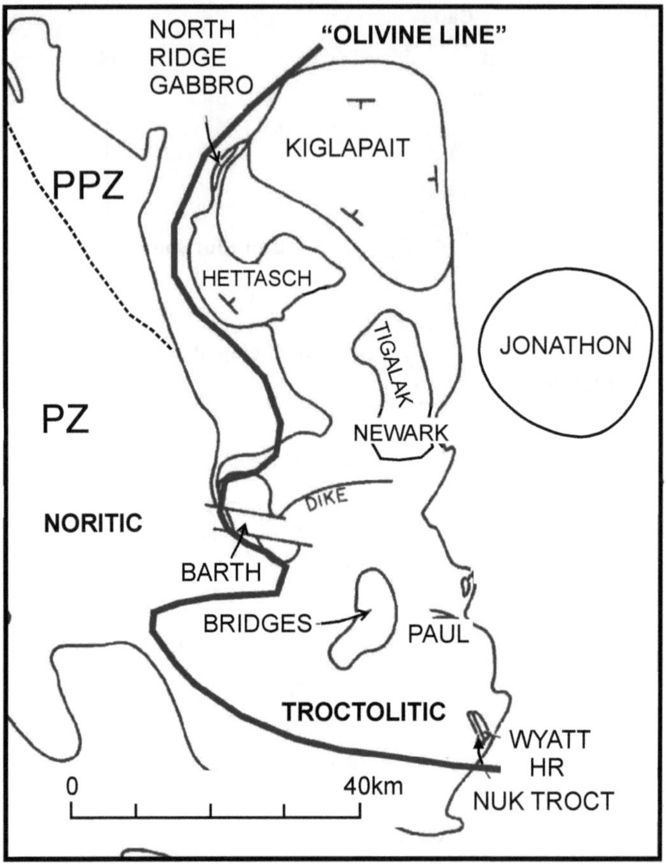

Fig. 13.3 Sketch map of the eastern part of the Nain Plutonic Suite (NPS; Ryan 1990) showing several layered intrusions and an "Olivine Line" roughly separating the troctolitic eastern intrusions from the noritic western intrusions. Adapted from Figs. 2 and 7 of Xue and Morse (1993), where further identification of the intrusions is given. The principal intrusions east of the "Olivine Line" are the Kiglapait (Morse 1969), Hettasch (Berg 1980), Tigalak, Newark Island (Wiebe 1988), and Jonathon Island (Berg 1994) intrusions. Other layered intrusions are listed in Morse (1977). Most of the rocks to the west of the olivine line are the ~2.2 Ga Paleoproterozoic (PPZ) and ~1.4 Ga younger Proterozoic intrusions of the Nain anorthosite, undifferentiated on this map

been reported by Wheeler (1942) on the north coast at Kiglapait Harbour. That is, until we saw the air photos and the air photo mosaic, and then we knew there was something big afoot. In my first week in camp there arrived a wild and wonderful cook, Jack Saunders, a big cook tent, the Provincial Geologist, the Company Manager, the Field Chief, and a helicopter. That gave us a first two traverses N-S and E-W and a good sense of the structure. There was a lot of troctolite.

I mention these matters because in due course the Lunar program came along and troctolite became considered a minor accidental product of basaltic magma. The evidence for such melts in the Nain Plutonic Series was stark. The partly adjacent

Hettasch Intrusion (Figs. 13.1 and 13.3) was found to have been invaded at an early stage by a melatroctolite magma that became supercooled in the local magma and then created sequences of dunite layers followed by 10-cm plagioclase "snowflakes" that were products of rapid growth from single centers of nucleation (Berg 1980). The well-sampled chilled margin of the Hettasch Intrusion was found to be troctolitic and close to Kiglapait in bulk composition (Morse 1981). Then with the discovery of the Newark Island Intrusion by Wiebe (1988) it was found that meter-scale pillows of troctolite magma were quenched by invading granite magma, again showing the undeniable existence of troctolite magmas. They exist, and they are hot.

The Kiglapait Working Hypothesis

The assumption has always been the that there was one prolonged emplacement of magma that then crystallized slowly in place, creating layered rocks at the base and unlayered ones at the roof simultaneously. Accordingly, the total volume of magma was estimated from cross sections using all the opposing dips of layering. Then the layered series was contoured by volume to give "PCS"—percent solidified—units assuming the hypothesis. The fractionation progression of the Upper Border Zone was later correlated to the Layered Series via the maximum An in the UBZ to the mean An in the LS. That solution gave the same An value for the arrival of apatite, Ap^+ at 94 PCS.

It was evident early on that some of the lowermost rocks were more evolved than those overlying them, and this feature was ascribed to the notion of leading—edge fractionated magma; i.e. magma fractionated in transit from the source to the site of emplacement. Work in progress on olivine compositions supports evidence that rocks representing new injections of magma are limited to the region 6–13 PCS.

The bulk magma composition was estimated by summation from the top down over all the layered series (Morse 1981).

Three major pieces of evidence bear on the working hypothesis. The first is that five very similar analyses of the chilled margin of the nearby Hettasch Intrusion, collected over a strike distance of 8 km, are very similar to the calculated Kiglapait bulk composition (Berg 1980; Morse 1981).

Secondly, the experimental study of Morse et al. (2004) showed that IF the estimated bulk composition is correct, then by the lever rule about 83 % fractionation would be required to reach saturation with augite, which is conventionally mapped at 84 PCS. That study also showed, paradoxically, that the coarse-grained Inner Border Zone rocks best resemble the bulk composition of the intrusion.

Third, the "slow arrivals" of the phases augite, sulfides, and Fe-Ti oxides, to be discussed under Petrography, absolutely require the residence of their components within the magma chamber over a very protracted history, that is then followed by the abrupt arrival of apatite, which is construed as evidence of violent stirring attended by the precipitation of the Main Ore Band.

If instead it is argued that a new, evolved magma suddenly arrived to form the Upper Zone then some other magma in some other place would have to have evolved to the same extent and then ascend to a higher level even while denser than the originally troctolitic magma. This is an untenable hypothesis because it is circular.

At this point the working hypothesis is still considered to be unfalsified. The gravity study, below, also supports the hypothesis.

Field Relations

The Kiglapait Intrusion was emplaced into a pair of Aphebian (ca. 1.8 Ga) sedimentary sequences known as the Snyder Group and the overlying Falls Brook Group (Williams 1985), remnants of which are found in the northwest corner of the intrusion (Figs. 13.4 and 13.5) and in xenoliths found in the southeast part of the intrusion, particularly at the west end of Partridge Point and on the shore north of Slambang Bay (Fig. 13.5). The evidence suggests the possible pre-existence of a local sedimentary basin. A feeder is presumed from field relations to lie near the west edge of the intrusion. The Falls Brook group is distinguished by its variety of clastic sediments, carbonates, banded iron formation, and a km-long slab of spinel harzburgite, all of which deserve more investigation in addition to the study of Schuh (1981). The spinel harzburgite is of intense potential interest because that lithology is the same as that of the inferred mantle source region of the Kiglapait magma near the spinel-garnet transition (McIntosh 2009). The Falls Brook Group is of special interest from its contact with the Kiglapait magma, leading to a contact metamorphic aureole from greenschist to sanidinite facies (Speer 1982; Berg 1977). From petrological studies a best estimate of the contact pressure is ~2.5 kbar (Berg and Docka 1983 and personal communications), suggesting a roof thickness of ~9.6 km using the modified PREM configuration of the Earth's crust (Stacey and Davis 2009).

Xenoliths in the Layered Series deserve special mention. Many of them come with cm-scale rinds of augite (Ball 1984; Owens 1986) that tend to preserve even some of the volatile contents of the country rocks. Of particular interest is the study of Owens and Kremser (2010) reporting the occurrence in a large xenolith of cuspidine, a mineral rich in fluorine and yet not demonstrably part of the carbonate—rich protolith. The authors infer that the source of the F was the magma itself, a possibility that might help explain the frequency of minute but very strongly reversed rims on plagioclase (Morse and Nolan 1984) if fluorine may widen the plagioclase loop as does water (e.g., Morse 2000).

Structure

The geologic map (Fig. 13.4.) shows the shape of the intrusion at the present erosion level, with color coding and labels defining the Lower Zone, Upper Zone, and Upper Border Zone within the intrusion, and some supracrustal country rocks in the northwest corner. The main lithology is comprised of the troctolitic Lower Zone,

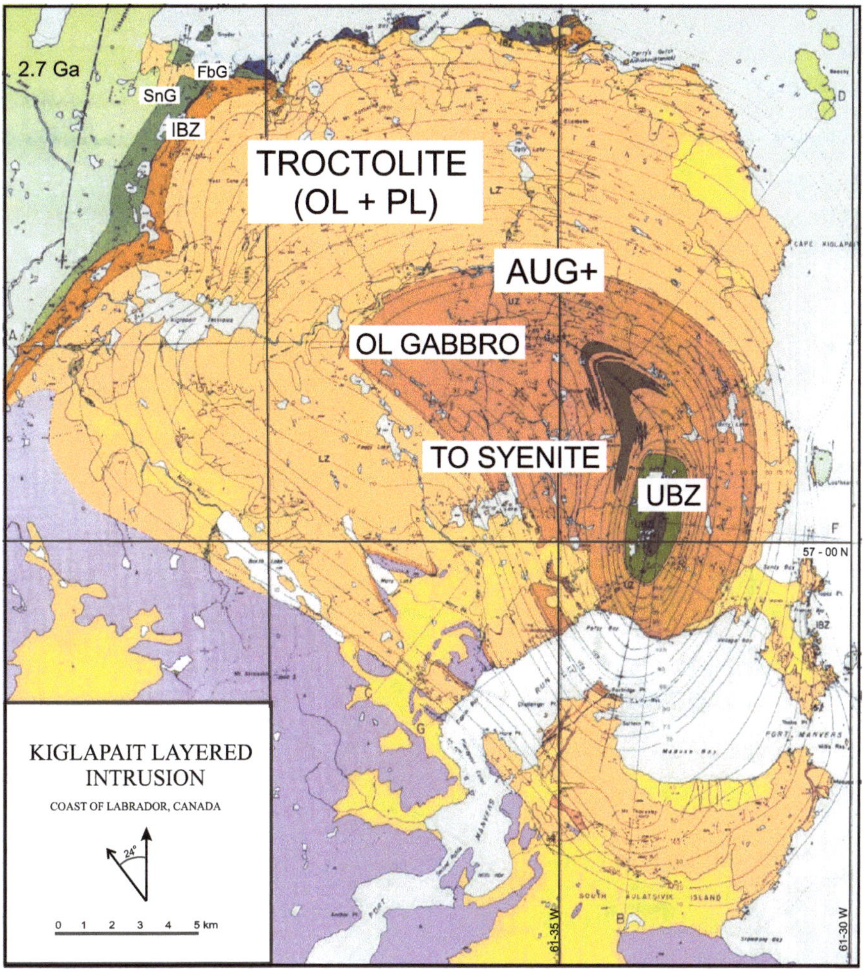

Fig. 13.4 Reduced image of the 1:50,000 geologic map of the Kiglapait intrusion (Morse 1969) showing the main zones and country rocks. In the northwest corner attention is called to the ~2.7 Ga tonalitic Archean rocks of the Nain Province, succeeded eastwards by the ~1.8 Ga Aphebian Snyder Group (*SnG, green*) and Falls Brook Group (*FbG, orange*), inside which is the Inner Border Zone of the Kiglapait intrusion, too narrow to show at this scale. The Falls Brook Group was earlier called the Outer Border Zone of the Kiglapait intrusion but was later found to be a supracrustal sequence with many mafic granulites similar to Kiglapait compositions. Symbols: *AUG*$^{+}$ Arrival of augite to mark the LZ-UZ boundary; *UBZ* Upper Border Zone

the olivine gabbroic to ferrosyenitic Upper Zone, and the Upper Border Zone with inverted stratigraphy and divided into a troctolitic central oval and a UZ-equivalent outer part in green.

Contours are drawn for percent solidified (PCS) levels, and these define the general strike of layering as it intersects the surface. The wider spacing of these contours in the western region of the intrusion signifies the flatter dips of layering in

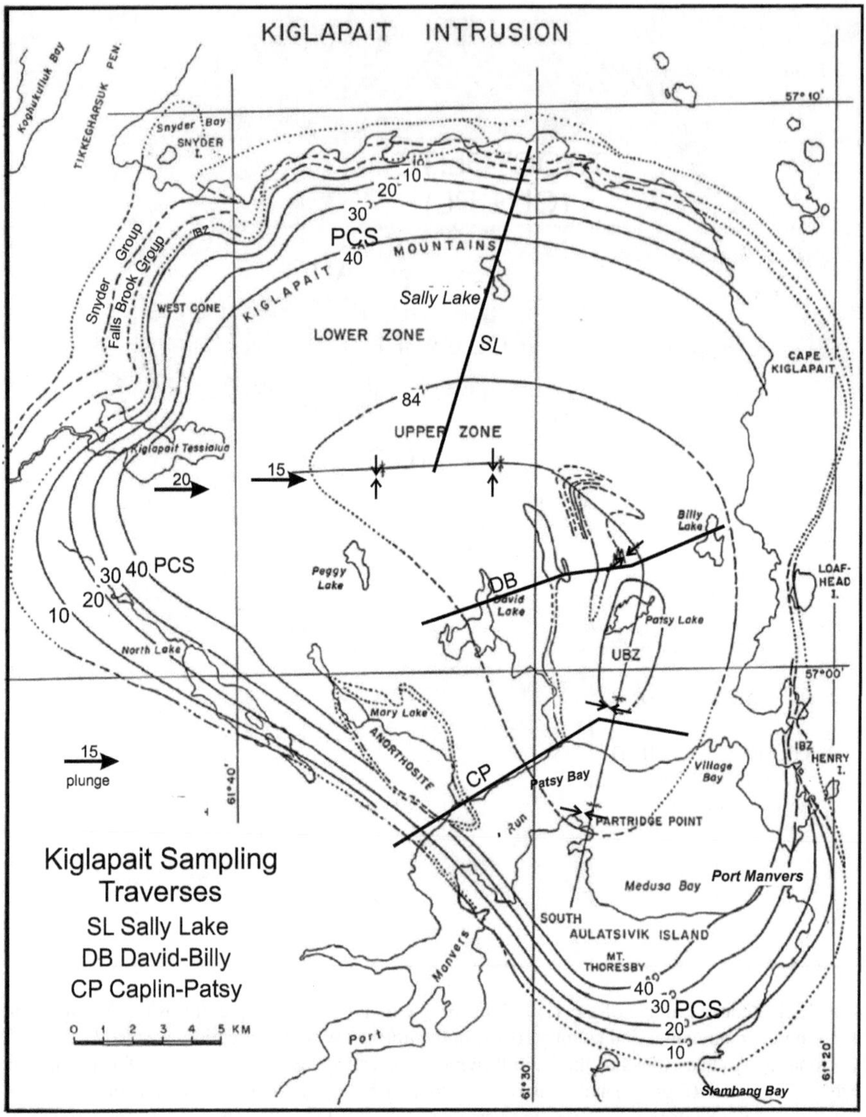

Fig. 13.5 Kiglapait sampling traverses. The map also shows the northwest supracrustal groups and the Inner Border Zone (*IBZ*), *PCS* contours 0–40, and the synclinal axis with plunge at its western edge

that region compared to the more steeply-dipping eastern region. A central synclinal axis is drawn in a light line from the western extremity of the UZ around the corner through the synclinal finer-grained rocks thought to represent foundered parts of the UBZ, thence curving around to pass through the median line of the UBZ and out through the LZ-UZ boundary at Partridge Point, better seen in the sketch map of Fig. 13.5.

In this figure Port Manvers Run is seen as an inland waterway leading north to the base camp site at Village Bay and out to the sea through Port Manvers (Fig. 13.5). The northwest shoreline of this Run (Fig. 13.5) is the site of the CP mapping and sampling traverse, the most accessible exposure of the Lower Zone and lower parts of the Upper Zone. The traverse extends to the Upper Border Zone contact. The anorthosite mass breaking the traverse at 40–50 PCS is a roof pendant, with a collar of olivine gabbro (UZ equivalent) quenched onto its lower surface.

Sampling Traverses

Detailed sampling traverses were mapped in Morse (1979b) and their location is given on the map of Fig. 13.5, which also shows the locations of the supracrustal groups in the northwest, PCS contours 10–40 PCS, and the central synclinal axis. At the upper left end of this axis are two plunge arrows showing the shallow dip of layering in this western region, reaching 15° at the LZ-UZ boundary. This shallow plunge of the axis is relevant to the origin of steeper dips in the bowl-shaped interpretation of the intrusion, to be discussed.

Analysis of the Intrusion Structure

Cross sections are shown in Figs. 13.6. and 13.7; a N-S and half of an E-W cross section. The eastern side of the intrusion has steeper dips than the western side. Dips of layering to the east of the UBZ range as high as 60°. The question arises as to whether these reflect sagging at the center or deformation from westward compression. Because the trend of the synclinal axis turns through 108°, any sagging or deformation is limited by the 15° plunge at the western lobe of the Upper Zone. That is to say, any consistent compression or sagging in the E-W direction could have steepened the E-W axis from a limit of zero (if it is not to be reversed) to 15°, but no more. The default interpretation is to exclude sagging or compression unless that interpretation can be falsified. It should be noted that many intrusions, particularly the younger syenitic cousins to the Kiglapait in the Gardar Province to the east, are exposed in cross section and show bowl-shaped forms (e.g., Upton 2013) that are interpreted as original.

Volume Estimates

The reasoning above suggests that the dips should be treated as primary, hence the local dip should be produced downward to a constant thickness of rocks and then back up on the other side. That is the hypothesis on which the cross sections were made. The volume of the intrusion was estimated (Morse 1969) by dividing the intrusion up into 13 sectors, of which five were rectangular and the rest arcuate, all centered on or near the synclinal axis (Morse 1969; Fig. 13.9). Then cross sections

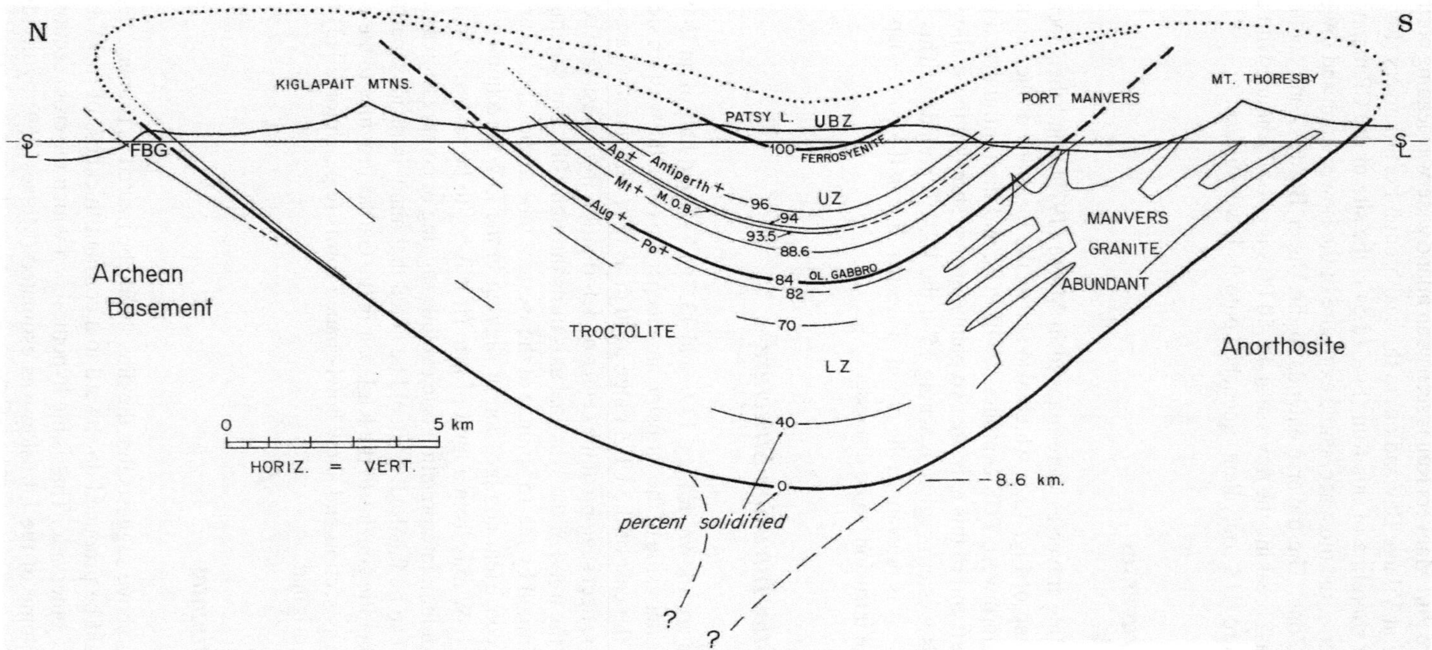

Fig. 13.6 Kiglapait North-South cross section. Symbols: *LZ* Lower Zone, *UZ* Upper Zone, *UBZ* Upper Border Zone, *FBG* Falls Brook Group, *M.O.B.* Main Ore Band (Fe-Ti oxide layer), *OL* olivine, *Po* pyrrhotite, *Aug* augite, *Mt* magnetite, + phase arrival

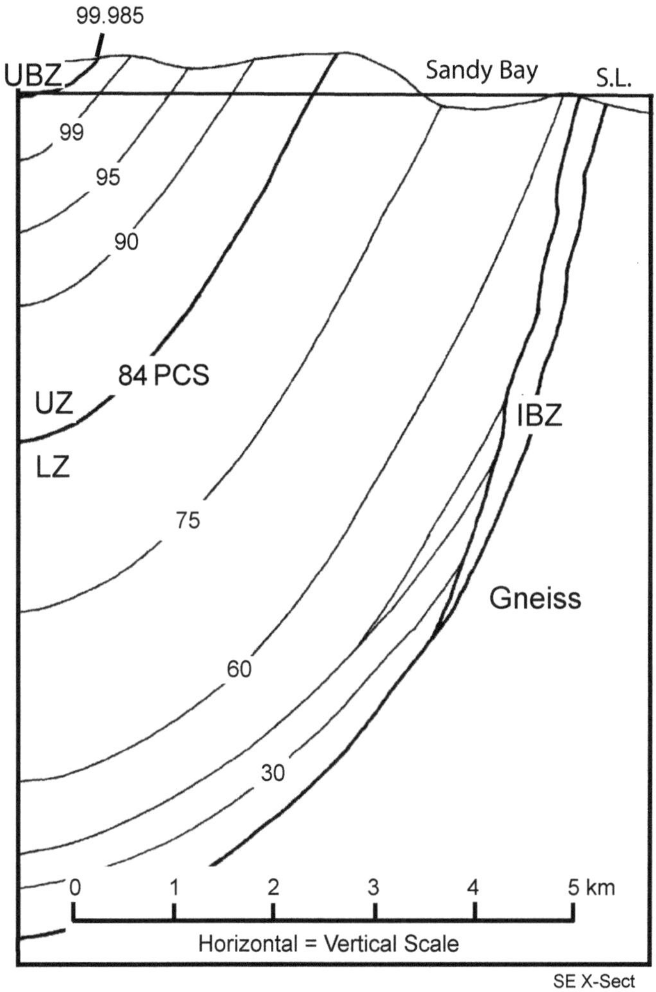

Fig. 13.7 Kiglapait cross section at the southeast margin, in the latitude of *Sandy Bay* and the south end of the Upper Border Zone (*UBZ*). Symbols: *S. L.* Sea level, *IBZ* Inner Border Zone (conceptual), *UZ* Upper Zone, *LZ* Lower Zone, *PCS* percent solidified

were drawn to a consistent central depth, and measured manually at fixed intervals of depth, thereby creating slices of volume. These were the summed over the whole intrusion, and then contoured in terms of total volume so that the volume percent values describe a set of nesting bowls. Layering was assumed to define time horizons. The total depth to the floor from the base of the UBZ was determined to be 8400 m.

The pre-erosion volume was estimated by continuing the cross sections into the air so as to converge just above the top of the UBZ. The resulting original volume was estimated at 3530 km^3 and the existing volume at 1870 km^3 (Morse 1969; Table 7). The existing mass is estimated at 5.61×10^{15} kg.

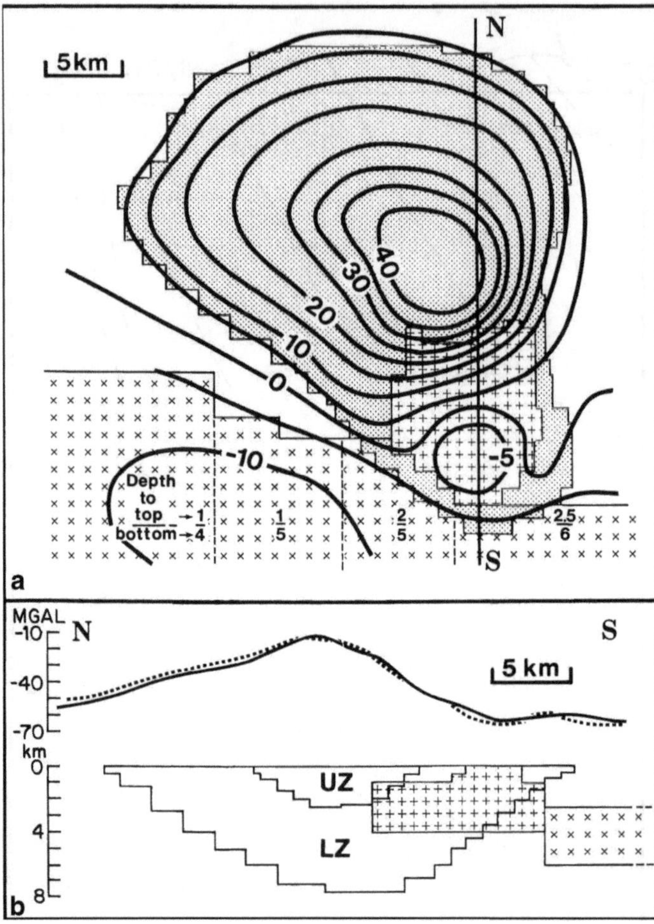

Fig. 13.8 **a** Kiglapait gravity model, Fig. 13.8 of Stephenson and Thomas (1979) courtesy the authors. Contours in mGal. The major (*northern*) portion of the intrusion matches the structural cross sections reasonable well, but the southern region has evidently been stoped off by the Manvers Granite (+symbols) and also affected by the proximity of anorthosite (X symbols), as shown in the cross section **b.** The curves in (**b**) show the measured gravity (*solid*) and the calculated model (*dashed*) assuming the low-density units shown

Gravity Studies

A detailed gravity survey by Stephenson and Thomas (1979) using many rock densities within and outside the intrusion for control is consistent with the geometrically-derived floor, the intrusion thickness, its bowl shape, and resultant mass of the main body of the intrusion (Fig. 13.8). The data require a surprisingly large volume of low-density material in the southeastern part of the intrusion.

Field evidence shows an abundance of the younger fluorine-rich Manvers Granite in dikes, pods, and small plutons in this portion of the intrusion. The granitic magma was emplaced along brittle fractures in the Kiglapait rocks, and thin (cm-dm-scale thickness) veins of pegmatitic granite have been seen to occupy outward-dipping joints in the Kiglapait rocks, extending for kilometers beyond known outcrops in the south. The Manvers Granite has a concordant age of 1300 ± 1 or 2 Ma (M.A. Hamilton, written communication 07 Dec 1995), and is thus ~7 Myr younger than the Kiglapait Intrusion. It is assumed that the low-density gravity signal in the southern part of the Kiglapait Intrusion is due to stoping and replacement of the intrusion rocks by the younger granite. The working hypothesis is that the original intrusion had the depth and form as calculated, before the invasive granite stopped some of it away.

Stratigraphy

Layered Series

The updated stratigraphic section of Fig. 13.9 describes the zones and subzones of the intrusion, shown in cross section in Fig. 13.6. An Inner Border Zone (IBZ) occupies a small (~200 -m) thickness at the base. It is a coarse-grained subophitic olivine gabbro that turns out on examination to be a best candidate for the bulk composition of the intrusion. It competes among summation liquids, chilled margins of nearby troctolitic intrusions, and experimental studies as described in detail by Morse et al. (2004) and listed in Tables 13.1 and 13.2 here. An earlier-interpreted Outer Border Zone of mafic granulites (Morse 1969), once considered to be a chill containing olivine-gabbroic granulites, is now relegated to the interlayered mafic compositions within the Falls Brook Series, which is clearly an older supracrustal edifice (Schuh 1981). The Lower Zone (LZ) is composed of troctolite, forming >82 % of the volume of the intrusion. The arrival of augite (first as oikocrysts but then as cumulus grains at 84 PCS) defines the base of the Upper Zone (UZ). This event is succeeded in turn by the arrivals of Ti-magnetite, sulfide globules (now mostly pyrrhotite, Po), apatite (Ap), and antiperthite blebs in plagioclase, followed by mesoperthite as the primary feldspar. The diagram also shows the main compositional variation of plagioclase (An) and olivine (Fo).

What the stratigraphic column cannot show is the uppermost feldspar complexity that involves wormy and blocky symplectite replacing oligoclase in the company of stringy remnants of mesoperthite, and many other curious textures. These are illustrated photographically in several plates in Morse (1969) and are described by Speer and Ribbe (1973).

Upper Border Zone

This roof zone occupies the structural center of the intrusion (Fig. 13.5 and the cross sections, Figs. 13.6 and 13.7). All the UBZ stratigraphy is inverted, with the most

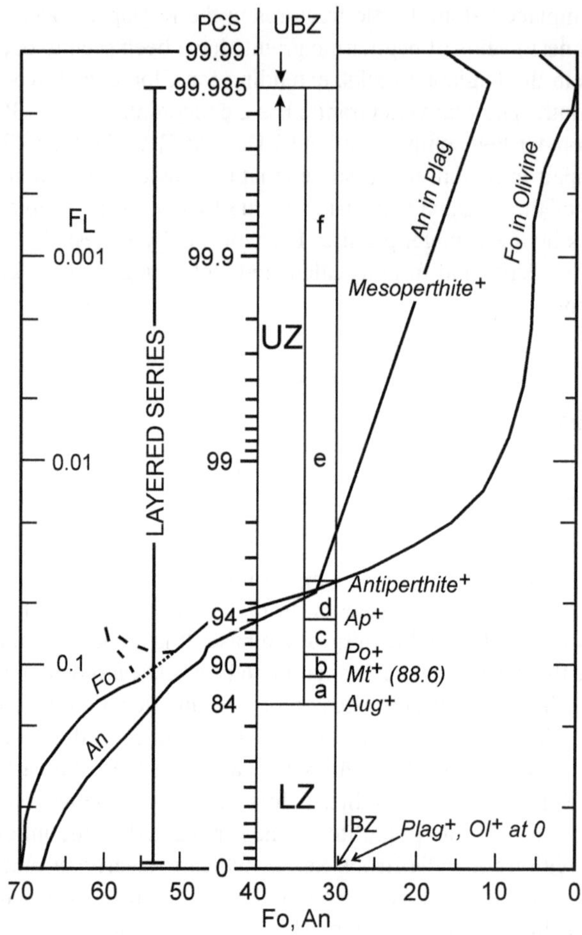

Fig. 13.9 Kiglapait stratigraphic column. The X-axis denotes mineral compositions for plagioclase and olivine. The zones and subzones are defined by their increasing mineralogical complexity, beginning with plagioclase and olivine in the Lower Zone (*LZ*) and the following phases defining the subzones of he Upper Zone (*UZ*): *Aug*: augite, *Mt*: titanomagnetite (and minor ilmenite), *Po*: pyrrhotite (near troilite) and other sulfides, *Ap*: fluorapatite; others as named varieties of feldspar. *PCS*: percent solidified. F_L: fraction liquid. The vertical scale (horizontal in succeeding figures) is required to be logarithmic in order to plot all the samples at high stratigraphic levels, given that the section is ~8400 m thick and there are still ~266 m of rocks above 99.9 PCS

primitive mineral compositions in the highest elevations of a central unit, UBZ-α. The most evolved compositions in the intrusion are found at the outer rim of the outer shell (UBZ-β) and in the adjacent UZf unit. The UBZ-UZ contact delineates the Kiglapait Sandwich Horizon.

The correlation from the UBZ to the LS is made from plagioclase compositions. The cores or maximum compositions of UBZ plagioclase correlate with the mean plagioclase composition of the Layered Series (Morse and Allison 1986). The same correlation has been found in the Skaergaard Intrusion (Salmonsen and Tegner 2013).

Table 13.1 Estimates of Kiglapait composition

	1		2	3	4	5	6	7	8	9
	MODE		HETTAS	JONATH	IBZAVG	KI1981	KI1986	SMAR	KI3763	JB94-4
SiO_2			47.94	48.07	47.98	47.46	47.46	49.83	49.78	49.52
TiO_2			1.24	0.68	0.76	0.81	0.81	0.30	0.24	0.33
Al_2O_3			18.95	19.16	18.42	19.46	19.46	19.58	19.76	19.21
Fe_2O_3			1.12	2.09	1.80	1.91	1.04	0.68	0.44	0.91
FeO			10.66	8.90	10.12	9.36	10.14	8.30	8.72	7.38
MnO			0.14	0.13	0.16	0.15	0.15	0.13	0.13	0.17
MgO			7.67	8.48	7.92	8.04	8.04	8.17	8.65	8.03
CaO			8.60	8.60	8.19	9.27	9.27	9.46	8.90	8.86
Na_2O			3.21	3.14	3.43	3.13	3.13	3.10	3.18	3.88
K_2O			0.40	0.30	0.29	0.22	0.22	0.25	0.24	0.96
P_2O_5			0.13	0.30	0.12	0.11	0.11	0.04	0.03	0.07
Total			100.06	99.85	99.20	99.92	99.83	99.84	100.07	99.32
	Vol%					Oxygen Norms				
AP	0.21	Ap	0.29	0.65	0.26	0.25	0.25	0.07	0.07	0.11
TIMT	1.09	Mt	1.01	1.87	1.65	1.74	0.97	0.62	0.38	0.82
ILM	0.36	Ilm	1.68	0.92	1.04	1.09	1.09	0.43	0.32	0.43
		Or	2.32	1.73	1.75	1.16	1.16	1.71	1.71	5.77
		Ab	30.12	29.37	32.10	28.94	29.00	28.55	29.08	28.26
FSP	73.20	An	37.54	38.58	35.78	40.12	40.20	39.68	39.81	33.27
AUG	5.74	Di	4.56	2.81	4.38	5.21	5.22	6.21	3.85	9.08
HY	0.36	Hy	2.71	6.80	3.50	2.17	0.43	7.39	6.95	0
OL	18.76	Ol	19.76	17.28	19.55	19.32	21.68	15.34	17.82	18.24

Table 13.1 (continued)

	1		2	3	4	5	6	7	8	9
	MODE		HETTAS	JONATH	IBZAVG	KI1981	KI1986	SMAR	KI3763	JB94-4
BIOT	0.16	Ne	0	0	0	0	0	0	0	4.04
TROIL	0.05	Q	0	0	0	0	0	0	0	0
PARG	0.07		Total Feldspar + Olivine + Pyroxene, FOP:							
		FOP	97.01	96.57	97.06	96.92	97.69	98.88	99.22	94.62
						Ternary Compositions				
FSP	74.6		72.1	72.2	71.7	72.5	72.0	70.7	71.2	71.7
AUG	5.9		5.6	3.5	5.4	6.5	5.8	7.5	4.7	9.4
OLHY	19.5		22.2	24.4	22.8	21.1	22.2	21.7	24.2	18.9
An	nd		55.5	56.8	52.7	58.1	58.1	58.2	57.8	52.4
Fo	nd		60.2	67.1	61.9	64.6	61.5	65.2	64.9	68.1

1. Mode constrained by 120 rock densities (Morse 1979a, b)
2. Hettasch Intrusion chilled margin, average of 5 analyses (Morse 1981)
3. Jonathon Intrusion chilled margin (Berg 1994)
4. Average of three Inner Border Zone analyses (KI 3708, KI 3567, KI 3623; Morse 1981)
5. Summation over 48 wet chemical analyses constrained by modes (Morse 1981)
6. Same with modified ferric iron (Nolan and Morse 1986)
7. Average South Margin composition (Nolan and Morse 1986)
8. Sample KI 3763 from the South Margin (Nolan and Morse 1986)
9. SRun 94-4 (Pl, Ol, Liq), 1208EC, 1 atm on KI 3763 (Blundy 1997); Ferrous ratio set at 0.9 Values for Ti, Mn, and P carried in from run 94-7

Table 13.2 Experimental bulk composition compared to other estimates

Oxide	1	2	3	4	Norms	1	3	4
SiO_2	47.46	48.96	48.55	48.07	AP	0.25	0.25	0.26
TiO_2	0.81	0.63	0.79	0.76	MT	1.74	1.31	1.65
Al_2O_3	19.46	18.58	18.43	18.46	IL	1.09	1.09	1.04
FeO	*11.08*	13.24	*13.13*	*11.76*	OR	1.16	1.74	1.75
MnO	0.15	0.15	0.15	0.16	AB	28.94	33.72	32.10
MgO	*8.04*	7.06	*7.00*	*7.94*	AN	40.12	34.77	35.78
CaO	*9.27*	8.03	*7.96*	8.21	DI	5.21	4.14	4.38
Na_2O	*3.13*	3.64	*3.60*	3.44	HY	2.17	3.27	3.50
K_2O	0.22	0.22	0.27	0.29	OL	19.32	19.69	19.55
P_2O_5	0.11	0.11	0.11	0.12	An/PL	*58.1*	*50.8*	*52.7*
Total	99.73	100.62	100.00	99.20	XMg(0)	*0.584*	*0.516*	*0.619*
					FSP	72.0	72.2	71.7
					AUG	5.8	5.1	5.4
					OLHY	22.2	22.7	22.8
					OLRAT	0.793	0.817	0.809

1. KI1981 summation, Table 13.1
2. Average of KIBC2 and KIBC3 glasses
3. Same, but with Ti and K from KI1981
4. IBZ average from Table 13.1

Fig. 13.10 Photo gallery of Kiglapait field and textural relations. **a** Basal *Lower* Zone troctolite at Hare Point on Port Manvers Run, 10 PCS. Mafic layers are olivine cumulates, felsic layers are plagioclase cumulates. **b** Graded layering, *Lower* Zone ~15 PCS, southeast coast. **c** Wall of layered troctolite, 15 PCS, southeast coast. **d** *Upper* Zone ferrodiorite near 95 PCS, southeast of *Upper* Border Zone. **e** Subophitic olivine in troctolite, 21 PCS, partly crossed polars in doubly-polished thin section. Length of field is 3 mm. **f** Normal and reverse-graded layering in *Upper* Zone ferrosyenite, southeast of *Upper* Border Zone, 99.8 PCS

Field and Laboratory Illustrations

In two Photo Galleries (Figs. 13.10 and 13.11) are shown images that may help to visualize the Kiglapait rocks as seen in the field and through the microscope. Most of the information needed will be found in the captions, but a few remarks may be helpful here. In Fig. 13.10a the dip is taken as primary. This and nearby outcrops

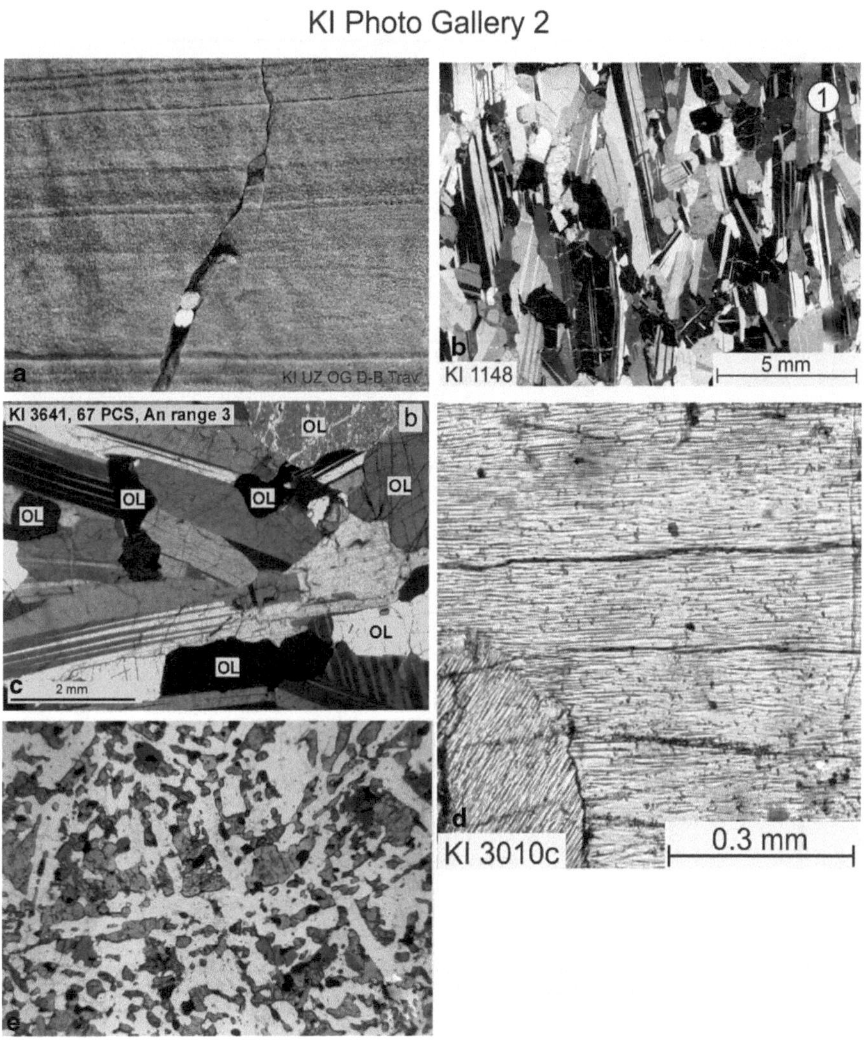

Fig. 13.11 Second photo gallery of Kiglapait field and textural relations. **a** Delicate layering in *Upper* Zone olivine gabbro, DB traverse. **b** Lamination in *Lower* Zone troctolite, Kiglapait Mountains, 38 PCS. **c** *Lower* Zone troctolite, 67 PCS, SL traverse, with triple junctions approaching textural maturity. **d** Mesoperthite in ferrosyenite, 99.86 PCS, CP traverse. **e** Roof chill of Kiglapait magma within the *Upper* Border Zone, sample KI 4085 at ~90 PCS equivalent. Bladed plagioclase (*white*) among fine-grained olivine, augite, and Fe-Ti oxide minerals

of the basal Lower Zone show structures and textures ranging from crystal mushes, dikes, and rolls to rigorously parallel layers with thicknesses from millimeters to meters. Graded layering as in Fig. 13.10b shows clear evidence of feedback, with the mafic base giving way to a very felsic top. The mesoperthite in Fig. 13.11b has spacings of only a few microns. Unless reacted with a melt, it retains its bulk composition of a Sr-rich sodic sanidine (Speer and Ribbe 1973).

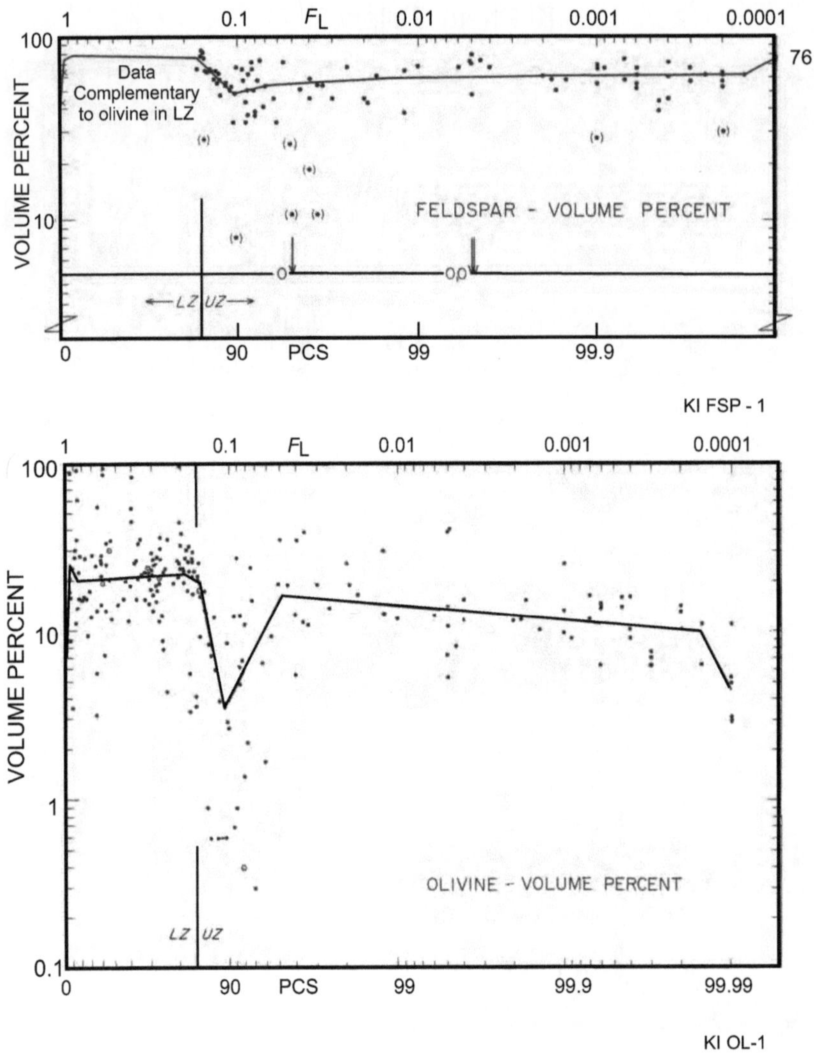

Fig. 13.12 Kiglapait modal data, modified from Morse (1979c). Plagioclase **a** and olivine **b** No data are plotted for LZ plagioclase because they are simply the complement of the olivine modes in this troctolitic zone. The olivine abundance falls off sharply near 90 PCS where augite and opaques become super-abundant

Petrography

Modal Compositions

Modal data for all cumulus minerals in the intrusion, and one non-cumulus (biotite) are reproduced here from the study of Morse (1979c) in Figs. 13.12–13.15, all plot-

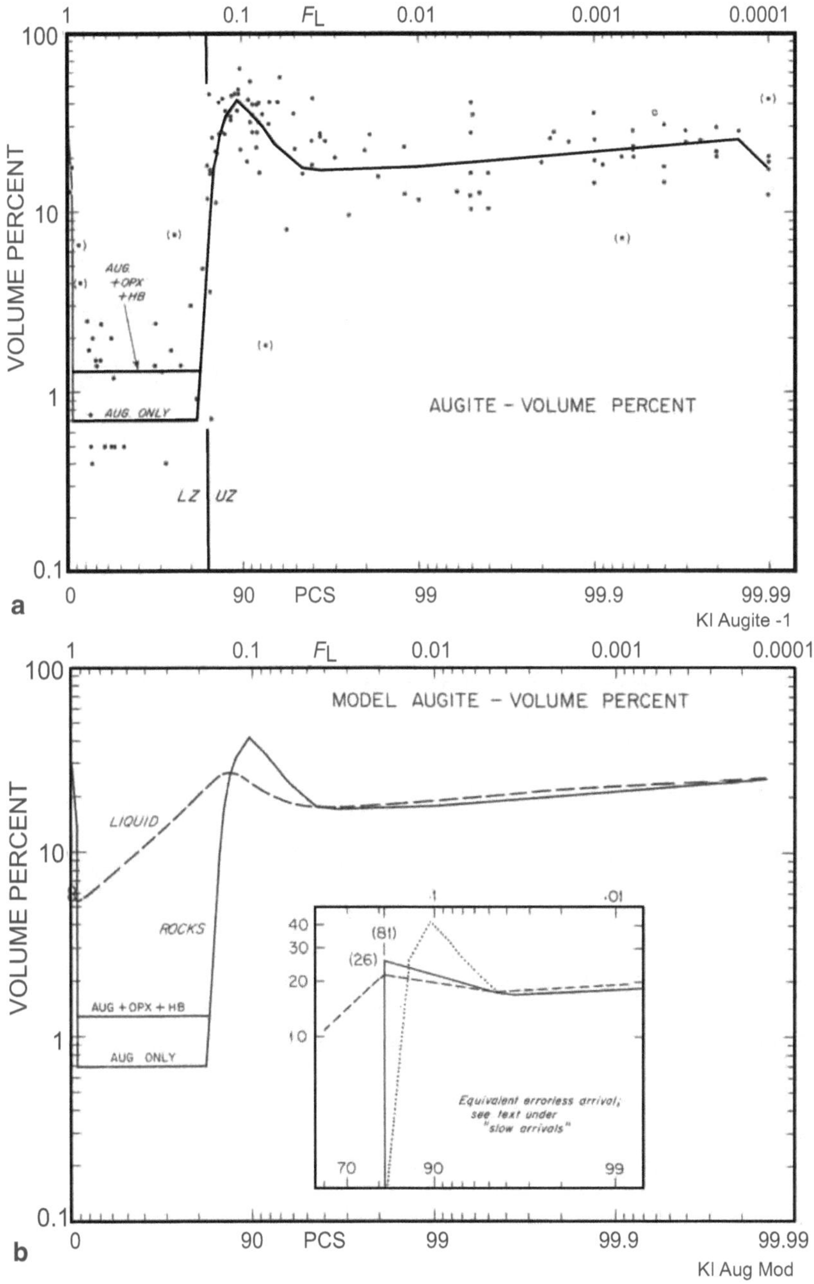

Fig. 13.13 Modal augite **a** and the companion summation model **b** that gives an estimate of the liquid composition by summation from the top down. Part (**a**) also gives data for minor interstitial Opx and hornblende. In Part (**b**) the inset shows the same area under the curve in (**a**) redistributed to map an "Equivalent errorless arrival" assuming that the slow arrival and overproduction of the actual modal quantities are errors due to poor stirring of the magma. The errorless arrival happens to coincide with the first (oikocrystic) arrival of augite crystals at 81 PCS. Its peak value is 26% rather than the 42% of the slow but overproduced trend

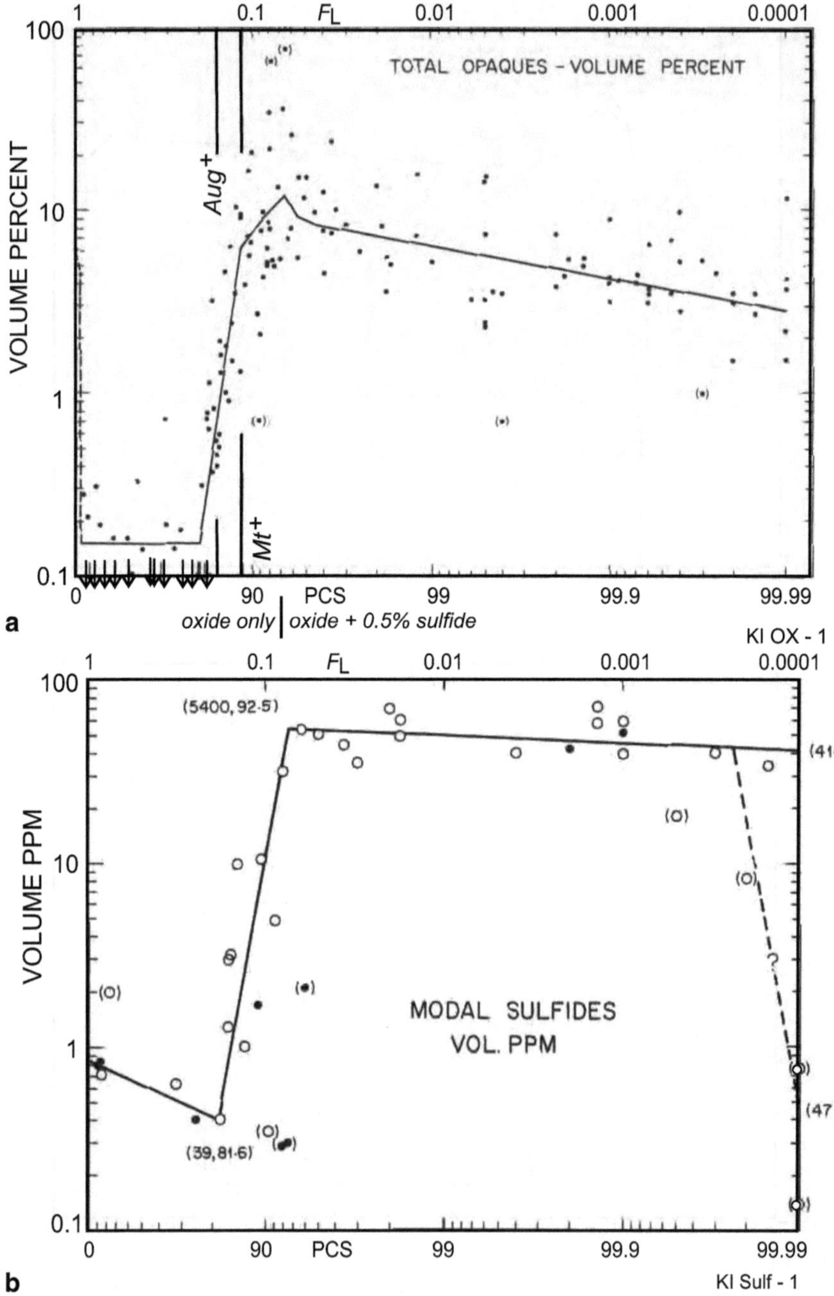

Fig. 13.14 Modal data for total opaques **a** and for sulfides **b**. Note that the sulfide data are given in volume ppm, hence they have little effect on the opaques plot

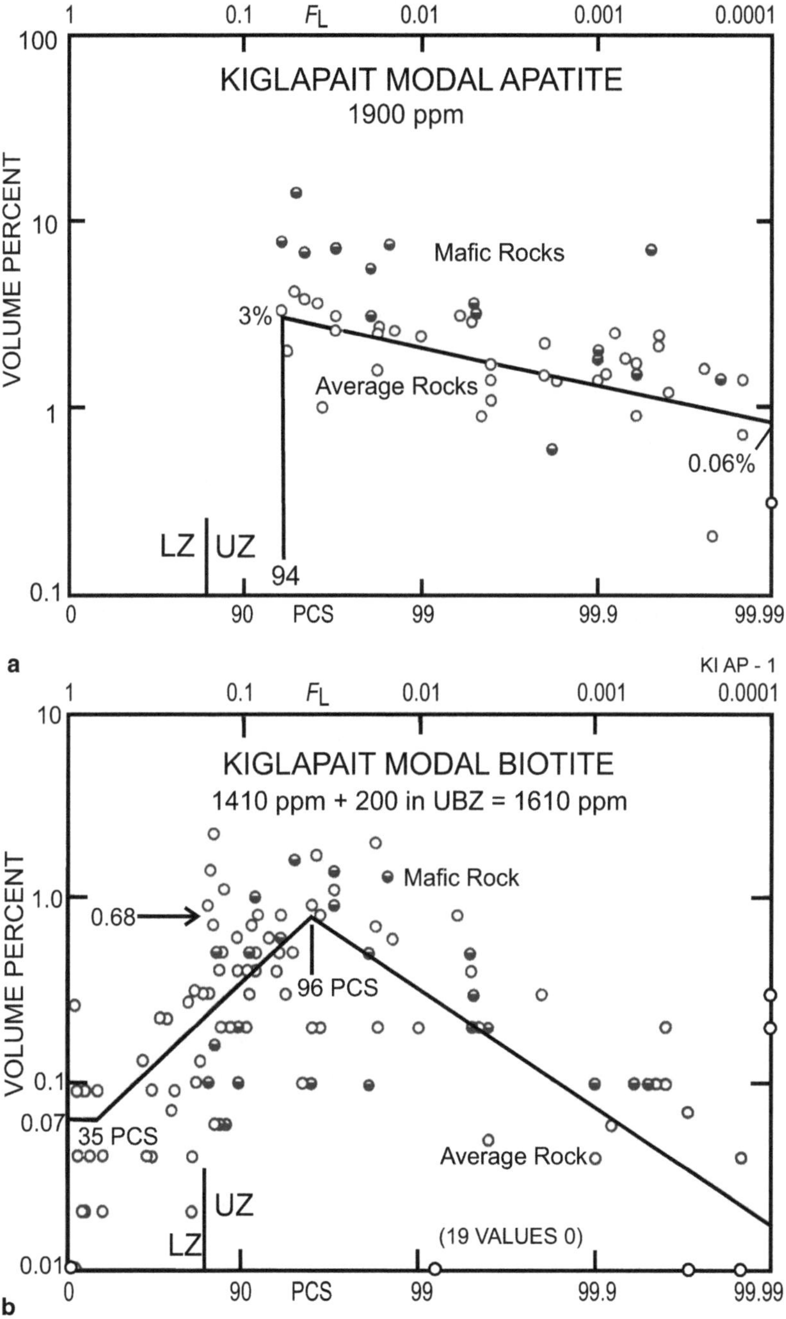

Fig. 13.15 **a** Modal data for apatite. Modes for *mafic rocks* tend to be significantly richer in apatite than for *average rocks*. **b** Modal data for biotite. The modes are not significantly different in *mafic rocks* from those in *average rocks*

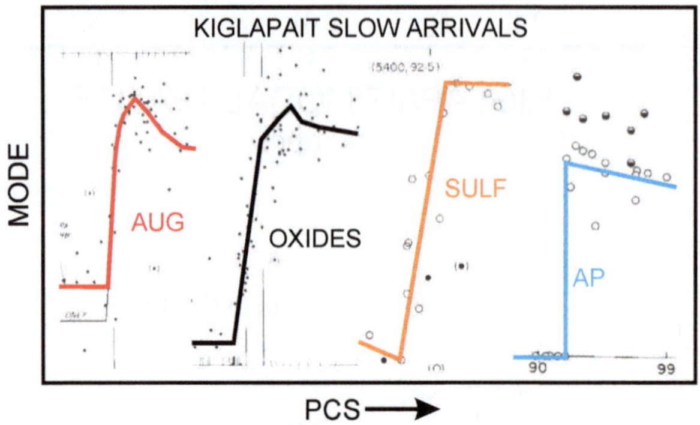

Fig. 13.16 Slow arrivals of augite, Fe-Ti oxide minerals, and sulfides plotted from Figs. 13.10, 13.11, 13.12 and 13.13. The first three seem to leak in rather than jumping in: they arrive over many thousands of years—oxides especially. But then just afterward, Ap arrives instantaneously—presumably because the magma got stirred by the sudden violent descent of oxide minerals at the Main Ore Band. The sulfides were originally immiscible liquid droplets, as reproduced frequently in experimental melts

ted against stratigraphic height expressed as PCS and the fraction of the magma present as liquid, F_L. The data need no description beyond that given in the captions, except to emphasize that the modal data for plagioclase in the LZ are omitted because in each case they are complementary to the olivine data in the figure below. The liquid composition for each mineral species can also be calculated by summation over all the overlying rock compositions, as illustrated in the augite model of Fig. 13.13b. These individual modal diagrams are followed by a comparison (Fig. 13.16) of the "slow arrivals" of augite, Fe-Ti oxide minerals, sulfide globules and grains, contrasted with the sudden arrival of apatite.

Residual Porosity

The modal data for the Lower Zone rocks permit the calculation of residual porosity based on the amount of an excluded component captured in the cumulate. For example, if a LZ troctolite contains no trace of augite it has by definition a residual porosity of zero because all the originally trapped liquid has been removed. By hypothesis, it was replaced by isothermal adcumulus growth in which the departing rejected solute was exchanged by incoming fresh magma. If, however, some grains of augite are found, they are attributed to the incomplete action of adcumulus growth and the residual porosity p_r is calculated as

$$p_r = D = C_s / C_L \tag{13.1}$$

where

D is another name to show that the residual porosity has the dimensions of a partition coefficient,
C_S is the concentration of that mineral component measured in the solid rock, and
C_L is the concentration of that mineral component in the calculated liquid.

When this is done for all four excluded components shown in Fig. 13.16 and calculated at 0 and 84 PCS (the limits of the Lower Zone), the combined stratigraphic result is given by

$$p_r = 0.139 * F_L^{0.91} \tag{13.2}$$

and using this calibration it then becomes possible to scale the calculated residual porosity against the *An range* of the observed individual samples when studied in grain mounts to give robust estimates of the within-sample compositional variation of plagioclase (Morse 1979c; 2012). This procedure succeeds because the plagioclase does not re-equilibrate within the cumulate in all of geologic time at high temperature (Morse 1984).

Recent results are shown in Fig. 13.17a, where the An range is mapped over the entire stratigraphic range of the intrusion, and then converted to residual porosity in Fig. 13.17b. The most important feature of the latter is the cluster of data near zero that proclaims the high frequency of near-perfect adcumulates. Adcumulates as defined by their An-range were solidified isothermally (Morse 1984), so they constitute impermeable barriers to the motion of intercumulus liquid across levels of stratigraphy. Their concentration in the Lower Zone was used (Morse 2012) to show that the high probability of occurrence of isothermally solidified adcumulates was high enough to falsify the hypothesis of compaction as an important agent of solidification.

Photomicrographs of Kiglapait feldspars with high and low values of the An range are shown in Fig. 13.18.

Apatite, Biotite, and Volatiles

Cumulus apatite arrives abruptly at 94 PCS in the Kiglapait stratigraphy; its appearance in the rocks has been located where predicted within a few tens of meters on the ground. Its occurrence, along with that of biotite, is reported with crystallographic, electron probe, infra-red and neutron activation analyses by Huntington (1979). In the Layered Series apatite is strongly dominated by F among the volatiles, with minor Cl and OH. Modal apatite ranges up to 14 % in mafic rocks lying above 94 PCS, but in average rocks declines steadily from ~3 to 1 % at the end of crystallization, where cm-scale crystals of apatite have been seen. The OH content is negligible and the Cl content is estimated at ~10 ppm for the intrusion.

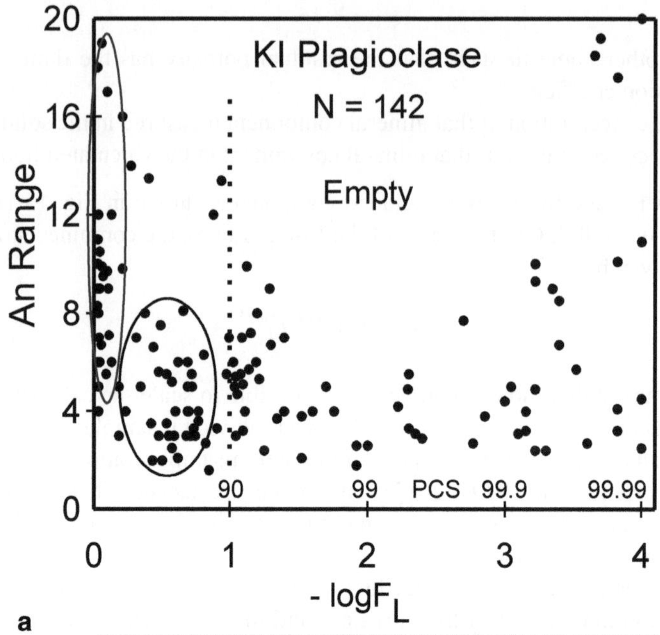

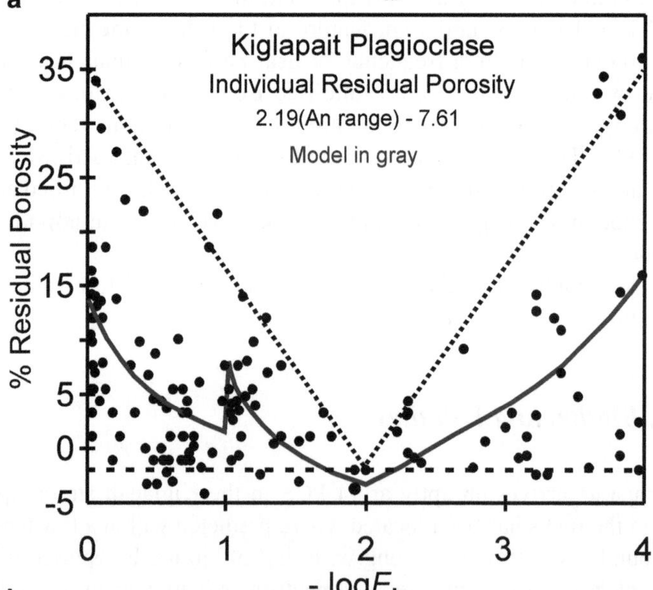

Fig. 13.17 **a** An range (maximum - minimum An in grain mounts) in the Kiglapait intrusion. The *X*-axis is given as the log scale indicated and as PCS (volume percent solidified). The ellipses call attention to two groups of interest, being very large at low PCS and moderate in the range 40–90 PCS. The dotted *vertical line* calls attention to the 90 PCS level in the stratigraphy. No data occur in the empty region. **b** Residual porosity of individual samples as calculated from the equations in Morse (2012), and with the model (solid curve) derived from the model from the An range in (**a**). The *dotted lines* approximate the *upper* limits to the data. The horizontal *dashed line* calls attention to the fact that at least 15 samples broadly distributed share the low values at or *below* those at 99 PCS

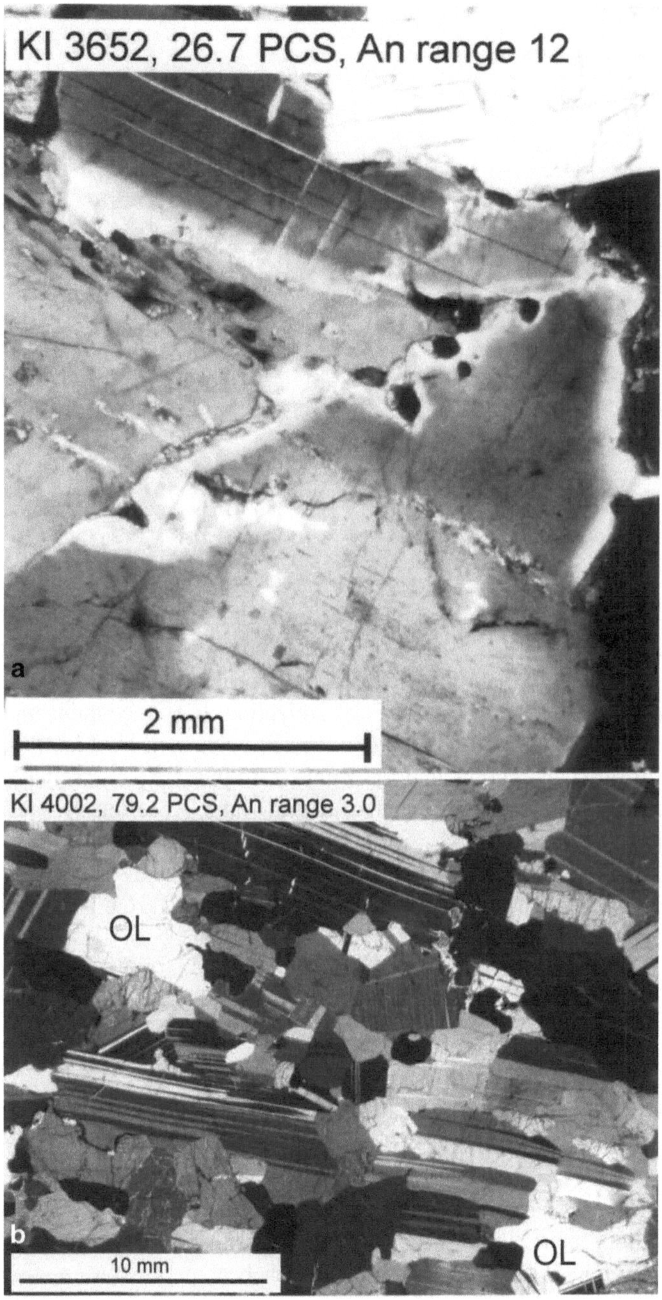

Fig. 13.18 Photomicrographs of Kiglapait cumulates, showing different amounts of zoning for different values of the An range. **a** From 26.7 PCS with An range = 12 mole % An showing normal zoning in adjacent grains of plagioclase. **b** From 79.2 PCS with An range = 3 mole % An showing little if any visible zoning in plagioclase. Untwinned blocky grains are olivine (*OL*). **c** Cumulate from the *Upper* Border Zone, 99.99 PCS, mottled feldspar intergrowths of oligoclase and orthoclase with An range = 20 mole % An

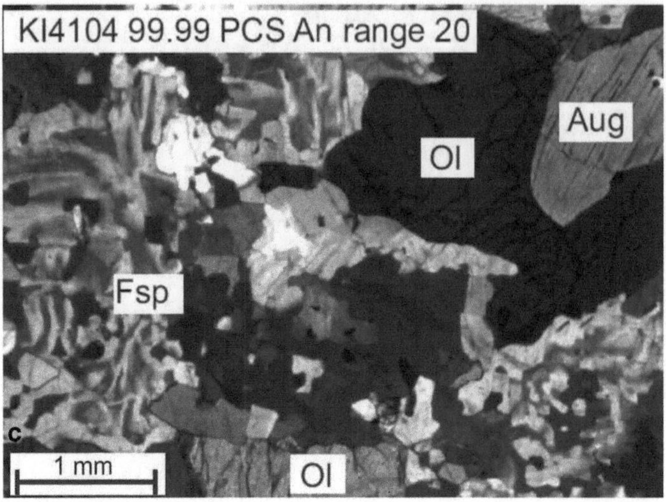

Fig. 13.18 (continued)

Red biotite is not a cumulus phase in the Kiglapait rocks but occurs mainly as rims on Fe-Ti oxide minerals, hence its modal abundance roughly tracks those minerals. It rises from 0.08 % in the Lower Zone to peak at nearly 1 % at 95 PCS, then declines to trace values at the end of crystallization. Biotite contains an average of 0.4 % F and 0.07 % Cl, and its mean content of calculated H_2O is 4 % by weight. However, infrared absorption analyses range from 0 to 1.5 ± 1.1 % by weight, leaving uncertainty about any significant structural water. Accordingly, these minerals are regarded as Ti-oxybiotites.

Summation of the volatile contents of apatite and biotite over the intrusion yields 165 ppm F, 10 ppm Cl, and the maximum content of H_2O from both phases is estimated at 68 ppm by weight, with a minimum of 8 ppm.

Gabbro Pegmatites

An abundance of distinctive pods of gabbro pegmatite occur at low stratigraphic levels of the Lower Zone. These objects contain major amounts of black augite and 10 cm tablets of plagioclase. The augite must signify undercooling to the point of augite saturation. The coarse grain size suggests the influence of either water or fluorine or both, confined to these pods, which range in size from very small to a few meters in long dimension. These local pods become scarcer up-stratigraphy. It is interesting to consider that these local, perhaps volatile-bearing pods may represent ingestion of volatiles from the country rocks, or merely concentrations of endogenous volatiles sequestered in the coarse-grained mafic pods. In principle, these could be dehydration or defluorination features, in which any local volatile components were captured, thus depleting the residual magma in volatiles.

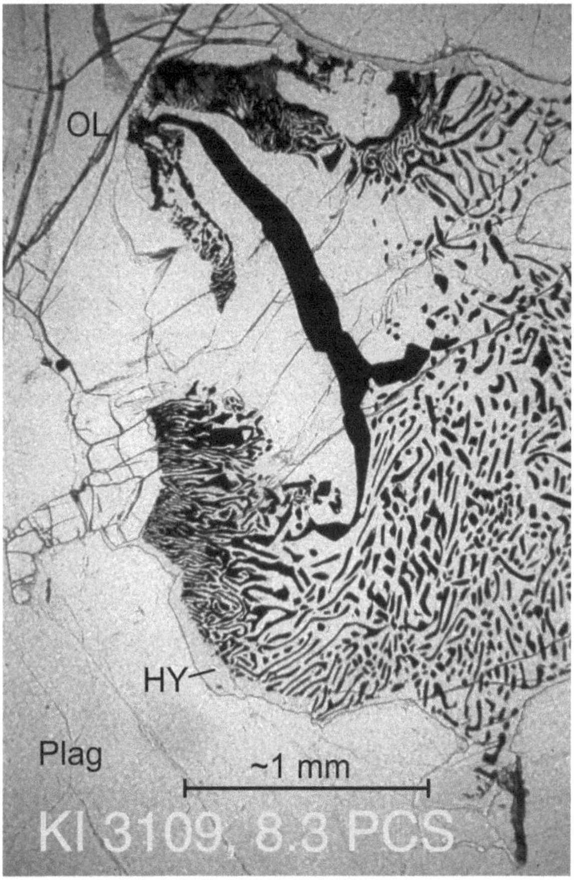

Fig. 13.19 Oxysymplectite from the Kiglapait *Lower* Zone at 8.3 PCS. The large grain of olivine in the *upper left* region has reacted with intercumulus liquid to make hypersthene and magnetite by the coupled end-member reactions Fa + O_2 = Mt + Q and Q + Fo = En (e.g., Morse 1979a). The main part of the intergrowth is a mosaic of hypersthene + magnetite, and there is an outer hypersthene rim. The reaction has exhausted all the available liquid, leaving abundant olivine intact

Fe–Ti Oxide Minerals

Opaque minerals in the Lower Zone occur only in small patches, in which evidence for trapped intercumulus liquid is unambiguously recorded by traces of magnetite, oxysymplectite, hypersthene, brown hornblende, or augite in local occurrences. Oxysymplectites (Fig. 13.19) are strictly local intergrowths (scarce in thin section) that consist of magnetite and hypersthene, partially invading crystals of olivine that have reacted with trapped liquid. While fundamentally caused by the oxidation of olivine with release of silica to form hypersthene, the bulk analyses of the phases also show the introduction of Fe, Ti, Al, and Si as well as oxygen, so the symplectites result from a late, local magmatic reaction above the solidus. The process eventually runs out of oxygen and stops eating olivine. The persistence of unaltered olivine exterior to the oxysymplectites defines a low activity of silica and oxygen in the rock generally.

Cumulus titanomagnetite and minor ilmenite first appear in the Upper Zone of the Kiglapait intrusion (Morse 1979c). They first occur as disseminated grains within gabbroic cumulates (Morse 1980). In the region above 90 PCS, oxide-rich layers a few centimeters thick begin to appear, and these become more numerous and thicker to a maximum at the 80 cm thick Main Ore Band (MOB), at 93.5 PCS. The oxide-rich layers and particularly the MOB represent rapid transit of roof crystals to the floor, and would have promoted stirring of the magma. The MOB event would have been profound enough to homogenize all the remaining magma body, then ~2 km thick. This event is inferred to have marked the end of slow arrivals and was almost immediately succeeded by the sudden arrival of apatite at 94 PCS. Above this level rare clusters of Ti-Mt+apatite occur to make local "nelsonitic" intergrowths seen in thin section, but these never reach the scale of observed hand specimens.

Mineralogy

Textures

Igneous lamination is prevalent as defined by the (010) faces of plagioclase. It is generally concordant with the layering. In sections cut parallel to the lamination, olivine is commonly seen as subhedral with stubby prisms that commonly lie either across the dip or down-dip. There is a notable, strongly laminated, large (2 m) boulder erratic from the Upper Zone near the mouth of Spar Brook west of Patsy Bay (Fig. 13.5). Here the grain size is coarse. What appear to be laths of plagioclase are commonly 2 × 8 cm in cross section. The dip slope is exposed, and the form of the close-packed plagioclase is oval, typically ~6 × 8 cm, without facets. The only face seen is the (010) that defines the lamination. Elsewhere, subsolidus maturity of textures is commonly seen, as in Fig. 13.11c.

Intrusion-Wide Olivine and Plagioclase Compositions

All compositional data for these two primary phases in the intrusion are shown together in one color-coded plot, Fig. 13.20a. In the Lower Zone the olivine compositions are ~10 mol percent richer in Fo than the plagioclase crystals are in An, but after the Main Ore Band they play second fiddle and end up at zero, which the feldspars cannot do because they terminate at a ternary minimum. Despite the usual scatter to be expected from blind sampling, there are clear parallel tendencies in

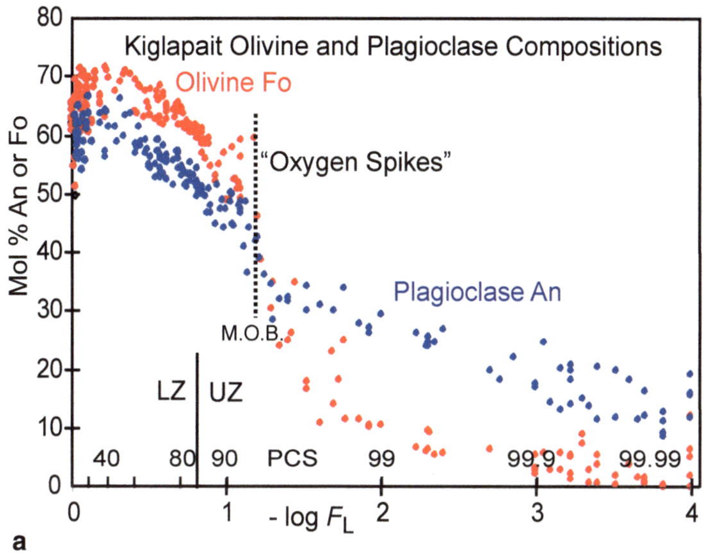

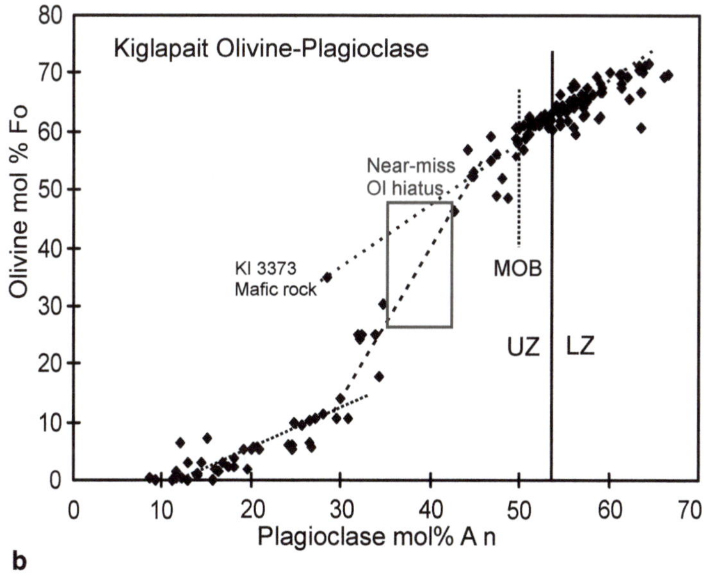

Fig. 13.20 **a** Compositions of olivine (*red*) and plagioclase (*blue*) with *PCS* and -log F_L. The "*Oxygen Spikes*" reflect the oxidation of the melt at high abundances of titanomagnetite leading up to the Main Ore Band (*M.O.B.*) at 93.5 PCS. This reaction diminishes the activity of fayalite and thereby enhances the activity of forsterite in the olivine, generating more Fo-rich olivine than normal for this stratigraphic height. The *An* and *Fo* values cross over at about 95.5 PCS; the Fo content goes to zero while the An content barely reaches 10–13. Symbols: *LZ* Lower Zone, *UZ* Upper Zone. **b** Olivine compositions plotted against plagioclase compositions. The main *LZ* trend stops at the *MOB* but continues metastably as far as the mafic sample *KI 3373* at 95 PCS. No data are found within the box labeled "*Near miss Ol hiatus*;" however a few samples do occur at that stratigraphic level, as found in the previous figure

the reverse variation 0–40 PCS and the near-parallel declining paths thereafter to 90 PCS.

However, many olivine compositions from about 89–95 PCS show an excursion to higher Fo values. As indicated in the figure, these aberrant values are taken to reflect the abundance of Fe-Ti oxide minerals being crystallized in this interval. These in turn are mainly titanomagnetite with a binary ratio near 55 % ulvøspinel: 45 % magnetite. Accordingly there is a major component of ferric iron in the liquid and that reduces the activity of Fe^{2+}, thereby favoring the Mg endmember of olivine. Even so, the Kiglapait magma is so rich in Fe^{2+} and so low in oxygen fugacity that it does not quite have an olivine hiatus, as does the Skaergaard intrusion (e.g., Morse et al. 1980).

Not quite: but as in the Fo-An plot of Fig. 13.20b, there is a near miss. From the Main Ore Band at ~ An_{42} to ~An_{36} there are no olivines. In terms of stratigraphic height, there are four samples (Fig. 13.20a), but in the Fo-An space they are absent. Denser sampling would probably turn up more olivines in this region, but the message is clear: the liquid was barely buffered with olivine in this region (see also the modal data in Fig. 13.12).

Ternary Feldspar

Kiglapait feldspars occupy perhaps the most highly evolved continuum known anywhere. The data for grain mounts and wet chemical analyses shown in Fig. 13.21 illustrate an interesting difference between the bulk compositions and individual grains. The two overlap from An 70 to about An 35, but then the bulk compositions continue around the corner to the ternary end point whereas the individual points show a dearth of data in the antiperthite range, with a renewed heavy population near the minimum. At the most evolved stage, the feldspars are at first mesoperthites, but eventually become complex intergrowths of microsyenite, mesoperthite, oligoclase, and perthite (Fig. 13.18c). Some assemblages show the odd reaction liquid = sanidine—oligoclase in which oligoclase is dissolved. The long tail of oligoclase toward An 10 (Fig. 13.21) is a product of sub-solidus exsolution. Further analysis of these details is in preparation for a detailed paper on the Kiglapait feldspars. Thermal studies suggest that the ternary minimum is hypersolvus, yet some of the reaction relations suggest otherwise, so the final history is one of low temperature subsolidus reactions.

Augite

This important phase throughout the Upper Zone has compositions that run nearly parallel to the 40 % diopside line in the pyroxene quadrilateral, being unaccompanied by low-Ca pyroxene (Morse and Ross 2004). That publication shows the composition range in detail, along with optical, X-ray diffraction, and chemical data. Here a summary diagram will be found under "Experimental" in which augite, olivine, and liquid are all plotted together.

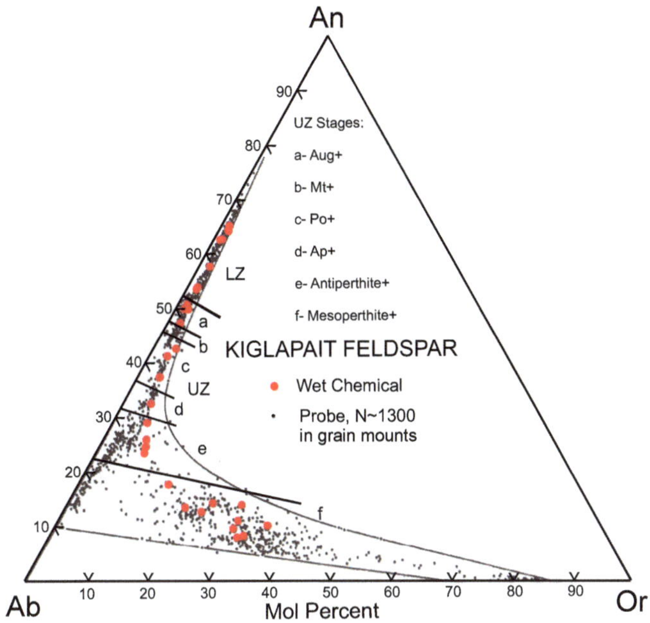

Fig. 13.21 Ternary feldspar plot for the Kiglapait intrusion. The subzones of the *Upper* Zone are shown as listed in the diagram. The bulk compositions define the overall trend of feldspar compositions but the individual points from grain mounts show a dearth of compositions in the transition from antiperthite to mesoperthite. There is also a considerable overlap in feldspar types in this region. The terminus of the main trend is a ternary end point near *Or* 33. The Or-poor trend from *An* 28 to *An* 14 is a subsolidus exsolution trend partnered by the *Or*-rich, *An* 0 group

Whole-Rock Compositions

Sources

Estimates of the Kiglapait bulk composition compared to nearby intrusions and to marginal rocks and one experimental composition are given in Table 13.1, from Morse et al. (2004). Further experimental compositions are listed in Table 13.2, to illustrate comparison with the Inner Border Zone composition, which is currently regarded as a best estimate. The Kiglapait and associated intrusions share low silica (48 %), high alumina (>18 %), high total iron (12 %), low Mg# (0.6), low K_2O (<0.3 %) and low P_2O_5 (0.12 %). They have abundant normative feldspar (72 %) and olivine (19.5 %) and very low augite (5.4 %). Here the protocols of Morse et al. (2004) involve assigning a HY component to the original diopside (i.e. Wo = 0.5) to make the augite found to have Wo ~0.4, and because the silica activity is too low to produce actual HY in the rocks, the normative HY (if any) is lumped with olivine to form OLHY. The norms also calculate the ratio of olivine to augite as a convenience to plotting in the ternary Ol-AUG-FSP.

Normative Compositions

Oxygen norms of the whole-rock analyses (Morse 1981) have an An content that is 1.5 An% lower than CIPW norms and which plots very close to the results for separated feldspars, Fig. 13.22a. The more abundant data of the separates show a steeper slope after 90 PCS, but in other respects the normative results are similar.

The abundance of normative feldspar, however, strikingly demonstrates the overall pattern of modal color index (Fig. 13.22b), beginning with a steep decline as mafic phases are added after 84 PCS, a trough centered near 94 PCS, a crest near 99.6 PCS, a second trough, and a felsic end point. A set of blue symbols and a sloping straight line show the samples nearest to a cotectic relationship, revealing that many samples are either more mafic, or more felsic, than the cotectic composition. Although most samples selected for analysis were collected in the field as locally "average" in color index, many were clearly not average. The diagram elegantly portrays the difference between haphazard modal variation (a mass-dependent, extensive variable) and the mass-independent (intensive) constraint of mutual saturation of liquid with two phases.

In hindsight, however, we may select those analyses that most closely follow the "nearest cotectic" line for a glimpse at the cotectic condition. These are listed in Table 13.3, and they furnish a useful characterization of the chemical fractionation in the intrusion.

The ternary plot of Fig. 13.23 shockingly reveals the extent of the modal diversity, from an abundance of highly felsic rocks to a somewhat less populated set of mafic rocks. The selected compositions from Fig. 13.22b and Table 13.3 are shown in blue. The troctolites (at ~zero AUG) form two separate groups straddling the cotectic composition; none are cotectic. Nor is the cotectic line of descent at all populated except near the triple point. There is, encouragingly, a true concentration near the triple point, with a cluster defining a trend of feldspar-unsaturated rocks. All of the normative rock compositions with more than a few percent AUG are from the Upper Zone. The grayscale plot of a rock at 99.9 PCS shows that there is little effect of composition on the mutual saturation point of the three corner phases.

Upper Border Zone

The Key in Fig. 13.24 shows the outline of the Upper Border Zone and the major sampling traverse that runs north-south (N-S). There are also two main east-west traverses not shown. The central highland oval in the UBZ is designated UBZ-α, corresponding to the Lower Zone, and the outer lowland region is designated UBZ-β, corresponding to the Upper Zone of the Layered Series. Unit α is coarser-grained, and has mineral compositions indicative of the Lower Zone, hence considered contemporary with it. Unit β is a finer grained "gray gabbro" in field terminology, with compositions consistent with being a mirror image of the Upper Zone.

Several compositional features of the UBZ will serve to show its reflection in most or all of the Layered Series. These attributes can be usefully illustrated in the

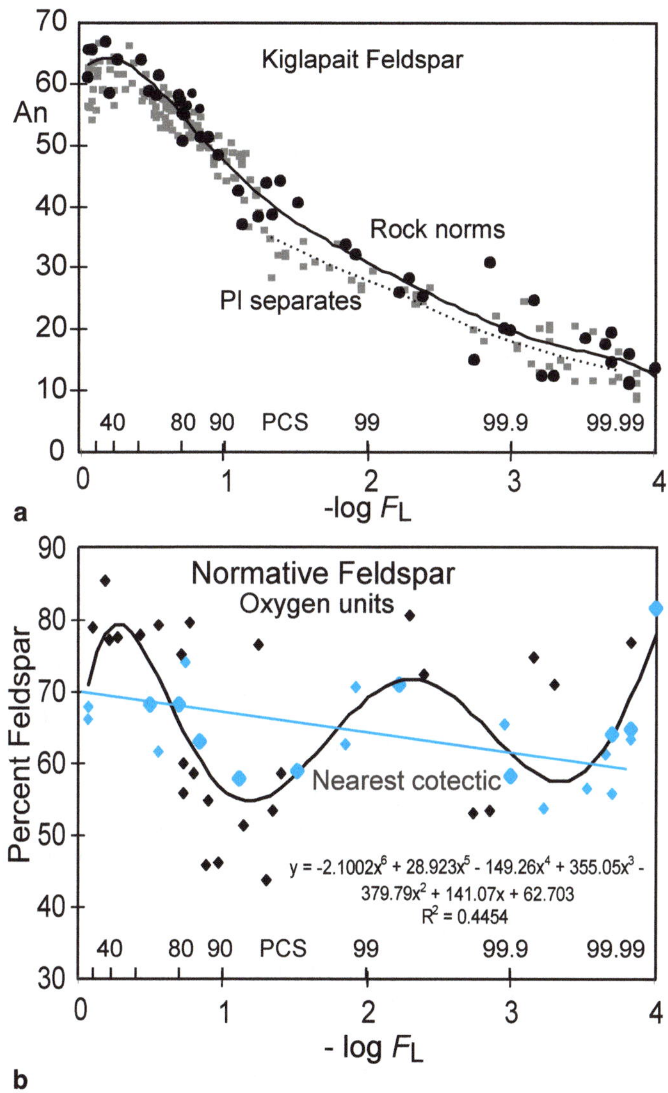

Fig. 13.22 **a** Oxygen–normative feldspar compositions from whole-rock analyses (Morse 1981) from the Kiglapait intrusion plotted over individual plagioclase compositions in grayscale. The curve is a sixth-order polynomial with $R^2 = 0.97$. **b** Normative feldspar amounts in the whole-rock analyses. The curve reflects known low values near the mafic overloads near 90 PCS and magnifies the variation seen less clearly in the modal plots of Fig. 13.13. The compositions nearest to cotectic estimates are shown in *blue* symbols and a straight *blue line*. The large *blue* symbols identify the analyses reported in Table 13.3

three panels of Fig. 13.24. Panel (a) shows coarseness as the average grain intercept in two orthogonal 17 mm traverses in thin section (Allison 1984). The general distinction of the α and β units is blurred by the presence of abundant trapped liquid. In addition, the northern limit of the β unit is as coarse as the α unit. In panel (b) the

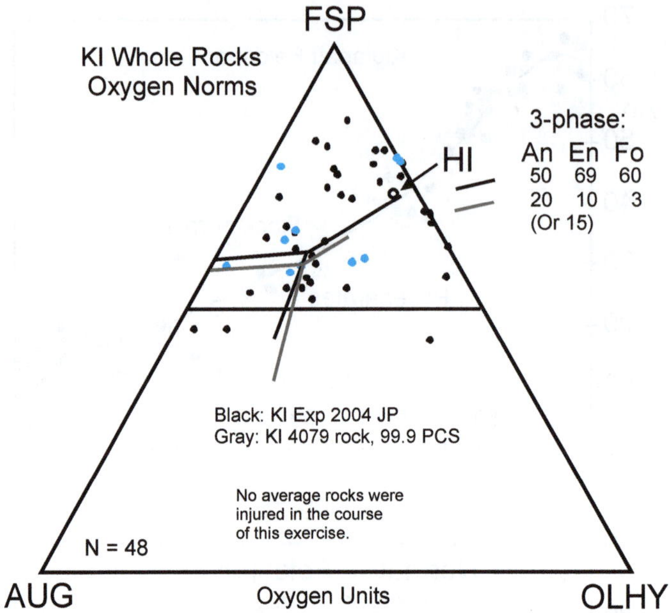

Fig. 13.23 Ternary plot of analyzed rocks using the normative conventions of Morse et al. (2004). The field boundaries are taken from the experimental study of that paper and the evolved field boundaries in grayscale are taken from a single evolved rock, *KI 4079*, at 99.9 PCS. The *Ol-FSP* cotectic trace is invariant with pressure to 13 kbar (McIntosh 2009). The small *circle* marked "HI" is the composition of the Hettasch intrusion chilled margin (Berg 1980). The slopes of the augite-feldspar and augite-olivine field boundaries are schematic but the triple points are precise. In the key, the *black* and *grayscale* lines are characterized by their triple-point compositions in feldspar, augite, and olivine. Analyzed samples from Fig. 13.22 and Table 13.3 are identified in *blue*

maximum An content is shown to vary from An_{69}, a value greater than any in the basal Lower Zone, to as low as An_{10}, equal to the limit in the LS. In panel (c) the Fo content of olivine is seen to be likewise similar to those values in the LS, reaching to as low as Fo_2. In all three panels the south end of the UBZ appears to be less fractionated than the north end, suggesting the possibility of a southern tilt to the unit.

The residual porosity of the UBZ rocks covers a much greater range than that of the LS; Fig. 13.25. As the value of porosity approaches 100 % the rock becomes essentially a quench of the local melt. Two such samples are conspicuous at 64 and 0 PCS. The sample with 87 % at 64 PCS is KI 3143, with fresh olivine, apatite interstitial to fresh, strongly zoned plagioclase, oikocrysts of augite, and oxy-symplectites with Opx, biotite, and magnetite. At 0 PCS and a residual porosity of 59 % the sample (KI 3133) contains major oikocrysts of augite, fresh, rounded olivine, biotite, and zoned plagioclase. One should take note that the red biotites common in the UBZ and occurring in the LS also are fluorine oxybiotites (Huntington 1979) not known to be hydrous.

The normative ternary composition of UBZ rocks (Fig. 13.26) is well-ordered compared to that of the LS shown in Fig. 13.19. With only three samples very rich in feldspar and one or two mafic rocks on the OL-AUG boundary, most of the rocks

Table 13.3 Selected wet-chemical analyses of near-cotectic Kiglapait rocks

Zone	LZ	LZ	UZa	UZc	UZe	UZe	UZf	UZf	UZf	UBZ
PCS	68	80	85.8	92.3	97	99.4	99.9	99.98	99.985	99.99
Spl NO	KI 3276	KI 3360	KI 3364	KI 4141	KI 3377	KI 4123	KI 4079	KI 3381	KI 4077	KI 4110
PCS	68.00	80.00	86.00	92.00	97.00	99.40	99.90	99.98	99.985	99.99
SiO2	46.85	47.52	51.10	49.42	40.92	48.91	46.81	50.91	50.26	54.91
TiO2	0.27	0.23	0.73	3.20	3.37	2.83	2.76	1.54	1.98	1.57
Al2O3	18.59	18.73	16.46	13.11	11.40	14.09	10.10	11.67	10.77	14.10
Fe2O3	1.80	1.27	1.23	3.52	3.40	0.22	2.14	1.42	2.84	3.02
FeO	10.48	11.00	7.35	8.85	21.79	17.76	22.71	18.96	19.15	11.92
MnO	0.16	0.16	0.14	0.19	0.41	0.33	0.58	0.48	0.58	0.35
MgO	11.31	10.58	7.31	6.07	3.33	0.96	0.80	0.45	0.08	0.01
CaO	7.65	7.47	12.15	11.92	9.96	7.85	7.86	7.10	6.69	5.70
Na2O	2.90	2.97	3.16	3.19	2.80	4.72	3.54	3.68	3.61	4.47
K2O	0.21	0.20	0.25	0.29	0.43	0.91	1.33	2.54	3.11	3.67
P2O5	0.02	0.03	0.02	0.01	2.51	1.38	1.08	0.55	0.43	0.01
Sum	100.24	100.16	99.9	99.77	100.32	99.96	99.71	99.30	99.50	99.73
Oxygen Norm[a]										
AP	0.07	0.07	0.07	0.00	5.75	3.07	2.46	1.22	0.96	0.00
MT	1.64	1.16	1.10	3.23	3.34	0.20	2.09	1.37	2.72	2.79
IL	0.38	0.33	0.97	4.41	4.93	3.99	4.05	2.23	2.89	2.15
OR	1.16	1.16	1.72	1.76	3.13	6.00	8.46	16.50	20.35	22.90
AB	27.17	27.77	29.28	29.97	28.19	44.38	35.66	36.05	35.77	42.28
AN	38.55	38.18	30.83	21.92	19.30	15.59	8.76	8.67	4.51	8.04
DI	0.65	0.00	23.47	30.41	13.39	12.37	20.41	19.71	21.74	16.52
HY	1.63	5.10	3.01	7.71	2.00	0.00	4.69	4.70	1.04	3.41
OL	28.76	26.18	9.55	0.59	19.97	13.79	13.14	9.55	10.02	1.91

Table 13.3 (continued)

Zone	LZ	LZ	UZa	UZc	UZe	UZe	UZf	UZf	UZf	UBZ
NE	0	0	0	0	0	0.60	0	0	0	0
Q	0	0	0	0	0	0	0	0	0	0
AN%	58.7	57.9	51.3	42.3	10.6	26.0	19.7	19.4	11.2	16.0
XMg^0	0.66	0.63	0.64	0.55	0.21	0.09	0.06	0.04	0.01	0.00
FSP	68.30	68.20	63.20	58.10	58.90	71.10	58.20	64.30	64.90	77.00
AUG	0.80	0.00	27.10	39.50	17.90	13.30	26.80	24.80	24.40	20.80
OLHY	30.90	31.80	9.80	2.40	23.20	14.90	15.00	10.80	10.70	2.10

[a] Oxygen Norm conventions and definitions as in Morse et al. (2004)

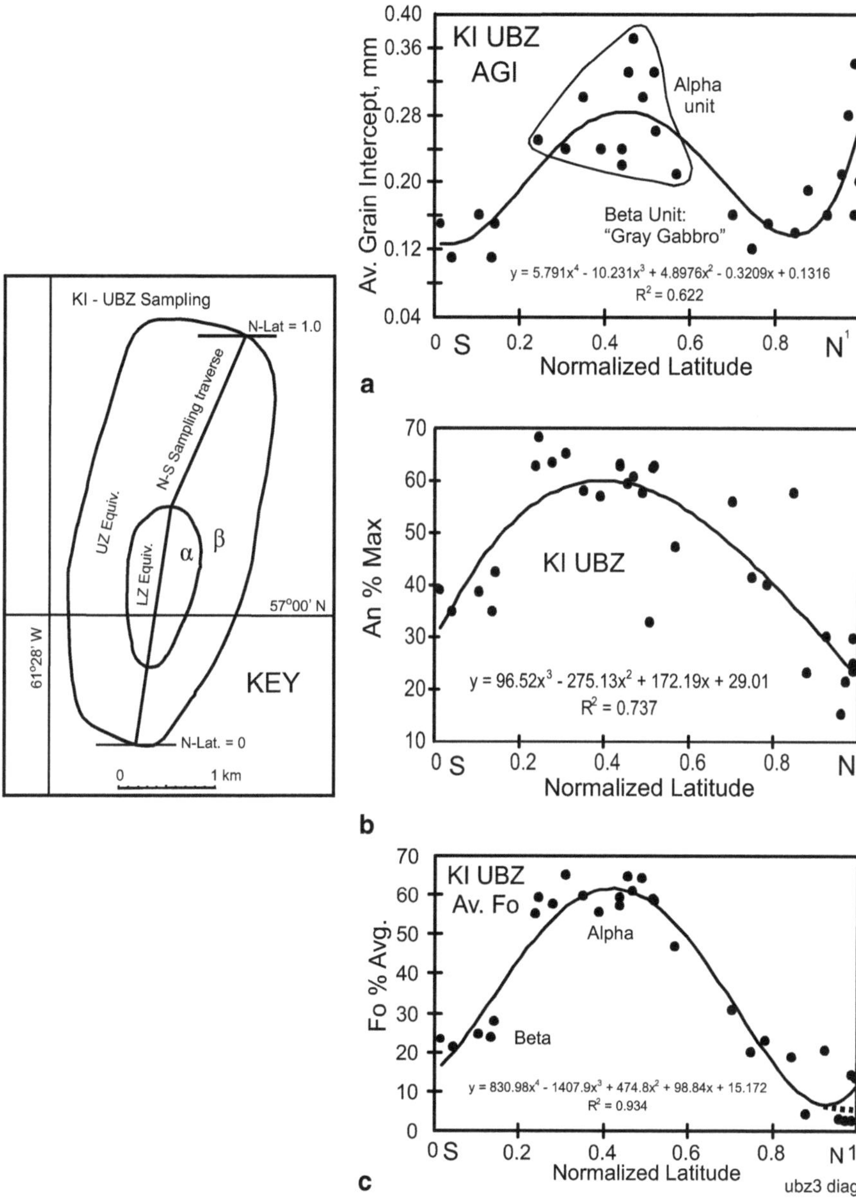

Fig. 13.24 North-South variation of **a** grain size as average grain intercept, **b** An_{max}, and **c** average Fo% in the *Upper* Border Zone. The X-axis is normalized latitude along the N-S sampling traverse shown in the Key map

plot very near the plagioclase—olivine cotectic and show the dominance of LZ variation. There is a weak tendency to group slightly to the olivine-rich side of the cotectic, which should be the case for rocks with large amounts of trapped liquid that is more mafic than the original cumulate.

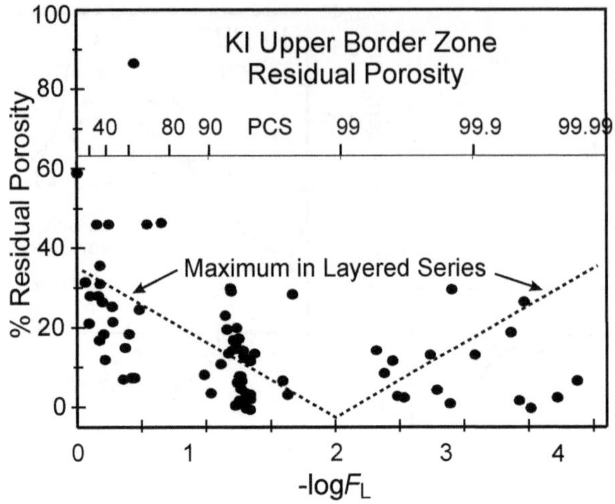

Fig. 13.25 Residual porosity (Morse 2012) in the UBZ as determined from the An range; data from Allison (1984). The PCS scale for the *Upper Border Zone* is determined from the maximum An of UBZ plagioclase scaled to the mean An of *Upper* Zone plagioclase at the arrival of apatite in each unit (Morse and Allison 1986). The high values are particularly significant in the range 0–80 PCS and at the mafic maximum near 95 PCS

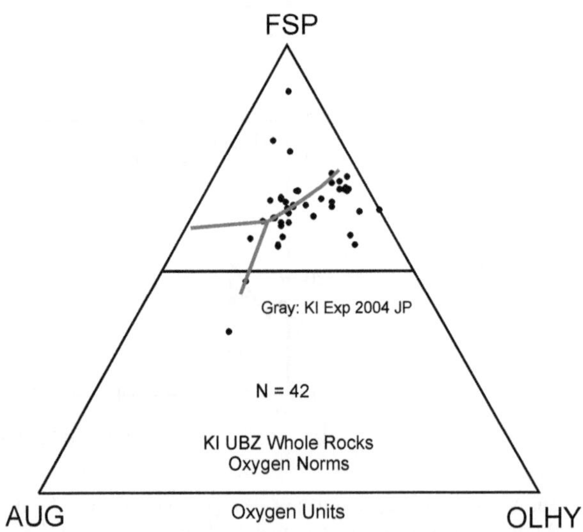

Fig. 13.26 Ternary plot of 42 UBZ whole rock analyses using the oxygen norm conventions of Morse et al. (2004), The Ol-FSP field boundary is from that experimental study (see Fig 13.23) and the other two boundary slopes are approximations. The UBZ samples cluster rather strongly near the cotectic but are slightly more mafic as expected from their higher residual porosity

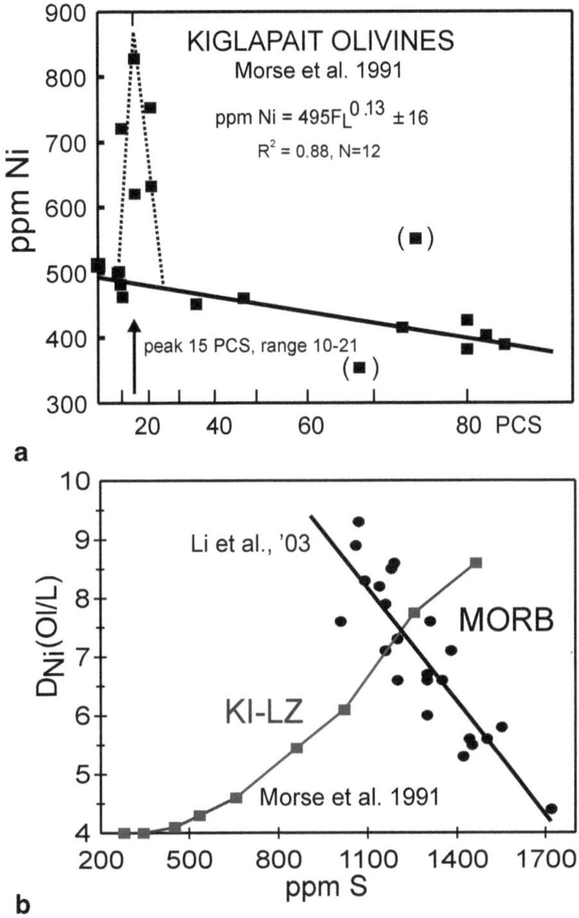

Fig. 13.27 **a** Nickel in Kiglapait olivine shows a distinct spike near 15 PCS interrupting a steady decline to 84 PCS, and suggesting a unique, short-lived, Ni-rich event. Diagram after Morse et al. (1991). **b** Plot of the partition coefficient (*KI-LZ*) of Ni in olivine divided by the calculated sulfur content of the Kiglapait liquid (*grayscale*) based on summation over the sulfur modes shown in Fig. 13.12. This result is compared with the contrasting partitioning in *MORB* from Li et al. (2003). If the *Ni* peak in (**a**) actually represents a Ni-sulfide saturation event now hidden at depth, there should have been a corresponding peak in sulfur, now lost to view. Hence the observed sulfide trend strictly records only the re-saturation in sulfur, as long speculated

Trace Elements

Nickel in Olivine: The Sulfur Problem

A persistent feature of interest in the characterization of magmas and the origin of ores is the abundance of nickel in olivine. As shown in Fig. 13.27a, the Kiglapait data show a smooth depletion trend except for a distinct spike at 10–25 PCS. This is interpreted as a pulse of fresh magma, rich enough in Ni to cause a 400-ppm spike but small enough in volume to effectively disappear by 25 PCS. The Ni spike offers some indirect evidence that there might have been an early deposit of Ni-rich sulfides from a nickel-rich magma, as at Voisey's Bay in troctolite some 60 km to the south (Lightfoot et al. 2011). If so, that would explain the long fractionation to a re-saturation of sulfides at 91 PCS, and also suggest that any ore deposit might lie >8 km beneath the present erosion surface.

The partition coefficient of Ni in olivine versus liquid was found to vary from $D_{Ni}^{OL/L}=4$ to 10 in the Kiglapait Lower Zone (Morse et al. 1991). These low values were thought to be explained by the low values of oxygen fugacity in the Kiglapait magma. An alternative explanation was suggested by the study of Li et al. (2003), who found that Ni could also be partitioned to sulfur in the melt, thereby lowering the activity of Ni available for olivine. These authors showed a strong negative correlation between the value of *D* and the sulfur content of the melt in MORB compositions, confirming the effect of S in generating low values of *D* (Fig. 13.27b).

However, when the observed values of *D* in the Kiglapait Lower Zone are plotted against the strongly rising sulfur content of the summation liquid, the correlation between *D* and S is strongly *positive*, Fig. 13.27b. Because of the late arrival of sulfides (Fig. 13.16), a strong enrichment of the liquid in S with PCS is clearly mandated, so there would seem to be little doubt that it must increase with PCS. Still, this account depends on the assumption that the only sulfur in the magma was what we see in the eventual sulfides. Perhaps what the low initial values of *D* are telling us is that there really was much more S in the magma, consistent with the Ni spike, and that indeed there was an early saturation in metal sulfide. The sulfur we see records only the re-saturation after an early precipitation of sulfides now lost to view.

Intensive Parameters

Oxygen Fugacity

Cumulus Ti-magnetite crystallized as ulvøspinel and was accompanied by ubiquitous but minor ilmenite. The Fe-Ti oxide mineral geothermometer—geobarometer yields a nominal history of oxygen fugacity beginning at the wüstite-magnetite (WM) buffer, thence moving on a shallow slope to a point about halfway between the WM and FMQ buffers at 86 PCS, the errorless arrival of Mt, then in the UZ trend between the buffers to the end of crystallization (Fig. 13.28, modified for temperature from Morse, 1980). After 86 PCS the probable path proceeds along the metastable extension of the Lower Zone trend to 93.5 PCS at the Main Ore Band, and then (dashed line in the figure) drops down onto the UZ trend.

A similar treatment for the Skaergaard intrusion was given by Morse et al. (1980) to illustrate the effect of oxidation and silication on the olivine hiatus in that intrusion. In contrast, the Kiglapait intrusion has no such hiatus and therefore has a lower

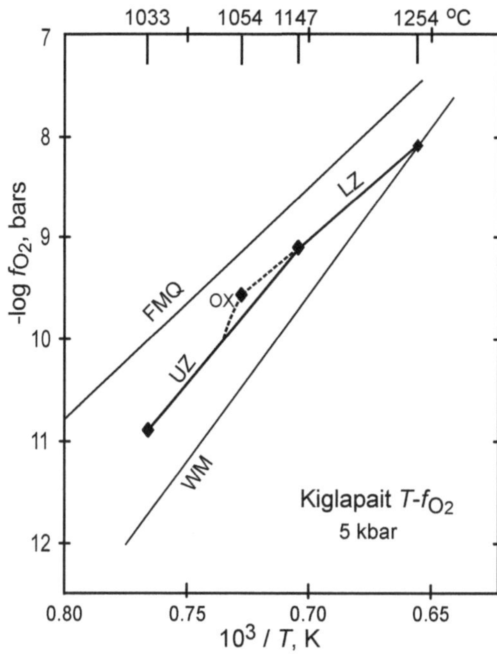

Fig. 13.28 Oxygen fugacity of the Kiglapait magma, modified from Fig. 12.10 of Morse (1980) using updated temperatures from experiments. The magma arrived at the *WM* buffer and then became slightly more oxidized within the *Lower* Zone as the oxide mineral components accreted in the liquid. The low-f_{O2} condition at the Main Ore Band where abundant *T-X* conditions are available for study is marked "*OX*" in the figure and it lies on the extension of the *LZ* path, so it is considered an appropriate fixed point. Temperatures are from this study

oxygen fugacity and silica activity. These two mass-independent parameters were combined into an "oxygen-silica activity ratio" by Morse (1980), who showed that this ratio could facilitate comparisons among magma types and occurrences.

Silica Activity

This intensive parameter, previously mapped by Morse (1980) and by Morse and Ross (2004) has been re-evaluated for the Lower Zone using the graphical method of olivine composition contours of log (a_{SiO2}) plotted against the Kelvin temperature (Fig. 1 of Morse 1979a). The protocol for 5 kbar was used here, with quartz as the base for calculation. The result is shown here as Fig. 13.29. This figure has been used to address the problem of plagioclase fractionation in the Kiglapait Lower Zone (Morse 2014), where it is discussed in detail. The main point here is to show that the silica activity in the Kiglapait magma is very low compared to more tholeiitic magmas such as Skaergaard. The latter has a distinct olivine hiatus owing

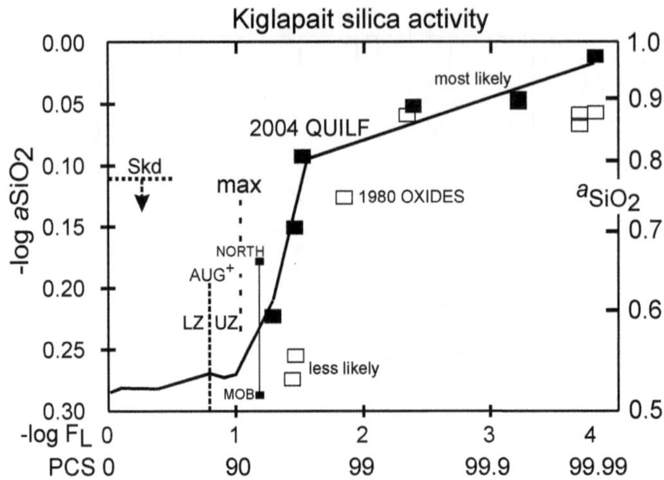

Fig. 13.29 Activity of silica in the Kiglapait liquids calculated from coexisting pairs of augite and olivine at high temperature. *Black rectangles* from the *QUILF* equilibrium of Andersen et al. (1993), calculated by D. H. Lindsley Pers Comm. 2004. *LZ-UZ* region calculated by Morse (2014). The vertical *dashed line* is located at the maximum in modal abundance of augite Morse (1979b). Open symbols represent augites with higher Ca, considered to have gained most Ca from their initial condition and hence less likely to give relevant results

to its higher silica activity (Morse et al. 1980) and the Kiglapait instead only has elevated Fo contents in olivines when accompanied by abundant Fe-Ti oxide minerals (Fig. 13.20). In this case the locally low activity of ferrous iron causes the higher Mg contents of the olivines over the range 90–95 PCS.

The Liquid Path in the Pyroxene Quadrilateral

A liquid path in another dimension is that of the pyroxene quadrilateral, Fig. 13.30. Here the augite series has been smoothed by calculation with the QUILF equilibrium (Morse 1996), and the Mg ratios and diopside content of the liquids also calculated from that equilibrium, as detailed in Morse and Ross (2004). The interesting feature of the liquid path is that it runs to low values of calcium until the Mt+ event, after which the crystallization of Fe-Ti oxide minerals drives the liquid up in Ca to the end point at zero Mg and at a binary eutectic near ferrohedenbergite. This end-member pyroxene is stable at the pressures of the Kiglapait intrusion, in contrast to the case of the Skaergaard intrusion, where the low-pressure stable phase at or near the liquidus is ferrowollastonite (e.g., Lindsley et al. 1969). In their Fig. 13.6, these authors experimentally show the pressure dependence of the green Wo phase at high temperatures, with mixtures of Wo and Hd at somewhat lower temperatures, and Hd only at high pressures and liquidus temperatures. The Kiglapait pressure at

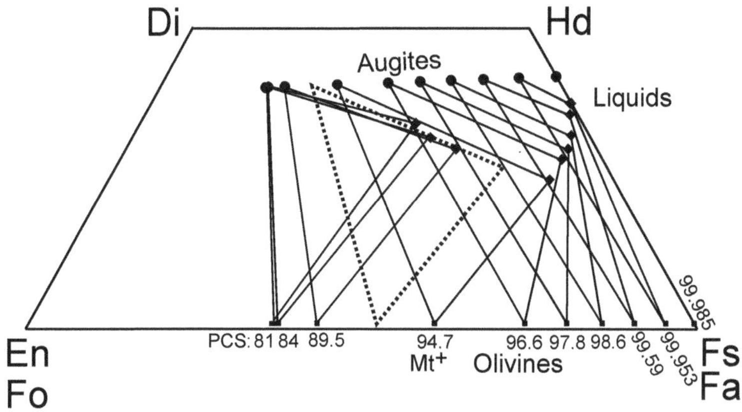

Fig. 13.30 Calculated liquid compositions in equilibrium with augite and olivine. The stratigraphic "PCS" values are shown along the base of the quadrilateral for reference to the liquid compositions, which are taken from Morse (1981). However, the most Mg-rich *triangle* at 81 PCS represents an experimental determination of the liquid composition at the augite saturation point at 5 kb in graphite (Morse et al. 2004). Figure from Morse and Ross (2004)

2.5–3 kbar and ~1000°C lies clearly in the Hd field. The nearly or quite silica—saturated Skaergaard magma at temperatures ≤980°C with ferrowollastonite is characteristic of pressures ≤1 kbar, in contrast to the conclusions of Larsen and Tegner (2006) who suggest an increasing pressure with time for that intrusion, based on fluid inclusions and silicic granophyres near the base of the intrusion.

Stable Isotopes

Oxygen

The Kiglapait layered intrusion was the first major intrusion found to have all whole rock and calculated liquid $\delta^{18}O$ values close to a normal uncontaminated gabbroic value of 6.0 (Kalamarides (1984). The $\delta^{18}O$ values of average whole rocks vary smoothly from 6.0 at the base of the Lower Zone to 6.3 at the top of the Upper Zone. Calculated liquid $\delta^{18}O$ values lie practically superimposed on the whole rocks trend. The effect of incoming cumulus phases such as magnetite and augite on the $\delta^{18}O$ of the liquid and rocks during fractional crystallization is negligible. A further study by Kalamarides (1986) describes the fractionation between liquid and the temperature-corrected mineral phases, obtainable owing to the closed-system behavior of the intrusion.

Sulfur

Sulfide minerals, mainly pyrrhotite crystallized from immiscible liquid globules, were studied in detail by Shirey (1975). They reach their modal maximum at 92 PCS (Fig. 13.16). Samples from 93 to 99.985 PCS were studied by ion probe for their isotopic ratio by M. Chaussidon (personal communication 1989). The isotopic composition of these grains and globules, $\delta^{34}S$, averages -1.0 ± 1.1 ‰. (1-SD) with a range from 1.0 to −3.6 ‰ on 19 measurements. Seventy-four percent of the data points fall in the range 0 to −2‰. There is no tendency to vary up-stratigraphy. The mean isotopic composition is very close to the meteoritic (bulk Earth) standard (0.0 ‰) and implies a largely mantle source for the sulfur.

Radiogenic Isotopes

Strontium

A preliminary study of Rb and Sr isotopic data by Simmons and Lambert reported by Morse (1982) showed that the initial Sr ratio was constant at Sr 0.7040 over the entire Lower Zone but then rose to 0.7054 at 96 PCS and then linearly on $-\log F_L$ to 0.7066 at the end of crystallization. The interpretation was that the radiogenic strontium-87 was endemic to the magma and retained in the alkali feldspar components of the magma and only precipitated in the actual feldspars after 90 PCS when the plagioclase composition began to fall below An_{50}. Although this is an outrageous hypothesis, subsequent investigations (Weis et al. 2004) have shown that isotopic disequilibrium among the coexisting mineral phases is ubiquitous in the Kiglapait and other layered intrusions and we have no hypothesis outrageous enough to explain this effect. The current tendency is to suppose that the late-stage Kiglapait magma might have become contaminated with the residues of the nearby Newark Island layered intrusion (Wiebe 1988). The entire isotopic investigation is now being repeated with new instruments to be determined at higher precision and more petrographic detail.

The study of DePaolo (1985) found approximately the same variation in Sr initial ratios, and concluded that the Lower Zone magma was pristine, whereas the Upper Zone magma was progressively affected by assimilation of country rocks. Examination of country rocks and xenoliths has not yet provided much support for this hypothesis (see Weis et al. 2004).

Geochronology

The age of the Kiglapait intrusion was first determined by DePaolo (1985) at 1305 ± 22 Ma from Sm-Nd chronometry. It was then determined by U-Pb zircon at

1306±2 Ma by T. E. Krogh (personal communication., 17 March 1989) and again by U-Pb baddeleyite at 1308±2 Ma by M. A. Hamilton (personal communication 18 June 1995). We take 1307±2 as the most likely description of the age, with a deep bow to the skills that got us to such a secure result.

Ar–Ar Cooling History

An Argon-Argon dating study by Yu and Morse (1992) showed that the country rocks cooled from 350°C (and > 2.5 kbar) at the time of emplacement, 1307 Ma, for about 9 Ma when the contact aureole and the mass of the intrusion had cooled to 560°C. Then the system cooled at a rate of ~25°C/Ma to 300°C at 1290 Ma, then at ~4°C/Ma to 230°C at 1286 Ma, at which point the intrusion and the country rock had equilibrated in temperature. Then the ensemble cooled at ~3°C/Ma to 185°C at 1256 Ma, and finally at ~1.5°C/Ma to 130°C at 1220 Ma, a cooling record lasting 87 million years (inaugurated by a magmatic history estimated to have lasted 1 Ma).

Experimental

Strontium Partitioning

Feldspar-liquid equilibria in experimental runs at 5 kbar are presented by Morse and Allaz (2013). The experimental partition coefficient $D_{fsp/L}$ for Sr in plagioclase is almost constant near 2.0 until the plagioclase composition reaches ~An_{40}, then it climbs strongly to a value of 6.5 at An_{15} where the host feldspar is mesoperthite. The partition coefficient is linear on the Or content of the feldspar with D=0.1006 (Or)+1.7657, R^2=0.992 where 'Or' is 100Or/(Or+Ab). Similarly, the partition coefficient for Ba is linear with Or as D=0.1273 (Or) −0.0339, R^2=0.9991. An Addendum to the cited paper demonstrates how the partition coefficient for Sr can be used to infer liquid compositions from the feldspar compositions (Morse and Allaz 2014) Similarly, the partition coefficient for Ba is linear with Or as D = 0.1273 (Or) -0.0339, R^2 = 0.9991.

Thermometry

The cotectic path for the Kiglapait Lower Zone was studied at 5 kbar by Morse et al. (2004) and that work contained useful thermal information for these compositions. It showed a concave-up curve of temperature against An content from 1253°C at 0 PCS to 1200°C at 84 PCS, from steep (~6°C/An%) at first to flat at the LZ-UZ boundary. The cotectic-up curve is a feature expected from the Wyllie (1963) prin-

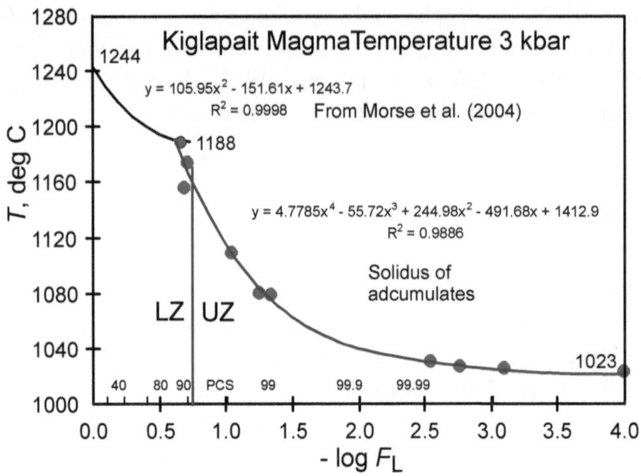

Fig. 13.31 Experimental temperatures of Kiglapait compositions referred to 3 kbar. The determinations were made in graphite at 5 kbar and temperatures were adjusted for the *Lower* Zone at −6.7°C/kbar, and for the *Upper* Zone at −5°C/kbar. The *Lower* Zone liquidus trend is taken from Morse et al. (2004), transformed to the F_L basis as shown by the third-order polynomial in the figure. The *UZ* data are based on the solidus of adcumulates, taken as the magma temperature at the closure of reactive exchange (Morse et al. 1980)

ciple that liquidus curves flatten at the arrival of each new phase (Morse 2011). This curve for the Lower Zone is brought up to date for a pressure of 3 kbar in Fig. 13.31.

The experimental study is now extended to the Upper Zone by means of finding the solidus of Kiglapait adcumulates. These adcumulates are chosen from the master list of the An range and residual porosity discussed in Morse (2012). The theory is that isocompositional growth of feldspar down to a residual porosity of ~0 must reflect isothermal crystallization, so therefore the beginning of melting of an adcumulate recovers the liquidus = solidus of the cumulate (Morse et al. 1980). For this document the temperatures were adjusted to 3 kbar using the adiabatic gradient of −6.7°C/kbar from McIntosh (2009) for the Lower Zone and −5°C/kbar newly determined for the Upper Zone.

The results are shown here in Fig. 13.31. The UZ curve is concave-up and flattened toward the end of crystallization on this semi-log scale, as shown in the figure. There is a sharp discontinuity at the LZ-UZ boundary, where the arrival of augite at the liquidus makes a jump in the fractional latent heat (Morse 2011). One might expect that this new evolution of latent heat would slow down the cooling of the magma, but instead the data show that a sharp drop in temperature is required to crystallize the over-abundant augite. The curve begins to flatten with the arrival and overproduction of Fe-Ti oxide minerals culminating in the Main Ore Band at 93.5 PCS, and the flattening continues to the end of crystallization.

The entire range of crystallization shown in Fig. 13.31 is plotted against plagioclase composition in Fig. 13.32, with the single regression as shown. The overall slope is 4.85°C/An%, compared with 4.8°C/An% earlier estimated by

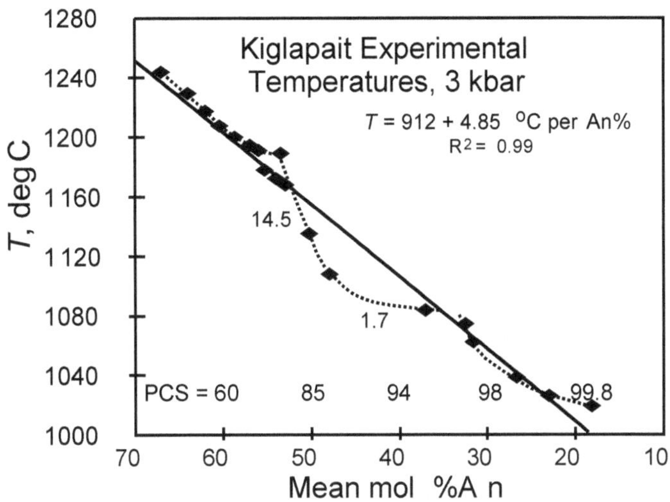

Fig. 13.32 Temperatures from Fig. 13.31 plotted against plagioclase compositions. The global dataset yields a slope of 4.85°C/An% compared with 4.8°C/An% estimated from Peterson (1999) by Morse (2008). However, there is a local excursion in the trend at ~80–93 PCS where the mafic maximum occurs and the latent heat release falls off from the peak associated with that overproduction of augite and Ti-magnetite (Morse 2011). Here as shown by the dotted curve, the slope steepens to 14.5 and then flattens to 1.7°C/An%. The array is characterized by three concave-up curves

Morse (2008a). However, there are three separate, concave-up shelves that can be interpreted as follows. The first is the Lower Zone shelf already discussed. There follows a steep drop at ~14.5°C/An% beginning with the arrival of augite at ~84 PCS and continuing with flattening to 1.7°C/An% at 95 PCS where the overshoot of augite and oxide minerals has ended. Then there is a last concave-up shelf to the end of crystallization. Each shelf represents a large drop in An per degree of temperature, and each steepening represents a smaller change in An per degree. The greatest change in An occurs at ~89–95 PCS at the peak of augite production (Fig. 13.13a), which can be visualized in the system Di-An-Ab as the range in which the liquid finds the cotectic and runs down toward Ab, directly away from the An—Di sideline.

The Magma Source Region

A study at high pressures in graphite by McIntosh (2009) showed that at 13 kbar the model Kiglapait bulk composition became saturated with spinel, Al-cpx, Al-opx, and garnet. The relevant phase space and compositions are shown here in Fig. 13.33 (inset). Of special interest in this ACF diagram is that augite gains alumina toward the sideline garnet composition and hence essentially parallel to the CA sideline, not toward CaTs or plagioclase. In the main diagram the phase relations show that the bulk composition of the intrusion ("BC3") lies outside the mantle source

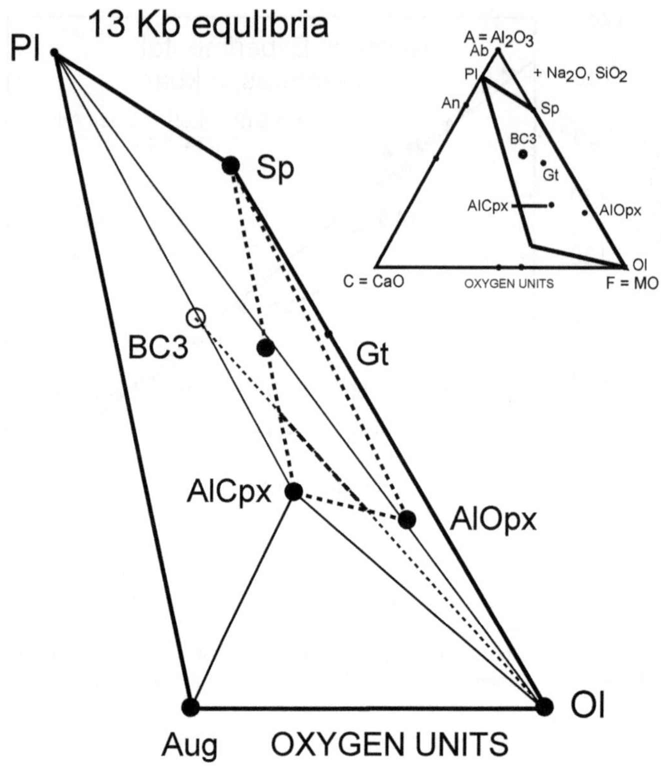

Fig. 13.33 ACF diagram from McIntosh (2009) showing the compositions of phases at 13 kbar in the starting composition *BC3* of the Kiglapait magma. The inset shows the full ternary plot from which the quadrilateral is taken. The plot shows that the source rock (harzburgite) is dominated by two pyroxenes + spinel and garnet which is not in the plane of the section. The bulk composition lies outside the source and hence must have lost *OL* on ascent. Addition of Fo-rich olivine to the assemblage would then approximate more closely to the true source rock, bringing the source composition into the mafic *triangle* as shown by the heavy *dashed line*

assemblage. For this composition to have originated from a harzburgitic source (Morse 2006) it must have dropped off a large component of olivine in transit, as shown by the dotted and heavy-dashed line within the triangle AlCpx—Gt—Sp—AlOpx. (The garnet is not in the AlCpx—Sp plane, so it is not a redundant phase.)

The theme of the experimental program is to extend the bulk composition backwards toward olivine to find the appropriate mantle temperature and pressure consistent with a plausible mantle source composition. This is done by adding olivine as Fo so as to achieve a mantle olivine composition relevant to a depleted mantle source, at least Fo_{90}. The bulk compositions achieved to date reach only to Fo_{85}, so are not yet at the target. Nevertheless, one fortuitous result brings us a long way toward the final goal, as described in another context by Morse (2013).

Several euhedral olivine crystals with melt inclusions occur in bulk composition BC 35 in experiment KIPM-2. One of these contains a central melt inclusion

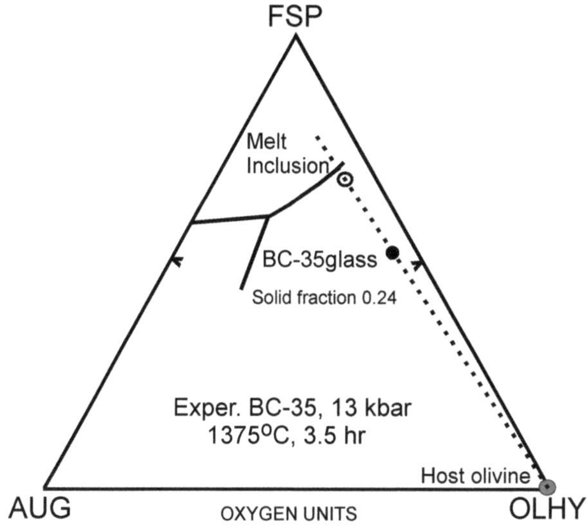

Fig. 13.34 Ternary plot of the experimental result with composition *BC-35*. The experiment was designed as part of a project to estimate the source temperature of the Kiglapait magma before its loss of olivine to the site of emplacement. Fortuitously the homogeneously nucleated olivine trapped a sample of the boundary layer resulting in the melt inclusion, which became further evolved by growth of the host during the experiment. As a result the melt inclusion became enriched in feldspar almost to the cotectic condition L=OL+FSP. The lever rule gives the solid fraction as 0.24, meaning that if 24 % of the original magma were to have crystallized olivine of the host composition, the residual liquid would have reached the position shown

that has been analyzed by electron probe, along with its host. The result is shown in the ternary plot of Fig. 13.34. The melt inclusion is evolved almost to saturation with plagioclase, with a solid fraction 0.24, so 24 % crystallized from the bulk composition. The criterion of plagioclase saturation would be reached with 26.3 % crystallized. This preliminary result is qualitatively consistent with the geometrical analysis in Fig. 13.33.

Not yet having quite the mantle composition wanted, these results can nevertheless be used to illustrate a plausible scenario for the origin of the Kiglapait magma; Fig. 13.35. The experimental data of McIntosh (2009) show that the reactions of plagioclase to spinel and then garnet occur very close to each other, as illustrated by the heavy dashed field boundaries terminating at a liquidus. The experiment with elevated Fo content is shown as a star, exactly on the CMAS lherzolite solidus of Presnall et al. (2002). It is isobaric with the garnet saturation at 13 kbar. If we add more Fo the temperature will rise, schematically as shown, with a small pressure increase to a hypothetical harzburgite solidus. From there, with an eventually enormous melt fraction, a liquid would rise to emplacement at 3 kbar, crystallizing copious amounts (i.e. more than perhaps 30 %) of olivine and repeatedly doing so for the few thousand years or so required to fill the magma chamber.

All this after already yielding the nearby Hettasch, Jonathon, and Newark Island Intrusions, and several others mentioned in Fig. 13.3. The troctolitic suite associated

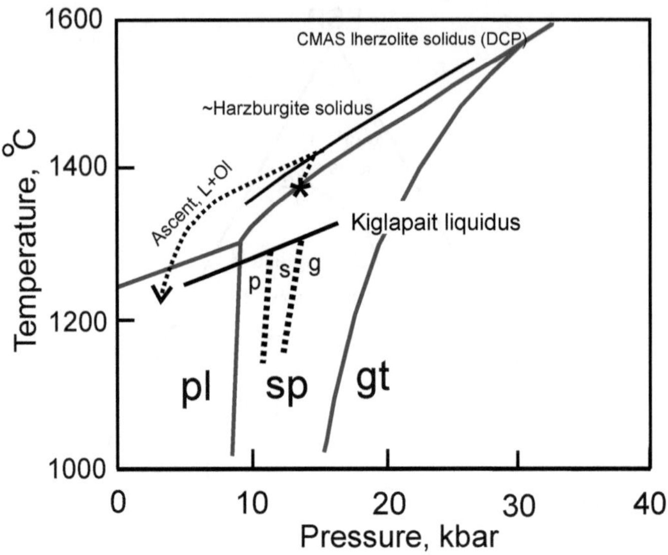

Fig. 13.35 The *lower* solid *black line* represents the experimental liquidus data (McIntosh 2009) from 5 to 13 kbar for the Kiglapait bulk composition. The fields of plagioclase (*pl*), spinel (*sp*) and garnet (*gt*) are shown for the generalized system CMAS (Presnall et al. 2002), along with the CMAS solidus. The corresponding fields for the Kiglapait composition are shown as bold broken lines, being precise only for the sp+gt+*L* equilibrium. Results from the recent experiment with added Fo (*at the asterisk*) are extrapolated to a plausible solidus at higher Mg# for a depleted harzburgitic mantle source of the Kiglapait magma (Morse 2006), followed by a curve of ascending magma shedding olivine enroute to the site of emplacement at ~3 kbar

with the Nain Plutonic Series represents a truly impressive amount of troctolitic magma generated at considerable depth and high temperature over some considerable time span, perhaps millions of years, from a seemingly inexhaustible source. Instead, it was more likely a renewable source of constant composition traveling over a durable hot spot. It should not be neglected to mention the very hot melatroctolite magma that quenched within the Hettasch magma to form the snowflake troctolite zone there (Berg 1980). That would precisely describe an early pulse from the source represented in Fig. 13.35 here, ascending too fast to shed all its olivine to a cotectic condition with plagioclase. The magma was hot, and at Hettasch we have a sample of it.

Cooling History

The physics and thermodynamics of a deep body of cooling magma with crystals at the roof and floor are straightforward to specify. What occurs in between the upper and lower boundary layers is a matter of hot debate, called hydrodynamics by some and confusion by others. But some limits on these histories can be described. The cooling of intrusive magma, like the freezing of a pond, is a Stefan problem (Turcotte and Schubert 2002) in which the loss of heat is almost immediately encumbered by the presence of boundary layers of crystal mush (or ice on the pond) that are themselves sources of latent heat that constitutes something like 80 % of the total loss of enthalpy (e.g., Morse 2011). These impediments constitute for us the fixed points in an otherwise unruly universe, so it is helpful to talk about them and their constraints.

At the start we need to consider the filling of the magma chamber, which then begins to heat the floor and roof dramatically. In the present case we take the initial country rock as at 350°C (Berg and Docka 1983) and the magma at 3 kbar as ~1,240°C, giving in the classic thermal approximation an instantaneous contact temperature of 1590/2=795°C. From the estimates of the maximum temperature in the metamorphic aureole we have 950°C (Berg and Docka 1983), so in due course the magma has heated up its floor and roof by at least 155°C. We may take this condition as a steady-state approximation. The initial mean roof gradient is then 1240/9.6=130°C/km.

Roof Cooling

Fig. 13.36 is a depth-temperature plot showing the thermal constraints at the roof cooling surface of a deep crystallizing magma. The durable presence of an Upper Border Zone is assumed. At the thermal buffer the temperature is constrained by the ambient phase assemblage (here taken as Plagioclase+Olivine+Liquid). The roof cumulate overlying the buffer interface must also contain this assemblage, but also, deeper inside it, the fractionated products of trapped liquid, which are chilled and evolved. Here the thermal gradient (heavy line) runs to lower temperatures upward until it encounters the solidification front above which no melt occurs.

Below the thermal buffer lies the magmatic part of the thermal boundary layer, of arbitrary thickness including zero. When zero, the roof has foundered and left cold cumulates in contact with hot magma, which is then quenched (Fig. 11e here and with composition shown in Fig. 20 of Morse et al. 2004). When non-zero, the thickness of the layer is unspecified but considered to be in the range of centimeters or perhaps meters. Over this thickness interval the thermal difference is estimated to be in the range 6–18°C, depending on the magma depth as assumed to be varying from 3 to >8 km. This difference is given by the known slope of the liquidus for Pl+Ol as indicated.

The dotted curve attached to the liquidus at the buffer is that of a two-phase plume of dispersed crystals plus melt that has nucleated in the nearby thermal region of the buffer (with or without supercooling). As the plume descends on a steep adiabat, it becomes increasingly supercooled relative to the liquidus, and crystals may grow en route to the bottom. The other steep dotted curve is that of the crystal-free return adiabat from the floor.

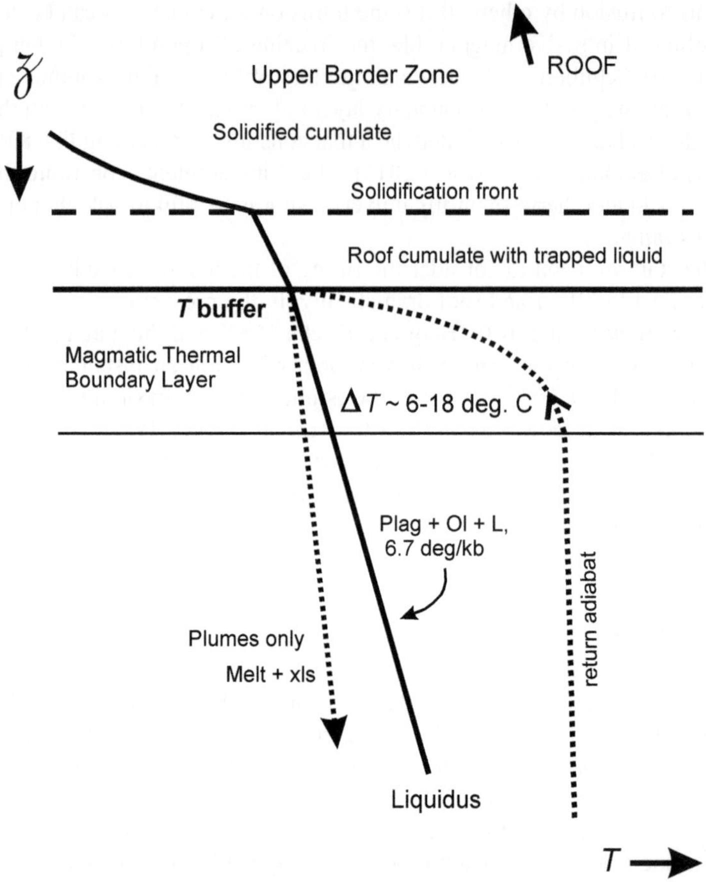

Fig. 13.36 Depth versus temperature diagram of the roof cooling of a large troctolitic intrusion with an *Upper Border Zone* (*roof cumulate*). The principal heat loss is through the roof and the lateral dimension is taken as effectively infinite. The work done is to heat the roof as well as cool the magma, and this combination is a Stefan effect (see text) owing to the diminishing thermal contrast between magma and roof. The situation considered here is one of temporal equilibrium at steady state. Crystals may nucleate in the magmatic thermal boundary layer and because they are all denser than in their liquid state, they will tend to sink. The downward path is a "wet adiabat," a liquid with crystals, and because that path is much steeper than the *liquidus PT* slope, the sinking packet is increasingly supercooled. The return liquid adiabat from the floor is a decompression liquid adiabat cooling at about 0.3°C/km and returning from a floor liquidus differential of perhaps 6–18°C, depending on magma depth, and hence cooling through (and defining) the *upper* magmatic thermal boundary layer. The liquidus slope is from experiments on the Kiglapait bulk composition (McIntosh 2009)

With a mean roof gradient ~125°C/km, a liquid adiabat making a short-circuit for advective heat transfer to the roof boundary layer, and a hot thermal buffer at the floor (see below) it is easy for the magma to cool through the roof. The main point of the figure is that crystal-poor plumes are always significantly cooler than the liquidus.

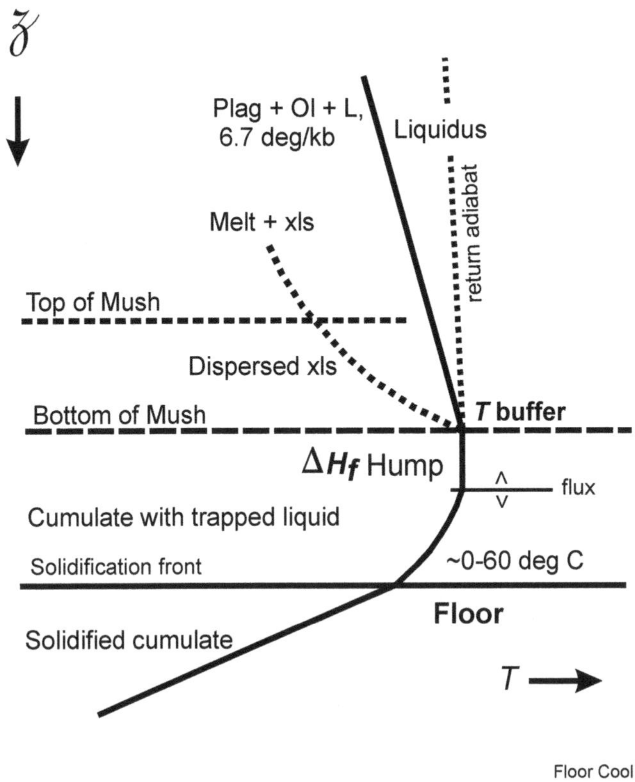

Fig. 13.37 Thermal constraints near the floor of a steady-state igneous cumulate, using the same coordinate system as in the previous figure. It is assumed that a dilute crystal-bearing packet arrives from *above*; it is metastably supercooled with respect to the *liquidus*. Within a boundary layer of increasing crystal population (a crystal mush) the temperature reaches stability at the liquidus. This crystal population approximates to the crystal-bridging criterion of ~50 %, leading to a porosity of ~50 %. This assemblage locks the ambient temperature to the crystal-liquid maximum, which is nominally the highest temperature in the intrusion assuming perfect stirring. For adcumulates, the solidification front lies at the base of the latent heat hump. For extreme orthocumulates, the temperature range may be as great as 60°C (Morse 2013)

Floor Cooling

The diagram of Fig. 13.37 shows, with a few quantities but schematically, the thermal relations near the floor of an igneous cumulate. It is assumed that 2-phase plumes of dispersed crystals plus melt arrive near the floor from the roof (where nucleation occurred) on a near-liquid adiabat, the crystal population being of order 10–100 ppm or perhaps percent. As they descend, the crystals may grow because the adiabat becomes progressively cooler than the liquidus. Lacking mutual contact, the crystals are not at saturation equilibrium with the melt. A mushy zone develops near the floor, and within that zone the crystals become closer packed with depth (dotted curve) until they reach thermodynamic equilibrium with melt, hence lie at the liquidus temperature. At the bottom of the mush it is assumed that the residual porosity is ~50 %, consistent with crystal bridging. This interface has the maximum temperature of the system, the floor liquidus, a thermal buffer. Assuming that permeability allows some interchange of liquid within the top of the edifice, the thermal buffer extends some distance downward, forming the "ΔH_f Hump" (Morse 1986) more elegantly called the latent heat boundary layer (Martin et al. 1987). Below this limit, the trapped liquid undergoes closed-system crystallization until it solidifies completely. Below this, the temperature profile goes to the perturbed geotherm.

If the cumulate crystallizes with perfect exchange to the overlying magma it forms an adcumulate within the latent heat boundary layer. If only partial or no exchange occurs, the result is an ortho- or meso- cumulate that will crystallize over a range of temperature. This range may most likely be less than 50–60°C (Morse 2013).

The magmatic system is closed at the bottom of the latent heat hump, and the rest of the solidification zone is assigned to the floor region and has nothing further to do with the magma. The return adiabat to the surface from the thermal maximum is shown dotted.

With a thermal maximum at the base of a mushy zone there is no cooling of the magma through the floor. All the rocks below the base of the thermal buffer belong to the floor regime. The magma cannot cool through the floor because that place is on the wrong side of the latent heat hump. Only the mushy zone can cool through the floor.

Between Roof and Floor

The invariant condition OL+PL+L moves upward in *T* with *P* at a slope of 6.7°C/kbar, as determined in our laboratories, and its olivine to plagioclase ratio is independent of *P* at least from 5 to 15 kbar (McIntosh 2009). This path is that of the liquidus.

The *P-T* liquidus is not generally populated with crystals, however, except in the extreme case of thin sills. Instead, crystals that nucleate and grow at the base of the UBZ (as they are found from the petrographic record to have done) may sink in two-phase plumes with dispersed crystals nearly along a liquid-only adiabat (e.g., the return adiabat of Figs. 13.36 and 13.37), which has a slope of a fraction of a degree per kbar, nearly vertical. This "damp" adiabat thus becomes supercooled relative to the liquidus, accelerating the nucleation and growth of crystals, significantly plagioclase, which goes along for the ride with olivine and cool liquid. If the accumulation rate is ~2 cm/yr then the mean population in the overlying 1-cm^2 column is of order tens of ppm, very dilute.

The entire liquidus then defines the coolest place in the magma body to be at the roof contact with the UBZ, and the hottest place to be in the boundary region of the magma at the floor, where the variance is likewise zero. If the roof is mushy for a small distance and porous enough for permeability, the upper thermal buffer may extend some small distance into the UBZ. If the floor is mushy enough for permeability, the lower, high-temperature thermal buffer may extend some small distance into the floor.

Many schemes of crystallization within the magma body have been ventured including wholesale suspension and mushy zones hundreds of meters thick. Evidence from the An range of plagioclase shows that such configurations are highly unlikely in the Kiglapait intrusion (Morse 2012).

Where is the Hottest Magma?

This issue has been carefully skirted in the discussion above, for the answer is probably NOT the thermal maximum at the floor boundary. It more likely resides in the deep interior of the magma body, of which we have, surprisingly, some evidence. In both Kiglapait (Morse 1979c) and Skaergaard (Holness et al. 2007; Tegner et al. 2009) intrusions there are modal overshoots where augite and oxide minerals become saturated in the magma but then increase in mode well beyond their equilibrium concentrations (Fig. 13.16) before returning to a stable amount (which defines the saturation condition). The components of these mineral phases have evidently been lying in wait somewhere within the magma body, and yet it is unlikely that the components of such dense phases existed as crystals in this hidden reservoir without sinking earlier. If not then crystallized, yet oversaturated, these components can only have been in solution at a temperature higher than their saturation temperature. These sites must, then, have been the hottest places in the magma, but in the interior, not at the boundaries.

These hot regions were identified as the cores of convection cells (Morse 1979c) and by implication are also identified as the hidden internal reservoirs of large magma chambers that may have influenced the fractionation path of major binary solutions during the crystallization of the intrusions (Morse 2008b).

Acknowledgements This work has involved the help and skills of John Brady, Deb McIntosh, Mike Rhodes, Mike Jercinovic, Julien Allaz, Mike Vollinger, Sean Regan, and many others, for which I am most grateful. Reviews of an earlier version by Lew Ashwal and Ed Mathez were most helpful in steering the work along the path of progress. I am grateful for the good work of the Editor, Bernard Charlier, in trimming these sails. This chapter is based on research supported by NSF under Award No. EAR 0948095.

References

Allison JP (1984) Petrography of the Upper Border Zone of the Kiglapait intrusion, Labrador. MS thesis, University of Massachusetts, 220 pp

Andersen DJ, Lindsley DH, Davidson PM (1993) QUILF: a Pascal program to assess equilibria among Fe–Mg–Mn–Ti oxides, pyroxenes, olivines, and quartz. Comput Geosci 9:1333–1350

Ball SS (1984) The partridge point area of the Kiglapait layered intrusion. MS thesis, UMASS, Amherst MA 98 p

Berg JH (1977) Dry granulite mineral assemblages in the contact aureole of the Nain Complex, Labrador. Contrib Mineral Petrol 64:32–52
Berg JH (1980) Snowflake troctolite in the Hettasch intrusion: evidence for magma mixing and supercooling in a plutonic environment. Contrib Mineral Petrol 72:339–351
Berg JH (1994) Jonathon Intrusion and associated rocks. In: Berg JH et al Anorthositic, granitoid and related rocks of the Nain Plutonic Suite. Internat Geol Correlation Programme IGCP Projects #290 and #315
Berg JH, Docka JA (1983) Geothermometry in the Kiglapait aureole, Labrador. Am J Sci 283:414–434
Blundy JD (1997) Experimental study of a Kiglapait marginal rock and implications for trace element partitioning in layered intrusions. Chem Geol 141:73–92
DePaolo DJ (1985) Isotopic studies of processes in mafic magma chambers: I. The Kiglapait Intrusion, Labrador. J Petrol 26:925–951
Hildreth W, Wilson CJN (2007) Compositional zoning of the Bishop Tuff. J Petrol 48:951–999
Holness MB, Tegner C, Nielsen TFD, Stripp G, Morse SA (2007) A textural record of solidification and cooling in the Skaergaard intrusion, East Greenland. J Petrol 48:2359–2377
Huntington HD (1979) Kiglapait mineralogy I: apatite, biotite, and volatiles. J Petrol 20:625–652
Kalamarides RI (1984) Kiglapait geochemistry VI: oxygen isotopes. Geochim Cosmochim Acta 48:1827–1836
Kalamarides RI (1986) High-temperature oxygen isotope fractionation among the phases of the Kiglapait Intrusion, Labrador, Canada. Chem Geol 58:303–310
Larsen RB, Tegner CT (2006) Pressure conditions for the solidification of the Skaergaard intrusion: eruption of East Greenland flood basalts in less than 300,000 years. Lithos 92:181–197
Li C, Ripley EM, Mathez EA (2003) The effect of S on the partitioning of Ni between olivine and silicate melt in MORB. Chem Geol 201:295–306
Lightfoot PC, Keays RR, Lamswood D-E, Wheeler R (2011) S Saturation history of Nain Plutonic Suite mafic intrusions: origin of the Voisey's Bay Ni–Cu–Co sulfide deposit, Labrador, Canada. Miner Depos 47:23–50
Martin D, Griffiths W, Campbell IH (1987) Compositional and thermal convection in magma chambers. Contrib Mineral Petrol 96:465–475
McIntosh DB (2009) High pressure liquidus studies of the inferred magma composition of the Kiglapait layered intrusion, Labrador, Canada. MS thesis, University of Massachusetts
Morse SA (1969) The Kiglapait layered intrusion, Labrador. Geol Soc Am Mem 112:1–198
Morse SA (1977) Catalog of layered intrusions in the Nain complex. In: Morse SA (ed) The Nain Anorthosite project, Labrador: field report 1976. Univ Mass Dept Geol Geog Contr No 29, pp 47–49
Morse SA (1979a) Reaction constants for En-Fo-Sil equilibria: an adjustment and some applications. Am J Sci 279:1060–1069
Morse SA (1979b) Kiglapait geochemistry I: systematics, sampling, and density. J Petrol 20:555–590
Morse SA (1979c) Kiglapait geochemistry II: petrography. J Petrol 20:591–624
Morse SA (1980) Kiglapait mineralogy II: Fe–Ti oxide minerals and the activities of oxygen and silica. J Petrol 21:685–719
Morse SA (1981) Kiglapait geochemistry IV: the major elements. Geochim Cosmochim Acta 45:461–479
Morse SA (1982) Strontium isotope fractionation in the Kiglapait intrusion. Science 220:193–195
Morse SA (1984) Cation diffusion in plagioclase feldspar. Science 225:504–505
Morse SA (1986) Convection in aid of adcumulus growth. J Petrol 27:1183–1215
Morse SA (1996) Kiglapait mineralogy III: olivine compositions and Rayleigh fractionation models. J Petrol 37:1037–1061
Morse SA (2000) Linear partitioning in binary solutions. Geochim Cosmochim Acta 64:2309–2319
Morse SA (2006) Labrador massif anorthosites: chasing the liquids and their sources. Lithos 89:202–221

Morse (2008) The internal magma reservoir of large intrusions revealed by multiphase Rayleigh fractionation. J Petrol 49:2081–2098
Morse SA (2008a) Toward a thermal model for the Skaergaard liquidus. Am Mineral 93:248–251
Morse SA (2008b) The internal magma reservoir of large intrusions revealed by multiphase Rayleigh fractionation. J Petrol 49:2081–2098
Morse SA (2011) The fractional latent heat of crystallizing magmas. Am Mineral 96:682–689
Morse SA (2012) Plagioclase An range and residual porosity in igneous cumulates of the Kiglapait Intrusion. J Petrol 53:891–918
Morse SA (2013) Solidification of trapped liquid in rocks and crystals. Am Mineral 98:888–896
Morse SA (2014) Plagioclase fractionation in troctolitic magma. J Petrol 55:2403–2418
Morse SA, Allaz J (2013) Experimental partitioning of Sr and Ba in Kiglapait feldspars. Am Mineral 98:2197–2200
Morse SA, Allaz J (2014) Addendum to above paper. Am Mineral 99:557–558
Morse SA, Allison JP (1986) Correlation between roof and floor cumulates of the Kiglapait intrusion, Labrador. Geophys Res Lett 13:1466–1469
Morse SA, Nolan KM (1984) Origin of strongly reversed rims on plagioclase in cumulates. Earth Planet Sci Lett 68:485–498
Morse SA, Ross M (2004) Kiglapait mineralogy IV: the augite series. Am Mineral 89:1380–1395
Morse SA, Lindsley DH, Williams RJ (1980) Concerning intensive parameters in the Skaergaard intrusion. Am J Sci 280-A:159–170
Morse SA, Rhodes JM, Nolan KM (1991) Redox effect on the partitioning of nickel in olivine. Geochim Cosmochim Acta 55:2373–2378
Morse SA, Brady JB, Sporleder BA (2004) Experimental petrology of the Kiglapait intrusion: cotectic trace for the Lower Zone at 5 kb in graphite. J Petrol 45:2225–2259
Myers JS, Voordouw RJ, Tettelaar TA (2008) Proterozoic anorthosite–granite Nain batholith: structure and intrusion processes in an active lithosphere-scale fault zone, northern Labrador. Can J Earth Sci 45(8):909–934
Namur O, Charlier B, Toplis MJ, Higgins MD, Liégeois J-P, Vander Auwera J (2010) Crystallization sequence and magma chamber processes in the ferrobasaltic Sept Iles layered intrusion, Canada. J Petrol 51:1203–1236
Nielsen TFD (2004) The shape and volume of the Skaergaard intrusion: implications for mass balances and bulk composition. J Petrol 45:507–530
Owens BE (1986) Xenoliths and autoliths in the Kiglapait intrusion, Labrador. MS thesis, UMASS, Amherst MA 149 pp
Owens BE, Kremser DT (2010) Åkermanite breakdown to a cuspidine-bearing symplectite in a calc-silicate xenolith, Kiglapait Intrusion, Labrador, Canada. Can Mineral 48:809–819
Peterson AL (1999) Quest for the liquid line of descent of the Upper Zone of the Kiglapait intrusion, Labrador, Canada: an experimental study. MS thesis, University of Massachusetts, 80 pp
Presnall DC, Gundfinnsson GH, Walter MJ (2002) Generation of mid-ocean ridge basalts at pressures from 1 to 7 GPa. Geochim Cosmochim Acta 66:2073–2090
Ryan AB (1990) Preliminary geological map of the Nain Plutonic Suite and surrounding rocks (Nain-Nutak, NTS 14 SW). Newfoundland Department of Mines and Energy, Geological Survey Branch, Map 90–44, scale 1:500,000
Salmonsen LP, Tegner C (2013) Crystallization sequence of the Upper Border Series of the Skaergaard Intrusion: Revised subdivision and implications for chamber-scale magma homogeneity. Contrib Mineral Petrol 165:1155–1171
Schuh ML (1981) Geology of the Avakutakh iron formation, Labrador. MS Thesis, Northern Illinois University, DeKalb, 192 pp
Shirey SB (1975) Sulfides and sulfur content of the Kiglapait layered intrusion, Labrador. MS Thesis, University of Massachusetts, Amherst, 75 pp
Speer JA (1982) Metamorphism of the pelitic rocks of the Snyder Group in the contact aureole of the Kiglapait layered intrusion, Labrador: effects of buffering partial pressure of water. Can J Earth Sci 19:1888–1909

Speer JA, Ribbe PH (1973) The feldspars of the Kiglapait intrusion, Labrador. Am J Sci 273-A:468–478
Stacey FD, Davis PM (2009) Physics of the earth, 4th edn. Cambridge University Press, Cambridge, UK, 532 pp
Stephenson R, Thomas MD (1979) Three-dimensional gravity analysis of the Kiglapait layered intrusion, Labrador. Can J Earth Sci 16:24–37
Tegner C, Thy P, Holness MB, Jakobsen JK, Lesher CE (2009) Differentiation and compaction in the Skaergaard intrusion. J Petrol 50:813–840
Turcotte DL, Schubert G (2002) Geodynamics, 2nd edn Cambridge University Press, Cambridge, UK, p 456
Upton BGJ (2013) Tectono-magmatic evolution of the younger Gardar Southern rift, South Greenland. Geol Surv Den Greenl Bull 29:124 pp
Weis D, Morse SA, Scoates JS (2004) Mineral-scale and regional isotopic heterogeneity within the Kiglapait Intrusion and other mafic intrusions of the ca. 1.3 Ga Nain Plutonic Suite, Labrador. EOS Trans Am Geophys Union 85(17):JA514, Abstr V52A-02
Wheeler EP (1942) Anorthosite and associated rocks about Nain, Labrador. J Geol 50:611–642
White SM, Crisp JA, Spera FJ (2006) Long-term volumetric eruption rates and magma budgets. Geochem Geophys Geosyst 7(3):Q03010
Wiebe RA (1988) Structural and magmatic evolution of a magma chamber: the Newark Island Layered intrusion, Nain, Labrador. J Petrol 29:383–411
Williams GL (ed) (1985) Lexicon of Canadian stratigraphy, VI, Atlantic region. Canadian Society of Petroleum Geologists, Calgary, Canada
Wyllie PJ (1963) Effects of the changes in slope occurring on liquidus and solidus paths in the system diopside-anorthite-albite. Minrol Soc Am Spec Pap 1:204–212
Xue S, Morse SA (1993) Geochemistry of the Nain massif anorthosite, Labrador: magma diversity in five intrusions. Geochim Cosmochim Acta 57:3925–3948
Yu Y, Morse SA (1992) Age and cooling history of the Kiglapait Intrusion from an $^{40}Ar/^{39}Ar$ study. Geochim Cosmochim Acta 56:2471–2485

Chapter 14
The Ilímaussaq Alkaline Complex, South Greenland

Michael A. W. Marks and Gregor Markl

Abstract The Ilímaussaq complex in South Greenland is a well-studied multiphase alkaline to peralkaline intrusion of Mesoproterozoic age. Most of the Ilímaussaq rocks are extremely enriched in alkalis, iron, halogens, high-field-strength elements (HFSE) and other rare elements, forming one of the most differentiated peralkaline rock suites known.

The major factors causing the extreme differentiation trends are low oxygen fugacity and silica activity as well as very low water activity in the melts. These inhibit the early exsolution of aqueous NaCl-bearing fluids and facilitate the enrichment of alkalis and halogens in the melts, thereby increasing the solubility of HFSE. The unusually long crystallization interval of these rocks and the suspected continuous transition from melt to fluid results in extensive (auto)metasomatism and hydrothermal overprint. Primary mineral assemblages are therefore partially resorbed in most rock units and replaced by secondary minerals to various extents.

The Ilímaussaq complex is a well-known example of magmatic layering in peralkaline plutonic rocks. Recent investigations of mineral chemical trends in the layered rocks permit better understanding of their formation. The mechanism of crystal mats formation in the cooling magma causing crowding effects during settling of the layering-forming minerals is believed to govern the formation of the Ilímaussaq layered sequence.

Despite more than 100 years of research and hundreds of publications on Ilímaussaq, important aspects on the origin of some Ilímaussaq rocks, the architecture and deep structure of the complex and the significance of hydrocarbons and bitumens present in the peralkaline rocks remain unclear. Thus, plenty of room for further studies on these unusual rocks exists and interdisciplinary research is needed to better understand the genesis of this unique magmatic complex.

Keywords Peralkaline rocks · Agpaitic rocks · Eudialyte · Oxygen fugacity · Halogens · Metasomatism

M. A. W. Marks (✉) G. Markl
Mathematisch-Naturwissenschaftliche Fakultät, FB Geowissenschaften,
Universität Tübingen, Wilhelmstrasse 56, 72074 Tübingen, Germany
e-mail: michael.marks@uni-tuebingen.de

B. Charlier et al. (eds.), *Layered Intrusions*, Springer Geology,
DOI 10.1007/978-94-017-9652-1_14

Table 14.1 Overview on the less common minerals found in the Ilímaussaq complex. Minerals for which the Ilímaussaq complex is the type locality are marked in italic, asterisks mark minerals, which have not been described from other localities. For a full list of minerals indentified in the Ilímaussaq alkaline complex, the reader is referred to Petersen (2001).

Aegirine	$NaFeSi_2O_6$
Aenigmatite	$Na_2Fe_5TiSi_6O_{20}$
Analcime	$NaAlSi_2O_6 \cdot H_2O$
Arfvedsonite	$Na_3Fe_5Si_8O_{22}(OH)_2$
Astrophyllite	$(K,Na)_3(Fe,Mn)_7Ti_2Si_8O_{24}(O,OH)_7$
Baddeleyite	ZrO_2
Catapleiite	$Na_2ZrSi_3O_9 \cdot 2H_2O$
Chkalovite	$NaBeSi_2O_6$
Djerfisherite	$K_6(Fe,Cu,Ni)_{25}S_{26}Cl$
Elpidite	$Na_2ZrSi_6O_{15} \cdot 3H_2O$
Epididymite	$Na_2Be_2Si_6O_{15} \cdot H_2O$
Eudialyte	$Na_{15}Ca_6Fe_3Zr_3Si_{26}O_{73}(O,OH,H_2O)_3(Cl,OH)_2$
Murmanite	$Na_2(Ti,Nb)_2Si_2O_9 \cdot 2H_2O$
Natrosilite	$Na_2Si_2O_5$
Natrophosphate	$Na_7(PO_4)_2F \cdot 19\ H_2O$
*Naujakasite**	$Na_6(Fe,Mn)Al_4Si_8O_{26}$
Neptunite	$KNa_2Li(Fe,Mg,Mn)_2Ti_2Si_8O_{24}$
Pyrochlore	$(Na, Ca)_2Nb_2(O,OH,F)_7$
Rinkite	$(Na,Ca)_3(Ca,REE)_4TiSi_4O_{15}F_3$
Sodalite	$Na_8(Al_6Si_6O_{24})Cl_2$
*Sørensenite**	$Na_4Be_2Sn(Si_3O_9)_2 \cdot 2\ H_2O$
Steenstrupine	$Na_{14}Mn_2(Fe,Mn)_2REE_6ZrSi_{12}O_{36}(PO_4)_7 \cdot 3H_2O$
Thermonatrite	$Na_2CO_3 \cdot H_2O$
Trona	$Na_3(HCO_3)(CO_3) \cdot 2H_2O$
Tugtupite	$Na_4(AlBeSi_4O_{12})Cl$
Ussingite	$Na_2AlSi_3O_8OH$
Villiaumite	NaF

Introduction

The Ilímaussaq igneous complex in South Greenland is a Mesoproterozoic composite intrusion that mainly consists of syenitic and nepheline syenitic rocks. It is the type locality of agpaitic rocks, which are peralkaline (molar $(Na + K)/Al > 1$) rocks containing complex Na–Ca–(Ti, Zr)–silicates, the most common being eudialyte-group minerals (Sørensen 1997; Marks et al. 2011, see Table 14.1 for mineral formulae). To date, it is believed that such rock types form by fractional crystallization processes of alkali basaltic and nephelinitic parental magmas derived from lithospheric mantle sources (Larsen and Sørensen 1987; Kramm and Kogarko 1994;

Marks et al. 2011; Schilling et al. 2011). In that sense, the Ilímaussaq rocks are some of the most extreme products of magmatic differentiation processes known. As a consequence, they comprise some of the most unusual magmatic rock types—both in terms of mineralogy and geochemistry (Table 14.2).

The agpaitic rocks of the complex are highly enriched in alkalies (Li, Na, Rb, Cs), halogens (F, Cl, Br, I), high field strength elements (HFSE) such as Zr, Hf, Nb, Ta, REE, U, Th, and otherwise relatively rare elements such as Be, Sn, Sb, W, Mo, As, Zn, Pb and Ga (Bailey et al. 2001). It is therefore unsurprising that the complex is the type locality for some thirty minerals, some of which like sodalite, arfvedsonite and eudialyte occur as major rock-forming components; about ten minerals have only been found in the Ilímaussaq complex so far (Petersen 2001; Table 14.1). Consequently the area remains a magnet for mineral collectors as well as petrologists, mineralogists and geochemists and has given rise to several hundred publications to date.

Despite its exotic petrology, the Ilímaussaq complex can serve as a text-book example for studying a range of magmatic and hydrothermal processes. Because of its exceptionally long crystallization interval and the major changes in the mineralogy of the successive rock units, the complex offers a unique possibility to study the stability relations and compositional evolution of rock-forming minerals in alkaline to peralkaline magmas and the parameters governing the enrichment of Na, Fe, halogens, HFSE and other rare elements in such systems (e.g., Larsen 1976; Markl et al. 2001; Markl and Baumgartner 2002; Andersen and Sørensen 2005; Graser and Markl 2008; Marks et al. 2011).

Some Ilímaussaq rocks have economically very attractive levels of REE, Zr, Nb, Be and U. The complex has therefore a long prospection and exploration history (Sørensen 1992, 2001). Recent developments on the REE market as well as political changes in Greenland has led to increased exploration activity in the complex and the Ilímaussaq complex hosts some of the largest deposits of REEs and U worldwide (Parsons 2012; GMEL 2015; TANBREEZ 2014).

Igneous layering is another characteristic of most Ilímaussaq rocks. In general, igneous layering is a common feature in large mafic and ultramafic intrusions. Some of the most famous localities are Skaergaard in Greenland, Rum in the UK, Stillwater in the USA, Kiglapait in Canada, Bushveld in South Africa and the Great Dyke in Zimbabwe (e.g., McBirney 1996; Emeleus et al. 1996; McCallum 1996; Morse 1969; Eales and Cawthorn 1996; Wilson 1996). Whilst magmatic layering in granitic complexes is relatively rarely described (e.g., Dolbel in Niger, Pupier et al. 2008; High Tatra in Poland, Gaweda and Szopa 2011) it is a phenomenon quite commonly encountered in syenitic intrusions. Well-known examples occur in the Kola peninsula, Russia (Vlasov et al. 1966; Sørensen 1968; Féménias et al. 2005; Galakhov 1975; Kramm and Kogarko 1994; Arzamastsev et al. 1998) as well as in the Gardar Province, South Greenland (see Upton et al. 1996 for a review) including the Ilímaussaq complex. In the latter, the most spectacular example is exhibited by the so-called kakortokite-lujavrite sequence (see Table 14.2 for the major Ilímaussaq rock units). However, the origin of this layered sequence and its relation to the other Ilímaussaq rocks is still not satisfactorily explained.

Table 14.2 Overview on the major rock types of the Ilimaussaq complex (Partly adapted from Sørensen 2006).

Rock type	Texture	Major minerals	Minor minerals
Augite syenite	Medium- to coarse-grained, hypidiomorphic to xenomorphic granular, massive or layered	Alkali feldspar, olivine, Ca-rich augite (diopside to hedenbergite), titanomagnetite, calcic amphibole (ferro-edenite-ferro-pargasite-hastingsite)	Nepheline, biotite, apatite, baddeleyite/zircon, pyrrhotite, chalcopyrite, sphalerite, galena
Pulaskite and foyaite	Medium- to coarse-grained, massive	Alkali feldspar, nepheline, olivine, hedenbergite to aegirine, sodic-calcic to sodic amphibole (katophorite to arfvedsonite)	Titanomagnetite, apatite, biotite, aenigmatite, fluorite, eudialyte, analcime, pyrrhotite, sphalerite, galena and others
Sodalite foyaite	Coarse-grained, foyaitic	Alkali feldspar, nepheline, sodalite, hedenbergite to aegirine, katophorite-arfvedsonite	Olivine, titanomagnetite, apatite, biotite, aenigmatite, fluorite, eudialyte, analcime, pyrrhotite, sphalerite, galena, djerfisherite and others
Naujaite	Coarse-grained to pegmatitic, poikilitic, massive and layered	Alkali feldspar, nepheline, sodalite, hedenbergite to aegirine, katophorite-arfvedsonite, eudialyte	Olivine, titanomagnetite, apatite, biotite, aenigmatite, fluorite/villiaumite, pectolite, pyrrhotite, sphalerite, galena, djerfisherite and others
Kakortokite	Mostly medium- to coarse-grained, mostly laminated and layered	Alkali feldspar, nepheline, eudialyte, sodic amphibole (arfvedsonite)	Sodalite, fluorite, aegirine, aenigmatite, rinkite, pyrrhotite, galena, sphalerite, native Sn and Pb and others
Lujavrite	Mostly medium- to fine-grained, mostly laminated	Albite, microcline, nepheline, sodalite, analcime, aegirine, arfvedsonite, eudialyte	Villiaumite, steenstrupine, naujakasite, sphalerite, galena, native Pb and many others. Depending on the respective Lujavrite type, some of these minerals (e.g., naujakasite) may be major constituents
Peralkaline granite and quartz syenites	Medium- to coarse-grained, hypidiomorphic granular	Alkali feldspar, quartz, arfvedsonite, aegirine	Aenigmatite, astrophyllite, elpidite, epididymite, zircon, ilmenite, fluorite, pyrochlore, sphalerite and others

Here we present an overview of the present understanding on the origin and evolution of the Ilímaussaq magmatic system. Furthermore, based on detailed mineral chemical investigations and structural data, we provide a new hypothesis on the origin of the kakortokite-lujavrite sequence of the complex. Finally, potential directions for further research aimed at a better understanding of this unique system are proposed.

Geological Setting of the Ilímaussaq Complex

The Ilímaussaq complex is part of the Gardar province in South Greenland, which represents an uplifted and eroded continental rift province of Mesoproterozoic age (see Upton et al. 2003 and Upton 2013 for recent reviews). During the early stages of rifting, between 1350 and 1140 Ma, a thick (~3500 m) sequence of lavas and sediments called Eriksfjord Formation accumulated (Poulsen 1964; Larsen 1977a; Halama et al. 2003). A large number of dykes and several mostly composite complexes intruded into a mainly Paleoproterozoic and subordinately Archean basement. The Paleoproterozoic basement country rocks for much of the Gardar Province consist of calc-alkaline granitoids composing the Andean-type Julianehåb batholith that developed within the Ketilidian orogenic belt. The batholith formed between 1850 and 1725 Ma and is considered to have originated following subduction of an oceanic plate beneath the Archean craton to the North (Garde et al. 2002). In the northwestern part of the province, Archean high-grade gneisses form the country rocks for some of the Gardar intrusives (Allaart 1976). Major dyke swarms were emplaced into the Julianehåb batholith mainly in the Tuttutooq-Ilimmaasaq-Narsarsuaq and the Nunarsuit-Isortoq zones. They span a wide compositional range from basaltic to trachytic, phonolitic, rhyolitic, lamprophyric and carbonatitic (e.g., Upton and Emeleus 1987; Goodenough et al. 2002; Halama et al. 2004; Köhler et al. 2009). Anorthositic xenoliths of up to 100 m size occur frequently in some of the mafic and intermediate intrusions (Bridgwater and Harry 1968). It is therefore assumed that a large anorthositic complex may be present at depth underlying the whole province (Bridgwater and Harry 1968; Upton et al. 2003). A detailed study of plagioclase zonation in megacrysts allied to the anorthosites indicates crystallization at about 10–12 kbar prior to a rapid ascent to upper crustal levels (Halama et al. 2002). Some of the dykes, referred to as giant dykes on account of their widths of up to 800 m, are largely composed of coarse-grained gabbro (Bridgwater and Coe 1970). Where they are composite, the dykes have gabbroic margins and syenitic to alkali granitic centres (Upton and Emeleus 1987; Halama et al. 2004).

Several larger central-type plutonic complexes intruded into shallow crustal levels. Their depth of intrusion was approximately 2–5 km and it is likely that many of them had surface expressions (Emeleus and Upton 1976; Upton et al. 1990). They are often composite and the most common rock types are quartz syenite, syenite and nepheline syenite with subordinate alkali granite, gabbro and carbonatite. Most of the complexes contain layered cumulates and give evidence for strong in

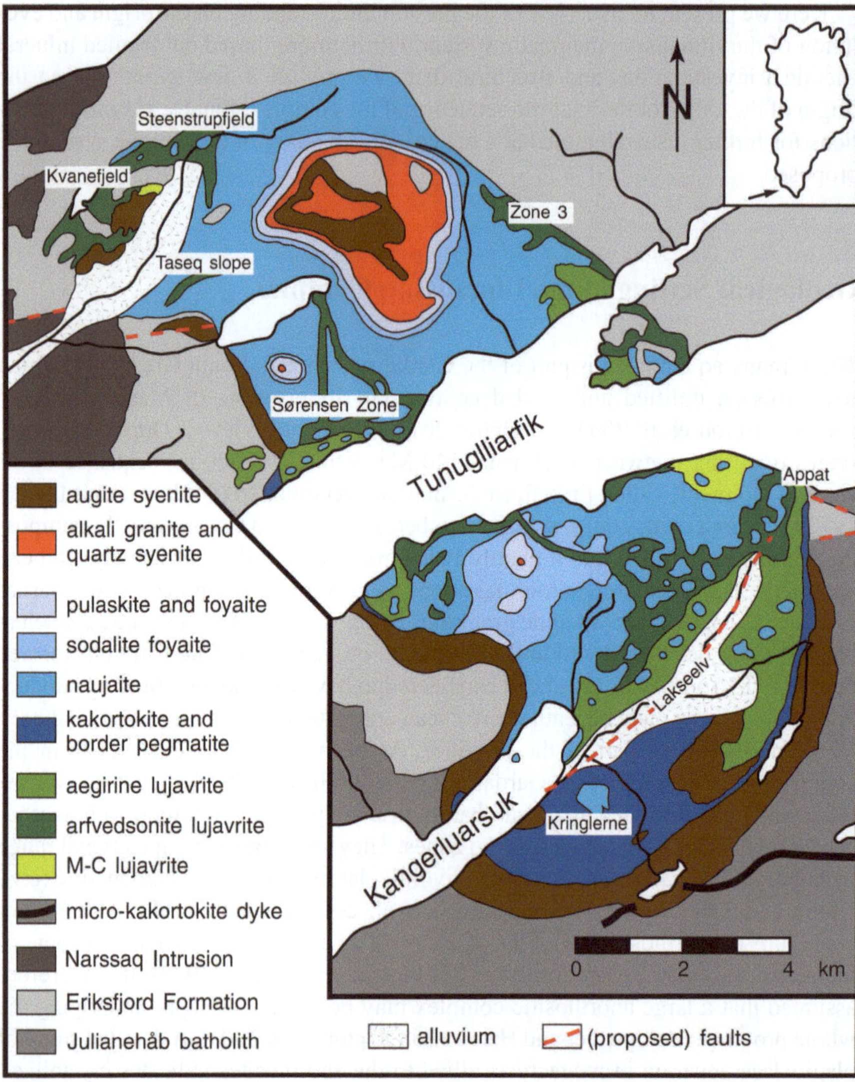

Fig. 14.1 Simplified geological map of the Ilímaussaq complex, South Greenland (Modified after Ferguson 1964 and Sørensen 2001).

situ fractionational crystallization (Upton et al. 1996). Some of the complexes are silica-undersaturated and some are wholly oversaturated, which was probably controlled by the amount of crustal contamination experienced by the parental magmas (e.g., Stevenson et al. 1997; Marks et al. 2003, 2004). Unlike most other Gardar intrusions, the Ilímaussaq complex comprises both quartz-bearing and nepheline-bearing rocks, although the latter largely dominate (Fig. 14.1).

Fig. 14.2 Overview photographs of the Ilímaussaq complex. **a** The northern part of the complex as seen from the Tunulliarfik fjord towards the north. On the right side of the image, the contact between the Ilímaussaq rocks (*mostly light gray*) and the basaltic country rocks (*green from vegetation*) is clearly visible. The horizontal *dark gray top* of the mountain *right* of the *center* of the image is a large xenolith of basalt (compare Fig. 14.1). The red weathered areas are zones of intense oxidation. **b** The southern part of the complex with the Kangerluarsuk fjord in the foreground, view to the south. Note the brownish rim of augite syenite, which separates the layered kakortokite sequence from the mountainous granitoid basement rocks in the background. The *light gray rocks* in the *left* part of the picture are naujaites, which are separated from the kakortokites to the right by slightly darker lujavrites.

Field Relations among Major Rock Types and General Structure of the Complex

The complex has an ovoid plan measuring 18 by 8 km. Its location was structurally controlled, and emplacement of the Ilímaussaq magma(s) is considered to have been by block subsidence and stoping (Ussing 1912; Ferguson 1964; Sørensen 1978). The northern part of the complex exposes the roof region of the Ilímaussaq magma chamber and intrudes the volcano-sedimentary succession of the Eriksfjord Formation. In contrast, a stratigraphically lower level is exposed in the southern part of the complex in which the contacts of the Ilímaussaq rocks are against the granitic country rocks beneath the unconformity with the Eriksfjord Formation (Figs. 14.1 and 14.2).

Several major intrusive phases can be distinguished in the complex: (1) metaluminous augite syenite, (2) peralkaline granite and quartz syenite and (3) peralkaline nepheline syenites. Most of the latter contain eudialyte-group minerals, mineralogically classifying them as agpaitic rocks (e.g., Ussing 1912; Sørensen 1997; Marks et al. 2011). The nepheline syenites largely dominate the complex and are subdivided into (i) coarse-grained roof series (pulaskite, foyaite, sodalite foyaite, naujaite) and floor series (kakortokite) cumulate rocks and (ii) several types of mostly fine-grained and melanocratic rocks (lujavrites), many of which possess strong magmatic fabrics. The lujavrites intrude the roof series rocks, but not the floor series cumulates. Detailed petrographic descriptions of the major Ilímaussaq rocks are presented in Ussing (1912), Ferguson (1964) and Hamilton (1964). Here we present only a brief summary, Table 14.2 and Fig. 14.3 give an overview of the textures and mineralogy of the major Ilímaussaq rocks.

The oldest unit of the complex is augite syenite. It consists of perthitic alkali feldspar, fayalitic olivine, Ca-rich augite (diopside to hedenbergite), Fe–Ti oxides (ulvöspinel-rich magnetite ± ilmenite) and calcic amphibole (hastingsite to ferropargasite to ferro-edenite), with minor amounts of nepheline, biotite, apatite, zircon/baddeleyite and various sulphides (Ferguson 1964; Larsen 1976; Karup-Møller 1978; Marks and Markl 2001; Fig. 14.3a). In the southern part of the complex, it occurs as a nearly continuous shell with variable thicknesses between about 10 and 1000 m (Fig. 14.1). The augite syenite is relatively fine-grained at its contacts towards the Julianehåb granitoids and its grain size increases towards the agpaitic rocks. An increase in the modal amount of interstitial nepheline as well as systematic changes in the composition of olivine and augite from the margins inwards imply that the augite syenite crystallized inwards perpendicular to this contact (Larsen 1976; Marks and Markl 2001). Rafts of augite syenite (up to tens of metres across) occur within the agpaitic rocks. In places these are surrounded by up to 2 m thick pegmatites and show variable signs of fluid-induced interaction with their host rocks (Schönenberger et al. 2006). In the northern part of the complex, the augite syenite envelope is largely missing except for a subhorizontal lid in the roof zone of the complex. Larger augite syenite masses (in parts strongly metasomatized) occur within the agpaitic rocks at the western and northwestern margins of the complex (Fig. 14.1). The field evidence implies that the augite syenite outcrops are relics of a once much larger rock body, which was disaggregated due to the intrusion of agpaitic magma.

Peralkaline granite and quartz syenite occur in several areas of the complex and it is believed that their original extent has been much reduced by erosion (Ferguson 1964; Steenfelt 1981). The largest of these occurrences is in the roof zone of the complex (Fig. 14.1). Alkali feldspar, quartz, sodic amphibole (arfvedsonite) and sodic pyroxene (aegirine) are their major constituents. The perthitic alkali feldspars contain high amounts of tiny aegirine needles, resulting in a green coloration of the rock. Minor components are zircon, astrophyllite, aenigmatite, pyrochlore, sphalerite, fluorite, elpidite, epididymite and others (Ussing 1912; Hamilton 1964; Ferguson 1964). The peralkaline granite and associated quartz syenites post-date the augite syenite but are older than the agpaites. They are regarded as products of an independent silica-oversaturated magma batch (Steenfelt 1981; Stevenson et al. 1997; Marks et al. 2004). The contacts between the different units are not always intrusive but are gradational in places, and it seems possible that assimilation and hybridization may have happened in-situ (Steenfelt 1981). No detailed mineralogical and geochemical study on the peralkaline granites and associated quartz syenites

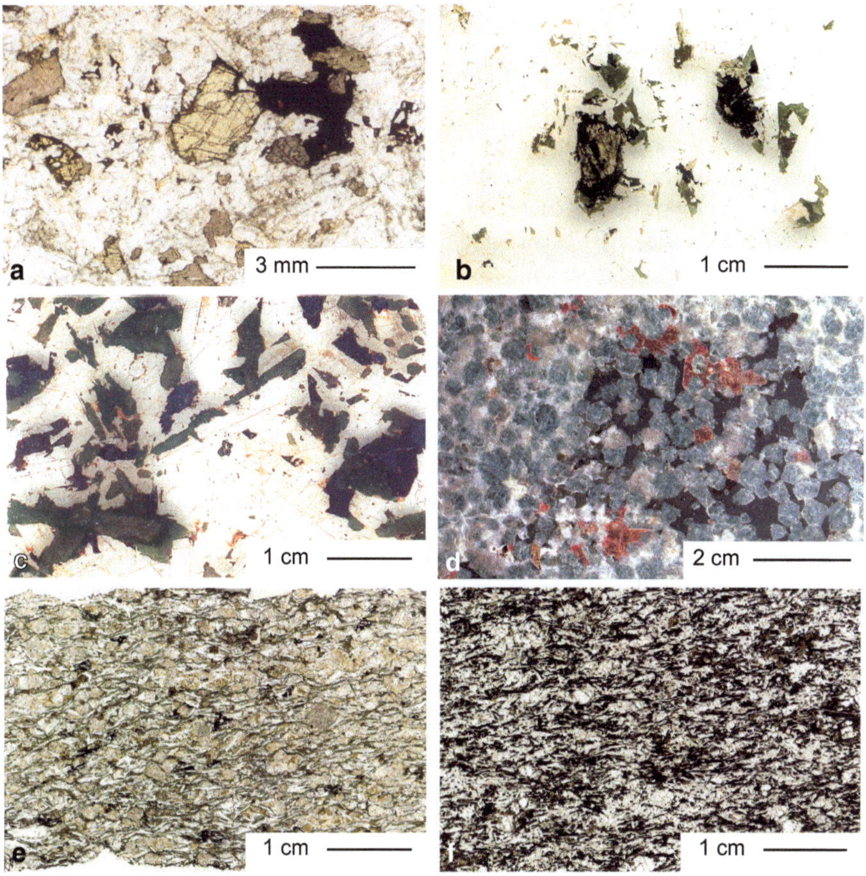

Fig. 14.3 Photographs and microphotographs of the major Ilímaussaq rocks. Note the very different scales. **a** Augite syenite with olivine (*yellow*), augite (*gray*) and ulvöspinel-rich magnetite (*opaque*) in a colorless and cloudy matrix of alkali feldspar with minor amphibole (*reddish brown*). **b** Typical texture in pulaskite with olivine (*yellow*) being replaced by arfvedsonite (*dark bluish*) and aegirine (*green*). **c** Typical foyaite with randomly orientated laths of alkali feldspar and interstitial aegirine, arfvedsonite and aenigmatite. Note the augitic core (*dark gray*) overgrown by green aegirine-rich pyroxene in the lower left corner. **d** Polished surface of naujaite showing euhedral sodalite (*green*) with interstitial arfvedsonite (*black*), eudialyte (*red*) and alkali feldspar/nepheline (*white*). **e** Typical aegirine lujavrite showing needles of aegirine (*green*) and laths of alkali feldspar (*colorless*) wrapping around eudhedral phenocrysts of eudialyte and nepheline (*colorless and cloudy*). Note growth of arfvedsonite (*black*) in cases in pressure shadow position of eudialyte. **f** Typical arfvedsonite lujavrite with arfvedsonite (*black*), minor aegirine (*green*) and eudialyte (*round colorless grains*) together with colorless laths of microcline and subhedral nepheline and albite.

is available to date and therefore, the genesis of these rocks, including their genetic relations to the augite syenite and the early agpaites is not well understood.

The peralkaline nepheline syenites represent the main volume of the complex and exhibit large scale stratification: The uppermost nepheline syenites (roof series) are preserved beneath the augite syenite roof and successively grade from top downwards from pulaskite over foyaite and sodalite foyaite to naujaite, with the latter largely dominating (around 70 % of the volume). These rocks are coarse-grained and contain small but appreciable amounts of the same mafic minerals (olivine, augite, magnetite) and apatite as the augite syenite, which are successively replaced by katophorite-arfvedsonite, aegirine-augite to aegirine and aenigmatite (Figs. 14.3b and 14.3c), accompanied by increasing amounts of nepheline and additional eudialyte. Sodalite appears first as an interstitial phase during the foyaite stage but becomes a cumulus phase in sodalite foyaites and naujaites, the latter contain typically around 40–50 vol.% of euhedral sodalite crystals several mm in size (Fig. 14.3d). Interestingly, some of these sodalite crystals occur as hexagonal prisms. They were thought to represent pseudomorphed nepheline (Hamilton 1964), but could also represent paramorphs after a high-pressure polymorph of sodalite (A. Finch, pers. comm.). Besides, fluorite and villiaumite occur in significant amounts as several sulfides (e.g., pyrrhotite, sphalerite, galena and djerfisherite) do. The roof series rocks essentially formed by downward solidification in a phonolitic magma. Pulaskite, foyaite and sodalite foyaite represent a border facies, whereas the naujaite formed by the flotation of sodalite less dense than the melt from deeper levels of the magma chamber, followed by compaction and the crystallization of intercumulus melt (Ferguson 1964; Engell 1973; Larsen and Sørensen 1987).

The lowest rocks exposed (floor series) consist of medium- to coarse-grained, mostly layered, agpaitic nepheline syenites called kakortokites and are divided into three sub-units (lower layered kakortokites, slightly layered kakortokites and transitional layered kakortokites). The latter grade upwards into typically fine-grained and melanocratic rocks (lujavrites), most of which possess a strong magmatic fabric (Ferguson 1964, 1970; Bohse and Andersen 1981; Andersen et al. 1981, Pfaff et al. 2008; Ratschbacher et al. submitted; Fig. 14.4).

The lower layered kakortokites (LLK) are at least 200 m thick with an unknown continuation at depth. They mainly consist of alkali feldspar, nepheline, arfvedsonite and eudialyte with minor amounts of sodalite, aegirine and aenigmatite together with fluorite, rinkite, pyrrhotite, galena, sphalerite, native Sn and Pb and other accessory minerals. Their layering is defined by the recurrence of on average 8 m thick units (numbered from −11 to +17; Bohse et al. 1971) with pronounced thickness variations between layers (Fig. 14.5a). Most of these 29 units consist of three layers: A basal arfvedsonite-rich layer of black appearance, which is in many cases (but not always) followed by a red eudialyte-rich layer. The upper white layer is rich in alkali feldspar and nepheline and is generally much thicker than the black and red layers (Figs. 14.5b and 14.5c). Within a given three-layer unit, the transitions between the layers are gradual, while the boundaries to consecutive units are often (but not always) sharp (Fig. 14.5b). Generally, the units dip gently (with about 10–20°) towards the center of the intrusion (Fig. 14.5d), with a steepening towards the margin (up to 50°) resulting in a bowl-like geometry (Bohse et al. 1971). One specific unit (+3) contains several hundred meters large inclusions of earlier rock types (augite syenite, naujaite and sodalite foyaite), which were detached from the overlying rocks by a proposed roof collapse (Bohse et al. 1971). The overlying

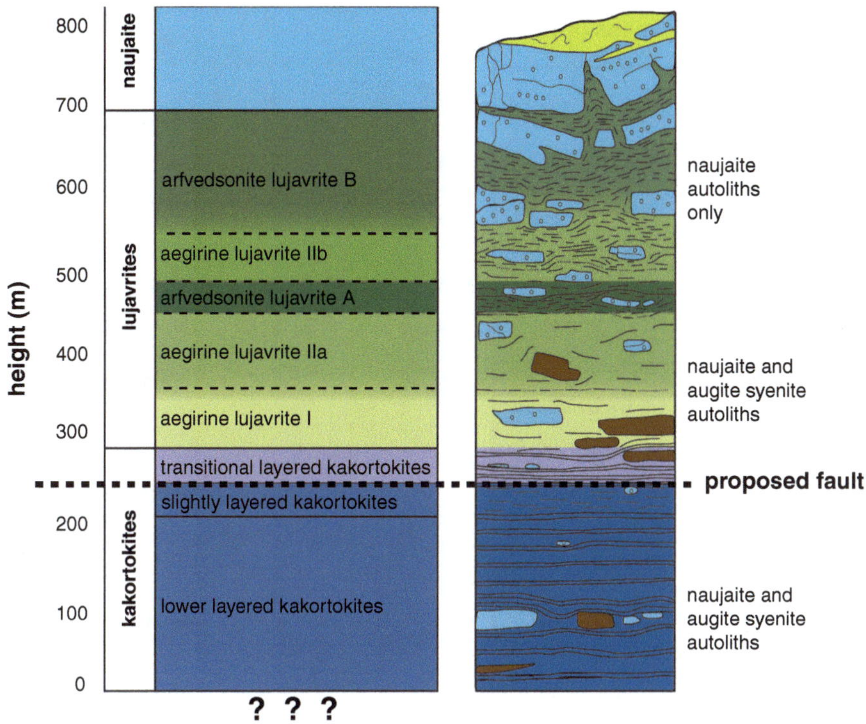

Fig. 14.4 Stratigraphy of the layered kakortokite-lujavrite sequence in the southern part of the complex with major sub-units and the proposed Lakseelv Fault (Modified from Andersen et al. 1981).

kakortokite units envelope these autoliths, and beneath them the layering of the underlying units is depressed (Fig. 14.5e). Other sedimentary structures (i.e. trough banding and current bedding) are mainly confined to the intrusion's margins or occur close to the autoliths (Upton and Pulvertaft 1961; Bohse et al. 1971). An about 50 m thick sequence of finer grained rocks called slightly layered kakortokites (SLK) conformably overlies the LLK (Fig. 14.4). This part of the kakortokites is virtually unlayered and is proposed to have the composition of an average kakortokite (Bohse and Andersen 1981).

A marginal pegmatite zone separates the agpaites from the surrounding augite syenite and country rocks. It forms a rim around the southern part of the complex and is largely restricted to the lower part of the complex made up of kakortokites and lujavrites (Figs. 14.1 and 14.6). The marginal pegmatite zone consists of a mixture of a fine- to medium-grained kakortokite-like matrix cut by a network of pegmatite veins (Fig. 14.8a). The fine- to medium-grained parts of these rocks are considered to represent equivalents of the kakortokite-forming melt (Sørensen 2006). In places, strongly metasomatized autoliths of augite syenite are present in this zone.

On the south shore of the Kangerluarsuk fjord, medium- to fine-grained and melanocratic rocks occur, which where interpreted as representing slumped masses

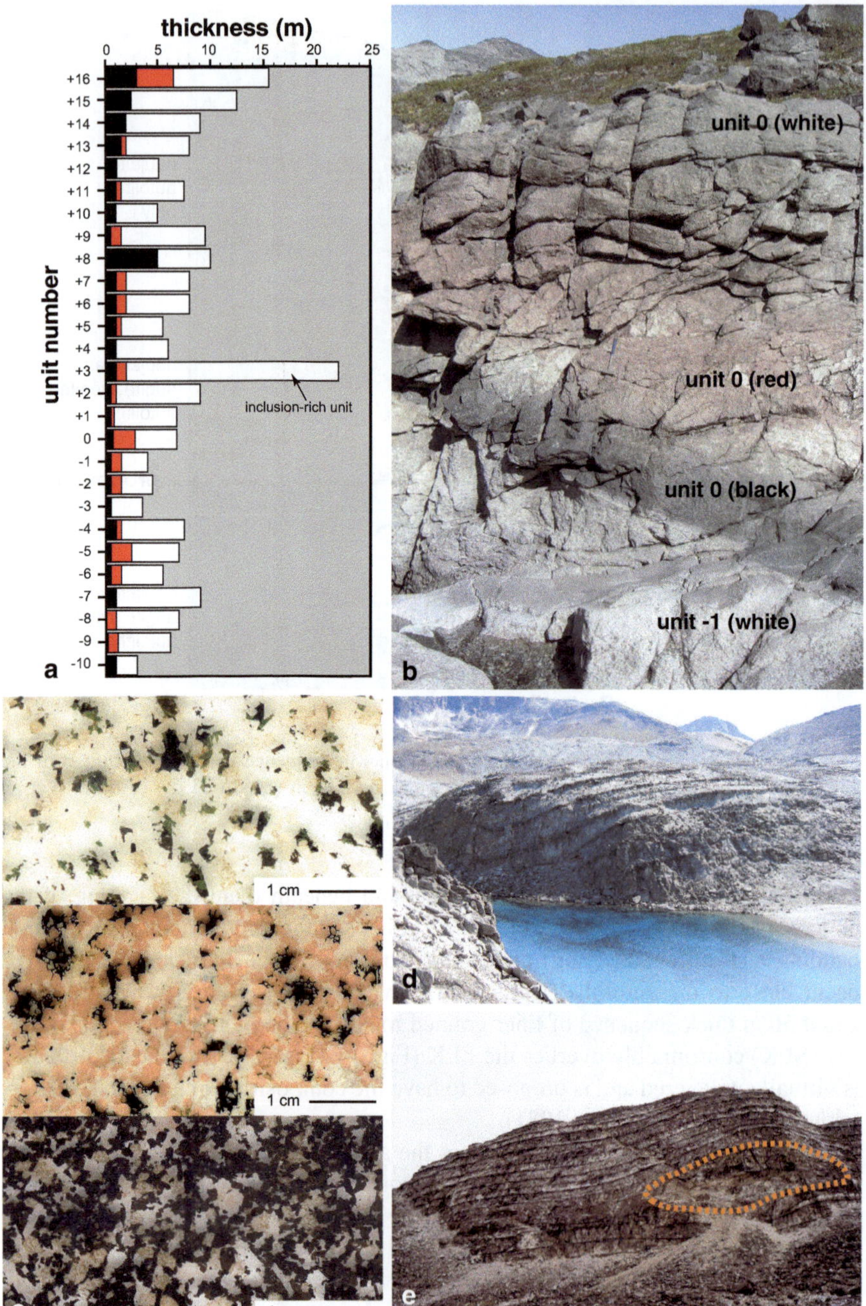

Fig. 14.5 Field appearance of the lower layered kakortokites. a Variation of the thickness of black, red and white layers in the different units along the stratigraphy established by Bohse et al. (1971) using their published data. **b** A close-up of unit -1 (*white layer*) to 0 (*black, red and white layer*) as exposed in the Laksetværelv valley, the main valley running approximately NW-SE through the LLK sequence (compare Fig. 14.6). Note the sharp contact between the base of unit 0 (*black layer*) and the

of unconsolidated "kakortokite mush" (Bohse et al. 1971). In places, however, their field appearance and their relationship towards the LLK may also be explained by an intrusive character of these rocks or by mingling (see e.g. Fig. 8 of Bohse et al. 1971). The rocks are mostly arfvedsonite-rich (even compared to the black layers of the LLK) and further contain microcline, albite, aegirine, rinkite, fluorite, sodalite and up to several cm large eudialyte, which often forms clusters in albite-rich schlieren. Their main occurrence is just south of the proposed Lakseelv fault (Fig. 14.6). How important these rocks are volumetrically is unknown and except for their field occurrence, no further work has been published on them. Future work on drill cores throughout the SLK-LLK sequence is required to reveal their significance and to provide data on the hidden rock succession.

Immediately north of the proposed Lakseelv fault, an about 40 m thick sequence of transitional layered kakortokites (TLK) occurs (Fig. 14.6). These show again distinctive modal layering but their dip of foliation is much steeper (45–75° towards NW) than that of the LLK (see above). Unfortunately, the field relations against the other kakortokite units are obscured by alluvial deposits in the Lakseelv valley (Fig. 14.6). Bohse et al. (1971) proposed the presence of a fault in the Lakseelv valley, with a north-down offset that increases from zero at Appat to several hundreds of metres in the Kangerluarsuk Fjord (Figs. 14.1 and 14.6). Based on this, Bohse and Andersen (1981) assigned the TLK sequence to the structurally highest level of the kakortokites in general. The lowest units of the TLK are mineralogically and texturally similar to the LLK, but their grain-size very significantly decreases upwards through the sequence. Furthermore, the modal ratio of aegirine to arfvedsonite increases strongly, continuously developing into the lower parts of the overlying lujavrites.

Lujavrites comprise several varieties of mostly fine-grained and melanocratic agpaitic nepheline syenites, which often show a strong magmatic fabric. A large number of publications deal with the geology, mineralogy and geochemistry of these unusual rocks (e.g., Andersen et al. 1981; Bohse and Andersen 1981; Bailey and Gwodz 1994; Bailey 1995; Rose-Hansen and Sørensen 2002; Sørensen et al. 2003, 2006, 2011; Bailey et al. 1993, 2006; Ratschbacher et al. submitted). Their largest occurrence is in the south-east of the complex to the north and east of the Lakseelv valley (Fig. 14.1). Here the sequence is about 350 m thick and occurs intermediate between kakortokites and naujaites (Figs. 14.4 and 14.6). The lower

top of unit −1 (*white layer*) but the gradual contacts between the black, red and white layers within unit 0. Hammer for scale in the centre of the picture. The viewing direction is approximately towards the east. c Scanned thin sections of a white, red and black unit of the kakortokites. The white one is dominated by alkali feldspar and nepheline with minor amounts of eudialyte (*very pale purple*) interstitial arfvedsonite (*black*) and aegirine (*green*). The red layer is characterized by euhedral eudialyte (*red*) with alkali feldspar, nepheline and interstitial arfvedsonite. In black layers, euhedral to subhedral arfvedsonite dominates with minor amounts of eudialyte (*light brownish*), alkali feldspar laths and nepheline. d Gently dipping layering in the lower layered kakortokites (units +4 to +10) at the SE tip of the Blå Sø, the small lake close to the naujaite autolith as shown in Fig. 14.6. Viewing direction is approximately southwards. e The steep cliffs of Kringlerne as seen from the south shore of the Kangerluarsuk Fjord. The viewing direction is towards the south (compare Figs. 14.1 and 14.6). The cliff is about 110 m high and exposes layers +1 to +13, including unit +3, which contains several larger autoliths of augite syenite, naujaite and foyaite, as marked with the orange dotted line.

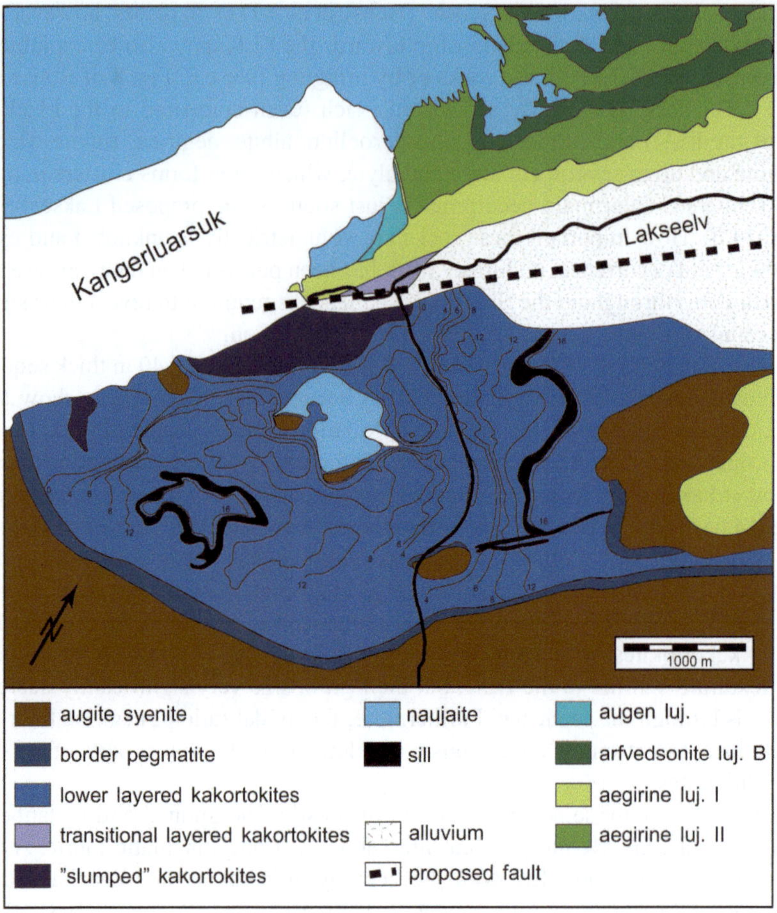

Fig. 14.6 Map of the southwestern part of the Ilímaussaq complex (modified from Bohse et al. 1971 and Andersen et al. 1988). The proposed Lakseelv Fault separates the lower layered kakortokites and the so-calledof the complex made up of kakortransitional layered kakortokites, which grade into the overlying lujavrite sequence. Numbers and thin lines are outcrop traces of black layers of respective lower layered kakortokite units. The lake in the center of the image is the same lake as in Fig. 14.5d The abundant inclusions of naujaite within the lujavrite are omitted for clarity.

part of the sequence gradually evolves from the underlying transitional layered kakortokites (see above) into green aegirine-rich lujavrites (Fig. 14.3e), whereas the upper part is dominated by black arfvedsonite-rich lujavrites (Bohse and Andersen, 1981; Fig. 14.3f). In between, aegirine- and arfvedsonite-dominated sheets of variable thickness alternate. In places these are separated by several dm-thick layers of felsic rocks rich in feldspar and nepheline with smaller amounts of eudialyte, arfvedsonite and aegirine. The lujavrites intrude and disintegrate the naujaites both horizontally and vertically, resulting in a zone where many naujaite blocks of variable size (up to hundreds of meters) occur in a lujavrite matrix, which wraps around them (Fig. 14.7). Mineralogically, lujavrites are similar to the kakortokites

Fig. 14.7 Two photographs of the contact zone between lujavrites and naujaites. **a** Lujavrites intrude and disintegrate the naujaites, resulting in a zone where naujaite blocks are floating within a fine-grained lujavrite matrix. **b** One of the larger naujaite autoliths (about 25 m high), which is split in two parts by intruding aegirine lujavrite I.

except for the high amounts of aegirine in the green lujavrites and the occurrence of microcline and albite instead of perthitic alkali feldspar and abundant interstitial analcime and natrolite in some lujavrites (e.g., Bohse and Andersen 1981). Most of the lujavrites show fluidal textures with a strong alignment of crystals, especially in the aegirine-dominated units (Fig. 14.3e). The initially steep dip (50–85°) towards the NW in the lower part of the sequence first overturns towards SE and within less than 300 m following the Lakseelv valley upwards, the dip rapidly shallows to about 20° towards the NE (Ratschbacher et al. submitted). Accordingly, the area may be interpreted as a steep feeder zone opening upwards in a fan-like manner. The steeper part is composed of aegirine lujavrite but this flattens out over a short distance whilst grading upwards into arfvedsonite lujavrite B. It is also intersected by approximately horizontal sheets (sills) of arfvedsonite lujavrite A (Ratschbacher et al. submitted). The field evidence suggests that the lujavrite sequence did not form through continuous closed-system in situ differentiation from a single magma batch but evolved at least partly through melt replenishment from deeper levels.

Further aegirine and arfvedsonite lujavrite occurrences of variable size occur throughout the complex but their detailed genetic relationships to the lujavrites of the Lakseelv area are not known in detail. A number of locally restricted and special lujavrite types are known as well (e.g., M-C lujavrites, augen lujavrite, murmanite lujavrite, villiaumite lujavrites and others; e.g., Sørensen et al. 1969; Bohse and Andersen 1981; Figs. 14.1 and 14.6). Some of them may have crystallized from independent magmas whereas others could be products of hybridization and metasomatism with other Ilímaussaq rocks and fluids (e.g., Bohse and Andersen 1981; Sørensen et al. 2011). It is beyond the scope of this review to discuss their genesis in detail.

The lujavrites of the Kvanefjeld area, situated in the northwestern corner of the complex, however, are of special interest (Fig. 14.1). This area represents the roof zone of the complex, where lujavrites are in contact with the volcanic host rocks (e.g., Sørensen et al. 1969, 1974, 2011; Nielsen and Steenfelt 1979). The geology of this area is exceptionally complex as different Ilímaussaq units, country rocks, gabbros and anorthosites occur as inclusions in a matrix of at least two distinct types of lujavritic rocks, with strong metasomatic overprint and hydrothermal activity (see below). A number of late lamprophyric dyke rocks intersect the area (Sørensen et al. 1969, 2011). The area was recently interpreted to represent an off-shoot from the arfvedsonite lujavrite magma through a fracture zone along parts of the north contact of the complex (Sørensen et al. 2011). One of the latest lujavrite members of this area is rich in naujakasite and steenstrupine. These rocks probably represent some of the final residual melts of the long lasting evolution of the parental magma and are termed hyper-agpaitic rocks (e.g., Khomyakov 1995; Sørensen and Larsen 2001; Andersen and Sørensen 2005). Given their exceptionally high contents of REE, they are of major economic interest.

Kvanefjeld differs from many other emerging REE projects in that it is a multi-commodity project that is also expected to produce U and Zn. Several new target areas in the northern part of the complex (Steenstrupfjeld, Sørensen Zone and Zone 3; Fig. 14.1), each associated with lujavritic rocks, have also been identified as potential economic deposits. To date, these deposits are owned by Greenland Minerals and Energy Ltd (GMEL) and have an estimated resource of 1010 MT containing 593 Mlbs U_3O_8, 11.13 MT RE_2O_3 and 2.3 MTZn (GMEL 2015).

The agpaitic rocks contain abundant pegmatites (Fig. 14.8). They occur as thick layers within a given rock type (e.g., in the naujaites), at the boundaries between different rock types (e.g., the marginal pegmatite; Fig. 14.8a), as discordant dyke-like bodies (Fig. 14.8b) and as irregularly formed lenses enclosed in their host rock (Fig. 14.8c; e.g., Sørensen 1962; Ferguson 1964; Bohse et al. 1971; Müller-Lorch et al. 2007). Throughout the complex, hydrothermal veins containing partly very uncommon mineral assemblages are abundant but are especially concentrated in the Kvanefjeld, on the Taseq slope and some other areas. They show a large range in textures and mineralogical composition (e.g., Engell et al. 1971; Markl 2001a; Markl and Baumgartner 2002; Graser et al. 2008; Hettmann et al. 2014). The most common types include (i) aegirine-dominated veins with variable amounts of albite and analcime (Fig. 14.8d), (ii) albite- analcime- and natrolite-dominated veins, in cases with galena, sphalerite and various Be- and Nb-minerals (e.g., tugtupite, chkalovite, pyrochlore; Fig. 14.8e) and various silico-phosphate and phosphate minerals (e.g., steenstrupine, natrophosphate; Petersen et al. 2001) and (iii) zoned veins with aegirine-rich rims and arfvedsonite-rich cores with or without sodalite, nepheline, albite and analcime (Fig. 14.8f). Furthermore, bitumens are known from some Ilímaussaq rocks (Petersilie and Sørensen 1970; Konnerup-Madsen et al. 1988; Laier and Nytoft 2012). They occur finely dispersed (not visible with the naked eye) in most Ilímaussaq rocks and are also found as milimetre-sized droplets (Konnerup-Madsen et al. 1988). The latter are not well studied to date, but seem especially common in late-stage pegmatites and veins from the Kvanefjeld and Taseq slope rich in Be- and Nb-minerals (H. Friis, per. comm.).

Fig. 14.8 Field photographs of various pegmatites and hydrothermal veins from the Ilímaussaq complex. **a** Close up of the marginal pegmatite area in the southern part of the complex. **b** Discordant arfvedsonite-alkali feldspar pegmatite intruding naujaite. **c** Irregularly formed pegmatite lens enclosed in naujaite. **d** A fine-grained aegirine-albite-analcime vein. **e** Albite- and analcime-rich vein containing purple tugtupite and other Be-silicates like e.g. chkalovite. **f** Zoned vein with a fine-grained aegirine-rich rim and a medium- to coarse-grained core rich in sodalite, nepheline, albite and euhedral arfvedsonite.

A variety of fine-grained sills and dykes occur within the Ilímaussaq complex and its adjacent country rocks (Ferguson 1964; Allaart 1969; Sørensen et al. 1969; Bohse et al. 1971; Rose-Hansen and Sørensen 2001). One of them is an agpaitic and Fe-rich phonolite (A.I. =1.44; FeO_{tot} =12 wt.%) called micro-kakortokite (Fig. 14.1), which was intensively studied (e.g., Larsen and Steenfelt 1974; Marks and Markl 2003; Giehl et al. 2012; 2014). Mineralogically, geochemically and petrologically it shows striking similarities with the various Ilímaussaq rocks and the intruded magma may have had a composition similar to that which produced the agpaitic part of the complex.

Time Constraints

Over the past decades, several attempts to date the Ilímaussaq rocks have been undertaken (e.g., Blaxland et al. 1976; Paslick et al. 1993; Marks et al. 2004; Waight et al. 2002; Krumrei et al. 2006). The results of these studies largely differ in precision but point to the formation of the Ilímaussaq complex at around 1160 Ma. This shows that the Ilímaussaq magmatic system belongs to the youngest of the Gardar intrusive complexes (Upton et al. 2003).

The detailed geochronological study of Krumrei et al. (2006) provides important insights into the duration of intrusion, fractionation and cooling. Baddeleyite from the augite syenite gives an age of 1160 ± 5 Ma, indistinguishable from a U–Pb age of zircon (1166 ± 9 Ma) from the subsequently intruded peralkaline granite (Heaman and Upton, cited in Upton et al. 2003). Amphibole from all major Ilímaussaq rocks (including amphibole from a late-stage pegmatite of the youngest lujavrites) was dated with the Ar–Ar technique and gave a spread in age data which was interpreted to reflect cooling ages (Krumrei et al. 2006). These data and the application of a thermal cooling model led Krumrei et al. (2006) to conclude that the fractionation and cooling history of the complex was very short-lived, probably on the order of 500–800 ka.

Source Constraints and Crustal Contamination

The Ilímaussaq rocks are interpreted to be derived from a parental (transitional) olivine basalt magma that fractionated in a deep-seated magma chamber and was successively tapped by cauldron subsidence and/or stoping (e.g., Larsen and Sørensen 1987; Stevenson et al. 1997). Neodymium and oxygen isotope compositions of mafic minerals from the major rock types are relatively homogeneous (εNd = − 0.9 to − 1.8 and $\delta^{18}O_{V\text{-}SMOW}$ = + 5.2 to + 5.7 ‰) with no significant difference between the early augite syenite and the agpaites or between early and late agpaites. These data imply that most of the Ilímaussaq melts were derived from an isotopically homogeneous mantle source (Marks et al. 2004). Because of the slightly negative εNd values, a lithospheric or an OIB-type mantle source is likely but probably accompanied by small amounts of interaction with the host rock during storage in a deeper magma chamber. Trace element data (e.g., enrichment of LILE, LREE, Sr and F) and isotopic signatures of other Gardar dyke rocks and volcanics suggest considerable heterogeneity. It is proposed that these magmas were generated within lithospheric mantle that was variably metasomatized because of earlier subduction processes during the Ketilidian orogeny (e.g., Upton and Emeleus 1987; Halama et al. 2003; Goodenough et al. 2002; Upton et al. 2003; Köhler et al. 2009; Upton 2013). Indeed, potential fragments of the Gardar lithospheric mantle enclosed as xenoltihs in an ultramafic lamprophyre were described by Upton (1991) from the small island of Illutalik, just a few of kilometers southwest of the Ilímaussaq complex.

A slightly positive εNd value of + 0.7 for the above-mentioned micro-kakortokite dyke may imply less contamination with low εNd material and thus, this dyke rock may represent the closest approach to the composition of the magma from which the Ilímaussaq rocks crystallized (Marks et al. 2004).

At the margins of the complex, however, crustally contaminated rocks occur (Stevenson et al. 1997) and show that significant assimilation of country rocks occurred on a local scale, probably during emplacement, consistent with field observations of peralkaline granite forming around quartzite xenoliths enclosed in the augite syenite (Ferguson 1964). The peralkaline granite unit has a lower εNd (−3.1) and was interpreted to reflect contamination with lower crustal rocks probably at the time of storage (Marks et al. 2004). However, no systematic geochemical investigations on the various occurrences of peralkaline granite and associated quartz syenites are available to date, giving room for further studies. In the following sections, these rocks are therefore not considered further.

Major Mineralogical and Mineral Chemical Changes During the Magmatic to Hydrothermal Evolution of the Ilímaussaq System

Detailed investigations of mineral assemblages and their compositional evolution trends within and between the various rock types provided deep insight in the magmatic evolution of the Ilímaussaq system (e.g., Larsen 1976, 1977b, 1981; Markl et al. 2001; Marks and Markl 2001; Marks et al. 2004; Müller-Lorch et al. 2007; Schmid 2008; Pfaff et al. 2008; Rønsbo 1989, 2008; Zirner et al. submitted; Lindhuber et al. in press; Ratschbacher et al. submitted). The major results of these studies are briefly summarized below.

From the augite syenite stage to the agpaitic stage the mafic mineral assemblage changes from olivine + augite + Fe–Ti oxides to arfvedsonite (± aegirine) + eudialyte + aenigmatite. This is illustrated well by disequilibrium textures found in the agpaitic roof series, where cumulus fayalite, hedenbergite and ulvöspinel are resorbed and replaced by arfvedsonite, aenigmatite and aegirine (Figs. 14.3b and 14.3c). Due to extensive fractionation of olivine and augite, X_{Fe} in mafic minerals increases from intermediate values towards compositions almost devoid of Mg (Fig. 14.9). Besides increasing X_{Fe}, a strong enrichment of Fe_{tot} and depletion of Si in the melts result in increasing Fe/Si ratios. Simultaneously, the relative amount of Na with respect to Ca increases (as mirrored by pyroxene and amphibole evolutionary trends; Fig. 14.9a and b), eventually stabilizing a Na- and Fe-rich mafic mineral assemblage.

In the early agpaites, amphibole generally dominates over pyroxene which, in most cases, is later than amphibole. In most lujavrites, however, primary aegirine and arfvedsonitic amphibole are in textural equilibrium (except for arfvedsonite lujavrite B, where primary aegirine is often replaced by arfvedsonite) and in green

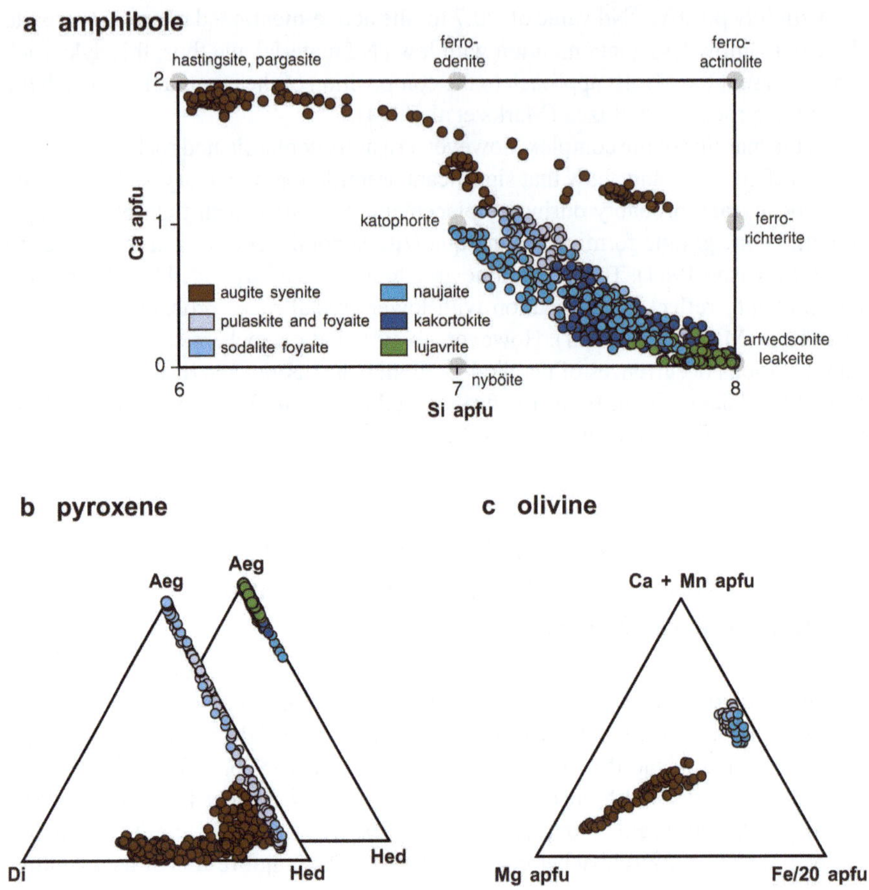

Fig. 14.9 Compositional variation of mafic minerals of the Ilímaussaq rocks: **a** amphibole, **b** pyroxene and **c** olivine (apfu = atoms per formula unit).

lujavrites, aegirine is largely dominating over arfvedsonite (e.g., Rose-Hansen and Sørensen 2002; Ratschbacher et al. submitted). This was interpreted to reflect changes in water activity and/or redox conditions in the lujavritic melts (e.g., Sørensen 1962; Markl et al. 2001; Rose-Hansen and Sørensen 2002). Slight differences in Na/Si and Na/Fe ratios in the lujavritic melts may also play a role, but the details of the relative stability of aegirine compared to arfvedsonite are to date not quantitatively constrained.

Felsic minerals during the augite syenite stage are alkali feldspar and minor amounts of nepheline, the latter increasing in abundance during differentiation. Ternary feldspar (with up to 20 mol% An) is restricted to the most primitive augite syenite samples from the relatively fine-grained border facies (Larsen 1981; Marks and Markl 2001). In the agpaitic rocks, alkali feldspar, nepheline and sodalite are very abundant and analcime forms during the late-magmatic stage. In some lujavrites, albite and microcline form separate crystals and alkali feldspar is not present anymore.

The formation of large amounts of sodalite (and eudialyte) in the early agpaites clearly reflects the unusual Cl-rich composition of the agpaitic melt and both minerals buffer NaCl activity during differentiation. High levels of F are indicated by the early magmatic occurrence of fluorite and in some agpaitic rocks villiaumite occurs. Because of the high solubility of Ti, Zr and other HFSE in such alkali- and halogen-rich melts (e.g., Watson 1979; Linnen and Keppler 2002), their concentration reaches high values and eudialyte and aenigmatite are stabilized as magmatic phases. These minerals do not occur in the augite syenite, where the HFSE elements are mainly contained in accessory baddeleyite and zircon (Marks and Markl 2001).

Agpaitic pegmatites studied by Müller-Lorch et al. (2007) offer detailed insights into the final stages of agpaitic melt evolution. Mineralogically, they are largely similar to their host rocks. They mainly consist of variable amounts of sodalite, nepheline, microcline, albite and analcime and the dominant mafic mineral is mostly arfvedsonitic amphibole accompanied by aegirine, eudialyte and aenigmatite. Besides others, minor and accessory minerals include fluorite, galena, sphalerite, rinkite, biotite, monazite and in some cases even zircon and various alteration products like natrolite, catapleite, hematite, chlorite, biotite, astrophyllite, aegirine, fluorite and others (Müller-Lorch et al. 2007). However, the generally observed increase of Na/K ratios in early magmatic arfvedsonite is reversed in some analcime-rich pegmatites. These differences are probably related to the buffering of Na and K activities by two coexisting alkali feldspars, which favors Na over K in cooling fluids and the ceasing of the buffer capacity when albite is replaced by analcime as a stable phase (since analcime does not incorporate significant amounts of K). Arfvedsonite in such assemblages is enriched in K (Müller-Lorch et al. 2007). In the very late stages, virtually K-free assemblages in analcime-aegirine veins are present and support the existence of Na-dominated aqueous fluids, augmented by molar Na/K ratios up to about 40 as analyzed by ion chromatography (crush-leach technique) from some late-stage hydrothermal veins (Graser et al. 2008).

Some lujavrites and agpaitic dyke rocks show ocelli-like and spheroidal structures, which have been interpreted as liquid-liquid immiscibility causing the separation of a Na–Al-rich melt from an Fe–Mn–Nb-REE-rich one (Markl 2001b; Sørensen et al. 2003). In some of the latest lujavrites, in fenitized volcanic rocks in the roof of the complex (Kvanefjeld) and in some pegmatites and hydrothermal veins, so-called hyper-agpaitic mineral assemblages formed, which mark an even more evolved stage compared to the agpaites in general (Khomyakov 1995; Sørensen and Larsen 2001; Andersen and Sørensen 2005). The most important mineralogical changes compared to the earlier agpaites include the formation of naujakasite and steenstrupine at the expense of nepheline and eudialyte. They often show complex textural relationships between different mineral assemblages and contain a wealth of otherwise extremely rare and unusual minerals. In these assemblages Na-rich aluminosilicates and phosphates with different amounts of other volatile species (F, Cl and OH) dominate and these rocks recently attracted increased interest as they contain the highest concentrations of REE. Besides REE, important metals that are extremely enriched during these stages include Be, Nb and Sn, resulting in the formation of various Be-silicates (e.g., tugtupite and chkalovite), Nb-minerals (pyrochlore) and the extremely rare Sn–Be silicate sørensenite. In these final products of the Ilímaussaq evolution several water-soluble minerals like villiaumite, trona, thermonatrite and natrophosphate are stable (Sørensen and Larsen 2001). Recently, even natrosilite was found in some of these rocks (Andersen and Friis submitted) as was expected, since it was described from similar assemblages in the Khibina and Lovozero complexes in Russia (Khomyakov 1995; Sørensen and Larsen 2001). Experimental data further imply the co-existence of two immiscible melts during final agpaite evolution: an ionic melt of almost pure NaF and an aluminosilicate melt (Kogarko and Krigman 1970; Kogarko and Romanchev 1983).

Estimation of Crystallization Conditions

Pressure

Based on the observed overburden of the Eriksfjord formation and on fluid inclusion data the emplacement pressure for the Ilímaussaq rocks is around 0.1 GPa corresponding to a depth of about 3–4 km (Konnerup-Madsen and Rose-Hansen 1984). The depth of initial crystallization of sodalite, however, might be significantly deeper, since primary fluid inclusions in sodalite from naujaite imply a pressure of 0.3–0.4 GPa (Markl et al. 2001; Krumrei et al. 2007). This indicates that sodalite crystallized at least partly from a rising magma at considerable depth (around 10–12 km), which requires a conduit towards deeper crustal levels. Based on heatflow data (Sass et al. 1972), however, the agpaitic rocks of Ilímaussaq have a total thickness of not much more than 1 km. Sparse gravity measurements indicate that a dense body extending deep into the crust underlies the Ilímaussaq intrusion (Forsberg and Rasmussen 1978) and the aeromagnetic map of the region shows a magnetic low in the Ilímaussaq area (Thorning and Stemp 1997). To date, however, detailed geophysical investigations such as 3-D density modeling (e.g., Arzamastsev et al. 2001) are lacking and thus the deeper structure of the Ilímaussaq area remains open for speculation until such work will be performed.

Temperature and Silica Activity

Overall, the Ilímaussaq rocks cover an extremely long crystallization interval from >950 °C to a solidus of around 500–450 °C (Larsen 1976; Markl et al. 2001). The highest temperatures were derived by ternary feldspar and nepheline thermometry in augite syenite samples and reach about 1000 °C. In the same rocks, near-solidus temperatures deduced from mafic minerals reach temperatures as low as about

650 °C. Silica activity in the augite syenite was initially at about 0.8 and closed system fractionation produced nepheline-bearing assemblages at silica activities of around 0.4 (see Marks and Markl 2001 for details).

Nepheline in sodalite foyaite, naujaite and kakortokites indicates slightly lower temperatures of 950–900 °C and equilibria among nepheline, sodic pyroxene and alkali feldspar indicate temperatures between 800 and 700 °C at silica activities between 0.5 and 0.3. Late-stage aegirine rims around magmatic arfvedsonite indicate temperatures of about 500 °C at silica activities of around 0.25. The highest temperatures in some lujavrites were estimated to 750–800°C with solidus conditions around 450–500 °C (see Markl et al. 2001 for details) with silica activity dropping from around 0.6 to around 0.15, the latter probably reflecting hydrothermal conditions.

Redox Conditions

During the augite syenite stage redox conditions are constrained by olivine—augite—Fe–Ti oxide assemblages. During fractional crystallization they decrease from about ΔFMQ= −1 to extreme values around or even below ΔFMQ = −4, reaching the theoretical stability of native iron (Marks and Markl 2001). These values represent some of the most reduced conditions reported for terrestrial plutonic rocks. Similar redox conditions (ΔFMQ = −2 to −4) were derived from preserved relics of the same assemblage in pulaskite and sodalite foyaite, which shows that the initial stages of agpaite formation were similarly reduced (Schmid 2008). These estimates are in agreement with the presence of ulvöspinel-rich magnetite as well as with experimental evidence (Giehl et al. 2012).

The quantification of redox conditions during the main agpaite stage is difficult due to the absence of suitable mineral assemblages for which reliable thermodynamic data exist. Nevertheless, the presence of primary methane-dominated and water-free fluid inclusions, as found in most agpaites (e.g., Krumrei et al. 2007), and low sulfate/sulphide ratios found in sodalite from naujaites, again indicate relatively reduced conditions (Hettmann et al. 2012). The high amount of Fe^{3+} in clinopyroxene of the agpaitic rocks does not contradict these observations, since the Fe^{2+} /Fe^{3+}ratio in silicate melts depends not only on oxygen fugacity, but is largely influenced by peralkalinity, the presence of water and other factors (Botcharnikov et al. 2005; Giuli et al. 2012; Markl et al. 2010).

During late-magmatic to hydrothermal stages, relatively oxidized conditions (around and above the HM buffer) were estimated for the formation of some skarn-like rocks as well as for some hydrothermal veins (Graser and Markl 2008; Karup-Møller 1978). Similar evolution trends starting with highly reduced conditions at magmatic stages evolving towards relatively oxidized during hydrothermal stages are know from other peralkaline rock suites and seem to be a typical feature of peralkaline magmatic systems (e.g., Marks et al. 2003; Mann et al. 2006; Markl et al. 2010).

Fluid Evolution

Numerous studies investigated the nature and composition of fluids trapped in minerals of the Ilímaussaq complex, mostly focusing on the agpaitic rocks and their associated hydrothermal veins (e.g., Sobolev et al. 1970; Petersilie and Sørensen 1970; Konnerup-Madsen 1980, 1984, 2001; Konnerup-Madsen and Rose-Hansen 1982, 1984; Konnerup-Madsen et al. 1979, 1981, 1985, 1988; Markl and Baumgartner 2002; Krumrei et al. 2007; Graser et al. 2008). These studies revealed the complex fluid evolution in Ilímaussaq and the unusual fluid composition: Fluid inclusions in the Ilímaussaq rocks are dominated by hydrocarbon gases (mostly methane) and hydrogen.

Presumably primary fluid inclusions in early magmatic sodalite from the roof and early magmatic sodalite and eudialyte in the floor cumulates are predominantly either methane-dominated (with minor amounts of hydrogen) or pure methane with generally no detectable water (not even with Raman techniques; Konnerup-Madsen et al. 1985; Krumrei et al. 2007). This was interpreted to result from strongly reduced conditions and low water activities during the early stages of agpaite crystallization and is in accordance with mineralogical observations and petrologic calculations (see above). Fluid inclusions of secondary origin in the same rocks are either methane-dominated (containing traces of higher hydrocarbons like ethane and propane) or aqueous, with high salinities of up to about 22 wt.% NaCl equiv. (e.g., Konnerup-Madsen et al. 1988; Markl et al. 2001; Krumrei et al. 2007). Krumrei et al. (2007) presented a two-stage model for the evolution of the Ilímaussaq fluid phases due to the much higher solubilities of water than hydrocarbons in silicate melts. At the beginning of sodalite crystallization, a hydrocarbon-rich fluid was in equilibrium with the agpaitic melt. The water activity in the agpaitic melt at this stage was very low (Markl et al. 2001). NaCl activity in the melt decreased with further sodalite fractionation from initial values of about 0.4 down to values well below 0.1, while the water activity successively increased. During late-magmatic conditions, analcime became stabilized instead of albite and nepheline (Ussing 1912; Markl et al. 2001; Müller-Lorch et al. 2007) and eventually, an aqueous fluid was expelled from the melt. Because NaCl strongly partitions into the aqueous fluid, the large-scale formation of sodalite ended with the exsolution of this aqueous fluid, which was then trapped only as secondary inclusions. In lujavrites, however, two types of apparently primary fluid inclusions occur: firstly again methane-dominated ones and secondly aqueous fluid inclusions with relatively low salinities (around 3 wt.% NaCl equivalent; Markl et al. 2001), pointing to the entrapment of two co-existing and immiscible water-rich and methane-rich fluid types (Konnerup-Madsen 2001).

Primary and secondary fluid inclusions in late-magmatic to hydrothermal veins are highly variable ranging from again hydrocarbon-dominated via hydrocarbon–water–NaCl mixtures to aqueous fluid inclusions with highly variable salinities with or without traces of hydrocarbons (e.g., Konnerup-Madsen and Rose-Hansen 1982; Graser et al. 2008). Detailed investigation on some of the late-stage agpaites and associated hydrothermally overprinted rocks revealed a large temperature range (from about 600–200 °C) for their formation, which marks the transition from the late-magmatic to the hydrothermal, now hyper-agpaitic stage (see above; Markl

2001a; Sørensen and Larsen 2001). The aqueous agpaitic fluids are characterized by a high pH (up to 10–12) enabling the formation of ussingite and the above-described hyper-agpaitic mineral assemblages (Markl and Baumgartner 2002). Kogarko (1977) and Khomyakov (1995) proposed a gradual transition from melt to hydrothermal fluid in such systems. Indeed, today it seems clear that the Ilímaussaq rocks are an exceptional archive for such a transition and the detailed study of these final products of agpaite evolution may reveal further details of this process, which to date is only fragmentally understood.

Methane-rich fluid inclusions with minor amounts of hydrogen and higher hydrocarbons and the occurrence of bitumen are also known from other agpaitic complexes (especially from the Kola peninsula) and may be a general characteristic of such rocks (e.g., Nivin 2002). Despite many studies, there was and is considerable debate whether such fluids are of a magmatic origin and represent primary high-temperature conditions (mantle gas theory) or are the result of various secondary processes (e.g., Konnerup-Madsen 2001; Potter and Konnerup-Madsen 2003; Krumrei et al. 2007; Graser et al. 2008; Beeskow et al. 2006; Nivin et al. 1995, 2001, 2005; Potter et al. 1998, 2004; Potter and Longstaffe 2007), but scientists agreed on the abiogenic origin of the methane-rich fluids found in the Ilímaussaq rocks and in similar agpaitic complexes in Russia (see above). This was, however, largely questioned by Laier and Nytoft (2012) who argued that the hydrocarbons of Ilímaussaq are of biogenic origin and migrated to Ilímaussaq from deep oil seeps offshore west of Greenland during the Cretaceous. Laier and Nytoft (2012) investigated the biomarker distribution in finely dispersed bitumens in the Ilímaussaq rocks. They concluded that a biogenic origin would be consistent with published stable isotope data for the bitumen, and even re-interpreted the published stable isotope characteristics of hydrocarbon gases from Ilímaussaq and from the agpaitic Khibina and Lovozero complexes in Russia mostly as of biogenic origin. Loss of hydrocarbon gases via diffusion (as reported from Khibina and Lovozero by Nivin et al. 2001, 2005) may significantly change their isotopic composition (Zhang and Kross 2001) and this process is invoked by Laier and Nytoft (2012) to be the reason that hydrocarbon gases from Ilímaussaq, Khibina and Lovozero could be mistaken as abiogenic gases. This theory is in stark contrast to numerous detailed fluid inclusions studies (see above) and has to be regarded as dubious until it can be documented that such fluids have also been trapped in minerals outside the Ilímaussaq complex as one would expect, if they migrated from deep oil seeps offshore west of Greenland. Hydrocarbons in the Ilímaussaq rocks are in agreement with the predicted behavior of the C–O–H system under the extremely reduced crystallization conditions reported for the Ilímaussaq magmas (e.g., Holloway and Jakobsson 1986; Scott 2004; Zhang and Duan 2010; Mysen and Yamashita 2010). So far the idea of Laier and Nytoft (2012) has to be regarded as highly unlikely and it is the responsibility of Laier, Nytoft and their followers to present the supporting evidence that the application of biomarker distributions to such magmatic to hydrothermal rocks is a reliable source of information. Obviously, the source and origin of the hydrocarbons and especially that of the so-far little studied bitumen in Ilímaussaq (and other agpaitic complexes) is an exciting topic which needs further investigation.

Metasomatic and Hydrothermal Processes

The fluids associated with the Ilímaussaq magmas caused intense metasomatism and hydrothermal overprint within the complex, at its margins, and in the surrounding country rocks, resulting in highly variable secondary mineral assemblages (e.g., Ferguson 1964; Sørensen et al. 1969; Graser and Markl 2008; Derrey 2012). Furthermore many varied types of hydrothermal veins (see above) and several hundreds of square meters large albitized and analcimized areas as well as oxidized and hematitized zones throughout the complex demonstrate the role of intensive late-stage fluid circulation both pervasive and channelized (e.g., Ferguson 1964; Engell et al. 1971).

In most rock types, the primary mineral assemblages are resorbed and replaced by secondary minerals to variable extents (see above): The olivine-augite-ulvöspinel assemblage is replaced by katophorite-arfvedsonite, aenigmatite and aegirine-augite (Fig. 14.3, e.g., Larsen 1976; Markl et al. 2001); primary apatite (up to 21 wt.% REE_2O_3) became unstable, was dissolved and recrystallized to several new generations of apatite with partly high Na- and REE contents (up to 2.9 and 27 wt.%, respectively; Rønsbo, 1989, 2008; Zirner et al. submitted) and felsic minerals (alkali feldspar, sodalite and nepheline) are commonly replaced by pure albite, analcime, secondary sodalite, nepheline, natrolite and other zeolithes (e.g., Ferguson 1964; Markl et al. 2001). Eudialyte is often altered to complex mixtures of albite, aegirine, microcline, catapleite, fluorite, monazite, apatite, neptunite, rarely zircon and a wealth of other REE-rich minerals (e.g., Rose-Hansen and Sørensen 2002; Karup-Møller et al. 2010; Karup-Møller and Rose-Hansen 2013). Furthermore, autoliths of various Ilímaussaq rocks within younger members of the complex are overprinted to variable degrees causing similar mineralogical changes (e.g. Schönenberger et al. 2006).

Although there exists a large variety of different kinds of hydrothermal veins in the complex, only the Be silicate-bearing ones were investigated in greater detail (Engell et al. 1971; Markl 2001a; Markl and Baumgartner 2002). Their field relations imply that they (at least partly) replaced earlier Ilímaussaq rocks and their microtextures and mineral assemblages point to strongly basic pH values for the responsible fluids (or melts?) governing the formation of minerals like ussingite and tugtupite (Markl and Baumgartner 2002) during hyper-agpaitic conditions (see above).

At the margins of the complex, Ca-rich skarn-like rocks with ilvaite-grossular-, and epidote-bearing assemblages occur. Their formation was interpreted in a model in which seawater was the metasomatizing fluid which entered the surrounding basaltic lavas, reacted with them (spilitization) and infiltrated Ca along fractures into the metasomatized rocks (Graser and Markl 2008). Further evidence for the infiltration of seawater along the margins of the complex is provided by Li and B isotope studies of Marks et al. (2007) and Kaliwoda et al. (2011).

Metasomatism in the country rocks of the complex was first described by Ferguson (1964). Detailed work of Derrey (2012) showed that in the granitic country rocks along the South coast of the Kangerluarsuk fjord metasomatism occurred up to at least 120 m away from the contact. Several effects on the granitic country rocks

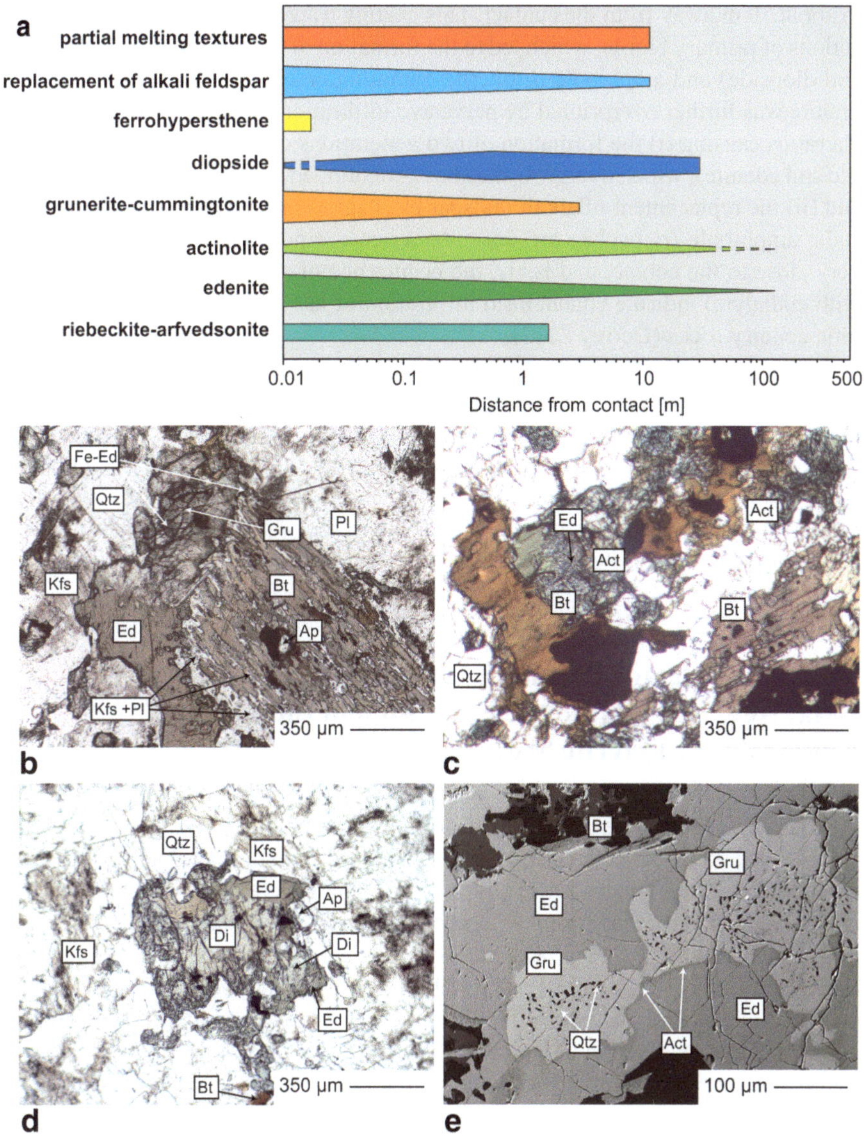

Fig. 14.10 Mineralogical and textural effects on the granitic country rocks caused by the emplacement of the Ilímaussaq complex (modified from Derrey 2012). **a** Overview on the textures and minerals found in the country rock against distance from the contact. **b** Breakdown of biotite with newly formed grunerite that is overgrown by edenite. **c** Intergrowth of actinolite and edenite replacing biotite. **d** A diopside core mantled by edenite. **e** BSE image of grunerite with tiny quartz inclusions, rimmed by a thin actinolite seam, which is overgrown by edenite.

caused by the emplacement of the Ilímaussaq complex can be distinguished and a detailed textural analysis allows the assessment of their relative timing (Fig. 14.10). First of all, the granite was heated to about 670 °C causing partial melting of the granite as recorded by granophyric intergrowth textures of alkali feldspar and quartz up to about 10 m away from the contact. This heating triggered several dehydration reactions of primary biotite, which led to the formation of pyroxene (ferrohypersthene and diopside) and amphibole (grunerite-cummingtonite series compositions). The granite was further overprinted by pervasive infiltration of fluids with variable Ca/Na ratios causing (i) the formation of two generations of calcic amphibole (actinolite and edenite), which overgrow the above-mentioned biotite-breakdown products and (ii) the replacement of alkali feldspar by plagioclase and quartz. Subsequently, sodic amphibole (riebeckite-arfvedsonite series compositions) formed in samples very close to the contact and lastly, the occurrence of aegirine-rich veins (in cases with eudialyte) indicate channelized infiltration of late agpaitic fluids into the granitic country rocks (Derrey 2012).

In all, the products of intensive late-stage fluid-assisted processes within and around the Ilímaussaq complex are plentiful and variable. These aspects of the Ilímaussaq evolution need to be investigated in detail, in order to better understand the physical and chemical characteristics of agpaitic fluids and their capability for transporting and concentrating HFSE and other unusual elements (e.g., Be, Sn) to sub-economic or economic interest. Clarifying the potential role of hydrocarbons and bitumen in such systems is another important task to achieve.

Combining Detailed Mineral Chemical and Structural Data: New Insight Into the Formation of the Layered Kakortokite-Lujavrite Sequence

Igneous layering has been described in almost all of the major Gardar plutonic complexes and from some of the larger dykes, covering a remarkably wide compositional spectrum of rocks (see Upton et al. 1996 for a review). In the Ilímaussaq complex itself, different forms of igneous layering are present in all rock types (e.g., Sørensen and Larsen (1987), but have been mostly studied in the kakortokites. Since their first description by Ussing (1912) their origin has been widely discussed, resulting in a number of very different models for their formation. These include intermittent crystallization of a multiply saturated magma (Ussing 1912; Sørensen, 1968), convective overturns (Upton and Pulvertaft 1961; Bohse et al. 1971), crystallization from a multiply layered magma chamber (Larsen and Sørensen 1987; Bailey 1995; Bailey et al. 2006), recurrent magma recharge (Pfaff et al. 2008) and changes in vapor pressure (Ussing 1912; Pfaff et al. 2008). A satisfactory model for the layering observed in the kakortokite sequence needs to explain at least the following features:

1. The formation of three-layer units with internal mineral grading, but with sharp upper and lower contacts between units of the lower layered kakortokites and the absence of clear layering in the overlying slightly layered kakortokites.
2. The recurrence of these rhythmic units of at least 29 times and their relative continuity over several kilometers.

Furthermore, the transitional layered kakortokites evolve gradually (structurally and mineralogically) into the lujavrite sequence. Their genetic relationship towards slightly layered kakortokites and lower layered kakortokites, however, are unclear (see above).

The Formation of Three-Layer Units

Textural evidence implies that amphibole, eudialyte, and alkali feldspar + nepheline crystallized more or less contemporaneously (e.g., Sørensen 1968; Pfaff et al. 2008; Lindhuber et al. in press), hence there must have been an effective mechanism which separated these minerals from each other after crystalization. Given the significant density contrast between these phases ($\rho_{arfvedsonite} > \rho_{eudialyte} > \rho_{alkali\ feldspar} \sim \rho_{nepheline}$), the modal grading within a given unit is believed to be due to gravity settling (e.g., Sørensen 1968; Larsen and Sørensen 1987; Pfaff et al. 2008). Field observations support this assumption: Arfvedsonite is strongly concentrated at the base of each unit (black layer), eudialyte is enriched in the middle (red layer) while upper white layer is rich in the light minerals (alkali feldspar + nepheline), with gradual transitions between the three layers of a given unit (Fig. 14.5b). This process is convenient to explain the formation of the modally graded three layer-units.

The overlying slightly layered kakortokites do not show clear modal layering and are finer grained than the lower layered kakortokites. Obviously, gravitational sorting processes were not effectively working in this upper sequence and these rocks may therefore represent an "upper border facies" of the kakortokites, similar to the pulaskite-foyaite-sodalite foyaite succession for the roof series rocks. However, detailed studies on the slightly layered kakortokites are needed in order to clarify their origin and to integrate these rocks into a comprehensive model.

Crystal Mat Formation and the Origin of the Layered Kakortokite Sequence

The recurrence of the three layer-units for at least 29 times (this is the number of three-layer units exposed today, with an unknown continuation at depth), however, is much less clear and cannot be explained by gravitational sorting of sinking minerals in a melt alone. An overview study on the compositional changes in eudialyte from black layers throughout the kakortokite sequence revealed surprisingly small indication of fractional crystallization upwards through the lower layered kakortokite sequence with only slightly decreasing Fe/Mn ratios in eudialyte (Fig. 14.11; Pfaff et al. 2008). A recent study by Lindhuber et al. (in press) focused on two sampling sites within the kakortokite sequence covering one and two three-layer units, respectively, including samples from the overlying base (black layer) of the next unit as well as from the top (white layer) of the underlying unit (Fig. 14.12). At

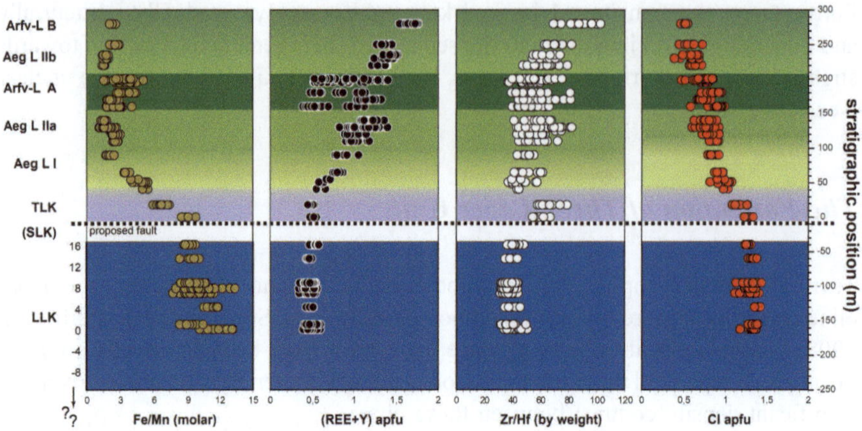

Fig. 14.11 Variation of eudialyte composition throughout the kakortokite-lujavrite sequence (Data from Pfaff et al. 2008; Lindhuber et al. in press and Ratschbacher et al. submitted). In lower layered kakortokites (LLK) only data for black layers are shown, for slightly layered kakortokites (SLK), no such data exist to date.

both sampling sites, the Fe/Mn ratio of eudialyte continuously decreases from the base (black layer) towards the top (white layer) of a given unit, which is interpreted to reflect fractional crystallization within a three layer-unit. The base of the overlying units (black layer), however, starts with a slightly higher Fe/Mn ratio (and thus less evolved), generally decreasing upwards through the sequence, resulting in a "christmas-tree pattern". This is indication for relatively little but detectable fractional crystalization within the lower layered kakortokites sequence. The black layer of unit seven, however, is less evolved than that of the underlying unit six (Fig. 14.12). This reset of the Fe/Mn ratio is associated with a small-scale fold at the contact between units + 6 and + 7, which may indicate post-cumulus movement within a semi-rigid crystal mush.

The model of mat formation and crystal settling in a magma chamber as introduced by Lauder (1964) and later invoked for the Skaergaard intrusion by Nielsen and Bernstein (2009) and Namur et al. (2015) can explain the recurrence of the kakortokite layering and is consistent with the observed changes in eudialyte composition. In a non-convecting magma chamber (as was proposed for the kakortokites by Sørensen (1968) and Sørensen and Larsen (1987)), cumulate minerals nucleate and grow throughout the magma. The heavier minerals (amphibole and eudialyte) would sink, while light minerals (alkali feldspar) would start to float. One possibility to produce a melt dense enough for feldspar to float is to crystalize large amounts of light minerals (sodalite) during an early stage. As cooling of the magma continues, amphibole, eudialyte and feldspar start to crystallize and the sinking amphibole and eudialyte interfere with the bouyont feldspar causing "traffic jams" (Bons et al. 2015). These crowding effects may result in amphibole-rich mats underlain by alkali feldspar-rich horizons (the later black and white layers). With time, the amphibole mats would further grow by the addition of more sinking amphibole crystals,

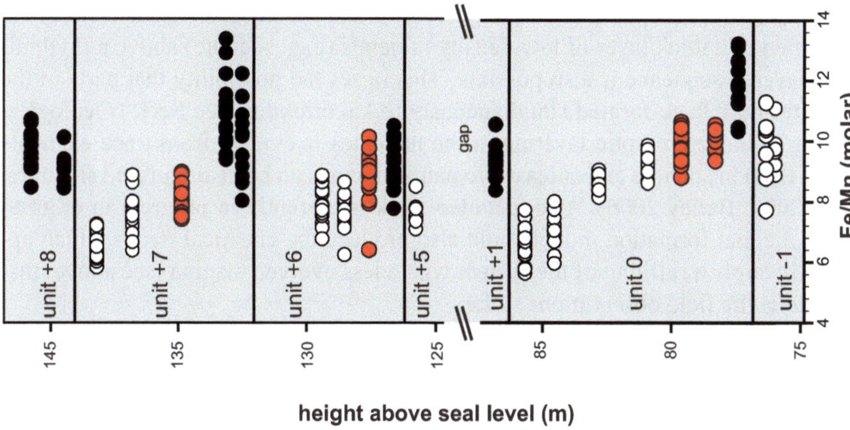

Fig. 14.12 Compositional variation of eudialyte in eudialyte from units −1 to +1 and +5 to +8, color-coded for *black*, *red* and *white* layers for the respective units (Data from Lindhuber et al. in press).

by intercumulus growth of grains already contained in the mat and by crystallization of melt pockets within the crystal mats. Hence, effective barriers would build up over time and the remaining melt and crystals between two of such amphibole barriers would eventually evolve as quasi-closed systems (Fig. 14.12). In such a scenario, the lowest mat within a given magma column would contain the earliest crystallized eudialyte with relatively high Fe/Mn ratios, because early crystallized eudyalite crystals may have been dragged down by amphibole and hence enclosed in the amphibole-mats before they became impermeable—consistent with the data of Pfaff et al. (2008) and Lindhuber et al. (in press). Relatively similar Fe/Mn ratios in eudialyte from black layers in the upper part of the lower layered kakortokites sequence (Pfaff et al. 2008) may imply that these were traped in the amphibole-rich mats and were therby protected from further equilibration with the evolving liquid and from exchange over unit boundaries. Thus, relatively primitive eudialyte grains should be present within all black layers, independent of their stratigraphic height. Later eudialyte crystallized from the evolving melt between two amphibole mats should then show decreasing Fe/Mn ratios upwards the sequence with distinct resets of the ratios between different units—again, consistent with the data of Lindhuber et al. (in press).

In all, the structural and mineral chemical features of the lower layered kakortokites can be explained by the proposed mechanism of mat formation associated with crowding effects. With time, amphibole-rich mats would sink downwards and a partly consolidated (and layered) kakortokite mush would accumulate from below. The presence of autoliths of other Ilímaussaq rocks within a distinct horizon of the lower layered kakortokites shows that such quasi-solid floors existed, whilst further upwards such mats were still forming.

A quantitative treatment of crowding effects and mat-forming processes in the lower layered kakortokites is given by Baur (2012) and Bons et al. (2015) and further implies that (depending on melt viscosity and a number of other factors) the formation of a thick layer of low-density minerals (e.g., sodalite) above a rhythmically layered sequence is also possible. This raises the possibility that parts of the naujaites may have formed simultaneously to kakortokites (see Sect. "Geological Setting"). Indeed, cryptic layering in the naujaites is evident from trace elements (Br, I, B, S) in sodalite and at least two naujaite units can be distinguished (Krumrei et al. 2007; Bailey 2006). Also, detailed work is currently in progress in order to test if the mat formation model could also produce the chemical trends which apparently imply a refilling of the system with a less evolved magma (see above) that contradict the field observations so far.

Surprisingly, the review by Naslund and McBirney (1996) on mechanisms of formation of igneous layering does not include the mat formation process proposed by Lauder (1964), although potentially representing a valid explanation for the formation of rhythmic layering in general. Recently, however, this process was invoked to explain layering features observed in the Triple Group of the Skaergaard intrusion (Nielsen and Bernstein 2009; Namur et al. 2015) and it remains to be tested if the process of mat formation during crystal settling can explain the reccurrence of rhythmic layering in other multiply saturated (anchieutectic) layered intrusions as well.

The Magmatic Evolution and Architecture of the Lujavrite Sill Complex

Pfaff et al. (2008) and Ratschbacher et al. (submitted) performed detailed studies on the compositional variation of eudialyte (which is a component of the early cumulus assemblage) throughout the transitional layered kakortokite-lujavrite sequence (Fig. 14.11). An almost continuous trend with strongly decreasing Fe/Mn and increasing (REE + Y) throughout the sequence is believed to reflect fractional crystallization (Pfaff et al. 2008; Ratschbacher et al. submitted). The simultaneous decrease of Cl in eudialyte may reflect an increase of water activity during lujavrite evolution, consistent with petrological calculations and fluid inclusion data (see above). In arfvedsonite lujavrite A, however, a reset to less evolved melt compositions (lower REE + Y and higher Fe/Mn in eudialyte) may have resulted from an influx of less evolved melt into the lujavritic magma bodies (Ratschbacher et al. submitted). This is consistent with observations of intrusive contacts between aegirine lujavrite IIa and arfvedsonite lujavrite A and suggests that the latter is a separate sheet intruded into aegirine lujavrite IIa instead of being an in situ differentiate from it.

Ratschbacher et al. (submitted) presents mineral chemical evidence indicating that at least two melt batches were responsible for the formation of the lujavrite sequence in the Lakseelv area, challenging the classical view of a continuous closed system evolution for the lujavrites (Sørensen 2006). The lujavrite sequence including the TLK is probably a distinct unit of the Ilímaussaq complex and may represent the remnants of a steep feeder zone once forming a sill complex consisting of at least two magma batches (see Sect. "Field Relations among Major Rock Types and General Structure of the Complex"). They intruded the naujaites both horizontally and vertically (Fig. 14.7). The texturally and structurally distinct M-C lujavrites west of Appat (Fig. 14.1) are probably products of another independent magma body.

A direct genetic relationship between lujavrites of the Lakseelv area and the LLK is questionable as is the significance of the spatially associated "slumped kakortokites" (Bohse et al. 1971). We speculate that the Lakseelv valley may be the trace of a shallow northwards dipping flexure or fault, which juxtaposed the layered kakortokites of Kringlerne against their once overlying naujaites. This triggered and channeled the intrusion of a feeder system for lujavrite-forming melts. Based on preliminary structural data on the lujavrite sequence, including foliations, lineations and shear-sense indicators, such movements were active during super-solidus and sub-solidus conditions of the lujavrites (Ratschbacher et al. submitted). However, detailed structural follow-up work, including the interpretation of microstructures in the lujavrites and in the "lujavrite-naujaite inclusion zone" are needed to clarify such important structural details. Other modern methods like AMS measurements (e.g., O'Driscoll et al. 2006) are promising tools to answer such questions about the architecture of plutonic systems. Also, detailed structural investigations in the closer and broader vicinity of the complex could probably improve the still sketchy picture of the tectonic evolution of the Gardar rift system and may help to understand how the emplacement of the Ilímaussaq magmas was influenced by tectonic structures (e.g., Ussing 1912; Ferguson 1964).

What is the parental magma of the Ilimaussaq complex?

The present model for the petrogenesis of the Ilímaussaq complex involves a deep-seated magma chamber, which was successively tapped resulting in several magma batches showing variable degrees of differentiation. These were successively emplaced at a shallow crustal level (e.g., Sørensen and Larsen 1987). Tracking and reconstructing the detailed geochemical evolution of the Ilímaussaq melt(s) is complicated by the fact that most rock types are the result of crystal fractionation and cumulate formation, and there is appreciable ambiguity about the detailed genetic relations between the different rock types and the number of magma batches involved in the formation of the complex (see above). Thus, when searching for a potential parental melt composition of the Ilímaussaq magmatic system, the first question must be: parental to which stage of the evolution? Parental to the augite syenite stage, the agpaitic stage, or parental to the whole complex?

New Experimental Evidence for the Augite Syenite—Agpaite Relations

There has been considerable debate about the detailed genetic relationships between the augite syenite and the early agpaites (e.g., Engell 1973; Sørensen 1978; Bailey et al. 1981). To date, it is not entirely clear if agpaites evolved continuously from augite syenites via fractional crystallization at depth or if they were formed from a distinct batch of mantle magma.

Based on geochemical considerations, about 80–95 % crystallization of augite syenite magma must have occurred in order to reach the sodalite foyaite stage, which would require a huge underlying magma chamber (Engell 1973). Similarly, but even more extreme, Bailey et al. (1981) estimated about 99 % fractionation of augite syenite required to reach the lujavrite stage. However, both calculations are based on the assumption that the trace element contents of the border facies of the augite syenite represent that of an actual melt, which is questionable.

Various dyke-rocks, some lujavrites and the relatively fine-grained rocks of the marginal pegmatite were proposed to represent the agpaitic stage of the Ilímaussaq magma (e.g., Larsen and Steenfelt 1974; Bailey et al. 1981; Sørensen 2006). The perhaps most promising candidate is the above-mentioned micro-kakortokite dyke (Larsen and Steenfelt 1974). Despite being highly evolved (A.I. = 1.4; X_{Fe} = 0.98) compared to the augite syenite (A.I. = 0.8; X_{Fe} = 0.77), this rock records the mineralogical evolution of the various Ilímaussaq units (for details see Marks and Markl 2003). Recent experimental work of Giehl et al. (2012, 2014) used the composition of this dyke as a model for the Ilímaussaq complex and simulated the liquid line of descent for peralkaline iron-rich phonolites. This work demonstrates that even strongly peralkaline melts can fractionate significant amounts of ol, aug, mag and alkfsp (representing augite syenite, a metaluminous and not peralkaline rock type), questioning the earlier assumption that augite syenite may represent a melt composition. We speculate that augite syenite could thus be a cumulate rock formed during the early evolutionary stages of a peralkaline melt. In some experiments, up to about 20 % of this assemblage formed, corresponding roughly to the amount of augite syenite relative to agpaites we see today in the field. When the residual melts reach A.I. values of about 2.4 or higher and are rich in chlorine (≥0.2 wt.%) and ZrO_2 (≥0.7 wt.%), aenigmatite, aegirine, eudialyte and sodalite become stable phases at about 800 °C (Giehl et al. 2014). Thus, this dyke rock may represent a melt composition that would be able to produce the phase assemblages seen in the Ilímaussaq rocks, allowing for the earlier fractionation of some olivine and augite. If this rock, however, separated from the Ilímaussaq source pool before, during or after the augite syenite stage remains questionable.

A Genetic Link Between Kakortokites and Naujaites?

Equally unclear are the detailed genetic relationships between naujaites and kakortokites. Naujaite contains large amounts of euhedral sodalite embedded in interstitial alkali feldspar, nepheline, arfvedsonite, eudialyte and a number of accessories. Kakortokites are mineralogically very similar, but texturally very different. Eudialyte, nepheline, feldspar and arfvedsonite are euhedral cumulus phases and sodalite is only present in minor amounts in the lowermost part and diminishes upwards the kakortokite sequence (e.g., Sørensen and Larsen 1987).

A straightforward and intriguing idea would thus be that (parts of) the naujaites resemble the counterparts of the kakortokites with both rocks forming from the same magma batch (as Ussing already proposed in 1912). Since sodalite has the least density of the four minerals, it would float in a melt fastest. If sodalite saturation of the melt was achieved first and feldspar, nepheline, arfvedsonite and eudialyte would join the crystallizing assemblage slightly later, but still during an early magmatic stage, no amphibole-rich crystal mats (see above) acting as barriers would be present at this stage of magma evolution. Therfore, sodalite would not be hindered to float (except for maybe the lowermost crystallized sodalites, which may be captured in the lowermost exposed kakortokites, see above). A change from solely sodalite to a sodalite + feldspar + nepheline + eudialyte + arfvedsonite ± accessories cumulus assemblage could also account for the two geochemically distinct naujaite layers (see above; Krumrei et al. 2007). Large autoliths of naujaite are present in one specific unit of the kakortokites, demonstrating that parts of the naujaites were already consolidated when parts of the kakortokites were still forming (Bohse and Andersen 1981). This does not contradict the model above since both naujaites and kakortokites probably formed over a considerable time interval, giving enough time for such a process.

Larsen (1976) suggested that kakortokites crystallized from a magma that was distinct from the naujaite parent magma. This was based on the fact that amphiboles in the kakortokites are more magnesian and less calcic than those of the naujaites. As far more data are available today (Pfaff et al. 2008; Marks et al. 2004; Lindhuber et al. in press), the apparent difference in X_{Fe} is not present anymore. Although amphibole compositions from naujaite and kakortokite show considerable overlap, those from naujaites tend to be higher in Ca and Si (Fig. 14.9a). This, however can also be explained by the fact that in naujaites, amphibole is not a cumulus phase, but crystallized from interstitial melt captured in the rising sodalite mush and these small interstitial melt pockets are probably not representative of the whole magma column.

Preliminary numerical simulations (Baur 2012; Bons et al. 2015 show that it is possible to produce an uppermost light layer (sodalite-dominated), followed downwards by a rhythmic layered sequence (rich in alkali feldspar, eudialyte and amphibole), which is underlain by an amphibole-dominated unit (potentially represented by the "slumped kakortokites"?) out of one magma volume. Indeed, no unequivocal evidence for the detailed genetic relationships between naujaites and kakortokites is available to date and the final answer to this question is maybe hidden below the exposed lower layered kakortokites and critically depends on the interpretation of the proposed fault in the Lakseelv valley (Fig. 14.6; see Sect. "The Magmatic Evolution and Architecture of the Lujavrite Sill Complex"). Also, detailed knowledge on the spatial distribution and frequency of the early fayalite-hedenbergite-ulvöspinel assemblage (see Sect. "Major Mineralogical and Mineral Chemical Changes During the Magmatic to Hydrothermal Evolution of the Ilímaussaq System") throughout the whole naujaite sequence (which is more than 600 m thick) could potentialy help to test this hypothesis, but such a study has not been performed yet.

In all, there is no direct information on the actual melt compositions coexisting with the observed mineral assemblages—neither for the Ilímaussaq complex nor for the Gardar province in general, since to date, no melt inclusion study on the Ilímaussaq rocks or other Gardar rocks has been performed. Such a study would reveal many important details on the geochemical evolution and the liquid line of descent of the fascinating Ilímaussaq magmatic system.

Acknowledgements The funding of various Ilímaussaq- and Gardar-related projects provided by the Deutsche Forschungsgemeinschaft is gratefully acknowledged. Also, we highly appreciate the work of numerous student assistants, Diploma students, and PhD students, which together contributed to our mineralogical, petrological and geochemical knowledge on the Ilímaussaq area over the past 15 years. Special gratitude is owed to Thomas Wenzel. He provided critical support during microprobe work and brought many of the past Ilímaussaq projects to a success, because of his expertise, his steady interest in this work and his constructive comments on an earlier version of this manuscript (as well of several others in the past). Many constructive and fruitful discussions with numerous colleagues enhanced our understanding on the Gardar Province and the Ilímaussaq area. Amongst others we would like to thank Brian Upton, Henning Bohse and Paul Bons. We are also grateful to Salik Hard and co-workers of the Narsaq tourist office, Harry Andersen, Peter Lindberg, Helgi Jonasson and Stefán Magnússon, who provided logistic support during various field campaigns in South Greenland. We also thank Lotte Melchior Larsen, Tom Andersen and Troels F.D. Nielsen for their very detailed and constructive reviews, which helped us to improve the present work.

References

Allaart JH (1969) The chronology and petrography of the Gardar dykes between Igaliko Fjord and Redekammen, South Greenland. Rapp Grønl Geol Under 25:26

Allaart JH (1976) Ketilidian mobile belt in South Greenland. In: Escher A, Watt WS (eds) Geology of Greenland. Grønlands Geologiske Undersøgelse :121–151

Andersen T, Sørensen H (2005) Stability of naujakasite in hyperagpaitic melts, and the petrology of naujakasite lujavrite in the Ilímaussaq alkaline complex, South Greenland. Mineral Mag 69:125–136

Andersen S, Bohse H, Steenfelt A (1981) A geological section through the southern part of the Ilímaussaq intrusion. Rapp Grønl Geol Undersøgelse 103:39–42

Andersen S, Bohse H, Steenfelt A (1988) Geological map of Greenland 1:20,000. The southern part of the Ilímaussaq complex, South Greenland. Grønlands Geologiske Undersøgelse/Geodaetisk Institut, Denmark

Andersen T, Friis H (submitted) The transition from agpaitic to hyperagpaitic magmatic crystallization in the Ilímaussaq alkaline complex, South Greenland. J Petrol

Arzamastsev A, Arzamastseva L, Glaznev VN, Raevskii A (1998) Petrologic-geophysical model for the structure and composition of deep levels of the Khibina and Lovozero complexes, Kola Peninsula. Petrology 6:434–450

Arzamastsev AA, Bea F, Glaznev VN, Arzamastseva LV, Montero P (2001) Kola alkaline province in the Paleozoic: evaluation of primary mantle magma composition and magma generation conditions. Russ J Earth Sci 3:1–32

Bailey JC (1995) Cryptorhythmic and macrorhythmic layering in aegirine lujavrite, Ilímaussaq alkaline intrusion, South Greenland. Bull Geol Soc Den 42:1–16

Bailey JC (2006) Geochemistry of boron in the Ilímaussaq alkaline complex, South Greenland. Lithos 91:319–330

Bailey JC, Gwodz R (1994) Li distribution in aegirine lujavrite, Ilímaussaq alkaline intrusion, South Greenland: role of cumulus and post-cumulus processes. Lithos 31:207–225

Bailey JC, Rose-Hansen J, Løvborg L, Sørensen H (1981) Evolution of Th and U whole-rock contents in the Ilímaussaq intrusion. Rapp Grønl Geol Undersøgelse 103:87–98

Bailey JC, Bohse H, Gwodzdz R, Rose-Hansen J (1993) Li in minerals from the Ilímaussaq alkaline intrusion, South Greenland. Bull Geol Soc Den 40:288–299

Bailey JC, Gwozdz R, Rose-Hansen J, Sørensen H (2001) Geochemical overview of the Ilímaussaq alkaline complex, South Greenland. Geol Greenl Surv Bull 190:35–53

Bailey JC, Sørensen H, Andersen T, Kogarko LN, Rose-Hansen J (2006) On the origin of microrhythmic layering in arfvedsonite lujavrite from the Ilímaussaq alkaline complex, South Greenland. Lithos 91:301–318

Baur A (2012) Modellierung von Schichtung in einer Magmenkammer durch Gravitationssortierung. Bachelorthesis, Universität Tübingen

Beeskow B, Treloar PJ, Rankin AH, Vennemann TW, Spangenberg J (2006) A reassessment of models for hydrocarbon generation in the Khibina nepheline syenite complex, Kola Peninsula, Russia. Lithos 91:1–18

Blaxland AB, van Breeman O, Steenfelt A (1976) Age and origin of agpaitic magmatism at Ilímaussaq, South Greenland: Rb–Sr study. Lithos 9:31–38

Bohse H, Andersen S (1981) Review of the stratigraphic divisions of the kakortokite and lujavrite in southern Ilímaussaq. Rapp Grønl Geol Undersøgelse 103:53–62

Bohse H, Brooks CK, Kunzendorf H (1971) Field observations on the Kakortokites of the Ilímaussaq intrusion, South Greenland, including mapping and analyses by portable x-ray fluorescence equipment for zirconium and niobium. Rapp Grønl Geol Undersøgelse 38:43

Bons PD, Baur A, Elburg MA, Lindhuer MJ, Marks MAW, Soesoo A, van Milligen BP, Walte NP (2015) Layered intrusions and traffic jams. Geology. 43:71–74

Botcharnikov RE, Koepke J, Holtz F, McCammon C, Wilke M (2005) The effect of water activity on the oxidation and structural state of Fe in a ferro-basaltic melt. Geochim Cosmochim Acta 69:5071–5085

Bridgwater D, Coe K (1970) The role of stoping in the emplacement of the giant dikes of Isortoq, South Greenland. Geol J Special issue 2:67–78

Bridgwater D, Harry WT (1968) Anorthosite xenoliths and plagioclase megacrysts in Precambrian intrusions of South Greenland. Medd om Grønl 185:243

Derrey I (2012) Element transport and mineral replacement reactions during alkali contact metamorphism: effects on the Julianehåb granite induced by the Ilímaussaq intrusion, SW-Greenland. Diplomathesis, Universität Tübingen

Eales HV, Cawthorn RG (1996) The Bushveld Complex. In: Cawthorn RG (Ed) Layered intrusions. Elsevier Science Ltd, Amsterdam, pp 181–230

Emeleus CH, Upton BGJ (1976) The Gardar period in Southern Greenland. In: Escher A, Watt WS (eds) Geology of Greenland. Grønlands Geologiske Undersøgelse :152–181

Emeleus CH, Cheadle MJ, Hunter RH, Upton BGJ, Wadsworth WJ (1996) The rum layered suite. In: Cawthorn RG (ed) Layered intrusions. Elsevier Science Ltd, Lausanne, pp 403–440

Engell J (1973) A closed system crystal-fractionation model for the agpaitic Ilímaussaq intrusion, South Greenland , with special reference to the lujavrites. Bull Geol Soc Den 22:334–362

Engell J, Hansen J, Jensen M, Kunzendorf H, Løvborg L (1971) Beryllium mineralization in the Ilímaussaq Intrusion, South Greenland, with description of a field beryllometer and chemical methods. Rapp Grønl Geol Undersøgelse 33:40

Féménias O, Coussaert N, Brassiness S, Demaiffe D (2005) Emplacement processes and cooling history of layered cyclic unit II-7 from the Lovozero alkaline massif (Kola Peninsula, Russia). Lithos 83:371–393

Ferguson J (1964) Geology of the Ilímaussaq alkaline intrusion, South Greenland. Bull Grønl Geol Undersøgelse 39:82

Ferguson J (1970) The significance of the kakortokite in the evolution of the Ilímaussaq intrusion, South Greenland. Bull Grønl Geol Undersøgelse 89:193

Forsberg R, Rasmussen KL (1978) Gravity and rock densities in the Ilímaussaq area, South Greenland. Rapp Grønl Geol Undersøgelse 90:81–84

Galakhov A (1975) Petrology of the Khibina alkaline massif. Nauka, Leningrad, 296 pp

Garde AA, Hamilton MA, Chadwick B, Grocott J, McCaffrey KJW (2002) The Ketilidian orogen of South Greenland: geochronology, tectonics, magmatism, and fore-arc accretion during Palaeoproterozoic oblique convergence. Can J Earth Sci 39:765–793

Gaweda A, Szopa K (2011) The origin of magmatic layering in the high tatra granite, Central Western Carpathians-implications for the formation of granitoid plutons. Trans Royal Soc of Edinb 102:129–144

Giehl C, Marks MAW, Nowak M (2012) Phase relations and liquid lines of descent of an iron-rich peralkaline phonolitic melt: an experimental study. Contrib Mineral Petrol 165:283–304

Giehl C, Marks MAW, Nowak M (2014) An experimental study on the influence of fluorine and chlorine on phase relations in peralkaline phonolitic melts. Contrib Mineral Petrol 167:977

Giuli C, Alonso-Mori R, Cicconi MR, Paris E, Glatzel P, Eeckhout SG, Scaillet B (2012) Effect of alkalis on the Fe oxidation state and local environment in peralkaline rhyolitic glasses. Am Mineral 97:468–475

GMEL (2015) Greenland minerals and energy Ltd. http://www.ggg.gl. Accessed 15 Feb 2015

Goodenough KM, Upton BGJ, Ellam RM (2002) Long-term memory of subduction processes in the lithospheric mantle: evidence from the geochemistry of basic dykes in the Gardar Province of South Greenland. J Geol Soc London 159:705–714

Graser G, Markl G (2008) Ca-rich ilvaite-epidote-hydrogarnet endoskarns: a record of late-magmatic fluid influx into the persodic Ilímaussaq complex, South Greenland. J Petrol 49:239–265

Graser G, Potter J, Köhler J, Markl G (2008) Isotope, major, minor and trace element geochemistry of late-magmatic fluids in the peralkaline Ilímaussaq intrusion, South Greenland. Lithos 106:207–221

Halama R, Waight T, Markl G (2002) Geochemical and isotopic zoning patterns of plagioclase megacrysts in gabbroic dykes from the Gardar Province, South Greenland: implications for crystallisation processes in anorthositic magmas. Contrib Mineral Petrol 144:109–127

Halama R, Wenzel T, Upton BGJ, Siebel W, Markl G (2003) A geochemical and Sr–Nd–O isotopic study of the Proterozoic Eriksfjord Basalts, Gardar Province, South Greenland: reconstruction of an OIB-signature in crustally contaminated rift-related basalts. Mineral Mag 67:831–853

Halama R, Marks MAW, Brügmann GE, Siebel W, Wenzel T, Markl G (2004) Crustal contamination of mafic magmas: evidence from a petrological, geochemical and Sr–Nd–Os–O isotopic study of the Proterozoic Isortoq dike swarm, South Greenland. Lithos 74:199–232

Hamilton EI (1964) The geochemistry of the northern part of the Ilímaussaq intrusion, S.W. Greenland. Bull Grønl Geol Undersøgelse 42:104

Hettmann K, Wenzel T, Marks MAW, Markl G (2012) The sulfur speciation in S-bearing minerals: new constraints by a combination of electron microprobe analysis and DFT calculations with special reference to sodalite-group minerals. Am Mineral 97:1653–1661

Hettmann K, Marks MAW, Kreissig K, Zack T, Wenzel T, Rehkämper M, Jacob DE, Markl G (2014) The geochemistry of Tl and its isotopes during magmatic and hydrothermal processes: the peralkaline Ilímaussaq complex, southwest Greenland. Chem Geol 366:1–13

Holloway JR, Jakobsson S (1986) Volatile solubilities in magmas: transport of volatiles from mantles to planet surface. J Geophys Res 91:D505–D508

Kaliwoda M, Marschall H, Marks MAW, Ludwig T, Altherr R, Markl G (2011) Lithium, beryllium, boron and boron isotope systematics in the peralkaline Ilímaussaq intrusion (South Greenland) and its granitic country rocks: interplay between magmatic and hydrothermal processes. Lithos 125:51–64

Karup-Møller S (1978) The ore minerals of the Ilímaussaq intrusion: their mode of occurence and their conditions of formation. Bull Grønl Geol Undersøgelse 127:51

Karup-Møller S, Rose-Hansen J (2013) New data on eudialyte decomposition minerals from kakortokites and associated pegmatites of the Ilímaussaq complex, South Greenland. Bull Geol Soc Den 61:47–70

Karup-Møller S, Rose-Hansen J, Sørensen H (2010) Eudialyte decomposition minerals with new hithero undescribed phases from the Ilímaussaq complex, South Greenland. Bull Geol Soc Den 58:75–88

Khomyakov A (1995) Mineralogy of hyperagpaitic alkaline rocks. Scientific Publications, Clarendon, Oxford, p 222

Kogarko LN (1977) Problems of the genesis of agpaitic magmas. Nauka, Moskow, p 294

Kogarko LN, Krigmann LD (1970) Phase equilibria in the system nepheline-NaF. Geochem Int 7:103–107

Kogarko LN, Romanchev BP (1983) Phase equilibria in alkaline melts. Int Geol Rev 25:534–546

Köhler J, Schönenberger J, Upton B, Markl G (2009) Halogen and trace-element geochemistry in the Gardar Province, South Greenland: subduction-related metasomatism and fluid exsolution from alkalic melts. Lithos 113:731–747

Konnerup-Madsen J (1980) Fluid inclusions in minerals from igneous rocks belonging to Precambrian continental Gardar rift province, South Greenland: the alkaline Ilímaussaq intrusion and the alkali acidic igneous complexes. PhD Thesis, University of Copenhagen 140 pp

Konnerup-Madsen J (1984) Composition of fluid inclusions in granites and quartz syenites from the gardar continental rift province (South Greenland). Bull Minéral 107:327–340

Konnerup-Madsen J (2001) A review of the composition and evolution of hydrocarbon gases during solidification of the Ilímaussaq alkaline complex, South Greenland. Geol Greenl Surv Bull 190:159–166

Konnerup-Madsen J, Rose-Hansen J (1982) Volatiles associated with alkaline igneous rift activity: fluid inclusions in the Ilímaussaq intrusion and the Gardar granitic complexes (South Greenland). Chem Geol 37:79–93

Konnerup-Madsen J, Rose-Hansen J (1984) Composition and significance of fluid inclusions in the Ilímaussaq peralkaline granite, South Greenland. Bull Minéral 107:317–326

Konnerup-Madsen J, Larsen E, Rose-Hansen J (1979) Hydrocarbon-rich fluid inclusions in minerals from the alkaline Ilímaussaq intrusion, South Greenland. Bull Minéral 102:642–653

Konnerup-Madsen J, Rose-Hansen J, Larsen E (1981) Hydrocarbon gases asociated with alkaline igneous activity: evidence from compositions of fluid inclusions. Rapp Grønl Geol Undersøgelse 103:99–108

Konnerup-Madsen J, Dubessy J, Rose-Hansen J (1985) Combined Raman microprobe spectrometry and microthermometry of fluid inclusions in minerals from igneous rocks of the Gardar Province (South Greenland). Lithos 18:271–280

Konnerup-Madsen J, Kreulen R, Rose-Hansen J (1988) Stable isotope characteristics of hydrocarbon gases in the alkaline Ilímaussaq complex, South Greenland. Bull Minéral 111:567–576

Kramm U, Kogarko LN (1994) Nd and Sr isotope signatures of the Khibina and Lovozero agpaitic centres, Kola Province, Russia. Lithos 32:225–242

Krumrei TV, Villa IM, Marks MAW, Markl G (2006) A $^{40}Ar/^{39}Ar$ and U/Pb isotopic study of the Ilímaussaq complex, South Greenland: implications for the ^{40}K decay constant and for the duration of magmatic activity in a peralkaline complex. Chem Geol 227:258–273

Krumrei TV, Pernicka E, Kaliwoda M, Markl G (2007) Volatiles in a peralkaline system: abiogenic hydrocarbons and F–Cl–Br systematics in the naujaite of the Ilímaussaq intrusion, South Greenland. Lithos 95:298–314

Laier T, Nytoft HP (2012) Bitumen biomarkers in the mid-Proterozoic Ilímaussaq intrusion, Southwest Greenland-a challenge to the mantle gas theory. Mar Petrol Geol 30:50–65

Larsen LM (1976) Clinopyroxenes and coexisting mafic minerals from the alkaline Ilímaussaq intrusion, South Greenland. J Petrol 17:258–290

Larsen JG (1977a) Petrology of the late lavas of the Eriksfjord formation, Gardar Province, South Greenland. Medd om Grønl 53:31

Larsen LM (1977b) Aenigmatites from the Ilímaussaq intrusion, South Greenland: chemistry and petrological implications. Lithos 10:257–270

Larsen LM (1981) Chemistry of feldspars in the Ilímaussaq augite syenite with additional data on some other minerals. Rapp Grønl Geol Unders 103:31–37
Larsen LM, Sørensen H (1987) The Ilímaussaq intrusion-progressive crystallization and formation of layering in an agpaitic magma. In: Fitton JG, Upton, BGJ (eds) Alkaline igneous rocks, Geological Society of London, Special Publication 30:473–488
Larsen LM, Steenfelt A (1974) Alkali loss and retention in an iron-rich peralkaline phonolite dyke from the Gardar Province, South Greenland. Lithos 7:81–90
Lauder W (1964) Mat formation and crystall settling in magma. Nature 202:1100–1101
Lindhuber M, Marks MAW, Bons PD, Wenzel T, Markl G (in press) Crystal mat-formation as an igneous layering-forming process: textural and geochemical evidence from the 'lower layered' nepheline syenite sequence of the Ilímaussaq complex, South Greenland. Lithos
Linnen RL, Keppler H (2002) Melt composition control of Zr/Hf fractionation in magmatic processes. Geochim Cosmochim Acta 66:3293–3301
Mann U, Marks MAW, Markl G (2006) Influence of oxygen fugacity on mineral compositions in peralkaline melts: the Katzenbuckel volcano, Southwest Germany. Lithos 91:262–285
Markl G (2001a) Stability of Na–Be minerals in late-magmatic fluids of the Ilímaussaq alkaline complex, South Greenland. Geol Greenl Surv Bull 190:145–158
Markl G (2001b) A new type of silicate liquid immiscibility in peralkaline nepheline syenites (lujavrites) of the Ilímaussaq complex, South Greenland. Contrib Mineral Petrol 141:458–472
Markl G, Baumgartner L (2002) pH changes in peralkaline late-magmatic fluids. Contrib Mineral Petrol 144:331–346
Markl G, Marks MAW, Schwinn G, Sommer H (2001) Phase equilibrium constraints on intensive crystallization parameters of the Ilímaussaq complex, South Greenland. J Petrol 42:2231–2258
Markl G, Marks MAW, Frost BR (2010) On the controls of oxygen fugacity in the generation and crystallization of peralkaline melts. J Petrol 51:1831–1847
Marks MAW, Markl G (2001) Fractionation and assimilation processes in the alkaline augite syenite unit of the Ilímaussaq intrusion, South Greenland, as deduced from phase equilibria. J Petrol 42:1947–1969
Marks MAW, Markl G (2003) Ilímaussaq 'en miniature': closed-system fractionation in an agpaitic dyke rock from the Gardar Province, South Greenland. Mineral Mag 67:893–919
Marks MAW, Vennemann TW, Siebel W, Markl G (2003) Quantification of magmatic and hydrothermal processes in a peralkaline syenite- alkali granite complex based on textures, phase equilibria, and stable and radiogenic isotopes. J Petrol 44:1247–1280
Marks MAW, Vennemann T, Siebel W, Markl G (2004) Nd-, O-, and H-isotopic evidence for complex, closed-system fluid evolution of the peralkaline Ilímaussaq intrusion, South Greenland. Geochim Cosmochim Acta 68:3379–3395
Marks MAW, Rudnick R, Vennemann T, McCammon C, Markl G (2007) Arrested kinetic Li isotope fractionation at the margin of the Ilímaussaq complex, South Greenland: evidence for open-system processes during final cooling of peralkaline igneous rocks. Chem Geol 246:207–230
Marks MAW, Hettmann K, Schilling J, Frost BR, Markl G (2011) The mineralogical diversity of alkaline igneous rocks: critical factors for the transition from miaskitic to agpaitic phase assemblages. J Petrol 52:439–455
McBirney AR (1996) The Skaergaard intrusion. In: Cawthorn RG (ed) Layered intrusions. Elsevier Science Ltd, New York, pp 147–180
McCallum IS (1996) The Stillwater complex. In: Cawthorn RG (ed) Layered intrusions. Elsevier Science Ltd, Oxford, pp 441–484
Morse SA (1969) The Kiglapait layered intrusion, labrador. Geol Soc Am Memoir 112:204
Müller-Lorch D, Marks MAW, Markl G (2007) Na and K distribution in agpaitic pegmatites. Lithos 95:315–330
Mysen BO, Yamashita S (2010) Speciation of reduced C–O–H volatiles in coexisting fluids and silicate melts determined in-situ to ~ 1.4 GPa and 800 °C. Geochim Cosmochim Acta 74:4577–4588

Namur O, Abily B, Boudreau A, Blanchette F, Bush JWM, Ceuleneer G, Charlier B, Donaldson CH, Duchesne JC, Higgins MD, Morata D, Nielsen TFD, O'Driscoll B, Pang KN, Peacock T, Spandler C, Toramaru A, Veksler I (2015) Igneous Layering in Basaltic Magma Chambers. In Charlier et al. (Editors) Layered Intrusions, Springer Geology, Dordrecht, in press. (DOI 10.1007/978-94-017-9652-1_2)
Naslund HR, McBirney AR (1996) Mechanisms of formation of igneous layering. In: Cawthorn RG (ed) Layered intrusions. Elsevier Science Ltd, Shannon, pp 1–44
Nielsen TF, Bernstein S (2009) Chemical variations in the triple group of the skaergard intrusion: insights for the mineralization and crystallization process. AGU Fall meeting 2009, Abstract V21A 1956
Nielsen BL, Steenfelt A (1979) Intrusive events at Kvanefjeld in the Ilímaussaq igneous complex. Bull Geol Soc Den 27:143–155
Nivin VA (2002) Gas concentrations in minerals with reference to the problem of the genesis of hydrocarbon gases in rocks of the Khibiny and Lovozero complexes. Geochem Int 40:883–898
Nivin VA, Devirts AL, Lagutina YP (1995) The origin of the gas phase in the Lovozero massif based on hydrogen-isotope data. Geochem Int 32:65–71
Nivin VA, Belov NI, Treloar PJ, Timofeyef VV (2001) Relationships between gas geochemistry and release rates and the geochemical state of igneous rock massifs. Tectonophysics 336:233–244
Nivin VA, Treloar PJ, Konopleva NG, Ikorsky SV (2005) A review of the occurence, form and origin of C-bearing species in the Khibiny alkaline igneous complex, Kola Peninsula, Russia. Lithos 85:93–112
O'Driscoll B, Troll VR, Reavy RJ, Turner P (2006) The Great Eucrite intrusion of Ardnamurchan, Scotland: reevaluating the ring-dike concept. Geology 34:189–192
Parsons I (2012) Full stop for mother earth. Elements 8:396–398
Paslick CR, Halliday AN, Davies GR, Mezger K, Upton BGJ (1993) Timing of proterozoic magmatism in the Gardar Province, southern Greenland. Bull Geol Soc Am 105:272–278
Petersen OV (2001) List of minerals identified in the Ilímaussaq alkaline complex, South Greenland. Geol Greenl Surv Bull 190:25–34
Petersilie IA, Sørensen H (1970) Hydrocarbon gases and bituminous substances in rocks from the Ilímaussaq alkaline intrusion, South Greenland. Lithos 3:59–76
Pfaff K, Krumrei TV, Marks M, Wenzel T, Rudolf T, Markl G (2008) Chemical and physical evolution of the 'lower layered series' from the nepheline syenitic Ilímaussaq intrusion, South Greenland: implications for the origin of magmatic layering in peralkaline felsic liquids. Lithos 106:280–296
Potter J, Konnerup-Madsen J (2003) A review of the occurence and origin of hydrocarbons in igneous rocks. In: Petford N, McCaffrey KJW (eds) Hydrocarbons in crystalline rocks, Geological Society Special Publication 214:151–173
Potter J, Longstaffe FJ (2007) A gas-chromatograph continuous flow isotope ratio mass-spectrometry method for ^{13}C and D measurement of complex fuid inclusion volatiles: examples from the Khibina alkaline igneous complex, Northwest Russia and the South Wales Coalfelds. Chem Geol 244:186–201
Potter J., Rankin AH, Treloar PJ (2004) Abiogenic Fischer-Tropsch synthesis of hydrocarbons in alkaline igneous rocks: fluid inclusion, textural and isotopic evidence from the Lovozero complex, N.W. Russia. Lithos 75:311–330
Potter J, Rankin AH, Treloar PJ, Nivin VA, Ting W, Ni P (1998) A preliminary study of methane inclusions in alkaline igneous rocks of the Kola igneous province, Russia: implications for the origin of methane in igneous rocks. Eur J Mineral 10:1167–1180
Poulsen V (1964) The sandstones of the Precambrian Eriksfjord formation in South Greenland. Rapp Grønl Geol Under 2:16
Pupier E, Barbey O, Toplis M, Bussy F (2008) Igneous layering, fractional crystallization and growth of granitic plutons: the Dolbel Batholith in SW Niger. J Petrol 49:1043–1068

Ratschbacher B, Marks MAW, Wenzel T, Bons P, Markl G (submitted) Emplacement and geochemical evolution of highly evolved nepheline syenites in the composite alkaline to peralkaline Ilímaussaq complex, SW Greenland. Lithos
Rose-Hansen J, Sørensen H (2001) Minor intrusions of peralkaline microsyenite in the Ilímaussaq alkaline complex, South Greenland. Bull Geol Soc Den 48:9–29
Rose-Hansen J, Sørensen H (2002) Geology of the Lujavrites from the Ilímaussaq alkaline complex South Greenland, with information from seven bore holes. Medd om Grønl Geosci 40:58
Rønsbo JG (1989) Coupled substitutions involving REEs and Na and Si in apatites in alkaline rocks from the Ilímaussaq intrusion, South Greenland, and the petrological implications. Am Mineral 74:896–901
Rønsbo JG (2008) Apatite in the Ilímaussaq alkaline complex: occurrence, zonation and compositional variation. Lithos 106:71–82
Sass JH, Nielsen BL, Wollenberg HA, Munroe RJ (1972) Heat flow and surface radioactivity at two sites in South Greenland. J Geophys Res 77(32):6435–6444
Schilling J, Wu F-Y, McCammon C, Wenzel T, Marks MAW, Pfaff K, Jacob DE, Markl G (2011) The compositional variability of eudialyte-group minerals. Mineral Mag 75:87–115
Schmid C (2008) Die Pulaskite der Ilímaussaq Intrusion in Südgrönland. Diplomathesis, Universität Tübingen
Schönenberger J, Marks M, Wagner T, Markl G (2006) Fluid-rock interaction on autoliths of agpaitic nepheline syenites in the Ilímaussaq intrusion, South Greenland. Lithos 91:331–351
Scott HP, Hemley RJ, Mao HK, Herschbach DR, Fried LE, Howard WM, Bastea S (2004) Generation of methane in the Earth's mantle: in situ high pressure-temperature measurements of carbonate reduction. PNAS 101:14023–14026
Sobolev AV, Bazarova TY, Shugorova NA, Bazarov LS, Dolgov YA, Sørensen H (1970) A preleminary examination of fluid inclusions in nepheline, sorensenite, tugtupite and chkalovite from the Ilímaussaq alkaline intrusion, South Greenland. Bull Grønl Geol Under 81:32
Sørensen H (1962) On the occurrence of steenstrupine in the Ilímaussaq massif, Southwest Greenland. Bull Grønl Geol Under 32:251
Sørensen H (1968) Rhythmic igneous layering in peralkaline intrusions: an essay review on Ilímaussaq (Greenland) and Lovozero (Kola, USSR). Lithos 2:261–283
Sørensen H (1978) The position of the augite syenite and pulaskite in the Ilímaussaq intrusion, South Greenland. Bull Geol Soc Den 27:15–23
Sørensen H (1992) Agpaitic nepheline syenites: a potential source of rare elements. Appl Geochem 7:417–427
Sørensen H (1997) The agpaitic rocks-an overview. Mineral Mag 61:485–498
Sørensen H (2001) Brief introduction to the geology of the Ilímaussaq alkaline complex, South Greenland, and its exploration history. Geol Greenl Surv Bull 190:7–24
Sørensen H (2006) The Ilímaussaq alkaline complex, South Greenland-an overview of 200 years of research and an outlook. Medd om Grønl Geosci 45:10–31
Sørensen H, Larsen LM (1987) Layering in the Ilímaussaq alkaline intrusion, South Greenland. In: Parsons I (ed) Origins of igneous layering. D. Reidel, Dordrecht, pp 1–28
Sørensen H, Larsen LM (2001) The hyper-agpaitic stage in the evolution of the Ilímaussaq alkaline complex, South Greenland. Geol Greenl Surv Bull 190:83–94
Sørensen H, Hansen J, Bondesen E (1969) Preliminary account of the geology of the Kvanefjeld area of the Ilímaussaq intrusion, South Greenland. Rapp Grønl Geol Under 18:40
Sørensen H, Rose-Hansen J, Nielsen BL, Løvborg L, Sørensen E, Lundgaard T (1974) The uranium deposit at Kvanefjeld, the Ilímaussaq intrusion, South Greenland. Rapp Grønl Geol Under 60:54
Sørensen H, Bailey JC, Kogarko LN, Rose-Hansen J, Karup-Møller S (2003) Spheroidal structures in arfvedsonite lujavrite, Ilímaussaq alkaline complex, South Greenland-an example of macroscale liquid immiscibility. Lithos 70:1–20
Sørensen H, Bohse H, Bailey JC (2006) The origin and mode of emplacement of lujavrites in the Ilímaussaq alkaline complex, South Greenland. Lithos 91:286–300

Sørensen H, Bailey JC, Rose-Hansen J (2011) The emplacement and crystallization of the U–Th-REE-rich agpaitic and hyperagpaitic lujavrites at Kvanefjeld, Ilímaussaq alkaline complex, South Greenland. Bull Geol Soc Den 59:69–92

Steenfelt A (1981) Field relations in the roof zone of the Ilímaussaq intrusion with special reference to the position of the alkali acid rocks. Rapp Grønl Geol Under 103:43–52

Stevenson R, Upton BGJ, Steenfelt A (1997) Crust-mantle interaction in the evolution of the Ilímaussaq complex, South Greenland: Nd isotopic studies. Lithos 40:189–202

TANBREEZ (2014) TANBREEZ Mining Greenland A/S. http://www.tanbreez.com. Accessed 2 Aug 2014

Thorning L, Stemp RW (1997) Projects Aeromag 1995 and Aeromag 1996. Results from aeromagnetic surveys over South Greenland (1995) and South-West and southern West Greenland (1996). Dan Grønl Geol Under Rapp 11:44

Upton BGJ (1991) Gardar mantle xenoliths: Iglutalik, South Greenland. Rapp Grønl Geol Under 150:37–43

Upton BGJ (2013) Tectono-magmatic evolution of the southern branch of the Gardar rift in the late Gardar period. Geol Surv Den Greenl Bull 29:124

Upton BGJ, Emeleus CH (1987) Mid-Proterozoic alkaline magmatism in southern Greenland: the Gardar Province. In: Fitton JG, Upton BGJ (eds) Alkaline igneous rocks, Geological Society of London, Special Publication 30:449–471

Upton B, Pulvertaft TCR (1961) Textural features of some contrasted igenous cumulates from South Greenland. Medd om Grønl 123:1–29

Upton BGJ, Martin AR, Stephenson D (1990) Evolution of the tugtutôq central complex, South Greenland; a high-level, rift-axial, late-Gardar centre. J Volcanol Geotherm Res 43:195–214

Upton BGJ, Parsons I, Emeleus CH, Hodson M (1996) Layered alkaline igneous rocks of the Gardar Province, South Greendland. In: Cawthorn RG (ed) Layered intrusions. Elsevier Science Ltd, Tokyo, pp 331–364

Upton BGJ, Emeleus CH, Heaman LM, Goodenough KM, Finch A (2003) Magmatism of the mid-Proterozoic Gardar Province, South Greenland: chronology, petrogenesis and geological setting. Lithos 68:43–65

Ussing NV (1912) Geology of the country around Julianehaab, Greenland. Medd om Grønl 38:426

Vlasov K, Kuzmenko M, Eskova E (1966) The Lovozero alkali massif. Oliver and Boyd, Edinburgh, p 627

Waight T, Baker J, Willigers B (2002) Rb isotope dilution analyses by MC-ICPMS using Zr to correct for mass fractionation: towards improved Rb–Sr geochronology? Chem Geol 186:99–116

Watson EB (1979) Zircon saturation in felsic liquids: experimental results and applications to trace element geochemistry. Contrib Mineral Petrol 70:407–419

Wilson AH (1996) The Great Dyke of Zimbabwe. In: Cawthorn RG (ed) Layered intrusions. Elsevier Science Ltd, pp 365–402

Zhang C, Duan Z (2010) A model for C–O–H fluid in the Earth's mantle. Geochim Cosmochim Acta 73:2089–2102

Zhang T, Kross BM (2001) Experimental investigation on the carbon isotope fractionation of methane during gas migration by diffusion through sedimentary rocks at elevated temperature and pressure. Geochim Cosmochim Acta 65:2723–2742

Zirner ALK, Marks MAW, Wenzel T, Jacob DE, Markl G (submitted) Rare earth elements in apatite as a monitor of crystallization and metasomatic processes: the Ilimaussaq intrusion, South Greenland, as a type example. Lithos

Chapter 15
Ophiolitic Magma Chamber Processes, a Perspective from the Canadian Appalachians

Jean H. Bédard

Abstract Oceanic/ophiolitic magma chambers differ from continental layered mafic-ultramafic intrusions because magmatism is synchronous with extensional tectonism in a submarine environment. Because oceanic ridges continuously extend, new magma formed by decompression melting of the upwelling mantle constantly arrives beneath the ridge axis. Arriving magma commonly ponds at the base of the crust, or forms sills where favourable crustal structures (faults, shear zones, older sills) are encountered. A sheeted sill architecture for the middle and lower oceanic crust is probably common. Many monomineralic facies (anorthosite, chromitite, pyroxenite) in ophiolites form as reaction rims between newly emplaced primitive magma and evolved host cumulates as a result of incongruent dissolution or mixing across phase boundaries. When deformation is broadly distributed through the crust (Bay of Islands ophiolite), many previously-emplaced rocks experience high-temperature ductile shear that straddles the solidus. Consequently, modal cumulate layering is not always produced by sequential crystallization/accumulation or crystal sorting against a cooling surface or floor, but may form by transposition and tectonic repetition of partly-solidified intrusions, hosts and reaction facies. Syn-magmatic deformation triggers and activates mixing between intra-cumulate intrusions and incompletely consolidated host rocks to create a range of hybrid facies, few of which have cotectic phase proportions. Cumulates affected by penetrative deformation tend to have lower trapped melt fractions (5–10 %) than those unaffected by shear (20–30 %), suggesting that shear pumping actively expells pore melt from the deforming matrix. Percolation of primitive to residual melt through a deforming cumulus framework has the potential to mobilize incompatible elements and transform chemical signatures (Annieopsquotch ophiolite). Cumulates in the Betts Cove ophiolite are not penetratively deformed, and show well-developed size-graded cumulate beds, some with basal load structures, indicating an origin as gravity deposits. These types of cumulates may form in subsiding, fault-bounded 'trap-door' chambers. Graded harzburgitic cumulate beds are intercalated with bedding-parallel pyroxenite sheets that merge with discordant pyroxenite dykes, suggesting that they are bedding-parallel melt segregation veins that fed residual melt into fault-guided conduits, allowing expelled pore melt to be evacuated efficiently from within the thick pile of compacting cumulates.

J. H. Bédard (✉)
Geological Survey of Canada, 490 de la Couronne, Québec QC G1K 9A9, Canada
e-mail: jbedard@nrcan.gc.ca

B. Charlier et al. (eds.), *Layered Intrusions*, Springer Geology,
DOI 10.1007/978-94-017-9652-1_15

Keywords Ophiolite · Cumulate · Deformation · Percolation · Residual porosity

Introduction

Concepts derived from continental layered mafic-ultramafic intrusions are often applied to oceanic magma chambers. However, although oceanic and continental basalts crystallize the same phases, they are emplaced into fundamentally different environments, which influences the mechanisms by which magma chambers are created, replenished and solidify. Since oceanic magma chambers are not easily accessed, ophiolitic analogues have been extensively investigated to gain insights into lower crustal processes. This contribution does not attempt an exhaustive review of the literature on oceanic/ophiolitic magma chambers, since there are recent comprehensive overviews (Thy and Dilek 2000, 2003). Instead, I present facets of three Paleozoic Appalachian ophiolites from Newfoundland, Canada, that shed light on controversial issues (cumulate deposition by gravity, degree of melt entrapment in cumulates, mechanisms of post-cumulus melt expulsion, oceanic crust formation models), or that document important petrogenetic processes that have not received much attention (intra-cumulate hybridization, formation of monomineralic reaction rims, igneous metasomatism). The principal differences between continental mafic-ultramafic and oceanic/ophiolitic magma chambers are that most well-studied examples of the former intrude near-static (non-orogenic) environments within continents, whilst the latter are emplaced into active extensional submarine environments. Apart from local slumps and flattening fabrics related to compaction, most continental layered intrusions show few signs of penetrative syn-magmatic deformation except for small-scale normal shear zones related to differential compaction (e.g. Nicolas 1992). In contrast, oceanic ridges are continuously extending, with basaltic magma constantly arriving from beneath the ridge axis. I first review the general rheological structure of the oceanic/ophiolitic lithosphere so as to provide a general framework through which to understand how melt rising from the mantle moves through oceanic crust; after which I briefly review existing concepts of how magma chambers develop there. I then present three ophiolite case studies (Fig. 15.1), emphasize the petrogenetic implications of these observations, and discuss how the processes exhibited by the ophiolitic rocks might apply to continental magma chambers.

Although most ophiolites formed near subduction zones and so do not represent pure ocean ridge environments (eg. Pearce 2003; Hawkins 2003), the magma chamber mechanisms should be very similar, since most ophiolites are dominated by fairly dry tholeiitic magmas that do not differ radically from MORB (mid-ocean ridge basalt). Some supra-subduction zone (SSZ) ophiolites differ more significantly from true spreading ridges because of the important involvement of boninitic magmas. Nonetheless, the interplay of magmatism, deformation and hydrothermal circulation will be very similar. Boninites are more primitive than MORB, with higher SiO_2 and water contents leading to the early appearance of orthopyroxene

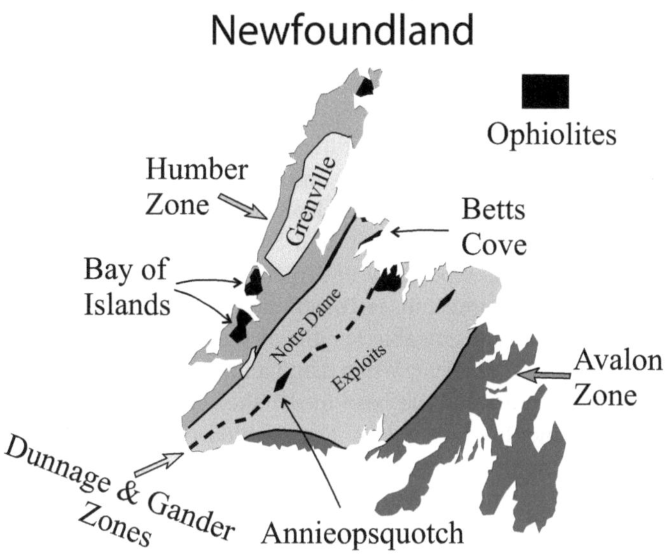

Fig. 15.1 Location map of the Newfoundland ophiolites discussed in this paper. The Tectonic zones are from Williams (1979)

and late appearance of plagioclase (e.g. Crawford 1989; Bédard 1999). Boninite-dominated ophiolites show many similarities with the Bushveld and Stillwater intrusions since boninites resemble the primitive 'U-type' magmas (eg. Barnes 1989) that show early orthopyroxene saturation.

Rheology of the Oceanic Crust

The rheological structure of actively-spreading oceanic lithosphere is principally controlled by the interplay between hydrothermal cooling, extension rate and the rate/volume of melt delivery (Harper 1985; Agar 1990, 1994; Chevenez et al. 1998; Manning et al. 2000; France et al. 2009; Koepke et al. 2011). Cooling of the submarine crust is mediated through forced circulation of seawater through a brittle, permeable carapace, with extensional tectonism facilitating circulation of water by increasing porosity. Typically, the upper basaltic crust is pervasively fissured and fractured, is commonly dissected by extensional normal faults, and is affected by hydrothermal/seafloor metamorphism (e.g. Wright et al. 1995; MacDonald 2001; Alt and Teagle 2003). The black smokers observed at many ridges represent the focussed efflux of the fluids that refrigerate the crust (Kelley et al. 1993, 2001; Humphris and Tivey 2000). The ridge axis is typically a graben at slow-spreading ridges, where extensional faults occur in a wide zone, and may exhume lower crust and mantle as core complexes (Karson 1990; Cannat et al. 1991; MacDonald 2001; Ohara et al. 2003; Smith et al. 2006). In contrast, at fast spreading ridges the axis is a broad dome elongated parallel to the ridge with only a shallow graben-like depression at the axial summit, and normal faults (often dipping away from the ridge) first appear on the dome flanks (MacDonald and Leyendyk 1986; Bohnenstiehl and Kleinrock 1999; MacDonald 2001).

When magma budgets are high the brittle domain near the ridge axis is generally restricted to the uppermost few km (lavas and sheeted dykes). High magma budgets are typical of fast-spreading ridges, but also occur episodically at slow-spreading ridges (Sinton and Detrick 1992; Cann and Smith 2005). When magma budgets are high the lower crust is repeatedly underplated or traversed by new batches of ascending melt; so its temperature is constrained to remain perched between the solidus and liquidus for considerable lengths of time (see France et al. 2009 for a discussion of process timescales). When magma budgets are high, the middle crust resembles the lower crust and the base of the sheeted dyke unit separates brittle rocks above from ductile rocks below. The position of this interface may fluctuate depending on small-scale phenomena (e.g. France et al. 2009; Koepke et al. 2011). For example, magmatic offshoots from beneath intrude, reheat, melt and assimilate rocks previously cooled to ambient (greenschist grade or lower) conditions; whereas pauses in magmatic activity result in hydrothermally driven crack propagation into the underlying gabbroic rocks, which causes hydrothermal metamorphism and metasomatism as it siphons heat out of the magma beneath to drive further downward solidification (Lister 1983; Lowell and Germanovich 1994; Manning et al. 2000). When magma budgets are low, these hydrothermal cooling cells can propagate down to the mantle and rigidify the entire crust. At the slow-spreading Mid-Atlantic Ridge, for example, earthquakes occur directly beneath the ridge axis at >10 km depth (Toomey et al. 1988) indicating that during a magmatic hiatus the hydrothermal front can penetrate to the mantle within a few thousand years, even while the crust is almost directly above the spreading ridge axis.

At active ridges mantle melt is nearly continuously being delivered to the base of the crust, such that the older rocks are repeatedly re-injected by new magma, providing common opportunities for reworking (Michael and Schilling 1989; Bédard et al. 2000a; Danyushevsky et al. 2004; Saal et al. 2007; Lissenberg and Dick 2008; Kvassnes and Grove 2008). This occurs until the segment that was located above the ridge axis has drifted far enough away from the melt delivery zone that it ceases to be affected. Consequently, an oceanic/ophiolitic crustal segment is typically subjected to multiple magmatic/deformational/hydrothermal events, the oldest of which will be most poorly preserved. In some ways the study of the lower oceanic crust is like the study of orogens, with evidence of the oldest events only surviving in strongly metamorphosed and deformed fragments. The final events seen by distal segments of crust that have drifted away from the melt delivery zone are likely to be structurally-controlled, sill-like offshoots from the ridge axis melt upflow zone, hydrothermal overprints and brittle-ductile shear zones (e.g. Harper 1985; Agar 1994; Alt and Bach 2006).

Since most of the oceanic middle and lower crust is perched at or near the solidus when magma budgets are high, the resulting crustal ductility implies that imposed tectonic extension could lead to large scale necking. When large volumes are hot, necking may be by distributed extensional deformation (Bédard 1991, 1993; Cannat et al. 1991), but it is common to see a progressive focusing of extensional structures into discrete shear zones that overprint early, more penetrative, higher-temperature fabrics (e.g. Cannat et al. 1991; Agar 1994; Giguère et al. 2003; Schroetter et al. 2003; Abily et al. 2011). These extensional shear zones may root into the crust/mantle transition (Moho), which also commonly represents a shear zone, typically a response to differential crust/mantle motion (Rabinowicz et al. 1984; Nicolas et al. 1988, 1994; Jousselin et al. 2012). This basal shear commonly imposes a dominant subhorizontal grain to the lowermost crust, typically with mineral stretching lineations oriented perpendicular to the ridge axis, and with a preferential fracture plane parallel to the ridge axis/sheeted dykes. The transpositional layering created by such ductile crustal deformation can easily be misidentified as depositional cumulate layering (see below).

When magma budgets are low, the crust alternates between magmatic and amagmatic extension phases (Harper 1985; Karson 1990; Cannat et al. 1991; Sinton and Detrick 1992; MacDonald 2001). When the bulk of the crust is cold and rigid, extensional strain increasingly focusses into shallowly-dipping (30–45°) normal faults that can exhume lower crustal and even mantle rocks to the surface as core-complexes (e.g. Karson 1990; Blackman et al. 2002; Schroetter et al. 2003, Ildefonse et al. 2007; Pagé et al. 2008; Tucholke et al. 2008; Miranda and Dilek 2010).

Oceanic/Ophiolitic Magma Chambers

Seismic data at fast-spreading ridges have revealed the common occurrence of long-lived melt-dominated pockets or lenses near the top of the ductile layer (Axial Melt Lense or AML), with regions beneath having lower porosities but still retaining small melt fractions (Sinton and Detrick 1992; Singh et al. 1998). It has been proposed that the bulk of the cumulate crust represents the solidified products of such AML that foundered into the lower crustal mush, and that these foundered cumulates are entrained towards the ridge flanks as a ductile gabbro 'glacier' (Sleep 1975; Henstock et al. 1993; Quick and Denlinger 1993). Notwithstanding the tremendous difficulties involved in transporting primary melts through many km of ductile lower crustal cumulates to refill the AML (intruding nowhere else!), there is also a serious mass balance problem involved if all of the lower crust is to be generated in this way. Another hypothesis is that the AML rarely sees primary melts, but rather represents accumulations of pore melt expelled from the underlying gabbroic mush, with melt pooling along the capping structural discontinuity (cf MacLeod and Yaouancq 2000; Lissenberg et al. 2004). In such a scenario, gabbro glaciers may indeed form beneath the high-level AML, but they will only characterize the uppermost cumulates located immediately below the sheeted dykes, and may be overprinted by through-flowing melt.

The principal alternative hypothesis for the genesis of large volumes of mushy lower oceanic crust is the small, multiple magma chamber model (Vine and Moores

1972; Smewing et al. 1975; Nisbet and Fowler 1978; Malpas 1990; Thy et al. 1989; Smith and Cann 1992) and its recent variant, the sheeted sill model (Bédard 1988, 1991, 1993; Boudier et al. 1996; Korenaga and Kelemen 1997; Lissenberg et al. 2004). A historical perspective on the development and evolution of ophiolitic magma chamber models can be found in Thy and Dilek (2003).

Multiple sill emplacement in the mid-lower crust is a predictable consequence of the crustal rheological structure. When new mantle-derived melt reaches the crust, the tendency is for intrusions to follow existing structural discontinuities and so abundant Moho-parallel sills are common (e.g. Garmany 1989; Bédard 1991; Boudier et al. 1996; Crawford et al. 1999; Dunn et al. 2000; Natland and Dick 2009). Because the hot lower oceanic crust is commonly ductile, melt-driven fractures (dykes) will have difficulty propagating upward and would commonly be arrested if a pre-existing melt-rich zone or shear zone should ever be intersected. If rising magma encounters obliquely-dipping structures, then obliquely dipping sheet-like intrusions would tend to form. There will be a dynamic interaction and feedback between deformation and intrusive activity, since new intrusions focus strain, being the weakest part of the crust.

Since many lower and middle oceanic/ophiolitic crustal rocks are layered adcumulates with low trapped melt fractions (e.g. Natland and Dick 2001), then an efficient mechanism is required to extract pore melt from solidifying intra-cumulate sills, or else the crust would instead be composed of orthocumulates with high trapped melt fractions. Korenaga and Kelemen (1997) proposed that the residual pore melt escaped lower crustal sills in the Oman ophiolite mainly through ephemeral cracks that closed up and healed afterwards. Dick et al. (1991) discussed the possible role of shear pumping to expel pore melt from oceanic cumulates, with shear zones guiding long-distance flow of residual melt. Field and geochemical evidence for oceanic/ophiolitic magma chamber dynamics will be presented in the next sections, as well as evidence bearing on the mechanisms by which melts are extracted from partially solidified intra-cumulate sills.

The Annieopsquotch Ophiolite

Description

The Annieopsquotch ophiolite (481–478 Ma Dunning and Krogh 1985) of central Newfoundland (Figs. 15.1 and 15.2) has been recently remapped and studied (Lissenberg et al. 2004, 2005). Pillow basalts (>500 m) grade down into a well-developed sheeted dyke unit (~1.5 km thick), that is rooted in the gabbroic crust beneath (>2.7 km thick). The base of the exposed crust is tectonically bounded and mantle rocks are absent. Most dykes and lavas are tholeiitic basalts with supra-subduction trace element signatures and most gabbros share this affinity (Fig. 15.3; Lissenberg et al. 2004, 2005). Lavas show slight LILE-depletion (LILE = large ion lithophile elements) in comparison to dykes (Fig. 15.3b), possibly as a result of seafloor weathering/hydrothermal alteration.

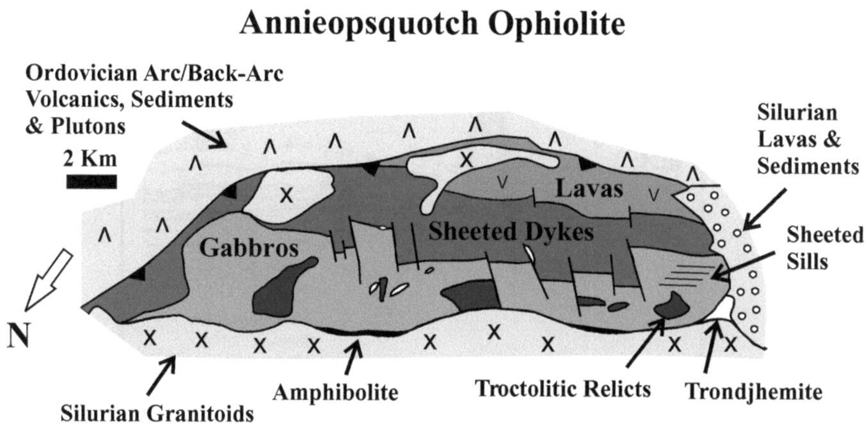

Fig. 15.2 Simplified map of the Anniopsquotch ophiolite adapted from Lissenberg et al. (2004). Map has been rotated so that the stratigraphic up direction is towards the top of the page

The gabbroic zone is divided into three units (Lissenberg et al. 2004). The highest level (~440 m thick) is composed of texturally heterogeneous gabbros that are geochemically almost identical to the overlying dykes and lavas (heavy dashed line in Fig. 15.3a). This implies that the upper gabbros are essentially a slowly-cooled, evolved (5–6 % MgO) multiphase basaltic melt, segregated into a high-level holding chamber that is possibly equivalent to the Axial Melt Lense (AML) of fast-spreading oceanic ridges. The high MgO content of many Annieopsquotch lavas and dykes implies that most of these magmas did not erupt from the 'high-level' AML, however, but were extracted from deeper, middle crustal reservoirs of the central gabbroic zone (Lissenberg et al. 2004).

The central gabbroic zone (~1.5 km thick), is constructed of palaeo-horizontal gabbroic bodies, typically 10–30 m thick, that are interpreted to represent sills (Lissenberg et al. 2004; their Figure 3). Upper and lower contacts of individual gabbroic sills commonly have finer grain sizes than their cores, indicating that the intrusions cooled from both top and bottom. The absence of hard chills implies that emplacement took place before previous sheets had cooled completely. Vein-like gabbroic apophyses (~20 cm) locally branch off from the upper contacts of gabbroic bodies and show inward growth structures from both sides (Lissenberg et al. 2004; their Figure 3). The local presence of crescumulate zones, up to 2 m thick and consisting of 5–10 cm long dendritic plagioclase and clinopyroxene branching down from the upper contacts of sills (Lissenberg et al. 2004; their Figure 3) clearly imply solidification from the top down. The sills are composed of weakly- or undeformed plagioclase + clinopyroxene +/− olivine cumulates. Trace element modelling implies that the magmas from which these cumulates formed had compositions very similar to the overlying sheeted dykes and basaltic lavas (Fig. 15.3c), with residual porosities (trapped melt fraction) most commonly between 15 % and 25 %. These high residual porosities are consistent with what is expected from compaction-driven melt expulsion in the absence of an imposed shear stress (eg. Schmidt

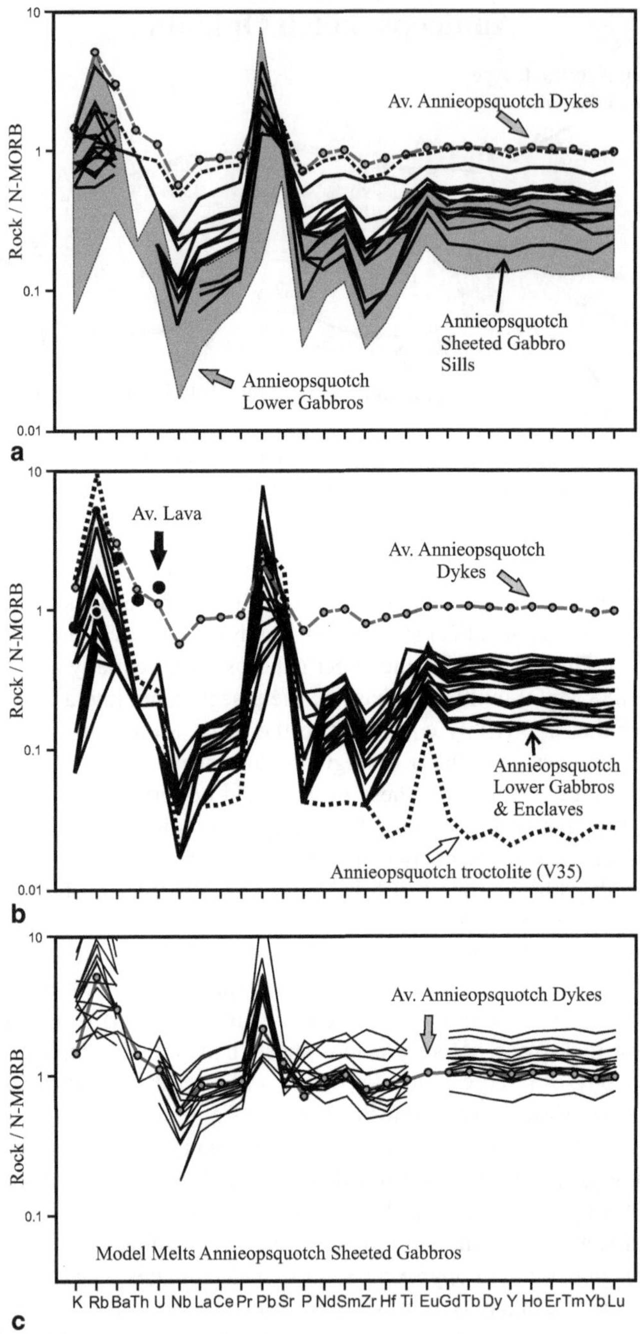

Fig. 15.3 Normalized extended trace element plots of Annieopsquotch rocks and model melts. The average of the Annieopsquotch sheeted dykes is shown on all panels as *grey circles* linked by a *grey dashed line*. **a** Gabbros from the sheeted sills (central crust), compared to the field of the lower crustal gabbros from panel b. A gabbro sampled immediately below the sheeted dykes is shown as a *black dashed line*, and represents a quasi-liquid composition. **b** Rocks from the lower

Fig. 15.4 Annieopsquotch field photographs of lower gabbroic crust. **a** Large enclave of paleo-vertically-dipping layered olivine gabbro. **b** Small plutonic enclave embedded in Annieopsquotch lower gabbros. **c** Larger olivine gabbro body that is veined by gabbro and pyroxenite (55 mm lens cap). **d** Pervasively veined 'background' gabbro (55 mm lens cap)

et al. 2012), in accord with the absence of evidence for superimposed penetrative deformation in the Annieopsquotch gabbros. The modelling also implies that the sheeted sills represent part of the plumbing system from which the Annieopsquotch lavas erupted.

The lowest exposed crust (~480 m thick) is characterized by heterogeneous gabbro, gabbro-norite, and minor websterite containing ~15 % layered olivine gabbro to troctolite enclaves. The enclaves range in size from 0.5 m to tens of metres (Figs. 15.4a and b; and Lissenberg et al. 2005 their Figure 2), with lithologies ranging from anorthositic leuco-gabbro to olivine mela-gabbro and troctolite. Modal and phase layering is restricted to the enclaves, and is on a 10 cm to 1 m scale (Fig. 15.4a). Layering orientations in the enclaves are variable, but many have paleo-vertical dips and some show macroscopic fold closures. Most enclaves contain euhedral to amoeboid clinopyroxene, some of which appears to replace an older clinopyroxene (Fig. 15.5). The enclaves typically have diffuse edges, marked by outwardly increasing proportions of pyroxene-rich veins, which eventually 'merge' with the

part of the crust. Spidergrams from enclaves and hosts are identical and are not distinguished on the plot. The boninitic troctolite (*V35*) enclave is shown as a *dashed black line*. The average basaltic lava is identical to the average sheeted dyke, except for LILE elements, where the average basalt composition is shown as a *black circle*. **c** Model melts from the mid-crustal zone sheeted gabbros (from Lissenberg et al. 2004). These models have trace element profiles that are parallel to the average sheeted dyke. See Lissenberg et al. (2004) for model constraints. Profiles normalized to the N-MORB composition of Sun and McDonough (1989)

Fig. 15.5 Photomicrographs of neoblastic (?) clinopyroxene-bearing enclaves from the lower gabbroic crust of Annieopsquotch. **a** Layered gabbro V37A. Note the amoeboid shape of clinopyroxene (*cpx*) and the scalloped pyroxene/plagioclase (*plag*) contacts. Crossed Nicols. **b** Layered gabbro V47A. Near-euhedral acicular clinopyroxene with relics of an older clinopyroxene within. *Ol* olivine, *opx* orthopyroxene. Crossed Nicols. **c** Scan of thin section V47A showing the euhedral clinopyroxene prisms. The *box* shows the location of the micrograph shown in **b**

dominant, more massive, lower crustal gabbros that surround them (Figs. 15.4c and d). The field and petrographic evidence implies that the 'background' lower crustal gabbros are partly intrusive into, and partly replacements of an older, heterogeneous cumulate crust, of which the layered enclaves are the vestiges.

Clinopyroxene-bearing layered enclaves have trace element signatures that are almost indistinguishable from those of surrounding lower crustal gabbros (Fig. 15.3b). In comparison to gabbroic rocks from the sheeted sills, the lower crustal gabbros (and pyroxene-bearing enclaves) are more depleted, with systematically lower L/HREE (Fig. 15.3). Figs. 15.6a–c show inverse trace element models for three of these pyroxene-bearing layered olivine gabbros, calculated using the equilibrium distribution method of Bédard (1994), with partition coefficients constrained as in Lissenberg et al. (2004, 2005). The trapped melt fractions of these

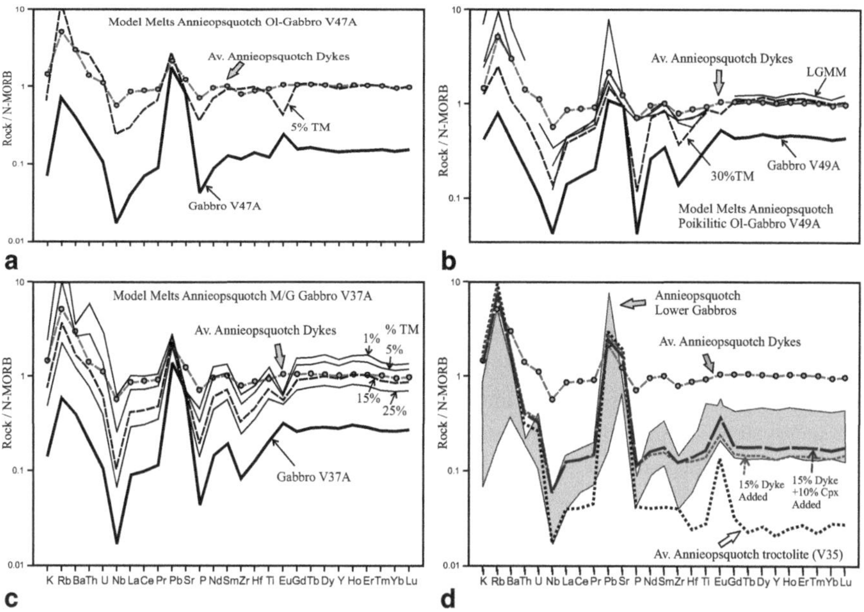

Fig. 15.6 Normalized extended trace element plots of Annieopsquotch rocks and model melts. **a–c** Three clinopyroxene-bearing enclaves from the lower gabbros (*heavy lines*) and their model melts (*heavy dashed lines*). The trapped melt fraction (TM) was varied to achieve a close fit to the HREE-content of the average sheeted dyke. Note that the LREE-contents of the model melts of all three enclaves are systematically depleted compared to the average Annieopsquotch dyke. **b** This panel shows three model melt profiles (*thin lines* labelled 'LGMM') from the lower gabbros taken from Lissenberg et al. (2004). **c** Inverse model melts corresponding to different assumed trapped melt fractions are also shown. Note that whatever TM fraction is assumed, the resulting model melts all show relative LREE depletion. **d** Mixing models attempting to reproduce the composition of lower crustal gabbros from Annieopsquotch. The *grey* field shows the compositions of lower crustal gabbros and enclaves from Fig. 15.3b. The *heavy black dotted line* is troctolite V35. The *grey dashed line* shows the result of adding 15 % average dyke to V35. The *black dashed line* shows result of adding an additional 10 % metasomatic clinopyroxene with a composition in equilibrium with the average dyke. Profiles normalized to the N-MORB composition of Sun and McDonough (1989)

models were adjusted to produce a reasonable HREE (Heavy Rare Earth Element) fit to the average of Annieposquotch dykes. The systematic difference in rock chemistry between the sheeted sill gabbros and lower crustal gabbros propagates into a systematic difference between model melts, with model melts of lower crustal gabbros showing LREE-depletion relative to dykes, lavas and model melts from sheeted sill gabbros (Fig. 15.6). Note that this difference is not an artifact of the modelling assumptions (Fig. 15.6c). This implies that unlike the gabbros in the sheeted sill zone, none of the lower crustal gabbros are simple accumulations of equilibrium crystals + trapped melt from the dominant tholeiites. On the other hand, the overall similarity of the trace element geochemical profiles suggests that the Annieposquotch tholeiites were involved somehow in the genesis of these rocks.

One of the most pristine of these layered enclaves (sample V35, a clinopyroxene-poor troctolite) differs considerably from clinopyroxene-rich layered enclaves. Sample V35 is strongly depleted in trace elements (Figs. 15.3b and 15.6d) and inverse modelling implies accumulation from a strongly depleted, possibly boninitic magma (Lissenberg et al. 2005). The absence of pyroxene and abundance of plagioclase is unusual for boninites, but the trace element signatures are typical of Appalachian low-Ti boninites. Lissenberg et al. (2005) suggested that troctolites like this one are relics of an older oceanic basement that was invaded by tholeiites during a later seafloor-spreading event. Figure 15.6d shows an attempt to reproduce the lower crustal gabbro geochemistry by adding 15 % average Annieopsquotch dyke to depleted troctolite V35. The mixtures show obvious differences from the target gabbros, with overly high L/HREE ratios. Thus, the enrichment process recorded in these rocks was not simple bulk magmatic impregnation followed by closed-system solidification. Two possible ways to explain the differences are: (1) loss of a late-stage melt enriched in the most incompatible elements (cf. Meurer and Meurer 2006), and (2) crystallization of a HREE-compatible phase (e.g. clinopyroxene) during percolative magma through-flow. Figure 15.6d shows the result of adding 15 % metasomatic clinopyroxene to the previous model 'impregnated' troctolite. This secondary model clinopyroxene was assumed to have formed in equilibrium with the average dyke composition. Although adding secondary clinopyroxene in this way decreases the final L/HREE to some extent, it does not exactly reproduce the lower crustal gabbro geochemical profiles. More complex through-flow/reaction/chromatographic models are probably needed (cf. Lissenberg et al. 2013), with formation of clinopyroxene from more evolved, more enriched pore melt. Late-stage loss of LREE-enriched evolved melt may also have occurred, but these hypotheses need further testing and refinement.

Implications

The well developed structures of the middle gabbroic zone at Annieposquotch provide unambiguous evidence of its origin as a sheeted sill complex emplaced into a stiff crust that was not affected by penetrative ductile deformation. Individual bodies (sills) have intrusive upper contacts and show evidence for cooling (crescumulates) from the top down. The absence of deformation allowed significant retention of trapped melt (15–25 %). Escape of some residual melt is recorded in the vein-like gabbroic apophyses emanating from upper sill contacts, but these are probably not sufficiently abundant to account for all the missing melt. However, unlike the Hole 735B drillcore (Dick et al. 1991), the density of sampling at Annieopsquotch is not adequate to determine whether intercalated bodies of more evolved gabbro might represent the full budget of missing evolved segregated melt. Consequently, it is not yet possible to know whether some evolved pore melt made its way back into the AML to mix and perhaps erupt, or whether it solidified as evolved veins within a larger mass of primitive cumulates.

The possibility of wholesale metasomatic replacement of cumulates is rarely considered, despite compelling evidence from some intrusions (e.g. Muskox: Irvine 1980; Rum: Bédard et al. 1988). The field, petrographic and geochemical evidence presented above implies that much of the Annieopsquotch lower crust has a metasomatic origin, representing impregnation and reequilibration of older cumulates by tholeiitic basalts associated with seafloor spreading. The combination of impregnation and metasomatic introduction of 'porphyroblastic' clinopyroxene into an older troctolitic cumulate (Figs. 15.4 and 15.5) was able to transform the whole-rock trace element chemistry to the point where it became nearly indistinguishable from that of surrounding 'intrusive' gabbros. Recent studies of the lower oceanic crust at both slow- and fast-spreading ridges (Lissenberg and Dick 2008; Lissenberg et al. 2013) also implies that percolation of residual basaltic melt through a thick gabbroic pile has the potential to remobilize and transform the chemistry of cumulate rocks on a large scale.

The Bay of Islands Ophiolite Complex

Description

The geology of the cumulate crust of the Latest Cambrian/Ordovician (~489 Ma, Kurth et al. 1998) Bay of Islands Ophiolite Complex (Fig. 15.7, BOIC) in the North Arm Mountain (NAM) Massif was outlined in a series of papers (Bédard 1988, 1991, 1993; Bédard and Hébert 1996, 1998; Bédard et al. 2000a; Giguère et al. 2003); providing a view that differs considerably from previous 'big-chamber' interpretations (Casey et al. 1983; Elthon et al. 1984; Komor et al. 1987; Komor and Elthon 1990). The plutonic crust of the BOIC at NAM can be divided into an uppermost, laterally-discontinuous unit (Unit IV, not distinguished from Unit III in Fig. 15.7) of isotropic near-liquid gabbro, anatectic trondhjemite, diorite and hornblendite (cf. Koepke et al. 2007). The 3–5 km thick underlying gabbro-dominated crust (Unit III) is laterally and vertically heterogeneous and will be discussed in more detail in the next section. The sheeted dykes (Rosencrantz 1983) are rooted in these two upper units. There is a laterally discontinuous unit at the base of the crust (Unit II, $\leq$1.5 km) that is dominated by pyroxenites and peridotites. These were previously interpreted to represent high pressure tholeiitic cumulates (Elthon et al. 1984), but are reinterpreted on the basis of detailed field mapping, petrology and geochemistry as cumulates extracted from a younger, underplating, syn-kinematic boninitic sill complex (Bédard 1991, 1993; Bédard and Hébert 1996). The transition to the underlying mantle rocks (Unit I) is typically marked by mylonitic shear zones, many of which branch up and dissect crustal blocks. The map shown in Fig. 15.7 represents the southern, steeply-dipping, homoclinal limb of a synclinal structure. As such, Fig. 15.7 is essentially a cross-section through the ophiolitic lithosphere.

Unit III.—A preliminary facies map was published in Giguère et al. (2003), and a modified version is presented here as Fig. 15.7b. The thicker western section has

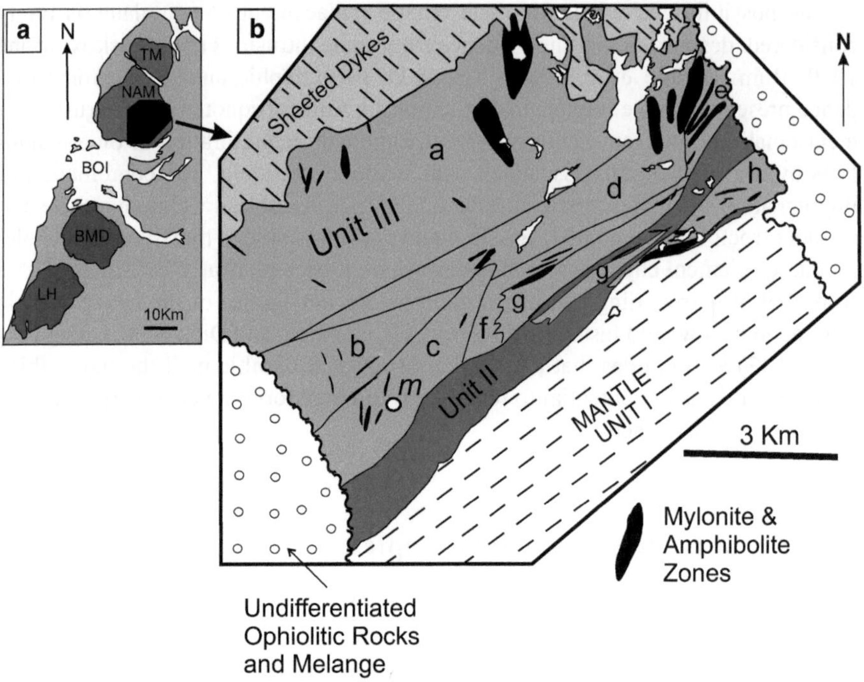

Fig. 15.7 Bay of Islands Ophiolite Complex. A) Four massifs of the Bay of Islands Ophiolite Complex on the west coast of Newfoundland. TM = Table Mountain, NAM = North Arm Mountain, BMD = Blow-me-Down, LH = Lewis Hills. BOI = Bay of Islands. B) Simplified facies map of the southern part of the North Arm Mountain Massif, adapted from Giguère et al. (2003, their fig. 2). Unit IV has been grouped into Unit IIIa. Unit IIIa: Olivine gabbro and leucogabbro, trondhjemite, diorite, hornblendite. IIIb: Micro-ferrogabbro, coarse gabbro, olivine gabbro. IIIc: Olivine gabbro, melagabbro, leucogabbro, anorthosite, picrite. IIId: Olivine leucogabbro. IIIe: Porphyritic ferrogabbro. IIIf: Ferrogabbro. IIIg: Olivine gabbro. IIIh: Partly hornblenditized olivine gabbro. The white circle marked m in domain IIIc is the location of the detail map of (Fig. 15.8)

abundant melagabbro to peridotitic intra-cumulate sills that were emplaced within a heterogeneous, strongly deformed, layered gabbro/melagabbro/leucogabbro (GML) complex (Unit IIIc). A detailed map of part of Unit IIIc is presented as Fig. 15.8 and will be discussed in more detail in the following. The olivine-rich sills gradually disappear up-section in Unit III, whilst intrusive ferro-gabbroic and late, less-deformed pegmatitic gabbro bodies increase in abundance up-section. In the east, the most primitive melagabbros and troctolites of Unit IIIc are absent, large crustal domains are composed of ferrogabbro, and there are prominent, high-temperature mylonite zones with steep paleo-dips (Fig. 15.7b). In fact, shear zones and mylonites are not restricted to these corridors, but occur throughout Unit III crust; such that the entire crust is dissected into 100–500 m scale phacoidal lenses by mylonites and younger brittle-ductile shear zones (Giguère et al. 2003). There is no clear spatial or depth zonation with respect to metamorphic grade, since granulite grade mylonites (orthopyroxene neoblasts) have been observed immediately beneath the sheeted

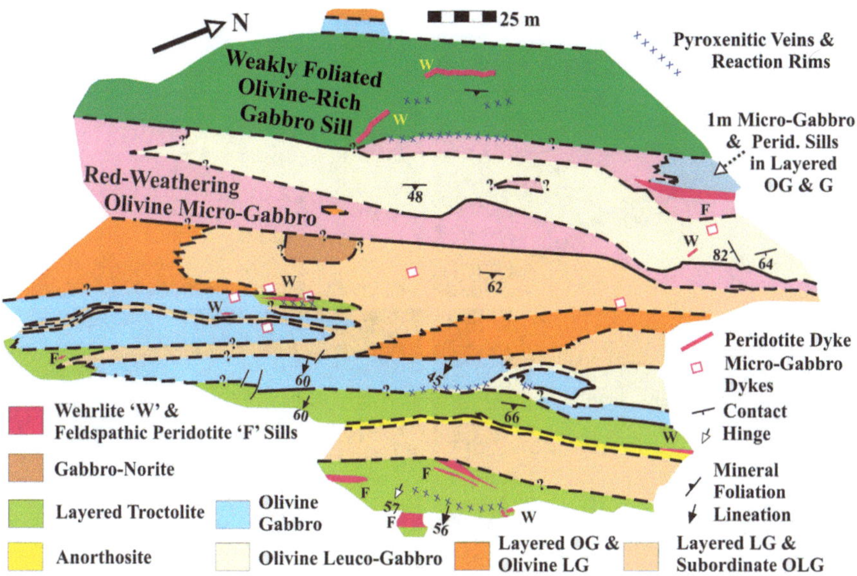

Fig. 15.8 Detailed map of a typical layered gabbro sequence from Unit IIIc in the BOIC at NAM; adapted from Bédard and Constantin (1990). The youngest units on the map are cross-cutting microgabbro and peridotite dykes. The large undeformed body of red-weathering olivine microgabbro and peridotite sills in the upper center of the map are either contemporaneous with these, or only slightly older. The weakly foliated olivine-rich gabbro sill in the upper part of the map has experienced only moderate deformation and is slightly older still. All other map units predate these intrusions, and have experienced various extents of deformation, often with isoclinal or sheath folds (cf Figs. 15.9 and 15.9b). Note the uniform orientation of fold hinges and stretching lineations. OG = olivine gabbro, LG = leuco gabbro, LOG = leuco olivine gabbro.

dykes, whilst greenschist grade mylonites occur throughout. Rather, it seems as though most began as broad, high grade shear zones (amphibolite to granulite) that cooled and became more focused with time. Many high-grade shear zones have narrower, greenschist-grade cores that are surrounded by hydraulic breccias, suggesting a transition from ductile to brittle-ductile behavior. Oxygen isotope data (Giguère et al. 2003) show that the brittle-ductile shear zones allowed seawater to penetrate to the vicinity of the Moho.

The nature and kinematics of these shear zones has only partly been resolved, but several features are noteworthy. First, many shear zones are associated with concentrations of hornblende-Fe-oxide rich microgabbro, or oxide-rich lenses that may represent accumulations from Fe-rich melts. This suggests that these zones of focused strain played a role in the movement of residual melts, as also proposed by Dick et al. (1991) and Natland and Dick (2001) for the Hole 735B gabbros. Second, on a larger scale the eastward lateral gradation from melagabbro to ferrogabbro-dominated crust (Fig. 15.7) suggests that dense, Fe-rich residual melt may have migrated laterally in response to pressure gradients, rather than ascending through low-density feldspar-rich cumulates. Third, in the eastern area the mylonitic zones

Fig. 15.9 Field photographs from the BOIC. **a** Glaciated pavement in Unit IIIc detail map area (Fig. 15.8) showing typical strongly deformed olivine gabbro with numerous anorthositic boudins. The *red arrows* show the trend and plunge of the stretching lineation, visible in the cleft just above the *arrow* on the *left*. The *orange strike* and *dip symbol* shows the orientation of the high-temperature trans-solidus foliation. Geologist's feet for scale. **b** Down-plunge view of anorthositic boudins, showing their strongly deformed nature. Glove for scale. **c** Folded olivine-rich vein in gabbro, with narrow anorthositic reaction rim. From complex zone of intercalated rocks of Units II and IIIg. Lens cap is 55 mm wide. **d** Pyroxenite reaction rim. The material at the bottom of the image is a typical Unit II gabbro-norite, which grades towards anorthositic gabbro as the contact with the pyroxenite is approached. The *bright green* clinopyroxene-rich websterite reaction rim fingers into the gabbroic substrate and thickens in the fold nose. The *khaki-coloured* rock at the *top* of the image is a lherzolitic intrusion of boninitic affinity

steepen and broaden, such that the paleo-vertical zone of deformation is almost 2 km wide, with sheared porphyritic ferrogabbros grading up into sheeted dykes that are oriented parallel to the mylonitic fabric. This suggests that deformation-enhanced melt migration may have played a role in channeling melt up through low-density gabbros towards an eruption centre.

Part of unit IIIc was mapped on a 1:100 scale (Fig. 15.8) to document the geometry of the gabbro/melagabbro/leucogabbro (GML) complex there. This coherent mass of near-continuous outcrop is bounded on all sides by coeval to younger shear zones. In the 'less-deformed' core zone, the basic pattern is of a near-concordant set of 1–30 m scale layers with contrasting mineral modes. Obvious outcrop-scale fold closures (Figs. 15.8, 15.9 and 15.10), the axial planar mineral foliation, and a consistent stretching lineation that is parallel to fold hinges, show that most of the small scale layering (Fig. 15.10d, cf. Bédard 1993, his Figures 2 and 3) represents isoclinally folded, transpositional repetitions of existing lithological heterogeneities; not

Fig. 15.10 Field photographs from the BOIC. **a** Intra-cumulate hybrid mela-troctolite from Unit IIIc detail map area (Fig. 15.8). The olivine-melagabbro matrix contains a variety of gabbroic enclaves, some angular (*upper left*), some more amoeboid (*center and center-right*), probably reflecting the relative degree of prior lithification of the heterogeneous host gabbro prior to injection of a picrite, which seems to be mixing with the gabbro. Hammer handle is 1 m long (from Bédard 1983, reproduced courtesy of the Geological Society of America). **b** Same location as previous photo. The *lower* part of the outcrop is a late wehrlitic sill with a fingered *upper* contact against a strongly deformed peridotite/gabbro/anorthosite complex. The peridotitic core to the swirls are inferred to be intrusions, whereas the *white* anorthositic bands are interpreted to be deformed reaction rims. **c** Shear hybridization in Unit III. The *lower* half of the photo shows an olivine gabbro with an orthopyroxene-rich 'pod' of uncertain origin. Just above the finger is a sharp contact with a strongly-foliated gabbro-norite with schlieren of orthopyroxenite. This is interpreted to represent the edge of a granulite grade trans-solidus high-strain zone, where orthopyroxenite and olivine gabbro are mechanically mixed during deformation, with olivine being lost through reaction. **d** Typical isomodal transpositional layering in Unit III. About 1.2 m of section shown. Note how the pyroxenite layer on the *left* side terminates towards the top, presumably as a fold closure. The gabbro has a strong foliation with extremely elongated olivine grains. The layering is interpreted to represent transpositional repetition of lithologies, as well as shear-related hybridization. Modal variations are not related to cotectic solidification. Instead, each individual layer is considered to represent a mechanical mixture of different proportions of intrusive peridotite, disaggregated partially molten host gabbros, and whatever reaction products may have developed between these two systems

a sequence of depositional cumulate layers. The most deformed and presumably oldest GML packages are cross-cut by slightly less deformed and more massive gabbroic sheets (the 'Weakly Foliated Olivine-Rich Gabbro Sill' on Fig. 15.8). Contacts of this body are shallowly discordant and clearly transect older foliations and layering in the host rocks. These slightly less sheared gabbroic sills are themselves cross-cut by younger, essentially undeformed ferrogabbroic sills and dykes (the 'Red Weathering Olivine Micro-Gabbro' of Fig. 15.8) and by peridotitic dykes and sills that appear to mix with partly consolidated gabbroic cumulates (Figs. 15.9a, b, 15.10a). The field relationships (Figs. 15.8, 15.9 and 15.10) demonstrate that these rocks were emplaced as a series of syn- to late-kinematic shallowly-dipping sill-like intrusions, and led to the original proposal of a sheeted sill model for crust formation in the BOIC (Bédard 1988, 1993). Similar intrusive contacts, folds and fabrics can be observed all the way to the base of the sheeted dykes, which implies that essentially the entire plutonic crust was behaving as a ductile medium. Structural data (Berclaz et al. 1994, 1998) suggest that this phase of deformation was dominantly sinistral extensional, as shown by the geometry of trondhjemite-filled pull-aparts developed in the coeval granulite- to amphibolite-grade shear zones.

The margins of late olivine-rich dykes are locally marked by 1–10 cm wide anorthositic reaction rims, interpreted by Bédard (1993, his Figures 3, 4 and 7) to have formed at intrusive contacts in response to pore-scale mixing between invading olivine-rich (picritic) primitive magma (mol. Fo up to 90.6, NiO up to 0.17 wt%) and partial melts of the olivine-poor gabbro host. Typical, massive, foliated gabbroic sheets of the GML complex often contain anorthositic boudins (Figs. 15.9a and b) that are interpreted to represent dismembered, stretched and refolded anorthosite reaction rims (Fig. 15.10b, and Bédard et al. 2000a, his Figures 2 and 3). If the anorthosite boudins are indeed disaggregated reaction rims, then the typical olivine gabbros and leucogabbros that host them, and which dominate this part of the crust (Figs. 15.9 and 15.10, cf. Komor et al. 1987) must be mixtures between: (a) the injected picritic peridotite, (b) wallrock reaction products, and (c) partly-disaggregated, partly-molten wallrock cumulate gabbros.

Unit II—Detailed mapping of Unit II (see Bédard 1991, his Figures 2 and 11) shows that this unit is also a syn-kinematic sill complex (Fig. 15.9c); but was in a dextral compressional régime (Berclaz et al. 1994, 1998). This underplating event drove intense seawater circulation through the already-cooled, brittle overlying crust (Units III and above), and created small hydrothermal sulphide bodies at the Unit II/III interface. In the NAM Massif, Unit II is dominated by peridotites and pyroxenites, with subordinate chromitite, olivine gabbro, ferrogabbro and gabbro-norite. Unlike the dominantly tholeiitic rocks of Unit III, the pyroxenites and peridotites of Unit II of the NAM Massif are all cumulates from melts of boninitic affinity (Bédard and Hébert 1996; Bédard 2005, 2007). Boninitic melts typically co-saturate in pyroxene prior to plagioclase, leading to the formation of abundant cumulate dunite, wehrlite, lherzolite, harzburgite, orthopyroxenite and websterite (Bédard and Hébert 1996). Because of the near-pervasive deformation (Fig. 15.10c), cross-cutting and enclave-host relationships are only rarely preserved (but see Bédard 1991, his Figures 9, 12, 13 and 15), such that the dominant transpositional layering was previously misidentified as depositional cumulate layering.

Unit II thins to a few tens of metres of thickness in the next massif to the NE (Table Mountain, TM, Fig. 15.7a), where its presence is inferred on the basis of rare feldspathic wehrlite layers along the Moho. In contrast, the massif to the SW (Blow-me-Down, BMD) has a very thick section of basal dunite (>3 Km), with few pyroxenites or cogenetic gabbroids. Geochemical data and modelling (Stern et al.

2011a, b; Stern 2012) suggest that some of the BMD dunites have boninitic trace element signatures. This regional-scale petrologic gradation from thick dunite at BMD, to a NE-thinning package of mixed peridotites, pyroxenites and gabbroids at NAMM (Fig. 15.7b), to a very thin ultramafic underplate at the base of the gabbroic TMM crust, implies that the underplating boninitic magmas propagated and fractionated towards the NE.

Inverse trace element modelling implies that Unit III cumulate rocks retained very low trapped melt fractions, typically 5–10 % (Bédard 2005, 2007; and unpublished results). Given the obvious evidence for pervasive syn-magmatic deformation at NAM, such low trapped melt fractions are plausibly interpreted as resulting from shear-enhanced melt segregation. Only a few, small, *in-situ* gabbroic veins that could represent discrete segregation structures have been identified, however. These may have been more abundant at an early stage, and could have been obliterated by contact-parallel shear. Alternatively, melt escape may have taken place by homogeneous percolation through the deforming cumulates, leading to a cryptic down-gradient evolution that cannot be resolved with the existing sample set.

Unit II (like Unit III) shows considerable field evidence for shear-enhanced hybridization between feldspar-rich hosts and olivine-rich intrusions (forming olivine melagabbros and feldspathic wehrlites), and for mechanical incorporation of reaction rims into invading olivine-rich sills (Bédard 1991). There are two common types of reaction rims formed in Unit II: chromitite and pyroxenite. Chromitite rims are proposed to have formed through incongruent dissolution of pyroxenite (high Cr# chromite) or feldspar-rich gabbros (low Cr# chromite) into intruding primitive boninite magma (Bédard 1991; Bédard and Hébert 1998; Bédard et al. 2000a). Outcrop-scale evidence (Bédard 1991) implies that the 1–10 cm scale pyroxenite rinds (Fig. 15.9d) that commonly separate peridotite intrusions from gabbroic host rocks also represent reaction rims. It was proposed that the pyroxenite reaction rims formed by pore-scale hybridization at the contacts between invading olivine-saturated primitive melt and partially molten gabbroic host rocks, since the phase topology constrains hybrid melts to lie in the pyroxene-only field (Bédard 1991, his Figs. 23 and 24). The pyroxenite rims are typically separated from distal host gabbroids by thin (1–10 cm) leucogabbroic haloes that contain high-An plagioclase and are extremely depleted in all but plagioclase-compatible elements (Sr, Eu), suggesting that they represent feldspar-rich restites from extensive melting of a plagioclase-rich gabbroic protolith, in accord with the proposed model. Field evidence implies that ongoing deformation actively stripped pyroxenitic reaction products from their gabbroic substrate and dumped them into the invading sills, thus exposing new reaction surfaces and accelerating reaction progress (Bédard et al. 2000a).

Implications

Essentially the entire cumulate crustal section of the BOIC is a syn-kinematic sheeted sill complex. If only the tholeiitic rocks of Unit III are considered, then the image that emerges is of distributed, channeled migration of primitive and evolved

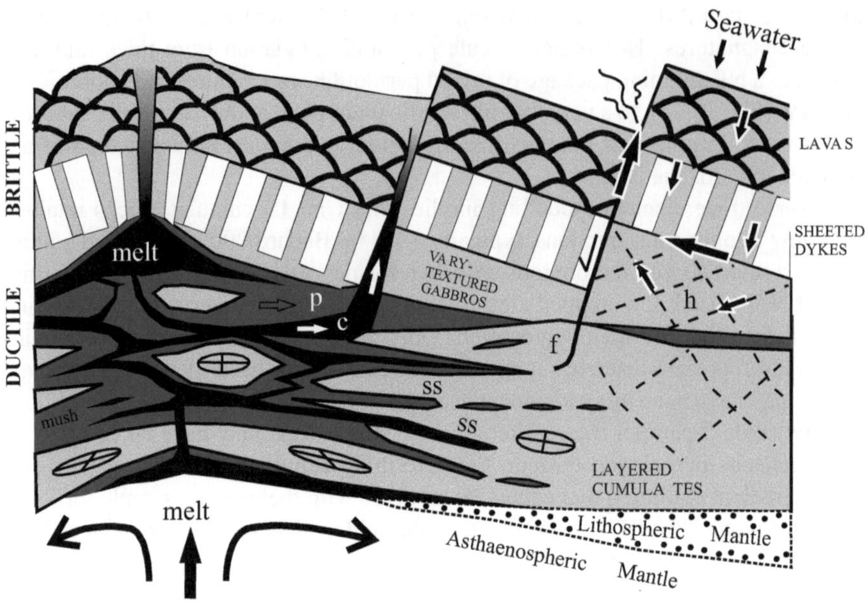

Fig. 15.11 Schematic cross-section of a ridge experiencing distributed deformation during a magmatic pulse. *Strain circles* are shown for the case where mantle is flowing faster than crust. At the ridge axis the melt from the mantle pools at the Moho and may spread out as underplating sills. As melt wends its way up through a tortuous feeder system, it evolves by assimilation-fractional crystallization as it reacts with solid products of previous magmatic pulses and hybridizes with partially consolidated host cumulates. Evolved melts from this process feed the axial melt lens. Towards the ridge flanks, the environment is cooler, mainly due to evacuation of heat by seawater circulation, and magma crystallizes against the roof to form the vary-textured upper gabbros. As distance from the axial hot zone increases, the brittle layer thickens as seawater penetrates deeper (*h* on figure) along brittle-ductile shear zones (*dashed lines*) that dissect the crust. During phases of high axial melt flux from the mantle, overpressured magma would be injected into the cooler, partly consolidated distal cumulate rocks to form the sheeted sill zone (*ss*). Evolved melt located in the pores of partly consolidated gabbros would migrate towards low-pressure sinks either through percolation flow (*p*), or where shear zones develop, through channelled flow (*c*), and could feed evolved off-axis volcanism

melts throughout a ductile crust that necks fairly homogeneously on a large scale (Fig. 15.11), with deformation progressively focusing into shear zones (Giguère et al. 2003). This focusing presumably records overall cooling and rigidification on a crustal scale. These shear zones appear to have guided movement of residual Fe-rich melt both laterally (towards a low-pressure structural sink) and vertically (to feed sheeted dykes). Shear-pumping may have helped move dense melt upward through low-density feldspar-rich gabbros; and is probably responsible for the low residual porosities.

Deformation in the BOIC plutonic crust played an important role in generating the observed lithological diversity by mixing incoming primitive melts with pre-existing, partly consolidated host rocks to create a range of hybrid facies. This accounts for the fact that most (transpositional) layers in the BOIC are isomodal

(Figs. 15.10c and d), with near-constant mineral proportions in each layer, but a wide range of mineral mode from layer to layer. In other words, few BOIC rocks are simple cotectic cumulates, most being variable mixtures of three components (new magma, old host rocks, reaction products), with subsequent loss of residual pore melt. Deformation also played an essential role in thickening reaction zones by cross-contact shear (mechanically juxtaposing reactants), stripping off reaction rims to expose new reaction surfaces, and adding reaction products to invading magmas (Bédard et al. 2000a).

The Betts Cove Ophiolite

Description

The Betts Cove Ophiolite (~489 Ma, Dunning and Krogh 1985) is roughly coeval to the BOIC, but is dominated by boninitic melts. Most of the sheeted dykes and 1.3 km of pillow lavas (Fig. 15.12) are low-Ca low-Ti boninite (Bédard 1999; Bédard et al. 2000b). Trace element inversion models (Fig. 15.13a) imply that peridotitic cumulates at Betts Cove formed from melts identical to typical Betts Cove boninite lava, with the best inverse solutions implying extensive melt entrapment (20–30 % trapped melt). The ubiquitous boninitic magmas constituting the plutonic, hypabyssal and extrusive crust at Betts Cove led Bédard et al. (1998) to suggest its formation through fore-arc seafloor spreading; but see van Staal et al. (2013) for the most recent palaeo-tectonic interpretation of the area.

The Betts Cove crust is partly dissected by oblique extensional faults (D on Fig. 15.12) that locally expose mantle harzburgite to the seafloor as core complexes (Bédard and Escayola 2010). Late gabbros were emplaced along some of these oblique faults and conjugate graben-bounding normal faults (Gb on Fig. 15.12). The late gabbros yield model melt solutions (also with 20–30 % trapped melt, Fig. 15.13b) that match a specific lava sequence (called intermediate-Ti boninites by Bédard 1999) found in the upper part of the seafloor lava stratigraphy, and which also occur as dykes in the sheeted dyke complex. It is inferred that the same faults that guided ascent of the intermediate Ti- boninites played a role in ascent of the low-Ti boninites.

The layered cumulate sequence was previously studied by Upadhyay (1973) and Riccio (1976). Although block faults are locally prominent, there are well-exposed coherent sections up to 1 km thick; but there is little lateral continuity because of overprinting events. A well exposed and preserved section was examined in more detail (a–d on Fig. 15.12). Near the base of this section, there are 10 cm to 2 m thick layers of cumulate dunitic harzburgite, harzburgite and lherzolite with weak bedding-parallel foliations (Fig. 15.14a). Static pseudomorphism allows microtextures to be deciphered, despite a near-pervasive greenschist facies hydrothermal metamorphic overprint (Fig. 15.14b). Olivine (serpentine) and chromite are always cumulus, orthopyroxene (talc, bastite) occurs most commonly as cumulus prismatic

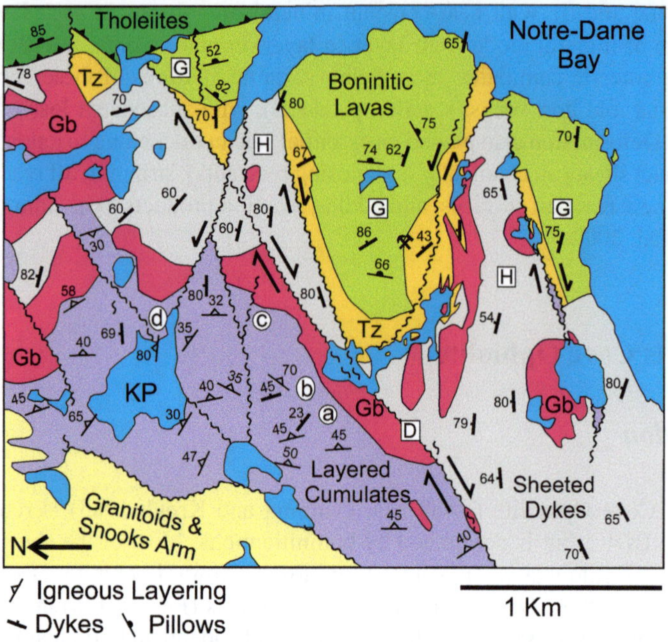

Fig. 15.12 Simplified map of the southern part of the Betts Cove ophiolite (adapted from Bédard et al. 1999), rotated so as to have stratigraphic way-up towards the *top* of the image. This steeply-tilted segment of the ophiolite is bounded to the NW by younger Cape Brulé intrusions and fault-bounded panels of Snooks Arm Group lavas and sediments. The Betts Cove boninitic lavas and sheeted dykes are dissected into a horst-and-graben (*H* and *G*) structure by syn-magmatic extensional faults. *TZ* zones of mixed lavas and dykes, with 10 and 50 % dyke representing the map unit thresholds. The major structure extending from the *top left* towards the *bottom right* labelled *D* is interpreted to be a décollement along which lower crustal cumulates were exhumed, and which also guided the ascent of a generation of late gabbroic intrusions (*red Gb*). These gabbros also exploited the steep graben-bounding faults to ascend further, and induced a hydrothermal cell that generated the Betts Cove Cu-Zn Mine (*hammer and pick symbol*). The tholeiites at the top of the map (Mount Misery Formation, base of Snooks Arm Group) were deposited above the boninites, but were thrust north-westward during obduction (Tremblay et al. 1997). The *circled letters* (*a–d*) are referred to in the text. *KP* Kitty Pond

prisms oriented parallel to bedding, but also forms anhedral poikilitic grains, whilst clinopyroxene and plagioclase are always post-cumulus in peridotites. Near the top of this exposure is a spectacular sequence of normally size-graded olivine + orthopyroxene-cumulate lherzolite to harzburgite beds, each about 1–1.5 m thick (Fig. 15.14c). Euhedral orthopyroxene prisms (replaced by bastite) are oriented with long axes parallel to bedding, with a weak alignment towards the NE. Orthopyroxene grades from ~ 30–40 % large (≤4 cm) primocrysts at the base of each bed to ~ 15 % smaller (~ 0.5 cm) primocrysts at the top (Fig. 15.14c). Discrete cumulus olivine (serpentine) grains are 2–5 mm in size, but the cumulus orthopyroxene prisms contain smaller (mm-sized) ovoid serpentine pseudomorphs after olivine, perhaps indicating co-precipitation of olivine and orthopyroxene. Small euhedral chromite occurs as inclusions in all other phases (Fig. 15.14b).

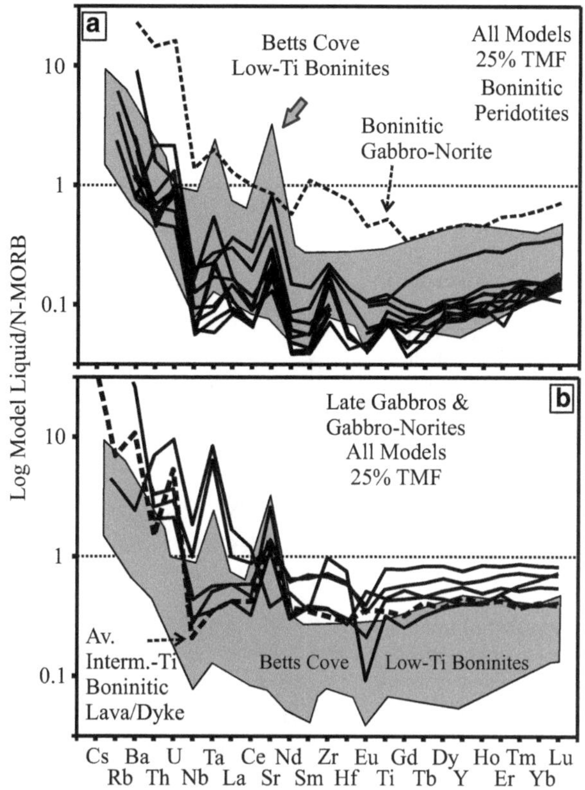

Fig. 15.13 Results of inverse trace element models applied to Betts Cove plutonic rocks using the method of Bédard (1994). All models assume 25 % trapped melt. D values were calculated using the parameterizations of Bédard (2005, 2006, 2007, 2014). **a** Model melts in equilibrium with peridotitic cumulates, compared to field of low-Ti boninites from Betts Cove. A model melt (*dashed line*) from the Kity Pond layered gabbro-norite occurrence (Fig. 15.16) has higher abundances, but a similar shape to the peridotitic model melts, indicating a cosanguineous relationship. **b** Model melts in equilibrium with the late gabbroic intrusions (Gb on Fig. 15.12) have flatter, more enriched trace element profiles that resemble those of the intermediate-Ti boninites (*heavy dashed line*) which occur in the *upper* part of the boninitic lava stratigraphy

The ultramafic cumulate layers are interbedded with cm-m sized pyroxenite layers (orthopyroxenite, websterite or clinopyroxenitic websterite) that may show micro-rhythmic modal layering on a cm scale. These bedding-parallel pyroxenites were interpreted to form part of the cumulate stratigraphy by Upadhyay (1973), representing melt evolution past the olivine/orthopyroxene peritectic reaction and eventual co-saturation in clinopyroxene. Although many of these pyroxenite layers are completely conformable with the peridotite beds, pyroxenites also form discordant metre-thick dyke-like intrusions (Fig. 15.15a and b). Pyroxenite dykes can cross thick sequences of layered peridotite-pyroxenite, or be restricted to individual peridotite layers sandwiched between concordant pyroxenites. In the specific case illustrated in Figs. 15.15b–d, the short discordant legs are restricted to an individual size-graded peridotite bed.

Bedding parallel orthopyroxene-rich websterite layers generally exhibit prominent flattening fabrics, with randomly-oriented prisms of orthopyroxene lacking a clear lineation (Fig. 15.15d). These strongly compacted pyroxenite layers (~5–10 cm thick) contrast with the short bedding-perpendicular segments of discordant pyroxenite (~1 m long and 30 cm wide) that appear to merge with them (Figs. 15.15b and c). The discordant pyroxenites are coarser-grained and contain much more clinopyroxene (concentrated in the cores of these bodies, Fig. 15.15c)

Fig. 15.14 Betts Cove layered ultramafic cumulates. **a** Typical sequence of rhythmic lherzolite-pyroxenite layering. Hammer handle is 30 cm long. This outcrop is from site **a** on Fig. 15.12. **b** Photomicrograph of typical cumulate lherzolite, crossed Nicols. Serpentinized olivine (*ol*) pseudomorphs are included in poikilitic clinopyroxene (*cpx*), as are chromite micro-phenocrysts (*red arrows*), which also occur included in olivine. **c** Olivine + orthopyroxene cumulate bed from site (*b*) of Fig. 15.12. Note the abundance of coarse, flow-foliated orthopyroxene primocrysts at the base of the bed (*bottom right*), and the steadily upward decrease in their abundance and size

Fig. 15.15 Pyroxenite segregation veins at Betts Cove (location b on Fig. 15.12). **a** Regular rhythmic peridotite-pyroxenite layering is dipping about 45° away from the viewer (see *orange strike and dip symbol*). Layering is truncated by a steeply dipping pyroxenite dyke (*left side* of picture under notebook). The pyroxenite in the dyke is continuous with the bedding-parallel pyroxenite, suggesting that melt was draining from the layered cumulates into the dyke. Notebook is 17 cm long. **b** Layered peridotite-pyroxenite. The bedding is striking up towards the *top right*, and dipping steeply towards the *lower right* (*strike and dip symbol*). The size-grading in the primocrystic orthopyroxene is just visible in the main body of the *tan-coloured* peridotite (*centre of picture*), which is cut by a short pyroxenite dyke (*circled c*). The circled *c* and *d* mark the locations of the next two figures, with the *red arrow* at *c* marking the view direction. **c** Short, zoned, websterite cross-dyke (*c* on Fig. 15.15b), with orthopyroxene-rich edges (*o*) against peridotite (*p*), and a coarser, clinopyroxene-rich core (*c*). **d** View onto bedding plane of concordant orthopyroxene-rich websterite (*d* on Fig. 15.15b). Note the random orientation of prismatic orthopyroxene pseudomorphs (*brown*), and the coarser clinopyroxene oikocrysts (*green*)

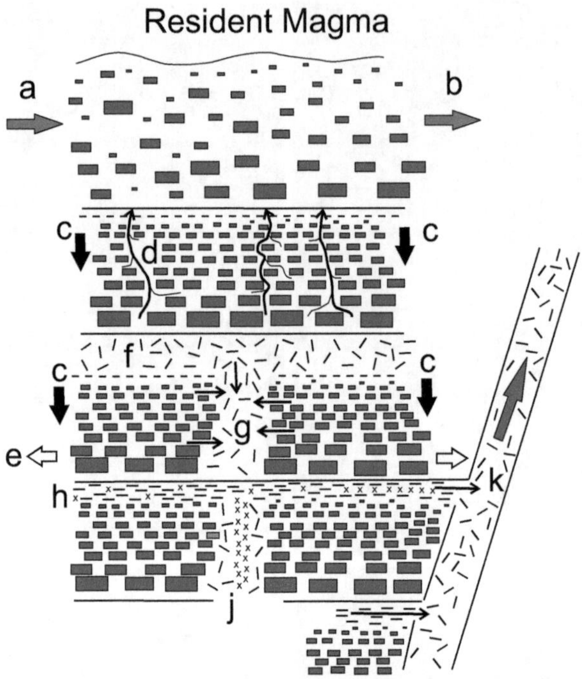

Fig. 15.16 Cartoon illustrating accumulation, compaction and melt segregation mechanism inferred for Betts Cove. At the *top* is primocryst slurry newly erupted at the magma crystal mush interface. Only the orthopyroxene primocrysts (*grey rectangles*) are shown. As magma flows from *a* to *b*, primocrysts get size graded. Each deposit is buried by new slurry replenishments. Eventually, the weight of the overlying cumulus pile drives limited compaction (*c*, *black arrows*). Pore melt expelled by compaction segregates into bedding planes (*squiggly arrows marked d*). With continuing compaction, the rigidifying peridotite extends (*e white arrows*) and boudinages, allowing pore melt to segregate into tension gashes (*g*). Because the pore melt has been mechanically separated from the olivine in the peridotitic cumulate, it crystallizes only orthopyroxene in the bedding-parallel layers (*f*) and tension gashes (*g*). A major fracture develops at *k*, and melt residing in the bedding-parallel segregation veins (*h*) drains into the fault and ascends as a dyke to erupt at the magma/mush interface above. By this time clinopyroxene has co-saturated, and some bedding-parallel pyroxenites contain rhythmic orthopyroxenite-clinopyroxene-rich websterite micro-rhythmic layering. In contrast, the melt residing in the tension gashes was protected from compaction by the fairly stiff peridotite, and so melt was not efficiently extracted from these sites, such that abundant clinopyroxene crystallizes in the core of the normally-zoned cross-dyke (*j*)

than the bedding parallel pyroxenites, suggesting less compaction and higher trapped melt fractions.

In many places the bedding-parallel concordant pyroxenites merge with the intrusive ones (Figs. 15.15a and b), suggesting either that some of the bedding parallel pyroxenites are intrusive sills, or that bedding-parallel pyroxenites might have formed from segregations of residual melt that crystallized some pyroxene and then migrated out into structurally-controlled conduits/dykes (Fig. 15.16). The absence of impregnations emanating from these bedding-parallel pyroxenites (compare with Fig. 15.17) suggests they are not intra-cumulate sills and an origin as cumulus grains and/or segregation veins is favoured.

Fig. 15.17 **a** Clasts of peridotite cumulate in a pyroxenite stockwork (from site c, Fig. 15.12). The sledgehammer handle is 1 m long (*arrow*). **b** Close-up of contact between peridotite clasts (*p*) and orthopyroxenite to websterite dykes (*o*) of previous photo. Note the impregnation (*i*) of the host peridotite (*p*) by the intrusive pyroxenite (*o*)

The relationships described above are interpreted as follows. The size-graded beds (Fig. 15.14c) suggest that a thick sequence (>100 m) of olivine + orthopyroxene cumulates was emplaced in an open chamber (of unknown size) as a series of crystal slurries that experienced size-grading during flow and deposition (Fig. 15.16). Compaction expelled residual intercumulus melt into the bedding planes between harzburgitic flow units (Fig. 15.16). Compaction was inefficient, as shown by the high residual trapped melt fractions in these peridotites (20–30 %). At some stage, bedding-perpendicular extension of the more rigid, partly-compacted peridotite beds led to boudinage, with the voids filling with residual melt. Because of the olivine to orthopyroxene reaction relationship, melts segregating from olivine + orthopyroxene cumulates crystallize orthopyroxene only, generating orthopyroxenitic cumulates in these segregation conduits. Further compaction and melt expulsion of this pooled residual melt developed strong flattening fabrics in the bedding-parallel pyroxenites. In contrast, the melt filling the tension gashes in the boudinaged peridotite beds were protected from compaction by the rigid peridotite and retained more trapped melt, so explaining their coarser textures, more abundant clinopyroxene and absence of flattening fabrics. The ongoing extension of the rigidifying mass allowed steeply-dipping faults to form and propagate, and these high-porosity low-pressure structures concentrated the pooled, partly solidified residual melt residing in the bedding-parallel pyroxenite layers, and channelled them to higher stratigraphic levels.

A few 100 m up-section to the NE (site c on Fig. 15.12; on the other side of an intrusive gabbro) is another spectacular outcrop where large blocks (5–10 m) of layered peridotite are separated by a stockwork of pyroxenite dykes (10 cm–3 m) that impregnate the edges of peridotite blocks (Fig. 15.17). I would suggest that the ascending evolved melt extracted from the underlying thick cumulate peridotite section fluidized this level of cumulate beds and disrupted it. An alternative hypothesis is that the brecciation resulted from faulting, with the ascending melt simply filling open spaces. Either way, it is clear that large volumes of residual melt expelled from the underlying section transited through this disrupted level.

A few 100 m up-section to the NE (site d on Fig. 15.12), on the flanks of Kitty Pond, an outcrop face over 30 m high is composed of 5 cm–3 m sized beds of rhythmically layered orthopyroxenite, clinopyroxene-rich websterite and gabbro-norite. The layered gabbro-norites are geochemically distinct (low TiO_2) from the late intrusive gabbroids (Gb) and have REE profiles that more closely resemble low-Ti boninitic magmas (Fig. 15.13a). Consequently, the layered gabbro-norites are interpreted to represent evolved facies of the boninitic cumulate suite. Many rhythmic sequences are incomplete, but there are several well developed metre-scale units (Fig. 15.18) grading from coarser orthopyroxenite at the base (cm-sized orthopyroxene) up to finer-grained orthopyroxenite (mm-sized orthopyroxene), followed by clinopyroxene-rich websterite (bedding-parallel clinopyroxene prisms) and then gabbro-norite (bedding-parallel plagioclase foliation). The basal orthopyroxenite displays lobate loading structures with a 10 cm amplitude and 30 cm wavelength (Fig. 15.18). The size-graded orthopyroxenite bases of these beds suggests that the replenishing magma was carrying a crystal cargo of variably-sized orthopyroxene, not unlike the graded beds of the underlying ultramafic cumulates. This orthopyroxene cargo was size-graded as it was deposited. This orthopyroxene accumulation was dense enough to trigger the formation of load structures as the orthopyroxene-rich cumulate subsided into the underlying, unconsolidated gabbronoritic mush. This is demonstrated by the fact that plagioclase laths wrap around the pyroxenite lobes. The regularity of the orthopyroxene to orthopyroxene + clinopyroxene to orthopyroxene + clinopyroxene + plagioclase sequence in metre-sized rhythmic beds implies the action of small-scale fractional crystallization. This further implies that the melt from which these layers formed evolved as a near-closed system during crystal fractionation, and that the melt layer thickness was small (2–3 m). While it is possible to suggest that the roof was jacked up by a few metres each time one of these small replenishments occurred, there is a notable absence of roof xenoliths, or of out-of sequence shallowly-discordant injections that should characterize such a scenario. It seems more likely that the arriving orthopyroxene-charged melt was injected into an open chamber and ponded beneath a long-lived body of less-dense basaltic melt, much as suggested by Huppert and Sparks (1980). Whether the transition from clinopyroxene-rich websterite to gabbro-norite in each rhythmic layer reflects closed-system differentiation of the ponded replenishing melt, or breakdown of the interface and renewed deposition of 'background' gabbro-norite cumulates, has yet to be determined.

Fig. 15.18 Rhythmic orthopyroxenite/clinopyroxene rich websterite/gabbronorite layering (from site d on Fig. 15.12). The orthopyroxenitic base of the layer is lobed against the subjacent foliated gabbro-norite, with the feldspar foliation wrapping around the lobes. The orthopyroxenite base is coarser than the top. **a** Close up of base of orthopyroxenite. **b** General view. Coin (*arrow*) is 18 mm wide. *W* websterite, *gn* gabbro-norite, *m/g o* and *f/g o* are medium-grained and fine-grained orthopyroxenite, respectively

Implications

The outcrop-scale structures preserved in the Betts Cove cumulate rocks provide unambiguous evidence for gravity-driven accumulation, since this is the most straightforward way to explain size-grading of primocrysts and the development of load structures at the base of ultramafic beds. Although gravity-driven processes have become unfashionable, the Betts Cove cumulates seem to provide unambiguous examples of layers formed this way.

The Betts Cove occurrences also record: segregation of melt from compacting peridotites into a variety of veins/dykes; ascent of these segregated melts through an overlying cumulate sequence; and injection of these segregated melts as small replenishments into the base of a higher-level basaltic magma chamber. Extracting pore melt from thin sequences of compacting cumulates deposited at the base of an open chamber is not difficult, since flow paths are short and escaping melt easily remixes with the larger body of supernatant magma. The question of how pore melts can be extracted from deeply buried cumulate layers, or from intra-cumulate sills emplaced beneath thick gabbroic successions, is more difficult to answer. The Betts Cove examples illustrate one mechanism by which this can occur. Initial formation of bedding-parallel and bedding-perpendicular segregation veins during compaction concentrates residual melts into channels, with the development of throughgoing faults in the rigidifying cumulate pile allowing these pre-concentrated melts to escape to higher levels. In contrast to the BOIC layered cumulates, the cumulate rocks at Betts Cove show no evidence for penetrative shear deformation, and most planar fabrics (except near faults) are consistent with a compaction origin, with only a weakly-developed primocryst lineation that formed during emplacement. Unlike the BOIC, it seems as though extensional strain at Betts Cove was almost completely partitioned into faults and décollement structures (Fig. 15.12). This geometry of extension might have allowed trap-door subsidence of accumulating cumulate piles (Fig. 15.19), such that new magmas ascending along faults would have had the

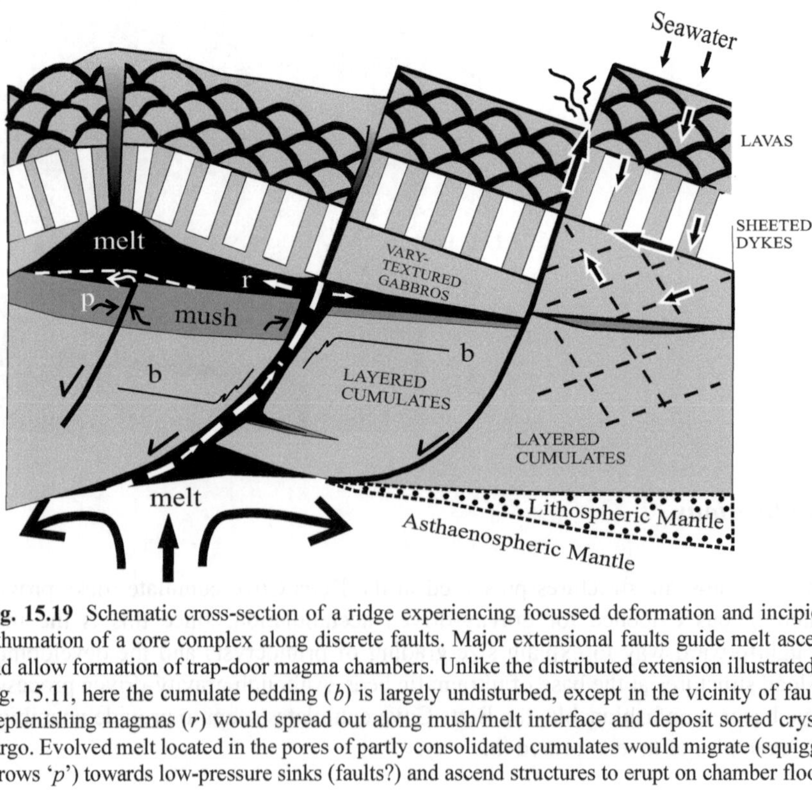

Fig. 15.19 Schematic cross-section of a ridge experiencing focussed deformation and incipient exhumation of a core complex along discrete faults. Major extensional faults guide melt ascent, and allow formation of trap-door magma chambers. Unlike the distributed extension illustrated in Fig. 15.11, here the cumulate bedding (*b*) is largely undisturbed, except in the vicinity of faults. Replenishing magmas (*r*) would spread out along mush/melt interface and deposit sorted crystal cargo. Evolved melt located in the pores of partly consolidated cumulates would migrate (squiggly arrows '*p*') towards low-pressure sinks (faults?) and ascend structures to erupt on chamber floor

chance to pond beneath summital magma pools (AML?) and add cumulates to the top of the previously-formed sequence. Further increments of extension along these same structures at a later stage could have led to exhumation of the cumulate layers and their juxtaposition against the lavas (Fig. 15.12).

Discussion and Implications for Continental Layered Intrusions

Gravitational Cumulate Deposition and Stratified Magma Chambers

In the model originally proposed by Wager and Brown (1968), crystals formed within or at the margins of cooling chambers were carried to the floor by gravity-driven mechanisms. Although a variety of arguments (many valid) have been advanced against this model, in some cases it seems clear that magmas entered chambers carrying abundant crystals that were deposited mechanically (Ferrar Sills: Marsh 2004; Bédard et al. 2007; Duke Island: Irvine 1974, 1987; Franklin Sills: Bédard et al. 2012). The field data from Betts Cove provide fairly unambiguous examples of gravity-driven sorting and deposition of orthopyroxene and olivine + orthopyroxene cumulates. In the case illustrated in Fig. 15.18, it seems likely that the orthopyroxene cargo was entrained as the evolved basaltic magma migrated up through open, dyke-like channels developed in a thick cumulate pile (Figs. 15.16 and 15.17), and that these small replenishing pulses ponded beneath a larger mass of basalt (cf. Huppert and Sparks 1980).

Formation of Monomineralic and Hybrid Facies

The common presence of monomineralic lithologies where replenishing primitive melts come into contact and react with older, more evolved host rocks in the BOIC provides important insights for how analogous facies can form in continental layered intrusions. The creation of monomineralic rocks, especially those composed of phases appearing 'late' in the crystallization sequence, represents a challenge for conventional cumulate-forming models. For example, conventional accumulation scenarios require that the feldspar constituting anorthositic layers must have separated by floatation (Namur et al. 2011) or by dynamic channeled segregation (Bédard et al. 2007) from co-crystallizing mafic phases. This would imply a genetic link between anorthositic and underlying pyroxenitic layers, and suggests there should be a near constant ratio of paired anorthosite/pyroxenite macro-rhythmic layer thicknesses. Data from the Bushveld provide some support for such an interpretation (e.g. Cawthorn 2002; Cawthorn and Ashwal 2009; Maier et al. 2013). Alternatively, if the evidence from the BOIC is considered, then monomineralic rocks could represent reaction products, formed when components of disparate systems were brought into contact (Bédard 1991; Bédard et al. 2000a). Thin anorthosite+chromitite sequences from Rum occur near major intrusive peridotite/host troctolite contacts, leading Bédard et al. (1988) to propose an origin by partial assimilation and reaction between intrusion and host; a suggestion validated by the detailed work of O'Driscoll et al. (2010). Chromitites, pyroxenites and anorthosites are commonly observed near contacts between major sequences in other layered intrusions (e.g. Bushveld and Stillwater: Wager and Brown 1968; Cawthorn 1996) and assimilation reactions may also have been involved in their genesis.

Field and textural evidence imply that many BOIC cumulates are hybrids (Figs. 15.9 and 15.10), representing mixtures of three components (new magma, old host rocks, reaction products), with subsequent loss of pore melt. Deformation played an important role in mixing incoming primitive melts with pre-existing, partly consolidated host rocks, and in expelling pore melts afterwards. Given the rarity of syn-magmatic penetrative shear deformation in continental layered intrusions, formation of abundant magmatic mechanical hybrids would require special circumstances.

Igneous Metasomatism and Recrystallization

The possibility that 'igneous' textured rocks might represent high-temperature metasomatic replacements is rarely considered (e.g. Irvine 1980; McBirney and Sonnenthal 1990); but Bédard et al. (1988) argued that extensive tracts of gabbro in the Eastern Layered Intrusion of Rum formed as replacements of pre-existing troctolite cumulates. Others have suggested instead that the Rum layered gabbros are being replaced by troctolite (Holness et al. 2007), an interpretation which seems less consistent with the field evidence (Bédard et al. 1988; Bédard and Sparks 1991). Whichever interpretation prevails, the implication is that large volumes of adcumulate-textured rocks at Rum have metasomatic origins, and so cannot retain 'primary' textures and mineral compositions. The evidence from Annieopsquotch presented above implies that significant thicknesses of the lower crust there formed as replacements/impregnations of older layered cumulates. A very similar process has been quantified by Lissenberg et al. (2013) for oceanic cumulates. It should not be assumed *a priori* that textures and chemical signatures in cumulate rocks are primary.

Segregation of Melt from Thick Cumulate Piles/Sheeted Intra-Cumulate Sills

The evidence from Betts Cove suggests that residual melt originating from the deeper parts of cumulate piles unaffected by penetrative shear segregates in two steps. (1) Diffuse percolation through a near-homogeneous cumulate matrix towards structural discontinuities. These are commonly bedding planes, but extensional fractures in compacting layers (boudinage/tension gashes) were also involved. (2) Segregation of pooled melts into throughgoing brittle or brittle/ductile structures and ascent to higher levels. In an oceanic magma chamber, extensional faults have a high probability of occurrence, but similar structures could also form in continental intrusions as a result of differential compaction between 'chilled' flanks and strongly-compacted, slow-cooling intrusion cores (eg. Nicolas 1992; Scoates et al. 2010), or due to regional-scale externally-driven tectonism (e.g. Long Loch fault at Rum: Emeleus et al. 1996). In régimes where most deformation is partitioned into discrete corridors (Betts Cove or Annieopsquotch) large trapped melt fractions (20–30%) are typically retained in the compacted but unsheared cumulates. A trap-door model (Fig. 15.19) is proposed to explain thick sequences of depositional cumulate beds, the absence of penetrative shear deformation, together with the existence of synmagmatic exhumation structures (Fig. 15.12).

A completely different set of processes seem to characterize the BOIC cumulates. The field data from NAM in the BOIC imply that almost the entire thickness of the plutonic crust is constituted of intra-cumulate syn-kinematic sills (Fig. 15.8). All BOIC cumulates modelled to date have low trapped melt fractions (5–10%), suggesting that shear deformation efficiently squeezed pore melt out of the matrix, all the way down to the percolation threshold. However, late evolved dykes or veins that could have channelled escaping melt (as seen at Betts Cove) are uncommon in the BOIC. It is possible that segregation veins were once present and that the intensity and penetrative nature of deformation obliterated the outcrop-scale evidence of their existence. It is also possible that some of the more evolved 'sills' in the sequence (Fig. 15.8) are the solidified products of these pooled segregated melts. A third alternative is that shear deformation enhanced the efficiency of percolation to the extent that discrete segregation veins rarely formed. The concentration of ferrogabbroic facies along major shear zones at NAM suggests that structures did indeed help channel the movement of evolved residual melt, but more detailed studies are needed to confirm whether the pore melt from the dominant gabbro/melagabbro/leucogabbro cumulates of Unit III all segregated into channels, and to document where it ended up. On a larger scale, the lateral gradation from olivine-rich gabbros in the west to ferrogabbroic cumulates in the east at NAM (Fig. 15.7b), suggests that much of the Fe-rich residual melt in the BOIC migrated laterally towards low-pressure structural sinks, with extrusion towards the surface (also structurally mediated) being offset by several km from their conjugate cumulates. This family of processes may be more common in the early stages of oceanic cumulate maturation than generally thought; but is not always recognized because transpositional layering is commonly misinterpreted as undeformed depositional layering. Most continental magma chambers probably do not experience such penetrative deformation and the attendant shear-pumping melt segregation. This is probably why most cumulates in layered intrusions retain fairly high trapped melt fractions.

Conclusions

Oceanic/ophiolitic magma chambers show significant differences from continental layered intrusions, mostly related to the common occurrence of syn-magmatic shear deformation in the former. This may result in lower trapped melt fractions because of shear pumping, and the more common occurrence of intra-cumulate hybridization and reaction. Conversely, there are also similarities. The examples documented in this paper show that formation of depositional (gravity-driven) cumulate beds is possible, and that segregation of residual melt from compacting cumulates can occur through a variety of mechanisms, most notably by segregation into open channels. Conversely, more homogeneous percolation of residual melt through a cumulate section can completely transform it by metasomatic reaction and impregnation. Monomineralic facies (anorthosite, chromitite, pyroxenite) may form at the contacts between intrusions of primitive replenishing magma and evolved host rocks as a result of the reactions that must occur when systems that are out of equilibrium are brought together. In most cases, juxtaposition of chemical systems, formation of layering, and segregations of melt from cumulates are controlled by physics, and occur at higher rates than those required for chemical equilibration, such that physics trumps chemistry. Conversely, mechanical mixing of reactants accelerates chemical reactions beyond what slow diffusion rates should allow.

Acknowledgments The geological relationships outlined in this paper result from 25 years of work on ophiolites with many colleagues and students, all of whom are thanked for their hard work, illuminated discussions, diligent hammering, and general good company while in the field. Constructive reviews from Georges Ceuleneer, Johan Lissenberg and Alex Zagorevski, and additional comments from Lyal Harris and Bruce Marsh have helped clarify the presentation. GSC staff are also thanked for technical, logistical and IT support. This is NRCAN/ESS/GSC Contribution # 20130156.

References

Abily B, Ceuleneer G, Launeau P (2011) Syn-magmatic normal faulting in the lower oceanic crust: evidence from the Oman ophiolite. Geology 39:391–394

Agar SM (1990) Fracture evolution in the upper oceanic crust: evidence from DSDP hole 504B. In: Knipe RJ, Rutter EH (eds) Deformation mechanisms, rheology and tectonics, vol 54. The Geology Society, Special Publication, London, pp 41–50

Agar SM (1994) Rheological evolution of the ocean crust—a microstructural view. J Geophys Res 99:3175–3200

Alt JC, Bach W (2006) Oxygen isotope composition of a section of lower oceanic crust, ODP Hole 735B. Geochem Geophys Geosyst 7:Q12008

Alt JC, Teagle DAH (2003) Hydrothermal alteration of upper oceanic crust formed at a fast-spreading ridge: mineral, chemical, and isotopic evidence from ODP Site 801. Chem Geol 201:191–211

Barnes SJ (1989) Are Bushveld U-type parent magmas boninites or contaminated komatiites. Contrib Mineral Petrol 101:447–457

Bédard JH (1988) Magma chamber dynamics and recycling of crustal cumulates by the mantle: evidence from the Bay of Islands ophiolite. EOS Trans Am Geophys Union 69:1476

Bédard JH (1991) Cumulate recycling and crustal evolution in the Bay of Islands ophiolite. J Geol 99:225–249

Bédard JH (1993) Oceanic crust as a reactive filter: synkinematic intrusion, hybridization, and assimilation in an ophiolitic magma chamber, western Newfoundland. Geology 21:77–80

Bédard JH (1994) A procedure for calculating the equilibrium distribution of trace elements among the minerals of cumulate rocks, and the concentration of trace elements in the coexisting liquids. Chem Geol 118:143–153

Bédard JH (1999) Petrogenesis of boninites from the Betts Cove Ophiolite, Newfoundland, Canada: identification of subducted source components. J Petrol 40:1853–1889

Bédard JH (2005) Partitioning coefficients between olivine and silicate melts. Lithos 83:394–419

Bédard JH (2006) Trace element partitioning in plagioclase feldspar. Geochim Cosmochim Acta 70:3717–3742. doi:10.1016/j.gca.2006.05.003

Bédard JH (2007) Trace element partitioning coefficients between silicate melts and orthopyroxene: parameterizations of D variations. Chem Geol 244:263–303

Bédard JH (2014) Parameterizations of Ca-rich clinopyroxene—melt trace element partition coefficients. Geochem Geophys Geosyst 15:303–336. doi:10.1002/2013GC005112

Bédard JH, Constantin M (1991) Syn- and post-kinematic intrusions of troctolite, gabbro, and peridotite into layered gabbroic cumulates in the Bay of Islands ophiolite: genesis of anorthosite by partial assimilation, and troctolite by hybridization. Geol Surv Can, Curr Res Pap 91–1 part D, 79–88

Bédard JH, Escayola M (2010) The advocate Ophiolite mantle, Baie Verte, Newfoundland: regional correlations and evidence for metasomatism. Can J Earth Sci 47:237–253

Bédard JH, Hébert R (1996) The lower crust of the Bay of Islands ophiolite, Canada: petrology, mineralogy, and the importance of syntexis in magmatic differentiation in ophiolites and at ocean ridges. J Geophys Res 101:25105–25124

Bédard JH, Hébert R (1998) Formation of chromitites by assimilation of crustal pyroxenites and gabbros into peridotitic intrusions: North Arm Mountain massif, Bay of Islands ophiolite, Newfoundland, Canada. J Geophys Res 103:5165–5184

Bédard JH, Sparks RSJ (1991) The structure and petrogenesis of the Trallval and Ruinsival areas of the Rhum ultrabasic pluton—comment. Trans R Soc Edinb 82:389–390

Bédard JH, Sparks RSJ, Renner R, Cheadle MJ, Hallworth MA (1988) Peridotite sills and metasomatic gabbros in the Eastern Layered Series of the Rhum complex. J Geol Soc Lond 145:207–224

Bédard JH, Lauzière K, Tremblay A, Sangster A (1998) Evidence for forearc seafloor-spreading from the Betts Cove ophiolite, Newfoundland: oceanic crust of boninitic affinity. Tectonophysics 284:233–245

Bédard JH, Lauzière K, Boisvert É, Sangster A, Tellier M, Tremblay A, Dec T (1999) Geological map of the Betts Cove Ophiolitic Massif and its cover rocks. Geol Surv Can Map 1969A, scale 1:20 000

Bédard JH, Berclaz A, Hébert R, Varfalvy V (2000a) Syntexis and the genesis of the oceanic crust. In: Dilek Y, Moores EM, Elthon D, Nicolas A (eds) Ophiolites and Oceanic Crust: new Insights from Field Studies and the Ocean Drilling Program, vol 349. The Geological Society of America, Special Paper, pp 105–119

Bédard JH, Lauzière K, Tremblay A, Sangster A, Douma SL, Dec T (2000b) The Betts Cove ophiolite and its cover rocks. Geol Surv Can Bull 550:76

Bédard JHJ, Marsh BD, Hersum TG, Naslund HR, Mukasa SB (2007) Large-scale mechanical redistribution of orthopyroxene and plagioclase in the basement sill, Ferrar dolerites, McMurdo dry valleys, Antarctica: petrological, mineral-chemical and field evidence for channelized movement of crystals and melt. J Petrol 48:2289–2326

Bédard JH, Naslund HR, Nabelek P, Winpenny A, Hryciuk M, Macdonald W, Hayes B, Steigerwaldt K, Hadlari T, Rainbird R, Dewing K, Girard É (2012) Fault-mediated melt ascent in a Neoproterozoic continental flood basalt province, the Franklin sills, Victoria Island, Canada. Bull Geol Soc Am 124:723–736. doi:10.11301/B30450.1

Berclaz A, Bédard JH, Hébert R, Varfalvy V, Melançon B, Yao KR, El Bilali L (1994) Tectonic evolution of the plutonic section of the Bay of Islands ophiolite, Newfoundland. EOS Trans Am Geophys Union 75:650–651

Berclaz A, Hébert R, Bédard JH (1998) Predominance and significance of deformation processes in the genesis of deep oceanic crust at a slow-spreading ridge: evidence from the Bay of Islands ophiolite, Newfoundland. Geol Assoc Can/Mineral Assoc Can Meeting 23:A15–A16

Blackman DK, Karson JA, Kelley DS, Cann JR, Früh-Green GL, Gee JS, Hurst SD, John BE, Morgan J, Nooner SL, Ross DK, Schroeder TJ, Williams EA (2002). Geology of the Atlantis Massif (Mid-Atlantic Ridge, 30°N): implications for the evolution of an ultramafic oceanic core complex. Marine Geophys Res 23:443–469

Bohnenstiehl DRR, Kleinrock MC (1999) Faulting and fault scaling on the median valley floor of the trans-Atlantic geotraverse (TAG) segment, 26°N on the Mid-Atlantic Ridge. J Geophys Res 104:29351–29364

Boudier F, Nicolas A, Ildefonse B (1996) Magma chambers in the Oman ophiolite: fed from the top and the bottom. Earth Planet Sci Lett 144:239–250

Cann JR, Smith DK (2005) Evolution of volcanism and faulting in a segment of the Mid-Atlantic Ridge at 25°N. Geochem Geophys Geosystems 6, Q09008

Cannat M, Mével C, Stakes D (1991) Stretching of the deep crust at the slow-spreading Southwest Indian Ridge. Tectonophysics 190:73–94

Casey JF, Karson JA, Elthon D, Rosencrantz E, Titus M (1983) Reconstruction of the geometry of accretion during formation of the Bay of Islands Ophiolite Complex. Tectonics 2:509–528

Cawthorn RG (1996) Layered intrusions. Elsevier Science Publication, Dev in Petrol 15:531p + map

Cawthorn RG (2002) Delayed accumulation of plagioclase in the Bushveld Complex. Min Mag 66:881–893

Cawthorn RG, Ashwal LD (2009) Origin of anorthosite and magnetitite layers in the Bushveld Complex, constrained by major element compositions of plagioclase. J Petrol 50:1607–1637

Chevenez J, Machetel P, Nicolas A (1998) Numerical models of magma chambers in the Oman ophiolite. J Geophys Res 103:15443–15455

Crawford AJ (ed) (1989) Boninites and related rocks. Unwyn Hyman, London

Crawford WC, Webb SC, Hildebrand JA (1999) Constraints on melt in the lower crust and Moho at the East Pacific Rise, 9°48′N, using seafloor compliance measurements. J Geophys Res 104:2923–2939

Danyushevsky LV, Leslie RAJ, Crawford AJ, Durance P (2004) Melt inclusions in primitive olivine phenocrysts: the role of localized reaction processes in the origin of anomalous compositions. J Petrol 45:2531–2553

Dick HJB, Meyer P, Bloomer S, Kirby S, Stakes D, Mauwer C (1991) Lithostratigraphic evolution in an in situ section of oceanic layer 3. In: von Herzen RP, Robinson PT, et al., Proc Ocean Drilling Progr, Sci Results 118:439–538

Dunn RA, Toomey DR, Solomon SC (2000) Three-dimensional seismic structure and physical properties of the crust and shallow mantle beneath the East Pacific Rise at 9°30′N. J Geophys Res 105:23537–23555

Dunning GR, Krogh TE (1985) Geochronology of ophiolites of the Newfoundland Appalachians. Can J Earth Sci 22:1659–1670

Elthon D, Casey JF, Komor S (1984) Cryptic mineral-chemistry variations in a detailed traverse through the cumulate ultramafic rocks of the North Arm Mountain Massif of the Bay of Islands ophiolite, Newfoundland. In: Gass IG, Lippard Sj, Shelton AW (eds) Ophiolites and Oceanic lithosphere, vol 13. Geological Society, Special Publications, London, pp 83–97

Emeleus CH, Cheadle MJ, Hunter RH, Upton BGJ, Wadsworth WJ (1996) the rum layered suite. In: Cawthorn RG (ed) Layered intrusions, vol 15. Elsevier Science Publication Dev in Petrol, pp 403–439

France L, Ildefonse B, Koepke J (2009) Interactions between magma and hydrothermal system in Oman and in IODP Hole 1256D: fossilization of a dynamic melt lens at fast spreading ridges. Geochem Geophys Geosystems 10, Q10019, doi:10.1029/2009GC002652

Garmany J (1989) Accumulations of melt at the base of young oceanic crust. Nature 340:628–632
Giguère E, Hébert R, Beaudoin G, Bédard JH, Berclaz A (2003) Hydrothermal circulation and metamorphism in crustal rocks of the Bay of Islands ophiolite complex, Newfoundland, Canada: evidence from mineral and oxygen isotope geochemistry. In: Dilek Y, Robinson PT (eds) Ophiolites in earth history, vol 218. Special Publication Geological Society, London, pp 369–400
Harper GD (1985) Tectonics of slow spreading mid-ocean ridges and consequences of a variable depth to the brittle/ductile transition. Tectonophysics 4:395–409
Hawkins JW (2003) Geology of supra-subduction zones—Implications for the origin of ophiolites. In: Thy P, Dilek Y (eds) Ophiolite concept and the evolution of geological thought, vol 373. Geological Society of America, Special Paper, pp 227–268
Henstock TJ, Woods AW, White RS (1993) The accretion of oceanic crust by episodic sill intrusion. J Geophys Res 98:4143–4161
Holness MB, Hallworth MA, Woods A, Sides RE. (2007) Infiltration metasomatism of cumulates by intrusive magma replenishment: The Wavy Horizon, Isle of Rum, Scotland. J Petrol 48:563–587
Humphris, SE, Tivey MK (2000) A synthesis of geological and geochemical investigations of the TAG hydrothermal field: Insights into fluid-flow and mixing processes in a hydrothermal system. In: Dilek Y, Moores E, Elthon D, Nicolas A, Ophiolites and the Oceanic Crust: new insights from field studies and the ocean drilling program, vol 349. Geological Society of America, Special Paper, pp 213–235
Huppert HE, Sparks RSJ (1980) The fluid dynamics of a basaltic magma chamber replenished by influx of hot, dense, ultrabasic magma. Contrib Mineral Petrol 75:279–289
Ildefonse B, Blackman DK, John BE, Ohara Y, Miller DJ, MacLeod CJ (2007) Oceanic core complexes and crustal accretion at slow-spreading ridges. Geology 35:623–626
Irvine TN (1974) Petrology of the Duke Island ultramafic complex, southeastern Alaska. Geol Soc Am Mem 138:240 pp
Irvine TN (1980) Magmatic infiltration metasomatism, double-diffusive fractional crystallization and adcumulus growth in the Muskox Intrusion and other layered intrusions. In: Hargraves RB (ed) Physics of magmatic processes. Princeton University Press, Princeton, pp 325–383
Irvine TN (1987) Layering and related structures in the Duke Island and Skaergaard intrusions: similarities, differences, and origins. In: Parsons I (ed) Origins of igneous layering, vol 196, D Reidel Dordrecht. NATO ASI series C: mathematical and physical science, pp 185–245
Jousselin D, Morales LFG, Nicolle M, Stephant A (2012) Gabbro layering induced by simple shear in the Oman Ophiolite Moho transition zone. Earth Planet Sci Lett 331:55–66
Karson JA (1990) Sea floor spreading on the Mid-Atlantic Ridge: implications for the structure of ophiolites and oceanic lithosphere produced in slow-spreading environments. In: Malpas J, Moores EM, Panayiotou A, Xenophontos C (eds) Ophiolites, Oceanic Crustal Analogues. Proc Symp “TROODOS 1987”, Nicosia, Cyprus Geol Surv Department, p 547–555
Kelley DS, Gillis KM, Thompson G (1993) Fluid evolution in submarine magma-hydrothermal systems at the Mid-Atlantic Ridge. J Geophys Res 98:19579–19596
Kelley DS, Karson JA, Blackman DK, Früh-Green GL, Butterfield DA, Lilley MD, Olson EJ, Schrenk MO, Roe KK, Lebon GT, Rivizzigno P (2001) An off-axis hydrothermal vent field near the Mid-Atlantic Ridge at 30°N. Nature 412:145–149
Koepke J, Berndt J, Feig ST, Holtz F (2007) The formation of SiO_2-rich melts within the deep oceanic crust by hydrous partial melting of gabbros. Contrib Mineral Petrol 153:67–84
Koepke J., France L, Muller T, Faure F, Goetze N, Dziony W, Ildefonse B (2011) Gabbros from IODP Site 1256, equatorial Pacific; insight into axial magma chamber processes at fast spreading ocean ridges. Geochem Geophys Geosystems 12, Q09014, doi:10.1029/2011GC003655
Komor SC, Elthon D (1990) Formation of anorthosite-gabbro rhythmic phase layering: an example at North Arm Mountain, Bay of Islands Ophiolite. J Petrol 31:1–50
Komor SC, Elthon D, Casey JF (1987) Petrology of a leucogabbroic interval within basal layered gabbros at North Arm Mountain, Bay of Islands Ophiolite. Contrib Mineral Petrol 95:278–300

Korenaga J, Kelemen PB (1997) Origin of gabbro sills in the Moho transition zone of the Oman ohiolite: Implications for magma transport in the oceanic crust. J Geophys Res 102:27729–27749

Kurth M, Sassen A, Sühr G, Mezger K (1998) Precise ages and isotopic constraints for the Lewis Hills (Bay of Islands Ophiolite): preservation of an arc-spreading ridge intersection. Geology 26:1127–1130

Kvassnes AJS, Grove TL (2008) How partial melts of mafic lower crust affect ascending magmas at oceanic ridges. Contrib Mineral Petrol 156:49–71

Lissenberg CJ, Dick HJB (2008) Melt-rock reaction in the lower oceanic crust and its implications for the genesis of mid-ocean ridge basalt. Earth Planet Sci Lett 271:311–325

Lissenberg CJ, Bédard JH, van Staal CR (2004) The structure and geochemistry of the gabbro zone of the Annieopsquotch ophiolite, Newfoundland: implications for lower crustal accretion at spreading ridges. Earth Planet Sci Lett 229:105–123

Lissenberg CJ, van Staal CR, Bédard JH, Zagorevski A (2005) Geochemical constraints on the origin of the Annieopsquotch ophiolite belt, Newfoundland Appalachians. Geol Soc Am Bull 117:1413–1426

Lissenberg CJ, MacLeod CJ, Howard KA, Godard M (2013) Pervasive reactive melt migration through fast-spreading lower oceanic crust (Hess deep, equatorial Pacific Ocean). Earth Planet Sci Lett 361:436–447

Lister CRB (1983) On the intermittency and crystallization mechanisms of sub-seafloor magma chambers. Geophys J R Astron Soc 73:351–365

Lowell RP, Germanovich LN (1994) On the temporal evolution of high-temperature hydrothermal systems at ocean ridge crests. J Geophys Res 99:565–575

Macdonald KC (2001) Seafloor spreading: mid-ocean ridge tectonics. In: Steele J, Thorpe S, Turekian K (eds) Encyclopedia of ocean sciences. Academic Press, p 1798–2001

MacDonald KC, Luyendyk BP (1986) Investigation of faulting and abyssal hill formation on the flanks of the East Pacific Rise (21°N) using Alvin. Mar Geophys Res 7:515–535

MacLeod CJ, Yaouancq G (2000) A fossil melt lens in the Oman ophiolite: Implications for magma chamber processes at fast spreading ridges. Earth Planet Sci Lett 176:357–373

Maier WD, Barnes SJ, Groves DI (2013) The Bushveld complex, South Africa: formation of platinum-palladium, chrome- and vanadium-rich layers via hydrodynamic sorting of a mobilized cumulate slurry in a large, relatively slowly cooling, subsiding magma chamber. Mineral Depos 48:1–56

Malpas J (1990) Crustal accretionary processes in the Troodos ophiolite, Cyprus: evidence from field mapping and deep crustal drilling. In: Malpas J, Moores E, Panayiotou A, Xenophontes C (eds) Ophiolites, Oceanic Crustal Analogues. Proc Symp "TROODOS 1987", Nicosia, Cyprus Geol Surv Department, pp 65–74

Manning CE, MacLeod CJ, Weston PE (2000) Lower crustal cracking front at fast spreading ridges: evidence from the East Pacific Rise and the Oman Ophiolite. In: Dilek Y, Moores E, Elthon D, Nicolas A (eds) Ophiolites and the oceanic crust: new insights from field studies and the ocean drilling program. Geol Soc Am Spec Paper 349:261–272

Marsh BA (2004) Magmatic mush column Rosetta Stone: the McMurdo Dry Valleys of Antarctica. EOS Trans Am Geophys Union 85:497

McBirney AR, Sonnenthal EL (1990) Metasomatic replacement in the Skaergaard Intrusion, East Greenland: preliminary observations. Chem Geol 88:245–260

Meurer WP, Meurer MES (2006) Using apatite to dispel the "trapped liquid" concept and to understand the loss of interstitial liquid by compaction in mafic cumulates: an example from the stillwater complex, Montana. Contrib Mineral Petrol 151:187–201

Michael PJ, Schilling J-G (1989) Chlorine in mid-ocean ridge magmas: Evidence for assimilation of seawater-influenced components. Geochim Cosmochim Acta 53:3131–3143

Miranda EA, Dilek Y (2010) Oceanic core complex development in modern and ancient oceanic lithosphere: gabbro-localized versus peridotite-localized detachment models. J Geol 118:95–109

Namur O, Charlier B, Pirard C, Hermann J, Liegeois JP, Van der Auwera J (2011) Anorthosite formation by plagioclase flotation in ferrobasalt and implications for the lunar crust. Geochim Cosmochim Acta 75:4998–5018
Natland JH, Dick HJB (2001) Formation of the lower ocean crust and the crystallization of gabbroic cumulates at a very slowly spreading ridge. J Volcanol Geotherm Res 110:191–233
Natland JH, Dick HJB (2009) Paired melt lenses at the East Pacific Rise and the pattern of melt flow through the gabbroic layer at a fast-spreading ridge. Lithos 112:73–86
Nicolas A (1992) Kinematics in magmatic rocks with special reference to gabbros. J Petrol 33:891–915
Nicolas A (1994) The genesis of oceanic crust—Magma injection, hydrothermal circulation, and crustal flow—comment. J Geophys Res 99:12029–12030
Nicolas A, Reuber I, Benn K (1988) A new magma chamber model based on structural studies in the Oman ophiolite. Tectonophysics 151:87–105
Nisbet EG, Fowler CMR (1978) The Mid-Atlantic Ridge at 37° and 45°N: some geophysical and petrological constraints. Geophys J R Astron Soc 54:631–660
O'Driscoll B, Emeleus CH, Donaldson CH, Daly JS (2010) Cr-spinel seam petrogenesis in the Rum Layered Suite, NW Scotland: cumulate assimilation and in situ crystallization in a deforming crystal mush. J Petrol 51:1171–1201
Ohara Y, Fujioka K, Ishii T, Yurimoto H (2003) Peridotites and gabbros from the Parece Vela backarc basin: unique tectonic window in an extinct backarc spreading ridge—art. Geochem Geophys Geosyst 4:8611 doi:10.1029/2002GC000469
Pagé P, Bédard, JH, Schroetter JM, Tremblay A (2008) Mantle petrology and mineralogy of the Thetford Mines Ophiolite Complex. Lithos 100:255–292
Pearce JA (2003) Supra-subduction zone ophiolites: The search for modern analogues. In: Thy P, Dilek Y (eds) Ophiolite concept and the evolution of geological thought, vol 373. Geological Society of America, Special Paper, pp 269–293
Quick JE, Denlinger RP (1993) Ductile deformation and the origin of layered gabbro in ophiolites. J Geophys Res 98:14015–14027
Rabinowicz M, Nicolas A, Vigneresse JL (1984) A rolling mill effect in asthenosphere beneath oceanic spreading centers. Earth Planet Sci Lett 67:97–108
Riccio L (1976) Stratigraphy and petrology of the peridotite-gabbro component of the western Newfoundland ophiolites. Unpublished PhD Thesis, University of Western Ontario, 264 p
Rosencrantz E (1983) The structure of the sheeted dikes and associated rocks in the North Arm massif, Bay of Islands ophiolite complex, and the intrusive process at oceanic spreading centers. Can J Earth Sci 20:787–801
Saal AE, Kurz MD, Hart SR, Blusztajn JS, Blichert-Toft J, Liang Y, Geist DJ (2007) The role of lithospheric gabbros on the composition of Galapagos lavas. Earth Planet Sci Lett 257:391–406
Schmidt MW, Forien M, Solferino G, Bagdassarov N (2012) Settling and compaction of olivine in basaltic magmas: an experimental study on the time scales of cumulate formation. Contrib Mineral Petrol 164:959–976
Schroetter J-M, Pagé P, Bédard JH, Tremblay A, Bécu V (2003) Forearc extension and seafloor spreading in the Thetford Mines Ophiolite complex. In: Dilek Y, Robinson PT (eds) Ophiolites in earth history, vol 218. Geological Society, Special Publication, London, pp 231–251
Scoates JS, Lindsley DH, Frost BR (2010) Magmatic and structural evolution of an anorthositic magma chamber: the Poe mountain intrusion, Laramie Anorthosite Complex, Wyoming. Can Mineral 48:851–885
Singh SC, Kent GM, Collier JS, Harding AJ, Orcutt JA (1998) Melt to mush variations in crustal magma properties along the ridge crest at the southern East Pacific Rise. Nature 394:874–878
Sinton JM, Detrick RS (1992) Mid-ocean ridge magma chambers. J Geophys Res 97:197–216
Sleep NH (1975) Formation of oceanic crust: some thermal constraints. J Geophys Res 80:4037–4042
Smewing JD, Simonian KO, Gass IG (1975) Metabasalts from the Troodos Massif, Cyprus: genetic implications deduced from petrography and trace element geochemistry. Contrib Mineral Petrol 51:49–64

Smith DK, Cann JR (1992) The role of seamount volcanism in crustal construction at the Mid-Atlantic Ridge (24°–30°N). J Geophys Res 97:1645–1658
Smith DK, Cann JR, Escartin J (2006) Widespread active detachment faulting and core complex formation near 13°N on the Mid-Atlantic Ridge. Nature 442:440–443
Stern FG (2012) Geochemistry of the ultramafic rocks from the Bay of Islands Ophiolitic complex, Newfoundland. Unpublished MSc thesis, University of Ottawa-Carleton, 83 p
Stern FG, Hattori KH, Bédard JH (2011a) Partitioning of PGE between silicate melt and residual mantle peridotites in the Bay of Islands Ophiolite, Newfoundland. Geol Assoc Can/Mineral Assoc Can meeting, Ottawa
Stern FG, Hattori KH, Bédard JH (2011b) Mantle evolution recorded in Cr-spinel from the Bay of Island ophiolitic complex, northern Appalachians. Geological Society of America, National meeting
Sun S-S, McDonough WF (1989) Chemical and isotopic systematics of oceanic basalts: implications for mantle compositions and processes. In: Saunders AD, Norry MJ (eds) Magmatism in the ocean basins, vol 42. Geological Society, Special Publication, London, pp 313–345
Thy P, Dilek Y (2000) Magmatic and tectonic controls on the evolution of oceanic magma chambers at slow-spreading ridges: perspectives from ophiolitic and continental layered intrusions. In: Dilek Y, Moores E, Elthon D, Nicolas A (eds) Ophiolites and the oceanic crust: new insights from field studies and the ocean drilling program, vol 349. Geological Society of America, Special Paper, pp 87–104
Thy P, Dilek Y (2003) Development of ophiolitic perspectives on models of oceanic magma chambers beneath active spreading centers. In: Thy P, Dilek Y (eds) Ophiolite concept and the evolution of geological thought, vol 373. Geological Society of America, Special Paper, pp 187–226
Thy P, Schiffman P, Moores EM (1989) Igneous mineral stratigraphy and chemistry of the Cyprus Crustal Study Project drill core in the plutonic sequences of the Troodos ophiolite. In: Gibson IL, Malpas J, Robinson PT, Xenophontos C (eds) Cyprus Crustal study project Initial report, Hole CY-4, vol 88–9. Geological Survey Can Paper, pp 147–185
Toomey DR, Solomon SC, Purdy GM (1988) Microearthquakes beneath the median valley of the Mid-Atlantic Ridge near 23°N: tomography and tectonics. J Geophys Res 93:9093–9112
Tremblay A, Bédard JH, Lauzière K (1997) Taconian obduction and Silurian exhumation of the Betts Cove ophiolite, Canadian Appalachians. J Geol 105:701–716
Tucholke BE, Behn MD, Buck WR, Lin J (2008) Role of melt supply in oceanic detachment faulting and formation of megamullions. Geology 36:455–458
Upadhyay HD (1973) The Betts Cove ophiolite and related rocks of the Snooks Arm Group, Newfoundland. Unpubl PhD Thesis, Memorial Univ Newfoundland, 224 pp
Van Staal CR Chew DM Zagorevski A McNicoll V Hibbard J Skulski T Castonguay S Escayola MP Sylvester PJ (2013) Evidence of Late Ediacaran hyperextension of the Laurentian Iapetan margin in the Birchy Complex, Baie Verte Peninsula, Northwest newfoundland: implications for the opening of Iapetus, formation of peri-Laurentian microcontinents and Taconic—Grampian orogenesis. Geosci Can 40:1–25
Vine FJ, Moores EM (1972) A model for the gross structure, petrology, and magnetic properties of oceanic crust. In: Shagam R, Hargraves RB, Morgan WJ, Van Houten FB, Burk CA, Holland HD, Hollister LC (eds) Studies in earth and space sciences: a memoir in honor of Harry Hammond Hess, Boulder Col, vol 132. Geological Society of America, Memoirs, pp 195–205
Wager LR, Brown GM (1968) Layered igneous rocks. Oliver and Boyd, Edinburgh, 588 p
Williams H (1979) Appalachian orogen in Canada. Can J Earth Sci 16:792–807
Wright DJ, Haymon RM, Fornari DJ (1995) Crustal fissuring and its relationship to magmatic and hydrothermal processes on the east Pacific Rise crest (9°12′ to 54′N). J Geophys Res 100:6097–6120

Index

B. Charlier et al. (eds.), *Layered Intrusions*, Springer Geology,
DOI 10.1007/978-94-017-9652-1

C

M

MIX
Papier aus verantwortungsvollen Quellen
Paper from responsible sources
FSC® C105338

FSC
www.fsc.org

If you have any concerns about our products,
you can contact us on
ProductSafety@springernature.com

In case Publisher is established outside the EU,
the EU authorized representative is:
Springer Nature Customer Service Center GmbH
Europaplatz 3, 69115 Heidelberg, Germany

Printed by Libri Plureos GmbH
in Hamburg, Germany